Biology

Biology
THE WORLD OF LIFE

Sixth Edition

Robert A. Wallace
UNIVERSITY OF FLORIDA

HarperCollins*Publishers*

Sponsoring Editor: Glyn Davies
Developmental Editor: Meryl R.G. Muskin
Project Coordination, Text and Cover Design: Proof
Positive/Farrowlyne Associates, Inc.
Cover Photo: Richard K. Laval/Earth Scenes
Cover Inset Photo: Tom McHugh/Photo Researchers, Inc.
Photo Researcher: Lynn Mooney
Production Manager: Michael Weinstein
Compositor: Graphic World, Inc.
Printer and Binder: Arcata Graphics/Kingsport
Cover Printer: The Lehigh Press, Inc.

Biology: The World of Life, Sixth Edition

Library of Congress Cataloging-in-Publication Data

Wallace, Robert A.
Biology, the world of life / Robert A. Wallace.—6th ed.
p. cm.
Includes bibliographical references and index.
ISBN 0-673-46480-6 (student ed.)
ISBN 0-673-46625-6 (teacher ed.)
1. Biology. I. Title.
QH308.2.W35 1991
574—dc20 91-19036
CIP

93 94 9 8 7 6 5 4 3

From that night he crashed my Christmas party in New York, wearing a T-shirt and an eight-pound scarf, until he ran me down in Colorado after all those years, he unknowingly taught me a great deal about living. So with great affection, I dedicate this book to the mad Hungarian, Zsolt Megai.

Brief Contents

Contents vii
Preface xvii
About the Author 1

PART ONE
Life and the Flow of Energy 2
1 A Brief History and the Enchanted Isles 4
2 Scientists and Their Science 32
3 The Chemistry of Life 52

PART TWO
Cells and Inheritance 88
4 The Cell and Its Structure 90
5 Energy: The Dance of Life 120
6 The Cell and Its Cycles 151
7 Inheritance: From Mendel to Molecules 181
8 Advances in Genetics 209

PART THREE
The Diversity of Changing Life 226
9 The First Life 228
10 The Processes of Evolution 250
11 Diversity I: The Monera, Protista, and Fungi 281
12 Diversity II: The Plants 316
13 Diversity III: Animals 336

PART FOUR
Reproduction and Development 380
14 Plant Reproduction and Development 382
15 Animal Reproduction 407
16 Animal Development 439

PART FIVE
Systems and Their Control 474
17 Support and Muscular Systems 476
18 Homeostasis and the Internal Environment 497
19 The Respiratory, Circulatory, and Digestive Systems 516
20 The Immune System 550
21 Hormones and Nerves 573
22 The Nervous System 597
23 The Senses 637

PART SIX
Behavior and the Environment 656
24 Animal Behavior 658
25 Biomes and Communities 694
26 Population Dynamics 735
27 Human Populations 763
28 Resources, Energy, and Human Life 780
29 Bioethics, Technology, and Environment 816

Appendices A–C A-1
Glossary G-1
Selected Readings R-1
Credits C-1
Index I-1

Detailed Contents

Preface xvii

About the Author 1

PART ONE
Life and the Flow of Energy 2

1 **A Brief History and the Enchanted Isles** 4
The Voyage of the *Beagle* 6
The History of an Idea 8
The Beginnings of Biology 10
Time and the Intellectual Milieu 13
The Development of Darwin's Idea of Evolution 14
The Impact of Lyell 14
Puzzling Change and Variation 15
The Galapagos Islands
The Impact of Malthus 17
Natural Selection 25
Summary 29
Key Terms 30
For Further Thought 30
For Review 30

2 **Scientists and Their Science** 32
Inductive and Deductive Reasoning 33
Science Trends and Shifting Logic 33
Hypothesis, Theory, and the Scientific Method 35
The Scientist as Skeptic 38
Defining Life 39
The Trap of Teleology 42
Science and Phenomenology 42
What Biologists Do for a Living 43
Science and Social Responsibility 43
ESSAY 2.1: *The Controlled Experiment* 36
Summary 50
Key Terms 50
For Further Thought 50
For Review 51

3 **The Chemistry of Life** 52
Atoms and Elements, Molecules and Compounds 54
CHNOPS 54
The Structure of an Atom 54
How Atoms May Vary 55
Isotopes 55
Ions 56
Electrons and Their Behavior 56
Oxidation and Reduction 57
Electrons and the Properties of the Atom 57
Chemical Bonding 59
Ionic Bonds 59
Covalent Bonds 60
Hydrogen Bonds 61
Chemical Reactions 63
Energy of Activation 65
Enzymes 65
Energy Changes 67
The Molecules of Life 68
The Magic of Carbon 68

The Role of Water in Forming Large Biological Molecules 72
Carbohydrates 74
Lipids 75
Proteins 78
Amino Acids • Peptides • Protein Structure
Nucleic Acids 82
ESSAY 3.1: *Acids and Bases* 73
ESSAY 3.2: *Steroids, the Drug of "Champions"* 79
ESSAY 3.3: *Unraveling the Structure of Insulin* 83
Summary 84
Key Terms 86
For Further Thought 86
For Review 86

PART TWO
Cells and Inheritance 88

4 The Cell and Its Structure 90
Cytology and Technology 93
Prokaryotic and Eukaryotic Cells 93
Specialization in Eukaryotic Cells 96
Cell Components 99
Cell Walls 99
Plasma Membrane 100
The Cytoskeleton 101
Microtubules 102
Centrioles 102
Cilia and Flagella 103
Mitochondria 104
Ribosomes 104
The Endoplasmic Reticulum 105
Golgi Bodies 107
Lysosomes 108
Plastids 108
Vacuoles 109
The Nucleus 109
How Molecules Move 110
Passive Transport 110
Diffusion • Facilitated Diffusion • Osmosis
Active Transport 114
Facilitated Active Transport • Endocytosis and Exocytosis
ESSAY 4.1: *Microscopes and What They Can Do* 95
ESSAY 4.2: *The Endosymbiosis Hypothesis* 105
Summary 116
Key Terms 117
For Further Thought 118
For Review 118

5 Energy: The Dance of Life 120
The Laws of Thermodynamics 123
ATP: The Energy Currency 124
Photosynthesis: A Poet's View 125
Photosynthesis: A Closer View 126
The Light-Dependent Reactions • The Light-Independent Reactions • What Does a Plant Do with Its Glucose?
Food to Energy 133
Food to Energy: A Poet's View 136
Food to Energy: A Closer View 137
Glycolysis
Fermentation 140
Aerobic Respiration • The Krebs Cycle • The Electron-Transport System • Chemiosmotic Phosphorylation
A Brief Review 146
ESSAY 5.1: *Chlorophyll* 127
ESSAY 5.2: *The Visible Light Spectrum and Photosynthesis* 130
ESSAY 5.3: *A Notable Exception from the Sea* 134
Summary 147
Key Terms 148
For Further Thought 148
For Review 148

6 The Cell and Its Cycles 151
The Cell Cycle and Mitosis 152
Meiosis 157
The Double Helix 164
DNA Replication 168
How Chromosomes Work 168
Transcription 169
The Genetic Code 171
Translation 172
Continuing Problems in Cell Biology 176
ESSAY 6.1: *When Meiosis Goes Wrong* 165

Summary 177
Key Terms 179
For Further Thought 179
For Review 179

7 Inheritance: From Mendel to Molecules 181
Mendelian Genetics 182
A Kindly Old Abbot Puttering in His Garden 182
The Principle of Dominance 183
The Principle of Segregation 184
The Principle of Independent Assortment 187
The Testcross 188
Classical Genetics—Beyond Mendel 189
Sex Determination 190
Gene Linkage and Crossover 193
Sex-linked Traits
Chromosome Mapping 197
Mutations 197
Gene Mutations 197
Chromosome Mutations 199
Further Beyond Mendel 199
Dominance Relationships 199
Multiple Alleles 200
Gene Interaction 200
Polygenic Inheritance 201
Other Sources of Variation 201
Population Genetics 202
ESSAY 7.1: *The Attempt to Recreate Leonardo da Vinci* 191
ESSAY 7.2: *The Disease of Royalty* 196
ESSAY 7.3: *Sickle-cell Anemia* 198
Summary 204
Key Terms 206
For Further Thought 206
For Review 207
Genetics Problems 208

8 Advances in Genetics 209
Genetic Engineering: An Unfolding Story 210
The Players 210
Finding and Isolating the Gene 211
How DNA Is Chopped Up 211
Getting Genes from One Place to Another 212
Splicing and Cloning the Genes 213
Sorting Through the Recombinants 213
Is It Tampering with Genes or Genetic Engineering? 216
How Great a Promise? 217
Advances in Agriculture 219
Diagnosis and Decision in Humans 220
Gene Replacement Therapy 222
ESSAY 8.1: *Bacterial Sex* 214
ESSAY 8.2: *Crippling a Microbe* 217
ESSAY 8.3: *The Gene Machine* 218
Summary 223
Key Terms 223
For Further Thought 224
For Review 224

PART THREE
The Diversity of Changing Life 226

9 The First Life 228
Life from Nonlife 229
Mechanism vs. Vitalism 229
Life's Many Meanings 229
The Expanding Universe 231
Our Early Earth 232
Hypotheses Concerning the Origin of Life 233
The Formation of the Molecules of Life 235
Protobionts: The Ancestral Droplets 236
Reproducing Droplets 238
Successful Droplets 239
Heterotrophy and Autotrophy 241
Oxygen: A Bane and a Blessing 246
ESSAY 9.1: *The Oldest Life* 244
Summary 247
Key Terms 248
For Further Thought 248
For Review 248

10 The Processes of Evolution 250
Variation in Populations 252
Variation by Single-Gene and Polygenic Inheritance 254
How the Environment Maintains Variation 255
Factors Involved in Evolution 256
How Natural Selection Can Influence Variation 257
Modes of Natural Selection

Small Populations and Genetic Drift 260
Bottleneck Effect • Founder Effect
The Question of Species 264
Speciation 266
Allopatric Speciation 266
Sympatric Speciation 268
Divergent and Convergent Evolution 269
Small Steps or Great Leaps? 270
Death Stars and Dinosaurs: The Great Extinctions 272
ESSAY 10.1: *Sexual Dimorphism* 261
ESSAY 10.2: *Continents Adrift* 267
ESSAY 10.3: *Extinction and Us* 274
Summary 277
Key Terms 279
For Further Thought 279
For Review 279

11 Diversity I: The Monera, Protista, and Fungi 281
Five Kingdoms and Splitters and Lumpers 283
The Kingdom Monera and a Continuing Simplicity 285
Division Archaebacteria and Extreme Habitats 285
Eubacteria and a Sea of Life 286
The Kingdom Protista and Accelerating Evolution 292
The Animal-like Protists 292
Phylum Ciliophora and Tiny Predators 292
Phylum Sarcodina and a Flexible Simplicity 294
Phylum Sporozoa and a Parasitic Existence 295
The Plantlike Protists 295
Phylum Euglenophyta and an Evolutionary Puzzle 295
Phylum Chrysophyta and Glassy Corpses 297
Phylum Pyrrophyta and Sparkling and Dreaded Waters 298
The Funguslike Protists 299
The Kingdom Fungi and Varied Lifestyles 299
Division Oomycota and Irish Americans 301
Division Zygomycota and Preserving Our Bread 303
Division Ascomycota and Remarkable Partnerships 303
Division Basidiomycota and the Destroying Angel 307
Division Deuteromycota—The Fungi Imperfecti 310
Mycorrhizae and Associating with Plants 311
ESSAY 11.1: *Classification* 284
ESSAY 11.2: *Botulism* 287
ESSAY 11.3: *Antibiotics* 290
ESSAY 11.4: *Lichens* 304
ESSAY 11.5: *Truffles and Sex and Truffles* 306
Summary 311
Key Terms 314
For Further Thought 314
For Review 314

12 Diversity II: The Plants 316
Alternating Generations 317
The Evolution of Alternating Generations 319
Aquatic Plants 320
Division Rhodophyta and Many Colors 320
Division Phaeophyta and Undersea Forests 321
Division Chlorophyta: Ancestors of Land Plants 322
Nonvascular Land Plants 323
Division Bryophyta and the Invasion of the Land 323
Vascular Plants 325
Division Pterophyta and Leaves and Roots 326
Gymnosperms and Naked Seeds 327
Angiosperms, Covered Seeds, and Flowers 331
Summary 333
Key Terms 334
For Further Thought 334
For Review 335

13 Diversity III: Animals 336
The Animal Kingdom 337
Phylum Porifera: An Evolutionary Dead End 339
Phylum Cnidaria and the Evolution of Radial Symmetry 340
Phylum Platyhelminthes and the Evolution of Heads and Sides 344

Phylum Nematoda and the Evolution of a Body Cavity 346
Phylum Mollusca and the Evolution of the Coelom 347
Phylum Annelida and the Evolution of Repeating Segments 350
Phylum Arthropoda and the Evolution of Specialized Segments 352
Phylum Echinodermata and an Evolutionary Puzzle 356
Phylum Hemichordata and a Half Step Along 357
Phylum Chordata and the Advent of Freeways 357
Class Agnatha • Class Chondrichthyes • Class Osteichthyes • Class Amphibia • Class Reptilia • Class Aves • Class Mammalia
Summary 375
Key Terms 378
For Further Thought 378
For Review 378

PART FOUR
Reproduction and Development 380

14 Plant Reproduction and Development 382
Sexual Reproduction 383
Disadvantages of Sexual Reproduction 383
Advantages of Sexual Reproduction 383
Reproduction and Life Cycles in Various Plants 384
Green Algae 384
Mosses 386
Ferns 386
Conifers 388
Flowering Plants 390
How Flowers Work 392
Plant Development 395
Plant Hormones 398
Auxins 398
Gibberellins 401
Cytokinins 402
ESSAY 14.1: *The Coevolution of Pollen and Insects* 393
Summary 403
Key Terms 404
For Further Thought 405
For Review 405

15 Animal Reproduction 407
Asexual Reproduction 408
Binary Fission 408
Regeneration 410
Budding 410
Sexual Reproduction 410
Timing of Sexual Reproduction 413
Human Reproduction 415
The Male Reproductive System 415
The Female Reproductive System 417
The Menstrual Cycle
The Human Sexual Response 421
Conception 422
Sex and Society 424
Contraception 425
Historical Methods 425
The Rhythm Method 427
Coitus Interruptus 427
The Condom 427
Diaphragm 430
The Cervical Cap 430
Spermicides 430
The Intrauterine Device 430
The Birth Control Pill 431
Norplant 432
The Sponge 432
Sterilization 432
Abortion 434
ESSAY 15.1: *How They Do It* 409
ESSAY 15.2: *STDs or Chlamydia Is Not a Flower* 428
Summary 435
Key Terms 437
For Further Thought 438
For Review 438

16 Animal Development 439
Early Stages of the Embryo 440
Influences on Organization 442
The Flexible Fate 442
Sequencing 443
Stress 445
Induction 445

Types of Eggs 446
Early Development in the Frog 447
The First Cell Divisions 447
Gastrulation 447
Embryonic Regulation 450
Birds and Their Membranes 451
Human Development 453
The First Trimester 462
The First Month • The Second Month • The Third Month
The Second Trimester 465
The Fourth and Fifth Months • The Sixth Month
The Third Trimester 466
Birth 467
Miscarriage 469
ESSAY 16.1: *Can Humans Regenerate Missing Parts?* 443
ESSAY 16.2: *Wound Healing* 444
ESSAY 16.3: *Why Your Children Will Not Look Exactly Alike* 454
Summary 470
Key Terms 471
For Further Thought 472
For Review 472

PART FIVE

Systems and Their Control 474

17 Support and Muscular Systems 476
Support Systems 477
Types of Skeletal Systems 478
Types of Connective Tissue 479
The Human Skeleton 482
Joints 485
The Variety of Contractile Systems 486
Types of Muscles in Humans 488
Skeletal Muscles 488
Antagonist Muscles
Muscle Contraction 492
Summary 493
Key Terms 495
For Further Thought 495
For Review 496

18 Homeostasis and the Internal Environment 497
Homeostasis and the Delicate Balance of Life 498
Feedback Systems 498
Homeostasis and the Regulation of Temperature 500
Ectotherms 500
Endotherms 500
Homeostasis and the Regulation of Water 504
Solutions to the Water Problem 505
Invertebrate Excretory Systems 506
The Human Excretory System 508
The Formation of Urine • The Concentration of Urine
ESSAY 18.1: *Special Solutions to the Water Problem* 507
ESSAY 18.2: *Pressure Gradients in the Kidney Tubule* 511
Summary 513
Key Terms 514
For Further Thought 515
For Review 515

19 The Respiratory, Circulatory, and Digestive Systems 516
Respiratory Systems 517
The Various Ways Species Get Oxygen 517
Aquatic Animals • Insects • Mammals
Circulatory Systems 525
Circulation in Small Animals 525
Circulation in Larger Animals 526
The Vertebrate Vessels • Arteries and Blood Pressure • Blood • The Vertebrate Heart • Control of Heartbeat
The Lymphatic System 536
Digestive Systems 538
Digestive Arrangements 538
The Human Digestive System 540
ESSAY 19.1: *The Joy of Smoking* 523
ESSAY 19.2: *The Incredible Blood-Brain Barrier* 527
ESSAY 19.3: *Heart Attack* 535
ESSAY 19.4: *CPR* 536
ESSAY 19.5: *Heimlich Maneuver* 544

Summary 546
Key Terms 548
For Further Thought 548
For Review 549

20 The Immune System 550
The Nonspecific Responses 551
Specific Responses 555
The Antigen-Antibody Response 557
Programming the Lymphocytes 558
The Role of the Cytotoxic T-Cells 558
The Roles of the Helper T-Cells and the B-Cells 558
Tolerance and Autoimmunity 561
Interferon: An Exciting New Promise 562
AIDS: A Devastating New Problem 563
Mind and Body 568
ESSAY 20.1: *Lyme Disease* 555
ESSAY 20.2: *Allergy: An Overreaction* 556
ESSAY 20.3: *How African Crocodiles Defend Against Texas Rabbits* 561
ESSAY 20.4: *AIDS* 566
Summary 570
Key Terms 571
For Further Thought 572
For Review 572

21 Hormones and Nerves 573
Hormones 574
Invertebrate Hormones 575
Human Hormones 576
How Hormones Work 579
Second Messenger Systems (Peptide Hormones) • *Direct Gene-Activation Systems (Steroid Hormones)*
Feedback in Hormonal Systems 583
The Relationship of Hormones and Nerves 584
Nerves 584
Impulse Pathways 587
The Mechanism of the Impulse 587
The Synapse 591
ESSAY 21.1: *Hormones and Test-Taking* 579
ESSAY 21.2: *Human Pheromones* 581
ESSAY 21.3: *Prostaglandins, Endorphins, and Runner's High* 582
Summary 594
Key Terms 595
For Further Thought 596
For Review 596

22 The Nervous System 597
The Evolution of Nervous Systems 598
The Central Nervous System 600
The Spinal Cord and the Reflex Arc 600
The Vertebrate Brain 602
The Human Brain 604
The Hindbrain 605
The Medulla • *The Cerebellum and Pons*
The Midbrain 607
The Forebrain 607
The Thalamus and Reticular System • *The Hypothalamus* • *The Cerebrum*
Hemispheres and Lobes 609
Two Brains, Two Minds? 610
The Split Brain 614
The Peripheral Nervous System 614
The Autonomic Nervous System 616
Autonomic Learning 619
Mindbenders 621
Tobacco 623
Caffeine 624
Marijuana and Hashish (THC) 625
Alcohol 626
Opiates 628
Cocaine 628
Amphetamines 631
Barbiturates 631
Phencylidine 632
Methaqualone 632
Psychedelics 632
ESSAY 22.1: *Penfield's Mapping* 612
ESSAY 22.2: *The Hangover* 627
Summary 634
Key Terms 635
For Further Thought 635
For Review 635

23 The Senses 637
Thermoreceptors 639
Tactile Receptors 640

Auditory Receptors 640
Chemoreceptors 646
Proprioceptors 650
Visual Receptors 651
ESSAY 23.1: *Moth-Bat Coevolution* 642
ESSAY 23.2: *That Wonderful You* 646
ESSAY 23.3: *Fooling Mother Nature's Sweet Tooth* 649
Summary 653
Key Terms 654
For Further Thought 654
For Review 655

PART SIX
Behavior and the Environment 656

24 Animal Behavior 658
Ethology and Comparative Psychology 659
The Development of the Instinct Idea 660
Releasers 660
Innate Releasing Mechanisms 661
Fixed Action Patterns 661
Learning 662
Habituation 663
Classical Conditioning 665
Operant Conditioning 666
How Instinct and Learning Can Interact 667
Imprinting 669
Orientation and Navigation 670
Social Behavior 672
Aggression 673
Fighting 674
Cooperation 680
Altruism 683
Kin Selection 684
Reciprocal Altruism 687
Sociobiology and Society 688
ESSAY 24.1: *The Advantage of Forgetting* 664
ESSAY 24.2: *Biological Clocks* 674
ESSAY 24.3: *How to Recognize Kin* 686
Summary 690
Key Terms 691
For Further Thought 692
For Review 692

25 Biomes and Communities 694
Ecosystems and Communities 695
Habitat and Niche 695
Succession 698
The Web of Life 701
Extinction and the Web of Life 703
Extinction and Us 704
The Land Environment 707
Biomes 707
The Water Environment 717
Freshwater Bodies 717
Rivers and Streams • Lakes and Ponds
The Oceans 726
Coastal Areas 730
ESSAY 25.1: *Subdue the Earth* 706
ESSAY 25.2: *The Destruction of Tropical Forests* 711
ESSAY 25.3: *The Redwoods* 715
Summary 732
Key Terms 734
For Further Thought 734
For Review 734

26 Population Dynamics 735
Populations, Ethics, and Necessity 736
How Populations Change 739
Population Growth 739
Controlling Populations Through Reproduction 743
Tapeworms 745
Chimpanzees 746
Birds 747
Controlling Populations Through Mortality 749
Abiotic Control and Density-Independence 749
Biotic Control and Density-Dependence 752
Predation 752
Parasitism 754
Competition
Disease 756
The Advantage of Death 757
ESSAY 26.1: *What Have They Done to the Rain?* 751
ESSAY 26.2: *Greenpeace and the Bartenders* 754

Summary 760
Key Terms 761
For Further Thought 761
For Review 761

27 Human Populations 763
Early Human Populations 764
The Advent of Agriculture 768
Population Changes from 1600 to 1850 768
Population Changes After 1850 769
The Human Population Today 770
The Future of Human Populations 773
Population Structure 775
ESSAY 27.1: *Doubling Times* 774
ESSAY 27.2: *Why Do Women Live Longer?* 777
Summary 778
Key Terms 779
For Further Thought 779
For Review 779

28 Resources, Energy, and Human Life 780
Renewable Resources: Focus on Food and Water 781
Food and the Present Crisis 782
The Global Implications of Food Resources • Some Global Realities • Projections of Food Supply • Farming the Earth's Jungles • Domestic Animals as Food • Fishing • Solutions to the Hunger Problem
Water and the Coming Crisis 796
Nonrenewable Resources 799
Recycling—and Around It Goes 800
Energy 801
Energy from Fossil Fuels 802
The Geopolitics of Fossil Fuel 806
Energy from Water 806
Energy from the Wind 806
Energy from the Earth 807
Energy from the Sun 807
Active Solar Systems • Passive Solar Systems • Photovoltaics
Energy from the Atom 810
Encouraging Conservation and Increasing Efficiency 811
ESSAY 28.1: *How Hunger Kills* 784
ESSAY 28.2: *The Food Pyramid* 786
ESSAY 28.3: *The Great African Famine* 788
ESSAY 28.4: *Miracle Crops and the Green Revolution* 792
ESSAY 28.5: *The True Cost of Gasoline* 803
ESSAY 28.6: *The Greenhouse Effect* 804
Summary 813
Key Terms 813
For Further Thought 813
For Review 814

29 Bioethics, Technology, and Environment 816
The Ethics of Doormats 817
Environmental Pollution 819
Air Pollution 820
Carbon Monoxide • Nitrogen Oxides • Sulfur Oxides • Hydrocarbons • Particulate Matter
Water Pollution 824
Sewage • Chemical Pollution of Water • Heat Pollution of Water
Pesticides 828
Radiation 829
Pollution from Nuclear Reactors
Hidden Decisions 835
The Future 837
ESSAY 29.1: *Whose Rights Are They, Anyhow?* 818
ESSAY 29.2: *Temperature Inversions* 821
ESSAY 29.3: *Holes in the Sky* 822
ESSAY 29.4: *Sewage Treatment* 825
ESSAY 29.5: *Radiation* 831
ESSAY 29.6: *The Chernobyl Meltdown* 833
ESSAY 29.7: *Nuclear Winter* 836
Summary 838
Key Terms 839
For Further Thought 839
For Review 840

APPENDIX A: **Classification of Organisms** A-1

APPENDIX B: **For Further Thought Answers** A-5

APPENDIX C: **For Review Answers** A-9

Glossary G-1

Selected Readings R-1

Credits C-1

Index I-1

Preface

PRESENTING THE SIXTH EDITION

In all the years I've been writing, I've rarely, if ever, brought out an edition that has had more care lavished on it than this one. This is, in a sense, the kind of work I wish I had produced every time, but that just wasn't possible for a number of reasons. Things have to come together, and this time they did.

Because of a fortunate combination of events, and because of the swiftly changing situation, both environmentally and scientifically, I asked my new publisher, HarperCollins, for permission, this one time, to follow closely on the heels of an earlier edition. I had three arguments.

I had a unique set of advisers standing by.

I pointed out that this group of dedicated teachers and scientists was standing by to confer with me and that this might not be the case a year from now. A number of these people had used this book for years, through its long evolution, and were anxious to have their say in the next edition. Living with the book, as they had, gave them a particular insight on improving it, making it an even more effective teaching aid. (I view all books as teaching aids. The real teaching is done by those people at the front of the class—in the trenches, as it were. My role is to be of assistance to them.) I felt I could not pass up the opportunity to be guided by such a group of willing and interested experts.

Environmental issues have become pressing.

I argued that biology is among the most swiftly changing of all sciences and it was time for this book to reflect those changes. In particular, I wanted a stronger ecological–environmental emphasis. Part of my argument was based on the advice of the people in the trenches, who almost unanimously urged me to strengthen this message. I suppose I was sensitive to such advice after conducting my own research in the Amazon rain forests these past several years.

There has to be a better way of handling chemistry for non-majors.

Finally, I had heard a recurring theme from the beginning, and this seemed like a good time to respond to it. This book was written for people who do not necessarily have a strong scientific background and those who perhaps do not intend to go on in the sciences (at least when they start out; if we happen to change their minds along the way, so much the better). This book was written for them from the ground up; it is not a cut-down, chopped-up, condensed version of some book that got so long it began to crumble of its own weight. But, as many of us are aware, we don't invite them into the world of life by hammering them with chemistry. With my advisers, I devised a way to handle this. I would present the chemistry at two levels: *A Poet's View,* with just the essentials one would need to understand what's going on, and *A Closer View,* with the information presented in more detail. In fact, The Poet's View could be used to prepare the reader to better understand The Closer View. It all seemed to make a great deal of sense to my advisers and me. We would see how the publisher responded.

How the Publisher Responded

You're holding the book, so you can see how the publisher responded. In fact, they picked up the banner with such enthusiasm that the rest of us had to scramble to keep up with their expectations. We not only had to incorporate these new features, but we also had to make the usual improvements one finds in a new edition. In this case, we updated, better explained, added, and deleted to make the old material more effective. We also made some major changes, such as moving the discussion of the beginnings of life to the evolution section, and condensing and modernizing the discussions of animal behavior. While we coordinated and fine-tuned these changes, one of our greatest challenges was handed to us by the publisher. They felt so committed to this edition that they decided to develop an almost entirely new art program. So they pulled together some of the best artists in the business and we began. It meant long hours and endless discussions, but you can see the result.

All this has not been easy, but we hope you agree it has been worth it. Even as we close this project, though, there are rumors of new explanations of the great extinctions, new approaches to battling AIDS, unexpected water crises, new ways of understanding variation in populations, atrocities in the rain forests, promising conservation measures in Mexico, and on it goes. We're already eyeing each other and thinking about the next edition. But we promise, first, a breather—for both you and us.

ANCILLARIES

Just as authors need help, so do the books themselves and so the publisher has put together an effective package of ancillaries to help you in the classroom. (I'm particularly impressed with the laser disk. If you haven't seen it, ask the sales rep for a demonstration.)

Instructor's Manual—Newly written by David Fox, the manual includes, for each chapter, an overview, list of key topics, lecture outline, classroom discussion topics and activities, list of readings, list of resources (software, videos, films, etc.), and a section relating the individual themes of each chapter to the book as a whole.

Student Study Guide—Also new for this edition, this invaluable resource is written by Steven Muzos of Austin Community College, Rio Grande Campus. The guide provides students with objectives and outlines for each chapter, as well as review summaries of key concepts and terms, and self tests to help them prepare for examinations.

Laboratory Manual—The HarperCollins Laboratory Manual was written by Bill Tietjen of Bellarmine College. This lab manual can be used independently or coordinated with *Biology: The World of Life*. All labs have been carefully chosen and classroom tested. Each begins with an introduction followed by exercises. Students are alerted to hazards or items of particular importance. Special boxes demonstrate mathematical or statistical material, and many of the exercises are illustrated.

Laboratory Manual Instructor's Guide—This preparation guide includes hints for preparing labs, methods of building equipment, transparency masters, learning objectives, projected times for each exercise, sample problems, advice on caring for organisms, and supply sources.

Testbank—A testbank containing 1500 questions has been prepared by Jon Sperling of Queens College, City University of New York. Questions include multiple choice, fill-in-the-blank, matching, true-false, and short answer.

Testmaster—The testbank is available in a computerized form for IBM or MacIntosh.

The HarperCollins Biology Encyclopedia Laser Disk—Produced by HarperCollins in conjunction with Nebraska Interactive Video, Inc., the **Biology Encyclopedia Laser Disk** offers the latest in visual technology. It contains transparencies, micrographs, slides, and film and video footage. Over 1500 images were provided by Carolina Biological Supply. The laser disk allows instant access to any image or footage, controlled by a hand-held remote.

Acetate Transparencies—The publisher provides, free, a comprehensive set of 125 four-color acetates of art and photo micrographs from the text.

Harper Dictionary of Biology—Written by W.G. Hale and J.P. Margham, both of the Liverpool Polytechnic Institute, this reference contains 5600 entries, which go far beyond basic definitions to provide in-depth explanations and examples. Diagrams illustrate such concepts as genetic organization, plant structure, and human physiology. The dictionary covers all major subjects (anatomy, biochemistry, ecology, and so on) including biographies of important biologists.

Student Environmental Action Guide—The Earthworks Group and HarperCollins has joined with the Student Environmental Action Coalition to bring your students a handbook of the environmental movement on campuses around the country. It contains a series of strategies through real campus examples for approaching the administration, the community, political leaders, student leaders, and one's own personal habits to achieve positive change. Examples include: population control, transportation, water conservation, and publishing a newsletter.

Two Minutes a Day for a Greener Planet—Written by Marjorie Lamb, a veteran reporter on environmental affairs, this book provides easy, practical steps that all of us can take to save Earth. It gives suggestions for individual action, on a small scale, which can make a big impact on our planet's future.

The Botany Coloring Book, The Zoology Coloring Book, The Anatomy Coloring Book, The Physiology Coloring Book—An exciting, new approach to learning science, coloring has been shown to be a remarkably effective way to learn anatomy. The text accompanying each figure provides explanatory material and leads the reader through the various anatomical structures, step by step.

Writing About Biology—Written by Jan A. Pechnik of Tufts University, this brief guide includes sections on writing lab reports, essays, term papers, research proposals, critiques and summaries, and in-class essay examinations. It also includes special sections on effective notetaking, how to give oral presentations and how to prepare applications for summer and permanent jobs in biology. Appendices include commonly used abbreviations for lengths, weights, volumes and concentrations.

ACKNOWLEDGMENTS

No book is better than its editor, and what you see here reflects the character, imagination and determination of Glyn Davies, Life Sciences Editor. Glyn has a commitment to good books. He's tough-minded and we've disagreed a few times, but mainly we've had some laughs and some great times along the way to becoming friends and to producing what we consider a major success. I'm looking forward to our next project.

Marianne Russell, our editor-in-chief, inherited what must be one of the toughest jobs in the business—taking over a wide range of existing projects, each with its own set of problems, and making them somehow work under entirely different conditions. Publisher Susan Katz has the keen intelligence and hands-on enthusiasm that somehow seem to provide the momentum to move things past those inevitable sticking-points. Overseeing all of the battles below are Bob Biewen and, of course, George Craig, and it has been a particular pleasure working with them.

My Developmental Editor was Meryl Muskin, who somehow managed to keep track of everything. Ellen Pettengell, the book's designer, as diligent as ever, let nothing escape her and often provided solutions to some thorny problems. It has been a particular pleasure working with Andy Hess, of Proof Positive/Farrowlyne Associates, whose energy, intelligence, and background have made many a difficult problem seem easy. My Project Editor, Jean Dal Porto, moved the manuscript along, making just the right editorial changes. My photoresearcher, Lynn Mooney, worked very effectively with me to provide some of the remarkable images you see here. Randee Ladden, Teri J. McDermott, Sandra E. McMahon, Precision Graphics, Rossi & Associates, Kevin A. Somerville, and Sarah Forbes Woodward were the artists who produced most of the outstanding renderings that accent this book. We owe a great deal to their talents. Karen Morse, of the University of Massachusetts, Amherst, developed the learning objectives at the beginning of each chapter.

The people who bring my books out into the real world are indeed a dedicated group of professionals. Perhaps you've met Arnold Parker, Marilyn Moran, John Cross, Lucinda Turley, Peter Adams, Dave Fleming, Kevin Connors, Kelly Bell, Jim Northington, Pat Kelley, Frank Capek, Pat Quinlan, Jane Gunton, Phil Howrigan, Craig Gagstetter, Pam Fullerton, Kurt Reynolds, Clare Lynch, Jim Marshall, Suzanne Sitlington, John Young, Cyndi Keen, Cynthia Biron, Chuck Hickman, Nan Williams, Greg Odjakjian, Joan McKee, Erin Kelly or Lynn Butler.

Then there are those who played indirect, but no less important, roles in the development of this book. Russell Hurd, old friend and coauthor, is a remarkable physiologist who has provided endless encouragement and information over the years. Fulton Crews is a first-rate pharmacologist and a fearless and cheerful companion in the jungle. Don Silva has taught me a great deal about mountain life and, in fact, simply makes it worth living there. My old friend and long-time guide Caento Padilla is literally magic in the jungle and much of what I know about the place I learned from him. Jack May, through thick and thin, ensures there's never a dull moment. Scott Fehr, stalwart friend and a budding young computer scientist, is a monument to determination. Jim Cook has forgotten more about the biology of North Florida than I'll ever know. I stand amazed at Jerry Sanders, my long-time friend and partner, who seems to understand everything. Aaron, next door, is the funniest little kid on the block and sure can break up a long day.

As usual, my friends and colleagues at the University of Florida have been an enormous help to me, especially Frank Nordlie, Lincoln Brower, Lou Guillette, Carmine Lanciani, Dave Evans, and Brian McNab. The energy and intelligence of my fellow members of the Explorers Club is a constant source of encouragement. In particular, I acknowledge John Bruno, Charles Webb, Nicholas Sullivan, and Lew Scotton. I am greatly pleased that HarperCollins has become a Corporate Member of the Explorers Club, in keeping with their role as the Environmental Publisher.

Finally, I thank Jayne Austin, jungle explorer and handholder, who knows not only about the insights of Tennessee Williams, but the overhand right of Michael Moorer as well.

ADVISERS AND REVIEWERS

These are the people who helped me shape this book. I am very grateful for their encouragement and assistance. In particular, I thank Judith Goodenough of the University of Massachusetts, Amherst. She played a very large role in the development of this edition and her careful analysis led to numerous improvements. It is hard to find the kind of diligence and intelligence she brings to a project. I am indeed grateful to her. The reviewers include:

Daniel W. Benjamin, *Central Michigan University*
David J. Cotter, *Georgia College*
Terence B. Curren, *College of San Mateo*
David Fox

Harvey P. Friedman, *University of Missouri, St. Louis*
Judith Goodenough, *University of Massachusetts, Amherst*
David Hansen, *Pacific Lutheran University*
Brenda K. Johnson, *Western Michigan University*
Carmine A. Lanciani, *University of Florida*
Michael Lang-Moreland, *Diablo Valley College*
Tom Linder, *University of Washington*
Dennis Martin, *Pacific Lutheran University*
Thomas A. McGrath, *Corning Community College*
Neil A. Miller, *Memphis State University*
Pamela Monaco, *St. John's University*
F. Scott Orcutt, Jr., *University of Akron*
Wallace H. Orgell, *Miami-Dade Community College*
John T. Romeo, *University of South Florida*
Gary A. Smith, *Tarrant County Jr. College*
Donald E. Stearns, *Rutgers University*
Sandra B. Wilson, *Manatee Community College*
Anne E.K. Zayaitz, *Kutztown University of Pennsylvania*

We would also like to thank our reviewers from previous editions who helped form the book in its years of development and evolution:

Bonnie Amos, *Baylor University*
Vernon Avila, *San Diego State University*
John E. Butler, *Humboldt State University*
William W. Byrd, *Arkansas State University*
Galen E. Clothier, *Sonoma State University*
Charles Cottril, *Diablo Valley College*
Charles F. Denny, *University of South Carolina—Sumter Campus*
Gary E. Dolph, *Indiana University at Kokomo*
William Dunscombe, *Union County College*
William A. Emboden, *California State University, Northridge*
Wayland L. Ezell, *St. Cloud State University*
Douglas Fratianne, *The Ohio State University*
John L. Frola, *The University of Akron*
Laurence Fulton, *American River College*
William Gnewuch, *Sacred Heart University*
Garland F. Hicks, Jr., *Valparaiso University*
Ronald D. Humphrey, *Prairie View A & M University*
Ross E. Johnson, *Washburn University of Topeka*
Jack L. King, *The University of California at Santa Barbara*
Peter Klopfer, *Duke University*
Charles Leavell, *Fullerton College*
William H. Leonard, *Clemson University*
Lee McGeorge, *Duke University*
Teresa C. Minter-Procter, *Porterville College*
James Morrow, *Allan Hancock College*
Maria Nakamura-Chapin, *St. Petersburg Junior College (Clearwater Campus)*
John D. Pasto, *Middle Georgia College*
Peter Pedersen, *Cuesta College*
Richard G. Pendola, *Community College of Rhode Island*
Dennis K. Poole, *Gulf Coast Community College*
Robert C. Romans, *Bowling Green State University*
Gerald Sanders, *San Diego State University*
Orlando A. Schwartz, *University of Northern Iowa*
Dr. William R. Sigmund, *Slippery Rock University*
Joseph Stevenson, *St. Louis Community College at Florissant Valley*
James L. Strayer, *Washtenaw Community College*
William J. Thieman, *Ventura College*
Steven Vogel, *Duke University*
David E. Youker, *Sauk Valley College*

Robert A. Wallace

About the Author

Bob Wallace received his B.A. in Fine Arts with an emphasis on seventeenth-century painting techniques, his M.A. at Vanderbilt University where his research dealt with muscle histochemistry, and his Ph.D. at the University of Texas at Austin with an emphasis on the behavioral ecology of West Indian birds.

He has had a variety of roles in his travels around the world, including art instructor, carpenter (a very poor one, he says), portrait painter, longshoreman, nickel prospector, track and basketball coach, farmhand, karate instructor (he has a black belt in Tae Kwon Do), and scuba diver. His interests include cooking, horseback riding, marathoning, and fly-fishing.

His work currently includes helping develop a meaningful scientific education program in the United States and the promotion of good scientific writing. He is involved in helping teach science writing at the graduate level at the University of Florida where he is adjunct associate professor of zoology.

He is a fellow of the Explorers Club and has made explorations in the Arctic, the Indian Ocean, the West Indies, the Mediterranean, and the Amazon. His research involves traveling through the Amazon Basin where he visits the shamans and *curanderos* of remote tribes, particularly the Sacha Runa (Forest People) and the often-feared Waoranis (Human Beings) of Ecuador. He collects the medicinal plants used by these people, which are then tested in laboratories in the United States, particularly as potential agents in the fight against cancer and malaria.

Bob Wallace, when he and Jayne are not in Florida or South America, can usually be found at their mountain home in Steamboat Springs, Colorado, where he attempts to impress the local trout with the cleverness of his tied flies (he reports that the trout are so far not very impressed).

PART ONE

Life and the Flow of Energy

We begin our voyage through the sixth edition of *Biology, The World of Life,* with an introduction to Charles Darwin and his voyage aboard the *Beagle*, an episode that marked the entrance to modern biology. With the science introduced, we turn to the scientists and their roles, emphasizing, finally, the building blocks of life.

CHAPTER 1

A Brief History and the Enchanted Isles

Overview

The Voyage of the *Beagle*
The History of an Idea
The Beginnings of Biology
Time and the Intellectual Milieu
The Development of Darwin's Idea of Evolution
Natural Selection

Objectives

After reading this chapter you should be able to:

- Describe Charles Darwin's background and his activities on the voyage of the *Beagle*.
- Trace the path of scientific discovery prior to the eighteenth century and describe the contributions of some early scientists.
- Compare the pre- and post-Darwinian views on the changeability of the species.
- Describe Lamarck's views of how evolution occurs.
- Outline the ideas of Charles Lyell and Rev. Thomas Malthus, stating how each influenced Darwin.
- Describe the process of natural selection as put forth by Darwin and explain how his thoughts on natural selection were molded by his observations as a naturalist.

Struggling and writhing, the young man lay on his back, his arms and legs flailing at the cool Argentine air. It was a most curious sight. Yet, there he was, twisting convulsively on the ground, dust and bits of grass flying around him. Who could have guessed that this lad, only months earlier, had been walking his dog down an English lane on his way to have lunch with his wealthy uncle? Furthermore, who could have known that this same young man would, years later, throw the entire Western world into an anxious turmoil? And why was he now rolling on the ground?

Curious eyes watched from the Argentine pampas (Figure 1.1). They had never seen anything like this before. Some animals darted for cover, but one large bird, overcome with curiosity, stepped closer for a better look. As the bird drew nearer, young Charles Darwin, aglow with good health and a crack shot to boot, leaped to his feet, snapped his rifle to his shoulder, and dropped the curious bird on the spot. That night the rhea would be served aboard the good ship, *Beagle* (Figure 1.2).

This was not the most curious thing that Darwin had ever done. His shipmates, in fact, had come to expect strange things from this bright and energetic young man with his notable strength and amazing endurance. He had climbed hill after hill, just to be the first Englishman to view the magnificent beyond. He relentlessly collected all sorts of plants and animals and would go sleepless, working night and day, to retrieve some fossil before the ship again set sail. His shipmates had marveled at his strength, and their admiration increased when he once saved a group that had become stranded without supplies. He was happy and exuberant these days, but, in time, this was all to change.

But what was young Darwin doing in South America? After all, he was of the British upper class, given to riding and hunting, good food,

FIGURE 1.1

(a) The Argentine pampas. (b) Charles Darwin at the age of 29.

(a)

(b)

FIGURE 1.2

Rheas are known for their vile personalities as much as for their obsessive curiosity. The birds supplemented the menu on the *Beagle* when the boat visited South America.

and an occasional rousing game of blackjack. He was apparently not unusual in most respects and was well liked. But he had not always pleased his father.

Darwin's father was a huge, commanding man, a noted physician of great will and principle. He often wondered aloud about his son, though. He had tried to send Charles to medical school when the boy was 16. But Charles couldn't stand the sight of blood and almost fainted while first witnessing surgery. Then young Darwin tried the clergy. He spent three years at Cambridge studying theology, but even then he spent much of this time wandering around the countryside, adding rocks, insects, feathers, or whatever, to his collections (Figure 1.3). Indeed, he seemed to be more concerned with beetles than beatitudes. He was not alone in his interest, however. At that time, the English countryside was alive with amateur butterfly collectors, rock hounds, and plant fanciers whose position and wealth permitted them to indulge in such hobbies. Even so, Darwin's academic prowess had been so thoroughly unremarkable that at one point, his father had told his trifling son, "you care for nothing but shooting, dogs, and rat catching, and you will be a disgrace to yourself and all your family."

THE VOYAGE OF THE *BEAGLE*

It must be admitted that at this point in Charles Darwin's career there was little to suggest that his mind would come to be regarded as one of the most brilliant and inquisitive in history. In a short time, however, this diffident young man was to hear, through his friend the Reverend John Henslow, of an offer of free passage on a survey ship called the *Beagle* (Figure 1.4). A naturalist companion was needed for a voyage that was

FIGURE 1.3

The Cambridge botanical gardens as they appeared when Darwin was a student there.

FIGURE 1.4

The *Beagle*, a solid wooden ship about 25 meters long, was one of the Royal Navy's several vessels used to chart foreign waters and to carry explorers and scientists over the world in Britain's quest for greater power and increased trade.

to last five years. There would be no pay, and the person chosen would have to sleep in a hammock in the cramped chart room (although he would be permitted to share the captain's table). Armed with Henslow's recommendation, Darwin eagerly applied for the position, but was nearly rejected because of the shape of his nose. Captain Fitzroy, himself only 23, believed that the nose reflected the character of its bearer, and Darwin's nose just didn't show much character (Figure 1.5).

Darwin's family required some persuasion to accept this "madcap scheme," which they considered scarcely suitable for a prospective clergyman. And evidently, Charles had his own trepidations about such a momentous decision. In a letter to his sister Susan he wrote:

> Fitzroy says the stormy sea is exaggerated; that if I do not choose to remain with them, I can at any time get home to England; and that if I like, I shall be left in some healthy, safe and nice country; that I shall always have assistance; that he has many books, all instruments, guns, at my service. . . . There is indeed a tide in the affairs of men, and I have experienced it. Dearest Susan, Goodbye.

Ultimately all the arrangements were made, and in 1831, the H.M.S. *Beagle* set out from Devonport with Charles Darwin gazing, perhaps a bit apprehensively, at the slowly retreating shoreline of his homeland. As the heavy wooden vessel (about the size of a tugboat) creaked and groaned

FIGURE 1.5

Captain Fitzroy, the young captain of HMS *Beagle*.

FIGURE 1.6
Darwin's desk aboard HMS *Beagle*.

its way across the Atlantic toward South America, Darwin's worst fears were realized. To begin with, shipboard life was tougher than he expected, but worse yet, he tended to get seasick! But he made the best of it, sometimes spending days in his bunk, as the boat continued relentlessly, captained by tough, young Fitzroy, even then, one of the world's best navigators (Figure 1.6). At long last, the *Beagle* reached South America and headed down the coast on the first leg of its voyage—sailing past the coasts of Brazil and Argentina, weaving through the terrible pounding gales of Cape Horn (Figure 1.7), and finally turning northward along the desolate coasts of Chile and Peru.

Fortunately, there were periods of respite when the *Beagle* dropped anchor to put foraging parties ashore. Darwin wasted no time getting to land. After resting a bit and regaining his land legs, he was irresistably drawn deeper into these new places—places that harbored all manner of new and fascinating things. Darwin took copious notes on everything he saw and he brought back to the ship all manner of things, much to the amusement and occasional dismay of the crew. Since the ship sometimes remained anchored for months, he ventured far inland into the wild South American terrain. Darwin, an excellent rider himself, soon developed the greatest respect for the riding skills of the rugged gauchos who often accompanied him.

To understand just what this adventure meant to Charles Darwin (and would later mean to the world), we must stop at this point to consider the state of scientific thinking at that time. What was known of the biological world? What prejudices or beliefs did Darwin have? More important, what seeds of ideas? What hunches?

THE HISTORY OF AN IDEA

FIGURE 1.7
The infamous Cape Horn, indicating the almost perpetual violence of the place.

New scientific ideas have appeared throughout human history. Although many of these ideas have been erroneous, they often provided at least a framework for the expansion of our knowledge. Nevertheless, new ideas—and even new facts—have not always been met with enthusiasm. In the best of times, when an assumption did not fit what was known about the real world, the assumption was discarded. The notions that remained, then, were likely to be based to some extent on the available facts, and as new information came to light, our body of scientific knowledge would be expanded.

It would be satisfying somehow to say that scientific knowledge progressed stepwise from ancient times until the present, knowledge and understanding accumulating all the while until we reached the modern crescendo of scientific expansion, all to the good of humankind. Alas, this is not the case.

As far back as the early civilizations of Babylonia, Egypt, and Greece, scholars were trying to figure out the nature of the world in which they lived (Figure 1.8). They looked for answers in the heavens, on the earth, and in scripture and philosophy. And they debated endlessly over the various "truths" they held. These were, indeed, curious and intellectually vigorous people. However, after the fall of these ancient civilizations, scientific curiosity steadily declined. Then, for a period of more than ten centuries, between A.D. 200 and 1200, there were virtually no scientific advances at all. In fact, much of what had been known was forgotten. Although the Greek mathematician Eratosthenes had calculated the circumference of the earth to within 50 miles more than 200 years before the birth of Christ, in 1492 Columbus had trouble convincing anyone except a few Moorish astronomers that the world was round!

During these centuries, the writings of a few ancient scholars were preserved in dusty monasteries where they were dutifully pondered and transcribed from one parchment to another. But instead of active investigation to verify statements or ideas, people increasingly turned to religious authority for absolute answers, seeking some sort of order in a world that was beyond their understanding. As a result, science and religion became hopelessly intertwined, so the "scientific" statements that were otherwise testable became religious doctrine and were, therefore, not to be questioned. Thus, by a twist of fate and politics, the church became, on one hand, the seat of higher learning and, on the other hand, a formidable opponent of new ideas. The Polish astronomer Copernicus (1473–1543), who voiced the heretical theory that the earth was not the center of the universe, escaped retribution by dying shortly after his work was published. Later, when the great Galileo (1564–1642) produced detailed evidence that Earth did indeed revolve around the sun, his writings were banned in Rome and he was summoned before the Inquisition (Figure 1.9) and forced to recant his belief in the Copernican theory. (The Catholic Church recently forgave Galileo.)

FIGURE 1.8

The ancient scholars tried to discover the nature of the world with very little information to go on.

FIGURE 1.9

Galileo was persecuted by the Church for relying on evidence in the formation of his opinions.

Nevertheless, as the world became more predictable, and people began to feel less at the mercy of magical forces and unseen beings, there was a great surge of scientific thinking. In fact, the very year Galileo died marked the birth of one of the greatest scientists of all time—Isaac Newton (1642–1727). By the age of twenty-four, Newton had already formulated the idea of universal gravitation, and in 1685, he presented a set of carefully structured laws of motion that were to revolutionize the physical sciences.

THE BEGINNINGS OF BIOLOGY

Until the eighteenth century, science was largely limited to topics dealing with the inanimate, such as mathematics, astronomy, and physics. The study of living things was largely exempt from such investigation because of the philosophical and religious influences of that time. After all, it was one thing to search for the physical laws that described salts, stones, or stars, but quite another to probe at the essence of life. Implicit in the reverence for life was the notion that since people are living things, life must surely have some special purpose, some grand design. People simply weren't ready to see themselves as just another physical phenomenon.

In addition, most scientists of that day believed that all species, or kinds, of living things were created in their present form—in other words, that they had not changed during their time on earth. This was certainly the view of the Swedish botanist Carl von Linné (1707–1778), who

FIGURE 1.10

The Swedish naturalist, Carolus Linnaeus, the founder of the scientific method of naming living things. He lived and worked in the century before Darwin. Here, also, are two plant species named by Linnaeus: (a) Arizona century plant (*Agave americana*); (b) Ox-eye daisies (*Chrysanthemum levcanthemun*).

(a)

(b)

devised a system of classification for all living organisms, naming them in Latin. In his fondness for Latin, he even called himself Carolus Linnaeus (Figure 1.10).

Although few questioned the Church's dogma on creation, one small departure from this idea was suggested in 1753 by Linnaeus's French contemporary, George-Louis Leclerc de Buffon (1707–1788). In 1753, Buffon proposed that, in addition to those animals that had originated in the creation, there were also lesser families "conceived by Nature and produced by Time." He explained that changes of this kind were the result of imperfections in the Creator's expression of the ideal.

Interestingly, a decade later, someone else joined the doubters of the fixity of species. This was Erasmus Darwin (1731–1802), the grandfather of Charles. Erasmus was a peculiar fellow, not only a physician but an amateur naturalist who wrote about botany and zoology, often in rhyme. In his ramblings, he referred almost incidentally to certain relationships among animals that, strangely, heralded a number of important ideas that would later be embraced by Charles. Among these were the importance of competition in the formation of species, the effect of environment on changes in species, and the heritability of these changes. As you might expect, there has been a great deal of speculation on the extent to which Charles may have been influenced by Erasmus. However, the influence may not have been very great because Charles apparently never had much use for the musings of his grandfather.

Other people of that era were also beginning to toy with the notion of the changeability of species and the heritability of those changes. In France, Jean Baptiste de Lamarck (1744–1829), a protégé of Buffon's, boldly suggested that not only had one species given rise to another, but that humans themselves had arisen from other species (Figure 1.11). A passionate classifier, Lamarck held to the old notion that every organism has its position on the "scale of nature"—with humans, of course, firmly at the top, thereby revealed as the highest form of life. Lamarck also observed that the fossil animals found in older layers of rock seemed to be somewhat simpler than those in more recently deposited rock. This difference suggested to him that the older ones had gradually given rise to the more recent ones (or become "higher"). He couldn't immediately account for how such changes might have arisen, but he finally surmised that there was some "force of life" that caused an organism to generate new structures or organs to meet its biological needs. Once formed, such structures continued to develop through use, and their development in the parents was inherited by the offspring. In this way, he said, the structure became perfected in succeeding generations.

FIGURE 1.11

Today, biologists often associate the name Lamarck with error. Lamarck's idea of evolution, we know, *was* wrong, but we must keep in mind that, in Darwin's time, he was a major intellectual force.

Lamarck's most famous example of such change was his explanation of how the giraffe developed a long neck. He maintained that the long neck evolved as giraffes of each generation stretched their necks in an effort to reach the topmost branches of trees. He argued that this effort altered the animals' hereditary materials so that a longer neck was passed along to the offspring (Figure 1.12). This example did not bring him great respect. In fact, it brought guffaws of derision and even today it is cited as a classic case of scientific error.

Whereas Lamarck's arguments did little to persuade his lecture audiences, he did create some lively discussions in intellectual circles. (Certain

FIGURE 1.12

Lamarck believed that acquired traits were inherited. For example, he suggested that giraffes, by continually stretching for higher leaves, would tend to have offspring with yet longer necks. In time, he explained, all giraffes would have long necks, as we see in the column at the left.

Darwin, on the other hand, suggested that from among any population of giraffes, some are born with longer necks than others and that these would be able to reach more leaves. Better-fed animals would then most likely be able to rear their young, and thus longer necks would come to predominate in subsequent generations, as we see in the second column.

Lamarck's explanation **Darwin's explanation**

recent findings have revived the discussions in modern scientific circles.) Nineteenth century society at large, however, held to the firm conviction that each form of life had arisen through special creation. The matter seemed settled. The English were more interested in discussing the French Revolution. They were still discussing it when Charles Darwin was born.

TIME AND THE INTELLECTUAL MILIEU

The intellectual climate of England was far more conservative than that in France when Charles Darwin was born (1809). The English had been horrified by the brutality of the French Revolution, and any ideas held by the "French atheists" were usually either dismissed out of hand or viewed with extreme suspicion. Consequently, the church continued to hold strong sway over the sciences in England.

But as everyone knows, the English are an irreverent lot, and even a small group of Englishmen is likely to have at least *one* rebellious soul. And among the British scientific community there were undoubtedly those who simply didn't care *what* the French were doing, and who, furthermore, believed that life changed on a changing planet. It has even been suggested that this sentiment was by now so strong that it set the stage for Darwin's theory, that his idea could not fail in such an environment, that the time was right to publish (Figure 1.13).

FIGURE 1.13

Interior of the Old British Museum, Montague House, where some of the most influential scientists worked. It was into this environment that Darwin introduced his controversial ideas in the *Origin of Species*.

THE ORIGIN OF SPECIES

BY MEANS OF NATURAL SELECTION,

OR THE

PRESERVATION OF FAVOURED RACES IN THE STRUGGLE FOR LIFE.

BY CHARLES DARWIN, M.A.,

LONDON:
JOHN MURRAY, ALBEMARLE STREET.
1859.

However, this was apparently not the case. In fact, even after Darwin had published *Origin of Species* and his revolutionary idea was catching on, he observed:

> It has sometimes been said that the success of the *Origin* proved that the subject was in the air, or that men's minds were prepared for it. I do not think that this is strictly true, for I occasionally sounded out a few naturalists, and never happened to come across a single one who seemed to doubt about the permanence of species. . . . I tried once or twice to explain to able men what I meant by Natural Selection, but signally failed.

THE DEVELOPMENT OF DARWIN'S IDEA OF EVOLUTION

When Charles Darwin set out on the voyage of the *Beagle*, he had no quarrel with the prevailing notion that life had originated through special creation and that species were fixed in form. Furthermore, he was aware that many scientists felt that the goal of science should be twofold: first, to discover how nature worked and, second, to use the findings to demonstrate the wisdom of the Creator. Darwin hoped to join in the parade of scientists and clergy as they marched arm in arm to produce a better world. Instead, he was to reluctantly whistle the parade to a stop.

The Impact of Lyell

FIGURE 1.14
Charles Lyell, perhaps the greatest geologist of Darwin's time and a close friend of Charles's. Lyell said that there was time for species to have changed.

Darwin's questions about life did not leave him alone among scientists. The physical sciences were less hampered by religious tenets, and there were signs of rumblings and stirrings in these disciplines, disturbances that would one day make it easier for some to fall into step with Darwin as he slowly marched away from the parade. Among the physical scientists who were also marching to a different drummer and developing bold new ideas about the earth was Charles Lyell (1797–1875) himself only a few years older than Darwin (Figure 1.14). Lyell had set forth many of his new ideas in his book *Principles of Geology,* the first volume of which was published before the *Beagle* set sail. Darwin had acquired a copy and had asked to have the second volume sent to him en route. Lyell (who was to become a good friend of Darwin) had some rather startling things to say about the physical evolution of the earth. He said that the world was much older than anyone had imagined; that over long periods of time, continents and mountains rose slowly out of the sea; and that they just as slowly subsided again or were washed away. Most importantly, Lyell claimed that the very forces that had so changed the earth in the past were still at work and that the world was still changing.

Darwin's own observations of South American geology seemed to confirm Lyell's position at every hand. In his adventurous climbing of the Andes, he had found fossil clam shells at 10,000 feet. Below them, near an ancient seashore at 8000 feet, he found a petrified pine forest that had clearly once lain beneath the sea because it, too, was interspersed with seashells. In fact, the *Beagle* had arrived in Peru just after a strong local earthquake had destroyed several cities, in some places *raising the ground level by two feet*. The earth had changed and clearly was still changing.

Darwin was excited by his developing idea, but he kept the most revolutionary of his thoughts to himself because he was sometimes uneasy with his ideas and often full of doubts. After all, he had studied for the ministry and had believed in the literal truth of the Bible.

Puzzling Change and Variation

While Darwin continued to divide his time aboard the *Beagle* between reading and hanging over the rail, the heavy, wooden boat had creaked and groaned across the Atlantic and was now making its way up the west coast of South America.

Darwin, of course, darted ashore at every chance, watching, collecting, probing, and taking notes. He was particularly interested in the South American plant and animal life and the fossil beds he found there. Quite early in the voyage he was struck by how living things could vary so markedly from one place to the next. He collected shells from the Atlantic shore and found that they were not like those picked up on the Pacific beaches. He wrote about how birds and mammals differed from one place to the next. He noted that in some cases, species changed gradually, from one place to another, one type giving way to another almost unnoticeably. But in other cases, one kind of organism would suddenly disappear, another having appeared in its place. This natural variation that Darwin noticed would play an integral role in the theory that Darwin would develop—after a momentous visit to a small group of islands.

The Galapagos Islands

When the *Beagle* finally left the shores of South America, an impatient Captain Fitzroy, concerned with completing charts, measuring harbors, and preparing the way for British commerce, set the sails of his sturdy craft (a dog to windward) for a straight run to the Galapagos, a chain of islands lying about 580 miles off the coast of Ecuador. It was these remarkable islands that history would most closely associate with an unsuspecting Darwin.

When the anchor clattered into the shallow waters of St. Stephen's harbor of the island the British called Chatham, Darwin scrambled ashore as usual, but as soon as he had looked around, he was almost ready to leave. Chatham was rough, crude, and barren, and he didn't like it. But as he began to explore the place, he encountered a very strange, and even fascinating, assortment of animals. "A little world in itself," he wrote, "with inhabitants such as are found nowhere else." There were lizards three feet long, grazing on seaweed beneath the turbulent sea—in Darwin's words, they were "imps of darkness, black as the porous rocks over which they crawl." And he saw the giant tortoises that had for years been captured by seafarers to be stacked upside down on the decks of their boats where the hardy beasts could somehow survive for months, providing fresh meat on the long voyages.

Darwin was more interested in the plants than the animals and soon spent most of his time "botanizing" over the dry and barren islands. He

FIGURE 1.15

Darwin's finches include the peculiar woodpecker finch *(Camarhynchus pallidus)* that uses twigs, much as a woodpecker uses its beak, to extract insects.

was struck by the strange animals of the islands, however, and collected not only the marine lizards, but also their land-bound cousins further inland as well. He also collected various kinds of birds, many of which, he was convinced, were undescribed species. Among these were a motley group, that were not unlike the finches he had collected on the South American mainland (Figure 1.15).

Darwin later regretted that he had been storing all the bird species collected from the islands together on the boat. He came to recognize the importance of separating them according to where they were taken one day while he was examining a few of the finches. He noticed that two of them taken from different islands differed in the size and shape of their bills. This struck him as odd and possibly significant. The importance of his observation was driven home as he was walking the four miles to the settlement of political outcasts, banished from Ecuador, who had been sent to Charles Island. His companion that day was the acting British governor of the island who informed Darwin that he was able to tell from which island any of the tortoises came. He explained that they differed, for example, in the size and shape of the carapace (shell) and in the length of their extremities. Darwin wondered if each island was somehow producing its own forms of creatures, and from that day on, he carefully separated his collections from each island. This was to prove a critical decision once he was home in England. Years later, Darwin had his finch collections examined by a British specialist, and it was decided that there were 13 species, differing primarily in size and shape of their beaks. Darwin surmised that these birds must have come originally from the South American mainland, since the volcanic islands of the Galapagos would have been formed later than the continent. But why were these birds so different from those on the mainland, and why did the assortment on each island differ so much from one to the next? Years would pass before Darwin would conclude that the birds were descended from mainland stock that had accidentally been blown out to the islands, and that the various islands themselves had changed the populations over time.

As the voyage continued, Darwin continued not only his collecting, but also his questioning of how things came to be. His letters and observations were reaching England and had the scientific community anxiously awaiting more of his findings. In fact, he was told that upon his return, he would be invited to take his place among the British scientific establishment. The letters waiting at various ports of call encouraged him and sent him bounding into the hills of each new land, his hammer joyously ringing against the rocks.

When Darwin agreed to go on the expedition, he had expected to be gone about two years, but five years were to pass before his return. When the *Beagle* finally made its way back to England, Darwin was greeted enthusiastically by his family, the scientific community, and his dog. His reception among the scientists of the day was a warm one, and immediately the questions began. What had he seen? What had he brought back? A new phase of work had just begun. (Because of the focus on Darwin and British science, it sometimes seems that the rest of the world was on hold during this time, but Table 1.1 shows this was not the case.)

The Impact of Malthus

Darwin was grateful to be back among his friends, his new colleagues, and his books. After a flurry of activity, he married his cousin, Emma Wedgewood, and retreated to the country and began to enjoy the quiet mornings when he could find time to work.

In his reading, he came across an old essay by the Reverend Thomas Malthus (1766–1834) that was probably the first clear warning of the dangers of human overpopulation. In the essay, which appeared in 1798, Malthus pointed out that populations tended to increase in a geometric (exponential) progression, and that if humans continued to reproduce at the same rate, they would inevitably outstrip their food supply and create a teeming world full of "misery and vice" (Figure 1.16). Malthus's message

FIGURE 1.16

The dilemma of the English poor helped prompt Malthus's conclusions.

TABLE 1.1
Historical Milestones

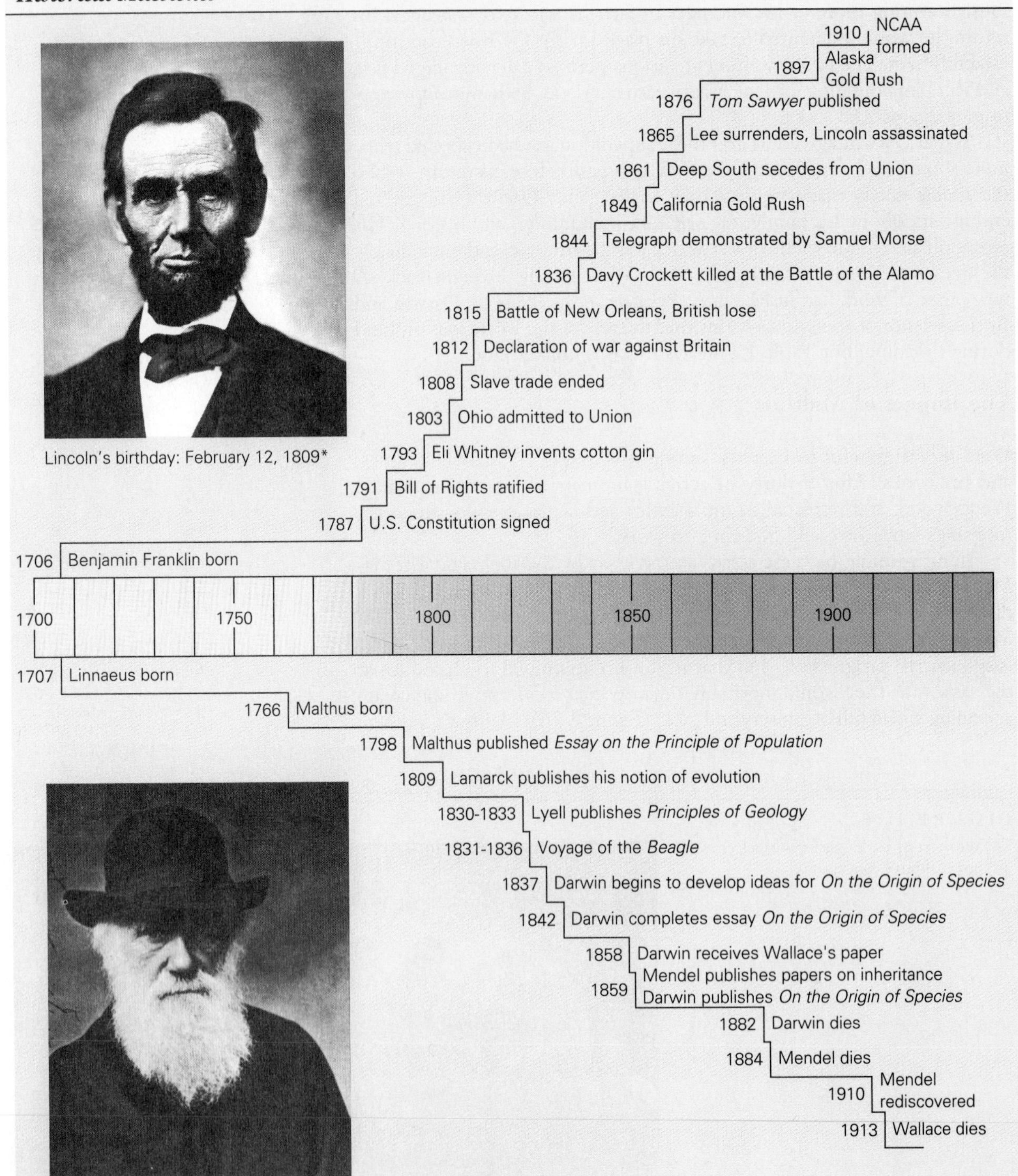

Lincoln's birthday: February 12, 1809*

Darwin's birthday: February 12, 1809

*You might find it a remarkable coincidence that both men were born on the same day.

The Galapagos Islands

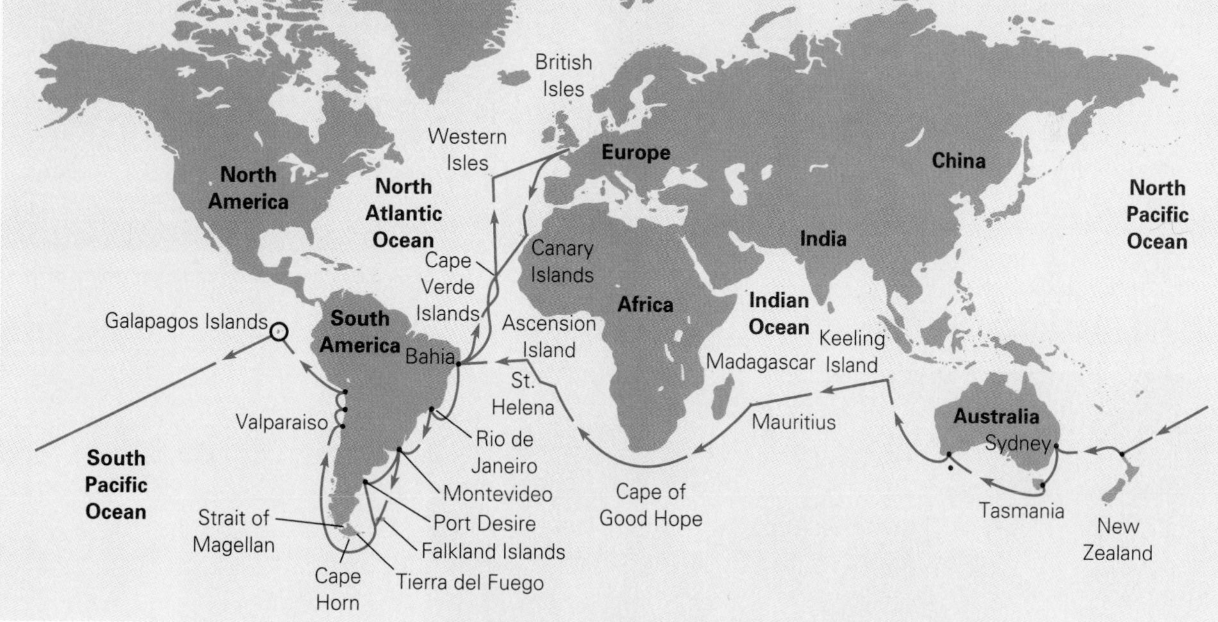

On Captain Fitzroy's charts, the Galapagos were only a group of islands to be visited as the *Beagle* left the coast of South America. Years later, his friendship with Charles Darwin would be severely strained largely because of what young Darwin would find in this forsaken place. Darwin at first thought the islands were strange and repugnant. But as he roamed about, he discovered places of indescribable beauty. He found plants and animals unlike those found anywhere else. More important, though, he found creatures somewhat like those he had seen on the mainland; a finding that would one day help launch his grand theory.

The volcanic origins of some of the islands are revealed by calderas, such as this one on Bartolome Island. A variety of life forms live in such sheltered places, many in close association with the mineral lakes found there.

The highland zone of Santa Cruz Island bears the heavy, varied vegetation typical of humid, tropical areas.

At the lowest level of the islands' vegetational zones are the remarkable mangroves, rooted in seawater and supporting a great variety of life.

Galapagos tortoises from different islands have distinctive appearances. Generally, the higher the arch over the neck, the farther the great reptile must reach for its favorite food, the *Opuntia* cactus. They are the islands' only native grazer.

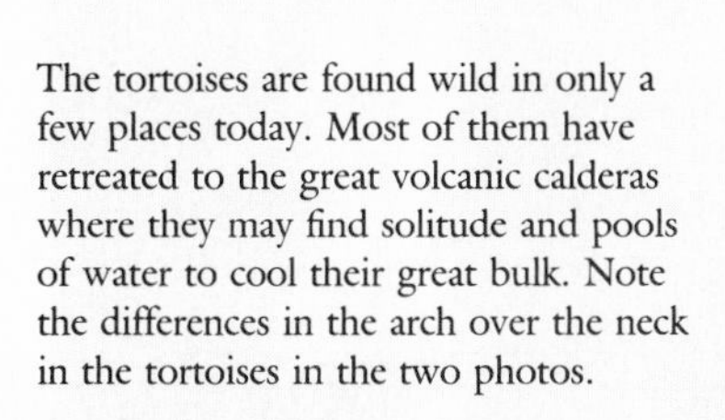

The tortoises are found wild in only a few places today. Most of them have retreated to the great volcanic calderas where they may find solitude and pools of water to cool their great bulk. Note the differences in the arch over the neck in the tortoises in the two photos.

Sally Lightfoot crabs lend splashes of brilliant color to the dull volcanic coastlines.

Among the most bizarre creatures on the islands are the marine iguanas, large lizards that graze on the seaweed just offshore (right). Their rarer cousins, the land iguanas (below), are found farther inland.

The waved albatross, one of the world's greatest fliers with a wingspan of 2.5 meters, nests only on Hood Island in the Galapagos.

The flightless cormorant has followed the evolutionary pattern of many island forms; it has lost the ability to fly. Its feathers are wettable to facilitate diving, and after each fishing trip, it must stand and dry itself. It has been suggested that flying increases the chances of being blown away from the safety of islands.

Whereas some birds nest only at certain times of the year, the blue-footed booby will rear a brood whenever conditions are good. If the fish supply should shift away from the islands, however, the blue-footed boobies do not hesitate to abandon eggs and young to follow the food.

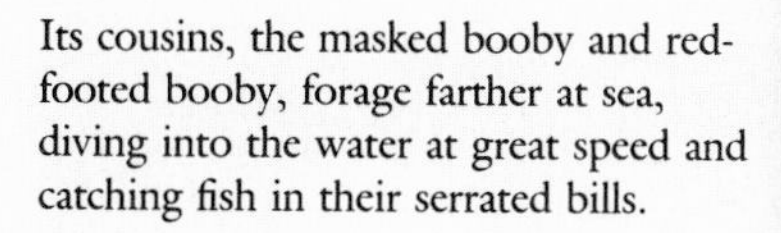

Its cousins, the masked booby and red-footed booby, forage farther at sea, diving into the water at great speed and catching fish in their serrated bills.

The Galapagos sea lions (above) are quite approachable and even friendly (except for protective bulls with harems). Fur seals (left), on the other hand, have been relentlessly hunted and are now quite skittish.

A male great frigate bird attempts to entice a white-throated female from a flock wheeling above the madly displaying males as a group of them inflate their red throat pouches and coo enticingly.

Swallowtail gulls breed in great numbers on volcanic sea cliffs of the Galapagos. They are the only nocturnal gull, catching squid and fish from the darkened sea.

The Galapagos hawk is an unusual species, with many traits of both hawks and vultures.

The Galapagos penguins are the northernmost members of the group. It is astonishing to watch the speed with which they move underwater in search of fish.

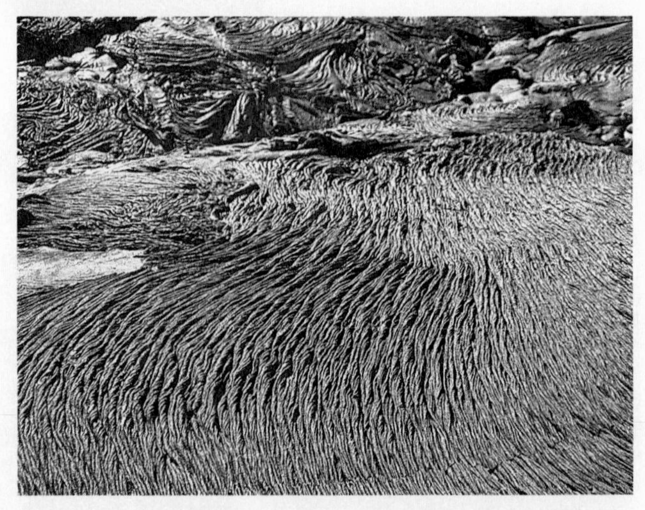

Areas of the islands are covered with ropy pahoehoe lava, or with sharp and spiny lava beds that are virtually impassable. The remarkable origins of these islands are largely responsible for the startling array of life there.

may have been theological, but Darwin applied the idea to his own work and concluded that species have a high reproductive potential, but that not all individuals reproduce, because of differences in their survival abilities. The idea was that populations are kept in check partly because not all animals survive long enough to reproduce.

Malthus's paper set Darwin to thinking. He calculated that even a pair of elephants, notoriously slow breeders, could produce 19 million progeny in only 750 years. Yet it seemed that through the years, the number of elephants on the earth stayed about the same. Something was obviously interfering with their reproductive output. But was that "something" exerting an equal effect on all elephants, or did individuals differ in their ability to reproduce? Were some less successful at leaving offspring? And if this were the case, Darwin wondered what factors determined which ones were to be successful.

NATURAL SELECTION

Darwin's answer came to him in part because of his background as a country gentleman. He was familiar with the principles of **artificial selection** (Figure 1.17); he knew that through careful selection of animals

FIGURE 1.17

Breeding programs based on artificial selection have produced plants and animals with extreme traits. Here, the obese mouse is two-and-a-half times heavier than normal mice. The rhino mouse (below) first develops a normal coat, but, after two weeks, gradually goes bald and wrinkled. Artificial selection has produced many kinds of flowers, crop plants, livestock, and other animals of particular interest to humans.

for mating, breeders were able to accentuate desired characteristics in the offspring. Breeders, then, determine the reproductive success of individuals. For example, by mating only the offspring of the greatest milk producers, breeders could develop high-yield dairy cattle. And by breeding only the offspring of good laying hens, they could eventually produce hens that were veritable egg-laying machines. The results of such artificial selection could be seen in only a few generations.

Darwin envisioned some sort of process of **natural selection** in which nature determines the reproductive success of individuals. It was analogous to the artificial selection imposed by the breeders. He remembered the variation he had observed in populations of plants and animals during his journey aboard the *Beagle*. Some of those variations, he reasoned, would give their bearer a competitive edge and increase that individual's chance of surviving and reproducing. Thus, nature would select the best of the individuals for mating. Natural selection would be far less efficient than artificial selection since individuals with only *somewhat* less-desirable characteristics might be able to produce at least *some* offspring, and thus their traits would take longer to disappear from the population. (Of course, those individuals with traits totally out of keeping with their environment would leave no offspring at all; hence, those traits would more quickly disappear from the population.) On the other side of the coin, the traits of those individuals with some reproductive *advantage* could be expected to *increase* through the generations.

Eventually, Darwin would express his idea of natural selection in his book, the *Origin of Species*:

> How will the struggle for existence . . . act in regard to variation? Can the principle of selection, which we have seen is so potent in the hands of man, apply in nature? I think we shall see that it can act most effectively. . . . If such [variations] do occur, can we doubt (remembering that many more individuals are born than can possibly survive) that individuals having any advantage, however slight, over others, would have the best chance of surviving and of procreating their own kind? On the other hand, we may feel sure that any variation in the least degree injurious would be rigidly destroyed. This preservation of favourable variations and the rejection of injurious variations, I call Natural Selection. . . .

Briefly, natural selection involves (1) overproduction of offspring, (2) natural variation within a population, (3) limited resources, and the struggle for survival, and (4) selection by the environment for those with traits that enable the individual to survive and reproduce.

Natural selection, then, came to be defined as the process through which certain types of organisms are more reproductively successful than other types, thereby disproportionately passing along those traits that led to their success.

Natural selection could account for many of Darwin's observations, such as the variety among the Galapagos finches. Their basic similarity suggested that they had all descended from the same ancestral stock. Yet the species differed from each other in critical ways, such as in color, bill size, foraging behavior, and food choice. By applying the principles of natural selection, Darwin could see how each island would have presented the original colonizers with its own array of opportunities and threats, and from among the variable offspring of any population, those with the traits best suited to a specific environment would best thrive and reproduce. In time, each environment would have molded the birds into populations so distinct that each became a species, distinct from the others.

Since the notion of natural selection depended on inequalities among members of a population, the source of the variation, of the inequalities, had to be accounted for. Why, then, are the individuals different? What brought about their inequalities? Darwin proposed that such variations (inequalities) appear randomly—that no driving force, no direction, and no design are necessary. He decided that if some new variation provided an advantage that increased the reproductive output of its bearer, it would spread through future populations. This meant that if the long neck of the giraffe is inherited and is helpful in acquiring food, then the giraffes with longer necks will be better nourished, and thus will be more likely to have the energy to leave offspring. Among these offspring, he suggested, some will have longer necks than others, and these, in turn, will be more successful than their shorter-necked brothers and sisters.

Since long-necked giraffes would leave more offspring, the result would be a general tendency for any generation to be composed of animals with longer necks than those of any preceding generation. (The differences in Lamarck's and Darwin's explanations of long necks in giraffes are shown in Figure 1.12.)

It is important to realize that Darwin developed his idea not only in the face of withering opposition, but also without hard proof. There was no experimental evidence he could offer, and to make things worse, he knew almost nothing about the field we now call genetics. If he had understood what was going on in genetics across the Channel in the garden of a monastery (see Chapter 8), he would have been able to save himself much grief. Because of such problems, he was reluctant to present his ideas. He felt he should be able to explain the mechanism by which variation appeared.

Then, in 1858, something happened that prompted the thoughtful Darwin into action. He received an unfinished paper from a young biologist working in Indonesia, Alfred Russel Wallace (1823–1913) (Figure 1.18). Wallace wanted Darwin's opinion of the merit of the paper that was, in effect, a sketchy outline of the principles of natural selection. By this time, Darwin had already planned to present his theory of natural selection to the Linnaean Society of London. Startled by the letter, and at the urging of the geologist Lyell, the botanist Joseph Hooker, and the scientific philosopher Thomas Huxley, he began to hasten his work.

Hooker and Lyell agreed to assist Darwin, whose young son had just died, by editing and condensing an earlier paper of his with a letter he had written the American botanist, Asa Gray, describing his developing principle of natural selection. Although Darwin feared he might be "scooped" (in science, being first often means being foremost), he asked that Wallace be permitted to present his paper first and receive credit for the idea, rather than have anyone think he had behaved in a "paltry spirit." The outcome was that the papers were read at the same meeting, with Darwin's presented first, in keeping with his much more substantial evidence. The papers were presented in July and published in August 1858. Darwin then went furiously to work, putting aside his idea of a huge

FIGURE 1.18
Alfred Russel Wallace, an English adventurer and field biologist who arrived at many of the same conclusions that Darwin had and pushed Charles into publishing, developed his own thesis on evolution by natural selection.

monograph describing his theory. Instead, he completed an "abstract" of the idea: the *Origin of Species*, which was published in 1859. The first edition sold out on the first day.

Darwin's carefully formulated ideas were greeted with enthusiasm in some quarters, but, needless to say, the response was not universal. He was forced to defend his idea of "descent with modification" not only against scientists who demanded hard evidence, but also against the attacks of philosophers, theologians, and a general public who thought the idea was heretical.

One difficulty from the start was the issue of human evolution. Then—as today—when the principles of evolution were applied to other species, humans were bound to find themselves under this great explanatory umbrella as well. A lot of people resented being placed in the shade with the worms and moles and did not believe they shared the expectations of other creatures. Moreover, the acquisition of such noble traits as a bipedal gait, thumb and hand dexterity, and extensive learning capacity was simply not to be attributed to a mere natural process. In the human desire to be set apart from other creatures, great ideological clashes began.

The battle grew and was soon full blown. Darwin himself was poorly equipped for such a fight. He had fallen ill upon returning to England after his journey, and he never recovered his health. He had become a dedicated family man, spending a great deal of time with his wife and children. He continued his experiments and observations, but because of weakness, he generally could only work from mid-morning until noon (Figure 1.19). His infirmities have been diagnosed time and again by

FIGURE 1.19

(a) An aged Darwin, often in ill health, about a year before his death. (b) Darwin's study at Down House.

(a)

(b)

medical historians who first suggested that they were psychosomatic, but some then suggested that Darwin may have contracted Chagas' disease by once allowing himself, as an experiment, to be bitten by a benchuga bug. This idea has now been largely discounted. So the source of Darwin's infirmity remains a puzzle to this day. (Darwin and Wallace had become good friends by this time and, as Wallace grew increasingly depauperate, Darwin arranged a government stipend for him; Wallace would later serve as Darwin's pallbearer.)

While Darwin was developing his theory, he became plagued by anxiety and self-doubt. He and Emma lived a rather quiet and somewhat reclusive life at Down House in the country, and Darwin became even more withdrawn when the noisy debate over natural selection started. However, he had formidable defenders. Many of the best minds of the time leaped to the defense of the grand idea—brilliant, hard-nosed, and combative souls who savored the taste of intellectual battle. Among his most brilliant defenders was his long-time friend, the great debater Thomas Huxley (Figure 1.20). (When Lord Wilburforce, a clergyman, facetiously asked Huxley in a debate if he was related to the apes on his father's or his mother's side, Huxley is reported to have muttered, "The Lord hath delivered him into my hands." He then won applause by saying that he would rather be related to an ape than to a man who refused to use his God-given powers of reason.) Of course, Darwin's defenders did not join the fray empty-handed. Here, after all, was a unifying concept, one that made sense of it all. It accounted for the observations. It was not to be rejected on any basis other than a better explanation. And there was none.

FIGURE 1.20
Thomas Huxley, a bright, energetic, and aggressive friend of Darwin's, who proved to be one of his greatest defenders.

SUMMARY

1. Charles Darwin served as a naturalist during a five-year voyage from England to many places around the world, including South America, aboard the H.M.S. *Beagle*. He collected specimens of plants and animals, and wondered about the variability among organisms he observed.
2. Darwin lived when the intellectual environment was dominated by the Church, which discouraged experimentation and resisted ideas that were in conflict with religious doctrine.
3. At this time, most scientists believed that God created all types of living organisms in their present form. A few scientists, such as Buffon, Erasmus Darwin (Charles's grandfather), and Lamarck suggested that species change through time. Lamarck believed that, when necessary, some "force of life" allowed organisms to generate new structures that could be developed through use and passed to offspring.
4. The ideas of Lyell and of Malthus were important to the development of Darwin's thoughts on how species might change over time. Lyell argued that the earth was much older than was thought. Therefore, there would have been time for the changes in living organisms that Darwin would come to suggest had occurred.
5. Malthus pointed out how quickly populations can grow. Darwin realized that population size remains under control only when some individuals do not survive and reproduce.

6. Darwin put Malthus's ideas together with what he knew about artificial selection and proposed the process of natural selection, a process in which nature selects the individuals who reproduce, leaving offspring possessing their parents' traits. As a result of natural selection, the traits of the successful breeders increase through successive generations, but the traits of less successful breeders become less common. He believed the variation appeared through random processes. Those traits of a species that allow an individual to leave more offspring than others would spread through future generations. Natural selection could account for many of Darwin's observations, including the variety among the Galapagos finches.
7. Darwin didn't publish his ideas until he realized that Alfred Russel Wallace had similar thoughts on natural selection. The papers of Darwin and Wallace were presented at the same scientific meeting in 1858, and Darwin's book, the *Origin of Species*, was published in 1859. It met with strong protest as well as grateful acceptance.

KEY TERMS

artificial selection (25)
natural selection (26)

FOR FURTHER THOUGHT

1. Describe how Darwin's *Origin of Species* might have been reviewed if it were introduced in the fifteenth century. Would it be accepted or rejected? Why?
2. How are the processes of artificial selection and natural selection similar? Which is the more efficient process?

FOR REVIEW

True or false?

1. ___ Charles Darwin set out as a naturalist on the *Beagle* in order to prove his theory of evolution.
2. ___ The *Origin of Species* was an abstract that supported the theory of fixity of species.
3. ___ Alfred Russel Wallace was a strong opponent of Charles Darwin and disputed his theory of species changeability.

Fill in the blank.

4. The works of ___ suggested to Darwin that the reproductive potential of a species is influenced by their survival abilities.
5. ___ is a process whereby the environment selects the animals that will reproduce.

Short answer

6. Describe the contribution of Carolus Linnaeus to the biological sciences.
7. Did Darwin's theory of evolution concur or conflict with the opinions of those in his era? Explain.

Choose the best answer.

8. In which of the following time periods was the pursuit of biological investigation most prevalent?
   A. A.D. 200 through 1200
   B. fourteenth century
   C. sixteenth century
   D. eighteenth century
9. The idea that great spans of time were available in which the earth and its inhabitants could change was suggested to Darwin by:
   A. The Rev. Thomas Malthus
   B. Charles Lyell
   C. Jean Baptiste de Lamarck
   D. Joseph Hooker
10. Which of the following was an erroneous concept formulated by Lamarck?
   A. Populations tend to increase in a geometric progression.
   B. The environment effects changes in species.
   C. The experiences of an individual can somehow change its hereditary material.
   D. Variations in populations occur randomly.

CHAPTER 2

Scientists and Their Science

Overview

Inductive and Deductive Reasoning

The Scientist as Skeptic

Defining Life

What Biologists Do for a Living

Science and Social Responsibility

Objectives

After reading this chapter you should be able to:

- Explain how deductive and inductive reasoning are employed to formulate a hypothesis.
- Describe the scientific method and differentiate between hypothesis and theory.
- Explain how a controlled experiment is used to test a hypothesis.
- List six features that characterize living things.
- Describe how the concepts of teleology and phenomenology affect the observations of scientists.
- Discuss some special social concerns that confront today's scientists.

Charles Darwin was well aware that his description of natural selection would offend a great many people, but he did not expect criticism of his methods. Yet it was his methods that drew some of the harshest attacks.

Darwin had both the fortune and misfortune to set forth his thoughts at a time when intellectual England had become interested in the philosophy of science, particularly in deciding how science should be done. The question was, what is the best way to gather valid information? There were two major schools of thought on the subject.

INDUCTIVE AND DEDUCTIVE REASONING

According to one school of thought, science should be done inductively. **Inductive reasoning,** in fact, had been promoted by none other than Isaac Newton (Figure 2.1). To use inductive reasoning, one first gathers simple, empirical data and, from the data, arrives at a generalization. Ideally, there is no goal, no premise, no hunch, and no preconception in the mind of the researcher. Inductive reasoning works because finally, from the sheer weight of unprejudiced evidence, some general statement emerges.

But there are other ideas about how good science should proceed. The Dutchman C. Huygens, a physicist and lens grinder (who set forth the wave theory of light in opposition to Newton's particle theory), was a firm proponent of **deductive reasoning.** In deductive reasoning, observation or insight should suggest some general idea from which specific statements can be *derived*. These specific statements can then be tested. In deductive reasoning, then, the generality is arrived at by some observation or insight, and the specific cases suggested by the general idea become the subject of experiments.

Scientists these days feel free to get at the truth by either method, and most would probably agree that perhaps the best science is done by employing both methods. For example, if a body of facts suggests some general principle (inductive), then specific propositions can be drawn from that principle (deductive) and tested. However, most scientists, from Darwin's day to this, probably rely more strongly on inductive evidence in developing scientific principles.

FIGURE 2.1

Isaac Newton, one of the truly important figures in the history of science. Newtonian physics was to dominate the world of the physical sciences until the twentieth century. It was Newton who, in 1687, provided final and irrefutable evidence for the Copernican doctrine of the solar system, using primarily inductive reasoning.

Science Trends and Shifting Logic

A problem in modern science is that it tends to be a bit trendy. In fact, this has been true for a long time. In Darwin's era, to be "scientific" meant to operate like the physicists, with fixed laws and great precision. In fact, to be sure that he was conforming with the rules of good science, Darwin sent a draft of his paper on natural selection to one of the great physical astronomers of the time, John Hershel. He then wrote to Lyell that he was most anxious to have Hershel's opinion. It was not long in coming. Darwin later wrote Lyell:

> I have heard, by a round-about channel, that Hershel says my book is the law of higgledy-piggledy. What this exactly means I do not know, but it is evidently very contemptuous. If true, this is a great blow and discouragement.

But later, Darwin was greatly pleased when another authority on how science is done, and one of the greatest philosophers of the time, John Stuart Mill, said that Darwin's reasoning was "in the most exact accordance with the strict principles of logic."

But which logic? In keeping with the times, Darwin tried to adhere to inductive reasoning, and he said as much when he referred to a "general truth" in a letter to Thomas Huxley:

> I have got fairly sick of hostile reviews. . . . I entirely agree with you, that the difficulties of my notions are terrific, yet having seen what all the reviews have said against me, I have far more confidence in the *general truth* of the doctrine than I formerly had.

In spite of Darwin's confidence in inductive reasoning, he, too, occasionally lapsed into philosophical inconsistencies. He wrote in a letter to Joseph Hooker (the first person in whom he had confided regarding his developing theory) that he looked upon "a strong tendency to generalize as an entire evil," but later admitted, "I cannot resist forming one [generalization] on every subject."

In fact, some of Darwin's most insightful work was done deductively. For example, in developing his compelling theory about how coral reefs are formed he once stated, "No other work of mine was begun in so deductive a spirit as this; for the whole theory was thought out on the west coast of South America before I had seen a true coral reef." Table 2.1 illustrates how the two methods might have been used in this case. The point, here, is that even the greatest scientists do not proceed by

TABLE 2.1
An Example of Inductive and Deductive Reasoning

INDUCTIVE REASONING

Observation: Coral atolls are usually composed of a circle of islands.

Observation: Coral atolls are formed from the deposits of living animals.

Observation: Animals without direct access to fresh seawater tend to die.

Observation: The interior of an atoll seems to be comprised of sunken coral.

Generality: Coral atolls are formed as the coral animals secrete deposits. The center, lacking nutrient-laden water, dies and sinks, leaving a ring that, in turn, breaks apart to form a circle of islands.

DEDUCTIVE REASONING

Generality: Coral atolls are formed as the coral animals secrete deposits. The center, lacking nutrient-laden water, dies and sinks, leaving a ring that, in turn, breaks apart to form a circle of islands.

Deduction: Coral atolls will have a sunken center.

Deduction: Coral animals need contact with fresh seawater.

Deduction: Seawater contains something that coral animals need.

Deduction: Coral atolls comprised of more nearly complete rings of land are probably recently formed.

cut-and-dried rules. They, too, may be unsure and at odds with themselves at times. Yet, even by using a variety of philosophical approaches to questions, and insisting on a certain leeway and elbow room, they eventually manage to get at the truth.

Hypothesis, Theory, and the Scientific Method

Darwin's theory of natural selection is occasionally criticized as being "only a theory." However, many critics seem to forget that *theory* in science is a relatively lofty position. What, then, does the term mean? Perhaps we can best describe a theory by showing how it is arrived at through the **scientific method.**

The scientific method has been notoriously difficult to define. Scientists use it every day, but most are at a loss to explain precisely what it is. So here, we will just say that the scientific method is the process of establishing facts. The key word here is *process*, and the process can be complex and fascinating, indeed, as it begins and ends with observations about the real world.

The initial observations usually reveal some problem, some question. Darwin, for example, observed the little Galapagos finches and wondered why they were so different from each other, yet all so similar to the mainland birds. (Most questions, by the way, stop right there, with the question. Scientists, though, by their nature, can't let things rest.)

What happens next is among the least understood and most critical aspects of science. It is a vaguely defined process and no one says much about it. It can happen while the scientist is mowing the lawn. It is probably best described as "mulling over." Scientists are always mulling over something.

If the process is to proceed, at some stage the scientist must make a **hypothesis** to explain the observation. A hypothesis is a tentative explanation of the observation that can be used to make predictions that can then be tested. (The tests are designed to test the predictions.) It is important, then, that the hypothesis leads to *testable predictions*.

Suppose the scientist suspects that a certain chemical has an influence on plant growth. This is the hypothesis. From the hypothesis, then, a prediction can be made. "If I add this chemical to the soil, the plant should respond in this way." Such thinking leads to the **experiment.** The experiment is a means to test the hypothesis.

The experiment must be set up so that there can be only one explanation for the observations to come. This usually involves the use of a **control**. A control is a procedure set up exactly like the experiment, except that the thing being tested by the experiment is left out. That sounds pretty convoluted, but the principle is a simple one. For example, if the scientist is trying to learn what effect some chemical has on plant growth, he or she needs to know that the effect is not due to something other than the chemical, such as handling the plant, or growing it under a certain kind of light, or whatever. So the experimenter sets up a second arrangement, called a *control*, just like the experiment except that, instead

ESSAY 2.1

The Controlled Experiment

Diver with right whale. The marked diving response of the whale also occurs as a vestigial behavior in humans.

Some evolutionists have argued that humans are descended from aquatic animals. One line of evidence is that we seem to have a vestigial *diving response*. To illustrate, if you submerge your face in a bowl of cold water you will not only gain the full attention of your little brother, but, in addition, your pulse rate is likely to drop. The slowed heart rate is associated with a decreased rate of metabolism. A diver with such a response could be expected to stay submerged longer.

Now, to test the reflex, the pulse of a subject could be taken normally, and then again with his or her face submerged. However, is it the water stimulating the receptors of the face that causes the reduced heart rate, or is the change due to something else? For example, could the bending over alone have caused the response? Controls can help answer the question in some cases. Here, a control might be to check the pulse while placing the face into an empty bowl. But how about the possibility that the lowered heart rate is due to either holding one's breath or holding one's breath while bending over? How could the experiment be altered to control these effects?

of adding the chemical to the soil, he or she adds plain water. Any changes in the experimental plant that do not appear in the control plant can be regarded as an effect of the chemical. The controlled experiment is reviewed in Essay 2.1.

Finally, the scientist may gain enough information to draw a **conclusion.** The conclusion may be tentative or firm, depending on the strength of the scientist's confidence. Any confidence the scientist has, though, may not be shared by other researchers. Their responses can vary from applause to hoots of derision, but someone is bound to repeat the experiment. If the hypothesis is supported, it may, in time, be elevated to a higher level and become a **theory,** which is a well-founded generalization supported by a considerable amount of evidence.

The theory itself, then, will point to yet other avenues of experimentation from which more conclusions will be generated, and in this way, science proceeds (Table 2.2).

The reception a theory receives may depend on the intellectual milieu into which it is introduced. The theory of evolution, for example, caused an initial shock in the Victorian world, but was assimilated by the intellectual community with surprising rapidity. The idea was accepted not only by intellectuals, but by much of the general populace as well. One possible reason for its widespread acceptance is that the deeply religious citizens of that era saw in evolution a preexisting program for change, a program authored by the deity.

The acceptance was not universal, of course. The theory had strong opposition then, even as now. There have always been those who pointed to the uncertainties in the evolutionary argument in an effort to deride the idea. This argument, however, is rather selectively applied. For example, no one has ever seen a hydrogen atom, and the behavior of hydrogen is indeed theoretical, but we can apply the theory and make water. The principles of evolution obviously strike stronger emotional chords than do the principles of chemistry.

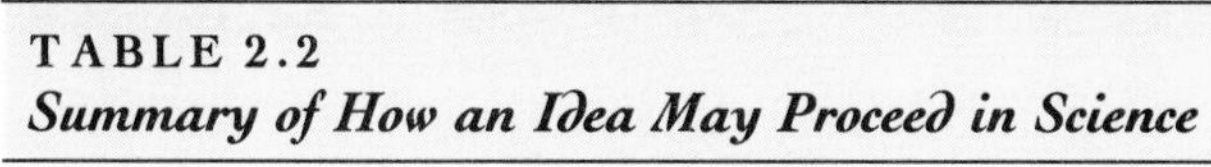

TABLE 2.2
Summary of How an Idea May Proceed in Science

Observation (or idea)
↓
Hypothesis ← Revision
↓
Experiment → Revision
↓
Conclusion

Scientists arrive at their conclusions in many ways, but one accepted route is through an observation or an idea leading to a hypothesis, which must then be tested through experimentation. The experiments may lead to the revision of the original hypothesis, which is then tested again by new experiments. Finally, the scientist is able to form a conclusion that supports or rejects the hypothesis.

FIGURE 2.2

Much of what we know in science is due to simple observation. In some cases, there is nothing else we can do, such as when we witness the birth of a volcanic island. Yet the information is valuable. Can you think of any problems using information derived from pure observation? In some cases, however, the observer can interact with the observed. Dian Fossey told us a great deal about mountain gorillas by being able to move among them. The arrogant, tough, and dedicated Fossey was instrumental in protecting these shy beasts from poachers (who sell gorilla hands to wealthy Europeans to be used for ashtrays) until she was killed by an assailant in her mountain cabin.

We should also add that scientific knowledge can be expanded not only through hypotheses and experimentation, but also by simple observation. For example, consider how much we learn from witnessing the birth of a volcano, from monitoring the results of an exploding star, or from simply watching wild animals (Figure 2.2).

THE SCIENTIST AS SKEPTIC

Scientists unrepentantly demand hard evidence. However, their rigor does not mean that they are eminently rational people, pristine, pure of heart, unemotional, and unfettered by personal prejudice. I certainly don't know any scientist who is like that, but, fortunately, such sterling character is not necessary. Important contributions have stemmed from hunches or personal prejudices, or from the work of someone with an ax to grind. In any case, for whatever reason those data are generated, the system works simply because most good scientists are skeptics. They are disbelievers at heart (and many love their hard-nosed image). Furthermore, because of their skepticism, any idea, hypothesis, or experiment is certain to be attacked by someone. You can rest assured that any new idea will be examined, the alternative hypotheses considered and weighed, and the experiments repeated. Even then, the conclusions may be met by a variety of responses from cheers to total rejection.

Because so many scientists love to flex their minds in what they may regard as the ultimate game (albeit an important one), we often end up with good science. The system works, and we are usually left with ideas that have withstood repeated attacks. In such an arena, simple observation has both a great disadvantage and a great advantage. The disadvantage is that a simple observation may simply be rejected, but not always by everyone and not always entirely. Pure observations are most likely to be taken seriously when the observer has an established reputation. (Unfortunately, one often hears the best stories from other sorts of observers—why don't those UFOs ever pick up a physicist?) In any case, though experimentation is fine, there are some very useful scientific ideas that have been developed simply because someone was there and paying attention.

Even with our carefully developed and well-groomed procedures for doing science, some of the most fundamental questions do not easily lend themselves to our probing. Perhaps the most basic and perplexing of these is the deceptively simple question, "What is life?"

DEFINING LIFE

It turns out that questions about the nature of life are often based as much in philosophy as in science (as we will see in Chapter 3). In fact, since the waters so quickly become muddied with disagreement, perhaps the best we can do is note a few of the most important characteristics of life. None alone can define life, but together, they form a composite that generally sets living things apart from the nonliving (Figure 2.3). With this in mind, we can say that (1) living things basically *have a cellular structure*. Cells, as we will see, are the smallest units of life that contain the necessary structural and chemical means to carry on the activities of life. (2) Many living things *show movement*. If you're watching something and it moves, your first assumption might be that it's alive. In fact, among animals, both the hunter and the hunted often use movement as an indicator of life. (3) Living things *metabolize and grow*. Generally, living things gain mass, to some degree, over time. The building material for the new mass comes from molecules that are manufactured (as with turnips) or acquired (as with wolves) and then reorganized into new kinds of structures in the living organism. (4) Living things *reproduce*. Since death is an integral part of life's experience, entire lines would die out with the deaths of its members had they not reproduced. But because they do reproduce, the lines go on. (5) Living things *respond*. Plants generally respond slowly, as when a potted petunia turns its leaves toward the lighted window. A more rapid response is seen by stepping on a cat's tail. The essential point is that

FIGURE 2.3

Living things generally exhibit six characteristics that, together, set them apart from the nonliving.

(a) Living things show movement, like the long powerful strides of a lynx chasing its prey.

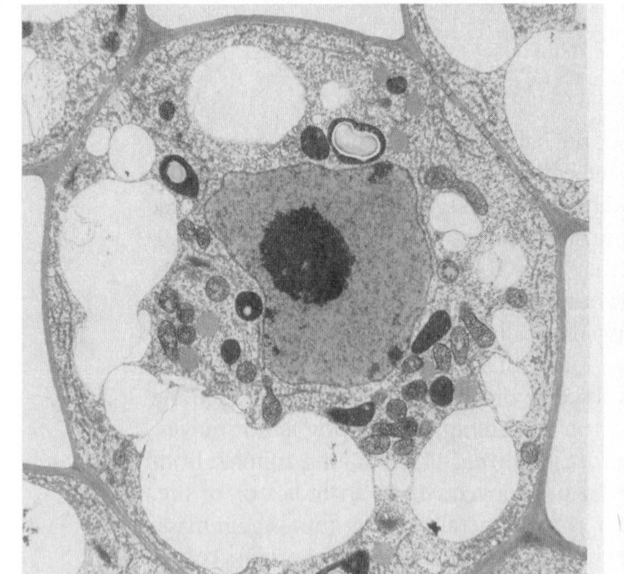

(b) Living things are essentially cellular.

(c) Living things metabolize and grow, as can be seen in the blossoming of this dwarf dahlia.

FIGURE 2.3 continued

(d) Living things reproduce, resulting in new generations, as in this group of elephant cows and calves.

(e) Living things evolve, as evidenced by this fossil insect, which represents a stage in the evolution of honey bees.

(f) Living things respond, as this cat illustrates with its ability to turn in midair to land on its feet.

organisms must have ways of reacting to the changing environment of an active world. (6) Living things *evolve and adapt*. Just as life responds quickly to immediate conditions, it also changes through the generations (evolution) to better adjust to long-term shifts in conditions (adaptation). Such changes are most easily detected in the geological record left by fossils.

You may have noted that there are some problems in defining life. After all, beehives are cellular, rivers move, oil droplets reproduce, crystals grow, a still pond is responsive, and mountains evolve. And what about viruses? They can crystallize. Can a crystal be alive? Nonetheless, should you encounter something that has even a few of these traits, it had best be considered alive—especially if it has teeth. Obviously, this list is by no means exhaustive. What other traits could be added? And what other exceptions to these can you think of?

The Trap of Teleology

The simple word "to," which implies "in order to," has caused a great deal of confusion in many areas of science. For example, consider the phrase, "Birds migrate southward to escape winter." The statement seems harmless enough, but if interpreted literally, it implies that the birds have a goal in mind, or that they are moving under the directions of some conscious force that compels them to escape winter. Philosophers have termed such assumptions *teleology*. (*Teleos* is Greek for *end* or *goal*.) It is commonly used in reference to ideas that go beyond what is actually verifiable and generally implies some inner drive to complete a goal or some directing force operating above the laws of nature.

Teleological statements are rampant in scientific literature today, partly from carelessness, but partly because scientists may feel free to communicate among themselves by using such phrasing, a kind of shorthand, when they are sure that their message will be interpreted correctly.

A biologist might say "Birds migrate southward to escape winter," (sentence 1) when he or she really means, (2) "by migrating southward, birds escape winter," or to be more precise, (3) "birds that migrate southward as winter approaches tend to escape the harsh seasonal weather of their northern summer homes." Do you see the several refinements in sentence (3) that make it more accurate than sentence (2)? Note primarily that both sentences avoid implying that birds flying south actually entertain conscious intentions of escaping cold weather.

Science and Phenomenology

If you've never had the experience of looking up a word in a dictionary and coming away more puzzled than you were to begin with, try *phenomenology*. So, with apologies to my philosopher friends, let me just say that phenomenology encompasses the notion of the observer's own interpretation of an observation or body of information. Even science, then, becomes personalized, as observations are described in terms of the scientist's own experience. And, if so, biology must be the most phenomenological of all the sciences. Part of the reason is that life is so variable and changeable. Hard and fast rules are few, and we find exceptions at every turn. Each time we draw up a new rule, some beastie lurking somewhere chuckles "Not me!" Since there are so many unusual cases in the realm of life, the biologist has the obligation and opportunity to exercise choices and to make decisions based, however subtly, on a personal set of values. The biologist is thereby forced to interpret much of what he or she sees, and the interpretation is always colored by one's own tendencies, wishes, hunches, or proclivities. Just as the distinct flora of the baker's hands flavor the bread, so does the biologist interject something of himself or herself into the science, and when biases are at odds with the findings, data tend to become suspect and arguments mount. It is just such arguments that help make biology fun and keep biologists honest.

WHAT BIOLOGISTS DO FOR A LIVING

The word *biologist* may cause some to conjure up the image of a balding little man with bad posture and a squeaky right shoe, padding through aisles of dusty books on the trail of some ancient description of an extinct lizard. Others may visualize a butterfly chaser with thick glasses—the one who couldn't get a date for the prom—net poised, leaping gleefully through the bushes. Others may think of biologists as bird-watchers in sensible shoes, peering through field glasses in the cold, wet dawn in hope of catching a glimpse of the rare double-breasted seersucker. Maybe such images do fit some biologists, but there are others who search for the mysteries of life in other places, such as in clean, well-lit laboratories amid sparkling glassware (Table 2.3, next page).

A few years ago it was said that a biologist is one who thinks that molecules are too small to matter, a physicist is one who thinks that molecules are too large to matter, and anyone who disagrees with both of them is a chemist. Many biologists, though, are in fact chemists, and vice versa. Perhaps it is true that the biologist whose scope is limited to molecules must periodically be convinced of the existence of the platypus, but no one else has been able to tell us how tiny hummingbirds are able to make it across the Gulf of Mexico. Fortunately, the sharp lines of division between such disciplines are becoming blurred and indistinct as scientists become more broadly trained and able to handle more kinds of ideas in their continuing effort to solve the "Great Puzzle."

SCIENCE AND SOCIAL RESPONSIBILITY

The question often arises, what is the social responsibility of informed scientists? Precisely what are their obligations to the rest of us, if any? Should their interests lie primarily in the continuing accumulation of more data, or are they morally obligated to help apply their findings to human

TABLE 2.3
Careers in Biology

AGRICULTURE CAREERS

agronomist
animal breeder
county extension agent
field crop manager
florist
greenskeeper
horticulturist
landscape contractor
nursery manager

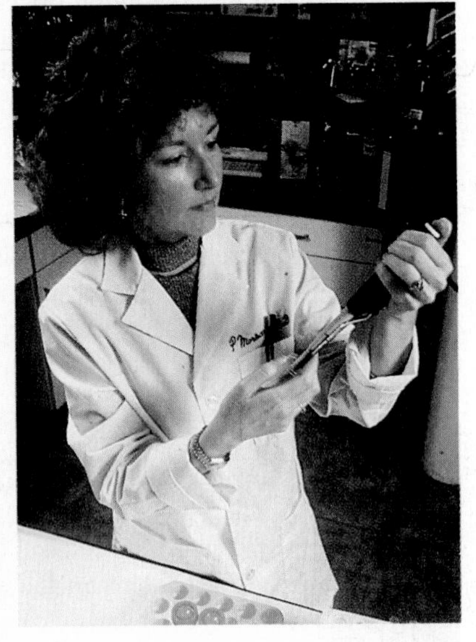

CONSERVATION CAREERS

fish and game manager
forester
range manager
wildlife manager
soil conservationist

MEDICAL CAREERS

dental hygienist
dentist
medical technician
midwife
mortician
nurse
nutritionist
optometrist
paramedic
pharmacist
physical therapist
physician
veterinarian
X-ray technician

RELATED CAREERS

curator—zoo, botanical garden, herbarium
occupational therapists—private industry
sales and service—biological supply firms and pharmaceutical companies
science writers, illustrators, photographers—publishing houses
teaching

FIGURE 2.4
Scientists have, in recent years, become more aware of the social implications of their work, partly because of the confrontational behavior of the activist groups both within and outside the scientific community.

problems (Figure 2.4)? The citizen might well ask, should researchers, funded with public money, pursue their own research interests, or should they be encouraged, by law if necessary, to direct their attention to solving our immediate problems? Who should decide where the money goes? You may one day have a voice in such decisions, even if by a simple vote. Should that day come, will you know enough? Furthermore, if you're of the adventurous sort, you may want to know what challenges remain. What's out there yet to be discovered? It may surprise you to learn that there are many places left on earth that have not yet been explored, and that there is an organization dedicated to worldwide exploration—so that we may know, in the end, the true nature of the planet. See Essay 2.2, "Exploring the Earth," on pages 47–49.

Another question is, should scientists with access to important information, "uncommon knowledge," put down the pen, shut off the computer, and step onto the soapbox? Are they obligated to tell us what they know about our problems and to suggest solutions? Or should they, as scientists, continue to generate more data and to seek yet more information? Is the information of greater value than the opinion? Furthermore, can concerned scientists assume that if the information is made available, an informed public will behave wisely (Figure 2.5)?

> How much radiation is safe? Can nuclear power plants explode? What is a meltdown? What sorts of meats reduce the risk of cancer? Of heart attack? Why is sunburn dangerous? What is the greatest risk of nuclear war? Do salt tablets help in summer? What is CPR? What exactly is AIDS? What is nuclear winter? What does El Niño have to do with food prices? What is the Ogalalla aquifer? What risks are associated with genetic engineering?

FIGURE 2.5

The problems associated with certain scientific technologies are being vividly driven home to more Americans each year. As the public becomes more aware of such problems, there is an increasing pressure on industry to become more socially responsible. Here, crews clean up dioxin.

Some people just don't want to concern themselves. After all, breakfast this morning was warm and on time, and the traffic lights worked. The room is comfortable, and our afternoon is planned. But keep in mind as we trip lightly through these pages that in spite of feelings of complacency or security, many of these concepts are not simply academic ones that may appear on some test. Some of the things we will encounter here may have a fundamental bearing on the quality of your daily life, and, perhaps, how long you will live. You may one day be forced to make some difficult, even excruciating, decisions. Of course, you may choose to hope for the best and to ignore questions of the long run, and who could blame you?

If one is interested in great problems these days, there is no reason to search farther than the human condition; it has the ring of a predicament. In fact, we hear that the people inhabiting this planet are, quite simply, in serious trouble. There is no need to recite the growing list of our problems. Why should we burden ourselves with the effects of burgeoning populations, hunger, resource maldistribution, nuclear wastes, human wastes, farming crises, dangerous air, fouled waters, dwindling mineral supplies, and international hostilities? Why should we even mention such things? After all, it's a nice day. There are things to do, people to meet, places to go . . . and perhaps an earth to save.

Exploring the Earth

Michael McBride observing whales up close and personal.

Science and Exploration–A Partnership

Exploration and science have long progressed together, one increasing the scope, as well as the dreams, of the other. In some cases, the exploration has been done entirely in the name of science, as when the great South American explorer Alexander von Humboldt set out in 1799 through the wilds of Venezuela and the Orinoco River, eventually bringing back some 12,000 botanical specimens. Or when the noted African explorer, Mary Henrietta Kingsley, led her expeditions up the Congo and Ogoone Rivers in the late 1800s, posing as a trader among the cannibals, later to describe her findings to an awed England.

In some cases, science has preceded the exploration, enabling the explorer to go further, to explore new realms that, without the science, would have remained inaccessible. Buzz Aldrin would never have stepped so boldly onto the surface of the moon, nor would Sylvia Earl and Bob Ballard have penetrated the ocean depths without a great deal of hard, thoughtful science being done first.

As both science and exploration have progressed, the horizons have been pushed back constantly, revealing more—much more—to be done. Even as our horizons expand, however, through new science and technologies, exploration continues the old-fashioned way. So, with information satellites hurtling overhead unseen, men and women, often bedraggled, hungry, or wet—yet determined—chart unmapped rivers of the Northwest Territories, discover new tributaries of the Amazon, track bizarre unknown species in Borneo, and wade ashore on undescribed islands of the Indian Ocean.

There is, indeed, an instinct to explore, to learn more. That instinct, we believe, is alive and well in the hearts of young people everywhere. Remember, this is your world and now is your time. By accepting the challenge, you join a host of courageous and well-trained people seeking to understand the ultimate truths of nature.

*John C.D. Bruno, Past President, The Explorers Club**

*The Explorers Club, New York City, is composed of world-class explorers and dedicated to the advancement of exploration everywhere in the furtherance of scientific understanding.

Charles Mazel, off Roatan, Honduras, photographs corals, finding that some absorb invisible light wavelengths and convert them to visible light through fluorescence. The significance of this ability is not known.

Charles Mazel.

The coral under visible light.

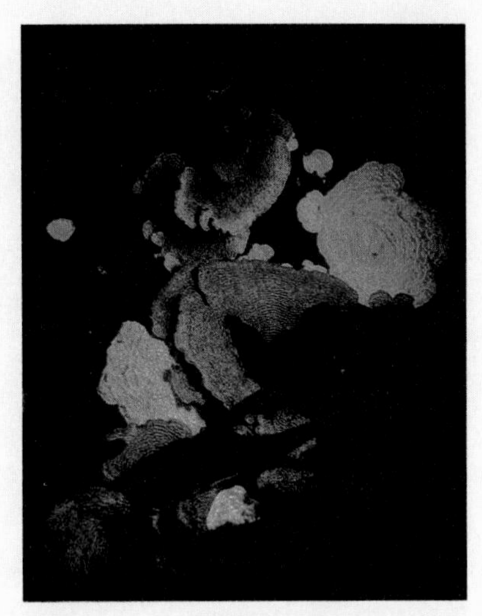

Fluorescing coral.

Explorers Club past president John C.D. Bruno recuperating aboard a research vessel near Tierra del Fuego after a second unsuccessful attempt to climb Chile's Cerro Hudson volcano.

And trekking in Patagonia, near where Charles Darwin worked.

The author, holding Explorers Club flag in the Ecuadoran Amazon.

Monarch butterflies, after a storm, festooning a tree in Sierra Chinqua, Mexico near an area being burned, are studied intensively by Lincoln Brower.

Nicholas DeVore III sharing a green sea turtle with the Satawalese of Micronesia, the famed celestial navigators.

SUMMARY

1. Both inductive and deductive reasoning are used in science. In inductive reasoning, data and observations lead to a general statement. In deductive reasoning, a general statement leads to specific statements that can then be tested by experimentation.
2. The scientific method begins with observations, usually ones that lead to a problem or question. A hypothesis is then proposed as a tentative explanation for the observation. The hypothesis must be testable through experimentation. In an experiment, the effects of the factor suspected to be important are determined by comparing what happens in the presence of that factor with what happens in its absence. This is done using a control, a second arrangement exactly like the experimental arrangement except that the factor being tested is left out.
3. There are several features that together characterize, but do not define, living things. Living things are made of cells. They move, metabolize, and grow. They reproduce and respond to change. They evolve and adapt through generations.
4. A statement is called teleological if it implies that the event is goal-oriented or driven by a force that is above the laws of nature.
5. Everyone, including a scientist, interprets an observation or body of knowledge based on his or her own experience and biases.
6. Biologists work in the laboratory as well as in the field. The lines that once separated physics, chemistry, and biology are becoming blurred.
7. Biologists must often make decisions about the extent of their responsibility to society.

KEY TERMS

conclusion (37)
control (35)
deductive reasoning (33)
experiment (35)
hypothesis (35)
inductive reasoning (33)
scientific method (35)
theory (37)

FOR FURTHER THOUGHT

1. Which of the following is a teleological statement? Suggest a refinement by rephrasing the sentence.
   A. Often, flowers that are pollinated by hummingbirds are red.
   B. Some birds engage in courtship behavior before mating.
   C. Sexually mature salmon swim upstream to spawn.
   D. Monarch butterflies migrate to coastal regions of California and Mexico.
2. In 1928, Alexander Flemming accidently contaminated a bacterial culture with the fungus *Penicillium* and later noticed that the fungus killed his bacteria. He subsequently theorized that the fungus was an antibacterial agent. State whether this discovery utilized the deductive method, the inductive method, or a combination of the two. Explain.

FOR REVIEW

True or false?

1. ___ In deductive reasoning, a series of observations leads one to a generality.
2. ___ Teleological statements express ideas that are generally unverifiable.
3. ___ A hypothesis is a possible explanation that can be tested.
4. ___ The hardest scientific evidence is obtained through repeated experimentation.

Short answer

5. A ___ is a crucial factor that is sometimes altered or left out of a controlled experiment.
6. List six qualities exhibited by living things.
7. Describe some ways in which scientific knowledge is expanded. Which method produces the least reliable scientific evidence?

Choose the best answer.

8. Phenomenology can best be described as:
   A. a compelling drive to complete a goal.
   B. an observation that is clouded by personal bias.
   C. a verified explanation of natural phenomena.
   D. an idea that can be tested.
9. Which statement best describes the inductive method?
   A. Some personal observation suggests an experiment.
   B. There is an ultimate goal.
   C. Evidence gathered is highly prejudiced.
   D. Generalizations are reached after gathering data.
10. Which of the following is a thoroughly tested idea that explains some natural phenomenon?
    A theory
    B. hypothesis
    C. teleological statement
    D. both B and C

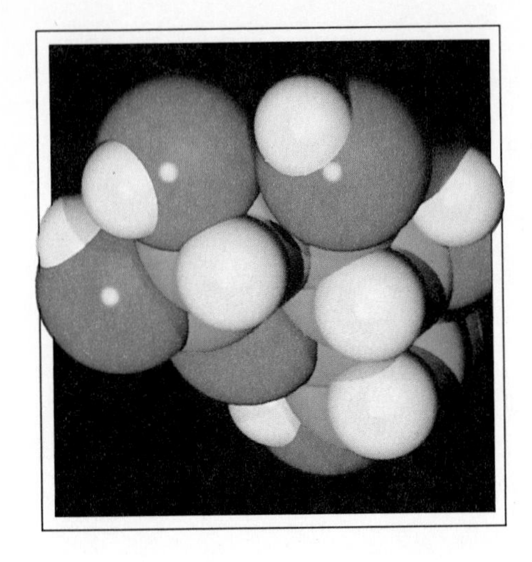

CHAPTER 3

The Chemistry of Life

Overview

Atoms and Elements, Molecules and Compounds
CHNOPS
How Atoms May Vary
Electrons and Their Behavior
Chemical Bonding
Chemical Reactions
The Molecules of Life

Objectives

After reading this chapter you should be able to:

- Distinguish between an atom and an element and list the six elements that comprise 99 percent of living matter.
- List the subatomic particles and relate the structure of an atom to its chemical properties.
- Describe how the organization of electrons in nuclear orbitals affects an element's reactivity.
- Name three types of chemical bonds and state how each is formed.
- Explain how elements combine in chemical reactions and describe endergonic and exergonic reactions.
- Describe the special bonding properties of carbon.
- Describe the mechanisms by which enzymes act as catalysts in chemical reactions.
- List four major groups of biological molecules and describe their components, structure, and properties.

So now we come to it. Chemistry, the part you've been dreading. You signed up for biology, you say, and now you're getting chemistry. Life is tough enough, anyway, and then this has to happen.

Before despondency sets in, though, let me say that we will cover only the basics, the material that will help you to better understand some of the discussions that follow. The material will be simple, and if you take it one step at a time, you will probably find it surprisingly easy, even, if I may say so, enjoyable. You also might keep in mind that the fact that we are led to chemistry at all underscores the unity of the sciences in explaining our natural world.

An understanding of the basic concepts of chemistry can also broaden one's view. An established and sound scientific view may well become increasingly important as time passes (Figure 3.1). For example, can automobile emission control devices be harmful if they destroy unburned hydrocarbons and allow nitrous oxides to pass into the air? Is all cholesterol bad? Is there any safe level of radiation? How can X rays cause cancer? Should nuclear power plants be built? Is the risk of nuclear accidents

FIGURE 3.1

Chemicals affect us in uncountable ways, as influences on our muscular development, our food chain (the *Valdez* oil spill), the air we breathe (urban smog), and rainfall (acid rain).

greater than the danger of the atmospheric pollutants produced by burning coal? Why do we age? How can chemicals cause hostility? Some of these questions will be answered here and some in later chapters, but in every case, our understanding of certain biological problems will depend, to some degree, on how much we know about basic chemistry. Chemistry, you see, influences our lives at a number of levels.

So here's the plan: We will begin with some basic definitions. Then we will take a look at the structure and components of life's building blocks and consider some of the ways they can interact. Later, we will see how certain combinations of these things are important to life. Keep in mind that the principles we will consider here will serve as a foundation for many of our following discussions.

ATOMS AND ELEMENTS, MOLECULES AND COMPOUNDS

FIGURE 3.2
Hydrogen and oxygen are gases, but when joined in a certain proportion, they form a liquid in which fish are found.

All matter, whether it exists in an alga or a distant star, is composed of elements. **Elements** are substances that cannot be divided into simpler substances by chemical means. **Atoms** are the smallest indivisible particles of these elements. Atoms *can* be divided into their component parts, but when this is done, the special qualities of that element are lost. Each element, or type of atom, has its own special chemical behavior by which it is identified. Oxygen, chlorine, and carbon are examples of elements.

Ninety-two elements occur naturally, and about 17 others have been synthesized in the laboratory. Each has been named and given a letter symbol. For example, the element called hydrogen is designated by the letter H. Sulfur is S. Oxygen is O. Of course, nothing is all that simple, so we find that sodium is designated by Na, from its Latin name *natrium*. The symbol standing alone usually designates one atom. For example, O refers to one atom of oxygen. The symbol 2O refers to two atoms of oxygen. But the symbol O_2 means that two oxygen atoms are joined together into one *molecule* of oxygen, the form in which oxygen usually exists in nature. **Molecules** are formed from combinations of two or more atoms of the same or different elements. A **compound** is two or more elements combined in a fixed proportion. Water, then, is a compound formed of hydrogen and oxygen atoms joined to form molecules that are designated by the symbol H_2O (Figure 3.2).

CHNOPS

CHNOPS is an elegant word that might help you to identify the six important elements that make up about 99 percent of living matter. However, if you drop the word casually at your next gathering, it is not likely that everyone will know you are referring to carbon, hydrogen, nitrogen, oxygen, phosphorus, and sulfur. If they do, you are in the company of organic chemists and should leave immediately.

The Structure of an Atom

Let's now consider what an atom is supposed to look like. We say "supposed to" because atoms cannot be seen, so all the evidence is circum-

TABLE 3.1
Components of the Atom

NAME	LOCATION	WEIGHT	CHARGE
Proton	Nucleus	One atomic unit	Positive
Electron	In orbitals outside nucleus	Negligible	Negative
Neutron	Nucleus	One atomic unit	No charge

FIGURE 3.3
The hydrogen atom and the helium atom. Note that hydrogen has one proton (this reveals it as hydrogen) and one electron (thus it is electrically balanced). The other atom has two protons (making it helium) and two neutrons in its nucleus. The neutrons bear no charges so the atomic number of helium is two, while the atomic number of hydrogen is one. Helium is also balanced in that its two positively charged protons in the nucleus exist with two negatively charged electrons in orbit around the nucleus.

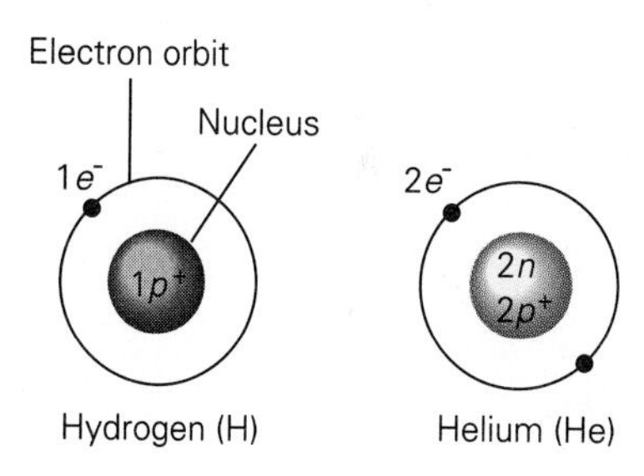

stantial. (But it is impressive. People have been hanged on less compelling circumstantial evidence.) Atoms, according to theory, are composed of a dense center around which small particles spin in orbit. The center, called the **nucleus,** consists of **protons** and, usually, **neutrons.** Each proton carries one positive charge, designated by a plus sign (+).

Neutrons, as the name implies, carry no charge. Protons and neutrons are similar in size and are said to weigh one atomic unit each. The particles in orbit around the nucleus are called **electrons.** They are much smaller than protons (having about 1/1835 the mass). Each electron carries one negative charge, designated by a minus sign (−). Table 3.1 lists the components of the atom. An atom of hydrogen, the simplest element, consists of one proton and one electron, with no neutrons (Figure 3.3).

Helium, the next simplest element, is composed of two protons, two neutrons, and two electrons. Since an atom's **atomic number** refers to the number of protons in the nucleus, the atomic number of hydrogen is one; of helium, two. It is important to know that elements differing by even a single proton may have vastly differing chemical properties.

You may have noticed in the diagrams of hydrogen and helium that the number of positively charged protons are balanced by the negatively charged electrons, resulting in the atoms having a net neutral charge. If they have equal numbers of protons and electrons, they are said to be electrically balanced. As a final note about our examples here, it should be mentioned that the figures are not drawn to scale. In nature, the electrons are nowhere near the nucleus. In fact, if the period at the end of this sentence were a nucleus, its nearest electron would probably be across the street somewhere.

HOW ATOMS MAY VARY

Isotopes

The atoms of any given element all have the same number of protons in their nuclei. If this number were to change the atom would, by definition, become a different element. However, the number of *neutrons* in the atoms of certain elements may vary. For example, most oxygen atoms have eight protons and eight neutrons, but some oxygen atoms may have nine, ten, or even more neutrons. Atoms that can vary in the number of their neutrons are called **isotopes** of an element. There are eight different neutron variants of oxygen, so oxygen has eight isotopes. Some elements have as

TABLE 3.2
Properties of Isotopes

ISOTOPES	NUMBER OF PROTONS (ATOMIC NUMBER)	NUMBER OF NEUTRONS
Hydrogen (H)	1	0
Deuterium (H^2)	1	1
Tritium (H^3)	1	2
Carbon 12	6	6
Carbon 13	6	7
Carbon 14	6	8

many as 20 isotopes. Usually, all the isotopes of an element have about the same chemical properties, but in some cases, isotopes of the same element may have vastly different chemical traits. Table 3.2 shows three isotopes each of hydrogen and carbon.

Ions

Ordinarily, each atom of an element has the same number of negatively charged electrons as it has positively charged protons; hence, it is electrically balanced. We know that the number of protons, with their positive charges, can't change without changing the element, but an atom *can* gain or lose electrons and, when this happens, it is no longer electrically balanced. If the atom loses an electron, it is left with a net positive charge, since its protons now outnumber its electrons. If it should gain an extra electron, it then has a net negative charge. Charged particles—atoms that have either a positive or a negative charge—are called **ions.** Many ions, as we will see later, are critical to the processes of life.

ELECTRONS AND THEIR BEHAVIOR

An electron moves around the atomic nucleus at almost unbelievable speed. Although its path is not as regular as a planet's orbit around the sun, there is an area of space, called an **orbital,** in which an electron is most likely to be found at any given moment. Actually, there are a number of potential orbitals in which an electron can move around a nucleus, and these are organized at various distances from the nucleus. Electrons in orbitals close to the nucleus are at lower energy levels than are those in orbitals farther away. Energy levels are also known as **shells.** The higher energy electrons found in the outer shell are in a position to best interact with other atoms.

It is possible for an electron to move from one shell to another. For example, if an electron is "excited" by some external energy source, such as heat, light, or electricity, it may jump outward to a higher energy shell. Should it later fall back into a more inner shell, the energy that had provided its boost is released. We will shortly see how this released energy can be used in living systems. Figure 3.4 will give you an idea of some energy relationships.

FIGURE 3.4

Electrons may exist in high or low energy states. An electron moving from an orbit near the nucleus to one farther out is moving from a lower to a higher energy level. "Exciting" an electron by raising it to a higher level by the input of energy is analogous to pushing a boulder up a hill. At the higher energy levels, they both increase their potential energy, which can be released if they return to their former states.

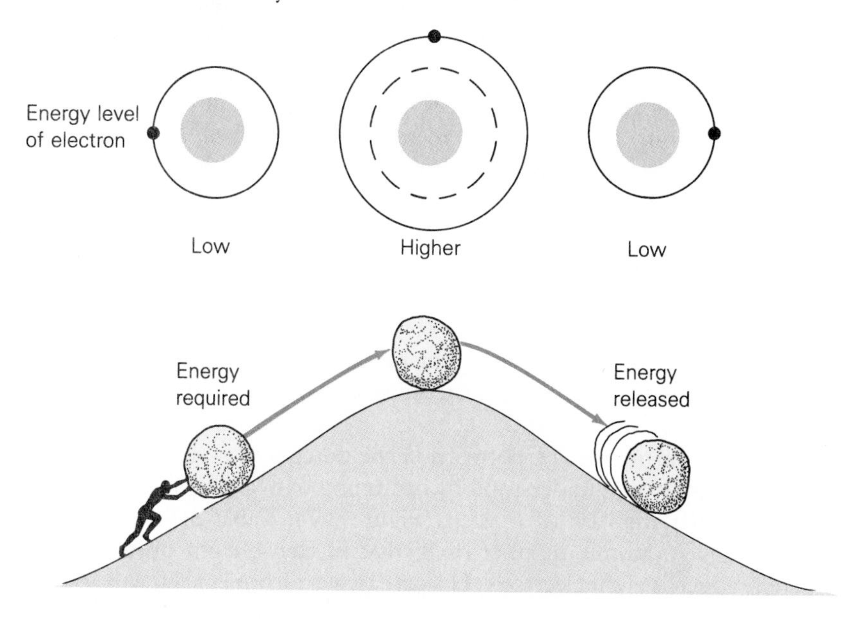

FIGURE 3.5

On May 6, 1937, Germany sought to impress the world with its technological achievements, so it sent the hydrogen-filled zeppelin *Hindenburg* across the Atlantic to the United States. We were impressed, indeed: the zeppelin blew up. The hydrogen, perhaps encouraged by a spark, explosively joined with oxygen (to make water). Thirty-six people died on the approach to Lakehurst, New Jersey. Miraculously, 61 people survived.

Oxidation and Reduction

In some cases, an electron may be so energized that it escapes from its atom altogether. The loss of electrons from an element is called **oxidation,** because the lost electron is often recaptured by an oxygen atom. In fact, some substances cannot lose electrons unless oxygen is available to accept them. The capture of such electrons by any element is called **reduction.** So, an atom losing an electron is *oxidized,* and an atom gaining an electron is *reduced.* This simple process of electron transfer has tremendous implications for life.

In some cases, as electrons are transferred in such reactions, they tend to travel with a proton. Together, of course, they form a hydrogen atom. So, oxidation and reduction can involve the transfer of hydrogen from one element to another. As an example, an atom of oxygen can be reduced by two hydrogen atoms, resulting in a molecule of water. Such reactions can sometimes proceed so rapidly that the effect is explosive (Figure 3.5).

Electrons and the Properties of the Atom

If you remember the numbers two, eight, and eight, you will have at your fingertips the answer to very few questions. But one such question would be, what is the maximum number of electrons that can exist in the first

FIGURE 3.6

Helium and neon are called inert atoms. They tend not to react with other elements because their charges are balanced and their electron shells are filled.

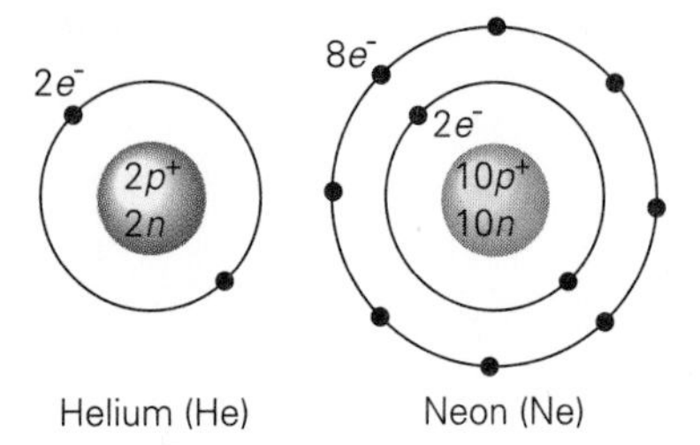

three shells of an atom?* Only a specific number of electrons can occupy any one electron shell; the first (the innermost) shell can accommodate only two, and the second and third hold only eight each. (The fourth can hold 32, and some atoms have even more shells.) However, no matter how many shells an atom has, its chemical properties are determined primarily by the outer shell. It is the one most likely to gain or lose electrons, and the atom tends to behave in such a manner as to keep this shell filled. When this outer shell is filled, the atom is said to be **inert.** It is almost impossible to make inert atoms react with any other element. The outer shells of helium and neon, for example, are full. In the case of helium, with only two electrons, there is only one shell, the inner one, and it has all the electrons it can hold. Neon, as you see, has only two shells, both completely occupied (Figure 3.6). These, then, are "inert" elements that generally do not react with other elements.

So we see that atoms tend to "satisfy" themselves in two ways. First, *they tend to balance the numbers of their protons and electrons,* and second, *they tend to keep their outer shells filled.* When these conditions are met, the atoms are inert, unreactive, and not of great interest to us here. With atoms, as perhaps with people, the most active and interesting ones are often the most "dissatisfied."

Why is the number of electrons in the outer shell so important? Perhaps this can best be understood by beginning with a couple of examples. Let's consider the case of oxygen (Figure 3.7). Table 3.3 tells us that oxygen has an atomic number of 8; that is, it has eight protons, and, therefore, needs eight electrons. However since the inner shell will accom-

*Specifically, in the lighter atoms; heavier atoms can hold 18 electrons in the third shell.

FIGURE 3.7

Oxygen is reactive because it has only six electrons in the outer shell. The atom is electrically balanced but its outer shell is unfilled.

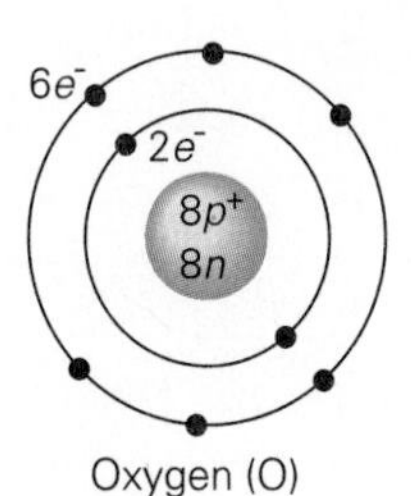

TABLE 3.3
Some Elements Found in Living Matter

ELEMENT	SYMBOL	ATOMIC NUMBER	ELECTRONS IN EACH SHELL 1	2	3	4	PERCENTAGE OF EARTH'S CRUST
Carbon	C	6	2	4	0	0	0.03
Hydrogen	H	1	1	0	0	0	0.10
Nitrogen	N	7	2	5	0	0	Traces
Oxygen	O	8	2	6	0	0	46.60
Phosphorus	P	15	2	8	5	0	0.10
Sulfur	S	16	2	8	6	0	0.05
Sodium	Na	11	2	8	1	0	2.90
Magnesium	Mg	12	2	8	2	0	2.10
Chlorine	Cl	17	2	8	7	0	0.05
Potassium	K	19	2	8	8	1	2.60
Calcium	Ca	20	2	8	8	2	3.60
Iron	Fe	26	2	8	8	8	5.00

modate two electrons, the outer shell is left with only six—two short of filling its shell. So, oxygen tends to accept two electrons from other atoms.

On the other hand, note that sodium (Figure 3.8), has an atomic number of 11. This means it has two electrons in its first shell, eight in the second shell, but only one in the third shell. Since it isn't energetically feasible for a sodium atom to gain seven electrons, it tends to fulfill its outer shell requirements by losing its single outer electron. This loss means that it now only has two shells, but at least both are satisfied.

Of course, the loss of the electron (with its negative charge) ionizes the sodium. That is, it leaves it with a positive charge. Sodium ionized in this way is written Na^{+}. Magnesium, which has 12 electrons, is sometimes written as Mg^{++} (do you see why?). The fact that chlorine has an atomic number of 17 may give you a clue to the reason salt (NaCl) is formed so easily. Take a minute here to consider the question.

FIGURE 3.8

Since sodium has 11 electrons, with ten existing in its first two shells, the third shell can harbor only one electron. Sodium can not rectify the situation by adding seven more electrons to the third shell; the resulting charge imbalance would be too great. Instead, it tends to give up its outermost electron. Thus, sodium is highly reactive with anything that accepts electrons.

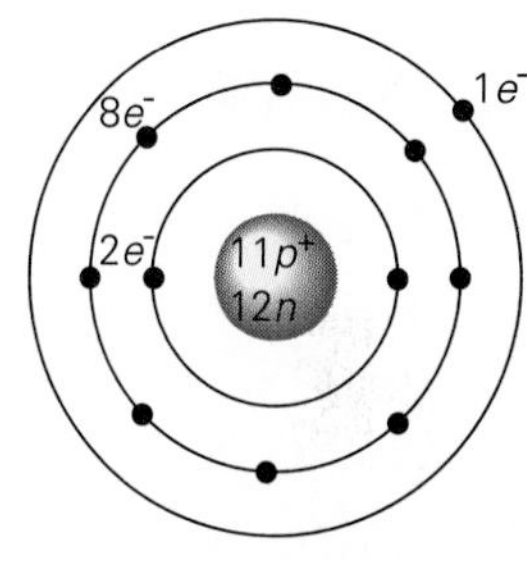

Sodium (Na^{+})

CHEMICAL BONDING

We've seen that atoms can join to form molecules, so let's take a look at how this can happen. (You already know enough to figure out that atoms can only join with certain other kinds of atoms.) By the way, it is interesting that the chemical properties of atoms may have little to do with the properties of the molecules that they form. For instance, both sodium and chlorine are deadly poisons, but together they form common table salt.

As we explore the ways in which atoms can join, you will be relieved to discover that even the most complex molecules are generally held together by only a few, and conceptually simple, kinds of bonds.

Ionic Bonds

First, let's reconsider how table salt is formed. We know that sodium has two electrons in its first shell, eight in its second, and only one in its third. Therefore, sodium has a tendency to give up its outer electron (at the expense of becoming ionized) in order to meet its outer shell requirements. We also know that chlorine has two electrons in its first shell, eight in its second, and seven in its third. Hence, it can fill its outer shell requirement rather easily by picking up a single electron. Not surprisingly, when a strong electron donor, such as sodium, encounters a strong electron acceptor, such as chlorine, there is a quick transfer of an electron from one atom to the other. This satisfies the outer shell requirements of both atoms, but as we know, it leaves the sodium, which has lost an electron, with a positive charge (Na^{+}). And, of course, the chlorine, with its additional electron, now has a negative charge (Cl^{-}).

Chemically (as well as socially, some would say), opposites attract, and the sodium and chlorine are now bearing opposite charges. As a result, they are electromagnetically attracted to each other. They draw together and join, forming the molecule NaCl, or table salt. This sort of reaction is called **ionic bonding,** and it involves the transfer of an electron from one atom to another and the subsequent union of the resulting oppositely charged ions.

FIGURE 3.9
When NaCl is placed in water, the salt molecule separates into its component sodium and chloride ions. Water has this effect on salt because of its uneven (polar) charge distribution.

In some cases, such bonding does not occur, even when the atoms bear opposite charges. For example, when table salt is placed in water, Na and Cl atoms dissociate into their ionic forms, Na^+ and Cl^- (Figure 3.9) and they stay in this condition because of the peculiar configuration of water molecules.

One of the reasons water has such special properties is because its molecules have unevenly distributed charges. In other words, water molecules are polarized, having a positive and negative end. Because of the polarization, the positively charged end of the water molecule is attracted to the negatively charged chlorine, and the negatively charged end of the water molecule is attracted to the positively charged sodium. As the water molecules cluster around these charged atoms, they effectively separate the two ions. We will learn more about the peculiar qualities of water when we discuss the topic of hydrogen bonding.

Covalent Bonds

In another type of bonding, called **covalent bonding,** atoms do not transfer electrons, but *share* them. We can illustrate with oxygen again. Oxygen, as we have seen, needs two electrons to satisfy its outer shell requirements. One way it can acquire them is by sharing electrons with two hydrogen atoms, thereby forming a water molecule (H_2O). This sharing satisfies the shell requirements of both the oxygen atom and the two hydrogen atoms (Figure 3.10).

When a covalent bond is formed between two different kinds of atoms, the shared electrons are usually attracted more strongly to one atom than to the other. The electron, of course, will be more powerfully drawn to the atom with the stronger positive charge. When water is formed from hydrogen and oxygen, for example, the shared electrons are pulled more strongly toward the more positively charged oxygen nucleus than they are toward the two hydrogen nuclei. It is this "imbalance" in the position of the electron that results in a polarized water molecule with negatively and positively charged ends.

In some cases, the electrons are about equally attracted to both elements, and, thus, the molecules are symmetrically charged, with very little polarity. But polarity is not an either-or situation. In fact, because the polarity of different kinds of molecules varies, there is no clear line of distinction between ionic and covalent bonding. Keep in mind that in ionic bonding, an electron from one atom is completely captured by another atom, and in covalent bonding, the electron is shared by two atoms, although perhaps being more strongly attracted to one than the other.

A word about notation may be in order here. Notation involves how chemical bonds can be written. A single bond, for example, can be indicated by a single line from one atom to another. Thus, when two hydrogen atoms join, the resulting hydrogen molecule (H_2) may be written as H—H. When an oxygen atom joins with two hydrogens, bonds can be shown as H—O—H. When two oxygen atoms join together, each of them requires two electrons, so their two shared electrons form a **double bond,** O = O. Look back at Table 3.3. Do you see why nitrogen can be written as N≡N?

FIGURE 3.10

Covalent bonds form as two hydrogen atoms move in to share their electron with an oxygen atom. The "unbalanced" structure of a water molecule results in a stronger negative charge on one side of the molecule. Another way to represent the water molecule is shown on the right:

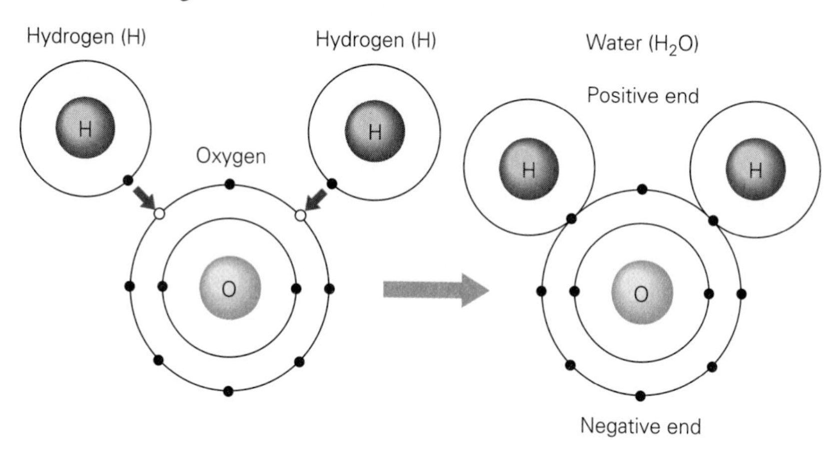

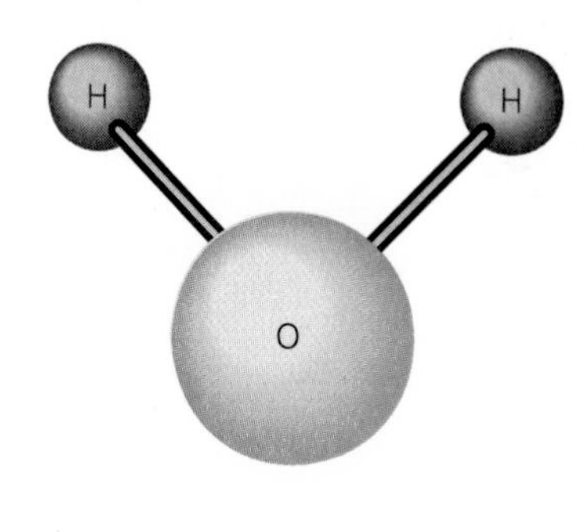

Hydrogen Bonds

Both covalent and ionic bonds are regarded as rather powerful forces, but there are a number of other, weaker types of bonds as well. Perhaps the most important of these is called the **hydrogen bond** (a weak bond between the positive part of a hydrogen atom that is covalently bonded to form one molecule and the negative part of another molecule). Strangely enough, hydrogen bonds are important to life precisely because of their weakness. It takes very little energy to form them, and they are easily broken. However, they are present in enormous numbers in biological molecules, and are essential to life processes. They are like the tiny threads that bound Gulliver in Lilliput Land: individually weak, but collectively strong.

We have seen that when a hydrogen atom is covalently bonded to another atom, the bond may be polar. The hydrogen atom bears a slight positive charge because the shared electrons are pulled toward the other atom (often oxygen or nitrogen). The weak attraction between a slightly positively charged hydrogen atom in a molecule and a slightly negatively charged atom in the same or another molecule is called a hydrogen bond. As an example, water molecules may be held together by hydrogen bonds. The bonds are formed because, as we have seen, the electrons are more strongly attracted to the oxygen than to the hydrogens. This means that the oxygen atoms, with their electrons drawn in close about them, become somewhat negative. This slight displacement leaves the hydrogen atoms (the protons from which the electrons were drawn) with a slight positive charge. They therefore have a tendency to attract negatively charged particles, and the nearest such particle is likely to be the oxygen atom of the next water molecule. Thus, water molecules tend to join in a loose, but highly structured and constantly changing latticework:

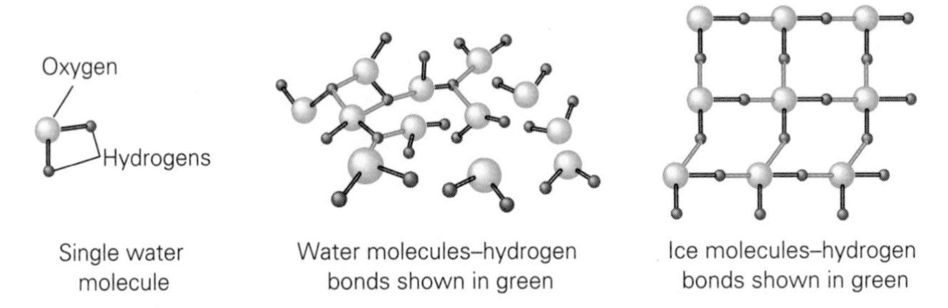

Hydrogen bonds last for only a very brief time—about 10 raised to the negative 11th power (1/100,000,000,000) second. But, in their short existence, they bestow upon water the unusual qualities of being very fluid and, at the same time, relatively stable. The stability of water is quite easily demonstrated by considering the time it takes for it to boil. (Legend has

FIGURE 3.11

The peculiar bonding properties of water allow it to take many forms in the pageant of life on Earth.

it that if the pot is watched, the data may be unreliable.) Why does water so strongly resist being changed to a gas? It is resistant because although the heat easily ruptures the bonds between the molecules, they form new bonds with dazzling speed.

The peculiarities of hydrogen bonding also explain why it is so difficult to freeze water. Ice is crystalline; that is, it is regular and repeating in its molecular structure. But the constantly shifting molecules of water don't hold still long enough to encourage such a regular structure. The molecules are joined so weakly that they continually break and form new bonds (Figure 3.11).

Water, then, because its molecules form hydrogen bonds so easily, is quite physically stable, a trait that has given it special importance in the processes of life. But, you may ask, just why is this sort of stability so important to life?

Actually, there are a number of reasons. For humans and other terrestrial air-breathers, for example, water's stability, or resistance to change, enables us to breathe in a wide range of temperatures and in very dry environments by retarding the evaporation of moisture from our lungs (Figure 3.12). In addition, we are resistant to freezing because our bodies contain so much water. (The formation of ice crystals, of course, can rupture delicate cell membranes.) Furthermore, the unequal charge distribution of water molecules, which permits hydrogen bonding in the first place, also makes water a powerful solvent. Water can form hydrogen bonds with any polar or ionic compound, thereby causing it to dissolve. When they are in solution, the molecules or ions can interact in new ways. In fact, all of the biochemical reactions in our bodies take place in water. Water's constantly changing structure also gives it a certain fluidity and movability that enables it to pass through our bodies' tissues and to seep deep into the earth's crust to reach the roots of the largest trees. Furthermore, because of water's powerful tendency to join, it can form "columns" and move from those roots to the highest leaves as these "columns" are drawn along as units.

I should add that hydrogen bonds are important to a number of critical biological molecules. For example, they hold the two long chromosomal molecules together, as we will see, and they are important in maintaining the configuration of other large molecules.

FIGURE 3.12

Humans have adapted to a wide variety of conditions—including deserts—due to mechanisms that retard the evaporation of water.

CHEMICAL REACTIONS

Most of the molecules associated with life tend to join, separate, shift around, and change their nature. They interact, however, in specific ways, according to the properties with which they are endowed. Such interactions between molecules are referred to as **chemical reactions.** Let's now consider a few common chemical reactions and see why they occur. We will see that, essentially, they simply involve the breaking and forming of bonds.

To begin, consider an example in which two elements simply join. The general statement can be written:

$$A + B \rightarrow AB$$

That is, A plus B yields AB. Suppose these elements are sodium and chlorine joining to make table salt. We can simply plug in their chemical symbols and we have:

$$Na + Cl \rightarrow NaCl$$

Hydrochloric acid (HCl) is formed in much the same way. Here, one ion each of hydrogen and chlorine forms one molecule of hydrochloric acid. This reaction, however, is a bit more complex to write since, in this case, we must show the number of atoms involved. This is because the ionized hydrogen and chlorine would ordinarily be in molecular form, that is, as H_2 and Cl_2, so their union would form two molecules of hydrochloric acid. Thus, the combination would be written as:

$$H_2 + Cl_2 \rightarrow 2HCl$$

Hydrogen gas is made up of molecules of hydrogen (H_2); that is, two atoms of hydrogen are bonded to form a molecule. Of course, only one atom of oxygen is needed to form H_2O, but oxygen also normally appears as a molecule (O_2). Here, we see the bonds of oxygen must be broken in order for new kinds of bonds to form. Then, since each oxygen atom requires two atoms of hydrogen, we get:

$$2H_2 + O_2 \rightarrow 2H_2O$$

That is, two molecules of molecular hydrogen and one molecule of molecular oxygen form two molecules of water. Note that the same number of hydrogen and oxygen *atoms* appear on both sides of the arrow. Thus, the equation is balanced. This is an important principle.

Another important point is that some reactions can go the other way. So far, the "yields" arrow has pointed only to the right, but many reactions are reversible. For example, water can be broken down into hydrogen and oxygen:

$$2H_2 + O_2 \leftarrow 2H_2O$$

As a convenience, if the reaction is more likely to go one way than another, it is shown by two arrows, the larger one indicating the general tendency of the reaction:

$$2H_2O \rightleftharpoons 2H_2 + O_2$$

In another type of chemical reaction, the **reactants** (the molecules involved in the reaction) may simply switch partners, forming the products. When molecules exchange components in this way, they form two completely different kinds of molecules. For instance, if hydrochloric acid (HCl) is combined with sodium hydroxide (NaOH), we get water (H_2O) and table salt (NaCl):

$$HCl + NaOH \rightarrow H_2O + NaCl$$

Notice, again, that the same elements, in the same quantities, appear on both sides of the reaction.

Energy of Activation

Some reactions occur very easily, but others only with great difficulty. In the easy cases, the reactions occur spontaneously if chemicals are simply mixed together. Dabbling amateur chemists sometimes blow up their labs this way. Reactions that proceed more reluctantly generally require an energy boost of some kind, just as a rock at the top of a hill may need a shove before it begins tumbling down. As a simple example, we know that a match can burn, but it can also just sit there in its box, doing quite nicely, thank you, without ever bursting into flame. To initiate the oxidation process one adds energy, generally by striking the match on the back of the leg, unless the tuxedo is rented. The heat of friction provides the boost and initiates the oxidation process. Once the reaction has started, the energy is provided by the heat of the flame, and the process will continue until it reaches your fingers or until the wood is depleted. The energy boost required to initiate a chemical reaction is called the **energy of activation.** (In spontaneous reactions, normal environmental conditions provide the energy of activation.)

Why, though, must a match be struck? How does the heat of friction provide the energy of activation? What does heat do? The addition of heat increases the **kinetic energy** of molecules; that is, it increases the speed of their random movements. (Molecules are in constant movement, even in solid substances.) The more active the molecules are, the greater the likelihood that they will bump into each other and it is in just such encounters that molecular bonds are altered. The oxidation that produces flame is based on such changes in bonds.

Another way of increasing the likelihood of molecular interactions (or speeding up chemical reactions) is to increase the concentrations of the two reactants. The more molecules there are, the greater the likelihood they will interact. The same principle can be achieved by subjecting a mixture of the reactants to pressure, thereby forcing the molecules closer together. We see, then, that the chemical reactions can be encouraged or accelerated by a number of processes, such as adding heat, increasing the concentration, or increasing the pressure.

Such methods are fine for encouraging reactions in the laboratory, but how about chemical reactions in living things? Imagine the effects of heat, pressure, or varying concentrations of solutes in a delicately balanced living body. Such drastic changes would be disruptive to say the least. Obviously, living things have other ways to encourage chemical reactions; ways that are, perhaps, more compatible with the delicate mechanisms of life.

Enzymes

If you tried to impress your little brother by mixing hydrogen and oxygen to form water, his attention span would be severely taxed and he would probably wander off, because nothing would happen. However, should you coax him into watching again, add a tiny piece of platinum to the mixture. He may never watch you do anything again, because this demonstration will blow his hat off. Furthermore, after you recover your composure, you can also recover your platinum. This is because platinum in this case served as a **catalyst.** A catalyst speeds up reactions between other elements without becoming part of the product. Thus, the catalyst emerged unchanged. Catalysts do not change the type of reaction that can take place or the direction of the reaction; they simply make it easier for the reaction to occur by lowering the necessary energy of activation. It is through such catalytic reactions that the "cold" chemistry of life can take place.

Special types of catalysts found within living things are called **enzymes** (defined by one memorable student as "naval officers"). Technically, they are among the molecular group known as **proteins** (about which we will say more later).

Enzymes are generally large, complex molecules with very specific shapes. Their shapes, in fact, are very critical to their functions. An enzyme's contorted configuration, produced as it loops and coils back on itself, forms a kind of groove or slot on the surface of the enzymatic mass. This slot is called the **active site** because of its role in the enzyme's chemical behavior. Lining the active site are certain functional groups (side groups with very specific chemical traits) arranged very specifically. These functional groups are able to interact with certain combinations of molecules with which they come into contact and to change them in some way. The

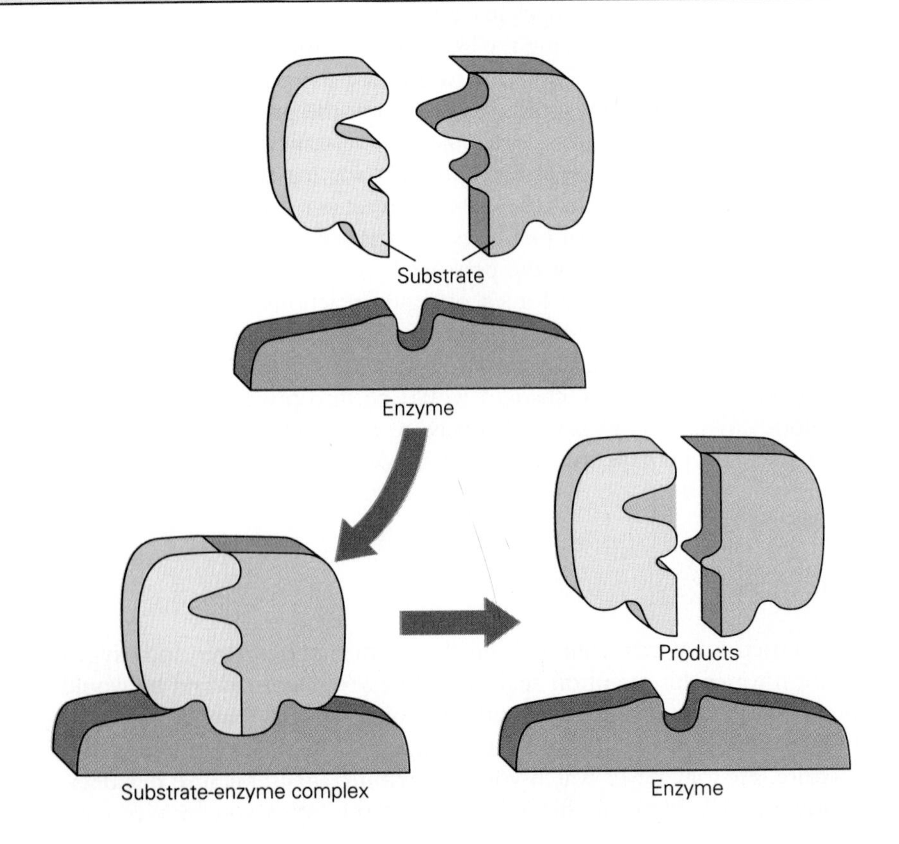

FIGURE 3.13

How an enzyme is believed to work. The enzyme, usually a complex protein, has an area that roughly fits the substrate upon which it will act. In this example, the enzyme is rearranging the component parts of two substrates. In other cases, enzymes may join smaller molecules to form larger ones. The presence of the substrate in the active site causes the enzyme to change shape slightly, making the fit between the substrate and the active site more precise. The molding of the enzyme molecule also stresses the substrate molecule, weakening bonds or exposing new regions to other reactants in the medium. As a result, the substrate molecule, which would otherwise have been quite stable, reacts. After the reaction, the substrate breaks away from the enzyme, leaving it unchanged and ready to initiate the same reaction with new substrates. Each type of enzyme reacts with only one kind of substrate.

molecules they act upon are referred to as **substrate.** The substrate binds to the enzyme's active site, forming a temporary enzyme-substrate complex. The molecular combination resulting from enzymatic activity is called the **product.**

Enzymes are specific. Each recognizes and interacts with only its own substrate because the shape of the active site is complementary to that of the substrate molecule. However, before the substrate slips into the active site, the fit between them is inexact. Once in place, the substrate induces a change in the shape of the enzyme that makes the active site embrace it snugly, like a hand grasping another in a handshake. This shifts the positions of the functional groups in the active site, enhancing their ability to work on the substrate. Most enzymes are believed to work by placing stress on the bonds of substrate molecules, weakening them. This lowers the activation energy, thereby aiding the formation of the product. When the interaction is completed, the product breaks away and the enzyme, unchanged, is ready to repeat the sequence (Figure 3.13). This whole process takes place so quickly that a single enzyme may go through several million such cycles each minute in a dazzling display of efficiency.

Enzymes are able to bring about several kinds of reactions. For example, they may encourage substrate molecules to *join* to other molecules. We see this as our bodies build large complex sugars from smaller subunits. Enzymatic action can also cause the substrate to be broken apart into its components. The disruption occurs when the altered substrate joins with H^+ or OH^- ions. Digestive enzymes, for example, function by such disruption.

Energy Changes

As the molecules of life square dance their way through their complex pageant, they must alternatively shift directions, change partners, stop, and go. But with each change, they alter their energy levels. It is important to keep in mind that chemical reactions generally result in a change in the energy level of the reactant molecules. In other words, the product of a chemical reaction will have either more or less energy than did the starting materials. Chemical reactions in which the product has more energy than the starting materials are called **endergonic** (or energy gaining); those in which the product has less energy than the starting materials is called **exergonic** (energy releasing).

It may have occurred to you that if a reaction results in a product with more energy than the starting materials, that energy would have come from somewhere—it would have to be *added* to the process somewhere along the line. Just as a boulder cannot roll up a hill without a push, neither can endergonic chemical reactions take place without energy somehow being expended. A general statement about endergonic reactions can be written as:

$$\begin{array}{c}\text{Energy} \\ \searrow \\ A + B \rightarrow AB\end{array}$$

(The addition or subtraction of some factor during a reaction is traditionally shown by a curved arrow at the step where it occurs.)

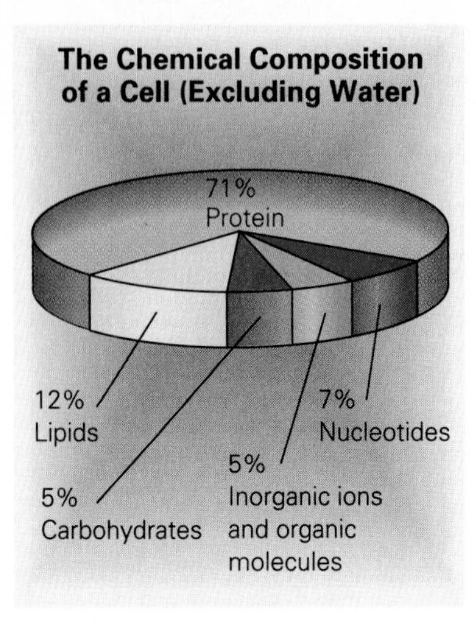

TABLE 3.4
The Macromolecules

MACROMOLECULE	COMPOSITION
Carbohydrates	Monosaccharides
Lipids	Glycerol and fatty acids
Proteins	Amino acids
Nucleotides	Sugar, phosphate group, and nitrogenous base

To continue our analogy, once we have gotten our boulder up the hill, it then holds considerable potential energy. Furthermore, if it should roll back down, that energy would be released along the way. The rolling down, then, is analogous to an exergonic reaction. When the product of a chemical reaction has less energy than the starting material (that extra energy being released during the reaction), the process can be summarized this way:

$$AB \rightarrow A + B$$

Energy

If the boulder should roll over you on its way down, the energy it releases might cause certain changes in your body. In this case, the boulder's energy would be doing *work* (causing changes). We'll see somewhat more useful ways energy can cause changes, but first let's review what sorts of molecules tend to wind their way through into this complex business we call life. We will consider only a few substances here, and they all have such familiar names that you may suspect that you already know what they are. For example, everyone knows what carbohydrates are, right? And fats? Let's see.

THE MOLECULES OF LIFE

It would seem that life might one day be understood in terms of the molecules that make up living things. Furthermore, in trying to retrace the development of life, we quickly become involved with the molecules of a raw and primitive earth. So although we still can't say exactly what life is and how it began, we are led to take a closer look at life's molecules and how they behave. (The molecules and their composition are summarized in Table 3.4).

The Magic of Carbon

Carbon has been described as the element most intimately associated with life, but you may have noticed that we haven't said much about it so far. Perhaps a look at Table 3.3 will show why. Carbon, you see, has six

FIGURE 3.14A

Methane is the simplest hydrocarbon, composed of one carbon and four hydrogen atoms covalently bonded. *Note that by sharing their electrons, the hydrogen atoms fill the shell requirements of both the carbon and the hydrogens.*

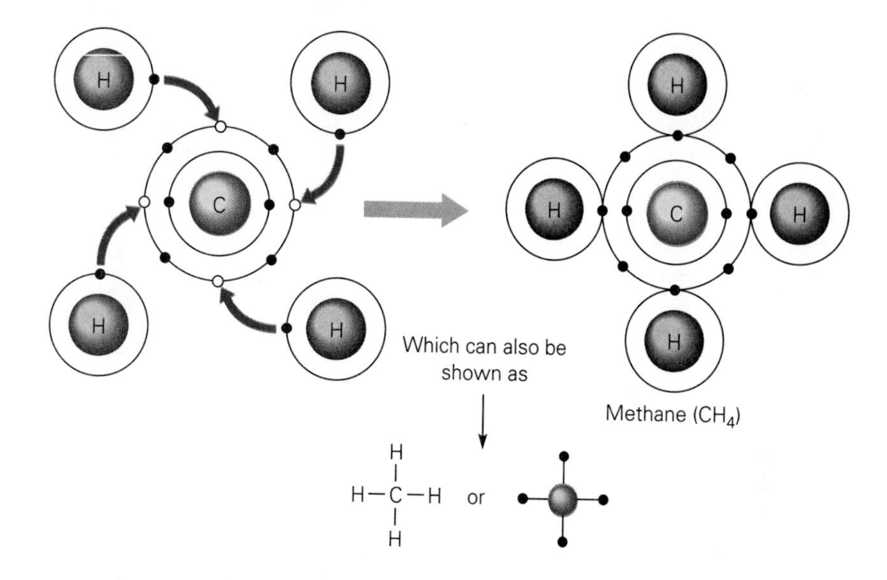

protons and six electrons, and thus it presents a special case. How might an atom like this be expected to react? After all, since two electrons will occupy its inner shell, it can only have four in its outer shell. In order to satisfy its shell requirements then, it must either give up these four electrons or gain four more. Either of these changes, however, would throw its electric charge too far out of balance. The only way carbon can resolve this dilemma is to share its four outer electrons with other atoms through covalent bonding.

The problem becomes interesting at this point because carbon can combine with either atoms that gain or atoms that lose electrons. For example, it can share the electrons of oxygen, nitrogen, sulfur, or phosphorus atoms, or it can share electrons with hydrogen atoms. If a carbon atom simply adds four hydrogens, the result is CH_4, the swamp gas called methane (Figure 3.14).

It should be pointed out that the four hydrogens do not simply stick out at right angles, as the diagram seems to indicate. Instead, they protrude three-dimensionally in such a way as to be as far from each other as possible. Thus, the structure would be more accurately represented as:

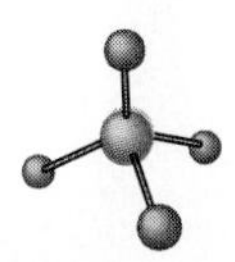

In ancient times, spikes on certain weapons were arranged in such a way because when they were thrown in the path of an advancing army, one spike always pointed up.

FIGURE 3.14B

Methane is a common compound of "swamp gas," which rises from mud, causing bubbles that make some people wonder what's down there. (Bacteria)

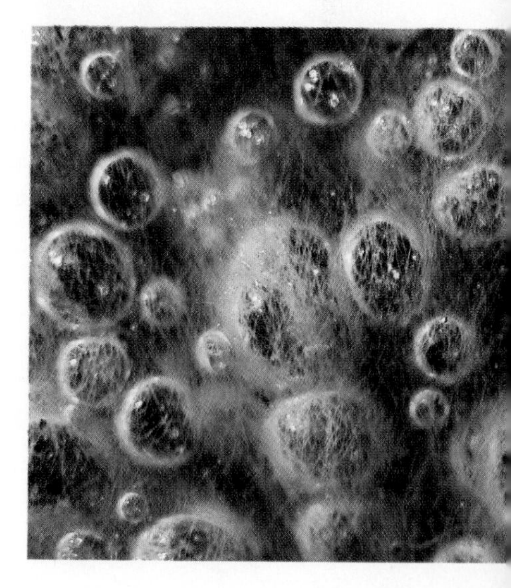

FIGURE 3.15
Some of the simpler carbon chains.

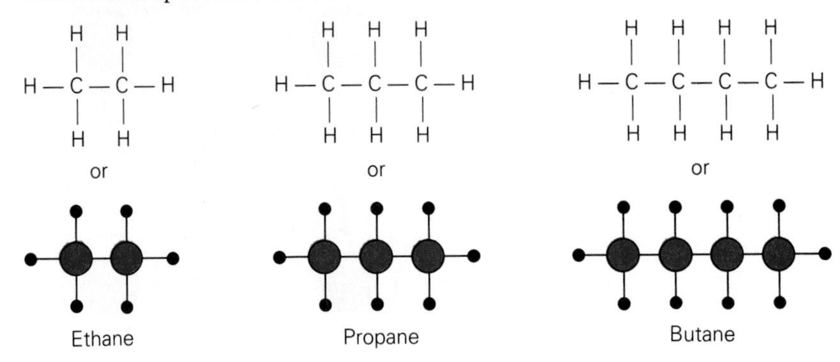

In addition, because of their peculiar bonding properties, carbon atoms can interact with each other as well. In this way, carbon atoms can join together to form complex molecules, as we shall see shortly. (Figure 3.15 shows some of the simpler carbon chains. It is interesting that of the some 2.2 million chemical compounds known, about 2 million contain carbon.)

Molecules produced by any combination of hydrogen and carbon atoms are called **hydrocarbons.** There are some very long hydrocarbon chains (Figure 3.16). You can see that in these kinds of molecules, the electron-shell requirements of every atom are satisfied; each carbon atom can form covalent bonds, and each hydrogen atom is satisfied through a single covalent bond with a carbon atom.

FIGURE 3.16
Among the longest hydrocarbons are those found in crude oil. Refinement generally involves shortening these huge molecules.

Because of the properties of carbon, some very complex configurations can be formed. For example, carbon chains can branch. A five-carbon branch can be written as:

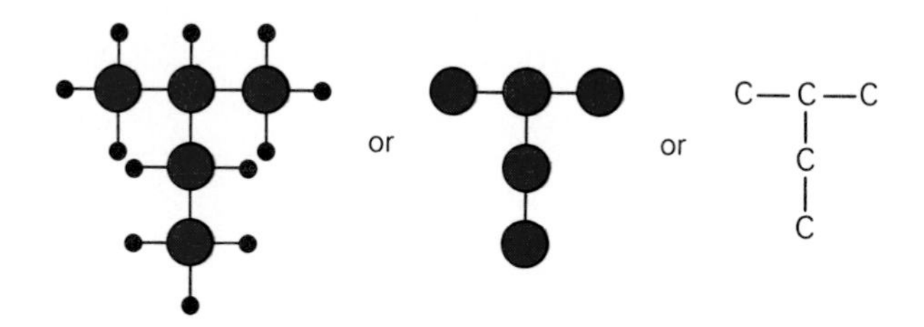

Carbon can also join ends to form rings. Five- or six-carbon rings are especially common:

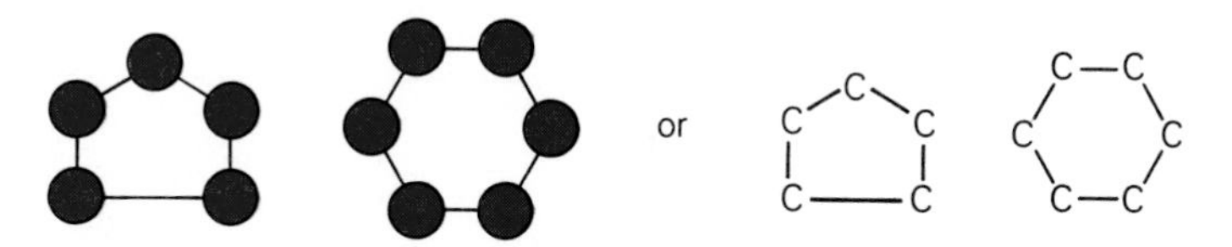

The notation of such molecules can be further simplified and written as:

You should be aware that if no atom is shown where one is expected along a carbon backbone, that atom exists and is hydrogen. In such a case it is assumed to be present. Other atoms attached to carbon are always noted.

Molecules with the same carbon backbone differ in their properties depending on the **functional group(s)** attached to the backbone. A functional group is a molecular side group that can enter into only specific chemical reactions. Among them are the **hydroxyl groups,** —OH, which combine with carbon to form alcohols (Table 3.5), the **amino groups,** —NH_2, which can form substances called amines (notations below), and

or — O — H or —N(H)(H)

the **carboxyl groups,**—COOH, (Essay 3.1) which form organic acids:

or —C(OH)=O

Functional groups tend to behave as a unit. That is, they move together as a group when the larger molecule of which they were a part becomes dissociated. If they break away from the molecule and exist as unattached charged particles, they can produce **free radicals.**

As we will see, the peculiar bonding of carbon not only permits the formation of large and complex molecules, but those sorts of molecules have precisely the traits that permit the complex processes of life. In fact,

TABLE 3.5
Functional Groups

FUNCTIONAL GROUP	STRUCTURE	OCCURS IN
Alcohol	Ⓡ—OH	Sugars
Acid	Ⓡ—C(=O)—OH	Sugars Fats Amino acids
Aldehyde	Ⓡ—C(=O)—H	Sugars
Amino	Ⓡ—N(—H)—H	Amino acids Proteins
Sulfhydryl	Ⓡ—SH	Amino acids Proteins
Phosphate	Ⓡ—O—P(=O)(—OH)—OH	Phospholipids Nucleotides Nucleic acids

Ⓡ = remainder of molecule

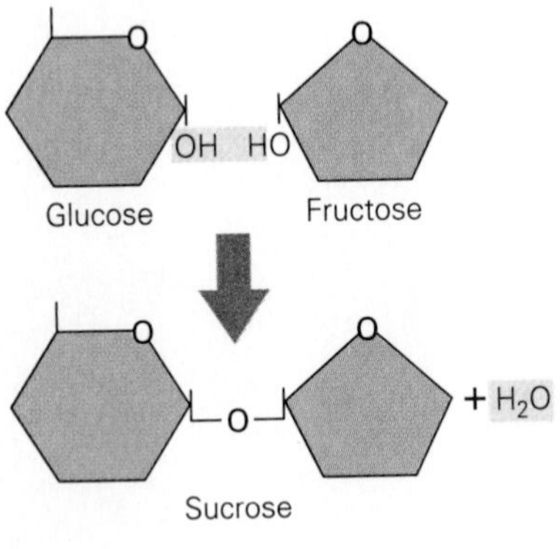

it can be said that because carbon is so intimately associated with the processes of living things, the chemistry of carbon is, in a sense, the chemistry of life.

The Role of Water in Forming Large Biological Molecules

We have seen that carbon atoms can be linked to form chains and rings. They can also form chains of rings and a host of other complex structures. Indeed, cells are composed primarily of large carbon-containing molecules, called macromolecules. In spite of their size, the structure of macromolecules is fairly easy to understand because each is comprised of repeating units. The units are called **monomers,** and the macromolecule is called a **polymer.** The monomers are joined together by **dehydration**—that is, the removal of the atoms that form water. On the other hand, when the components of water are *added* **(hydration)** to any molecule formed by dehydration, the molecule will be broken down into its component substances, a process called **hydrolysis** (Figure 3.17 illustrates simple examples of dehydration and of hydrolysis involving sucrose—table sugar.)

FIGURE 3.17
Sucrose, or table sugar, is a double sugar composed of both glucose and fructose. In the formation of sucrose, glucose loses an OH group that joins with an H from fructose to form water. The two simple sugars then join at the points at which these atoms were lost. The reaction does not proceed easily, and in living systems, an energy source must be provided. When the atoms of water are chemically added to sucrose, it breaks down into glucose and fructose, its component parts. Reactions involving loss of water are called *dehydrations;* those in which water is added are called *hydrations;* and they result in a chemical breaking apart known as *hydrolysis.*

ESSAY 3.1

Acids and Bases

Acids are recognized by their sour taste. An acid is any substance that donates positive hydrogen ions (H^+) to a solution. Since H^+ is hydrogen that has lost its electron, it is simply a proton. Thus, acids can be considered proton donors. Hydrochloric acid (HCl) is a "strong" acid because it has a strong tendency to dissociate, or to release H^+ ions into its environment. It cannot be neutralized by the negative chloride ions, Cl^-, that are released by the same action.

Bases taste bitter and flat. They reduce the amount of H^+ in a solution. In other words, bases are proton acceptors. The ion OH^- is a powerful base because it will readily capture free H^+ ions to form H_2O. For this reason, when an acid and a base combine, they form a compound plus water. To use a familiar example of neutralization:

$$NaOH + HCl \rightarrow NaCl + H_2O.$$

The acidity of a substance is measured on the pH scale. The term *pH* means "hydrogen power." The reference point on this scale is 7, the pH of pure water. Thus, substances with a pH value below 7 are acids and those with a pH value above 7 are bases. A substance with a pH of 7 is neutral. The scale is based on the concentration of hydrogen in one liter of water. For example, pH 3 means that there are 10^{-3} moles of hydrogen ions per liter of water, and pH 8 means that there are 10^{-8} moles per liter. A mole is the number of grams of a compound equal to its molecular weight. Almost all biological processes take place in a pH environment of 6 to 8, with a few important exceptions such as digestive processes in the stomach (which occur at about pH 1). The pH level in living systems is sometimes regulated by buffers, acid-base pairs that serve to "soak up" small amounts of excess acid or base ions.

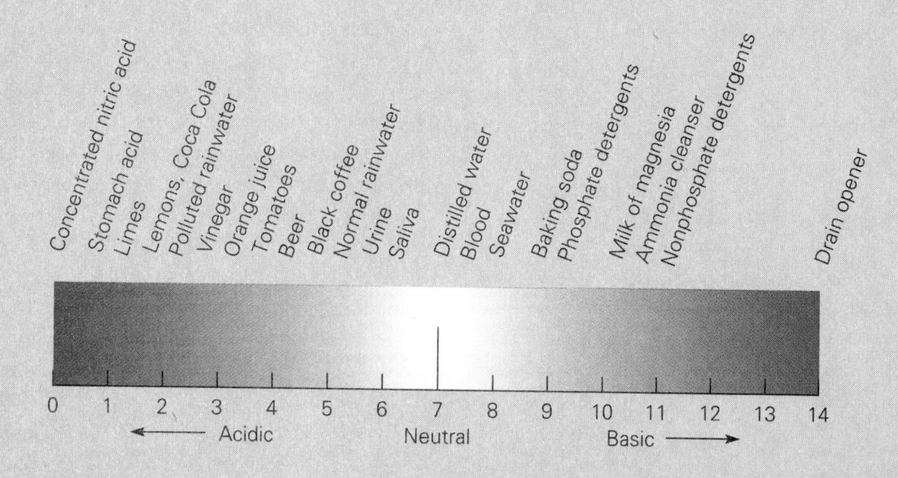

Carbohydrates

The first important group of molecules we will consider that are associated with life are the "familiar" carbohydrates. (Some have discovered that it's really not easy to be very familiar with a carbohydrate.) Let's begin with a simple definition. **Carbohydrates** are molecules that generally contain carbon, hydrogen, and oxygen in a proportion of one carbon to two hydrogens to one oxygen atom (the general formula for a monomer of a carbohydrate is thus CH_2O). (The simpler carbohydrates are also called **sugars,** and you may occasionally see the terms used interchangeably.)

In a nutshell, carbohydrates are important because they are the principal energy source for most living things and because they serve as the basic material from which many other kinds of molecules are built. Some carbohydrates are simple, such as glucose and fructose. Some carbohydrates, such as these two, may exist as either chains or rings. Other molecules of carbohydrates may be much larger, forming very long chains. The simpler carbohydrate molecules, those containing a single sugar group, are called **monosaccharides,** or *simple sugars* (Figure 3.18). A carbohydrate composed of two simple sugars is called a **disaccharide,** or double sugar. For example, table sugar (sucrose) is a disaccharide formed from the monosaccharides glucose and fructose. Long chains of monosaccharides are called **polysaccharides.**

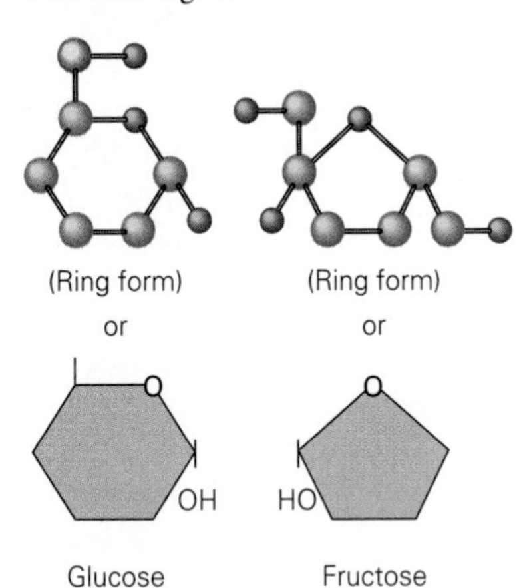

FIGURE 3.18
Glucose and fructose are important simple sugars. They are called simple sugars because they are not combined with other sugars.

Glycogen is a polysaccharide formed in many animals. It is essentially long branching chains of glucose units linked together by dehydration reactions (see Figure 3.17).

In mammals, glycogen is the principal form in which glucose is stored in the body, chiefly in the liver. When your body, for example, is in need of glucose, glycogen is hydrolyzed (breaking the sugar linkages) and the component glucose molecules are released into your bloodstream.

Plant cells store their glucose in a polysaccharide called **starch,** chains of glucose units that are less highly branched than those of glycogen. The glucose chains in starch coil and form starch granules (Figure 3.19).

FIGURE 3.19
Plants store their glucose in starch granules (upper inset). The pheasant, as in all animals, stores its starch in glycogen molecules (lower inset). The flight muscles of ground-feeders are composed largely of glycogen because it is readily converted to the energy they need to escape their predators.

FIGURE 3.20

Because of the bonding characteristics of cellulose (left), it is much less soluble than starch (right), although both are composed of units of glucose.

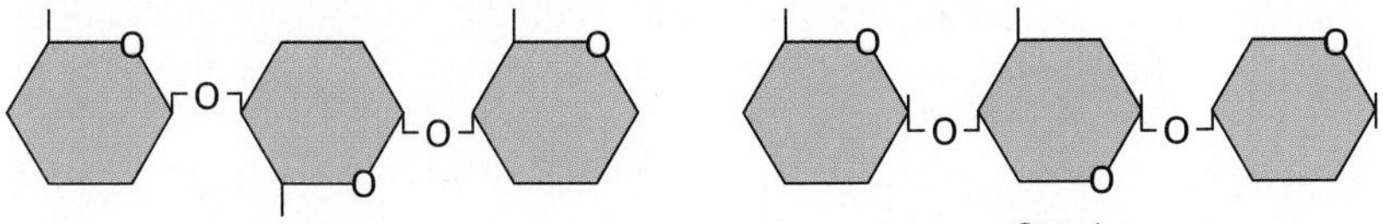

Cellulose

Starch
(segment of glycogen)

Cellulose, a relatively insoluble polysaccharide common in plants, is formed when glucose units are joined together by slightly different bonds than are found in glycogen or starch (Figure 3.20). Our digestive systems can readily break down starches into their component parts, but the linkages in cellulose cause the molecules to coil tightly around themselves so that the bonds are simply not accessible to enzymes. Thus, such molecules are rather resistant to the digestive enzymes of many animals. This is why people don't each much wood and rarely graze. Of course, termites eat wood and cows graze, but they have help; their digestive tracts harbor tiny organisms that can break down cellulose into its component sugars.

The hard covering, or exoskeleton, of insects is composed of **chitin,** which is made up of repeating six-carbon units like glucose, except these contain nitrogen. The units are bonded like those in cellulose, so that they are not easily digested by many species (Figure 3.21). However, on this point my confidence was once shaken as I followed a young man down a street in Austin, Texas, during one of the area's regular cricket invasions. He suddenly scooped a handful of the hapless insects from a wall and stuffed them into his mouth. I suppose I could have explained to him the nature of sugar bondings, but I walked into a trash can instead.

FIGURE 3.21

The hard skin of these crickets is made of a carbohydrate chitin related to cellulose.

Lipids

Like carbohydrates, **lipids** are composed of carbon, hydrogen, and oxygen. Lipids, however, have far less oxygen. Also, unlike carbohydrates, lipids are generally insoluble in water.

There are a number of types of lipids, but **fats** are the best known. The fats are important because they store a great deal of energy. In fact, they contain twice as much energy for their weight as do carbohydrates or proteins (Figure 3.22). You would think, then, that fat people would be bursting with strength and vigor. You'll see as we go along why this is usually not the case. But first, let's take a look at the composition of fats.

Each molecule of fat consists of two types of compounds, **glycerol** and **fatty acids.** Glycerol is a chain of three carbon atoms, each with a hydroxyl (OH) group. Fatty acids are generally chains of carbon and hydrogen that end in carboxyl groups. Fats are formed by the hydroxyl

FIGURE 3.22

Fat molecules (inset) are high in energy, so we find that the flight muscles of long-distance migrators have high concentrations of fat.

groups (OH) of the carboxyl (COOH) joining with the hydrogens of glycerol after dehydration. The process is outlined in Figure 3.23.

Fatty acid chains can be highly variable. Most are 16 to 18 carbon units long, but some are shorter and some longer. Also, not only may the three fatty acid components be of different lengths, but fats may differ in the way the carbons are bonded together. For example, in some fats, the carbon chains carry all the hydrogens possible, while in others, the carbon atoms have fewer hydrogens because they are joined by a double bond (Figure 3.24). A fat with its carbon chains filled with hydrogens is said to be **saturated.** Guess what one with double bonds—that is, with room for more hydrogens—is called. (The answer is **unsaturated,** but you knew that.) The biological properties of fats are largely determined by the length of the chains and the degree of saturation.

Most animal fats are saturated. The result is they tend to solidify at room temperatures. Examples are lard, grease, and butter. On the other hand, plant fats are, in most cases, liquid at room temperature—as in

FIGURE 3.23

A fat molecule is composed of a glycerol molecule joined to fatty acids. Three molecules of water are lost in the bonding of three fatty acid molecules to a single molecule of glycerol. The fatty acids form long hydrocarbon chains, the ends of which link with the glycerol molecule. The key questions that determine the nature of the fat are: How long are the chains? and, Are the fatty acids saturated or unsaturated? (as we see in the next figure).

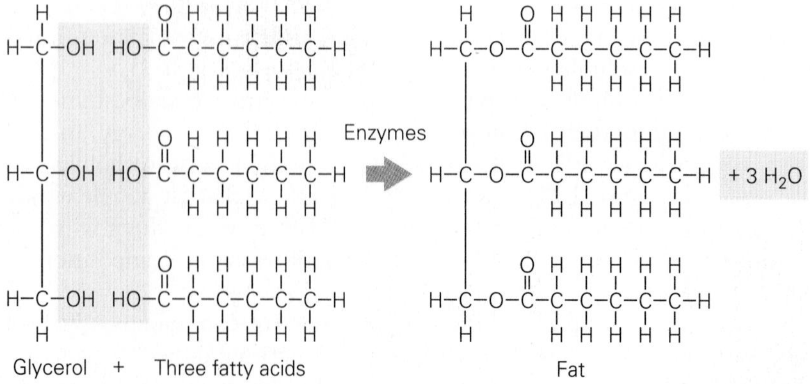

```
   H        O H H H H H               H   O H H H H H
   |        ‖ | | | | |               |   ‖ | | | | |
 H-C-OH  HO-C-C-C-C-C-C-H           H-C-O-C-C-C-C-C-C-H
   |          | | | | |               |     | | | | |
   |          H H H H H               |     H H H H H
   |                                  |
   |        O H H H H H   Enzymes     |   O H H H H H
   |        ‖ | | | | |               |   ‖ | | | | |
 H-C-OH  HO-C-C-C-C-C-C-H   ==>     H-C-O-C-C-C-C-C-C-H  + 3 H2O
   |          | | | | |               |     | | | | |
   |          H H H H H               |     H H H H H
   |                                  |
   |        O H H H H H               |   O H H H H H
   |        ‖ | | | | |               |   ‖ | | | | |
 H-C-OH  HO-C-C-C-C-C-C-H           H-C-O-C-C-C-C-C-C-H
   |          | | | | |               |     | | | | |
   H          H H H H H               H     H H H H H

 Glycerol  +  Three fatty acids              Fat
```

FIGURE 3.24

Saturated and unsaturated fats. Carbon atoms linked by double bonds form unsaturated fats because they are able to form new bonds with other atoms. Unsaturated fats are generally oily, like olive oil, while saturated fats, like lard, are usually solid at body temperature.

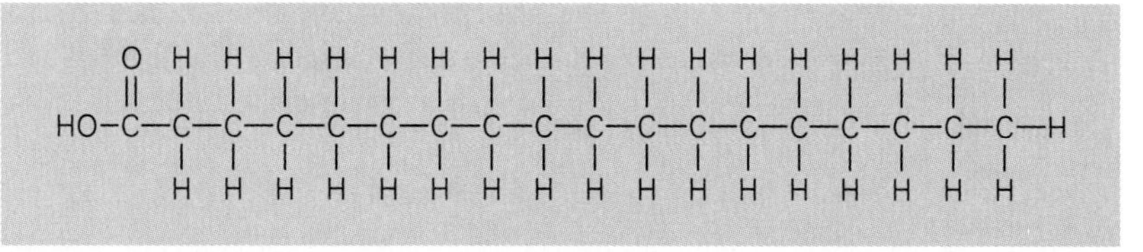

A saturated fat

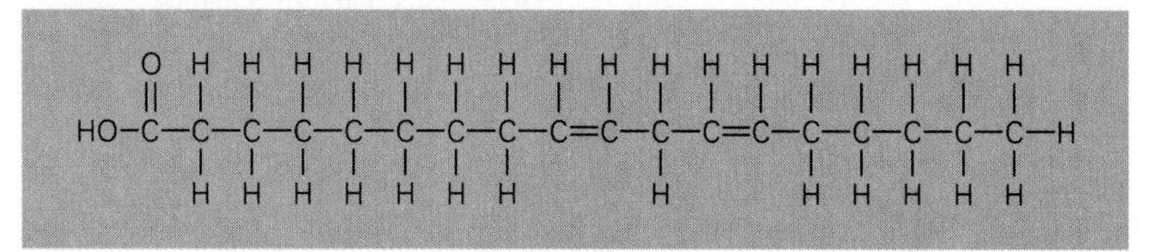

An unsaturated fat

vegetable oils and olive oil. The reason is, the double bonds cause the vegetable fat molecules to form irregular shapes that prevent them from packing together close enough to solidify. When hydrogens are artificially added to such fats (as we see in "hydrogenated vegetable oils"), the result is fat that is solid at room temperature. The vegetable fats in margarine and peanut butter are treated this way to prevent the oils from separating out. As we will see in Chapter 19, animal fats can contribute to the cardiovascular disease, atherosclerosis, in which "plaques" (deposits) accumulate inside blood vessels, damaging them and impeding blood flow.

The lipids also include the **waxes,** molecules with very long and saturated fatty acid chains combined with long-chain alcohols or with carbon rings. These are usually solid at room temperature. And then there are the **phospholipids,** in which the third hydroxyl group of glycerol is attached to a molecule containing phosphorus. Phospholipids are important in the structure of living membranes, as we shall see.

The last lipid group we will consider is the **steroids.** The steroids are fatlike substances with a different structure than anything we have considered. Steroids are formed from four interlocking carbon rings with many types of side groups (Figure 3.25). For example, cholesterol is a

FIGURE 3.25

Two steroids. All steroids are composed of four basic interlocking rings, but a variety of side groups may be attached to those rings. Cholesterol is abundant in foods such as milk, butter, and eggs. Cholesterol circulates in the bloodstream and, if present at high levels, it may be deposited in the arteries, causing arteriosclerosis and possibly a heart attack. Testosterone, another steroid, is synthesized in the body and contributes to the male characteristics, such as musculature, beard, and deep voice.

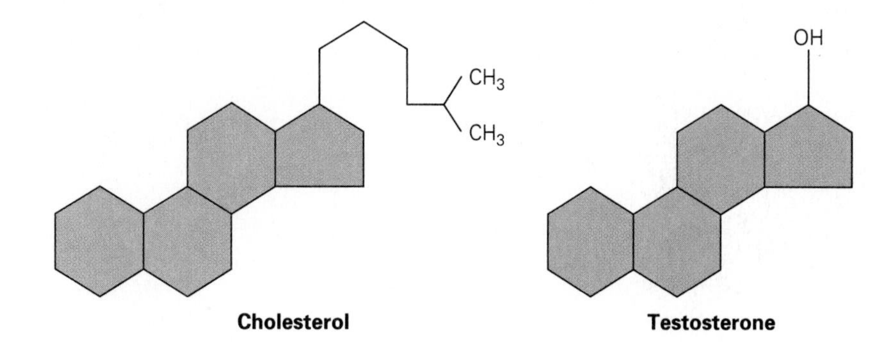

steroid. Some people develop a greater interest in steroids when they learn that sex hormones are included in this group. Certain synthetic male hormones have been abused by athletes seeking greater bulk and strength. The unfortunate side effects include physical, mental, and emotional problems, as is described in Essay 3.2.

Proteins

The general structure of proteins is easy to describe, but the simplicity is deceptive. Proteins are actually immense molecules with incredibly complex arrangements. Let us not be deterred, however, as we eagerly plunge ahead into new realms of knowledge.

Amino Acids

Proteins are made up of chains of smaller nitrogen-containing molecules called **amino acids.** As you may guess from the name, amino acids contain at least two kinds of functional groups, amino groups (NH_2) and acids ($COOH$). Both these groups are attached to the same carbon. The rest of the molecule contains carbon that may take a variety of forms, including chains, branches, or loops. So, for simplicity, we'll just call the group attached to the carbon that holds the amino and the acid group, R, for radical (which means "root," as in "radish"). For a molecule to qualify as an amino acid, it need only have the configuration shown at the left.

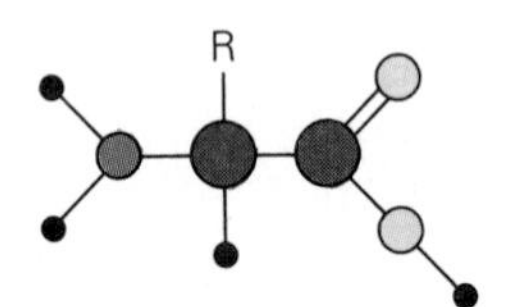

There are many amino acids, but only 20 are found in proteins. Some of these are shown in Figure 3.26. As you can see, the carbon group R determines the identity of the compound.

FIGURE 3.26

Some of the amino acids found in proteins. Note the variety of structures among amino acids, but also note that they all have two things in common: an amino group and an acid, or carboxyl, group. Proteins contain only 20 different kinds of amino acids, and the human body can manufacture 12 of these from other amino acids. It is essential, however, that the other eight be provided in the diet, so they are called the "essential" amino acids. It seems paradoxical that the amino acid element that is probably available from the fewest sources is nitrogen. And yet, it is this same nitrogen that provides us with such problems as its disposal, as we will see in our discussion of the kidney. The importance of amino acids, of course, lies in their role in the formation of proteins.

Glutamine

Glycine

Serine

Threonine

Phenylalanine

Alanine

Valine

Aspartic acid

Cysteine

ESSAY 3.2

Steroids, the Drug of "Champions"

In 1988, at the Seoul Olympics, sprinter Ben Johnson of Canada burst out of the starting blocks to win the 100 meter dash, leaving American Carl Lewis in the dust. Three days later the gold medal was awarded to Lewis because traces of stanozolol were found in Johnson's urine. Stanozolol use is banned in international sports because it is an anabolic (tissue-building) steroid. Anabolic steroids burst upon the athletic scene in Olympic competition in the 1950s, when it was learned that Soviet weightlifters were using them to increase their strength. The Americans quickly followed suit and the drugs were soon being used in any sport that required strength (as does virtually every sport). At first steroids were used primarily by elite athletes looking for the final edge that might put them over the top. Soon, however, they found their way into local gyms where body builders of mediocre abilities and grand egos were wolfing down pills in hopes of looking good around the pool.

The major benefit of steroids seems to be to allow muscles to recover more quickly, so that the athlete can train harder. As athletes were reporting remarkable results with steroids, the medical community began testing the effects of the drugs and the drugs were soon banned. Medical researchers reported a variety of remarkable side effects of steroid use including liver cancer, heart disease, and kidney damage. One problem with the drugs is that, along with tissue building, they also "masculinize." The masculinization is particularly acute for women who may grow facial hair as their voices deepen and their breasts decrease in size. They may, indeed, gain muscle mass, but the masculinizing effects may be impossible to reverse. In adolescents, steroids hasten maturation and may cause growth to stop and the loss of hair in boys. Strangely, in men, the high levels of steroids in the body may cause the body's own production of male hormones to cease, resulting in enlarged breasts and shrunken testes.

Anabolic steroids also can cause behavioral changes by increasing the aggressiveness of the user. The aggressiveness may be manifest by more vigorous training but also by hostile and sometimes violent social interaction. Interestingly, much of what we know about steroid use comes from observations of abusers since the medical community cannot administer such doses for athletes because of ethical considerations.

Peptides

A chain of up to 300 amino acids may join together to form long peptide chains or **polypeptides** (*poly,* many). Peptide chains, sometimes simply called **peptides,** are smaller than proteins, technically. However, they can join to make up a protein. The amino acids in polypeptides or proteins need not be of the same type, but they are always linked together in the same way: the OH part of the carboxyl group of one molecule joins with the H of the amine group of another by dehydration, as shown in Figure 3.27. The linkage between amino acids is called a **peptide bond.**

It is the *sequence* of amino acids in the chains that determines the biological character of the protein molecule. It has been discovered that a change in the position of even a single amino acid in a sequence of thousands can alter or destroy the activity of the protein. Imagine the incredible variety of proteins that might be possible by simply varying the number and sequence of 20 amino acids. It turns out, in fact, that although all proteins utilize only those 20 amino acids as building blocks, almost every species on earth has some proteins that are peculiar to it alone.

Protein Structure

A protein molecule may be described in several ways. For example, its amino acid sequence determines its **primary structure.** (For the story of the discovery of one such sequence, see Essay 3.3.) The **secondary structure** of a protein is due to hydrogen bonds between every fourth amino acid. Some proteins form a right-handed coil, called an alpha helix. Others form "pleated sheets." A few proteins that have structural functions in the body, such as the keratin in hair, nails, horns, and bird feathers, can be described only by their primary and secondary arrangements.

Other, more complex proteins fold back on themselves in what is described as a **tertiary structure.** The folding is very elaborate and highly specific, giving the protein the characteristic shape that is usually so important to its function. These folded proteins form dense clumps, also called globules (Figure 3.28). Enzymes, for example, are usually globular proteins with complex tertiary structures. Any such protein maintains its tertiary structure by means of hydrogen bonds, hydrophobic (water-

FIGURE 3.27

The carboxyl group of one amino acid can join with the amine group of the next amino acid by each contributing to the formation of water. The result is a peptide bond. Polypeptides then, and their larger cousins, proteins, are chains of amino acids joined by peptide bonds.

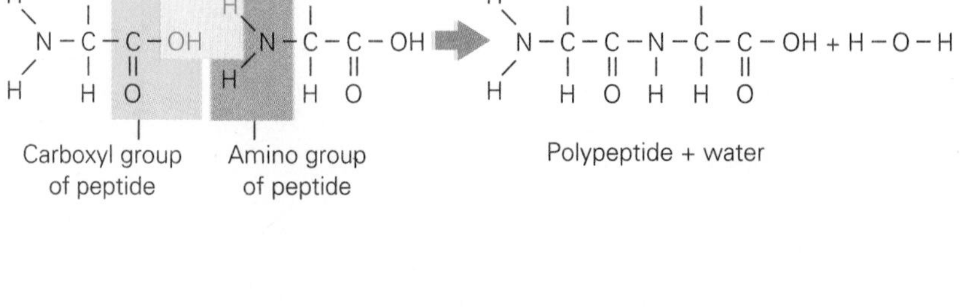

H_2O

Carboxyl group of peptide

Amino group of peptide

Polypeptide + water

FIGURE 3.28

Levels of protein structure. Proteins can be described first according to their primary structure (the sequence of specific amino acids). Their secondary structure describes how this long molecule spirals or folds along its length (largely due to the hydrogen bonds that occur at specific intervals). Their tertiary structure is determined by how that molecule folds back on itself to form a globular molecule. The most complex proteins form a quaternary structure by intertwining with other molecules and interacting with them in specific ways.

(a) Primary structure

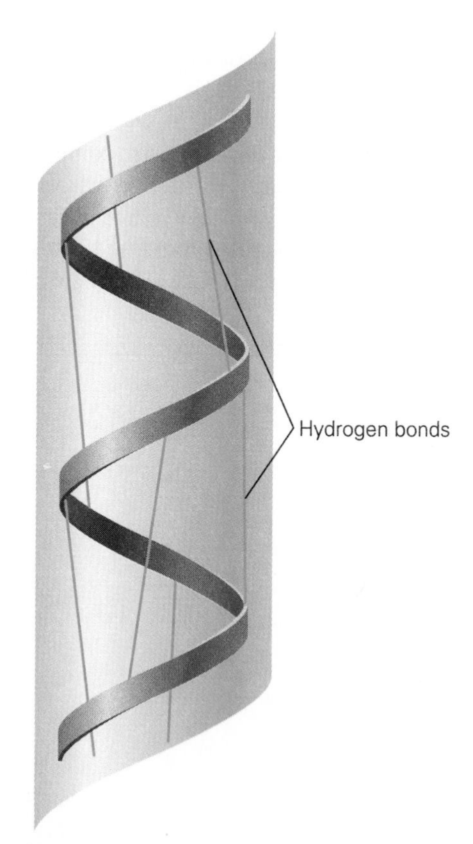

(b) Secondary structure (alpha helix with hydrogen bonds between carboxyl and amino group of amino acids)

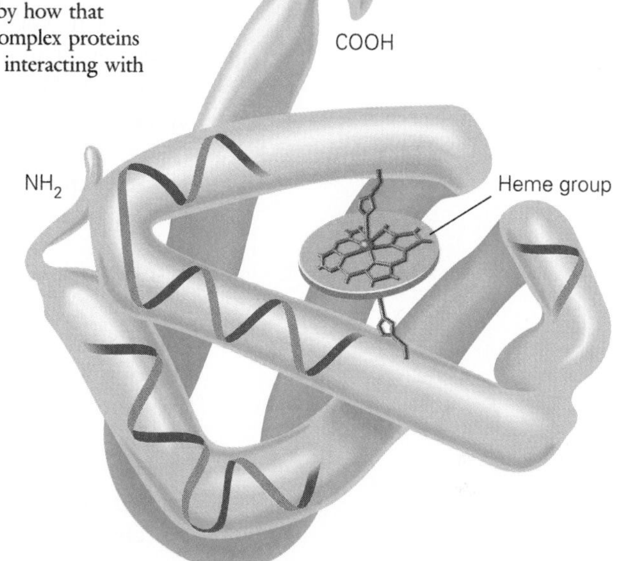

(c) Tertiary structure (myoglobin, a muscle protein)

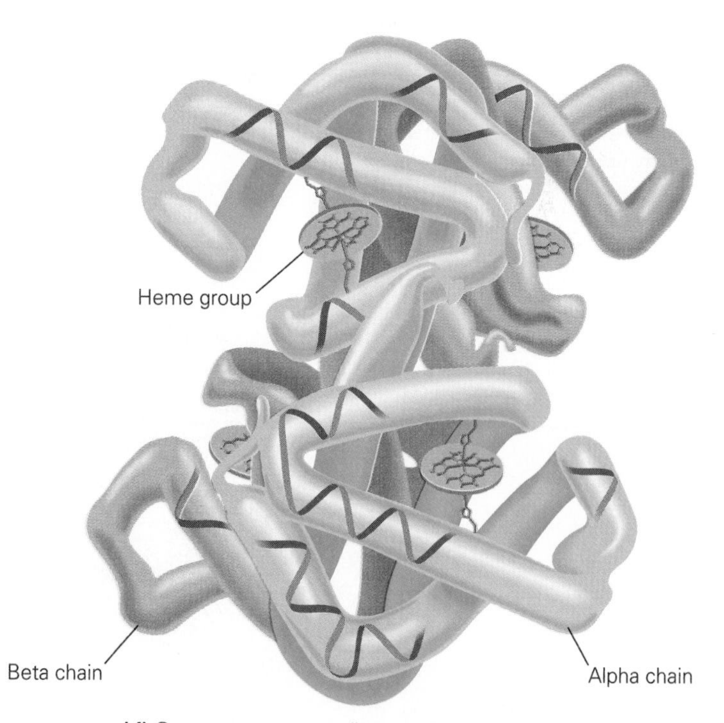

(d) Quaternary structure (hemoglobin, a blood protein)

repelling) forces, and disulfide linkages (bonds between the sulfur groups of certain amino acids along the chain). Tertiary bonds such as these are weak and can be broken easily, by heat or acid, for example. Once such tertiary bonds are broken, the molecules become fibrous (part of a process called **denaturation**). Denaturation involves altering a protein in such a way that its chemical and biological properties are destroyed, usually irreversibly. This is what happens when you cook the protein-rich white of an egg. The irreversibility is apparent if you have ever tried to unfry an egg.

Finally, two or more globular proteins may intertwine with each other to form a **quaternary structure.** Hemoglobin, the blood's oxygen-carrying molecule, has four protein chains arranged in such a configuration.

Nucleic Acids

Our final category in the molecules of life includes the **nucleic acids, RNA (ribonucleic acid)** and **DNA (deoxyribonucleic acid),** the largest of all biological macromolecules (see Chapter 7). In most organisms, the hereditary units called genes are composed of DNA. DNA molecules are immensely long polymers.

As in all of the large molecules of life, DNA and RNA are composed of repeated subunits, in this instance known as **nucleotides.** Nucleotides contain a five-carbon sugar, a phosphate group, and one of five different ringlike **nitrogenous bases** (Figure 3.29). Despite the immense length of nucleic acids, the organization of their nucleotides is surprisingly simple.

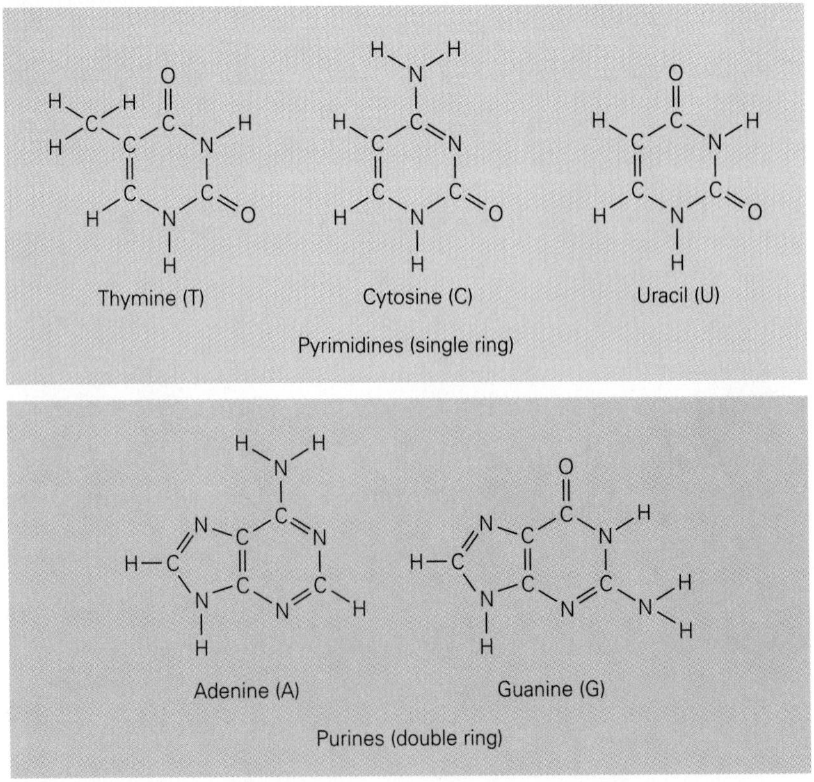

FIGURE 3.29
The five nitrogenous bases that are important in heredity: adenine, thymine, guanine, cytosine, and uracil.

Thymine (T)

Cytosine (C)

Uracil (U)

Pyrimidines (single ring)

Adenine (A)

Guanine (G)

Purines (double ring)

ESSAY 3.3

Unraveling the Structure of Insulin

In 1954, after ten years of painstaking research, Frederick Sanger and his colleagues at Cambridge University in England were able to tell us the full sequence of the amino acids in insulin, a rather small protein.

This work, for which Sanger was awarded a Nobel Prize, consisted of breaking the insulin molecule into smaller pieces of protein and then analyzing the pieces. First, he subjected the entire molecule to strong acid to break it down into its component amino acids. He was then able to identify each of these amino acids by a process called chromatography. Chromatography involves filtering the product down through a column of an absorbent material for which various kinds of molecules have different affinities. Each amino acid thus adheres to the material at a different rate and so is stopped at a specific point as it moves down the column. By comparing his results with the known absorbing point of each acid, Sanger determined which amino acids were present in insulin. He was also able to learn something about their relative amounts. To discover the sequence of the amino acids in the insulin molecules, Sanger used more specific hydrolyzing agents that break only bonds between certain molecules in the long chain.

Pepsin, for example, hydrolyzes bonds only between tyrosine or phenylalanine and other amino acids. To determine which end of the segment was free and which was attached, he added a chemical (dinitrofluorobenzene) that attaches to any terminal amino group and turns the compound a bright yellow.

After analyzing many such short segments, Sanger was eventually able to work out their structures, and where the terminal amino acids of these groups were identical, he could overlap them to reconstruct longer chains. For example, an overlap of the segments

A=C=X=N=O=F
N=O=F=G=O=I

yielded the longer segment

A=C=X=N=O=
F=G=O=I.

Later, he was able to determine how the two component molecules of insulin were joined by bonds between sulfur atoms of cysteine molecules. Thus, after years of patiently tracking down small bits of information, one step at a time, Sanger was eventually able to put all of the pieces together and tell us for the first time the structure of a relatively simple protein.

Insulin molecule (computer-generated model).

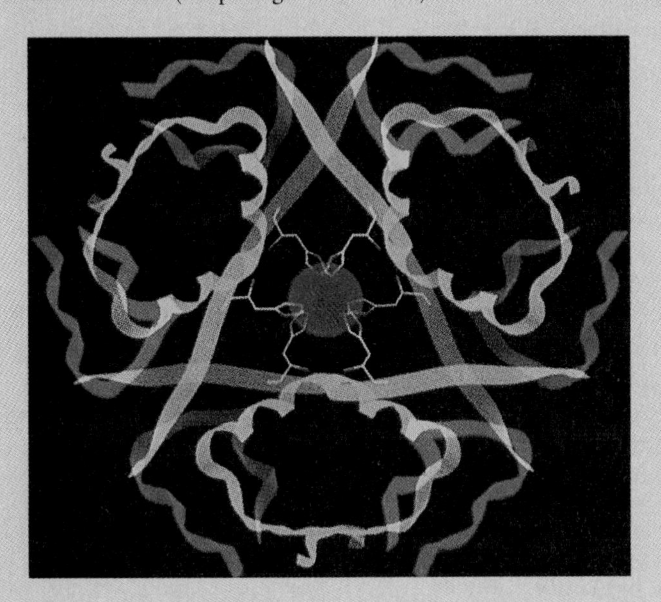

FIGURE 3.30
The double-helix configuration of DNA has been likened to two staircases winding around each other.

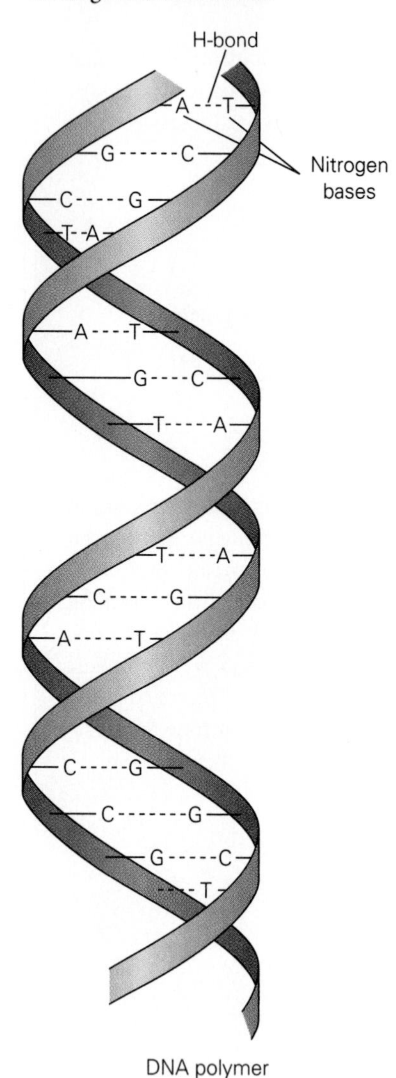

In fact, it was DNA's deceptive simplicity that, until the early 1950s, kept biologists in the dark as to the chemical nature of the gene. Although proteins can contain as many as 20 different amino acid subunits, there are only four different subunits (nucleotides) in DNA and four in RNA. These nucleotides are very much alike in most ways.

In the assembly of a DNA polymer, the nucleotides are linked into two opposing strands. The two strands are wound around each other in such a way as to form the well-known **double helix** (Figure 3.30). This double helix will shortly be our staircase leading into the world of genetics, but first we will wend our way through the life of the cell.

SUMMARY

1. An element is a substance that cannot be simplified by ordinary chemical means. The smallest indivisible particle of an element is an atom. Combinations of atoms form molecules. A compound is two or more elements combined in a fixed proportion. There are 92 naturally occurring elements and about 17 that have been synthesized in the laboratory. The elements carbon, hydrogen, nitrogen, oxygen, phosphorus, and sulfur (CHNOPS) make up 99 percent of living matter.
2. An orbital is the area of space where an electron is most likely to be found as it circles the nucleus. There are a number of potential orbitals at different energy levels, called shells. Higher energy electrons, found in the outer shell, are in the best position to interact with other molecules Isotopes of an element differ in the number of neutrons. An ion is a charged particle that results when an atom gains or loses an electron.
3. The gain of an electron is reduction, whereas the loss of an electron from an element is oxidation.
4. Each electron shell holds a specific number of electrons. A "satisfied" atom has the same number of electrons and protons and has its outer shell filled.
5. Bonds hold atoms together, forming molecules. An ionic bond, due to the attractive force between two ions of opposite charge, is formed when electrons are transferred from one atom to another. A covalent bond is formed when atoms share electrons. In some molecules shared electrons are more strongly attracted to one of the atoms, polarizing the molecule. A hydrogen bond is a weak bond formed when the positive end of a hydrogen atom that is covalently bonded to one molecule is attracted to the negative end of another polar molecule. Hydrogen bonding between water molecules gives water some of its unusual characteristics.
6. Chemical reactions, interactions between molecules, involve the making and breaking of bonds. An energy boost needed to initiate a reaction is called the energy of activation. Chemical reactions can be encouraged by adding heat, by increasing the concentration of reactants, or by increasing pressure.

7. A catalyst speeds up a reaction by lowering the energy of activation without changing the nature or direction of the reaction. Enzymes, catalysts found in living organisms, are proteins. The substrate fits into a region of the enzyme called its active site. Enzymes usually stress chemical bonds, thereby lowering the energy of activation. Enzymes emerge from such reactions unchanged.
8. An endergonic reaction, one in which the product has more energy than the reactants, requires the addition of energy. A reaction in which the product has less energy than the reactants is called exergonic.
9. Carbon is the element most intimately associated with life. When reacting with other atoms, a carbon atom shares the four electrons in its outer shell. Hydrocarbons are combinations of carbon and hydrogen. Carbon can bind to carbon in chains or rings. Carbon may bind to functional groups, molecular side groups that can enter into only certain reactions.
10. The molecules comprising living organisms include carbohydrates, lipids, proteins, and nucleotides. Most of these are large macromolecules that are comprised of repeating units called monomers linked together to form chains called polymers. The monomers are linked by dehydration, the removal of the atoms that form water. The polymers can be broken down by the addition of water molecules in a process called hydrolysis.
11. Carbohydrates contain carbon, hydrogen, and oxygen. Simple sugars (monosaccharides) are built into disaccharides or polysaccharides by dehydration. Animals store carbohydrates as glycogen by connecting many units of glucose. Plants store carbohydrates in branched chains of glucose called starch. Plants form cellulose by connecting glucose units in a different type of linkage than is found in glycogen or starch.
12. Lipids are composed of carbon, hydrogen, and oxygen. Lipids include fats, waxes, phospholipids, and steroids. Fats are energy-rich molecules comprised of glycerol and fatty acids. The carbon chains of saturated fats are filled with hydrogens. Animal fats are usually saturated and are solid at room temperature. Plant fats, or oils, are usually liquid at room temperature. Saturated fats can have adverse effects on the heart and blood vessels. Waxes are long-chain saturated fatty acids combined with long-chain alcohols or with carbon rings. Phospholipids, important components of membranes, are created when one of the hydroxyl groups of glycerol is attached to a phosphorus-containing molecule. Steroids are lipids formed of four interlocking carbon rings with many side groups. Cholesterol and the sex hormones are examples of steroids.
13. Proteins are long chains of nitrogen-containing amino acids linked by peptide bonds. Peptides, chains of up to 300 amino acids, can be joined to form proteins. About 20 amino acids are found in proteins. The sequence of amino acids forms the protein's primary structure and determines its character. Hydrogen bonds between amino acids cause some proteins to form a stable, regular shape called its secondary structure. Some protein chains fold back on themselves, forming a clump. This folding is the tertiary structure. The quaternary structure is formed when two or more twisted protein chains combine.
14. Nucleic acids, DNA and RNA, are long polymers of nucleotides, each of which contain one of five nitrogen bases, a sugar, and a phosphate.

KEY TERMS

active site (66)
amino acid (78)
amino group (71)
atom (54)
atomic number (55)
carbohydrate (74)
carboxyl group (71)
catalyst (66)
cellulose (75)
chemical reaction (63)
chitin (75)
compound (54)
covalent bonding (60)
dehydration (72)
denaturation (82)
deoxyribonucleic acid (DNA) (82)
disaccharide (74)
double bond (60)
double helix (84)
electron (55)
element (54)
endergonic (67)
energy of activation (65)
enzyme (66)
exergonic (67)
fat (75)
fatty acid (75)
free radical (71)
functional group (71)
glycerol (75)
glycogen (74)
hydration (72)
hydrocarbon (70)
hydrogen bond (61)
hydrolysis (72)
hydroxyl group (71)
inert (58)
ion (56)
ionic bond (59)
isotope (55)
kinetic energy (65)
lipid (75)
molecule (54)
monomers (72)
monosaccharide (74)
neutron (55)
nitrogenous bases (82)
nucleic acids (82)
nucleotide (82)
nucleus (55)
orbital (56)
oxidation (57)
peptide (80)
peptide bond (80)
phospholipid (77)
polymer (72)
polypeptide (80)
polysaccharide (74)
primary structure (80)
product (67)
protein (66)
proton (55)
quaternary structure (82)
reactant (64)
reduction (57)
ribonucleic acid (RNA) (82)
saturated (76)
secondary structure (80)
shell (56)
starch (74)
steroid (77)
substrate (67)
sugar (74)
tertiary structure (80)
unsaturated (76)
wax (77)

FOR FURTHER THOUGHT

1. Phosphorus has an atomic number of 15. How many electrons are contained in its outer shell? How many protons are contained in its nucleus?
2. Atom X has eight electrons in its second and outermost shell. Atom Y has seven electrons in its second and outermost shell. Which atom is inert? Explain.
3. Which of the following is a valid equation of a chemical reaction? Why?
   1. $HCl + NaOH \rightarrow H_2O + NaCl$
   2. $HCl + NaOH \rightarrow HO + NaCl$

FOR REVIEW

True or false?

1. ___ When an atom loses an electron it is said to be reduced.
2. ___ A covalent bond is formed when an electron is transferred from one atom to another.
3. ___ Carbon can combine with atoms that tend to gain electrons and those that tend to lose electrons.
4. ___ Polysaccharides are composed of long peptide chains.

Short answer

5. Atoms tend to satisfy themselves (make themselves electrically neutral) by balancing their numbers of protons and ___ .
6. ___ are charged particles that result from the loss or gain of an electron.
7. Catalysts that are found in living things are called ___ .
8. ___ are composed of glycerol and fatty acids.
9. Explain how the qualities of hydrogen bonding make water somewhat resistant to freezing and boiling.

10. Choose the correct statement.
    A. Atoms contain negatively charged neutrons and positively charged electrons.
    B. Molecules are the smallest indivisible particles of elements.
    C. Compounds are formed of two or more elements in a fixed position.
    D. Isotopes of a certain element vary in the number of protons they contain.

Choose the best answer.

11. The properties of carbon allow it to form all but which of the following?
    A. ionic bonds
    B. complex branching molecules
    C. bonds with other carbon atoms
    D. long hydrocarbon chains
12. Chemical reactions are:
    A. all irreversible, one-way processes.
    B. accelerated by catalysts.
    C. called endergonic when their products have less energy than their starting molecules.
    D. all of the above.
13. The primary structure of a ___ is determined by specific sequences of amino acids.
    A. protein
    B. lipid
    C. carbohydrate
    D. polysaccharide

PART TWO

Cells and Inheritance

Focusing on the smallest units of life, the cells, we see how cells acquire and handle energy and how they so faithfully reproduce themselves and so unfaithfully distribute genetic material to the next generation. We are introduced to the principles of genetics, which we then expand into the intriguing modern area of genetic engineering, replete with its dire threats and grand promises.

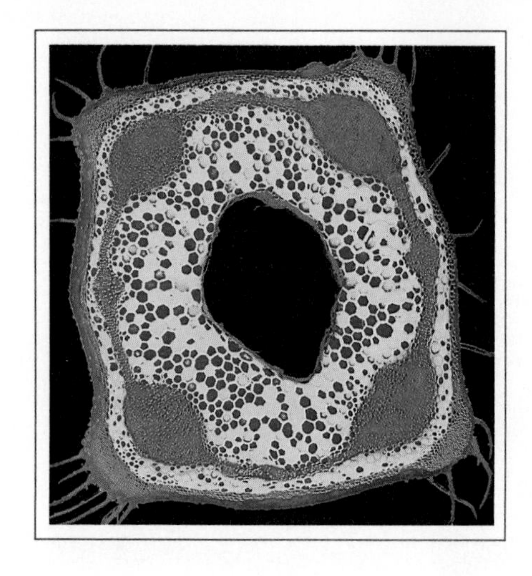

CHAPTER 4

The Cell and Its Structure

Overview

Cytology and Technology

Prokaryotic and Eukaryotic Cells

Cell Components

How Molecules Move

Objectives

After reading this chapter you should be able to:

- List the tenets of cell theory and describe the emergence of cytology as a science.
- Differentiate between eukaryotic and prokaryotic cells and describe the advantages of specialization in eukaryotic cells.
- Describe the components and physical properties of cell walls, plasma membranes and the cytoskeleton.
- List the cellular organelles, indicating the functions of each.
- Explain how the fluid mosaic model describes the structure of the plasma membrane.
- Describe the types of active and passive transport, explaining how each functions to move substances in biological systems.

In the mid-1600s, the prestigious Royal Society of London appointed Robert Hooke as their new curator of experiments. Hooke felt honored to have been chosen for such a prestigious position, but with the position came a problem. At the group's weekly meetings, the curator had to provide them with some sort of entertaining scientific demonstration. The problem was, the society was comprised of the elite scientists of that day—a rather jaded lot to be sure. They were indeed difficult to impress and almost impossible to entertain. Hooke pondered over what to show them. What would they like to see?

The scientific world was abuzz in those days with talk of lenses, bits of curved polished glass that could be used not only to restore books and letters to weakened old eyes, but to see the distant heavens more clearly. Hooke himself was interested in lenses, particularly in those that could bring into view things too small to be seen with the naked eye. In fact, he had built one of the first microscopes (Figure 4.1). He decided he would show them something with his microscope. They were, indeed, impressed and quite happy to peer through the remarkable device at meeting after meeting. Hooke was kept busy wondering what next to show them. One night, while tinkering in his laboratory, he cut a very thin slice of cork with his penknife and placed it under his lens. Cork was interesting because people wondered how so seemingly solid a substance could float. So he focused his lens, hoping to finally solve this vexing old problem. And what do you think he saw? Nothing! That's because reflected light

FIGURE 4.1

Hooke's primitive microscope consisted of two convex lenses at either end of a six-inch tube. The object for study was stuck on a pin attached to the base of the microscope. A flame served as the light source for the microscope.

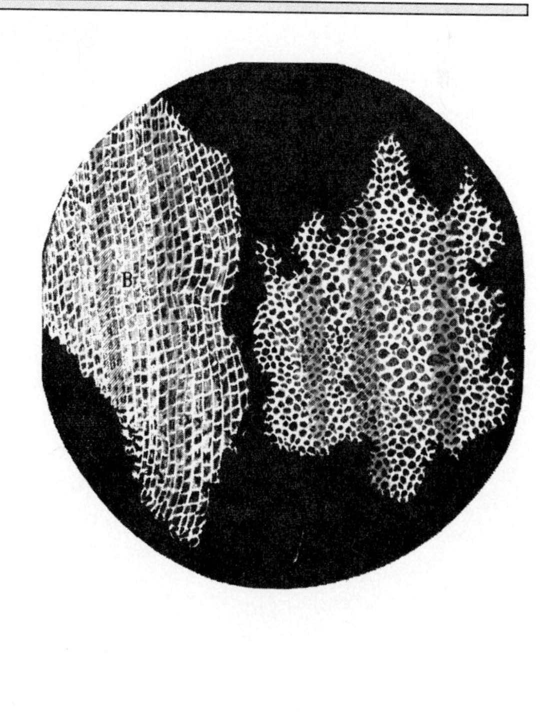

FIGURE 4.2

Diversity in plant and animal cells. These cells are from multicellular organisms, and each is well adapted to a specific role. (a) Compact bone cells (b) Buttercup root cells (c) leaf epidermis cells (d) Human fat cells

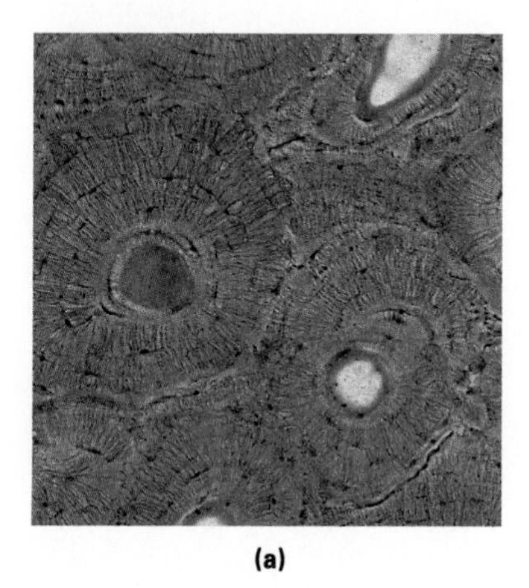

(a)

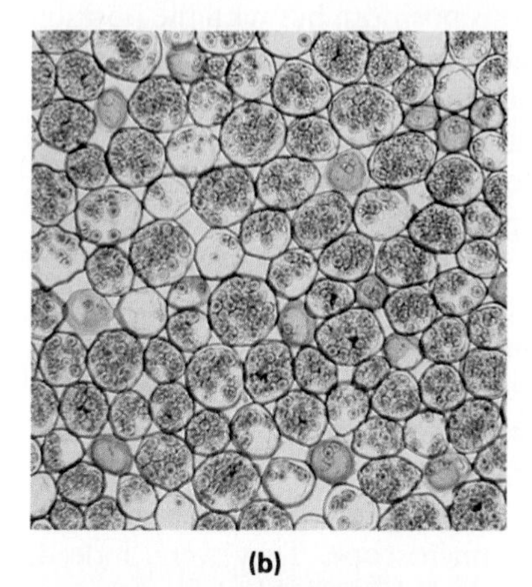

(b)

(c)

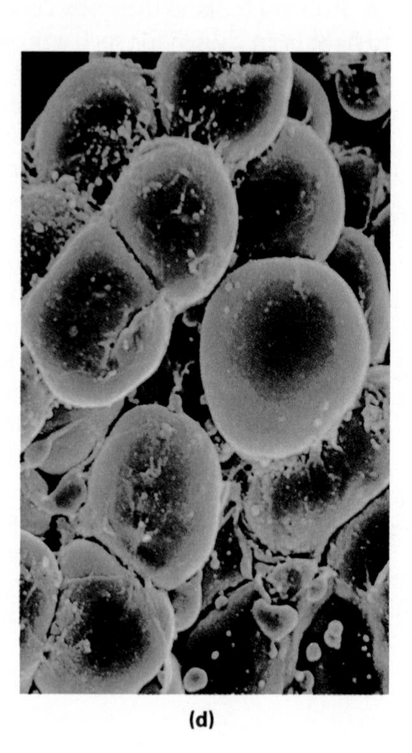

(d)

does not provide the illumination necessary to see objects of that nature. So he arranged to have the light pass from underneath, *through* the cork. This time Hooke did see something, and it surprised him. He wrote in his notes that the cork seemed to be composed of tiny "boxes" or "cells." He showed his cells to the society, the society was duly impressed, and a new field of science was born.

This field, however, was a bit slow abirthing. In fact, about a century was to pass before the scientific world would begin to grasp the extreme importance of Hooke's "little boxes." Actually, it wasn't until about 1805 that the German naturalist, Lorenz Oken, formally stated what has become known as the **cell theory:** "All life comes from cells and is made of cells." The idea was buttressed in 1839 by two other Germans, botanist Matthias Jakob Schleiden and zoologist Theodor Schwann, who independently concluded that all living things—from oak trees to squids, from tigers to humans—are made up of cells (Figure 4.2). Another 50 years were to pass before another German, Rudolf Virchow, devised the rather catchy statement, *omnia cellula e cellula,* "all cells from cells." About this time people began to take the whole concept somewhat more seriously. **Cytology** (*cyte,* cell), the modern study of cells, had indeed gotten off to a shaky start.

CYTOLOGY AND TECHNOLOGY

Once biologists began to see the importance of understanding cells, the field of cytology grew rapidly and sporadically as one fascinating discovery was jostled aside by yet another. For example, by the early 1900s, most of the cell structure visible through the light microscope had been described. Then, with the advent of the transmitting electron microscope (TEM), a whole new dimension of the cell, its ultrastructure, was revealed. We could see things we never imagined existed. Even with the new microscope, however, we didn't know much about the cytoplasmic fluid in which floated organelles, tiny structures within the cell, each with a specific role in the life of the cell. We could only assume that the organelles were suspended in some sort of "cytoplasmic matrix" or "ground substance," labels that seemed to convey information, but actually didn't. The problem was that in order to view anything with a TEM (Essay 4.1), electrons had to pass through the substance, so it had to be sliced into ultrathin sections, and this procedure demolished any cytoplasmic detail that might have been present.

But technology saved the day again. With the high-voltage electron microscope we could see a whole new realm of cytoplasmic structure. It was no longer necessary to open the cell to see its contents. Furthermore, we could now see cell contents in three dimensions. This was indeed a remarkable step forward, but even newer technologies continued to emerge, and with each new device, we learned more. Much of what we learned turned out to be quite unexpected.

Today, cytologists have become extremely sophisticated as they seek to answer increasingly specific questions about life's little boxes. Nonetheless, the basic definition of a cell has not changed much since cells first began to be taken seriously. Cytologists now define a cell as the smallest unit that can exist as an independent organism, and that it has, at some point in its development, everything necessary for life. There are organisms that consist of only a single cell, but here we will concern ourselves primarily with multicellular species, keeping in mind that many of the principles we will discuss also apply to single-celled organisms.

With all the attention given to cells these days it would seem that describing one should be a simple task. But this is not the case, partly because a living cell is a hotbed of activity, constantly changing with dramatic speed. But describing cells is also difficult because there are so many kinds of them.

PROKARYOTIC AND EUKARYOTIC CELLS

Before we become involved with the description and behavior of cells, we should first be clear about what kinds of cells we will be considering. We will *not* be talking about the bacteria and cyanobacteria. These are one-celled organisms that differ in several basic ways from all other cells. They are called **prokaryotic cells** (*pro,* early; Figure 4.3) to differentiate them from the **eukaryotic cells,** our primary concern here. The differences between the two types of cells are summarized in Table 4.1. The comparison will mean more to you after we wend our way through the rest of the chapter.

FIGURE 4.3

Compared to the typical eukaryotic cell, the prokaryotic cell is quite small. Further, it lacks the membrane-bound organelles of the eukaryote, including the organized nucleus. Prokaryotic DNA is naked and circular, lacking the protein complex of eukaryotic chromosomes. A dense cell wall, quite different chemically from eukaryotic cell walls, surrounds the membrane and is often itself surrounded by a slimy sheath. Prokaryotes may have tubelike, cytoplasmic projections known as pili. Where flagella appear, they are solid, rotating entities, anchored in the cytoplasm and cell wall, and quite unlike those of eukaryotes.

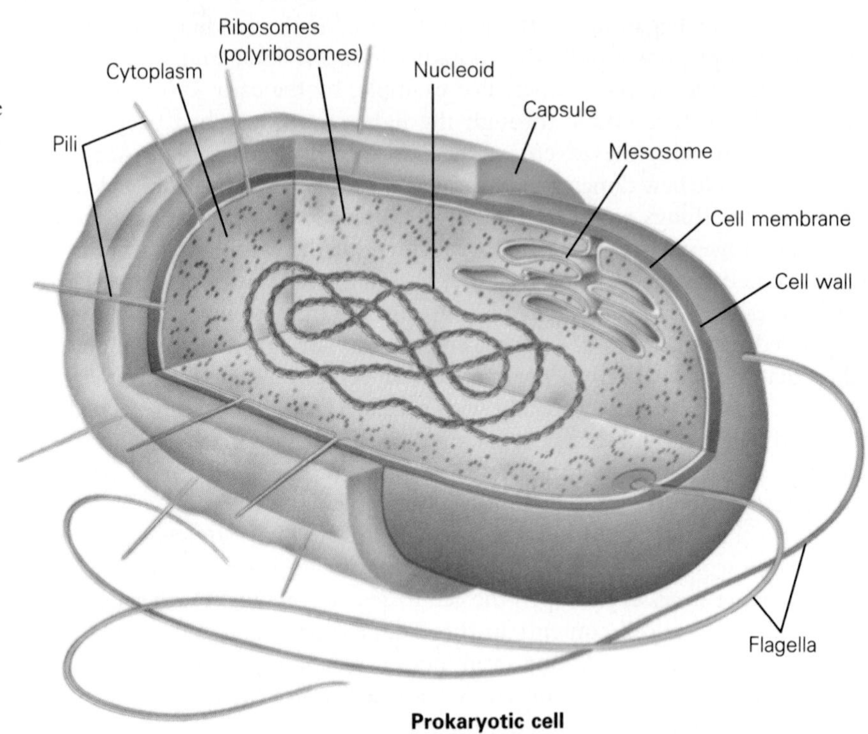

TABLE 4.1
Differences Between Prokaryotic and Eukaryotic Cells

STRUCTURE	PROKARYOTIC CELLS	EUKARYOTIC CELLS
Nuclear membrane	Absent	Present
Membrane-bound organelles (mitochondria, Golgi bodies, etc.)	Absent	Present
Flagella	Rotating and lacking 9 + 2 structure	Fixed with 9 + 2 structure
Chlorophyll	Not in chloroplasts	In chloroplasts
Cell wall	Contains amino acids and sugars	Does not contain amino acids and sugars (when wall is present)
Pili	Present	Absent
Chromosomes	Composed entirely of nucleic acids	Include nucleic acids and proteins

ESSAY 4.1

Microscopes and What They Can Do

The ordinary **light microscope (LM)** was once the mainstay of scientific magnification. There have been many variations on its theme, but the basic principle has remained the same. And so have the limitations of the technique—limitations imposed by the LM's low powers of resolution.

Resolution refers to the smallest distance by which two points can be separated and still be distinguished as separate points. If they are separated by a distance smaller than the resolving power of the microscope, they will be seen as a single point.

With the light microscope, the image is seen as contrasting light and dark areas as light passes through the specimen from below. The problem of resolution is therefore compounded by the inherent limitation of light diffraction, the bending of light rays as they pass through an object. The length of the light wave itself (that is, the color of the light) is a critical factor here. As the wavelength increases, so does the angle of diffraction, and its resolution is thus reduced. Under ideal conditions, the best light microscopes have a resolution of about 0.25 micrometers with ordinary white light. Resolution has recently been increased by the electron microscope. Ultraviolet light is shined on the subject, which absorbs the light energy and reemits it as electrons that are then magnetically focused.

The resolving power of the **transmitting electron microscope (TEM),** however, is far greater yet. Thus, any object can be seen more clearly with the electron microscope, and yet other objects can be seen that are not visible at all with the light microscope.

The electron microscope, which became available in the 1930s, works on an entirely different principle. Excited electrons are drawn from a heated filament and are then directed, or focused, by magnetic fields. As the electrons pass through an object, they are deflected or absorbed to various degrees by differences in the density of the matter. They then produce an image on a special film below the object. Electrons are easily deflected, so the specimen must be cut exceedingly thin and the operation must be carried out in a vacuum.

The wavelength of an electron beam is only about 0.005 nanometers (a nanometer is one billionth of a meter), so, theoretically, resolutions of this power are possible. A major disadvantage is that heavy metal stains must be employed, and these are usually highly poisonous to living things. Because of this and the necessity for extremely thin sections, there is no way living material can be examined. Thus, we can't say with certainty that what we see is not due to the method of preparation.

The **scanning electron microscope (SEM)** has far less resolving power than the standard electron microscope, but it produces the fascinating appearance of three dimensions. In SEMs, a thin beam of electrons is passed back and forth over an object, thereby scattering the electrons. They are then captured on a positively charged plate, causing it to generate a small current. The current is then amplified and fed into a cathode ray tube that projects the image caused by the reflected electrons onto a television screen.

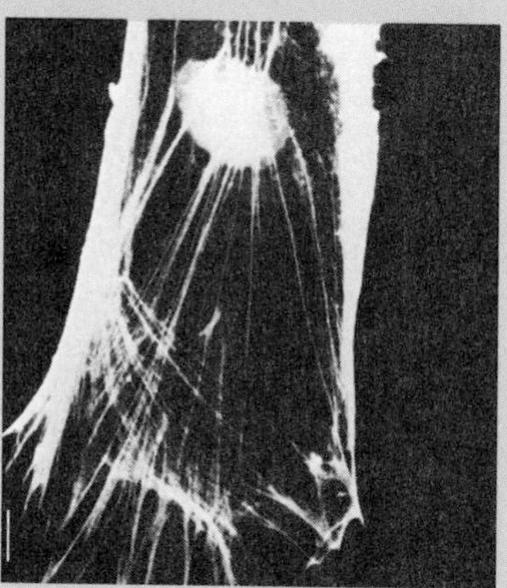

(a)

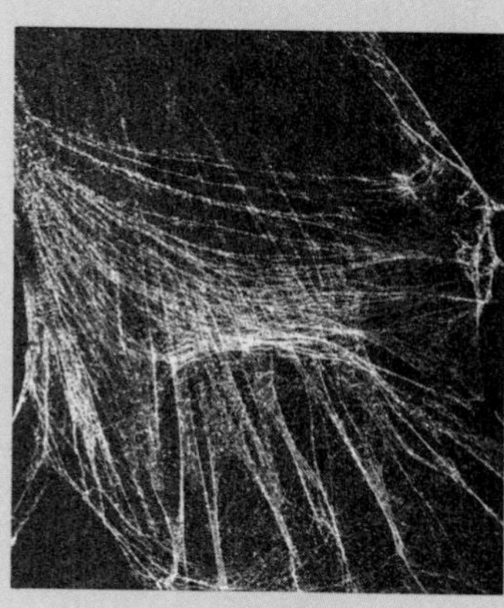

(b)

These two micrographs of actin in a rat embryo cell illustrate the dramatic improvements in microscopes. The image in (a) was produced using a fluorescent light microscope and is magnified 2000 times. The image in (b) was produced using the electron microscope and is magnified 3000 times. The electron microscope offers a more detailed and clearly focused image.

Specialization in Eukaryotic Cells

Eukaryotic cells combine in a variety of sizes and shapes, each with its own particular limits and abilities. Not only do the cells of various organisms differ, but the cells of a single organism may be quite different from each other. For example, the tiny boxlike cells beneath the skin of your hand are quite unlike the delicate threadlike nerve cells that stretch from the base of your spine down to your foot. And neither of these resemble the peculiar, pulsating cells that make up your heart.

So the body is composed of various kinds of cells with different abilities. But cells that differ in appearance may also behave quite differently, as would be expected. To illustrate, some are highly irritable—that is, they are sensitive to environmental changes. Yet others are contractile, and others secrete fluids, while still others have long tails and can swim (Figure 4.4). This sort of variety suggests a great advantage of cellularity that might well have made the condition worth developing through evolution. That advantage is **specialization,** where cells become different as they assume different roles. In other words, an organism living in a complex

FIGURE 4.4
Human cells vary widely, permitting great specialization and division of labor within a single body. (a) Human cardiac muscle (contractile cells) (b) Human cartilage cells (connective cells) (c) Neuromuscular junctions (between nerve [neuron] and muscle) (d) Human sperm (motile cells)

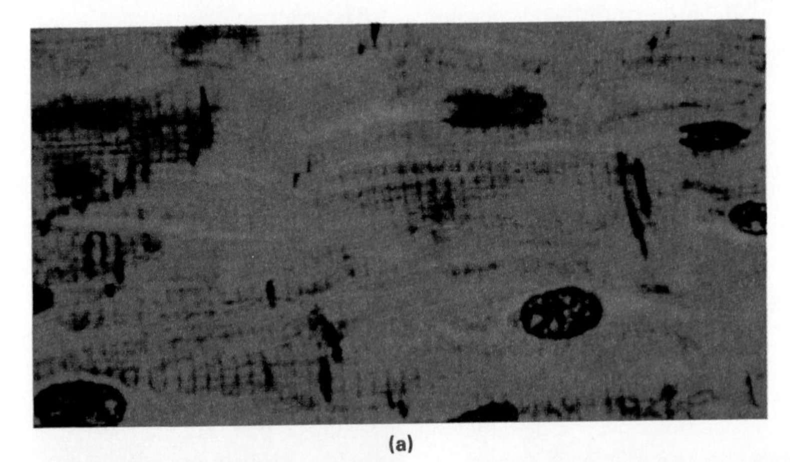

(a)

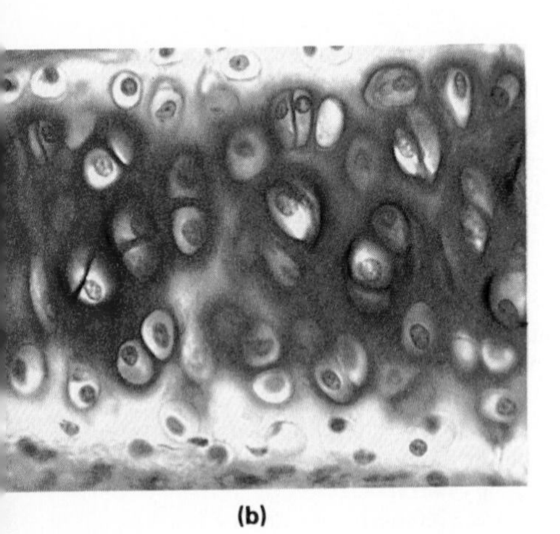

(b)

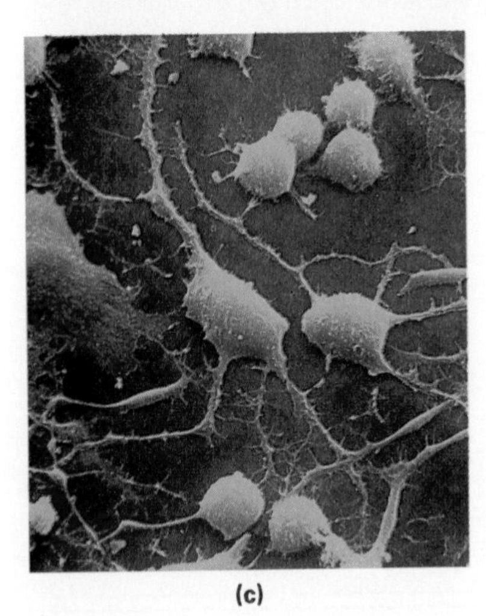

(c)

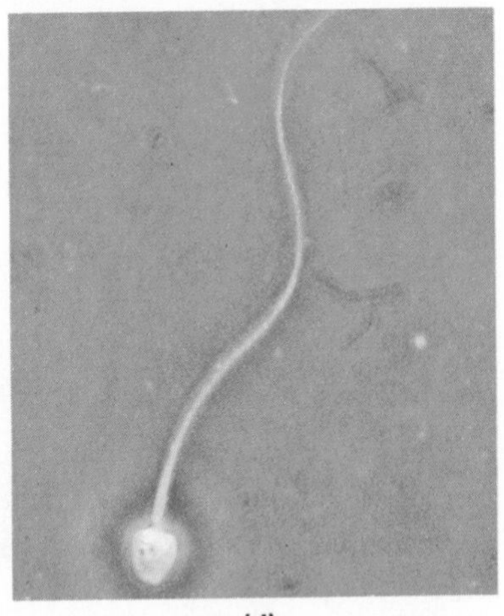

(d)

FIGURE 4.5

Both "cells" have the same volume, but the subdivided one has a far greater surface area. A large organism composed of one cell would be limited in its behavior and vulnerable to its environment. Having none of the advantages of cellular specialization, it would not survive.

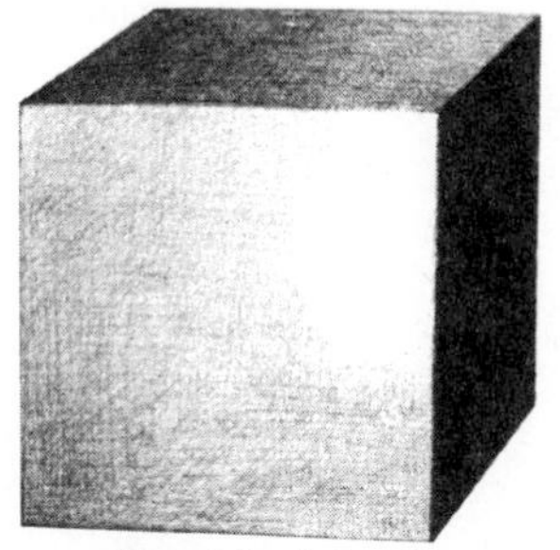

One 3-inch "cell"
Surface area of membrane
54 square inches

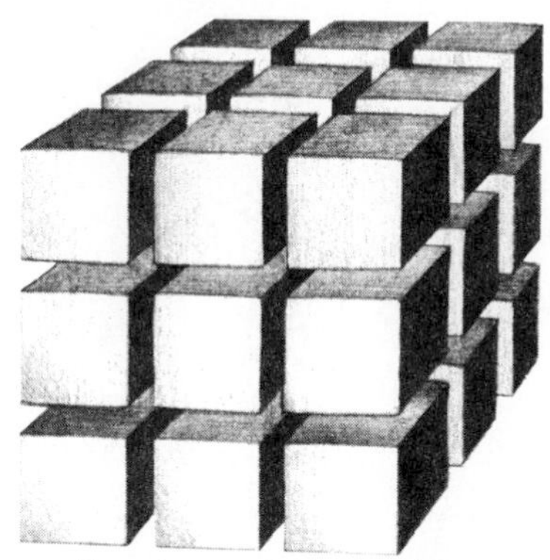

Twenty-seven 1-inch "cells"
Surface area of membrane
162 square inches

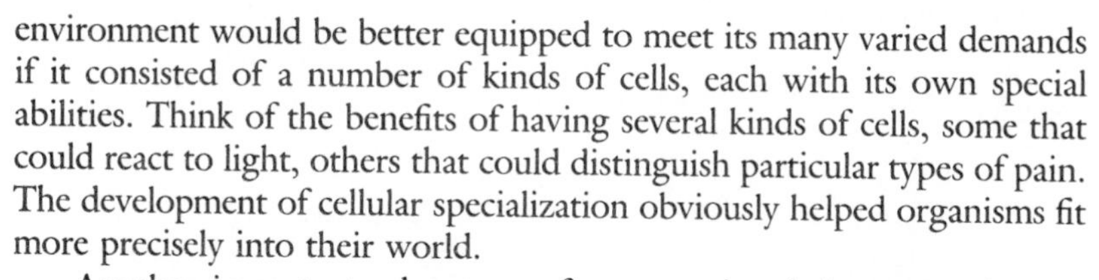

environment would be better equipped to meet its many varied demands if it consisted of a number of kinds of cells, each with its own special abilities. Think of the benefits of having several kinds of cells, some that could react to light, others that could distinguish particular types of pain. The development of cellular specialization obviously helped organisms fit more precisely into their world.

Another important advantage of an organism being comprised of many cells instead of one large cell is that the organism would have greater control over the interactions between itself and the environment. Such control would be possible because of the nature of cell membranes, the thin, living coverings of most cells. Such membranes are highly selective about what passes into, out of, and through the cells they cover. Because of this selectivity, cellularity would increase the level of control over what passes between the delicate cellular interiors and the outer environment, since, with living matter broken up into cells, the relative membrane area would increase. As you can see in Figure 4.5, a 2000-pound cell would have less surface area (and hence, less regulatory membrane) than a 2000-pound organism composed of many cells. In fact, this is why there is no such thing as a 2000-pound cell.

Before we discuss some of the major aspects of cells, it would be nice to pull it all together and have a look at a representative cell and its components. The problem is, as has been mentioned, that there is no such thing as a "representative" cell. Cells are too diverse, and they change too rapidly to be perfectly described in a simplistic diagram. But with this small embarrassment in mind, you might take a look at Figure 4.6. Cell components are also reviewed in Table 4.2.

FIGURE 4.6
Representative plant and animal cells.

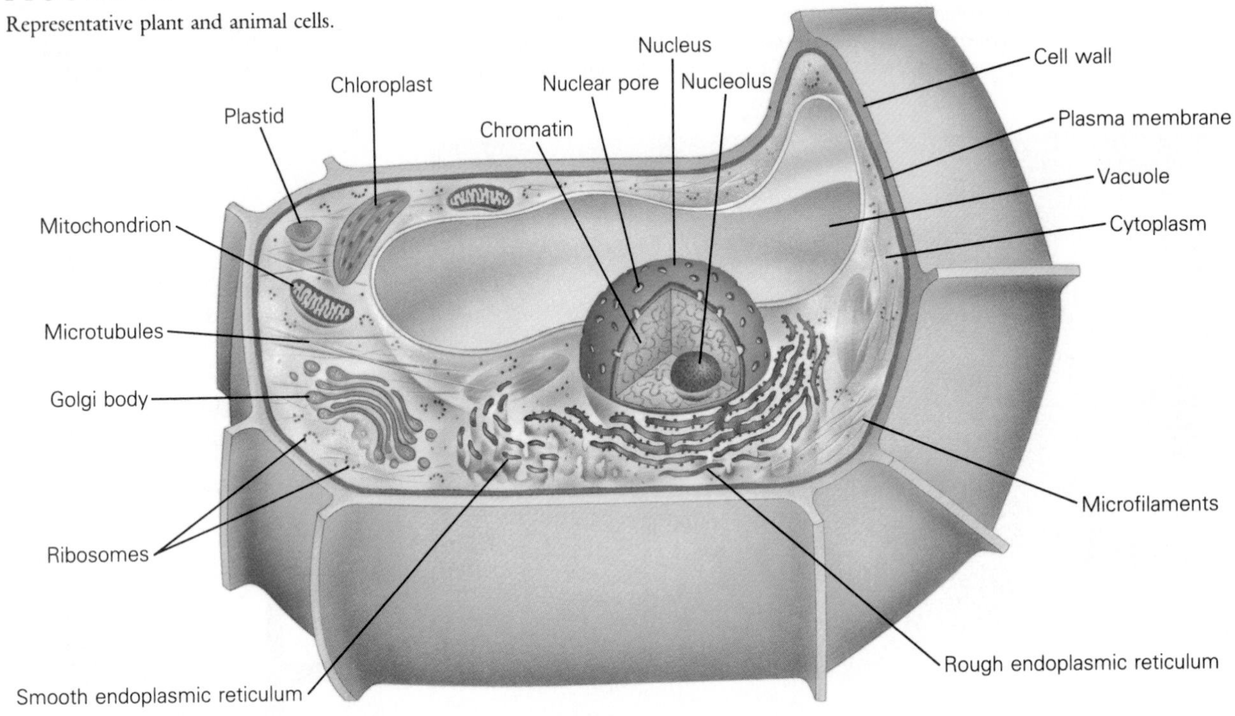

Plant cell

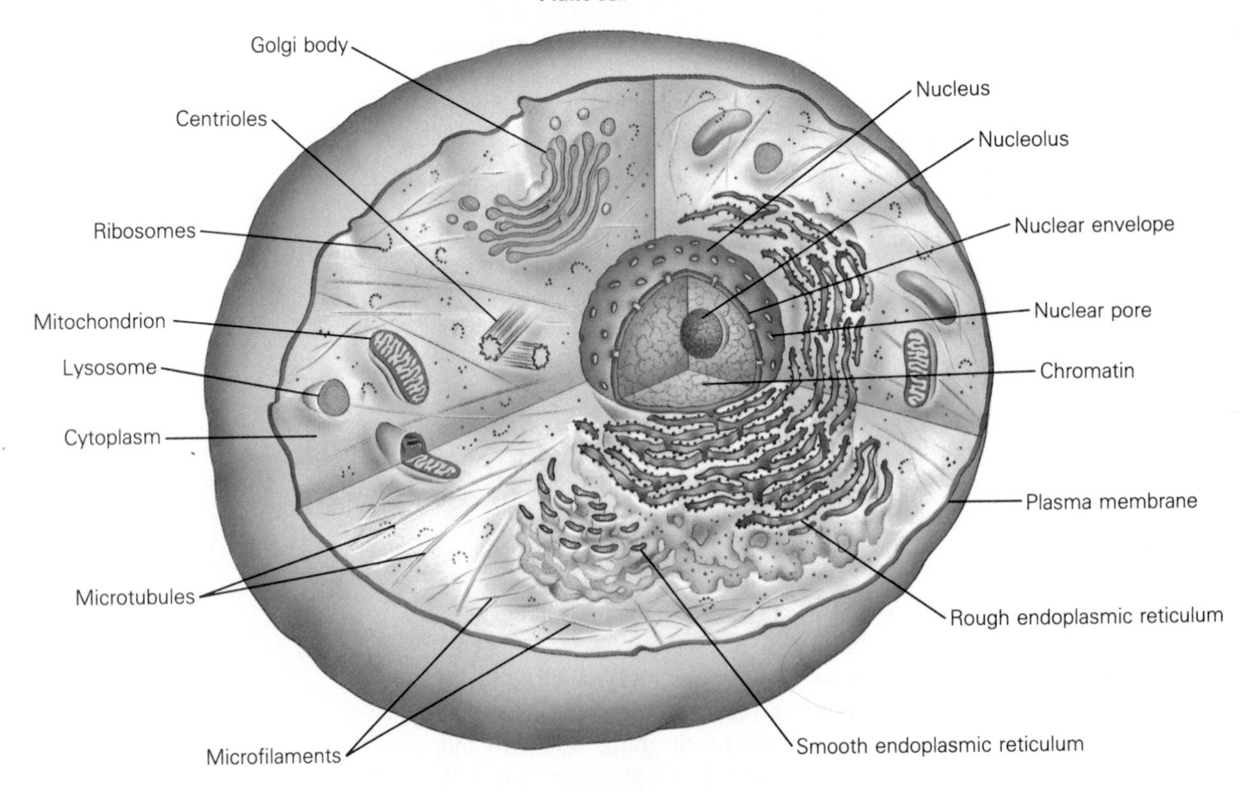

Animal cell

TABLE 4.2
Cell Components

ORGANELLE	DESCRIPTION	FUNCTION
Ribosomes	Tiny bodies consisting of two subunits	Protein synthesis
Cilia and flagella	Hairlike cell extensions	Movement of cells and substances
Mitochondria	Tiny round, elongate, or threadlike structures	Energy production
Cytoskeleton	Network of microtubules, microfilaments, and intermediate fibers	Internal framework, cell shape, and motility
Endoplasmic reticulum	Highly variable network of membranes	Protein and membrane synthesis
Smooth ER	Ribosomes absent	Transfer of rough ER products
Rough ER	Ribosomes present	Membrane and protein manufacture
Golgi body	Stacks of flattened vesicles	Packaging and storage center
Lysosomes	Spherical packets	Digestive enzyme storage
Plastids	Double-membrane sac	Plant pigment or nutrient storage center
Vacuoles	Fluid-filled sacs	Storage vesicles
Microtubules	Tiny tubes	Forms core of spindle fibers, cilia, and motile cells
Centrioles	Cylindrical bodies	Cell divisions and movement
Nucleus	Double-membraned prominent cell structure	Cell regulation and reproduction

CELL COMPONENTS

Our descriptions of the components (organelles) of eukaryotic cells must be a bit artificial because of the complexities of such structures and the fact that many of them can change over time. Just keep in mind that neither these descriptions nor these illustrations are likely to depict the situation for all cells, nor for any cell all the time.

It should be made clear that the inside of a living cell is a seething, roiling mass of viscous, grainy fluid that contains many types of tiny bodies—minute structures that are constantly growing, extending, moving, multiplying, appearing, and disappearing. Such activity perhaps should not be surprising since we know that most cells must constantly maintain and repair themselves and divide and grow. We also know that the cell's sensitive cytoplasm responds to specific environmental stimuli and that as those stimuli change, so does the behavior of the cell. The vigorous activity inside living cells should serve to remind us that the goings-on within cells are incredibly complex, and that we really cannot hope to understand everything that is happening there. At least not yet.

Cell Walls

Plant cells have cell walls; animal cells do not. **Cell walls** are nonliving, rather inflexible, and highly permeable. The main component of a cell wall is cellulose. The cellulose molecules are wound together to form strong "cables" (Figure 4.7). Plant cells that play supporting or mechanical roles also have lignin in their cell walls. The lignin adds rigidity to the cell walls. This is why trees are not limp.

FIGURE 4.7
A micrograph of the matlike fibers in a cell wall.

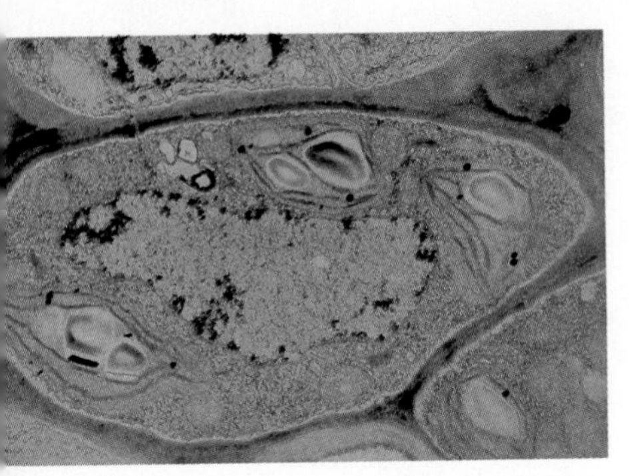

The cell walls of plants are commercially important in a number of ways. For example, we count on cell walls to hold up our own walls as we frame our houses with wood. Also, it is for the cellulose in cell walls that we have leveled vast areas of our forests in response to wheedling commercials designed to increase our demand for "disposable" paper commodities. The cellulose of plant cell walls is also valuable as a major component of celluloid, rayon, cotton, and hemp. (Hemp was once provided by legally cultivating a plant called *Cannabis sativa,* later known as "killer weed" or reefer.) Lignin was long considered a totally useless by-product of paper manufacturing. However, researchers worked hard to find ways to alter it so that it could be sold somehow, and they cleverly managed to discover uses for it in the manufacture of synthetic rubber, vanillin, and adhesives.

Plasma Membrane

All cells are bounded by a delicate structure called the **plasma membrane,** or cell membrane. The membrane is so thin that its presence was postulated on circumstantial evidence alone until the advent of the electron microscope. Modern research techniques have revealed that the membranes are composed of two layers of lipid-phosphate molecules (called phospholipids) in which are scattered various proteins. It seems, however, that the molecular arrangement of cell membranes may vary widely. For example, membrane thicknesses may be quite different from cell to cell, and the ratios of different lipids may vary from one type of cell to another.

A generalized plasma membrane may best be summarized in what is called the **fluid-mosaic model,** the currently accepted model of a plasma membrane in which "mosaics" of proteins are imbedded in a bilayer of phospholipids (Figure 4.8). At normal biological temperatures, the plasma membrane behaves like a thin layer of liquid, about the consistency of light oil, surrounding the cell. The phospholipids are polarized so that the **hydrophilic** ("water loving") heads are closest to the outer sides, and **hydrophobic** ("water fearing") tails project inward. Some of the globular proteins, imbedded here and there in the membrane, cross the membrane.

There may also be other kinds of molecules associated with membranes. For example, various sorts of carbohydrates may be found attached to the outer side of the membrane, specific carbohydrates that determine the cell type. The underside of the membrane may be attached to a sort of internal support (skeleton) for the cell, called the *cytoskeleton* (discussed later).

Plasma membranes are **selectively permeable;** that is, they permit certain molecules to cross them while prohibiting the passage of others. Some of the membrane proteins form tiny channels, protein-lined holes, through which certain materials may cross. Other substances may cross the membrane by moving through the lipid bilayer. Membranes may also actively assist some kinds of molecules in their passage, as we will see. Thus, this membrane regulates the constituents of the cell's interior, helping to protect it from an essentially hostile environment.

In a word, plasma membranes are more than sacs filled with fluid. They are living, responding structures. Thus, the more membrane area an organism possesses, the greater its control over its internal environment.

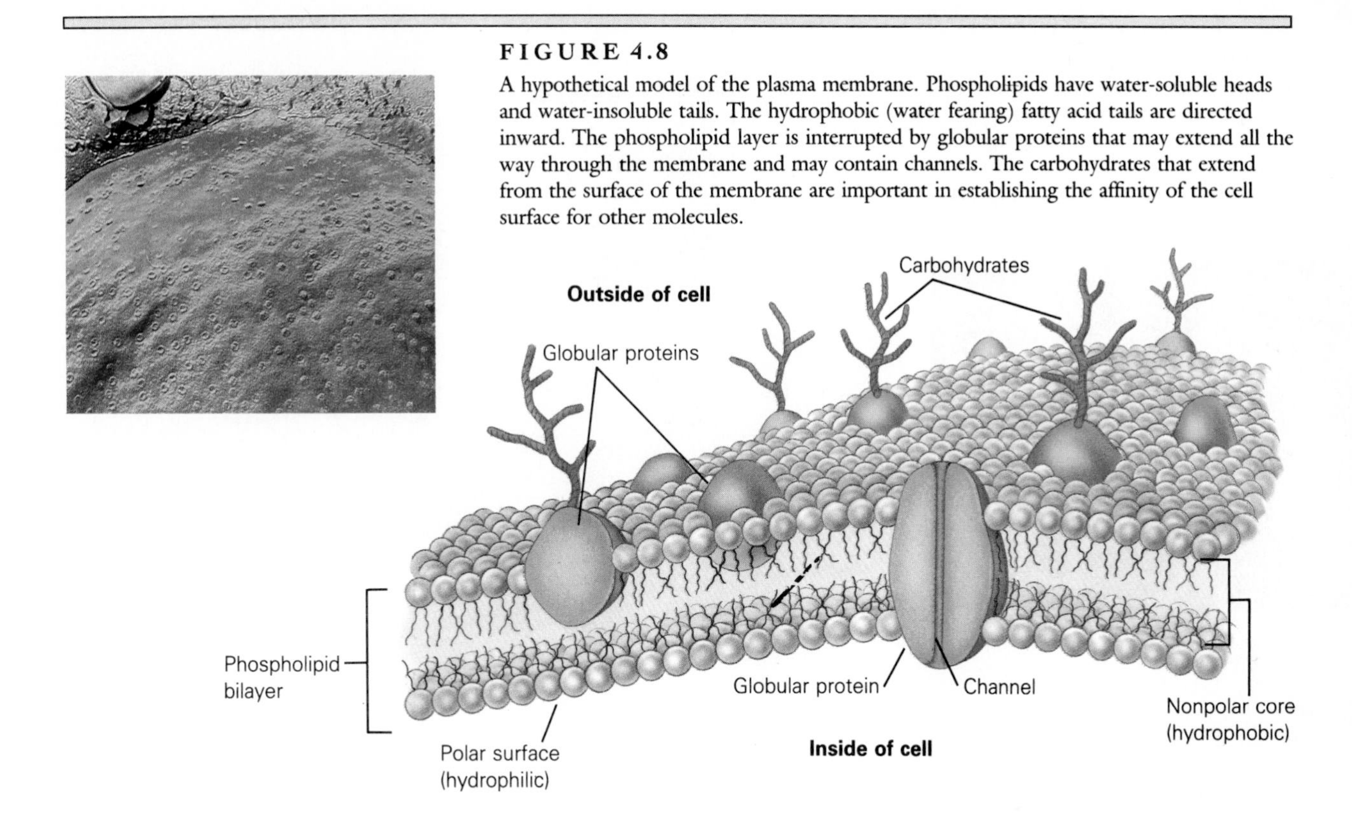

FIGURE 4.8

A hypothetical model of the plasma membrane. Phospholipids have water-soluble heads and water-insoluble tails. The hydrophobic (water fearing) fatty acid tails are directed inward. The phospholipid layer is interrupted by globular proteins that may extend all the way through the membrane and may contain channels. The carbohydrates that extend from the surface of the membrane are important in establishing the affinity of the cell surface for other molecules.

The Cytoskeleton

It was recently discovered that the cell's organelles may not float freely in an amorphous cytoplasm. Instead, they are held in place by a complex bridgework called the **cytoskeleton** (Figure 4.9), a weblike system of fibers that helps maintain cell shape and suspends organelles in a three-dimensional web.

There is also important evidence that the cytoskeleton holds even enzymes in place. It has been suggested that precise spatial arrangement of enzymes would increase their efficiency by encouraging a specific sequence of interactions. For example, enzyme B might be held near enzyme A, so that it might more easily interact with the products of A. Enzyme C would be near B, and so forth. Such structural organization presumably would be an improvement over random enzyme movement through the cell.

In addition, the cytoskeleton functions in cell motility. Parts of the cytoskeleton cause cilia and flagella to beat, while other parts cause muscle contraction. Materials and structures within the cell may move to their destinations along tracks formed by the cytoskeleton.

There are at least three types of fibers forming the cytoskeleton. The thinnest ones are the **microfilaments,** which are also called actin filaments. A variety of fibers with slightly larger diameters are called **intermediate fibers.** Those with the largest diameters are the **microtubules.**

FIGURE 4.9
The cytoskeleton is an extensive network of tubular elements that extends throughout the cell. The major organelles of the cell are suspended in this lattice. It also helps the cell maintain its shape and functions in cell motility.

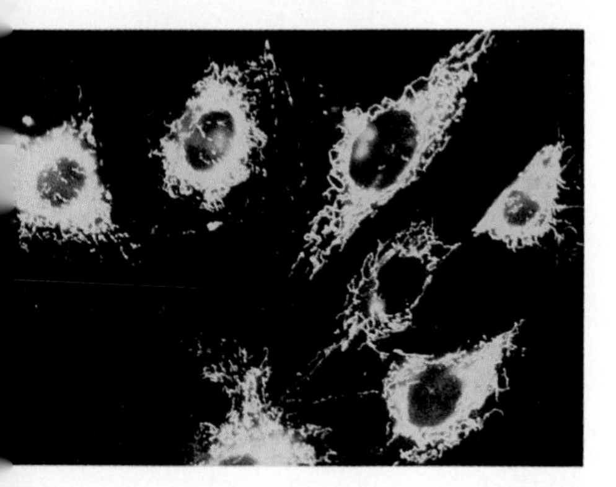

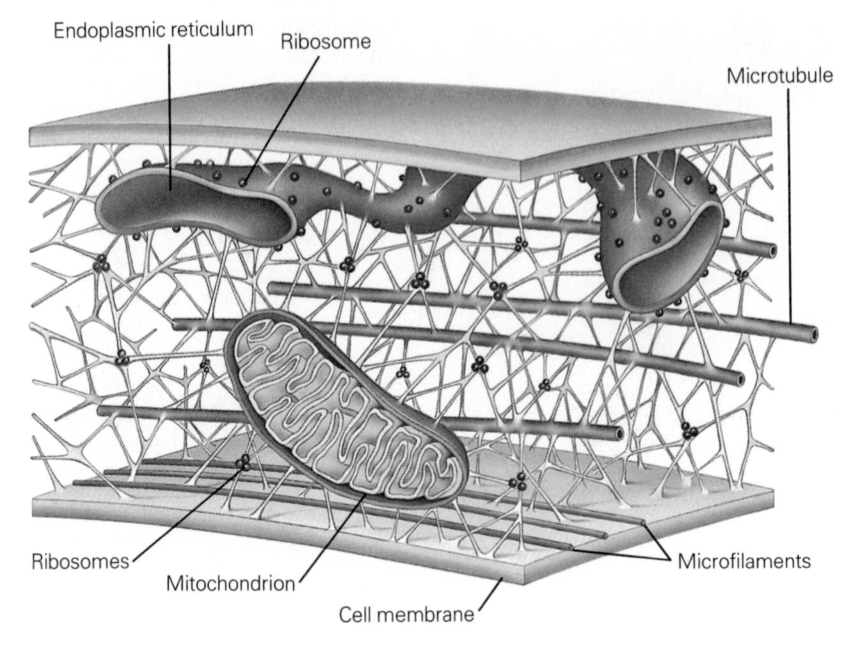

FIGURE 4.10
Microtubules are composed of a tight, spiral arrangement of tiny tubulin spheres.

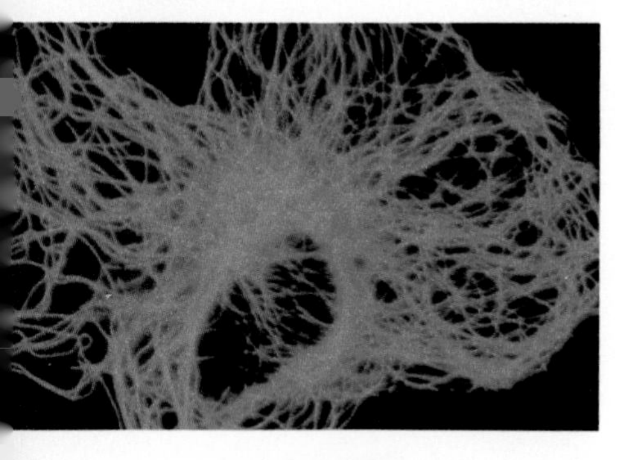

Microtubules

Microtubules, as you might expect, look like tiny tubes (Figure 4.10). Their constituent protein, called **tubulin,** occurs in doublet spheres whose arrangement forms the tubes or cylinders. Microtubules form the core, not only of flagella, cilia, and sperm tails, but also of the starshaped "asters" and the spindle fibers that appear at mitosis.

Centrioles

Centrioles are small cylindrical bodies, barely visible under a light microscope, that lie just outside the nucleus in an area of specialized cytoplasm. They are normally found in the cells of animals, algae, and some fungi; they are absent in the cells of flowering plants.

Centrioles have a characteristic structure of nine sets of microtubules running lengthwise just below their surface (Figure 4.11). Usually centrioles are paired, each lying at right angles to the other, forming a kind of "T."

They are associated with structures that function in organizing microtubules that serve in cell division and cell motility. For example, in some cells centrioles may play a role in organizing the microtubules of spindle fibers that serve to separate the copies of genetic information (chromosomes) during cell division. However, cells that lack centrioles, those of flowering plants for instance, are still able to form spindle fibers.

The role of centrioles in cell movement is to give rise to basal bodies, short cylinders of microtubules that are arranged in the same manner as those in the centrioles. In turn the basal bodies direct the formation of cilia and flagella.

FIGURE 4.11

Centrioles. A section across the central axis of one centriole. Note the nine sets of triplet microtubules.

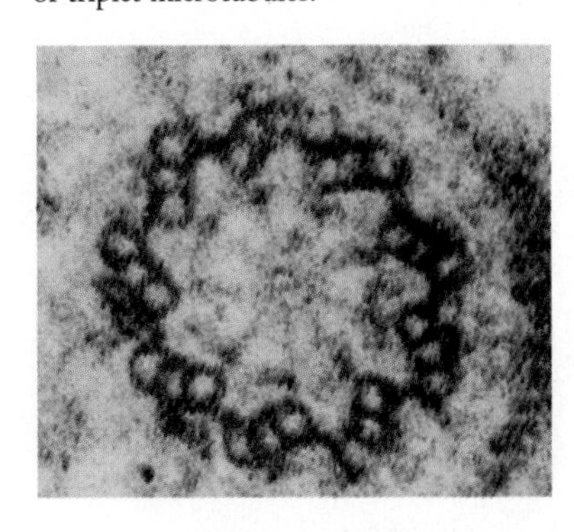

Cilia and Flagella

Cilia and **flagella** (Figure 4.12) are hairlike extensions that project from the surfaces of certain kinds of cells. They differ only in length; cilia are about 10 to 30 micrometers, whereas flagella may extend to thousands of micrometers. Both make "beating" movements, and both may function in moving the cell along through some fluid. Cilia, in addition, move substances across the surface of a stationary cell. As examples, a sperm cell swims by beating its whiplike flagellum, and the beating cilia that line your breathing passages help sweep away airborne debris (unless you have killed them by smoking). Both cilia and flagella contain microtubules that form a 9 + 2 arrangement. Basically, the structure involves a circle of nine pairs of microtubules surrounding two single microtubules. (The arrangement in centrioles is similar, but they lack the central pair.)

The 9 + 2 arrangement of cilia and flagella has generated much speculation about their origin. For example, the basal body may act as a template that organizes the microtubules within a cilium or flagellum. (Do you see how the structural arrangement of a cilium suggests this origin?)

FIGURE 4.12

An electron microscope view (a) of a flagellum cross-section and (b) a cutaway diagram of a flagellum. Note the 9 + 2 arrangement of the microtubules.

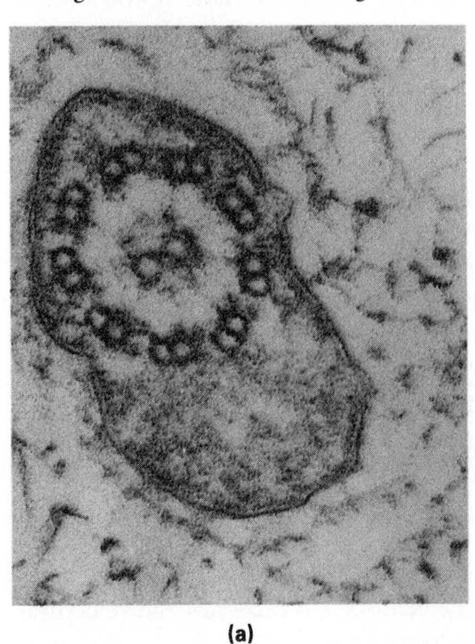

(a)

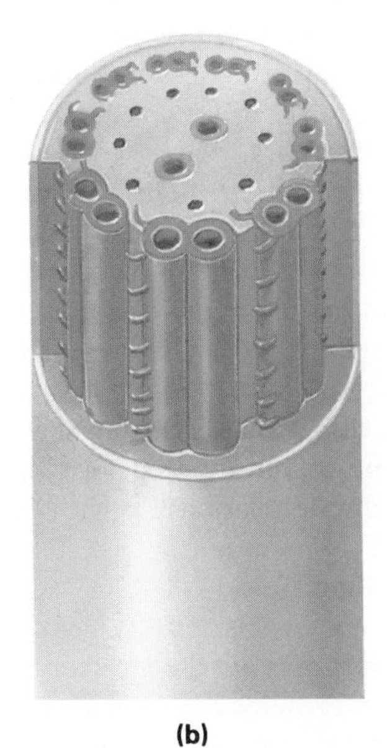

(b)

FIGURE 4.13

A cutaway view of a mitochondrion (a), and (b) micrograph of mitochondria.

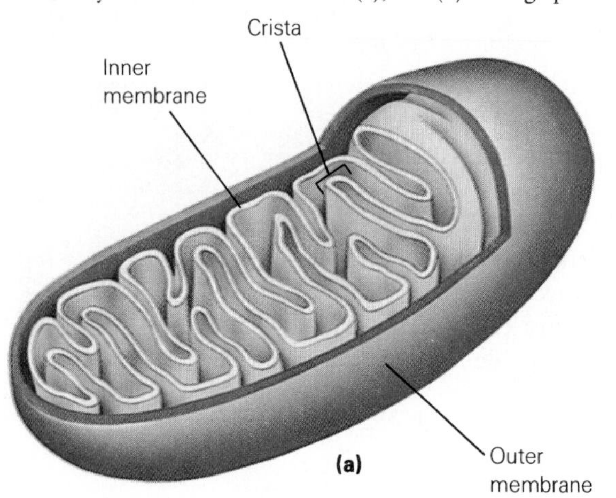

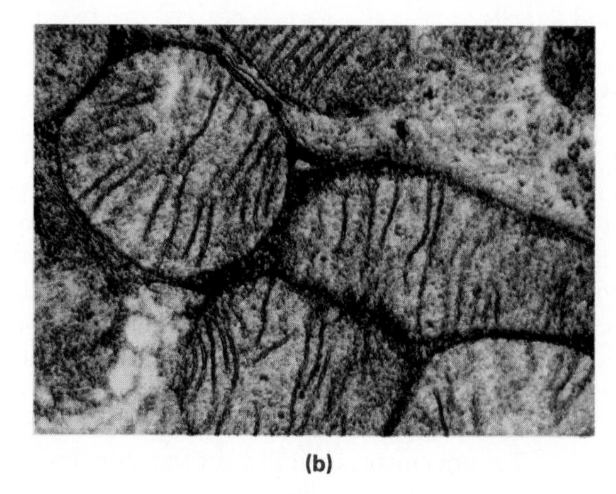

Mitochondria

Mitochondria are tiny structures (about the size of bacteria) that occur in almost all types of eukaryotic cells. They appear in a variety of shapes—round, elongate, and even threadlike. In some cells they seem to squirm around and move through the cytoplasm, while in other cells they appear more stationary or even immobile. They are not randomly distributed, but are more numerous in places where work is going on—places where the greatest amount of energy is required. There are more mitochondria in liver cells, for instance, than there are in less active cells, such as cartilage. Not only are they more common in more active types of cells, they also accumulate in the most active part of any cell. The concentration of mitochondria at places where work is being done is not coincidental; they are the "powerhouses" of the cell, with integral roles in energy production.

A mitochondrion is double membraned, with the inner membrane commonly folded to produce shelflike **cristae** that extend inward into the fluid **matrix** in which a number of reactions necessary for the production of energy occur (Figure 4.13). There has been vigorous discussion over just how mitochondria and certain other organelles came to be included in cells. A major argument is that many such organelles were once free-living organisms that long ago invaded the early cells and developed a symbiotic relationship with them (Essay 4.2).

Ribosomes

Ribosomes are the structures where amino acids are assembled into proteins. Proteins that remain within the cell, such as enzymes and structural proteins, are usually synthesized on ribosomes that are free in the cytoplasm. Proteins that are exported from the cell, hormones and digestive

ESSAY 4.2

The Endosymbiosis Hypothesis

The origin of cell organelles largely remains a mystery. How did they come to be there? How did they evolve? And from what?

One rather interesting idea has been well supported by Lynn Margulis, who contends that those energy-harnessing organelles, chloroplasts and mitochondria, may be the descendants of free-living prokaryotic organisms that invaded other cells early in life's evolutionary history. This idea is known as the **endosymbiosis hypothesis,** and it is continually gaining support. Eventually, she suggests, the chloroplasts and mitochondria became mutually dependent on their host cells and are unable to exist outside them. In time, the "living-together" arrangement evolved into a permanent marriage.

Lynn Margulis, a stimulating scientist who often takes on the establishment and wins, as when her endosymbiosis hypothesis—once unpopular—gained general acceptance.

There are several lines of supporting evidence for the idea. For example, some cell organelles such as mitochondria and chloroplasts are capable of reproducing themselves, but today their reproduction is often precisely synchronized with the reproduction of the cell in which they reside.

There are also other lines of supporting evidence, including the fact that both chloroplast and mitochondria are surrounded by a double membrane, very similar to that of prokaryotic cells from which eukaryotic cells are thought to arise. Also, both of these organelles contain their own genetic material, DNA (with slight differences in code from the DNA in chromosomes). This DNA occurs in simple, circular strands, as does the DNA of prokaryotes. Furthermore, chloroplasts have their own ribosomes, small bodies associated with energy production (discussed shortly). Interestingly, the ribosomes of these two organelles are smaller than those of the eukaryotic cell; in fact, they are about the size and shape of bacterial ribosomes (and bacteria, you recall, are prokaryotic).

The list goes on, forming a compelling body of evidence that the invasion of ancient cells by the ancestors of prokaryotes yielded many of today's cellular organelles. So far, no single alternative theory has been able to replace this encompassing scenario.

enzymes for example, and most proteins that become incorporated into cellular membranes, are generally manufactured on ribosomes that are bound to a membranous compartment called the endoplasmic reticulum.

The Endoplasmic Reticulum

The **endoplasmic reticulum (ER)** is a highly variable membranous structure whose job is the synthesis, modification, and transport of certain materials (Figure 4.14). The endoplasmic reticulum is a series of membranous channels that form a transportation network within the cell.

FIGURE 4.14

(a) Rough endoplasmic reticulum (ER). (b) The smooth endoplasmic reticulum lacks the intimate association with ribosomes seen in the rough endoplasmic reticulum. It functions primarily in the assembly and storage of lipids and carbohydrates.

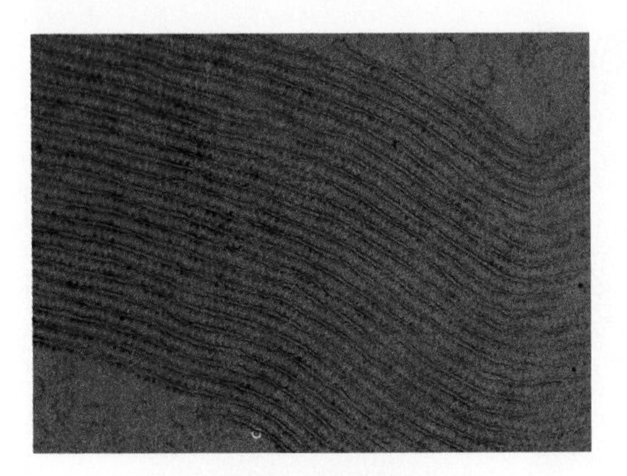

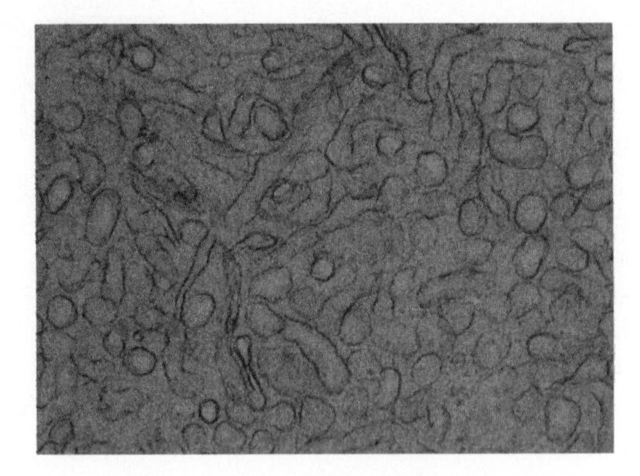

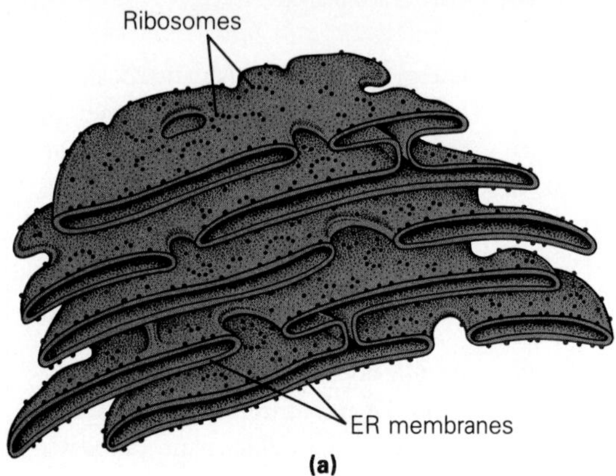

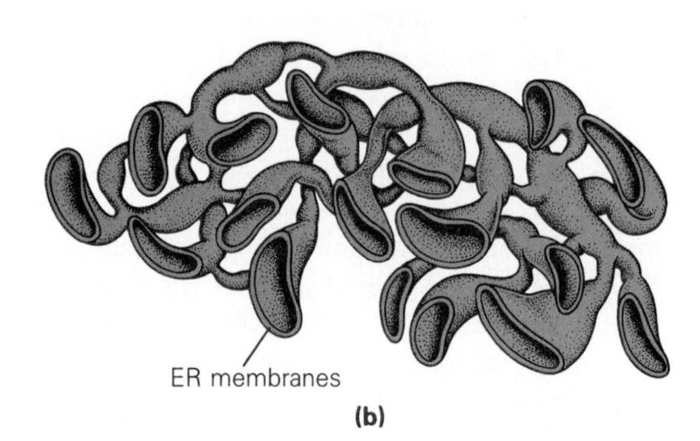

Under the microscope, the ER appears as light space surrounded by membranes. Some of these membranes are studded with ribosomes, some are not. These with ribosomes are called **rough ER,** and those without, **smooth ER.**

The proteins produced by the ribosomes on the rough ER pass into the channels formed by the membranes of the endoplasmic reticulum. Within these channels, the protein may be modified, or moved to other organelles for packaging, modification, storage, or export. Many proteins manufactured on the rough ER form new cell membranes. (Membranes are constantly forming, disintegrating, and coalescing in the active cell.) The smooth ER transfers the products formed by the rough ER to other locations by budding off tiny product-laden spheres that can then move easily through the cell.

Smooth ER is found primarily in cells that synthesize fats or steroids. The smooth ER, particularly in liver cells, is also important in detoxifying certain drugs.

Electron microscope studies suggest that the ER is connected to the outer border of the nuclear membrane. Furthermore, there is evidence that it is also confluent with the plasma membrane. This arrangement would result in an open channel from the nucleus to the outside of the cell. The cell might thus be able to easily transport products manufactured in the nucleus directly to the outside. But more important, with such a connection, the nucleus might be able to react quickly to changes in the cell's environment.

Golgi Bodies

In 1898, Camillo Golgi, an Italian cytologist, was experimenting with some cell-staining procedures and discovered that when he used certain stains, such as silver nitrate, peculiar "bodies" appeared in the cells. These structures had never been noticed before, but when other workers looked for them using the same stains, they turned up in a variety of cells. However, because they could not be seen in living cells, there was a great argument over whether they really were cell structures, or were just artifacts or debris produced by the staining process itself.

The electron microscope resolved the debate. Indeed, these strange bodies did exist, and, appropriately enough, they were named **Golgi bodies.** It was found that they had a characteristic and identifiable structure no matter what kind of cell they were found in. In every case, they appeared as a group of tiny flattened sacs, lying roughly parallel to each other, somewhat like pancakes. (Figure 4.15).

Even after their existence was confirmed, an argument continued over their function. What did they do for a living? We now know that they

FIGURE 4.15

Golgi bodies appear as flattened sacs in the cytoplasm, which serve as a kind of packaging center, enclosing various materials in a membrane.

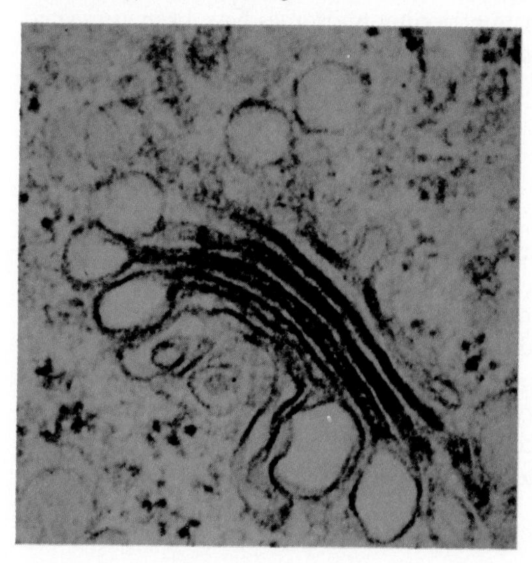

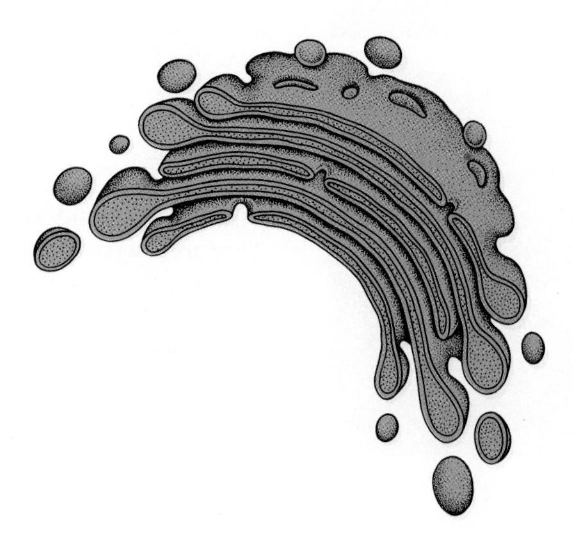

FIGURE 4.16
Lysosomes hold powerful digestive enzymes.

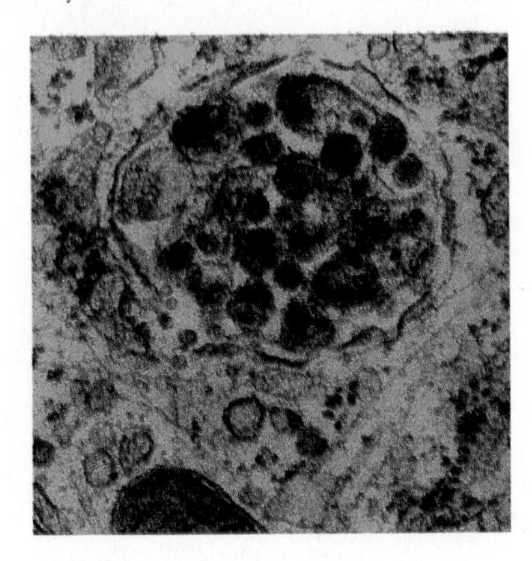

FIGURE 4.17
Plastids include colorless leucoplasts as well as green chloroplasts, such as those within plant cells.

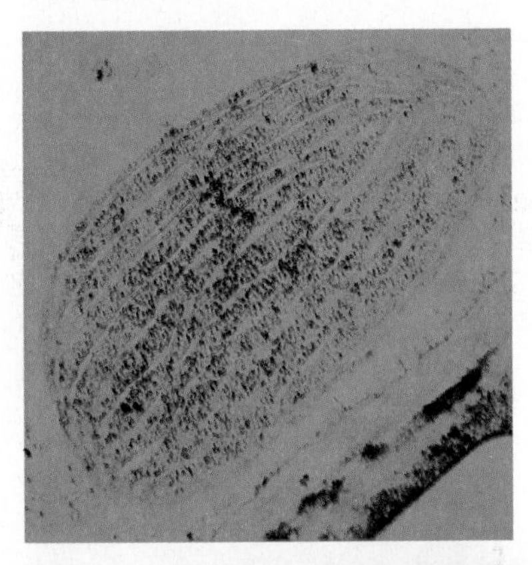

serve as a sort of packaging center for the cell. They have been likened to manufacturing, warehousing, and shipping centers (as well as Swiss finishing schools). Their role is indeed complex. Products formed by the endoplasmic reticulum are stored (and, in some cases modified) in the Golgi complex. The complex also manufactures many polysaccharides, including ones that will be secreted by the cell. Enzymes and other proteins, as well as certain carbohydrates, are collected in these bodies and packaged into sacs or vesicles. In this way, they are kept apart from the rest of the cell. In some cases, the packages break away from the Golgi complex and move to the plasma membrane where the enclosed molecules are released from the cell.

Lysosomes

Lysosomes are somewhat spherical, about the size of mitochondria, and are, in general, distinctly unimpressive bodies (Figure 4.16). Their appearance, however, belies the rather startling role they play in the history of a cell. Lysosomes are packets of digestive enzymes that are synthesized by the cell and packaged by the Golgi bodies. The packaging is important because if these enzymes were floating free in the cell's cytoplasm, the cell itself would be digested. Christian de Duve, who first discovered the lysosomes, called them "suicide bags," and the dramatic description is not entirely unwarranted, since they can actually destroy the cell that bears them.

The question arises, why would cells have ever developed such a risk to themselves? In some cases the destruction of cells is beneficial to the organism. For example, the cells could be old and not functioning well, or they might be in a part of the body that was undergoing reduction as a part of a normal developmental process, such as in the webbing between the fingers of a developing embryo. (They are generally not found in plants.)

Lysosomes may also help dispose of unwanted mitochondria, red blood cells, or bacteria. (Fragments of all these have been found within lysosomes.) Interestingly, malfunctioning lysosomes have been associated with a number of human diseases, including cancer. Rupturing lysosomes have also been accused of contributing to the aging process.

Plastids

Plastids are found in plant cells, but not in animal cells (Figure 4.17). They are interesting bodies, similar in some respects to mitochondria. For example, they are double-membraned and they have their own DNA. Strangely, their DNA may differ somewhat from that of the cells in which they reside.

There are two major types of plastids—colored and colorless. The colorless ones are called leucoplasts. The best known of the colored plastids are the green chlorophyll-containing chloroplasts. Chloroplasts may also contain yellow or carrot-colored pigments, which, logically enough, are called carotenoids. In addition to pigments, the chloroplasts contain enzymes and other molecules that capture the energy of sunlight and use it in the production of food (see Chapter 5). Other plastids contain pigments that give flowers and fruits their color. The colorless plastids, or leucoplasts, are sites where starch, or sometimes other substances such as oils or proteins, may be stored.

FIGURE 4.18

Vacuoles are fluid-filled sacs, here seen as the large space occupying almost the entire interior of a plant cell.

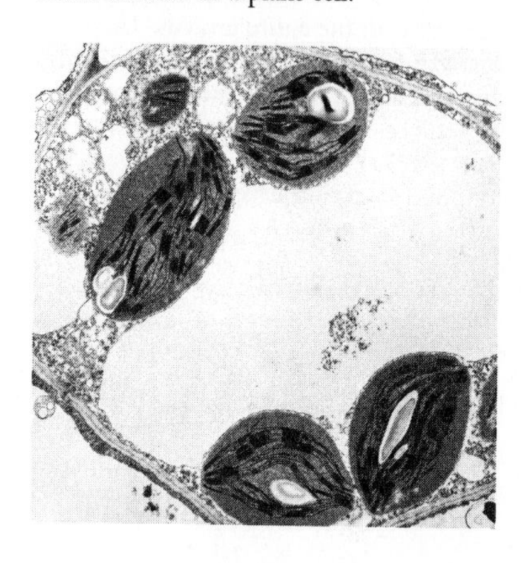

Vacuoles

Vacuoles are fluid-filled sacs found in the cells of both plants and animals, as well as in microscopic organisms called protists. It is in plant cells, however, where they reach their greatest development; in fact, they may be the most conspicuous bodies in the plant cell (Figure 4.18). There are a number of types of vacuoles, each with a different function. Some, for example, are highly active in the cell's metabolism, while others are simply storage vessels.

In plants, the vacuoles are filled with a "cell sap" that can change volume through osmosis. In fact, it is the pressure of the swelling vacuole that forces the plasma membrane against the cell wall and makes the plant tissue firm. A plant wilts when there is not enough fluid to keep its huge vacuoles filled. Besides water, the vacuole sap may contain sugars, proteins, pigments, and organic acids. It is these acids that give oranges and lemons their tart taste.

Some single-celled organisms (such as the paramecium, as we shall see) contain **contractile vacuoles** that enable them to squeeze excess water out through the plasma membrane. Such vacuoles may appear and disappear in response to the organism's needs, or they may be relatively permanent structures. The latter help us distinguish one species of these tiny organisms from another.

The Nucleus

FIGURE 4.19

The nuclear membrane is pitted with pores that allow easy movement of materials across the membrane (scanning EM).

Perhaps the most prominent structure in most cells is the **nucleus.** It also plays a central role in the life of the cell in that it is involved in critical processes, such as reproduction, and the regulation of virtually everything the cell does. The nucleus exerts its control by directing the synthesis of proteins that serve as enzymes, chemical messengers, or other active molecules.

The nucleus is surrounded by a double membrane, a fact that suggests two hypotheses regarding the origin of the nucleus. The **nuclear membrane** (also called the nuclear envelope) is perforated by **nuclear pores** (Figure 4.19). Such pores facilitate the movement of molecules between the nucleus and cytoplasm.

A number of rather straightforward experiments have illustrated the critical role of the nucleus. For example, if an amoeba is cut in half, the half with the nucleus will survive and continue to function as before. The

FIGURE 4.20

An experiment showing the importance of the nucleus. A living amoeba (a single-celled protist) was cut in half, leaving one half with the entire nucleus. The deprived half died, and the nucleated half survived. When the nucleus was removed from the surviving half and placed into the dying half, the latter recovered and continued a normal existence.

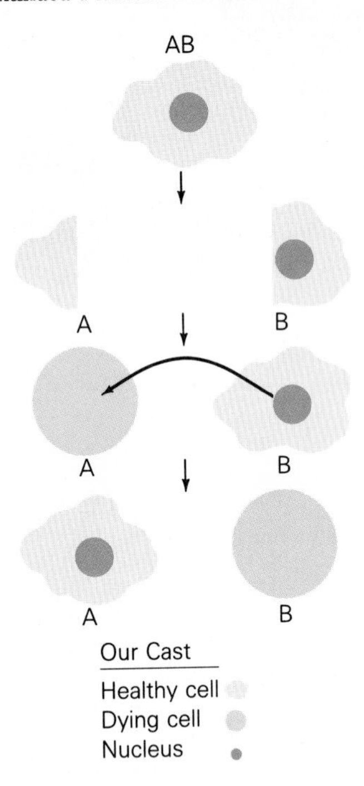

half without the nucleus is unable to move about or capture food and soon dies. However, if the nucleus is surgically restored to this half within a few days, it recovers and is soon as good as new (Figure 4.20).

Inside the nucleus is a fluid matrix in which a number of different types of bodies are suspended. When cells are appropriately stained, a netlike structure becomes visible within the nucleus. This material is called chromatin (from Greek: *chromo,* color) because of its affinity for the stains. At a certain period in the cell's cycle, the chromatin shortens and thickens and forms chromosomes, the structures that include sequences of genes. (As a highly educated person, you are undoubtedly aware that genes have a role in heredity. But just in case, we'll go over it later.)

Within each nucleus is a small structure called the **nucleolus,** a dark-staining body containing RNA, DNA, and proteins. Most nuclei contain only one nucleolus, but some kinds have more. The job of the nucleolus is to synthesize ribosomes. An actively growing cell can produce an astounding 10,000 ribosomes per minute.

HOW MOLECULES MOVE

The large multicellular organisms of the earth are faced with a number of general problems. One of these is how to move materials around through large, bulky bodies. Many of these materials must laboriously pass into and out of cells by crossing the plasma membranes. There are two broad categories of such movement across membranes: **passive transport** and **active transport.** Table 4.3 summarizes molecular movement.

Passive Transport

Passive transport is conceptually quite simple because it does not involve the expenditure of cellular energy. Instead, it requires only the random movement of molecules. However, as you may have suspected, in biology, even something so simple is not all *that* simple. We will consider three kinds of passive transport: diffusion, facilitated diffusion, and osmosis.

Diffusion

Diffusion is the random movement of molecules from areas of their higher concentration to areas of their lower concentration. Physicists tell us that the molecules of any substance—gas, liquid, or solid—are constantly in motion, bumping into each other, rebounding, and changing their direction of movement. The principle would become apparent if you were to open a vial of perfume (or if you're the baser sort, try hydrogen sulfide) in a corner of a room. Soon the smell will be detectable in all parts of the room. If the solution is allowed to evaporate completely, all parts of the room will eventually smell the same (Figure 4.21). Until that happens, however, there will be a concentration gradient, with a greater concentration of perfume molecules nearer the vial.

The molecules shift away from their center of concentration because they bump into each other with greater frequency where they are more densely packed. When they collide, of course, they rebound and change

TABLE 4.3
Movement of Molecules

TYPE	DESCRIPTION
PASSIVE TRANSPORT (NO CELLULAR ENERGY REQUIRED)	
Diffusion	Movement of molecules from an area of higher concentration to one of lower concentration; due to constant motion of molecules.
Facilitated diffusion	Molecules are carried across membrane by a carrier (permease) from an area of higher to one of lower concentration.
Osmosis	Diffusion of water across a semipermeable membrane from an area of lower solute concentration to one of higher solute concentration.
ACTIVE TRANSPORT (CELLULAR ENERGY REQUIRED)	
Facilitated active transport	Carrier molecule carries substance across membrane; can be toward an area of higher concentration.
Endocytosis	
Pinocytosis	Liquids are moved into cell by being surrounded by plasma membrane and pinched off into vacuole.
Phagocytosis	Solids are moved into cell by being surrounded by plasma membrane and pinched off into vacuole.
Exocytosis	The cell expels materials as the membrane surrounding those materials fuses with the plasma membrane after which both membranes rupture.

direction. However, their chances of collision are lessened when they are moving outward toward areas of lesser concentration. So there is a tendency for the *net* movement to be outward, *away* from the highest concentration gradient.

The rate at which molecules or ions diffuse depends on many factors. Increased heat produces faster molecular movement, so the diffusion rate is higher in a warmer system. This is why sugar diffuses more quickly in a cup of hot coffee than in a glass of iced tea. The size of the particle in motion also influences the rate of diffusion. Large particles diffuse more slowly than smaller ones.

Although molecules and ions diffuse through the fluid environment of living cells with relative ease, they may not be able to move as easily from one cell to another. This is because the membranes of living cells are selective. Some substances are allowed across, some are escorted across, and some are not allowed to cross at all. Some substances diffuse through the lipid portion of the membrane and some through channels formed by membrane proteins. The two most important factors influencing the rate of diffusion of those substances that cross the lipid portion of the cell membrane are solubility in lipids and size.

Certain molecules are too large to get through. Other molecules though, such as carbon dioxide, oxygen, and water, are able to diffuse rather freely across almost any plasma membrane. Some of the substances that cannot pass through the lipid portion of the membrane can pass through channels formed by membrane proteins. When these are open, they allow specific substances, usually ions of the correct charge and size, to diffuse through.

FIGURE 4.21
Molecules of perfume escaping from a bottle will soon permeate an entire room. At first, the molecules are concentrated near the bottle, but if evaporation is complete, in time, the random movements of the molecules will disperse them evenly throughout the room.

It should be stressed that in spite of its mechanical simplicity, diffusion is an important means by which small molecules move across cell membranes. Furthermore, simple diffusion works according to strict physical laws; that is, it works the same in both living and nonliving systems.

Facilitated Diffusion

Facilitated diffusion is similar to simple diffusion in that the molecules are stimulated to move by environmental heat. Also, the movement is away from areas of higher concentration to areas of lower concentration. However, facilitated diffusion differs from simple diffusion in that some kinds of molecules are moved more easily than others. This is because the movement of certain molecules is enhanced by specific carrier proteins embedded in the cell membrane. Although we are not yet certain how the carrier proteins work, it is thought that the carrier protein spans the membrane and changes shape to allow the substance through (Figure 4.22).

Osmosis

Osmosis is one of those words that has been borrowed from science and then twisted and misused beyond recognition. (Do any others come to mind? How about "instinct" or "ecology"?) The next time you hear the word *osmosis* in cocktail conversation, ask the speaker to define it. You will make a lifelong friend when you smirk and say, "No, osmosis is the diffusion of water across membranes, from an area of lower solute concentration to one of greater solute concentration." You may have to explain that a solute is a substance in solution. Say further, that it doesn't really matter what kind of ion or molecule is in solution: it could be sugar or salt, or any other soluble substance that cannot cross the membrane. Then say, "You see, when two solutions are separated by a selectively permeable membrane—that is, a membrane that allows only certain substances, such as water, to pass while being impermeable to the molecules in solution—the water will move from the solution with the lower solute concentration

FIGURE 4.22

A model of active transport across a membrane. A carrier molecule (purple) picks up ions such as sodium on one side of the membrane, then rotates and releases them on the other side, where it picks up other ions, such as potassium, for the reverse journey. The pump in this case is called a sodium-potassium pump.

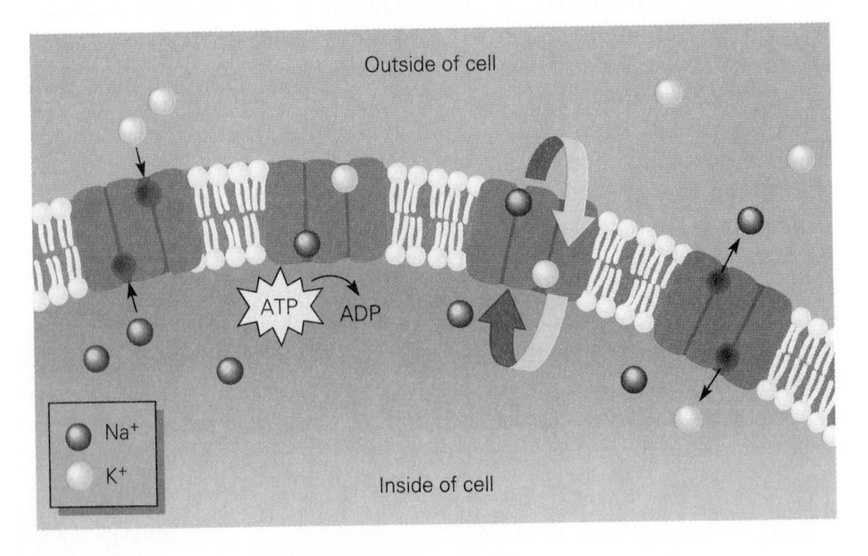

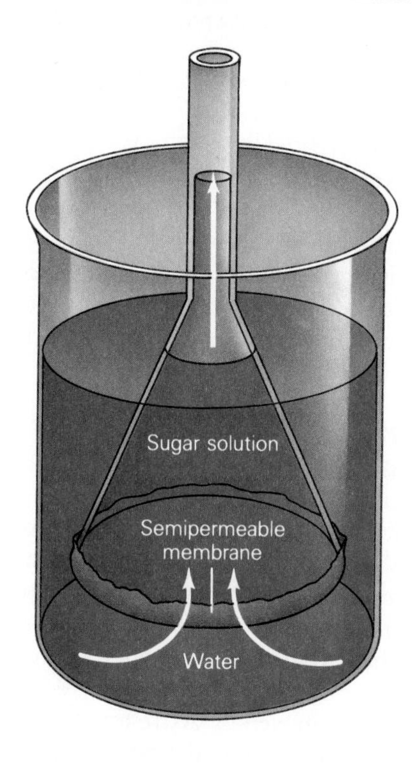

FIGURE 4.23
Osmosis will cause water molecules to move through a semipermeable membrane from a place where the solute concentration is higher to where it is lower. In the illustration, the inverted funnel is immersed in pure water, but inside the funnel, sugar is in the water. Because the membrane covering the funnel will not allow the sugar molecules to pass, we are interested only in the movement of water molecules. The resulting movement in this case will cause the sugar solution to become more dilute, thus raising its level in the funnel.

through the membrane to the solution with the higher solute concentration." The people there will greatly appreciate your discourse and you are sure to be invited back every February 30th.

If you look around and everyone is still there, you should be prepared to answer questions, since osmosis is one of those simple principles that, at first, may seem hard to understand. For example, one might wonder, how long does osmosis go on? Theoretically, it continues until the water concentration on both sides of the membrane is equal. However, this equilibrium rarely occurs because molecular concentrations in living systems may rapidly change, and because other physical factors may influence the movement of the water molecules. Pressure, for example, may influence osmotic movement. In the sort of setup we see in Figure 4.23, the rising column of solution will eventually become so heavy that its pressure finally stops the net flow of water molecules. At this point, the *net* movement of water stops. (When osmotic movement apparently stops, the water actually continues to move across the membrane, but equally in both directions, so that there is no gain or loss of volume on either side.)

Osmosis sometimes poses problems of water balance for cells. If a cell is placed in a solution with a solute concentration greater than that within the cell, it will lose water and shrink. Conversely, when a cell is placed in a solution with a solute concentration less than that of the cell, it will tend to gain water. Animal cells cannot tolerate excessive uptake of water. Since they lack a rigid cell wall, they may burst if too much water enters the cell. Plant cells, however, have a cell wall that will prevent them from rupturing under such conditions. When they are in an environment with a solute concentration less than that of the fluid within the cell, the soil in which most land plants grow for instance, water will enter the cell and increase the outward pressure against the cell wall. This pressure against the cell wall, called turgor pressure, creates a force opposing the one created by solutes within the cell that draws water in (Figure 4.24).

Active Transport

In some cases, a living cell can move molecules "uphill" across its membrane from a place of lower concentration to a place of higher concentration. Such a chore is uphill because it goes against the normal direction of movement of the molecules and thus requires energy.

Facilitated Active Transport

We see, in a process called **facilitated active transport,** that carrier molecules are able to move ions across a membrane. Some cells are able to maintain proper concentrations of sodium and potassium by pumping sodium out of the cell and potassium in. The mechanisms of such "pumps" are not completely understood, but it is generally assumed that some sort of carrier molecule is involved. It is visualized as a membrane protein that changes its configuration, thereby allowing the sodium and potassium to move through it (see Figure 4.22 on page 112). A variety of molecules can be moved by active transport in either direction, into or out of cells, but always with an expenditure of energy.

FIGURE 4.24

Turgor pressure helps give cells their shape as we see in single-celled paramecia (a), the water plant *Elodea* (b), and red blood cells (c). In land plants, loss of turgor pressure causes wilting (d).

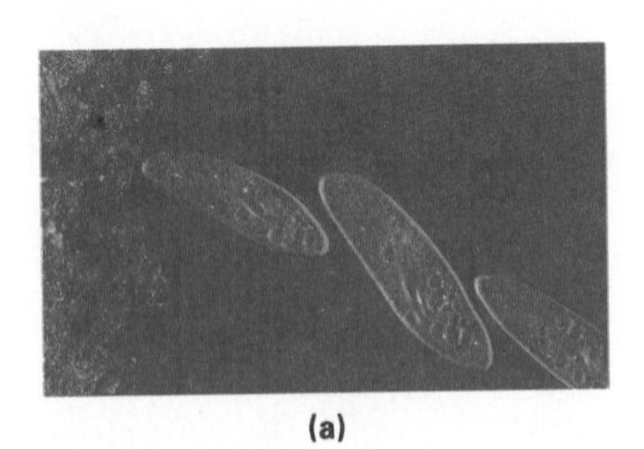

(a)

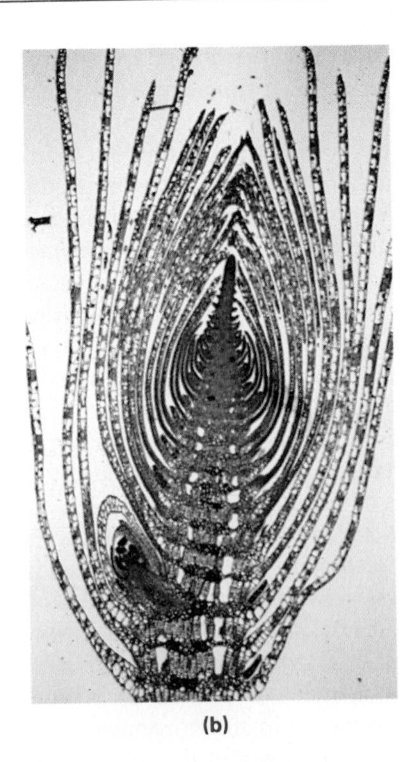

(b)

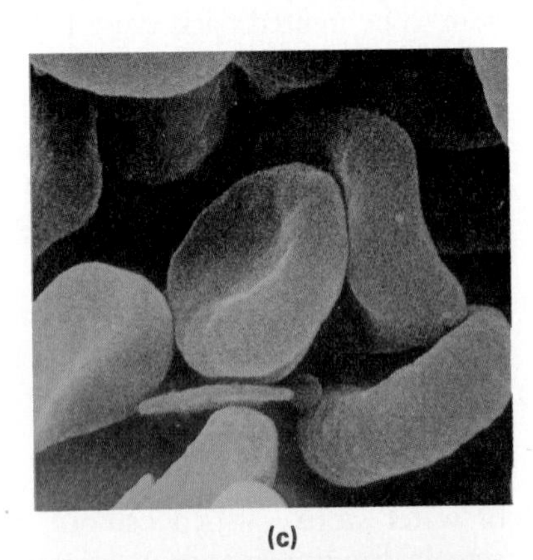

(c)

(d)

Endocytosis and Exocytosis

Some forms of active transport are so active that they are visible under microscopes. The best example, perhaps, is **endocytosis,** a kind of active transport in which material is surrounded by the plasma membrane and then pinched off into a vacuole. It was first observed in the feeding behavior of the amoeba. When an amoeba detects a particle of food, its surrounding membrane buckles inward, forming a deepening cup around the food. The *rim* of the cup eventually draws together, and the cup pinches off from the surface, forming a membrane-bound sac, or vacuole, within the amoeba. The food that was outside the cell is now enclosed by the vacuole. Digestive enzymes are secreted into the vacuole and the food is broken down. The resulting nutrient molecules are passed through the vacuole membrane and out into the amoeba's inner fluids.

Endocytosis, by the way, is a general term that includes two processes. If the vacuole engulfs solid material, the process is called **phagocytosis;** if it engulfs dissolved or fluid materials, it is called **pinocytosis.**

The opposite process is equally intriguing. It is called **exocytosis,** the process by which cells *expel* materials (Figure 4.25). The material to be

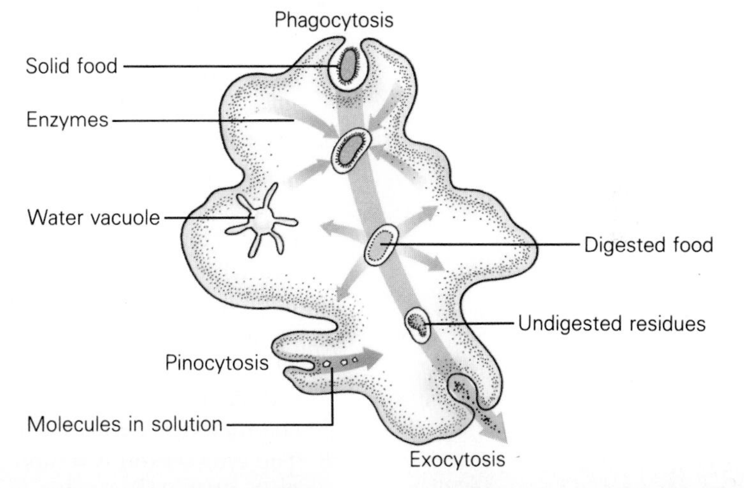

FIGURE 4.25

Active transport and vacuoles. In the drawing, a very busy amoeba is demonstrating both endocytosis and exocytosis in all their variations. At the upper region, the amoeba is engulfing a small ciliate protozoan by phagocytosis. At the left, a channel has surrounded a solution of large molecules by pinocytosis. At the lower side of the amoeba, the undigested residue from a food vacuole is being expelled by exocytosis.

Below, a dramatic sequence of micrographs demonstrates phagocytosis in the amoeba *Chaos carolinense,* as it captures *Pandorina morum.*

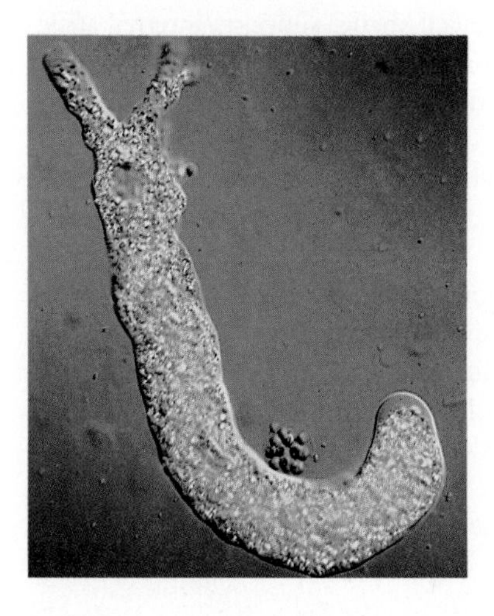

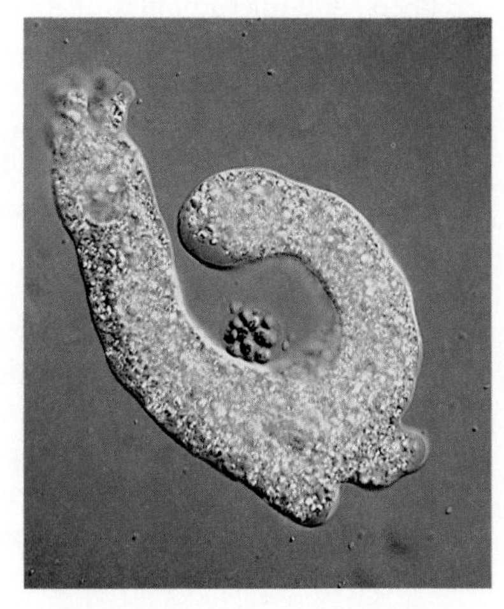

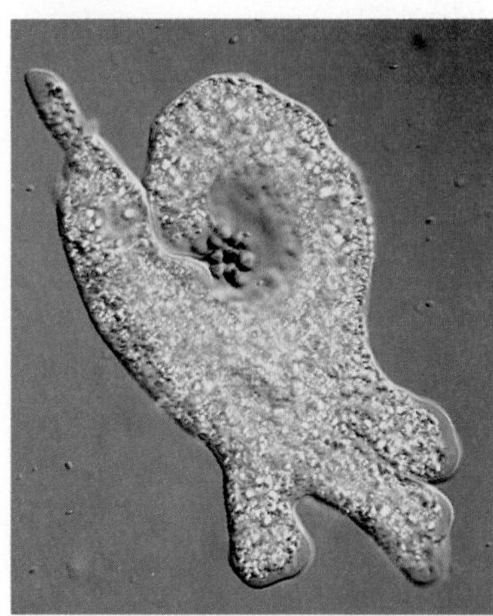

expelled becomes enclosed in a vacuole that eventually fuses with the plasma membrane. At the point where the membrane of the vacuole joins the plasma membrane, both membranes rupture and the contents of the vacuole are ejected from the cell.

We have, indeed, learned a great deal about Hooke's "little boxes" since he first looked at that cork. Of course, many problems remain unsolved. Interestingly, some of these problems were unveiled as we came to know more about cells. But other puzzles are almost as old as the field of cytology itself. These enduring questions have remained as monuments to life's intricacies. Even now, though, a host of dedicated researchers are determined that these monuments, too, shall fall.

SUMMARY

1. The cell theory states that all life comes from cells and is made of cells, the smallest units of life. The two types of cells are prokaryotic (bacteria and cyanobacteria) and eukaryotic.
2. Eukaryotic organisms composed of many cells have at least two advantages over single-celled organisms: an ability to meet a greater variety of environmental demands due to cell specialization and an increase in the degree of control over interactions with the environment due to the increased area of regulatory membrane covering the body surface.
3. Plant cells have a rigid, nonliving, permeable cell wall composed of cellulose and other compounds.
4. The cell is bounded by a living, selectively permeable cell (plasma) membrane. The fluid-mosaic model describes this membrane as composed of two layers of phospholipid molecules oriented with their hydrophilic heads outward and their hydrophobic tails inward. Various proteins scattered among the lipids form a mosaic. Some proteins form channels that allow certain materials to cross the membrane. Other molecules, such as carbohydrates, may be attached to the outer surface of the membrane, and the cytoskeleton may be attached to the inside surface.
5. The cytoskeleton is composed of a web of microfilaments, intermediate fibers, and microtubules. It gives the cell shape, supports internal structures and enzymes, and functions in motility.
6. Microtubules, made of the protein tubulin, form mitotic asters, spindles, and the core of cilia and flagella.
7. Centrioles are associated with cell division and movement. They are found in cells of animals, algae, and some fungi, but not in cells of flowering plants.
8. Cilia and flagella are hairlike extensions whose beating movements move the cell or move substances over the cell surface. They are composed of nine pairs of microtubules surrounding a central pair of microtubules.
9. Mitochondria have an important role in energy production. The most active types of cells, and active parts of each cell, have relatively more of these double-membraned structures.
10. Ribosomes are structures on which amino acids are synthesized into proteins.

11. The endoplasmic reticulum (ER) is a membranous structure. The rough ER is studded with ribosomes and manufactures proteins and membranes. The smooth ER packages the products of the rough ER into small sacs for distribution within the cell. The ER may provide a direct path between the nucleus and the cell's environment.
12. The Golgi bodies are stacks of flattened sacs that may modify, store, and package materials produced elsewhere in the cell and produce their own polysaccharides.
13. Lysosomes are small packets of digestive enzymes that control the process of digestion within the cell.
14. Plant cells contain plastids. The colorless plastids (leucoplasts) are storage bodies. Colored plastids contain pigments. Green plastids (chloroplasts) contain chlorophyll and some carotenoids, which capture light energy to make food.
15. Vacuoles are fluid-filled sacs. Some are involved in metabolism and others are involved in storage. The contractile vacuoles of single-celled organisms rid the cell of excess water.
16. The nucleus is important in the regulation of cellular activities and reproduction. It contains the genetic material (chromatin) and one or more nucleoli, which contain DNA, RNA, and protein. The nucleolus synthesizes ribosomes.
17. We have considered three types of passive transport. Diffusion, the movement of molecules from an area of high concentration to one of lower concentration, occurs because molecules are always in constant motion. Facilitated diffusion is similar to diffusion except that the substances are transported across the membrane by carrier proteins. Osmosis is the diffusion of water across a semipermeable membrane from an area of lower solute concentration to an area of higher solute concentration.
18. Active transport requires the expenditure of energy. One type of active transport involves a carrier molecule. In endocytosis either solid substances (phagocytosis) or liquids (pinocytosis) are actively moved into the cell as they are surrounded by cell membrane and eventually pinched off into a vacuole. The reverse process, exocytosis, actively moves substances out of the cell.

KEY TERMS

active transport (110)
cell theory (92)
cell wall (99)
centriole (102)
cilia (103)
contractile vacuole (109)
cristae (104)
cytology (92)
cytoskeleton (101)
diffusion (110)
endocytosis (115)
endoplasmic reticulum (ER) (105)
endosymbiosis hypothesis (105)
eukaryotic cell (93)
exocytosis (115)
facilitated active transport (114)
facilitated diffusion (112)
flagella (103)
fluid-mosaic model (100)
Golgi body (107)
hydrophilic (100)
hydrophobic (100)
intermediate fibers (101)
light microscope (LM) (95)
lysosome (108)

matrix (104)
microfilaments (101)
microtubule (101)
mitochondria (104)
nuclear membrane (109)
nuclear pore (109)
nucleolus (110)
nucleus (109)
osmosis (112)
passive transport (110)
phagocytosis (115)
pinocytosis (115)
plasma membrane (100)
plastid (108)
prokaryotic cell (93)
ribosome (104)
rough ER (106)
scanning electron microscope (SEM) (95)
selectively permeable (100)
smooth ER (106)
specialization (96)
transmitting electron microscope (TEM) (95)
tubulin (102)
vacuole (109)

FOR FURTHER THOUGHT

1. A certain cell has a large number of mitochondria. What can we conclude about the energy requirements of this cell?
2. What transport mechanism is at work when a burning incense stick fills a room with scent? Explain the changes in concentration gradients that occur.
3. Two solutions are separated by a semipermeable membrane. The solute concentration of solution A is 25%. The solute concentration of solution B is 50%. Will water diffuse across the membrane from A to B or from B to A?

FOR REVIEW

True or false?

1. ___ An organism composed of many cells has greater control over the interactions between itself and the environment than does a single-celled organism.
2. ___ Golgi bodies are protein synthesizing structures located on the endoplasmic reticulum.
3. ___ Colored plastids are pigment-containing bodies found only in plant cells.
4. ___ Carrier proteins assist the movement of water molecules across cell membranes in a process called osmosis.
5. ___ The active transport mechanism moves molecules across cell membranes from regions of low concentration to regions of high concentration.

Short answer

6. How does the fluid mosaic model describe the arrangement of phospholipids and proteins in a cell membrane?

Fill in the blank.

7. Cellular components are held in place by intricate webs of microtubules called the ___ .
8. The hairlike extensions of motile cells are composed of ___ .
9. The ___ is surrounded by a double membrane and contains the genetic material called chromatin.

10. Solid and liquid materials are engulfed by cells in an active transport process called ____ .

Choose the best answer.

11. A cell membrane is a structure that
    A. contains carrier molecules.
    B. is permeable to all molecules.
    C. is strengthened by cellulose fibers.
    D. occurs only in plant cells.
12. Which of the following correctly matches a cellular component with its function?
    A. ribosome/imparts strength and gives shape to the cell
    B. lysosome/stores digestive enzymes
    C. centriole/stores pigments and plant nutrients
    D. plastid/synthesizes proteins
13. Which of the following organelles is the primary production site for cellular energy?
    A. Golgi body
    B. lysosome
    C. vacuole
    D. mitochondria
14. Diffusion can best be described as
    A. the mass movement of fluids across semipermeable membranes.
    B. the movement of molecules against concentration gradients.
    C. the net movement of molecules away from areas of high concentration to areas of low concentration.
    D. none of the above
15. Which of the following is not a type of passive transport?
    A. diffusion
    B. facilitated diffusion
    C. exocytosis
    D. osmosis

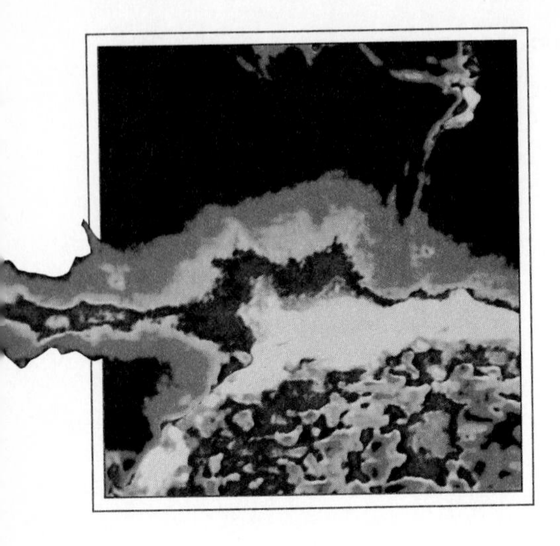

CHAPTER 5

Energy: The Dance of Life

Overview

The Laws of Thermodynamics
ATP: The Energy Currency
Food to Energy
A Brief Review

Objectives

After reading this chapter you should be able to:

- Define two energy states and list their various forms.
- State the tenets of the laws of thermodynamics and their implications for living systems.
- Describe how the ATP molecule is formed and its role in energy transfer.
- Summarize the photosynthetic process by tracing the steps in the conversion of light energy to chemical energy.
- Relate the structure of a chloroplast to its function.
- Describe four major processes in cellular respiration, stating where each takes place and listing their products.
- Compare and contrast the role of the electron transport system in the chloroplast and the mitochondrion.

As I sit on my porch swing writing this, the sun filters through the leaves above to mottle my paper. From here I can see the tracks of a possum that wandered by last night probably looking for any remains of Rachel's dog food (fat chance). That possum is safe from both me and fat old Rachel, but I suspect I may have eaten a few of its relatives as I was growing up in Arkansas. I find that this admission is best not made in certain social circles these days, but I did once mention it to a friend who was visiting. He was chagrined. (I'm sure he thought that eating possum is a good way to get run over.) On the other hand, he believed that monkeys were fine eating. He was a member of the often-feared Waorani tribe of Ecuador's Amazon. We had often hunted together there and had eaten monkeys, as well as capybara, caiman, and piranha. The experience had underlined for me that food can indeed come in a variety of forms, but I also know that whatever the immediate source of one's food, the ultimate source is that same sun that now dabbles this page.

We will now consider how energy available in the form of sunlight is handled by living things—how it is captured by leaves and then how it is passed to the bodies of plant eaters and the bodies of the animals who eat those herbivores. We will see that there are common themes among all the diverse forms of life on the planet. Also, because the chemistry of all this can get pretty complicated, we will take two looks at the processes, "the poet's view" for those of you who never intend to take another

FIGURE 5.1

The water behind a dam is a source of potential energy that can be released as kinetic energy.

biology course as long as you live, and "the closer view" for those of you who maybe aren't sure, or who may just want to know a little more (or those whose instructor says they have to). In either case, what you will see is that as simpler molecules are joined to form more complex molecules, energy is stored; and as complex molecules are changed to simpler molecules, energy is released.

First, though, what is energy? Energy is defined as the ability to do work (that is, to move matter). Energy exists in many forms, and at many levels, and it is on the move. It enables and it destroys, it binds and it releases, and it ebbs and flows in countless directions throughout the universe. As it shifts about, it can exist in two states: potential energy or kinetic energy. **Potential energy** is stored energy. Although it isn't doing anything, it has the capacity to do work. Matter has potential energy as a result of its location or spatial arrangement. Water behind a dam has potential energy. When potential energy is released, it becomes **kinetic energy,** the energy of motion. Kinetic energy is in the process of doing work. As water passes through the dam, its potential energy becomes kinetic energy, which can turn electrical generators (Figure 5.1). A person holding a jar of nitroglycerine would probably be obsessed by a single hope: that the molecules will *not* rearrange themselves to a lower energy level. We can say, then, that a jar of nitroglycerine has potential energy (or free energy). When potential energy is released, it becomes kinetic energy (Figure 5.2). The jar of nitroglycerine and a rock at the top of a hill both have potential energy that can be converted to kinetic energy.

FIGURE 5.2

Dynamite (a) contains nitroglycerine, an oily compound whose molecules can rearrange suddenly, releasing a blast of energy. Explosions (b) result from an unbridled and unchanneled release of energy.

(a)

(b)

FIGURE 5.3

The energy that is released as a boulder rolls down a hill is proportional to the work it took to get it up there in the first place.

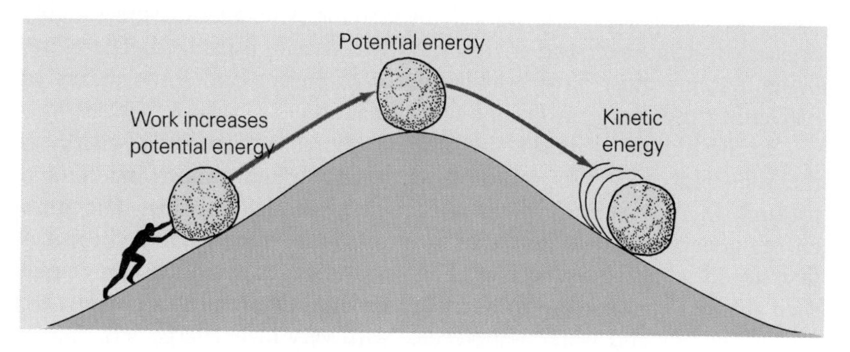

Both potential and kinetic energy can take many different forms. For example, a car battery has potential **electrical energy.** (We might also refer to it as chemical energy.) When the electrical energy is released to turn the starter, it becomes **mechanical energy.** As the parts of the starter move, friction causes some of the initial energy from the battery to be dissipated as **heat energy.** Thus, we see that not only can energy exist in different forms, it can also be converted from one form to another.

THE LAWS OF THERMODYNAMICS

We are about to consider a very impressive topic, as you can see by the heading. However, some of the mystique may be lost when you see the simplicity of the basic laws of thermodynamics that govern the behavior of all energy, regardless of its form.

According to the **first law of thermodynamics,** energy can be neither created nor destroyed, but only changed from one form to another. This means that the energy released as a boulder topples down the hill (Figure 5.3) is directly proportional to the amount of energy it took to push it up the hill in the first place. It also means that the heat energy that is "lost" as the tumbling boulder produces friction and releases heat is not actually lost, it is merely unavailable for use in that system; that is, this heat can't be used to do work.

The **second law of thermodynamics** tells us that in any energy transformation, not all the energy is transferred. Some energy will be "lost" simply because no transformations are 100 percent efficient. Thus, after each energy transformation, the energy available to do useful work in a system will be slightly lower than before. Thus, as a battery turns a starter, the starter produces less mechanical energy than was available in the chemical energy of the battery. Also, in chemical reactions, the product will contain less energy than did the reactants, unless energy is somehow added. All systems involving energy transformations, then, tend to fall to their lowest possible energy level.

The second law tells us that all processes tend toward increased **entropy**. Entropy is a measure of disorder, which means that as the available energy decreases in a system, the system "winds down" or becomes increasingly disorganized. Energy, then, must be added to a system to keep it in an organized state.

With a little imagination, you can see that these laws of energy have some interesting implications for living systems. After all, since life is highly organized, it must expend energy simply to remain organized. Therefore, living systems must have both a source of energy and some means of converting the energy from one form to another.

In the biological world, the ultimate source of virtually all energy is the sun. This energy is captured, changed, shifted about, and filtered through living systems, where it is used to reduce entropy. The sun's energy generally begins its useful progression in green plants, where it is first stored in certain molecules. The various stages in life's cycles extract their share of energy from these molecules until finally all that remains are carbon dioxide and water—molecules with very little energy left. Essentially, what we will learn now is that all life runs on sugar and a little electricity. (Keep that sentence in mind.) We will begin by considering a few of the chemical reactions that result in your having the energy to do whatever it is you would rather be doing.

ATP: THE ENERGY CURRENCY

The standard energy currency of cells is a fairly simple molecule called **adenosine triphosphate,*** or **ATP** (Figure 5.4A). ATP has been referred to as *energy currency* because it must be "spent" in order for work to be done. It is spent by splitting off the last phosphate group, releasing inorganic phosphate (or P_i), and converting the molecule to **adenosine diphosphate,** or **ADP.** The energy that was used to form the bond that attached phosphate is released, when the bond is broken. In other words,

$$ATP \rightarrow ADP + P_i + \text{energy}.$$

FIGURE 5.4A
ATP model.

As everyone keeps trying to tell you, in order to have something to spend, work must be done, and so it is with the chemistry of life (Figure 5.4B). In order to produce ATP, energy must be expended (work must be done) somewhere along the line. That energy is used to put a molecule of inorganic phosphate back onto an ADP molecule, a place it really doesn't "want" to go. It has to be forced back onto the ADP molecule by the expenditure of energy. Thus,

$$ADP + P_i + \text{energy} \rightarrow ATP.$$

So, ATP provides the energy necessary for life's functions, and energy is necessary for the reconstruction of ATP. Where does the energy necessary to put the phosphate on the ADP come from in the first place? Right! From the sun. What we will see now is how green plants use sunlight to form ATP, which they then use to make food (glucose), and

*It may have occurred to you that *tri* refers to three and *di* to two (dissect, cut in two). ATP has three phosphate groups, ADP has two.

FIGURE 5.4B

ATP is actually adenylic acid (comprised of adenine, ribose, and a phosphate) attached to two or more phosphates. When the bonds of the last two phosphates are broken, large amounts of energy are released—energy that is available to do work.

how animals, by eating plants, or by eating plant eaters, use that stored food to form ATP of their own (Figure 5.5). In simplified terms, we will see how plants use sunlight to make food (sugar) and how that food is then used to fuel the processes of life as it is broken down and its energy released by shifting electrons at each step along the way.

Photosynthesis: A Poet's View

In essence, **photosynthesis** simply involves using the energy of sunlight to make glucose (food) and oxygen from carbon dioxide and water. The energy is necessary because carbon dioxide and water exist at very low energy levels and have little inclination to interact on their own. In a sense, then, with the energy of sunlight, they are rearranged to form glucose. The energy that is used to join them is stored in the bonds of the glucose molecule. Glucose, then, contains a large amount of energy that can be released later as the molecule is broken down, for example, as in digestion.

The process of photosynthesis can be written as:

$$6CO_2 + 6H_2O \xrightarrow{\text{Sunlight}} \underset{\text{(glucose)}}{C_6H_{12}O_6} + 6O_2.$$

Photosynthesis takes place in two steps (Figure 5.6). The first, the **light-dependent reaction,** as you might expect, requires light. The glucose is not made at this step, though. Instead, the energy of sunlight is stored in two kinds of molecules that are manufactured at this stage, ATP and NADPH. It is in the second step, the **light-independent reaction,** in

FIGURE 5.5

The energy of sunlight is used by plants to make glucose. The energy stored in glucose is then released to make ATP, which can be broken down to ADP and inorganic phosphate (P_i) with the release of energy. That energy is then available to do work.

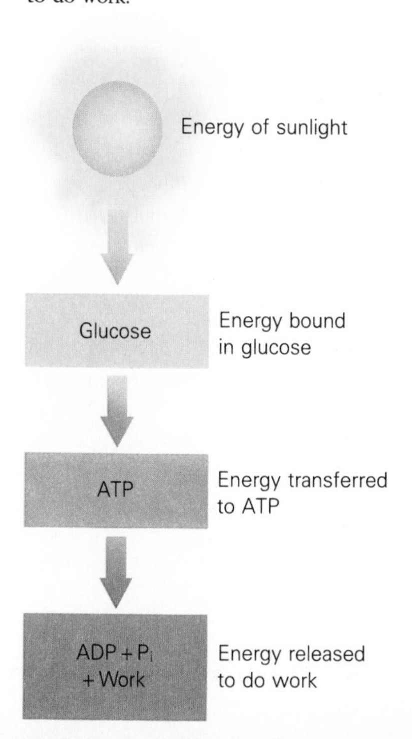

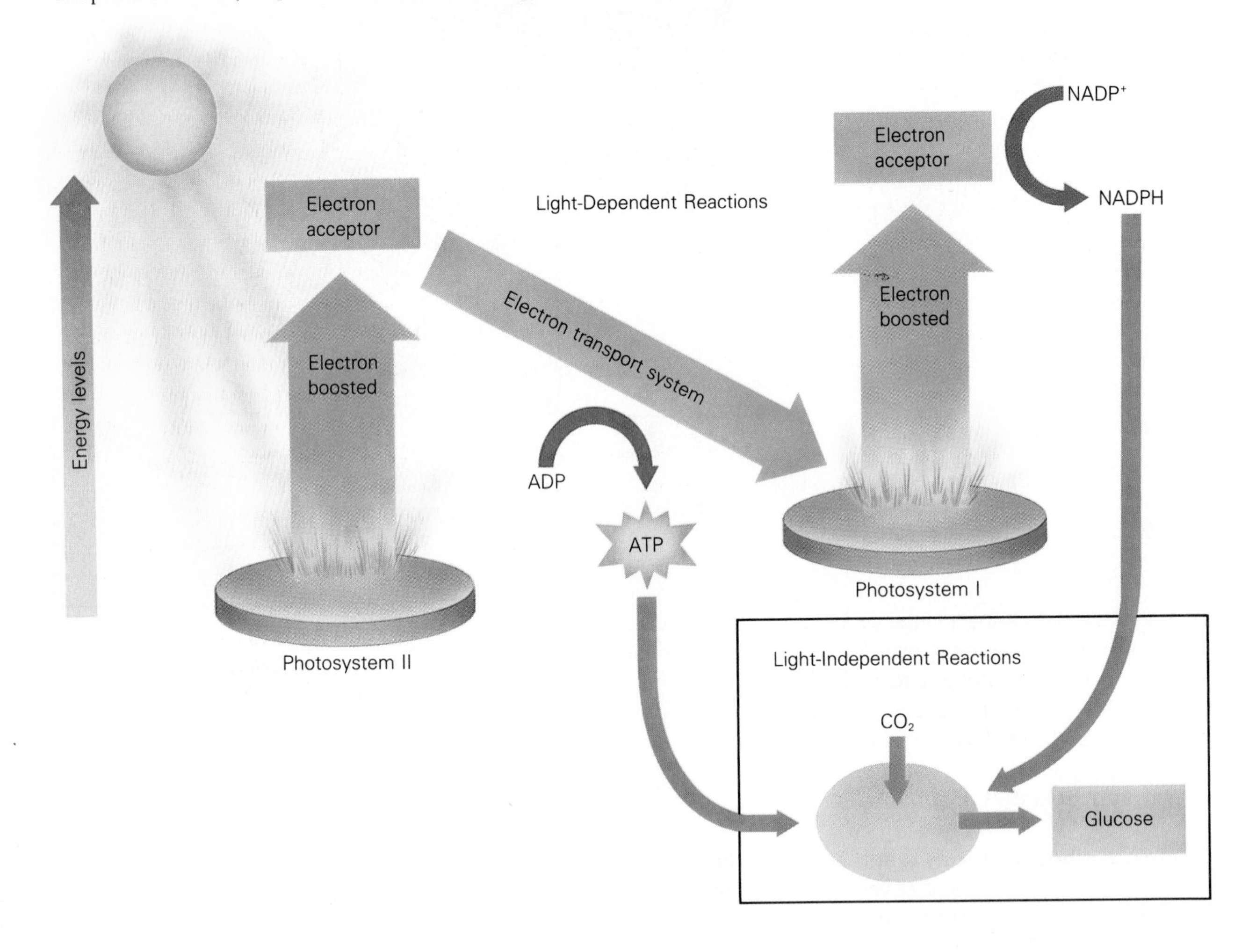

FIGURE 5.6
In the light-dependent reactions, the energy of sunlight produces ATP and NADPH, which then enter the light-independent reactions where glucose is formed. In the light-independent reactions, CO_2 and NADPH, with the energy from ATP, form glucose.

which light is not required, that glucose is formed. Here, the ATP and NADPH molecules provide the energy to form the glucose, using carbon dioxide, CO_2, as the carbon source. That's the basic idea; the details can grow a bit more complex.

Photosynthesis: A Closer View

Now for those who insist on the details, let's go over it again, being more specific, beginning with the role of light.

The Light-Dependent Reactions

Photosynthesis begins when light strikes a pigment, such as chlorophyll. Actually, there are several types, green plants, for instance, have chloro-

ESSAY 5.1

Chlorophyll

Chlorophyll is a pigment molecule found in the chloroplasts of all green plants. The green color you see in such plants is actually caused by light rays from the center portion of the spectrum that are reflected back to your eye by the molecular structure of the chlorophyll. The colors you do not see, the reds and purples at either end of the spectrum, are the ones that are absorbed. This scheme is not coincidental. The red and violet light rays are the ones that contain high levels of light energy, energy that the chlorophyll uses in the vital process of converting light energy from the sun into usable energy for all living things. Energy in plants is generally stored in the form of glucose, which is made available to animals that eat the glucose in one form or another.

Chlorophyll is usually a mixture of two different compounds known as **chlorophyll a** and **chlorophyll b.** Most green plants have more chlorophyll a, although a few, including some algae, primarily utilize chlorophyll b or some related compound. Each cell may contain as many as eighty chloroplasts—small bodies that hold the chlorophyll molecules in stacks of thylakoid disks (the grana). Because of this arrangement, the largest possible reactant surface is made available for the chemical reactions involved in photosynthesis.

phylls *a* and *b* (Essay 5.1). Orange pigments called **carotenoids** capture energy in certain wavelengths that chlorophyll cannot. When light hits the pigment, it initiates the light reaction.

The chloroplasts are tiny, double-membraned bodies within the photosynthetic cells. Inside the chloroplast is a jellylike **matrix** called the **stroma,** in which are embedded stacks of disks called the **thylakoids.** Each stack is called a **granum.** The thylakoids of different grana are joined by membranous **lamellae** (Figure 5.7). Embedded in the outer thylakoid membrane are tiny bumps, each penetrated by a minute channel leading from inside the disk out to the stroma.

Photosynthesis begins when the chlorophyll within the chloroplasts is activated by the energy of light (see Essay 5.2). This causes two of chlorophyll's electrons to move out into a more distant orbital; that is, to a higher energy state. From here, they can leave the chlorophyll entirely and be captured by a molecule called an **electron acceptor** (a molecule that can accept one or more electrons, thereby becoming reduced as it oxidizes the donor molecule). The chlorophyll replaces its lost electrons by literally tearing two of them from water molecules. When water mol-

FIGURE 5.7

The location of chloroplasts within the cells of a green leaf. Tissues within the leaf contain vast numbers of photosynthetic cells, each with numerous chloroplasts. Within the chloroplasts are membranous grana, stacks of thylakoids, in which the sunlight's energy first interacts with the biological realm. Surrounding the thylakoids are the clearer regions of the fluid stroma, where the final, carbohydrate-synthesizing events (light-independent reactions) occur.

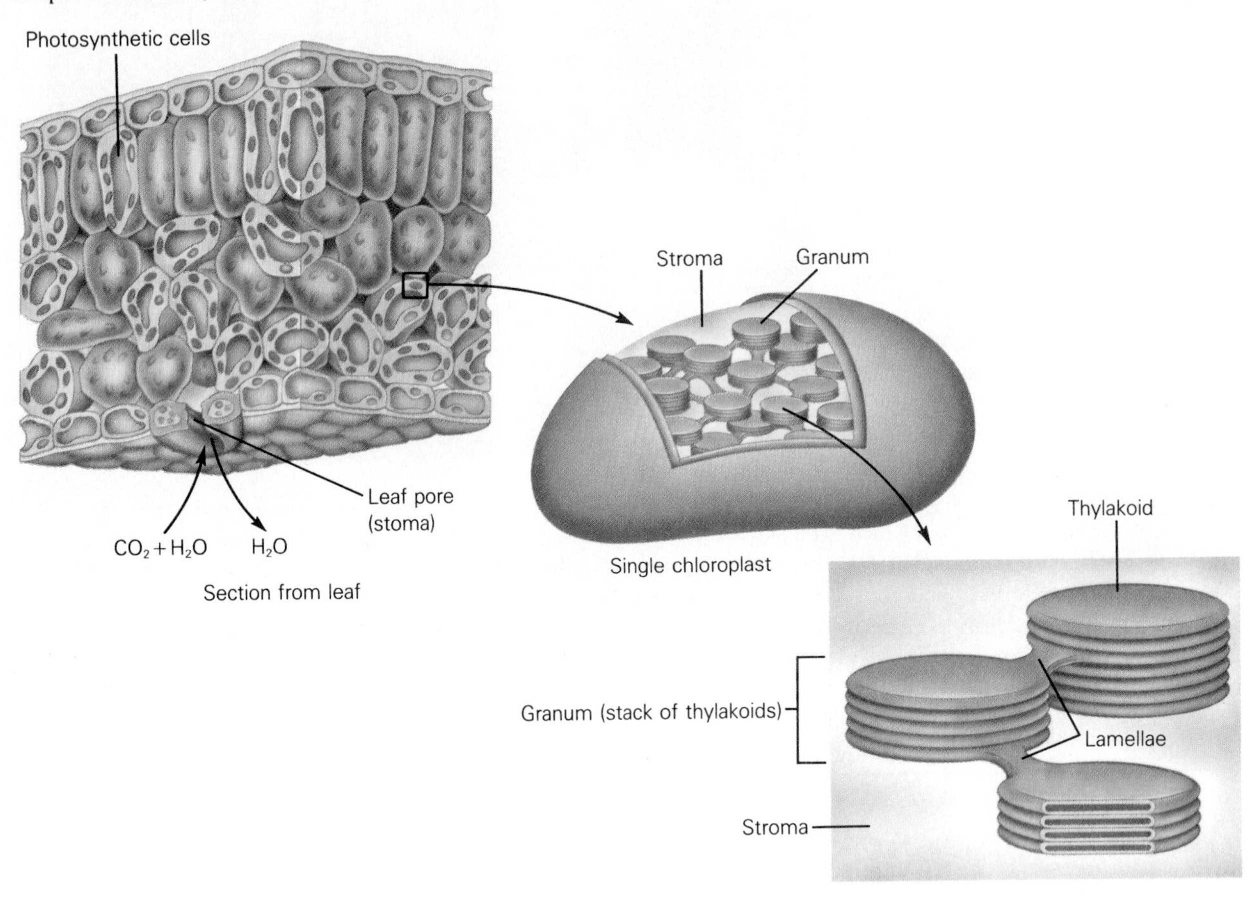

ecules are broken apart in this way, the oxygen is released as a gas. These are the essential events that occur in the pigment system known as **photosystem II.** You can follow along in Figure 5.8.

From the electron receptor in photosystem II, the electrons are passed along to the beginning of an **electron-transport system.** The system consists of a series of other electron-acceptor molecules, each at a lower energy level than the one preceding. Electrons progressively lose their energy as they are passed "downhill" from one molecule to the next. Some of the protons from which the electrons are stripped follow them because they have opposite charges. As the electrons are passed to increasingly lower energy states, giving up energy at each step, that energy is used to pump some of those accompanying hydrogen ions into the interior of the thylakoids (the lumen). We will soon see why this is an essential step, but first let's follow the electrons through their sequence.

FIGURE 5.8

Summary of the events of photosynthesis. In the light-dependent reactions, sunlight boosts electrons from water along photosystem II to an electron acceptor which passes them along a series of molecules at increasingly lower energy levels to photosystem I, which boosts them to another electron acceptor. They are then passed to $NADP^+$, forming NADPH, a high energy molecule that, with ATP, powers the light-independent reactions, where CO_2 is used to make glucose. The ATP is formed as protons, which have been pumped into the thylakoid lumen by the energy released as electrons move down the electron-transport system, move out of the lumen into the stroma. They must pass through enzyme-charged particles that are able to join ATP and inorganic phosphate (P_i) to make ATP.

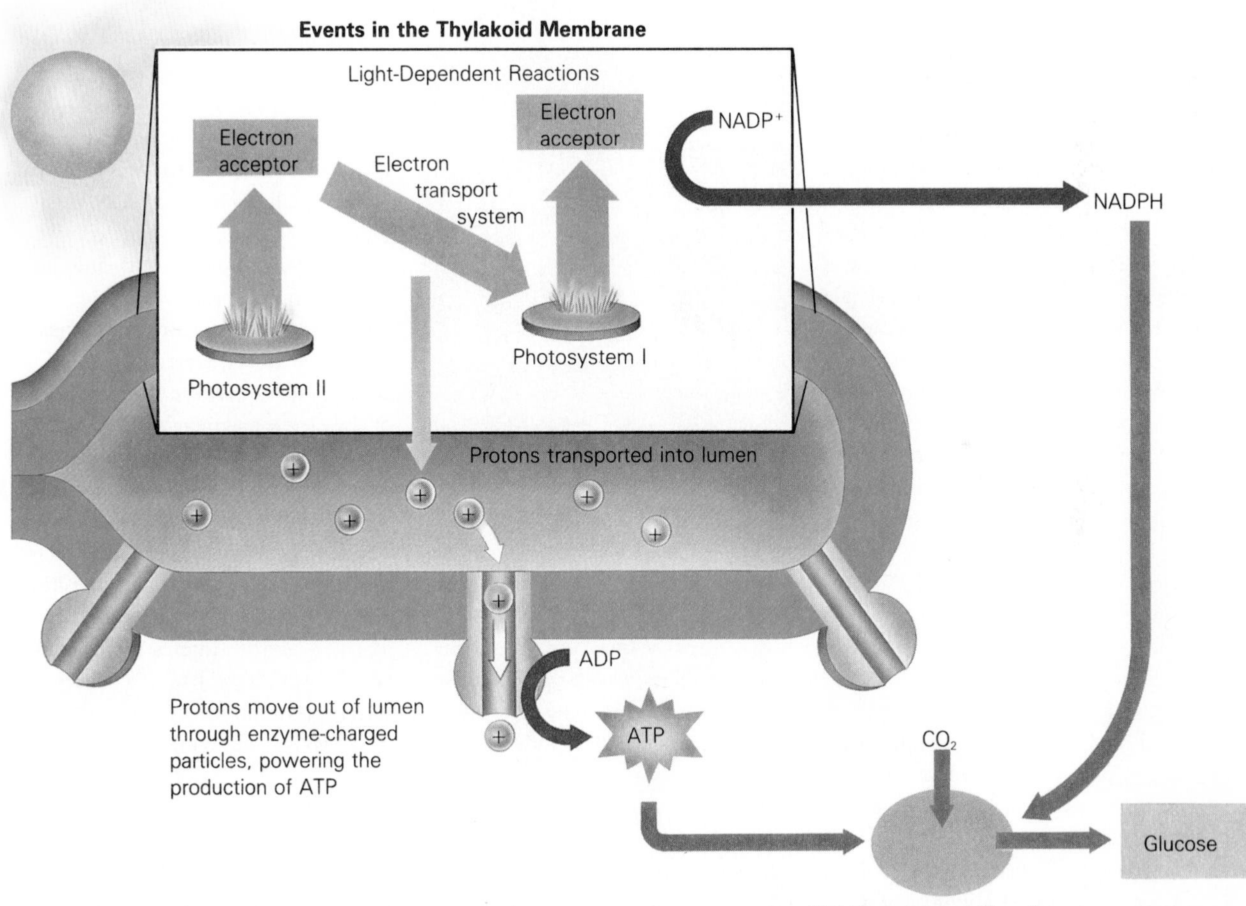

When the electrons have reached the lowest energy level of the electron-transport system in photosystem II, they are transferred to another chlorophyll pigment system called **photosystem I.** (It is designated as I because it is believed to have evolved first.) The electrons that enter photosystem I do not languish long. Light energy again boosts them to another electron acceptor, this one at an even higher energy level than the first acceptor in system II. From this acceptor, the electrons, some accompanied by their protons, are passed along to a nucleotide of adenine, a complex molecule known as $NADP^+$. The $NADP^+$ is thus reduced to NADPH. You will notice in Figure 5.8 that the electrons accepted by $NADP^+$ are still at a higher energy level.

ESSAY 5.2

The Visible Light Spectrum and Photosynthesis

The earth is constantly bathed in radiation emanating not only from the sun, but from a host of other celestial bodies. Part of the radiation that reaches us is visible light. Visible light, however, is only part of an **electromagnetic spectrum** that includes (in increasing order of energy) radio waves, microwaves, infrared radiation, visible light, ultraviolet radiation, X rays, and gamma rays.

The longest waves contain the lowest levels of energy, very short waves the highest, and in between is **visible light.** Visible light is visible because it interacts with special pigments (light-absorbing molecules) in our eyes. It also interacts with pigments such as chlorophyll, one of the molecules that can absorb the energy of light and provide power of photosynthesis.

Physicists have described light in two ways: as particles (called photons) or as **waves**. Arguments over which concept best accounts for the behavior of light have raged for years. So let's finesse it a bit and say that light is composed of **photons** and moves like waves. This side step will at least help us consider the energy of a photon in terms of its wavelength.

The various wavelengths of light are critical to life in a number of ways. Visible light, we have seen, interacts with the retinas of our eyes and provides the energy that enables green plants to grow. The longer infrared wavelengths can be reflected and dissipated as heat, and we know that warmth, of course, to some degree (no pun), is important to all life. At the other end of the spectrum, the shorter, more energetic wavelengths are usually too powerful for most forms of life to utilize since they tend to disrupt molecules, especially proteins and DNA. Ultraviolet light (UV), for example, burns our skin and damages the retinas of our eyes. The more penetrating radiation, such as gamma rays and X rays, is called ionizing radiation, because it can ionize, or break up, molecules within cells. The pieces of the disrupted molecules are very reactive and tend to enter into random and potentially harmful reactions. If such molecular fragments should disrupt the genetic material of a cell, the result may be a mutation. (We will see why shortly.) Fortunately for the life on earth, most of the more energetic and dangerous wavelengths of the sun are filtered out by ozone molecules in the stratosphere.

So, infrared warms us and ultraviolet burns us, and the wavelengths we can see lie in between. Obviously, though, there are no sharp dividing lines between visible and invisible rays. Both reds and violets become dully visible as they enter our range of sensitivity. It is interesting that different organisms may see different parts of the electromagnetic spectrum. For instance, many insects cannot see red. Red flowers appear to them as gray. But they may see deep violet (ultraviolet) in flowers that appear white to us. And whereas insects can see ultraviolet, but not red, the reverse is true for many birds. Interestingly, many people who have had the lenses of their eyes removed because of cataracts can see ultraviolet light that is invisible to the rest of us. As we will see in Chapter 23, the reason we see colors at all is because light of different wavelengths interacts in specific ways with the color-sensitive cells in our eyes. Various animals are able to perceive different parts of the light spectrum because of physiological differences in those receptors.

The process leaves photosystem I short two electrons, but these are replaced when two more from photosystem II are boosted by light energy (Figure 5.8), and, as we've seen, photosystem II will recover its lost electrons from water. Because the accompanying protons (H^+) are left behind in the lumen, their concentration there builds to over a thousand times that of the stroma. At the same time, other protons are pumped into the lumen of the thylakoid by the energy released from the electron transport systems. So, protons end up inside the thylakoid from two sources, from the disruption of water that occurs within the thylakoid (these are simply left there) and from protons being pumped in from the stroma.

The result of this buildup is that the hydrogen ion concentration inside the thylakoid is much greater than in the stroma surrounding the thylakoid. Because of the charges on the ions, this accumulation sets up a strong chemical and electrical imbalance across the thylakoid membrane. This imbalance establishes a great reservoir of potential energy.

The imbalance produces potential energy because there is a strong tendency for the positive charges inside the thylakoid and the negative charges outside to join and reestablish an equilibrium. Remember, though, that the outside of the thylakoid membrane that separates them is studded with particles that are penetrated by tiny channels extending into the lumen of the hydrogen-saturated thylakoid. It is through these channels that the hydrogen ions (protons) can pass outward and begin to reestablish an electrical and chemical equilibrium.

However, this is not a free ride: there is a toll for their passage. Within the particles is a battery of phosphorylating enzymes. As the protons pass through, the energy of their movement is used to phosphorylate ADP; that is, to reattach inorganic phosphate (P_i) to ADP, thereby forming high-energy molecules of ATP.

The use of an electron-transport system to pump hydrogen ions across a membrane, resulting in a concentration difference that can be used to make ATP, is known as **chemiosmosis.** We will see it again when we look at the role of mitochondria in respiration. (The process was first described by a maverick scientist named Peter Mitchell who largely supports his research by revenues from his British dairy herd.)

Now we will see that the energy in the newly formed molecules of ATP, along with the high energy molecules of NADPH, formed earlier, provide the necessary energy to juggle molecules in the light-independent stage. It is here that the nutrient molecules, such as glucose, are formed.

The Light-Independent Reactions

The light-independent reactions, we know, derive their name because light energy is not necessary to drive this part of photosynthesis. In this case, the energy comes from ATP and high-energy electrons held by the NADPH that was produced in the light reaction. The major biochemical pathway in this reaction, sometimes called the **Calvin cycle,** is outlined in Figure 5.9.

The light-independent reactions consist of 14 or 15 complex steps, which, in spite of your great disappointment, we will not go into here. Just keep in mind that a carbon dioxide molecule is added to a 5-carbon sugar molecule already present in the cell. This 5-carbon sugar is ribulose biphosphate, or RuBP.

FIGURE 5.9

The Calvin cycle. The light-independent cycle takes place outside the grana and is an extremely complex process. This is where plants will utilize CO_2 from the air to form glucose. First, RuBP is formed by adding ATP to a molecule called RuP; RuBP is the molecule that the CO_2 joins to form an unstable 6-carbon intermediate that immediately splits into two 3-carbon molecules. With the addition of yet more ATP and NADPH, two 3-carbon molecules called PGAL are formed. Some PGAL then forms glucose and other molecules, but most PGAL forms more RuP and continues the cycle.

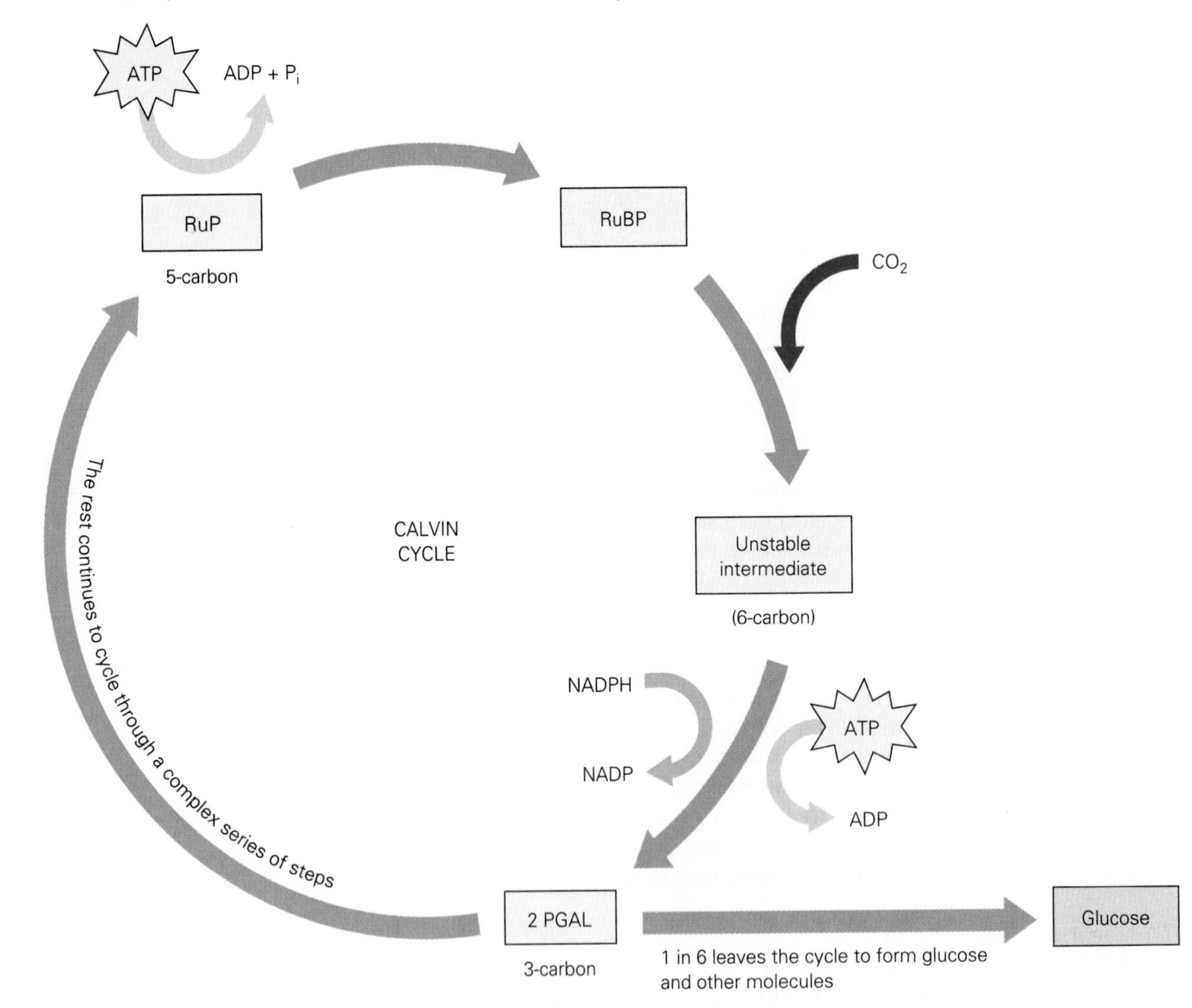

The additional carbon results in a very unstable 6-carbon sugar that quickly splits into two 3-carbon sugars. There follows a complex series of events in which the bonds between carbon atoms are broken and reformed, eventually producing two molecules of phosphoglyceraldehyde, or PGAL. The PGAL is the end product of photosynthesis.

PGAL may then enter into any of a number of metabolic pathways. If the plant is in need of food, PGAL can be used as is. It can also be converted to glucose and then to starch for storage, or it may be converted to proteins or fats. Most of the PGAL, however, is changed back to RuBP, with an input of energy from ATP. It is then ready to enter the cycle again.

What Does a Plant Do with Its Glucose?

The photosynthetic process is indeed functional, but it requires sunlight, and sunlight is not always available. Thus, many plants store some of the glucose they produce in photosynthesis. (In a sense, perhaps, saving it for a rainy day.) Most plants do this by simply linking the glucose molecules together and storing them as starches—either in the cells that produced them or in special storage areas. The potato plant, for example, stores starch in swollen underground nodules that humans have been known to eat. Other plants may convert glucose to other kinds of sugars, such as sucrose or fructose. Sugar beets store sucrose in underground stems. Many fruits are sweet because they are laden with fructose.

Such storage is obviously a temporary measure. Eventually the plant will utilize the glucose it has made. Some of it will go into the construction and repair of cell walls, but most of it will go to make ATP. We will see how this is done shortly, in our discussion of cellular respiration.

Let's close this part of our discussion by noting that photosynthesis is an energy-costly process. The reason is simply because its molecular reactants, water and carbon dioxide, are in a low, free-energy state—that is, they are rather stable and, thus, reluctant to participate in any chemical events. They can only be induced to react by an expenditure of some sort of energy. Fortunately for life as we know it, sunlight is plentiful and evolution has provided some organisms with the ability to use it as an energy source as part of an endless cycle of chemical events that permit the existence of most forms of life (Essay 5.3).

FOOD TO ENERGY

The great variety of living things on the planet plays many roles in the extemporaneous skit of life, but all life requires energy. We have seen that the most obvious storehouses of energy are the earth's green plants. We have also seen that after the energy of sunlight has been shuffled, juggled, and channeled through the photosynthetic processes, finally to end up in the bonds of glucose, it can take any of a number of routes. The plant may utilize that bond energy immediately, or it may store it in its body. It is this stored food that the heterotrophs are after, whether the heterotroph is an herbivore that eats the plant directly or a carnivore that seeks the food that was once stored in plants but now is stored in some other animal's body. Not only must heterotrophs have ways of acquiring energy, either from the plant or from each other, once they get it, they must have ways to use it.

The process by which cells drain high-energy molecules of their energy (to form ATP) is called **cellular respiration.** The process is usually gradual and tightly controlled so that the energy is released slowly and in a stepwise manner. If it were to be released all at once, there would be the crackle of exploding protoplasm followed by the cool silence of death.

ESSAY 5.3

A Notable Exception from the Sea

It was once suggested that all life on the planet draws its energy, in one way or another, from the sun. However, in 1977, geologists surveying the ocean bottom near the Galapagos Islands made a startling discovery. A group of excited scientists in a specially outfitted exploratory submarine found an entire community of animals drawing their energy, not from the sun, but from hot, sulfur-laden waters spewing from open vents on the seabed. A number of such vent communities have now been found. The waters there, it has been learned, are heated by the molten rock that lies beneath the earth's crust, the same phenomenon that produces volcanic action.

As shown in the accompanying figure, water seeping into cracks in the sea bottom is heated and sent upward into the frigid water of the deep ocean. Bacteria utilize the hydrogen sulfide in this water to begin an autotrophic food chain. The hydrogen sulfide is believed to provide electrons, much as water does in photosynthesis.

Here, the food-making process is referred to as chemosynthesis (making organic compounds from the energy derived from inorganic chemical reactions). Clouds of these microscopic species set up the beginning of remarkable food chains.

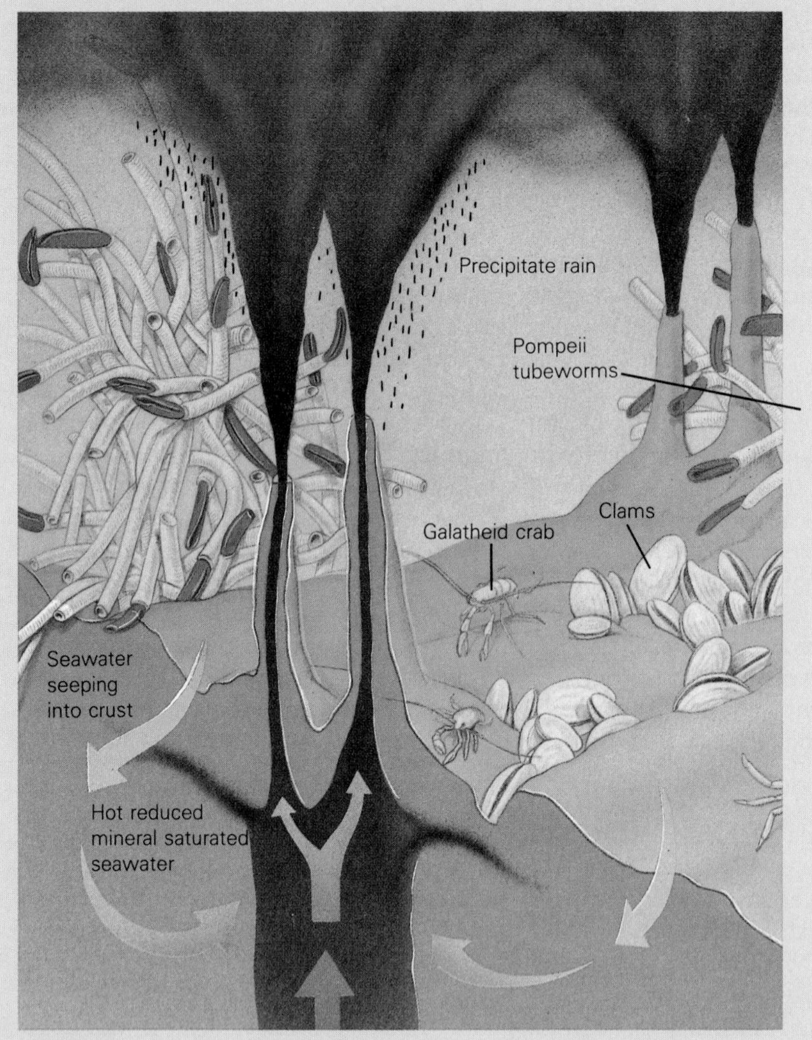

A diagram of the energy source of chemosynthetic communities beneath the sea.

Among the various kinds of animals living crowded around the vents are previously undescribed species of giant blood-red tubeworms, nearly 3 meters long, agile white crabs, mussels, barnacles, and even leeches.

The chemosynthetic communities of the sea are host to a wide variety of life. At left, a tubeworm bed in the Galapagos Rift area. In the lower right corner, galatheid crabs and mussels. These creatures draw energy not from the sun, but from the mineral-laden water that spews from the "smoker" chimneys (above). The discovery of these mysterious undersea communities was made possible through the use of small submarines such as *Alvin* (left). These submarines can carry out a variety of tasks despite the extreme pressure of the ocean depths.

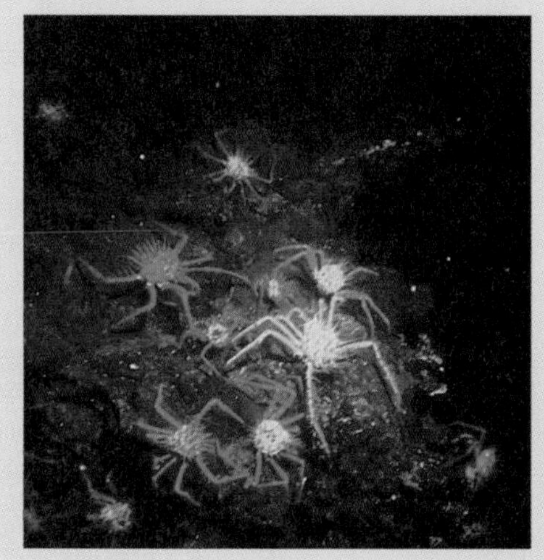

Living things have developed a variety of ways to handle such high-energy molecules, essentially through four major processes: glycolysis, the Krebs cycle, electron-transport, and chemiosmotic phosphorylation.

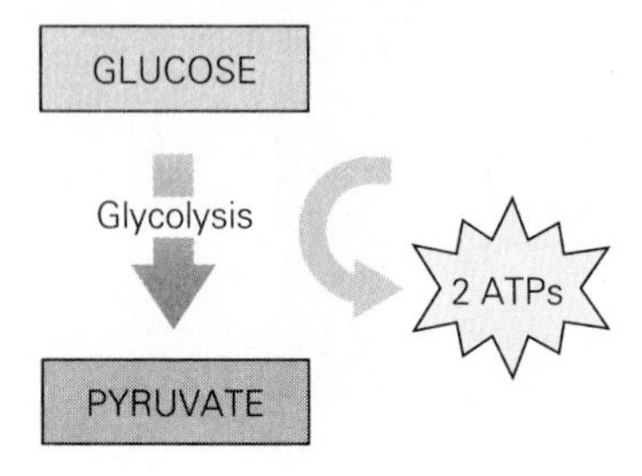

FIGURE 5.10
Glycolysis is the process that forms pyruvate from glucose.

Food to Energy: A Poet's View

The first thing to know about cellular respiration is that it begins with the breakdown of glucose through a process called glycolysis. The 6-carbon glucose molecule is broken down into two 3-carbon molecules of **pyruvate.** This happens in the cytoplasm and does not require oxygen. (Since it does not require oxygen, it is called **anaerobic**—which means no oxygen. Clever, eh?) The process produces two molecules of ATP (Figure 5.10).

The pyruvate can then enter into fermentation reactions or aerobic respiration, depending on the kind of cell in which it exists and that cell's immediate needs. If the pyruvate exists in certain yeasts (and bacteria), it will enter into fermentation, because that's their entire means of energy production. It may also enter into fermentation in an animal cell that contains no oxygen (such as in a fatigued muscle). The end products of fermentation in yeasts are quite different from those in animal cells. In yeasts and bacteria, the pyruvate forms alcohol (ethyl alcohol) and CO_2. This is how yeasts make booze. In animals, when no oxygen is present, the pyruvate is broken down into lactate (lactic acid). (The accumulation of lactic acid helps to make your muscles sore when you continue your effort after you're out of breath [Figure 5.11].)

In cells in which there is oxygen, the pyruvate can take other routes. For example, it can move into a mitochondrion where it enters the Krebs cycle. Here, the pyruvate will be drained of its energy, and broken down to simple CO_2 and H_2O while producing two molecules of ATP for each original glucose molecule. The Krebs cycle removes numerous electrons from the pyruvate molecules as the molecules move through the cycle and these are passed to an electron carrier at the top of the electron transport system, embedded in the mitochondrial membrane. The electrons are then

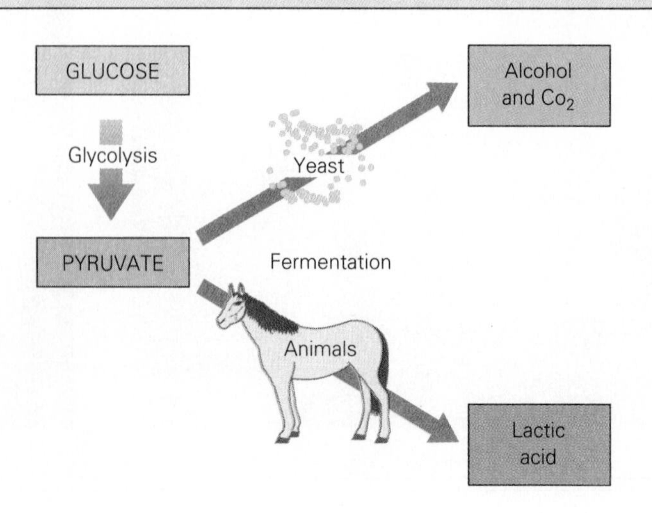

FIGURE 5.11
In yeasts, the pyruvate can form alcohol; in animals, it can form lactate.

FIGURE 5.12

Pyruvate can be altered to enter the Krebs cycle, which produces two ATPs and provides electrons for the electron transport system, where 32 ATPs are produced.

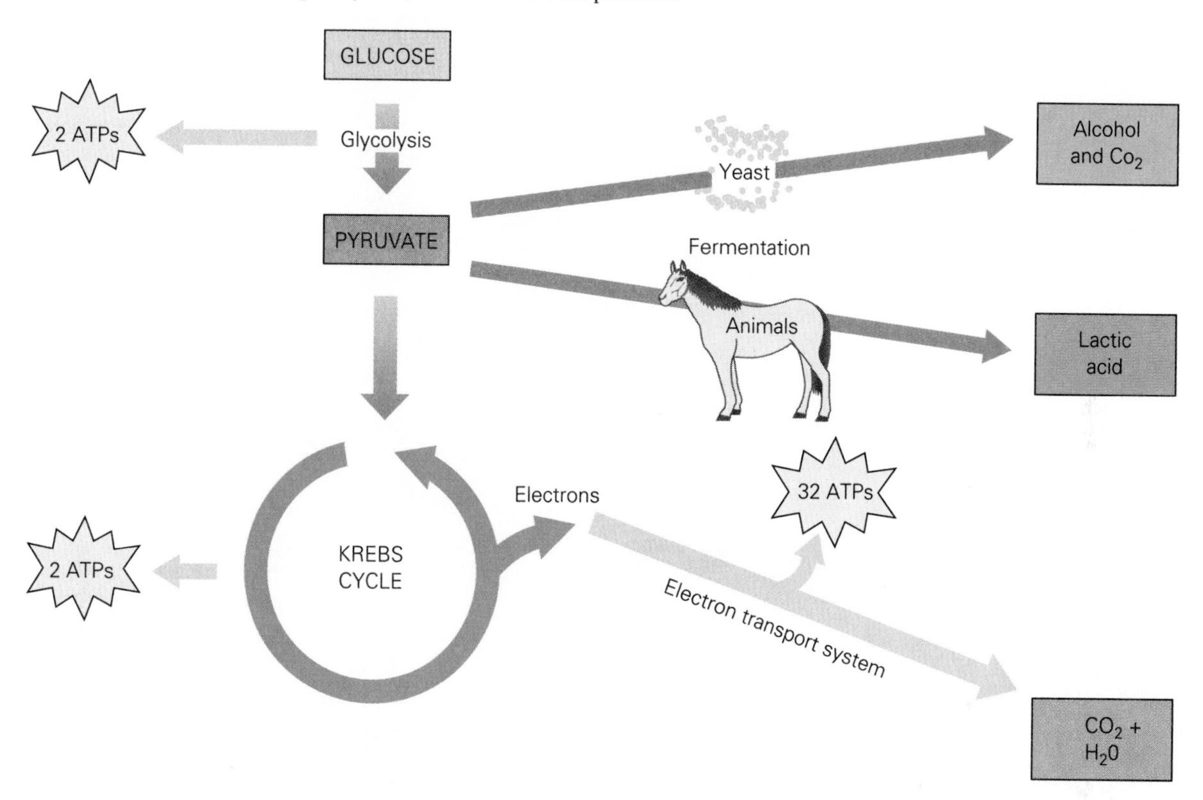

passed down the electron transport system with energy being released each time an electron is passed to the next lowest level. That energy is used to make a total of 32 ATP molecules per original glucose molecule, a very efficient system as you can see (Figure 5.12).

Thus, for every glucose molecule being broken down, 36 ATP molecules are produced. The ATP molecules, of course, contain the energy necessary to run the processes of life.

Food to Energy: A Closer View

Now for a closer look at the processes of cellular respiration. We will take the processes one at a time, beginning, as all life does, with glycolysis.

Glycolysis

Glycolysis (*glyco*, sugar; *lysis*, breakdown) is the anaerobic breakdown of glucose, a process that takes place in the cytoplasm. Glycolysis is the first step toward converting the energy within a glucose molecule to the usable energy of ATP. As we will see, though, glycolysis alone doesn't produce very much ATP.

In glycolysis, a 6-carbon glucose molecule is first destabilized by the addition of two phosphate ions in separate reactions. These ions come

FIGURE 5.13
The pathway from glucose to the Krebs cycle.

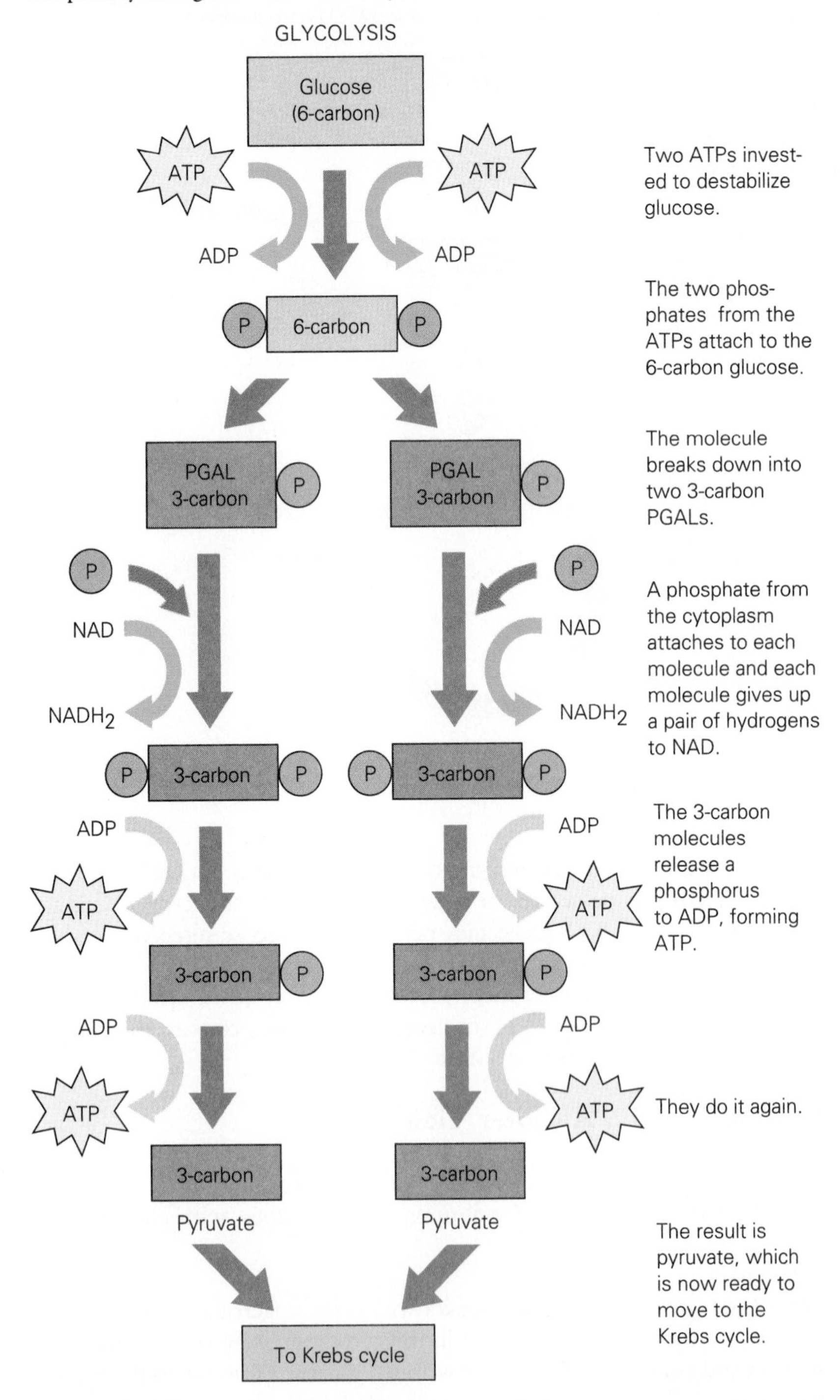

from two molecules of ATP, changing them to ADP. Although glycolysis actually produces four ATP molecules, it also uses up two ATP molecules at the beginning of each sequence, "priming the pump" as it were. So, the net gain is only two. In any case, the phosphorylated molecule quickly breaks down into two 3-carbon fragments. (The details are shown in Figure 5.13.) The two fragments are then repeatedly rearranged from one kind of molecule to another until, finally, two molecules of pyruvate (or pyruvic acid) are formed. Pyruvate is the end product of glycolysis.

Along the way the 3-carbon fragments are partly oxidized; that is, they give up electrons. Two electrons, with their protons, are removed from each fragment. These then reduce (or join with) two molecules called NAD^+ (nucleotides of adenine), thereby forming two molecules of NADH.

Once the pyruvate is formed, it may enter any of several different metabolic pathways. The particular route at this point may depend upon whether oxygen is present. If oxygen is present, in some species the pyruvate may be converted to a 2-carbon molecule and then enter the Krebs cycle, as we shall see shortly, and produce more ATP. If oxygen is not present, the pyruvate will enter a sequence called **fermentation,** an anaerobic process that usually produces CO_2 and alcohol or lactic acid. As we will see, though, it makes little ATP (Figure 5.14).

FIGURE 5.14
The process of glycolysis and respiration.

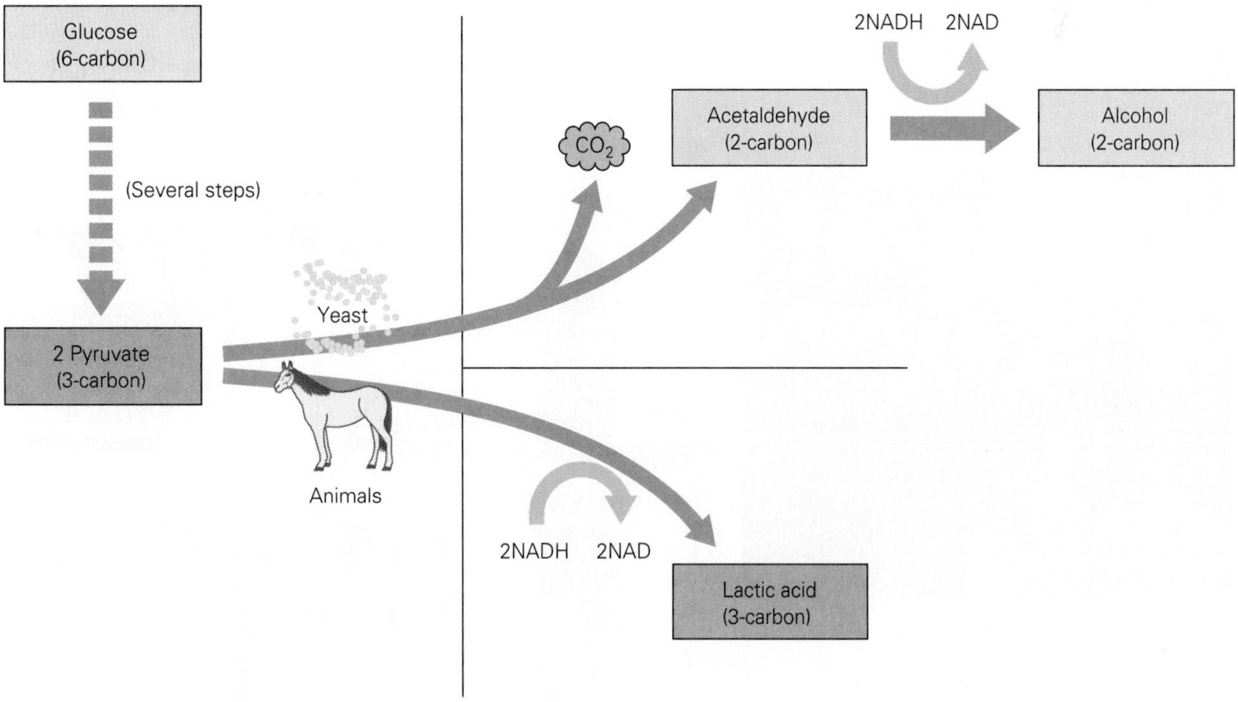

Fermentation

One such anaerobic pathway takes place in yeast and is called **alcoholic fermentation**. Here, as we see in Figure 5.15, the product is carbon dioxide and alcohol. In yeast, the glucose is only partially broken down and forms ethanol (the ethyl alcohol present in spirits). The two molecules of NADH produced in glycolysis give up their hydrogens to help form the ethanol and then return as NAD (more precisely written as NAD^+ since it bears a net positive charge) to pick up two more hydrogens from a glucose molecule. We have known about the practical aspects of fermentation for a long time because early man, confronted with all the great problems of the universe, at once set about learning how to make booze.

In animals, on the other hand, a process called **lactate fermentation** produces, not alcohol, but lactate (Figure 5.16). It is not surprising that anaerobic animals (those that do not utilize oxygen in their energy production), such as internal parasites, are fairly sluggish. They just don't have the ATP molecules to be frisky. But even aerobic animals—those, like us, that require oxygen for most respiration—under certain conditions may also derive energy from anaerobic processes. For example, if we are unable to get enough oxygen during exercise, our respiration may partly

FIGURE 5.15
The little old winemaker (a) knows that the anaerobic conditions in alcoholic fermentation are maintained by the careful exclusion of oxygen. Humans have an old acquaintance with the production of ethanol from pyruvic acid in yeast (b).

(a)

(b)

FIGURE 5.16

Humans also have an acquaintance with the production of lactic acid from pyruvic acid in the muscles.

shift to glycolysis (aerobic and anaerobic processes can occur simultaneously). When this happens, lactate accumulates in the muscles, and they weaken with fatigue. If enough lactate accumulates, the muscles will fail to contract at all, or they may contract in an uncoordinated manner. This is why a runner who is not in condition is usually gasping for air at the end of a race and may stagger across the finish line. A conditioned runner is likely to have developed the heart and lungs along with the muscles so that sufficient oxygen is supplied to prevent the buildup of lactate.

In mammals, lactate is washed from the muscles by the circulating blood and carried to the liver, where it is reconstituted to glucose. The glucose units are then linked together to form glycogen, which is stored in the liver. The glycogen can be broken down into its glucose components when glucose is needed, but the transformation is uphill—that is, it requires an expenditure of some energy. However, because the energy level of lactate is fairly high, the boost to glucose is a short one.

In aerobic animals, the pyruvate formed in glycolysis may take a different route. It may go through a series of reactions (the Krebs cycle, also called the citric acid cycle) in which pyruvate is broken down aerobically to make large quantities of ATP. In this sequence, in fact, almost all its bond energy is extracted to make new ATP molecules (many more than can be formed in glycolysis). Here, the energy-laden electrons of pyruvate are passed along to lower energy levels until they end up as mere components of humble water, as we will now see.

Aerobic Respiration

We have found that in the absence of oxygen, a glucose molecule provides an aerobic organism with only two molecules of ATP. If oxygen is present, it's a whole new ball game. The pyruvate will proceed to a series of reactions that generate many more ATP molecules in the mitochondrion (Figure 5.17). This happens in a series of reactions that are often divided into those of the Krebs cycle (named for its British discoverer, Sir Hans Krebs) and the electron transport chain, both of which take place in the mitochondria, tiny structures within the cell.

If oxygen is present, the pair of 3-carbon pyruvate molecules produced by glycolysis are prepared to enter the Krebs cycle. First, they become oxidized—with NAD^+ molecules again accepting the two hydrogens, becoming NADH. The pyruvic acid molecules each also lose one carbon in the form of carbon dioxide. The remaining 2-carbon fragment, **acetate,** quickly unites with a small organic molecule called *coenzyme A*, or *CoA*, to form a molecule called **acetyl-CoA.**

The Krebs Cycle

The **Krebs cycle** is a cyclic series of reactions in which electrons and protons (H^+) are stripped from intermediates. Carbons are also removed from the intermediates and are released as CO_2. In addition, each turn of the cycle produces one ATP molecule. Let's see how this happens.

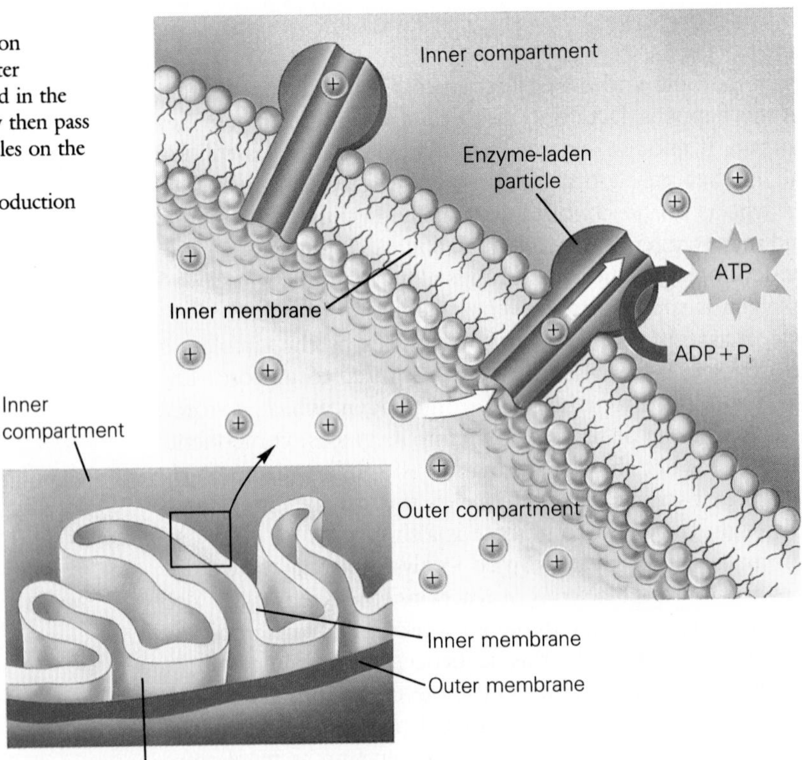

FIGURE 5.17
In chemiosmotic phosphorylation protons are moved into the outer compartment by energy released in the electron-transport system. They then pass through enzyme-charged particles on the inner membrane to the inner compartment, powering the production of ATP as they pass.

It is actually the acetyl-CoA that enters the Krebs cycle (Figure 5.18). It should be stressed that this molecule can be derived not only from glucose, but from fats and some amino acids as well (although by less direct routes).

Once the acetyl-CoA enters the Krebs cycle, the acetyl group separates from its CoA enzyme and combines with a 4-carbon molecule called **oxaloacetate** or oxaloacetic acid. The result is a 6-carbon compound called **citrate** or citric acid (the common acid of lemons and limes). The citrate is then oxidized, followed by the splitting off of two carbons in the form of carbon dioxide, with the remainder reverting to oxaloacetate. This is then recycled to pick up another acetyl group from the next acetyl-CoA molecule. The cycle continues as long as acetyl-CoA is available. Details of this process are illustrated in Figure 5.18. As you can see, the stored energy in the acetyl group has been used to convert three molecules of NAD^+ to NADH, one molecule of FAD (another electron acceptor, more precisely written as FAD^+ since it bears a positive charge), to $FADH_2$, and one molecule of ADP to ATP. The glucose molecule has by this time been completely oxidized. As we will see next, its products will go on to provide the cell with yet more energy after they are passed along to the electron-transport chain.

Note that the electron acceptors NAD^+ and FAD^+ are modified nucleotides, and each of these substances readily accepts and releases two electrons at once. Do you see a certain parsimonious efficiency in a system that makes multiple use of nucleotides in continuous oxidation and reduction processes?

The Electron-Transport System

The **electron-transport system** of respiration is a series of electron acceptors in which energy is released with each transfer of electrons to the next acceptor. Here, electrons that had been accepted by the NAD^+ and FAD^+ molecules are passed along to a series of electron carriers similar to those we mentioned in photosynthesis. Here again, each acceptor is at a lower energy level than the preceding one, and again the energy held in the electrons is gradually released in a stepwise process.

Chemiosmotic Phosphorylation

As we now know, the mitochondrion has two membranes, an outer one and an inner one. The outer membrane is the rather simple "covering," but the inner membrane forms a number of shelflike folds (cristae) that project into the mitochondrial matrix. The electron-transport system of the mitochondrion is embedded in the inner membrane. This membrane is also studded with stalked particles that appear to be similar to particles of the chloroplast, except that their arrangement is essentially reversed: the particles of the mitochondrial membrane project into the inner compartment from the surface of the inner membrane. The energy released in the electron-transport system is used to pump positively charged hydrogen ions into the mitochondrion's outer compartment. As they move to rejoin negatively charged hydroxyl ions (OH^-, formed by the splitting of water) in the inner compartment, they must pass through the particles where batteries of enzymes use the energy of their passage to form ATP from

FIGURE 5.18
Overview of the progression of pyruvate to the Krebs cycle.

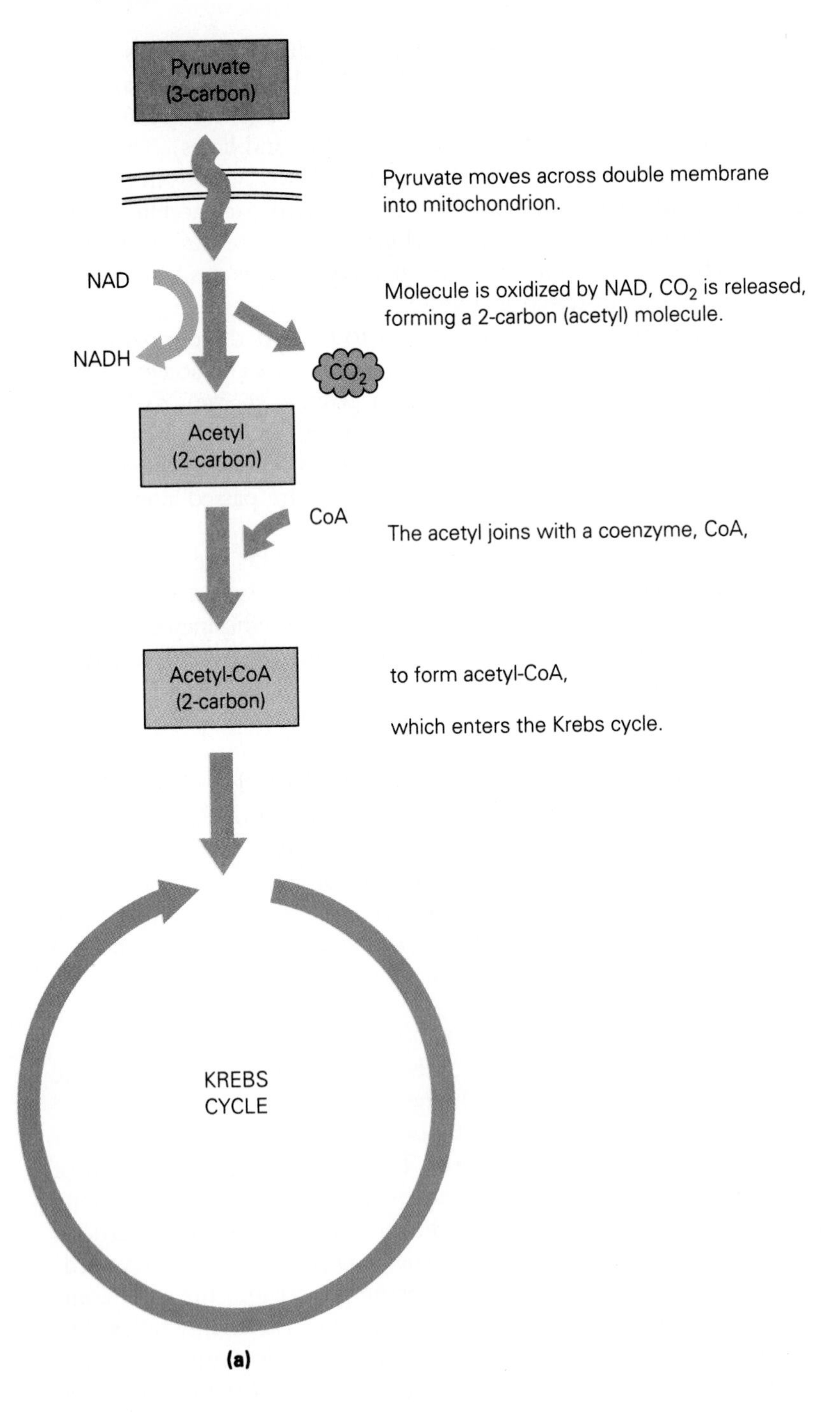

FIGURE 5.18 continued
Details of the Krebs cycle.

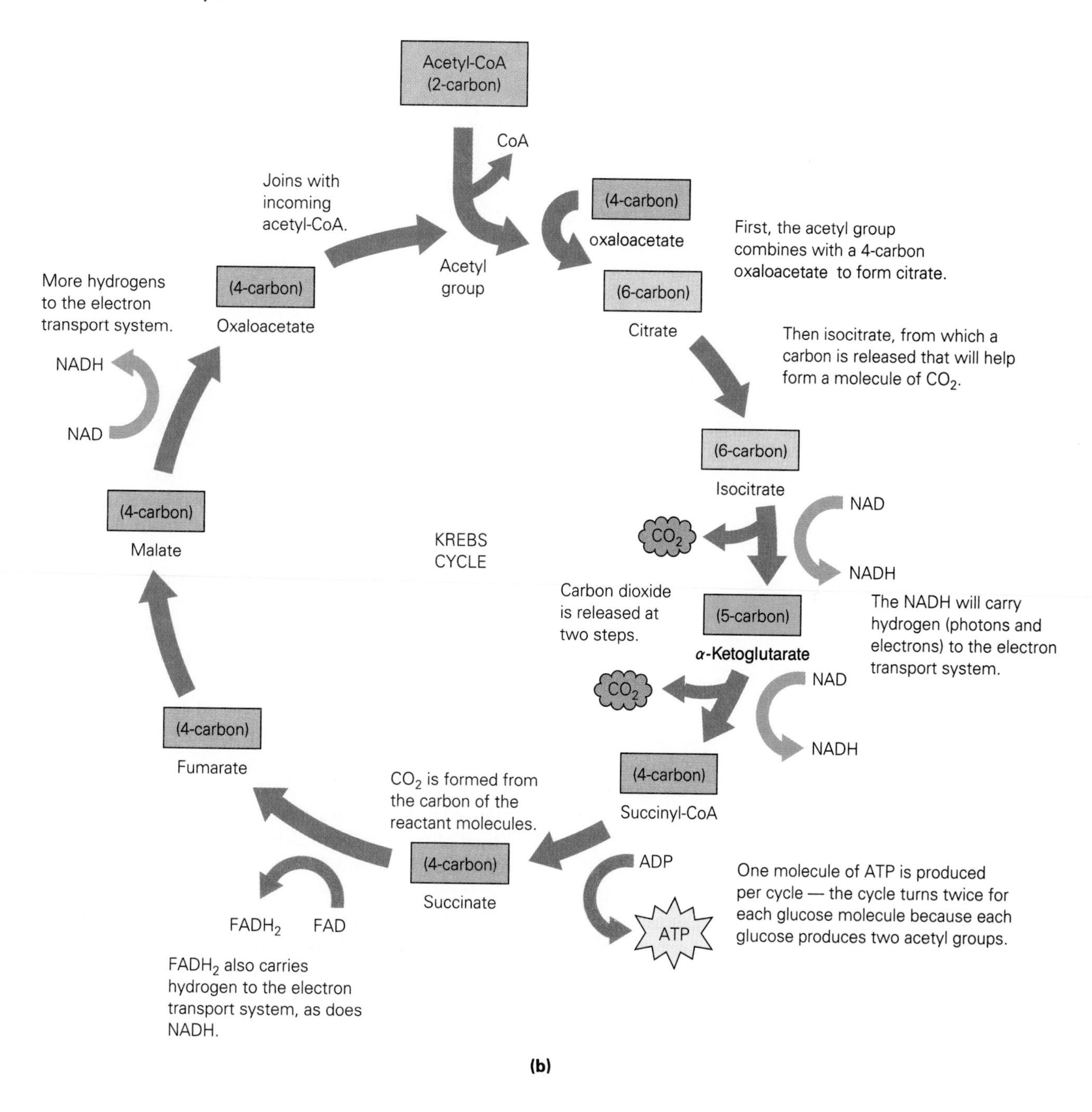

(b)

ADP and P_i (phosphorylation). The production of ATP using the energy of protons passing across a membrane through special enzyme-laden channels is called **chemiosmotic phosphorylation.** (The process is shown in Figures 5.8 and 5.17.)

A BRIEF REVIEW

By way of summary, in both the chloroplast and the mitochondrion, the energy released in the electron-transport system is used to move hydrogen ions (protons). However, they are moved in opposite directions within the two bodies. In the chloroplast, they are pumped into the lumen of the thylakoid (where they join those left by the splitting of water). As they pass outward into the stroma, they must move through the enzyme-laden channels, where they power the transformation of ADP and inorganic phosphate (P_i) into ATP. In respiration, on the other hand, hydrogen ions are transported by carriers across the inner membrane into the outer compartment of the mitochondrion, resulting in a concentration gradient between the outer and inner compartments. These hydrogen ions (protons) pass inward, reestablishing an equilibrium, only by passing through channels in the particles. Energy is extracted during this passage and then used to enable phosphorylating enzymes to form ATP from ADP and inorganic phosphate (P_i).

At this point, let's review the major points of respiratory reactions, beginning with glycolysis. During glycolysis, the 6-carbon sugar, glucose, is split into two 3-carbon molecules. These molecules are then rearranged to pyruvate, and, in the process, a small quantity of ATP is produced. In yeasts, pyruvate is converted to carbon dioxide and alcohol, while in aerobic animals, in the absence of oxygen, it is converted to lactic acid. In aerobic animals, if oxygen is present, the pyruvate is converted to acetate, which combines with coenzyme A. The resulting acetyl-CoA then enters the Krebs cycle, where carbon dioxide is produced, along with some ATP, and electrons are transferred to FAD^+ and NAD^+ molecules. The resulting $FADH_2$ and NADH then release these electrons to the electron-transport chain, and the electrons travel down the chain to increasingly lower energy levels. The remaining electrons are finally deposited in an "electron dump"—that is, combined with oxygen to form water. Less than half of the energy in the original glucose molecule ends up in ATP. The rest is released, primarily as heat. (As a demonstration, reach over and feel the body heat of the person next to you. Explain that it's an assignment.) As the electrons are passed along to lower energy levels, energy is released that is used to pump protons out of the inner compartment of the mitochondrion, establishing a concentration gradient. As the protons move inward, attracted by the net negative charge there, they pass through enzyme-laden particles. The particles use the energy of their passage to manufacture ATP from ADP and P_i through a process called chemiosmotic phosphorylation. The ATP produced in this way powers the innumerable activities associated with life.

SUMMARY

1. Energy is the ability to do work. It can exist in two states: potential energy (stored energy) and kinetic energy (energy of motion). Energy can exist in several interchangeable forms, including chemical, electrical, mechanical, and heat.
2. The laws of thermodynamics govern the behavior of all energy. The first law states that energy cannot be created or destroyed, but can change from one form to another. The second law states that when energy changes form, some will become unavailable to the system. As a result, processes move toward disorder (entropy). Therefore, life must expend energy to remain organized.
3. The sun's energy is stored in certain molecules by green plants, thereby becoming available to support all living systems.
4. ATP (adenosine triphosphate) is the cell's energy currency. Energy is put into the molecule to add the third phosphate bond to ADP. When this bond is broken, the released energy can fuel cellular work.
5. Photosynthesis is the process by which plants and algae use the sun's energy to make organic compounds. The sun's energy is captured by pigments such as the green pigments chlorophyll a and b and the orange pigments called carotenoids. The overall result is that carbon dioxide and water are converted to glucose and oxygen.
6. The light-dependent reactions of photosynthesis begin when light is captured by a pigment such as chlorophyll. Pigments are found in chloroplasts, which are double-membraned structures containing disks (thylakoids) and arranged in stacks called grana. Thylakoids of different grana are connected by membranous lamellae. Embedded in the thylakoid membrane are particles through which enzyme-laden channels connect the inside of the disk to the jellylike stroma surrounding the thylakoids.
7. When light strikes a chlorophyll molecule in photosystem II, two of chlorophyll's electrons move to a higher energy state where they can be captured by an electron acceptor and passed along an electron transport chain to another chlorophyll pigment system, photosystem I. After passing along a second transport chain, the electrons reduce $NADP^+$ to NADPH. The electrons taken from the chlorophyll in photosystem I are replaced by electrons from photosystem II and those, in turn, are replaced by electrons taken from water, forming oxygen.
8. Protons become more concentrated within the thylakoid because they are left when the electrons are removed and passed along photosystem II and because they are pumped in from the stroma. The proton accumulation within the thylakoid sets up a chemical and an electrical gradient. In response to these gradients, protons move through enzyme-laden channels in the membranes. As they do, the energy of their movement is used to produce ATP from ADP. This process is called chemiosmosis.
9. The ATP and NADPH formed in the light-dependent reactions are used to form nutrient molecules in the light-independent reactions. The major biochemical pathway of the light-independent reactions is the Calvin cycle. The basic events are the addition of carbon dioxide to a 5-carbon sugar,

ribulose biphosphate (RuBP) to form a 6-carbon sugar that quickly splits into two 3-carbon molecules of phosphoglyceraldehyde (PGAL), which can be converted to glucose, starch, or RuBP.

10. In a process called cell respiration, cells slowly release the energy in high-energy molecules and store it in ATP. The first step in this process, glycolysis, takes place in the cytoplasm and does not require oxygen. As each glucose molecule is converted to two molecules of pyruvate, there is a net gain of two ATP, two NADH, and two H^+. If oxygen is absent, pyruvate enters a pathway called fermentation, which produces alcohol in yeast plus carbon dioxide, and produces lactate (lactic acid) in animals. In the presence of oxygen, the 3-carbon pyruvate may be converted to 2-carbon acetate, combined with coenzyme A, and enter the Krebs cycle.
11. In the Krebs cycle, the energy in each of the two pyruvates formed from a glucose molecule is used to reduce three molecules of NAD^+ to NADH and one molecule of FAD^+ to $FADH_2$, and to form one ATP. Carbon dioxide is produced.
12. In the electron transport chain, electrons that had been accepted by NAD^+ and FAD^+ are passed along a series of electron acceptors, each at a lower energy level than the one before, releasing energy gradually. At the end of the chain, electrons join with protons to form hydrogen and then join with oxygen to form water.
13. ATP is produced by chemiosmotic phosphorylation as protons pass out of the mitochondrion through enzyme-laden channels in the particles that are embedded in the mitochondrial membrane.

KEY TERMS

acetate (142)
acetyl-CoA (142)
adenosine diphosphate (ADP) (124)
adenosine triphosphate (ATP) (124)
alcoholic fermentation (140)
anaerobic (136)
Calvin cycle (131)
carotenoids (127)
cellular respiration (133)
chemiosmosis (131)
chemiosmotic phosphorylation (146)
chlorophyll (127)
chlorophyll a (127)
chlorophyll b (127)
citrate (143)
electrical energy (123)
electromagnetic spectrum (130)
electron acceptor (127)
electron-transport system (128)
entropy (123)
fermentation (139)
first law of thermodynamics (123)
glycolysis (137)
granum (127)
heat energy (123)
kinetic energy (122)
Krebs cycle (142)
lactate fermentation (140)
lamellae (127)
light-dependent reaction (125)
light-independent reaction (125)
matrix (127)
oxaloacetate (143)
mechanical energy (123)
photon (130)
photosynthesis (125)
photosystem I (129)
photosystem II (128)
potential energy (free energy) (122)
pyruvate (pyruvic acid) (136)
second law of thermodynamics (123)
stroma (127)
thylakoids (127)
visible light (130)
wave (130)

FOR FURTHER THOUGHT

1. Describe how the role of glucose differs in the processes of photosynthesis and cellular respiration.
2. Do the light-independent reactions of photosynthesis derive their name because they can only take place in the dark? Explain.
3. Describe how glycolysis might be viewed as an inefficient cellular respiration process.

FOR REVIEW

True or false?

1. ___ The second law of thermodynamics states that energy transformations are usually 100 percent efficient.
2. ___ ATP and NADPH are produced during the light-dependent reactions of photosynthesis.
3. ___ The Krebs cycle takes place in the cytoplasm of a cell.
4. ___ Fermentation is an aerobic cellular respiration process.
5. ___ The formation of ATP via the use of an electron transport system is called chemiosmosis.

Fill in the blank.

6. ___ is defined as the ability to do work.
7. Energy transformation processes tend toward a measure of disorder called ___.
8. During photosynthesis, photosystem II recovers its missing electrons from ___ while photosystem I recovers its missing electrons from ___.
9. ___ is a process in which ADP and inorganic phosphate combine to form ATP.
10. ___ is the first step in the breakdown of glucose.

Choose the best answer.

11. In the light-independent reactions of photosynthesis
    A. electrons from photosystem II are passed to the beginning of the electron transport chain.
    B. $NADP^+$ is reduced to NADPH.
    C. PGAL is produced.
    D. ADP is phosphorylated.
12. When no oxygen is present, human cells convert pyruvate to
    A. citrate.
    B. acetyl-CoA.
    C. lactic acid.
    D. oxaloacetate.
13. Which of the following is a possible path for the pyruvate molecules that were formed in glycolysis?
    A. alcoholic fermentation
    B. Krebs cycle
    C. lactic acid fermentation
    D. all of the above

14. Which of the following correctly matches a cellular respiration process with the net number of ATP molecules it produces for each molecule of glucose?
    A. fermentation; 36 ATPs
    B. glycolysis; 2 ATPs
    C. Krebs cycle; 4 ATPs
    D. none of the above
15. The electron-transport system for respiration lies in the
    A. cellular fluid.
    B. thylakoid.
    C. mitochondria.
    D. stroma.

CHAPTER 6

The Cell and Its Cycles

Overview

The Cell Cycle and Mitosis
Meiosis
The Double Helix
DNA Replication
How Chromosomes Work
Continuing Problems in Cell Biology

Objectives

After reading this chapter you should be able to:

- Describe how cells reproduce by listing the events that occur during the stages of the cell cycle.
- List the stages of meiosis and describe how meiotic events halve chromosome numbers during gamete formation.
- Describe the components and spatial arrangement of the DNA molecule as proposed by Watson and Crick.
- Explain how the DNA molecule replicates.
- Describe how chromosomes direct the production of proteins via transcription and translation.
- Compare RNA's properties to those of DNA.
- Explain what is meant by the genetic code.

You may not be surprised to hear that when carrot cells reproduce, they form more carrot cells, and oyster cells form more oyster cells. After all, an oyster cell is very good at being an oyster cell, so when it reproduces, its best bet lies in making a very close approximation of itself. If a well-functioning oyster cell were to generate new kinds of cells, there would be no guarantee that those new cells would work—and, in fact, they probably wouldn't. So, there is an advantage in cells giving rise to new cells that are very similar to themselves. Such accurate duplication is assured by the precise mechanisms of cell reproduction.

We will now see how cells go about reproducing themselves so precisely. First, we will have a look at some cellular processes; then we will focus on what happens at the molecular level.

THE CELL CYCLE AND MITOSIS

One of the first questions we might ask is, Why do cells reproduce at all? That shouldn't be too hard to figure out. After all, in an active body, cells die or are worn away and must be replaced. In addition, simple growth is often based on the proliferation of cells. Many forms of life, then, demand that cells reproduce as part of maintenance and growth. So now, on to the process itself.

Before dividing, a **parent cell** makes copies of all of its chromosomes and many of its organelles. Then, the nucleus divides in a process called **mitosis** in which the duplicate copies of the chromosomes are carefully separated and distributed evenly, one to each daughter nucleus. Almost immediately after mitosis, the cytoplasm divides to form two **daughter cells,** each containing a nucleus formed by mitosis. The process of cytoplasmic division is called **cytokinesis.** After mitosis and cytokinesis, each daughter cell has a full chromosome complement that is identical to that of the parent cell, and the cytoplasm of the parent cell has been distributed about equally between the daughter cells.

The life of a cell, the cell cycle, can be divided into cell division (which includes both mitosis and cytokinesis) and interphase (Figure 6.1).

For purposes of discussion, mitosis is divided into four stages: prophase, metaphase, anaphase, and telophase. When the cell is not dividing, it is said to be in interphase. Interphase was once called the "resting stage," but we now know that this description is inaccurate. Interphase is a busy time, indeed, because it is then that many of the cell's components must double prior to the next mitotic division. I should add that the distinction between one phase and another is arbitrary, since there is a continuous progression from one phase to the next. Those who find comfort in labeling sometimes add further breakdowns, such as late prophase or early anaphase, so we will occasionally encounter these terms. Let's now review each of these stages in a bit more detail (Figure 6.2, pages 154–155).

Interphase is the time when materials needed for cell growth and division are synthesized. The animal cell at this stage almost doubles in size and generally becomes rounder as the cytoplasm becomes heavier and more viscous. Growth includes protein synthesis, organelle construction,

FIGURE 6.1

The cell cycle of liver cells grown in culture, showing the time spent in each phase.

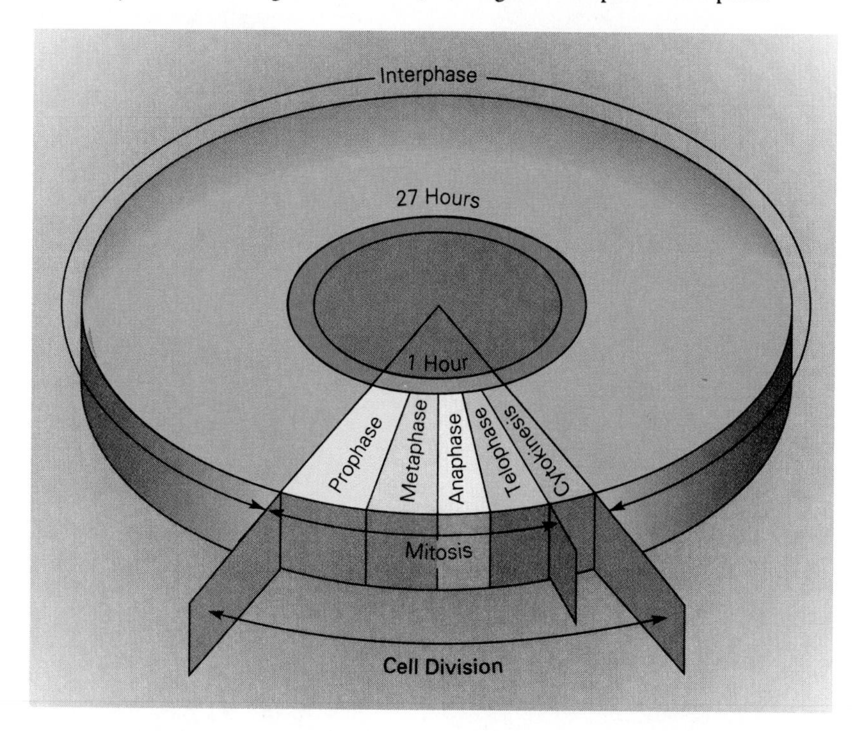

and storage of materials. The chromosomes are duplicated during interphase and the copies of each chromosome are bound to one another by a **centromere.** As long as the copies are attached by a centromere, each is called a **chromatid** (Figure 6.3).

During **prophase,** the chromatin begins to coil and those coils form larger coils. The coiling makes the chromatin strands thicker. As a result, the chromosomes become visible in a light microscope. While the chromosomes are condensing, the nuclear membrane gradually vanishes and a spindle begins to form in the cytoplasm. The **spindle** is a structure that consists of microtubules radiating across the cell from pole to pole.

As the nuclear membrane disappears, the chromosomes move toward the center of the cell and attach, one to each spindle fiber, marking **metaphase.** Although the chromatids lean out in all directions, the centromeres are held in position at the midpoint of the spindle fibers.

Anaphase begins with the division of the centromeres and the separation of the chromatids, each one now referred to as a **chromosome.** Put another way, chromosomes can be composed of two chromatids that, when separated, are themselves referred to as chromosomes. Subsequently, each of these single chromosomes (formerly, chromatids) moves along the spindle lines to opposite poles of the cell. It appears as if the thin micro-

FIGURE 6.3

Chromosomes become apparent in the nuclei of the cells for the first time at the beginning of prophase. The diffuse chromatin net has condensed and become twisted into strands. It is not always apparent, but chromosomes are composed of two parts called chromatids, which are joined at a centromere. When two chromatids that are joined in such a way become separated, each former chromatid is then called a chromosome. Thus, in this figure, there are two chromatids and one chromosome.

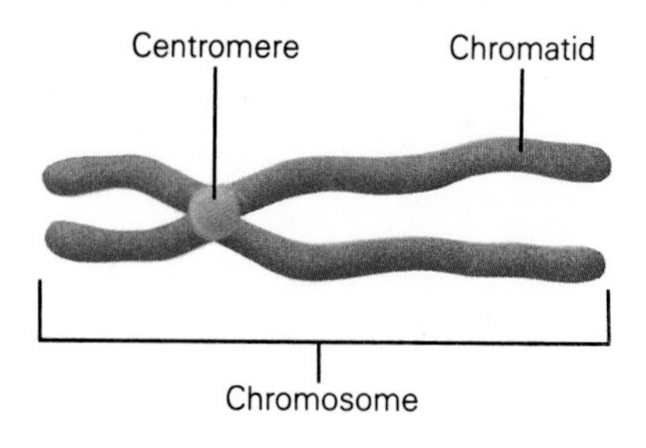

FIGURE 6.2
Cell cycle in the endosperm of a lily.

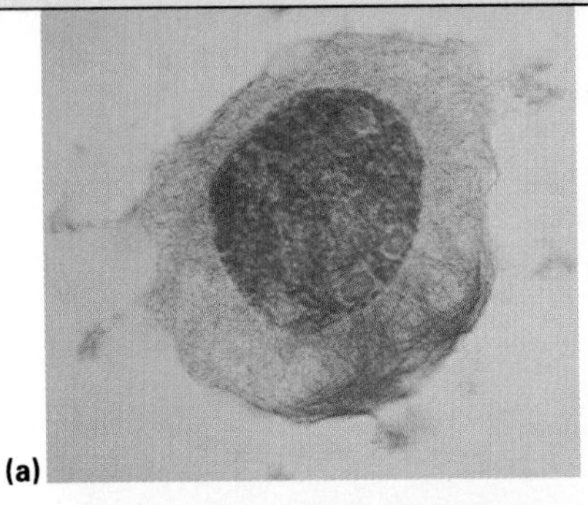
(a)
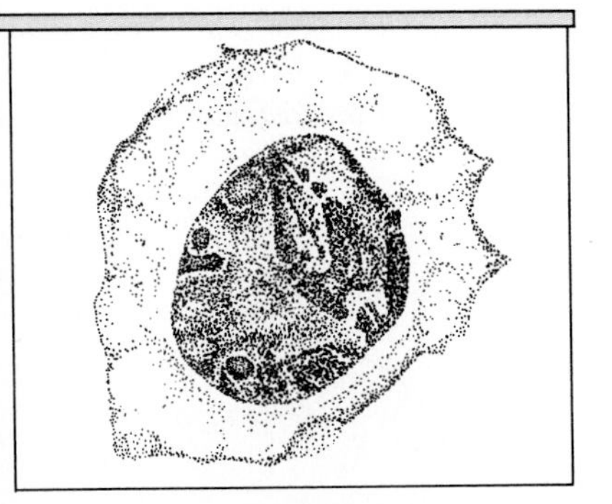

(a) Interphase
The chromatin is diffuse and spread thinly throughout the nucleus. The nucleolus is prominent.

(b)
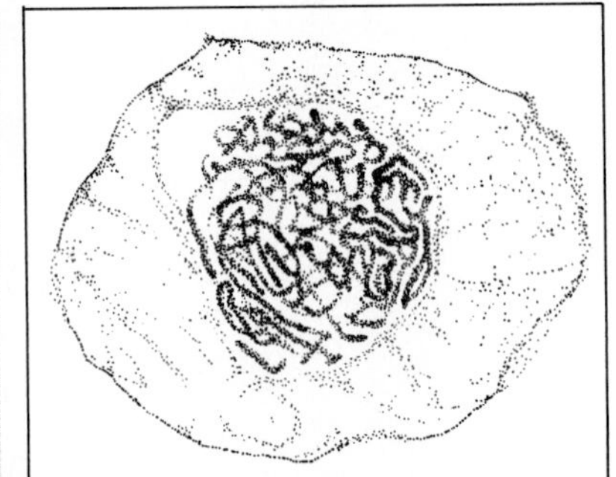

(b) Early prophase
The chromosomes begin to shorten and thicken.

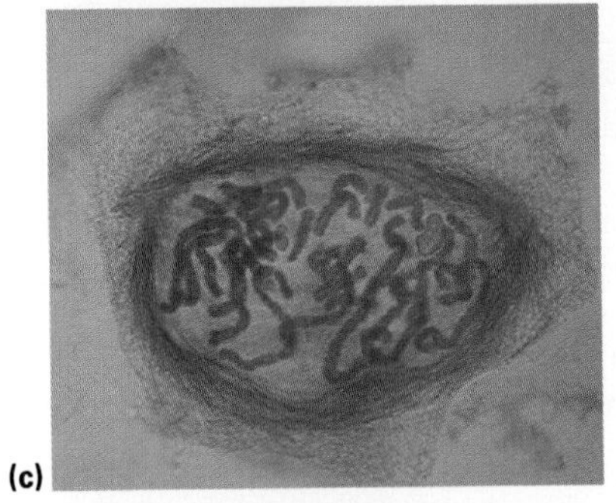
(c)
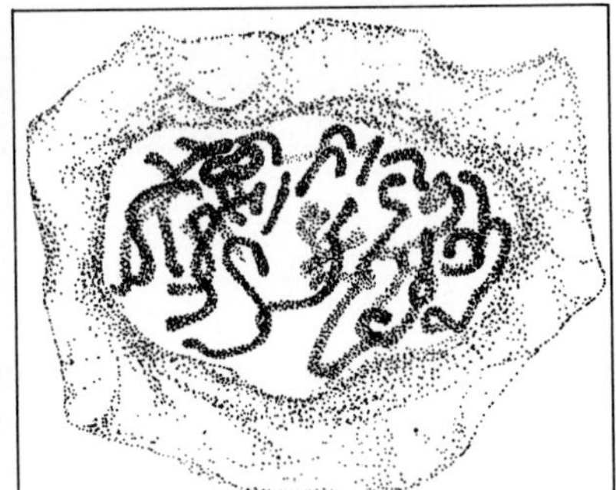

(c) Early prophase
The nucleolus is disappearing as the chromosomes are becoming distinct.

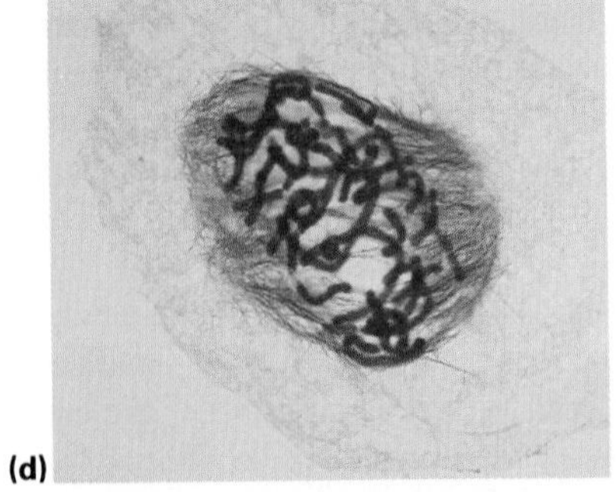
(d)
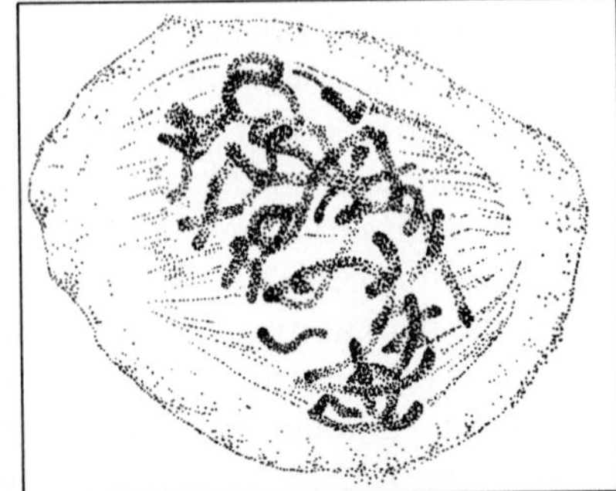

(d) Mid prophase
The nuclear membrane breaks and spindle appears.

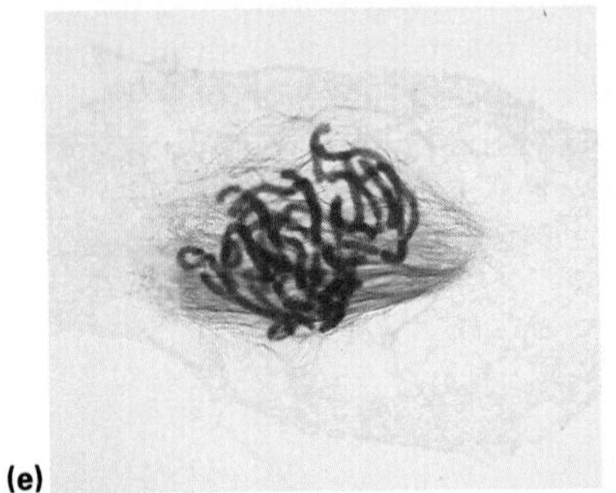
(e)
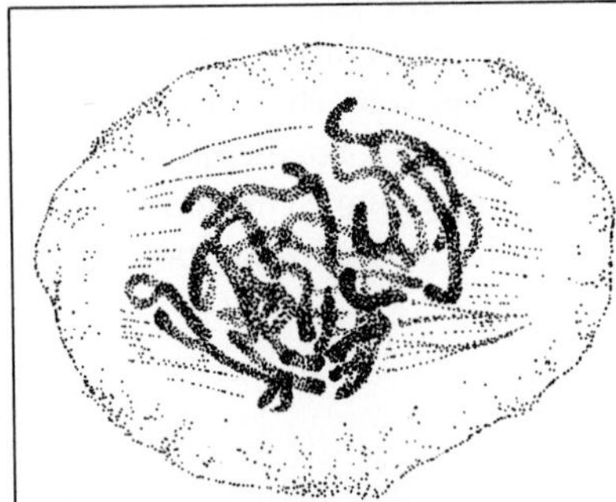

(e) Early metaphase
Chromosomes move toward the center of the cell.

FIGURE 6.2 continued

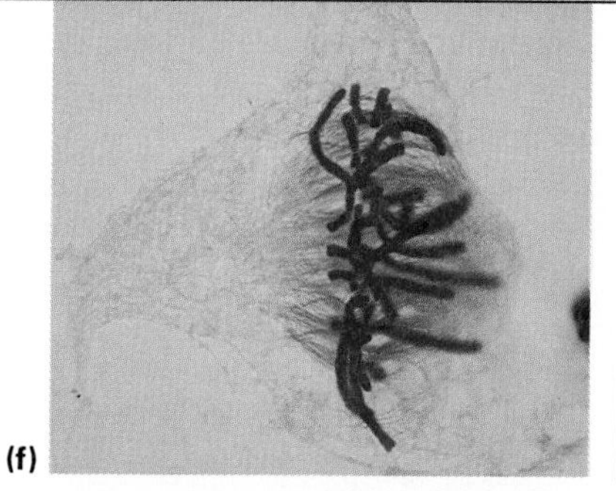
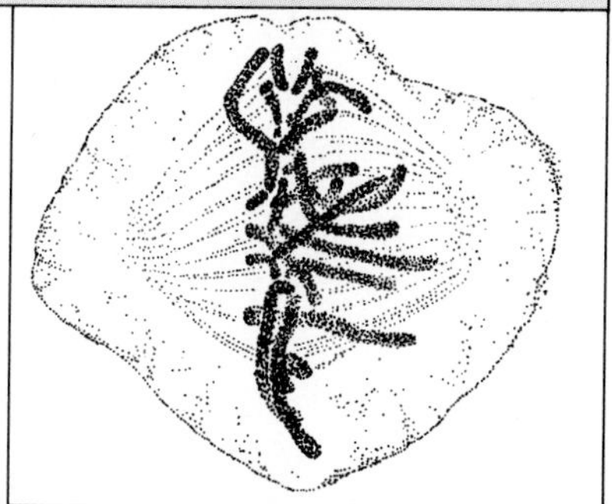

(f) Metaphase
Chromosomes line up at the equator and attach to spindle fibers.

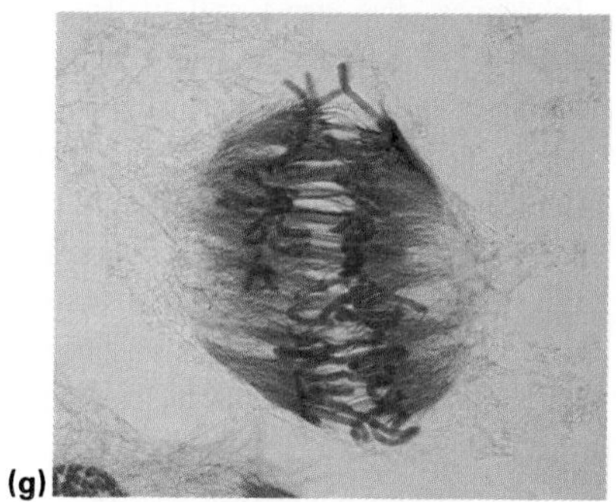
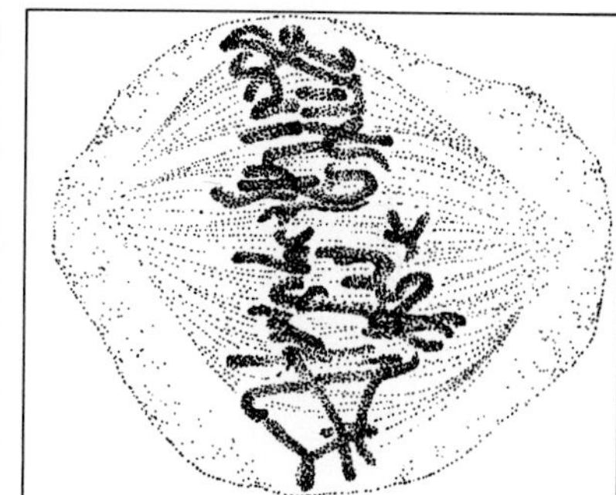

(g) Early anaphase
The chromosomes begin to separate.

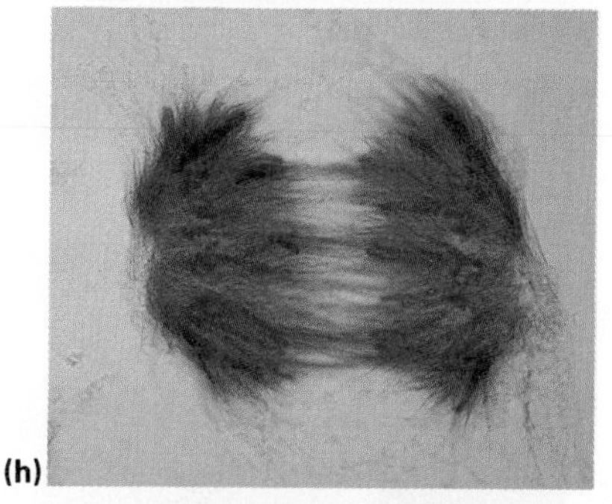
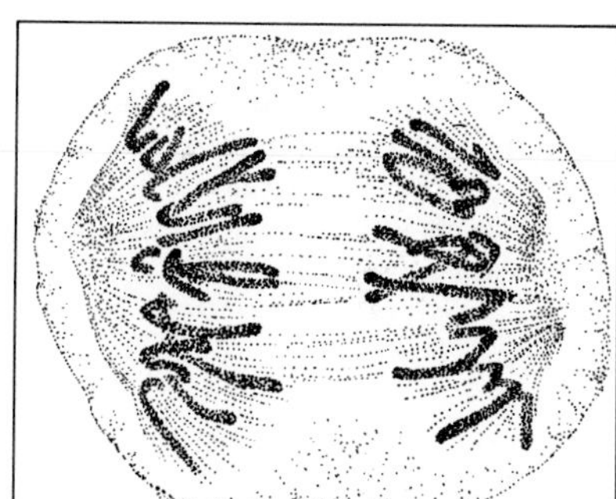

(h) Late anaphase
The chromosomes draw nearer to the ends of the spindle.

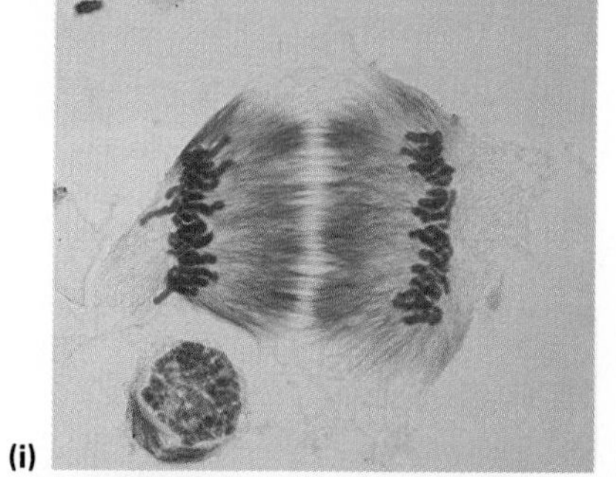
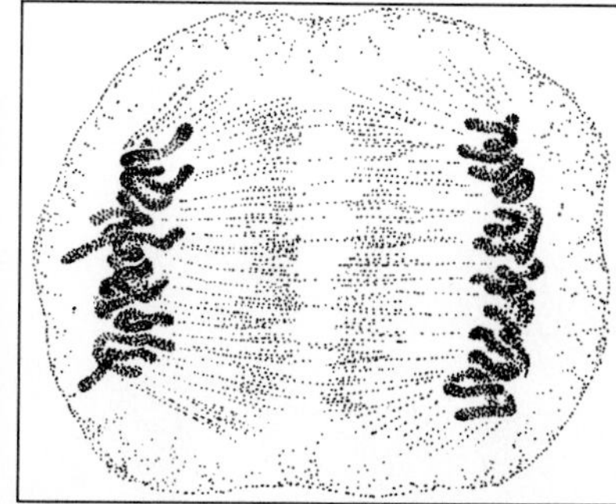

(i) Early telophase
New nuclei are becoming evident.

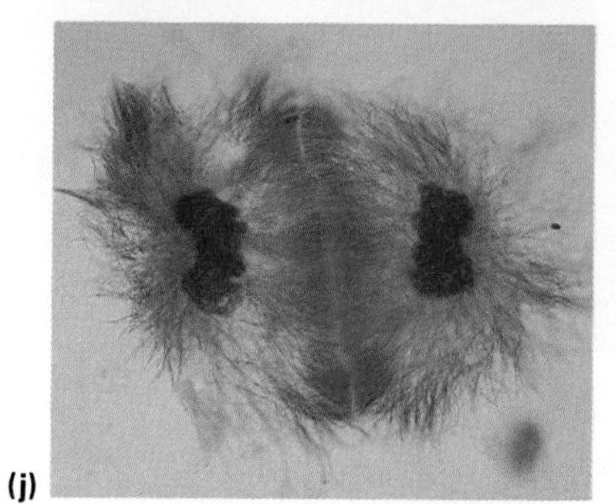
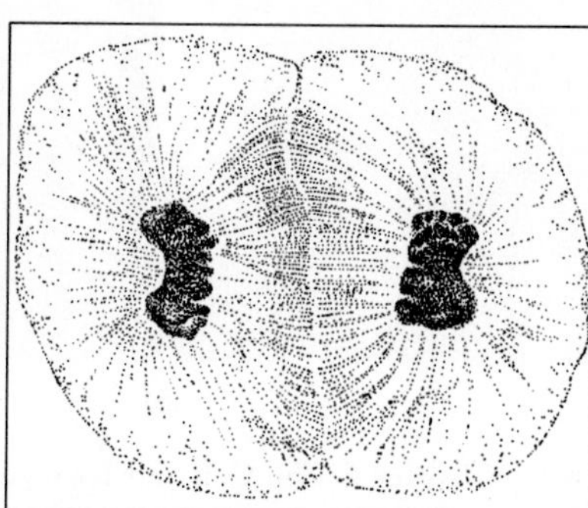

(j) Telophase and cytokinesis
The aggregation is complete and the two new cells are separating.

tubules that comprise the spindle have attached to the centromeres of the single chromosomes and are pulling each member of a pair in different directions. In any case, the chromosomes that converge at one end of the spindle are identical to those at the other end. When the chromosomes have formed two groups, one at each end of the cell, anaphase ends.

During **telophase,** the chromosomes unwind and stretch out once again. As this happens the nuclear membrane begins to form around each cluster of chromosomes.

In animals, cytokinesis begins as a furrow along the cell's equator. The furrow pinches in between the two clumps of chromosomes, finally closing together and roughly dividing the parental cytoplasm (cytokinesis). In plants, the rigid cell walls cannot pinch in. Instead, a new cell wall begins to form in the cytoplasm, roughly halfway between the divided chromosomes. In both animals and plants, each daughter cell may at first be somewhat smaller than the mother cell because it has less cytoplasm, but such differences are quickly reduced by water intake and the formation of new cell organelles. The daughter cells, by the way, may also be of unequal size since the cytoplasmic division may not be precise, but the important point is that the genetic component of each daughter cell will be identical to that of the parent cell (Figure 6.4).

FIGURE 6.4

(a) Cytokinesis in an animal cell. The deep furrow (closeup in b) marks the beginning of the two-cell stage.
(c) Cytokinesis in a plant cell, marked by the formation of a new cell wall.

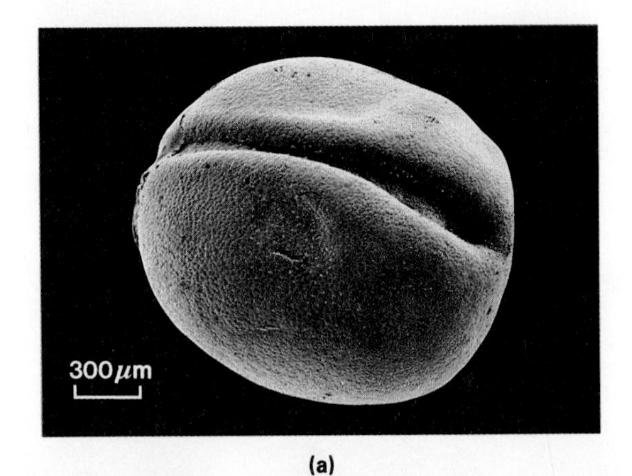

(a)

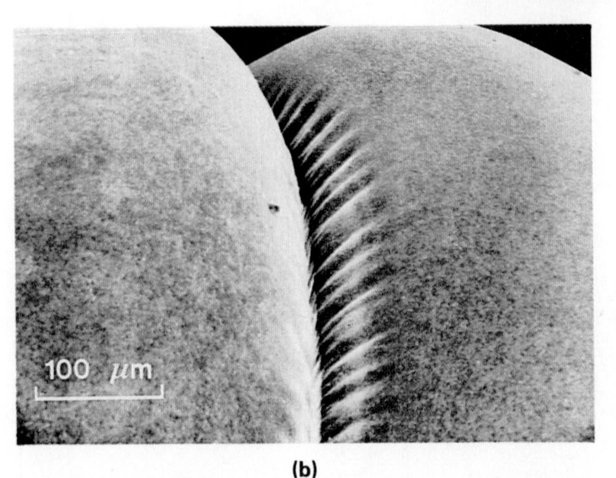

(b)

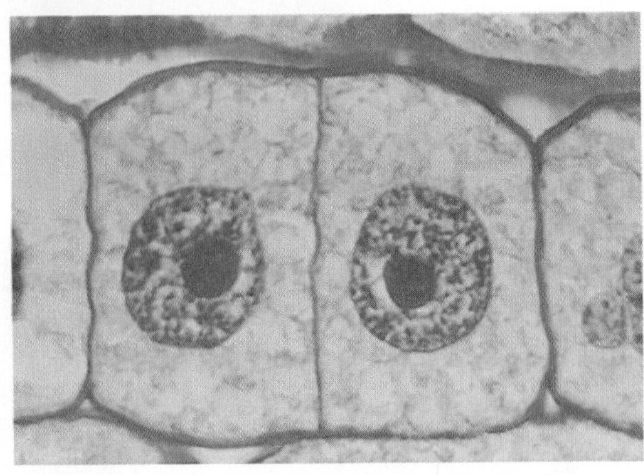

(c)

Before we take a look at this process at the molecular level, let's consider another process—one that may, at first, seem similar to mitosis. Keep an eye peeled, though, because the differences are critical. This one, called meiosis, is important to reproduction.

MEIOSIS

Following the steps of meiosis may answer some questions for you if you have thought much about the reproductive process, especially if you are aware that eggs and sperm are simply specialized cells. One basic problem unfolds this way: at fertilization, the nuclei of an egg and a sperm are joined, and from the subsequent divisions of the fertilized egg, new cells of the developing individual are produced which are genetically identical to the parent cell, and to one another. This means that all the cells of the developing organism have their full complement of chromosomes. Since eggs and sperm are both cells, when they combine at fertilization, why don't the resulting offspring have twice as many chromosomes as their parents?

In answering this question, we can begin by noting that the number of chromosomes is identical in virtually every cell of the body, since all the cells of any organism are genetically identical to that first fertilized egg. We say "virtually" because there are exceptions; there must be. For the union of an egg and a sperm to produce the correct number of chromosomes in the cells of the offspring, each of them must have exactly *half* the chromosome complement of the other cells, so that the *sum* of their chromosomes results in the full complement. As it turns out, each egg and sperm has exactly half the number of chromosomes characteristic of the species. Thus, when the egg and sperm are joined, the result is a primal cell with the "normal" number of chromosomes. Then, through millions of divisions of that cell (in conjunction with cell specialization, which we'll worry about later), a new individual is formed—resplendent in its own demands, hang-ups, and voting tendencies.

The halving of chromosomal complements is possible because, as you see in Figure 6.5, chromosomes come in pairs. Every kind of chromosome in the nucleus has a counterpart—another chromosome like itself in size, shape, and the kind of hereditary information (genes) it contains. These matched pairs of chromosomes are called **homologues;** a homologue is either of the two members of a pair of chromosomes. Chromosomes exist in homologous pairs, as you may have guessed, because the egg and sperm contribute identical *kinds* of chromosomes to the new individual. Thus, for any pair of chromosomes in the body, one can be considered the descendant of a paternal chromosome (from the father's sperm) and the other the descendant of a maternal chromosome (from the mother's egg). The result is that each parent contributes to every genetically determined characteristic of the offspring (with special exceptions, as we shall see). So, chromosomes come in pairs—for example, 23 pairs in humans, 9 pairs in cabbages, and 47 pairs in goldfish. (The numbers of chromosomes for some other species are shown in Table 6.1.) Thus, if we have a gene for some trait from one parent, we will have another gene for that trait from the other parent.

FIGURE 6.5

Pairing chromosomes of a female *Drosophila*. Note that each chromosome in a cell has its counterpart, one having come from the mother fly, the other from the father.

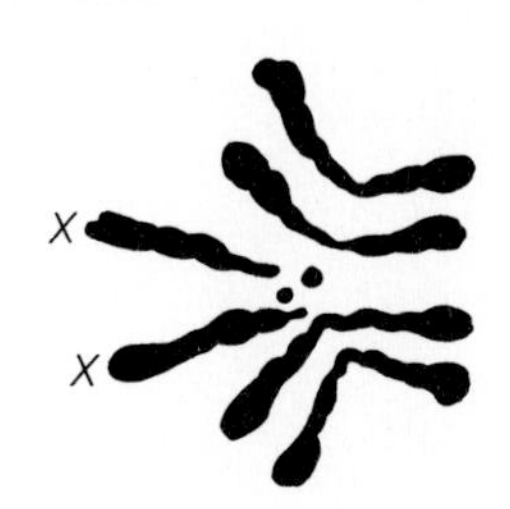

For virtually any species, then, eggs and sperm must have half the number of chromosomes as do other cells. But they don't start out that

TABLE 6.1
The Number of Chromosomes in a Variety of Species

CHROMOSOME NUMBERS*

Alligator	32	Garden pea	14	*Penicillium*	2
Amoeba	50	Goldfish	94	Planaria	16
Brown bat	44	Horse	64	Redwood	22
Carrot	18	House fly	12	Rhesus monkey	42
Cat	32	Human	46	Rose	14, 21, 28
Cattle	60	Lettuce	18	Sand dollar	52
Chicken	78	Magnolia	38, 76, 114	Starfish	36
Chimpanzee	48	Marijuana	20	Tobacco	48
Dog	78	Onion	16, 32	Turkey	82, 81
Fruit fly	8	Opossum	22	White ash	46, 92, 138

*There is no apparent significance to chromosome number as far as biologists can determine. If you feel good about your 46 (see human), check the turkey, amoeba, and cattle. Do you find it significant that the number of human chromosomes falls somewhere between that of a housefly and a sand dollar? What do you make of the tobacco and chimpanzee numbers? Note the variation in some plant and animal species. Plants sometimes undergo spontaneous doubling and tripling of chromosome number.

way. They start out with a full (double) complement of chromosomes, and somewhere along their developmental route, half the chromosomes are lost.

Cells that contain a full complement of chromosomes, that is cells with both members of each homologous pair, are called **diploid.** Cells that contain only one member of each homologous pair of chromosomes, eggs and sperm for instance, are called **haploid.** Eggs and sperm don't start out haploid. They develop from cells with a full (double) complement of chromosomes, and somewhere along their developmental route, half the chromosomes are lost. However, since the chromosomes contain the genetic information that governs cellular structure and function, the egg and sperm cannot contain just a random assortment of chromosomes. They must contain one complete set of instructions for forming and running a cell, one member of each homologous pair.

Meiosis, then, is the process by which cells with two complete sets of chromosomes halve their chromosomal number so that they contain only one complete set. We've seen that meiosis is important in the formation of **gametes**—eggs and sperm. However, in some organisms, certain plants for instance, meiosis may result in the formation of spores.

Now, let's follow the development of gametes as they begin their meiotic divisions in the **gonads** (ovaries or testes) of animals. Remember, it may seem at first that we are retracing mitosis, but this will change.

First, note that the meiotic process is divided into parts, meiosis I and meiosis II (Figure 6.6, pages 160–161). But before meiosis begins at all—that is, during the cell's interphase—each chromosome is duplicated. The copies (chromatids) remain attached to each other by a centromere, just as they do in mitosis. Then the chromatin net shortens and thickens so that the chromosomes become visible, and the spindle starts to form about the time the nuclear membrane begins to disappear (all this should sound familiar so far).

The process continues as, in early prophase I, each pair of homologous chromosomes comes together. The chromosomes themselves consist of two identical chromatids (so each chromosomal group is made up of *four* identical chromatids at this point). As the chromosome pairs line up side by side, an important event may take place: the chromosomes may exchange parts. This exchange is called **crossing over,** technically, an exchange of DNA segments during meiotic prophase (Figure 6.7). The crossovers take place at various points along the chromosomes, and as chromosomal parts are exchanged, new combinations result. (We will discuss later the role of crossovers in producing variation and see why such variation is so important in evolutionary processes.)

In metaphase I (as you might expect if you recall our discussion of mitosis), the chromosomes line up across the developing spindle. In meiosis, however, both members of a homologous pair attach, side by side, to the center of the same spindle fiber.

During anaphase I, the members of each pair separate and move toward opposite spindle poles. Note that the centromeres do not divide, as they do in mitosis. Thus, the chromatids of each chromosome now travel together to the spindle poles. At telophase I, the end of the first stage of meiosis, the chromosomes begin to elongate and to again take on a netlike appearance. A nuclear membrane may now form as the cell enters interphase II. Each nucleus now contains only half the number of chromosomes of the parent cell, although each chromosome is two-stranded. Not only are there now fewer chromosomes in each cell, but they are probably qualitatively different from any of the original chromosomes as a result of crossovers.

Interphase II differs from the interphase preceding meiosis I in an important way: this time the chromosomes do not replicate themselves. Aside from this, however, the second meiotic division proceeds in a way similar to mitotic events.

In prophase II, the chromosomes shorten and thicken, and the nuclear membrane (if it is present) begins to disappear. The spindle fibers begin to form. At metaphase II, the chromosomes line up and attach to the midpoint of the spindle fibers, one to each spindle fiber. Now, at anaphase II, in contrast to anaphase I, the centromere divides in two, each half bearing a centromere, and the single chromosomes (earlier called chromatids) move toward opposite poles. At telophase II, a nuclear membrane forms around each set of chromosomes. The cytoplasm is divided as new cell membranes appear, thus separating each group into a different cell. So where there was one cell with two copies of each chromosome, there are now four cells, each with half the chromosomal complement (Figure 6.8, pages 162–163).

If the meiotic process has been taking place in a testicle, each of four resulting cells will undergo further changes, grow a tail, and become a sperm. However, if the meiosis was in an ovary, three of the resulting

FIGURE 6.7

While homologous chromosomes are closely associated in prophase I of meiosis, they may exchange parts. Their chromatids may cross (a) and then break at the point of their crossing. Each separated part may then fuse to the other chromosome at the area of the breakage (b). New genetic combinations are created in this way.

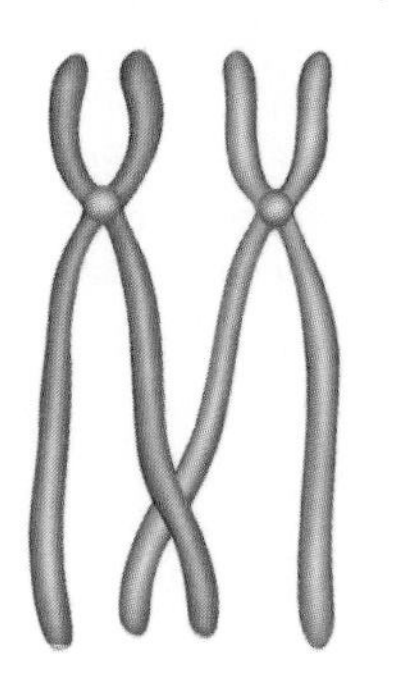

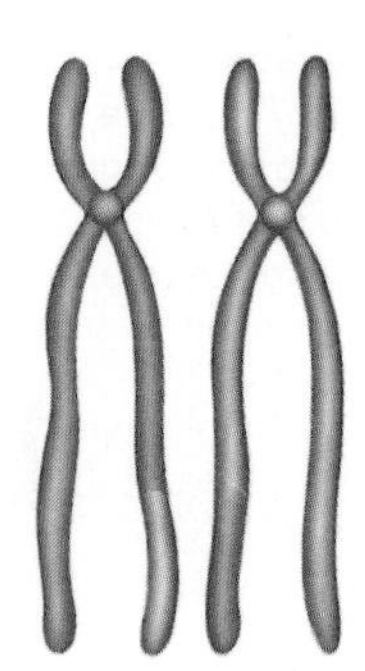

FIGURE 6.6

Meiosis in the anther of an Easter lily *(Lilium longiflorum)*. The meiotic process is divided into two parts: meiosis I and meiosis II.

Meiosis I.

(a) Prophase I

The homologous pairs of chromosomes come to lie next to one another. Each chromosome consists of two chromatids attached by a centromere.

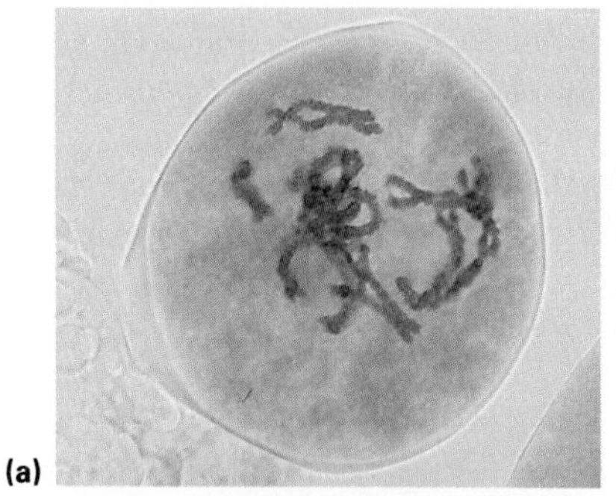
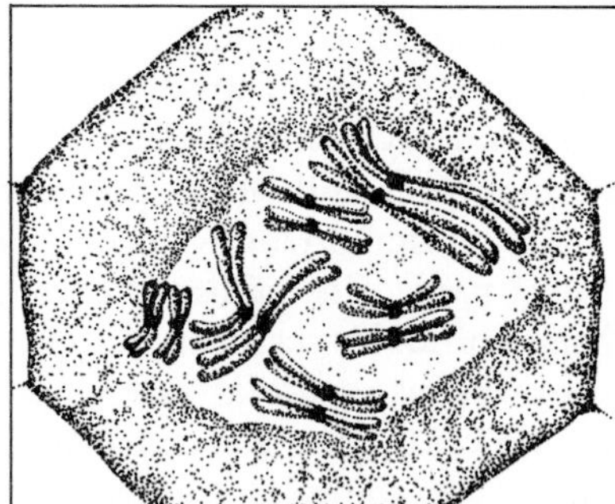

(a)

(b) Metaphase I

The homologous pairs of chromosomes attach to the same spindle fiber.

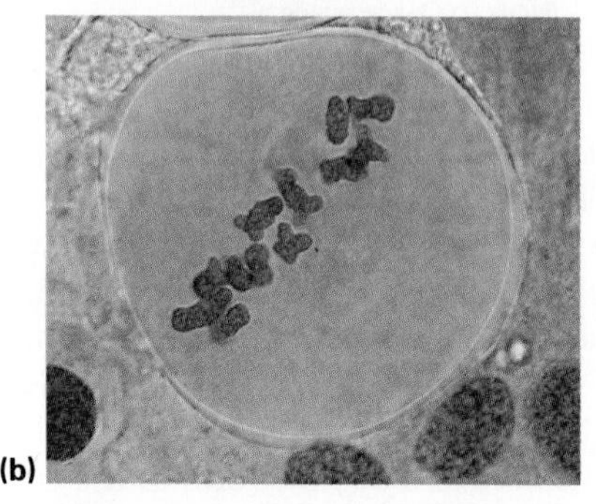
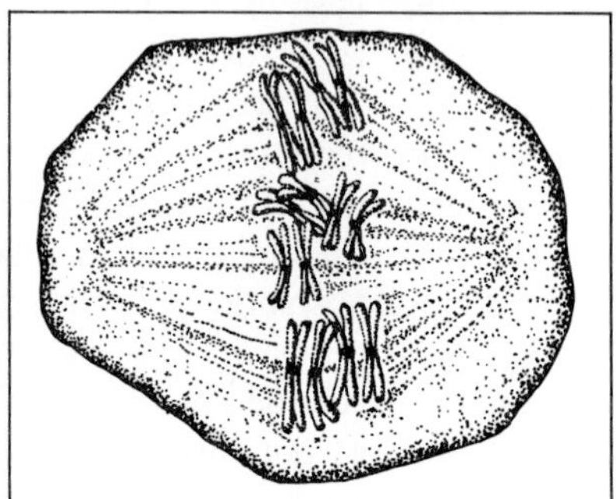

(b)

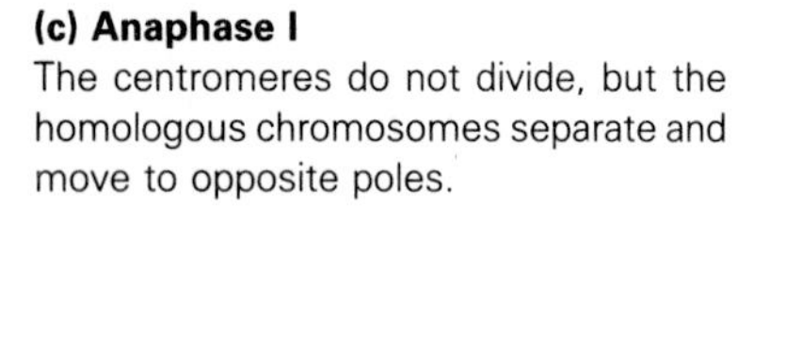

(c) Anaphase I

The centromeres do not divide, but the homologous chromosomes separate and move to opposite poles.

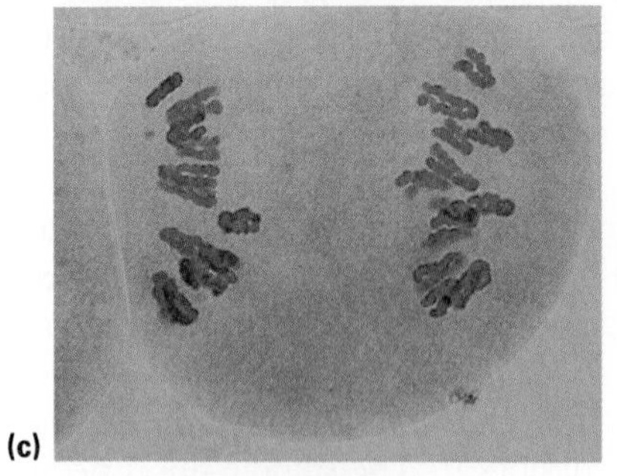
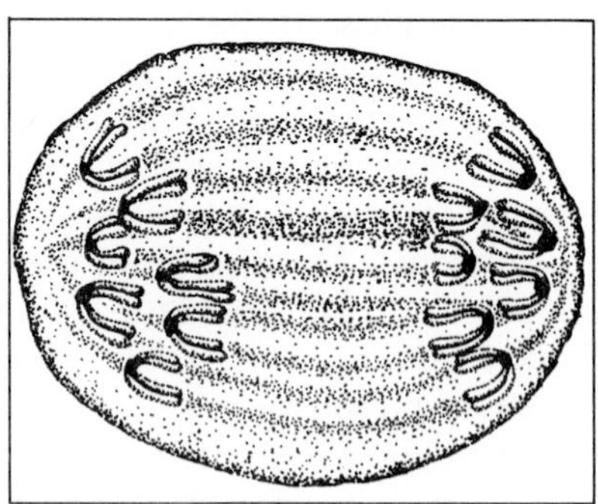

(c)

(d) Telophase I

New nuclei are forming at each pole.

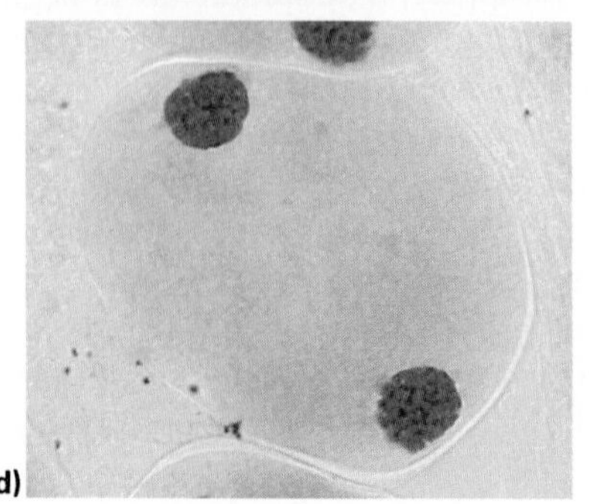
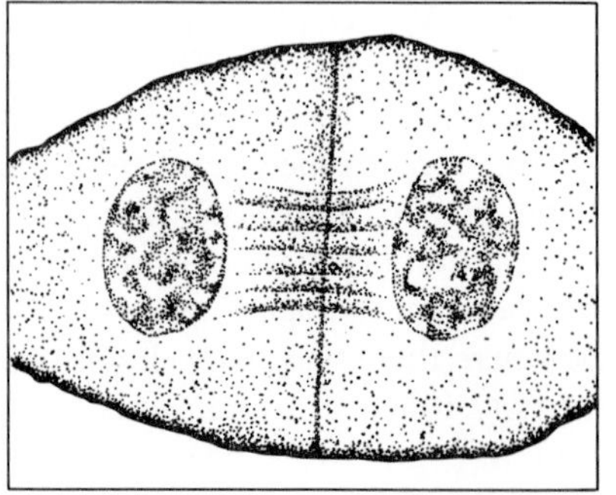

(d)

FIGURE 6.6 continued

Meiosis II.
(e) Early prophase II
The diffuse chromatin is beginning to condense.

(e)

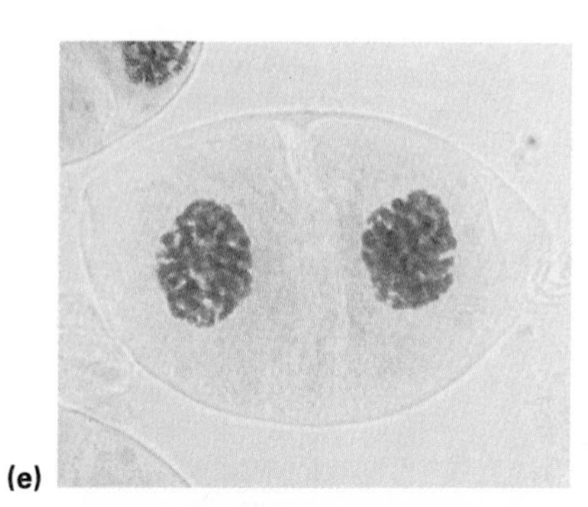

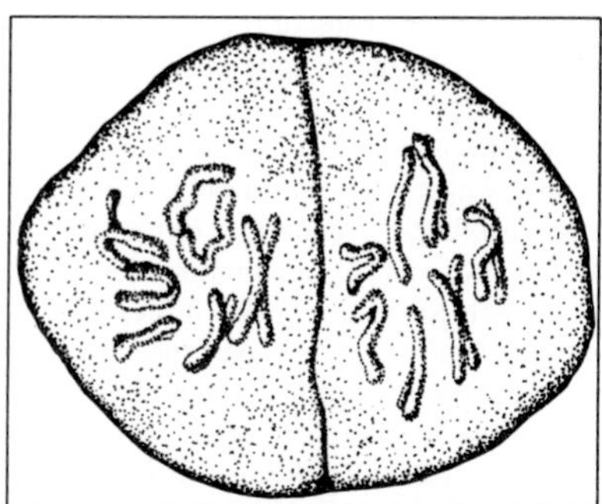

(f) Metaphase II
Distinct chromosomes line up, one of each pair to a spindle.

(f)

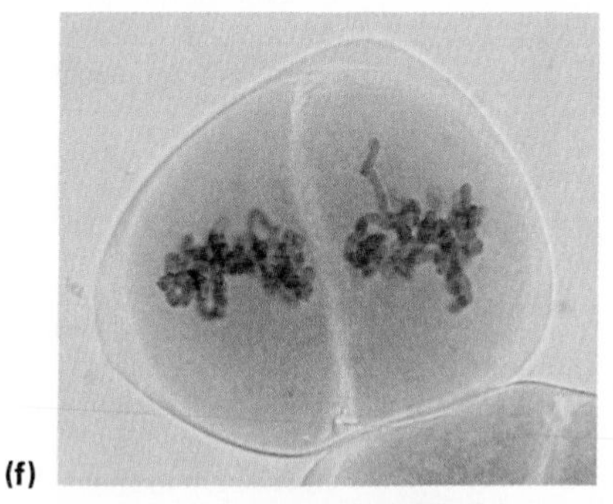

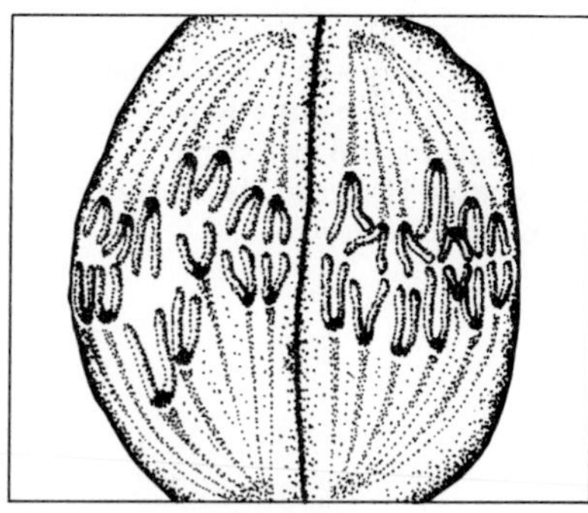

(g) Late anaphase II
The centromeres divide. Single-stranded chromosomes move to opposite poles.

(g)

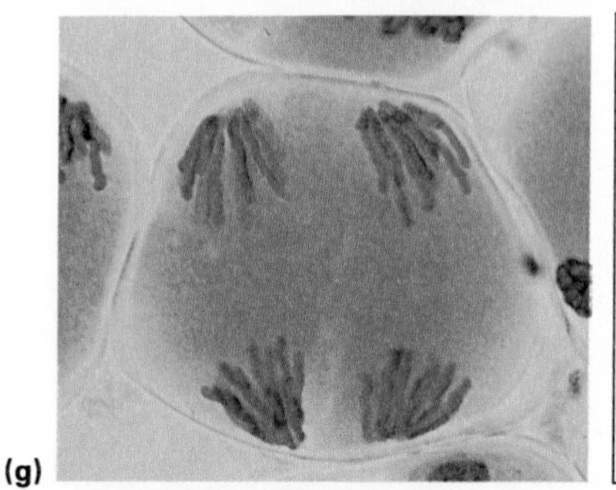

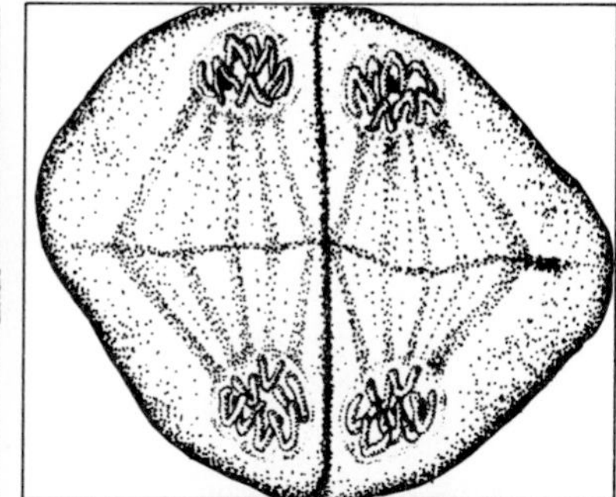

(h) Telophase II
New nuclei and new cell walls are beginning to form, resulting in four haploid cells. (Photos: Carolina Biological Supply Co.) Keep in mind that cytokinesis is a bit different in animals because no cell wall forms; instead, the cells pinch inward, finally closing together, and two new cells are formed.

(h)

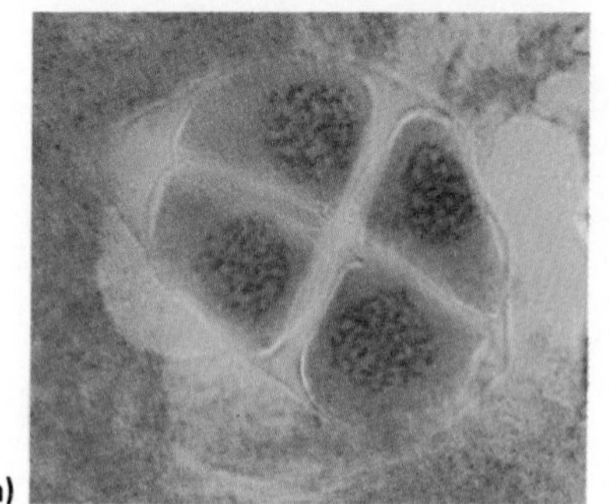

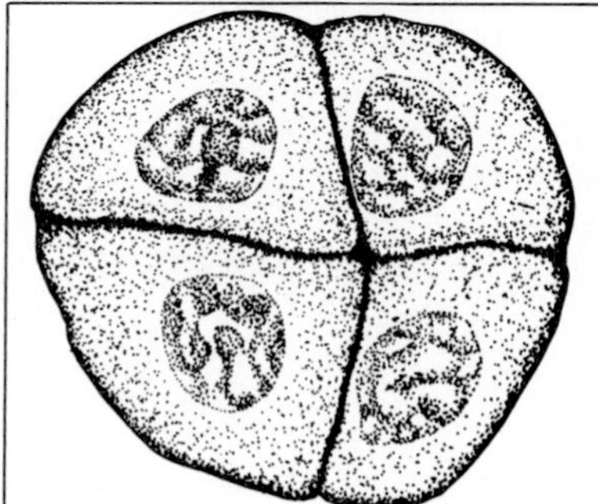

FIGURE 6.8

By way of review and comparison, the outline of mitosis and meiosis in animal cells. For convenience, we have here an animal with *two* pairs of chromosomes. Also, to make things conceptually easier, paternal and maternal chromosomes are identified by their darkness. It doesn't matter which you want to call paternal or maternal, but it is important to notice the relative number of chromosomes from each parent that makes it into the daughter cells. Notice in particular the different behavior of the chromosomes in metaphase and anaphase of the two processes. Also keep in mind that meiosis is essentially two divisions. Finally, compare the chromosome number in mitotic telophase with that of telophase in meiosis II. Note that chromosomes are most easily counted by counting centromeres. Here, also, the behavior of the centrioles is shown.

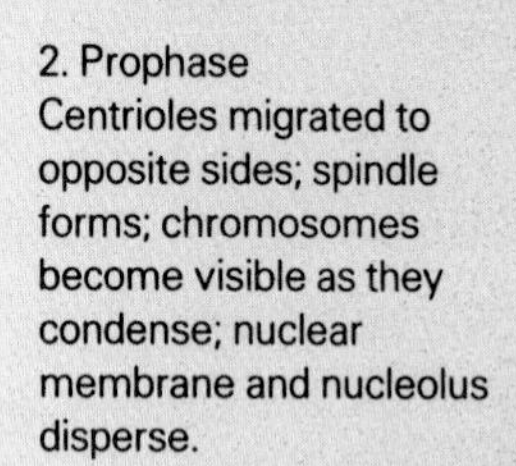

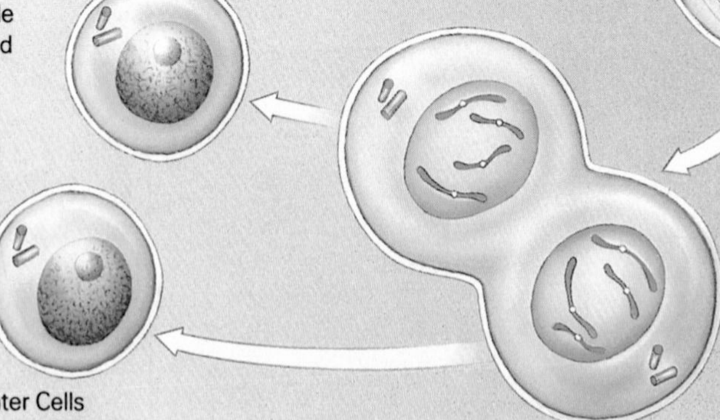

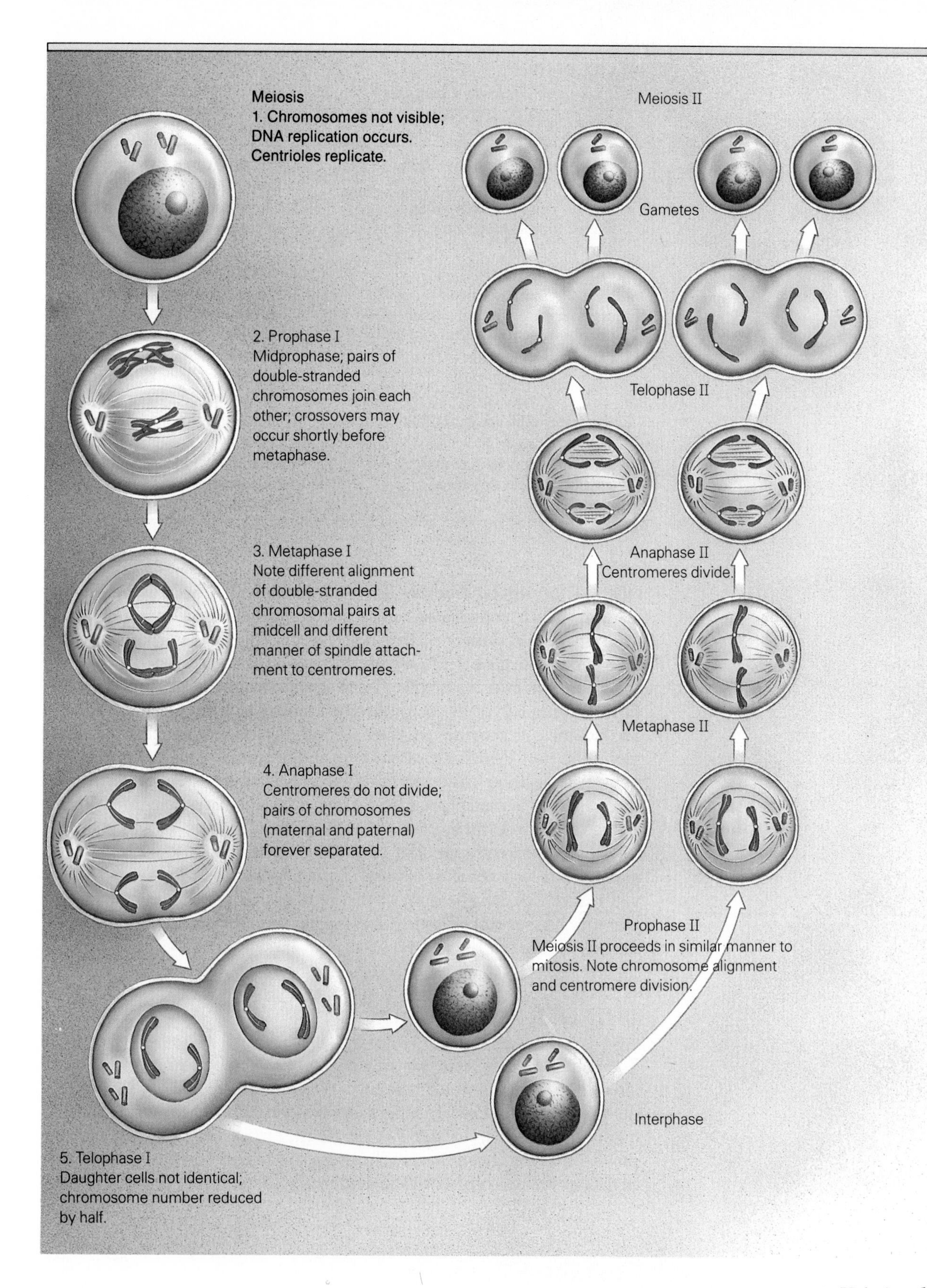
Meiosis
1. Chromosomes not visible; DNA replication occurs. Centrioles replicate.
2. Prophase I
Midprophase; pairs of double-stranded chromosomes join each other; crossovers may occur shortly before metaphase.
3. Metaphase I
Note different alignment of double-stranded chromosomal pairs at midcell and different manner of spindle attachment to centromeres.
4. Anaphase I
Centromeres do not divide; pairs of chromosomes (maternal and paternal) forever separated.
5. Telophase I
Daughter cells not identical; chromosome number reduced by half.
Interphase
Prophase II
Meiosis II proceeds in similar manner to mitosis. Note chromosome alignment and centromere division.
Metaphase II
Anaphase II
Centromeres divide.
Telophase II
Gametes
Meiosis II

TABLE 6.2
Mitosis and Meiosis Compared

MITOSIS	MEIOSIS
1. Cell undergoes one division.	1. Cell undergoes two divisions: nuclear and cytoplasmic.
2. Has no genetic exchange between homologous chromosomes.	2. Chromosomes exchange parts in crossovers.
3. Produces two cells.	3. Produces four cells.
4. Keeps amount of chromosomal material constant and equally distributes it to daughter cells.	4. Halves chromosomal material before distributing to daughter cells.
5. Occurs in most cells.	5. Occurs only in the formation of eggs and sperm (or spore formation).
6. Daughter cells are genetically similar.	6. Daughter cells are genetically dissimilar.
7. Mitotic products may undergo further mitotic divisions.	7. Meiotic products cannot participate in further meiotic divisions.

cells will form **polar bodies**—tiny, nonfunctional cells. Most of the cytoplasm, which contains materials that could nourish a young embryo, will remain in the one cell that becomes the egg. It is not known what determines which of the four cells this will be.

There are two major take-home lessons from the story of meiosis. First, meiosis halves the chromosome number to form eggs and sperm. Second, meiosis provides a means of shuffling and reorganizing the chromosomes, thus increasing variation in the offspring. This reorganization takes place in three major ways: (1) in the apparently random lineup of chromosomes at metaphase I, so that paternal and maternal chromosomes are mixed by the time a cell enters meiosis II; (2) in crossovers; and (3) in the apparently random selection of polar bodies during egg formation. You may come to have an abiding appreciation for meiosis when you realize that it is because of this process that you don't look exactly like your little brother. Table 6.2 compares several features of mitosis and meiosis. (Also, see Essay 6.1.)

THE DOUBLE HELIX

The search for the structure of hereditary material is one of the most interesting stories in modern science. By about the middle of this century, the role of chromosomes in inheritance was well established. But a growing question was, What are they? What are they made of? Chemical analysis showed that the chromosomes themselves were composed of protein and an acid called **deoxyribonucleic acid,** or **DNA** (a nucleic acid containing

ESSAY 6.1

When Meiosis Goes Wrong

Meiosis is an extremely complex process, so it should not be surprising that frequently something goes wrong. In fact, about one-third of all fetuses spontaneously abort in the first two or three months of pregnancy, and it turns out that most of them have the wrong number of chromosomes. Failure of the chromosomes to separate correctly at meiosis is termed *nondisjunction*.

Not all failures of meiosis result in early miscarriage. There are late miscarriages and stillbirths of severely malformed fetuses. About one liveborn human baby in 200 has the wrong number of chromosomes, often accompanied by severe physical and/or mental abnormalities.

About one baby in 600 has three copies of tiny chromosome 21. The syndrome is known both as trisomy-21 and Down's syndrome, after the nineteenth century physician who first described it. Characteristics of the syndrome are general pudginess, rounded features, a rounded mouth, an enlarged tongue which often protrudes, and various internal disorders. Often there is a peculiar fold in the eyelid. In the past this was erroneously equated with the characteristic eyefold of Asians (thus the earlier name "mongoloid"). Trisomy-21 individuals also have a characteristic barklike voice and unusually happy, friendly dispositions.

Trisomy-21 occurs most frequently among babies born to women over 35 years old. At that age, the incidence is about two per 1000 births. By age 40, this climbs to about six, and by age 45, 16 children with Down's syndrome are born for each 1000 births. The age of the father apparently has little, if any, effect. It has been suggested that since all the eggs in the ovary, thousands of them, remain frozen in prophase I until receiving the hormonal signal to begin development, this permits the chromosomes to perhaps interact abnormally over time, setting the stage for nondisjunction. We can guess that the much-prolonged prophase I of the human oocyte might have something to do with this.

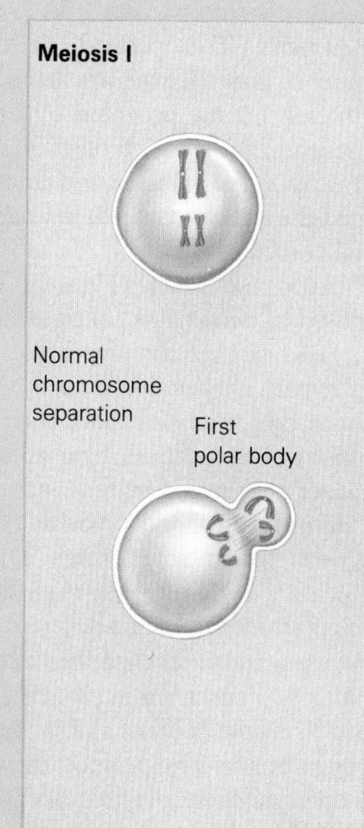

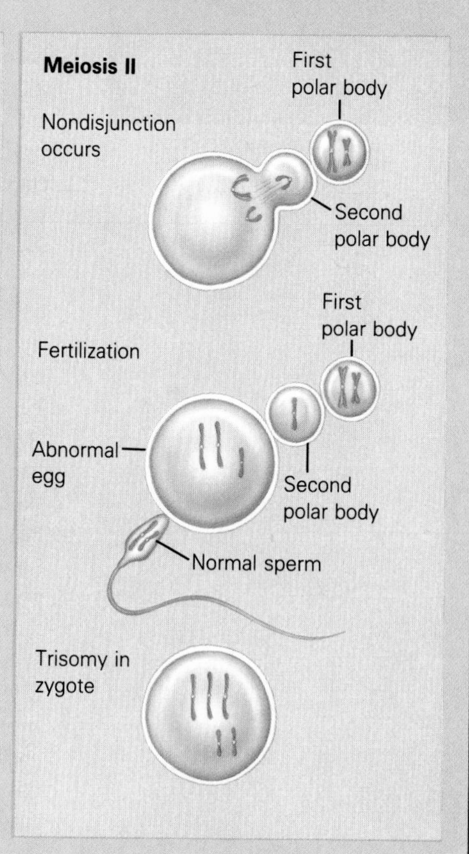

FIGURE 6.9
James Watson (right) and Francis Crick, who, with a little help from their friends, won the race with the great Linus Pauling to discover the structure of DNA.

the sugar deoxyribose). By 1950 it was generally accepted that the hereditary material itself was the DNA component of the chromosomes. But how was the DNA arranged in the chromosome? What did it look like? How did it duplicate itself before mitosis? And, in particular, how could it direct protein synthesis? By the middle of the century, several intensive research efforts were well under way in the United States and Britain. The prevailing notion was that there were three chains in each DNA molecule. Linus Pauling, the great American chemist, and Rosalind Franklin of England both favored this idea, but they disagreed over such details as whether the nitrogen bases stuck outward from the molecule or inward.

Then a brilliant and confident young American postdoctoral fellow ran across an ebullient and talkative English graduate student in biophysics at Cambridge University. The results of what transpired were later described by the American, James Watson, in his controversial best-seller, *The Double Helix*.

The partnership, it turned out, was a most fortunate one. The Englishman, Francis Crick, had noted that when the DNA portion of chromosomes was photographed by x-ray diffraction techniques, a peculiar image appeared—one which could only be produced by molecules arranged in a helical, or spiral, pattern. Watson's biological intuition told him that only two strands were involved. (After all, reproduction usually involved only two parents, and both chromosomes and genes generally come in pairs.) They pressed relentlessly at the question and constructed a number of unusual experimental devices. On the basis of x-ray diffraction information mainly provided by colleague Rosalind Franklin, they even built several DNA models of sticks and metal in their laboratories. Finally, they hit upon the solution and immediately published their findings (and scooped the scientific world) in a simple one-page article in the prestigious journal *Nature*.

Watson and Crick (Figure 6.9) announced in 1953 that DNA is composed of two chains coiled around each other in the form of a double helix. (You can get the idea if you visualize a ladder twisted so that the rungs remain perpendicular to the sides.)

According to their model, the two longitudinal sides of the ladder are composed of alternating sugar and phosphate molecules. The "rungs" of the ladder are formed by four nitrogenous bases—**adenine** (A), **guanine** (G), **thymine** (T), and **cytosine** (C). Adenine and guanine are **purines,** bases with two nitrogen rings. Thymine and cytosine are **pyrimidines,** bases with a single nitrogen ring. These bases, with their sugar and phosphate components, are nucleotides (as we saw in Chapter 3).

Watson and Crick described a chain with one base per sugar-phosphate and with two bases joining together to form each rung. The bases extend inward from the ladder's sides, they said, and are held together by weak hydrogen bonds. The purines, they pointed out, could not lie adjacent to each other because they are so large they would overlap. They simply couldn't fit, and the two pyrimidines, with their single nitrogen-containing rings, were too short to reach across. The arrangement could work only if a pyrimidine paired with a purine to form the rungs (T to A and C to G) (Figure 6.10). Furthermore, according to Watson and Crick's deductions, the nucleotides along one of the strands of the double helix could follow any sequence. For example, one might "read" along a strand and

FIGURE 6.10

In DNA, the backbones are composed of sugar and phosphate. Replication proceeds as the two strands of DNA unwind and then "unzip." Their bases are exposed to the nuclear fluid that contains free nucleotides of thymine (T), adenine (A), cytosine (C), and guanine (G). The exposed bases on the chains, then, join with their complementary free-floating nucleotides. Thus, each strand manufactures a strand that is identical to the one from which it had separated, and the familiar double helix is reestablished.

In (b), geometric figures represent each base and the specificity of base pairing becomes obvious.

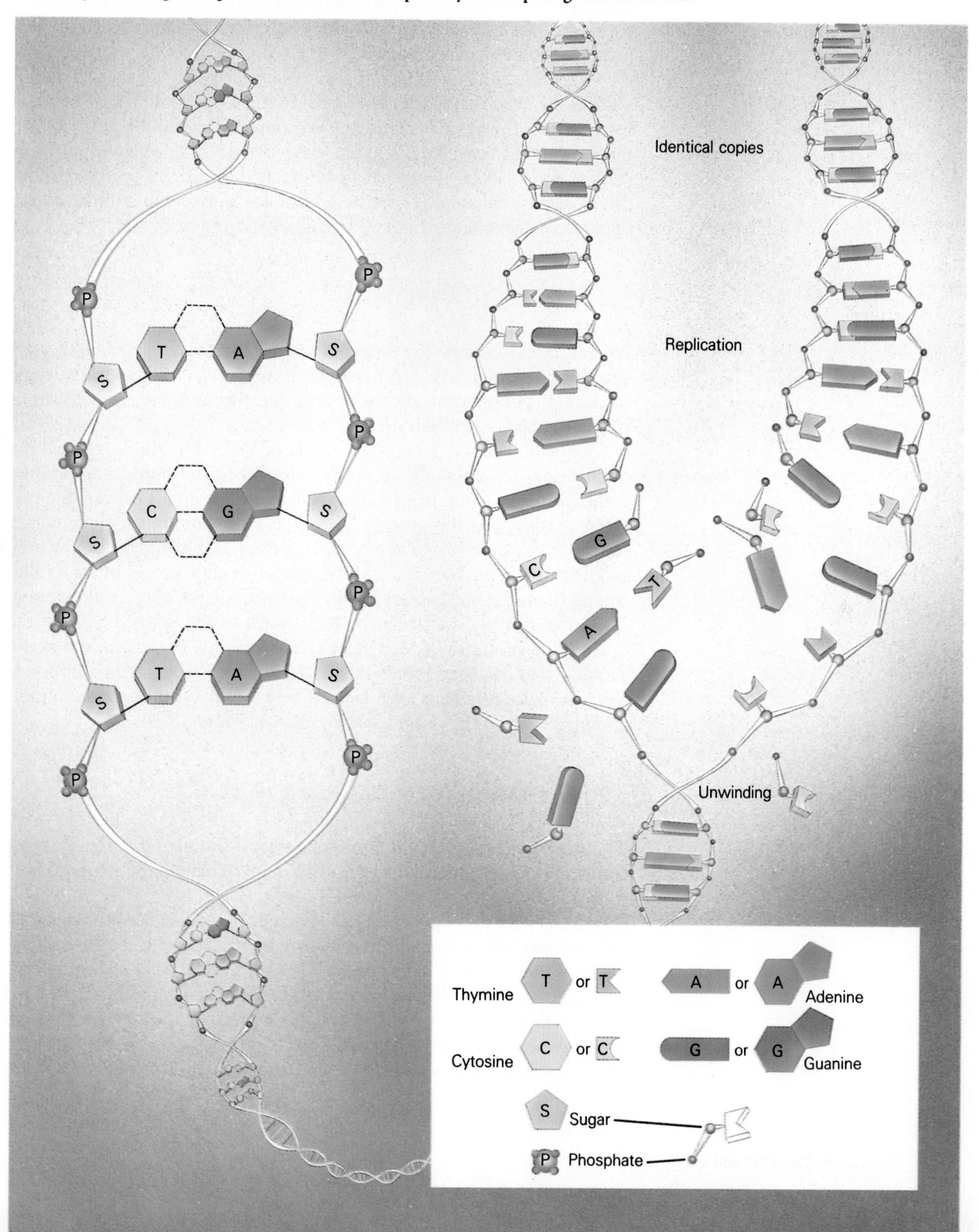

find ATTCGTAACGCGT in one segment and something quite different along another segment of the same strand. Also, the strands were found to be very long, so that the necessary complexity for life could be provided by the possible variations in the sequence and the great numbers of nucleotides in any chain. In fact, if the DNA from a single cell were laid out in a straight line, it would be about five feet long! Furthermore, there are about 10 billion nucleotide parts in the 46 chromosomes of a human cell, so maybe you can (and then maybe you can't) imagine the number of possible variations in the sequences.

According to this model, if the sequence of one strand of DNA is known, the sequence of the other strand can be deduced. This is because, for example, the purine adenine (A) of one strand will always pair with the pyrimidine thymine (T) of the other strand, and wherever cytosine (C) exists, it will be paired with guanine (G) on the other strand. With this in mind, what would be the nucleotide sequence of the complementary segment of the strand described in the preceding paragraph?

DNA REPLICATION

As you recall from our discussion of mitosis, before the cell divides, each chromosome makes an exact copy of itself (replicates) so each chromosome consists of two chromatids. How does the chromosome, with its intertwining strands, go about it?

Actually, Watson and Crick's model of DNA structure suggested the mechanism—conceptually, a simple one. When the time comes for the chromosome to replicate itself, the weak hydrogen bonds holding the two DNA strands together break. Then the two strands simply unwind, "unzip," and open up. The strands' purines and pyrimidines are now exposed to the nuclear sap, which is loaded with a variety of molecules, among them the nucleotides of adenine, guanine, cytosine, and thymine. These free-floating molecules can then attach to the exposed bases of the open DNA strand (T to A, C to G). The original strand and the newly forming one can then bind together by forming weak hydrogen bonds along their length. Thus, step by step, new DNA is produced, each molecule identical to the original (refer again to Figure 6.10).

HOW CHROMOSOMES WORK

The extreme precision of the mitotic process suggests that the chromosomes are rather important to the cell's development and function. How, then, do they carry out their critical roles?

It will help to keep two things in mind. First, cells become *specialized* into muscle cells or nerve cells or whatever because of differences in the proteins they produce. Some of these proteins are directly involved in the special job of the cell, for instance, the one that carries oxygen in red blood cells or those that allow a muscle cell to contract. Other proteins are enzymes. Enzymes, you recall, are biocatalysts, and their role is to facilitate certain, very specific biochemical events. It is through differences in the summation of such events that cells become specialized. The second thing to remember is that protein production is the result of chromosomal action.

So, what we are really asking here is, How do chromosomes direct the production of proteins such as enzymes? The answer can be summarized simply: DNA makes RNA. RNA makes protein. **RNA,** or **ribonucleic acid,** differs from DNA in that the sugar component of its nucleotides, ribose, has one more oxygen atom than the sugar in DNA, and in place of thymine there is uracil. Also, most RNA exists as a single strand, not as a double helix. In a process called **transcription,** the sequence of bases in a portion of DNA is used as a template to form an RNA molecule with a complementary sequence of bases. Next, the language of RNA (its sequence of bases) is translated into the language of proteins (the sequence of amino acids). The process is called **translation.** Keep in mind the flow of information depicted in the following diagram as we fill in some of the details.

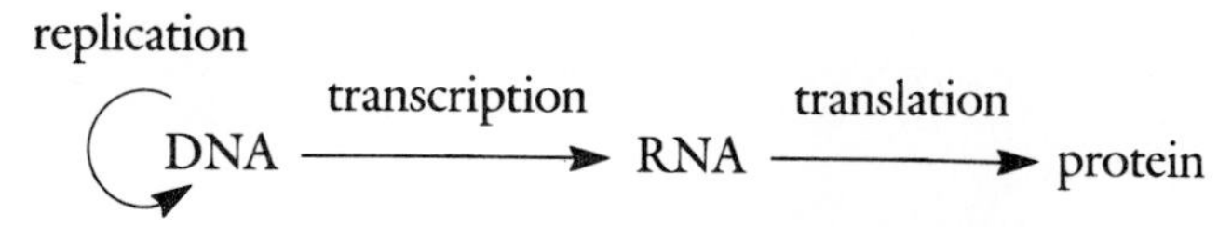

Transcription

When a chromosome begins to direct the formation of proteins, it must first unwind, as in the replication of DNA. Now, however, only certain sections along the chromosome unwind (Figure 6.11). In order for the unwinding to occur, the hydrogen bonds between purines and pyrimidines must be broken, exposing these nucleotide bases to the nuclear fluid. As a result, free nucleotides begin to attach to the exposed bases along the DNA strands. As before, cytosine attaches to exposed guanine, guanine to exposed cytosine, and adenine to the exposed thymine. This time, however, where adenine lies exposed along a DNA molecule, it is attached, not to thymine, but to **uracil,** another nitrogenous base in the cytoplasm. The nucleotides attach to each other by the usual sugar-phosphate connections, and the result is the formation of long strands, similar to the original DNA. Here, though, the DNA strands are not building complementary strands of DNA, as they do when they replicate themselves in mitosis. They are building RNA.

There are three types of RNA (Table 6.3). **Messenger RNA (*m*RNA)** carries the information from DNA, its message for producing a protein, to the cytoplasm where the protein will be manufactured. **Ribosomal RNA (*r*RNA)** together with certain proteins form ribosomes, the workbenches on which proteins are assembled. **Transfer RNA (*t*RNA)** transfers amino acids, the building blocks of proteins, from the cytoplasm to the ribosome and positions it properly in the growing protein chain.

We return to our story as the DNA unwinds at various sections along its length and directs the formation of *m*RNA. The *m*RNA then separates from the DNA strand and moves from the nucleus into the cytoplasm, each segment carrying a "message" encoded in its nucleotide sequence. Here it directs the formation of proteins.

The *m*RNA is a rather long and, as we will see, somewhat unstable molecule that contains guanine, cytosine, adenine, and uracil in sequences directed by the DNA. The sequencing of its nucleotide bases is very precise and very critical because the sequence of bases in RNA forms what is called the genetic code.

FIGURE 6.11

DNA manufactures RNA, including *m*RNA in much the same way as it duplicates itself. The DNA, *then, determines* the sequence of bases along the *m*RNA. However, as *m*RNA is made, thymine is replaced by uracil. As long *m*RNA molecules break away from their DNA template, they move into the cytoplasm to direct protein formation. Obviously, entire DNA strands do not unwind when forming *m*RNA. If they did, since each cell contains the same chromosome complement, they would all make the same enzymes and no specialization could occur. One of the enduring questions in biology concerns how the manufacture of *m*RNA is directed. How does one part of a chromosome come to actively make *m*RNA while other parts lie dormant, their DNA strands tightly interwound? And what causes long dormant segments to suddenly become active?

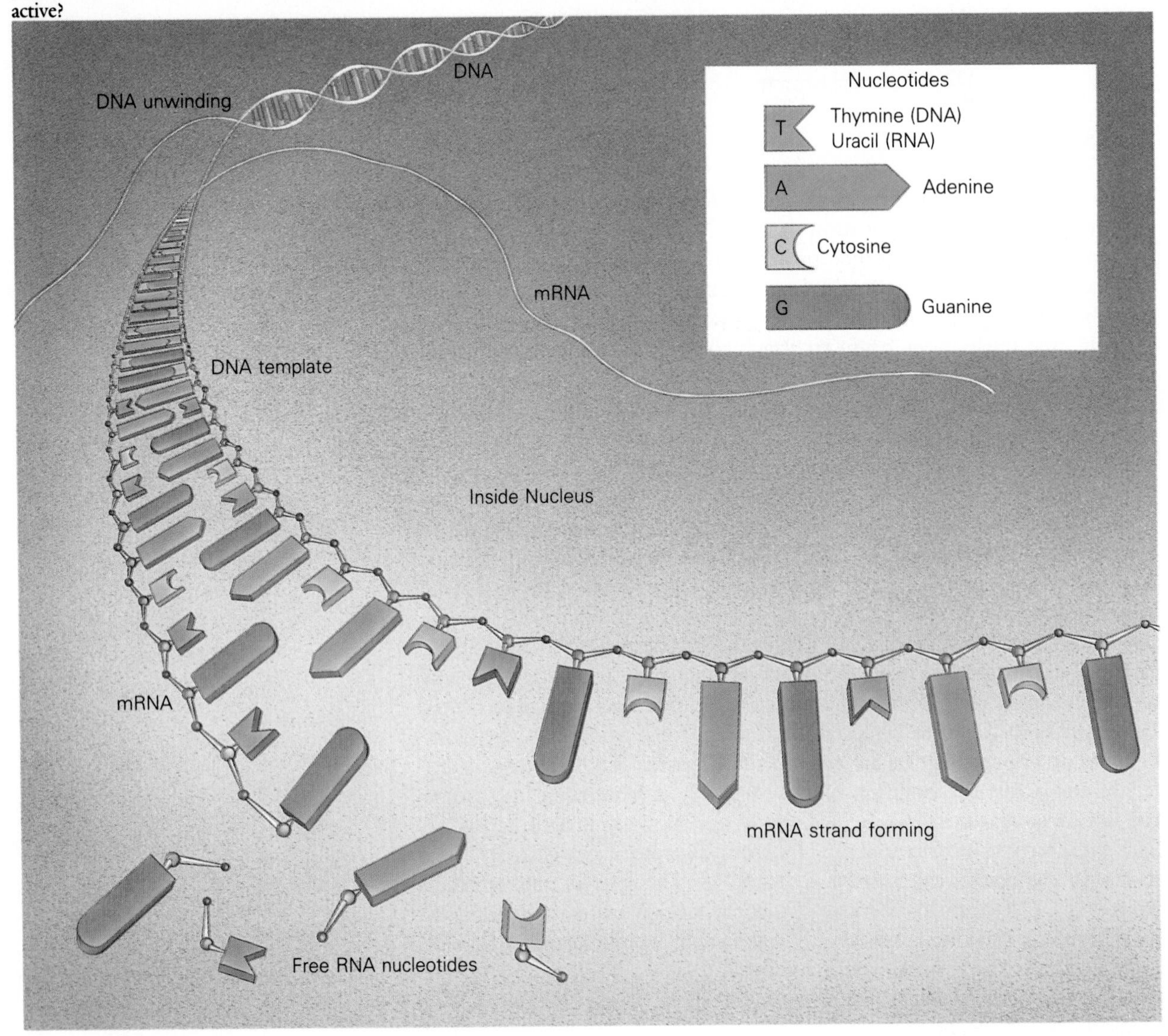

TABLE 6.3
Nucleic Acids

ACID	STRUCTURE	FUNCTION
DNA (deoxyribonucleic acid)	Double twisted strand; deoxyribose (sugar), phosphate, thymine, cytosine, adenine, and guanine (bases).	Holds genetic information in its base sequence.
RNA (ribonucleic acid)	Ribose (sugar), phosphate, uracil, cytosine, adenine, and guanine (bases).	
Messenger RNA (*m*RNA)		Carries DNA's message in its sequence of bases (codon) to cytoplasm.
Transfer RNA (*t*RNA)		Bears anticodon; adaptor molecule that brings the correct amino acid to the *m*RNA molecule.
Ribosomal RNA (*r*RNA)		Combines with protein to form ribosomes.

The Genetic Code

Before we see what *m*RNA does in the cytoplasm, let's learn something about how its message is carried—the mysterious **genetic code.** The genetic code is the "language of genes" that is used to convert the linear sequence of DNA into the sequence of amino acids in proteins. It may, at first, seem strange that only four nucleotide bases are sufficient to code for the thousands of proteins that make up the highly complex structures of living organisms. How, then, is the complexity achieved with only a four-unit code (or "alphabet")?

Let's go through it stepwise. First, you recall that although proteins are very large and complex, they are made up of amino acids—of which there are only 20 kinds. The question, then, is how can four "letters" specify for 20 amino acids? The problem can be reduced to simple mathematics. If each nucleotide specified one amino acid, only four amino acids could be coded for. A two-base combination for each amino acid would provide four times as many possible combinations ($4 \times 4 = 16$), which is still not enough.

So, early in the search for the genetic code it was suggested that a three-base combination was necessary to code for each amino acid. At first this was just a guess; the verifying data came later. It was reasoned, however, that with four nucleotide bases, a combination of any three would provide 64 possible combinations ($4 \times 4 \times 4 = 64$). But there are only 20 amino acids. So, according to this scheme, more than one nucleotide triplet can code for the same amino acid (Table 6.4).

A nucleotide triplet on *m*RNA that codes for the insertion of one amino acid in a protein is called a **codon.** Of the 64 codons, it is known that 61 code for specific amino acids. What about the other three, specifically, UAA, UAG, and UGA? These particular codons, interestingly enough, serve as periods in the language of life; they signal the end of a message. In addition, the codon AUG apparently doubles as an initiator, indicating the start of a message.

TABLE 6.4
The Genetic Code, or the "Language of Life"***

FIRST LETTER	SECOND LETTER: U		C		A		G		THIRD LETTER
U	UUU	PHE	UCU	SER	UAU	TYR	UGU	CYS	U
	UUC		UCC		UAC		UGC		C
	UUA	LEU	UCA		UAA	stop	UGA	stop	A
	UUG		UCG		UAG	stop	UGG	TRP	G
C	CUU	LEU	CCU	PRO	CAU	HIS	CGU	ARG	U
	CUC		CCC		CAC		CGC		C
	CUA		CCA		CAA	GLU	CGA		A
	CUG		CCG		CAG		CGG		G
A	AUU	ILE	ACU	THR	AAU	ASN	AGU	SER	U
	AUC		ACC		AAC		AGC		C
	AUA	MET	ACA		AAA	LYS	AGA	ARG	A
	AUG (start)		ACG		AAG		AGG		G
G	GUU	VAL	GCU	ALA	GAU	ASP	GGU	GLY	U
	GUC		GCC		GAC		GGC		C
	GUA		GCA		GAA	GLU	GGA		A
	GUG		GCG		GAG		GGG		G
		Phenylalanine Leucine Leucine Isoleucine Methionine Valine		Serine Proline Threonine Alanine		Tyrosine Histidine Glutamine Asparagine Lysine Aspartic Acid Glutamine		Cysteine Tryptophan Arginine Serine Arginine Glycine	

*There are 64 triplet combinations. Each triplet is shown with the amino acid or the signal for which it codes.

Translation

When the newly formed *m*RNA leaves the nucleus and enters the cytoplasm, it diffuses through the cell fluid until it encounters one of the cell's many ribosomes. It joins temporarily with a ribosome to form a ribosome-*m*RNA complex. Amino acids are brought to the complex by *t*RNA.

There are at least as many kinds of *t*RNA molecules as there are amino acids, and each kind of *t*RNA molecule can attach to only one type of amino acid. The specificity of the *t*RNA molecule is determined by the three nucleotides (forming an **anticodon**) that are found at one end of the molecule. For example, if these three anticodon bases all happen to be adenine, an AAA sequence, then the *only* amino acid that the *t*RNA molecule can attach to is one called *phenylalanine*. Now, when a *t*RNA with this nucleotide arrangement with its phenylalanine encounters the *m*RNA-ribosome complex, the phenylalanine will be inserted only at a place along the *m*RNA where there are three complementary uracils (UUU). So both the *kind* of amino acid and its *position* in the developing protein are dictated by the DNA back in the nucleus.

FIGURE 6.12

Protein synthesis. Messenger RNA is assembled along strands of DNA. The rather large *m*RNA molecules then move through the nuclear membrane into the cytoplasm. In the cytoplasm, they are joined by ribosomes that "read" the base sequence of the molecule. A sequence of three bases (a triplet) codes for a particular amino acid. As a triplet is read, a smaller RNA molecule, transfer RNA, or *t*RNA, moves to the site with an amino acid in tow.

The *t*RNA has an anticodon sequence that fits the triplet's codon. When the appropriate *t*RNA locks onto the codon, it releases its amino acid to a developing chain of amino acids and then unlocks and moves away. Any *t*RNA molecules can carry only one kind of amino acid. After the *m*RNA has directed a number of amino acid sequences, it breaks apart. Long sequences of amino acids form proteins, and some of these are enzymes. Enzymes determine the nature and behavior of the cell.

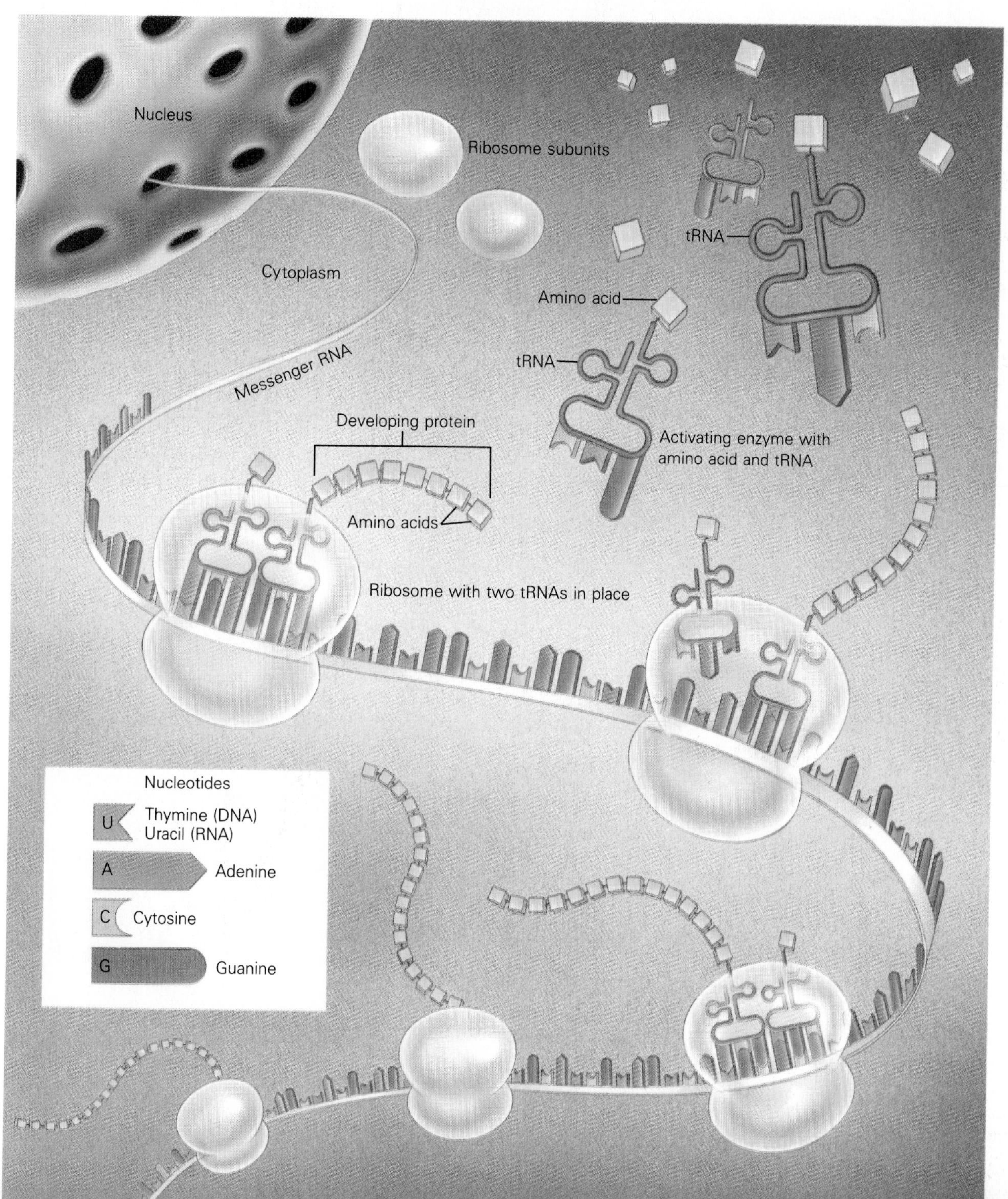

ESSAY 6.2

Aging

It somehow seems highly unlikely to the young, but under the best of circumstances, they will grow old—old and ugly, some say. The changes will be almost imperceptible, normally just a slow accumulation of small changes, a wrinkle here, a backache there. Perhaps the best barometer of the change is meeting an old friend after several years. The compliments on your appearance will be effusive and directly proportional to your state of deterioration. What brings about these changes? Why do our bodies let us down? What's happening anyway?

Several things are happening. Primarily, though, cells are becoming less efficient. They haltingly carry out their tasks of removing wastes, destroying poisons, repairing genes, and manufacturing proteins. As cells weaken, so do the bodies they serve.

There are a number of factors that may contribute to the problems of aging cells. There is an accumulation of free radicals (atoms with only a single electron in their outer shells). They may satisfy their shell requirements by taking electrons that were in the process of meeting some cellular need, such as generating ATP. Certain pigments may also accumulate and pose a threat. Lipofusion, a brown pigment produced by the metabolism of fat, may choke cells and impair their delicate functions. Even glucose can be a problem. For example, in time, glucose can damage proteins, including collagen, one of the primary supportive materials of the body. Glucose can even clog the DNA helix and halt the orderly formation of proteins.

The causes of such changes remain a mystery, but there are two major theoretical approaches to the problem: (1) the time bomb theory and (2) the wear-and-tear theory. The time bomb theory is based on the idea that we are genetically programmed to fade out. The mechanisms include a genetic clock that causes healthy cell lines to die out after about 50 doublings (cancer cells do not). A hormone clock has also been suggested. It could operate, for example, if the pituitary gland began to secrete a "killer hormone" after puberty. (Aging has been retarded in some animals by removing

the pituitary and dosing them with hormones artificially.) There may also be a group of clocks, all running at different rates, but interdependent, so that when one slows (perhaps due to some environmental change or signal) the others slow as well. (*Environmental,* in this sense, implies anything that is not internally based.)

The wear-and-tear theory suggests a genetic mechanism in which random mutations can be expected eventually to alter a few important genes that control an array of other genes or that manufacture certain vital proteins. The mutation rate, of course, can be increased by environmental abuses. The level of free radicals can also build up due to environmental abuses (and can also be reduced environmentally, such as by ingesting vitamins E and A). Finally, cells may give way to the continual assault from a number of sources, such as the constant (and increasing) barrage of radiation and chemicals that can cause an accumulation of small mutations that slowly destroy us.

The rate at which we age, no matter what the causes, seems to be largely genetic. Some people remain vigorous and alert into their eighties, while others show the effects of time much earlier and walk a degenerative and painful path to death. It has been suggested that the best way to live long is to choose an elderly set of grandparents.

The mechanisms of aging remain a mystery, but we are keenly aware that humans change over time in that inexorable march to their fate. Some traits, however, do not change. Can you see here some that do, and some that don't?

Review Figure 6.12 to see the process by which a new protein is synthesized. The ribosome attaches to one end of the *m*RNA strand and moves along its length, stopping to read off each nucleotide triplet of the *t*RNAs, which, with their amino acids, have diffused to the ribosome-*m*RNA complex. When the ribosome pauses at an *m*RNA triplet, it waits for a *t*RNA molecule with the proper complementary sequence to drift to that position. The *t*RNA briefly plugs into the *m*RNA codon and then moves away, leaving its amino acid behind. Meanwhile, the ribosome has moved along to the next codon site, where another *t*RNA falls into place, leaving its amino acid attached to the previous one. Thus, amino acids are joined one by one in the proper sequence for the formation of a specific protein. As soon as the *m*RNA is read and has directed a sequence of amino acids, it begins to break down.

Chains of amino acids formed in this way eventually form polypeptides or larger molecules of proteins. The resulting protein may be structural material or it may be an enzyme, depending on the *m*RNA sequence that directed its formation. An enzyme formed in this way will participate in the biochemical activities of that cell and, hence, help to determine the cell's very nature (that is, answer the questions about what it looks like and what it does—perhaps the same things one might wonder about a blind date). We know that the chromosomes are identical in every cell, but how can the same DNA component result in cells of entirely different types?

We know that the fate of cells is partially directed by their enzymes. We also know that enzyme production is directed by the DNA of the chromosomes as it determines the sequence of nucleotides along *m*RNA strands, which then move into the cytoplasm where they direct protein formation. Since there are many kinds of proteins within any cell, we can assume that many kinds of *m*RNA molecules may be formed along a single DNA chain. This happens because the *m*RNA molecules are formed along only short sections of the long DNA molecule. Obviously, it wouldn't do for a DNA molecule to continually transcribe *m*RNA along its entire length since this would result in only one kind of *m*RNA, and every protein would be the same. So we see that parts of chromosomes are shut down while other parts are unwound and busily forming *m*RNA.

Cells, then, take on their specific characters as DNA directs the synthesis of different enzymes. This explanation of cell differences, as we will see, simply places the question of control one step further back. We are now faced with the question of what determines which section of a chromosome will be active. At this point, the answers become rather varied and complex, but one line of evidence suggests that the cell's location may be critical. The idea is that the immediate environment of a cell may have an effect on its chromosomes.

CONTINUING PROBLEMS IN CELL BIOLOGY

Complex flowcharts, stroked chins, and remarkable electron microscope photographs of cell interiors may lead some people to believe that the "great questions" have been answered, and all we need to do now is to tie up a few loose ends. Nothing could be further from the truth. In fact, we have barely touched the hem of the garment (an old Arkansas phrase) in many areas of cell biology (see Essay 6.2).

In spite of what may have seemed great assurance in our foregoing description, we actually know very little about such matters as mitochondrial replication, spindle formation, cell-environment interaction, cell specialization, cell growth, chromosome structure, or the permeability of membranes. In fact, we know precious few of the simplest enzymatic pathways. It seems that the intense research efforts in this area are currently overmatched by some very fundamental questions that, so far, have simply refused to yield answers. But new minds, vigorous and well trained, are continuing to focus on the questions—often from odd angles—even as new technologies are developed. We can, therefore, expect a continuing stream of revelations as these people go about their work, confident that few secrets are held forever.

SUMMARY

1. Cells reproduce mitotically for maintenance and growth. In mitosis, a parent cell divides, forming two daughter cells that are each about the same size and have about the same number of organelles as the parent cell, and have a full chromosome complement identical to that of the parent cell.
2. During interphase, the stage when the cell is not dividing, most cellular components, including the chromosomes, make copies of themselves in preparation for division. The chromosomes unwind into fine strands (chromatin).
3. Mitosis has four stages: prophase, metaphase, anaphase, and telophase. During prophase, the nuclear membrane disappears, the spindle forms, and the chromatin condenses, making the chromosomes visible. Each chromosome is composed of two chromatids connected by a centromere. In metaphase, the chromosomes line up in the center of the spindle. During anaphase, the centromeres split and the chromatids (now called chromosomes) separate and move along the spindle toward opposite ends of the cell. During telophase, the chromosomes unwind and form strands, and the nuclear membrane re-forms. Then cytokinesis (division of the cytoplasm) occurs. Animal cells pinch in, forming a membrane between the groups of chromosomes. In plant cells, a new cell wall forms between the groups of chromosomes.
4. Most cells are diploid. That is, their chromosomes exist in homologous pairs. Cells that have one copy of each homologous pair of chromosomes are haploid. Meiosis is the process by which certain cells halve their chromosome number on the way to becoming gametes (eggs, sperm, or spores). The gametes contain one member of each homologous pair, whereas body cells typically contain matched pairs of chromosomes, called homologues. In animals, gametes are produced in the gonads (ovaries or testes). The typical chromosome number is restored when the egg and sperm fuse forming a cell that will develop into a new individual.

5. During the interphase preceding meiosis, the chromosomes replicate and, therefore, consist of two chromatids connected by a centromere. Meiosis involves two cell divisions. As in mitotic prophase, during prophase I of meiosis, the chromatin forms chromosomes, the nuclear membrane disappears, and the spindle begins to form. Unlike mitotic prophase during prophase I of meiosis, the homologous pairs of chromosomes, each chromosome consisting of two chromatids, line up next to one another. DNA segments may be exchanged between the homologues in a process called crossing over. This produces genetic variation. In metaphase I, unlike mitotic metaphase, the chromosomes line up on the spindle in homologous pairs. In anaphase, the members of homologous pairs separate and move toward opposite ends of the cell. During telophase I, the chromosomes elongate and the nuclear membrane re-forms. During interphase II, the chromosomes do not replicate. The events in meiosis II are similar to those in mitosis. At the end of meiosis II, four cells, each with half the chromosomal complement, have been created. In the testicle, meiosis produces four sperm. In the ovary, meiosis produces one egg and three nonfunctional polar bodies.
6. Meiosis has two important consequences: it creates gametes containing half the chromosome number and it increases genetic variation.
7. A chromosome is composed of protein and deoxyribonucleic acid (DNA), a nucleic acid containing the sugar deoxyribose. DNA is composed of a chain of nucleotide pairs in a ladderlike arrangement. The sugar and phosphate of the nucleotides form the sides of the ladder and the bases form the rungs. Each rung contains one purine (adenine or guanine) and one pyrimidine (thymine or cytosine). The sequence of bases along each strand is variable, but adenine is always paired with thymine and guanine is always paired with cytosine. The entire ladder is twisted in the form of a double helix.
8. When DNA replicates, the two strands of the DNA molecule unwind and open up so that the bases are exposed to the nuclear sap. Bases of nucleotides in the nuclear sap attach to the appropriate bases of the open strand (A to T and C to G), producing two new molecules exactly like the original.
9. Cells become specialized by differences in their proteins. To make a protein such as an enzyme, first RNA (ribonucleic acid) is formed as directed by a section of DNA (transcription). The DNA unwinds and comes apart between the bases as in replication. RNA differs from DNA because its sugar has one more oxygen and uracil replaces thymine. Messenger RNA (*m*RNA) separates from the DNA and carries the message of DNA in its sequence of bases to the cytoplasm.
10. The genetic code is the language of genes that is used to convert the linear sequence of DNA into a sequence of amino acids in proteins. A sequence of three nucleotides, called a codon, codes for an amino acid. Sixty-one of the possible 64 codons code for specific amino acids. One (AUG) signals the start of a message. Three others, (UAA, UAG, and UGA) signal the end of a message.
11. RNA directs the synthesis of protein (translation). In the cytoplasm, *m*RNA forms a complex with a ribosome. (A ribosome consists of ribosomal RNA, *r*RNA, and protein.) Amino acids are brought to the complex by a transfer RNA (*t*RNA). At one end of each type of *t*RNA

molecule is a sequence of three nucleotides called the anticodon that specifies a particular amino acid. The *t*RNA positions an amino acid only where it encounters three complementary bases (a codon) in the *m*RNA molecule. When the amino acid has attached to the growing protein chain, the *t*RNA moves away. The sequence of bases on *m*RNA is read three at a time and directs the sequence of amino acids in a protein that may be structural or an enzyme.

KEY TERMS

adenine (166)
anaphase (153)
anticodon (172)
centromere (153)
chromatid (153)
chromosome (153)
codon (171)
crossing over (159)
cytokinesis (152)
cytosine (166)
daughter cell (152)
deoxyribonucleic acid (DNA) (164)
diploid (158)
gamete (158)
genetic code (171)
gonad (158)
guanine (166)
haploid (158)
homologue (157)
interphase (152)
meiosis (158)
messenger RNA (*m*RNA) (169)
metaphase (153)
mitosis (152)
parent cell (152)
polar body (164)
prophase (153)
purines (166)
pyrimidine (166)
ribonucleic acid (RNA) (169)
ribosomal RNA (*r*RNA) (169)
spindle (153)
telophase (156)
thymine (166)
transfer RNA (*t*RNA) (169)
transcription (169)
translation (169)
uracil (169)

FOR FURTHER THOUGHT

1. A segment in a DNA strand is thymine, guanine, and adenine. What nucleotide bases are contained in its complementary strands of DNA; of RNA?
2. If an organism has 36 chromosomes, how many chromosomes will its daughter cells have after mitosis; after meiosis?

FOR REVIEW

True or false?

1. ___ Cytokinesis is a process whereby cells replicate their chromosomes.
2. ___ Cells halve their chromosome numbers in a process called mitosis.
3. ___ The backbone of the DNA molecule is composed of alternating sugar and phosphate molecules.
4. ___ Codons are triplet combinations of nucleotide bases on *m*RNA.

Short answer

5. List the four stages of mitosis in order of their occurrence after interphase.
6. Matched pairs of meiotic chromosomes are called ___ .
7. List three ways in which meiosis produces variations in offspring.
8. The two strands of the DNA molecule easily separate during replication when their ___ are broken.

Choose the best answer.

9. Which of the following occurs during the telophase stage of mitosis?
   A. Chromosomes are clustered at opposite poles of the cell.
   B. Spindle fibers first appear in the cytoplasm.
   C. Chromosomes line up along the spindle fibers.
   D. The nuclear membrane disappears.
10. Meiosis occurring in a testicle cell produces
    A. four functional cells.
    B. one functional egg cell.
    C. daughter cells with 46 chromosomes.
    D. three polar bodies.
11. Amino acids are carried to the ribosome by
    A. DNA.
    B. *t*RNA.
    C. *m*RNA.
    D. *r*RNA.
12. ___ is a nucleotide base found in RNA but not in DNA.
    A. Adenine
    B. Guanine
    C. Uracil
    D. Thymine

CHAPTER 7

Inheritance: From Mendel to Molecules

Overview

Mendelian Genetics

Classical Genetics — Beyond Mendel

Chromosome Mapping

Mutations

Further Beyond Mendel

Population Genetics

Objectives

After reading this chapter you should be able to:

- List Mendel's three principles and fully describe how his experimental crosses illustrate each.
- Use a Punnett square to predict phenotypic and genotypic ratios of genetic crosses.
- Show how the behavior of chromosomes during meiosis explains Mendel's principles.
- Describe how sex chromosomes determine gender and list some sex-linked traits.
- Describe the chromosome mapping technique.
- List the types of mutations that occur at the gene and chromosomal level.
- Describe the mechanisms that account for deviations in the ratios predicted by Mendel's principles.
- State how the Hardy-Weinberg principle accounts for constant allele frequencies within a population.

Charles Darwin was one of the best geneticists of his time. Charles Darwin was a terrible geneticist. That tells us something about the state of genetics in the Victorian world. But interestingly, one of the very few people in Darwin's time who knew more about genetics than Darwin did was one of the best geneticists who ever lived. In fact, some of his findings would have answered some of Darwin's most vexing questions.

MENDELIAN GENETICS

The answers to many of the questions about heredity that plagued Darwin were published by a contemporary, a monk named Gregor Johann Mendel (1822–1884) who was a member of an Augustinian order in Brunn, Moravia (now Czechoslovakia) (Figure 7.1).

FIGURE 7.1
Gregor Johann Mendel.

How close did Darwin come to learning about Mendel's findings? The journal in which Mendel's work was published was found in Darwin's library. However, whereas the adjacent article is heavily marked with Darwin's scrawl, Mendel's paper was not touched. Apparently, Darwin saw the paper but was unable to grasp its implications. Keep in mind that Mendel was probably the first mathematical biologist and that Darwin was not a good mathematician. While in school, he once wrote that he was "mired firmly in the mud of mathematics" and that there he would remain.

A Kindly Old Abbot Puttering in His Garden

The abbot, Gregor Mendel, was a rather remarkable man in his own right, although he is usually depicted as a kindly old man of the cloth who, while puttering around in his monastery garden, somehow stumbled upon important laws governing genetic transmission. Alas, our penchant for building myths to embellish the memory of notable people has led us astray again. The fact is that early in his life, Mendel began training himself in the sciences, and he became a rather competent naturalist. To support himself during those years, he worked as a substitute high school science teacher. The professors at the school, noting his unusual abilities, suggested that he take the rigorous qualifying examination and become a regular member of the faculty. Mendel took the test and did surprisingly well, but he failed to qualify. The conditions for teaching high school were rigorous indeed.

However, the monastic order to which Mendel belonged was confident of his abilities and, in 1851, sent him to the University of Vienna for two years of concentrated study in science and mathematics. When Mendel returned to the monastery, he decided to put aside his teaching duties for a time in order to begin his plant hybridization studies in earnest. He was noted for his intelligence and vigor, and he applied these qualities to the study of plant breeding. He worked hard, developed new varieties of fruits and vegetables, kept abreast of the latest developments in his field, and became active in community affairs.

In 1865, while Darwin was still puzzling over the enigma of heredity, wondering how traits are actually transmitted, Mendel presented a single paper at a meeting of the Brunn Natural Science Society. His talk was politely applauded. Then the group burst into a lively discussion of the hot topic of that day—the idea of natural selection. Mendel's paper was published in the society's proceedings the following year. However, no one had a clue as to what Mendel was going on about, and the work for which he was to become so famous was met with resounding silence. But why? What had he said? Why was it so hard to understand?

Put in simple terms, Mendel said that (1) when parents differ in some trait, usually the offspring will show only one parent's trait, not a mixture; (2) the offspring of parents differing in some trait will produce eggs or sperm, of which half will bear the trait of one parent, half of the other; (3) the fact that an organism carries the genetic material for one trait does not mean that it carries the genetic material for any other specific trait as well.

At this point, you are probably as mystified as was Mendel's audience, but let us go on and rephrase all this in scientific terms. Perhaps the best way to clarify things is to see exactly how Mendel arrived at his conclusions.

The Principle of Dominance

To begin with, Mendel based his information on a carefully planned series of experiments and, what is more important, on a statistical analysis of the results. The use of mathematics to describe biological phenomena was a new concept. Clearly, Mendel's two years at the university had not been wasted.

The care with which Mendel planned his projects is reflected in his selection of the common garden pea as his experimental subject (Figure 7.2). There were several advantages in this choice. Pea plants were readily available, fairly easy to grow, and Mendel had already developed some 34 pure strains. These strains differed from each other in very obvious ways, so there would be little difficulty in classifying the results of a given

FIGURE 7.2

Pink garden pea flowers. The petals of the garden pea usually ensure self-pollination; that is, the plant fertilizes itself. For his cross-pollination studies, Mendel had to open the young flower and remove the pollen (the plant's "sperm"). The pollen was then transferred to the female part (stigma) of another flower, from which the male, pollen-producing parts have been removed.

experiment. Mendel chose to study seven characteristics, each of which could be expressed in only two different forms.

1. Seed form—round or wrinkled
2. Color of seed contents—yellow or green
3. Color of seed coat—white or gray
4. Color of unripe seed pods—green or yellow
5. Shape of ripe seed pods—inflated or constricted between seeds
6. Length of stem—short (9 to 18 inches), or long (6 to 7 feet)
7. Position of flowers—axial (along the stem), or terminal (at the end of the stem)

Mendel's approach, a novel one at that time, was to cross two true-breeding strains that differed in only one characteristic, such as seed color. Peas ordinarily self-fertilize, so for this cross it was necessary to transfer the pollen by hand. Mendel called this original parent generation P_1 and designated their first-generation offspring the F_1 (first filial) generation. When the F_1 plants were allowed to self-pollinate, so that they crossed with each other at random, the offspring resulting from this cross were called the F_2 generation, and so on.

Now, when Mendel crossed his original P_1 plants, he found that the characteristics of the two parents didn't blend, as prevailing theory said they should. When plants with yellow seeds were crossed with plants that had green seeds, their F_1 offspring did not have yellow-green seeds. Instead, all of them had yellow seeds. Mendel termed the form of the trait that appeared in the F_1 **dominant** and the one that wasn't seen as **recessive.** The expression of one factor preventing the expression of the other one is called the **principle of dominance.** But Mendel was left with a vexing question. What had happened to the factor for the recessive trait? (See Table 7.1 for some dominant and recessive traits in humans.) After all, it had been passed along through countless generations so it couldn't have just disappeared.

The Principle of Segregation

In the next stage of his experiments, Mendel allowed the F_1 plants to randomly self-pollinate, and, lo and behold, the missing recessive trait reappeared in some of their F_2 offspring! Also, the ratio of F_2 plants with recessive traits to those with dominant traits was fairly constant, regardless of the particular characteristics involved. The exact results for each of the seven pairs of characteristics Mendel used in his tests are shown in Table 7.2 on page 186.

Finally, in the third year of the experiments, Mendel allowed the F_2 plants to self-pollinate. He found that all those with recessive traits produced only recessive F_3 offspring. However, the F_2 plants that showed a dominant trait produced both types of F_3 offspring. One-third of them produced *only* dominant offspring. The other two-thirds produced both dominant and recessive offspring, but they produced three times as many offspring with the dominant trait as they did with the recessive trait. In

TABLE 7.1
Some Human Traits That Are Inherited

	DOMINANT	RECESSIVE
Hair, skin, etc.	Dark hair	Blond hair
	Nonred hair	Red hair
	Curly hair	Straight hair
	Abundant body hair	Little body hair
	Early baldness (dominant in males)	Normal
	Normal	Absence of sweat glands
Eyes and facial features	Brown	Blue or gray
	Hazel or green	Blue or gray
	Congenital cataract	Normal
	Normal	Nearsightedness
	Farsightedness	Normal
	Astigmatism	Normal
	Free earlobes	Attached earlobes
	Broad lips	Thin lips
	Large eyes	Small eyes
	Long eyelashes	Short eyelashes
	High, narrow bridge of nose	Low, broad bridge
Skeleton and muscles	Dwarfism	Normal
	Normal	Midget
	More than five fingers or toes	Normal
	Webbed fingers or toes	Normal
	Short fingers or toes	Normal
	Normal	Progressive muscular atrophy
Some conditions affecting systems	Hypertension	Normal
	Normal	Hemophilia (sex-linked)
	Normal	Sickle-cell anemia
	Normal	Congenital deafness
	Migraine headaches	Normal

Source: Adapted from C. Villee, *Biology,* 4th ed. (Philadelphia: Saunders, 1962), and Evelyn Morholt, Paul F. Brandwein, and Alexander Joseph, *A Sourcebook for the Biological Sciences* (New York: Harcourt, Brace & World, 1966, p. 170).

other words, all the F_2 recessives and one-third of the F_2 dominants bred true (that is, passed their visible traits on to all offspring). The other two-thirds of these dominants produced mixed offspring—but in the same 3:1 ratio of dominant to recessive as in the plants their F_1 parents had produced.

TABLE 7.2
Mendel's Pea Plant Experiment

DOMINANT FORM	NO. IN F_2 GENERATION	RECESSIVE FORM	NO. IN F_2 GENERATION	TOTAL EXAMINED	RATIO
Round seeds	5474	Wrinkled seeds	1850	7324	2.96:1
Yellow seeds	6022	Green seeds	2001	8023	3.01:1
Gray seed coats	705	White seed coats	224	929	3.15:1
Green pods	428	Yellow pods	152	580	2.82:1
Inflated pods	882	Constricted pods	299	1181	2.95:1
Long stems	787	Short stems	277	1064	2.84:1
Axial flowers	651	Terminal flowers	207	858	3.14:1

Mendel's experiments led him to conclude that every trait is controlled by two factors, one inherited from each parent. Today we call Mendel's unit of inheritance a **gene** and the alternative forms of a gene, **alleles.** Alleles of a gene exist at specific locations, or loci, (singular, **locus**) on a chromosome. In other words, we might talk about a gene for seed color having an allele for yellow seeds and an allele for green seeds. We might also speak of a gene for seed form having an allele for wrinkled seeds and an allele for round seeds. Since one allele is dominant over the other, we can distinguish between an individual's **genotype,** that is, the assortment of alleles in its genetic makeup, and its **phenotype,** or its expressed traits.

Mendel realized that the 3 (dominant):1 (recessive) pattern of inheritance he observed when he bred the F_2 plants could only have arisen if the allele pairs separated during the formation of gametes (eggs and sperm). This realization led him to formulate his **principle of segregation:** When gametes form, the two alleles for a genetic trait separate so each egg or sperm carries only one allele. The paired condition is restored by the random fusion of gametes at fertilization.

Let's consider a specific example of a cross between two pure lines—plants that breed true when they self-pollinate and are identical in all characteristics except one. In this case, suppose one strain always produces round seeds and the other always produces wrinkled seeds. We have already seen that the round form is dominant, so we'll designate that form as **R** and the recessive wrinkled form as **r.** Each plant has two genetic components, or alleles, one derived from each of the two parents. Since both plants are true breeding, we can assume that in each plant the two alleles are identical. That is, in one plant, the components are **RR,** and in the other, they are **rr.** Plants such as these, in which both alleles code for the same form of the trait, are said to be **homozygous** for that particular characteristic; that is, having identical alleles for that trait. As a result, all the gametes they produce will be the same.

Remember that the plants are identical in all characteristics except one, so we only need to diagram the differing characteristics. Since half the genes in the F_1 generation come from each parent, the genotype of any F_1 plant would have to be **Rr.** Hence, these plants would be **heterozygous** for the characteristics in question: that is, having different alleles for that trait. Now, when these plants produce gametes (eggs and sperm), each gamete will carry either an **R** allele or an **r** allele, and the combination that occurs when two gametes come together at fertilization will determine the genetic characteristics of the resulting F_2 plant.

The results we should expect in the F_2 generation can be diagrammed in a form called a **Punnett square,** a gridlike device that helps one predict the outcome of simple genetic crosses. The possible gametes produced by one parent are indicated along one side of the grid and the possible gametes of the other parent are indicated along an adjoining side. The possible combinations are indicated in the boxes within the grid. A Punnett square predicting the outcome of the cross we have been discussing is shown in Figure 7.3.

If two homozygous parents were represented by **RR** and **rr** (as for seed form), their offspring would have to be heterozygous **Rr,** since one component would have come from each parent; and when this **Rr** individual produced gametes, half would contain an **R** component and the other half would contain the **r** component. A cross between two **Rr** individuals could then be expected to yield offspring with the genotypes **RR, rR, Rr,** and **rr,** or a genotype ratio of 1:2:1. What would be the genotypes and the phenotypes of the offspring if an **Rr** crossed with a homozygous recessive **rr** instead?

FIGURE 7.3

Arriving at the expected ratio in the F_2 generation of plants that differ in only one character (here, seed shape). The parent (P_1) generation is a cross between a pure-breeding plant with round seeds **(RR)** and a pure-breeding plant with wrinkled seeds **(rr).** Obviously, a plant that had received an **R** gene for round seeds from each parent could only pass along **R** genes to its progeny, and the same for a plant receiving only **r** genes from its parents. The F_1, then, must be **Rr** since one gene for seed shape must come from each parent. If two individuals from the F_1 are crossed, the expected ratio of their offspring can be calculated by use of a Punnett square. Remember, though, that these are only *expected* results, and this ratio would probably be approached only if a great number of crosses were made.

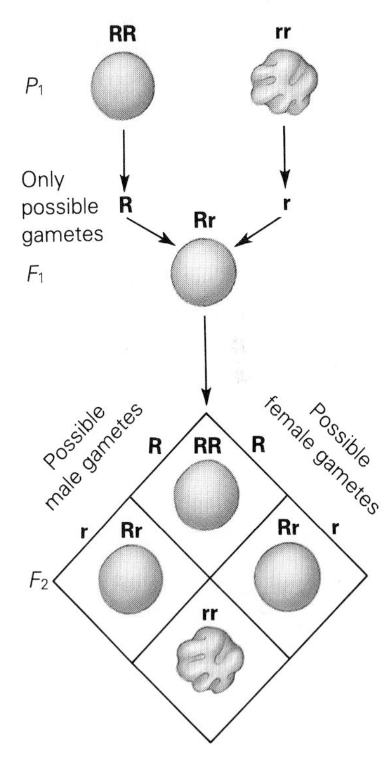

The Principle of Independent Assortment

In his next set of experiments, Mendel crossed **dihybrids,** plants differing in two characteristics. In one such experiment, plants with round **(R),** yellow **(Y)** seeds were crossed with a strain that had wrinkled **(r),** green **(y)** seeds. We saw in Table 7.2 that round and yellow are both dominant, so we shouldn't be astounded to learn that all the F_1 plants had round, yellow seeds. Since one P_1 parent was **RR YY** and the other was **rr yy,** all the F_1 plants would have to be **Rr Yy** or heterozygous for both characteristics. When these plants were allowed to self-pollinate, we would expect a random and independent assortment of the various characteristics to produce the F_2 generation shown in Figure 7.4.

It was apparent to Mendel from the results of this dihybrid cross that the segregation of genes for one characteristic was not affected by the segregation of genes for the other characteristic. This led him to formulate the **principle of independent assortment,** which states that each pair of alleles segregates independently during gamete formation. It should be pointed out here that if you were to run a single experiment to test Mendel's results, you might just end up with all wrinkled, green seeds in the F_2

FIGURE 7.4

In this case, we have the expected F_2 ratio from a dihybrid cross (a cross between plants that differ in only two characters—here, seed shape and seed color). One pure-breeding parent in the P_1 generation bore both dominant traits; the other, both recessive traits. Obviously, they could pass along only dominant and recessive genes, respectively. This means the F_1 is necessarily heterozygous for both traits. At meiosis, however, the traits assort independently. This means that when the F_1 is crossed with itself, we must account for all possible combinations in order to determine the F_2 ratio. Only one possible combination is labeled. If you were to work this out, you would get nine genetic combinations. Calculate the genotypes and phenotypes and then check page 191 to see how wrong you were. Again, keep in mind that we're only dealing with probabilities. This means that all the F_2 *could* be **rryy** (double recessives), but the odds are very much against it.

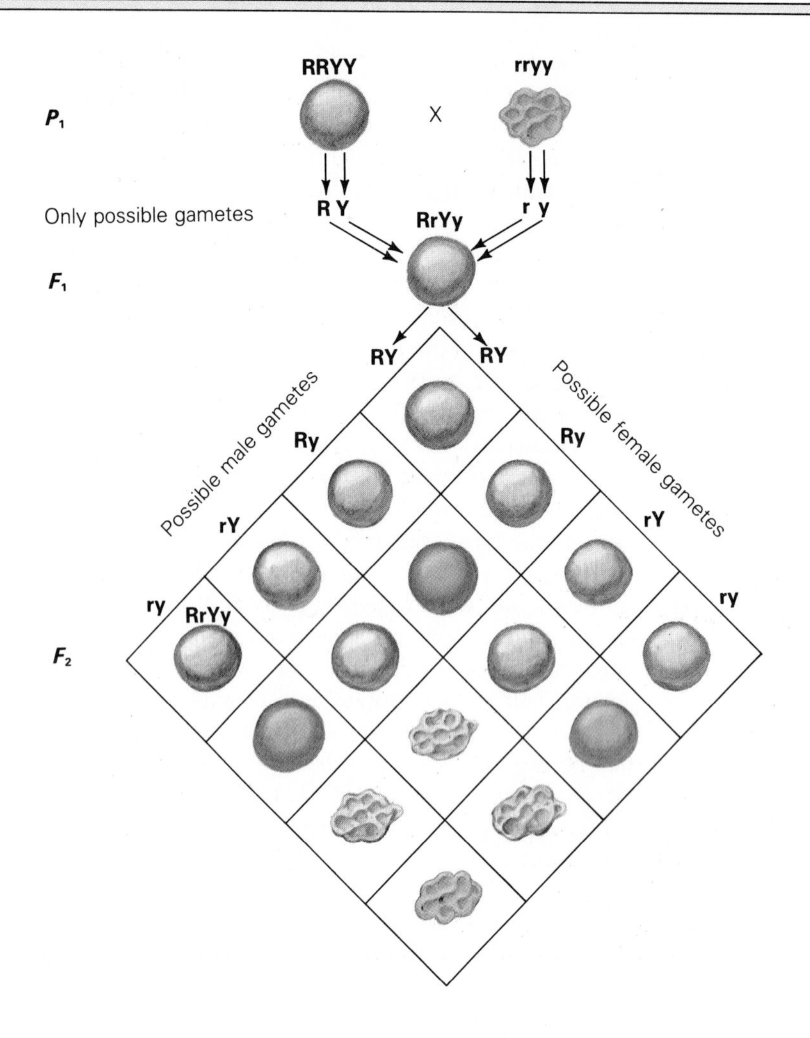

generation—just as when you toss a coin ten times, it *might* come up tails each time. You could expect the results shown in Figure 7.4 only if you ran enough tests for the law of averages to apply.

The Testcross

Although Mendel had carried out numerous progeny tests for determining whether an individual who showed the dominant form of a trait was homozygous or heterozygous, he soon devised a much simpler procedure, the **testcross.** The subject was simply crossed with an individual who was recessive for the trait in question. The recessive form of a trait can only be shown if the individual is homozygous for that trait. Therefore, that individual could produce only one type of gamete—one containing the factor for the recessive form. Let's use the traits round and wrinkled as an example:

1. If the dominant round individual in question is homozygous, then the testcross becomes **RR × rr,** and all of the progeny will be round **(Rr).**
2. If the dominant round individual is heterozygous, then the testcross becomes **Rr × rr,** and, statistically, half the offspring will be heterozygous round **(Rr)** and half will be wrinkled **(rr).**

The testcross is often applied today to test the pedigrees of plants and animals in agriculture (Figure 7.5).

FIGURE 7.5
The genotype (that is, whether homozygous or heterozygous) of an individual showing a dominant phenotype (here, round seeds) can be distinguished by the results of a cross with homozygous recessive **rr.** The homozygous **RR** produces all round offspring. The heterozygous **Rr** produces 1(round) : 1(wrinkled).

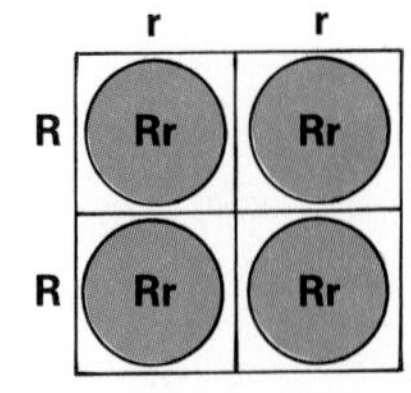

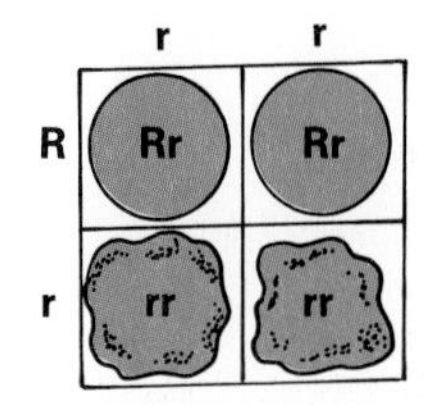

CLASSICAL GENETICS – BEYOND MENDEL

It is unfortunate that those nineteenth-century biologists who did read Mendel's reports were unimpressed with his findings. It is also unfortunate that no one saw enough significance in them to bother repeating his laborious experiments. Early in the twentieth century, however, some 17 years after Mendel died in obscurity, his work was rediscovered by three researchers who independently arrived at essentially the same conclusions. They were each elated and anxious to be the first to tell the world their findings. By now, though, enough was known for such information to be appreciated, and, in fact, it was met with great applause. The applause rang a bit hollow over the mossy grave of old Mendel, though, the man who had beaten them all.

Mendel's rediscovery set the stage for extensive research in what is now considered **classical genetics,** the modern study of genetics based essentially on Mendel's techniques. With Mendel's ideas as a foundation, the chromosomal theory of inheritance was soon established.

By the beginning of this century it was assumed that chromosomes of any organism were of identical makeup, and that meiosis was simply a way of dividing them up so that the gametes would receive the same number. It was also believed that during meiosis, the maternal and paternal chromosomes separated and that each set traveled together at anaphase, so that any gamete would contain chromosomes from one parent or the other, but not both. In the early 1900s, however, it was found that the chromosomes are not identical, but come in a variety of sizes and shapes, and that they also come in pairs (members of a pair are now called homologues—Chapter 6). In addition, it was found that meiosis divides chromosomes not only quantitatively but also qualitatively, so that each gamete receives both the same *number* and *kind* of chromosomes.

A young graduate student at Columbia University, Walter S. Sutton, began to try to relate Mendel's findings to what was known about meiosis. His results were surprising. For example, he found that when chromosome pairs line up at meiosis, there was no way to determine which way a paternal or maternal chromosome would go at anaphase. In one case, some

FIGURE 7.6

Polydactyly, the condition of having more than five digits on the hands or feet, is a dominant trait in humans. In the left photo, two girls and their brother show variation in the condition. The girl at left has six toes on both feet, her sister has six toes on one foot, and her brother has the normal complement of five toes on each foot. In the right photo, a pitcher for the Montreal Expos system, Antonio Alfonseca, shows that polydactyly need not be a disadvantage.

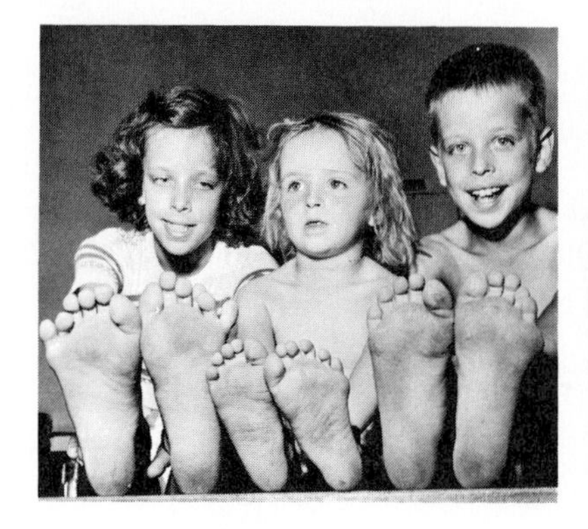

maternal chromosomes moved to one pole and some to the other, but there was no way to determine their direction. At that time, it was not known that chromosomes were the carriers of genes. The independent assortment of chromosomes, however, confirmed a hunch Sutton had that hereditary units are associated with chromosomes. Also, since there were far more hereditary characteristics than there were chromosomes, he reasoned that each chromosome must be responsible for many genetic traits. (See Table 7.1 and Figure 7.6 for examples of human traits known to be determined by genes, and Essay 7.1 for a description of an early attempt to breed for certain traits.)

In 1910, after Sutton's report, an innocuous little insect entered the world of genetics. Thomas Hunt Morgan, also at Columbia, began a program of breeding experiments with the fruit fly (or vinegar fly), *Drosophila melanogaster* (Figure 7.7). These little flies are ideal subjects for genetic studies. They are easy to maintain in the laboratory, they mature in 12 days and reproduce prolifically, and they have a variety of features that can be readily identified. They also have giant chromosomes in their salivary glands. Specific areas, or loci, of the chromosomes are readily distinguishable. In addition, they have only four pairs of chromosomes. The insects have little economic importance, and they don't seem to bother humans much when the two species come in contact. In fact, it has been said that God must have invented *Drosophila melanogaster* just for Thomas Hunt Morgan.

FIGURE 7.7

The fruit fly, *Drosophila melanogaster*.

Sex Determination

One of the first things Morgan did was attempt to cause genetic changes, or **mutations,** in his flies. He subjected them to cold, heat, X rays, chemicals, and radioactive matter. At first, he found nothing, but then he noticed among a group of normal red-eyed flies that there was a single male with strange *white* eyes. Morgan was convinced that this was a mutation (a subject we will take up later). He carefully nurtured his little white-eyed specimen and crossed it with several of its red-eyed sisters. In these crosses, he found that the F_1 generation was all red-eyed, just as you would expect

ESSAY 7.1

The Attempt to Recreate Leonardo da Vinci

Leonardo da Vinci was the illegitimate son of Piero, a notary from the town of Vinci, and a peasant girl named Caterina. The third wife of Leonardo's father later bore a son named Bartommeo who idolized Leonardo although he was 45 years younger. After the death of the legendary Leonardo, Bartommeo attempted an amazing experiment. He studied every detail of his father's relationship with Caterina. Then Bartommeo, himself a notary by family tradition, returned to Vinci and found another peasant girl who seemed similar to Caterina, according to all Bartommeo knew. He married her and she bore him a son whom they called Piero.

Strangely, the child actually looked like Leonardo and was brought up with encouragment to follow in the great man's footsteps. The boy became an accomplished artist and was becoming a talented sculptor when he died, thus ending the experiment.

A red chalk self-portrait done by da Vinci circa 1512.

if white eyes were recessive. Simple enough so far—but then the story became a bit more complicated.

When the F_1 flies were allowed to interbreed, the result was 3470 red-eyed and 782 white-eyed flies. This is not the simple 3 : 1 ratio you might have expected according to Mendelian principles. Also, there was another troubling factor. All the 782 white-eyed flies were males. The red-eyed flies, however, were of both sexes: 2459 females and 1011 males. At this point, an obvious answer was suggested: only males are white-eyed. Alas, when the original white-eyed male was crossed with its F_1 red-eyed daughter, the result was 129 red-eyed females, 132 red-eyed males, 86 white-eyed males, and 88 white-eyed females! If you have a penchant for puzzles, can spare a few minutes, and think you are smart, try to figure out how this could have happened. As a clue, keep in mind that sex is determined by chromosomes. The rest of us will trudge along.

Morgan surmised that although chromosomes may assort independently, the genetic determiners (or genes) may not. It then occurred to him that the sex-determining factor and the eye color factor may be somehow linked together. Hence, as one goes at anaphase, so must go the other.

By this time, Morgan knew something about the chromosomes that determine the gender of an offspring. He knew that of the four pairs of chromosomes in *Drosophila,* only one pair was responsible for gender. He even knew which ones were the sex chromosomes and which were the

ANSWERS TO FIGURE 7.4

You should get:

Genotype	*Phenotype*
1 **RRYY**	9 round yellow
2 **RRYy**	3 round green
1 **RRyy**	3 wrinkled yellow
2 **RrYY**	1 wrinkled green
4 **RrYy**	
2 **Rryy**	
1 **rrYY**	
2 **rrYy**	
1 **rryy**	

FIGURE 7.8

The chromosomes of male and female *Drosophila*. Note that each chromosome of one sex is represented in the genotype of the other, except for the twisted Y chromosome of the male. Because there are few chromosomes and they are rather easily identifiable, *Drosophila* was chosen for intensive research.

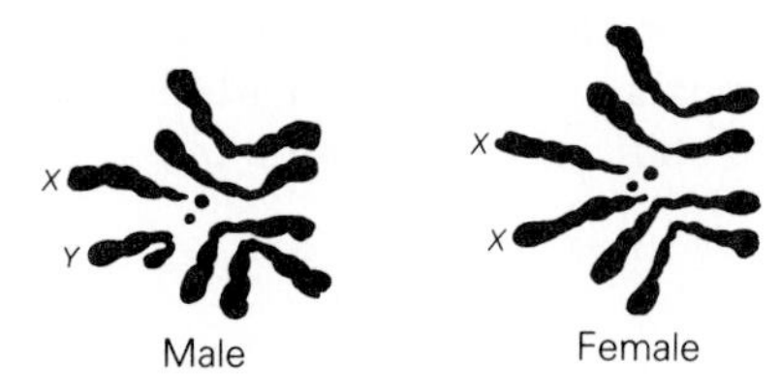

autosomes, the chromosomes responsible for characteristics other than sex. In females, the sex chromosomes are a rodshaped pair of chromosomes, for some reason called X chromosomes. In males, the sex chromosomes consist of only one X chromosome and a J-shaped chromosome called a Y chromosome (Figure 7.8). The X chromosomes are usually larger and contain far more genes than the Y chromosome. This difference in size will be important to the relative genetic contributions of the two kinds of chromosomes, as we will see.

In the production of sperm, the X and Y chromosomes, while not homologous, line up together at meiosis, so that half the gametes will contain an X and half will contain a Y. Since females have only X chromosomes, every egg will contain an X. This, then, sets the basis for gender determination in offspring. If a Y-bearing sperm reaches the egg first and fertilizes it, the offspring will be XY, and thus male. If an X-bearing sperm fertilizes the egg, the offspring will be XX, a female. (This is how gender is determined in many animals including mammals, but in other organisms it may be determined differently.)

An interesting human application of the findings on gender inheritance is gender testing. The cells in women have tiny specks in their nuclei called **Barr bodies.** Also, white blood cells in women show a small "drumstick" attached to the nucleus (Figure 7.9). Both the Barr bodies and the drumsticks are believed to represent condensed and nonfunctional X chromosomes in females. Cells from normal men lack both the Barr body and the drumstick. There was a stir at the Olympic games not long ago when a European woman who showed remarkable strength in the field events was found to be genetically abnormal, lacking the Barr bodies and drumsticks in some of her cell nuclei. Her strength was believed to be due to the "male" characteristics of her cells. For certain athletic competitions these days, sex tests are mandatory.

An imbalance of X and Y chromosomes causes abnormal sexual development. For example, certain females may have cells with no Barr bodies or drumsticks. In such cases, there is only one X chromosome instead of two. The physical characteristics that accompany this condition, resulting

FIGURE 7.9

Electron micrographs of the Barr body (a) and drumstick (b) from the cells of a woman.

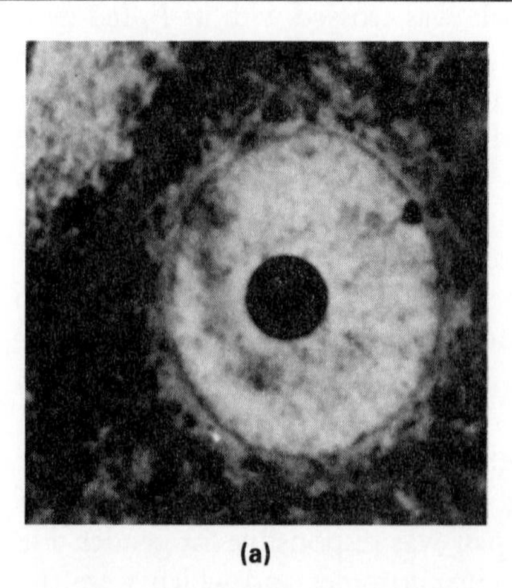

(a)

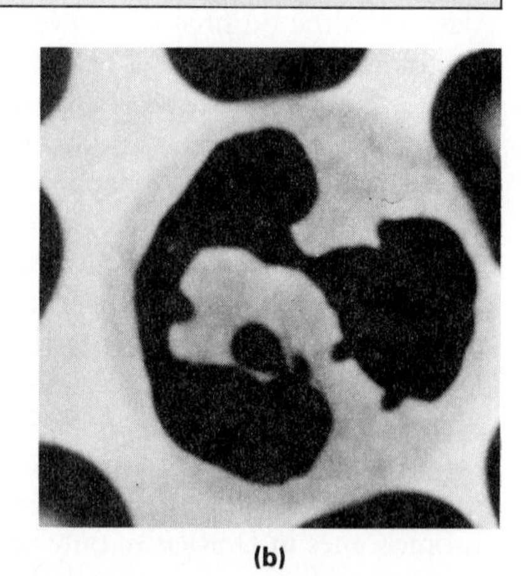

(b)

in what is called Turner's syndrome, are female genitalia, but underdeveloped breasts and tiny ovaries— and for some reason, short stature. The presence of an extra X chromosome in males (XXY) results in certain anomalies called Klinefelter's syndrome. Such males usually have sparse body hair, some breast development, and cells that show Barr bodies and drumsticks.

A study relating to sex chromosomes that began in 1968 raised some critical social issues. The goal was to identify XYY males at infancy and then to study their behavior. The reason was that XYY males are not only unusually tall, but they are 20 times more likely to be placed in a criminal institution than are genetically normal males. The study was to determine whether XYY males are genetically prone to antisocial behavior. (Some researchers argued that they were more likely to be institutionalized because the genetic condition was often associated with subnormal intelligence.) In any case, the announcement caused a furor in some circles. It was argued that identifying such males would result in their being treated differently, and that this might, itself, result in abnormal behavior. Other people, perhaps for political or philosophical reasons, also objected to studying the role of genetics on human behavior. In any case, the program was stopped in 1975.

Gene Linkage and Crossover

Recall from our discussion of meiosis (Chapter 6) that chromosomes assort independently. Thus, the fate of genes on one chromosome of a pair would have to be independent of those on the homologous chromosome. If two genes are on the *same* chromosome, however, it is likely that they will be passed together to the gamete. The occurrence of two or more hereditary units on the same chromosome is called **gene linkage.**

In *Drosophila,* for example, the recessive alleles for a black body **(b)** and a curved wing **(c)** (Figure 7.10) are carried on the same chromosome. The dominant alleles are for a beige body **(B)** and a normal wing **(C).** So if we diagram a cross between a homozygous normal fly, **BB CC,** and a homozygous recessive fly, **bb cc,** we will get the results shown in Figure 7.11. The only gametes the beige normal-winged fly can produce are **BC,** and the only gametes the black, curved-wing fly can produce are **bc.** Hence, their F_1 offspring would all be heterozygous for both characteristics—**Bb Cc**—and would, of course, show beige bodies and normal wings.

Now, if our **Bb Cc** fly were crossed with a homozygous recessive, **bb cc,** you would expect half of their offspring to be beige with normal wings and half to be black with curved wings (the reason for this is shown in the Punnett square). This is not what happens, however. In all likelihood, this cross would produce about 37 percent beige, normal-winged flies, about 37 percent black, curved-wing flies, 13 percent *black,* normal-winged flies, and 13 percent beige, *curved*-wing flies. This kind of assortment, it turns out, can best be accounted for if two conditions are met: (1) the genes for both body color and wing shape would have to be on the same chromosome, with their alternative alleles on homologous chromosomes, and (2) the chromosomes would have to be able somehow to exchange genes.

FIGURE 7.10

Curved wings in *Drosophila.* The curved wing is one of the obvious traits, due to a single gene mutation, that makes this small insect so easy for geneticists to work with. The curved wing trait is recessive, as is black body color, the normal color being beige. It was discovered later that the genes for wing shape and body color lie on the same chromosome and thus do not follow the principle of independent assortment.

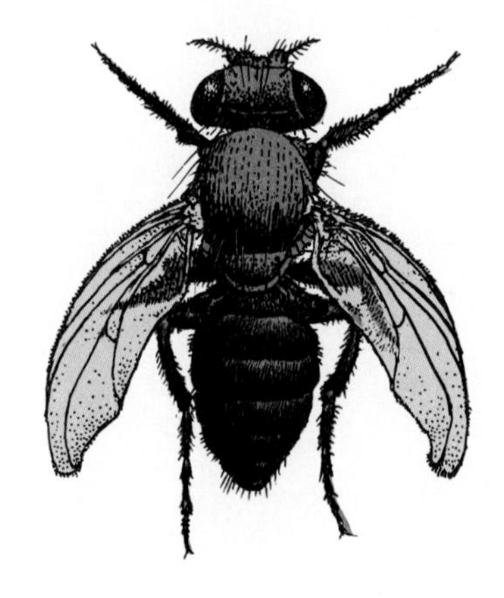

FIGURE 7.11

When genes are linked by existing on the same chromosome, you would expect these results from the cross shown here. However, the linkage may be disrupted by crossovers.

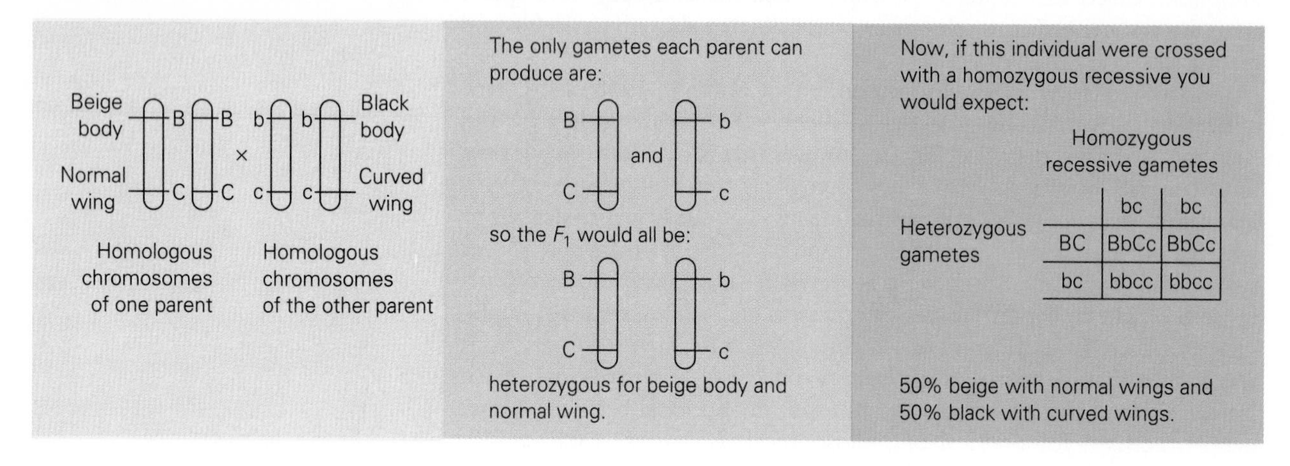

	bc	bc
BC	BbCc	BbCc
bc	bbcc	bbcc

You can probably believe condition 1 easily enough on the basis of our previous discussions, so let's consider how we might get condition 2. Recall that in meiosis, the homologous chromosomes line up during metaphase I, and as anaphase begins, the homologues of each pair move toward opposite poles. While they are lined up side by side in prophase, however, the homologues may stick together at various places and then exchange parts of their chromatids. The opposite chromatids of the homologue cross each other and then break off at their point of contact. Each broken part then fuses with the opposite portion of its homologue (Figure 7.12).

You might wonder how the chromatids manage to break at *exactly* the same place on each one, so that one chromosome does not end up with more genes than the other chromosome. Geneticists investigating

FIGURE 7.12

In meiosis, homologous chromosomes come to lie side by side in early prophase. During this time, their arms may cross (a), break at the point of crossing, and the parts rejoin (b) in such a way that the chromosomes below the cross are exchanged.

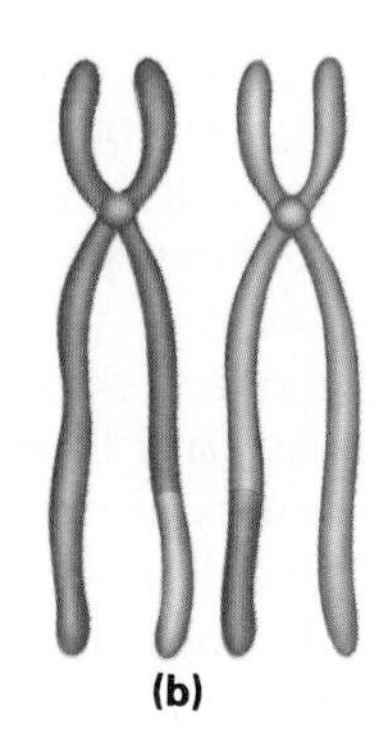

crossovers are asking the same question—but they have no answer yet. We do know, however, the basic mechanism by which crossovers occur, and we also know that some crossovers take place at virtually every meiosis.

Sex-linked Traits

The sex chromosomes carry some genes other than those involved in gender determination. Such traits are called **sex-linked traits.** An example is the allele for white eye color in *Drosophila.* Morgan showed that this allele is located on the X chromosome and that it is recessive. Therefore, females will have white eyes only if both eye-color alleles are for white eyes. In males, however, when a white-eye factor turns up on the X chromosome, there is no dominant counterpart on the Y chromosome, so the white eye color will be expressed (Figure 7.13).

Red-green color blindness in humans is caused by a recessive allele on the X chromosome. Thus, a woman who is heterozygous for this condition will show no symptoms of color blindness. Among her children, however, she can expect half her sons to be color blind and half her daughters to be carriers of color blindness (as she is). In her sons, the recessive gene on the X chromosome is not overridden because the tiny Y chromosome bears no allele for color discrimination. Figure 7.14 shows the expected occurrences of color blindness in this woman's grandchildren, the F_2 generation.

There are a number of other phenomena associated with sex-linked genes that have great impact on people's lives. For example, one type of muscular dystrophy is caused by a recessive gene on an X chromosome. The effects of this abnormality are so pronounced, however, that males carrying the gene usually die in their teens before reproducing. As a result, females rarely have the disease, since the probability of receiving the allele from both parents is very low. However, they can carry a single gene for the condition, so that when they reproduce, they may pass it along to their offspring. Another such disorder is hemophilia, which is also a recessive gene carried on an X chromosome (Essay 7.2).

FIGURE 7.13
In *Drosophila,* the gene for white eyes (a recessive gene) exists on the X chromosome. In females, the gene can be masked by a dominant gene for red eyes on the other X chromosome. But males carry no such dominant gene on their Y chromosome, so if a male's X should bear the gene for white eyes, it will be expressed.

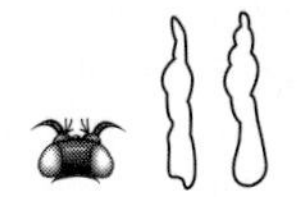

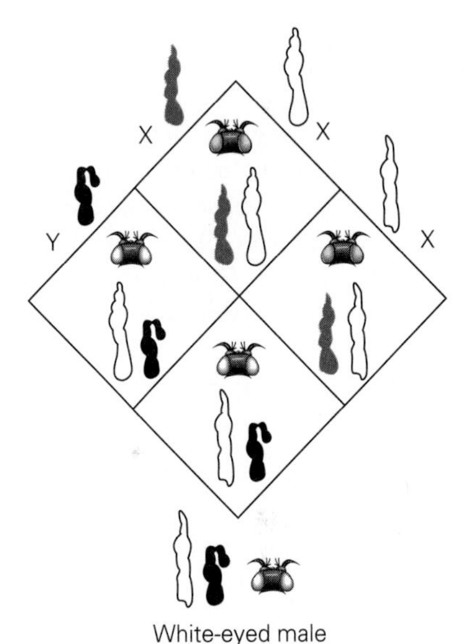

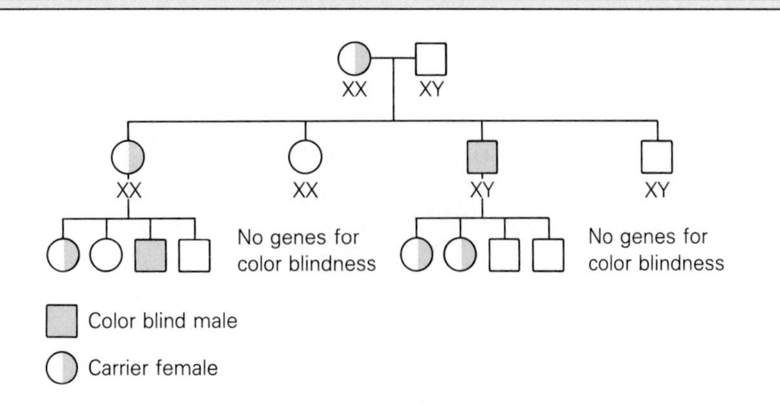

FIGURE 7.14
How color blindness is inherited. The original mating is between a carrier female and a normal male. Spouses (not shown) are presumed to be free of the color-blindness gene. The squares represent males, the circles females, and the condition (heterozygous or homozygous) is represented by color (which may seem odd, but it's convenient).

ESSAY 7.2

The Disease of Royalty

Queen Victoria, Prince Albert (center, seated), and their large family.

Because it was once the practice of ruling European monarchs to consolidate their empires through marriage alliances, hemophilia came to be transmitted throughout the royal families. Hemophilia is a sex-linked recessive condition in which the blood does not clot properly. Any small injury can result in severe bleeding and, if the bleeding cannot be stopped, in death. Hence, it has been called the "bleeder's disease." Hemophilia has been traced back as far as Queen Victoria, who was born in 1819. One of her sons, Leopold, Duke of Albany, died of the disease at the age of 31. Apparently, at least two of Victoria's daughters were carriers, since several of their descendants were hemophilic. Hemophilia played an important historical role in Russia during the reign of Nikolas II, the last Czar. The Czarevich, Alexis, was hemophilic, and his mother, the Czarina, was convinced that the only one who could save her son's life was the monk Rasputin—known as the "mad monk." Through this hold over the reigning family, Rasputin became the real power behind the disintegrating throne.

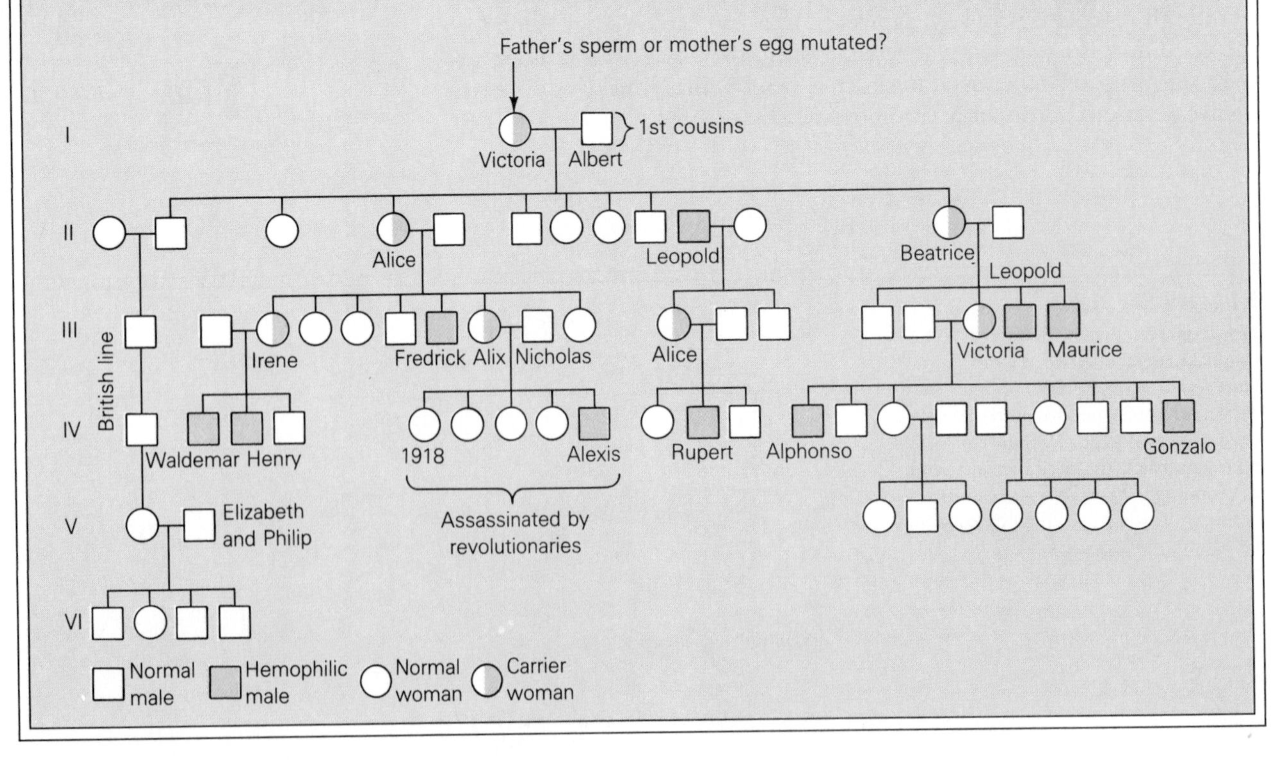

CHROMOSOME MAPPING

If genes are located on chromosomes, and if specific alleles are precisely exchanged through crossovers, then the genes for certain characteristics must lie at specific points along each chromosome. But how do we know which genes lie where? Finding the answer involves "mapping" the chromosome—a technique that is largely based on information from observations of crossovers.

Thomas Morgan and A. H. Sturtevant, one of the brilliant young graduate students attracted to Morgan's laboratory in those days, hypothesized that if genes were arranged linearly along the chromosome, then those lying closer together would become separated by crossovers less often than those lying farther apart. Genes lying closer together would thus have a greater *probability* of being passed along as a unit. Stated another way, the percentage of crossover is proportional to the distance between two genes on a chromosome—so *percent crossover* is the number of crossovers between two genes per 100 prophase opportunities.

As an example, suppose two characteristics, which we can call A and B, show 26 percent crossover. We can assign 26 crossover units to the distance between the two genes. Then, if some characteristic C turns out in breeding experiments to have 9 percent crossover with B and 17 percent crossover with A, it would be located between A and B at a point 9 units from B and 17 units from A. After the information from many such experiments has been compiled, a chromosome map can be constructed that indicates the position along the chromosome of the genes that code for certain characteristics. Figure 7.15 shows chromosome maps developed by this technique for chromosomes in the salivary gland of *Drosophila melanogaster*.

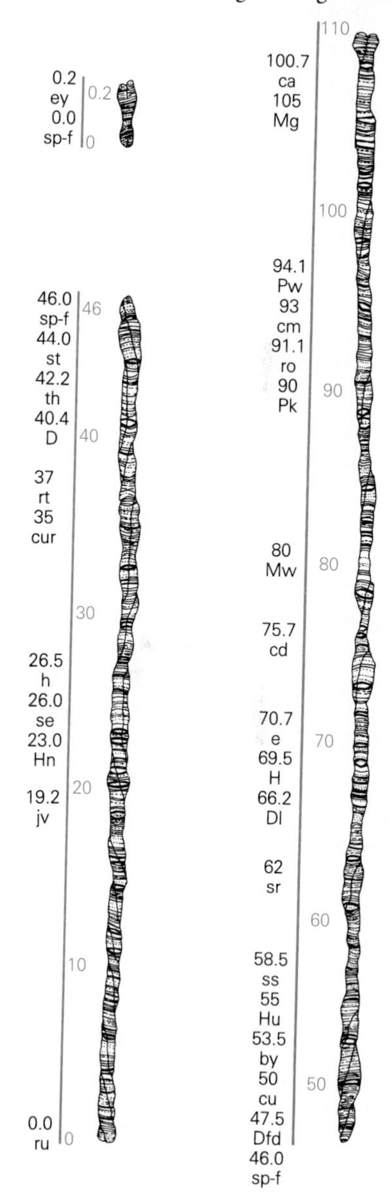

FIGURE 7.15
A map of chromosomes from the salivary gland of *Drosophila*. The units are shown in the numbered column at right. The relative positions of specific genes (such as one called jv, lower left) are shown by the numbers above the gene designation.

MUTATIONS

Mutations, or changes in the hereditary material, can be caused by X rays, chemicals, ultraviolet radiation, and a number of other agents. In fact, even now, as you sit peacefully reading this fascinating account, your body is being bombarded with tiny subatomic particles from outer space—a form of radiation. If one of these particles should strike you right in the gonad and alter a nucleotide in a developing gamete, the genetic instructions carried by that gamete could be changed. Recall that a small change in a nucleotide sequence can specify a different amino acid and consequently result in the production of a different protein. If an altered gamete happens to enter the reproductive process, whatever change has taken place will be passed along to the next generation, unless the change is lethal or otherwise prevents reproduction, or unless it is reversed through a second mutation, a highly unlikely event.

Gene Mutations

Mutations can occur at two levels, the level of the gene and the level of the chromosome. Mutations involving alterations in the genes are called **gene mutations** (or point mutations). Gene mutations occur during the duplication of DNA. They may involve the omission of nucleotide units,

ESSAY 7.3

Sickle-cell Anemia

Sickle-cell anemia is a recessive condition characterized by fragile red blood cells that collapse into a sickle shape when the oxygen concentration is low. These cells may clog the blood vessels and break down more easily than normal cells. Persons who have the disease suffer painful symptoms and usually die at an early age. The cause of sickle-cell anemia has been traced to a mutation in a single gene, resulting in an alteration of one nucleotide in a codon. The result is that in part of the complex polypeptides that comprise hemoglobin, a single amino acid is changed—and so is the capability of the entire molecule.

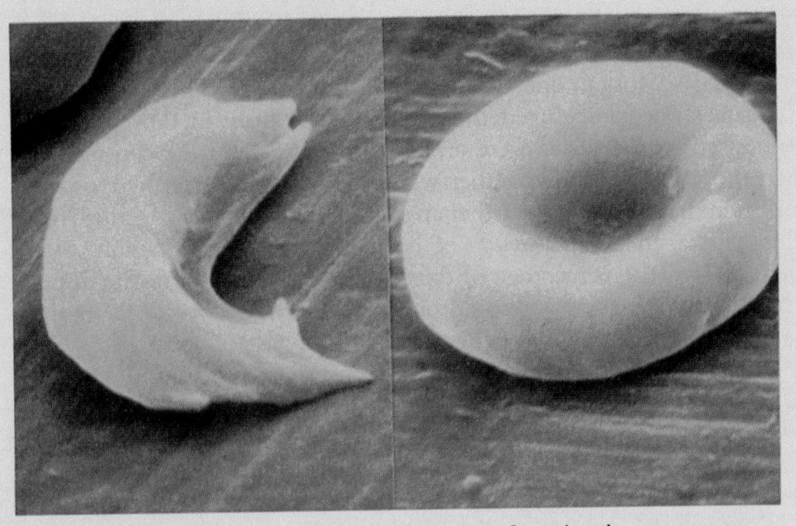

Sickled cell (left) showing the ragged and collapsed configuration that may cause them to clog blood vessels, thereby depriving tissues of oxygen. Normal at right.

In persons homozygous for this condition, all of the hemoglobin molecules are affected. These people do not normally survive long. In heterozygous persons, only some of the molecules are affected, and the red blood cells are able to function under normal oxygen requirements. Under conditions in which the blood's oxygen concentration is usually low, as during strenuous exercise or at high altitudes, these blood cells are likely to collapse. Nevertheless, the heterozygous condition also carries a compensatory advantage in certain environments. Persons who are heterozygous for sickle-cell anemia are more resistant to malaria than those with normal hemoglobin. Thus sickle-cell anemia is maintained in malarial areas from Africa to India.

the repetition of nucleotide units, or mistakes in bonding between these units. They may also involve the substitution of one nucleotide for another—either the substitution of one purine or pyrimidine for another or the substitution of a purine for a pyrimide or a pyrimidine for a purine. In some cases, gene mutations can have far reaching effects, as we see in sickle-cell anemia (Essay 7.3).

Not all such changes result in a change in the phenotype, the observed trait. For example, the sequence GAC codes for the insertion of aspartic acid into a protein, and a change from GAC to GAU causes no alteration in the code. Mutations can also be masked in some cases by other mutations that compensate for the effect of the first change.

The evolutionarily important mutations, of course, are those with some observable result. Often, in fact, these small changes in genes can have enormous effects. For example, the addition or deletion of a single nucleotide could change the entire amino acid product. This is because the codons are read in groups of three nucleotides. For example, ACG—GCC—GGA codes for threonine, alanine, and glycine. If another G should be inserted after the first one, the sequence would read instead ACG—GGC—CGG—A, and the amino acids that would be inserted into the protein would be threonine, glycine, and arginine. Thus, this one added nucleotide could change every amino acid in the sequence from that point on.

Chromosome Mutations

The second level of mutations involves changes in the structure of chromosomes. These, logically enough, are called **chromosome mutations.** Such changes may occur when the chromosomes are moving about or undergoing changes—for example, during meiosis. At such times, a part of a chromosome may be broken off and lost (a **deletion**), or it may break off and reattach at its other end (an **inversion**), or a segment may be duplicated, doubling the number of certain genes (a **duplication**) (Figure 7.16). The most serious of these mishaps is a deletion, since the absence of a large number of genes is likely to result in death.

In some cases, mutations may have little effect on the phenotypes of the individual. For example, a slight difference in hair color as the result of a mutation might scarcely be discernible. Some mutations may also be highly beneficial in some environments, but not in others. A change that produces blue eyes makes them more sensitive to light, but this sensitivity ceases to be an advantage in environments subject to glaring sun. Mutations are not necessarily bad when viewed in the context of life's big picture. We will see later how increased variation in a population can be important to natural selection, but it should be clear that random changes are not likely to be helpful to a population that has become finely attuned to its habitat through countless generations. Suppose you raised the hood on a sports car, and, knowing nothing about motors, you made an adjustment. It may be that the change you made was just the one needed to make the car run even better, but the odds are a lot greater that you screwed something up. The odds with random mutations are about the same.

FIGURE 7.16

Chromosomal mutations and their epistolary counterparts, showing the normal message with deletions, inversions, and duplications (or additions) to indicate how such changes can garble the message.

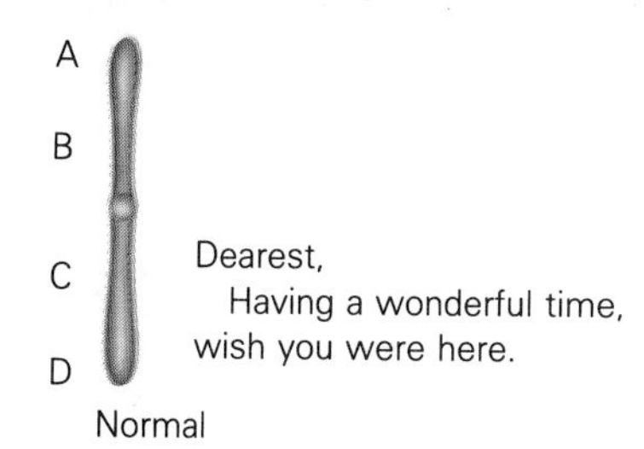

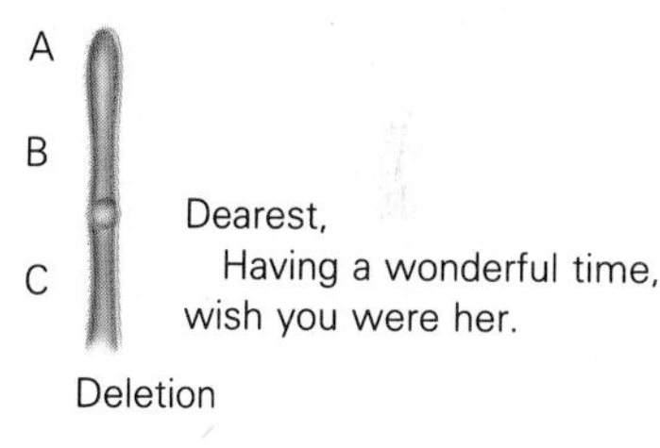

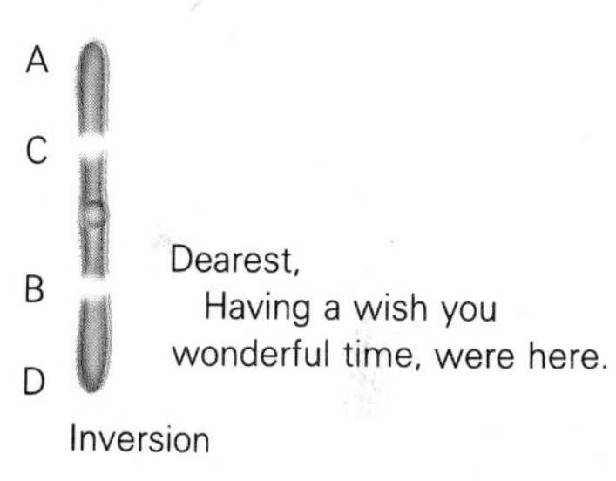

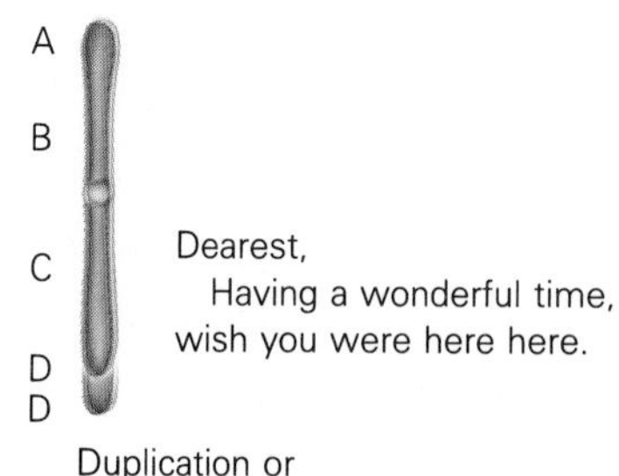

FURTHER BEYOND MENDEL

Why don't we find Mendel's classical 3:1 ratio everywhere? Obviously certain factors are at work confounding his elegant predictions. We have already seen two of these: linkage and crossing over. (Think now, do you see how they could have confused Mendel by introducing elements that he could have known nothing about?) It didn't take the classical geneticists of this century long to figure out these two complicating factors. But they still couldn't account for all the variation in their test program. As time went on, though, they began to unveil other mechanisms that helped explain the ratios they were getting.

Dominance Relationships

Mendel considered only **complete dominance** in which the effects of one allele completely masks the effect of the other. Actually, some homologous alleles show **incomplete dominance,** that is, neither is dominant and they produce intermediate effects.

An example of incomplete dominance is the inheritance of flower color in snapdragons. The presence of two red alleles produces a red flower and two white alleles result in a white flower. However, unlike the results of Mendel's crosses of garden peas, neither allele is dominant. Heterozygotes, individuals with one red allele and one white one, have pink flowers.

Multiple Alleles

We have generally considered only two alleles at each locus. Actually, there may be many traits that are the product of **multiple alleles,** three or more alleles of each gene. (Keep in mind that in this situation, each individual still receives just one pair of alleles.)

The most familiar example of multiple alleles is found in human ABO blood group systems, a discussion of which follows. Let's consider an example of multiple alleles from the ABO blood group system where there are three alleles possible at a single locus. There are several alleles responsible for blood types, with A, B, and O being the most common.

Persons who are homozygous for the **A** allele produce one kind of cell-surface antigen and are said to be **type A.** Persons who are homozygous for the **B** allele produce a different kind of cell-surface antigen and are said to be **type B.** Heterozygotes, **AB,** carry both alleles, producing both kinds of cell-surface antigens, and are said to be **type AB.** Persons who have neither the A antigen nor the B antigen belong to a fourth type—**type O.** Type O is the most common blood type. It seems that O is a third allele at the same gene locus, one that has neither enzymatic activity. It is recessive when paired with either B or A. With three different possible alleles, there are six different possible genotypes (Table 7.3).

Although **A** and **B** are codominant with respect to each other, they are dominant with respect to the **O** allele. Note that although there are three alleles of this gene locus in the human species, no one individual can have all three alleles at once. Diploid organisms like ourselves are limited to two at a time.

Gene Interaction

The genes at one locus may interfere with the expression of genes at another locus. The best known examples are due to **epistasis,** where one gene masks the effects of another. Mouse coat color, for example, is determined by genes for the dominant black and the recessive brown. The inheritance

TABLE 7.3

GENOTYPE	A ANTIGEN	B ANTIGEN	BLOOD TYPE
AA	present	absent	type A
AO	present	absent	type A
BB	absent	present	type B
BO	absent	present	type B
AB	present	present	type AB
OO	absent	absent	type O

of coat color proceeds without surprises unless a mouse is homozygous for an albinism gene at yet another locus. If this is the case, certain critical pigments are not provided for coat color and the mouse is white regardless of whether the coat-color gene codes for black or brown.

Polygenic Inheritance

Many different genes may have an additive effect on the same trait, a condition called **polygenic inheritance** (also see the discussion in Chapter 11). A number of genes acting on the same trait produces a wide range of variation for that trait, as we see in height, nose size, foot length, and weight. Let's see how polygenic inheritance might affect bill size in a population of woodpeckers.

Instead of an either/or situation where the birds have either long or short beaks controlled by genes for "long" or "short," we see a condition where several genes influence bill length. Some birds, by chance, will bear only alleles for long bills and so their bills will be long, indeed. Those with a preponderance of alleles for short bills will have to make do with a short beak. The laws of chance, though, result in most birds having a mixture of genes for long and short bills and the odds are the genes will be present in roughly equal numbers. If you toss a coin ten times (representing, here, ten genes for bill length), you could get any kind of result, even ten tails in a row (say, a very long bill). But if you made the ten tosses a thousand times (the number of birds in the population), you would, by chance—since the odds of getting heads or tails are equal—get closer to a 50:50 ratio of heads and tails (representing birds with average bills). And that, in fact, is what we get (Figure 7.17).

Other Sources of Variation

Other factors that may contribute to variation in populations are *environmental interactions* (the same genotype may be expressed differently in different environments), *sex influences* (the same trait may be expressed to

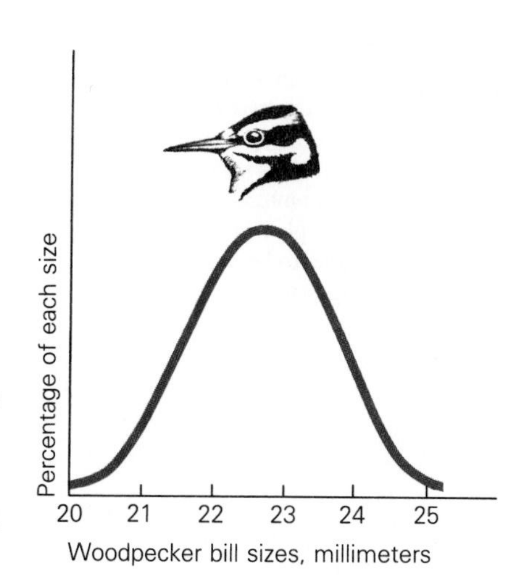

FIGURE 7.17
The bills of populations of woodpeckers vary. Some bills are long, some short, but the majority are of an intermediate length. In the population shown in this figure, there are few birds with bills 21 millimeters long and virtually none with bills of 20 millimeters. Nor do any have bills longer than 25 millimeters. Practically every woodpecker you see, though, has a bill within the 22 to 24 millimeter range. When a population shows a continuous distribution like this, instead of distinct classes, one can assume that the trait is controlled by several genes. Of course, a bird genetically disposed to have a long bill might not achieve that status if it has been subjected to a poor diet.

different degrees in males and females), and *effects of aging* (some genes exert their effects at only certain times, such as adulthood or old age).

POPULATION GENETICS

At one time, a genetic question making the rounds was: If brown eyes are dominant over blue eyes in humans, why doesn't everyone have brown eyes by now?

We now know the genetics of eye color to be more complicated than the question suggests, but even so, the answer is not a simple one. In fact, at the beginning of the century, many geneticists raised this point to argue against Mendel's principles. Actually, however, Mendel's laws held up quite nicely under this attack.

We can appreciate the argument by considering how allele frequencies can change in populations. We have already been using the term *population*, and you may have taken it to mean a group of individuals. You were not wrong, but at this point, we can give it a more precise meaning. In biology, a **population** is a group of interbreeding or potentially interbreeding individuals. With this in mind, we can consider the ratios of different alleles in a population for a specific characteristic, such as eye color.

First, imagine a population of only two individuals in which one sex is homozygous for dominant trait **A** and the other sex is homozygous for recessive trait **a,** as shown in Figure 7.18. Now, we know that all their F_1 offspring will be heterozygous for that characteristic. The Punnett square

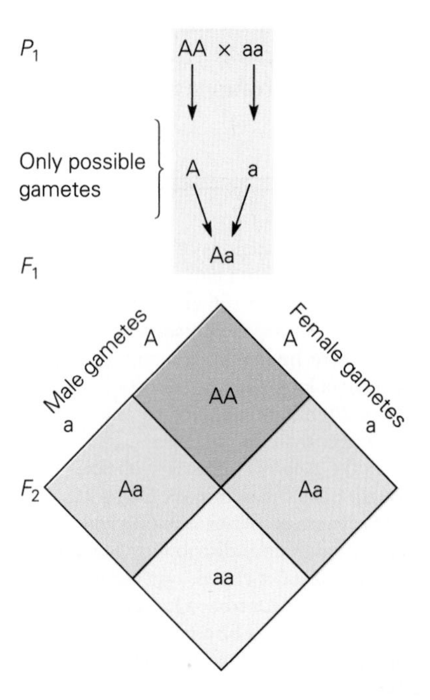

FIGURE 7.18

Here, one of the parents bears the dominant trait and one the recessive, but in the F_1, only the dominant phenotype exists, and in the F_2, the recessive is outnumbered 3 to 1. So it would seem that the recessive trait is disappearing from the population. However, an examination of the genotype shows that the ratio of dominant to recessive genes is precisely the same in each generation. The ratio will remain constant through the generations if mating is random, unless it is disrupted by some exterior influence such as immigration, selection, or drift. The principle here formed the basis of what came to be known as the Hardy-Weinberg principle.

then shows us that in the F_2 generation, three out of four individuals will show the dominant trait, and only one will show the recessive trait. So it might appear that we are on the way to eliminating the recessive allele from the population. However, if we now plot the F_3, F_4, F_5, and so on, generations, we will find that the proportion of dominant and recessive alleles in the population has not changed at all. In fact, look again at the F_1 and F_2 generations. There is a 1:1 ratio of the two alleles even at these stages.

Of course, a population of two is unrealistic, but the example illustrates a principle that also operates in larger populations. We can show by means of Punnett squares that the frequency of alleles for any characteristic will remain unchanged in a population through any number of generations—unless this frequency is altered by some outside influence.

The principle came to light during a discussion at Cambridge University in 1908. R. C. Punnett, the young Mendelian geneticist, was having lunch with his older friend, G. H. Hardy (Figure 7.19). Punnett said he had heard an argument critical of the Mendelian approach that he couldn't answer. According to the argument, if the allele for short fingers were dominant, and the allele for long fingers were recessive, then short fingers ought to become more common in each generation. Within a few generations, the critics said, no one in Britain should have long fingers.

Punnett disagreed with the conclusion, but he couldn't give a good reason for his argument. Hardy said he thought the answer was simple enough and jotted down a few equations on a napkin. He showed the amazed Punnett that given any particular frequency of alleles for normal or short fingers in a population, the relative number of people with short or long fingers ought to stay the same as long as the population was not subject to natural selection or other outside influences that could lead to changes in allele frequency. (Allele frequency refers to the ratio of different kinds of alleles in a population.) Punnett talked a reluctant Hardy into publishing the idea somewhere other than on a napkin (Hardy thought the notion was too trivial to be published). Others were developing the same notion, including a German physician named Wilhelm Weinberg, so the idea came to be known as the Hardy-Weinberg principle. (The principle was actually developed even earlier by an American named W. E. Castle, but for convenience and convention, we will sacrifice patriotism and propriety.) The principle is stated today as: In the absence of forces that change allele ratios in populations, when random mating is permitted, the frequencies of each allele (as found in the second generation) will tend to remain constant through the following generations. We will consider forces that change allele ratios in populations later, but for now let's examine the basis for the principle.

First of all, if we calculate the allele frequencies in the F_2 generation from Figure 7.18, we see in the F_2 generation that one-fourth of the population is **AA,** one-half is **Aa,** and one-fourth is **aa.** So, to determine what fraction of the F_3 generation will be offspring, say, of **AA** and **Aa** crosses, we simply multiply ¼ by ½ and get ⅛. In the third generation, then, assuming random mating, we can expect one-eighth of the population to have the genotypes resulting from this cross. What part of this generation will be the offspring of **AA** and **AA** crosses? Of **Aa** and **aa** crosses?

FIGURE 7.19
The English mathematician G. H. Hardy.

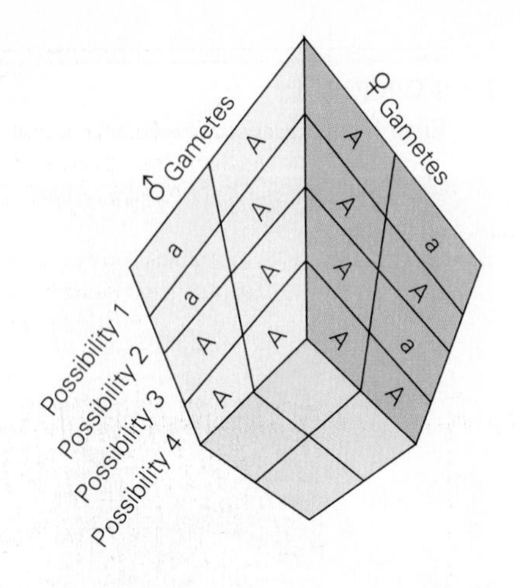

Now, there are four possible types of crosses that could produce **AA** individuals in the F_3 generation. You can work out for yourself what they are or conserve your energy and read the answers at left. (For any of these four mating possibilities, the boxes there can be filled in accordingly.)

Each of these combinations can be expected to produce one-eighth of the F_3 generation, so what part of the population will be heterozygous, or **Aa**? By counting the frequency of this type of gamete among the possible combinations, we see that $1/8 + 1/8 + 1/8 = 1/2$. And what was the fraction of **Aa** individuals in the F_2 population? One-half. If you work out the results for all possible combinations for the F_3 generation, or for any combination in any generation after the F_3 generation, you will see that the ratio of genetic components in a population remains stable. This is why we continue to have blue eyes in our population, even though they are a recessive trait.

Considering the genetic stability suggested by the Hardy-Weinberg principle, how can allele ratios in populations change? After all, evolution is essentially just changes in allele ratios. If there were some great advantage in having brown eyes that gave brown-eyed people a better opportunity to survive and reproduce, then the ratio of brown eyes to blue eyes could shift. It could also change if only people with eyes of a certain color were considered sexually attractive, or, if mutations were more likely to occur in one direction than another—say, from **a** to **A** rather than from **A** to **a.** If the population were small, shifts in gene frequencies could occur rapidly, since in a small group only a few accidental deaths could result in a great change in the overall ratios. Also, gene frequencies could change as a result of the flow between populations through immigration and emigration. Any of these things can occur in human populations, so the Hardy-Weinberg model is to some extent an artificial one. However, it helps to explain the relative constancy we see in populations around us.

SUMMARY

1. Mendel's principle of dominance states that when parents differ in one trait, their offspring, the F_1 generation, will display only one parental trait, the dominant trait. The recessive form of the trait is not seen in the hybrid.
2. Mendel's unit of inheritance, today called a gene, exists in alternative forms called alleles. An individual's assortment of alleles is its genotype. The phenotype is the expressed traits.
3. Mendel's principle of segregation states that when gametes form, the two alleles for a genetic trait separate so each egg or sperm carries only one allele.
4. An individual who is homozygous for a trait has two identical alleles for that trait. An individual who is heterozygous for a trait has two different alleles for the trait. When two heterozygous individuals mate, they generally produce three offspring showing the dominant form of the trait for every one offspring showing the recessive form (i.e., their offspring show the trait in a 3 : 1 ratio).

5. Mendel's principle of independent assortment states that each pair of alleles segregates independently during gamete formation. As a result, the occurrence of any characteristic in the next generation will be independent of the occurrence of any other characteristic.
6. A testcross in which a recessive individual is crossed with a dominant one can be used to determine whether a dominant individual is heterozygous or homozygous for the trait. If any recessive offspring result, the dominant individual must have been heterozygous.
7. Many of Mendel's findings can be related to the behavior of chromosomes during meiosis.
8. Gender is determined by a pair of sex chromosomes. Autosomes are chromosomes responsible for characteristics other than sex. In some organisms, such as the fruit fly *Drosophila,* those of males are an X and a Y chromosome, whereas the sex chromosomes of females are two X chromosomes. During sperm production, the X and Y chromosomes pair and separate so that half the sperm produced carry an X chromosome and the other half bear a Y chromosome. Since females are XX, all eggs have an X chromosome. If a Y-bearing sperm fertilizes an egg, the offspring will be XY and male. If an X-bearing sperm fertilizes the egg, the offspring will be XX and female.
9. Cells of human females have dark specks called Barr bodies, and the nucleus of a white blood cell in human females may have a structure called a drumstick. Both are believed to be inactive X chromosomes. Barr bodies and drumsticks are not found in the cells of human males.
10. An imbalance of sex chromosomes results in abnormal development of sexual characteristics. When only one X chromosome is present, the female will have Turner's syndrome. Klinefelter's syndrome occurs in XXY males.
11. Gene linkage is the occurrence of two or more genes on the same chromosome. Linked genes will be inherited together unless crossing over occurs during prophase I of meiosis.
12. The X chromosome carries some genes not concerned with gender determination. In fruit flies, a gene for eye color is on the X chromosome and in humans, the X chromosome may have an allele for red-green color blindness, multiple sclerosis, or hemophilia. Recessive alleles on the X chromosome will be expressed in males because they lack a dominant counterpart on the Y chromosome, but will only be expressed in females who are homozygous recessive for the trait.
13. The position of genes along a chromosome can be mapped using information from the observations of crossovers. The percentage of crossover is proportional to the distance between two genes along a chromosome.
14. Mutation is a change in genetic material. It can be caused by a number of agents including X rays, chemicals, and ultraviolet radiation. Mutations that alter a gene (gene mutations) occur when a nucleotide is omitted or repeated, or the wrong one added during DNA replication. Mutations involving the structure of chromosomes (chromosomal mutations) may involve deletions, inversions, or duplications. Most mutations are harmful.
15. There are several reasons why the 3 : 1 ratio in the appearance of a trait among offspring may not result from a cross between parents differing in the expression of a trait. Rather than showing complete dominance, in which one allele completely masks the other, some alleles show incomplete dominance and produce intermediate effects. Some genes have mul-

tiple (three or more) alleles. In epistasis, one gene masks the expression of a gene at another locus. For example, in mice the gene for albinism masks the expression of the gene determining coat color. In polygenic inheritance, different genes have an additive effect on a trait, producing variation in expression. Variation in gene expression may be caused by factors including environmental conditions, gender, and age.

16. A population is a group of interbreeding (or potentially interbreeding) individuals. The frequency of alleles for any characteristic will remain unchanged in a population through all generations unless it is altered by an external factor such as better survival and reproduction by those individuals bearing a particular allele, nonrandom mating, mutation, or accidents affecting a small population. Evolution is a change in allele ratios (allele frequencies).

KEY TERMS

allele (186)
autosomes (192)
Barr body (192)
chromosome mutation (199)
classical genetics (189)
complete dominance (199)
deletion (199)
dihybrid (187)
dominant trait (184)
duplication (199)
epistasis (200)
gene (186)
gene linkage (193)
gene mutation (197)
genotype (186)
heterozygous (187)
homozygous (186)
incomplete dominance (199)
inversion (199)
locus (186)
multiple alleles (200)
mutation (190)
phenotype (186)
polygenic inheritance (201)
population (202)
principle of dominance (184)
principle of independent assortment (187)
principle of segregation (186)
Punnett square (187)
recessive trait (184)
sex-linked trait (195)
testcross (188)

FOR FURTHER THOUGHT

1. The ability to roll one's tongue into a U shape is controlled by a dominant gene **(R)**.
   a. What genotypes could be produced if a homozygous individual for tongue rolling were mated with a recessive individual?
   b. What percentage of F_1 offspring will be able to roll their tongues?
2. Genes **A, B, C,** and **D** reside on the same chromosome. Genes **A** and **B** are 25 units apart. Genes **C** and **D** are 50 units apart. Would crossovers be more likely to occur between **A** and **B** or between **C** and **D**? Why?

FOR REVIEW

True or false?

1. ___ Genotypes are expressed, observable characteristics.
2. ___ A testcross is a cross between two dominant individuals.
3. ___ Gene linkage refers to the occurrence of two or more genes on the same chromosome.
4. ___ Autosomes are chromosomes that are responsible for determining gender.

Short answer

5. Mendel's principle of ___ states that when a heterozygote reproduces, half its gametes will carry a dominant trait and half will carry a recessive trait.
6. ___ is a condition in which several genes act on the same trait to produce variation.
7. Name the technique in which crossover events are used to determine a gene's location on a chromosome.
8. ___ mutations occur during DNA replication and may include repetition of nucleotide units.
9. The ___ principle offers a possible explanation of how gene frequencies operate within populations.

Choose the best answer.

10. A monohybrid cross such as **Bb** × **Bb** could produce individuals that are:
    A. homozygous dominant
    B. homozygous recessive
    C. heterozygous
    D. all of the above
11. Choose the correct statement.
    A. Sex-linked traits are always attached to the Y chromosome.
    B. Y chromosomes carry more genes than X chromosomes.
    C. Male offspring are produced when a Y-bearing sperm fertilizes an egg.
    D. None of the above.
12. During meiosis, if part of a chromosome breaks and is lost, it results in a ___ called a ___.
    A. gene mutation; deletion
    B. point mutation; deletion
    C. chromosome mutation; duplication
    D. chromosome mutation; deletion
13. ___ is a condition whereby genes at one locus mask the expression of genes at another locus.
    A. incomplete dominance
    B. inversion
    C. epistasis
    D. polygenic inheritance
14. Which of the following would not induce a change in the allele ratios of a population?
    A. random mating
    B. flow between populations (immigration and emigration)
    C. sexual selectiveness
    D. mutations

GENETICS PROBLEMS

1. A. Suppose Mendel had carried out the pea plant cross shown below. The characteristic under study is flower color, and each parent is derived from a pure-breeding line.

P_1	Purple	×	White
F_1	100% purple		
$F_1 \times F_1$	Purple	×	Purple
F_2	Purple		White
	705 plants		224 plants

   Identify each phenotype shown in the P_1, F_1, and F_2 generations as either homozygous or heterozygous.

   B. A pea-plant cross carried out by Mendel is shown next. The characteristic under study is seed color, and each parent is derived from a pure-breeding line.

P_1	Yellow	×	Green
F_1	100% yellow		
$F_1 \times F_1$	Yellow	×	Yellow
F_2	Yellow		Green
	6022 seeds		2001 seeds

   Identify the type or types of gametes produced by each type of individual in the P_1, F_1, and F_2 generations.

2. For each genotype, give the type or types of gametes produced and the frequencies.
   A. **aaGG**
   B. **AaGG**
   C. **AaGg**
   D. **Aagg**
3. The allele for yellow fruit color, **Y,** is dominant to the allele for orange, **y,** in certain species of squash. A heterozygous plant with yellow fruit is crossed with a plant with orange fruit.
   A. Identify the genotypes of both parent plants.
   B. Identify the kinds of gametes produced by each parent.
   C. Give the genotypes and phenotypes of the progeny and the expected percentage of each type.
4. The allele for red feather color in pigeons, **R,** is dominant to the allele for brown feathers, **r.** A red pigeon who had a red parent and a brown parent is mated with a brown pigeon.
   A. Give the genotypes of the two pigeons being mated.
   B. Identify the gametes produced by each of the pigeons being mated.
   C. What proportion of the F_1 progeny would be expected to have brown feathers?
5. The color of tomatoes is under genetic control, with the allele for red (**R**) dominant to the allele for yellow (**r**). A plant with red tomatoes is crossed with a plant with yellow tomatoes, and the F_1 consists of 42 plants with red tomatoes and 37 with yellow tomatoes. Based on this outcome, what can be concluded about the genotype of the red-fruited parent plant?

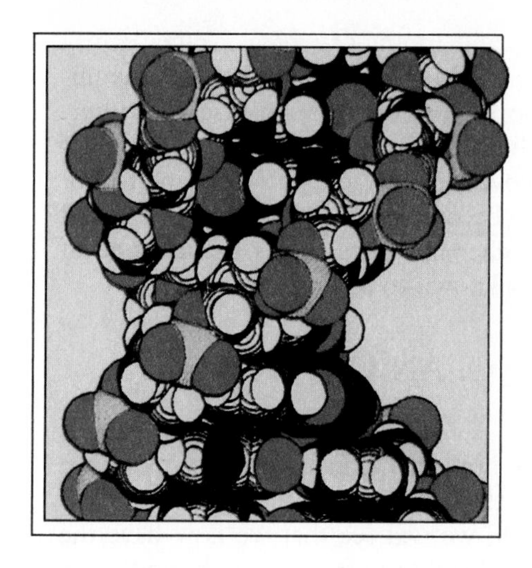

CHAPTER 8

Advances in Genetics

Overview

Genetic Engineering: An Unfolding Story

Is It Tampering with Genes or Genetic Engineering?

How Great a Promise?

Objectives

After reading this chapter you should be able to:

- Briefly outline the steps in recombinant DNA technology.
- Describe in detail the role of restriction enzymes.
- Explain how vectors and plasmids are used to clone a gene.
- Discuss the advantages and possible problems associated with genetic engineering.
- List ways genetic engineering techniques are used in agriculture and medicine.
- Describe the advances made in gene replacement therapy.

The field of genetics is presently undergoing sweeping changes largely because of a simple introduction: Madison Avenue has met the microbe. Already we find that the most conservative of financiers are interested in the mating habits of a lowly bacterium that is found in the bowels of everyone (including their employees). How did this all come to be? How did a generally harmless germ become the focus of such unexpected attention? Perhaps because, with other microbes, it has become one of the potentially most powerful tools known to science.

GENETIC ENGINEERING: AN UNFOLDING STORY

In this chapter, we will learn just how tiny, innocuous microbes have suddenly presented scientists with the ability to quite literally alter the course of humankind. In a nutshell, we will see that we now have the ability to find specific genes, to cut them away from their chromosomes within their natural organisms, to insert them into the chromosomes of other species, and then, after they have been duplicated countless times along with the normal genes of those species, to harvest them or their protein product.

To illustrate, in general terms, we can now excise from human chromosomes the genes that make human growth hormone. We can also open up certain bacterial chromosomes and insert the human gene into the bacterial chromosome. There, as the bacterial genes go about their business of making bacterial products, the human product (growth hormone) is made right along with them. Furthermore, each time the bacteria reproduce, they reproduce the human gene along with their own. At some point, the human product can be extracted in great quantities from millions of altered bacteria.

The Players

Before we focus on the pageant associated with placing genes in new environments, let's meet the players:

The **recombinant DNA** is a hybrid of DNA from two sources, the source DNA (later it will be called *foreign* DNA) and the host DNA.

The **source DNA** contains the "gene of interest," that is, the gene that will be added to a new chromosome.

The **host DNA** is the DNA that will be opened up to allow the insertion of the source DNA (that includes the "gene of interest").

The **vector** (a biological means of transfer) is the organism that will be used to transfer the source DNA into the host DNA.

The **restriction enzymes** are the enzymes that cut both the source DNA and the host DNA at specific places so that the former can be inserted into the latter.

A **clone** is a group of identical organisms that are descended from a single ancestor. Once a particular gene is inserted into a bacterium, that bacterium is then cloned and the gene product can then be recovered in great amounts from the clone.

All of these components and the processes associated with them in DNA technology are simply an extension of natural parts of biological systems.

Finding and Isolating the Gene

Obviously, the first thing to do in cloning a particular gene is to find that gene. This first step is rather simple. If an organism produces a certain substance, a protein, one can be sure that it has a gene for producing that protein. The trick is in rummaging through its chromosomes to find just where that gene is so that it can be excised and reproduced.

One way to approach the problem is to subject the cells containing the gene to restriction enzymes. These, as we will see next, will chop the chromosomes apart, cutting them in specific places and leaving fragments that can then be inserted into the DNA of host chromosomes.

In this way, a pool of cells is produced, but only some cells carry the gene of interest in their chromosomes. From among this pool then, the researcher must isolate those cells containing that gene. This can be done by identifying those cells that manufacture the protein produced by the gene. (Specific proteins interact with certain molecules called antibodies. If that reaction is detected, the gene is present.)

How DNA Is Chopped Up

Every species of bacteria produces at least one specific restriction enzyme. They don't go to all this trouble just to assist recombinant researchers, though. The enzymes are a defense against invasion by viruses that insert their own chromosomes into bacteria, thereby taking over the genetic machinery of the bacteria, causing them to make new viruses. So any viral DNA is likely to be chopped up and rendered harmless by bacterial restriction enzymes.

Restriction enzymes operate at very specific places along DNA strands. The restriction enzyme produced by the intestinal bacterium ***E. coli,*** for example, recognizes the following nucleotide sequences and cleaves the two complementary strands at the places indicated here by arrows:

```
. . .G↓A A T T C. . . .
. . .C T T A A↑G. . . .
```

So, the sequence is broken, wherever it appears, in a specific way:

```
. . .G      A A T T C. . . .
. . .C T T A A      G. . . .
```

The result is two molecular sequences with uneven, or staggered, ends. Because they are staggered, these ends are "sticky." That is, they tend to join with other molecules with a complementary sequence of nucleotides.

In the laboratory, fragments of DNA from different organisms, even from different species, can be joined together at their sticky ends (in the presence of an enzyme called ligase), thereby producing recombinant DNA. Remember, the restriction enzymes don't care about the source of the DNA; wherever this specific target sequence of nucleotides exists, they cut it. And whenever complementary sticky ends encounter each other, they join.

So the next step, after ascertaining the existence of a particular gene within the cells of an organism, is to disrupt those cells and subject their chromosomes to a specific restriction enzyme. This results in a pool of DNA fragments with very specific sticky ends, that is, with a particular sequence of exposed, single-stranded nucleotides.

Getting Genes from One Place to Another

The next step is to get those fragments, some carrying the gene of interest, into a bacterial population where they can be cloned. This is done by using a vector, a biological means of transfer. A common vector is the bacterium *E. coli.*

E. coli, and many other bacteria, contain a large, circular chromosome and a small circle of DNA called a **plasmid** (Figure 8.1 and Essay 8.1). Some plasmids contain a few genes, including those conferring immunity against **antibiotics,** substances that kill infectious organisms (*anti:* against; *bios:* life).

FIGURE 8.1
E. coli, (a), with a circular chromosome and smaller plasmid. Photomicrograph is shown in (b). Diagram (c).

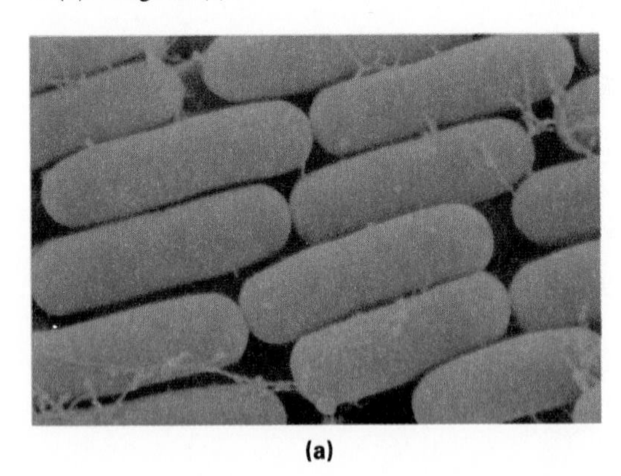
(a)

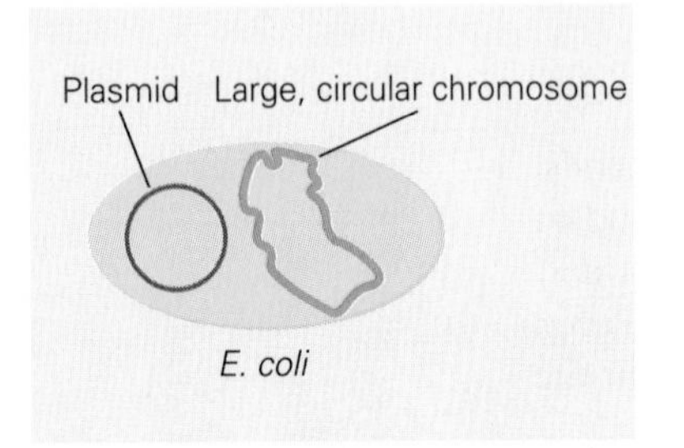

(c)

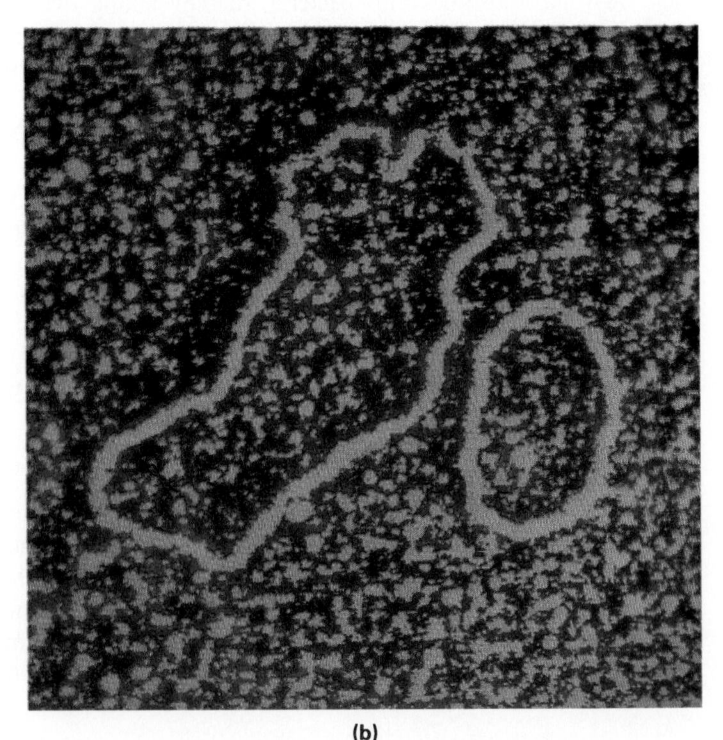
(b)

Splicing and Cloning the Genes

By using the same restriction enzyme that broke apart the source DNA, the plasmid rings can be broken apart at those places where the appropriate nucleotide sequence occurs. If the source fragments and the opened plasmids are then mixed together, the source fragments can be spliced into the plasmid molecule, as their complementary sticky ends fuse (again in the presence of ligase).

Now the plasmids, with their new DNA insertions (some carrying the gene of interest), can be inserted into host bacterial cells. This can be done by subjecting the new host cells to calcium chloride, which renders their cell walls permeable to the altered plasmids. Once the altered plasmids are inside their new host cells, they will reproduce themselves and their new DNA sequence, each time the host cell reproduces. In this way the recombinant DNA is cloned, making many copies that can later be retrieved. The process is straightforward, as we see in Figure 8.2.

Sorting Through the Recombinants

At this point the researcher has a population of bacteria, some of which contain plasmids and some of which don't. The problem is, the process is not particularly efficient, so the next thing is to find those cells that now contain plasmids and to separate these from the rest. This is actually not as difficult as it sounds because most plasmids used in cloning experiments

FIGURE 8.2

Using bacteria to make human hormones. The plasmids are first separated out from ruptured bacteria and subjected to a disruptive enzyme that breaks the circle at specific points. These opened loops are then mixed with a sequence of DNA from another source, say the gene sequence that makes a human hormone. The foreign sequence is inserted at the broken ends, reforming the circular plasmid, but now with new human genes inserted. As the bacterial genes direct the formation of bacterial substances, the human genes direct the formation of human hormones. The bacteria continue to reproduce in great numbers until finally they are chemically ruptured and quantities of human hormone can be retrieved.

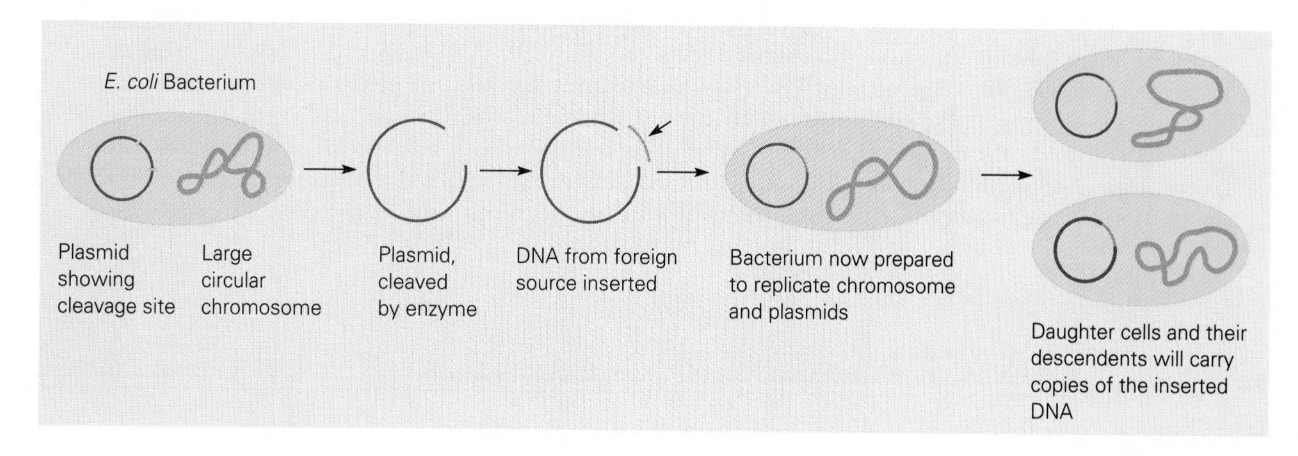

ESSAY 8.1

Bacterial Sex

Now let's consider a topic you may not have thought much about: how bacteria have sex. Many kinds of bacteria, including *E. coli,* normally reproduce through a procedure quite similar to mitosis. That is, they reproduce their genetic component and then pinch in two, each half receiving identical genes. But *E. coli* also has ways of recombining its genes in a process somewhat akin to sex.

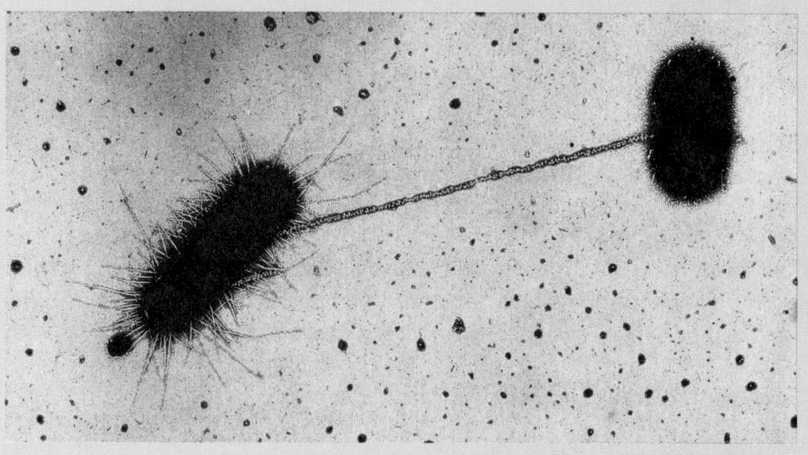

Two bacteria involved in the transfer of genetic material from one organism to another through a conjugation tube.

According to those who have seen bacterial sex, *E. coli* doesn't really have the hang of it (but then, who does?). However, in about one case in a million, two bacteria will undergo something called **conjugation.** In conjugation, one bacterium forms a long, slender tube that penetrates the body of an adjoining bacterium. Then, a type of genetic material moves from the bacterium that produced the bridge through the bridge into the body of the second bacterium. In this way, genes from one organism can join the genes of another organism, and as the genes are shuffled about, recombination can take place.

Now the story gets even stranger. In a nutshell, maleness has turned out to be a kind of disease. It is now known that populations of *E. coli* are composed of two "mating types" that were (a bit hastily) labelled "male" and "female." Each type will conjugate only with the other type. The label "male" was given to the donors of genetic material during conjugation, and the label "female," to the recipients. In conjugation, once the tube-like bridge between the two mating types has formed, certain of the "male's" chromosomal rings break, and the linear results slowly begin to move across the tube from the male into the female.

But what's this about maleness being a kind of disease? (Some people have suspected as much for years.) Experiments have shown that when male and female strains are mixed, the males remain males, but about a third of the females also become male. On the surface it seems that maleness is, indeed, somewhat contagious.

The reason for this contagion was a great puzzle until it was learned that maleness is due to the presence of plasmids (see Figure 8.1), a small circle of DNA found inside bacteria. Some kinds of plasmids contain only a few genes, and their activity is intensely directed toward their own reproduction. In *E. coli,* the plasmid not only contains the genes that confer "maleness," but it also directs the bacterium to form a conjugation tube. As the tube forms, the plasmid reproduces, forming linear copies of itself. These copies pass through the bridge into the female. The female is, in this way, changed to a male and rendered impermeable to penetration by any other male. (This accounts for the growing number of males in the mating group.) Scientists by now knew enough to discard the old "male" and "female" labels. The males simply had an F plasmid and the females didn't. So the males were considered F^+ and the females F^-.

Conjugation begins with an F⁺ cell developing a conjugation tube and penetrating a receptive F⁻ cell. The DNA of the plasmid then begins to replicate, remaining circular as the copied strand, produced in linear form, breaks away and passes through the tube and into the F⁻ cell. Once the entire sequence of DNA is inside the recipient and replication is completed, the strand becomes circular, and a new male (F⁺) is formed. Now you know about the private life of a common bacterium. And remember, you learned it here.

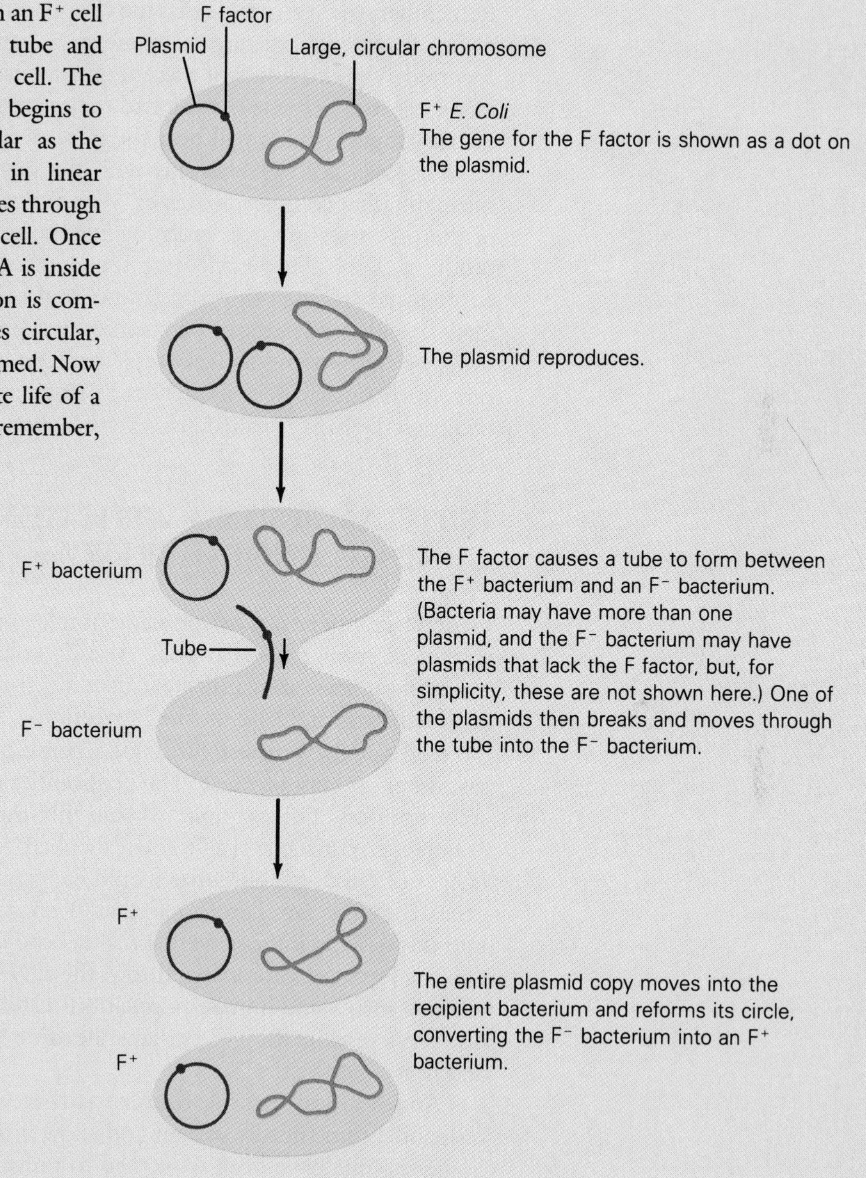

confer antibiotic resistance to their hosts. So the host population is simply subjected to antibiotics. Those that die didn't have the plasmid. The rest are cloned.

The next problem is more difficult. Now, the researcher must somehow find those cells carrying plasmids that have the gene of interest. Remember, the restriction enzyme chopped the source DNA into a number of fragments, cutting it wherever a particular nucleotide sequence occurred. Various kinds of these fragments could have been incorporated into a plasmid because they all had the appropriate sticky ends. The result is that some plasmids will bear the gene of interest and some won't.

The cells bearing plasmids with the gene of interest are sorted out through rather complex processes. We won't go into details here, but some of the processes involve screening for the protein the gene is known to produce. Also, if the nucleotide sequence of the gene is known (or can be deduced by analyzing the amino acid sequence of its protein), then nucleic acid segments can be chemically synthesized and used to identify the gene by bonding to its complementary sequences. Some laboratories carry such nucleic acid sequences in stock and the molecules can simply be ordered. (This is called "cloning by phone.")

IS IT TAMPERING WITH GENES OR GENETIC ENGINEERING?

Genetic engineering, as the name implies, involves manipulating genes to achieve some particular goal. Already you can see what sorts of objections might arise. But it doesn't take much imagination to see what new achievements might be on the horizon.

Perhaps the greatest threat of recombinant techniques, some would say, lies in its very promise. The possibilities of such genetic manipulation seem limitless. For example, we can mix the genes of anything, say, for example, an ostrich and a German shepherd. This may only bring to mind images of tall dogs, but what would happen if we inserted cancer-causing genes into the familiar *E. coli* that is so well adapted to living in our intestines? What if the gene that makes botulism toxin, one of the deadliest poisons known, were inserted into the DNA of friendly *E. coli* and then released into some human population? One might ask, "But who would do such a terrible thing?" Perhaps the same folks who brought us napalm and nerve gas.

Another, less cynical, concern is that well-intended scientists could mishandle some deadly variant and allow it to escape from the laboratory. Some variants have been weakened to prevent such an occurrence (Essay 8.2), but we should remember that even after smallpox was "eradicated" from the earth, there were two minor epidemics in Europe caused by cultured experimental viruses that had escaped from a lab. One person died of a disease that technically didn't exist.

Such things, of course, are the stuff of nightmares, and when gene splicing became a real possibility (Essay 8.3, p. 218) there was immediate furor. Many of the scientists involved met together in Asilomar, California, in February 1975 and, in some rather heated sessions, forged a set of

ESSAY 8.2

Crippling a Microbe

At one of the conferences on the risks of recombinant studies, geneticist Roy Curtiss III volunteered to produce a weakened mutant—one that could not survive outside the laboratory.

Given the go-ahead, he first produced an *E. coli* with a defective gene that prevented it from manufacturing its protective coat. The material that the gene normally made had to be provided artificially. But microbes mutate on their own, so some were soon back to producing the normal gene. Curtiss then deleted another gene necessary for coat production. But he was outfoxed by the crafty germs; they reproduced anyway. Dennis Pereira, a graduate student working with Curtiss, found that they were manufacturing a sticky substance called colanic acid, which acted as a kind of protective coat. So Curtiss and Pereira produced a microbe that couldn't make colanic acid. Finally they had a germ that depended on scientists for its livelihood. As an unexpected bonus, this new bug was sensitive to ultraviolet light. Ordinary sunlight would kill it.

This fermenter is a biochemical cauldron for growing gene-spliced microorganisms.

One problem remained. Even dying *E. coli* can conjugate with normal *E. coli*, so an escaped germ might be able to pass its dangerous gene to a healthy colleague. Curtiss, however, altered the gene of the dependent bacteria that makes thymine. Thymine, therefore, had to be supplied, and without it, how could DNA be made? Perhaps now the bug was helpless enough.

operating guidelines to keep such potential catastrophes from ever becoming a reality. These guidelines were self-imposed and have evolved into quite firm rules at this point. But not everyone was satisfied. Some communities even banned such research in their areas. Then something happened. No one is quite sure why the change came about, but much of the opposition to such research was quietly abandoned. It all happened at about the time recombinant techniques showed promise of great commercial feasibility.

HOW GREAT A PROMISE?

There were people who feared the results of gene splicing, but no one could deny that the promise was great and that the successes of the technique were beginning to mount. The beneficial aspects of recombinant DNA research were first proposed in the late 1970s, when a number of laboratories announced preliminary results suggesting that gene splicing could be a practical solution to a great many human problems. As these various small laboratories, often staffed by maverick graduate students,

ESSAY 8.3

The Gene Machine

New techniques based on plasmid transfer now make it possible to determine the exact nucleotide sequence of any isolated section of DNA or RNA. Already, the entire base sequences of DNA in bacteriophages, plasmids, polio viruses, and human mitochondria have been determined. They produce pages and pages filled with A, T, C, and G. (There are exactly 5315 nucleotides in the case of the first bacteriophage sequence to be discovered.) Your own DNA sequence of some 6.4 billion nucleotide pairs would fill a thousand volumes the size of this one. And, incredible as it sounds, the task is already well under way. Tom Maniatis, a Harvard molecular biologist, has chopped human DNA into fragments a few tens of thousands of nucleotides long and inserted them into the DNA of bacteriophages. His laboratory is virtually a DNA library of three million separate clones of cultured human DNA.

And now we have the gene machine. This is a device with Madison Avenue overtones, but it is available for only about $30,000, and it does just what it says. An operator can create any DNA sequence simply by typing it out. The computerized machine does the rest. For example, one might know a protein amino acid sequence but be unable to isolate the corresponding *m*RNA. If the investigator knows the sequence of even a short segment of the gene, it is typed into the gene machine, and a probe locates the rest of the gene from a DNA library consisting of cloned chromosome segments of the species in question. If all one has is a tiny amount of protein, a highly sensitive protein-sequencing accessory is also available. Protein samples placed in the automatic sequenator are dismantled chemically, one amino acid at a time from one end of the polypeptide. These fragments are indentified automatically and the sequence is printed out.

So, once a probable nucleotide sequence is determined, it is typed into the gene machine, which then goes through the steps of that DNA's synthesis. In about a day's time, a small quantity of the desired DNA is produced. Sections are cloned in a bacterial plasmid, which is then taken up by a receptive bacterium. Standard biological gene cloning methods then take over. Potentially, then, any enzyme the experimenter can imagine or any other kind of protein or gene can be made to order.

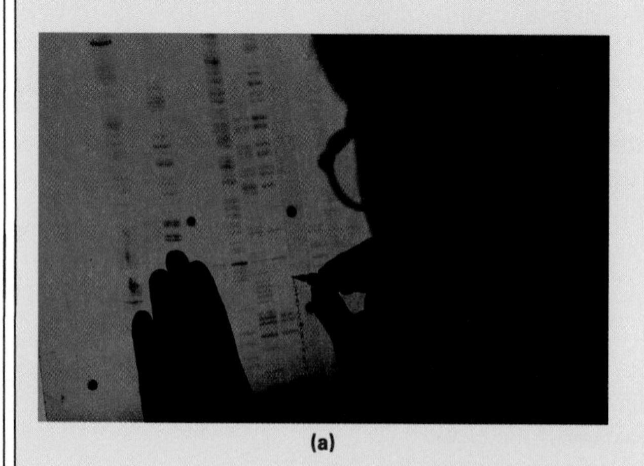

(a)

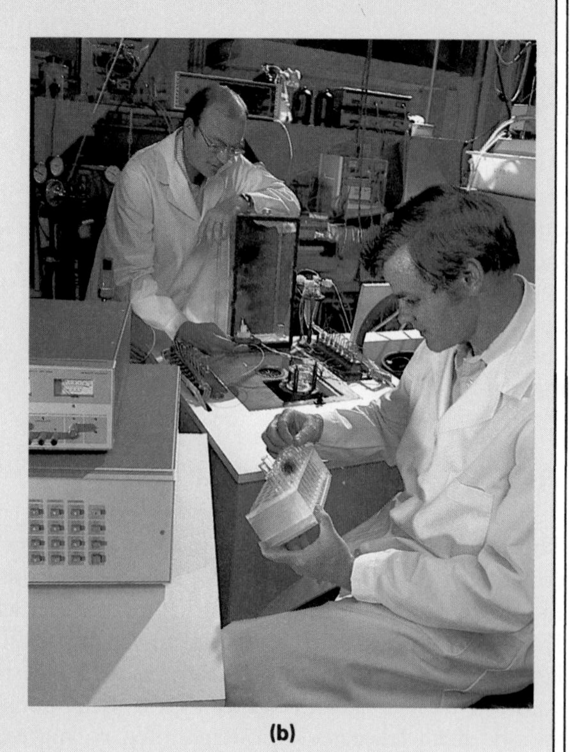

(b)

In radionucleotide sequencing (a), scientists can label each protein in a human gene to look for congenital disease. With a gene machine (b), scientists can create any DNA sequence simply by typing it out—the machine does the rest.

announced their findings, Wall Street reacted with great verve and fervor, and millionaires were made overnight (Figure 8.3). The hooks and ladders of financial gain tempted a number of dedicated scientists to escape the ivory towers of academia.

These people, imaginative, highly trained and highly motivated, saw all kinds of possibilities in recombinant DNA research. They realized that the major ability of genetic engineering is to simply produce great amounts of rare DNA. As we've seen, it is possible to take a human gene, insert it into a plasmid, grow the bacteria that harbor the plasmid, and then recover the products of the human DNA in vatloads.

Take, for example, insulin. Insulin has traditionally been harvested from human tissue, or from slaughtered pigs and cows (which yielded a similar kind of insulin that worked in most humans, but caused allergic reactions in others). Unfortunately, very little insulin could be extracted by such methods. Now, though, a gene that codes for human insulin has been isolated, inserted into a plasmid, and cloned, so the source for authentic human insulin is essentially unlimited. Thus, great quantities of insulin are now available that can be used to treat people with diabetes mellitis, who do not produce sufficient amounts of insulin. A sample of therapeutic drugs now manufactured through recombinant DNA is found in Table 8.1.

FIGURE 8.3
Two of the more successful early genetic engineers, Robert A. Swanson (left) and Herbert Boyer (right), co-founders of Genentech. Their research firm conducted the breakthrough gene-splicing study in which insulin was manufactured by *E. coli* bacteria.

Advances in Agriculture

Some of the greatest promises of recombinant research lie in agriculture. For centuries farmers have been waging war with plant pests and diseases, infertile soil, and an increasing need to produce more food for the world's burgeoning human population. Genetic engineering offers new hope as it raises the possibility of making plants resistant to insects, bacteria, fungi, and unsuitable growing conditions. Researchers are also working to simply increase the size of the edible parts, such as roots, seeds, or fruits. Further, plant foods can now be made more nutritious through alterations in their amino acid content.

There are both advantages and disadvantages in doing recombinant research on plants. The good news is that several important species (carrots,

TABLE 8.1
Genetically Engineered Products

PRODUCT	FUNCTION
Human growth hormone	Promotes growth in children with hypopituitarism
Interferon	Helps cells resist viruses
Interleukin	Stimulates the proliferation of white blood cells that take part in immune responses
Insulin	Treats diabetes by enabling cells to take up glucose
Renin inhibitor	Decreases blood pressure

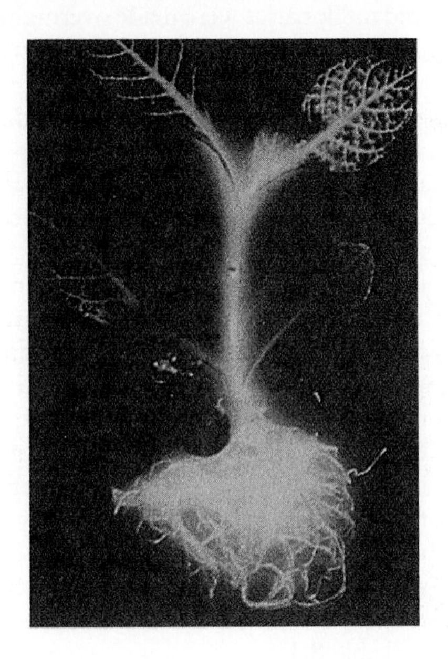

FIGURE 8.4

The gene for the enzyme luciferase, responsible for the firefly's glow, was placed into the genome of the tobacco plant at left through a remarkable feat of genetic engineering, causing it to glow as well (right). Such accomplishments are showing wide application for a pressing array of human needs.

cabbage, citrus, and potatoes) can be grown from single cells. So once a gene is introduced to a cell, a clone of that cell can produce countless altered progeny. The bad news is that most plant characteristics that need improvement (such as growth rate, size of edible parts, and amino acid balance) are polygenic—controlled by many genes. The problem is we haven't identified most of the genes responsible for such traits, and even if we had, the replacement of even five or ten genes would be terribly difficult.

Still, there have been some impressive feats of genetic engineering in plants (Figure 8.4). Some of the work has shown great potential. For example, not long ago, it was discovered that the herbicide glycophosphate could not be used to kill the weeds that grew among the crops because it killed all green plants—weeds and crops alike. A bacterial gene was discovered that could serve as a weapon against the herbicide. This gene has now been cloned and transferred into a number of crop varieties, causing these crops to be resistant to glycophosphate. Now the cropland can be sprayed to control the weeds without killing the crops.

Diagnosis and Decision in Humans

At least 2000 genetic errors in human metabolism are known. Some, such as Tay-Sachs disease and sickle-cell anemia occur primarily in certain races. Others show no racial or ethnic bias and plague all groups about equally. Some disorders appear in the young and kill them or mark them for life. Yet others, though, appear later in life, perhaps after the individual has reproduced and passed on the defective gene. In the past, we could only look at family trees to predict the odds of someone turning up with one of these diseases.

FIGURE 8.5

Rita Hayworth, one of America's most glamorous and successful actresses in the nineteen forties and fifties, was to fall victim to Alzheimer's disease at a time when her career could have continued.

Now, though, we are able to look at the genes themselves, at least in certain cases. If a mutation results in a specific disease, and that mutation occurs in a restriction site (and there are many such mutations, so the odds are pretty good), then the corresponding restriction enzyme will no longer work there. So if the enzyme is added, and the site remains unbroken, a mutation has occurred. It is thus possible to tell people if they are heterozygous (carriers) for a condition (if only half of the restriction sites are broken). It is also possible to tell pregnant women whether their fetuses bear certain inborn errors of metabolism.

In some cases, such information can result in our having to make very hard decisions. As examples, Alzheimer's disease (Figure 8.5) and Huntington's disease (Figure 8.6) are among the "delayed onset" disorders. If the gene for the disease is present, an awful fate awaits the bearer of the gene in his or her middle or late years. There may be two reasons for someone wanting to know if he or she bears the gene for these terrible

FIGURE 8.6

Woody Guthrie, folk hero who wrote "This Land Is Your Land," was one of the most influential figures in American folk music until he was gradually stricken with Huntington's disease. He was angry much of the time, partly as a symptom of the disease and partly because he knew he had it.

conditions. One reason is to decide whether or not to have children. Huntington's disease, for example, is due to a dominant gene, so if a person bears the gene, there is an even chance that the offspring will be afflicted. Another reason is that, with a dismal and dependent future ahead, one might like to prepare for the worst, even while living each day with such knowledge.

Whereas most people would want to protect any children they might have, many do not want to know about their own fate. Think about it; would you want to be told that sometime after your thirty-fifth birthday you will begin a period of rapid neural deterioration leading to an early death (Huntington's disease), or that by age 45 you will probably begin to lose your memory and soon after most of your mental abilities (Alzheimer's)? (A recent study at Johns Hopkins revealed that 60 percent of us don't want to know.)

But what if your condition could be treated by providing you with the normal gene product? Would you change your mind? Such treatment is becoming increasingly likely. Or how about a much more direct treatment. That is, simply replacing the defective gene. We are, in fact, on the threshold of being able to replace defective genes with normally functioning genes.

Gene Replacement Therapy

Hypogonadism is a relatively common recessive condition in mice and humans. Homozygotes have underdeveloped gonads, are sterile, and (in mice anyway) seem to have no idea of how to go about mating. The problem is, they are unable to produce a particular hormone (called gonadotropin-releasing hormone, the messenger that stimulates the pituitary to prompt the formation of gametes and sex hormones by the gonads). The condition was cured in mice by injecting eggs with the normal gene. In about 20 percent of the cases, the egg's chromosomes incorporated the normal genes and later, when fertilized and reimplanted into a surrogate mother, the altered embryos grew into normal fertile mice.

FIGURE 8.7
Two of the gene therapy researchers, Drs. Blaese and Anderson, on the eve of performing their historic procedure.

Replacing defective genes with normal genes is called **gene replacement therapy.** This treatment is the newest application of genetic engineering and treating human problems with altered genes has just begun.

In 1990 the first human gene-therapy was performed. Researchers at the National Institute of Health (Figure 8.7) infused gene-altered immune system cells into a 4-year-old girl with an inherited immune deficiency. Other researchers corrected a cellular defect that caused up to 75 percent of cystic fibrosis cases by inserting a gene that produces normal protein into cells bearing a defective gene. The engineered cells then produced the normal protein and corrected the abnormality. With such treatments having crossed the legal, ethical, and scientific hurdles, scientists can be expected to attack human problems on a broad front—using genetically engineered cells.

We see then, that the potential of genetic engineering is immense. There probably isn't a genetic defect that can't be reversed in some way, a protein that cannot be made in bacterial cultures, or a gene that cannot be replaced. The only question is whether technology will outrace our abilities to use such powerful tools wisely and well.

SUMMARY

1. It is possible to remove specific genes from the chromosomes of their natural bearer and insert them into the chromosomes of other species who will duplicate the gene as they reproduce, producing enormous quantities of the gene product that can then be harvested.
2. When a source DNA, which contains a gene of interest, is inserted into host DNA, the product is recombinant DNA.
3. A first step in producing recombinant DNA is to fragment the chromosomes containing the gene of interest. A restriction enzyme cuts DNA at specific places, leaving uneven ends called sticky ends (because their unpaired bases tend to join with other fragments having a complementary sequence of nucleotides).
4. Each chromosome piece is inserted into a host chromosome of a different cell. The cell that produces the gene product contains a DNA fragment with the gene of interest. The fragments are transferred to bacterial cells where many exact copies can be made by cloning. (A clone is a group of identical organisms with a single common ancestor.)
5. The transfer requires a vector, which is an organism used to transfer the source DNA to the host DNA. A common vector is the bacterium *E. coli,* which contains a circular chromosome and a small circle of DNA called a plasmid. DNA fragments are spliced into plasmids using a restriction enzyme. The plasmids carry the inserted DNA to new host bacterial cells for cloning. Since plasmids also contain genes for resistance to antibiotics, plasmid-containing cells are selected by exposing all cells to antibiotics.
6. The cells containing the gene of interest are identified by screening for the gene product or by matching up the base sequence in the DNA.
7. Although some people feared that recombinant DNA techniques might be misused and be dangerous, genetic engineering has many practical applications. It has been used to make large quantities of useful proteins such as insulin. It raises the possibility of making plants resistant to pests, more hardy, or with larger edible parts. It is useful in diagnosis of certain human metabolic disorders. Genetic engineering can detect the presence of the gene for delayed onset disorders such as Alzheimer's and Huntington's disease. In gene replacement therapy, defective genes are replaced with normal ones.

KEY TERMS

antibiotics (212)
clone (210)
conjugation (214)
E. coli (211)
gene replacement therapy (222)
genetic engineering (216)
host DNA (210)
plasmid (212)
recombinant DNA (210)
restriction enzyme (210)
source DNA (foreign DNA) (210)
vector (210)

FOR FURTHER THOUGHT

1. Describe the difficulties one would encounter in trying to clone DNA if the plasmids of bacteria should suddenly lose their resistance to antibiotics.
2. Describe a major concern you might have if a recombinant DNA research center were constructed in your neighborhood.
3. Why would a disease that is controlled by many genes be a poor choice for recombinant DNA research?

FOR REVIEW

True or false?

1. ___ Source DNA can only be inserted into bacterial cells that lack plasmids.
2. ___ Vectors transfer host DNA into source DNA.
3. ___ Restriction enzymes protect bacteria from viral invasions.
4. ___ In order for the cloning to be successful, the host and source DNA must be from the same species.

Fill in the blank.

5. The small circle of DNA in addition to the chromosome in a bacterial cell is called a ___ .
6. ___ is a genetic engineering process that substitutes normal genes in place of defective genes.
7. ___ such as *E. coli* carry genes of interest to bacterial populations.
8. A group of identical organisms descended from a common ancestor is called a ___ .

Choose the best answer.

9. Genes of interest are contained in:
   A. restriction enzymes
   B. host DNA
   C. source DNA
   D. all of the above
10. Restriction enzymes:
    A. are produced by bacteria
    B. operate at specific sites along DNA strands
    C. inactivate viral DNA
    D. all of the above
11. Which of the following can serve as a recombinant DNA vector?
    A. plant cells
    B. bacterial cells
    C. any cells
    D. only animal cells
12. The first step in the recombinant procedure is:
    A. Bacterial populations are subjected to antibodies.
    B. Source DNA is cut by enzymes.
    C. Plasmid rings are subjected to restriction enzymes.
    D. Bacterial cells are cloned.

13. Recombinant DNA is cloned at the point when:
    A. the restriction enzyme is produced.
    B. the source DNA replicates.
    C. The vector reproduces.
    D. the host cell reproduces.
14. One *disadvantage* in recombinant research on plants is that:
    A. important plant characteristics are controlled by many genes.
    B. important plant species grow from single cells.
    C. bacterial DNA cannot be cloned by plant host cells.
    D. the food value of plants cannot be increased.
15. Gene replacement therapy uses recombinant DNA to:
    A. harvest genes of interest.
    B. replace defective genes.
    C. diagnose "delayed onset" disorders.
    D. create beneficial bacteria.

PART THREE

The Diversity of Changing Life

The beginnings of life are traced through various scenarios, focusing on the modern evidence of that distant process. We then see how life changes on a changing planet to produce the grand diversity of living things existing today, paying particular attention to the evolutionary and adaptive themes influencing all life from the ancient bacteria to humans.

CHAPTER 9

The First Life

Overview

Life from Nonlife

The Expanding Universe

Our Early Earth

Hypotheses Concerning the Origin of Life

The Formation of the Molecules of Life

Protobionts: The Ancestral Droplets

Heterotrophy and Autotrophy

Oxygen: A Bane and a Blessing

Objectives

After reading this chapter you should be able to:

- State the mechanists' and vitalists' views on the origins of life.
- Describe how the hypothetical Big Bang accounts for the formation of the universe and describe the environmental conditions of early earth.
- Explain how the experiments of Stanley Miller and Sidney Fox are consistent with current scientific thinking on how life might have arisen.
- Trace the steps by which coacervate droplets may have evolved into living cells.
- Describe the evolution of early autotrophs and heterotrophs.
- Describe how the presence of oxygen changed the world and its early life forms.

Our best attempts at defining life seem to have a way of grinning back at us, making us feel a bit unsure of ourselves. Furrowed brows, cleaned spectacles, cleared throats, and all the other trappings of great knowledge do not diminish the simple reality that no one knows what life is. We have, though, listed some of the special qualities associated with life. You may have noticed that we didn't *define* life, we only *characterized* it, pointing out some of its common traits.

LIFE FROM NONLIFE

Questions regarding the origin of life have given rise to two major schools of thought, each with its own underlying assumptions. One is called *mechanism,* the other *vitalism.* It may seem, as we begin to define our terms, that any such arguments can only be sterile and academic, and, indeed, to consider the concepts in these terms is a historical convenience. But if we are to discuss how life might have begun, we should begin by focusing our quest on one of these two principles (Figure 9.1).

Mechanism vs. Vitalism

It will soon become apparent, if you have one ounce of poetry in your soul, that the mechanistic explanation of life is the less romantic of the two. This is because **mechanism** implies that life is the result of, and is subject to, the same laws of the universe as any other physical entity. It reduces the wondrous qualities of life to the simple interactions of mindless molecules. **Vitalism,** on the other hand, is based on the premise that living things are more than the result of molecular interaction, that life inherently possesses some undefinable and unmeasurable quality—an essence that some have called the "life force." The mechanists respond that if any such force exists, it too must be physical in nature, an argument that drives the vitalists right up the wall. Since the "life force" of the vitalists is generally not considered to be measurable, the explanation does not present the sort of questions that can be answered by scientific investigation. As a result, we will explore the alternative, the mechanistic view. In particular, let's see if there are ways that life might have arisen based on what we know of physical and chemical principles. (We have already considered such principles in some detail; here we will deal only with a very general overview.) Actually, the distinction between these venerable old ideas has been blurred by those philosophers who argue that the specialness of life is due to its being greater than the sum of its parts, that the interplay of its components yields a product that is greater and more momentous than can be accounted for by simply examining its constituents.

Life's Many Meanings

It seems that those who object to the mechanistic view, so far, have little to worry about. Up to this point, one cannot be very encouraged about our ever understanding the "essence of life" at the molecular level. The

FIGURE 9.1

Signs of life. Arguments can rage over how life began and whether it includes rocklike forms or crystals or whatever, but other forms seem to celebrate their existence, such as (a) the vibrant hummingbird, (b) the elegant hibiscus, (c) the clownfish nestled among the poisonous tentacles of the anemone, and (d) the totally focused jumping spider.

(a)

(b)

(c)

(d)

problem has perhaps best been expressed by one of our most capable scientists, Albert Szent-Gyorgi (Figure 9.2), who wrote:

In my hunt for the secret of life, I started research in histology. Unsatisfied by the information that cellular morphology could give me about life, I turned to physiology. Finding physiology too complex I took up pharmacology. Still finding the situation too complicated I turned to bacteriology. But bacteria were even too complex, so I descended to the molecular level, studying chemistry and physical chemistry. After twenty years' work, I was led to conclude that to understand life we have to descend to the electronic level, and to the world of wave mechanics. But electrons are just electrons, and have no life at all. Evidently on the way I lost life; it had run out between my fingers.

—Personal Reminiscences

FIGURE 9.2
Albert Szent-Gyorgi.

As Szent-Gyorgi lamented, the "meaning of life" will probably never be grasped as a pure, crystalline gem of truth. It is just not likely to be that simple. Many dedicated biologists have concluded that life's many meanings, if they exist, apparently lie somewhere in the very complexity that Szent-Gyorgi so valiantly worked through.

THE EXPANDING UNIVERSE

As we continue our search for the theoretical beginning of life, let's take a big leap. Let's go back to the real beginning—back to the instant that the entire universe came into being. (About now you may notice just the tiniest bit of speculation about a few very minor points here and there.) The universe appeared, we are now told, some 13–20 billion years ago. And before that? Theoretical physicists, working with the most powerful mathematical tools, tell us that there was nothing before that but a lump, that all the matter in the universe was compressed into that lump, no larger than a walnut. (Are you still with us?) And within this lump, we are told, was all the stuff of the heavens. In fact, every bit of matter in your body right now, at this very instant, was once in that lump. Furthermore, the stuff that comprises your house, your shoes, and your dog was there too, as was the air you now breathe. So were the other planets and all the far-flung galaxies. It was all there (Figure 9.3).

FIGURE 9.3
One part of the vast and mysterious realm we call the universe. At present, we are aware of life on only one small watery planet.

Then the lump exploded (or began to explode, since some say it is still exploding). At the moment of the explosion, the particles that composed the lump flew in every direction, bumping, colliding, and interacting. In the course of such chance collisions, some of the particles joined. Others formed clusters. The more unstable ones immediately fell apart, but some had been changed by their temporary associations. Some were now able to interact in new ways with yet other kinds of particles. It may have been just such chance changes and aggregations that finally formed the great celestial array of clusters that we call our universe. Then, from amidst all this, we are told, our solar system was born. And here, on the third planet from the sun, there is hope that intelligent life may one day exist in harmony with its home.

So, now you know how the universe came to be. Even this simplified scenario, only one of several explanations, may be a bit much, so I won't mention that astrophysicists also tell us that time didn't exist before the explosion, called the *Big Bang,* and that even all the empty space of the universe was also in that walnut.

Nonetheless, let's move along and assume that the earth now exists and that the celestial events that brought it about were quite impressive. It is time to focus our attention on the development of the *life* that somehow managed to come into being amidst such cataclysmic events. Obviously, we will lean a great deal on intuitive or circumstantial evidence.

OUR EARLY EARTH

In order to imagine how life may have appeared here, we must visualize quite a different kind of earth from the one we know today. For example, our air is composed of about 78 percent molecular nitrogen, 21 percent molecular oxygen, 0.03 percent carbon dioxide, and traces of other gases such as argon and helium. But it seems that this was not always so. In 1936, the Russian scientist, A. I. Oparin, proposed that the atmosphere of the early earth contained little or no oxygen and much more hydrogen than it does now. The implications of this assumption are enormous. You will recall from our earlier discussion of chemistry that hydrogen tends to join with all sorts of elements, such as nitrogen, oxygen, and carbon. Therefore, Oparin believed that the early atmosphere of the earth consisted of methane (CH_4), ammonia (NH_3), water vapor (H_2O), and hydrogen gas (H_2). Geological evidence has forced us to modify Oparin's view of the primitive atmosphere somewhat. Currently, it is thought that the early atmosphere was similar to the atmosphere today, except that it lacked free oxygen. The lack of molecular oxygen in the early atmosphere made all the difference. In the presence of O_2, organic molecules would have broken down before more complex structures could be formed.

Perhaps the most difficult thing to imagine is that the earth—this earth, our home—was lifeless. There were no, nor had there ever been, ears to hear or eyes to see. There were no crusty lichens or slimy molds. Nothing lurked anywhere. The place was a dead, unchronicled ball of matter covered by a very thin layer of hot swirling gases. The surface itself was hot, molten, and volcanic. In time, the surface began to cool and solidify, but the piercing shrieks and groans and deep rumblings from

FIGURE 9.4
The theoretical early earth; the birthplace of life (as far as we know).

below startled no one. Heavy billowing clouds, miles thick, surrounded the darkened sphere. The murky blackness was continually split by bright, spewing gapes in the earth and thunderous lightning from above (Figure 9.4). Water vapor condensed and fell to the sterile earth, immediately exploding from the heat to be lifted skyward with a crackling hiss. It is difficult to imagine that this seething place would, in time, give rise to something called life.

HYPOTHESES CONCERNING THE ORIGIN OF LIFE

The first life appeared on earth somewhere between 4.1 billion years ago (when the earth's crust began to cool) and 3.5 billion years ago (when the planet came to be populated by bacteria). The question then arises, where did they come from? The answer leads us directly to the origins of life.

Conjecturing about how life began seems both fruitless and safe; one can never be proven right and it is impossible to be proven wrong. Nonetheless, a number of very reputable scientists have dedicated their careers to the question. For us, it is a good intellectual exercise and it can help us to understand scientific reasoning. So let's ignore all the problems associated with the question, gird our loins, and tread into an arena littered with dead guesses.

FIGURE 9.5

As the earth cooled, various life forms could have originated in the still-hot waters. Here, life exists in the hot springs of Yellowstone.

We can begin by noting that the molecules that form living things today are apparently far more complex than the kind that existed when the earth was young. Thus, one can assume that the sort of molecules associated with today's life was formed by simpler molecules that somehow managed to join. The trouble was that those early molecules would have had almost no tendency to join; they could have coexisted quite nicely, side by side, without ever interacting at all. As time went by, however, things changed. The earth became a different kind of place and the complex molecules of the planet became more likely to interact (Figure 9.5).

According to our scenario, as the mountainous and volcanic earth continued to cool, pools of condensed water collected in the valleys. Then, as the atmosphere itself cooled, this water was lifted by evaporation, only to be returned in torrential rains of incredible proportion. The constantly falling water rushed over the earth's exposed surfaces sweeping away the salts and minerals embedded in the cooling rocks. These mineral-laden waters poured into the cracks and crevices of the parched earth, turning pattering streams into pounding rivers, and finally forming great bodies of still waters. As the waters filled the deepest valleys and greatest canyons of the planet, the oceans were born. It was in such places that molecules of all sorts were brought together, bumping, jostling, and interacting.

Interacting? Why would they interact? Such interaction requires energy to make it go. Where did the energy come from? It turns out (fortunately for our story) that there are a number of sources of such energy. For example, the sun produces a variety of types of radiation. Another source of energy is heat. Remember, the earth's surface was still far from placid; the searing landscape was constantly jarred with violent eruptions, and molten or hardening rock covered much of the surface. Lightning continually streaked earthward from the dense, heavy clouds, jolting the molecules of the earth below. Energy, we see, was abundant.

We had the essential elements and possible energy sources to promote interactions. Importantly, there was no free oxygen to destroy molecules that may have formed. But was all this enough? Could a mixture of water, ammonia, and simple molecular compounds, such as methane, form the complex molecules of life?

THE FORMATION OF THE MOLECULES OF LIFE

In 1953, a graduate student at the University of Chicago provided a partial answer to the question of how complex molecules could have arisen on the early earth. The student, Stanley L. Miller, built an airtight apparatus through which four gases—methane, ammonia, water vapor, and hydrogen (the gases thought at that time to compose the atmosphere of the early earth)—could be circulated past electrodes (Figure 9.6). He permitted the gases to circulate together in his chamber for a week, subjecting the mixture to intermittent jolts of electricity. Then he simply analyzed the contents of the chamber. His findings were indeed startling. The experiment had produced a number of organic compounds, including amino acids, the building blocks of proteins. We have seen some of the important ways proteins function in the complex pageant of life.

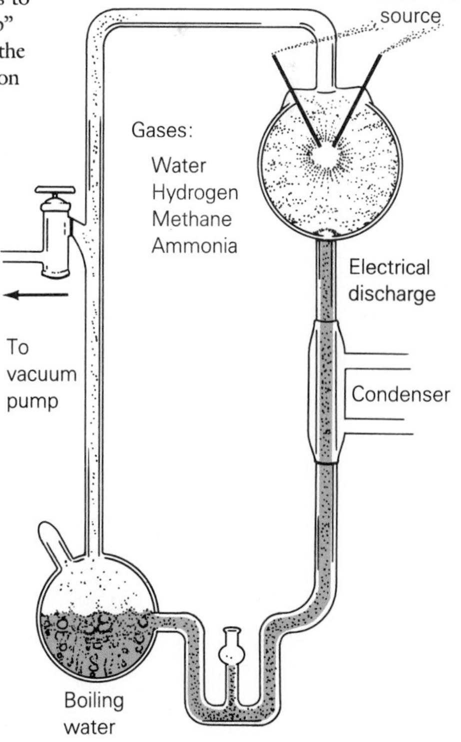

FIGURE 9.6

In 1953, Stanley L. Miller, then a graduate student, subjected a mixture of methane, ammonia, water vapor, and hydrogen to a series of electrical charges. He imagined this to be a rough duplication of conditions on the primitive earth when the "primordial soup" was subjected to bolts of lightning. The result justified his expectations. After a week, the simple molecules had joined to form amino acids. Here, he continues his experiments on other facets of the problem. His original apparatus is illustrated at right.

FIGURE 9.7

A coacervate droplet. Notice its organization. Such droplets have often been mistaken for living things under the microscope.

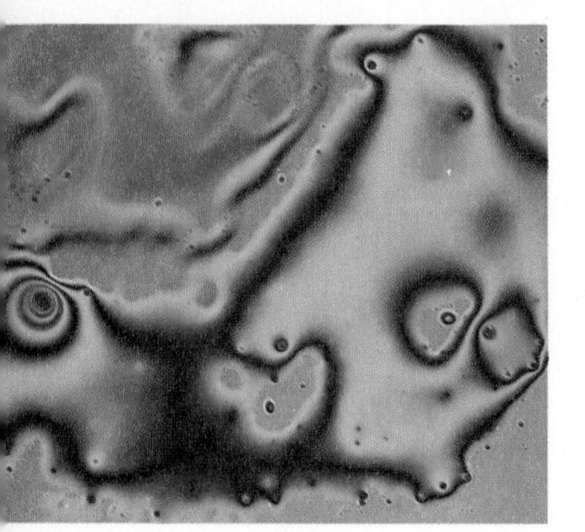

Miller was already a careful scientist and he was well aware that this experiment was too important to botch. So he also ran two controls. In one case, he sterilized the gas mixture at 130° C for 18 hours before he subjected it to the electrical power source. The yield of complex molecules was the same as in his first test. Then he ran the test exactly as before, but without the energy supplied by the electrodes. This time there was no significant production of the critical molecules. As a result, Miller felt safe in concluding that the formation of organic molecules was not due to the presence of contaminants, but was the result of smaller molecules joining together when energized.

Miller's experiment has since been duplicated and extended many times. Some experiments used different proportions of starting gases and others used different energy sources. In addition to the biological molecules that Miller obtained, these experiments have produced yet other molecules generally associated with life. They include purines, pyrimidines, and sugars, all of which are critical to the processes of life.

Sidney Fox, of Southern Illinois University, has taken the next step. He has shown how complex molecules called polypeptides can routinely form under lifeless conditions. (Polypeptides, as we know, are strongly associated with the processes of life.) Essentially, Fox showed that the simple kinds of molecules believed to have existed on the early earth could be readily joined to form more complex molecules. He dripped dilute solutions of the simpler molecules onto hot sand, rock, or clay, thereby vaporizing the water and concentrating the molecules on the earthen substrate. The result was the formation of complex molecules he called **proteinoids.** Fox imagined the torrential rains of the early planet splashing dilute solutions of simple molecules onto the hot earth, where the molecules would have joined, forming complex molecules that were then washed into the ancient seas where they could interact.

It has also been shown that clay surfaces favor the formation of polymers from their simpler building blocks. Many of the building blocks of complex biological molecules adhere to clay. As the clay dries out and is warmed, the building blocks link together to form polymers.

PROTOBIONTS: THE ANCESTRAL DROPLETS

We have now traced a theoretical pathway whereby large, complex organic molecules could have been produced and maintained when the earth was young. Obviously, though, we have not even begun to answer the question about how *life* came to be. It is indeed a long step from complex molecules to the simplest forms of life. Even the simplest living thing is composed of a fantastically ordered arrangement of complex molecules. How, then, could life have arisen from some ancient hot, chemical soup?

That's a good question (which is another way of saying we don't really know). Indeed, while there are no firm answers, we do have some ideas that bear mentioning. One of the most compelling was set in motion by

A. I. Oparin, who pointed out that when polymers such as proteins and carbohydrates, or proteins and nucleic acids are mixed in water, as might have occurred in the primeval oceans, the polymers tend to clump together into increasingly complex masses. These may be held together by electrostatic (positive and negative) forces and thus form **coacervate droplets** (Figure 9.7). Each such droplet consists of an inner cluster of colloidal molecules surrounded by a shell of water. The molecules of this water shell are arranged in a specific manner in relation to the colloid center. As a result, there is a clear demarcation between the colloidal protein mass and the water in which it is suspended. Such coacervate droplets may have led to the first **protobionts** *(proto:* first; *biont:* life form). These are specialized droplets with internal chemical characteristics that differ distinctly from those of the external environment.

The peculiar orientation of the water molecules around a coacervate droplet causes the water shell to act as a sort of membrane, a shell that allows some molecules through while excluding others. In the meantime, the molecules within the droplet tend to arrange themselves in an orderly manner, due to a variety of molecular interactions, such as those due to electrostatic charges along the molecules.

The selectivity of the droplet's absorption from its surroundings is also due to its internal molecular structure. Just as the water shell only allows the passage of certain molecules, the inner mass of the droplet itself is very selective regarding which molecules it will absorb into its mass. In fact, it may absorb almost all of some kinds of molecules from the medium while taking almost none of the others. As a result, its internal structure constantly changes, all the while becoming more different from its chemical surroundings. At the same time, because of its continuing absorption, the droplet grows, always becoming more specific, more precisely composed.

As our developing "cell" continues its changes, the molecules at the surface of the droplet rearrange themselves into a membranelike structure just under the water layer. This new "membrane" is even more selective than the water layer. It allows even fewer kinds of molecules to pass through. Thus, with increasing selectivity in the kinds of molecules that the mass will accept, the structure tends to become more regularly organized. At this stage, the complex droplets begin to show many of the properties of living things. (Even experienced microbiologists have, on occasion, attempted to identify the species of such droplets.)

Sidney Fox has suggested an even more rapid route to the internal organization of the developing cell. He has shown that when proteinoids are mixed with cool water, the proteinoids will self-assemble into small droplets he calls **microspheres** (Figure 9.8). The microspheres grow by absorbing proteinoids until they become so large and unstable that they fall apart, each fragment forming a new "daughter" with constituents similar to those of the "parent." Another kind of droplet forms when molecules called phospholipids are placed in water (Figure 9.9). These readily form membranes, similar to those in living systems, which can then break apart forming **liposomes** *(lipo:* fat; *soma:* body). So we see that coacervates, microspheres, and liposomes could have all been involved in the formation and development of the first protobionts.

FIGURE 9.8

Photomicrograph of an individual proteinoid microsphere.

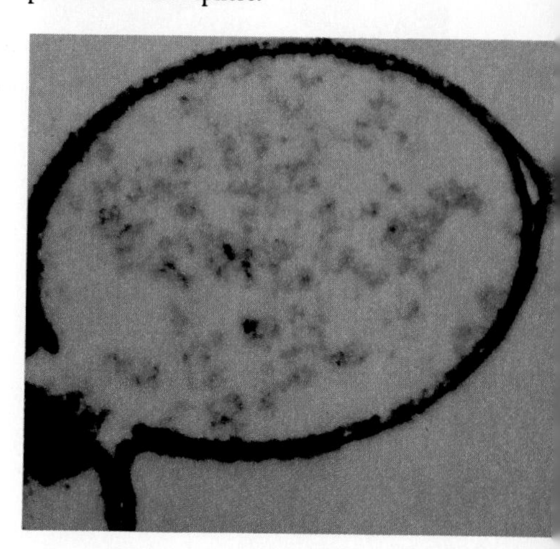

FIGURE 9.9
Photomicrograph of a droplet that forms when phospholipids are placed in water.

Reproducing Droplets

FIGURE 9.10
Soap bubbles, like oil droplets, fragment when they reach a certain size.

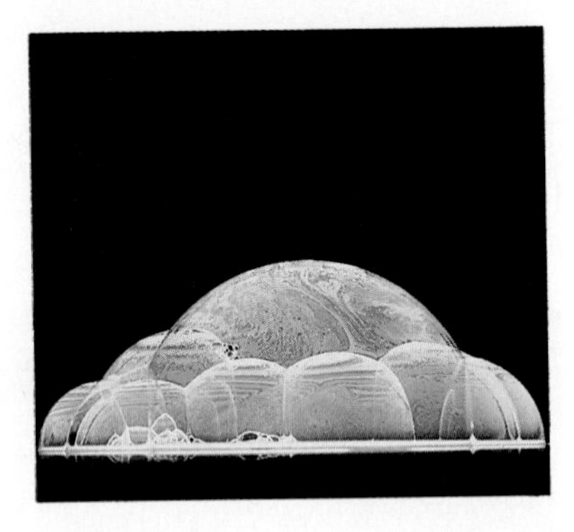

Now let's take another conceptual leap and see if we can account for how such droplets were able to reproduce. This intellectual exercise is critical if we are to accept these simple droplets as ancestral to life. Again, we must rely on a good healthy imagination—held in check by circumstantial evidence. We should first be aware that *most* of those early protobionts did *not* reproduce. But since the earth undoubtedly harbored a wide variety of droplets, some would have been more stable than others, and it was these that had the greatest chance of eventually reproducing.

The simplest form of reproduction has already been mentioned: fragmentation. Here, surviving droplets would have continued to grow as new molecules entered their masses. Finally, the mass would have become so large and unwieldy that the droplet would then fragment as a simple result of its large size, as we saw with microspheres. In a sense, the droplet breaks apart under its own weight (Figure 9.10). Each fragment, already blessed with stable parental organization, then begins its own growth. This, obviously, is a very simple form of reproduction, but reproduction nonetheless. Simple fragmentation, though, has its limits as a reproductive mechanism. With each fragmentation, the "daughter" cells would be increasingly unlike that first successful parent, their internal makeup constantly diluted as new kinds of molecules entered their interiors. Obviously, some other mechanism would be required to ensure the passage of successful traits through the generations.

So let's consider a more complex scenario, one that results in a very orderly and highly regulated descendancy. This type of reproduction is possible because of the great numbers of complex molecules that were generated in that early atmosphere.

The large, complex molecules within the early protobionts were certainly highly varied. Among them, some would have existed as "chains" of simpler molecules. These chains could have attracted other molecules floating free in the fluid interior. Each segment along the chain's length, of course, could have attracted only certain kinds of those free-floating molecules. The makeup and sequence of the new chain developing along the first chain's length, then, would have been directed by the makeup and sequence of the first chain. Now suppose the second chain breaks away from the first. It, then, could begin to attract certain molecules from the droplet's interior, and thus begin the formation of a third chain along its length. Because each segment along any such complex molecule can only attract (and bind) to certain other kinds of molecules, the third chain would, of necessity, be very similar to the first. Thus, a mechanism develops that would ensure the descendancy of the molecular makeup of certain successful droplets, as we see in Figure 9.11.

This scenario is, of course, highly simplistic. However, the basic scheme is supported by impressive evidence that is beyond our scope here. We encountered the basic idea in our discussions of the functions of molecules called RNA and DNA in genetics.

No matter how the droplets reproduced, the point is that there are ways in which even the most complex and precise ones *could* have come to reproduce at least some of the molecules within them. It should be apparent that the most "successful" droplets would be those that could produce the closest approximations of themselves since, by their very existence, they would have effectively demonstrated the stability of their molecular makeup.

Successful Droplets

According to our scenario, as the less successful droplets disappeared, the survivors could have continued to change, becoming ever more refined until they developed into the first cell-like bodies. Thus, the ratio of the more stable forms in the population of developing cells would have continued to increase. In addition, the components of the lost and decomposing droplets would have been released and made available to the increasingly efficient early "cells." All the while, the more efficient types would continue to accumulate. Any existing "cells," then, would have been the result of the most fortuitous of the molecular combinations in the untold number of trials over the millions of years. But finally, through such processes, according to this scenario, actual cells appeared. (We have seen what "actual" cells are; use your own judgment to compare the "organized droplets" we are studying here.) The point at which those droplets became cells is not important, but we can be sure that there was no sudden and dramatic appearance of "life" on the earth. No trumpets blared; no flags waved. There were only a few peculiar droplets drifting mindlessly in the warm seas, droplets a little more organized than the rest.

You are undoubtedly aware that any description of how life appeared on earth will not be accepted by everyone. Some scientists, for example, contend that protobionts probably did not give rise to complex molecules that could reproduce themselves. Instead, they say, it is more likely that complex molecules first appeared and then built something like a cell around themselves. Other people, of course, argue that there is an essential distinction between life and nonlife, that life is not merely a serendipitous accumulation of molecules, and that life—or at least human life—was

FIGURE 9.11

A suggested mechanism for replication of early, stable molecules.

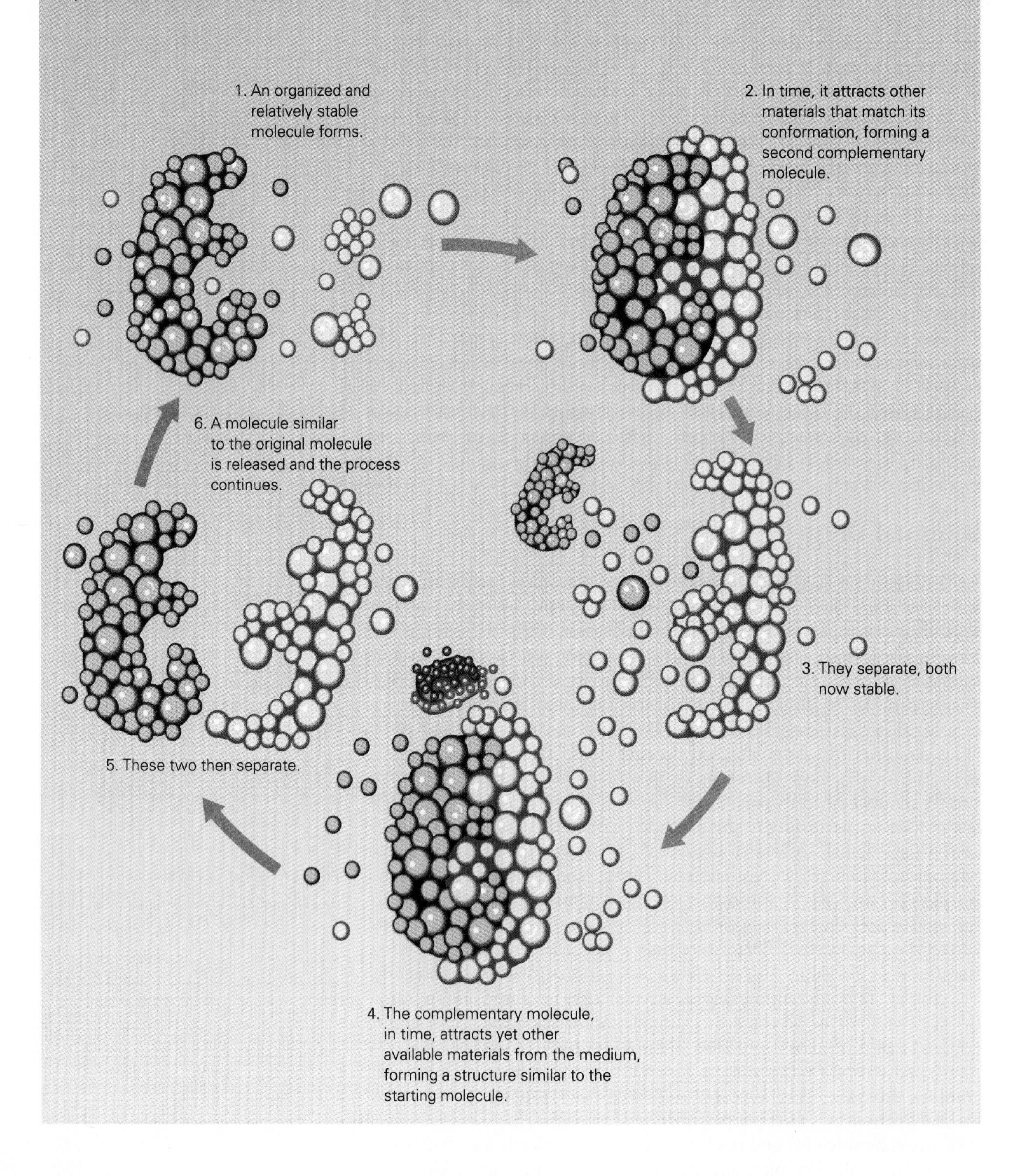

created by some greater power. And then there are a host of lesser hypotheses—including one recently suggested by a few well-known scientists who propose that life was brought here by travellers from outer space. They note a certain "cosmic unity" in the chemistry of galactic gases that are similar to that of living things. We will now leave this intellectual exercise, having set the stage for the processes called natural selection. We will simply assume that life appeared on the earth, and then consider the more approachable question of how it might have made its living.

HETEROTROPHY AND AUTOTROPHY

The molecular interactions necessary to support even the simplest forms of life are so complex that we must wonder how such interactions ever started. What triggered their development? Perhaps the impetus was partly food shortages. The earliest cells are considered to be **heterotrophs** (other-feeders) because they absorbed molecules from their surroundings for their maintenance, growth, and reproduction. These molecules would have been the amino acids, polypeptides, sugars, and nucleic acids that had spontaneously formed as we have described.

Certain kinds of molecules in the early environment would have been important for the processes of growth and reproduction, and as populations of life forms increased, some such "food" molecules would have fallen into short supply. In time, competing cells would have run up against "food barriers" and some would have lost the race to acquire these precious molecules. These cells would have had to develop some alternative or perish. One alternative might have been not to find food, but to *manufacture* it. Let's see if this route might have led to increasingly complex and efficient means of competing in a newly competitive world.

Among the small changes that were continually arising by chance in early cells, some would have resulted in the cells producing various sorts of molecules, some perhaps unusual. If a cell began to produce, say, one of the rare "food molecules" (or a reasonable substitute), then the pressure to compete for that molecule would be relieved. We can call this molecule nutrient A. Let's also say that it was manufactured from a more abundant nutrient B. This would mean that cells that previously had been held in check by the scarcity of nutrient A were then able to grow and multiply—that is, until nutrient B fell into short supply. At this point, there perhaps appeared, by chance, some other slight change that enabled a cell to manufacture B from another abundant molecule, say nutrient C, thereby increasing the supply of both nutrients B and A. In time, with competition increasing, and with one molecule after another falling into short supply, long series of interactions involved in the manufacture of needed molecules would arise. Obviously, those cells best able to carry out such chains of chemical reactions would be more likely to survive.

Any such system would require an efficient energy source to drive the process. Thus, it is interesting that a chemical called ATP, which is critical in the production of energy in today's cells, is one of the molecules that appears in those laboratory experiments that simulated the conditions of the early earth.

The cells that were able to chemically manufacture their own food are now called **autotrophs** (self-feeders). As time passed, the autotrophs became quite efficient at manufacturing their food. The biochemical chains became long, complex, and refined. Today, the result is most evident in our silent partners on the planet, the green plants. As we already saw, they actually use the energy of sunlight to power their intricate food-making machinery. They obviously had an advantage in the early days, and so their numbers would have multiplied. The first autotrophs were the bacteria that appeared about 3.5 million years ago. Their fossilized remains formed mounds called *stromatolites* (Essay 9.1). The food-laden bodies of these early autotrophs would not have gone unnoticed by other life forms. Other kinds of cells, the new heterotrophs, soon developed means of robbing the food makers.

Today the earth abounds with heterotrophs. Some eat autotrophs (these are called **herbivores**), while some (the **carnivores**) eat the species that eat the autotrophs. Most species, however, eat both autotrophs and heterotrophs (and are called **omnivores;** *omni:* all). (Figure 9.12 shows a modern representative of each.) The presumed sequence in the development of life is shown in Table 9.1.

TABLE 9.1
Possible Steps in the Origin of Life

Big Bang
(origin of the universe)
↓
Formation of the Earth
↓
Formation of Primitive Atmosphere
(contains abundant free hydrogen gas but no free oxygen gas; contains the elements carbon, hydrogen, oxygen, and nitrogen)
↓
Formation of Small Biological Molecules
(amino acids, glucose, nucleotides)
↓
Polymerization to Form Larger Biological Molecules
(proteins, carbohydrates, nucleic acids)
↓
Formation of Protobionts or Ancestral "Cells"
(coacervate droplets, microspheres, liposomes)
↓
Cell Reproduction and Genetic Material
↓
Heterotrophs
(absorbed molecules from surroundings for nourishment; anaerobic metabolism since no free O_2)
↓
Autotrophs
(photosynthetic cells give off O_2)
↓
Animal-like Heterotrophs with Aerobic Respiration
(herbivores, carnivores, omnivores)

FIGURE 9.12

Autotrophs (self-feeders) use the energy of the sun to manufacture their own food. Autotrophs come in many shapes and sizes, including (a) *Volvox*. Heterotrophs (other-feeders) can be divided into herbivores (which eat autotrophs), such as the meadow vole (b), and carnivores (which eat species that eat autotrophs), such as snakes (c), and omnivores, such as humans (d), which eat virtually anything.

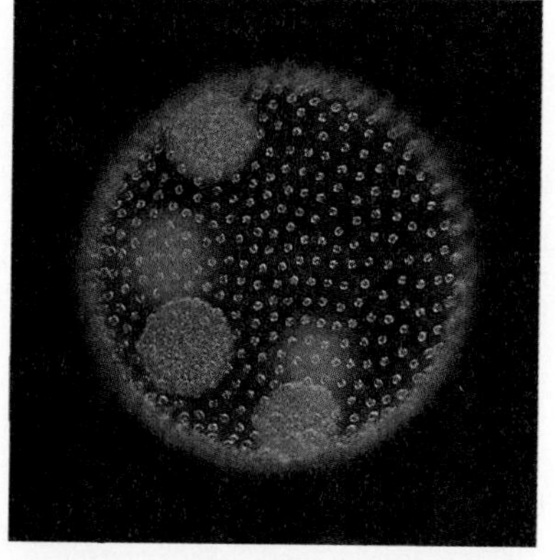

(a)

(b)

(c)

(d)

ESSAY 9.1

The Oldest Life

There is an area of Western Australia so forbidding that only the most hardened of miners and scientists go there. The miners are there for the usual reasons; the scientists go to dig for some of the oldest known rocks on earth. The rocks, called stromatolites, are interesting precisely because of their age. In 1980, a surprising announcement was made: the rocks contained fossilized remains of life that existed about 3.5 billion years ago—that is, a billion years before any other known life.

Furthermore, these rocks contained at least five different life forms. The fossils were tiny, to be sure. Most of the forms were elongated, or strandlike, and the cells lacked any sign of a nucleus, as do bacteria today. Furthermore, the organisms probably lived under a thin layer of warm water, and, since they used carbon dioxide, they were

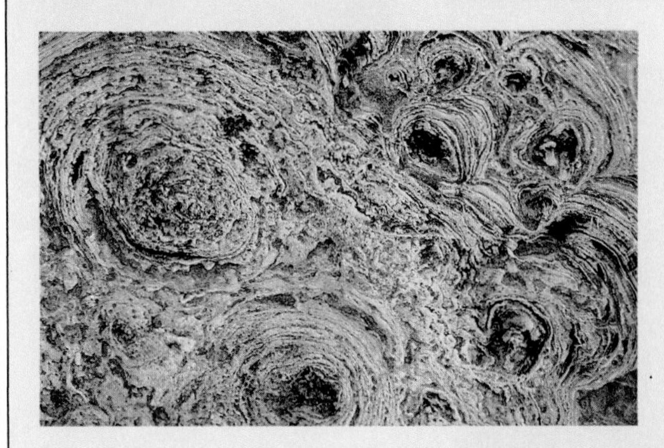

A stromatolite fossil embedded in a 3.5-billion-year-old rock from Western Australia.

Scientists have used data from fossils and rocks to help date the origins of the earth and life on earth. Some scientists believe that if the physical conditions permitted, life may have appeared on some other planets in our immense universe. Here, the surface of Mars awash in the sun's rays.

probably photosynthetic. It appeared that scientists had found one of the earth's earliest pond scums.

The oldest known rocks on earth are 3.8 billion years old, and some scientists think that they may contain evidence of life also. If this is so, since the earth is estimated to be only some 4.5 billion years old, life may have appeared very suddenly after the earth's surface cooled. According to some scientists, this means that there is an increased likelihood of life appearing wherever physical conditions permit. This, in turn, suggests to some that there is a greater likelihood that life has appeared on other planets in the vast, far-flung universe.

Stromatolites found in the shallow water of a lake at Isla Angel de la Guarda, in Baja California, Mexico. Life is believed to have evolved in shallow, warm waters such as these.

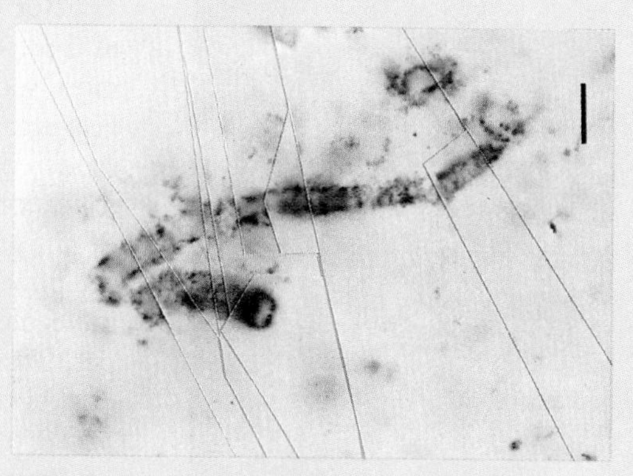

A fossil of the earliest life form known. Note the visible cell walls. Life appeared suddenly (in geologic time) after the earth became cool enough, some 3.5 billion years ago.

OXYGEN: A BANE AND A BLESSING

FIGURE 9.13
Many of the seemingly most indestructible structures on earth slowly yield before the devastating power of oxygen. This neglected machinery gives mute testimony to the destructive effects of the gas.

In the photosynthesis in green plants, as the energy of sunlight falls on the green pigment in the leaves, carbon dioxide and the hydrogens of water are used to make food, and water and oxygen are released. The release of oxygen by those first photosynthesizers was a critical step in the direction of life's development. In a sense, the production of oxygen falls into the "good news—bad news" category. It's good news for us, of course, since we need oxygen, but as oxygen began to replace hydrogen as the most prevalent gas in the atmosphere, it sounded the death knell for many of the early organisms. This is because oxygen is a disruptive gas, as seen in the process of rusting (Figure 9.13). So, in the early days of life on the planet, many life forms were destroyed by the deadly and accumulating gas.

But we must keep in mind that there were many forms of cellular life in those days, and that life was constantly changing in all directions (with most such "hopeful experiments" doomed to failure). We can assume, however, that some of those cells would have been changed in ways that rendered them less vulnerable to the dangers of oxygen. In fact, in time, some forms would have gone one step further and begun to *utilize* the deadly gas. These cells developed aerobic respiration, the ability to harness the power of oxygen to extract the maximal amount of energy from food molecules. This ability, obviously, would have given them a great competitive advantage over the forms that were unable to do so.

As oxygen collected in the upper atmosphere, it would have formed ozone (the union of three oxygens), and further altered the development of life on earth. This is because ozone acts as a shield against the sun's ultraviolet rays. Thus, the earth under the ozone shield would be protected against the terrible energy of raw sunlight (Figure 9.14). Before the ozone formed, the early cells probably took refuge five to ten meters beneath the

FIGURE 9.14
Raw sunlight is extremely destructive, with its powerful ultraviolet rays largely filtered out by a sparse layer of ozone in the upper atmosphere.

surface of the earth's shallow seas, where the force of the ultraviolet rays was diminished. But now, with the development of a protective layer of ozone, dry land became safe to inhabit. Thus, the shorelines were soon invaded by strange, primitive life forms.

We must keep in mind that the rise of the oxygen-rich atmosphere meant that life could no longer arise in the way it once had. In other words, as life developed, it destroyed the very mechanism that had produced it.

SUMMARY

1. There are two schools of thought on the nature of life. Mechanism implies that all physical entities, including living organisms, must follow the same chemical and physical laws. Vitalism implies that life is more than its molecular interactions because it has some unmeasurable "life force."
2. It is thought that the universe began 13–20 billion years ago with the Big Bang, an explosion and expansion of a small lump of matter containing all the material present today.
3. The atmosphere of the early earth had gases containing the elements essential to life (carbon, hydrogen, oxygen, and nitrogen). However, it contained little or no free oxygen (O_2). Available energy sources included the sun's radiation, volcanic heat, and lightning.
4. Stanley Miller built an apparatus in which an atmosphere thought at the time to be similar to that of the early earth (methane, ammonia, water vapor, and hydrogen) could be sterilized and circulated past electrodes. After a week, organic compounds such as amino acids had been formed.
5. Sidney Fox demonstrated that polypeptide molecules, called proteinoids, could form when dilute solutions of simple molecules splashed onto the hot earth. The proteinoids would have accumulated in the ancient seas. Clay surfaces may have also promoted the formation of polymers.
6. Coacervate droplets, microspheres, and liposomes could all have been steps in the formation of the first life forms (protobionts). A coacervate droplet, which is comprised of proteins held together by electrostatic forces and surrounded by water molecules, selectively absorbs materials from the surroundings, has an orderly arrangement of molecules within it, and grows. Microspheres are small droplets formed by self-assembling proteinoids. They grow until they break apart, forming daughter droplets. Phospholipids placed in water form membranes similar to those in living cells, which fragment and form liposomes.
7. Successful droplets would have had to have some way to reproduce. The simplest form of reproduction is fragmentation. Some droplets may have contained chains of molecules that could copy themselves, keeping the molecular makeup of the daughter droplets similar to that of the successful parent droplet.
8. Less successful droplets disappeared while successful droplets accumulated and were gradually refined into the first cell-like bodies.
9. At first, all cells depended on an external food source (heterotrophy). As the molecules needed for growth and reproduction were absorbed from

the environment, some of these molecules became rare. Complex biochemical pathways might have developed as some cells changed in ways that allowed them to produce rare "food" molecules (autotrophy). In time, other organisms would have become food for heterotrophs.

10. Oxygen, produced by photosynthetic autotrophs, began to accumulate in the atmosphere and killed many life forms. Some cells developed ways to escape from the damaging oxygen. Lines of cells less vulnerable to oxygen developed the ability to utilize it (aerobic respiration). Oxygen formed a layer of ozone in the upper atmosphere that shielded the earth from the damaging ultraviolet rays of the sun and made the dry land safe to inhabit.

KEY TERMS

autotroph (242)
carnivore (242)
coacervate droplet (237)
herbivore (242)
heterotroph (241)
liposome (237)
mechanism (229)
microsphere (237)
omnivore (242)
proteinoid (236)
protobiont (237)
vitalism (229)

FOR FURTHER THOUGHT

1. Could the results of Stanley Miller's experiment be duplicated by using a sample of earth's current atmospheric gases? Why or why not?
2. Explain how the emergence of early autotrophs proved fatal for some early organisms.

FOR REVIEW

True or false?

1. ___ The principle of mechanism states that the nature of life is determined by interactions of molecules.
2. ___ The earliest life forms left fossilized remains called microspheres.
3. ___ Coacervates are complex living cells.
4. ___ Heterotrophs extract food from other life forms.

Short answer

5. Name an important organic compound that was produced in Stanley Miller's experiment and explain why its formation was significant.

Fill in the blank.

6. By concentrating simple molecules, Sidney Fox created complex polypeptides called ___ .
7. ___ droplets are clusters of colloidal protein molecules surrounded by a water shell and held together by electrostatic forces.

Choose the best answer.

8. The atmosphere of the early earth contained no (or very little) free:
   A. ammonia
   B. oxygen
   C. helium
   D. none of the above
9. Green plants are classified as:
   A. omnivores
   B. heterotrophs
   C. herbivores
   D. autotrophs
10. Complex coacervate droplets
    A. may have developed into protobionts.
    B. show some properties of living things.
    C. are surrounded by a water shell that acts as a membrane.
    D. all of the above
11. Oxygen in the atmosphere of the early earth
    A. was consumed by photosynthesizers.
    B. destroyed the protective layer of ozone.
    C. acted as a deadly gas to some early organisms.
    D. Both A and B

CHAPTER 10

The Processes of Evolution

Overview

Variation in Populations

The Question of Species

Speciation

Small Steps or Great Leaps?

Death Stars and Dinosaurs: The Great Extinctions

Objectives

After reading this chapter you should be able to:

- State how the modern synthesis accounts for evolution within populations.
- Describe how variations are produced and environmentally maintained in a population.
- Explain how populations produced by stabilizing, directional, and disruptive selection influence variation within a population.
- Describe two mechanisms that bring about genetic drift in small populations.
- Explain some difficulties encountered in defining the term *species*.
- List several reproductive isolating mechanisms.
- Explain how species arise by allopatric and sympatric speciation.
- Recount several hypotheses for species extinction during the Cretaceous period.

Here we are, the dominant life forms of the third planet from a rather average star. As we look around us we see other forms of life, some common (and only a little different than other common forms), and some so disparate that we are shocked when we encounter them. Yet all forms of life are the result of the earth's various threats and blessings which cause that life to adjust or die out. The adjustment produces the kinds of changes that are encompassed by the term *evolution.*

The concept of evolution is, in fact, one of the more important ideas in modern history, so it is a bit peculiar that it is so little understood in what is undoubtedly one of the best-educated and most sophisticated societies of all time. Why is the idea so comfortably accepted and routinely utilized by some, while driving others right up the wall? Why is any such "cornerstone" the object of so many critics' chisels (Figure 10.1)? Probably for two reasons—reasons at quite different intellectual levels. One is based on what some people perceive as the theory being at odds with religious dogma. Matters of faith, however, lie outside the realm of scientific inquiry, and arguments aimed by either side clearly and inevitably miss their targets. The second reason for some of society's uneasiness with the concept of evolution is based on a simple misunderstanding of its principles.

So let's see if we can clear up some of these misunderstandings while building a foundation that will help to clarify much of what will follow. First, of course, a definition is in order. In the simplest terms, **evolution** involves change through time (or—as Darwin phrased it—descent with modification). It is important to keep in mind, though, that individuals do not evolve; populations evolve. So here we will refine our definition of evolution to be *the change in populations through time.*

As you may recall from Chapter 1, Darwin suggested that the mechanism of evolution is natural selection. Critical to the concept of natural selection are the ideas that members of a population have inherited differences, or variations, and that some of these variations result in certain individuals leaving more offspring than others. Darwin suggested that nature "selects" those individuals who will reproduce. So some forms of the inherited traits in a population improve the individual's chance of surviving and reproducing under the existing environmental conditions. The offspring of those individuals are likely to inherit the adaptive traits and they, too, will leave more offspring than individuals who lack the traits. Thus, the frequency of the adaptive traits will tend to increase in the population with each new generation.

Although Darwin's concepts regarding evolution were quickly accepted by the scientific community of his time, critical questions arose immediately. In particular, people asked, just how could variation be inherited? What are the mechanisms? The answer, of course, lay directly in the area of genetics that Mendel, not far away, had so elegantly described. But Darwin wasn't prepared to delve into anything of a mathematical nature and simply did not grasp Mendel's conclusions. Then, after the rediscovery of Mendel's principles in the early part of this century, there followed a flurry of interest in Mendelian genetics. People began to see that the Mendelian principles provided the mechanism of variation, so important to natural selection, that had eluded Darwin and other evolutionists for so long. By the 1940s, Darwin's original theory was firmly

FIGURE 10.1

News services reported that certain respected journals recently refused to hire Forest M. Mims III as a science writer because of his professed belief in divine creation.

integrated with Mendelian genetics, forming what is now called the modern synthesis.

The modern synthesis combines what we know about the inheritance of traits today with Darwin's ideas on how populations can change through time. In simpler terms, it is the integration of genetics and evolution; it is the meeting of Darwin and Mendel.*

We can ask, then, when a population evolves what changes? According to the modern synthesis, evolution is a change in the **gene pool,** which is the total of all the alleles of all the genes of all the individuals in an interbreeding population (Figure 10.2). If the frequencies of some alleles change, the population also changes, or evolves.

Here then, we will focus on the modern synthesis. We will begin at the beginning—we will ask about the variation that exists in populations.

VARIATION IN POPULATIONS

As you watch your little brother out of the corner of your eye, you may wonder how you could have the same parents. After all, you're a sophisticate, and he can't even be trusted not to eat what he finds on his shoe. Furthermore, if you come from a large family, you may have noticed that none of the kids are very much alike. They seem to vary, not only in appearance, but in such traits as attitudes, friendliness, and even agility. If this sort of variation can be found among the children of the same parents, it is not surprising that in an unrelated group of people (Figure

*Mendel, you recall, proposed that traits are controlled by "factors" that we now call genes. He suggested that each factor exists in alternative forms, which we call alleles today. For instance, the color of seeds in garden peas may be yellow or green depending on the alleles inherited.

FIGURE 10.2
The wildebeest herds that roam the African plain may be composed of many thousands of individuals, yet they are one population.

FIGURE 10.3

These Turkish men watching a cockfight show great variation although they are descendents of people who have for generations lived in the same area.

10.3) no two people will look alike—at least not much. There is, indeed, a great deal of variation among humans.

There is also great variation in other species. We can usually tell dogs apart, even without resorting to the crass and overly personal means they employ among themselves. Of course, there are species that, from our view, don't seem to vary much. They all seem to be rather alike. For example, standing on the seashore, it is difficult to tell one gull from another. We know, however, that they can recognize individuals among themselves. (Perhaps we all look alike to them.) The point is, variation among individuals is a common theme among the populations that inhabit the earth, although the differences may not always be apparent to us.

What is the source of the variation in a population? Some of it is caused by the environment. For example, a plant will grow taller if it receives adequate water and sunlight. Others less fortunate may be shorter. However, this variation is unimportant to evolution because it cannot be passed to offspring. Natural selection can only act on inherited (genetic) variation. As we saw in previous chapters, genetic variation can arise in a number of ways such as through meiosis, crossing over, and the chance combination of gametes. This sort of variation, however, involves the shuffling of alleles already present in the population. So the question becomes, how do new kinds of alleles arise? Even the best gamblers must draw new cards. In living things, gambling through evolution, new kinds of genes must arise from mutation, the random alteration of DNA (the genetic material).

Since variation is so common, one might ask if it has some fundamental advantage. Indeed it does—if not for the individuals themselves, then for the population in general. The advantage of a variable population is this: In a variable population, there are likely to be some individuals with traits that will ensure their success under a wide range of conditions. Put another way, no matter how the environment should change (within limits), in a variable population there are likely to be some individuals that will fare well. Those that do, by surviving and reproducing, will cause subsequent generations to follow along new paths that are more in keeping with the new conditions. Of course, if the environment should shift again, there would be some individuals capable of surviving those conditions, too (Figure 10.4).

The inherited variation in a population provides the raw material on which natural selection acts. So let's take a look at how variation may be influenced.

Variation by Single-Gene and Polygenic Inheritance

One factor that influences variation is the number of genes involved in the expression of the trait. Some traits are due to the expression of a single gene. Figure 10.5 illustrates some such traits. These are usually manifested as either-or conditions. Thus, some variation is produced by the interactions of rather simple genetic systems.

Most genetic differences among individuals in a population are caused by numerous genes with small individual effects. Polygenic inheritance is the control of a single trait by multiple genes, as we saw in Chapter 7. Height in humans is influenced this way. Because of polygenic inheritance for height, people are not either tall or short. Instead, there is a gradual variation that enables us to generate a smooth, gradual, and continuous bell-shaped curve as each gene adds its own small influence. It can thus be assumed that if there are a number of genes that contribute to tallness,

FIGURE 10.4

There may be great variation even among individuals with the same parents. (Recall the genetic shuffling that takes place in meiosis and crossing over.) From among such variation, some individuals will be more successful than others (the one on the end already looks pretty successful).

FIGURE 10.5

Only a few traits in humans are known to be controlled by single genes. Two such traits are the ability to roll one's tongue (a) and unattached earlobes (b) versus attached (c).

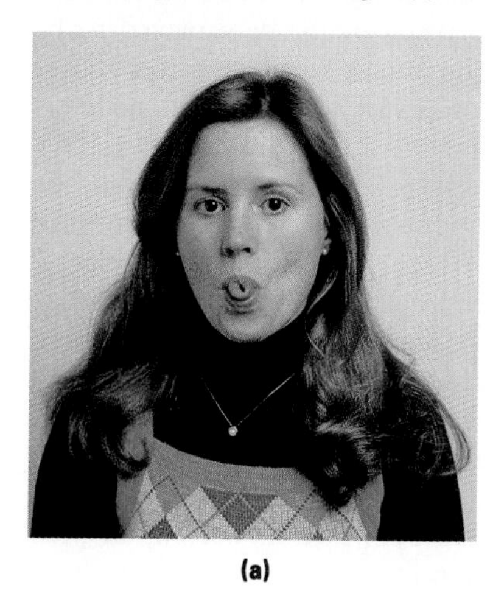

(a)

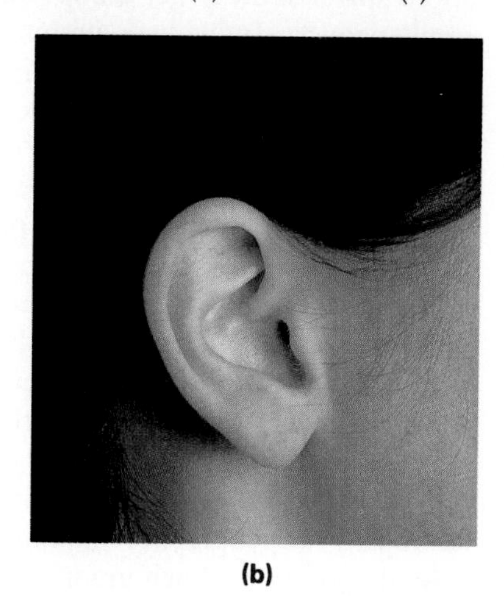

(b)

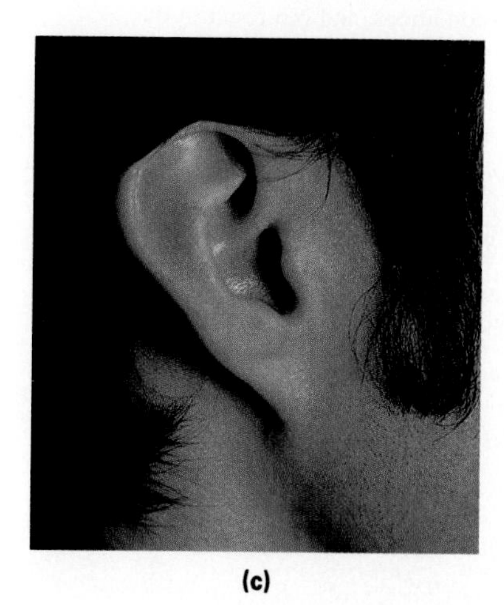

(c)

while others produce shortness, taller individuals are likely to possess a disproportionate number of the "tall" genes. On the other hand, those people with a preponderance of "short" genes will be short. By the same token, this is why parents of average height can produce exceptionally tall or short offspring. Let's say there are only seven genes that influence height (T for tall and S for short). If the parents are **TSTTTSS** × **STTTSST**, they might yield offspring that are **TTTTTTS** (very tall), **SSSTSSS** (very short), **TSTSTST** (intermediate), or any number of other possible combinations and heights.

How the Environment Maintains Variation

We have noted some of the ways in which genetic mechanisms can increase variation in a population. Here, let's consider a few of the ways in which the environment can interact with populations in such a way as to maintain or to further increase that variation.

In our first example, we might imagine what would happen if the environment were to have the tendency to shift wildly or unpredictably. Obviously, populations would have difficulty "tracking" it—that is, changing to conform to the requirements set by existing conditions. For example, if taller animals do best under certain environmental conditions, the population will become taller with each generation as it approaches that optimum. But if the environment should suddenly change and begin to favor short individuals, the frequency of short individuals would begin to increase. If, before the shift to the optimum was complete, the environment were to suddenly change, favoring tallness again, the populations would be unable to stabilize around a mean. Each time the environment changed, the population would begin to shift. If the optimum continually changed, never allowing the population to stabilize, the result would be a population with high variation.

FIGURE 10.6
Patchy environments, such as this tundra, cause populations to adapt to a variety of conditions, and can result in the population being highly variable.

High variation can also result if the environment is "patchy"—that is, comprised of markedly different kinds of places (Figure 10.6). Because a patchy environment presents a number of kinds of habitats, subpopulations (parts of the larger group) may adapt to a number of environments simultaneously. Individuals within each patch might be expected to be rather similar as they conform to the requirements set by local conditions, but the overall population might well be quite variable. As long as members of these different subpopulations continue to interbreed, however, they will tend to "dilute" each other's adaptations. Thus, no group would be able to fully adapt to its kind of patch. One might envision a group of permanent ponds dotted across the landscape, some deeper or cooler or with grassier bottoms than the others. The frogs that inhabited those ponds might develop different traits, each adapting to the particular features of its own watery home. Gene transfer would keep speciation from occurring as the frogs occasionally visited other ponds and mated.

Environmental characteristics often vary gradually over large geographic expanses. Regular variations in one or more traits commonly accompany the geographic variations because populations become adapted to local conditions. Such gradual variation in a trait or group of traits is called a **cline.** For example, the eastern United States changes somewhat gradually from North to South. Some animals, such as mice and frogs, range from, say, Vermont to Louisiana, and over this range they vary gradually from one place to the next (Figure 10.7). We noted earlier that individuals at the extremes of such a range may be so different as to be unable to successfully interbreed.

These, then, are some of the ways that the environment can influence variation. We might also keep in mind that they operate with genetic influences on variation, such as meiosis and crossing over, processes that at a different level encourage variation among individuals in populations.

Factors Involved in Evolution

The variation we have been discussing is largely due to the frequencies of different alleles within the population. In Chapter 7, we discussed the Hardy-Weinberg law, which says that allele frequencies will remain constant from generation to generation under certain conditions. If this is true, how can allele frequencies change and populations evolve? Obviously, the conditions necessary for the Hardy-Weinberg law to hold must be violated. The following factors invalidate the Hardy-Weinberg law and bring about evolutionary change:

1. Mutation. Changes in the genetic material (DNA)—that is, mutations—can add new alleles to the gene pool.
2. Migration. Allele frequencies may also change if new individuals enter the population (immigration), or if individuals leave the population (emigration).
3. Natural selection. If individuals with certain alleles leave more offspring than others, their alleles will increase in frequency over succeeding generations.
4. Small population size. Random changes in allele frequencies due to chance alone will have their greatest effect on small populations.

Because of their greater importance, we will consider more carefully the last two of these factors, natural selection and small population size, to see how they can act on genetic variation and cause evolutionary change.

How Natural Selection Can Influence Variation

We saw earlier that Charles Darwin recognized the importance of variation as he uncovered its role in the processes of evolution. Variation, in a sense (and as he implied), provides the "raw material" upon which natural selection can act. Natural selection, as we saw, is the process by which natural forces favor some traits over others, resulting in the accumulation of those traits in subsequent generations. Thus, some of the members of a population will possess inherited traits that enhance reproductive efforts, and so those traits will be carried into future generations. As they are carried along, they will replace less successful alternative traits.

Some nineteenth-century journalist is said to have coined the term "survival of the fittest" and it came to refer to selection that operates in this way. The acceptance of the term was unfortunate because many people have taken it to refer to the continued success of the best specimens. Actually, though, in the cool arithmetic of evolution, "fittest" refers solely to reproductive success. It has nothing to do with appearance or abilities unrelated to reproduction. Thus, scroungy little animals that outreproduce more beautiful specimens have a greater fitness. Evolutionarily, then, we can expect the greatest success from the best reproducers. Thus, each generation will be composed primarily of the offspring of the best reproducers from previous generations, and we can expect those traits that led to reproductive success to increase in frequency from one generation to the next. Nonetheless, variation will remain in the population for virtually any trait, including those directly related to fitness.

FIGURE 10.7

Species that cover vast ranges encounter a variety of environmental influences. Thus, they may vary from place to place. Where the environment varies gradually, the population may change gradually from one place to the next, so that individuals at the extremes of the range may be quite dissimilar, as we see here in these leopard frogs, which range from northern states to southern states.

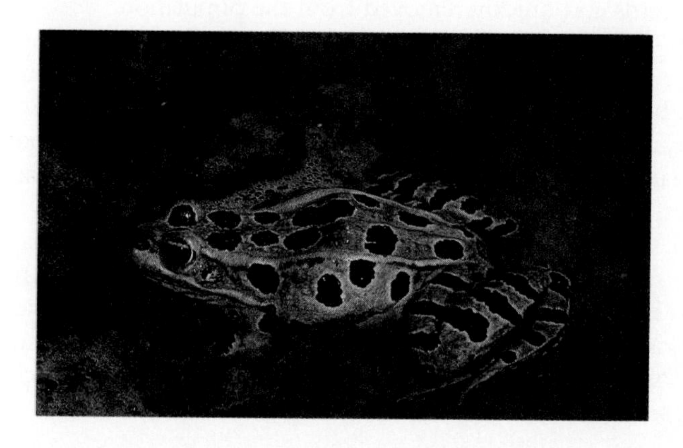

Modes of Natural Selection

Variation for any trait in a population can be described graphically, as we see in Figure 10.8A. Let's say this curve describes height in a group of people. We see, then, that most individuals in the population are of intermediate height, and that there are increasingly fewer taller or shorter people as they diverge from that intermediate height. Any trait distributed this way in a population produces the familiar bell-shaped curve.

The shape of any such curve, of course, can change. For example, it can be taller and thinner, it can widen, it can be skewed to one side or the other, or it can be "bumpy." Let's examine the implications of such shapes.

First, if being of intermediate height is especially important, then those people who are taller or shorter will suffer some increased disadvantage. Because of such *strong selection* (strong selection against departure from the optimum), in time, there will be fewer of them, and the curve will become narrower. If height is not so important, shorter and taller people will tend to survive and reproduce, and because of *weak selection* (weak selection against departure from the optimum), such a population will produce a wider, more relaxed curve.

So, when the environment is stable over a long period of time, as is assumed in these cases, the population tends to cluster around a single type, with divergent forms being less successful to some degree. The result is called **stabilizing selection.** In stabilizing selection, then, the population is clustered around a mean, or average, condition that is assumed to be optimal for the prevailing conditions. Yet, conditions can change. Suppose, as we've just seen, taller becomes better, and increased height yields a new kind of advantage (remember the giraffes). The population then shifts toward the high end of the height scale, as in Figure 10.8B. Such a process is labelled **directional selection** (changes due to populations tracking a new optimum).

A clear example of directional selection is the change in coloration of peppered moths that accompanied the environmental changes caused by the industrial revolution in England. In the 1700s, the British countryside was as green as its poets said. The trees were numerous, and many were covered with a light, gray-green lichen. Furthermore, if one looked carefully at the lichen coverings, it was sometimes possible to discern a light-colored moth sitting concealed upon the bark. One might also occasionally spot a black moth perched rather conspicuously on the light-colored lichen. The black moths were rather rare, largely because they were easily spotted by hungry birds and quickly removed from the population.

But "progress" will have its way, and England's entrance into the industrial revolution was marked by the construction of huge, coal-burning factories. These factories belched forth their billowing, black smoke, and soon the English countryside quietly submitted to its cloak of soot. As the trees darkened, the more common white moth became increasingly easier to see, while the darker moth became less visible. Thus, birds began to eat more of the light moths and their "light genes" began to disappear from the population. Finally, by the late 1840s, 98 percent of the moths were dark (Figure 10.9). The transition, then, from the light to the dark form was virtually completed in only 50 years. Interestingly, the use of pollution-control devices in British manufacturing has produced cleaner air, lighter trees, and a resurgence of the lighter moth.

FIGURE 10.8A

Normal distribution, a statistical term that is graphically depicted as a bell-shaped curve. The thin blue line represents the mean. The arrows (A) show the selective pressures against those individuals that are not at the means for some trait. In (B) there is strong selection against those not at the mean, resulting in most individuals being close to the mean, and in (C) the weak selective pressures allow the curve to expand.

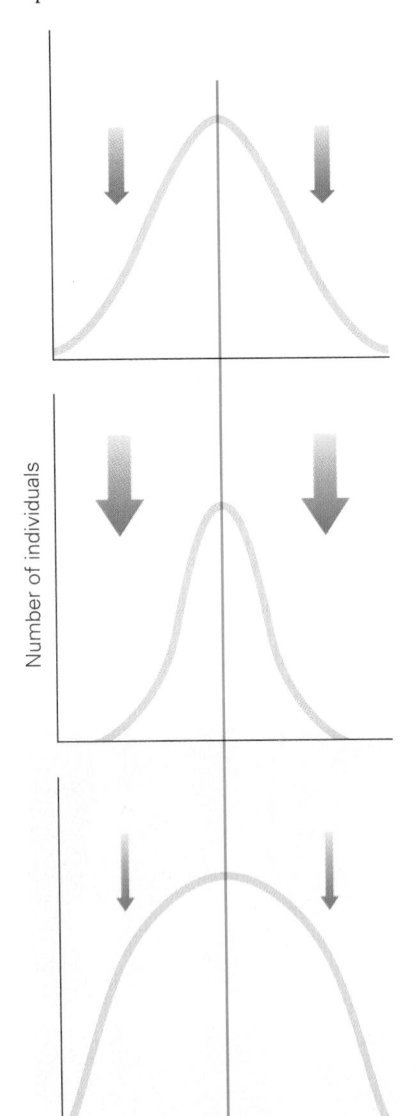

Not all populations evolve toward one optimal condition. In some cases the population may have two or more "desired" characteristics. In this case, there may be selection against the mean condition, resulting in an accumulation of individuals at the extremes. This is called **disruptive selection.** For example, consider a population of marine skates—large, flat fish that escape predators either by lying flat on the sea bottom covered with sand or by rapidly fleeing. Suppose most of the skates try concealment for a time and then begin to flee when a shark draws near. Others, though, perhaps bolder or more lethargic, do not move, remaining hidden, as the shark swims over. Yet others are very skittish and flee at the first sign of danger. In this population, then, we might expect selection against the average behavior as the sharks caught those that tried to flee while the predators were near. If the others tended to escape more frequently, the population would tend to be composed increasingly of both lethargic and skittish fish ("hiders" and "runners"). If escape tendencies were plotted on a graph, the population would produce a **bimodal** (two-humped) **curve,** as in Figure 10.8C (page 260). (Such a bimodal curve can also be

FIGURE 10.8B

Here, Y marks the optimum value for some trait, such as height. Those taller or shorter would therefore not do as well. Their numbers would be reduced by ecological pressures (shown by X and X′). If it should become better to be taller, then Y, the optimum, would shift. Individuals at X would suffer, while those at X′ would thrive. The result would be directional selection, producing a taller population until the population stabilized again around the new optimum.

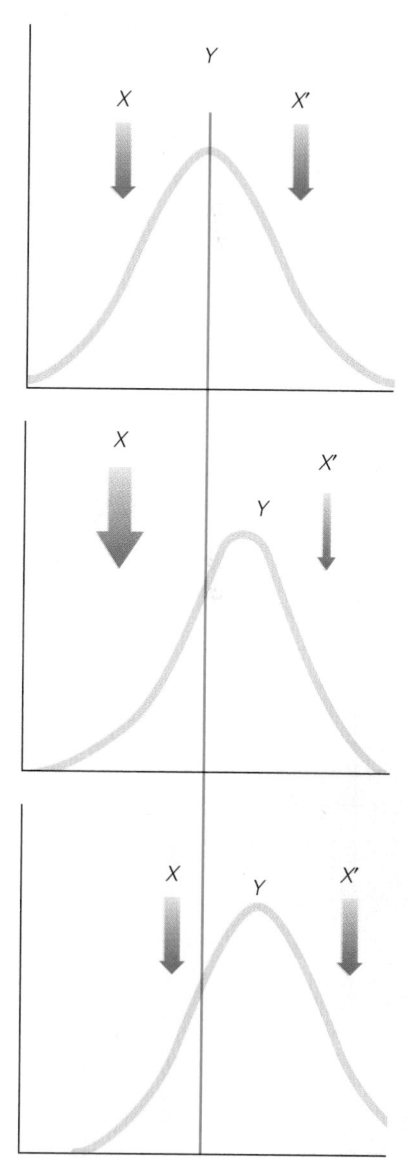

FIGURE 10.9

Two forms of *Biston betularia,* the peppered moth, on a lichen-covered background. The dark form became predominant as soot blackened the lichen, exposing the lighter form to predatory birds. With the advent of pollution-control devices, the background is becoming lighter and the light form is returning.

FIGURE 10.8C

Disruptive selection can produce a bimodal (two-humped) curve. Bimodal curves are produced in populations in which it is advantageous to be at one of the two extremes of a curve since the intermediate condition is selected against. Here, a continuously distributed population produces a bell-shaped curve (top) with individuals with intermediate traits being favored (Y). Then selection against the intermediate form begins to favor individuals with extreme traits (Y and Y′) until a bimodal distribution is achieved.

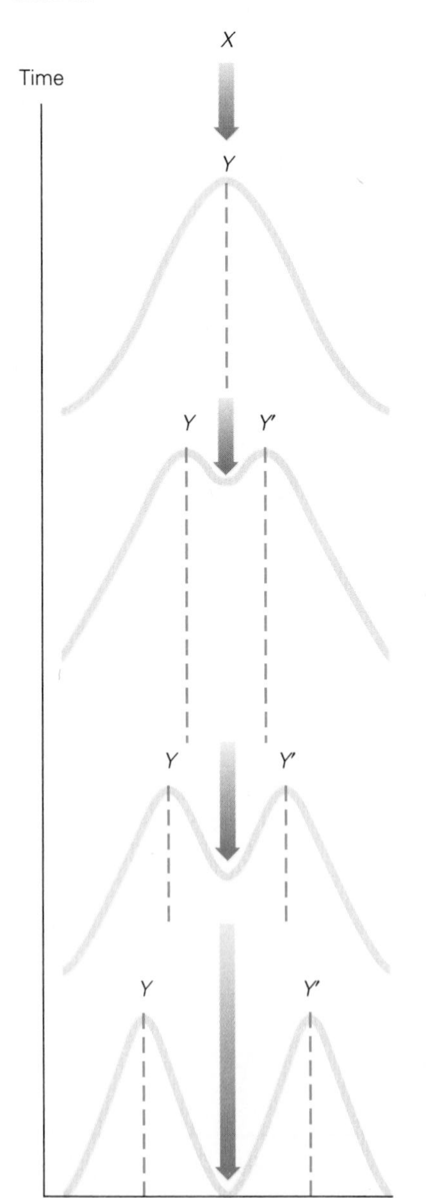

a result of sexual dimorphism as is described in Essay 10.1). Populations clustering around two phenotypes are called **dimorphic** (*di:* two; *morph:* shape). Other populations may produce more than two types and are called **polymorphic** (*poly:* many). We find a fascinating example in the brittle stars, spindly sea stars that may have any of several distinctly different appearances. This diversity seems to confuse predators by not allowing them to build a single image of the sea stars that would aid the predators in their search (Figure 10.10).

Small Populations and Genetic Drift

So far we've been discussing large, randomly mating populations that have been molded by the powerful effects of natural selection. In these groups, as adaptive traits are selected, others are relentlessly winnowed out as the populations track the environment by directional selection. But what about small populations? Do they change in the same ways? Or are they under different sorts of pressures? In other words, do they shift in response to different kinds of factors?

Small populations, indeed, may be molded by different factors than those operating in larger populations. For example, in small populations, **genetic drift** may be important. Genetic drift is a change in the frequency of genes in a population due to simple chance. Changes of this sort are *not* produced by natural selection. Genetic drift is important in small populations because they would, by definition, have a smaller gene pool (genetic reservoir). This means that any random appearance or disappearance of an allele would have a relatively large impact on allele frequencies. In large populations, any random change would have little impact since it would be swamped by the sheer numbers of alleles in the gene pool of the population. There are two important ways that small populations can set the stage for such random events.

FIGURE 10.10

Brittle stars vary in their appearance. Thus, it is difficult to find one, generalize about its appearance, and use that information to find another one.

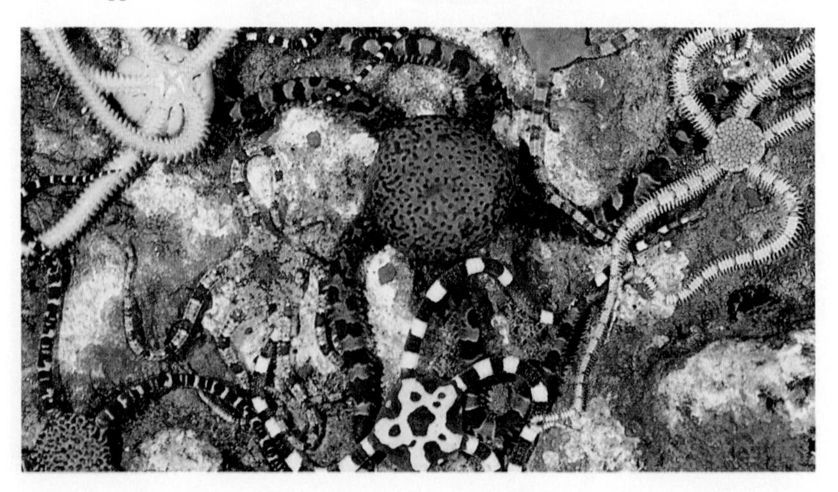

ESSAY 10.1

Sexual Dimorphism

You may have noticed that boys look different than girls and that it's easy to tell a rooster from a hen. In some deep-sea fish, the sexes are quite dissimilar; the male may even be an appendage, fused to a knob on the head of the female. On the other hand, it may be difficult to tell a male mouse from a female, and the sexes of seagulls are almost identical in appearance. **Sexual dimorphism**—or the different appearance of the sexes—varies widely among animals.

Among leafhoppers, the male is much smaller than the female. Among peafowl, the peacock is far more garish and conspicuous than the brown and drab peahen.

There are interesting evolutionary implications of sexual dimorphism. How did it start? With respect to vertebrates, one idea is that as they came under increasingly severe competition for commodities such as food, the males increased in size and strength in order to compete more successfully. The females were restrained from such changes by the demands of gestation (which placed a premium on a certain body size) and the fact that they were more closely associated with the young (which made it important that they be physically and behaviorally less conspicuous). Males that were best able to gain commodities might then have been disposed to share those with females for reproductive purposes. This meant that females might then be more likely to choose the more vigorous males as mates.

This tendency would have driven males toward increasingly garish appearances and bold, aggressive, and conspicuous behavior in order to display their vigor and to intimidate other males in disputes over commodities or females. Exceptions exist, of course, but generally today, males of most species, especially among birds and mammals, are larger, stronger, more aggressive than females and are dominant over them in social interactions. It provides some entertainment, and a lot of argument, to try to determine how this general condition could have arisen.

An interesting relationship between sexual dimorphism and domestic duties exists among some species. Consider an example from birds. The sexes of song sparrows look very much alike. The males have no conspicuous qualities that immediately serve to release reproductive behavior in females. Thus, courtship in this species may be a rather extended process as pairbonding (mating) is established. Once a pair has formed, both sexes enter into nest building, feeding, and defense of the young. The male may only mate once in a season, but he helps to maximize the number of young that reach adulthood carrying his genes. He is rather inconspicuous, so whereas he doesn't turn on females very easily, he also doesn't attract predators to the nest.

The peacock, on the other hand, is raucous and garish. He displays madly and frequently and is successful indeed. Once having seduced an awed peahen, he doesn't stay to help with the mundane chores of child rearing, but instead disappears into the sunset looking for new conquests. The peahen he has just left may be an inferior bird, not likely to be able to rear young successfully. He could have discovered this by a more careful process of mate selection. Just as he attracts females, he also attracts tigers, and his flurry of sexual activity may end in a flurry of feathers. Instead of carefully rearing one brood each season for several seasons, he maximizes his reproductive success through another means.

Think of other species of birds and mammals. Which evolutionary route have they taken? What about humans?

Bottleneck Effect

Certain kinds of disasters can reduce a population unselectively, that is, without favoring one genetic type over another. Among these are earthquakes, floods, fire, and drought. The result can be a genetic bottleneck. The **bottleneck effect** results when a population is drastically reduced, but the remaining population is not genetically representative of the original population. By simple chance, some alleles will be overrepresented, some underrepresented, and some eliminated entirely. The principle is demonstrated in Figure 10.11. Populations of elephant seals and cheetahs (Figure 10.12) are remarkably similar, since it is believed that both groups have recovered from very low populations at one time.

Founder Effect

Another influence operating on small populations is called the **founder effect.** Occasionally, a few organisms will be separated from the main population and establish their own breeding group. For example, a few birds might be blown from the mainland to some remote island where they become the founders of a new population, or a storm might wrench a large tree loose, allowing it to float out to sea. Such a raft could bear small creatures and eventually deposit them on some new land, perhaps an island, where they could begin to take their own evolutionary direction (Figure 10.13).

Any such small group is unlikely to bear all the alleles found in the main population, nor could the alleles of the small group be expected to appear in the same frequencies as those of the larger group. Certain alleles might not be represented at all in the small groups, and other alleles might be disproportionately common. As the organisms breed and begin propagating their own genetic ratios, it is soon apparent that they have departed from the evolutionary course of the parent population. Furthermore, if the founders must respond to different selective pressures in their new habitat, they may diverge even more rapidly from the parent population through natural selection.

FIGURE 10.11

The bottleneck effect. A large population may have a certain gene frequency (say light to dark individuals). A small part of that population may have a different gene frequency. If such a smaller population then serves as the breeding stock for the next generation, that generation will have a different gene frequency from the original population.

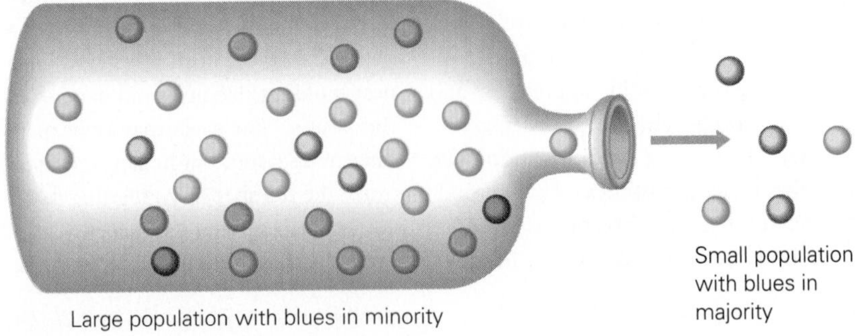

FIGURE 10.12

The founder effect is believed to be responsible for the remarkable genetic similarity of all cheetahs today. Apparently at one time their numbers dropped drastically, and today's population is descended from those few survivors.

It is because of the potential importance of seemingly insignificant changes that small populations can change, or evolve, so rapidly and, by the same token (their resulting lack of genetic variability), why they may be expected to become extinct quite easily.

There is another factor that places small populations at high risk of extinction. As a population grows smaller, genes that interfere with re-

FIGURE 10.13

The isolation on islands has allowed the evolution of some remarkable forms, as each species—bearing its own sets of genes—adapted to new opportunities. This is the world's smallest chameleon, ***Brokosia psyriaras,*** found on the island of Madagascar.

productive processes tend to accumulate. Such interference, for example, is given as a reason for the extinction of the American heath hen in 1930 (Figure 10.14). And, even now, the wisent is at dangerously low levels and may not be able to recover. (What is a wisent?*)

FIGURE 10.14
The last American heath hen on Earth died in 1930. It was probably due to an accumulation of deleterious genes in what had been a small population.

THE QUESTION OF SPECIES

Here's a simple question: What is a species? It may seem like a simple question, but it's not. Biologists find it quite difficult to answer, and, in fact, it is said that there is no record of any two biologists ever agreeing on an answer. If they seem to agree, it's generally because each one has stifled what he or she knows about this exception or that. We shall not be daunted by lack of consensus; we shall admit a certain ignorance and plunge right in.

The question of species seems easy because almost all of us can tell a dog from a fern (the leash slips off the fern), and we know an oyster from a horse (stirrups drag on the oyster). These, then, are clearly different species. Their differences in appearance tell us so. However, appearance can be deceiving. For instance, in a small, narrow area in Texas, we find a vexing situation in the form of the red-bellied woodpecker and the golden-fronted woodpecker. At first sight, they are difficult to tell apart (Figure 10.15). Their voices, feeding habits, and nest sites are also very similar. Yet, they are quite different species and obviously have no trouble telling each other apart. They do not attempt to interbreed, and no hybrids have been seen. However, far more dissimilar types than these (at least to the human eye) are known to interbreed.

As an example of interbreeding between distinct types, we find that coyotes and dogs freely interbreed. American ranchers have often wit-

*A European buffalo.

FIGURE 10.15
The golden-fronted woodpecker (a) looks remarkably similar to the red-bellied woodpecker (b). The range of the golden-fronted extends through dry areas from Honduras northward to central Texas, where it rather abruptly stops and that of the red-bellied begins. The red-bellied's range extends to the eastern coast, where it lives primarily in wooded habitats. The birds overlap in a narrow range in Texas, but in spite of strong similarities in appearance, vocalization, and behavior, they do not interbreed.

(a)

(b)

nessed their favorite hound giving birth to a litter of grouchy, furtive offspring with long noses. So, are the dog and coyote of different species? Since coyotes and dogs interbreed, even while differing greatly in appearance and behavior, some scientists treat them as a single species. Others argue that the *ability* to interbreed implies the *opportunity* to interbreed and that dogs and coyotes are sufficiently isolated that interbreeding between them is rare enough that they can be considered different species. However, when they do interbreed, the offspring are healthy and *able to breed among themselves* (Figure 10.16).

So, we see, a common criterion for animal species (but not necessarily for plant species) is the ability to interbreed and leave fertile offspring. Horses and donkeys can produce mules, but the mules are sterile, so the parents are considered to be of different species. How about the grackles of Texas and Puerto Rico? Are these noisy blackbirds of the same species? They are identical in virtually all respects except for the smaller size of the Puerto Rican birds. They obviously don't interbreed because they are geographically distant, but some researchers believe they could and would interbreed if they ever met. There are also species that cover a large area, changing gradually across the range. The subgroups interbreed all along the range, but individuals at either end of the range are so different as to make breeding impossible. Are those individuals of the same species?

These, then, are some of the problems with defining **species.** So, we will accept the definition of the renowned zoologist Ernst Mayr: *"A species is a group of actually or potentially interbreeding populations that is reproductively isolated from other such groups."*

Reproductive isolation refers to the inability of members of one group to successfully interbreed with another. Such isolation ensures that the genes from one group will not be combined with those from another group, thus each can accumulate its own genetic distinctions. There are several ways that groups can be reproductively isolated:

1. Geographic isolation. Obviously, if the groups can't reach each other, they can't interbreed. They can be geographically isolated in a number of ways, such as by the appearance of some sort of barrier (such as a canyon or a body of water) between them, or by a small group being transported away from the parent population (such as birds or insects being blown to some remote area by a storm).
2. Behavior isolation. The behavior of two groups may keep them from interbreeding, even where their ranges overlap. For example, they may not recognize each other's mating signals, or they may utilize different aspects of the environment in such different ways that their paths do not cross.
3. Mating isolation. Some species, especially insects, have very complex "lock and key" genitals. They may not be able to mate simply because their reproductive apparatuses do not fit.
4. Genetic isolation. In cases where individuals of two species do interbreed, the embryo may fail to develop because the mismatched chromosomes impede the normal pairing processes that must occur before development can proceed.
5. Hybrid isolation. In rare cases, offspring can be produced, but these are sterile. The best-known example is the cross between a horse and a donkey that produces a sterile mule.

FIGURE 10.16

A coyote-dog mix. This animal is able to breed with either dogs or coyotes. So are coyotes and dogs of different species?

SPECIATION

One of the most interesting and challenging questions in biology involves **speciation,** the formation of new species. The question is, how do they arise? As we will see, the answer is strongly dependent on geography.

Allopatric Speciation

Most species arise through **allopatric speciation** (*allopatric:* other land), the formation of new species after the geographic separation of once continuous populations. The process involves a population being somehow divided and each subgroup taking a different evolutionary route until they have diverged so much that interbreeding is no longer possible, even if they should rejoin. (Essay 10.2).

Populations can be divided in two basic ways. One is a small group of individuals (or even a seed or a pregnant female) being separated from the parent population and the descendants therefore becoming established in a new place. An example is Darwin's finches that are believed to be the descendants of birds that were swept to the Galapagos Islands, perhaps by some great storm. They not only became genetically isolated from the parent population, but also from each other as the various islands gave rise to their own distinct forms.

Populations have also been divided by geological events, perhaps as some great barrier appeared, separating a larger group into two smaller groups that then went their own evolutionary ways. An example is seen in the Abert and Kaibab squirrels. These distinct species are believed to be the descendants of a single species that was separated by the formation of the Grand Canyon (Figure 10.17).

(a)

(b)

FIGURE 10.17
The Abert squirrel (a) and Kaibab squirrel (b) are two distinct species that live on opposite sides of the Grand Canyon. It's believed that they were once one population that was divided as the great chasm developed. The two populations followed their own paths of evolution and now are quite different and unable to interbreed.

ESSAY 10.2

Continents Adrift

In 1912, Alfred Wegener published a paper that was triggered by the common observation of the good fit between South America's east coast and Africa's west coast. Could these great continents ever have been joined? Wegener coordinated this jigsaw-puzzle analysis with other geological and climatological data and proposed the theory of continental drift. He suggested that about 200 million years ago, all of the earth's continents were joined together into one enormous land mass, which he called Pangaea. In the ensuing millennia, according to Wegener's idea, Pangaea broke apart, and the fragments began to drift northward (by today's compass orientation) to their present location.

Wegener's idea received rough treatment in his lifetime. His geologist contemporaries attacked his naiveté as well as his supporting data, and his theory was neglected until about 1960. At that time, a new generation of geologists revived the idea and subjected it to new scrutiny based on recent findings.

The most useful data have been based on magnetism in ancient lava flows. When a lava flow cools, metallic elements in the lava are oriented in a way that provides permanent evidence of the direction of the earth's magnetic field at the time, recording for future geologists both its north-south orientation and its latitude. From such maps, it is possible to determine the ancient positions of today's continents. We now believe that not only has continental drift occurred, as Wegener hypothesized, but that it continues to occur today.

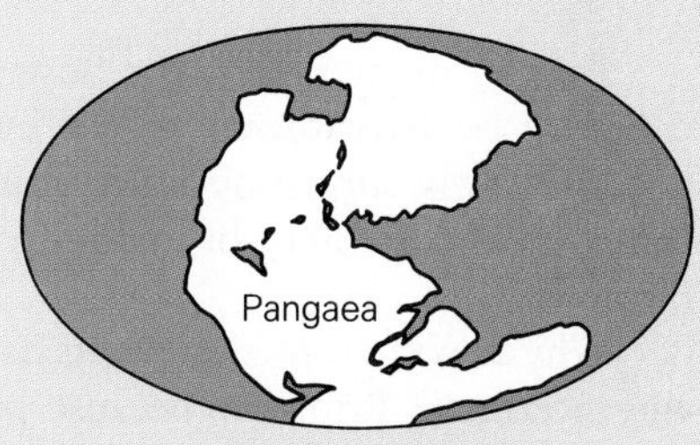

Lower Triassic Period
225 million years ago

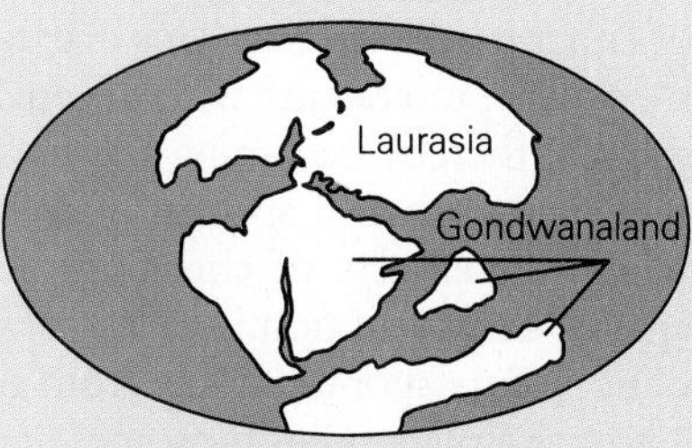

Lower Cretaceous Period
90 million years ago

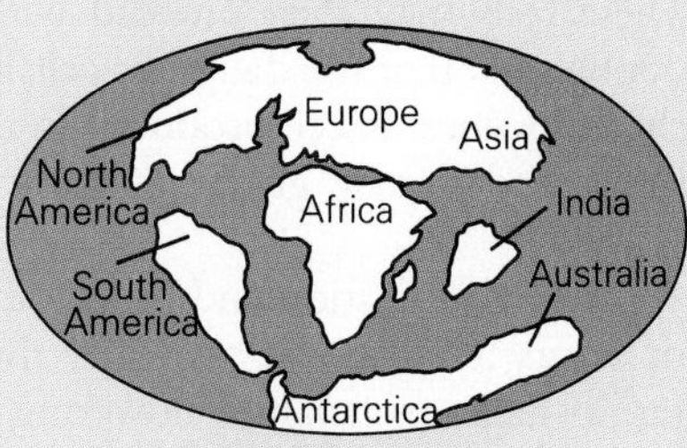

Paleocene Epoch
65 million years ago
End of Mesozoic Era

Geologists have long maintained that the earth's surface is a restless crust, constantly changing, sinking, and rising because of incredible, unrelenting forces beneath it. These constant changes are now known to involve large, distinct segments of the crust known as plates. At certain edges of these masses, immense ridges are being thrust up, while other edges sink lower. Where plates are heaved together, the buckling at the edges has produced vast mountain ranges. When such ridges appear in the ocean floor, water is displaced and the oceans expand. (Astoundingly, precise satellite studies reveal that the Atlantic Ocean grows 5 cm wider each year.)

An understanding of continental drift (or plate tectonics) is vital to the study of the distribution of life on the planet today. It helps to explain the presence of tropical fossils in Antarctica, for example, and the unusual animal life in Australia and South America. Continental drift provided just the sort of separation of subpopulations that would permit widespread speciation, forming the basis for widely diverging groups of primitive organisms.

As the composite maps indicate, the disruption of Pangaea began some 230 million years ago in the Paleozoic era. By the Mesozoic era, the Eurasian land mass (called Laurasia) had moved away to form the northernmost continent. Gondwanaland, the mass that included India and the southern continents, had just begun to divide. Finally, during the late Mesozoic era, after South America and Africa were well divided, what was to be the last continental separation began, with Australia and Antarctica drifting apart. Both the North and South Atlantic Oceans would continue to widen considerably up to the Cenozoic era, a trend that is continuing today. So we see that although the bumper sticker "Reunite Gondwanaland" has a third-world, trendy ring to it, it's an unlikely proposition.

FIGURE 10.18

Populations tend to influence each other according to how greatly they overlap. Allopatric populations normally do not interact. Sympatric species must minimize competition by subdividing a shared habit.

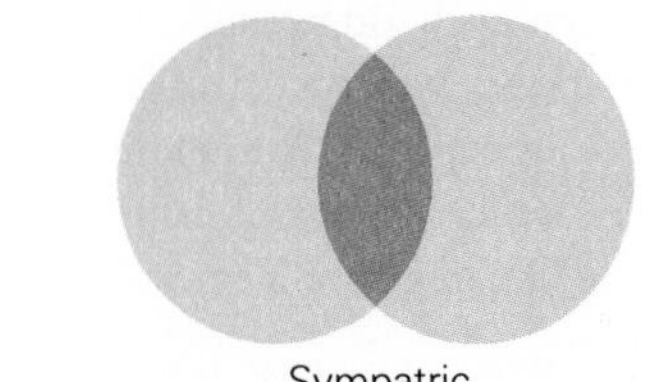

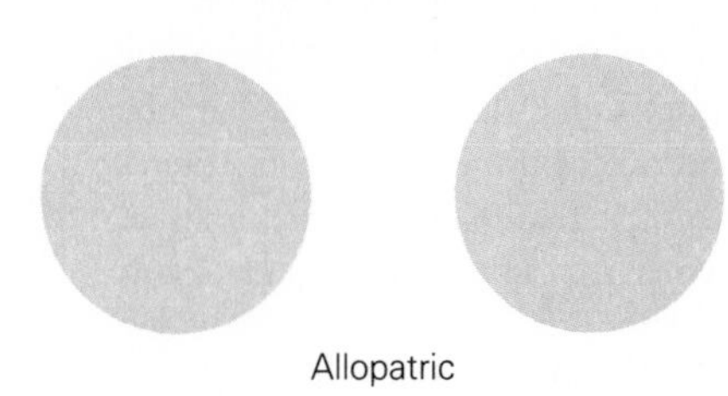

Sympatric Speciation

Sympatric speciation (*sympatric:* together land) is a less common event, involving the formation of two species from one, continuously interbreeding population. It is, by far, more common in plants than animals (Figure 10.18).

Among many plants, the flowering plants in particular, new species can arise by **hybridization,** the interbreeding of existing species. This discovery was surprising since animal hybrids are usually infertile due largely to the inability of dissimilar chromosomes to line up properly at metaphase.

Why don't hybridizing plants have the same problem with mismatched chromosomes? Primarily because plants with very different appearances may be very similar genetically. Where such genetically similar species overlap, there may be extensive hybrid populations. Surprisingly, such hybridization doesn't seem to result in the breakdown (through merging) of either parent species. This may be because hybrid populations find their own niche, interacting with the environment differently than do the parent groups, thereby becoming truly a new and distinct species.

Plants can also speciate successfully in a quite dramatic way, one in which whole sets of chromosomes become doubled (or even doubled again). The condition is called *polyploidy* (*poly:* many). Chromosome doubling occurs spontaneously from time to time in the mitotic divisions of a growing plant. Chromosomes normally reproduce in preparation for cell division; for some reason, however, the cell itself occasionally fails to divide while it continues through its normal sequences as if nothing were unusual. These cells may then proceed with normal mitosis. The result of such doubling is that the daughter cells will end up with four complete sets of chromosomes. A cell produced in this way is called a *tetraploid* cell (*tetra:* four), and it may go on to form tetraploid tissue and even tetraploid flowers.

So hybridization and polyploidy can create instant, sympatric species of plants, ready to be tested by the forces of natural selection. This versatility helps explain how flowering plants arose so abruptly in evolutionary history and how they then so quickly spread out over the landscape to help create the incredible diversity of plant species that so beautifully accent our world today.

Sympatric speciation in animals is a much rarer event, but a few rather clear-cut examples have been found. One of the best-known cases involves flightless grasshoppers. The grasshoppers appear to be of one species across their range, but chromosomal analysis reveals them to be of two different species, each occupying a particular part of the range. Apparently, a random genetic change had occurred in the parent population that allowed the descendants of those carrying that change to better adapt to a part of the parent population's former range. The change was apparently great enough to preclude further mixing of genes of the two populations, thereby giving rise to a new species in the midst of an existing one.

As an example of sympatric speciation in progress in an animal, consider the maggot fly. The fly has long parasitized the North American hawthorn trees, and when apple trees were introduced into North America in the nineteenth century, the fly began to infect them too. Now the fly has become divided into two groups of specialists, one line infecting hawthorn trees, the other preferring apple trees. The two lines can still interbreed, but they differ in some important respects, such as in several genes and in their maturation time.

Keep in mind that, although we have devoted more attention to the special cases of sympatric speciation, allopatric speciation is considered to be far more important in the evolution of life on earth.

Divergent and Convergent Evolution

In determining just how and when speciation has occurred, it is important to realize that similarities may not be a reflection of evolutionary relationships. Just as different environments cause populations to diverge, a process called **divergent evolution,** sometimes similar environments can cause species to become more similar through a process called **convergent evolution** (Figure 10.19). Convergent evolution can cause a great deal of

FIGURE 10.19

Examples of convergent evolution. (a) Upper, a tropical Asian cyprinid; lower, an African characin. (b) Upper, an Australian agamid; lower, an American horned toad. (c) Upper, an African civet; lower, an American weasel. (d) Left, an American meadowlark; right, an African yellow-throated longclaw. Each pair lives in ecologically similar environments.

confusion for those trying to work out evolutionary histories, since physical similarity does not necessarily indicate relationship (Figure 10.20).

SMALL STEPS OR GREAT LEAPS?

A cornerstone of Darwin's explanation of evolution as "descent with modification" was the notion that evolution is essentially the gradual accumulation of small changes over a long period of time. (The time available for life to have evolved is shown in Figure 10.21. See also Table 10.1 on page 272.) Such changes, he believed, can account for the sweeping changes that result in adaptation and speciation.

Recently, however, Niles Eldredge and Stephen Jay Gould, of the American Museum of Natural History, suggested that the fossil record does not show a gradual transition of one life form to another. Rather, they have reported that one life form persists in the record for long periods of time and then suddenly is replaced by a different life form apparently descended from the older one. They say that the story of evolution is marked by long periods of stabilization in which life forms changed little, occasionally punctuated by episodes of rapid change (Figure 10.22).

They argue that the stable periods often last about 3 million years, before the onset of a sudden change. The new version appears quickly, leaving little evidence of intermediate forms. Eldredge and Gould labeled evolution that proceeds by such fits and starts **punctuated equilibrium.**

FIGURE 10.20

Divergent evolution occurs when species divide into subgroups that then become increasingly different from each other. Convergent evolution involves the development of similar traits in dissimilar species of different stocks.

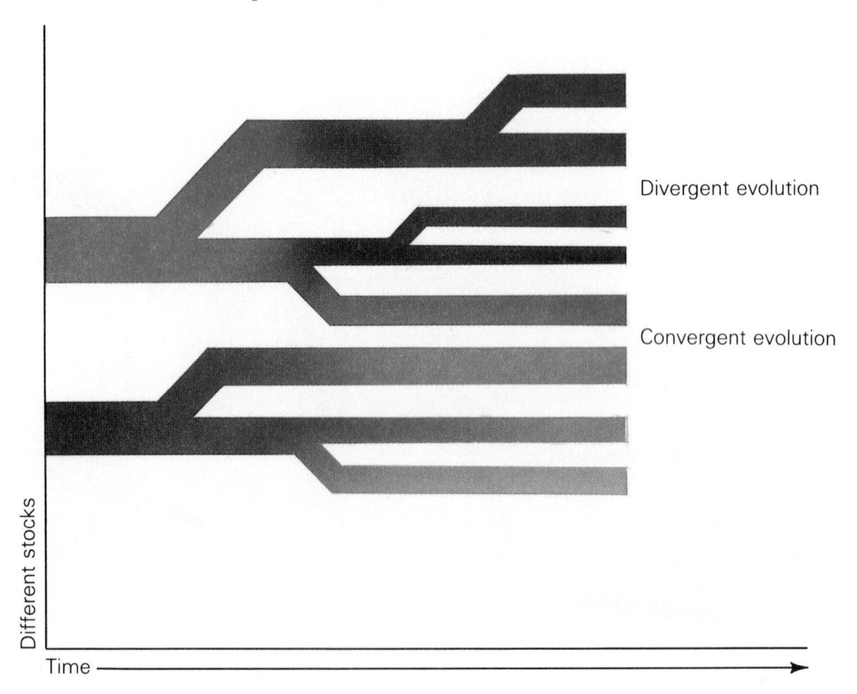

FIGURE 10.21

An evolutionary clock. Here, the earth's history is shrunk into a period of 12 hours—from midnight to noon. The events are in chronological order and the spacing indicates lapsed time.

Evolutionary Clock

12 Midnight —	Earth forms
3:00 A.M. —	First undisputed life
3:00 A.M.–9:15 A.M. —	Prokaryotes
9:15 A.M. —	First eukaryotes
10:45 A.M. —	Primitive animal phyla evolving
10:54 A.M. —	First terrestrial plants
11:00 A.M. —	First vertebrates
11:30 A.M. —	Age of dinosaurs
11:50 A.M. —	Age of mammals
11:59:00 —	First hominids
11:59:40 —	First humans
11:59:59 —	All of human history
12:00 Noon —	Present

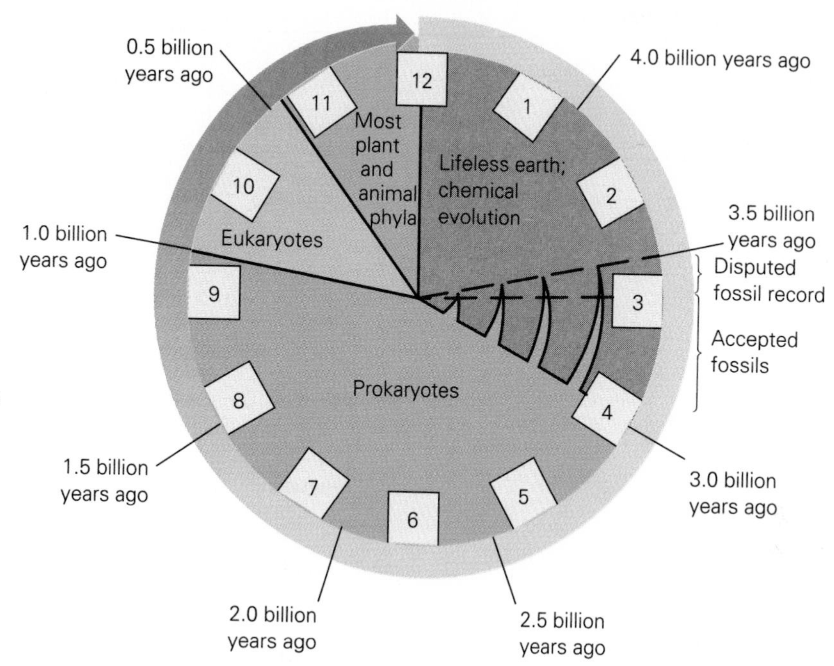

FIGURE 10.22

Comparison of the theory of gradual evolution in which one form slowly changes to another, and the theory of punctuated equilibrium in which forms change abruptly and then continue for long periods unchanged.

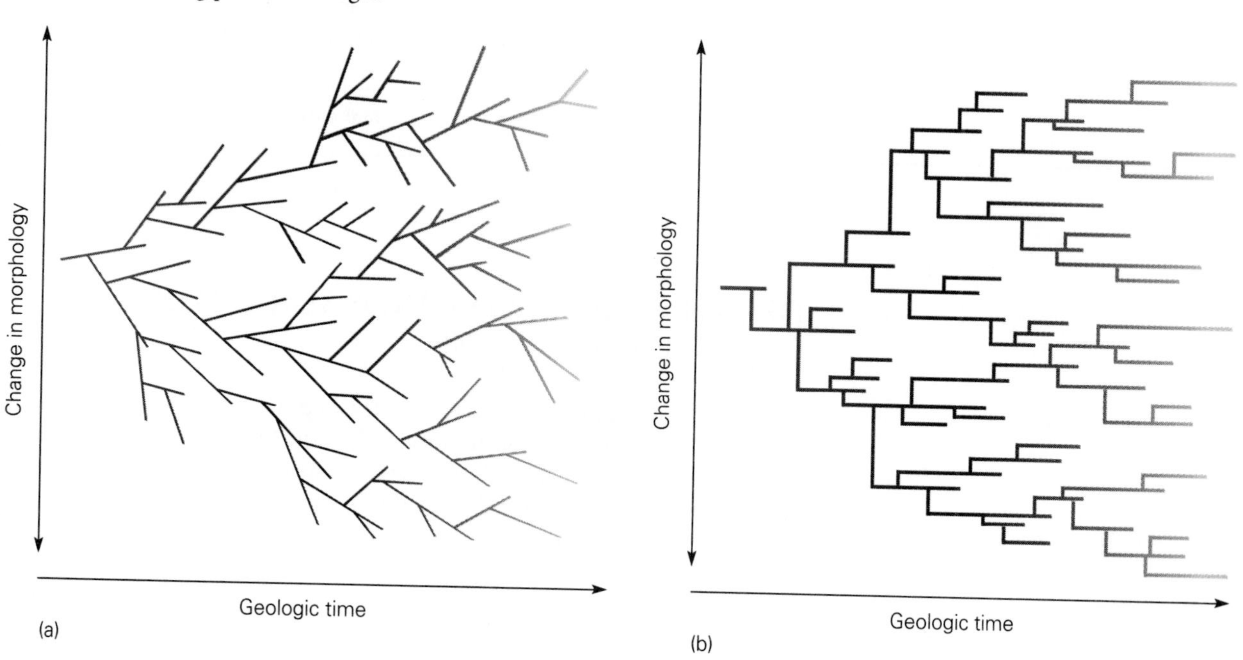

TABLE 10.1
Geological Timetable

ERA AND DURATION	PERIOD	EPOCH	YEARS BEFORE PRESENT (IN MILLIONS)	PRINCIPAL EVENTS OF THE ERA
CENOZOIC 65 Million years	Quaternary	Recent	(11,000 yrs)	Age of man; end of last ice age; warmer climate.
		Pleistocene	1	First human societies; large scale extinctions of plant and animal species; repeated glaciation.
	Tertiary	Pliocene	11	Appearance of man; volcanic activity; decline of forests; grasslands spreading.
		Miocene	25	Appearance of anthropoid apes; rapid evolution of mammals. Formation of Sierra Mountains.
		Oligocene	36	Appearance of most modern genera of mammals and monocotyledons; warmer climate.
		Eocene	54	Appearance of hoofed mammals and carnivores; heavy erosion of mountains.
		Paleocene	65	First placental mammals.
MESOZOIC 160 Million years	Cretaceous		135	Appearance of monocots; oak and maple forests; first modern mammals; beginning of extinction of dinosaurs. Formation of Andes, Alps, Himalayas, and Rocky Mountains.
	Jurassic		181	Appearance of birds and mammals; rapid evolution of dinosaurs; first flowering plants; shallow seas over much of Europe and North America.
	Triassic		220	Appearance of dinosaurs; gymnosperms dominant; extinction of seed ferns; continents rising to reveal deserts.

DEATH STARS AND DINOSAURS: THE GREAT EXTINCTIONS

It may be quite difficult to account for the beginnings of new species, but it is often easy to see how some species died out. We have built dams and knowingly doomed small pockets of isolated species. We have hunted other animals to extinction, as we did the dodo, the carrier pigeon, and the last common ancestor of the horse and zebra. However, other species have

TABLE 10.1
Geological Timetable (cont.)

ERA AND DURATION	PERIOD	EPOCH	YEARS BEFORE PRESENT (IN MILLIONS)	PRINCIPAL EVENTS OF THE ERA
PALEOZOIC 360 Million years	Permian		280	Widespread extinction of animals and plants; cooler, drier climates; widespread glaciation; mountains rising; atmospheric carbon dioxide and oxygen reduced.
	Carboniferous		345	Appearance of reptiles; amphibians dominant; insects common. Gymnosperms appear; vast forests; great life abundant. Climates mild; low-lying land; extensive swamps; formation of enormous coal deposits. Many sharks and amphibians; large-scale trees and seed ferns; climate warm and humid.
	Devonian	Paleocene	405	Appearance of seed plants; ascendance of bony fishes; first amphibians; small seas; higher, drier lands; glaciations.
	Silurian		425	Atmospheric oxygen reaches second critical level. Explosive evolution of many forms of life over the land; first land plants and animals. Great continental seas; continents increasingly dry.
	Ordovician		500	Appearance of vertebrates, but invertebrates and algae dominant. Land largely submerged. Warm climates worldwide.
	Cambrian		600	Atmospheric oxygen reaches first critical level. Explosive evolution of life in the oceans; first abundant marine fossils formed; trilobites dominant; appearance of most phyla of invertebrates. Low-lying lands; climates mild.
PRECAMBRIAN 3500 Million years			2700	Life confined to shallow pools; fossil formation extremely rare. Volcanic activity, mountain building, erosion, and glaciation. Photosynthetic life.

passed into extinction for reasons that continue to puzzle us. In particular, it is difficult to account for the massive, large-scale extinctions in which many species passed from the earth at once. For example, why did so many species die out with the great dinosaurs at the end of the Cretaceous (Figure 10.23, page 275)?

One hypothesis advanced to account for the die-off at the end of the Cretaceous is based on data suggesting that the temperature of the earth dropped drastically about that time. The dinosaurs and the others, some

ESSAY 10.3

Extinction and Us

There is disconcerting evidence that we are apparently in the midst of another, truly devastating mass extinction, and this time there is little question of the cause. As Pogo once said, "We have met the enemy, and he is us." Indeed, the extinction rate is presently estimated by a number of experts to be between a thousand and several thousand species per year. If we continue on this course of environmental destruction, by the year 2000 the extinction rate could be about 100 species per day. Within the next several decades, we could lose a quarter to a third of all species now alive. This rate of loss is unprecedented on the planet. Furthermore, the problem is qualitative, as well as quantitative. Not only are we losing *more* species, but we're also losing different *kinds* of species. The earlier mass extinction involved only certain groups of species, such as the cycads and the dinosaurs. The other species were left more or less intact. At present, though, species are dying out across the board. That is, the current extinction affects all the major categories of species. Of particular note, this time the terrestrial plants are involved. In the past, such plants provided resources that the surviving animals could use to launch their comeback. With the plants also devastated, any comeback (marked by a period of rapid expansion and speciation) by animals will be greatly slowed.

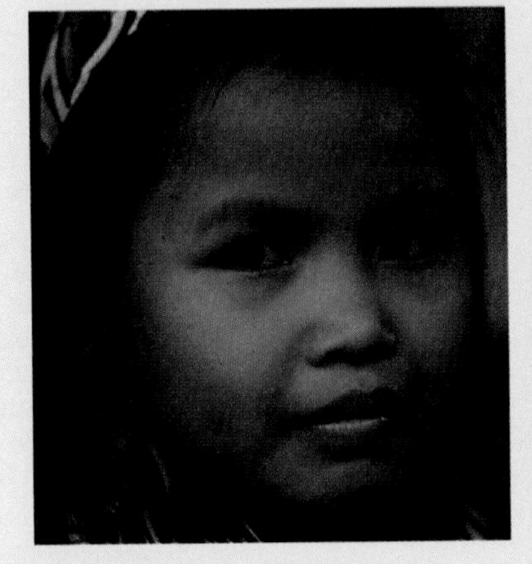

Indigenous peoples of Borneo, such as the Penans and the Dayaks, are watching almost helplessly as the timber industry destroys their forest home, undoubtedly causing the extinction of many potentially useful species at the same time.

The present great extinction will also prove to deter a resurgence of species for another reason. This time we are killing the systems that are particularly rich in life, including tropical forests, coral reefs, salt water marshes, river systems, and estuaries. In the past these systems have provided genetic reservoirs from which new species could spring and replenish the diversity of life on the planet. In effect, we are drying up the wellspring of future speciation.

In the past, extinctions have been due to two major processes, environmental forces (such as climatic change) and competition. For the first time, a single species (our own) has had the opportunity to cause mass extinctions at both levels. Much of the habitat destruction has been due to humans needing the land where other species lived, forcing them into extinction as they failed in their competition with us. And now we find ourselves in the remarkable position of being able to alter environmental forces. As the Amazon basin is destroyed, the trees are no longer available to cycle water back to the atmosphere, causing many experts to predict sweeping changes in the weather.

We are continuing to interact in new ways with the environment. As we continue to release chlorofluorocarbons into the atmosphere, the ozone holes grow larger and the delicate veil of life on earth becomes increasingly bathed in destructive radiation. And we are learning that the oceans have become perilous places for many forms of life because we continue to use the great waters as dumps for dangerous or unknown chemicals.

There are those who say that our course, by now, is irreversible, that the damage is done and now we must sit back and prepare to reap what we have sown. Others, though, argue that there is still time to alter our course, to salvage much of what is left and to protect it from further damage. The danger is in accepting the first alternative if the second is really the case.

say, simply died of the chilling effects of hypothermia. However, others have argued that the cooling of the earth unbalanced the sex ratios of many species. They note that the sexes of many kinds of animals, such as alligators, amphibians, and some fish are temperature dependent. That is, eggs raised in environments below a certain temperature will give rise to animals of one sex, and above that temperature, to the other sex. The argument is that as the earth cooled below a certain critical point, all the hatchlings of some species would have been of one sex and doomed to roam the earth without ever knowing the joys of parenthood. How awful.

Another explanation of the great extinctions of the Cretaceous also involves a cooling episode, but this one accounts for its origins. According to the **Alvarez hypothesis,** the earth was struck by a great asteroid that raised a cloud of dust that blocked the sun for many months and effectively impeded photosynthesis. Any such disruption of the food chain would have led to the demise of a great many species. The evidence here is circumstantial but solid. The best line of evidence is the discovery that a rare element called iridium was deposited in a fine layer over the earth at about the end of the Cretaceous. Iridium is uncommon in the earth's crust, but quite common in asteroids (Figure 10.24).

Other researchers have suggested that the earth was struck by some heavenly body that was not an asteroid, but perhaps three or four comets. (Ready for this?) The comets, they say, were from the great belt of 100 million or so comets that lazily circle the sun far beyond the reaches of the solar system. Occasionally, though, about every 26 to 30 million years, great numbers of the comets are jerked toward our sun by the passage of a companion star to the sun. The companion star has not been located or identified, but it has been named: Nemesis. Journalists, with their flair for high drama, have taken to calling it the Death Star.

FIGURE 10.23

These sauropod footprints are a poignant reminder of the great extinctions that occurred at the end of the Cretaceous Period, when so many land species and ocean-dwellers died.

FIGURE 10.24

Luis and Walter Alvarez point to a layer of iridium crucial to their explanation of the last great extinction.

Any such companion star, if it exists, could be any of the hefty little "black stars," dense bodies with a gravitational pull so strong that not even light can escape. It has been suggested that the star has an extremely elliptical orbit that takes it far into the celestial realm—until its next deadly loop through the belt of comets. Geologists have found that the earth has indeed been peppered every 30 million years or so by celestial objects big enough to form craters, and that the cycles roughly match the great extinctions that paleontologists tell us have taken place on our planet. Again, there is great disagreement about the existence of Nemesis and, in fact, about the periodic extinctions themselves.

Finally, there are those who say that about every 20 to 30 million years, the earth passes through severe galactic storms of dark clouds and gas as the sun takes us through the plane of the Milky Way, and that these storms are responsible for the great surges of death on the planet (Figure 10.25).

However, on a brighter note, the sun is now passing through a clear zone, the pristine aftermath of an exploded star. So all should be clear sailing for a while . . . or so they say.

In summary, we have seen that, as Darwin suggested, variation in natural populations results in some organisms being better reproducers than others and that their kinds of alleles tend to increase in populations. The best reproducers, of course, will tend to be those most in harmony with the environment. Thus, populations tend to track their environment through adaptation. Evolution, then, as we understand it today, is simply a function of basic arithmetic. Those alleles that promote successful reproduction will increase in frequency.

Life on earth is subject to countless pressures as it continues striving for its very existence. It must constantly react to the nature of its situation (or its predicament) and it must change. The world is a variable and

FIGURE 10.25

Although the nearest star to our sun is light years away, some researchers believe that such celestial bodies have exerted a powerful influence on life on earth. Countless numbers of these stars, by the way, have planets. Do any of them have life? Are we the most intelligent form of life in the universe? Do you find that idea depressing?

changeable place, and different life forms have evolved that are uniquely able to utilize one aspect of the earth or another in countless ways. Put simply, life must change in order to take advantage of that part of the world available to it. In the next chapter, we will begin to explore this vast array of life and see just what the processes of natural selection have wrought.

SUMMARY

1. Evolution can be defined as change in a population through time. The integration of Darwin's ideas on evolution with Mendelian genetics is called the modern synthesis.
2. There is likely to be a great deal of variation within populations. Variation is largely a reflection of the frequencies of different alleles. Evolution is a change in the allele frequencies in the gene pool (all the alleles in an entire population).
3. Inherited variation in a population arises by meiosis, crossing over, and the chance combination of gametes. New alleles are added by mutation.
4. If the environment changes, variation is advantageous because it increases the chances that some individuals survive and reproduce, leaving offspring that also fare well under the new conditions.
5. The inherited variation in a population provides the raw material on which natural selection acts. Variation is influenced by the number of genes involved in the expression of a trait. Traits that are manifested in an "either-or" condition are generally due to the expression of a single gene. In polygenic inheritance, traits are produced by the expression of multiple genes and vary more gradually and continuously in the population. An example of polygenic inheritance is human height.
6. Variation in the environment is a factor that maintains variation in populations. Shifting environmental conditions maintain variability in a population if the population lacks sufficient time to adjust to the new set of conditions. An environment consisting of patches of different habitats or with gradual changes across the habitat also maintains variability because subgroups adjust to local conditions.
7. The Hardy-Weinberg law says that allele frequencies will remain constant through generations under certain conditions. Factors that may change allele frequencies are mutation, migration, natural selection, and small population size.
8. Variation is the raw material for natural selection. The fittest individuals are those who leave the most offspring. The traits of the best reproducers increase in the population, while the traits of those leaving fewer offspring decrease. As a result, the allele frequencies change.
9. When presented graphically, variation in a trait such as human height produces a bell-shaped curve. Strong selection pressure eliminates individuals with extreme forms of the trait, producing a narrower curve. When selection pressure is weak, more individuals with extreme traits survive, and therefore the curve is wider.
10. As a result of stabilizing selection, which occurs when the environment is stable, the population is clustered around an average condition that is assumed to be optimal for the existing conditions.

11. When environmental conditions change, directional selection leads to changes in the population as the population conforms to the new conditions. An example of directional selection is the change in coloration of the peppered moth that accompanied the changing environment during the industrial revolution in England.
12. Disruptive selection (selection against the mean condition) results in the accumulation of individuals showing more extreme forms of a trait and a bimodal curve. Disruptive selection may produce dimorphic or polymorphic populations. Dimorphic populations cluster around two phenotypes. Polymorphic populations have more than two phenotypes.
13. Small populations may evolve through genetic drift, which is the change in frequency of genes in a population due to chance. When a disaster strikes a population, drastically reducing its size, the gene frequencies of the survivors may not be representative of those in the original population (the bottleneck effect). Gene frequencies of small populations may also be changed through the founder effect in which a few organisms become separated from the main population and establish their own breeding group. The gene frequencies of the new group change further in response to selection pressure in the new environment.
14. Small populations may become extinct easily because they lack genetic variability and because genes that interfere with reproductive processes may accumulate.
15. A species is a population whose members actually or potentially interbreed, producing offspring that are able to breed among themselves. Two problems with using the ability to interbreed as a definition of species include the difficulty of determining whether individuals without the opportunity to mate could actually do so, and deciding whether individuals at the extremes of a range, who are unable to successfully interbreed, are of the same species when subgroups interbreed all along the range.
16. When one group is unable to interbreed with another they are reproductively isolated. Groups may become reproductively isolated due to geographical barriers or distance, differences in certain behaviors such as mating signals, mismatched reproductive apparatuses, differences in genes that prevent pairing and embryological development, or the production of sterile hybrid offspring.
17. Speciation is the formation of new species. Among animals, the most common type of speciation is allopatric speciation, which occurs when subgroups become isolated from the parent group. When separation occurs, genetic differences accumulate between the groups that would prevent interbreeding if the groups were rejoined.
18. Sympatric speciation, common among plants, is the formation of new species from groups living within the same area. Sympatric speciation may occur by hybridization, particularly among flowering plants. New plant species may also form through polyploidy, which is the doubling or quadrupling of whole sets of chromosomes when the reproduced chromosomes fail to separate during mitosis.
19. Divergent evolution increases the differences between species living in different environments. Convergent evolution increases the similarities of species living in similar environments.
20. Darwin believed that evolution occurred by the gradual accumulation of small changes over long time periods. In contrast, Eldredge and Gould

argue that evolution is characterized by punctuated equilibrium, long periods in which life forms change little, followed by occasional periods of rapid change.

21. There are several hypotheses to explain the massive extinctions of species, including the demise of the dinosaurs at the end of the Cretaceous period. The decrease in temperature at this time may have caused extinction because organisms died from the cold or, in some cases, because the change in temperature altered the sex ratio so that one sex predominated. According to the Alvarez hypothesis, the dust cloud caused by an asteroid colliding with the earth blocked the sunlight needed for photosynthesis and interrupted the food chain. Other researchers suggest that the earth was struck by comets that were pulled toward our sun by an as yet undiscovered companion star, Nemesis. Others have suggested that the extinctions occurred when the earth passed through severe galactic storms of dark clouds and gas.

KEY TERMS

allopatric speciation (266)
Alvarez hypothesis (275)
bimodal curve (259)
bottleneck effect (262)
cline (256)
convergent evolution (269)
dimorphic (260)
directional selection (258)
disruptive selection (259)
divergent evolution (269)
evolution (251)
founder effect (262)
gene pool (252)
genetic drift (260)
hybridization (268)
polymorphic (260)
punctuated equilibrium (270)
reproductive isolation (265)
sexual dimorphism (261)
speciation (266)
species (265)
stabilizing selection (258)
sympatric speciation (268)

FOR FURTHER THOUGHT

A population of African grasshoppers exhibits several phenotypes. Most individuals are brown, about 30% are green and about 3% exhibit either purple heads or purple stripes.

1. Is this a dimorphic or polymorphic population?
2. If the distribution of phenotypes in this population were graphed, would they produce a bimodal or bell-shaped curve?
3. What type of environment would produce this type of population?

FOR REVIEW

True or false?

1. ___ Animal speciation occurs most often in sympatric populations.
2. ___ The ability to interbreed is sometimes used as a criterion for determining species.

3. ___ Populations living in similar environments tend to exhibit convergent evolution.
4. ___ Erratically shifting environments tend to produce highly varied, unstabilized populations.
5. ___ The founder effect is a type of genetic drift.

Fill in the blank.

6. ___ and ___ are two routes of sympatric speciation exhibited by plants.
7. ___ populations produce only two distinct types of individuals or phenotypes.
8. ___ is the type of reproductive isolating mechanism in which two species interbreed but fail to produce living offspring.
9. ___ is the kind of natural selection in which a population maintains a mean or average condition.
10. The ___ is a type of genetic drift in which some natural disaster randomly eliminates some alleles from a population.

Choose the best answer.

11. ___ isolation is a reproductive isolating mechanism in which two different species produce a sterile offspring.
    A. Genetic
    B. Mating
    C. Hybrid
    D. Geographic
12. Reproductively successful traits
    A. are carried into future generations.
    B. replace less successful traits over time.
    C. increase proportionately in future generations.
    D. all of the above
13. ___ results in a selection against a mean condition and produces a population of extreme types.
    A. Directional selection
    B. Disruptive selection
    C. Stabilizing selection
    D. none of the above
14. Genetic drift
    A. only occurs in large populations.
    B. is a natural selection process.
    C. changes gene frequency in a population.
    D. produces two genetically identical populations.
15. The Alvarez hypothesis suggests that extinctions in the Cretaceous era resulted from
    A. unbalanced sex ratios.
    B. the disruption of photosynthesis.
    C. the gravitational pull of black stars.
    D. the increased temperature of the earth.

CHAPTER 11

Diversity I: The Monera, Protista, and Fungi

Overview

Five Kingdoms and Splitters and Lumpers

The Kingdom Monera and a Continuing Simplicity

The Kingdom Protista and Accelerating Evolution

The Animal-like Protists

The Plantlike Protists

The Funguslike Protists

The Kingdom Fungi and Varied Lifestyles

Objectives

After reading this chapter you should be able to:

- Name the five kingdoms into which living things are grouped.
- List the different types of monerans and describe their diverse habitats.
- Describe the features of the animal-like, plantlike, and funguslike protists.
- Describe representatives in the four divisions of fungi.
- List ways in which fungi are beneficial and detrimental to humans.

Photo: Live foraminiferan.

At this point, we must begin to tread softly. Although it is well known that biologists are eminently agreeable people, when it comes to how living things are to be classified, they can be a bit testy. In fact, it may be difficult to find any two biologists who can agree on how any living things should be classified. They may even argue over what an *animal* is. That may seem strange because, of course, anyone can recognize an animal. If something has two eyes and large teeth and is after us, we are not going to be easily convinced that it's a plant. We're going to scream that "an animal" is chasing us. If we're lucky, our rescuer won't be a biologist, who would immediately demand, "Quick! Is it a pinniped? Good grief, what *order* is it in?"

Of course, there are many ways to distinguish an animal or a plant or a fungus. A toadstool is a fungus. The little elves that sit under it are animals. So where does the disagreement arise? The disagreement over classification is essentially of two sorts. Experts may disagree over just how closely species are related, and they may disagree over whether an organism is, say, a plant or an animal at all.

The latter is the more fundamental question. As an example, there is great disagreement over just what group sponges are in. And then, just because an organism bears chlorophyll, is it a plant? Maybe not. We will shortly consider the perplexing *Euglena,* which has chlorophyll and behaves like an animal. And what can we say of an organism that can turn into a crystal, as viruses do? What about tiny organisms, such as amoebas, composed of only a single cell that creep around and devour other tiny creatures? Are they animals? If they aren't plants, aren't they animals? Not necessarily. Today, most scientists believe that living things should be divided into five kingdoms. The amoeba, then, is removed from the animals and placed within a group of tiny one-celled species, like itself, called Protista (Table 11.1).

TABLE 11.1
Characteristics of the Five Kingdoms

CHARACTERISTICS	MONERA	PROTISTA	FUNGI	PLANTAE	ANIMALIA
1. Nuclear membrane?	No	Yes	Yes	Yes	Yes
2. Mitochondria? (involved with ATP production in the cells)	No	Yes	Yes	Yes	Yes
3. Ability to photosynthesize?	Some do	Some do	No	Yes	No
4. Motility? (ability to move)	Some have it, some do not	Some have it, some do not	Primarily nonmotile	Primarily nonmotile	Yes
5. Form?	One-celled	Most are one-celled	Molds, multicellular and yeasts, single-celled	Multicellular, some algae	Multicellular

FIVE KINGDOMS AND SPLITTERS AND LUMPERS

We will adhere to the five-kingdom approach in our discussion because it helps to simplify things and because many evolutionists tell us that it makes sense (Essay 11.1 describes the various biological classifications). They say that there are five major groups that are rather closely related and that are different from the other groups in important ways (Figure 11.1). However, a newly discovered group of ancient monerans, the *archaebacteria* (*archae:* old), are clearly not closely related to other bacteria, called the *eubacteria* (*eu:* true), and some would put them in their own kingdom. We will take the simpler approach and lump them together until the jury returns on that one.

The task of separating the various species into their proper groups is not easy. One problem is that there are simply so many species. Whereas there are only about 1500 species of bacteria, there are about 8500 species of birds. The carrot family can boast of about 3000 species, and there are about 15,000 species of wild orchids. The beetles weigh in with about 300,000 species, about 30,000 of which live in North America. (These, by the way, are minimal numbers. There are undoubtedly many more species, and many of these are disappearing from the earth before we even get to know them.)

It is a great pleasure to walk the countryside with a first-rate naturalist. These people are storehouses of information regarding the flora and fauna of their regions. They seem to know every living thing you come across. Occasionally, though, they can't name something. They may thoughtfully

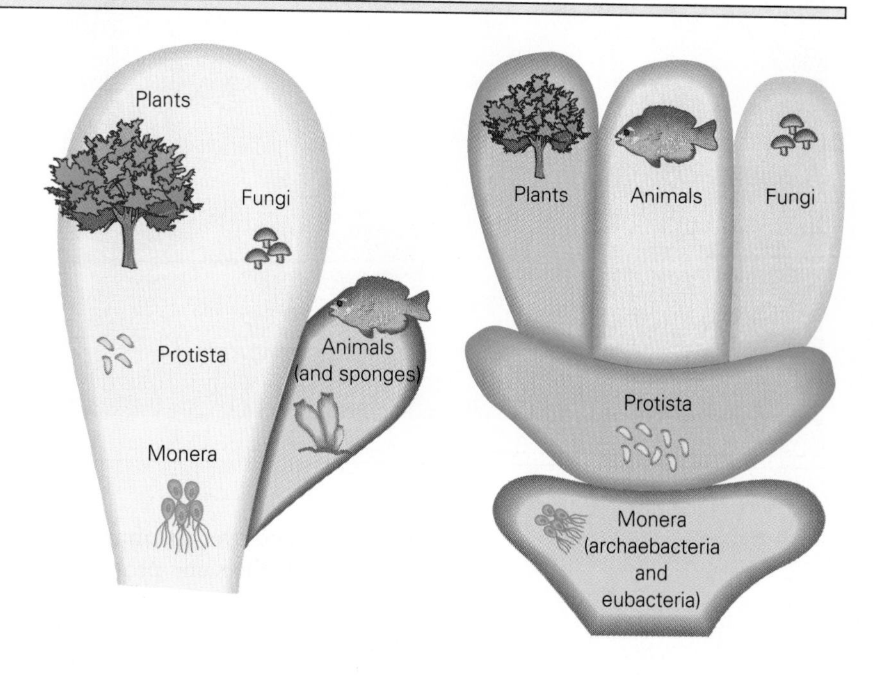

FIGURE 11.1
The two- and five-kingdom schemes of living things.

ESSAY 11.1

Classification

Living things are divided, for convenience, into increasingly smaller groups. One such division separates living things into kingdoms. Thus, an organism in the animal kingdom could be expected to have more in common with another organism in its own kingdom than one from the plant kingdom. Among animals, certain ones share a number of traits in common, and these may be placed together into a phylum. Thus, all animals that have rodlike structures called notochords at some period in their development are placed in one phylum, chordata. As finer and finer divisions are made among plants and animals, one finally arrives at a group of organisms that have so many things in common that they normally interbreed. These organisms are said to be of the same species. The taxonomic breakdown is:

Classifications

Kingdom
Phylum (zoology),
Division (botany)
Class
Order
Family
Genus
Species

Memory Aid

King
Philip
David
Came
Over
From
Greece
Singing

Members of two genera, then, won't share enough things in common to be able to interbreed, but they are still all within the same family.

Members of one family have more things in common with other members of the order than with families from another order. The lower a grouping is, the more closely related are the individuals within it.

Kingdom: Animal
Phylum: Arthropoda
Class: Insecta
Order: Coleoptera
Family: Scarabaeidae
Genus: *Dynastes*
Species: *D. granti*

turn it over and over and mumble something about what it *might* be. This, in essence, is the same problem confronting **taxonomists,** those people interested in naming and classifying living things. Not only are there too many kinds of things to be handled easily, but taxonomists take different approaches to naming things.

There are two main philosophical camps among taxonomists, the "splitters" and the "lumpers." The splitters tend to place species with similar, but distinct, traits into different groups. The lumpers tend to ignore slight differences and to place similar species into the same group. It is because of such fundamental differences that we have so many disagreements regarding biological relationships (Figure 11.2).

We will follow the road most taken, then, and try to avoid the controversies. We will begin by describing the largest groups, the kingdoms, and then discuss various representatives at lower levels in the taxonomic scheme (see Essay 11.1: "Classification").

The five kingdoms as generally accepted by biologists are:

Monera: The monera are prokaryotic cells (*pro:* before; *karyote:* nucleus). Neither the nuclear material nor the cell organelles are bound by membranes. The sole members of this kingdom are the bacteria—the archaebacteria and the eubacteria. The rest of the life on the planet is eukaryotic (that is, with a true nucleus).

Protista: These include all eukaryotes that are not clearly plants, animals, or fungi. Most are single-celled organisms, but some are multicellular.

Fungi: These include the oomycetes, zygomycetes, ascomycetes, basidiomycetes, and the deuteromycetes.

Plants: These include the red, brown, and green algae; the mosses; and the vascular plants (such as ferns, club mosses, horsetails, and seed plants).

Animals: These include a host of familiar species from sponges to mammals.

FIGURE 11.2

Both Dr. Lumper and Dr. Splitter agree that there are eight species of plants being classified. Dr. Lumper, however, places all eight in one genus, believing that the species all descended from a single species that flourished at time (a). Dr. Splitter disagrees and calls for four genera. She concludes that the eight species descended from four distinct groups that already had been established at time (b). By the looks of things, both may be right, depending on where one draws the line—literally.

THE KINGDOM MONERA AND A CONTINUING SIMPLICITY

The **monerans,** being composed of bacteria, are not only extremely small, but they are structurally the simplest of all living things. They have apparently changed little since they first appeared on the earth. In fact, forms resembling today's monerans can be found in the oldest known fossil-bearing rocks. Their simplicity is reflected in the fact that they lack nuclear membranes, membrane-bound organelles, and flagella with microtubules having the characteristic 9 + 2 structure. Because they are simpler, some people believe that they are ancestral to the eukaryotes (the other four kingdoms).

Division Archaebacteria and Extreme Habitats

The **archaebacteria,** the oldest known monerans (Figure 11.3) are rather mysterious life forms, and as we learn more about them, the more intriguing they become. They seem to be very primitive, somewhat rare, and closely related to the earliest bacteria. They are obligate anaerobes—that is, they cannot survive in the presence of oxygen. However, they live in a number of extreme environments, such as hot springs (including highly acidic hot springs such as we find in Yellowstone National Park), ammonia-

FIGURE 11.3

The archaebacteria are among the oldest known forms of life (scanning electron micrograph).

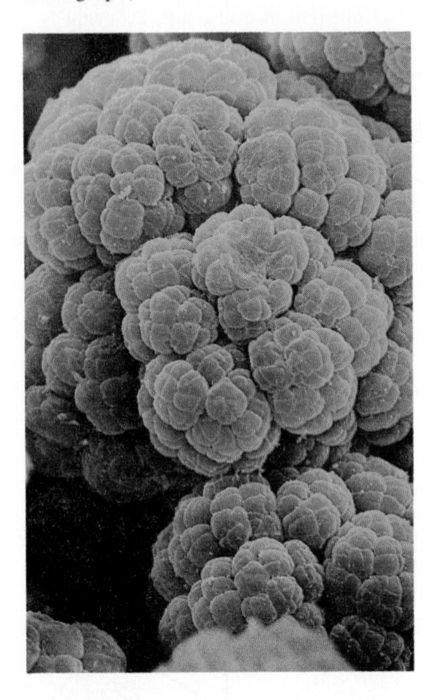

rich habitats (like cattle feedlots), and salty habitats such as salt marshes. Many are also common in a number of other lovely places, such as large intestines and swamp bottoms.

Eubacteria and a Sea of Life

Eubacteria, the true bacteria, are extremely small (most are only a few micrometers in length) (Figure 11.4). Their cell walls and cell membranes have a different composition from those of the Archaebacteria. They exist in vast numbers virtually everywhere on earth. They range from the upper atmosphere to hot springs, polar ice caps, raw petroleum, the deepest reaches of the sea, and animal guts. Some surfaces are free of the creatures since most of them dislike acid, high temperatures, and dryness. Nonetheless, it is believed that their mass outweighs that of all plants and animals combined.

Most bacteria require oxygen, but a few can live without it (including *Clostridium botulinum,* which manufactures one of the deadliest poisons known (Essay 11.2). Oxygen actually inhibits the growth of some bacteria such as *Clostridium tetani,* which lives deep in soil or in deep puncture wounds where there is little oxygen. Thus, puncture wounds from an object lying on the soil can bring about deadly lockjaw or tetanus.

Some bacteria can synthesize their food (Figure 11.5), but others cannot. Some take their nutrients from living things, such as the sorts that

FIGURE 11.4

To give you an idea of the size of bacteria, at left you see the tip of a pin magnified 85 times. Bacteria are visible at 440X (center) and clearly shown at 11,000X (right).

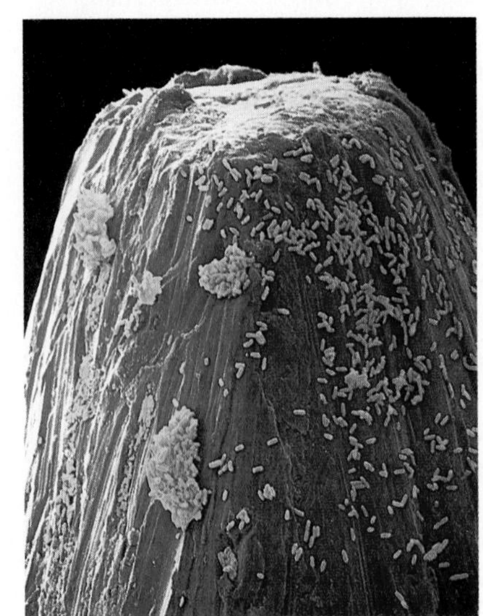

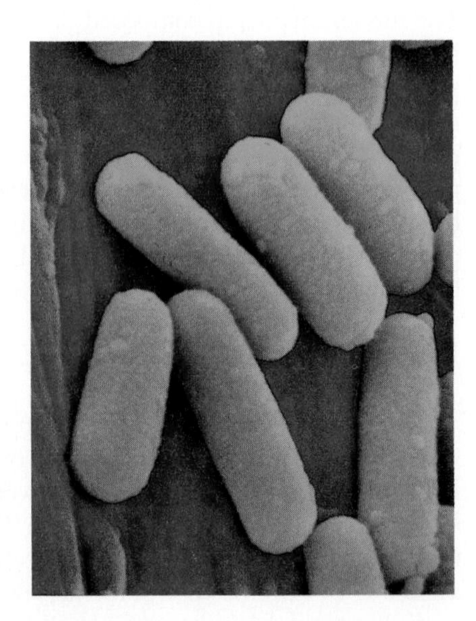

ESSAY 11.2

Botulism

The toxin produced by the bacterium *Clostridium botulinum* is among the most deadly poisons known. Only 60 billionths of a gram will kill a human. The endospores (shrunken, dehydrated, and highly resistant forms of the organism) are found as a common inhabitant of the soil around the world (so we are literally surrounded by an incredibly dangerous life form). Fortunately, however, the spores grow only under very specific conditions, basically in an anaerobic and nutrient-rich medium. Canned food that has not been properly sterilized is an ideal home for the microorganisms.

If canned food has been contaminated, however, it will produce a telltale foul smell and gas (which may cause cans to swell). This is why it can be suicidal to taste suspicious food. The symptoms of botulism begin about three days after eating poisoned food and include vomiting, constipation, and paralysis of the eyes and throat making breathing difficult. The poison functions by interfering with the release of acetylcholine, a neurotransmitter. Recovery is possible if antitoxins are administered within the first 24 hours.

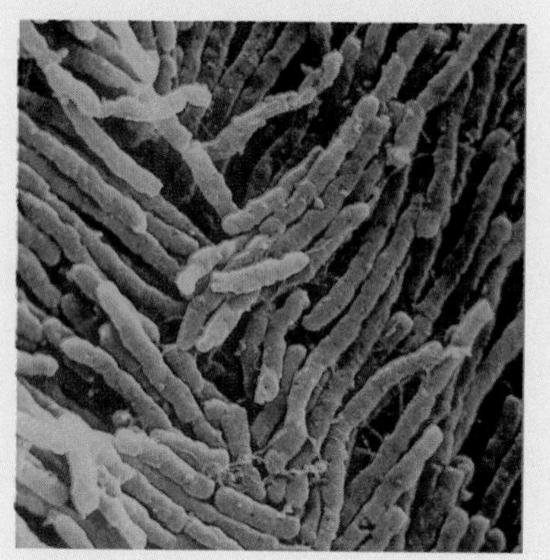

Clostridium botulinum bacteria.

This swollen can warns of botulism poison in its contents. Tasting this food is a bad idea.

invade our bodies and make us sick (Figure 11.5). Others take their nutrients from material that is no longer living (Figure 11.5). These are the decomposers that often cause us such problems as they quickly attack our unprotected food stuffs. Decomposition does have its beneficial effects, though. Without it, the earth would be littered with corpses, a situation

FIGURE 11.5

Bacteria are a highly varied group with very distinct life styles. In (a), chemosynthetic bacteria from the surface of a mussel at a deep water volcanic vent. They derive their energy not from the sun, but from the chemical energy of the molten core of the earth. In (b), *Neisseria gonorrhoeae,* the organism of gonorrhea. In (c), *Pseudomonas aeruginosa* bacterium, a denitrifier that releases nitrogen gas.

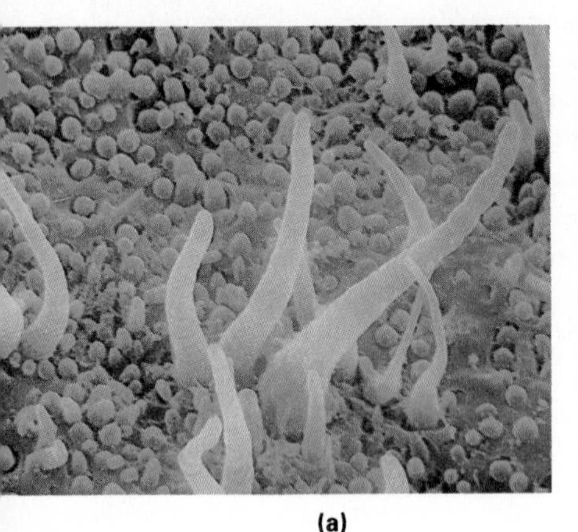

(a)

(b)

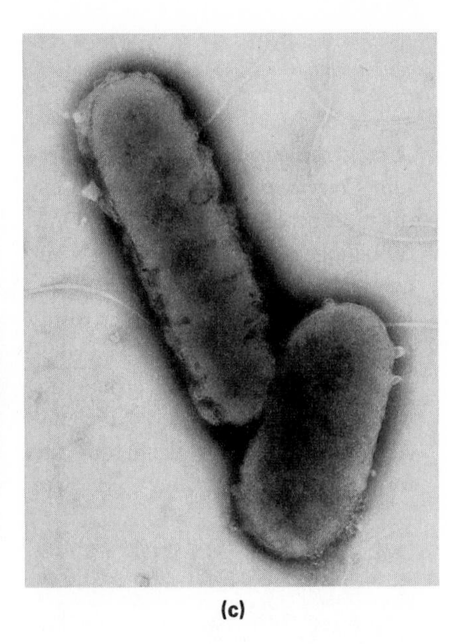

(c)

that would clearly interfere with dancing and marching. Also, decomposition means that the essential mineral nutrients locked in the corpses of dead organisms are available to be recycled by new generations of organisms.

Bacteria generally exist in one of three shapes: the rodlike **bacillus,** the spherical **coccus,** and the long, twisted **spirillum** (Figure 11.6).

Most bacteria reproduce asexually (without sex) after the circular chromosome duplicates itself (Chapter 9). As the two loops of DNA separate, the cell pinches in two, one complete strand of DNA going to each daughter cell. The process is far more simple than is the mitosis of eukaryotes. The division can be accomplished in 20 minutes, which is one reason infections can spread so fast. Also, we saw in Essay 8.1 that some bacteria can mate.

Do you like the smell of freshly turned earth? It is not the earth that you smell, but a special group of bacteria called *Actinomycetes* that live in the soil. The Actinomycetes are of particular interest because they are the source of many antibiotics (Essay 11.3), such as streptomycin, chloramphenicol, and the tetracyclines. However, the same group of bacteria produces leprosy (perhaps the least contagious and most slowly developing of all diseases since the organisms divide only once every 12 days).

One of the problems in controlling bacterial infection is that bacteria have some rather effective defense mechanisms. For example, some bacteria can form **endospores,** a kind of living kernel made from material inside the bacterial body. After an endospore is produced, the living material outside it is simply sloughed away. The resistant endospore may enable the organism to survive harsh conditions. (Some endospores can survive in boiling water for several hours.)

FIGURE 11.6

The three basic shapes of bacteria: the rounded coccus, the rodlike bacillus, and the corkscrew-shaped spirillum.

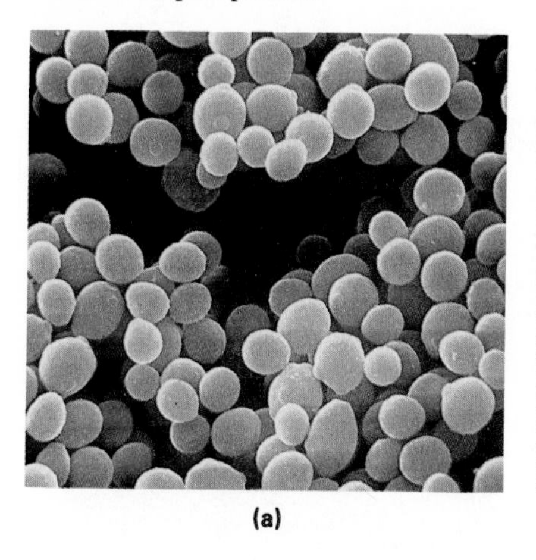

(a)

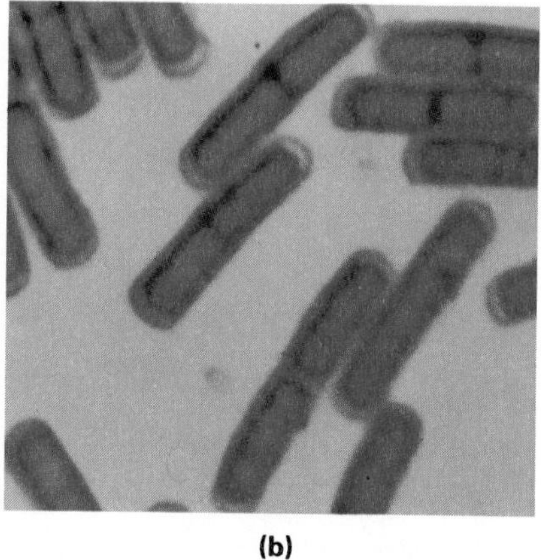

(b)

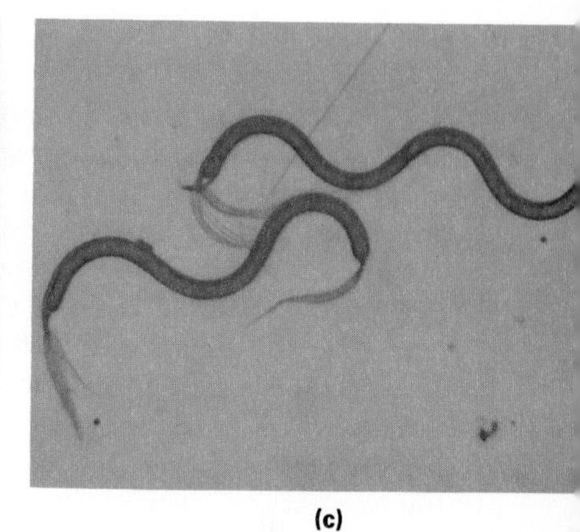

(c)

Some bacteria are photosynthetic, the best-known being the cyanobacteria. The cyanobacteria (Figure 11.7) are prokaryotes, but they differ in several respects from other eubacteria. Like other photosynthetic bacteria, the cyanobacteria (or blue-green algae as they used to be called) lack chloroplasts, their light-capturing pigments being scattered over those

FIGURE 11.7

Cyanobacteria largely form what is poetically called pond scum. They may also live in hot springs and oceans.

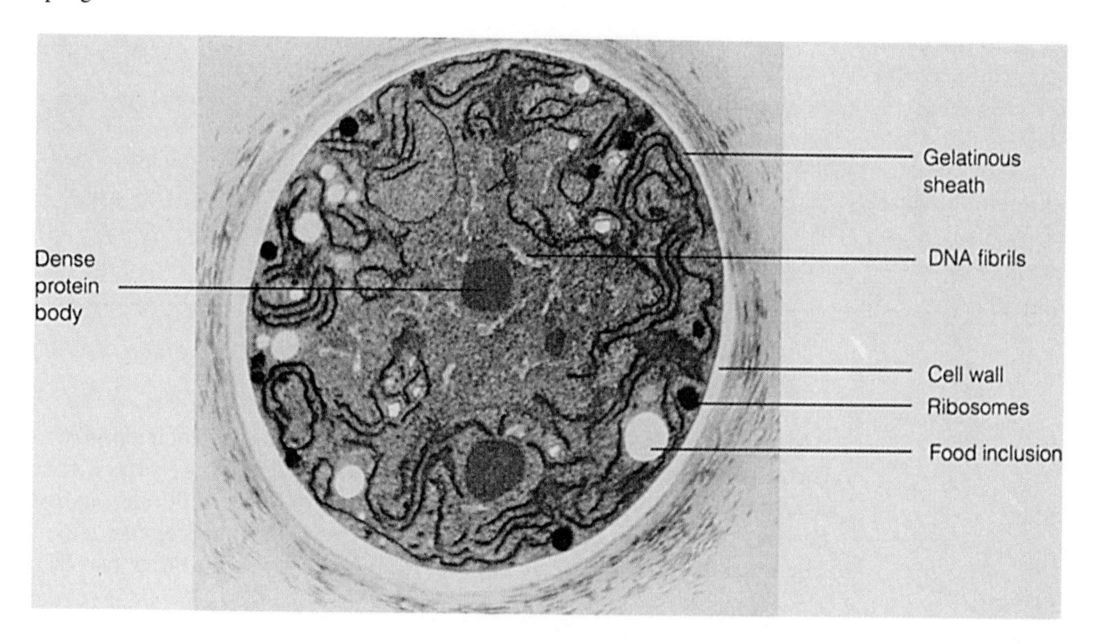

ESSAY 11.3

Antibiotics

For centuries, people have been looking for ways to heal themselves from bacterial infection or to keep from falling to bacterial attack in the first place. One of the efforts involved simple cleanliness (an unusual idea among many of our forebears). This led to efforts to perform surgery under antiseptic conditions. Then it was found that various chemicals could impede bacterial growth. The problem here was that the treatment was often as unpleasant or dangerous as the infection.

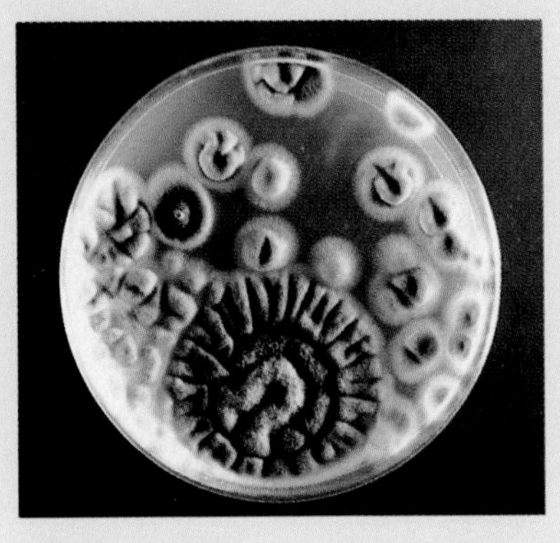

Penicillium culture.

Researchers plotting the molecular structure of an antibiotic. Such research has led to great advances in the development of more effective antibiotics.

The real revolution in bacterial control, however, was spawned by those first attempts, centuries ago, to apply molds to wounds in order to control infection. By the nineteenth century, Louis Pasteur and other scientists of the time were well aware of the ability of certain microorganisms to kill other microorganisms. However, the importance of a new group of antibacterial agents, called antibiotics, was not really established until 1927, when Alexander Fleming isolated the active, bacteria-killing agent from a mold called *penicillium.*

Antibiotics are, in most cases, naturally occurring chemicals produced by microorganisms, chemicals that kill or inhibit the growth of other microbes. However, scientists have now learned to augment the production of the drug by encouraging the growth of the microorganisms that produce it, and in some cases, they are able to produce antibiotics entirely artificially by chemical means.

Almost all natural antibiotics are produced by three groups of microorganisms: actinomycetes (the branching chains so common in soil), the rod-shaped bacilli, and molds. Some of the most potent antibiotics have been found in some rather peculiar places. As examples, bacitracin was isolated from bacteria found on the skinned knee of a little girl named Tracy; cephalosporin was derived from a mold taken from the sea near a sewage outlet; and streptomycin was isolated from bacteria found in the throat of a chicken. Given this, where is the next logical place to look for a new antibiotic?

folded inner membranes known as thylakoids. However, their photosynthetic mechanism differs from that of other bacteria. Like plants, they derive their required electrons by breaking apart water molecules. (Most bacteria get their electrons from other sources, such as hydrogen sulfide.) The cyanobacteria share an important ability with some other bacteria,

FIGURE 11.8

Nitrogen-fixing bacteria, shown glowing by special photographic technique, live in the roots of some plants and are able to incorporate the nitrogen found in the air into a water-soluble molecule that the plant can use. The plant, in turn, provides the bacteria with a suitable environment in which to live. This is one of the clearer examples in nature of a mutualistic symbiotic relationship, in which two species live in intimate association to the good of both.

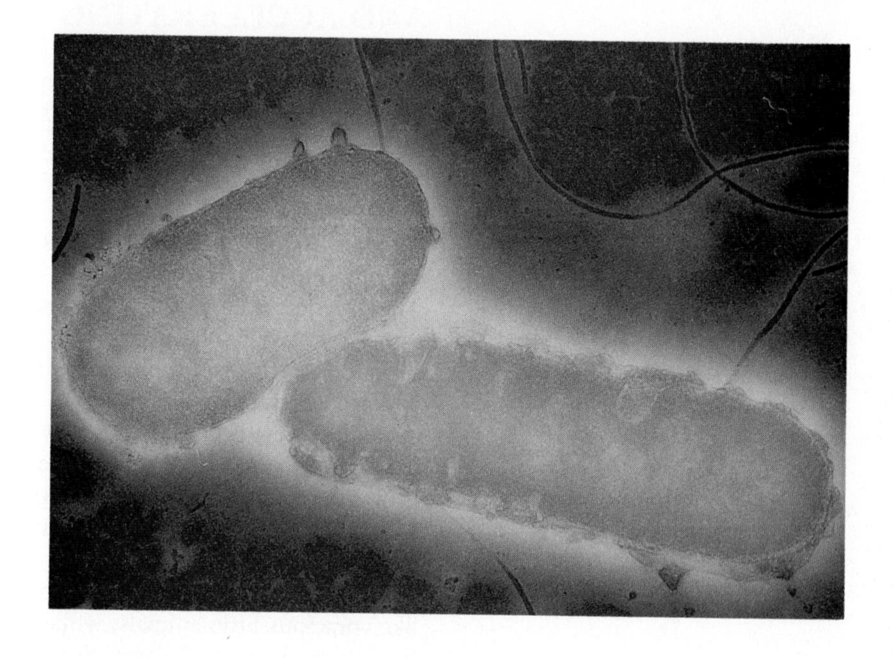

however—the ability to "fix" nitrogen. In fact, the earth's living systems indeed depend strongly on the ability of these organisms to incorporate ("fix") atmospheric nitrogen into the molecules of life (Figure 11.8). Nitrogen fixation by bacteria is of major importance to the movement of nitrogen through the earth's living systems. It is in the cyanobacteria that we see the first, primitive "division of labor," considered to be an evolutionary advancement, as different cell types assume different roles—one type specializing in photosynthesis, the other in nitrogen fixation.

Most species of cyanobacteria contain blue-green, light-absorbing pigments (green chlorophyll and blue phycocyanin). Others, though, are reddish in color. (As an aside, the Red Sea is named for its occasional eruptions of red cyanobacteria.) Cyanobacteria may exist in long chains, but each cell is physically separated from the others without the cytoplasmic bridges that commonly link plant cells. Whereas cyanobacteria are abundant in the soil, they do best in water. In fact, they may become so abundant in water that it not only develops a bad taste but can make one quite ill.

Cyanobacteria are a rather hardy lot. They can survive quite nicely in places where most other species cannot exist, such as in hot springs, salty lakes, and glaciers, and on parched desert rocks. One reason for their success is that they require only sunlight, carbon dioxide, water, and a few minerals. If conditions are unfavorable, some species form thick, tough walls or a jellylike protective coat. Under favorable conditions, they reproduce by simply dividing—and they do it rapidly. It is suspected that they have ways of mixing their genes, but they have never been caught in the act of mating.

THE KINGDOM PROTISTA AND ACCELERATING EVOLUTION

We now move to the eukaryotes, quite a different matter, as we shall see. The simplest of these (and the first to have evolved) form the kingdom **Protista.** The protistans possess a membrane-bound nucleus and other typical eukaryotic structures. Most live alone as single-celled organisms, but some form rather specialized colonies, and a few are multicellular. They are a complex and diverse group of organisms that are placed together in a grab-bag of eukaryotes that are neither fungi, plants, nor animals. Their predecessors on the planet were prokaryotic autotrophs. These were rather successful creatures, probably able to rapidly multiply when conditions were good and to become dormant during hard times. Eukaryotes appeared between a billion and 2 billion years ago. These new kinds of organisms were bound by complex and efficient membranes, and they had the ability to exchange genetic material in a kind of sexual union. They evolved rapidly in a few rather distinct directions and eventually gave rise to today's dazzling variety of plants, animals, and fungi, as well as the protists.

The protistans are, today, quite a varied group. Some prowl the earth like voracious little animals, while others lie quietly and make their own food in the manner of plants. Yet others take up the molecules of the dead, as do today's fungi.

Because of this diversity within the group, it is assumed that many of the protistans are not very closely related. In fact, their closest common ancestor probably existed among the first life forms. Nonetheless, because of important similarities, they are, for convenience, placed in the same kingdom.

THE ANIMAL-LIKE PROTISTS

The single-celled, animal-like protists are called **protozoans** (which means "first animals," a misnomer). They exist all around us in a variety of habitats, but most go unnoticed because they are microscopic, ranging from about 5 to 100 micrometers. However, some are up to several millimeters long—large enough to threaten small insects. Table 11.2 describes six of the protistan phyla.

Phylum Ciliophora and Tiny Predators

If you should find yourself with a microscope and unable to resist examining the water of a scummy pond, you would likely find great numbers of tiny protozoans. Some of these would be covered with tiny, hairlike **cilia** that propel them through the water. Protists that bear cilia at some stage in their life cycles comprise the phylum **Ciliophora.** The most familiar of these are the **Paramecia** (Figure 11.9). Paramecia (singular, Paramecium) are recognizable by their slipper shape, which is maintained by their outer thickened membrane, the **pellicle.** The pellicle, while holding its form, is flexible enough to enable the paramecium to bend around objects as it furiously swims. Behind its rounded anterior (or front) end

TABLE 11.2
Selected Protists

PHYLUM AND EXAMPLE	HABITAT	FOOD SOURCE	LOCOMOTION
Ciliophora (paramecia)	Pond water (surface)	Cilia sweep food into oral groove digested in food vacuole	Cilia
Sarcodina (amoeba)	Pond water (bottom)	Engulfs food with pseudopods	Amoeboid movement using pseudopods
Sporozoa (*Plasmodium*)	In animals	Parasitic upon host animal	None
Euglenophyta (*Euglena*)	Pond water (surface)	Photosynthesis (chlorophyll *a* and *b*)	Flagellum
Chrysophyta (diatoms)	Fresh and salt water	Photosynthesis (chlorophyll *a* and *c*, carotenoids)	None
Pyrrophyta (dinoflagellates)	Salt water	Photosynthesis	Two flagella

lies a deep **oral groove** into which food is swept by other cilia. The food is then forced through a mouthlike pore (**cytostome**) at the end of the groove and into a bulbous opening, which will break away and move into the cytoplasm as a **food vacuole.** Digestion takes place as enzymes enter the vacuole, the products passing out into the cytoplasm. Undigested food passes out where an opening appears at a fixed place in the pellicle. The

FIGURE 11.9
The paramecium is an active, spinning ciliate that inhabits pond water.

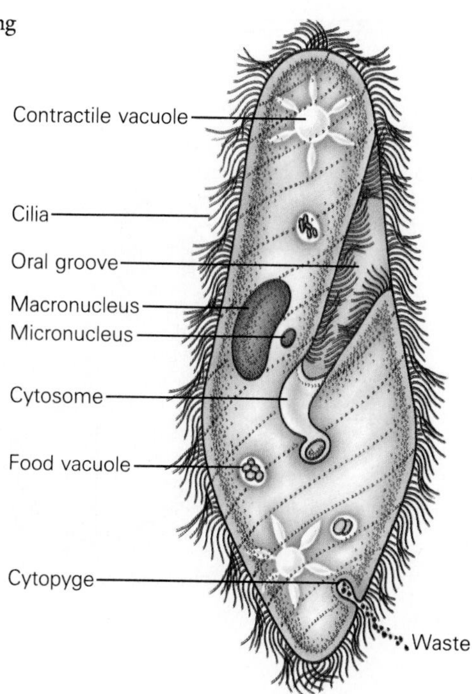

opening is called the **cytopyge.** Excess water entering the cell is expelled by a star-shaped **contractile vacuole.** In addition, each organism contains two nuclei of unequal sizes.

Paramecia are relentless hunters. They not only sweep tiny organisms such as bacteria into their oral groove, but they ensnare larger prey by explosively discharging long, barbed shafts with trailing threads, called **trichocysts,** from beneath the pellicle. The trichocysts are also used in defense.

Paramecia normally reproduce by simply replicating their genetic material and then dividing. They may also conjugate. During **conjugation,** a kind of sexual reproduction, two paramecia join at their oral grooves and exchange genetic material. Conjugation begins when the larger of the two nuclei, called the **macronucleus,** disintegrates while the smaller **micronucleus** divides twice, halving their chromosome number and then exchanging some of the haploid products with the mating partner.

Phylum Sarcodina and a Flexible Simplicity

Imagine a tough, well-trained army on the move through some primitive land, set on conquering everything in its path. Then imagine that army brought to a dead halt as soldiers drop their weapons, clutch at their bellies, and dart for the bushes. It has happened. Entire armies have been stopped by a tiny protist called *Entamoeba histolytica,* the cause of amoebic dysentery and a member of the phylum **Sarcodina,** the amoeboid protozoans. Amoeboid refers to the way they move through their watery surroundings. To illustrate, the most famous member of the phylum is the common amoeba (Figure 11.10) that creeps about on the bottoms of

FIGURE 11.10

The Sarcodina include the amoebas, small organisms that move and feed by extending a part of their body and then flowing into that extension.

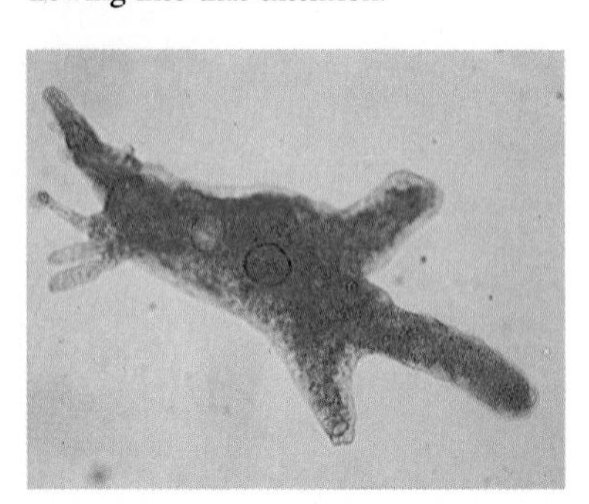

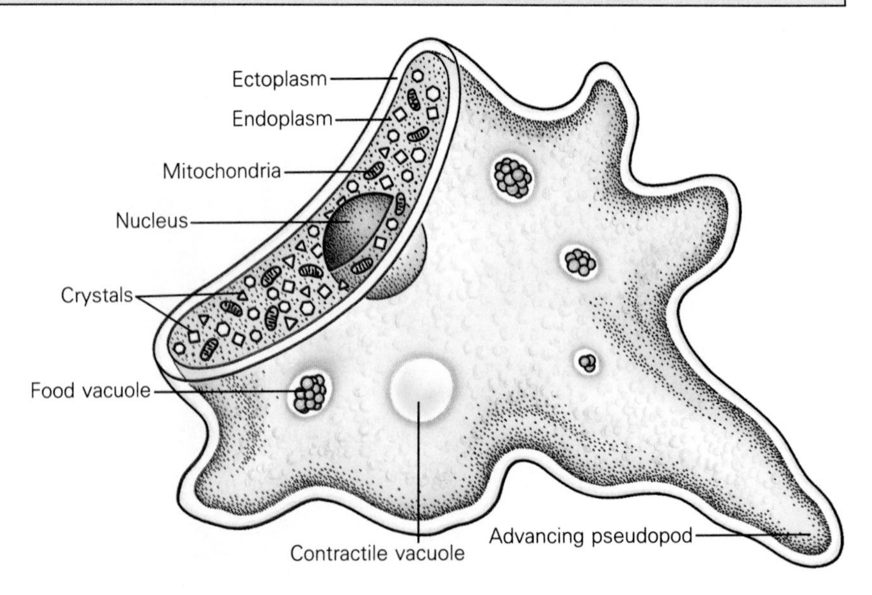

still ponds by sending out **pseudopods** ("false feet"), which are extensions of the main body, and then flowing into these outstretched arms. It also uses the pseudopods to engulf food materials, which then become contained in food vacuoles. Interestingly, amoeboid movement relies on the same contractile proteins as do many other species, including vertebrates. Such findings are considered evidence of the common heritage of living things. Some relatives of the amoeba have not only pseudopods, but also flagella (singular, *flagellum*), long, whiplike structures used in locomotion. Still others secrete a shell or cover themselves with sand grains.

Sarcodines employ rather simple means of regulating water and waste in their bodies. Amoebas, for example, take in oxygen and excrete metabolic wastes, such as ammonia, through diffusion. Excess water that flows in through the thin membrane by osmosis collects in one area and becomes encircled by a membrane, forming a contractile vacuole. Later, the vacuole draws near the cell membrane, then contracts and squirts out the water by temporarily rupturing the membrane.

The cytoplasm of the amoeba is divided into a clear, thin, outer layer, the **ectoplasm,** and a granular inner **endoplasm** that contains numerous crystals and mitochondria. Some sarcodines reproduce both sexually and asexually, but asexual reproduction is more common. Here, mitotic division of the nucleus is followed by the cytoplasm simply pinching in two.

Phylum Sporozoa and a Parasitic Existence

The phylum **Sporozoa** is a diverse group. Some of them are so different from each other that they are probably more closely related to other phyla of protists than they are to other sporozoans. Thus, it is difficult to place the group in any evolutionary line. They are placed together because they are all parasitic, living within animals. In fact, because of their reliance on other species, many of them have lost the ability to survive independently. One group has been critically important to the human species—the genus *Plasmodium,* which is the causative agent of malaria (Figure 11.11).

THE PLANTLIKE PROTISTS

The other major group of the protists are called the algae, "plantlike." These are photosynthetic plantlike protists. They are extremely widespread, living in all aquatic habitats and a few terrestrial ones. We will briefly consider the three major phyla.

Phylum Euglenophyta and an Evolutionary Puzzle

If, for some reason, you were to examine your scummy drop of pond water again, you might find another kind of organism, a green one lacking cilia but possessing one or more *flagella.* This organism is a member of the phylum **Euglenophyta.** One of the best-known is *Euglena* (Figure 11.12). *Euglena* has presented an enduring puzzle to taxonomists. They behave like animals, thrashing their way through the waters, while photo-

FIGURE 11.11

Malaria is caused by a protistan, called a *Plasmodium* (right), that develops a form called a sporozoite within the mosquito. When it is injected into the human body, it enters the liver and multiplies repeatedly there, releasing the rapidly increasing agents into the bloodstream, where they enter red blood cells. As they multiply within the cells, they are sporadically released into the blood, thereby infecting new cells. These releases mark times of high fever, followed by chills. Some of these released agents occasionally produce male and female forms. When a mosquito draws blood from an infected person, it may consume these male and female gametes, which join in the mosquito's body to form new sporozoites.

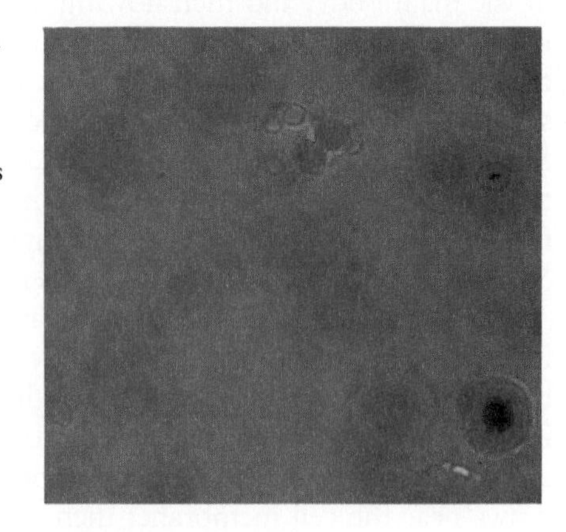

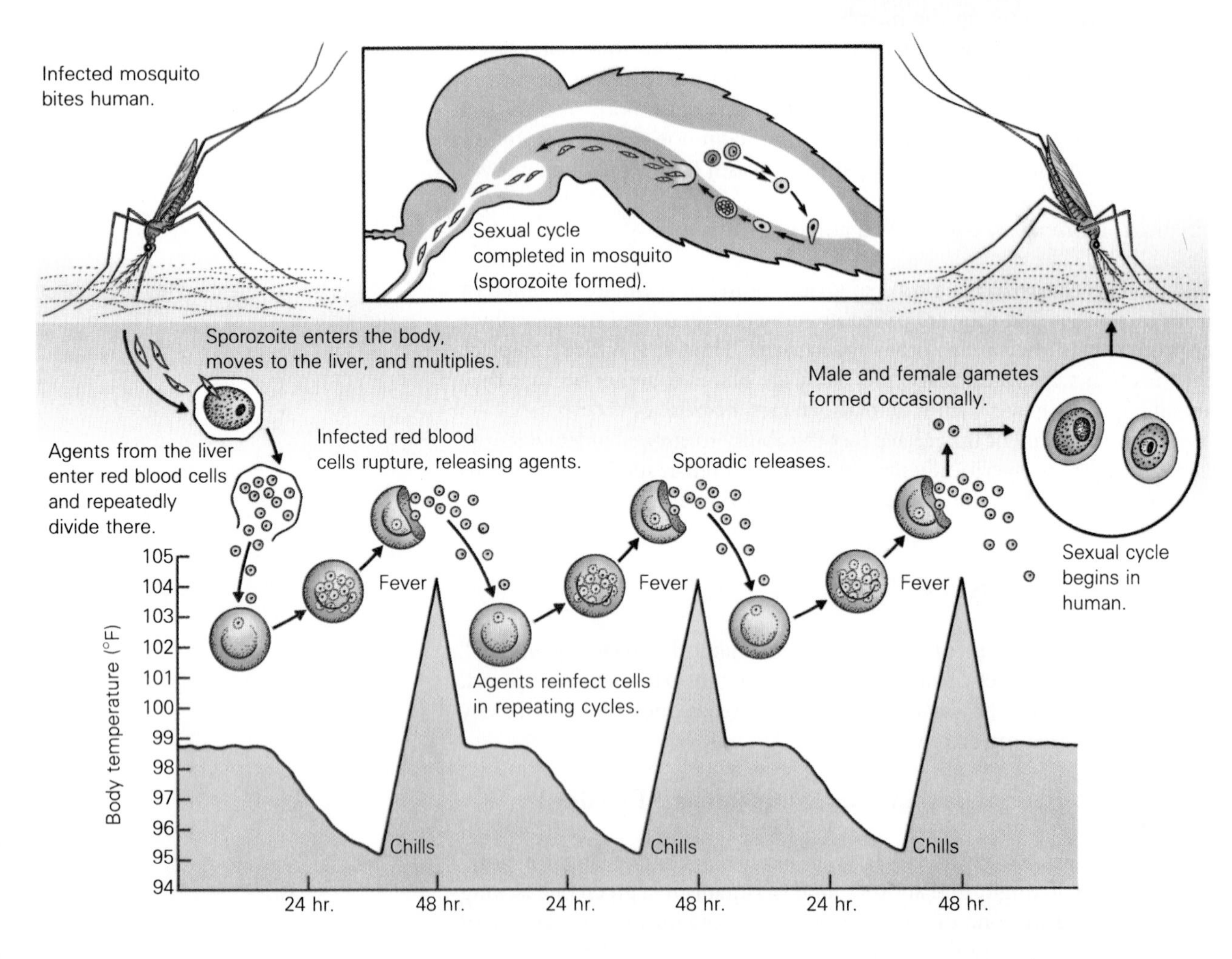

FIGURE 11.12

Euglena has presented an enduring puzzle to taxonomists, having characteristics of both plants and animals.

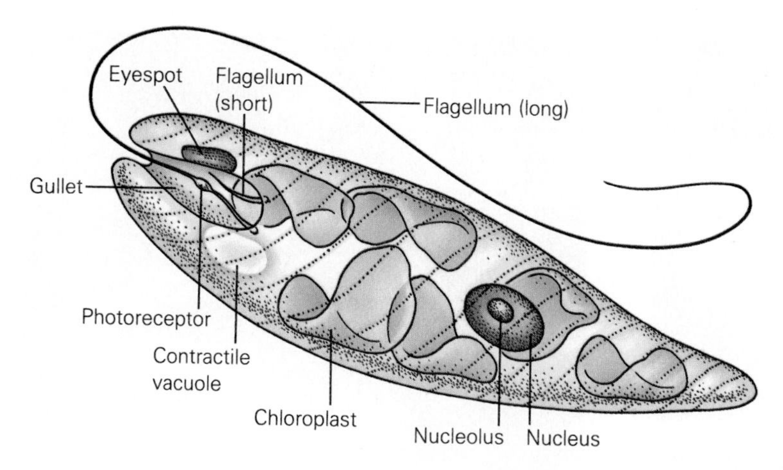

synthesizing with the use of chlorophyll within chloroplasts. *Euglena* has a long appendage protruding from the **gullet** at its rounded anterior end, leading to a saclike reservoir. The gullet in this group probably serves only as an area of attachment for the flagellum. (A pulsating contractile vacuole continually expels excess water.) *Euglena* contains chloroplasts and can manufacture its own food through photosynthesis. Near the gullet lies a light-sensitive structure, or photoreceptor, that is shielded on one side by the **eyespot,** so that it can only be stimulated by light coming from a certain direction. In this way, the organism can determine the direction of that light and move toward it, enabling the chloroplasts to make food.

Interestingly, it swims *toward* the moderate levels of light that penetrate its waters, but avoids the destructive direct rays of the sun. However, even with abundant light, *Euglena* cannot make all its food; it must absorb some vitamins from the environment. In the absence of light, it may take all its food from the environment. In fact, if it is kept in the dark for long periods, the chloroplasts may disappear.

Under adverse circumstances, the *Euglena* can simply lose water, reduce its internal (metabolic) activity, and become rounded and dormant. This **cyst** may persist until the return of better times.

The *Euglena* reproduces by dividing after mitosis. It can also dehydrate, form cysts, and divide at this stage. It has never been seen engaging in any sexual activity whatsoever.

Phylum Chrysophyta and Glassy Corpses

The **Chrysophyta** ("golden plants") comprise the yellow-green and golden brown algae and are named for their yellow carotenoid pigments, although they also possess chlorophyll *a* and *c*. This group includes the beautiful

FIGURE 11.13
Diatoms are unicellular or colonial algae that live both in fresh and salt water. They possess two shells, or valves, that in some species fit together like the halves of a pillbox.

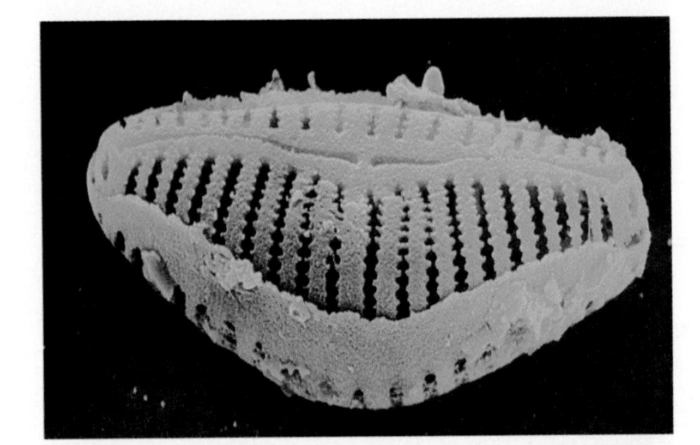

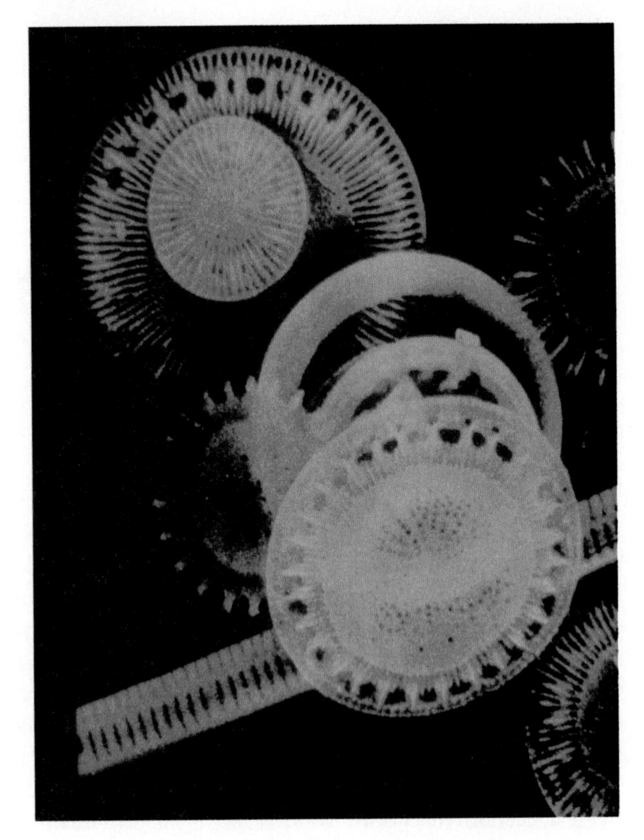

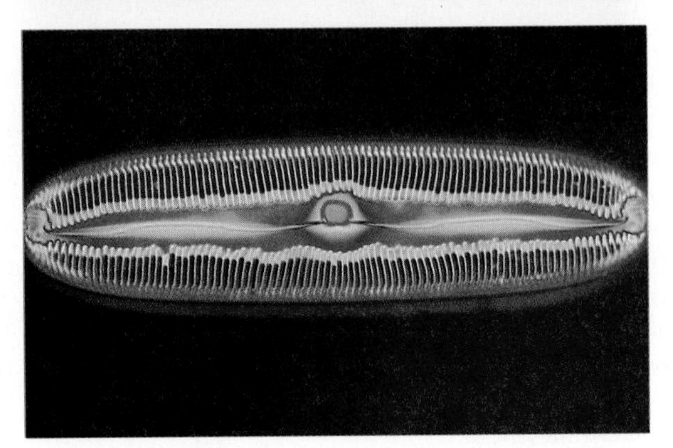

FIGURE 11.14
The dinoflagellate *Gymnodinium,* one of the dinoflagellate that produce red tide.

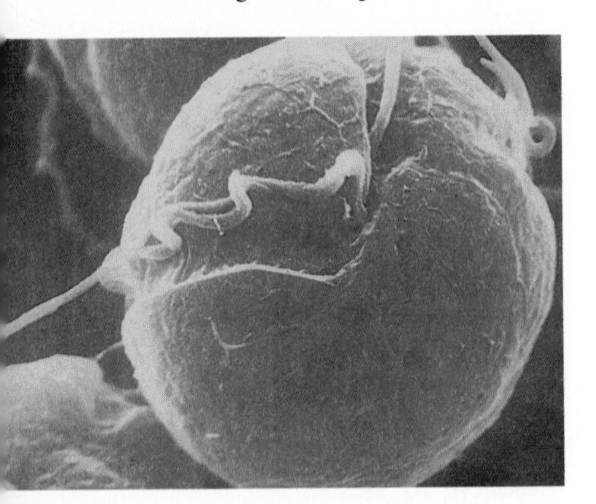

diatoms (Figure 11.13). Their cell walls are comprised largely of silicon, and their glassy corpses have slowly rained down upon lake and ocean bottoms for millions of years, producing very thick layers of ***diatomaceous earth,*** an extremely fine-grained material that is often used as filters. Some reproduce sexually, but most simply undergo cell division.

Diatoms are abundant and they can store fat, so they comprise a very basic and critical link in aquatic food chains. It is the fat of diatoms, by the way, that imparts that beloved fishy taste to fish. This is why some predatory fish that are not part of the diatom chain, such as trout, do not have that taste.

Phylum Pyrrophyta and Sparkling and Dreaded Waters

The **Pyrrophyta** ("fire-colored plants") are microscopic, photosynthetic algae with two tinsellike flagella. These are the **dinoflagellates.** They, too, are an important basic link in the sea's food chain (Figure 11.14). This group has enchanted sailors for as long as men have gone down to the sea. When they are mechanically disturbed, as in the wake of a ship or a churning oar, dinoflagellates emit a bright flash of light. One of the most fascinating sights I've seen was the trail of sparkling lights trailing my skiff, accented by flashes from the oar placement, in the waters of the Indian Ocean as I rowed back to the ship anchored offshore one June night.

Other dinoflagellates are not so charming. Some reproduce wildly at times, turning the sea a deep red and producing the dreaded "red tide" (Figure 11.15). Their sheer numbers may seriously deplete the water of its oxygen and thereby suffocate other species. Some also produce a powerful nerve poison that can kill vertebrates. These poisons accumulate in the bodies of the animals that eat the dinoflagellates and are ultimately consumed by vertebrates somewhere along the food chain. Thus, not only are fish killed by the thousands, but the lives of humans who eat tainted shellfish, for example, are threatened as well.

FIGURE 11.15
Red tides can wipe out other marine species by depleting the water of oxygen and producing poisons.

THE FUNGUSLIKE PROTISTS

Although they are unrelated and dramatically different, the two groups of slime molds, cellular slime molds and acellular slime molds have something in common: they both lead double lives.

In one portion of their lives, cellular slime molds multiply and grow over their food supply (often a rotting log), each cell behaving like an amoeba. But when their food source dries up, they change drastically. The individual cells crawl together and coalesce, forming a single body called a slug. The slug wanders blindly about like a mindless worm. Eventually, it stops and vertical stalks begin to arise from its mass. The stalks bear fruiting bodies that give rise to airborne spores, as any mold might.

Acellular slime molds form huge multinucleate masses, forming what looks like a giant amoeba—large enough to cover an entire log. The mass, however, may be only a millimeter thick. As it creeps along, it engulfs bits of organic matter and shows some sensitivity in that it avoids obstacles. When its habitat dries out, it produces slender stalks that will release haploid spores. These are released into the environment to later fuse and begin dividing, producing the diploid, amoebalike mass again.

As we leave the world of tiny monerans and protists, we should remind ourselves of two things. One, they are still with us, a testament to the resilience of their seemingly frail bodies; and two, they set the stage for many other forms that followed. The protists, in particular, set the stage for many of the groups that followed by developing a nucleated cell with a complex and efficient genetic apparatus, and by developing systems of handling energy by (1) creating energy-laden molecules (for example, through photosynthesis), and (2) developing (or acquiring) structures (like mitochondria) that can utilize that energy efficiently.

THE KINGDOM FUNGI AND VARIED LIFESTYLES

A peaceful walk in a damp forest may bring you upon a most enchanting sight: a ring of delicate mushrooms, their little caps pushing up through the woodland floor (Figure 11.16). The beautiful little caps may be of various colors and different shapes, and some have bold markings. Mushrooms can be beautiful indeed. We must keep in mind that a mushroom is a fungus. A fungus? Athlete's foot is a fungus. So what is it we put on steaks?

FIGURE 11.16

A fairy ring. Mushrooms are charming, if occasionally deadly, organisms. Sometimes they seem to spring forth in your yard almost overnight. This usually happens during a wet period and is only the final stage of a long developmental sequence. They form hyphae so quickly that, under ideal conditions, a single mushroom can produce over half a mile of these filaments in a single day (laid end to end).

We do indeed decorate our steaks with fungi, but, one would hope, a different variety than we find on our feet. (We rarely find truffles there.) Obviously, the **fungi** are a variable group, but they are all characterized chiefly by an absence of chlorophyll and by absorbing organic molecules from living or dead organic matter. Some have certain remarkable characteristics. For example, the cell walls of most species are not made of cellulose but of chitin, the protective covering of insects. Whereas a few, such as those on our feet, are parasitic on living things, the majority are saprophytic, drawing nutrients from the bodies of dead organisms. They do this by secreting digestive enzymes into and around the dead bodies and then absorbing the digestive products. Other species actually trap animals and digest them on the spot (Figure 11.17). Table 11.3 describes the characteristics of the fungi.

Fungi were once placed with the true plants and are still largely considered in the domain of botanists, but they are quite different from plants

FIGURE 11.17

The fungus *Arthrobotrys dactyloides* lassoing a roundworm. As the worm crawls through the loop, the noose will tighten and the fungus will begin to secrete digestive enzymes.

TABLE 11.3
Fungi

DIVISION	EXAMPLES	NUTRITION	NOTES
Oomycota	Water molds; white rusts; downy mildews	Saprophytic (decaying plants or animals), parasitic	Similar to algae; some crop damage; motile zoospores
Zygomycota	Black bread mold	Saprophytic (dead plants or animals), some symbiotic, some parasitic	Zygospores produced sexually; asexual reproduction by spores borne on hyphae; some form association with plant roots called mycorrhizae
Ascomycota	Pink bread mold; yeast; morels, truffles	Saprophytic; some parasitic	Asexual reproduction via conidia on conidiophores; sexual reproduction with ascospores in ascus; largest division; symbiotic association with algae produces lichens
Basiodiomycota	Mushrooms; puffballs; bracket fungus	Soil; decaying plants	Can be edible or deadly; sexual reproduction via basidiospores
Deuteromycota	*Penicillium*	Saprophytic, parasitic	No known sexual phase

in a number of ways. For example, they grow by producing long, thin filaments, or **hyphae,** that join together to produce a spongy, cottony mass, the **mycelium.** (Those fungi that produce cottony or fuzzy growths are sometimes referred to as molds.) In this group, as in the plants, phyla are called *divisions.*

Division Oomycota and Irish Americans

The **oomycetes** (Figure 11.18) are usually referred to as water molds, but they also include white rusts and downy mildews. The name means "egg

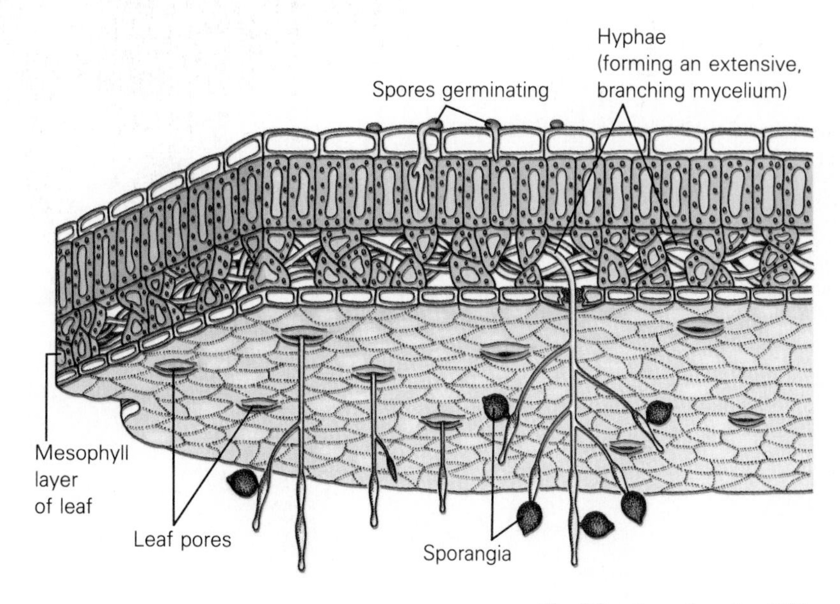

FIGURE 11.18
In the oomycete that causes potato blight, an airborne spore lands on the leaf of a potato plant and germinates. As it grows, long tubelike hyphae penetrate the mesophyll cells of the host's leaves. From these, sporangia arise, each sporangium producing many spores. The potato plant dies, but not before the fungus completes this asexual, spore-forming cycle. The sporangia will rupture, and the airborne spores will drift away, each to give rise to a new organism. Because the hyphae penetrate the interior of the leaf, the fungus is impermeable to most fungicidal sprays.

FIGURE 11.19

Asexual reproduction in black bread molds occurs as haploid spores germinate and produce extensive hyphae. Sexual reproduction occurs in bread molds by a form of conjugation. The hyphae are of two types, plus and minus. If the two contact each other, bridges will join them near their tips. The nuclei at the ends of the converging bridges will form the gametangia, which contain the gametes. When the bridges join, a zygote is formed around which develops a thick-walled zygospore. During the germination of the zygospore, meiosis occurs, resulting in the formation of a sporangium that releases haploid spores. Each spore can generate new hyphae. When it germinates, new hyphae are formed.

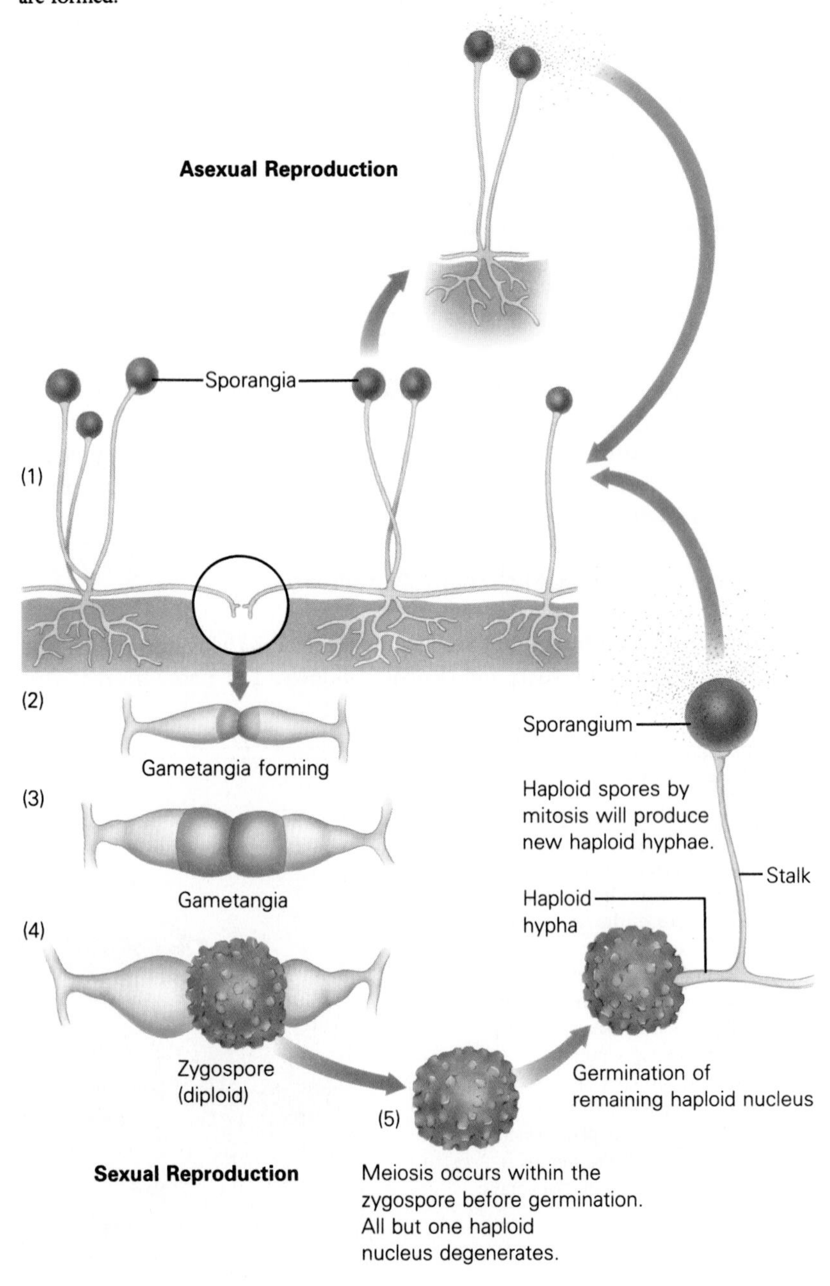

fungi" because a large egg is fertilized by a much smaller sperm. Some are saprophytic and others are parasites. The oomycota have some traits more in common with algae than with the other fungi (causing some people to place them with the algae). For example, their cell walls are made of cellulose (like the algae) and not of chitin (like the other fungi).

The oomycota have had an extraordinary impact on humans. For example, they can grow on the injured gills of our aquarium fishes and around 1880, the form that causes downy mildew on grapes almost wiped out the French wine industry until a copper-containing treatment was developed. The most serious impact on humans (as if losing our French wines wasn't serious enough) resulted from a form that causes late blight in potatoes. This fungus attacked the potato crop in Ireland hard between 1845 and 1847 when the country went through an unusually wet, cool period. In the summer of 1846, the entire potato crop was wiped out in one week. Over 1 million people died as a result of the ensuing famine. Many of those who survived began an exodus from Ireland, most heading for the United States and Canada. Thus, if you are of Irish descent, your citizenship may be due to a water mold. Interestingly, it is believed that the remarkable tonal qualities of the violins made by Stradivari were due partly to the growth of water mold on the raw soundboard wood.

Division Zygomycota and Preserving Our Bread

As you can see in Figure 11.19, the **zygomycetes** reproduce by both asexual and sexual means (but primarily asexually). The name is derived from the production of the **zygospore** by sexual means. The zygospore is a tough, resistant structure produced normally in response to harsh conditions, such as drying.

The zygomycetes are entirely terrestrial, most living on dead plant and animal material in the soil. They are a diverse group, and like the oomycetes, they have some characteristics in common with the algae. Virtually everyone, though, places the zygomycetes squarely in the fungi.

The most common zygomycete is the common bread mold, which circumvents all our preservatives and finally claims the result of our baking efforts. The life cycle of the very common black bread mold, *Rhizopus,* is shown in Figure 11.19.

FIGURE 11.20
Morels, often mistaken for mushrooms, have a delicate flavor and are highly prized by gourmets. Unfortunately, the morel season is short—less than one month in the spring.

Division Ascomycota and Remarkable Partnerships

The **ascomycetes** are the largest division of fungi, with some 30,000 free-living species and about 18,000 species forming a symbiotic association with algae to produce lichens (Essay 11.4). Other species form associations with the roots of oaks to form edible morels (Figure 11.20) and truffles (Essay 11.5). Most species are saprophytic but some important ones are parasitic, including the group responsible for Dutch elm disease, which has almost depleted our eastern forests of the magnificent elm.

Members of the division can reproduce both asexually and sexually. In asexual reproduction, spores are produced (called **conidia,** Greek for "dust".) The conidia develop on the tips of special aerial hyphae called

ESSAY 11.4

Lichens

We will consider the lichens separately because stretching our system of classification in this case is asking a bit much, even for the lumpers. One reason is that lichens are comprised of two different kinds of organisms, a fungus (usually an ascomycete or a member of the Fungi Imperfecti) and an alga (usually a moneran or one of the true green algae described later). The relationship is believed to be mutually beneficial in that the fungus, with its probing filaments, can cling tightly to rocks, and the alga, through photosynthesis, manufactures food molecules, some of which are used by the fungus.

High arctic lichen growing on a rock on the rugged Ellesmere Island, the focus of numerous expeditions by the early explorers who attempted to eat anything they could find growing.

The foliose, or leaflike lichens, are joined to the rock by a thin stalk that makes them rather easy to remove. All lichens are extremely sensitive to air pollution and this one quickly withers and falls off in polluted air.

Lichens grow extremely slowly, and although they may be an early and critical phase of the soil-building process as they attach to barren rocks, they have attracted relatively little research attention. In fact, only recently have scientists been able to grow the alga and fungus separately and then join them to form a lichen.

Lichens are found clinging to all sorts of exposed surfaces including bare rock, soil, and tree bark. Vast areas of the arctic and subarctic regions are covered by the famed "reindeer moss" of the lichen Cladonia.

Not all lichens grow in polar regions. Some grow in deserts, some actually inside rocks, and others in the tropics. The flowered beard lichen grows in Kilauea, Hawaii.

The British soldier lichen, like all lichens, grows very slowly. This is probably because lichens lack the means to store water.

C. rangiferina, the reindeer moss of the North, is an important food for many arctic animals such as caribou and reindeer, and humans as well.

ESSAY 11.5

Truffles and Sex and Truffles

Truffles are the fruiting bodies of mycelia living in a mycorrhizal association with tree roots (especially oaks). There are many species of truffles, boasting a variety of smells and tastes ranging from nuts and cheese to sewer gas and rancid bacon grease. Those falling into the latter two groups are not particularly prized. Only a few of some 70 European species are sought after by gourmets, and none of the over 50 American species is remotely tolerable.

European truffle hunters have traditionally used pigs to sniff out truffles, sometimes buried three feet deep. The pigs work assiduously to dig out the truffles. (They must be muzzled or intimidated to keep them from devouring the expensive delicacies.)

The pigs work so hard that one wonders, are the pigs' tastes so exquisitely refined that they appreciate the delicate taste of truffles so much? Actually, it turns out that they root them out for another reason. Truffles contain androstenol, a pig sex attractant. The hormone is released by boars and is exceedingly appealing to sows who then dig out the fungi, and by eating them, break open the ascocarp and scatter the spores.

It also turns out that men produce androstenol, and after either men or women are subjected to the smell, they rate photographs of women as more attractive then they do without the smell (make of that what you will).

conidiophores. They can then be dispersed by air currents. When the conidia land in a suitable environment, they begin to grow and to form new hyphae. These hyphae may reproduce asexually by conidia. In sexual reproduction, the hyphae of different mating strains fuse, forming hyphae composed of cells containing two nuclei. These hyphae form a cuplike structure called an **ascocarp.** Within the ascocarp are specialized saclike cells called *asci* (singular, **ascus**) (Figure 11.21). The two nuclei within an ascus fuse, forming a diploid cell called a zygote that divides by meiosis to form four haploid **ascospores.** Each ascospore then divides mitotically. As a result, each ascus contains eight ascospores. Each ascospore may form new hyphae that can reproduce asexually or sexually.

In seventeenth century Salem, Massachusetts, a number of people were put to death for being "witches." Historians have decided that many were simply the unfortunate victims of poisoning by an ascomycete. One strain can infect rye grain, forming a hardened growth called ergot. If bread from the grain is eaten, the ergot poisoning produces hallucinations, convulsions, and tissue damage in the extremities, accompanied by an intense, painful burning (called St. Anthony's fire). Ergot, by the way, is the source of the hallucinogen, LSD.

FIGURE 11.21

Yeasts (a), ascomycetes that do not produce mycelia, reproduce by budding. The ascomycetes also include the sac fungi (b), morels, truffles, the agents of Dutch elm disease, and a host of other plant ailments. In many species of ascomycetes, a fleshy cup is lined with specialized cells, forming the ascus, that contain spores called ascospores. These ascospores are the result of the union of two different types of nuclei that then undergo meiotic divisions.

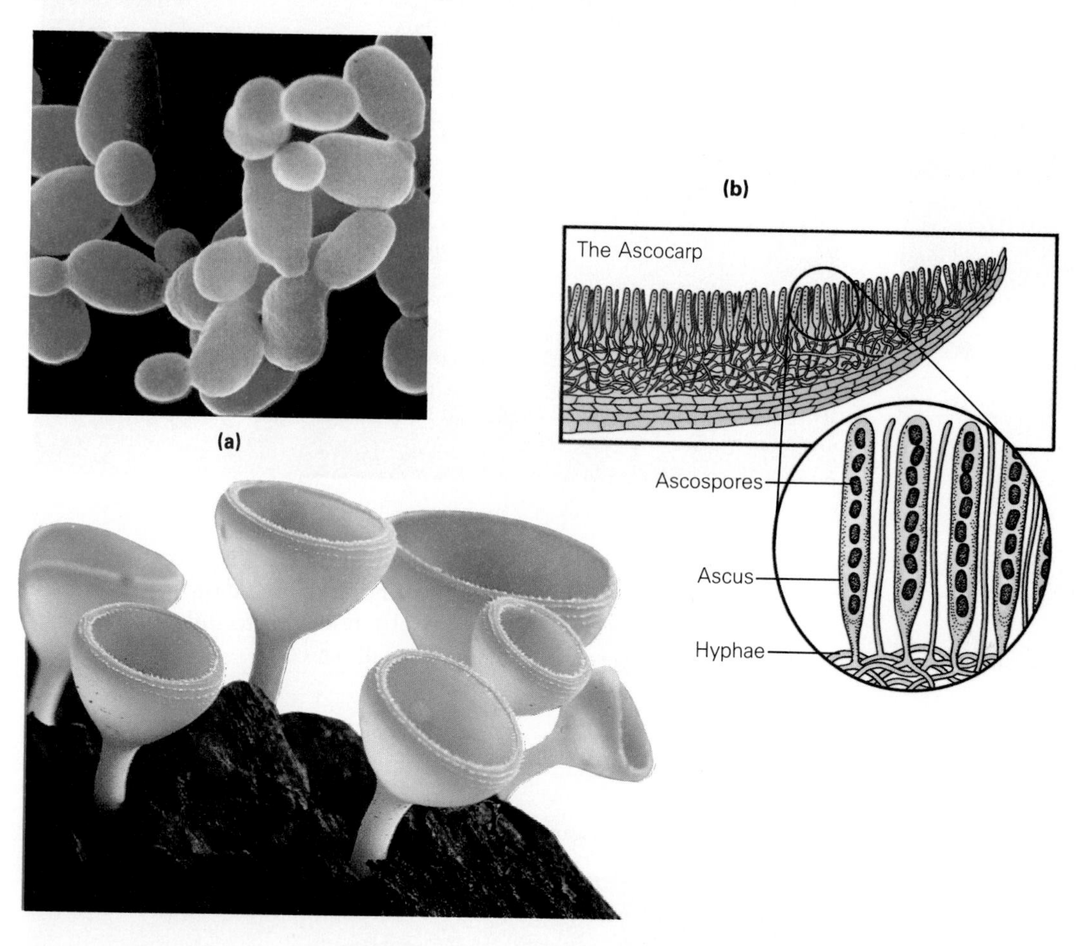

Division Basidiomycota and the Destroying Angel

The **basidiomycetes,** the group that includes the mushrooms, puffballs, and bracket fungi, produce the most spectacular fruiting bodies of all the fungi. The puffballs, for example, may be over half a meter wide (Figure 11.22), and the giant bracket fungi may be found growing on damp logs (Figure 11.23). Although the basidiomycetes include a variety of edible mushrooms, it also includes the dangerous, even deadly, forms. For example, the delicately hued *Amanita phalloides* is called the destroying angel—and with good reason since it not only kills annually many scavenging dogs (dachshunds are particularly susceptible), but also many inexpert

FIGURE 11.22

Puffballs are among the club fungi, perhaps the most advanced of the fungi. They are filled with spores that on a dry day may fill the air around the puffball (especially if the puffball is prodded by a foot). They are in the same class as mushrooms and toadstools, with such elegant species as smuts, rusts, and stinkhorns.

FIGURE 11.23

The club fungi comprise the class Basidiomycetes. The basidium is covered with tiny projections that contain haploid spores. Most basidiomycetes are saprophytic, but a few, such as the smuts and rusts, are important parasites of plants.

human mushroom hunters as well (Figure 11.24). Wheat rust and smut fungi that attack cereal grains are also among the basidiomycetes and are so devastating to crops that they have been known to influence international relations as certain food supplies fall scarce.

FIGURE 11.24

Amanita phalloides, the "destroying angel," usually fatal if eaten.

Most basidiomycetes form visible fruiting bodies. For example, the familiar mushroom is the fruiting body of a tangled mass of hyphae interwoven through the soil below. The part of the fungus that can be seen above the ground, such as the visible "mushroom," is actually a dense mat of hyphae called a **basidiocarp.** Extending from the gills on the mushroom's underside are countless club-shaped *basidia* (singular, **basidium)** (Figure 11.25), specialized reproductive cells, in which, as we will soon see, nuclear fusion and meiosis occur. In most species, each basidium bears

FIGURE 11.25

Life cycle of the common edible mushroom, a basidiomycete. Spores are dispersed by wind. (a) If they land in a suitable place, they will germinate into primary mycelia, which are comprised of tubelike hyphae. If two hyphae of appropriate types come together, they will join by extending their walls toward each other. These extensions contain nuclei, which end up in the same cell at fertilization. (b) A secondary dikaryotic mycelium (containing cells with two unfused nuclei) now spreads through the soil, and some may eventually form the visible mushrooms. In sexual fusion, the diploid nucleus undergoes meiosis to form four haploid nuclei. Each nucleus is expelled (with some cytoplasm) through an extension at the tip of the basidium, a clublike structure (for which the club fungi are named). A tough wall forms around it and it becomes a spore.

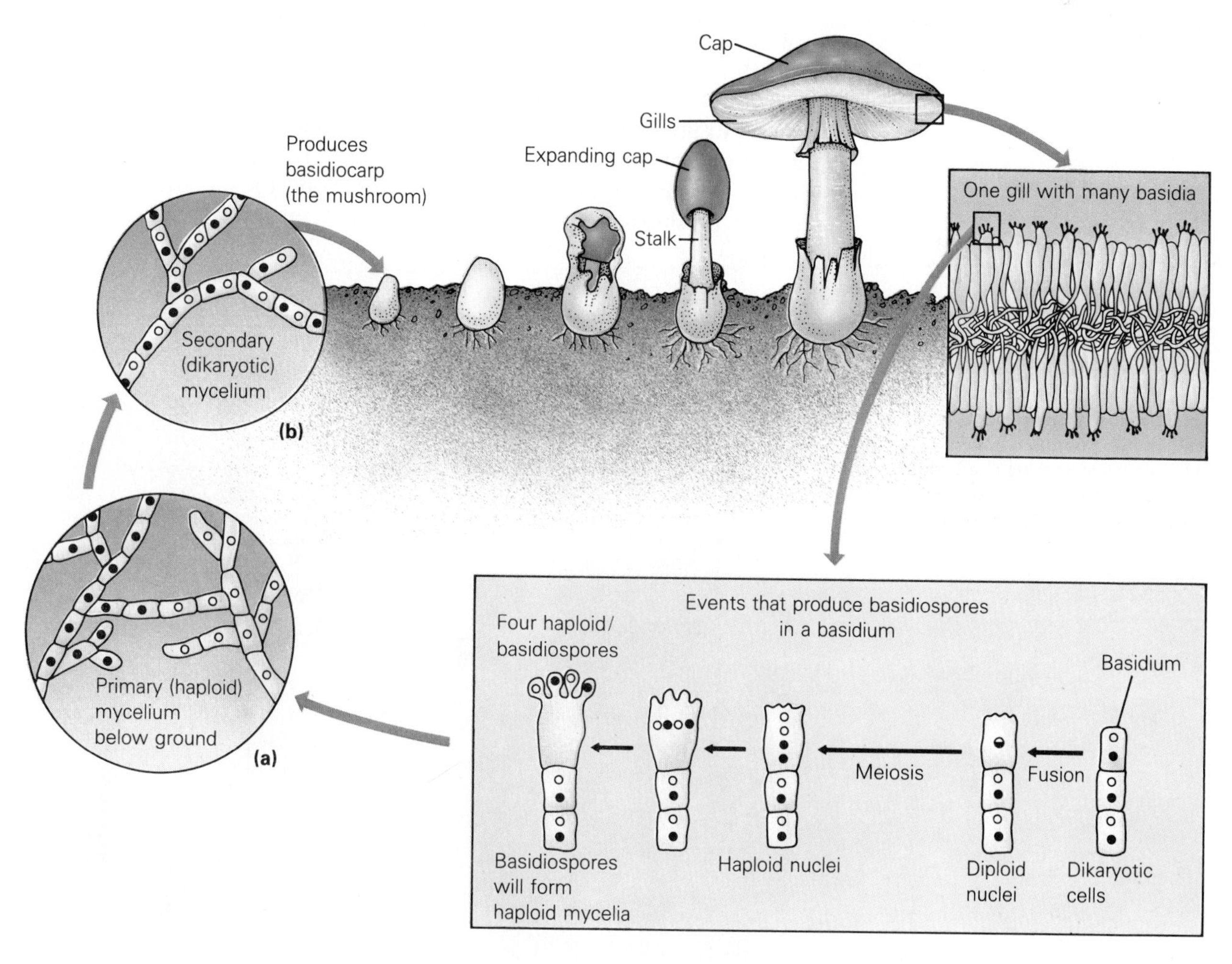

four haploid **basidiospores.** The basidiospores are released and carried, usually by air currents, to some new location where each produces a haploid mycelium. When two mycelia of different mating types encounter each other, the hyphae fuse, forming cells that contain two haploid nuclei and initiating sexual reproduction. These hyphae give rise to the basidiocarp (mushroom). In the basidia of the mushroom, the nuclei fuse, becoming diploid. Then, through meiosis, they produce the haploid basidiospores.

When new mushrooms form, they may appear overnight. Because they form from mycelia that have grown out from an original basidiospore, they often appear in a ring, called a fairy ring (see Figure 11.16). Even after the mushrooms die, the mycelia keep advancing outward at a rate of about 30 cm per year. Some huge fairy rings may be hundreds of years old.

Division Deuteromycota – The Fungi Imperfecti

The fungi that are not known to reproduce sexually are lumped together in an orphanage called **deuteromycetes,** or the Fungi Imperfecti. The imperfection, however, lies in our knowledge—not in the fungi. In fact, some of these may have sexual stages that have not yet been discovered. In other cases, sexual stages may have been present once, but have been lost through evolution.

Although some of the fungi imperfecti are parasites, others are beneficial. Examples of those that are parasites on humans are those that cause athlete's foot and ringworm. On the other hand, *Penicillium* (Figure 11.26), another of the fungi imperfecti, is not only the source of penicillin; it also produces the remarkable flavor in Camembert and Roquefort cheeses.

FIGURE 11.26

Penicillin was discovered in 1928 when *Penicillium* contaminated a bacterial culture being nurtured by Alexander Fleming while working in a London hospital. The fungus wiped out the bacteria and a new antibiotic was born.

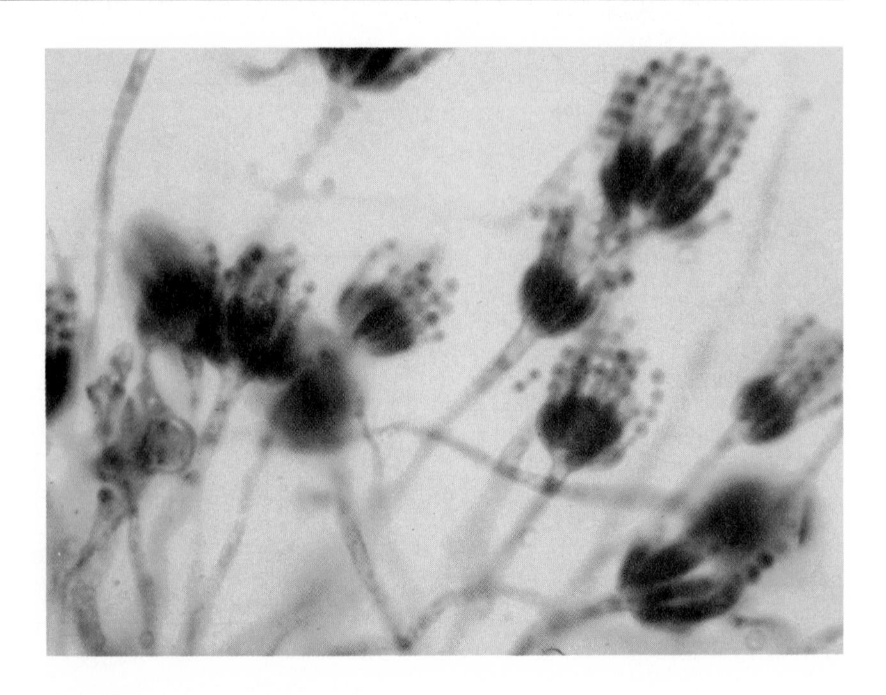

Mycorrhizae and Associating with Plants

Certain species of fungi develop a symbiotic relationship with most of the earth's plants. The association formed by roots and fungal filaments are called **mycorrhizae.** This relationship is beneficial to the plants because as the tubelike filaments spread out from the plant root, they add an enormous absorptive area. Minerals and water that move into the filaments can be tapped by the plant roots. The fungus, in return, is supplied with food manufactured by the plant's photosynthesis. Some of the earliest fossil plants show mycorrizal associations, thus fungi may have played an important role in the evolutionary success of plants.

SUMMARY

1. Many taxonomists support organization of the vast number of organisms on earth by grouping those with similar characteristics into one of the five kingdoms. Monera (prokaryotic cells including the archaebacteria and the true bacteria), Protista (mostly unicellular eukaryotes), Fungi (oomycetes, zygomycetes, ascomycetes, basidiomycetes, and deuteromycetes), Plants (red, brown, and green algae; mosses; and vascular plants), and Animals.
2. The monerans are structurally the simplest and the oldest organisms. They lack nuclear membranes, membrane-bound organelles, and flagella with the 9 + 2 microtubular arrangement. Kingdom Monera includes the archaebacteria and eubacteria.
3. Archaebacteria are the oldest known monerans and are primitive, rare, and closely related to the earliest bacteria. They cannot live in the presence of oxygen.
4. Eubacteria (true bacteria) exist virtually everywhere on earth. Although oxygen is required by most, the growth of some types is inhibited by oxygen. Some species make their own food. Others take in nutrients from living or from dead organisms. When nutrients are obtained from dead organisms, the corpses decompose and the nutrients recycle for use by other organisms. There are three basic shapes: rodlike (bacillus), spherical (coccus), or long and twisted (spirillum). Bacteria typically reproduce asexually, but sexual reproduction is possible. Some bacteria (particularly the Actinomycetes) produce certain antibiotics. Some bacteria can form resistant endospores that enable them to survive harsh conditions.
5. The cyanobacteria carry out photosynthesis with the light-capturing pigments chlorophyll and phycocyanin that are scattered over membranes called thylakoids. Cyanobacteria are important because they are able to incorporate molecules of atmospheric nitrogen into molecules of life, making nitrogen available for other living systems. Division of labor through cell specialization is an evolutionary advance seen in cyanobacteria. They may exist in long chains of separate cells. They live in both water and soil. In unfavorable conditions, some species form tough walls or jellylike coats. Reproduction is asexual.

6. Members of Kingdom Protista are mostly single-celled eukaryotes. Some form colonies and a few are multicellular.
7. Protozoans are single-celled, animal-like organisms. Three phyla are considered: Ciliophora, Sarcodina, and Sporozoa.
8. Members of Phylum Ciliophora bear cilia at some stage in their life cycle. Paramecia are ciliophorans with a thick but flexible outer pellicle. Food is swept into the oral groove, through the cytostome, and encapsulated within a food vacuole where digestion occurs. Undigested material passes out through the cytopyge. A contractile vacuole ejects excess water. They have two nuclei of different sizes. Paramecia have trichocysts, which are long, barbed shafts with trailing threads, for capturing food and defense. Paramecia may reproduce asexually by division or sexually by conjugation.
9. Members of Phylum Sarcodina are amoeboid protozoans that move and capture food using pseudopods. Food is contained in a food vacuole. Amoebas excrete ammonia, take in oxygen by diffusion, and expel excess water with a contractile vacuole. The cytoplasm of the amoeba is divided into a thin outer ectoplasm and a granular inner endoplasm. Sarcodines reproduce primarily asexually, but some also reproduce sexually.
10. Members of Phylum Sporozoa are all parasitic. An example is *Plasmodium*, which causes malaria.
11. Members of Phylum Euglenophyta move by means of one or more long flagella. *Euglena* is photosynthetic with a contractile vacuole that expels excess water and a photoreceptor that allows it to determine the direction of light and move toward moderate levels of light. *Euglena* can form a cyst under unfavorable conditions. It reproduces by dividing.
12. Two phyla of plantlike photosynthetic protists, the Chrysophyta and the Pyrrophyta, are considered.
13. The yellow-green and golden brown algae (Phylum Chrysophyta) possess carotenoid pigments as well as chlorophyll *a* and *c*. The diatoms have cell walls containing primarily silicon. Diatoms are important in aquatic food chains.
14. Members of Phylum Pyrrophyta, the dinoflagellates, are protists with two flagella. They are important in the sea's food chain. *Gonyaulax* is a dinoflagellate that causes "red tide" and produces a poison that accumulates in organisms as it is passed along the food chain.
15. Funguslike protists include the cellular and acellular slime molds. They resemble amoebae in their feeding stages and fungi in their asexual reproductive stages. During reproduction they give rise to stalks with spore-forming bodies at their tips.
16. The protists set the stage for the evolution of the groups that followed by developing a nucleated cell with a complex genetic apparatus, creating energy-rich molecules, and developing structures for efficiently using energy.
17. The fungi lack chlorophyll. Parasitic fungi obtain nourishment from living organisms and the more common saprophytic fungi obtain nutrients by secreting digestive enzymes into dead organisms and absorbing the products. The cell walls of most fungi are made of chitin. Fungi grow by producing long hyphae that join to form a mycelium.

18. There are five divisions of fungi: Oomycota, Zygomycota, Ascomycota, Basidiomycota, and Deuteromycota.
19. Members of Division Oomycota (water molds, white rusts, and downy mildews) develop from a large egg fertilized by smaller sperm. They are sometimes placed with the algae. Unlike other fungi, they have cell walls made of cellulose. They may be parasitic or saprophytic. Some cause diseases in plants or animals.
20. Members of Division Zygomycota reproduce primarily asexually. However, when conditions are harsh they may reproduce sexually, producing a tough resistant zygospore. All zygomycetes are terrestrial. There are parasitic and saprophytic species. Bread mold is a zygomycete.
21. The largest division of fungi is Division Ascomycota. Many ascomycetes form a symbiotic relation with algae to produce lichens. Most ascomycetes are saprophytic but some, such as the one that causes Dutch elm disease, are parasitic. They can reproduce asexually by forming spores called conidia at the tips of aerial hyphae called conidiophores. In sexual reproduction, two different strains of hyphae meet and ascospores are produced within a sac called an ascus. One ascomycete infects rye grain producing ergot, which can produce hallucinations if consumed by humans.Truffles are another ascomycete.
22. Division Basidiomycota includes the mushrooms, puffballs, and bracket fungi. Some mushrooms are edible, but others are poisonous. Wheat rust and smut fungi are basidiomycetes that destroy crops. The mushroom we see is the fruiting body (basidiocarp) of a mass of hyphae in the soil below. From the gills of the basidiocarp extend the many club-shaped basidia. Each basidium bears four haploid spores. The basidiocarp forms from the fusion of two mycelia of different mating types. Mycelia grow outward from a basidiospore, often forming a ring of basidiocarps called a fairy ring.
23. Fungi that are not known to reproduce sexually are called the Fungi Imperfecti (division Deuteromycota).
24. Symbiotic associations formed between fungi and plant roots are called mycorrhizae. The plant benefits from mycorrhizae by an increase in absorptive area, and the fungus obtains food from the plant.

KEY TERMS

archaebacteria (285)
ascocarp (306)
ascomycetes (303)
ascospores (306)
ascus (306)
bacillus (288)
basidiocarp (309)
basidiomycetes (307)
basidiospores (310)
basidium (309)
Chrysophyta (297)
cilia (292)
Ciliophora (292)
coccus (288)
conidia (303)
conidiophore (306)
conjugation (294)
contractile vacuole (294)
cyst (297)
cytopyge (294)
cytostome (293)
deuteromycetes (310)
dinoflagellate (298)
ectoplasm (295)
endoplasm (295)
endospore (288)
eubacteria (286)
Euglenophyta (295)
eyespot (297)
flagellum (295)
food vacuole (293)
fungi (300)
gullet (297)
hypha (301)
macronucleus (294)
micronucleus (294)
moneran (285)
mycelium (301)
mycorrhizae (311)
oomycetes (301)
oral groove (293)
Paramecium (292)
pellicle (292)
Protista (292)
protozoan (292)
pseudopod (295)
Pyrrophyta (298)
Sarcodina (294)
spirillum (288)
Sporozoa (295)
taxonomist (284)
trichocysts (294)
zygomycetes (303)
zygospore (303)

FOR FURTHER THOUGHT

1. Into which kingdom (Monera, Protista, or Fungi) would you place an organism that is single-celled, photosynthetic, and eukaryotic?
2. In what ways are the activities of bacteria ecologically important?
3. Surgical instruments that must be sterile and bacteria free are often autoclaved—subjected to intensely high temperatures of steam under pressure for long periods of time. Why?

FOR REVIEW

True or false?

1. ___ All monerans are prokaryotes.
2. ___ Kingdom Protista is composed entirely of multicellular organisms.
3. ___ Some cyanobacteria fix nitrogen.
4. ___ Red tide is produced by a dinoflagellate known as *Gonyaulax*.
5. ___ Ascomycetes reproduce asexually by spores called conidia.

Fill in the blank.

6. The ___ are a group of newly discovered ancient monerans.
7. The ___ are commonly called water molds.
8. Two paramecia may reproduce by exchanging genetic material in a form of sexual reproduction called ___ .
9. The symbiotic association formed between roots and fungal filaments are called ___ .
10. Penicillin is produced by a fungus of the division ___ , the largest division of fungi.

Choose the best answer.

11. All bacteria:
    A. are eukaryotic
    B. require oxygen
    C. reproduce slowly
    D. are one-celled
12. Which of the following is considered a plantlike protist?
    A. slime mold
    B. diatom
    C. amoeba
    D. *Paramecium*
13. Members of Phylum Ciliophora use ____ as a means of locomotion.
    A. a flagellum
    B. a mycelium
    C. pseudopods
    D. cilia
14. ____ are fruiting bodies produced by some basidiomycetes.
    A. ascospores
    B. conidiophores
    C. zygospores
    D. mushrooms
15. Which of the following is not characteristic of at least some members of Kingdom Fungi?
    A. presence of chloroplasts
    B. cell walls made of chitin
    C. saprophytic means of nutrition
    D. sexual reproduction

CHAPTER 12

Diversity II: The Plants

Overview

Alternating Generations
Aquatic Plants
Nonvascular Land Plants
Vascular Plants

Objectives

After reading this chapter you should be able to:

- Describe the alteration of gametophytic and sporophytic generations in plants and explain why this cycle may have evolved.
- Give the characteristics of three algal divisions, indicating which division is the most likely ancestor of land plants.
- List three types of bryophytes and explain how they differ from vascular land plants.
- List three groups of vascular land plants and describe their special adaptations to terrestrial life.
- Describe the evolutionary advances characteristic of ferns.
- Differentiate between gymnosperms and angiosperms.
- List some evolutionary trends in flower development.

We seem to innately understand that we are firmly bound to the world of plants. We are attracted to them, and we somehow feel better when we are with them. We walk among them when we can get out, and we bring them to us when we can't. I even recall a conflicted New Yorker who had covered his lawn with asphalt and then painted it green. Apartment dwellers in concrete cities often carefully tend their beloved window plants in a poignant homage to our bond with that silent world.

Interestingly, the kind of plant that attracts us may depend on where we were raised. I grew up in the South, and I love the dark forests. I even live in one. Californians may feel claustrophobic in such places and crave the sight of distant vistas and low-lying plants. They may be solaced by the sight of a Joshua treee silhouetted against a desert sky. Others, living in the North, may roam hardwood forests and mark the seasons by the color of the leaves. No matter what our preferences in such matters, though, we seem to find plants compelling and comforting, a feeling that may stem from our long association and profound interdependence with them.

But what exactly are plants? If you asked an old man watering an African violet what a plant is, he is not likely to say, "According to the five-kingdom scheme, it is a multicellular, eukaryotic, photosynthetic organism." If he did, his answer would exclude the photosynthetic cyanobacteria because they are not eukaryotes. He would rule out fungi because they are not photosynthetic. He might also mutter something about the problems of classification and then say that a more precise definition is: Plants are usually photosynthetic, have cell walls, are multicellular, and have a life cycle involving alternating generations.

Using that definition, we will consider the plants in two separate discussions. Here, we will look at their diversity with particular attention to the evolutionary significance of each group's special traits. Then, in Chapter 14, we will consider in more detail the reproduction and development of each group.

ALTERNATING GENERATIONS

There are a great many species of algae and plants with both sexual and asexual phases. Often the phases alternate in what is called an **alternation of generations,** essentially a continuing sequence of haploid and diploid forms (Figure 12.1A). For example, in some plants, spores grow into organisms that do not release more spores, but rather gametes ("eggs" and "sperm"). When a male and female gamete unite, the fertilized egg will grow into an adult individual that looks entirely different from the one that had grown from the spores. When it matures, this organism will then produce spores. Gamete-producing generations are called **gametophytic,** and spore-producing generations are called **sporophytic.**

The spores are produced by meiosis; hence, they are haploid and have only half the complement of chromosomes characteristic of the other phase and grow into individuals of the gametophyte generation, which then

FIGURE 12.1A

Some plants show a pronounced alternation of generations. The sporophyte generation begins when the eggs and sperm are joined in fertilization. The mature sporophyte then produces spores (male or female) whose subsequent growth comprises the gametophyte generation. The gametophytes then produce gametes ("eggs" or "sperm"), which join in fertilization to form a new sporophyte.

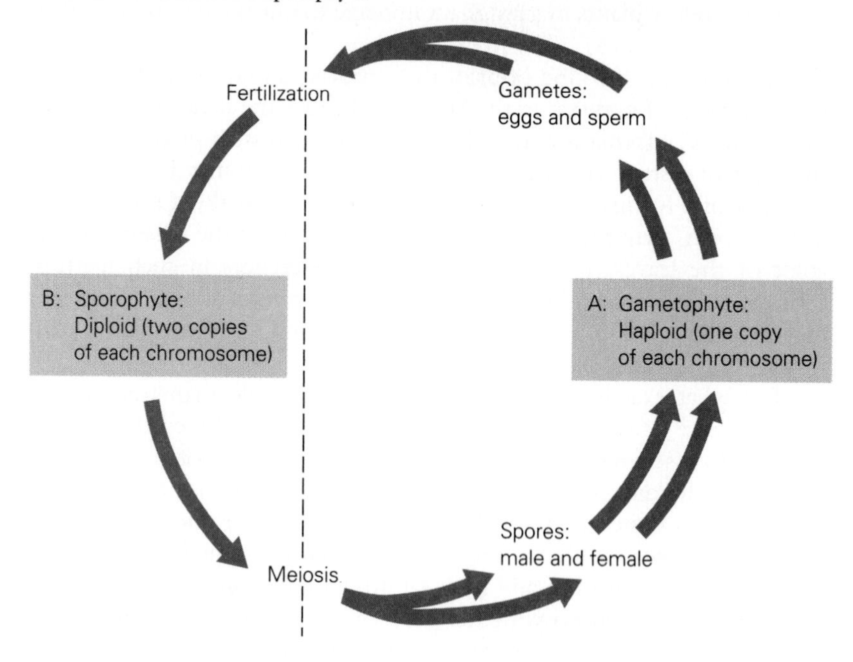

release gametes. Of course, both the gametes are haploid. When these join, they begin the sporophyte generation, restoring the full chromosomal complement. The sporophyte generation releases spores, initiating the gametophyte generation again. Nearly all plants go through an alternation of generations. However, one generation may be much more prominent than the other, depending on the species. The idea is diagrammed in simplified form in Figure 12.1B.

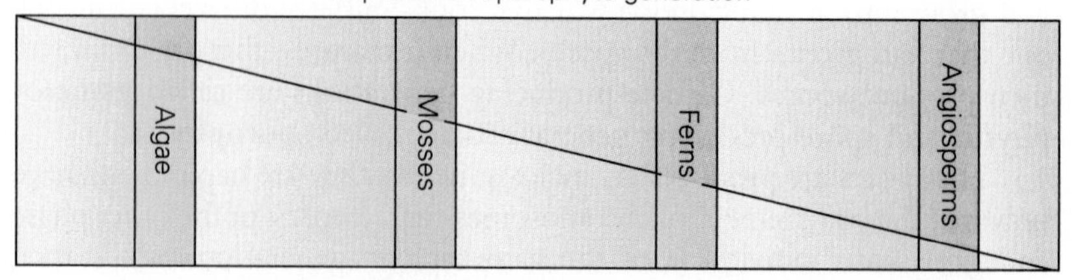

FIGURE 12.1B

The relative importance of diploidy in the generations of various groups of plants. We see, then, that the most conspicuous generation in algae is the gametophytic (haploid), while in angiosperms it is sporophytic (diploid). What about mosses and ferns? (There are exceptions to this trend within each group.)

The Evolution of Alternating Generations

Alternation of generations adds a complex dimension to life histories, so one may well wonder why it even evolved to begin with. We can never be sure what prompted its evolutionary development, but alternating generations may have evolved in response to regularly recurring stressful periods in the organism's life history. It has been suggested that such stress could have prompted the development of a gametophytic stage on two bases: (1) the subsequent fertilization could produce a mix of genes that might be better able to survive the stressful conditions; and (2) the often smaller, metabolically less expensive gametophytic stage might be better able to survive stressful times by simply needing less from the environment.

The relationship between stress and reproduction is an interesting one. For example, in some species, sexual reproduction increases in periods of stress. In earlier days, steel girders were placed around surviving trees in burned-over areas. The idea was that as the tree grew and came to exert pressure against the girder, the stress would cause the tree to produce more seeds. Interestingly, many species of animals, after being subjected to some sort of harsh conditions that reduces their numbers, tend to undergo a rapid surge of reproductive activity (which, one could argue, might be a response to either stress or low numbers). Stress has even been suggested as the stimulus of those baby booms that usually follow great wars (but the effect of so many randy returning servicemen cannot be discounted).

As we begin our survey of plants, the first few **divisions** (the botanical name for phyla) we will look at are loosely called **algae.** The term once referred to aquatic plants, but has largely lost any precise scientific meaning. We will find that the red and brown algae are quite different from the rest of the plants. We will also see that although the green algae are similar to the protista in many respects, biochemically they are similar to other green plants and probably gave rise to them. Table 12.1 lists the plant divisions we will discuss. Table 12.2 presents the important characteristics of each division.

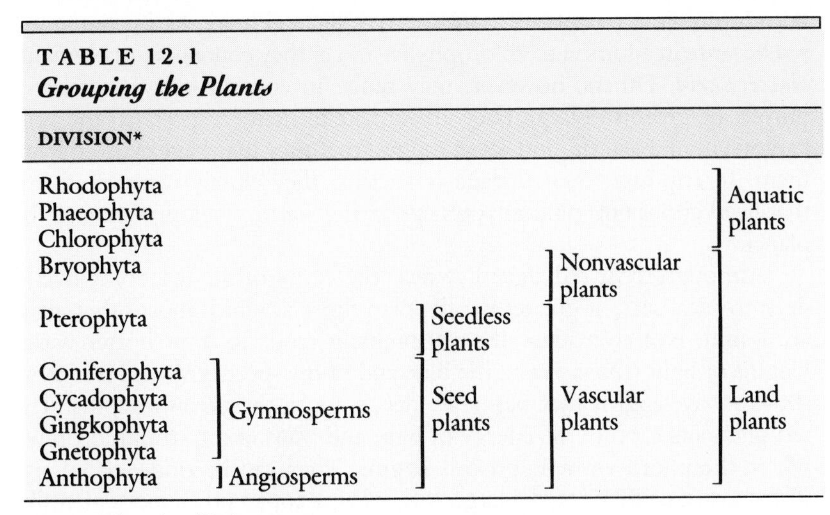

TABLE 12.1
Grouping the Plants

DIVISION*				
Rhodophyta				Aquatic plants
Phaeophyta				
Chlorophyta				
Bryophyta			Nonvascular plants	Land plants
Pterophyta		Seedless plants	Vascular plants	
Coniferophyta	Gymnosperms	Seed plants		
Cycadophyta				
Gingkophyta				
Gnetophyta				
Anthophyta	Angiosperms			

*A somewhat simplified arrangement.

TABLE 12.2
Plant Characteristics

DIVISION	HABITAT	LIGHT-GATHERING PIGMENT	BODY FORM
Aquatic Plants			
Rhodophyta (red algae)	Deep, tropical seas; some fresh-water	Phycoerythrin; chlorophyll *a* and *d*	Single celled or filamentous and branched; nonvascular
Phaeophyta (brown algae)	Coastal waters	Fucoxanthin, chlorophyll *a* and *c*	Nonvascularized; "stems" (stalks) and "leaves" (thalli)
Chlorophyta (green algae)	Mostly freshwater; some live in the sea	Chlorophyll *a* and *b*	Filamentous, bladelike, or live in colonies; nonvascular
Nonvascular Land Plants			
Bryophyta (mosses, liverworts, hornworts)	Moist habitats	Chlorophyll *a* and *b*	Nonvascular stems, leaves, and "roots" (rhizoids)
Vascular Plants			
Pterophyta (ferns)	Moist soils of woods and swamps	Chlorophyll *a* and *b*	Vascular leaves, roots, and rhizomes
Gymnosperms ("naked seeds" consist of conifers, cycads, gingkos)	Diverse climates	Chlorophyll *a* and *b*	Woody, vascularized trees; seed-bearing cones
Angiosperms (flowering plants; monocots and dicots)	Diverse climates; some aquatic	Chlorophyll *a* and *b*	Vascularized trees and plants with flowers

AQUATIC PLANTS

Division Rhodophyta and Many Colors

Rhodophytes ("red plants") are the red algae (Figure 12.2). Some are red because in addition to chlorophyll *a* and *d,* they contain a red pigment, *phycoerythrin.* Others, however, may range in color from green to red, purple, or greenish-black. They are similar in many respects to the prokaryotic cyanobacteria, and some suggest that they may have evolved from them. In any case, their lineage is ancient; they closely resemble forms that lived about 600 million years ago in the warm, shallow waters of the planet.

Although most red algae live near the edges of the sea, some live in deep tropical seas, at greater depths than those at which most other algae are found. Not coincidentally, red pigment tends to trap shorter wavelengths of light (those towad the blue end of the spectrum), and it is these shorter wavelengths that penetrate deepest into the ocean's depths. The red pigments capture the energy of light and transfer it to the chlorophyll where the photosynthetic process begins. They can be single-celled and filamentous, or they may be large with what is apparently a more plantlike body. Even the larger species, though, are composed of mats of filaments, much in the manner of fungi. Alternation of generations is common in

FIGURE 12.2

Fauchea, a red alga. There are a few freshwater forms, but most are marine.

red algae. They have incredibly complex and diverse reproductive lives that, in spite of your pleading for more information, are a bit beyond the scope of our discussion here.

Division Phaeophyta and Undersea Forests

The **phaeophytes** ("dusky plants") are the brown algae. If you have spent much time prowling rocky ocean shores, you have undoubtedly encountered them (Figure 12.3). They are the shiny brown plants with stemlike and leaflike structures. However, their "stems" and "leaves" are really not stems and leaves at all. The stems lack conducting tissues, and the leaves

FIGURE 12.3

Fucus, or rockweed, is a brown alga found in cold, temperate waters. It is commonly seen by walking along rocky areas when the tide is out.

have no veins. They possess chlorophyll *a* and *c* and a brown pigment called *fucoxanthin*. Unlike flowering plants, their cells contain centrioles, as do animal cells.

The brown algae show alternation of generations. In the simpler, smaller species, the gametophyte and sporophyte may be very similar. In the larger species, the sporophyte generation is dominant.

The brown algae are among the most conspicuous of marine plants. After all, the common rockweed, *Fucus*, is in this group, as are the giant kelps along the Pacific coast. (These are among the world's largest plants, reaching 100 meters in length.) The blades or **thalli** (singular, *thallus*) are anchored to rocks by **stalks** that are attached to **holdfasts.** The thalli are lifted toward the energizing light at the surface by air-filled bladders. The great "forests" often tempt divers who risk becoming entangled in the dense growth. The famous Sargasso Sea is a thick matlike raft, comprised of *Sargassum*, a brown alga (Figure 12.4). This remarkable body covers over two million square miles of the Atlantic Ocean and is comprised largely of plants that have grown in shallower waters, broken off, and floated to sea.

FIGURE 12.4
Sargassum, the alga that comprises the infamous Sargasso Sea.

Division Chlorophyta: Ancestors of Land Plants

The **chlorophytes** are the green algae (Figure 12.5). They are mostly freshwater organisms, but a few species live in the sea. They are believed to be directly in the evolutionary line that gave rise to the land plants—

FIGURE 12.5
The green algae are tiny plants, some of which float aimlessly in the seas, while others live in fresh water ponds and cover them with scum. *Acetabularia* (a) is very unusual, seemingly off on its own evolutionary pathway. Each thallus is comprised of a single wineglass-shaped cell. *Codium* (b) lacks cell walls that would separate its many nuclei. Its branching structure gave rise to its romantic common name, "dead man's fingers."

(a)

(b)

FIGURE 12.6

A group of *Volvox* colonies. In a *Volvox* colony, a few cells begin to become distinguished from the others early in their development. They will grow noticeably larger than the rest and will become reproductive cells. During asexual reproduction, each of these will divide and begin to form a daughter colony, even while it is inside the parent colony. When the divisions are complete, the new colony will depart through the wall of the parent colony. The parent colony then dies. Under certain conditions, the reproductive cells can also form eggs and sperm instead of new colonies.

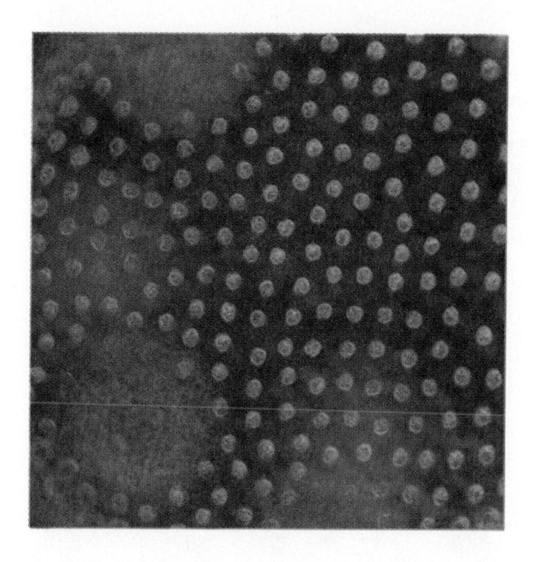

the mosses and vascular plants (ferns and seed plants). They share three important traits with the land plants: (1) they store carbohydrates such as starch; (2) the cell walls are comprised mainly of cellulose; and (3) they contain chlorophyll *a* and *b*. In spite of such common traits, green algae are extremely diverse. Many exist as single cells, while others are filamentous or even bladelike, and some form aggregates or specialized colonies (Figure 12.6). Their fragile beauty has captured the imagination of botanists for centuries. It is almost paradoxical that a close look at green scum from a still pond can provide such enchanting sights. The green algae, by the way, may be important for more than aesthetic reasons since a great deal of research attention has been focused on them as a potential source of food from the sea. Let's now leave the algae and move on to the next most complex and organized group of plants, the mosses. As usual, we will reserve the discussion of the life cycle for Chapter 14.

NONVASCULAR LAND PLANTS

Division Bryophyta and the Invasion of the Land

The **bryophytes** are the mosses, liverworts and hornworts. They are an ancient lineage, among the oldest forms of land plants. Bryophytes are widely diverse and grow in a variety of places. You probably can conjure up an image of a moss, but what about a hornwort? A liverwort? (See Figure 12.7). For simplicity's sake, let's focus on the more familiar mosses. Mosses, as you know from popular folklore, always grow on the north sides of the trees, except when they don't. It is true that in the northern hemisphere, they are more *likely* to be found on the north sides of trees, but that's because the midday sun shines from the south. Mosses grow

FIGURE 12.7

Hornworts (a) and liverworts (b) are bryophytes, ancient but inconspicuous species. Humans may live haughtily among such ignominious species without ever recognizing them, but when we die, they may grow ignominiously on our graves.

(a)

(b)

best where its shady, cool, and wet. (One reason they need the moisture is that the motile sperm of mosses must swim to the egg, a trait abandoned by more recently evolved plants.) Mosses also grow on rocks and bare soil. Some species grow in the cracks of concrete sidewalks, almost totally ignored by city dwellers. Many species are very specific about where they live. For example, some grow only on reindeer bones while others grow only on large insects and yet others on carnivore droppings.

In a sense, mosses are midway between green algae and the vascular plants that we will consider next. *Vascular* refers to plants with conducting vessels that transport food, minerals, and water. Bryophytes differ from the vascular plants in one important way—they lack conducting vessels. Thus they lack true leaves, stems, and roots (since all of these contain vascular conducting tissue). They do have structure similar to leaves, stems, and roots, however. The "roots" are really tiny cellular threads (rhizoids) that in most species serve only as anchors; they do not absorb water. Water enters through the flattened outgrowths that resemble leaves. The "leaves," just a few cells thick, curl up under dry conditions, thereby reducing water loss. Mosses lack the conducting vessels that support vascular plants, and they often grow crowded together, supporting each other and trapping water between their tiny bodies (Figure 12.8). Apparently, mosses did not give rise to vascular plants because fossil vascular plants have been found that are older than any fossil moss yet discovered.

FIGURE 12.8

Mosses are rather common bryophytes. They lack vascular systems and even roots. They survive so well partly because of their small size. (Can you see why?) The moss gametophyte (foreground) consists of a central upright strand of cells with leaflike growths where most photosynthesis occurs. These outgrowths are not true leaves, nor is their "stem" a true stem since both lack conducting tissue. The plant is attached to the soil by threadlike rhizoids. Rhizoids also lack conducting tissue. The brown stalks are the sporophytes, their enlarged tips containing spores.

VASCULAR PLANTS

Now we reach the more familiar vascular plants. In fact, when we think of green plants, we usually have in mind the vascular plants. They appeared some 400 million years ago, and within only 50 million years or so, they had, through divergence, formed the major evolutionary lines of land plants. Many have become extinct, but others have persisted to grace our landscape today.

The vascular plants include the ferns, the gymnosperms (including the conifers—cone bearers such as pine, spruce, and fir), and the angiosperms (the flowering plants). Vascular plants, then, usually are relatively large land plants, and thus they have had to solve the problems of large size and terrestrial life. This means (1) they had to be able to support themselves as they reached up toward the sunlight, (2) they needed a way to stand anchored in the soil, (3) they had to develop ways to withstand the deadly drying air and to conserve their water, (4) they needed a system to conduct water upward from the ground and food downward from the leaves, (5) they needed a way to exchange vital gases with the atmosphere, and (6) they needed a way to achieve fertilization in the dry air. Thus, we find land plants with strong fibers, conducting vessels, roots or rootlike structures, openings in the leaves or stems, waterproof and airproof coverings, and tough, resistant pollen.

Among the earliest evidence of vascular plants are fossils of *Rhynia* from Scotland (Figure 12.9). These ancient Devonian plants were preserved remarkably well, and we can clearly see features found in many plants today, such as a waxy covering, openings in the leaflike structures

FIGURE 12.9

Ancient 400-million-year-old fossils have enabled us to reconstruct an early organism called *Rhynia*. They were gone long before our first shuffling ancestors appeared.

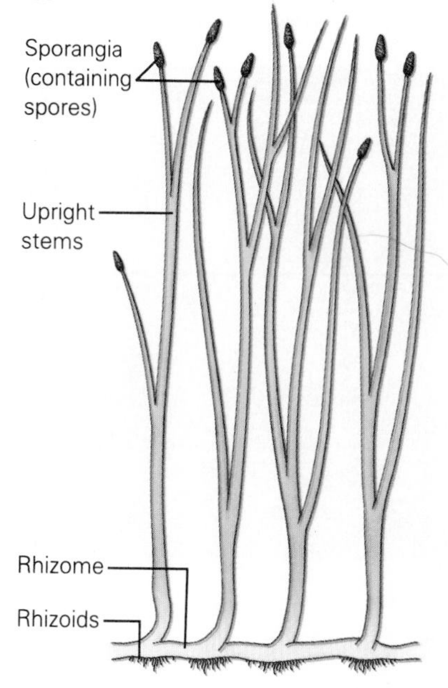

FIGURE 12.10

The horsetails (a) and club moss (b)—not real mosses—are ancient species with true vascular systems. Their ancestors were towering figures on the earth's landscape, but today they are represented by a few species that are likely to be found in areas abandoned by humans, such as vacant lots. These are both in the sporophyte generation.

(a)

(b)

FIGURE 12.11

Ferns are among the oldest vascular plants. Fossil ferns, found in deposits from the Middle Devonian, grew from the Arctic to the tropics.

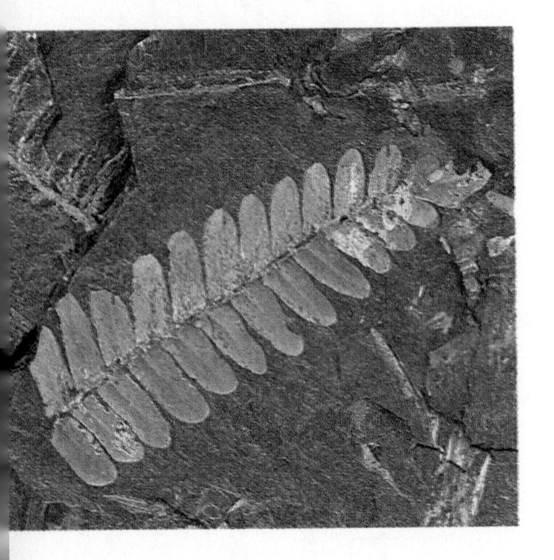

(through which gases could pass), rootlike structures called **rhizomes** (that are actually underground horizontal stems), and tiny, hairlike **rhizoids** through which water could pass into the plant. They had no leaves and no roots. Their most advanced feature seemed to be a tendency to branch, to form a two-pronged fork. Probably none of these gave rise to more advanced plants. Certain other primitive vascular plants have managed to survive until now, as we see in Figure 12.10.

Division Pterophyta and Leaves and Roots

The **pterophytes** are the ferns and their relatives (Figure 12.11). In this class we see true leaves for the first time. (**Leaves,** we can now say, are flattened photosynthetic structures that grow from a stem and have true vascular tissue.) Fern leaves are traditionally called **fronds.** With the ferns we also see the development of true **roots,** structures usually found below ground that function in anchoring and absorbing water and minerals from the soil. The stem exists in the form of a rhizome. Ferns and fernlike plants do not produce seeds, a trait that sets them apart from the next two classes.

The most familiar ferns are small plants that live in the moist soils of woods and swamps. Their delicate beauty enhances a walk in the dark woods, especially if we find them while they are still pale "fiddleheads," uncurling shyly from the damp earth. Other ferns are not so small and may grow to 30 meters as they wind their way up tall trees. Still others

FIGURE 12.12
Tree ferns are impressive plants that seem to be vestige of some bygone era of the earth. Here we see the sporophyte stage. Spores are produced in clusters under the fronds. Such spores travel easily and are often among the first colonizers of new areas, such as oceanic islands formed by volcanic activity.

may stand alone as tree ferns, rising to 20 meters in height (Figure 12.12). Ferns lack the strength of woody plants, however, so they usually grow in protected ravines or under the forest's canopy of leaves, a tendency that only heightens their aura of subtle drama. Thoreau was once moved to say "God made ferns to show what he could do with leaves."

Gymnosperms and Naked Seeds

The **gymnosperms** are the plants with "naked" seeds. That is, the seeds are not surrounded by a fleshy ovary, a trait characteristic of the next class we look at. There are four living divisions of gymnosperms. The **coniferata** (cone bearers) include the pine, spruce, fir, cedar, and their relatives. The other divisions are the **cycadophyta** (the cycads), the **gingkophyta** (the gingkos), and the **gnetophyta,** the peculiar gnetophytes.

The gymnosperms are among the most beautiful and impressive of all plants. After all, the redwoods are gymnosperms (Figure 12.13). The tallest living thing in the world is a redwood—over 100 meters tall—that was saved from lumbering interests. (A taller tree was discovered by lumbermen and quickly cut down and sawed into boards before anyone could prevent it.) The tree with the greatest bulk is another gymnosperm, a giant sequoia that reaches 80 meters in height and has a girth of 20 meters at its base. It is about 4000 years old, but it is a youngster compared to a bristlecone pine, almost 5000 years old, that was found in the mountains of eastern Nevada. Other gymnosperms include the hemlocks of the North, the cypress of the southern swamps, the wiry ephedra of the deserts (Figure 12.14).

FIGURE 12.13

The redwoods are among the most beautiful and overwhelming treasures of the plant kingdom. The trees are fire-and insect-resistant and they once covered vast areas of the West Coast. They have now been decimated by public demand for insect-resistant tomato stakes and lawn furniture that does not require paint. It would be nice to say that we have at last come to our senses, but the great trees continue to fall daily, except perhaps on Sundays.

FIGURE 12.14

The cypress family is distributed worldwide. Most are trees (b) but a few are shrublike. In all species, the leaves are small and scalelike like those of the primitive *Ephedra* (a) from whose ancestors they may be descended.

(a)

(b)

FIGURE 12.15

The gingko, *Gingko biloba,* is represented by only one species. Many species thrived in the Mesozoic era, but today it owes its survival to Eastern religionists who, for thousands of years, have planted them around temples. The plant is highly resistant to insects, disease, and air pollution. Thus, it is planted as a "street tree" throughout the world.

FIGURE 12.16

At first glance, you might mistake a cycad for a palm tree, but they are not closely related at all. Cycads are unmasked as gymnosperms in that they bear cones. Their leaves are covered by a waxy cutin that retards water loss, and their spines probably help protect them from foraging animals. They are very slow to reproduce, and only a few species remain, mostly limited to the warmer temperate zones and to semiarid regions of the tropics.

The gymnosperms first appeared during the Paleozoic era, but they didn't reach their heyday of diversity until the Mesozoic, a time when great cycads shaded the dinosaurs (Figure 12.16). There were several species of gingkos (Figure 12.15) in those days, but only one hardy species remains today. There are about 70 species of gnetophytes, but surely the most remarkable is the *Welwitschia* (Figure 12.17).

FIGURE 12.17

The splitting, twisted, and tortured foliage seen here belongs to *Welwitschia mirabilis,* a Southwest African desert-dwelling gnetophyte. The leaves would extend for many meters if they had not split and become gnarled.

FIGURE 12.18

The angiosperms are a diverse group. In spite of the different appearances of these species, they are all flowering plants. Water lily (a), bamboo (b), saguaro (c), and pink thistle (d). What other characteristics might you expect these four plants to share?

(a)

(b)

(c)

(d)

Not only are the conifers (the cone-bearing gymnosperms) abundant today, but they dominated the Mesozoic landscape as well. Such long-term success undoubtedly lies in their remarkable adaptations in fertilization, seed protection, and seed dispersal; traits that we will consider in more detail in Chapter 14.

Angiosperms, Covered Seeds, and Flowers

The **angiosperms** comprise the Division Anthophyta. These plants are among the most spectacular of all living things—the flowering plants. They are also among the most evolutionarily successful forms of life, and they include more than what we think of as flowers. For example, violets are flowering plants, but so are grasses and palms (Figure 12.18). In fact, more than 80 percent of all plant species on earth are angiosperms.

There are about 230,000 species of angiosperms, ranging in size from tiny, floating duckweeds (about 1 millimeter long) to the gigantic eucalyptus that can reach 100 meters in height (Figure 12.19). In the 100 million years or so that they have been on the planet, the angiosperms have managed to invade a remarkable diversity of habitats from deserts to

FIGURE 12.19

Diversity among the angiosperms. All the duckweed plants shown here could be held in the palm of your hand. The towering eucalyptus, strangely enough, is in the same division. The angiosperms are indeed a diverse group.

rain forests. No matter what their appearance or life style, though, all angiosperms bear seeds surrounded by a protective ovary. Also, all angiosperms bear a special trademark, the flower. Technically, the flower is a specialized leaf shoot with a variety of roles in reproduction, as we will see in Chapter 14.

We know little about the evolution of flowering plants; most of our information has been pieced together from a comparison of living plants, which we arranged from primitive to advanced. One fossil form has been found in what is now central Kansas, as we see in Figure 12.20. (A Kansas friend said they probably perished out of sheer boredom, but at that time central Kansas was a subtropical coastal plain with dinosaurs roaming the water's edge.)

There are four major evolutionary trends in the flowers themselves:

1. Flowers with many parts have evolved to form flowers with fewer parts.
2. Parts of the flower have become fused into single structures.
3. The ovary has dropped to a position in the flower, closer to the stem.
4. The regular, radial symmetry has become irregular and bilaterally symmetrical.

Of course, it is important to remember that any flower today can be made up of parts at various stages of evolutionary advancement.

FIGURE 12.20

Archaeanthus linnenbergeri is the only discovered member of an extinct family of Angiosperms that lived about 95 million years ago. It was apparently a small tree or shrub with a strong trunk and stout flowers. It may have resembled a magnolia (right) in some ways.

Although we don't know how the flowering plants evolved, we have an idea about how they reached a dizzying crescendo of success in a very brief period. Their sudden emergence about 127 million years ago may have had something to do with a climatic change, since the worldwide climate about this time became dramatically cooler and drier. Angiosperms in those days didn't have to worry much about being eaten by a grazing dinosaur, however, because the period mysteriously marked the extinction of these great beasts as well as many conifers and cycads.

No brief introduction to the world of plants could reveal how tightly our lives are bound to theirs. In fact, the bond between animals and plants is so strong that probably it cannot yet be entirely understood. The interplay between these life forms has had a profound effect on the evolution of both, and they remain inextricably linked in countless ways today. It may be more accurate to try to understand, emotionally as well as intellectually, that we are simply different but interdependent parts of the incredible and wonderful phenomenon called life.

SUMMARY

1. Plants are eukaryotic organisms that are usually photosynthetic and multicellular with cell walls. Their life cycle involves alternation of generations.
2. Alternation of generations is a repeating cycle of haploid and diploid forms. The haploid gametophytic generation produces gametes by mitosis. The gametes fuse and grow into an adult of the diploid sporophytic generation, which produces haploid spores by meiosis. The spores grow into new gametophytes. Alternation of generations may have evolved in response to regularly recurring periods of stress in the life history. This may have evolved because fertilization of gametes may produce a mix of genes able to survive stress better than the old mix and because the smaller gametophyte may be able to survive stress better because it requires less from the environment.
3. Members of Division Rhodophyta, the red algae, are not always red in color. The red pigment phycoerythrin, which is present in addition to chlorophyll *a* and *d,* gives a red color to some algae. Red pigment absorbs shorter wavelenghts of light, which are those that reach the deep water where most red algae live. The light energy is then transferred to chlorophyll for photosynthesis. Red algae are filamentous and can exist as single cells or matted into a large plantlike body. Reproduction is complex and diverse.
4. Division Phaeophyta, the brown algae, have leaflike and stemlike parts. They possess chlorophyll *a* and *c* and fucoxanthin (brown pigment). Their cells contain centrioles. The common rockweed *(Fucus),* the giant kelp, and sargassum are brown algae.
5. Members of Division Chlorophyta, the green algae, are thought to be ancestral to the land plants. Like land plants they store carbohydrates, have cellulose walls, and contain chlorophyll *a* and *b*. They may exist as single cells, filaments, blades, or colonies. They are a potential source of food from the sea.

6. The mosses, liverworts, and hornworts (Division Bryophyta) are among the earliest forms of land plants, but did not give rise to vascular plants. Their environment must be moist enough for sperm to swim to the egg. Mosses lack true leaves, stems, and roots as well as conducting vessels. They are anchored by tiny rhizoids and absorb water through leaflike outgrowths.
7. The vascular plants include the ferns, gymnosperms, and angiosperms. Adaptations of vascular plants to deal with the problems associated with life on land include: strong fibers, conducting vessels, roots or rootlike structures, openings in stems or leaves, waterproof and airproof coverings, and tough resistant pollen. One of the earliest vascular plants is a fossil, *Rhynia*.
8. The ferns and fernlike plants are the first group to have true leaves (fronds), which are structures with vascular tissue that grow from a stem. They are also first to have true roots: underground structures that are used for anchorage and absorption. They do not produce seeds and lack the strength of woody plants.
9. There are four groups of gymnosperms: conifers, cycads, gingkos, and gnetophytes. Their "naked" seeds are not surrounded by a fleshy ovary. The gymnosperms, particularly the conifers, have been successful plants since the Mesozoic Age because of their adaptations for fertilization, seed protection, and seed dispersal.
10. The angiosperms are the flowering plants. A flower is a specialized leaf shoot with a variety of roles in reproduction. Four evolutionary trends in flowers are: a reduction in the number of parts, a fusion of flower parts, a shift to a lower ovarian position in the flower, and a change from radial to bilateral symmetry. Existing flowers can represent any stage of evolutionary advancement. The emergence of angiosperms may be related to a change in worldwide climate to cooler and drier conditions.

KEY TERMS

algae (319)
alternation of generations (317)
angiosperm (331)
bryophyte (323)
chlorophyte (322)
coniferata (327)
cycadophyta (327)
division (319)
frond (326)
gametophytic (317)
gingkophyta (327)
gnetophyta (327)
gymnosperm (327)
holdfast (322)
leaf (326)
phaeophyte (321)
pterophyte (326)
rhizoid (326)
rhizome (326)
rhodophyte (320)
root (326)
sporophytic (317)
stalk (322)
thalli (322)

FOR FURTHER THOUGHT

1. A magnolia flower exhibits radial symmetry and has numerous parts that are separate from one another. Is this flower an example of an early or advanced stage of evolution?
2. In the absence of vascular tissue, bryophytes have evolved certain water conservation measures. What are they?

FOR REVIEW

True or false?

1. ___ In alternation of generations, it is the sporophyte that produces gametes.
2. ___ Vascular land plants evolved from the Rhodophyta.
3. ___ Vascular plants are most prevalent in the earth's oceans.
4. ___ The gymnosperms have "naked seeds."
5. ___ Red, brown, and green algae all contain chlorphyll *a*.

Fill in the blank.

6. The giant kelps of the Pacific Ocean are contained in the Division ___ .
7. Three traits shared by chlorophytes and vascular land plants are ___, ___, and ___ .
8. Angiosperms are ___ .
9. Cycads, gingkos, and gnetophytes are all ___ .
10. Ferns have true leaves called ___ .

Choose the best answer.

11. When a male and female gamete join they:
    A. develop into the gametophyte generation
    B. produce haploid gametes
    C. produce diploid gametes
    D. none of the above
12. ___ are characterized by the absence of vascular conducting tissue
    A. Ferns
    B. Bryophytes
    C. Angiosperms
    D. Tracheophytes
13. Ferns are plants that do not:
    A. possess fronds
    B. produce seeds
    C. inhabit moist environments
    D. possess true roots
14. Bryophytes:
    A. contain vascular tissue
    B. possess true leaves
    C. have motile sperm
    D. grow only on trees
15. Stalks and holdfasts are structures found in:
    A. angiosperms
    B. gymnosperms
    C. brown algae
    D. moss

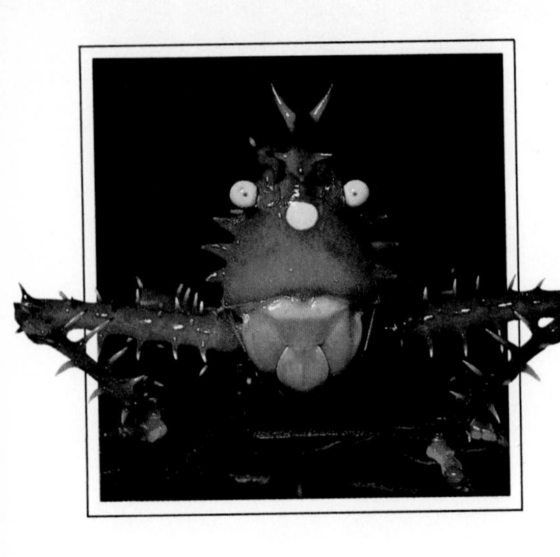

CHAPTER 13

Diversity III: Animals

Overview

The Animal Kingdom
Phylum Porifera
Phylum Cnidaria
Phylum Platyhelminthes
Phylum Nematoda
Phylum Mollusca
Phylum Annelida
Phylum Arthropoda
Phylum Echinodermata
Phylum Hemichordata
Phylum Chordata

Objectives

After reading this chapter you should be able to:

- Name the major phyla and their classes.
- Describe the body plan of each class and distinguish between simple and advanced characteristics.
- Describe two types of body symmetry and relate it to an animal's lifestyle.
- State the evolutionary importance of cephalization, segmentation, and presence of a coelom.
- Name the features common to all chordates.
- Prepare an evolutionary tree that shows the origins of each major phylum.

Here's an easy question: What is an animal? But perhaps by now you won't be taken in by any such "simple" question in biology. And right you are. In fact (and perhaps not surprisingly), there is even strong disagreement among biologists about such questions. For example, many biologists might be willing to say, "An animal eats and has a nervous system." But then we might hear a shout from the back of the room: "What about sponges?" It's true, sponges don't have digestive or nervous systems, so perhaps they should be placed in another kingdom, but who wants to deal with another kingdom? So, sponges uncomfortably reside with elk and grizzly bears in the animal kingdom. Thus, again we find ourselves stretching things a bit, compromising and making exceptions. Let's admit that we have an imperfect system, forge ahead, and see just where animals fit in our five-kingdom scheme, paying particular attention to the evolutionary processes that got them there.

THE ANIMAL KINGDOM

The first animals appeared between one and two billion years ago. We don't know much about them since the earliest ones were small and soft-bodied and didn't fossilize very well. About all they left of themselves were some burrows and tracks for us to ponder, and, as a final insult, a few droppings. But by the early Cambrian, roughly 600 million years ago, hard-bodied **metazoan** (multicellular) species had appeared and we have abundant fossils from that period, although we know almost nothing about the evolutionary processes that produced them.

We will use a simplified scheme to describe the animal kingdom, considering the major phyla (Figure 13.1), a few subphyla, and the major

FIGURE 13.1
The major modern phyla. As a general rule, in this scheme, the phyla that evolved earlier are shown closer to the right. Why are the echinoderms shown on the same branch as the chordates?

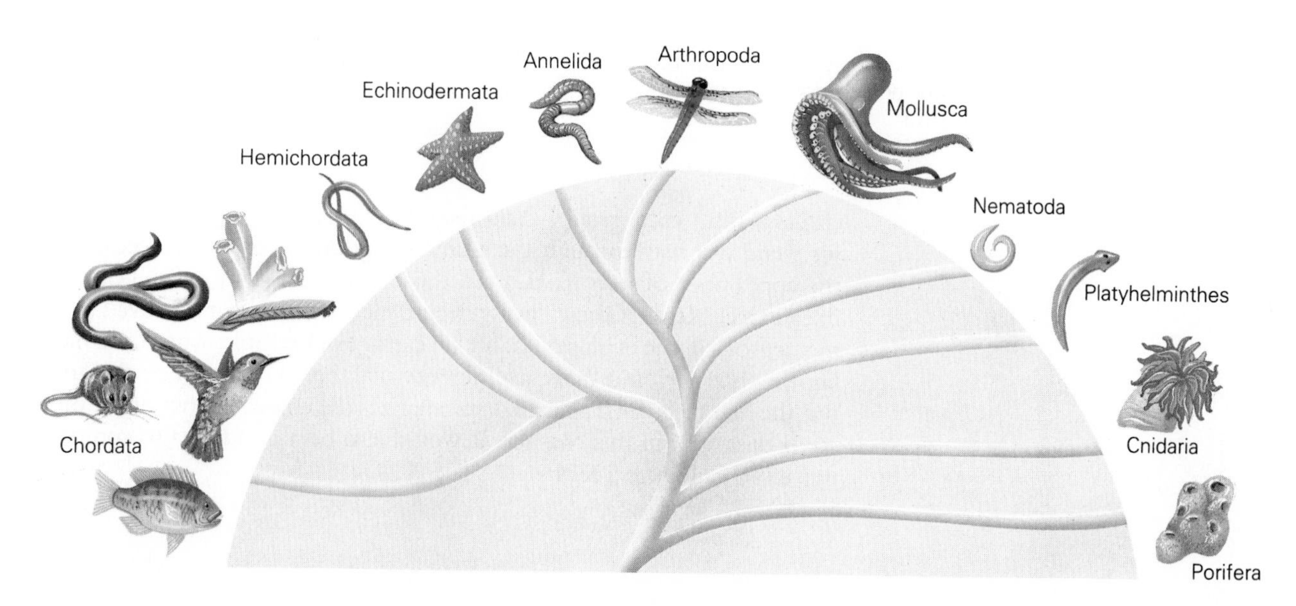

TABLE 13.1
Animal Diversity

PHYLUM	TYPICAL HABITAT	NERVOUS SYSTEM	DIGESTION	RESPIRATION	CIRCULATION	REPRODUCTION
PORIFERA	Salt water; some fresh water	None	Collar cells and trap food; Intracellular digestion	Diffusion across cell membrane	No system	Sexual with some asexual
CNIDARIA	Salt water; some fresh water	Nerve net	Saclike gut; one opening	Diffusion across cell membrane	No system	Sexual and budding
PLATYHELMINTHES	Salt + fresh water; other animals	"Nerve clumps" and nerves	Saclike gut; one opening	Across body surface	Via gut	Sexual and asexual
NEMATODA	Salt + fresh water; other animals; land	Nerve ring + nerve cords	Gut tube; two openings	Across body surface	None	Sexual
MOLLUSCA	Salt + fresh water; land	Brain; ganglia; nerve cords	Gut tube; two openings	Gills	Open; but closed in cephalopods	Sexual
ANNELIDA	Salt + fresh water; land	Ventral nerve cord with enlargements in each segment; "brain"	Gut tube; two openings	Across body surface	Closed; "pumping vessels"	Mostly sexual
ARTHROPODA	Salt + fresh water; land	Brain + ventral nerve cords	Gut tube; two openings	Gills; tracheal tubes; (book lungs)	Open	Sexual
ECHINODERMATA	Salt water	Nerve ring around mouth + radial nerves	Usually gut tube	Across body surface	No system	Sexual
CHORDATA	Salt + fresh water; land	Brain; single hollow dorsal nerve cord	Gut tube; two openings	Lungs + gills	Closed	Sexual

classes within each group. (You may want to refer to Appendix A.) As we wend our way through the animal kingdom, keep an eye peeled for the appearance of new traits, even those that may seem insignificant on the surface. Many of these "insignificant" new developments proved to be momentous in the evolution of life on earth. For example, who could have known that it would make a difference whether the mouth or the anus was the first to form. Table 13.1 summarizes the characteristics of the nine phyla discussed in this chapter. It would also be a good idea to compare this table to Figure 13.14.

FIGURE 13.2

The sponges are delicate animals that are different in so many critical respects from other animals that their placement among the kingdoms has been heavily debated.

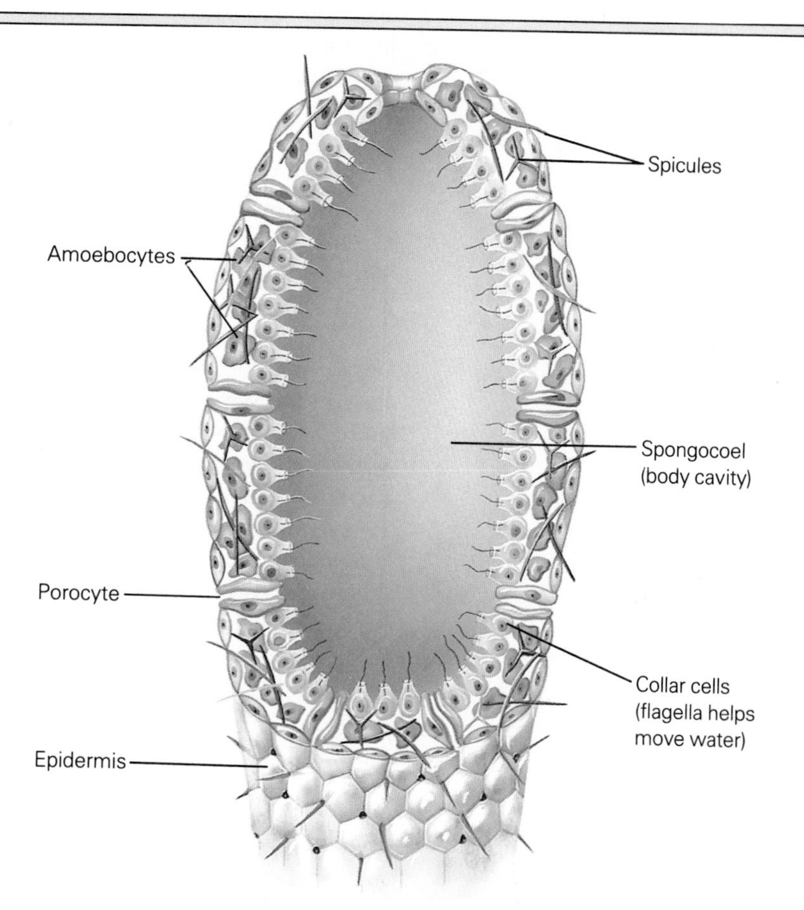

Phylum Porifera: An Evolutionary Dead End

Because they are so different from the other animals, sponges are sometimes called parazoa, which means "beside the animals." It is believed, in fact, that they arose independently from different protistan ancestors than those that gave rise to the other animals.

The **Porifera,** or sponges, are of three classes: calcareous (chalky) sponges, glass sponges, and bath sponges (Figure 13.2). The bodies of simple sponges are like vases with holes in the sides. Water enters these holes and is expelled out the top, having been relieved of some of its oxygen and food particles. The food particles are trapped by flagellated **collar cells,** that may pass them to **amoebocytes,** cells that can creep through the body wall, distributing the nutrients as they go. Water enters through the porocyte and is kept in motion by the flagellated collar cells. The body is supported by spicules (as we will see in Chapter 17). Sponges are extremely primitive animals, as demonstrated by the fact that their bodies can be shredded and passed through a sieve, and the cells will crawl back together and reassemble themselves into a functional animal. Sponges are considered to be evolutionary dead ends in that it is unlikely that their ancestors gave rise to other lines. So you needn't worry about one being in your family tree. In any case, they are successful enough, having survived on the planet for about a billion years.

Phylum Cnidaria and the Evolution of Radial Symmetry

The **Cnidaria** (mercifully, the *C* is silent) are the hydrozoans, jellyfish, sea anemones, and corals. The name *cnidarian* refers to the specialized stinging cells, called cnidocytes, that characterize these animals. These cells are responsible for the sting of a jellyfish and Portuguese man-of-war. They have **radial symmetry** (described in Figure 13.3), tentacles, a saclike gut, and a very primitive nerve net (these are the first organisms we've seen with true nerves, but the nerves are diffuse—they have not yet formed the aggregate at one end that will mark the "head" in other species). Essentially, their bodies are comprised of two layers, an outer **epidermis** (*epi:* outside; *dermis:* skin) one-cell thick and an inner **gastrodermis** (*gastro:* stomach) that lines the saclike gut, the **gastrovascular cavity** (Figure 13.4). A jellylike substance called **mesoglea** separates the two layers.

FIGURE 13.3

In radial symmetry, the body is essentially spherical, disk-shaped, or cylindrical. Any plane drawn through the center produces roughly equal left and right halves. Bilateral symmetry, on the other hand, means that the body can be equally divided by one plane only. This division produces mirror-image left and right halves that contain similar structures. Perfect examples of radial or bilateral symmetry are unusual. Can you think of structures in your own body that are asymmetric—that is, neither paired nor composed of mirror-image right and left halves?

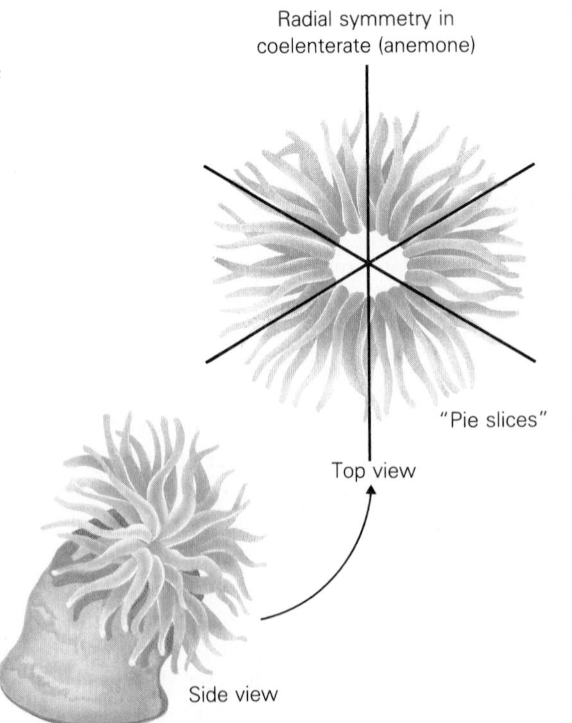

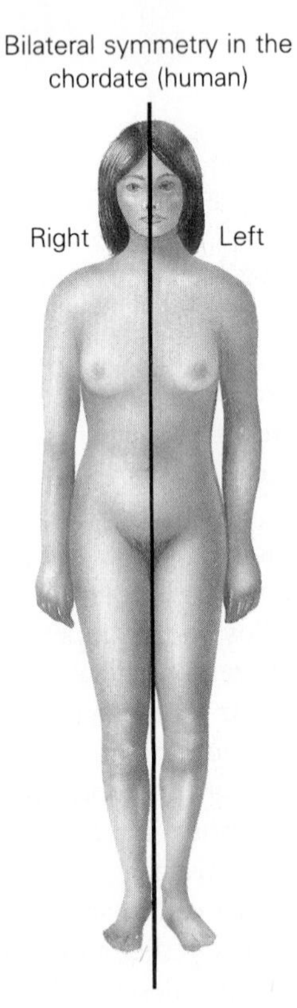

FIGURE 13.4

The *Hydra* is a cnidarian with only two cell layers and a radially symmetric body plan.

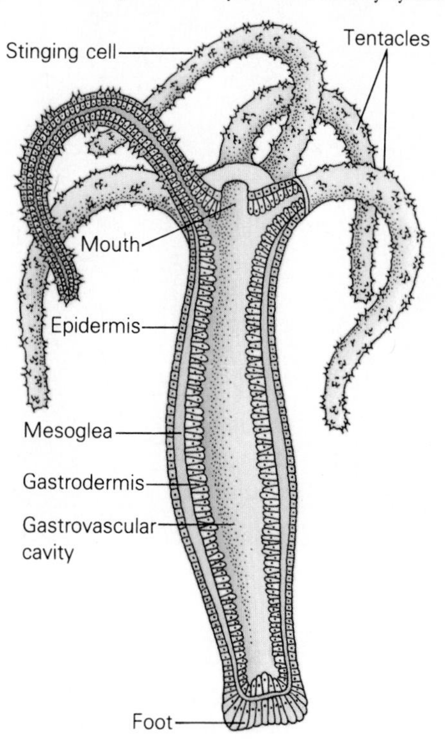

Some cnidarians have two distinct body types, and in the course of their existence, they may change from one phase to another. One body type, the **polyp,** is vaselike. The other form is bell-like and is called the **medusa.** The medusa is free swimming. The sedentary polyp stage predominates in the members of class *Hydrozoa,* examples of which include the *Hydra* (which incidentally, has *only* a polyp stage). The medusa is the predominate form in class *Scyphozoa* (the jellyfish) (Figure 13.5, next page). In the class *Anthozoa,* we find the sea anemones and corals (Figures 13.6 and 13.7, next page). Corals secrete hard, limy structures and come in many sizes, shapes, and colors.

Cnidarians have a most interesting life cycle that suggests the next crucial step in evolution—a change in body symmetry. The suggestion is based on the very different appearances of the adult and larval cnidarian. The larva is called a **planula,** and it has bilateral symmetry and resembles a flattened mass of cells covered by cilia. It crawls about on the bottom of its watery habitat until the time comes for its next stage. Then a most peculiar thing happens. It stands on one end and its body slowly begins to change. The free end becomes cuplike as tentacles develop around the rim of the forming cup. It is soon apparent that the little flattened planula is developing into a polyp with radial symmetry. This phase is very brief

FIGURE 13.5

Jellyfish can be among the most compelling sights of the sea. Their delicate bodies pulsate, ghostlike, through the depths as they trail their deadly tentacles.

FIGURE 13.6

A coral reef is a beautiful and exciting place. Corals are polyp forms, similar to the anemones, but generally smaller. They can be very diverse, as we see here.

FIGURE 13.7

The anemone has a large, fleshy body that is easily distinguished from the small polyps of hydrozoans. Many anemones are colorful, resembling flowers, but their beautiful tentacles may spell death for small animals that touch them.

in some groups, but very extended in others (Figure 13.8). The polyp stage begins to form layers of tentacled cups. The outer layers break away to become male and female medusae. When their eggs and sperm join, a new planula is produced.

FIGURE 13.8

The planula is the larval stage in the life cycle of the cnidaria. The planula usually has an outer layer of ciliated cells surrounding a rather compact mass of cells. It has a very simple nervous system, with most neurons aggregated at the anterior end. It lacks a mouth and must depend on whatever food it has stored. It attaches to some substrate and forms the polyp stage. Medusas are formed as cuplike buds break away from the polyp stage.

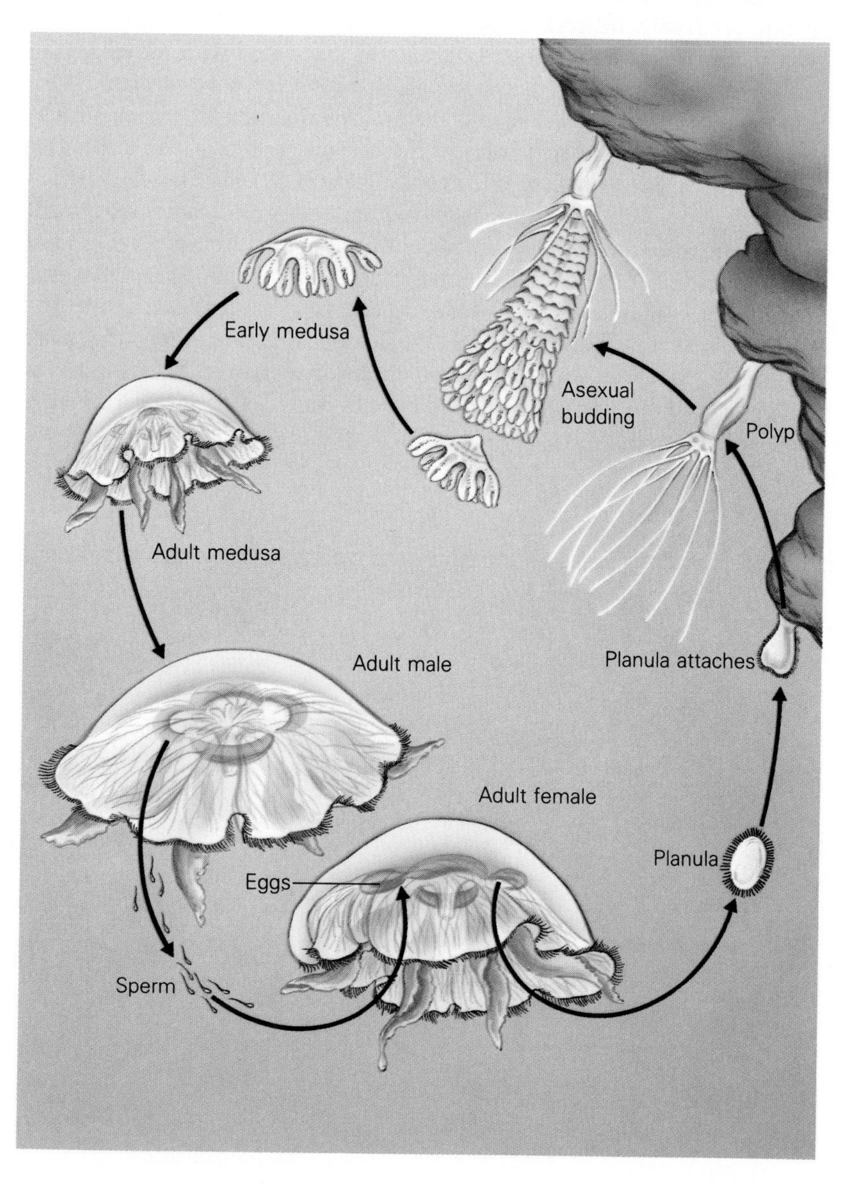

The evolutionary significance of the little planula is suggested by the fact that it looks a great deal like the next group we will discuss, the bilateral flatworms. Evolutionists have wondered what would happen if certain mutations caused the planula to continue in its wanderings, never to stand on end, never to become a "vase"? What if such an "eternal" planula should develop the ability to reproduce? Is this how the flatworms began?

Phylum Platyhelminthes and the Evolution of Heads and Sides

The **platyhelminths** are the flatworms. These are the first group we've encountered with bilateral symmetry as adults. The evolution of such a body plan would be important because it is associated with movement in a single direction; that is, with animals that tend to move forward. With forward movement would have come an accumulation of sensory mechanisms and integrating structures of the nervous system in the leading or "head" end, a development called **cephalization** (*cephalo:* head). Cephalization at any level, it is thought, could have triggered the development of a primitive brain. Such senses would have helped in discovering food and in avoiding danger in the environment into which the animal was moving. (It wouldn't do to go around backing in to new environments.) As animals swam, crawled, or glided in one direction, then, the stage would have been set for the next organizational change. One surface of the animal (the belly surface) would have become specialized for moving over the habitat, or perhaps feeding. The opposite surface (the back) would

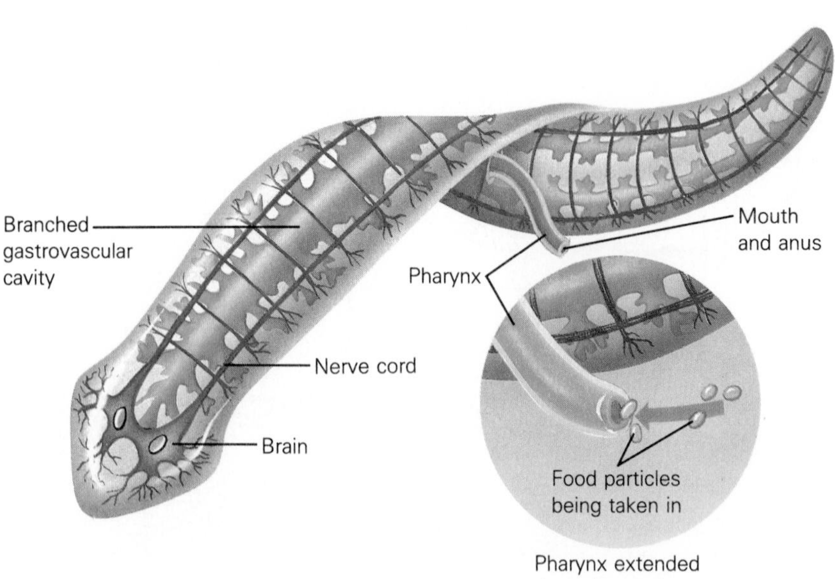

FIGURE 13.9
The planarian flatworm, with its extensive branched gastrovascular cavity, is able to evert its pharynx to ingest food.

have then become specialized perhaps for camouflage or defense. The two sides would not have had to differ much from each other, since they would have experienced much the same thing, and so they would have developed the organizational similarities that led to bilateral symmetry, or two-sidedness.

The flatworm nervous system shows cephalization and bilateral symmetry in that it has a clump of nerves at the head end and two large nerves running parallel down the length of the animal, linked by crossbridges. The muscular system is also well developed, the first we've seen. Flatworms live in both fresh and salt water, and as parasites. Their muscles, nerves, gonads, and digestive tract are specialized enough to contain true organs.* This is the first group we've encountered that has the three fundamental embryonic tissues (or "germ layers") as discussed in Chapter 16.

The best known flatworms are the planaria, free-swimming species of the class *Turbellaria.* Planaria are common experimental animals because they can regenerate after being mutilated and because they can learn simple tasks, such as turning a certain direction in response to a stimulus. The planarian is unlike some flatworms in that it has a protrusible ("stick-outable") pharynx that can be everted to encompass food outside the animal and draw it back into its gut (Figure 13.9). The saclike gut (gastrovascular cavity) is so extensive that it also serves to distribute digested materials to the cells.

Members of the class *Trematoda,* the flukes, are a grisly group of parasites that infect the various organs of many animals—including humans (Figure 13.10). They attach to these organs by suckers. The internal systems of trematodes are well developed and include the reproductive organs of both sexes. Their life cycles may be extremely complex, involving a number of hosts.

The class *Cestoda* includes the tapeworms. They are textbook examples of evolution moving toward simplification rather than complexity. With each simplification, they foreclosed new evolutionary options, so their ancestral line probably gave rise to no other animal today.

The tapeworm head, or **scolex,** anchors the worm in the gut of its host by a group of suckers and hooks (Figure 13.11). The body segments, or **proglottids,** contain both male and female reproductive structures. Mature proglottids are little more than increasingly swollen sacs of eggs that break off and pass out with the host's feces. Tapeworms lack a mouth and digestive tract, but these are hardly necessary since they live in a sea of digested food; they can simply absorb nutrients through their body walls. Tapeworms grow quite long and can rob the host of enough nutrients to cause extreme weakness. A common human tapeworm can grow to be over 5 meters long, but the fish tapeworm (which can also infect humans) can reach 20 meters in length!

Unlike the tapeworms, the other flatworms are evolutionarily important due to their remarkable success in giving rise to new species. In fact, long ago, an ancestral form diverged into two lines that would produce

*Animals have different "levels of organization." Some have achieved only cellular organization, whereas others have reached tissue, organ, and organ system levels. Tissues are organized groups of cells, organs are organized groups of tissues, and organ systems are organized groups of organs.

FIGURE 13.10

The liver fluke, a trematode, is a parasitic flatworm with rather well-developed systems.

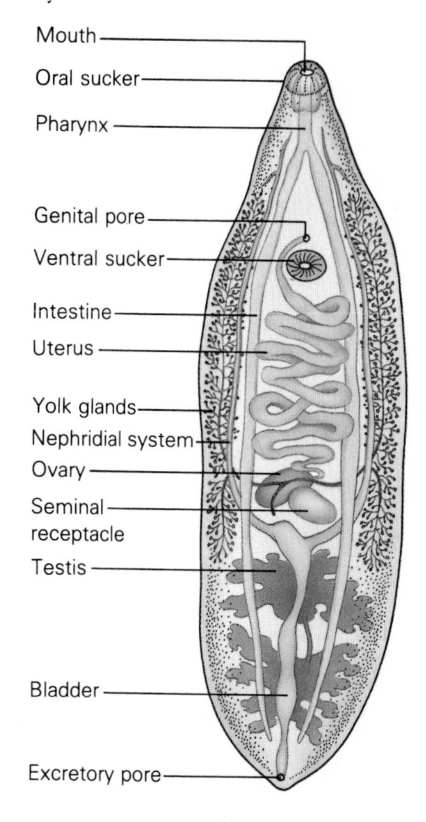

FIGURE 13.11

The head, or scolex, of a tapeworm is crowned with a circle of hooks by which it imbeds in the intestinal wall of its host. It then begins to form a long chain of segments called proglottids, each a veritable sack of eggs.

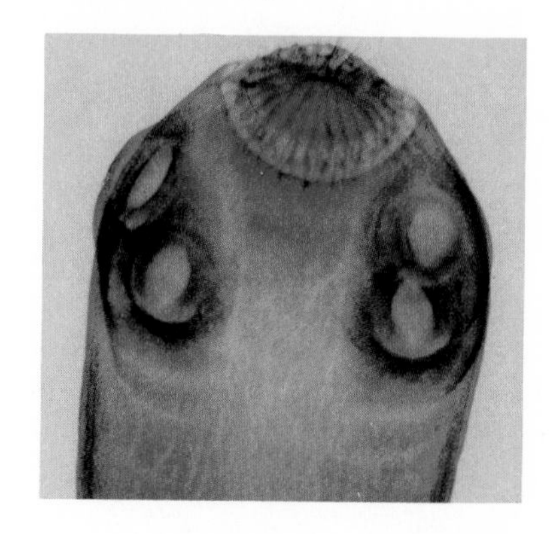

quite distinct groups of animals. The two great lines formed from a seemingly small difference. In one group, the first embryonic opening to form became the mouth; in the other, it became the anus. In today's **protostomes,** the first embryonic opening becomes the mouth; the anus forms later. This group includes the annelids, arthropods, and mollusks (see Figure 13.1). In the **deuterostomes** of today, the first opening forms the anus, the mouth appearing later. This group includes the echinoderms and the chordates. It is because of embryological evidence such as this that we place the humble echinoderms, such as the starfish, closer to ourselves than we do the highly intelligent mollusks or the very complex social arthropods.

Phylum Nematoda and the Evolution of a Body Cavity

The **nematodes,** or roundworms, are among the most startling of all living things in many ways. You may wonder how a roundworm could be startling—until you learn that you may eat thousands each day. Roundworms live practically everywhere, and those who study them are often given to

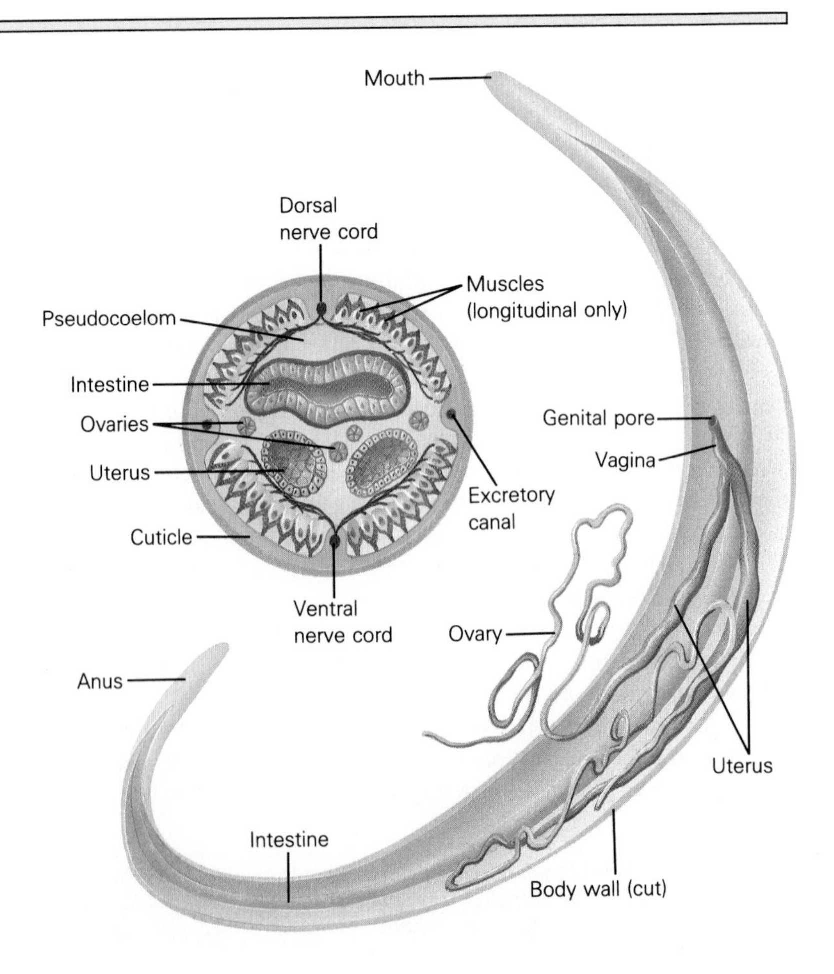

FIGURE 13.12

The body plan of a nematode, the female *Ascaris*. The body cavity is incompletely lined with mesoderm. Having only longitudinal muscles, the worm moves by thrashing back and forth.

FIGURE 13.13

The *Trichina,* or pork roundworm, can be transmitted to humans by eating pork that has not been properly cooked, as we can see here. Most roundworms, however, are harmless and live on decaying matter.

obsessive cleanliness, washing their food meticulously. Roundworms are distinctive in that they are the first group we've considered with a *complete* (tubular) *gut*—one in which food goes in one end and the waste goes out the other (decidedly more civilized than a saclike *incomplete* gut, the gastrovascular cavity of cnidarians and flatworms, that has a single opening through which food enters and undigested material exits).

With a tubular gut running through an elongated body, another important evolutionary advancement is achieved. Between the gut and the body wall now lies a body cavity (Figure 13.12). This general plan will mark the rest of the animals we will consider. In the nematodes and only a few other groups, the cavity is called a **pseudocoelom** (*pseudo:* false; *coel:* cavity). It gets this slightly derogatory term simply because the lining of the cavity is not entirely derived from an embryonic tissue called *mesoderm* (discussed in Chapter 16).

Most roundworms live harmlessly on decaying matter, but the best known roundworms are human parasites, such as *Trichina*—the worms that infect pork and then are passed to humans who eat incompletely cooked meat (Figure 13.13). Other notorious roundworms can cause monstrous deformities such as elephantiasis (resulting in blocked lymphatic ducts that cause grotesque swelling of the affected parts of the body). Another species, about a full meter long, winds its way through human tissue, pressing and squeezing its way through and between the cells of the body and sometimes causing great pain. Treatment has traditionally involved pulling this worm out through the skin by slowly winding it around a stick. Interestingly, although the roundworms have been remarkably successful in their own right, they haven't diverged much over the course of their evolution. That is, they haven't given rise to important new groups of animals, as did the early flatworms.

Phylum Mollusca and the Evolution of the Coelom

The **mollusks** are in the evolutionary line with earthworms and insects, since they are all protostomes. They did not give rise to the other protostomes, but they shared a common ancestor. In addition, they give us our first look at the **coelom.** The coelom is the cavity between the gut

FIGURE 13.14

An evolutionary tree showing hypothetical descendancy. The earliest animals are believed to have been protistlike. From these arose the radial creatures, then the bilateral. Coeloms appeared later. Note the strong division of the coelomates—the two groups form differently as embryos, as we will see in Chapter 16.

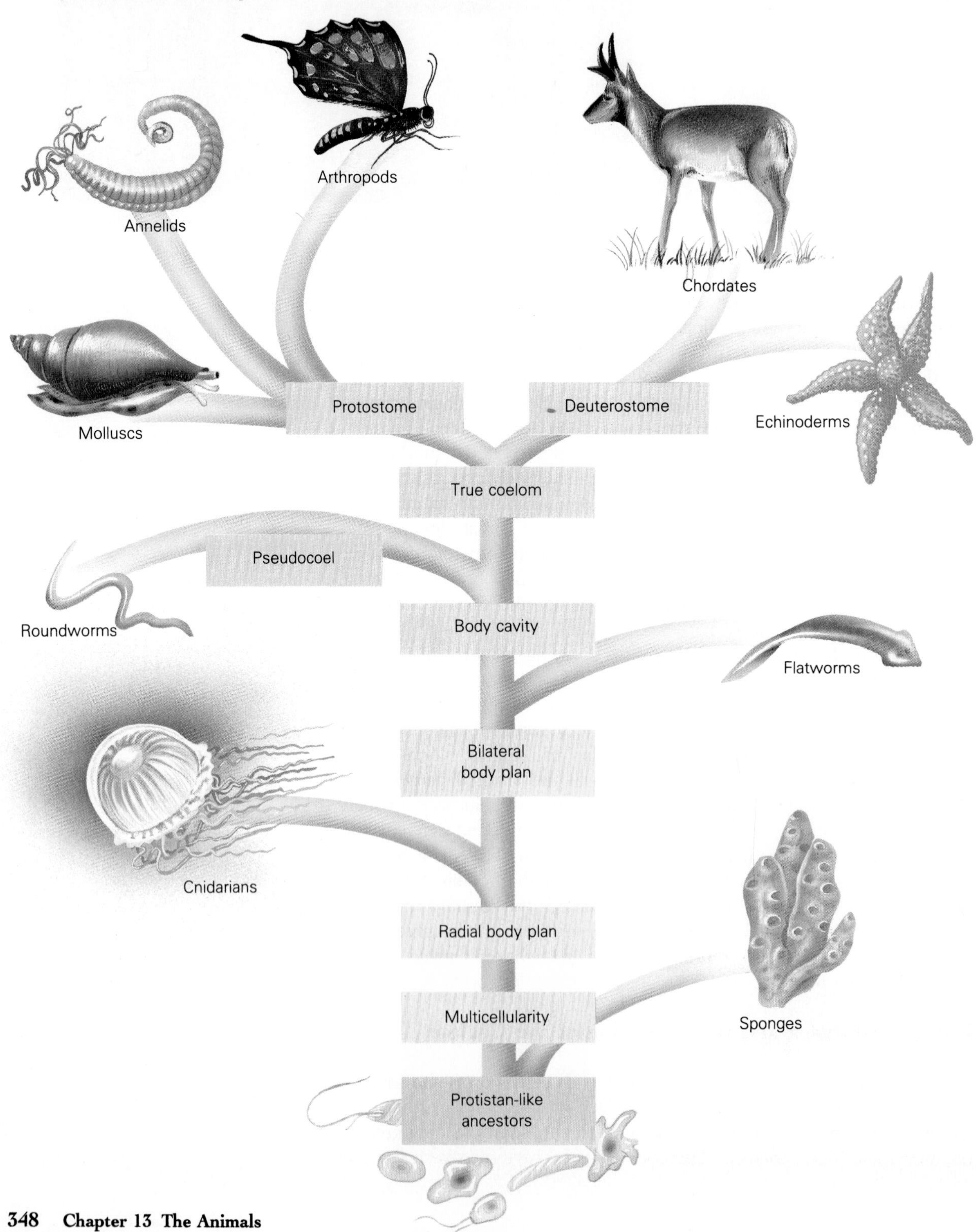

FIGURE 13.15

The internal structure of the clam. Clams, like other bivalve (two-shelled) mollusks, are filter feeders that strain food particles from water flowing across their gills. Note the relationship of the various systems and compare them to what you know about vertebrates. Do you see anything unusual about the relationship of the heart and gut? What part do we eat?

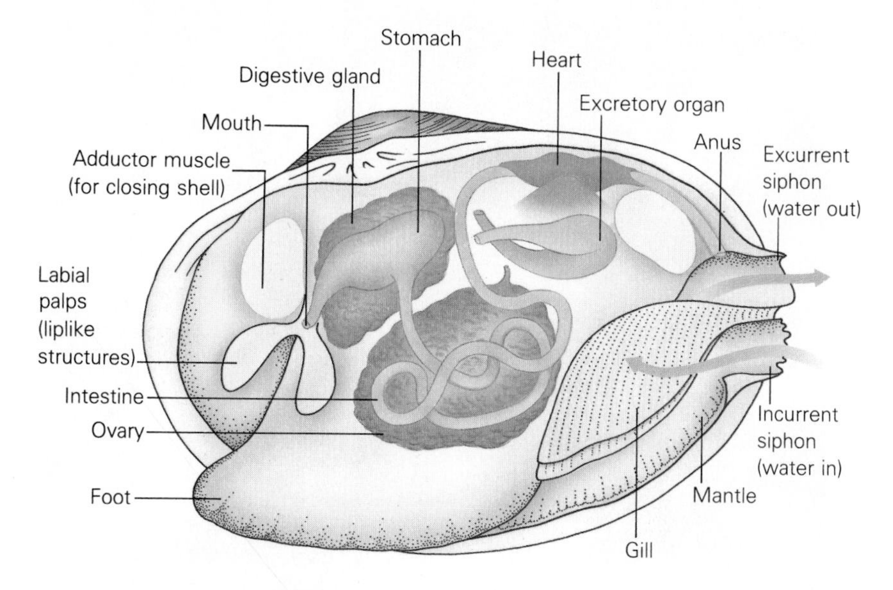

and the body wall, entirely lined with mesoderm. It gives the internal organs a certain freedom of movement and it provides an area into which other anatomical structures can protrude. For example, the liver is largely a specialized extension and elaboration of the gut that protrudes into the coelom. Figure 13.14 shows the major evolutionary developments in animals.

Mollusks are named for their soft bodies, although some are encased in hard shells and others have stiff shells inside their bodies (parakeets like to sharpen their bills on cuttlebones, the inner skeletons of cuttlefish). There are about 100,000 species of mollusks, and they first appeared about 500 million years ago.

All mollusks have a mass of muscle on one side of the body (called the **foot**), a fold of tissue over the body (called a **mantle**), and some (such as snails) have a rasplike, filing tongue (called a **radula**). They have well-developed systems (Figure 13.15). The various species may differ greatly in size. Some mollusks are almost microscopic, while one clam, *Tridacna,* may be a meter wide and weigh several hundred pounds (Figure 13.16). And let's not forget the giant squids of the North Pacific that have been known to attack small boats but that mainly terrorize people in movie theaters (Figure 13.17). In spite of the size of such giants, mollusks probably have more reason to be afraid of us than we do of them. After all, we lustily devour all sorts of oysters, clams, mussels, snails, squids, octopuses, and periwinkles. (Think back to your last periwinkle.)

FIGURE 13.16

The *Tridacna,* or giant clam, is a most beautiful and awesome creature when seen underwater. The mantle is a lovely, undulating blue. Divers have died by reaching a hand inside, only to have the massive shell close protectively.

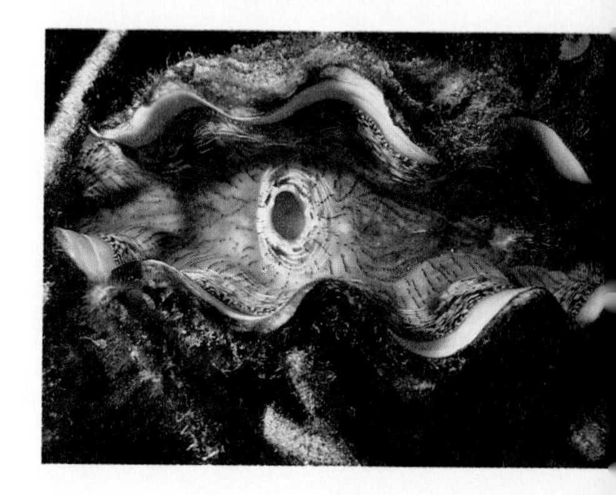

FIGURE 13.17
Little is known about giant squids, but they grow to great lengths and are known to be predators.

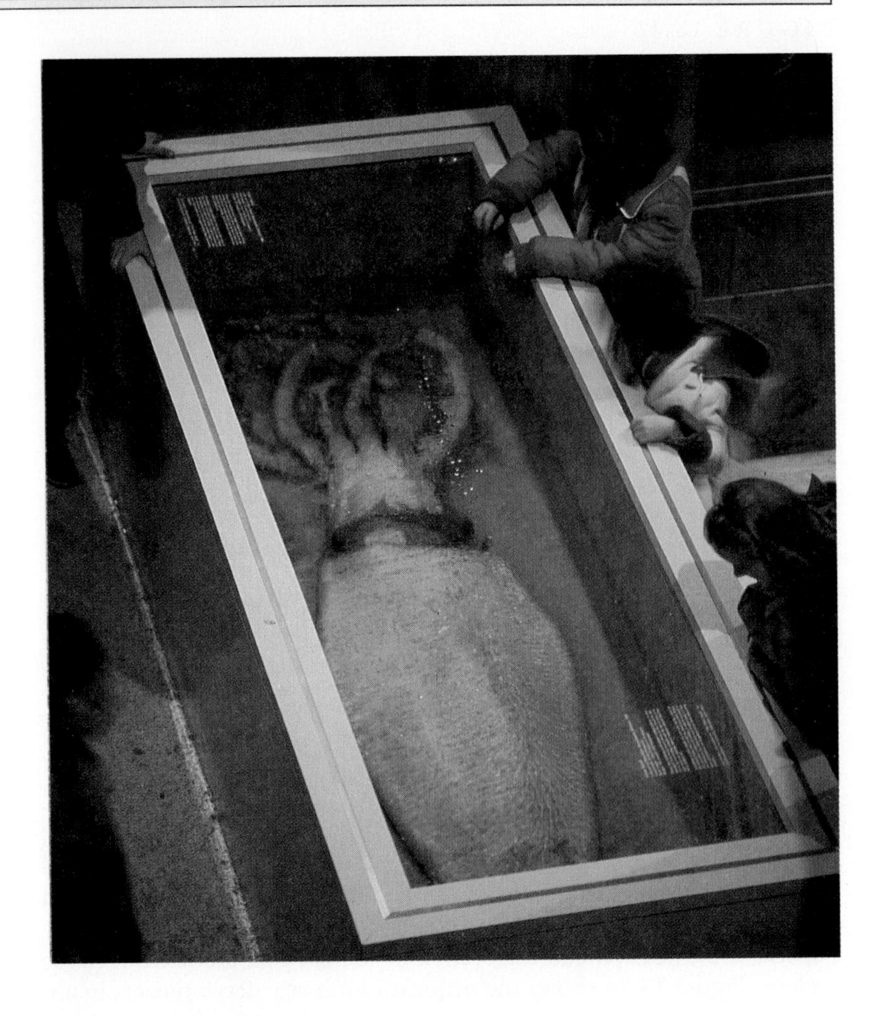

Phylum Annelida and the Evolution of Repeating Segments

With the **annelids,** we come to the truly segmented worms, the most familiar being the earthworms. The repeating units of segmented animals were an important evolutionary advancement (because of the specializations they made possible, as we will see in the next group). The annelids have a closed circulatory system; that is, the blood always remains within blood vessels. Thus, food from the gut does not have to diffuse through the coelomic fluids. Instead it can be carried by blood vessels, a much more efficient arrangement. The circulatory system set the stage for the development of larger bodies in which oxygen and nutrients could be distributed and waste removed rapidly. Five pairs of aortic arches extend from the dorsal (upper) vessel to the ventral (lower) vessel. The aortic arches serve as hearts, pumping the blood through the closed system of vessels.

Typical of many invertebrates, the annelids have a ventral nerve cord with enlargements in each segment. Two nerves encircle the beginning of the digestive system and join on the upper surface to form a "brain." Earthworms are in the class *Oligochaeta* ("few hairs") (Figure 13.18). They have bristlelike **setae** that help them to crawl through the earth and to brace against the tugging robin. There are some earthworms, though, that a robin might hesitate to tangle with, such as the 4-meter giants that rumble and gurgle their way through the Australian soil (Figure 13.19).

Earthworms reproduce in an interesting fashion. Although each worm contains both testes and ovaries, they do not self-fertilize. Instead, they exchange sperm by lying head to tail. Each worm releases sperm that swim into the special **seminal receptacle** of its friend. Later the ring, or **clitellum**

FIGURE 13.18

The body plan of an earthworm. The blood remains enclosed in vessels and is moved along primarily by a pulsating dorsal blood vessel. The digestive system is complete; that is, it includes a mouth and anus. Both circular and longitudinal muscles enable complex movements.

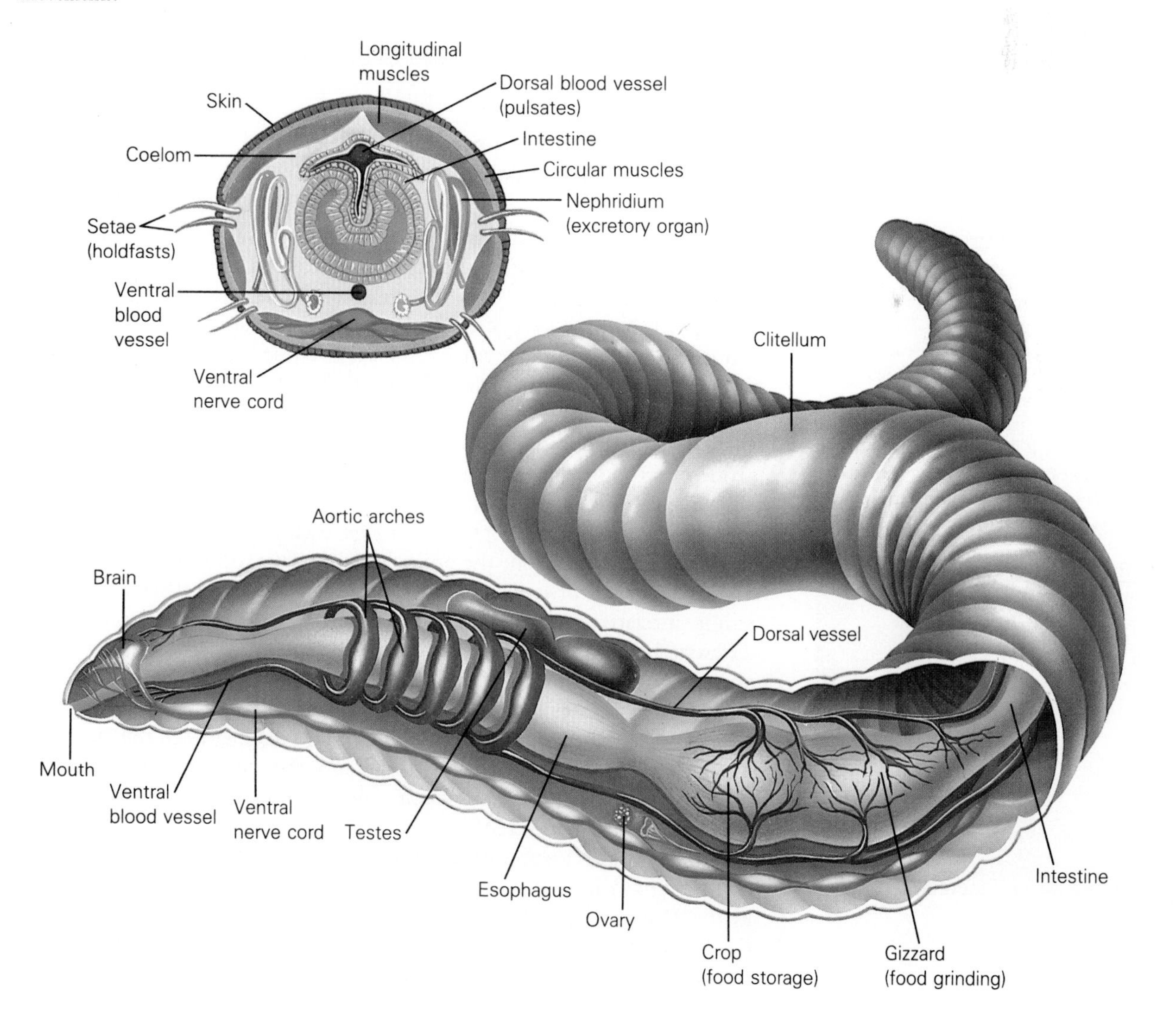

FIGURE 13.19
A giant earthworm (left).

(the band that you've seen around earthworms), secretes a layer of slime that begins to move along the worm. As it passes the egg-producing cells, the eggs move into it and are joined by sperm when the ring passes the pores of the seminal receptacles. The earthworm finally shrugs off its slimy ring, and the ends immediately seal. The "cocoon" then lies in the soil while the embryos develop inside its protective coat.

Another class, called *Polychaeta,* is interesting in that the segments of these worms are more specialized than those of the earthworm. The polychaetes have "legs," actually **parapodia,** that are used in both swimming and respiration, and they have heads with eyes, jaws, and tentacles (Figure 13.20). And they can bite.

The class *Hirudinea* includes the leeches, another group of beauties. Leeches, of course, suck blood. They do this by attaching with a suckerlike head that harbors three sharp teeth. After they attach, they inject an anticoagulant to keep the blood from clotting and then they suck until they are bloated, or until you discover them. After you regain consciousness, you may choose to interrupt their activities. Leeches were once used to reduce blood pressure and to take color out of the black eyes of the doctors who prescribed them. (In fact, the word "leech" originally referred to the physician, but came to be attached to the animal as well.)

Phylum Arthropoda and the Evolution of Specialized Segments

FIGURE 13.20
Polychaetes are segmented worms with rather marked specialization. They have well-developed heads and appendages on their segments. Whereas oligochaetes, such as earthworms, have only a few bristles per segment, polychaetes have many bristles. Polychaetes live almost entirely in salt water and eat a variety of foods, from algae and detritus to small animals.

Some have called the present period in history the age of man, but it might just as well be referred to as the age of **Arthropoda.** Members of this phylum have jointed legs and an exoskeleton (*exo:* outer). The hardened exoskeleton (made of chitin, see Chapter 3) functions in both protection and as an attachment for muscles. These creatures are abundant, diverse, and successful, indeed. The arthropods include millipedes, centipedes, crustaceans, and, most important, insects (Figure 13.21).

The arthropods are evolutionarily important because they are the first group to strongly capitalize on the segmented body plan. They do this by allowing various segments to specialize for different roles. Thus we find segments with antennae, or claws, or legs, each performing different tasks. With such specialization, the animals can behave more flexibly and meet a greater variety of environmental challenges. The specialized segments require a relatively complex nervous system to coordinate the activities of different body parts. There are two subphyla, *Chelicerata* ("claw horn ones") and *Mandibulata* ("jawed ones").

The chelicerates have clawlike fangs as their first appendages, no antennae, and the head and thorax are fused and distinct from the abdomen. Most species have four pairs of legs and they are represented by the horseshoe crabs (class *Merostomata*) and the spiders and scorpions (class *Arachnida*) (Figure 13.22).

The first appendages of the mandibulates are modified into antennae and they have jaws that work from side to side. There are more species in the Mandibulata than in any other group of any phylum. The class *Crustacea* includes the shrimps, crabs, lobsters, barnacles, and countless smaller creatures, most of which live in the water. The segmentation of

FIGURE 13.21

Arthropod diversity is remarkable, as we see in the crab spider (a), the painted grasshopper (b), the Hawaiian lobster (c), and the purple shore crab (d).

FIGURE 13.22

Tiny spiders have been known to strike terror in the hearts of the bravest men. It is probably no great consolation to know that only about 30 species are dangerous to humans. Like insects, they have tracheal tubes for breathing. A spider's respiratory surface is increased by an area of flaplike tissue called a book lung. After mating (when the female may capture and devour her suitor), the sperm are stored in a receptacle until eggs are released from the ovary. The chelicera (fangs) are the modified first appendages.

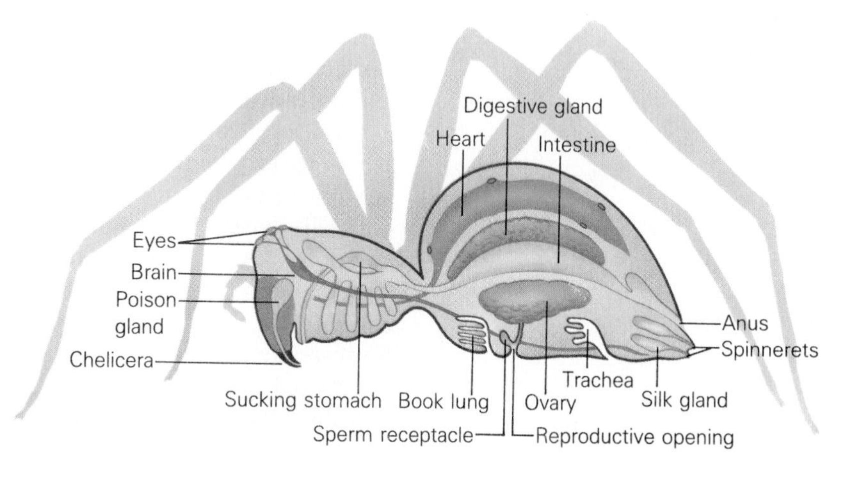

FIGURE 13.23

The anatomy of a male lobster. Blood travels only partway in vessels, then percolates through tiny openings in the tissue. In mating, a male forcibly turns over a female, inserting the first abdominal appendages into special grooves on the sides of her body, and filling her receptacle with sperm. Fertilization occurs later, whenever she releases the eggs. The offspring will develop after they become entwined in the swimmerets.

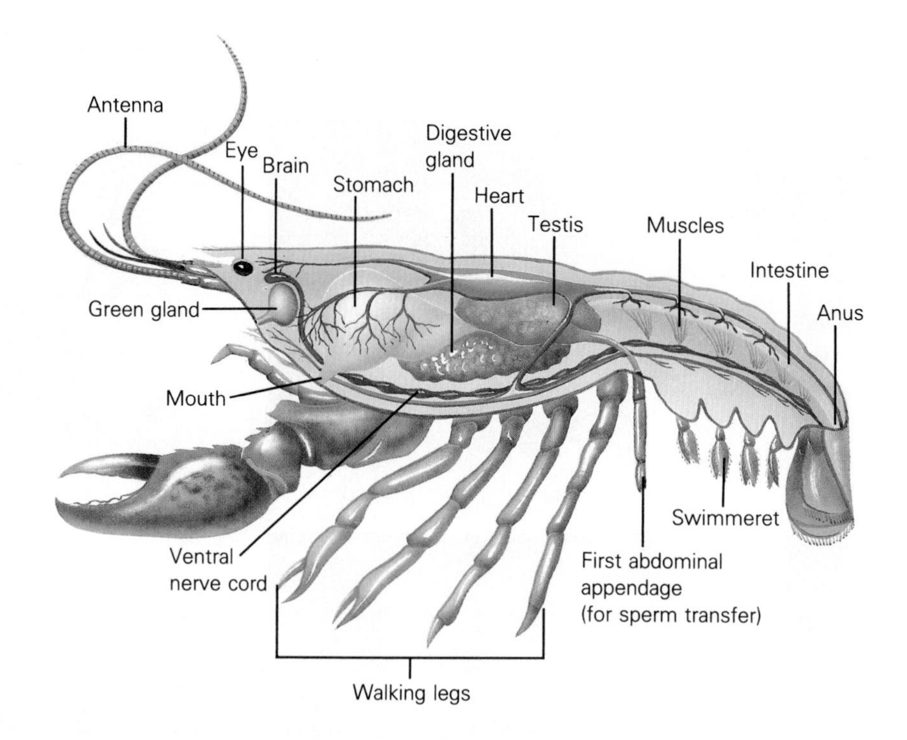

the arthropods may not be obvious, partly because much of the body is covered with a continuous, smooth chitinous shell (Figure 13.23). The classes *Chilopoda* and *Diplopoda* include the centipedes and millipedes, respectively. They are not economically important, except for tropical hotels losing an occasional guest when one crawls up the wall. The centipedes may have a mildly painful sting. Most species scavenge plant food, although a few will go after bedbugs and roaches. The centipedes have one pair of legs per segment, while the harmless millipedes have two (Figure 13.24).

Perhaps the most diverse and impressive class on earth is the *Insecta*. You probably won't be startled to learn that these are the insects. The planet is literally crawling with insects and so far about 900,000 species have been described, with more being added to the list almost daily. Perhaps you are one of those people who reflexively squashes any such small creature. If so, you may have slaughtered a member of some beautiful and unknown species "because it was a bug."

The astute observer may have noticed that not all insects are harmless. If you haven't noticed, go lie down on a grassy dune; the bite of some ants can be a religious experience. (In the Amazon, the sting of the Konga ant is like someone gripping your skin with pliers and twisting hard for two hours.) Also, insects pester and parasitize our livestock, eat our crops, walk on our food when we don't know where they've been, carry parasites and diseases, and threaten us in countless other ways. At the same time, however, they provide us with honey and silk and help pollinate our crops. (In the representative insect body in Figure 13.25, we see a head attached to a thick **thorax,** followed by a segmented **abdomen.**)

FIGURE 13.24

Centipedes (a) have only one pair of legs on each segment. Some species are quick and relentless hunters. Millipedes (b) have two pairs of legs per segment and are harmless herbivores.

(a)

(b)

FIGURE 13.25

The insect body plan (represented by the grasshopper). Note the breathing system. Air enters the body through the spiracles and diffuses through finely branching tubules, dead-ending in the tissues, as we will see in Chapter 19. How does this system differ from that of mammals? What aspects of the insect body are similar to those of mammals? How could such similarities have evolved? What are the major, general differences in insects and mammals?

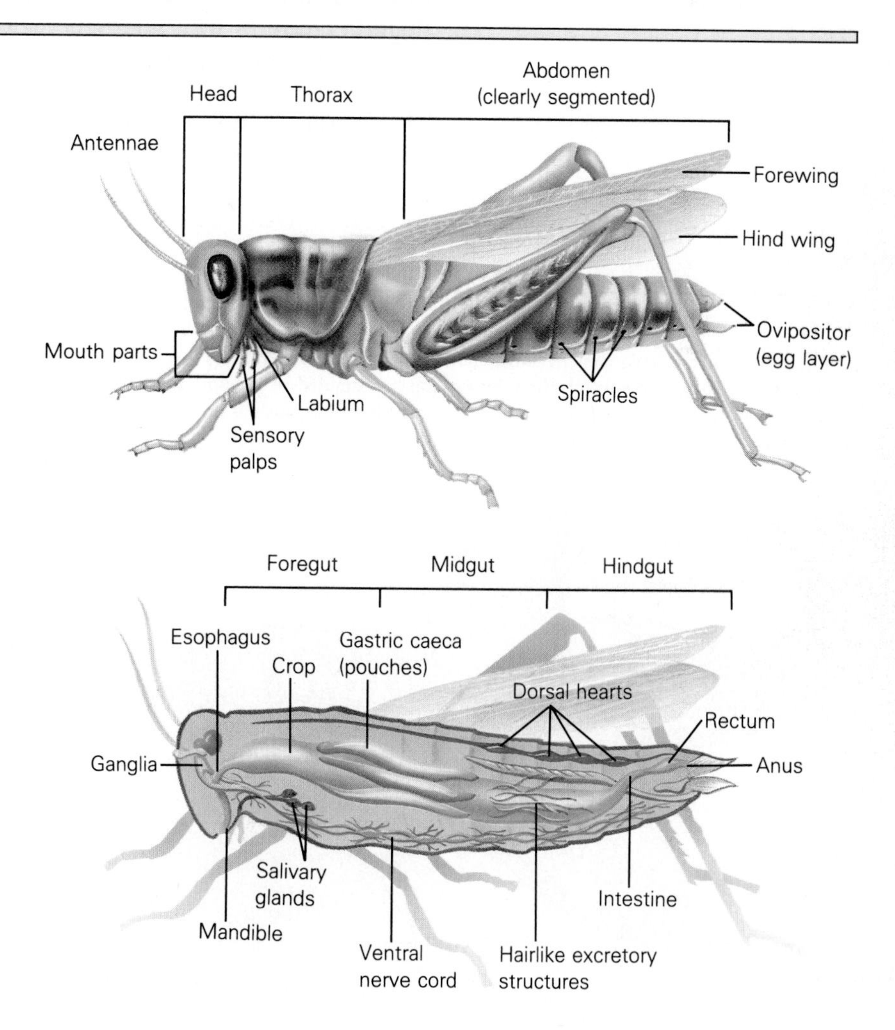

Phylum Echinodermata and an Evolutionary Puzzle

The **echinoderms** ("spiny skin") share the deuterostome line with our own group, the chordates. There are five living classes of echinoderms: The *Crinoidea* (primitive stalked sea lilies), the *Asteroidea* (sea stars), the *Ophiuroidea* (brittle stars), the *Echinoidea* (sea urchins and sand dollars), and the *Holothuroidea* (those great, limp sacs called sea cucumbers).

The evolution of the echinoderms is interesting because they apparently once developed a longitudinal and bilateral body plan and then reverted back to the more primitive radial plan, at least in the adult stage. Their bilateral heritage is today evident only in their free-swimming, ciliated larvae. Perhaps some ancestral form was forced, by competition or predation, back to the two-dimensional sea floor where the ability to move equally well in any direction was advantageous and the radial plan thus became superimposed on a bilateral one.

Whereas echinoderms seem to wear their skeleton on the outside, like the insects, their supportive skeletal elements are actually embedded just under the skin. They also have an unusual feature called a **water vascular system,** which is a kind of hydraulic pump that extends the soft, pouchlike tube feet, with their terminal suckers (Figure 13.26). Echinoderms are sluggish creatures with poorly developed nervous systems, but they are efficient and relentless predators. A sea star attacking an oyster wraps its arms around its prey and pulls tenaciously until the shell opens just a bit. Then it everts its stomach, squeezing it between the shell halves, and digests the oyster in its own shell, drawing the products into its own body.

FIGURE 13.26

The peculiar water vascular system of the starfish. Water enters and leaves through the sieve plate. By shifting the water, the tube feet can be extended by pressure from the inside. When the tube feet are pressed against something solid, such as an oyster shell, strong suction can be developed as water is moved from the foot.

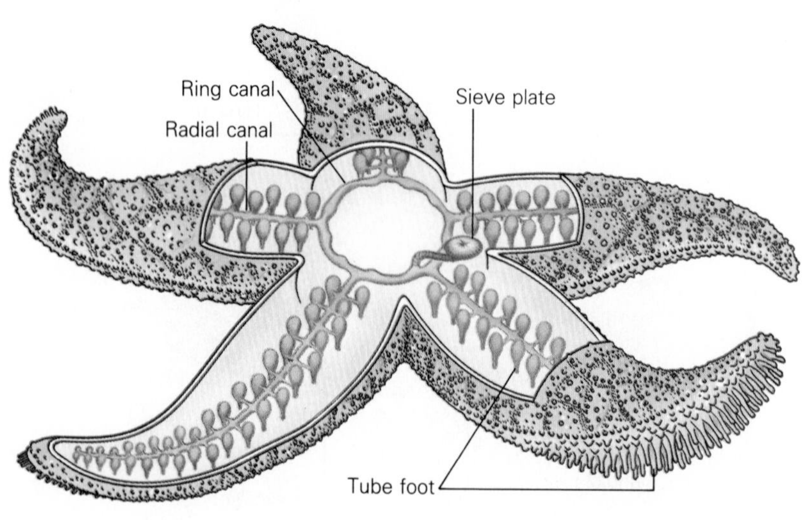

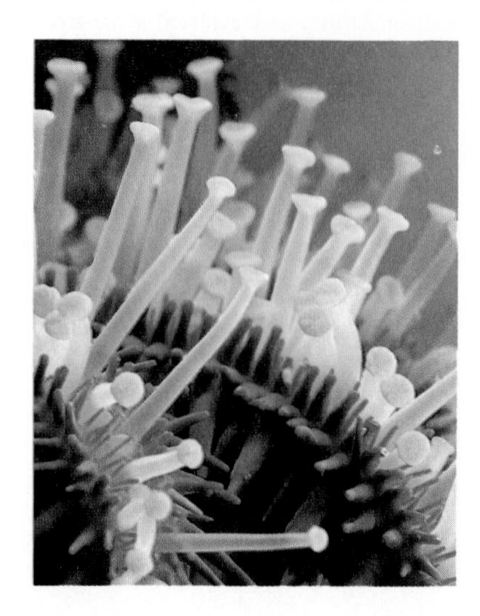

Phylum Hemichordata and a Half Step Along

Aside from a few biologists, not many people can muster much interest in the **hemichordates** ("half chordate"). Most are marine worms of the most uninspiring sort, and no one is likely to ask a hemichordate researcher how the day went. So why are biologists interested? Because the hemichordates occupy an important position in the evolutionary development of the vertebrates. Specifically, their embryonic development and a few fundamental aspects of their anatomy indicate that they are distantly related to the backboned animals. The relationship is revealed by the presence of gill slits and a hollow, dorsal nerve cord. They also have a stiff rod that acts as a supporting structure similar to the notochord of the chordates that we will see next.

Phylum Chordata and the Advent of Freeways

The **chordates** are named for possessing, at some stage, a **notochord,** or rigid, cellular rod covered with two layers of strengthening fibers. The chordates are quite a varied group, ranging from slimy little sea creatures to Las Vegas gamblers, from chipping sparrows to blue whales, from dingos to dugongs. Their evolutionary relationship is suggested in Figure 13.27 on the following page.

All chordates have three basic features in common. At some time in their lives they all have (1) a notochord, (2) a dorsal, hollow nerve cord that enlarges at one end to form the brain, (3) gill slits, which are openings in the throat region that lead from the digestive canal to the outside. Gill slits are used for respiration in fishes, but in air breathers, they disappear early in embryonic development. Chordates also generally share a few less distinctive traits. For example, there is almost always a tail that extends beyond the anus at some stage. (Some children are born each year with rudimentary tails that do not fit the expectations of their parents and the tails are quickly removed.) Chordates are also bilaterally symmetrical, with distinct heads, some segmentation, three embryonic germ layers (Chapter 15), and a well-developed coelom.

The three chordate subphyla include the *Urochordata,* the *Cephalochordata,* and the *Vertebrata,* the group that includes us. Taking them one at a time, we notice that the urochordates don't look like our close relatives (at least I hope not). Actually, they look more like sea squirts, and that's good, because that's what they are (Figure 13.28). In fact, they can only be identified as chordates when they are larvae. Once they change from their free-swimming larval stage, they lose their similarity to other chordates. The difference in adult urochordates and other chordates is underscored by the fact that the body walls of urochordates contain cellulose, a common substance of plant walls. Their strange larval history has led some scientists to speculate that the vertebrates arose from a line of urochordates in which the larval stages never matured physically, but became able to reproduce. It is suggested that this line could have given rise to animals like the lancelets, which would have then been ancestral to the *Agnatha* (jawless fishes). Evolutionists have little trouble guessing at the evolutionary route from the jawless fishes to other vertebrates.

FIGURE 13.27

The presumed family tree of the chordates. The fishes are believed to have developed heavy, moveable fins that enabled them to move over solid substrate. These became modified into increasingly efficient legs as airbreathing developed. Notice the relationship of birds and mammals to reptiles.

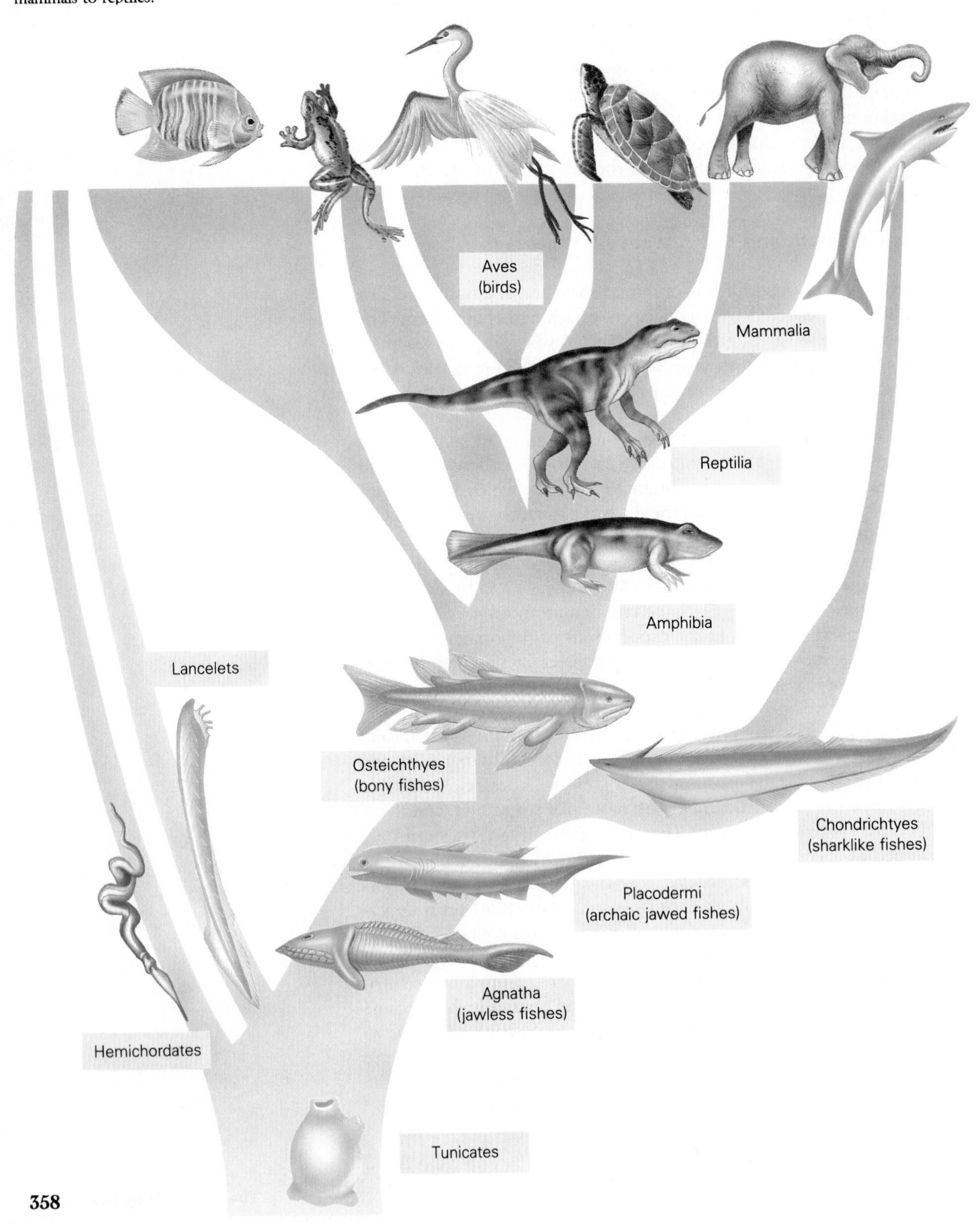

FIGURE 13.28

Sea squirts or tunicates. As larvae (b), they resemble tadpoles. At this stage, they are elongate and bilaterally symmetrical. The tail is stiffened by a rodlike notochord to which muscles are attached. Later they will settle down and attach, by their heads, to the ocean floor, where they will lose their brain, notochord, and tail, and develop a tougher outer coating (a).

(a)

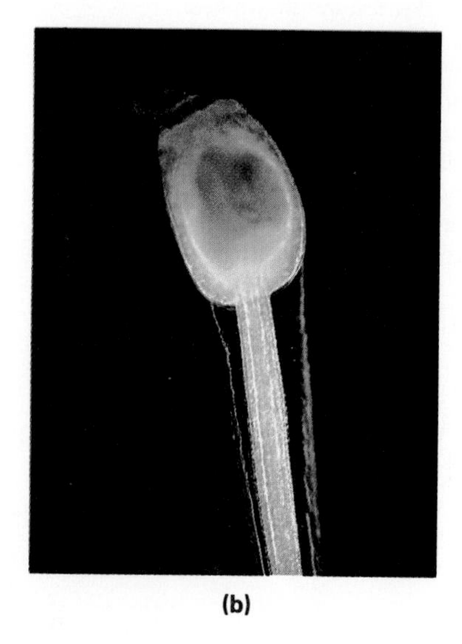

(b)

The lancelets, or *Branchiostoma* (formerly known as *Amphioxus*) are the best known cephalochordates (Figure 13.29), which some say are descended from the hemichordates. They spend much of their lives burrowed in the sandy bottoms of warm seas. Their entire bodies, 5 to 7 centimeters long, are thus protected, as only their mouths are visible. The

FIGURE 13.29

The *Branchiostoma*, or lancelet. Some zoologists believe that it is descended from larval tunicates that failed to metamorphose while becoming sexually mature. Adult lancelets are fishlike and they retain their notochord. They burrow into sandy bottoms, leaving only their mouth exposed. Food-laden water is drawn in and nutrient particles are swept along by cilia into the gut.

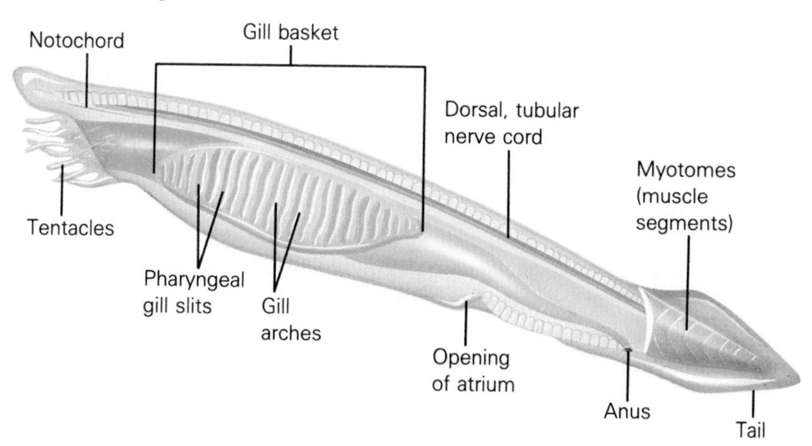

FIGURE 13.30

The general body plan of vertebrates, including humans, although humans are usually taller. All vertebrates have vertebrae (as one would hope), and they also have a dorsal nerve cord with a brain at one end. Other features shown here may or may not be present, or may be present only at some time in development. These include gill slits and the notochord. In vertebrates, the notochord disappears in the embryonic stage. Most vertebrates have some segmentation, bilateral symmetry, a head and tail, a coelom, and some way to get about.

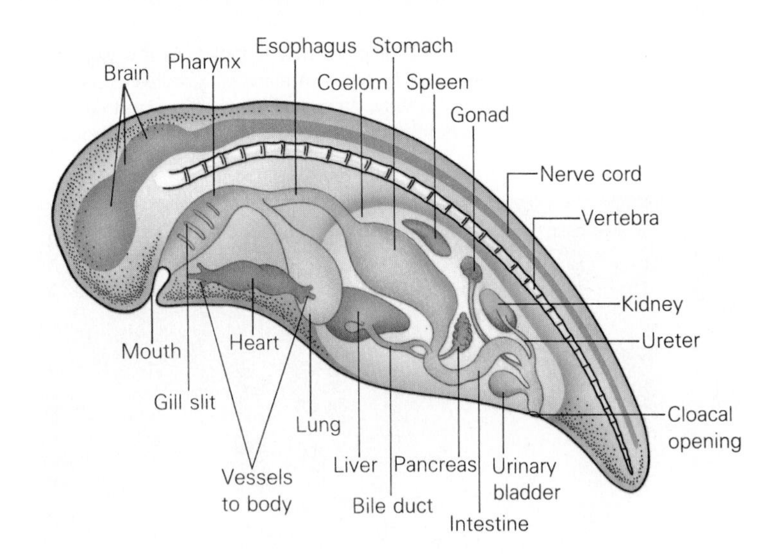

adults look a bit like larval sea squirts, but they are adult chordates with segmentation of muscles along their length.

The vertebrates are the best known and most varied of the chordates. (The general vertebrate body plan is shown in Figure 13.30.) They are distinct in that their dorsal, hollow nerve cords are protected by a series of bones called **vertebrae,** which together make up the **vertebral column** or *backbone*. This flexible backbone also serves as an attachment for a number of powerful muscles. In addition, the vertebrates have a complex skin (some with hair, glands, hoofs, feathers, scales, or fur, albeit not all on one animal). The heart is located ventrally (all we've seen so far are dorsal hearts) and pumps red iron-containing blood. Kidneys remove the waste from that blood. There is an endocrine regulatory system (Chapter 22), and the sexes are usually separate.

Class Agnatha

The members of class *Agnatha* ("jawless") are the jawless fishes. This group, which includes the lamprey and hagfish, is marked by round mouths without jaws. These fish are unusual in that they retain parts of their notochord through adulthood. Biology may introduce you to some ugly animals now and then, and these two are good examples (Figure 13.31). Their behavior also may not gain them great respect. The lampreys, in particular, are not admired because of their habit of attaching by their sucker mouths to the bodies of live or dead fish and sucking out their juices. Hagfish are scavengers and unusual in that they are **hermaphroditic** (with both male and female parts), but they produce eggs and sperm at different periods in their lives.

From the jawless fishes, we go to the jawed fishes, a group that includes some of the most fearsome creatures on earth. In large part, they are feared precisely because they have developed a jaw, a mouthpart with a hinge. With this hinge, a great number of fish have taken to biting.

FIGURE 13.31

Two beauties. The hagfish (a) and the lamprey (c). The notochord as a supportive structure is replaced by vertebrae. These fish have no jaws and therefore must feed by sucking the juices out of other animals (b). Jaws are believed to have developed from the support bars of the most anterior gill slit in some distant ancestor of these creatures. Think of how different the world would be if the simple ability to bite had never evolved.

(a)

(b)

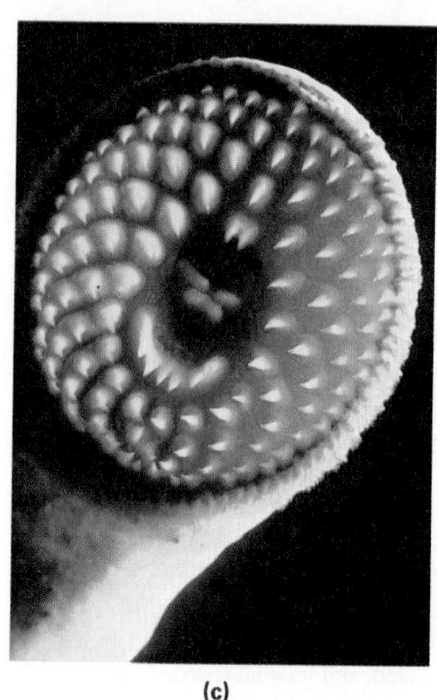

(c)

Once biting became possible, animals would have begun to use their new talent in increasingly efficient ways. Should they bite pieces out of leaves? Should they bite other animals? Would it be better to try to eat the whole animal, or to take a bite and run for it? What if it, too, had teeth? The ability to bite was undoubtedly an important development in predator-prey relationships (Chapter 25), and it forced certain evolutionary directions on the bearers of jaws. For example, some biting species would have had to develop great strength, visual acuity, and perhaps speed or stealth. Of course, new digestive abilities would be required with any change in diet. The development of jaws was a critical change, indeed, and one that triggered any number of other modifications.

Class Chondrichthyes

The class *Chondrichthyes* ("cartilage fish") includes the sharks, rays, and skates (Figure 13.32). They, of course, have jaws, and sharks have been known to bite. As their name implies, their skeletons are comprised of cartilage rather than bone. A great deal of attention is paid to sharks in the popular press, but the reports are not always accurate. It is generally safe to say that people who make blanket statements about shark attack behavior don't know much about sharks. (I generally ignore such statements, but when I spot a shark while diving, it all comes back and I am prepared to believe anything.)

All chondrichthyes have huge, oily livers that give them buoyancy (since they lack the air bladder of bony fishes). Thus, sharks don't have to spend much energy adjusting their depth (so they can spend that energy making life miserable for other creatures). They are heavier than water, though, and if they stop swimming, they sink. It was once thought that all sharks had to keep swimming to move water over their gills or they would drown, but they have been discovered sleeping on the ocean floor (Figure 13.33). Although most sharks are carnivores and able to tear flesh from prey, the largest are filter feeders that sift tiny crustaceans from the waters of the open oceans.

Some female sharks lay eggs, but others nurture their offspring within their bodies until the young can be born alive. The reputation of sharks is not helped by the knowledge that in some species, the first embryonic shark to develop will turn and devour its brothers and sisters while still in the uterus.

Class Osteichthyes

The class *Osteichthyes* ("bone fish") includes the bony fish (Figure 13.34). There are about 18,000 living species and they can be found from shallow

FIGURE 13.32

The shark is one of the most fabled, feared, and least understood large animals on earth. They are reported to have well-developed jaws and teeth, and to be totally humorless.

FIGURE 13.33

Sleeping shark, demonstrating that sharks do not have to keep moving in order to receive enough oxygen. Obviously, they are able to survive nicely on far less oxygen than we had believed.

FIGURE 13.34

The bony fish. How could the fins have given rise to legs? Do you see a relationship between swim bladders and lungs? What evidence is there that their ancestors could have invaded dry land? For such evidence, do we look to living or extinct species, or both?

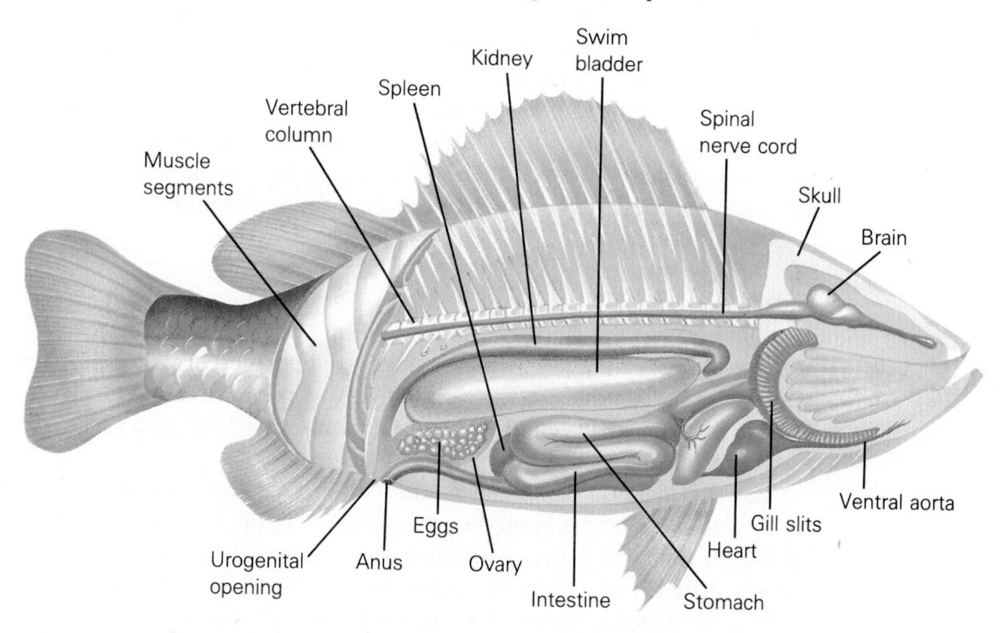

FIGURE 13.35
Some fish are extremely bizarre. Just by looking at this specimen, what can you deduce about its lifestyle? Does it live in lighted waters? What does it eat? What might you guess its social life is like? Does it swim in schools? Alone? Is it a prey animal? The enormous jaws indicate that the fish is prepared to eat just about anything it comes across.

ponds and rippling brooks to the darkest ocean depths (Figure 13.35). Bony fish have evolved a number of remarkable adaptations to survive in their unpredictable world. Some use their swim bladders as lungs and can survive for months encased in mud, and some catfish can walk clumsily from one pond to the next by using their fins as limbs. The biggest bony fish are probably the swordfish, which grow to 4 meters long.

In 1939, a fishing boat plying the waters between Madagascar and Africa hauled up a great, ugly fish that no one recognized. It turned out to be a coelacanth, or *Latimeria,* a member of a species that had long been believed to be extinct. The "living fossil" was particularly interesting because biologists believe that its heavy lobed fins were the evolutionary precursors of legs (Figure 13.36). The search for the beasts intensified and scores of these *Latimeria* have been caught and studied.

Class Amphibia

The class *Amphibia* ("double life") includes aquatic animals that have successfully invaded the land without completely relinquishing their watery heritage. One might ask why they would give up a lifestyle, to which they were well adapted, for even a part-time rigorous existence on land. Perhaps because the land had been exploited less, competitors and predators were fewer and oxygen was more readily available.

But the invasion of land also had its dangers. There was the risk of drying out and being subjected to drastic temperature fluctuations. Without the buoyancy of water, a land dweller also needed strong support and a sturdier frame, metabolically expensive structures. So, amphibians retained some dependence on water, especially in reproduction. Amphibians never developed skins that were impermeable to water and able to hold

FIGURE 13.36

The famed coelacanth. In 1939, one of these beasts was hauled from the waters off southeast Africa, reminding us once again that some fossils yet live. Fishermen were warned in a series of posters and bulletins to keep an eye peeled for them and not to throw them back or eat them. The fishermen ate some of them anyway. Notice the rudimentary bones similar to the bones of arms and legs.

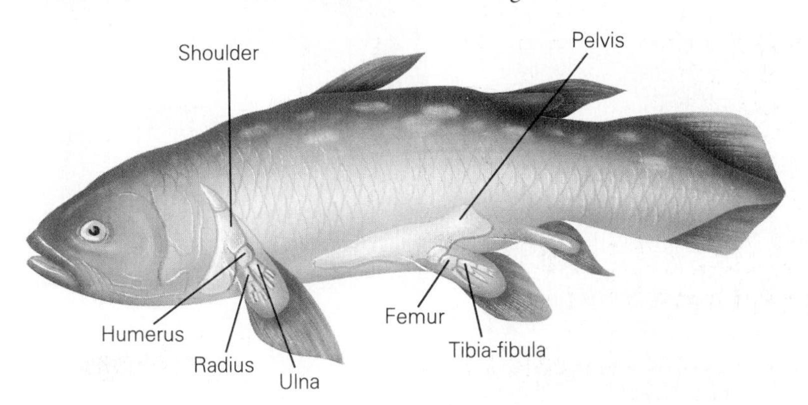

moisture in. But they did develop inpocketing lungs (instead of outpocketing gills) that held moisture inside, away from the drying air, thereby saving water. And they developed a strong skeleton and sturdy, muscular legs. Even the most advanced amphibians, such as frogs, toads, and salamanders, return to lay eggs in the water, after having spent their early lives there, and some show even greater specialization (Figure 13.37).

FIGURE 13.37

The amphibians have had to make many compromises as they sought to bridge the gap between water and land. The marsupial frog keeps its eggs safe in moist pouches on its back.

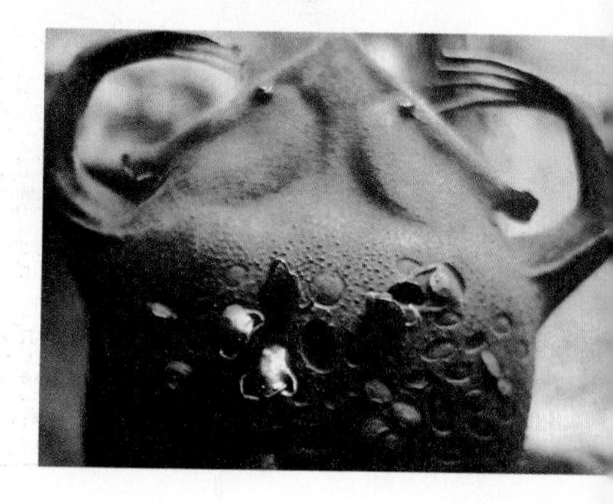

Class Reptilia

Reptiles (class *Reptilia,* "crawlers") probably evolved from an amphibian ancestor about 300 million years ago. Members of this class include the snakes, turtles, lizards, and crocodilians (Figure 13.38). They probably rank somewhere between spiders and bats in popularity. Snakes, in particular, often meet their fate by encounters with humans. (We are very hard on bugs and snakes.) We might remember, however, that the ancestors of those snakes, the dinosaurs, once ruled the earth while our ancestors were scurrying around under leaves.

With the reptiles came the complete transition to life on land. The reptilian skin is impermeable to water and keeps them from drying out on land. The scales that are so characteristic of reptilian skin help make it waterproof. Furthermore, reptiles no longer require water for reproduction. Fertilization is internal, so the sperm don't have to swim through water to reach the egg. The eggs of reptiles are covered by a waterproof, leathery sheath that allows air to enter while preventing water loss. Their eggs also contain a food supply and protective membranes that support development on land.

The reptiles apparently gave rise to both birds and mammals, but whereas birds and mammals have physiological ways to regulate their body

FIGURE 13.38

Reptile diversity, as seen in the gila monster (a), the long-nosed tree snake (b), the frilled lizard (c), and the alligator snapping turtle (d).

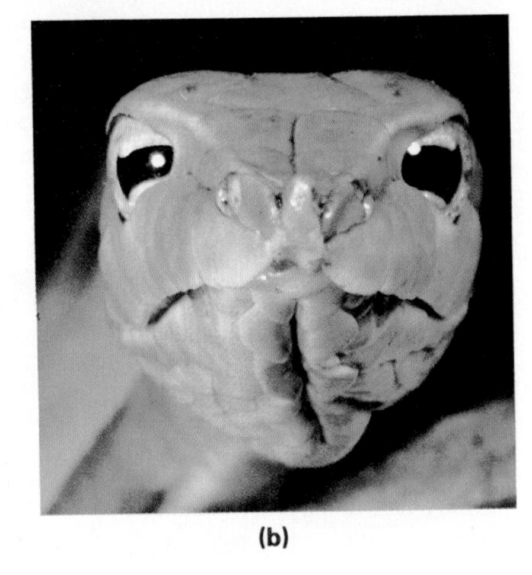

(b)

(a)

(c)

(d)

temperature, fish, amphibians, and reptiles do not. Thus, they are regarded as "cold-blooded" (see Chapter 18). Reptiles, however, do have behavioral methods of regulating their body temperature. They bask in the sun to raise their body temperature and seek shade when they become too warm.

Class Aves

Birds comprise the class *Aves* and are easy to identify: they have feathers (Figure 13.39). Feathers are not only suited to flight, they also provide insulation to help regulate body temperature. Birds also have scaly legs and feet, internal fertilization, and an egg well suited to development on land, characteristics that unmask their reptilian heritage. There are about 8600 species of birds, most of which can fly. Some species have secondarily

FIGURE 13.39

A rogues gallery of birds whose appearances reflect a variety of lifestyles: the king vulture (a), the long-eared owl (b), the bald eagle (c), the flamingo (d), the red-billed tropicbird (e), and the spoonbill (f).

become grounded; that is, they are the descendants of birds that could fly, but for some reason have lost the ability. Flightless birds include the ill-tempered rhea and the powerful ostrich (ostriches have been known to disembowel predators with a single kick from a huge, clawed foot). The much smaller flightless kiwi rummages through leaves, hunting insects by

FIGURE 13.40

Modifications for flight are seen in nearly every aspect of the bird's anatomy and physiology, from its streamlined form to its elevated metabolic rate. In spite of the demands placed upon it, the skeleton is extremely light. The frigate bird, for instance, has a wingspan of just over 7 feet, yet its skeleton weighs just 4 ounces. In general, the slender, hollow bones of birds have a deceivingly delicate appearance; in fact, however, they are strong and flexible, containing numerous triangular bracings within. Flight feathers, which can weigh more than the skeleton, owe their extreme strength and flexibility to numerous vanes. These have an interlocking arrangement of hooklike barbules.

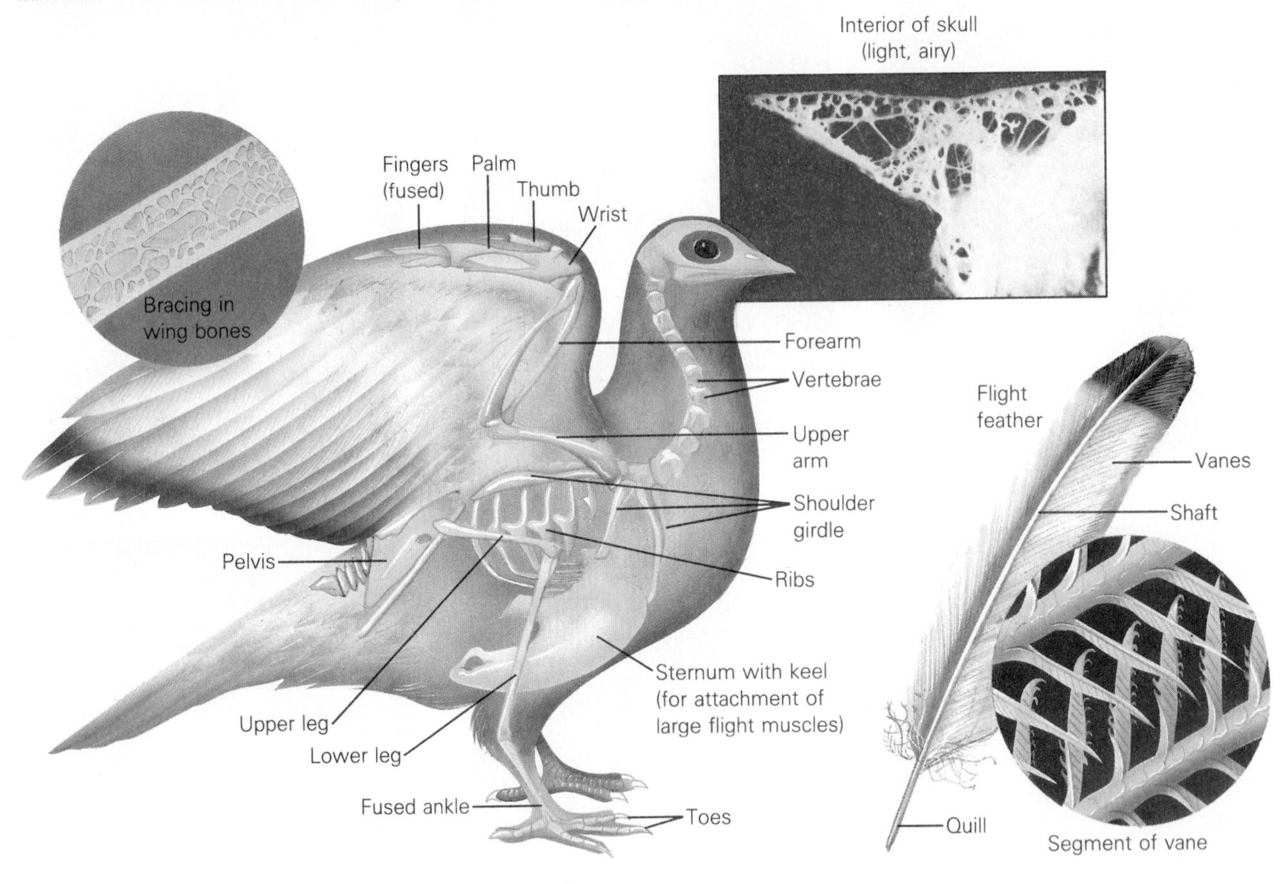

a remarkable sense of smell. And then there are the penguins, now suffering reduced numbers in some areas since Antarctic vacations became more popular. Nonetheless, even the flightless birds still possess many of the specializations for flight.

The structural specializations for flight are quite pronounced, but can vary depending on the role of flight in the bird's natural history. Some birds, such as quail, spend little time in the air, but can rise explosively from the ground when threatened. At the other extreme, some species, such as frigate birds and some swifts, rarely land, spending almost their entire adult lives in the air.

So we see that the anatomy of flying birds is geared not just for flight, but for different kinds of flight. Nonetheless, a heritage of flight has placed special demands on the body of all birds (Figure 13.40). For example, they all have a streamlined shape and feathers that can be tightly compressed against their bodies. They also have light, hollow bones, some of which are filled with long saclike extensions of the lungs. Many species have remarkably developed flight muscles, and some have even sacrificed the development of one gonad, thereby further lightening their body weight. These are indeed all rather marked changes from the reptilian forms that gave rise to birds.

Class Mammalia

Another descendant of reptiles are the mammals. The class *Mammalia* is comprised of hairy milkgivers (such as ourselves). Mammals emerged along with the birds to share dominance as the earth grew cooler during the Cenozoic. It is probably not coincidental that both groups are able to regulate their internal temperatures (through processes we will see in Chapter 18).

Hair helps regulate body temperature by providing insulation and their more efficient four-chambered heart enables mammals to maintain a high metabolic rate and high body temperature.

In the placental mammals, the young develop in the protective environment of the mother's body, where the embryo is nourished by the placenta. Females produce milk to nurse their young. As a result, females need not desert their helpless offspring to gather food for them.

The subclass *Prototheria* includes the egg-laying *monotremes,* such as the duck-billed platypus and the spiny anteater (Figure 13.41). Prototherian females lack breasts and secrete milk directly into the belly fur, where it is licked up by the infants.

FIGURE 13.41

The spiny anteater is a monotreme (a mammal that lays eggs and, when the eggs hatch, feeds its young with milk that oozes from the mammary glands). It has a remarkably sensitive nose and can detect insects that could have sworn they were well hidden.

FIGURE 13.42

Marsupials are rather primitive, pouched mammals. They are rather diverse, as we see here with Tasmania's pygmy-possum (a), this cuscus from the Solomon Isles (b), and this wallaby mother with her joey (c).

(a)

(b)

(c)

Another subclass, *Metatheria,* is comprised of the *marsupials*—such as opossums, kangaroos, wombats, and koala bears (Figure 13.42). The young are generally born in a relatively undeveloped state, and most species carry their offspring in protective pouches.

As an aside, no one knew precisely how kangaroos were born until one day when a zookeeper with not much to do was standing around looking at a kangaroo's vagina. To his surprise, he saw a tiny sluglike creature emerge and, using its forelegs in a kind of Australian crawl, begin to wend its way through the fur to its mother's pouch. It was an embryonic kangaroo, clearly not well developed. Subsequent research revealed that such tiny embryos attach to a nipple in the pouch and become fused to it for a time. Even while they nurse, the mother may become pregnant again, but her body can hormonally stop the development of the second embryo until the pouch is free.

FIGURE 13.43

Some mammals are not very well known at all, such as this star-nosed mole. Many kinds of mammals live among us, however, and escape our attention by being furtive or living on a different schedule. Flying squirrels, for example, are very common in many areas, but since they are nocturnal, few people have seen them.

The third subclass, *Eutheria,* is composed of the *placental mammals*. Perhaps because we are in this group, we know more about the placentals than the other mammals. The group is rather diverse, so let's review only a few of the orders found within our subclass.

The most primitive order of *Eutheria* is the *Insectivora* ("insect eaters"). These include moles and shrews—small but voracious animals that live on or under the ground and are constantly and frenetically foraging for morsels such as grubs, insects, or worms (Figure 13.43). The *Rodentia* ("gnawers") are the rats, mice, beavers, and other chiselers (Figure 13.44). The *Carnivora* ("meat eaters") include cats, bears, dogs, otters, and jackals (Figure 13.45). The vertebrate carnivores all have elongated canine teeth that enable them to catch game and tear its flesh, but their hunting patterns can be quite diverse. For example, they may hunt singularly or in groups, by stealth or by chase. (Actually, there are probably few true carnivores other than some sharks and flies. Almost all other species include some vegetable matter in their diets.)

FIGURE 13.44

The capybara is a water-loving animal and the world's largest rodent, sometimes reaching 100 pounds in weight. When chased, they may jump into a river and swim away underwater, a most disconcerting habit to the Indians hunting them.

FIGURE 13.45

Some animals are specialized in seeking out and eating the flesh of other animals. These are the carnivores. Since their prey may be quick and sensitive, or even dangerous, many carnivores have developed high intelligence and complex social systems. These lions have killed a wildebeest.

FIGURE 13.46

Przewalski's horse, a rugged native of Asia. This is one of the last wild horses known. (The mustang is feral, the descendant of escaped domesticated horses.) There is some indication that these tough little horses are similar to the ancestors of the domestic horse.

FIGURE 13.47

The vampire bat secretes an anticoagulant that keeps the blood flowing from tiny incisions it makes with its razorlike teeth. Most human victims are bitten at the hairline or on the thin skin of the ankles.

The *Artiodactyla* ("even toes") include the pigs, camels, sheep, goats, deer, cattle, hippopotamuses, and giraffes. They walk about on the tips of two toes (you may have noticed that two is an even number). Many artiodactyls can "chew their cud" by regurgitating plant food that they chew a second time to help break down tough cell walls, enabling more of the plant's nutrients to be extracted. The *Perissodactyla* ("odd toes") are the mammals that walk on a single, modified digit. These include the horses, tapirs, rhinoceroses, and zebras (Figure 13.46).

The *Chiroptera* ("hand wing") include the bats. These animals may not please your aesthetic sensibilities, but they are for the most part unfairly maligned. The reputation of bats stems largely from one group, the vampire bats (Figure 13.47). My friend Samuel Padilla, an Amazonian Indian, has awakened many times feeling tired and listless, only to have someone discover tiny hairline wounds, usually at the nape of his neck and heels. (When in a village we try to get cats to sleep with us.) Other bats, though, can be quite beautiful and graceful, such as the fruit-eating flying foxes (with a wingspan of over a meter). Most bats are helpful insect eaters (Figure 13.48). The *Lagomorpha* ("rabbit shape") include rabbits, hares, and pikas (Figure 13.49). The *Proboscidea* ("front feeder") are the elephants, the group with a nose that defies imagination, and long tusks that are actually modified incisors (Figure 13.50). Elephants have very complex social systems and interact with their environment in very intricate ways that we are only now beginning to understand. Finally, there are the

FIGURE 13.48

Bats are extremely specialized mammals that do not entirely arouse our admiration, perhaps because one species (see page 372) is specialized for gently fanning sleeping humans while they make a painless cut in the skin and then lap the blood. Most bats, however, are harmless or helpful. Many specialize in catching and eating insects, as does this spotted bat.

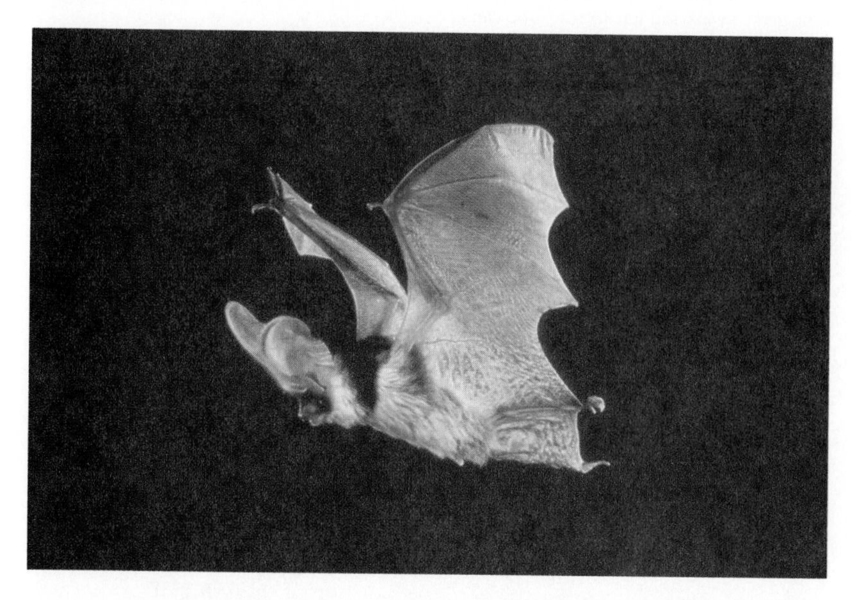

FIGURE 13.49

The pika is a familiar sight to many alpine hikers. They spend their brief, beautiful summers collecting plant material that they will live on during the harsher times.

FIGURE 13.50

Elephants are intelligent and highly social animals. They may occasionally fight viciously, and tuskers have been known to fatally wound an opponent. They may also be very helpful to each other. A sick or injured elephant may be helped along by two comrades on either side. Unfortunately, elephant country is being claimed by growing populations of humans, and in some cases, hunters have killed entire small herds illegally, for the sport. Their bodies, tusks and all, have been found in great heaps and left to rot. One can now lease land adjoining the great reserves for the privilege of killing an elephant that inadvertently crosses into the leased land, again for sport.

Primata ("first"). The primates include the monkeys, chimpanzees, gorillas, orangutans, baboons, humans, and even tree shrews, lemurs, and tarsiers (Figure 13.51).

There are obviously a great many things to be learned about our own kingdom, but time is growing short, in many cases, because our world and its denizens are changing at a dazzling and disconcerting rate. Many animals do not even live in the kind of world that they were born into. (We certainly don't.) Our studies and descriptions, then, must account for a past that will never exist again and a very near future quite unlike anything existing now.

FIGURE 13.51

Primates are fascinating creatures because of their high intelligence and physical dexterity. If they often remind us of humans, it may be because we are included in this group. We are only now discovering just how much we have in common with some of the other primate species. Here, a female patas monkey grooms another. Such behavior helps maintain strong social bonds.

SUMMARY

1. Members of Phylum *Porifera,* the sponges, have different protistan ancestors than other animals and are an evolutionary dead end. The three classes of sponges are calcareous, glass, and bath. As water flows into a sponge's vaselike body through holes in the sides and exits through an opening in the top, oxygen is extracted. Food particles are trapped by flagellated collar cells and passed to amoebocytes for distribution.
2. Members of Phylum *Cnidaria* (jellyfish, polyps, sea anemones, and corals) have radial symmetry, stinging cells, tentacles, a saclike gut called a gastrovascular cavity, and a nerve net. They have two cell layers, the epidermis and gastrodermis, with a jellylike mesoglea between them. Cnidarian life cycles usually include two distinct body types—a vaselike sedentary polyp and a free-swimming medusa. The three classes are *Hydrozoa, Scyphozoa* (jellyfish), and *Anthozoa* (sea anemones and corals). During development, the bilaterally symmetrical larva, called a planula, develops into a radially symmetrical adult.
3. Members of Phylum *Platyhelminthes* are flatworms. Two changes important in evolution are seen first in the flatworms: cephalization (the con-

centration of sensory and nervous structures at the head end) and bilateral symmetry (two-sidedness). Flatworms have organ systems and three germ layers. Members of Class *Turbellaria,* the free-swimming forms such as planaria, can learn simple tasks and have a protrusive pharynx for sucking food into the extensive gut. Class *Trematoda* contains the parasitic flukes, which attach to their hosts with suckers and have reproductive organs of both sexes. Their life cycles are complex. The tapeworms (Class *Cestoda*) are parasites that attach to their host by suckers and hooks on the scolex, have body segments called proglottids, and lack a mouth and a digestive system.

4. There are two evolutionary lines of animals. In the protostomes (annelids, arthropods, and mollusks), the first opening in the embryo becomes the mouth and in the deuterostomes (echinoderms and chordates) this opening becomes the anus.
5. The roundworms, which comprise Phylum *Nematoda,* are the first to have a complete tubular gut and a pseudocoelom (body cavity not lined with mesoderm). Some roundworms live on decaying material and others are parasites. Some cause human diseases.
6. The mollusks (e.g., oysters, clams, mussels, snails, squids, and octopuses) have a coelom, a cavity between the gut and the body wall lined with mesoderm. A coelom is important because it provides an area into which other structures can protrude. Mollusks have soft bodies, but some are encased in hard shells. The mollusk's body consists of a muscular foot, a fold of tissue over the body called the mantle, and, in some, a rasplike tongue called the radula.
7. Members of Phylum *Annelida* are segmented worms. Annelids have a circulatory system and a ventral bilateral nerve cord with enlargements in each segment. Although earthworms (Class *Oligochaeta*) contain both ovaries and testes, they don't self-fertilize. Members of Class *Polychaeta* have specialized segments called parapodia that are used in swimming and respiration and have heads with eyes, jaws, and tentacles. The leeches are in Class *Hirudinea.*
8. Members of Phylum *Arthropoda* have jointed appendages and a hardened exoskeleton made of chitin. Their body segments are specialized for different roles, allowing them to meet a greater variety of environmental challenges. The two subphyla are *Chelicerata* and *Mandibulata.* Chelicerates have clawlike fangs on their first appendage, no antennae, a fused head and thorax, and an abdomen. Mandibulates include class *Crustacea* (crabs, shrimp barnacles), *Chilopoda* (centipedes), *Diplopoda* (millipedes), and *Insecta* (insects).
9. The echinoderms and chordates are deuterostomes. There are five classes of echinoderms: *Crinoidea* (primitive stalked sea lilies), *Asteroidea* (starfish), *Ophiuroidea* (brittle stars), *Echinoidea* (sea urchins and sand dollars), and *Holothuroidea* (sea cucumbers). During evolution, echinoderms reverted from the longitudinal bilateral body plan, still seen today in their larvae, to radial symmetry. The skeletal elements are embedded in the skin. A water vascular system is used to extend the tube feet. They have poorly developed nervous systems. Digestive juices are released into prey and the products are absorbed.

10. At some point in their lives, all chordates possess a notochord (a rigid cellular rod covered with two layers of strengthening fibers), a dorsal hollow nerve cord that enlarges to form a brain at one end, gill slits (openings in the throat region connecting the alimentary canal with the outside), and usually a tail that extends beyond the anus. Chordates are bilaterally symmetrical with distinct heads, some segmentation, three germ layers, and a coelom. There are three subphyla: *Urochordata, Cephalochordata,* and *Vertebrata.*
11. The urochordates are sea squirts and can only be identified as chordates in their larval stage. Some scientists speculate that vertebrates may have arisen from a line of urochordates in which the larvae never matured.
12. The best known of the cephalochordates are the lancelets *(Branchiostoma)*.
13. In the vertebrates, the dorsal hollow nerve cord is protected by a series of bones called vertebrae, which make up the vertebral column (backbone) and serve as an attachment for powerful muscles. Vertebrates have complex skin, a ventrally located heart, iron-containing blood, kidneys to remove waste from blood, an endocrine regulatory system, and separate sexes. There are seven classes: *Agnatha, Chondrichthyes, Osteichthyes, Amphibia, Reptilia, Aves,* and *Mammalia.*
14. Class *Agnatha* are the jawless fishes, such as the lamprey and hagfish. They retain parts of the notochord through adulthood. Lamprey attach to living or dead fish and suck their juices. Hagfish are scavengers. Although hagfish are hermaphroditic (with both male and female parts), they produce eggs and sperm at different times.
15. Class *Chondrichthyes* includes sharks, rays, and skates. They have jaws that are made of cartilage rather than bone and have huge oily livers that provide buoyancy. Some female sharks lay eggs, while others bear live young.
16. Class *Osteichthyes* includes the bony fish.
17. Members of Class *Amphibia* (e.g., frogs, toads, and salamanders) reproduce in water and spend at least the early part of their lives there, but later move onto land. The move to land was advantageous because there was less competition, fewer predators, and oxygen was easier to obtain. Problems associated with life on land include the risk of drying out, dealing with temperature fluctuations, and support against gravity. Amphibians remain dependent on water for reproduction. Their skin does not hold moisture in, but their internal lungs prevent water loss.
18. Members of Class *Reptilia* (e.g., snakes, turtles, lizards, and crocodilians) are completely adapted to life on land. Their skin is impermeable to water. They have internal fertilization. Their eggs are water-filled and covered by a waterproof, leathery sheath, and contain food and protective membranes. They have no physiological methods to regulate body temperature. They probably evolved from an amphibian ancestor and apparently gave rise to both birds and mammals.
19. Birds (Class *Aves*) have feathers as well as scaly legs and feet. Most can fly, but some have lost this ability. Adaptations for flight include streamlined shape, feathers that can be tightly compressed against the body, light hollow bones, and well-developed flight muscles. Birds can regulate their internal temperatures.

20. Members of Class *Mammalia* have hair, produce milk, and can regulate their internal body temperatures. Subclass *Prototheria* are the egg-laying monotremes, such as the duck-billed platypus and spiny anteater. The marsupials (e.g., opossums, wombats, and koala bears), which comprise subclass *Metatheria,* bear relatively undeveloped young that are carried in pouches. The orders of subphylum *Eutheria* (the placental mammals) are: *Insectivora,* insect eaters such as moles and shrews; *Rodentia,* gnawers such as rats, mice, and beavers; *Carnivora,* meat eaters such as cats, bears, and dogs; *Artiodactyla,* animals with an even number of toes such as pigs, camels, sheep, and goats; *Perissodactyla,* animals that walk on a single modified toe; *Chiroptera* (bats); *Lagomorpha* (e.g., rabbits, hares, pikas); *Proboscidea* (e.g., elephants); and *Primata,* the primates, including monkeys, gorillas, humans, tree shrews, lemurs, and tarsiers.

KEY TERMS

abdomen (354)
amoebocyte (339)
annelid (350)
Arthropoda (352)
cephalization (344)
chordate (357)
clitellum (351)
Cnidaria (340)
coelom (347)
collar cell (339)
deuterostome (346)
echinoderm (356)
epidermis (340)
foot (349)
gastrodermis (340)
gastrovascular cavity (340)
hemichordate (357)
hermaphroditic (360)
mantle (349)
medusa (341)
mesoglea (340)
metazoan (337)
mollusk (347)
nematode (346)
notochord (357)
parapodia (352)
planula (341)
platyhelminths (344)
polyp (341)
Porifera (339)
proglottid (345)
protostome (346)
pseudocoelom (347)
radial symmetry (340)
radula (349)
scolex (345)
seminal receptacle (351)
setae (351)
thorax (354)
vertebrae (360)
vertebral column (360)
water vascular system (356)

FOR FURTHER THOUGHT

1. You have discovered a new animal that exhibits a segmented body plan. What assumptions can you make about the nervous system of this animal?
2. Into which phylum would you place an animal that is bilaterally symmetrical, has pharyngeal slits, and a well-developed coelom?
3. Describe some characteristics that are common to both humans and flatworms.

FOR REVIEW

True or false?

1. ___ The polyp and medusa stages represent two distinct body types exhibited by some cnidarians.
2. ___ Echinoderms have a notochord at some stage in their life.
3. ___ Animals adapted to a life that is spent partly on land and partly in water are called amphibians.

4. ___ Birds and mammals may have evolved from early reptiles.
5. ___ The jawless fishes are members of the class Chondrichthyes.

Fill in the blank.

6. ___ is a term that describes the accumulation of sensory mechanisms in the head region.
7. Body parts such as a foot, mantle, and radula would be found in members of the phylum ___.
8. The cavity between the gut and the body wall that is lined with mesoderm is called a ___.
9. Three features that are common to chordates are ___.
10. Individuals such as hagfish that contain both male and female parts are said to be ___.

Choose the best answer.

11. Members of the phylum ___ are characterized by jointed legs and hardened exoskeletons.
    A. Annelida
    B. Agnatha
    C. Arthropoda
    D. Nematoda
12. Which of the following incorrectly matches an animal with its appropriate body plan?
    A. jellyfish: radial symmetry
    B. planaria: bilateral symmetry
    C. adult echinoderm: radial symmetry
    D. sponges: bilateral symmetry
13. All of the following are chordates except:
    A. mollusks
    B. sharks
    C. penguins
    D. snakes
14. Which of the following incorrectly matches a phylum with its members?
    A. Platyhelminthes: planaria, tapeworms, flukes
    B. Mollusca: mussels, octopuses, clams
    C. Annelida: sponges, snakes, sea anemones
    D. Chordata: bears, bats, man
15. In the following list, which trait is least associated with evolutionary advancement?
    A. notochords
    B. coeloms
    C. nerve nets
    D. cephalization

PART FOUR

Reproduction and Development

The story of how living things reproduce and develop is woven together by the twin themes of evolution and adaptation. We follow our own development in detail, emphasizing social issues from contraception to STDs, and the philosophical notions of why we reproduce.

CHAPTER 14

Plant Reproduction and Development

Overview

Sexual Reproduction

Reproduction and Life Cycles in Various Plants

Plant Development

Plant Hormones

Objectives

After reading this chapter you should be able to:

- Discuss the advantages and disadvantages of sexual reproduction in plants.
- Describe the gametophyte and sporophyte stages of green algae, mosses, and ferns, and the reproductive cycle of each.
- Outline the process of seed production in conifers and flowering plants.
- List the structures of a typical angiosperm flower.
- Name and describe the plant tissues that are produced by differentiation.
- Describe the functions of the apical and lateral meristems.
- Name three plant hormones and state how they influence plant growth and development.

Reproduction is one of the most prevalent themes in the drama of life. Humans are keenly interested in it (or at least in the tawdry pageant that accompanies it), and while it is incessantly discussed, it remains one of the most regulated of controversial topics. The tension eases a bit, however, if the topic is *plant* reproduction. After all, what could be unseemly (or perhaps interesting) about plant reproduction? A book called *Sex in the Garden* would be perfectly acceptable in any circle once it was understood that the topic was plants.

However, perhaps plant reproduction is unfairly slighted when we bequeath our prejudices. The various processes involved in begetting plants can be quite interesting indeed. Furthermore, they can illustrate some fundamental aspects of reproduction in a simple form that can then be applied to more elaborate rituals in other species. So, keep an eye out for emerging general principles as we go, and perhaps we can come to some better understanding of just what's going on out there in the garden and other places as well.

SEXUAL REPRODUCTION

Disadvantages of Sexual Reproduction

We must keep in mind that evolution rewards *only* reproductive output. (Reread that sentence.) The genes of individuals who do not reproduce maximally will be replaced by the genes of those who do. There are no values to be attached to reproduction; it is neither good nor bad. There is only the relentless arithmetic of natural selection. Reproduction, of course, involves passing along replicas of one's genes into the next generation. Thus, arithmetically, if one wished to maximize one's genes in the next generation, it would be best to produce genetic *replicas* of one's self. That is, to pass *all* of one's genes into each of one's offspring. Yet many organisms dilute their genes by mixing them with those of a partner through sexual reproduction. But why? Why dilute one's genes by mixing them with a partner's? The question of how sex arose has produced a great deal of debate in scientific circles.

Aside from the question of gene dilution, there are a number of other problems associated with sexual reproduction. For example, sexual reproduction demands the production of enormous numbers of gametes, although only a few will ever enter into fertilization—a great waste. Also, for some species there is a real chance that the individuals might be so sparsely distributed that two individuals, or their gametes, might not be able to find each other. For example, a plant that releases pollen into the air must not be too far from its nearest neighbor or its effort will be wasted. In fact, the probability of breeding failure for members of some species is very high. Obviously, if you can reproduce all by yourself, you not only ensure that your offspring will bear *all* of your genes, but you eliminate the risks and wastes associated with sexual reproduction.

Advantages of Sexual Reproduction

Why, then, have sex? Your first response notwithstanding, one advantage of sexual reproduction should be apparent if you recall that the meiotic processes yield gametes that carry different combinations of alleles and normally there is no way to predict which of these gametes will eventually enter into fertilization. As a result, any individual's offspring are likely to be highly variable. This variation, we have seen, can be critical to evolutionary processes in a changing or unpredictable world.

You recall that under such circumstances, with a genetically variable population there is an increased likelihood of survival for some members of the group if environmental conditions change. Sexual reproduction also allows the union of superior new genes. Suppose, for example, two individuals bearing two newly mutated superior genes should mate. Those offspring bearing both mutations would be at a marked advantage indeed. Another advantage of sex is that with diploidy, recessive genes can essentially lie dormant in the population. (After all, if they aren't expressed, they can't be subjected to natural selection.) These genes, then, create a reservoir of potential variation should the alternative, dominant gene suddenly become unfavored.

In populations that reproduce without sex, a new "good" gene that arises in a population can quickly become fixed (reach 100 percent). This is good (for the gene) in a sense, but such an event can cause problems. Suppose an asexual population of pond dwellers is subjected to a period of hot, dry weather. A mutation that allows the population to survive this condition could quickly spread through the gene pool. If, however, things changed and cool, wet weather set in, the entire population might be lost. In sexually reproducing populations, the diploid condition (with masked recessive genes), and the high variation in such groups may help protect against such devastation.

A fascinating recent finding indicates a somewhat surprising advantage of sex. There is evidence that the same gene produces different protein products in males and females. This sexual influence has been called **genetic imprinting.** Any such effect, of course, would increase the level of variation in a population.

Even with such pronounced theoretical advantages to the evolution of sex, there are those who continue to effectively argue that these advantages simply do not offset the disadvantages, and that we have not yet come up with a suitable explanation for the evolution of sex.

REPRODUCTION AND LIFE CYCLES IN VARIOUS PLANTS

As we saw in Chapter 12, the life cycle of many species of plants involves an **alternation of generations** in which individuals of the haploid **gametophyte** generation produce gametes that fuse and develop into individuals of a diploid **sporophyte** generation. The sporophyte produces haploid spores that develop into members of the gametophyte generation (Figure 12.1A).

Green Algae

In most green algae, the haploid gametophyte generation is the dominant or most conspicuous phase (Figure 14.1). The plant undergoes a period of marked growth during this time. The diploid sporophyte stage is usually represented by only a single cell that seems to be waiting out adverse conditions.

Whereas the sporophyte is rather unspectacular, its development may have been of enormous evolutionary importance. It is believed to have evolved from a long line of algae that reproduced solely by asexual means. Its appearance in the distant past, then, as algae were beginning to give rise to land plants, would have set the stage for the evolution of the

FIGURE 14.1

Life cycle of *Ulothrix*, a green alga. (1) During the extremely brief sporophyte generation, the zygospore (the result of gametes joining in fertilization) undergoes meiosis and becomes filled with flagellated spores called zoospores. (2) The zoospores produce filaments that form either sporangia (spore-forming structures) (3), or gametangia (gamete-forming structures) (4). The gametangia produce flagellated isogametes (gametes of the same appearance; *iso*, same) (4) that can join in fertilization with gametes from another filament (5). This union results in a new zygospore. The sporangium, on the other hand, releases spores that simply settle to begin the growth of new filaments in a form of asexual reproduction. Under certain (usually adverse) conditions, these filaments may stop reproducing asexually and begin to form gametangia, leading to sexual reproduction.

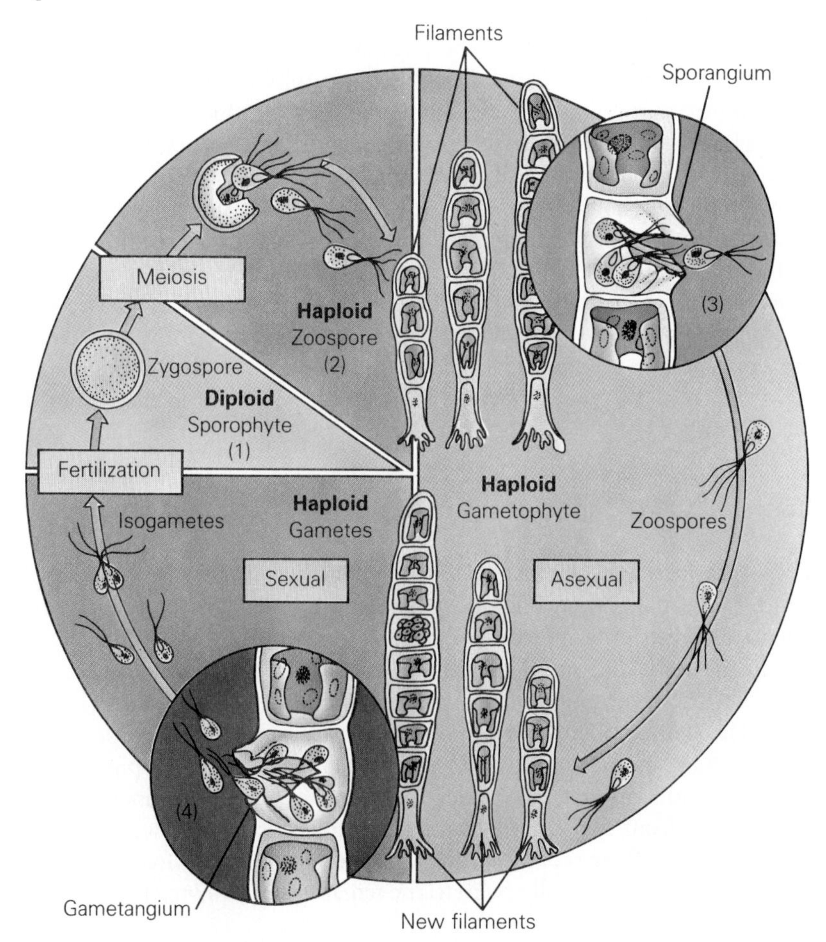

elaborate and conspicuous multicellular sporophytes that dominate our landscape today. As the sporophyte gained importance in the rigorous terrestrial realm, it could have become increasingly large and complex until it came to house and protect the gametophytes themselves, as we see in many land plants today. (Remember that most green algae exist as single cells of two different mating types. These could then join in fertilization to form the sporophyte stage.)

In more complex green algae, gametes of the two mating types are clearly distinct in their overall appearances. In these species, the larger gamete (the "egg") waits in a protective body for its smaller, motile partner (the "sperm"). Since the egg doesn't have to move about, it can grow large and save its energy to nurture the zygote. Specialized structures could then be expected to develop that would make fertilization of the egg and protection of the resulting zygote even more efficient. This trend sets the stage for a great many important evolutionary developments, as we will see when we consider the flowering plants.

Mosses

Mosses have made a reasonably successful transition to terrestrial life, but they still have some of the same requirements for reproduction as did the ancient green algae from which they descended. For example, there must be enough water available for the flagellated sperm to swim to the egg. The eggs and sperm are formed by the gametophyte generation, the leafy structure we think of as moss. Each gametophyte produces either the male **antheridium** or the female **archegonium,** or, in some types, both. Sperm develop within the antheridium and swim through a film of water to an archegonium on a nearby plant where it will fertilize the egg within. The resulting zygote develops into the sporophyte (which, we see, grows out of the female part of the gametophyte). Within the diploid sporophyte, spores are produced by meiosis. These spores are then released. The successful ones will grow into new gametophytes (Figure 14.2). With mosses, the gametophyte generation is dominant, whereas with the higher plants, as we will see, the gametophyte is reduced to living within the tissues of the sporophyte.

Ferns

The reproductive cycles of the vascular plants (plants with conducting vessels) are quite varied. One of the most ancient of these cycles is represented by the ferns (Figure 14.3).

In the ferns and some of their close relatives, such as horsetails and lycopods, the sporophyte stage is dominant and is well adapted to life on land. The gametophyte stage, however, is much smaller and more vulnerable to the inherent dangers of terrestrial life. Fern gametophytes usually lack conducting tissue, so they must remain very small since no cells can exist far from the surface that comes into contact with water. Water is necessary for their reproduction since the motile sperm from the male gametophytes must actually swim to the female gametophyte (Figure 14.4, page 388).

FIGURE 14.2

The moss life cycle. The haploid generation dominates. In fact, the diploid sporophyte grows out of the gametophytic tissue and depends on it for nourishment. The diploid sporophyte is essentially a stalked capsule that forms spores (1). These are released and borne by the wind to some suitable, moist place where they germinate and produce the slender thread, the protonema (2). This develops into the familiar carpet we think of as moss. Some tissues produce antheridia that form flagellated sperm, others produce archegonia that enclose a stationary egg (3). The union of the sperm and egg initiate the new sporophyte generation (4).

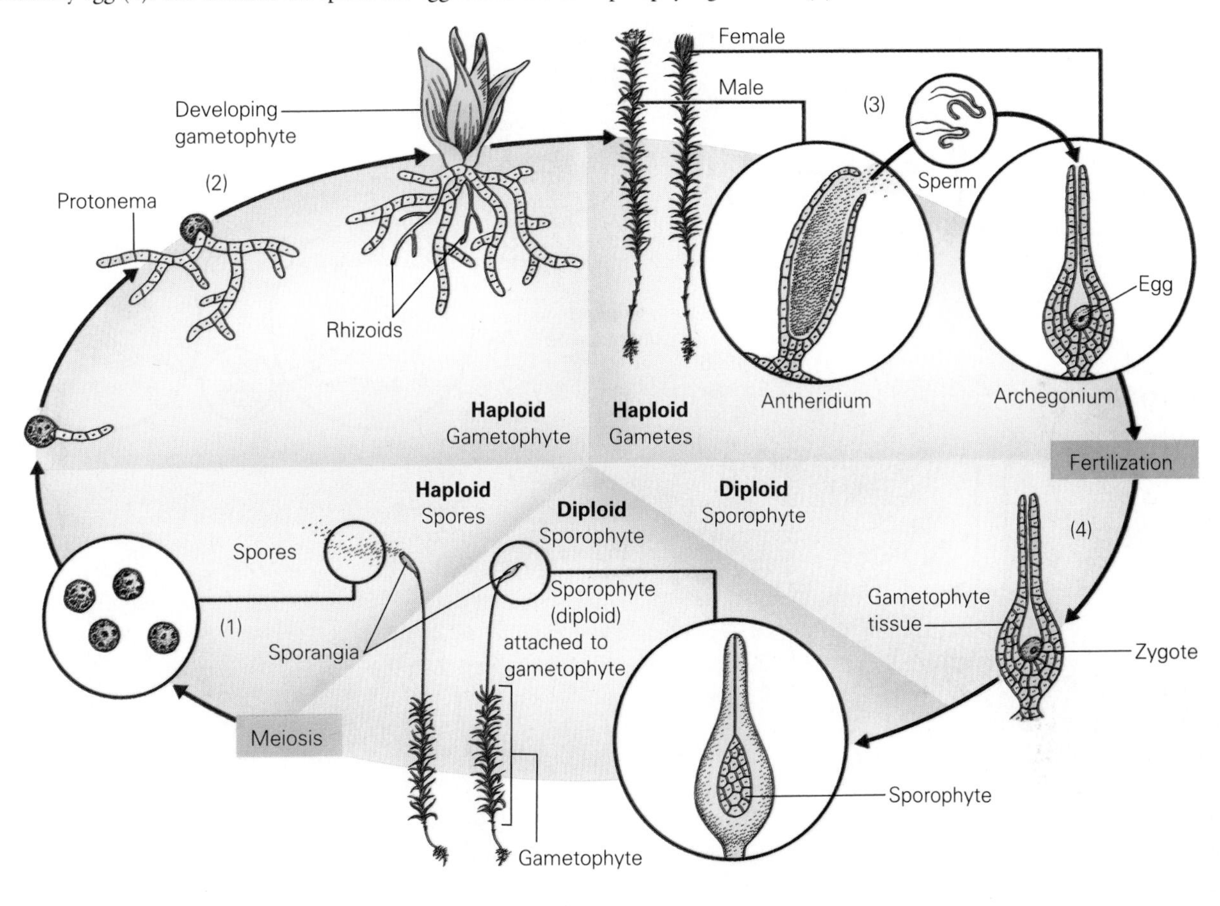

FIGURE 14.3

Many North American woodlands are enhanced by the presence of delicate ferns. In the spring, one may find them only blushing with a hint of the light green that will be their color, as they stand protectively curled, tentatively rising from the forest floor. Later, they will radiate outward, full grown, giving the forest a greenhouse atmosphere. Perhaps you have seen those peculiar brown dots on the "leaf" undersides. These are the spore-producing organs that will give rise to the next kind of generation.

FIGURE 14.4

Life cycle of the fern. In ferns, the sporophyte generation is highly dominant. After fertilization, the sporophyte begins to grow from the gametophytic archegonium (1). It soon produces its own roots and leaves, called fronds, as the old gametophyte withers and dies (2). Underneath the fronds form the dark spots called sori, within which are the spore-producing sporangia (3). The spores are released and, once settled, form flattened structures of the gametophyte generation, called prothalli (4,5). These form either sperm-producing antheridia or archegonia, in which develop the eggs. Fertilization initiates the next sporophyte stage (6).

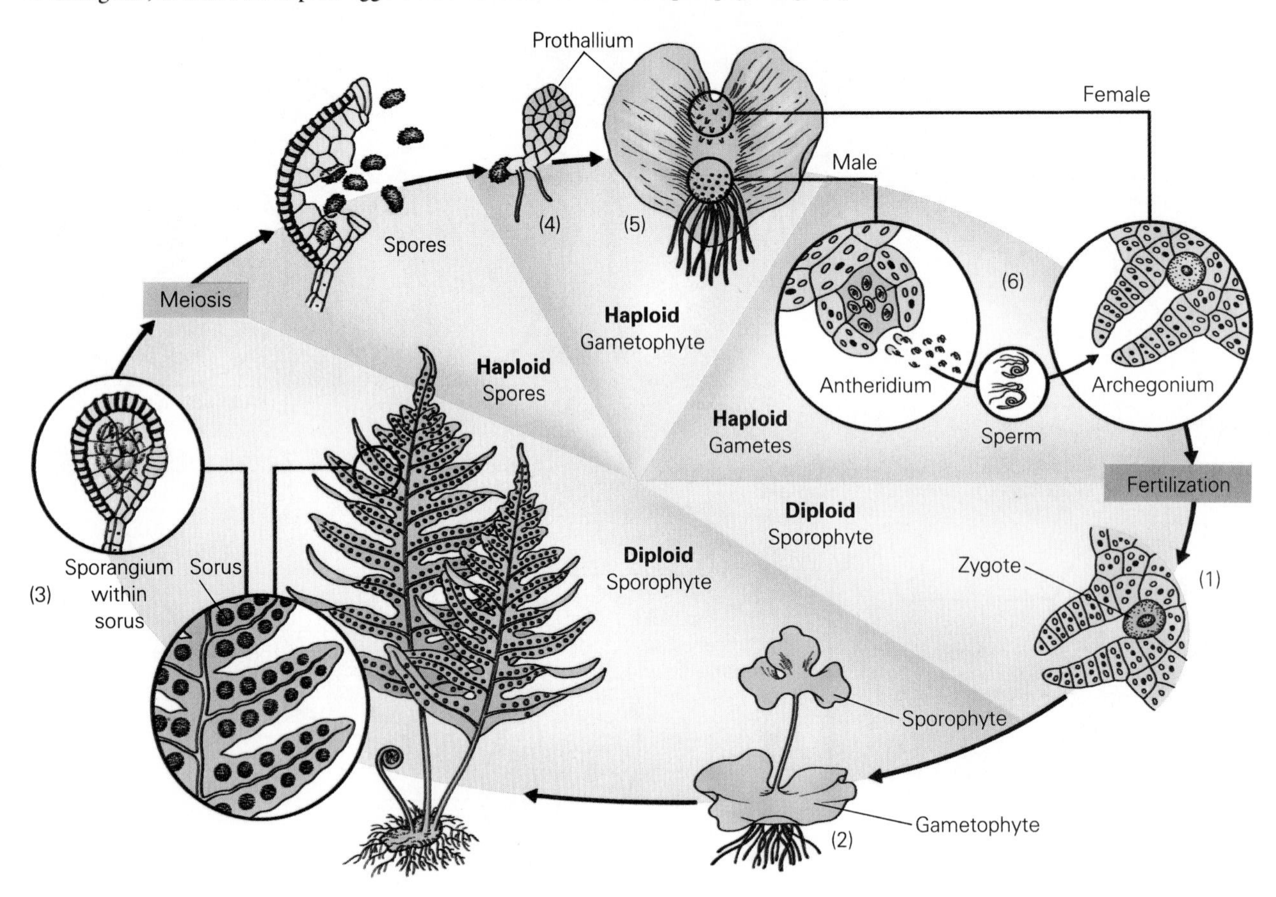

Conifers

Conifers are gymnosperms that, you will probably not be surprised to learn, bear cones. For example, pine trees bear pine cones, if all is going well. But precisely what is a pine cone? (Are you just a bit embarrassed because you really do not know what a pine cone is? You shouldn't feel badly, though, because the question is not as simple as it seems.) To begin with, there are two kinds of cones, male **(staminate)** and female **(pistillate)** (Figure 14.5). Early in their development, certain cells deep within the cones (the "mother cells": **microspore mother cells** in male cones, **megaspore mother cells** in female cones) undergo meiosis and produce spores. In the male cone, each spore will become a pollen grain (the male gametophyte), comprised of only four cells, two of which degenerate. These delicate pollen grains will eventually become enclosed in a thick,

FIGURE 14.5

Alternation of generations in a conifer. With the exception of cells hidden in their reproductive structures and an occasional yellow cloud of windborne pollen, all of the conifer we see is the sporophyte. The conifers have no truly separate gametophyte generation. Cells in the male cones, known as microspore mother cells, enter meiosis and form haploid microspores (1). These microspores form a winged pollen grain and, later, the male gametes (2). Cells in the female cone, the megaspore mother cells, enter meiosis and produce haploid megaspores. These produce the female gametophyte with two archegonia (3). When pollen (the male gametophyte generation) lands on the female cone, a pollen tube grows into the tissue surrounding the egg cell within an archegonium (4). Fertilization occurs when a sperm from a pollen tube reaches an egg cell (5).

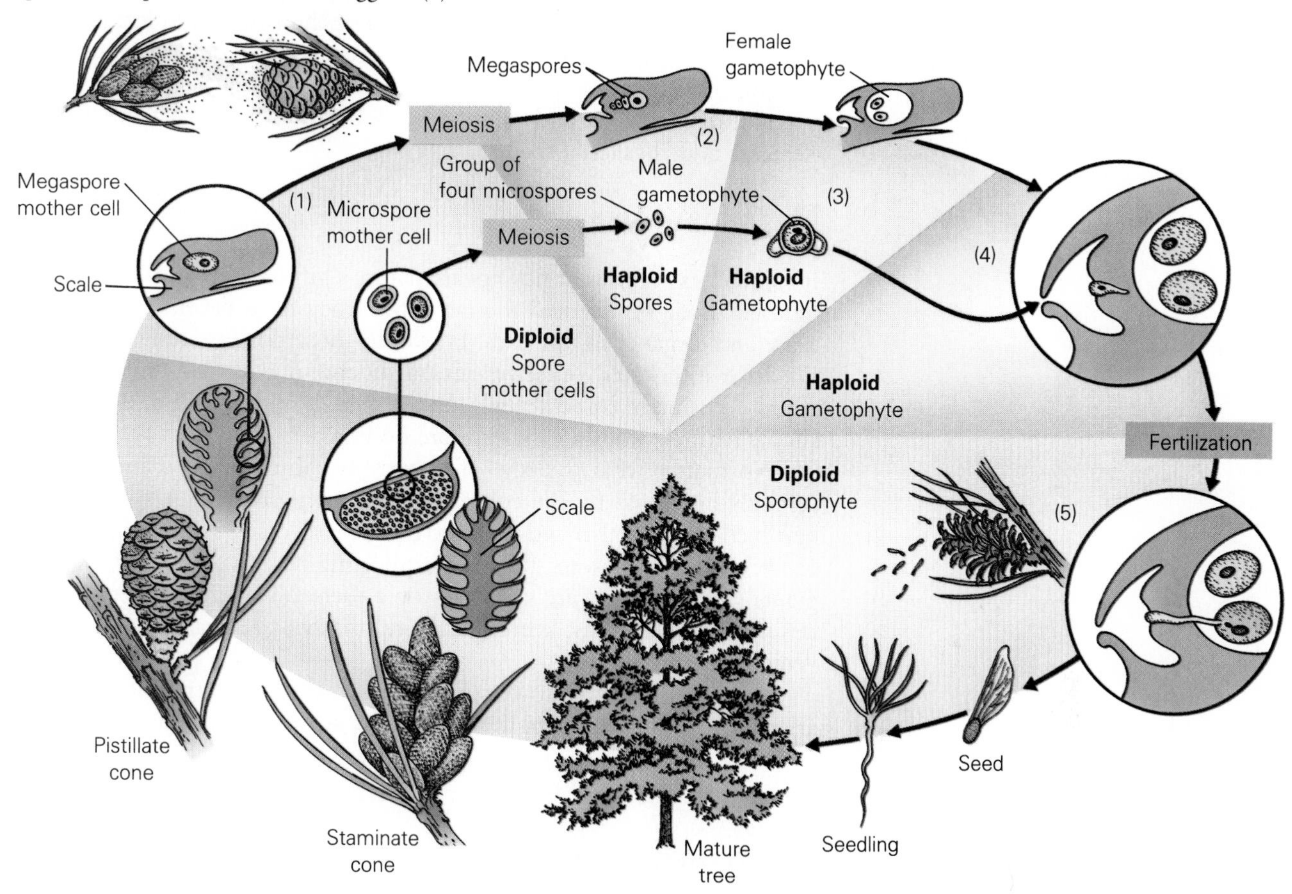

protective covering with thin, winglike extensions. When the scales of the male cone lift, the tiny pollen grains are released to be carried by the wind, perhaps eventually to settle on female cones. The evolution of tough, resistant pollen was an important step in completing the transition to land.

At the base of the female cones are two cells that will undergo meiosis, each producing four cells. Three of these will degenerate, leaving a single spore that will develop into the female gametophyte. Since there are two cells undergoing meiosis, two female gametophytes will be produced at the base of each scale. Then the nucleus of each repeatedly divides, producing gametophytes with multinucleated cytoplasm. Only one of these nuclei will become the **egg nucleus** (it is not known how one comes to be "chosen"). The egg nucleus will be fertilized by the sperm cell nucleus in a pollen grain.

In tracing the route of the pollen grain, we find that when a pollen grain reaches the base of a scale, it usually encounters a moist pool through which it can move until it reaches one end of the female gametophyte. There it germinates and produces a **pollen tube** that slowly elongates, reaching, finally, the archegonium of the female gametophyte. As the tube grows, the pollen grain begins nuclear division, producing the **sperm nucleus,** which will ultimately join with the egg nucleus. After the pollen tube has grown deep into the female tissue, the sperm nucleus is released and makes its way to the egg nucleus, where the two join in fertilization. The zygote, or fertilized egg, undergoes repeated mitotic divisions until it forms an embryonic plant that is surrounded by both protective and nutritive tissue. This entire structure is borne exposed (naked) to the elements and is called a **seed.**

Flowering Plants

It sometimes seems that flowers are nature's gift to humans. We make so much of their beauty and fragrance that they have become symbols of grace and elegance in our lives. However, those flowers are not "for" us. We have interjected ourselves into a functioning natural unit. Flowers appeal to us quite coincidentally. In the wild, it might be said, flowers are simply nature's way of making more flowers.

No matter what the role of flowers in nature, a woodland stroll is certainly enhanced by the colorful accents they provide. A spot of blue here, yellow there, then red attests to the great diversity in flowering plants. However, all the flowers, from tiny, quivering bells in a meadow to forest giants (Figure 14.6), are variations on a theme. Therefore, we can make certain generalizations, keeping in mind that we are dealing with a convenient fiction.

In our generalized flower (Figure 14.7) the **petals** (modified leaves) arise from a widened area, the **receptacle,** at the base of the flower. The

FIGURE 14.6

The incredible diversity of flowers is illustrated by the differences in sizes of these flowers: (a) the tiny ladies' tresses, (b) black-eyed Susan, and (c) one of the world's largest flowers, Rafflesia.

(a)

(b)

(c)

FIGURE 14.7

Overall structure of a flower and fertilization in a flowering plant. The development of the male gametophyte begins when a microspore mother cell (1) divides through meiosis (2), producing four microspores (3). Each will become a mature pollen grain when its coat develops and a mitotic event produces a pollen tube nucleus and a generative cell that (4) eventually divides to form two sperm. These will travel down the pollen tube (5), produced by the pollen tube nucleus, toward the ovary after the pollen lands on the stigma (pollination).

The female gametophyte, the embryo sac, begins its development when meiosis occurs in the megaspore mother cell (6), producing four haploid megaspore nuclei (not shown). Three of these degenerate and the surviving haploid cell divides mitotically three times to produce (7) the eight-celled embryo sac within the ovule. The endosperm cell contains two nuclei that will later fuse with a sperm cell nucleus to form the endosperm. Before entering the ovule, the sperm nucleus divides to form two sperm cells. Since one fertilizes the egg nucleus and the other fertilizes the endosperm nucleus, the result is a diploid zygote and a triploid endosperm.

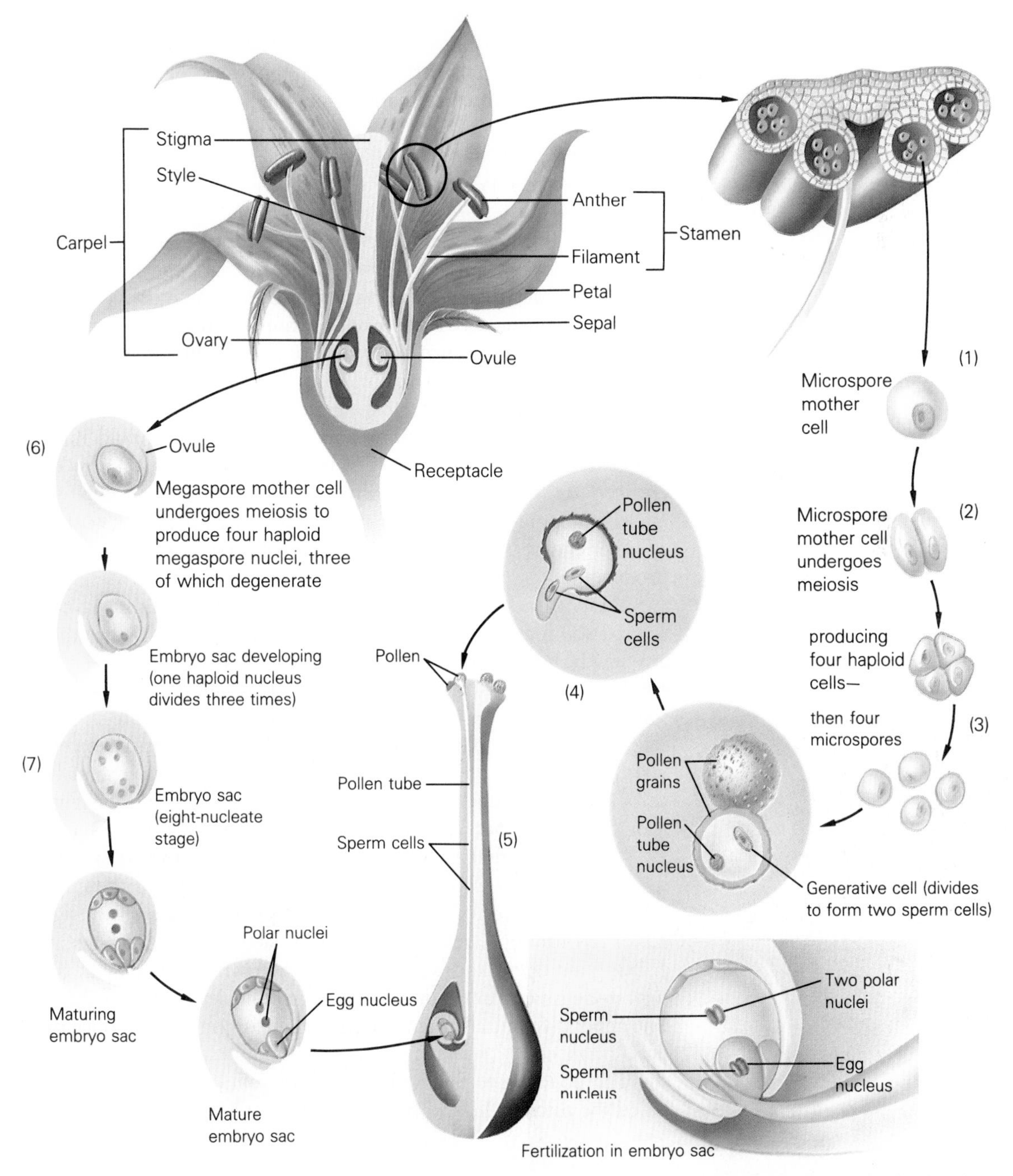

parts of the flower closest to the stem are the leaflike **sepals** that once covered the developing bud. Above these are the petals. In species that must attract animal pollinators (such as insects, birds, or even bats, see Essay 14.1), the petals may be very bright and attractive, some with just the qualities to attract certain species of pollinators. The petals surround the reproductive organs. The male part is the **stamen,** the female part, the **carpel.** Each stamen consists of a slender **filament** capped by an **anther** where, following meiosis, the male gametophyte is produced and released as **pollen.** Each carpel has three parts: the **ovary, style,** and **stigma.**

The ovary is the widened base of the carpel. It will later form the fruit. In its early developmental stages, the ovary contains the cells that will undergo meiosis to produce the female gametophyte. After fertilization, it will house the growing seed and then the developing embryo. The style is the slender stalk that arises from the ovary. At its tip is the stigma, a sticky or hairy structure that receives the pollen.

How Flowers Work

Within the flower's young ovary are the **ovules** (see Figure 14.7). Ovule technically means "little egg," but these are not eggs. They contain the sporophyte cells that will produce the female gametophyte. Each ovule consists of a megaspore mother cell surrounded by nutritive and protective tissues. The megaspore mother cell is diploid, but it will produce the haploid female gametophyte by meiosis. The meiosis results in four haploid cells, but three disintegrate, leaving a functional megaspore cell that goes through three rounds of mitosis to produce eight haploid nuclei. This eight-nucleate structure is the mature female gametophyte, now called the **embryo sac.** Cell walls then form in such a way that all but two of the nuclei (the **polar nuclei**) are isolated. There are now three isolated nuclei at either end of the embryo sac, and one of these begins changes that will produce the egg nucleus.

While this is going on, the male gametophyte is also developing in preparation for that glorious union of sperm and egg. Within the anthers are chambers called **pollen sacs** that contain numerous diploid cells. These are the flower's microspore mother cells. Through meiosis, each produces four microspores. Each of these then doubles by mitosis to produce two-celled male gametophytes. One of the cells is the **generative cell,** which will divide again to produce two sperm cells, the other, the **pollen tube cell.** The pollen tube cell then expands to enclose the generative cell. A tough, resistant coat forms over the two. This is the **pollen grain** (Figure 14.8). So we see that the gametophyte stage, so prominent in more primitive plants, is now represented only by the pollen grain and embryo sac.

Pollination occurs when the pollen grain lands on the stigma, but actual fertilization does not occur until later. On the moist stigma, the pollen grain opens and sends a long tube down the style, with the pollen tube nucleus near the tip and the generative nucleus lingering farther back. The generative nucleus then undergoes mitosis once to produce two identical sperm.

The pollen tube finally penetrates the ovule through a tiny opening called the **micropyle,** and the two sperm enter the embryo sac. Only one

ESSAY 14.1

The Coevolution of Pollen and Insects

We have seen the evolution of tiny, resistant pollen that requires no water and can even be carried about by light winds. This development, then, set the stage for the development of another means of pollen transport. As flowering plants were evolving on earth, so were other species. Many of these developing forms began to react to each other and to utilize one another in increasingly intricate and complex ways. As the insects were spreading out, invading new niches, taking advantage of what the environment offered, some began to exploit the food in flowers. These insects, taking food from one flower after the next, became vehicles for pollen transport. As time passed, some flowers developed an increasing dependency on the visits of pollen-bearing insects.

In time, plants began to compete for the attention of insects. They developed increasingly attractive colors that insects can see (such as those with shorter wavelengths, like violet), sweet nectar, and shapes that are somehow appealing to insects (such as broken margins—as are produced by petals). Those that attracted more insects left more offspring, and some lines of flowers became true specialists in attracting certain insects. Insects, on the other hand, became increasingly specialized so that they could tap the offerings of the specific sorts of plants that yielded the nutrients most closely fitting their specific needs. The insects and the flowers changed in ways that enhanced their interdependency and specializations in a steady progression toward finer attunement to each other. Such reciprocal influences on the development of interdependent species is referred to as *coevolution*.

of these will fertilize the egg nucleus. The resulting zygote then begins the next sporophyte generation. Meanwhile, the other sperm penetrates the large binucleate cell in the center of the embryo sac. The three nuclei fuse to form a special nutritive tissue called the **endosperm.** (The endosperm tissue also provides us with the flour and meal from wheat, corn, rice, rye, millet, and oats.)

The seed is well adapted to the terrestrial life, as evidenced by its tough, protective **seed coat.** Once the seed is released into the environment, it will remain dormant, germinating (beginning growth) only when triggered by the proper stimuli. This waiting period varies enormously among different kinds of seeds. For example, the seeds of the mangrove

FIGURE 14.8

Two evolutionary advances that helped the gymnosperms adapt to life on land. The pollen grain (right) carried the male gamete safely in a tough, resistant case, and the embryo came to be protected in an equally tough container of its own, the seed. The seed contains the embryo of the plant, some stored food, and a surrounding seed coat, which is hard and water-resistant. The protective seed coat makes it possible for the embryo to wait until conditions are right for growth.

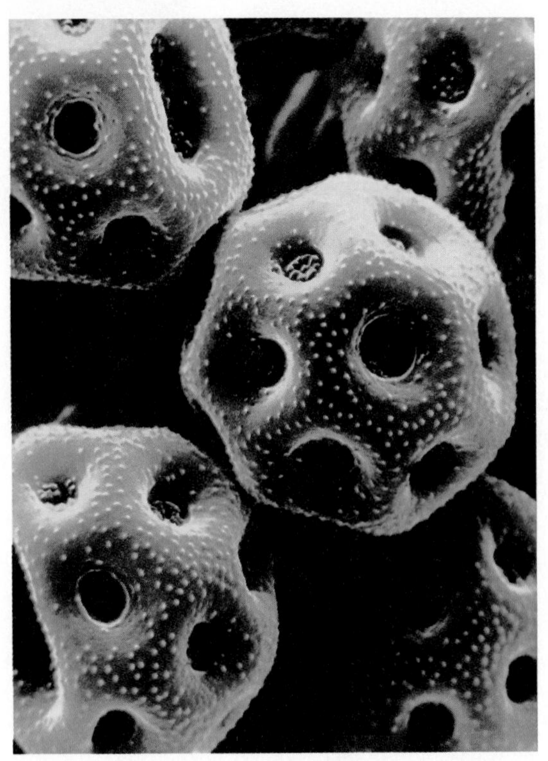

may start to germinate while the fruit is still on the tree, whereas Indian lotus seeds have been known to lie dormant for as long as 400 years, frustrating generations of gardeners.

Unlike the "naked" seeds of gymnosperms, the angiosperm seed develops within an ovary. As the seed develops, the ovary undergoes rounds of mitosis and grows to form a structure called the **fruit.** You may have an idea of what a fruit is, but technically, a fruit is any structure that develops from the ovary and encloses ovules or developing seeds. Fruits come in all sorts of sizes and shapes, ranging from grains and nuts (dried fruits) to pineapples and raspberries (clusters of small fruits) to fleshy tomatoes and apples. (You thought a tomato was a vegetable, didn't you?)

These, then, are the events of reproduction in plants. Reproduction is followed by a remarkable chain of developmental events that eventually produces the familiar kinds of plants that grace our world. As one might

FIGURE 14.9

Embryonic development in seeds of *Capsella,* or shepherd's purse. Note the uneven first division. The larger cell will grow and ultimately become the basal cell. The basal cell will form an elongated suspensor structure that probably functions in moving nutrients to the embryo while it is within the seed. The smaller cell will continue dividing linearly for a few divisions. Then it will begin dividing in every direction, forming a globular embryo. The embryonic tissue will grow increasingly specialized until, as we see in (g), several distinct tissues have formed.

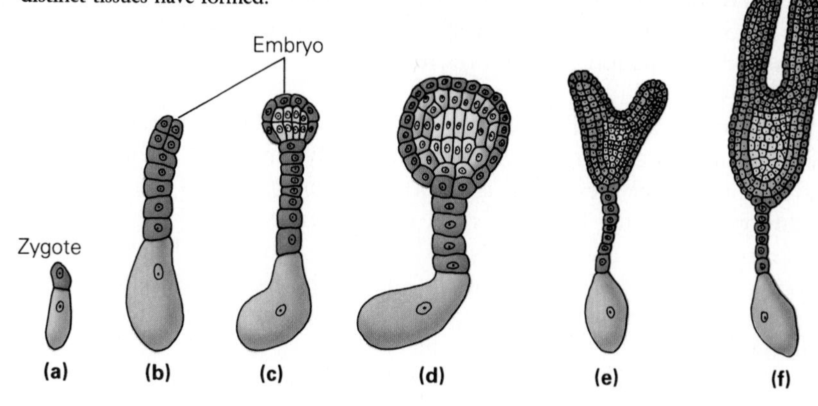

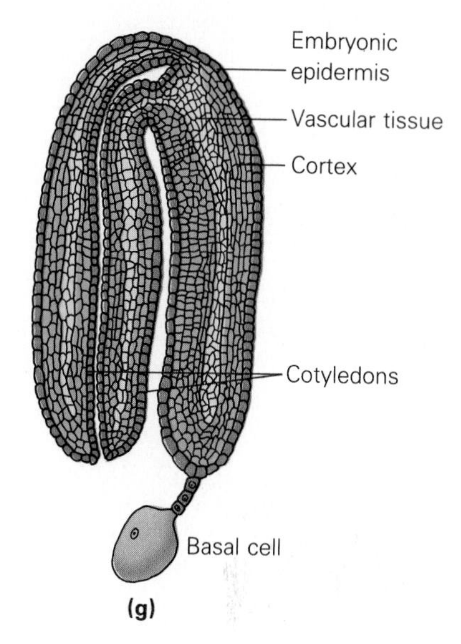

expect, different kinds of plants develop in different ways, but because of a common heritage, they all progress along a basic developmental theme.

PLANT DEVELOPMENT

Plants develop differently from animals in two basic ways. First, because of the rigid walls of plant cells, new cells are prevented from developing in the interior tissue, but this is not the case in animal tissue. Plants therefore grow by adding new cells to the periphery of the established cells. Second, whereas most animals stop growing at some point (maturity), plants continue to produce new cells and to grow throughout their lives.

Let's begin by considering the early development of the angiosperm, starting with the seed. Before the seed germinates (sprouts), one of the enclosed nuclei, the endosperm nucleus that will nourish the seedling, begins a period of rapid division followed by the mitotic divisions of the zygote itself. As we see in Figure 14.9, the first division of the zygote is uneven, so one cell is larger than the other. The smaller cell continues to divide, forming a chain, or stalk, of cells. Cells near the end of the stalk then begin to divide in every direction, producing a globular **embryo.** As the divisions continue, the cells in the embryo begin to change and become different from one another. Such a developmental process is known as **differentiation.** Some will eventually produce the **epidermis,** or protective "skin," typically one-cell thick and covered with a waxy layer; others will form the **vascular tissue,** which transports fluids, and still others will form the **cortex,** or structural tissue.

As differentiation continues, the globule flattens and elongates, forming the embryonic leaves, or **cotyledons.** Seeds of cone bearers, such as pines, have many cotyledons, but those of flowering plants have only one or two. Those with one are called **monocotyledons;** those with two, **dicotyledons.** In general, broad-leaved plants are dicotyledonous (or **dicots**) and narrow-leaved plants such as grasses are monocotyledonous (or **monocots**). Functional differences in monocots and dicots are shown in Figure 14.10.

There are areas of the plant in which the cells do not mature or differentiate; instead, they remain permanently embryonic. The division of cells in two of these areas is responsible for growth in the length of the roots and stems. They are called **apical meristems** and are located at the root tips and the stem tips. As the roots and stems lengthen, new embryonic tissue is formed near the tips as the older meristematic cells are left behind to mature and specialize. (The root meristem is seen in Figure 14.11.)

The apical meristem of the stem differs slightly from that of the root in that the stem tip lacks a protective cap and it may have leaves. The stem tip bears a **terminal bud,** which in most plants is safely shielded by highly modified leaves called **bud scales.** The bud scales fall off when stem growth begins.

In the trees and shrubs of the temperate zones, the terminal bud opens in the spring, giving rise to a new area of tissue. Then, in late summer or

FIGURE 14.10

Summary of differences between monocots and dicots.

Monocot

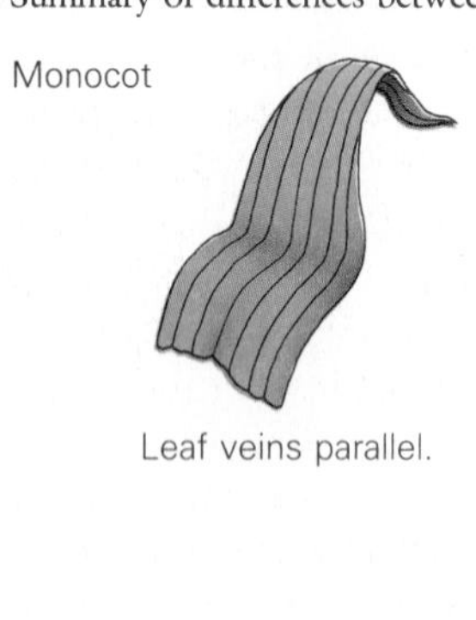

Leaf veins parallel.

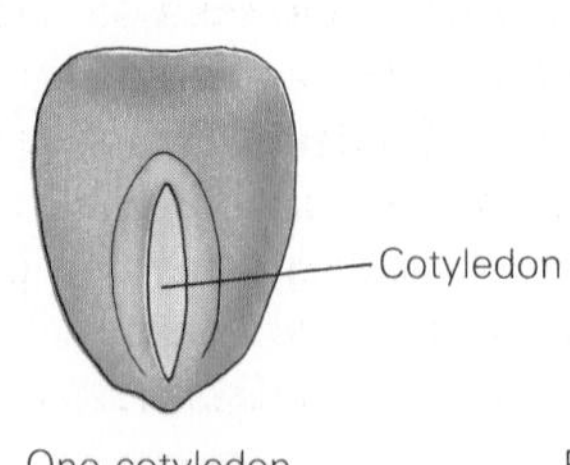

One cotyledon (as in corn).

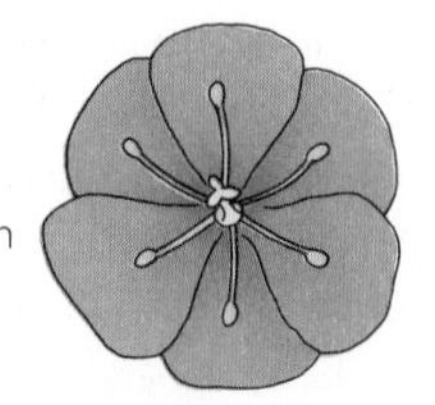

Flower parts in threes or multiples of three.

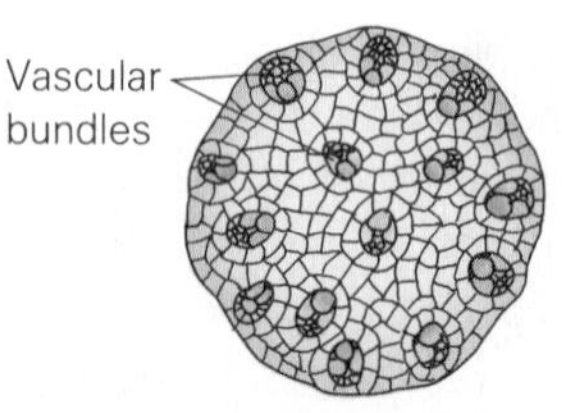

Vascular bundles in stem scattered. No vascular cambium.

Dicot

Leaf veins branching (netlike).

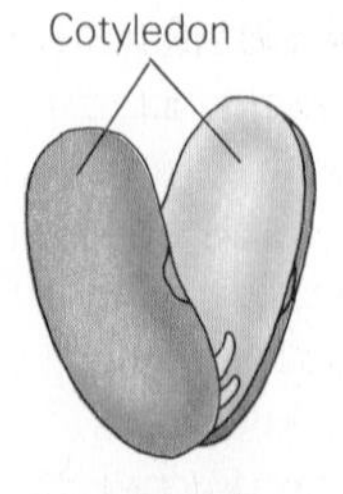

Two cotyledons (as in beans).

Flower parts in fours or fives or multiples of four or five.

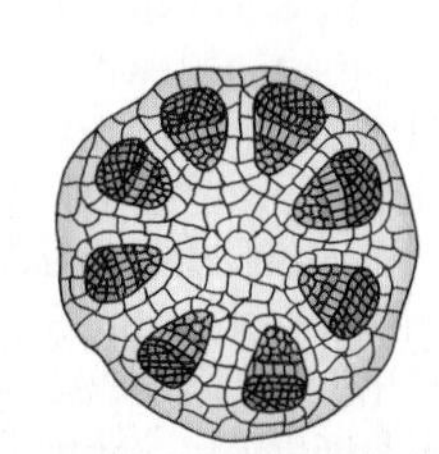

Vascular bundles in cylindrical arrangement in stem. Vascular cambium present.

FIGURE 14.11

The development of the root. The apical meristematic region is shielded by the large cells of the root cap (a). Such protection is necessary as the delicate root pushes its way between soil particles. The apical meristem is the site of intense mitotic activity, hence these cells, not yet mature, are smaller. As they mature, they increase their cytoplasm and cell organelles, eventually elongating. The tissues of the root (b), are basically arranged as two concentric cylinders separated by the endodermis. Here, the roles of the various tissues are indicated. Exterior cells may send out delicate root "hairs" through which water is readily absorbed. The profusion of root hairs can provide an enormous absorptive surface.

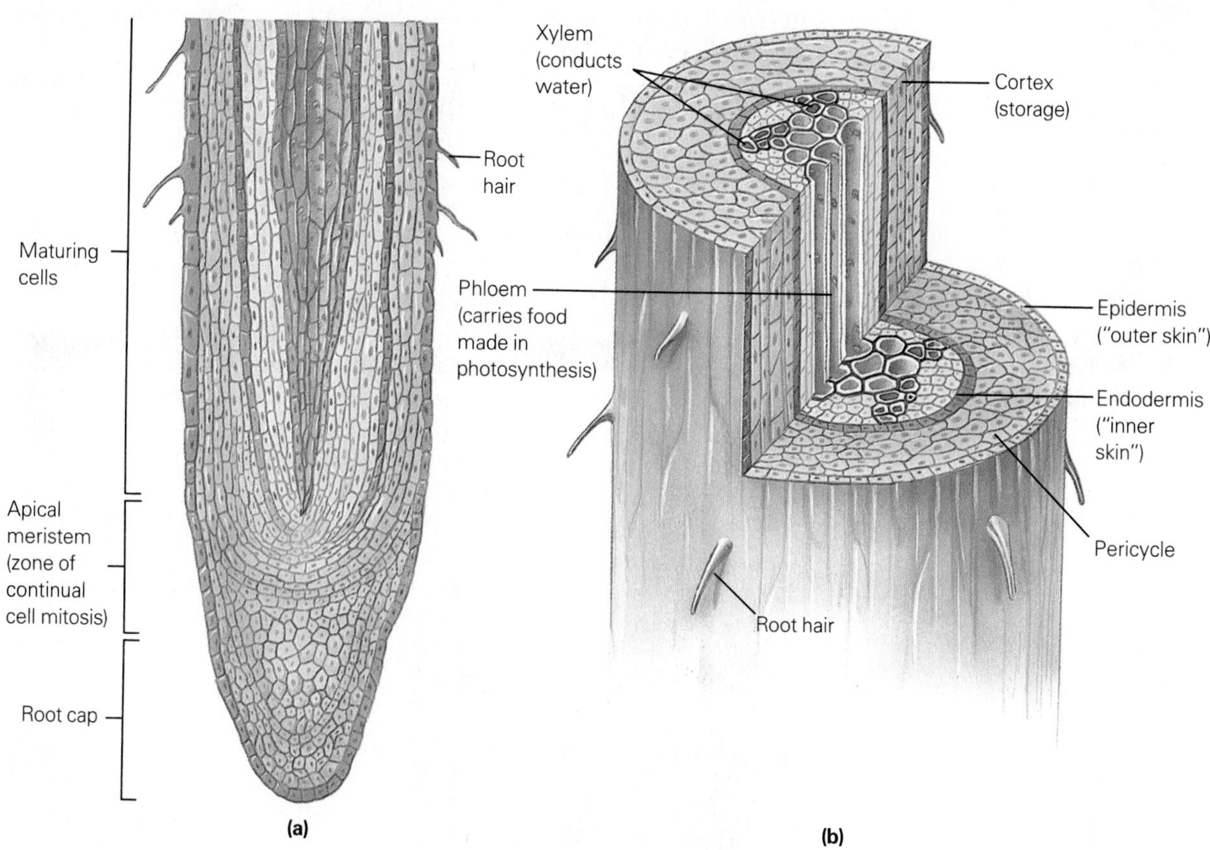

fall as growth ceases, a new terminal bud forms. This means that one can determine how much a stem has grown in any year by measuring the distance between the bud scale scars. If the weather was unfavorable in a certain year, the distance will be small (Figure 14.12).

The lateral buds on stems end up doing nothing if their growth is suppressed by hormones from the terminal bud at the tip. If the buds are not suppressed, they grow into branches. Such growth occurs, for example, if the terminal bud has been removed. Gardeners sometimes snip off the terminal bud in order to encourage a plant to be more bushy.

Just as apical meristems allow growth in length, **lateral meristems** (also called **cambia;** singular, cambium) are the tissues that enable growth in girth or diameter. This kind of growth, however, presents plants with a special problem. As a woody plant grows in diameter, the vascular system must be enlarged by adding cells to the **xylem,** which carries water up from the roots, and to the **phloem,** which carries food down from the leaves. Since these areas lie deep within the plant, it might be asked, how

FIGURE 14.12

The growth area of a stem. One can read a bit of the plant's history by noting the distance between bud scale scars, the distance being greater for good years when the stem has been able to grow rapidly.

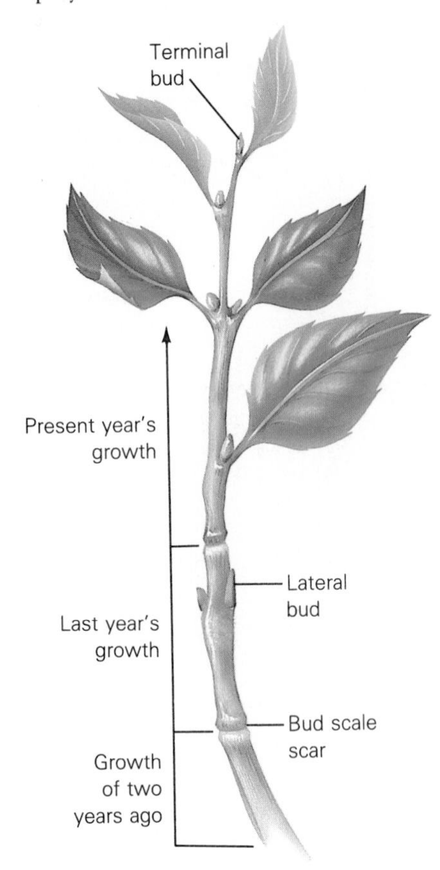

can new cells be added without rupturing the plant's surface? The answer is that there are two kinds of cambium. The **vascular cambium** produces new xylem and phloem and the **cork cambium** produces new **cork,** a protective tissue that functions as a kind of epidermis, or skin, which protects the tissue inside that would otherwise be exposed as growth caused the surface layers to rupture. As an aside, the cork in wine bottles does not occur naturally. Cork growers damage the bark of the cork oak, causing the tree to form new, smoother cork cambium. The tissue is then removed, cut into cylinders, and placed into wine bottles in such a way that they crumble at any attempt to remove them. The vascular cambium lies between the xylem and phloem, and as it produces new cells, those toward the inside of the stem form xylem and those toward the outside form phloem. In temperate regions, growth rings may result from seasonal differences in the size of the xylem cells that are formed. The larger cells are produced when more water is available in the moist spring. The smaller cells are produced in the drier summer (Figure 14.13). During the winter, the vascular cambium is dormant and no new cells are produced.

Plants are able to carry out a number of very specialized functions that belie their simple and seemingly inactive appearance. For example, the thick-walled tissues called **sclerenchyma** and **collenchyma** provide structural supports (Figure 14.14). **Parenchyma,** on the other hand, is composed of thin-walled cells with large vacuoles. Some parenchyma cells contain chloroplasts.

Plant development, we see, is very complex and coordinated. Obviously, there must be some means of regulating such activity. That regulation, we will see now, is largely due to the activities of special kinds of molecules to which the plant is particularly sensitive.

PLANT HORMONES

The seeming simplicity of plants sometimes leaves us unprepared for their complexities. As a case in point, researchers have for years been trying to unravel the mysteries of plant hormonal control. **Hormones** are substances formed in one part of an organism that are transported to other places where they cause changes. Many aspects of plant growth and development are under the control of these itinerant chemicals, and so the search is not only fascinating but increasingly important as we grow more reliant on these air-cleansing, green, intermediate links with our life-giving sun. Let's briefly consider three of the better known groups of plant hormones involved in regulating growth and development: *auxins, gibberellins,* and *cytokinins* (Table 14.1).

Auxins

Some of the first experiments on plant hormones were conducted by one Charles Darwin and his son, Francis, and reported in an 1881 publication.

After noticing that plants tend to bend toward light, the Darwins wondered if the plants were responding to some sort of signal from the stem tip. So they covered the shoot tips of growing plants and exposed them to light from the side. The plants now did not bend toward the light. However, they found that if the tips were capped with transparent

FIGURE 14.13

Section from the stem of a dicotyledonous plant. The woody xylem shows two years' growth, the winters being recorded in the areas of the smaller cells (and hence less growth). The vascular cambium produces xylem on its inner surface, phloem on its outer surface.

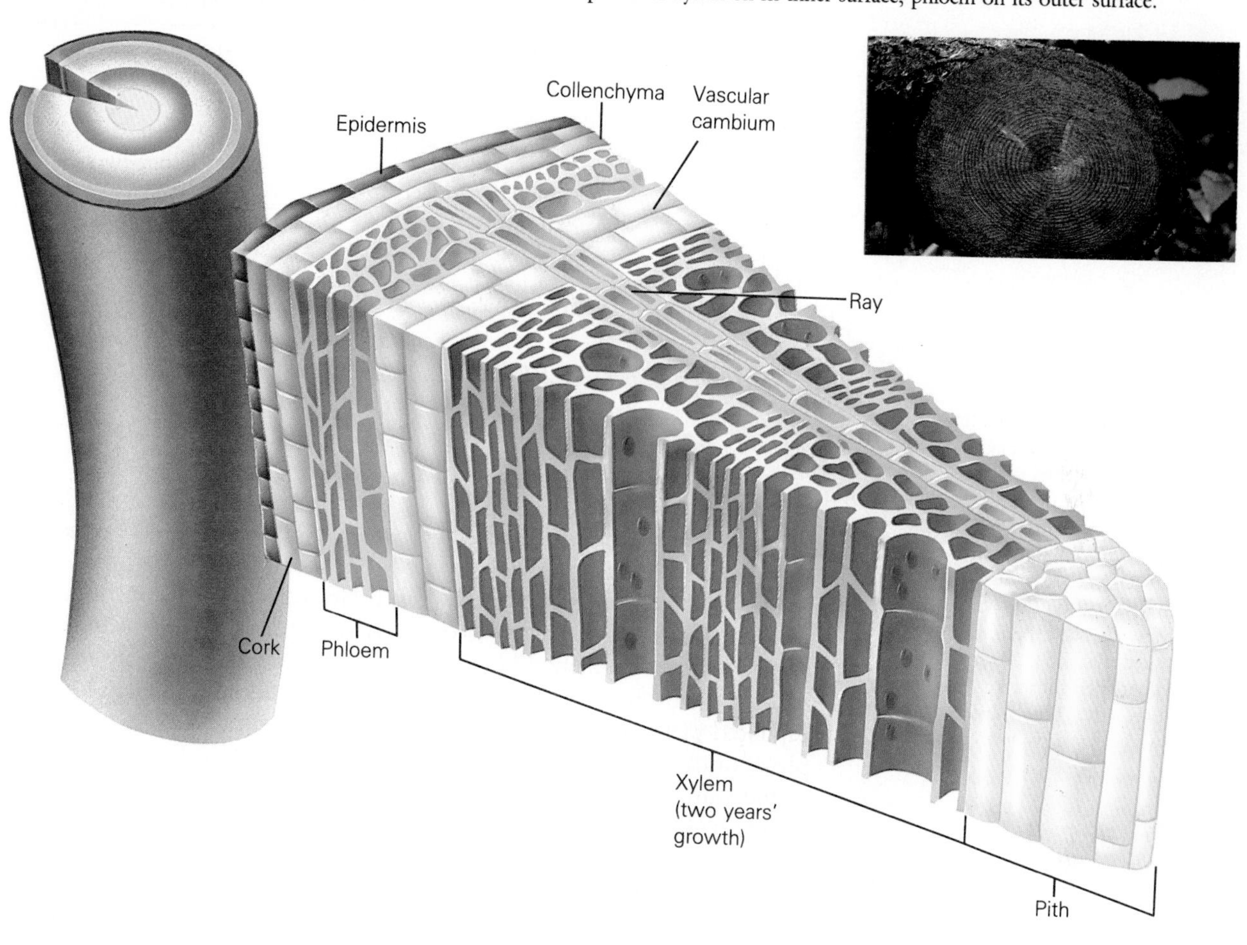

FIGURE 14.14

Sclerenchyma (a) and collenchyma (b). Both lend support to plants. The sclerenchyma dies at maturity, its thickened walls remaining. The collenchyma is alive when functioning, and its strong, supple walls are prevalent in growing plants.

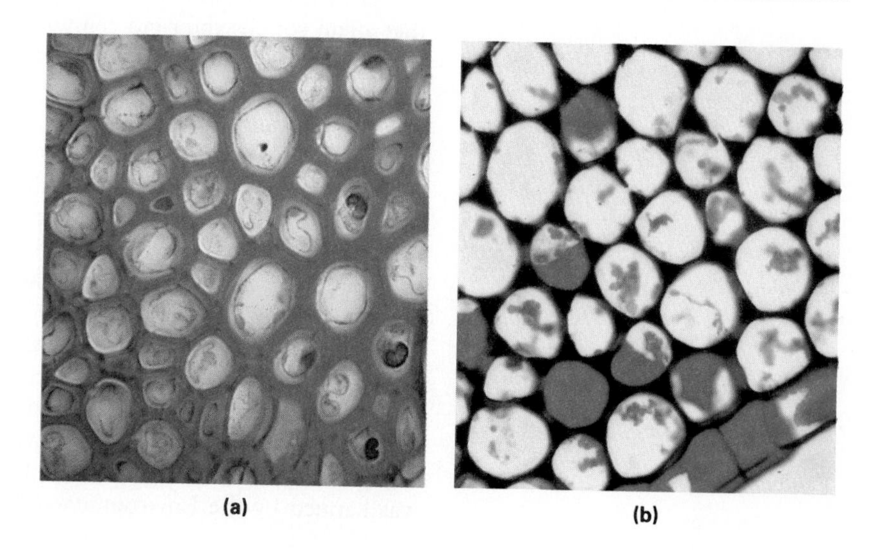

TABLE 14.1
Plant Hormones

GROUP	ACTIONS	LOCATION
Auxins	Control elongation; aid in fall leaf drop	Produced in stem tips; found in all parts of plants
Gibberellins	Stimulate cell division and elongation; aid in pollen germination, break seed dormancy	Found in all parts of plants, especially immature seeds
Cytokinins	React with auxins to produce growth	Found in actively dividing tissue areas: roots, seeds, fruit; also in sap

glass, normal bending occurred. They wrote, "We must therefore conclude that when seedlings are freely exposed to a lateral light some influence is transmitted from the upper to the lower part, causing the latter to bend."

In 1926, the Dutch physiologist, Fritz Went, discovered something else about that "influence." He cut off the tips of emerging oat seedlings and set them on a block of gelatinous agar for about an hour. The decapitated plants were kept in a dark place. Then they were removed and pieces of the agar on which the tips had rested were placed along one side of the stump. Within an hour, the plant began bending in the opposite direction (Figure 14.15). Thus, the "influence" was believed to be chemical and was called **auxin** (from the Greek *auxein,* to increase). It was later found that auxin is generally produced in the stem tips and seeps through the cells to the rest of the plant, instead of traveling in pipelines of xylem and phloem.

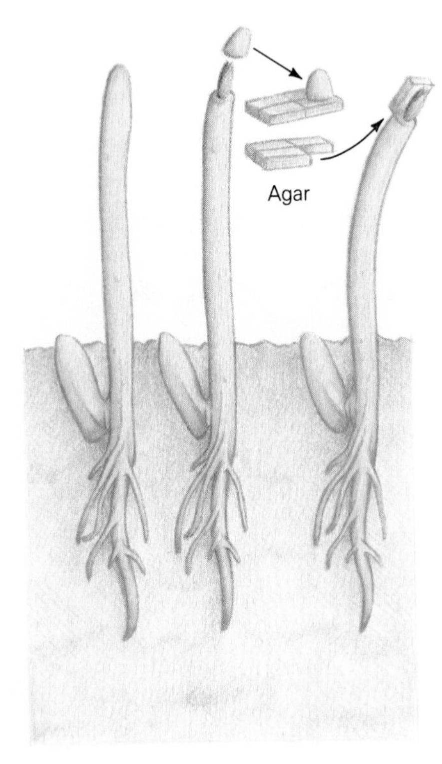

FIGURE 14.15
In this classic experiment, the tip was cut from an oat seedling and placed on an agar cube that absorbed any fluids that might seep from the tip. Then the block was placed off center on a decapitated seedling. As the seedling grew, it bent away from the side with the agar block (meaning that cells on this side were growing faster). The experiment demonstrated that the fluids from the plant tip contained substances that enhanced growth. Those substances, it was discovered, were plant hormones.

How does auxin control elongation in plants? It is currently thought that auxin softens the rigid cell walls of plants and allows water to swell the cells. Thus, plants get taller because of cell elongation. However, they are believed to get wider because auxins stimulate cell division in the vascular cambium.

Small amounts of auxin may stimulate the growth of roots, and although it doesn't affect the growth of leaves, it plays a role in the leaf drop and is often used as a defoliant. With the approach of fall, the plant draws certain ions, amino acids, and sugars from its leaves, and auxin helps to break down the cells that hold the leaf to the stem. The application of auxin also stimulates fruit to grow without the flower having ever been pollinated, thus growers can produce seedless varieties.

A synthetic form of auxin, called 2,4,5-T, is a powerful weedkiller (as is auxin at high concentrations) and was used as one ingredient in Agent Orange by the military in southeast Asia during the Vietnam War to expose enemy movement through the jungle (Figure 14.16). The use of 2,4,5-T was banned by the Environmental Protection Agency in 1979 for most

FIGURE 14.16

Agent Orange was sprayed over the jungles of Vietnam to defoliate trees and make trails more visible. It was later implicated in a host of severe health problems among veterans exposed to it.

uses in the United States, partly as a result of the controversy over its effects as a component of Agent Orange. Laboratory animals exposed to the chemical have developed birth defects, miscarriages, leukemia, and liver and lung diseases. The problem is apparently caused by trace amounts of a contaminant called dioxin, unavoidably produced in the manufacture of the defoliant. A number of military personnel exposed to Agent Orange have developed a wide range of medical problems, including cancer.

Gibberellins

While Went was performing his agar experiments, E. Kurosawa in Japan was looking for the cause of the "foolish seedling disease" of rice. The disease caused the plants to become spindly and pale before finally collapsing. It turned out that the culprit was a parasitic fungus that contained substances called **gibberellins.**

Gibberellins are found in most, if not all, plants, and they have some rather surprising properties. For example, they cause dramatic increases in stem length by stimulating cell division and cell elongation in both leaves and stems. I recall that in my first garden, the lettuce, to my astonishment, grew to about five feet tall. It was very embarrassing. However, it turns out that in some plants, certain weather patterns can produce effects that are similar to the effects of gibberellins (Figure 14.17), and the weather in Santa Barbara was right for tall lettuce.

Finally, gibberellins stimulate pollen germination and the growth of pollen tubes in a number of plant genera and can break the dormancy of many kinds of seeds that normally can be aroused only by cold or light.

FIGURE 14.17
Gibberellins have a dramatic effect on stem growth, as seen in these plants. The plant on the left was grown normally, while the one on the right was treated with gibberellins.

Cytokinins

In 1941, a Dutch physiologist, J. van Overbeek, found that coconut milk (which is really a liquid endosperm, or "seed food") contained a peculiar growth factor unlike anything known. The factor, whatever it was, not only accelerated the development of plant embryos, but it increased the growth rate of isolated cells in a test tube. Also, a drop of coconut milk could stimulate mature, nonmitotic cells to begin dividing again.

What, exactly, in the coconut milk was causing such changes? After all, coconut milk is rich in a variety of substances. Years of research efforts to isolate the growth factor proved futile. Then, researchers began looking for the factor in something besides coconut milk, something easier to work with. Somehow a graduate student discovered that old herring-sperm DNA could make tobacco cells divide. (One wonders what possessed him to investigate the effect of aging fish sperm on tobacco.) Others picked up on the line of investigation, and it was soon found that any stale DNA would provide the factor. Apparently, it was a breakdown product of nucleic acid. The product was finally isolated. It was called ***kinetin*** and was found to be in the chemical group called **cytokinins** (from "cytokinesis," cell division).

Cytokinins can react with auxin to produce a variety of growth effects, such as rapid cell division. They can also work alone to enhance germination once it has started. In addition, cytokinins can somehow keep leaves from turning yellow after they are removed from the tree. Apparently, they keep the DNA sequences that function in young, healthy leaves from ceasing activity. In general, however, the workings of cytokinins remain a mystery.

Reproduction and development in plants is highly regular, specialized, and coordinated. Cells divide, grow, and change in a silent ballet that moves strangely, and often in complex ways, to their own cellular choreography. Even when it seems plants are at rest, they are not. On a peaceful forest walk, those plants that shade us and soften our steps are not resting. They are working. They are alive.

SUMMARY

1. Disadvantages of sexual reproduction include gene dilution, gamete waste, and possible breeding failure. The diploid condition (with masked recessives) and the high variability of sexually reproducing populations help protect them from devastation if conditions change.
2. The haploid gametophyte generation is dominant in most green algae, whereas the diploid sporophyte stage is dominant among many land plants today. In complex green algae, a large, protected, immobile and energy-rich egg is fertilized by a motile "sperm."
3. In mosses, a flagellated sperm swims from an antheridium to an egg, which is held in an archegonium. The zygote develops into a sporophyte. Cells within the sporophyte undergo meiosis and form spores that will develop into gametophytes. The gametophyte will form either an antheridium or an archegonium, or both. The gametophyte generation is dominant.
4. Fern gametophytes lack conducting tissue so they are small. The sperm must swim through water to reach the egg within the female gametophyte. The sporophyte is dominant.
5. In conifers such as pine trees, a microspore mother cell in a male (staminate) pine cone undergoes meiosis to produce pollen grains, each comprised of two cells, a protective coat, and winglike extensions that carry the pollen on the wind. When it lands on a female cone, pollen germinates, producing a pollen tube that elongates and reaches the female archegonium. In female (pistillate) pine cones, megaspore mother cells undergo meiosis, each producing four cells. One of these will develop into a female gametophyte with a multinucleated cytoplasm. One of these nuclei, the egg nucleus, is fertilized by a sperm nucleus in a pollen grain.
6. A generalized flower is composed of leaflike sepals that once protected the bud; the petals that may attract pollinators; the stamen, which is the male reproductive organ consisting of the filament and anther; and the carpel, which is the female reproductive organ consisting of the ovary, style, and stigma. The male gametophyte is produced in the anther and is released as pollen. The widened base of the carpel, the ovary, contains cells that will produce the female gametophyte, houses the seed and growing embryo, and finally forms the fruit.
7. The ovules of the ovary consist of the megaspore mother cell and nutritive and protective tissues. The diploid megaspore mother cell undergoes meiosis to produce four haploid cells. Only one survives as a functional megaspore cell and divides mitotically to form a female gametophyte, now called the embryo sac, which consists of eight haploid nuclei in seven cells. The two nuclei sharing a cell are called the polar nuclei. The egg forms from one of the three haploid nuclei in cells isolated at one end of the ovary.

8. Each of the diploid microspore mother cells within the pollen sacs of the anther undergoes meiosis, producing four haploid microspores. Each microspore divides mitotically and produces the male gametophyte (pollen grain), which consists of a sperm cell and a pollen tube nucleus protected by a tough coat.
9. When the pollen grain lands on the stigma pollination occurs. The pollen grain sends a long tube down the style: the pollen tube nucleus is near the tip of the tube and the generative nucleus, which divides to produce two sperm, is farther back. When the pollen tube enters the ovary through the micropyle, the two sperm enter the embryo sac. One sperm fertilizes the egg and forms the zygote. The other sperm fuses with the two polar nuclei and forms a nutritive tissue (endosperm). The embryo develops one or more cotyledons that serve as the immediate source of food.
10. The embryo develops within a seed where it is protected by a tough seed coat until proper conditions trigger germination. The seed develops within the ovary, which divides mitotically forming the fruit.
11. Plant development differs from animal development in two ways: plants must grow by adding cells to the periphery and plants do not stop growing.
12. Before a seed sprouts, the endosperm nucleus divides rapidly. Then the zygote divides and forms a small cell that divides again to produce a stalk of cells. The end of the stalk divides to form the embryo. Cell differentiation (specialization) within the embryo produces cells that will become epidermis for protection, vascular tissue for transportation, or cortex.
13. The embryo elongates and forms embryonic leaves called cotyledons—one in monocots and two in dicots.
14. Some plant tissue (meristem) remains in an undifferentiated condition. The apical meristem is responsible for growth in length of roots and stems. Lateral meristem (cambium) allows growth in diameter.
15. Other specialized tissue includes the conducting tissues, xylem and phloem, thick-walled supporting tissues, sclerenchyma and collenchyma, and thin-walled parenchyma cells that contain many vacuoles and, in some, chloroplasts.
16. Hormones are substances produced in one part of an organism that are transported to other places where they cause changes. Three types of plant hormones involved in growth and development are auxins, gibberellins, and cytokinins. Auxins cause cell elongation, which enables plants to bend toward the light. Gibberellins increase the stem length by stimulating cell division and cell elongation. Gibberellins also stimulate pollen germination, the growth of pollen tubes in certain plants, and break the dormancy of many kinds of seeds. Cytokinins increase growth of plant embryos or plant cells in tissue culture and react with auxin to cause rapid cell division. Cytokinins enhance germination once it has started and keep leaves from turning yellow.

KEY TERMS

alternation of generations (384)
anther (392)
antheridium (386)
apical meristem (396)
archegonium (386)
auxin (400)
bud scale (396)
cambium (397)

carpel (392)
collenchyma (398)
cork (398)
cork cambium (398)
cortex (396)
cotyledon (396)
cytokinin (402)
dicot (396)
dicotyledon (396)
differentiation (395)
egg nucleus (389)
embryo (395)
embryo sac (392)
endosperm (393)
epidermis (395)
filament (392)
fruit (394)
gametophyte (384)
generative cell (392)
genetic imprinting (384)
gibberellin (401)
hormone (398)
lateral meristem (397)
megaspore mother cell (388)
micropyle (392)
microspore mother cell (388)
monocot (396)
monocotyledon (396)
ovary (392)
ovule (392)
parenchyma (398)
petal (390)
phloem (397)
pistillate (388)
polar nuclei (392)
pollen (392)
pollen grain (392)
pollen sac (392)
pollen tube (390)
pollen tube cell (392)
receptacle (390)
sclerenchyma (398)
seed (390)
seed coat (393)
sepal (392)
sperm nucleus (390)
sporophyte (384)
stamen (392)
staminate (388)
stigma (392)
style (392)
terminal bud (396)
vascular cambium (398)
vascular tissue (396)
xylem (397)

FOR FURTHER THOUGHT

1. If you wanted to transform your spindly houseplant into a fuller, more bushy specimen, would you remove the terminal bud, lateral buds, or lateral meristems? Explain.
2. When a pollen grain lands on a female stigma, may we correctly state that fertilization has occurred? Why or why not?
3. You wish to speed up the germination rate of a slow-growing plant. Would you treat it with auxins, cytokinins, or gibberellins?

FOR REVIEW

True or false?

1. ___ Asexual reproduction produces more genetic variation than does sexual reproduction.
2. ___ In mosses, eggs arise from the female archegonium.
3. ___ Vascular cambium produces xylem and phloem cells.
4. ___ Dicotyledons are plants that have only one embryonic leaf.
5. ___ Auxin is produced in plant roots.

Fill in the blank.

6. The male reproductive structures in flowering plants are collectively called the ___ and the female structure is called the ___.
7. ___ are the regions of growth at the tips of plant stems and roots.
8. In woody plants, the ___ carries water up from the roots and the ___ carries food down from the leaves.
9. Synthetic ___ is a plant hormone that is also a powerful weedkiller.

Choose the best answer.

10. In the conifers, the ___ produces the spores that will become pollen.
    A. microspore mother cell
    B. antheridium
    C. megaspore mother cell
    D. micropyle
11. The pollen grain of flowering plants
    A. contains the sperm cell.
    B. contains the pollen tube nucleus.
    C. is covered by a resistant coat.
    D. all of the above
12. The endosperm is best described as a ___ tissue.
    A. water conduction
    B. support
    C. nutritive
    D. meristematic
13. Which structure houses the seed in flowering plants?
    A. stigma
    B. style
    C. ovary
    D. none of the above
14. ___ are hormones that cause "foolish seedling disease" and stimulate cell division and elongation in plant stems.
    A. Gibberellins
    B. Cytokinins
    C. Auxins
    D. Synthetic auxins

CHAPTER 15

Animal Reproduction

Overview

Asexual Reproduction
Sexual Reproduction
Human Reproduction
The Human Sexual Response
Contraception

Objectives

After reading this chapter you should be able to:

- Describe ways in which animals reproduce asexually.
- Describe the reproductive strategies involved in internal and external fertilization.
- List the reproductive structures of the human male and female and describe the functions of each.
- Explain how the menstrual cycle is regulated.
- Describe the physiological changes that occur in males and females during each stage of the human sexual response.
- Describe how conception occurs.
- List the various means of contraception and their relative effectiveness in preventing pregnancy.

It struck me as strange a few years ago when I was writing a book called *How They Do It,* just how little we know about animal sex. It seemed strange because we are obviously fascinated by practically anything that has to do with sex, and the weirder the better. I learned there is indeed weird sex going on out there (Essay 15.1), but most of us really don't know much about it. One reason is because the information is not easy to find. I was forced to scour reams of dusty journals to dig out bits and pieces of information that had somehow failed to make its way to the public view. After the book was published, I learned that many people (even urbane, educated people) were somehow embarrassed to take the book to the cashier. (Many of the books, I was told, were *stolen!* People apparently would rather steal than have a cashier see that they were buying a book on sex.)

Why were people embarrassed? Why is the subject such a delicate one? You undoubtedly have your own ideas about such matters, but it cannot be denied that sex is a touchy subject, even the sex of other species. Nonetheless, this chapter is about animal reproduction, so it seems that we will be traveling over some hallowed ground. Of course, all due effort will be made to maintain our usual humble reverence.

Our plan will be to briefly survey the major means of reproduction in the animal kingdom, both asexual and sexual, and then to focus on the human as a representative animal. After all, this information may prove to be very important to you.

ASEXUAL REPRODUCTION

Some animals avail themselves of the great advantages of asexual reproduction that we discussed in the last chapter. As you might expect, these are generally rather primitive species, but the processes can involve some rather elaborate and complex patterns. We will consider three rather distinct cases here.

Binary Fission

A number of organisms reproduce by simply splitting in two, a process called **binary fission**. The planarian flatworm, for example (Figure 15.1), can reproduce itself in this way. (It can also reproduce sexually, but we'll ignore that small inconvenience for now.) In the process of fission, the planarian simply begins to constrict about midway along its length until it pinches in two. Cells from each part then move into the severed area and regenerate the missing parts. The result is two new planarians.

It is apparent that the cells that rebuild the missing structures in such cases must undergo some sort of fundamental reorganization. After all, these cells had already differentiated and specialized and were functioning in some specific role. Now they must somehow change and take on a new role. In some cases, cells can reverse their development; that is, after proceeding along one very specific developmental pathway, they reverse the process, revert to an earlier stage, and proceed along a different developmental pathway.

ESSAY 15.1

How They Do It

Many people seem to be under the impression that sex is sex, that basically all the species must do it the same way. Some animals do mate in ways that humans would recognize, but most of them don't. For example, bedbugs are often homosexual, however neither male nor female bedbugs can survive many matings because the male pierces his mate's back with his sharp penis and ejaculates directly into the body cavity. Special cells then capture and ingest many of the sperm as they roam the recipient's tissues. Thus, the recipient is nutritionally rewarded.

Animals mate in a variety of ways. Circumpolar bluets mate tail-to-thorax. And bighorn sheep rut in a fairly traditional way, if a bit precariously.

Other species may also surprise us. In some mites, brothers and sisters copulate before they are born, so the little females are born pregnant and undoubtedly disillusioned. Certain snails have an enormous penis just over the right eye. They are hermaphroditic (each possessing the organs of both sexes), but they can't exchange sperm until they have pierced each other with a chalky dart that usually acts as a sexual stimulant but can also kill. Other snails begin life as wandering males, but eventually settle down and become sedentary females. If a wandering male mounts a female, he must copulate before his masculinity fades. The female praying mantis sometimes devours the male even while he's copulating with her. In fact, after she eats his head and brain, his sex becomes more intense (but we'll try not to extrapolate from that).

Some snails, fish, and lizards change gender, so they may be a father at one time and a mother at another. In yet other fish, there are no males at all. The egg is stimulated to develop when the female mates with a male of a different species. His sperm only activates the egg; it doesn't join with it.

On the kinkier side, by our standards, we find geese that form ménages à trois (a threesome, usually with two males). Rape, or forcible sex, by the way, is common among lobsters, skunks, and orangutans. Furthermore, the female of many species must be subdued before they will allow mating. These include camels and rhinos. Rhino females are apparently stimulated by a good fight.

The reproductive structures of many animals can be somewhat surprising. Males of many insects have penises that look like instruments of torture with points, hooks, barbs, and impossible angles. That of each species is so distinct that inbreeding is usually not possible. Male opossums have split penises, with grooves instead of tubes, that match the divided vagina of the females. Pigs have a corkscrew-shaped penis that locks tightly into the female's vagina. Snakes and alligators can copulate with either of two penises that evert from the cloaca like turning the finger of a glove inside out. Each is covered with backwardly directed barbs, as are the male organs of cats and skunks. And, finally, as the old joke goes, how do porcupines do it? They, indeed, do it very, very carefully.

FIGURE 15.1

Planarian worms reproduce through binary fission. Basically, the body pinches in two roughly halfway along its length. Each end then regenerates the missing parts. Here, a worm demonstrates its powers of regeneration after its head was split longitudinally.

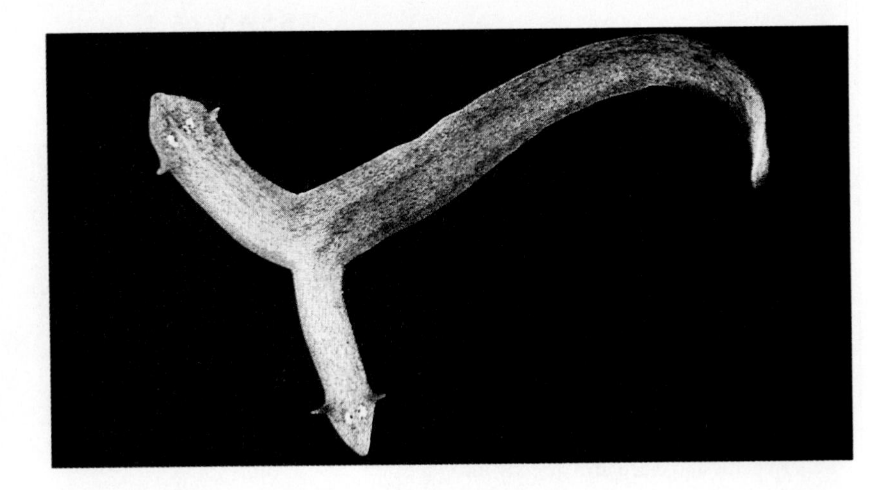

Regeneration

Regeneration, the replacement of missing parts, may be considered another form of asexual reproduction. For example, if a planarian is cut in half, each half can regenerate missing parts, presumably by the same reorganizational pathway that follows binary fission. In nature, body parts may be lost through injuries caused by predators. Sea stars (you may know them as starfish), for instance, can cast off arms to escape the grasp of a predator. Not only can the sea star replace its lost arm, the arm can regenerate into a whole new individual. Not long ago, before this fact was discovered, vengeful oystermen hauling up a sea star in a dredge would chop up the unfortunate creature and throw the pieces back in the water, quite sure that the sea star would eat no more oysters. However, they were wrong—an entire sea star can regenerate from a single small piece. As a result, the sea star population flourished, and the oyster catches dwindled, causing the oystermen to chop away with even greater diligence until biologists were able to convince them to axe their behavior instead of the sea star.

Budding

Another method of asexual production is **budding,** a process in which a new organism is produced as an outgrowth of the parent organism. The new appendange develops while attached to the parent, then pinches off and moves away on its own. Many of the simple invertebrates reproduce this way. An example is found in the tiny *Hydra,* a freshwater relative of the sea anemone (Figure 15.2).

SEXUAL REPRODUCTION

Sexual reproduction may not involve the intimacy that many of us associate with the process. This is because fertilization may be accomplished both *externally,* outside the body, as we see in frogs (Figure 15.3), as well as

FIGURE 15.2

The *Hydra* can reproduce by asexual budding. In this process, buds appear along the trunk of the body. They grow into tubelike structures complete with developing arms. Finally, the structure, resembling a tiny adult, breaks off and swims away.

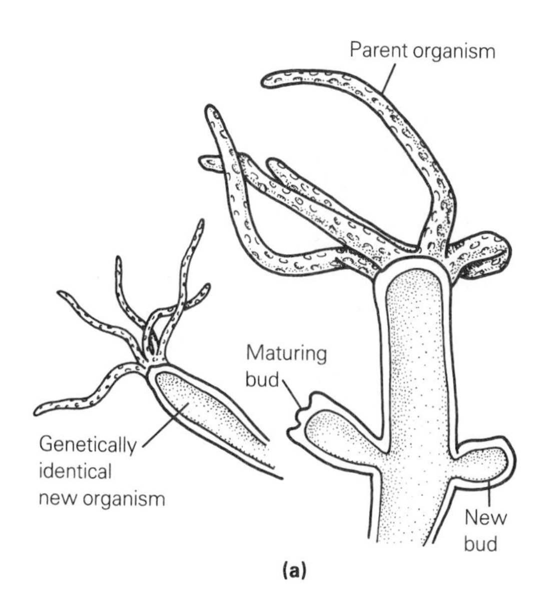

FIGURE 15.3

Insemination in chorus frogs. The male clasps the female from the back, gripping her behind her forelegs. As she releases her eggs he pours sperm over them. The eggs are then on their own, protected only by their gelatinous coating.

internally, as in mammals. The two methods involve quite different evolutionary strategies. In external fertilization, the eggs and sperm are scattered into the environment, leaving the rest to chance. External fertilization apparently evolved first, and whereas there is great wastage of gametes, there is little cost involved in the production of each.

In some cases, external fertilization can involve rather complex and even moving rituals. For example, the Caribbean fireworm spends its time burrowed in the bottom of bays in the West Indies. On the fifth night after an August full moon, the worms crawl out of their burrows and swim to the surface of the warm tropical seas. In a remarkable timing feat, they reach the surface after sunset, about an hour before the moon rises and in the blackness of night, so that nothing can interefere with what they are about to do. Then, for the only time in their lives, they become phosphorescent. The water sparkles and dances with greenish blinking lights. This is their time of reproduction. The water seems to be alive with their shimmering bodies. Then, suddenly, in a tiny phosphorescent explosion, clouds of eggs and sperm fill the water, and the ruptured bodies of the fireworms sink slowly to the bottom, leaving the gametes to play out a game of chance in the dark waters above.

The method obviously works for fireworms and for the salmon that shed their gametes in the northern streams each year (Figure 15.4). But, one might ask, isn't there a better way? (I'm sure you have an answer, but let's think in evolutionary terms for a moment.) Why risk having the sperm or eggs exposed to such a hostile environment? Why not just put the sperm inside the female where the eggs are? (I can hear the cheering.) With the sperm safe in her body, there would be less risk of unsuccessful matings and fewer eggs and sperm would have to be produced. Thus, internal fertilization arose. Probably at first, the reproductive orifices were simply pressed together as the male released sperm, as most birds do today (Figure 15.5). In time, however, many species would have developed specific

FIGURE 15.4

Salmon are born in freshwater streams. They then swim to the sea and several years later, at maturity, they swim back to the very stream where they were hatched, where the females release eggs and the males shed sperm over them. (This behavior not only gets eggs and sperm together, but it ensures a certain level of reproductive isolation among the various populations of salmon. Do you see any evolutionary advantage to such isolation?)

FIGURE 15.5

Birds copulate by the male rather clumsily placing his reproductive opening against that of the female.

reproductive structures and behavioral patterns that would have increased the probability of successful matings. The array of specialized male and female sex organs of today's creatures attests to the pervasiveness of this evolutionary direction. So today, sexual reproduction in many animals is internal, accomplished by **copulation,** the insertion of the penis into the vagina where sperm is deposited.

FIGURE 15.6

Female chimpanzees signal their readiness to copulate by a reddened rump that accompanies behavioral and physiological changes.

Timing of Sexual Reproduction

In many species, the timing of the copulation is critical to maximum reproductive output. For example, mammals copulate only at specific periods—usually when the female is physiologically and behaviorally receptive (in **estrus,** or heat). Such timing increases the likelihood of the young being born at the most opportune time of year, such as when the weather is warm or food is most abundant.

In many mammalian species, including dogs and horses, the timing is controlled by the receptivity of the female. Her condition, in turn, may be controlled by environmental conditions, such as the changing of the seasons. In many species, the female advertises receptivity through odors associated with vaginal discharge and changes in her behavior. In some primates, receptivity is also communicated by a reddened and enlarged area around the rump (Figure 15.6). (Researchers sometimes refer to receptive female chimpanzees as "pink ladies.") Obviously, sexual signaling can be accomplished through a variety of means.

In humans, theoretically, there is no period when the female is not sexually receptive or when she is not sexually attractive to males, your personal experience notwithstanding. Women are physiologically able to copulate at almost any time and use no particular signal to attract males. (Although several sure-fire ones may come to mind.) It has been suggested that the continual receptivity of women may have evolved as a means of enticing larger and stronger males to remain with a specific female, one that may be bearing his children. In such an association, the female might have greater reproductive success if she can elicit the male's assistance, and the male might also leave more offspring by assisting his reproductive partner, who is usually smaller and also burdened with the responsibility of nursing the young. Traditionally, the male has been useful both in protection and in bringing home protein-laden meat, while the female gathers plant material (she actually supplies more calories than the male among today's hunters and gatherers). Of course, the whole thing has not been a cold business arrangement (at least historically it hasn't been). It is simply the system that has produced the most offspring, and therefore it has become woven into our social fabric (Figure 15.7).

Considering such matters in scientific terms admittedly puts some people's teeth on edge. We should hasten to say that such theorizing does not neglect those more tender emotions of which we are so proud. A certain bonding, feeling, or attraction seems to develop between reproductive partners, emotions that serve to strengthen a union—even if it was originally established on a purely physical basis. Perhaps this kind of bond provided the basis for what is called "love," or perhaps such bonding *is* love. It's hard to say since no universally acceptable definition exists. In any case, the human bond often develops and, in fact, people seem to have a need for strong attachments with others. Attraction between people, as we all know, may be long term or short term. The resulting bonds

FIGURE 15.7

The family unit is remarkably similar across a range of cultural patterns. The Euro-American family (a) is organized much as is the stone-age Waorani group (b) of Amazonian Ecuador's Conanaco River.

(a)

(b)

may also be of intermediate length, or intermittent, and even retractable. Some long-term relationships among humans are sometimes formalized and referred to as "marriage," a system that, in our culture, discourages the breaking of the relationship. Interestingly, other species may also build long-term bonds. For example, lovebirds form devoted pairs (Figure 15.8) and geese often mate for life, and a member of a pair may never mate again if its mate is lost.

Copulatory behavior in many species is rather easy to predict. Naturalists are often generally aware of when the wild animals of their area will begin breeding, and animal breeders usually have a keen sense of when copulation is likely to occur between their charges. Humans, however, are highly variable in this regard. There is no specific point at which a man and woman are likely to initiate copulation. It can happen any time the coast is clear. Furthermore, the events preceding copulation in humans may vary widely. In certain other animals, specific signals always precede the copulatory act, such as a female bird "soliciting" by lowering her fluttering wings, or a female chimpanzee's "presentation" of her livid rear end to a male (although the blasé males may not mount the soliciting temptress).

FIGURE 15.8
Long-term bonds are built between the sexes of a variety of animals (although such bonds are relatively quite rare from species to species).

HUMAN REPRODUCTION

The human reproductive system is quite similar to those found in other mammals. So, as a representative example, we will focus our attention on human reproduction.

The Male Reproductive System

The male reproductive system includes the testes, scrotum, accessory glands, various ducts, and the penis (Figure 15.9, Table 15.1). The **testes** produce both sperm and male sex hormones. The two testes (testicles) lie in a pouch called the **scrotum**. Here, outside the abdominal cavity, the cooler conditions are more favorable for the development of sperm. Much of the testes consist of highly coiled **seminiferous tubules,** within which the sperm are produced. Between these tubules are the cells that produce male sex hormones. The seminiferous tubules lead into the **epididymis**, a highly coiled tube about 20 feet long that sits atop and along one side of the testes. The epididymis stores sperm as they mature to their final, functional stages, and at **ejaculation** (the forceful release of sperm), transports them to another tubule, the **vas deferens**. Waves of contraction of the vas deferens then propel the sperm over the pelvic bone to the back of the urinary bladder where the vas deferens from each testis joins. This tubule soon empties into the **urethra,** the tubule that runs through the penis and conducts urine and sperm (at different times).

The accessory glands include the **seminal vesicles,** the **prostate gland,** and the **bulbourethral glands** (also called Cowper's glands). The secretions of these glands together with the sperm comprise the semen. The secretions of both the seminal vesicles and the prostate gland activate and nourish the sperm. Shortly before ejaculation, the bulbourethral glands begin secreting a small amount of slippery fluid that may aid in penetrating

FIGURE 15.9
The male reproductive system.

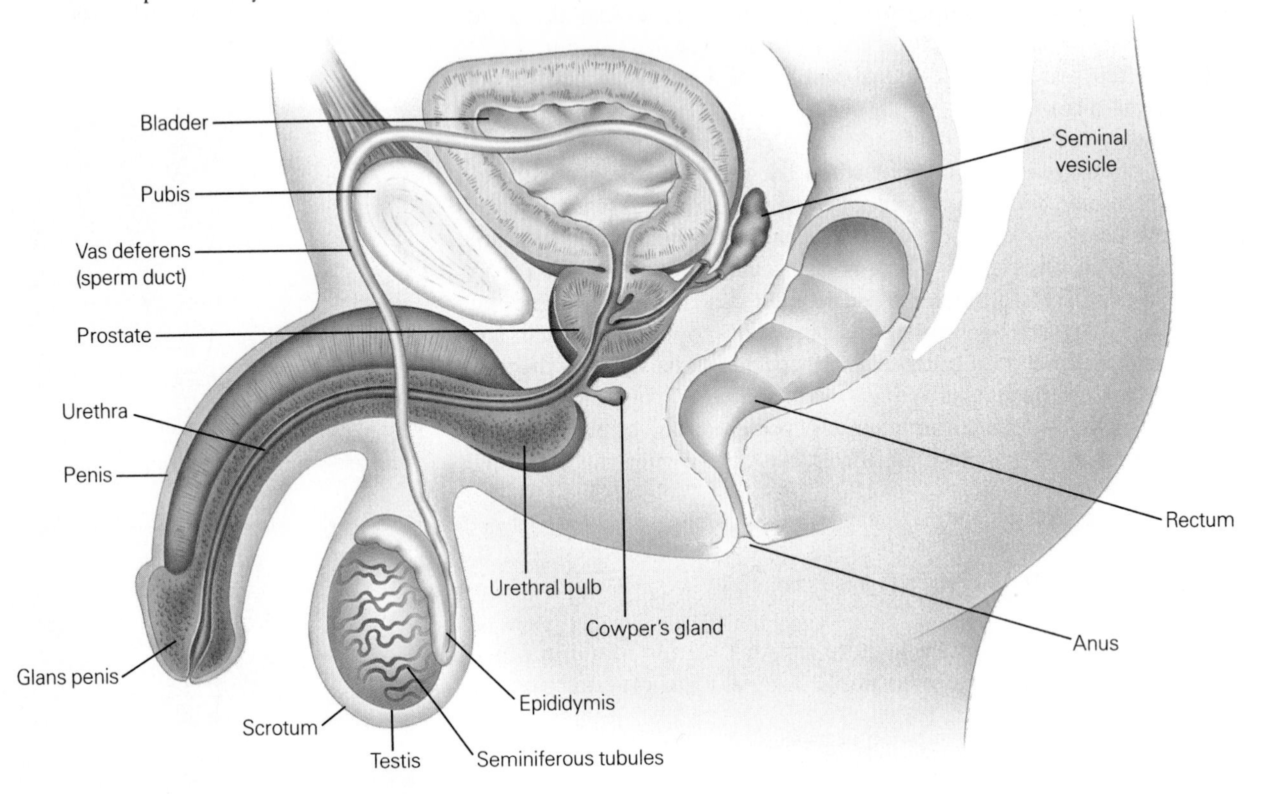

TABLE 15.1
Male Reproductive System

STRUCTURE	FUNCTION
Testes	
seminiferous tubules	Sperm production;
interstitial cells	production of male sex hormone
Epididymis	Storage and maturation of sperm
Vas deferens (sperm duct)	Tube that conducts sperm from epididymis to urethra
Urethra	Tube that conducts sperm (and urine) out of body
Penis	Transfer of sperm to vagina
Accessory glands	
seminal vesicles	Secrete most of volume of semen;
prostate gland	Adds secretion to semen;
bulbourethral glands (Cowper's glands)	Secrete small amount of fluid before ejaculation for lubrication and rinsing urine from urethra

the vagina as well as serving to rinse the urine from the urethra, making the pathway more hospitable for the sperm.

The **penis,** the organ that delivers the sperm to the female reproductive system, is composed of three long areas of spongy tissue bound together by fibrous tissue. One of these columns of spongy tissue is enlarged at the penis tip, forming the glans penis. This structure may be covered by the foreskin, the fold that is removed at circumcision.

The Female Reproductive System

The reproductive system of the human female includes the ovaries, oviducts, uterus, vagina, and external genitals (Figure 15.10, Table 15.2). The **ovaries** produce the eggs (ova) and the female sex hormones. When it is released from an ovary, an egg is caught by the open end of the closest **oviduct,** the tube that conducts the egg to the uterus. The oviduct (or fallopian tube) is lined with cilia that sweep the egg toward the uterus. There are two oviducts, one leading from each ovary to the uterus.

The **uterus** is pear-shaped, muscular, and thick-walled, with the lower end, the **cervix,** opening into the vagina. The vascular and glandular inner lining or **endometrium** of the uterus is the site where the embryo will develop. The muscular walls will help push the baby out of the uterus during labor.

FIGURE 15.10
The female reproductive system.

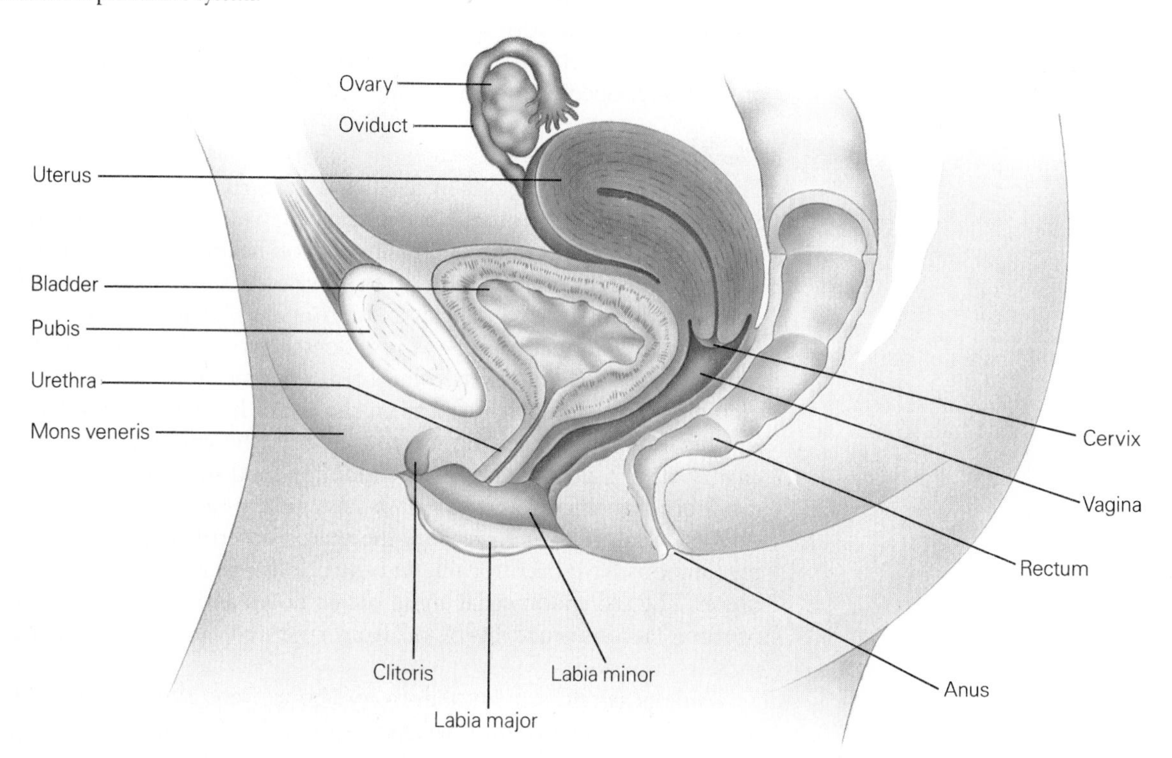

TABLE 15.2
Female Reproductive System

STRUCTURE	FUNCTION
Ovaries	Source of eggs, estrogens, and progesterone
Oviducts (fallopian tubes)	Catch and transport egg to uterus
Uterus	
endometrium	Houses developing embryo
cervix	Narrow lower region of uterus opening into the vagina
Vagina	Receives penis during copulation; birth canal

The **vagina** is a muscular tube 3 or 4 inches long that leads from the exterior of the body to the cervix. The vaginal muscles form a thin wall (vaginal wall) with a corrugated appearance. There may be a mucous membrane called the hymen stretched across the vaginal opening at some point.

The external genitals or **vulva,** are comprised of the *mons veneris,* a fatty mound that becomes covered with hair at puberty. The **labia majora** are thick folds that lie on either side of the vaginal opening and are also covered with hair. The **labia minora** are smaller folds of skin that lie between the labia majora and the vaginal orifice. Where the labia minora meet above the vaginal orifice is the **clitoris,** a small erectile organ that is highly sensitive and corresponds developmentally to the penis of the male. The *urethra,* which is part of the urinary system and conducts urine out of the body, opens between the clitoris and vagina.

The Menstrual Cycle

The menstrual cycle begins in females at puberty, ordinarily around the age of 12 to 14. (The onset has become progressively earlier in American girls, probably as a result of increasingly good health practices.) The cycles will continue for about the next 30 to 40 years of the woman's life, usually interrupted only by pregnancies and terminated finally in menopause.

What exactly is this cycle? How does it work? **Menstruation** is the result of the shedding of the blood-rich endometrium, the lining of the uterus, when the egg released from the ovary has failed to be fertilized. In preparation for receiving a fertilized egg, the endometrium becomes engorged with blood carrying life-sustaining food and oxygen. In addition, it has become thicker and its glands have begun secreting nutritious mucous material. If the egg reaches the uterus unfertilized, the endometrial preparations have been for nil, and the uterus reverts to its previous condition. The old endometrial lining breaks down and slides out the vaginal opening, accompanied by blood from ruptured vessels. This marks the menstrual period.

An egg is released from an ovary about every 21 to 32 days. If it is not fertilized, menstruation will begin about 14 days, give or take 2 days, after that release. The woman's "period" will last from three to six days.

However, there is a wide variation in the length of both the cycle and the menstrual period, and almost any pattern is considered normal as long as it recurs on a somewhat regular basis. Figure 15.11 describes the relationship of the ovarian, uterine, and hormonal cycles as they would occur during a 28-day cycle. *Hormones* (Chapters 14 and 21) are "chemical messengers," produced in certain parts of the body, that regulate events in other parts of the body.

At the end of a menstrual period, the cycle begins anew, deep within an ovary, where other follicles are developing. **Follicles** are rounded bodies, each containing an egg that will protrude inward from the follicle wall into a large fluid-filled space when the follicle reaches maturity. Their development is stimulated by rising blood levels of a hormone simply called the **follicle-stimulating hormone (FSH),** which interacts with another hormone called the **luteinizing hormone (LH).** Both are produced in a small but remarkable structure called the *pituitary gland* located at the base of the brain. (We'll look more closely at this gland later.)

As the follicles mature, they begin to secrete a hormone of their own called **estrogen,** which induces a thickening of the endometrium. For some reason, one bubblelike follicle begins to outgrow the others, and as soon as its ascendancy is established, the others stop developing. This larger follicle then moves to a position just below the surface of the ovary. The rapid rise in estrogen produced by the growing follicle, causes a surge in LH from the pituitary. The LH surge triggers **ovulation,** that is, the rupture of the follicle and the release of the egg from the ovary. In a 28-day-cycle, ovulation occurs on or about the thirteenth or fourteenth day after the follicle-stimulating hormone initiated the growth of the follicle. The egg is released, leaving behind follicle cells that will become transformed under the influence of luteinizing hormone into a yellowish structure, called the **corpus luteum** ("yellowed body"). However, the corpus luteum is far from being a corpse, for now, in addition to estrogen, it begins to secrete a second hormone, **progesterone.** Together, these hormones stimulate the uterus to begin final preparations for a fertilized egg. The endometrium thickens and becomes engorged with tiny blood vessels that will carry food and oxygen to the embryo in the event that fertilization occurs. The interaction of the progesterone and estrogen also inhibit the production of the pituitary's follicle-stimulating hormone at this time and, thus, prevents the start of a new cycle. These inhibiting properties of estrogen and progesterone are the basis for their use in birth control pills, as we will see.

The corpus luteum is maintained, at least initially, by the pituitary's luteinizing hormone, but the release of this hormone is inhibited by estrogen and progesterone. So, by producing its hormones, the corpus luteum may write its own death sentence. It may also be that the cells of the corpus luteum are programmed to die after two weeks. In any event, when the corpus luteum degenerates, the levels of estrogen and progesterone begin to fall and swollen uterine lining begins to shrink. Deeper within the uterus, blood vessels contract, reducing the supply of nutrients and oxygen to the endometrium. The tissue then breaks down, rupturing the vessels there, and bleeding begins. The blood flow carries bits of cell tissue out of the uterus and through the vaginal opening, marking menstruation.

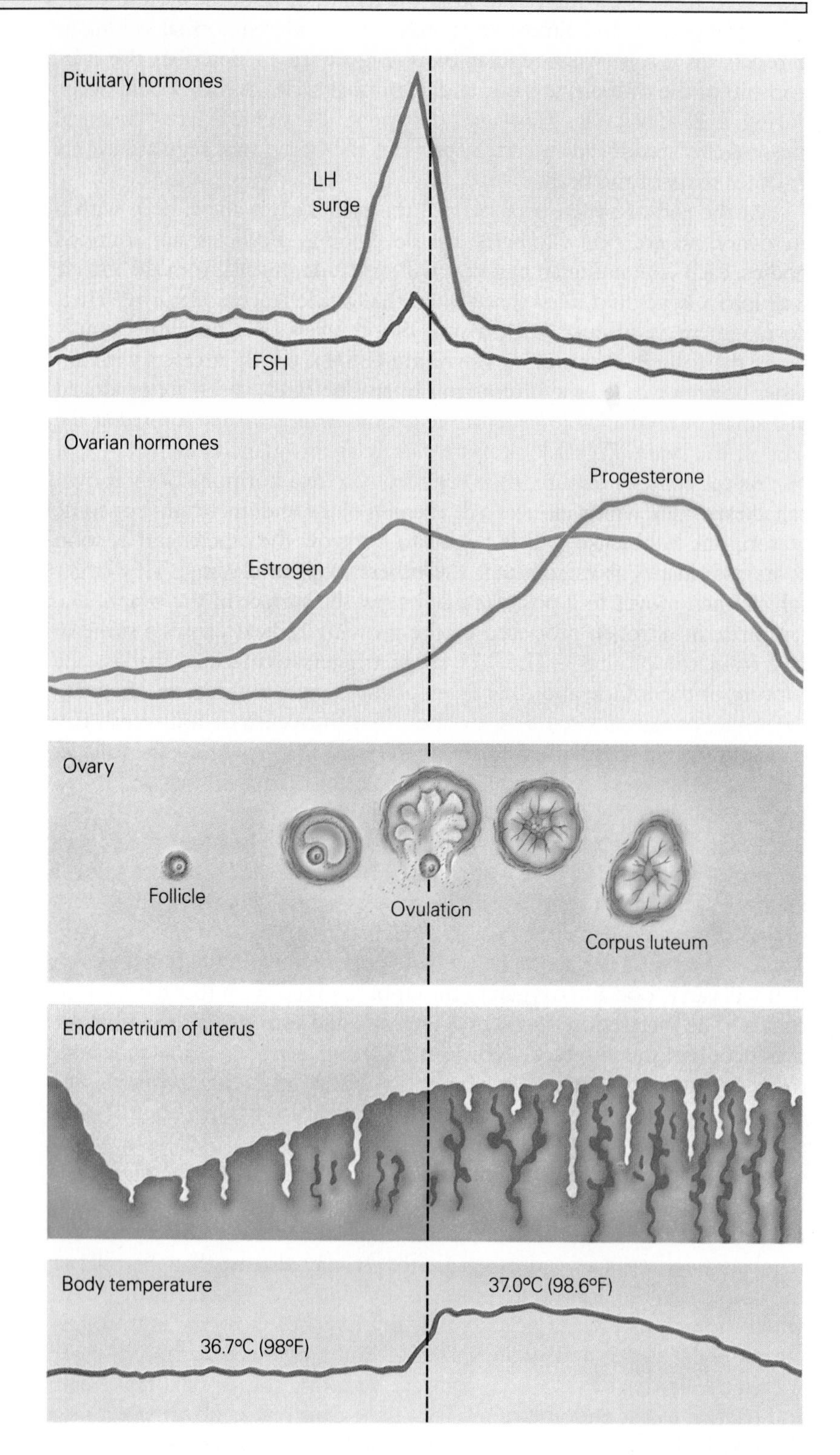

FIGURE 15.11

The relationship of female reproductive hormones and the events in the ovary and uterus during a 28-day menstrual cycle. Note that FSH and LH are released simultaneously from the anterior pituitary, but that LH is much more abundant. In the meantime, the maturing follicles are producing estrogen, but progesterone will not begin to rise until after ovulation, when the corpus luteum is formed. The sharp rise in LH and estrogen marks the rupture of the egg, but by this time the uterine endometrium is halfway through its period of development. Note that estrogen peaks just before day 14, about the time that progesterone levels begin to rise, peaking from day 21 to 24. The follicle reaches maturity at day 14, when ovulation occurs, leaving cells that develop into the corpus luteum, which continues secreting some estrogen, but particularly progesterone, for about 10 more days. The endometrium (lining) of the uterus undergoes growth and repair during the first 14 days, reaching full development about a week later. At day 14, the body temperature elevates slightly, a signal that ovulation has occurred.

Contractions of the muscles in the uterine wall are caused by chemical messengers called ***prostaglandins*** that are released from the endometrial cells as they are shed. If the level of prostaglandin is excessive, the normal contraction response may become painful spasms, called menstrual cramps. There may be additional pain resulting from the reduction in blood flow to uterine tissue as the muscle spasms constrict blood vessels. Women who suffer with severe cramping have higher levels of prostaglandins than women who do not have cramps.

The decline in the level of estrogen and progesterone that caused the loss of the endometrium also removes the inhibition on the pituitary's secretion of FSH. So, as the uterus is returning to its original state, the pituitary stimulates the development of more follicles, and the cycle, well, the cycle—begins again.

If copulation has been successful (depending on your point of view), and the egg is fertilized, the early embryo will implant in the uterine lining and begin its growth, taking its necessities from its mother's blood in a parasitic fashion. (Some say it's just the beginning of a long parasitic existence.) A hormone produced by the early embryo maintains the corpus luteum. As a result, the supply of estrogen and progesterone remains high and the endometrium is maintained.

THE HUMAN SEXUAL RESPONSE

In humans, under normal conditions the physical prerequisites of copulation or intercourse are simply an erect penis and a lubricated vagina. These conditions are usually achieved during the course of precopulatory sexual behavior, or foreplay. Foreplay, as well as the copulatory act itself, varies widely in its expression between cultures, from one individual to the next, and even between established sexual partners from one time to the next. Let's examine some of the changes that accompany sexual intercourse in our species.

The human penis is supplied with a great number of blood vessels, some of which open into large blood chambers. Parallel veins and arteries service the penis, and during sexual arousal, the arteries relax and enlarge, allowing an increased blood flow into the spongy tissue of the penis. The veins do not expand, however, and in fact are compressed as the spongy tissue fills with blood. Thus, the penis becomes filled with blood, causing it to grow stiff and lengthen, sometimes to a surprising degree—perhaps more surprising to some than others. The penis stands erect at such times and is said (by the more delicate among us) to be in the tumescent condition.

In women, foreplay also causes blood to accumulate in certain structures. One result of this is the vaginal walls seep fluid that serves as a lubricating substance which aids in the insertion of the penis. The lubricating may begin, however, well before she is emotionally ready for intercourse. With more extensive foreplay, the labia minora (small lips) and clitoris, a small fleshy, highly vascularized structure with an incredible density of sensory receptors, may enlarge and redden as blood rushes to those areas. About this time, the nipples may also harden and enlarge. As foreplay continues, the shoulders and chest may mottle and the breasts may enlarge and grow more sensitive.

Copulation involves the insertion of the erect penis into the vagina. After a few or many pelvic thrusts (depending on the species, and, in the case of humans, depending also on the individual, the degree of arousal, and one's schedule), the male ejaculates.

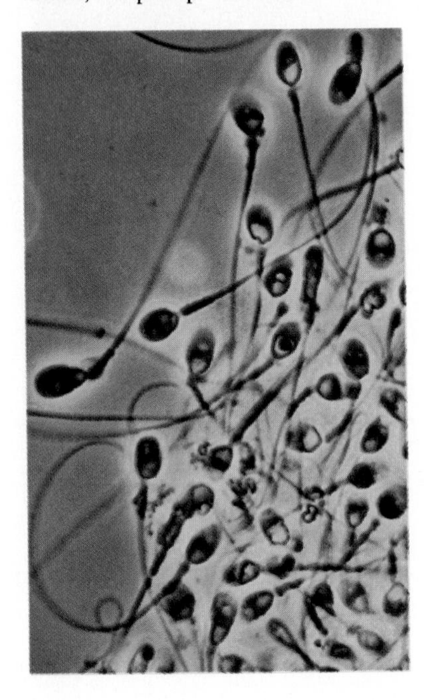

FIGURE 15.12
Human sperm, each bearing its own unique set of characteristics for physical, mental, and perhaps social traits.

Orgasm in males takes place in two stages. First, the entire genital tract contracts, including the epididymis, vas deferens, and seminal vesicles (see Figure 15.9). Fluid from the seminal vesicles flows into the urethra (the tube leading outside the body through the penis that is shared by the urinary system), where another fluid from the prostate gland is added. Thus the sperm that has been stored and matured in the epididymis is combined with the nourishing and activating secretions of the accessory glands, forming a fluid called *semen*. The bulbourethral glands have already begun secreting a small amount of a slippery fluid. As the urethra becomes filled with semen, the urethral bulb at the base of the penis expands to accommodate the influx. During the second stage, ejaculation, the semen is forcefully expelled from the penis. The second stage begins as a sphincter muscle (a ring of muscle under involuntary control) closes off the urethra at the bladder. A sphincter at the base of the prostate then relaxes, allowing semen to move into the urethral bulb and then into the urethra within the penis. This is followed by contractions of the muscles in the area behind the testicles, which pump the accumulated fluid out through the penis. Several contractions may follow, but most of the semen is expelled in the first. The semen usually ejects from the penis with considerable force in the first few contractions. Also, it is in this first fluid that the greater part of the sperm is concentrated (Figure 15.12).

Ejaculation, which culminates the orgasm, is the final stage of the copulatory act. The accompanying physiological changes include an increase in blood pressure and a quickened pulse. The postclimactic period is marked by a feeling of relaxation, sometimes to the point of drowsiness. The blood vessels return to their original state, and the penis again becomes flaccid or limp.

The physiology of orgasm in women is not as well understood. (The traditional advice to Victorian brides was said to be "lie still and think of England." Orgasm obviously was not expected.) In fact, the very existence of the female orgasm was argued until 1966, when a pioneer study of human sexual response was published by the physician-psychologist team of Virginia Masters and William Johnson. On the basis of many carefully monitored observations of various sexual activities, they demonstrated that the female orgasm, when it occurs, differs little from that of the male in that it involves contractions in certain reproductive structures. In women, the vagina and uterus contract rhythmically. Also, orgasm among women is not so universal a response as among men. Some women rarely or never reach climax, while others do so repeatedly and easily. The inability to reach orgasm, by the way, has absolutely no effect on the ability to reproduce (Figure 15.13).

Conception

Although normally only one egg is released during each menstrual cycle, the semen released at each ejaculation may contain millions of sperm, each

FIGURE 15.13

A human egg descending toward the uterus. The egg is released regardless of the woman's sexual behavior.

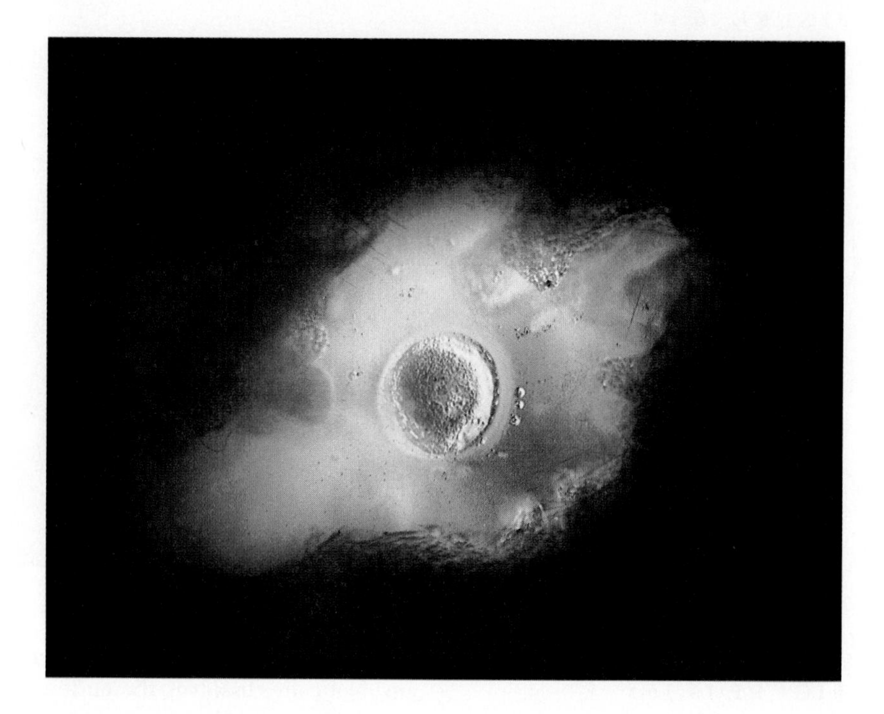

with its own genetic makeup. Of course, there is no way to predict which sperm will fertilize the egg. During copulation, the sperm are usually ejaculated into the upper reaches of the vagina, near the cervix. Each sperm cell has a whiplike "tail" that can propel it along at a rate of about ½ centimeter per minute.

The sheer number of sperm in each ejaculation is an adaptation to the extraordinary hazards they face in making their way to the egg. Some may die of "natural causes," as a result of their own physiological changes. Others may be rendered immobile by the acidic environment of the vaginal tract. Some are devoured by the woman's roaming white blood cells, which treat them as the foreign bodies they are.

In spite of the dangers, some sperm are able to make it into the relatively hospitable alkaline environment of the uterus. Through ways that largely remain a mystery, a few manage to find their way through the uterus and into the tiny pore leading into the oviduct and up the tube nearly to the ovary, where they may encounter an egg on its way down. **Conception** (fertilization) actually takes place, then, not in the uterus, but far up in the oviduct. Surprisingly, however, sperm may reach the egg only 1½ to 3 minutes after ejaculation. This is probably because they are assisted by currents created by uterine contractions (Figure 15.14).

Several sperm may reach the egg at once, and we now know that only one will penetrate the egg's membrane. Penetration by more than one sperm is prevented by dramatic changes in the membrane surrounding the egg (Figure 15.15). **Fertilization,** the union of the nuclei of the egg and sperm, follows.

Once fertilization has taken place, the zygote (fertilized egg) will continue down through the oviduct to be received by the ready uterus.

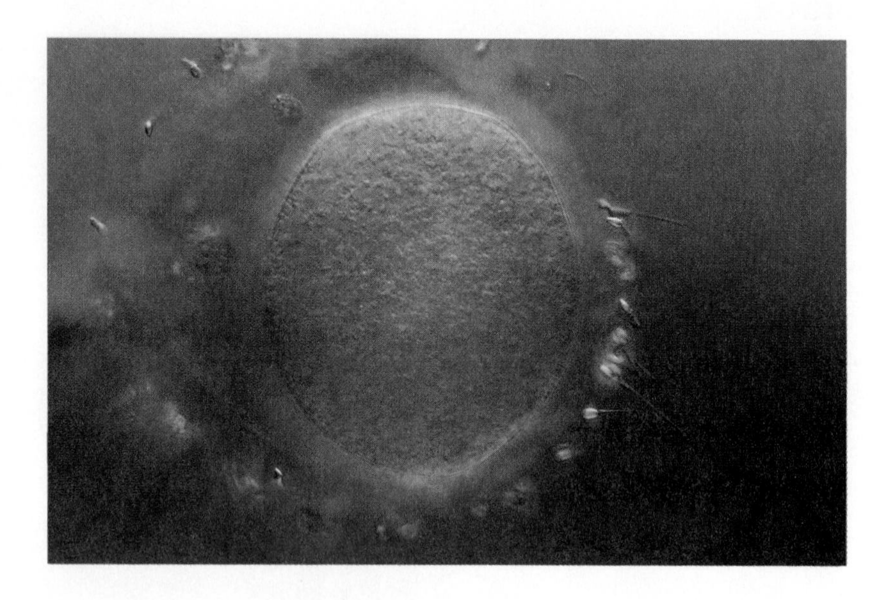

FIGURE 15.14
Uterine contractions have helped bring these sperm to the vicinity of the egg.

FIGURE 15.15
The sperm have reached the egg and have surrounded it. However, only one sperm will enter into fertilization. The size, personality, health, and tendencies of the offspring are largely determined at this point. (You would be a different person had another sperm penetrated your mother's egg just a millisecond sooner—a sobering thought, perhaps.)

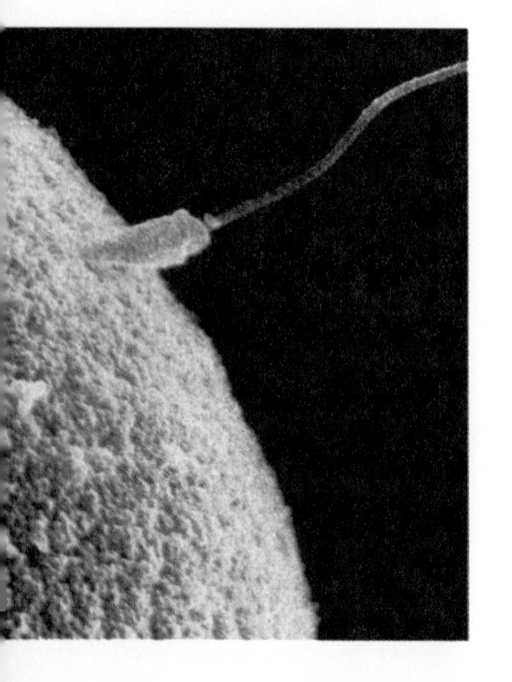

It will quickly begin to draw sustenance from the uterine fluids. Then, it enzymatically dissolves the endometrium and sinks deeper into the maternal tissues until it is completely embedded. It soon begins to receive sustenance as oxygen and nutrients filter through from the blood of the mother. In approximately 266 days, a new individual will join its relatives.

There has never been another individual quite like this one, nor will there be. The genetic combination cannot be duplicated, and neither can the experiences that will be superimposed on that constitution. In some cases, its existence will be brief, the time counted in days or even minutes. Most, though, will join the planet's teeming human population where it may become anything from a Sikh to a Republican. Whatever it will call itself, though, it will be unique.

Sex and Society

The act of copulation in many societies has been regulated by a number of religious, legal, and social restrictions. In fact, in our culture, "morality" has come to refer primarily to sexual matters. To many people, an "immoral" person is one who is sexually promiscuous, period.

It has been suggested that such values may stem from the idea that copulation must be reserved for childbearing activities. One argument is that in nature, copulation is reserved solely for the purpose of procreation and that copulation for any other reason is "unnatural," and therefore wrong. Logic aside, the premise is erroneous; copulation has many functions in nature. For example, among rhesus monkeys, subordinate males often present their rears to dominant males as an act of submission—an act that has the effect of reducing the dominant individual's aggression. The dominant male may follow up with a demonstration of his authority by mounting the subordinate and making several pelvic thrusts, although there is no actual penetration. Furthermore, among chimpanzees, a pink lady may copulate with every adult male in the group, as well as some adolescents. Since only one male is needed to impregnate the female, the act of copulation is believed to be a means of developing social bonds within the group. Among baboons, the females in the earlier phase of their estrus period copulate with males of any rank, but at the time they are most likely to conceive, they may accept only high-ranking males. Baboons are also highly social, and in this case, too, the earlier acts of copulation may serve to reinforce group bonds. Female monkeys have been known to present themselves to a male, distracting him long enough for her to swipe his bananas. This is a "natural" act, but not many people emulate it, at least in the better social circles.

If copulation is an important bond builder, it would help to explain our caution regarding what seems, on the surface, to be a rather mechanical process. In other words, it may not be advantageous to build bonds indiscriminately.

CONTRACEPTION

Contraception, or the avoidance of pregnancy, can be achieved in many ways. Some methods are employed by the female, others by the male, and some by both partners. Unfortunately, none, except abstinence and sterilization, is 100 percent effective or completely safe (Table 15.3, next page).

Historical Methods

Historically, contraception has been attempted in a variety of ways, some effective, some worthless. (One might wonder just how our ancestors must have reacted when they began to discover what causes pregnancies. It's still a bit hard to believe.) The ancient Egyptians, for example, blocked the cervix with leaves, cotton, or cloth. The encouragement of homosexuality among the ancient Greeks probably reduced the rate of conception in that culture, although the practice was undoubtedly not instigated for that reason. In the Middle Ages, condoms that fit, sheathlike, over the penis were fashioned from such materials as linen and fish skin. Douching has long been practiced as a means of flushing sperm out of the vagina. The finger has also been used to remove semen after it has developed a stringy quality through being exposed to the vaginal environment. The Old Testament refers to removing the penis just prior to ejaculation. This is unreliable for reasons we will see shortly. The most effective method, of course, has always been total abstention, a method that has met with little applause.

There are also a number of useless "folk" devices that are employed even today. These range from regulating intercourse according to the phases of the moon, to stepping over graves, to using cellophane sandwich wrappings as condoms, to douching with soft drinks (the carbonic acid and sugar are believed to be spermicidal, and a shaken drink provides the propulsion for the agents to reach far into the vagina). Since the effectiveness of these folk methods is so low, we might consider a few other contraceptives that we know more about.

TABLE 15.3
Summary of Contraceptive Methods

METHOD	ACTION	USER	FAILURE RATE (PREGNANCY PER 100 WOMEN PER YEAR)*	DISADVANTAGES
Douche	Washes away sperm in vagina	Female	40	None
Rhythm	Abstinence during fertile period	Male and female	24	Requires meticulous recordkeeping of woman's body cycle (e.g., temperature)
Coitus interruptus	Removal of penis from vagina before ejaculation	Male	23	Frustration, some sperm may be released before withdrawal
Condom	Traps ejaculated sperm	Male	10	Some loss of sensation in both male and female. Condom material may tear if not used properly
Diaphragm (with spermicide)	Prevents sperm from entering cervix; kills sperm	Female	19	Must be applied before intercourse and inserted correctly
Cervical cap	Prevents sperm from entering cervix	Female	13	Infection, allergic reactions
Spermicide (foam, jelly, aerosols, creams, suppositories)	Kills sperm	Female	18	May cause irritation. Must be applied before intercourse and left in for at least six hours
IUD	Prevents implantation	Female	5	Infection, menstrual discomfort, expulsion of device possible
"Combination" pill (estrogen and progestin)	Inhibits production of FSH and LH, preventing ovulation	Female	2	May produce unwanted side effects (e.g., risk of cardiovascular disease, water retention, nausea)
Minipill (progestin alone)	Thickens cervical mucus; inhibits sperm movement; prevents implantation	Female	2.5	None known
"Morning after" pill (50 × normal dose of estrogen)	Arrests pregnancy	Female	**	Physical discomforts increase (e.g., nausea, water retention)
Norplant	Thickens cervical mucus; inhibits sperm movement; prevents implantation	Female	1	None known
Vaginal sponge	Mechanical barrier to sperm; kills sperm	Female	10–20	May cause irritation or allergic reactions
Sterilization	Surgical control of sperm and egg movement	Male and female	0	May produce irreversible sterility

*There is a wide range; these figures are taken to be an average.

**Insufficient data.

The Rhythm Method

As an old joke goes, people who practice the rhythm method are called "parents," although the joke is funnier to some than others. Basically, the **rhythm method** involves periodic abstention. It is based on the notion that a woman's fertile period can be predicted or revealed by tests and that by refraining from intercourse during this time, pregnancy cannot occur.

Ovulation occurs 14 (plus or minus 2) days before the next menstrual period begins. Sperm is thought to live for two days after it has been released in the female's reproductive tract and an egg remains fertile for about a day following ovulation. The couple must therefore avoid intercourse at least two days before and one day after ovulation, although, for safety, this interval is usually increased. There are changes in the consistency of the cervical mucus accompanying ovulation and body temperature increases slightly on the days following ovulation. These changes can help identify the day of ovulation. The difficulty, of course, is knowing when ovulation will occur two days before the event. Although it requires the use of thermometers, calendars, paper, and pencils, the claim is made that this is the only "natural" means of birth control. In any case, the method is not particularly reliable (about 76 percent under ideal conditions), partly because the menstrual periods of many women are highly irregular. One way of interpreting these numbers is that for every 100 couples using this method for a year, 24 of the women are likely to become pregnant.

FIGURE 15.16
An advertisement encouraging the use of condoms as protection against sexually transmitted disease.

Coitus Interruptus

In spite of the obvious inconvenience of withdrawing the penis from the vagina immediately before ejaculation, **coitus interruptus** is one of the most widely used contraceptive practices in Europe today. There are little data on the effectiveness of this method, but it is probably not very reliable (probably about 70 percent), because the man must do precisely the opposite of what he wants to do. The practice is also risky because sperm may be present in the secretions of the bulbourethral glands and unobtrusively leave the penis before it is withdrawn. In addition, because of residual semen in the penis after the first ejaculation, there is higher risk of conception for those who are able, and have the time, to copulate again within a short time.

The Condom

The **condom,** or "rubber," is one of the most widely used contraceptive devices in the United States. Basically, it is a balloonlike sheath made of rubber, or other material, that fits tightly over the penis and traps the ejaculated sperm. There are many grades of condoms, ranging from the two-for-a-quarter specials sold in restroom vending machines to the expensive types that are made from animal membranes.

Condoms are only about 70 to 93 percent effective (partly because of differences in quality and the fact that they are often incorrectly used). The disadvantages of using condoms include the frantic effort to find one and put it on, and the reduced level of the man's sensation. It remains

ESSAY 15.2

STDs or Chlamydia Is Not a Flower

Not long ago, say a couple of decades past, frolicking Americans were warned of the two great hazards associated with promiscuity: gonorrhea and syphilis. Actually, there were other problems even then, but those were the two greatest risks by far. They were called VD (venereal disease*). The picture has changed a bit now, and today, the friendlier folk face a virtual battery of sexually transmitted diseases (STDs), to the expressed delight of certain evangelists. Let's briefly review a few of the threats. You may be surprised at what we're up against and how startling some symptoms can be. We'll begin with the old, familiar duo.

Gonorrhea, or clap, is caused by the bacterium *Neisseria gonorrhoeae*. It is transmitted by crossing mucosal surfaces and may affect the reproductive tract, the throat, or the rectum. Afflicted individuals, especially women, often show no symptoms. When a woman does show symptoms, they generally include bleeding and discharge between periods. Men show a discharge from the penis and severe burning on urination. (One suggested diagnosis is to place a nail between the man's teeth before he urinates. If he bites the nail in two, it's gonorrhea.) Symptoms usually develop within ten days of exposure. It generally subsides on its own after several weeks. Drug-resistant strains are known.

Syphilis afflicted Benjamin Franklin, Florence Nightingale, Isak Dinasen, and Napoleon and can rightly be said to have influenced the course of history. It is caused by the bacterium *Treponema pallidum,* and is particularly prevalent among the lower socioeconomic groups. There are three stages. (1) Primary stage: a dull red and painless crater-like bump, that may open anywhere on the skin, which then disappears. (2) Secondary stage: after a few weeks, a rash or discoloration over the body (a rash on the hands and feet is a strong sign of syphilis), a general malaise, hoarseness, fever, headache, muscle soreness, and possibly hepatitis and hair loss. These, after 2–10 weeks, disappear for 1–40 years (although there may be reoccurrences of these symptoms). (3) Tertiary stage: a range of symptoms from benign lesions over the body, sharp pains along the legs and trunks as large nerves degenerate, blindness, weakened blood vessels, and a changing brain accompanied by insanity. It can be treated with antibiotics in the early stages, with less effectiveness in later stages.

Chlamydia (you probably didn't know this) is the most commonly transmitted bacterial disease in America. It is particularly prevalent on college campuses. It is caused by a bacterium, *Chlamydia trachomatis.* It is easy to cure but may cause no symptoms. If left alone, it can persist for years. In men symptoms may mimic gonorrhea, but chlamydia also causes aching in the testicles and groin. Initially women may have no symptoms, but as the disease progresses there may be great discomfort, including pelvic pain, and pain in the lower left abdomen, often with vaginal discharge, especially intense during menstruation and deep contact during sexual intercourse. It can produce severe secondary complications. It is spread almost entirely by vaginal intercourse, and chances of infection increase dramatically with multiple sexual partners.

Venereal warts are more common than esthetics would demand. They appear on the genitals or rectal area and can reach the size of marbles. They are caused by a papilloma virus and are extremely contagious. Most appear within about a month after exposure. Young people are especially vulnera-

popular because, not only is it an effective means of birth control if used properly, it also provides the extra bonus of protection against sexually transmitted diseases. However, the various types of condoms have different degrees of effectiveness against conception and disease, and the two are ble. In women the virus may later cause cervical cancer.

Vaginitis is "simply" an infection of the vagina. It can be caused by a variety of organisms, such as fungi (yeast), bacteria, and trichomonas (a protistan). Yeast causes a discharge like cottage cheese; trichomonas causes a fishy smell, and a greenish, watery discharge. Bacteria produce similar, but more subtle, symptoms. The woman may also be asymptomatic but contagious. Vaginitis is not always due to sexual intercourse, but can be caused by anything that increases vaginal moisture, such as sitting around in a wet swimming suit. Douching and taking antibiotics kills off the vagina's normal protective flora and can contribute to infection. Vaginitis can also be caused by soaps, powders, and deodorants, involving no organisms at all.

Honeymoon cystitis is caused by unusually frequent and vigorous intercourse. Usually the culprit is the intestinal bacterium, *E. coli,* that is transferred from the anus to the urethra in women. The bacterium can then reach the bladder where it causes pain, and possibly even the kidneys where it can be dangerous. If not treated, it will reappear again and again, with increasing severity. The risk can be reduced by drinking a quart of water after each sexual episode (a moderating effect in itself), and for some reason cranberry juice seems to help.

Herpes is caused by the *Herpes simplex* virus. Type I usually causes fever blisters (cold sores) on the lips. Type II causes the same sort of blisters on the genitals. It is generally not dangerous but can be aggravating to the afflicted. It usually appears as small blisters in the genital area, accompanied by a "tingling" sensation. When it erupts, the blister is extremely sore, it then disappears in about two weeks. It is transmitted across mucosal surfaces, or openings in the skin. The virus never leaves the body ("What's the difference between love and herpes? Herpes is forever."), but simply retreats along nerves where it resides in the spinal cord, to blossom forth on your vacation. Its appearance can be triggered by stress, either emotional or physical. The average American male has a 40 to 50 percent chance of having been affected if he contacts an infected surface. (For some reason, 3 percent of nuns show antibodies.) In rare cases, the virus can inflame the brain and spinal cord. Herpes of the eye can cause blindness. People showing no symptoms *can* transmit the disease, but this is probably unusual.

Molluscum contagiosum sounds like a case of the snails, but it's actually known as "the hugging disease" since it is transmitted by skin-to-skin contact. It is not a serious condition, causing only a bump that reappears with annoying frequency. Young people are particularly susceptible.

AIDS stands alone as a threat to humanity and is discussed elsewhere.

One final note: Most of these problems can be largely avoided by a common sense approach to sex (if there is such a thing). Minimizing the number of partners is the safest bet, as well as developing an "index of suspicion," and avoiding unclean or promiscuous partners. Condoms, especially latex types, rather than those made of animal membranes, reduce the transmission of STDs.

*Venereal, from Venus, goddess of love.

not necessarily correlated. Animal membranes, for example, may allow a heightened sensitivity, and they may trap sperm effectively, but their relatively large pores allow the passage of viruses such as those that cause herpes and AIDS (Essay 15.2).

FIGURE 15.17

Top, diaphragm in place. Bottom, cervical cap in place. They both operate on the same principle—that is, blocking the cervix. However, the diaphragm is normally inserted each time before intercourse, while the cervical cap may be worn for up to two days. Both devices should be used with spermicidal cream or jelly.

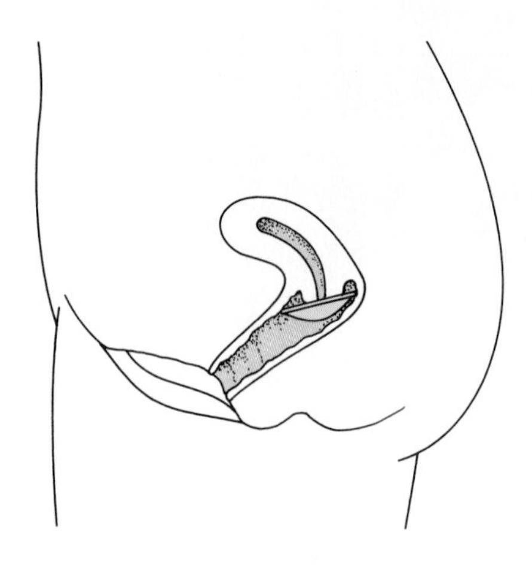

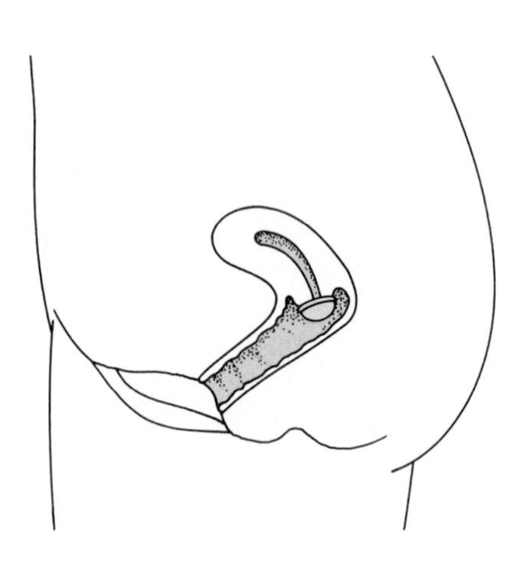

Diaphragm

The **diaphragm** is a dome-shaped rubber device with a flexible steel spring enclosed in the rim. It is inserted into the vagina and fitted over the cervix to prevent sperm from entering the uterus. Unlike the condom, the diaphragm must be individually fitted. It is partially filled with a spermicidal cream or jelly and should be left in place for at least six hours after intercourse. The device is about 80 to 98 percent effective if well-fitted and used correctly. With a certain element of anticipation on the part of the woman, it need not interrupt sexual activities (Figure 15.17).

The Cervical Cap

The **cervical cap** functions similarly to the diaphragm. It covers the cervix and prevents sperm from entering. Essentially it is a small rubber cap that fits tightly over the cervix and stays in place partly by suction. The cap should be partially filled with spermicidal cream or jelly. It should be left in place for at least six hours after intercourse, although one of its advantages is that it may be left in place for up to two days. It is about as effective as the diaphragm. One problem is that it can move out of position without being noticed.

Spermicides

A number of sperm-killing foams, jellies, aerosols, suppositories, and creams are available as contraceptives, or creams and jellies may be used to increase the effectiveness of other contraceptives such as the diaphragm or cervical cap. The **spermicides** are usually quickly and easily applied inside the vagina and require neither fitting nor prescription by a doctor. Used alone, they are 75 to 90 percent effective, depending on whether they are used properly.

The Intrauterine Device

The **intrauterine device,** or **IUD,** was recognized as an effective contraceptive for many years. The principle has apparently been utilized for a long time. In fact, it is said that Arab camel drivers on long, hard caravan marches prevented pregnancies in their charges by placing apricot seeds in the camels' uteruses. (One wonders how the practice got started.) Modern IUDs have been made from several materials and come in a variety of sizes and shapes. No one knows exactly how the IUD works beyond the fact that it seems that a foreign body in the uterus somehow prevents pregnancy. It is thought that the IUD prevents the implantation of the embryo, or that it may cause the egg to move too rapidly through the oviduct, and even that it may alter the condition of the uterine endometrium. In any case though, it is technically not a contraceptive device since it functions not by preventing conception, but by preventing implantation.

The IUD has several major advantages. Once it has been installed in the uterus, it requires little attention for one to several years. Also, it is inexpensive and about 95 percent effective.

The IUD has, in recent years, fallen into disrepute because of the many problems it causes. One problem is that the device may be expelled by the uterus in some women, perhaps without being noticed, leaving the woman unprotected against pregnancy. For some reason, IUDs are more likely to be rejected by the uterus in women who have never been pregnant. There may also be side effects just after insertion, including bleeding and pain. If these persist, the IUD must be removed. The IUD has also been associated with a number of ectopic pregnancies (those that occur in the body cavity, outside the uterus). More important, however, the IUD has been implicated in a number of serious disorders of the pelvic region, perhaps brought on as the string hanging from the uterus into the vagina serves as a "wick" for vaginal bacteria, permitting them to ascend into the uterus and on into the oviduct. These bacteria may then cause *pelvic inflammatory disease (PID)*, an infection of the female reproductive structures, which can result in reduced fertility or even sterility. PID may also increase the likelihood of the embryo implanting in the oviduct (an ectopic pregnancy), a condition that is very dangerous for the mother.

The Birth Control Pill

The combination **birth control pill** contains either natural or synthetic forms of the hormones estrogen and progesterone. Together these hormones inhibit the development and ovulation of the egg in the ovary. The "minipill" contains only synthetic progesterone. It is thought to work by increasing the thickness of cervical mucus, interfering with the movement of sperm, or making implantation more difficult. The great advantage of the birth control pill is its effectiveness. The combination pill, when used properly, may be more than 99 percent effective. The minipill is slightly less effective, about 97 percent, but it has fewer side effects.

Birth control pills may cause undesirable side effects similar to those of pregnancy, especially during the first few months of use. They include weight gain, fluid retention, nausea, headaches, depression, irritability, and increased facial pigment in certain areas. Another side effect may be an enlargement of the breasts. Some or all of these symptoms may diminish after the first few months of use.

Although the pill has produced no serious medical problems for the overwhelming majority of its users, some argue that it is too early to assess its long-term use. There are some rather serious side effects in a small percentage of pill users. The most serious are those that affect the circulatory system. For example, some users develop high blood pressure and this makes it more likely that they may have a heart attack or stroke. However, there are also some benefits of pill use. For instance, it is thought to decrease the risk of ovarian and endometrial cancer.

Massive doses of estrogen may be given when one suspects pregnancy may have occurred (the "morning after" pill). A great deal of research attention has been directed toward developing a contraceptive pill for men, but so far the research is unsuccessful.

Norplant

Norplant is a hormonal means of contraception that eliminates the major inconvenience of using the birth control pill—remembering to take one every day. Norplant implants consist of six thin 1 to 2 inch rods containing a form of progesterone that are inserted just under the skin through a tiny incision, usually in the upper arm. The hormone continuously leaches out through the walls of the rod and enters the bloodstream. It prevents pregnancy in much the same way as the minipill. The contraceptive effectiveness is thought to be significantly better than any other means of contraception, except for sterilization, and its effectiveness lasts for about five years. Although Norplant was only recently approved (at the end of 1990) for general use in the United States, new Norplant systems have already been designed that consist of fewer hormone-containing rods and are effective for a shorter period of time. The rods can be removed quickly if any complications develop or if the woman decides she would like to become pregnant.

The Sponge

A recently developed contraceptive that is gaining popularity is the **sponge** ("Held tightly between the knees," someone has said). The sponge is a small absorbent polyurethane sponge that is saturated with a spermicide. It is inserted into the upper area of the vagina so that it covers the cervix up to 24 hours before intercourse and is left there for about 8 to 10 hours afterward. Its success rate approaches 85 percent.

Sterilization

Sterilization, as you might guess, works, and it's 100 percent effective. Furthermore, either the male or female can be sterilized, although it is a much simpler matter in the male.

A **vasectomy** for a male normally takes about 15 minutes and can be performed in a doctor's office. A small incision is made in the side of the scrotum, a short section is cut from the vas deferens, and then the incision is closed (Figure 15.18). The operation can be reversed in 50 to 80 percent of the cases in those areas of the country where doctors have had the

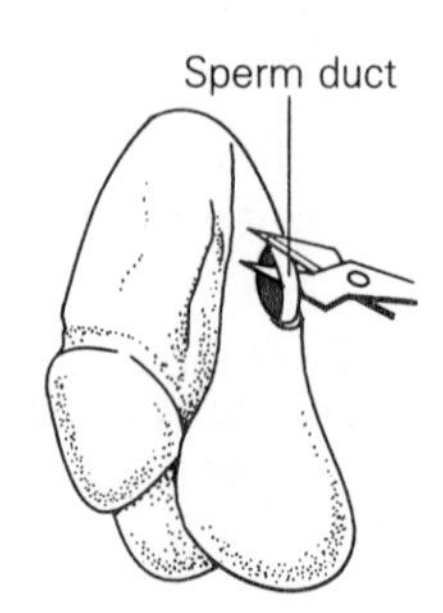

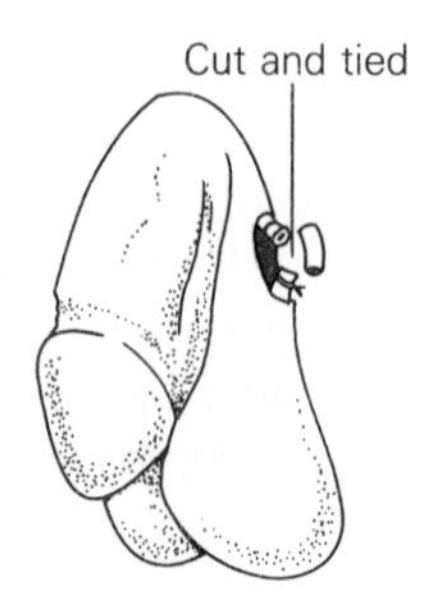

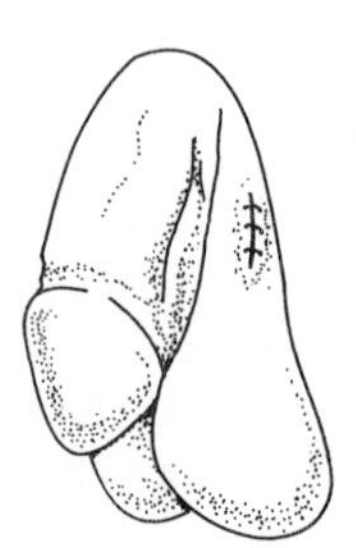

FIGURE 15.18
The simple procedure of a vasectomy. The operation has been described as similar to a visit to the dentist—with a few differences. Normally, under local anesthesia, the vas deferens is exposed by a small cut in the scrotum, and a section is removed from the tube. Both ends are tied in case they show a tendency to rejoin.

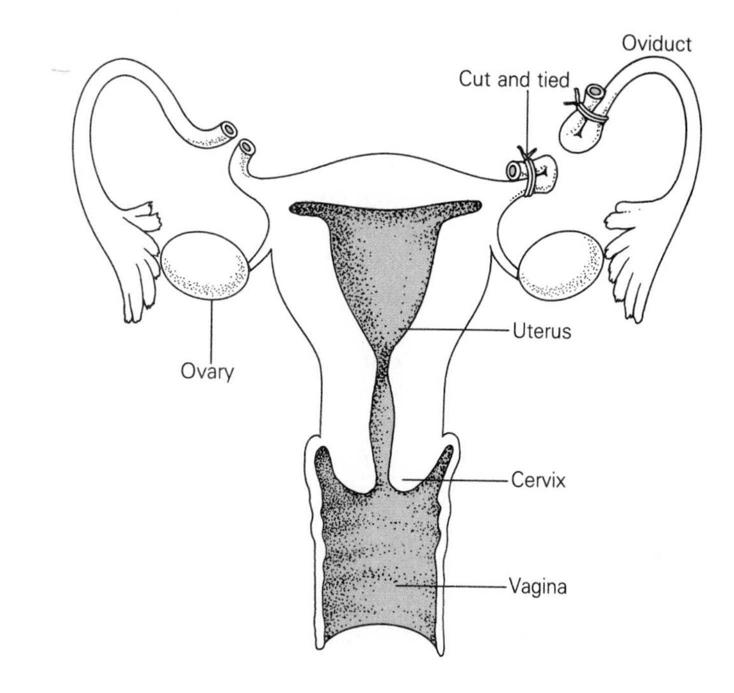

FIGURE 15.19

Tubal ligation, or "tying the tubes," is a means of sterilization in women. It is a more serious operation than the male's vasectomy. It has traditionally been accomplished by opening the body wall, cutting the oviduct, and tying back the ends. Newer methods, however, include tying with a device that is pushed up the oviduct from the uterus. Ligation does not affect the ovaries, thus they continue to produce their hormones, so there is no change in the physiology of the woman other than totally negating the possibility of pregnancy. Some women have reported increased sexual satisfaction with the knowledge that they cannot become pregnant, but such reactions are strictly a personal matter and cannot be predicted.

greatest experience with the operation. Another method, in which a removable plastic plug is inserted in the vas deferens, shows promise of being reversible in almost all cases.

Sperm continue to be produced after a vasectomy, but since they cannot be ejaculated, they are reabsorbed by the body. Semen is ejaculated as before, but it contains no sperm. Since the secretions of the accessory glands make up at least 80 percent of the normal ejaculate, and hormonal levels are not affected, there is no noticeable difference in the male's sexual performance.

Vasectomy is, in itself, a simple procedure, but other factors must be taken into account. For example, in many cultures, including that of the United States, men tend to identify rather strongly with the "male role." The realization that they are no longer able to impregnate women may have a rather marked psychological effect in some individuals. For this reason, the procedure is ordinarily advisable only for those men who are fairly secure in their own sexual identity.

Sterilization in women is usually accomplished by **tubal ligation** (Figure 15.19). The oviduct is cut and both ends are sealed. The oviduct can be cauterized, or seared, causing the opening to seal shut as it heals, or it can be pinched together with a tiny clip or rubber band. This operation is much more complicated than a vasectomy since the abdominal wall must be opened. In some cases, the abdomen can be entered through a small incision at the navel or just above the pubic bone. It is also possible to enter through the vagina, but this is riskier because of the increased possibility of infection. As with vasectomy, there is no evidence that, in humans, tubal ligation causes any change in hormone production, sex drive, or sexual performance. However, the psychological effects may be varied. Some women experience regret that they are no longer able to have children. Others feel liberated because they no longer have to worry about becoming pregnant.

Another means of sterilization, the use of a **fallopian plug,** is being tested in the United States. The process essentially consists of simply inserting silicone rubber plugs into the oviducts (fallopian tubes). The egg, unable to reach the uterus, simply withers and is reabsorbed. A fiber-optic device helps guide a tube to a place near the opening of the oviduct and liquid silicone, with a hardener, is then slowly pumped into the oviduct. The liquid hardens in about six minutes and X rays determine if the plug is in place in both oviducts. The woman is free to leave the physician's office right away, but she may be bothered by cramps at first. The process is easier on the woman than any surgical techniques and its effectiveness approaches 100 percent. At present, the process is considered irreversible, but new methods are being devised that permit retrieval of the plug.

Abortion

Abortion, the artificial termination of a pregnancy, has been one of the more common forms of birth control throughout recorded history. In the United States, restrictions on abortions have been relaxed, but this has not necessarily resulted in more abortions. Instead, it very likely means that *illegal* abortions may now be less common. When abortion was illegal, there was estimated to be one illegal abortion for every two live births in the United States.

Illegal surgical abortions and do-it-yourself methods have resulted in innumerable deaths and permanent physical and psychological damage as well. Some of the methods employed under illegal or "back alley" conditions have been as barbaric as they are hazardous. Furthermore, abortions by untrained people may not even be successful. Perhaps the most important point here, however, is that it is probably safer to deliver than to attempt to abort under these conditions.

When done properly, under medical supervision, an abortion during the first three months of pregnancy is a simple procedure. The most common method of aborting a fetus is called *dilation and curettage*, or *D and C*. The vagina and cervix are dilated, and the lining of the uterus is scraped with a curette, a steel loop at the end of a handle. Other methods are used quite successfully also, such as *vacuum curettage*, an increasingly popular technique in which a suction device pulls the embryo from the uterine wall. Another common procedure is called *salting out*. It involves the injection of salt solutions into the uterine cavity.

Abortion first became illegal in the United States in the nineteenth century. At that time, it was ruled dangerous to the prospective mother. Recently, courts have upheld the right of women to have abortions under certain conditions, often determined by state governments, but the decisions have caused heated debates, animosity, and even violence. The groups opposing abortion have coined the term "right to life," but others argue that pregnancy is a very personal matter and that the decision to accept this condition and the ensuing years of responsibility are primarily the woman's since she must bear the brunt of any pain, risk, and responsibility.

One argument has revolved around the question of when a developing embryo becomes a "person." Some believe that the fertilized egg is a human being and has a "right" to develop and be born. Others argue that abortion does not actually destroy "life" until later, such as when the heart (still a tubular muscle) begins to beat, or when the embryo begins to stir or "quicken," or the time when it becomes capable of surviving after birth.

The question basically boils down to opinions, and emotional ones at that. Logic has rarely found a place in these discourses, nor have decisions based upon the good of all. For example, many antiabortionists have stated their support for capital punishment or military enterprises that involve loss of life. It almost seems that life's sacredness is related to one's birthplace, ancestors, or voting record. Obviously, the question of abortion is not likely to be resolved on any rational or scientific basis (Figure 15.20).

FIGURE 15.20

People feel strongly about the rights of women to decide whether to have children. Some are very opposed to the right to end a pregnancy with an abortion, even embracing violence to make their argument.

It should be mentioned that abortion has various effects on different women. Some take abortion very lightly and measure it in physical or economic terms. Others, however, suffer aftereffects such as extreme guilt, a feeling of loss, or some other severe emotional trauma. People who do not want to be parents may be forced to make some very difficult decisions. Some are prepared for the effects, but others are surprised at their own anguish. Furthermore, the repercussions may be severe to both the man and the woman, a point often overlooked.

Abortion is also opposed on the assumption that it encourages sexual freedom, a possibility that is viewed with more alarm in some quarters than others. More than one legislator has publicly stated that if a woman has sex, it is only fair that she "pay the piper" and accept her punishment in the form of "compulsory pregnancy." Apart from the tacit assumption that moral responsibility should be expected only of women, children being considered a form of punishment is interesting, but probably rarely acknowledged by its *de facto* proponents. These are only a few of the arguments regarding abortion, but they may provide food for thought or perhaps launch a discussion or two. If you would like to hear some other arguments, just bring up the subject at your next family gathering.

As a final point of consideration, Swedish scientists attempted to find out what usually happened to unwelcome children whose mothers were denied abortion. Such a group was compared to a matched set of children who were wanted. They found that twice as many of the unwanted ones grew up labeled as "illegitimate," or in broken homes or institutions. Twice as many unwanted ones had records of delinquency, twice as many were declared unfit for military service, and twice as many had required psychiatric care. Finally, five times as many had been on social welfare programs even in their teens.

SUMMARY

1. Asexual reproduction among animals may occur by binary fission, in which the animal splits in half; by regeneration, in which missing parts are reformed; or by budding, in which the new organism forms as an outgrowth of the parent and pinches off to independent life.

2. Fertilization may be external or internal. In external fertilization, many gametes, each with little energy investment, are shed outside the body where fertilization occurs by chance. Few of the many gametes that are shed result in offspring.
3. Internal fertilization increases the chance of successful fertilization. Some species press their reproductive openings together. Others deliver sperm to the vagina through a penis (copulation). Reproduction may be timed to occur when the chances of offspring survival are greatest. Among some mammals, timing is controlled by estrus, a period when the female is physiologically and behaviorally receptive.
4. Human females are almost always physiologically able to copulate, an adaptation thought to have evolved as a means of maintaining a bond between mates that will ultimately result in successfully rearing more offspring.
5. The human reproductive system is quite similar to those found in other mammals. The male reproductive system includes the testes, scrotum, accessory glands, various ducts, and the penis. Their functions are given in Table 15.1. The reproductive system of the human female includes the ovaries, oviducts, uterus, vagina, and external genitals. Their functions are given in Table 15.2.
6. The menstrual cycle begins at puberty, usually between the age of 12 and 14, and continues for 30 or 40 years until menopause.
7. During each menstrual cycle, the endometrium, which is the lining of the uterus, becomes engorged with blood that will provide nutrients and oxygen for the early embryo. If fertilization does not occur, the endometrium is shed, marking the menstrual period.
8. At the beginning of the menstrual cycle, two hormones from the pituitary gland, follicle-stimulating hormone (FSH) and luteinizing hormone (LH), stimulate the development of egg-bearing follicles within the ovary. As the follicles mature, they secrete the hormone estrogen, which causes the endometrium to thicken. One follicle usually outgrows the others. LH triggers the release of the egg from this follicle. The corpus luteum from the follicle left behind in the ovary secretes the hormones estrogen and progesterone, which cause the endometrium to prepare for a fertilized egg and inhibit the production of FSH, thereby preventing the start of a new cycle. When fertilization does not occur, the corpus luteum stops its secretions so the endometrium breaks down and is lost. If fertilization does occur, the embryo implants in the endometrium and begins growth, obtaining necessities from the mother's blood.
9. Human precopulatory behavior (foreplay) is variable. In the male, foreplay causes the penis to swell with blood, making it tumescent and erect because more blood enters through enlarged arteries than leaves through the compressed veins. In women, the vagina becomes lubricated, the labia minora and clitoris enlarge and redden, the nipples and breasts enlarge, the breasts become more sensitive, and the shoulders and chest may have a mottled appearance.
10. During copulation the penis is inserted into the vagina. Pelvic thrusts result in ejaculation. Orgasm takes place in two stages. During the first stage, the epididymis, seminal vesicles, and prostate gland deliver their products to the urethra and a secretion of the bulbourethral glands (Cowper's glands) lubricates the penis for insertion into the vagina. During

the second stage, semen is first delivered to the urethral bulb and then forcefully ejected out of the body through the urethra. Ejaculation and an increase in blood pressure and pulse rate are physiological changes that occur at climax (orgasm). After orgasm, there is a period of relaxation. The physiology of orgasm in women is similar to that in men.

11. The sperm's tail and currents created by uterine contractions propel the sperm to the egg, which is in the oviduct near the ovary where conception occurs. Many of the millions of sperm released die due to natural causes or the unfavorable acidic environment of the female reproductive system, or are destroyed by the female's white blood cells. Although many sperm reach the egg, only one effects fertilization, perhaps because of changes in the egg's membrane that are triggered by the initial sperm penetration.
12. The embryo travels along the oviduct to the uterus. Initial nourishment is from the uterine fluids. Soon it embeds in the uterus and obtains nourishment from the mother's blood.
13. There is no means of contraception that is 100 percent effective in preventing pregnancy other than abstinence and sterilization. Methods of birth control are described in Table 15.3.
14. Sterilization is 100 percent effective. A vasectomy involves removing a small piece of the seminal duct or plugging the duct with plastic. A tubal ligation involves cutting the oviduct and sealing the ends.
15. Abortion is the artificial termination of a pregnancy. Methods include: dilation and curettage (D and C), in which the lining of the uterus is scraped; vacuum curettage, in which a suction device pulls the embryo from the uterine wall; and salting out, in which salt solutions are injected into the uterus.

KEY TERMS

abortion (434)
binary fission (408)
birth control pill (431)
budding (410)
bulbourethral glands (Cowper's glands) (415)
cervical cap (430)
cervix (417)
clitoris (418)
coitus interruptus (427)
conception (423)
condom (427)
contraception (425)
copulation (413)
corpus luteum (419)
diaphragm (430)
ejaculation (415)
endometrium (417)
epididymis (415)
estrogen (419)
estrus (413)
fallopian plug (434)
fertilization (423)
follicle (419)
follicle-stimulating hormone (FSH) (419)
intrauterine device (IUD) (430)
labia majora (418)
labia minora (418)
luteinizing hormone (LH) (419)
menstruation (418)
orgasm (422)
ovaries (417)
oviduct (417)
ovulation (419)
penis (417)
progesterone (419)
prostate gland (415)
regeneration (410)
rhythm method (427)
scrotum (415)
seminal vesicles (415)
seminiferous tubules (415)
spermicide (430)
sponge (432)
sterilization (432)
testes (415)
tubal ligation (433)
urethra (415)
uterus (417)
vagina (418)
vas deferens (415)
vasectomy (432)
vulva (418)

FOR FURTHER THOUGHT

1. If you used a spermicide as a method of birth control, would you choose one that is acid or alkaline? Why?
2. Why is the fallopian plug not the best choice of contraceptives for young college-aged women?

FOR REVIEW

True or false?

1. ___ In many mammalian species, copulation is most likely to occur when a female is in estrus.
2. ___ Progesterone and estrogen inhibit FSH production.
3. ___ Most sperm are expelled in the first contractions of an ejaculation.
4. ___ It is physiologically impossible for a woman to achieve an orgasm.
5. ___ The condom is a method of contraception that also provides some protection from sexually transmitted diseases.

Fill in the blank.

6. Two types of asexual reproduction are ___ and ___.
7. The ___ is the uterine lining that is shed during menstruation.
8. Large numbers of human sperm are required so that one may fertilize the egg, because ___.
9. The ___ method of contraception is one in which abstinence is observed around the time of ovulation.
10. Three medically supervised methods of abortion that may be employed during the first three months of pregnancy are ___, ___, and ___.

Choose the best answer.

11. The birth control pill acts by
    A. inhibiting the development and ovulation of the egg.
    B. preventing sperm from entering the uterus.
    C. increasing FSH production.
    D. preventing implantation.
12. In the following list, which method of contraception is least effective?
    A. IUD
    B. douche
    C. cervical cap
    D. tubal ligation
13. Conception takes place in the
    A. uterus
    B. vagina
    C. oviduct
    D. ovary
14. Fluids from the female ___ act as lubricating substances that aid in penis insertion.
    A. labia minora
    B. cervix
    C. vaginal wall
    D. clitoris
15. Choose the incorrect statement.
    A. LH is secreted by the pituitary gland.
    B. The corpus luteum secretes estrogen and progesterone.
    C. Estrogen induces a thickening of the endometrium.
    D. Progesterone stimulates follicle development.

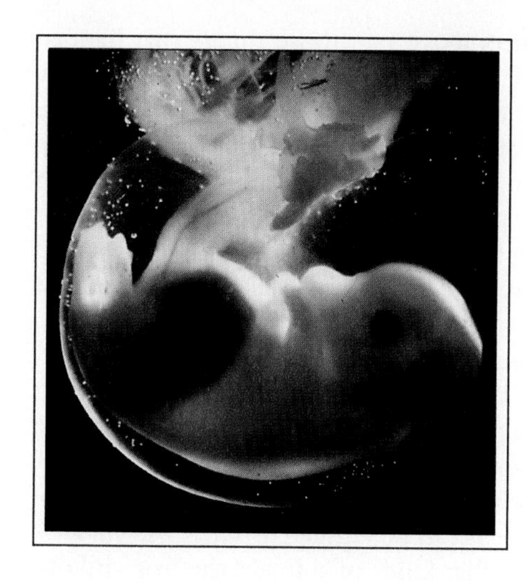

CHAPTER 16

Animal Development

Overview

Early Stages of the Embryo
Influences of Organization
Types of Eggs
Early Development in the Frog
Birds and Their Membranes
Human Development

Objectives

After reading this chapter you should be able to:

- Describe in general how an embryo develops, listing events of fertilization, cleavage, and gastrulation.
- Name three embryonic germ layers and the tissues formed from each.
- Describe four ways in which embryonic development is regulated.
- Explain how the amount of yolk in an organism's egg correlates to its life history.
- Fully describe the early developmental stages of a frog and a chick.
- Describe human developmental events during each of the three trimesters.

The story of development encompasses all those things that happen to us from the moment of conception (which most of us hardly remember) to that final indignity that mocks us all. In its broadest sense, then, it includes not only the events of the womb, but our birth, growth, aging, and perhaps even our death. (Some "primitive" tribes consider death just another phase of life, but that philosophy may take a bit of adjustment on our part.) Obviously, any such broad treatment of development would involve as much philosophy as science, so we will take a more modest approach.

We will focus primarily on the events from fertilization to birth. Furthermore, we will concentrate on humans. However, rather than jumping right in and considering the complex issues of human development, we will begin with a general overview of the early division of the embryo, the establishment of the three primary germ layers, and finally, some of the ways the embryonic development can be regulated. Then we will look briefly at the development in two other kinds of animals, the frog and our friend the chicken, with which we can most easily see just how the various membranes surrounding the embryo are formed. Keep in mind that the processes are different from one group to the next (for example, membrane formation in chickens may be notably different than in humans in certain ways). Yet by examining principles of development in three quite distinct groups, we can develop a reasonably good idea of how development proceeds in general.

EARLY STAGES OF THE EMBRYO

Development begins with **fertilization,** the fusion of the nuclei of the egg and the sperm (Table 16.1). Fertilization is important because it creates a new cell, called the **zygote,** which has two copies of each chromosome. In addition, it triggers a number of new metabolic reactions and begins a reorganization of the contents of the egg cytoplasm.

Next there is a rapid series of mitotic cell divisions called **cleavage**. Cleavage divides the zygote into many smaller cells of various sizes while causing no increase in the overall size of the embryo. Cleavage results in

TABLE 16.1
Early Stages of Development

STAGE	DESCRIPTION	RESULT
Fertilization	Fusion of nuclei of egg and sperm	Creates a cell (the zygote) with two copies of each chromosome; activates reactions in the egg
Cleavage	Rapid series of mitotic divisions	Divides the zygote into smaller cells of varying size, shape, and activity
Gastrulation	Migration of cells	Forms three primary germ layers (ectoderm, mesoderm, endoderm), each of which will give rise to specific tissues

TABLE 16.2
Structures Produced by the Three Germ Layers

GERM LAYER	DERIVATIVES
Ectoderm	All nervous tissue; epidermis of skin; parts of eyes and ears; hair; feathers; pituitary gland; adrenal medulla
Endoderm	Most epithelium of digestive system, of respiratory system, of ducts of reproductive system, of urethra and urinary bladder; thyroid, parathyroid and thymus glands; glands in liver and pancreas
Mesoderm	Muscle (all skeletal, most smooth, and all cardiac); cartilage; bone; blood; kidneys and gonads

the contents of the zygote being unequally distributed among the cells, so the newly created cells differ in their contents. This is important because gene activity may be altered by the cytoplasmic environment, thereby giving the cell a push along a particular developmental path.

During **gastrulation,** certain groups of the embryo's cells migrate, forming three distinct types of cells called **germ layers** (germ: beginning). They are the **ectoderm, endoderm,** and **mesoderm.** Each will contribute to specific structures as the body slowly forms (Table 16.2). The ectoderm lies on the outside, an inner endoderm is formed from cells of the yolky vegetal area, and the mesoderm is derived from the cells that rolled under from the outer layer. The mesoderm now lies between the ectoderm and endoderm.

Few structures of the vertebrate body are formed entirely from any one germ layer, but it is traditional, and simpler, to refer to them as if they are. For example, we will say that the ectoderm forms the outermost layer of the skin, the sense organs, parts of the head and neck, and the nervous system. In birds and mammals, the ectoderm is held responsible for hair and feathers. The mesoderm forms tissues associated with support, movement, transport, reproduction, and excretion (for example, muscle, bone, cartilage, blood, heart, blood vessels, gonads, and kidneys). The endoderm is also largely responsible for the structures associated with breathing and digestion (including the lungs, liver, pancreas, and other digestive glands).

Soon after gastrulation, some of the newly formed mesodermal cells form the **notochord,** a rodlike structure along the length of the developing embryo. The notochord will prompt the development of the brain and spinal cord (the central nervous system). Under the influence of the notochord, the overlying ectoderm begins to thicken, forming a **neural plate.** Soon tissue starts to build up along each side of the long, thickened plaque (Figure 16.1). After a period of enlargement, these two ridges, called the **neural folds,** lying along either side of the **neural groove,** begin to grow toward each other. The margins of the folds grow together until their tips touch and finally join and fuse. The result is a hollow tube lying along the dorsal surface of the embryo. This is the early central nervous system, and it is soon overgrown by ectodermal cells from the surrounding area so that it comes to lie beneath the skin.

FIGURE 16.1

Spinal cord induction. The rodlike notochord in this chick embryo lies just below the ectoderm. At some point, it begins to cause the overlying ectoderm to thicken into a neural plate. Then two ridges, the neural folds, form along its length on either side of the neural groove. These grow upward and inward toward each other and join, forming a tube over the notochord that will be the spinal cord. The anterior end of the spinal cord will develop into swollen bulbs that eventually become the brain.

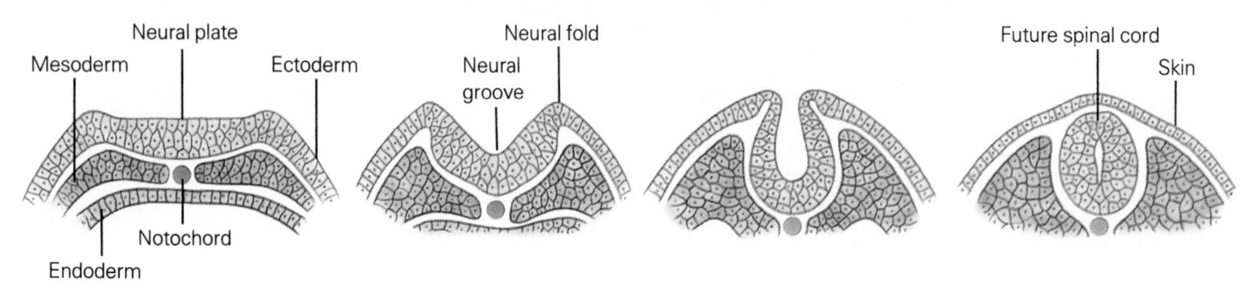

In most vertebrates, the formation of the central nervous system marks the partial or complete disintegration of the notochord that triggered it. Any supportive function it might have had is taken over by bones that first form largely from somites, segmented blocks of mesoderm that lie alongside the spinal cord. These will form the vertebrae (backbones) and eventually enclose the spinal column in protective bone.

INFLUENCES ON ORGANIZATION

At this point, we will consider a few general principles about what causes an embryo to take its final form. Then we will see how these principles apply to embryonic frogs, birds, and mammals.

The Flexible Fate

We are aware that the fate of some cells is never sealed. Otherwise, how could a flatworm that is cut in half regenerate its lost parts? Or how could a new sea star form from only a piece? (Essay 16.1). Furthermore, if sponge cells are separated by being passed through a sieve, they will crawl back together, ameboid fashion, and form a new sponge. (I suspect that if *my* cells were separated, they would take advantage of the situation and clear out.) Such cell flexibility is also present to a degree in many vertebrates. For example, in some kinds of salamanders, when a leg is amputated, the cells in that area of the cut despecialize and revert back to a more primitive state that is usually associated with embryos. Then they begin to specialize again, taking new developmental routes, moving about and changing until a new leg is formed.

What about mammals? If the leg of a mouse is amputated, it doesn't grow back. Cell differentiation in mammals is more permanent; the cells are less able to revert to an earlier condition. There are some interesting exceptions to this rule, however. For example, certain cells in the reproductive tract of rats are able to dedifferentiate and take a new route,

ESSAY 16.1

Can Humans Regenerate Missing Parts?

A few years ago, a young boy was admitted to the Children's Hospital in Sheffield, England. He had accidentally cut off the end of his finger. Ordinarily the part would have been reattached by a plastic surgeon in the hope that it would grow back and that the feeling would be restored. However, due to a clerical error, the stub was simply bandaged and the boy was ignored for several days. When the error was discovered, anxious physicians unwrapped the finger to see how extensive the damage had become. They were stunned to see that the finger was regenerating the missing part. They carefully monitored the progress of the finger over the next few weeks until the damage had, for all practical purposes, repaired itself and the finger was restored. The technique is now routinely used for young children, and there are several cases of regrown fingers, complete with nails and fingerprints. But how? This was contrary to all medical expectations. Should it have been so unexpected? After all, identical twins are formed when a developing human breaks apart at the two-cell stage, each part regenerating whatever was lost. So there are some regenerative powers in humans. The phenomenon is apparently associated with youth. It has not been found in adults, other than to a limited degree in wound healing. Subsequent research has shown that age does, in fact, have something to do with the regeneration. It seems that children under the age of 11 have marked powers of regeneration, and these rapidly dissipate after this time. As far back as the early 1970s, it was demonstrated that electrical charges surrounding the tissues in young people become reversed about the time that regenerative powers dissipate. Research attention is now focusing on maintaining particular (young) electrical fields around those areas in which regeneration is being attempted. The time when humans can regenerate lost limbs is probably not near, but scientists have new reasons for optimism.

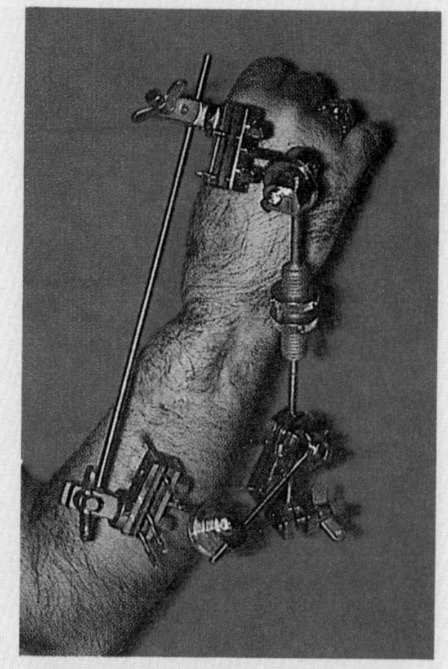

This arm was severed by an alligator. The arm was reattached and the above apparatus was used to stabilize it after a certain amount of healing had occurred.

becoming different sorts of cells, replacing those experimentally removed. We are aware that, in mammals, epidermal cells grow over a wound as the cells underneath are engaged in the healing processes (Essay 16.2). In general, though, among mammals, cellular deorganization and regeneration is very limited. In most cases, once a cell has developed along a certain route, its fate is sealed. So what determines that route? How is it regulated?

Sequencing

Embryonic regulation demands stringent controls on the sequence of development. For example, before those first tremblings in the developing heart begin, signalling the onset of a rhythmic beating, there must be some place for the blood to go. So great, irregular channels appear even earlier in development, formed along extensions of loose-knit, undifferentiated first-blood pools. One immediately wonders what sorts of mechanisms determine such sequencing.

ESSAY 16.2

Wound Healing

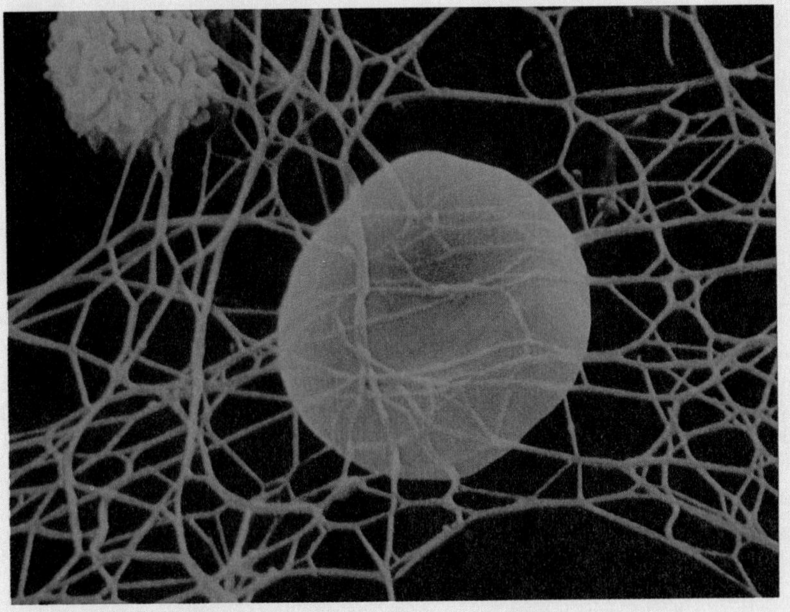

Here, a netlike structure called fibrin forms, holding red blood cells as a clot develops.

Centuries ago, Egyptians covered open wounds with raw meat, and today people are known to place raw steak on black eyes. Both habits may be regarded as a bit peculiar in some circles, but perhaps both have merit. In fact, it has recently been discovered that something in muscle tissue indeed promotes healing. Healing, in general, is a remarkably orchestrated sequence, especially when open flesh wounds are involved.

Flesh wounds are common and dangerous. After all, the protective covering around the body has been broken, leaving an opportunity for invasion by dangerous organisms. Thus, the body acts quickly by first sealing blood vessels in a process called *clotting*. Platelets—tiny bodies in the bloodstream—rush to the site where they interact with a blood protein called *fibrinogen* to form the clot. Some of the platelets disintegrate, releasing a substance called serotonin, which constricts the blood vessels in the area. The blood that has already escaped hardens and forms a protective scab. Once the blood flow has stopped, the body begins to respond to other chemicals (*pyrogens*) that have been released in response to the trauma.

If the wound is deep, the pyrogens will cause the body to become warm (perhaps too warm for invading bacteria) and the heart rate to increase. Blood vessels open up (bringing in not only white blood cells that will attack any invading bacteria but nutrients and oxygen for stepped-up metabolism) as new tissue forms. In the meantime, epidermal (skin) cells begin to multiply in the area of the wound, extending underneath the scab and growing together from all sides. Capillaries in the area begin to grow side branches and to penetrate the regenerating area, bringing in yet more fresh blood laden with oxygen and nutrients.

In the meantime, cells called *fibroblasts* rapidly multiply, forming "scar tissue," and filling in the depression caused by the wound. The scar tissue itself is inordinately strong because the fibroblasts produce cablelike fibers with the ability to stretch and contract. As they grow over the area and contract, they help pull the edges of the wound together. By now, nerve cells have begun to sprout side branches that invade the injured area. After a time, the scab loosens and falls away, exposing the layer of epidermis underneath. Below this, fibers of a connective tissue called collagen grow along the lines of stress that produced the injury in the first place, further strengthening the healed area.

The details of most such controls remain a mystery, but ultimately the control must reside in the chromosomes. In the tightly choreographed sequence of developmental processes, some cells slow down or even cease mitotic activities just as others begin a burst of reproduction, accompanied by a surge in the growth of the tissue they comprise. Some cells must even die as part of the developmental process. (It is essential that some cells go through the process of formation and then vanish, such as those that lie between what will be the fingers of a developing hand.)

Such timing can be accomplished because particular segments along the length of some chromosomes may be active for a shorter time than other segments. Thus, after an initial period of activity, one set of genes might cease its activity while another continues to turn out its enzymes. So the real questions of control involve the signals that turn genes on or off, and the challenge lies with identifying those signals and perhaps learning to control them.

Stress

Interestingly, stress can act as a regulatory factor in embryos. Just as karate practitioners build up the tissues around their knuckles by pounding makiwara boards, so an embryo builds up tissue along its lines of stress.

Embryos are stressed in a number of ways during normal development—for example, by bending and twisting that places tension on the back area and encourages the development of sturdy muscles in that area. This happens when a type of mobile, undifferentiated tissue called **mesenchyme** moves into the areas of tension or presssure, settles there, and forms new supportive tissue.

Induction

Perhaps the most fascinating form of regulation involves one type of tissue influencing the development of another, a process called **induction.** As an example, after the mesodermal layer is formed, the tissue just under the area that will ultimately become the spinal cord begins to thicken, forming that rodlike structure, the notochord. The notochord itself doesn't actually form the spinal cord; instead, as we have seen, it causes the overlying ectoderm to do so.

The role of the notochord as the **inducer** of the spinal cord has been demonstrated by transplanting segments of the notochord to other areas of the embryo. If this is done before the ectoderm has become too differentiated, it will form incipient spinal cords almost anywhere on the surface of the embryo. For example, a spinal cord could be induced to grow along the belly. Such freaks, of course, usually do not survive.

Sometimes different inducers operate together in a coordinated fashion. For example, in some cases, where two tissues have developed from the same germ layer, one may have induced the other. Perhaps the best example is the induction of the lens of the eye from ectodermal cells by underlying tissue also formed from ectoderm (Figure 16.2, page 446).

FIGURE 16.2

Lens induction in the vertebrate eye. By the fifth to sixth week (a), outpocketings of the brain (called optic vesicles) come to lie under the head ectoderm, inducing that ectoderm to thicken and roll, forming the lens of the eye by the eighth week (b) and (c). The underlying brain tissue will form elongated cells that become the light-sensitive retina of the eye. Thus, part of the eye is directly confluent with the brain and both the lens and the retina are ectodermal in origin. By the eleventh week (d), the cornea will have formed.

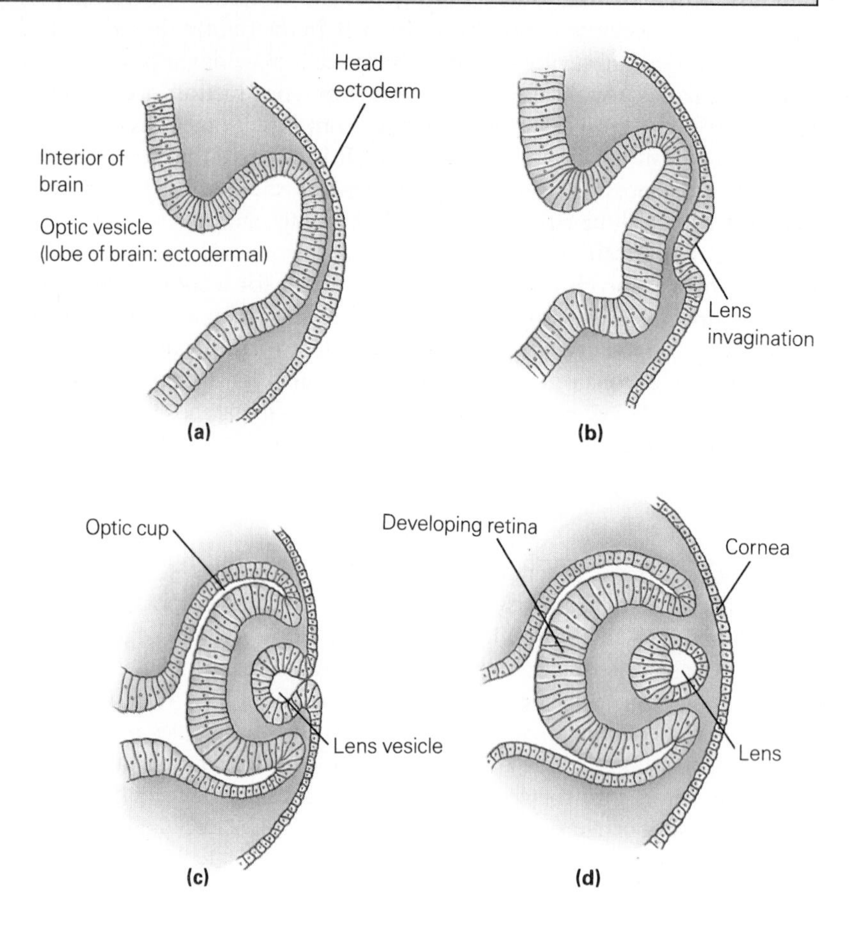

TYPES OF EGGS

The three basic types of animal eggs are roughly categorized according to the amount of **yolk** they have. The amount of yolk is critical since it is the embryo's food supply, at least for a time. In some species, the embryo needs only a small supply of yolk since it soon switches to nutrients derived from the mother's blood, as is the case with humans. We only need enough to last until the embryo has implanted in the wall of the uterus. In contrast, birds leave their mother's body at a very early developmental stage, and they must carry their entire embryonic food supply with them. So, whereas a human egg is smaller than the period at the end of this sentence, it would be hard to hide an ostrich egg with this whole book. Interestingly, both the young and the adults of these two species are about the same size.

Other kinds of animals have a moderate yolk supply. In these, the young must begin to find their own food long before they have reached final body organization. The frog is an example. The frog egg has just enough yolk to get the developing embryo to the tadpole stage; after that, the tadpole can survive on food stored in its tail for a time, but it must soon begin to eat on its own.

So human eggs have little yolk, frog eggs have an intermediate amount, and bird eggs have a great deal. Can you see how the relative amounts of yolk are correlated with life histories of the species?

EARLY DEVELOPMENT IN THE FROG

You may recall that at fertilization, the zygote receives all the genetic information it is ever going to get. No matter what manner of creature will ultimately be formed, all the genetic instructions are in that first cell. Let's see how those instructions tell a cell to build a frog.

The First Cell Divisions

In frogs, as in other animals, the egg begins drastic changes as soon as it is fertilized. First, the egg becomes unresponsive to other sperm, even as the two sets of chromosomes draw together for the first of countless mitotic divisions. The genetic material doubles, and virtually identical chromosomes move to opposite poles. Then the cell divides, forming two daughter cells. The embryo is on its way.

The first two cell divisions are usually at right angles to each other, along the same vertical axis; the third line of division is perpendicular to these. (Figure 16.3 compares the developmental sequence of frogs, chickens, and humans.) In the case of the frog, this third cleavage (division into smaller cells) is somewhat closer to one pole. The third division takes place closer to the more pigmented **animal pole** because the relatively large supply of yolk at the other, **vegetal pole,** resists dividing. The body axis is determined by the point at which the sperm penetrated the egg at fertilization.

As the cells of the embryo continue to divide, they form a cluster, or ball, of cells called a **morula** (Figures 16.3 and 16.4). A cavity, the **blastocoel,** appears within the ball, which is now called a **blastula.** Because of the unequal distribution of yolk, the cells are smaller at the animal pole.

Gastrulation

Early on the second day after fertilization, another important change occurs. At this time, certain cells migrate toward an area in the vegetal hemisphere. Here, they converge and then begin to roll under the surface of the embryo, marking the stage called gastrulation, during which the three germ layers form. Gastrulation, as we see in Figure 16.3, forms a cavity called the **archenteron,** which will become the digestive tract. The archenteron opens to the outside through the **blastopore.** The blastopore forms a curved line and eventually a circle over the yolky area at what will be the back, or posterior, side of the embryo. The involution continues, extending its margins, until the crescent finally becomes a circle. The vegetal area inside this circle is called the yolk plug and will eventually be enclosed by the expanding layer of the surface cells around it.

FIGURE 16.3

A comparison of the development of three vertebrates. In the early cell divisions (1,2,3), differences are obvious. The chick embryo, perched atop a massive and inert yolk, first undergoes incomplete cleavages. Frog and human begin similarly, but soon cell divisions lag in the larger, yolk-laden vegetal region of the frog, resulting in larger cells there. All three embryos produce a hollow blastula stage (4). Frog and human blastulas are spherical with prominent cavities. In the chick embryo, a streak of cells rises up slightly, forming only a minute cavity above the yolk. Gastrulation (5) produces new cavities and the three germ cell layers. From these layers, future tissues and organs will be molded. About this time, the human embryo implants in the uterine wall, its meager yolk reserves depleted. Its chorion grows into the uterus, seeking nourishment. The bird, too, will produce membranes that assist in bringing food from the yolk. The frog soon incorporates its food supply into its body.

In (6), all three produce the rudiments of a nervous system, ectoderm rising up in folds to outline the system. Subsequently, discrete pockets of cells contribute to organs as systems are built. (7,8) Heart and blood vessels in the chick and human form early. In each embryo, mesodermal blocks called somites contribute to vertebrae, muscle, dermis, and other structures. Interestingly, each species forms pharyngeal gill pouches, a primitive vertebrate feature. In land creatures, these contribute to structures other than gills.

With continued refinement of form (9,10,11), the species become easier to recognize. The tadpole will be eating long before the chick has used up its food supply, and the chick will be an adult by the time the human is born.

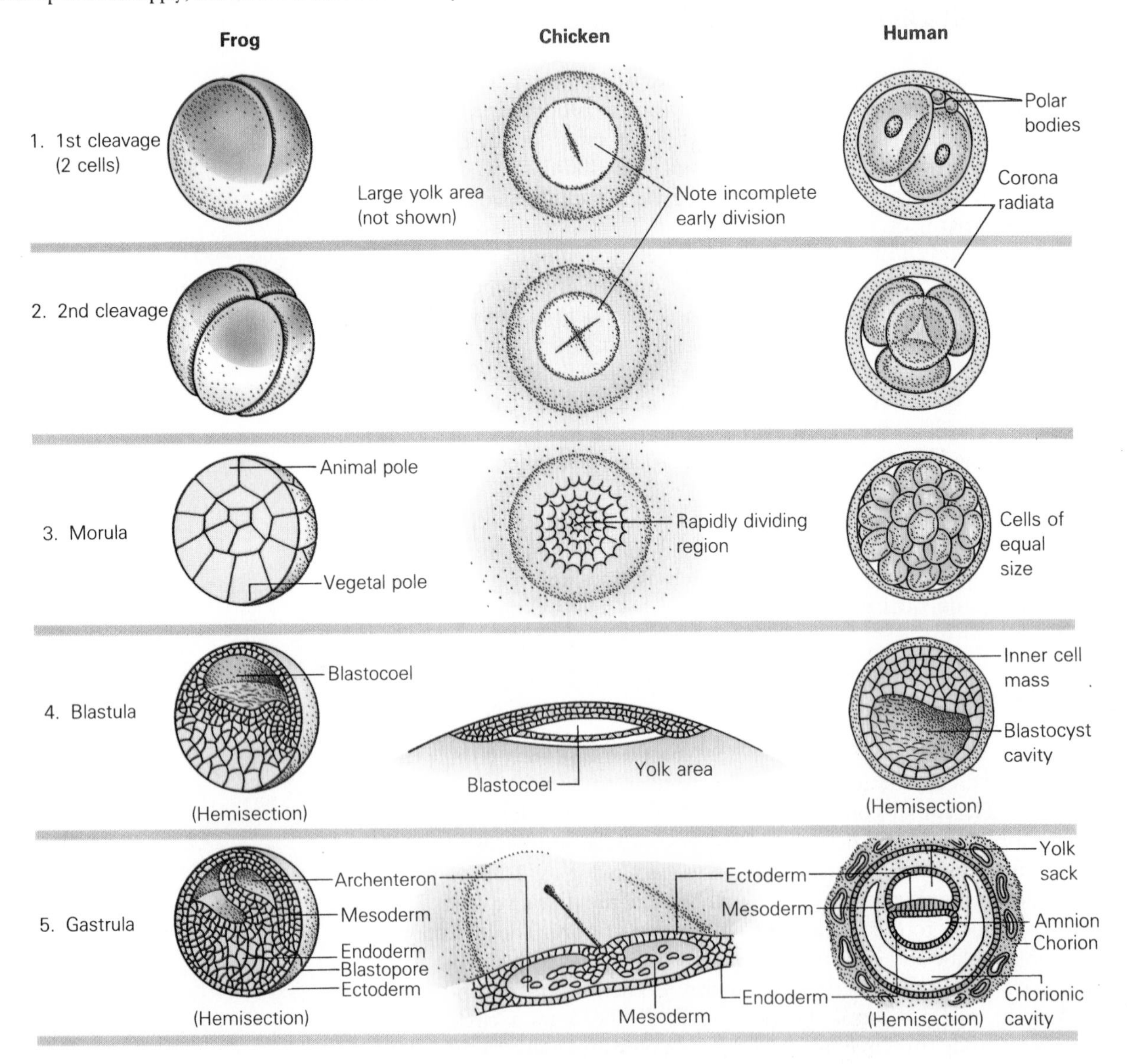

FIGURE 16.3 continued

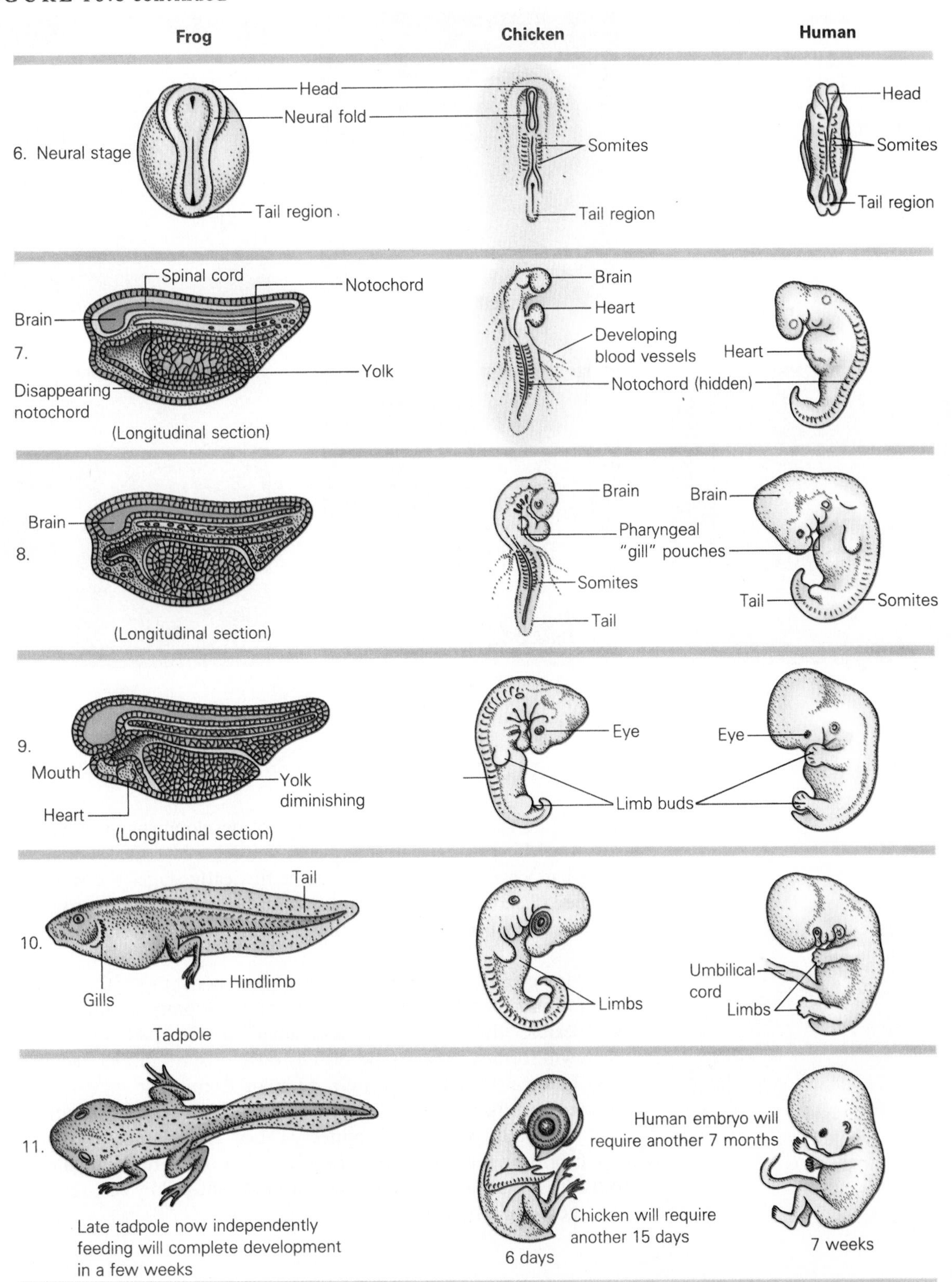

FIGURE 16.4

The gradual development of a frog embryo from the first cleavage to the morula stage. Note the smaller cells at the animal pole due to more rapid cleavage in this area that is less encumbered by yolk.

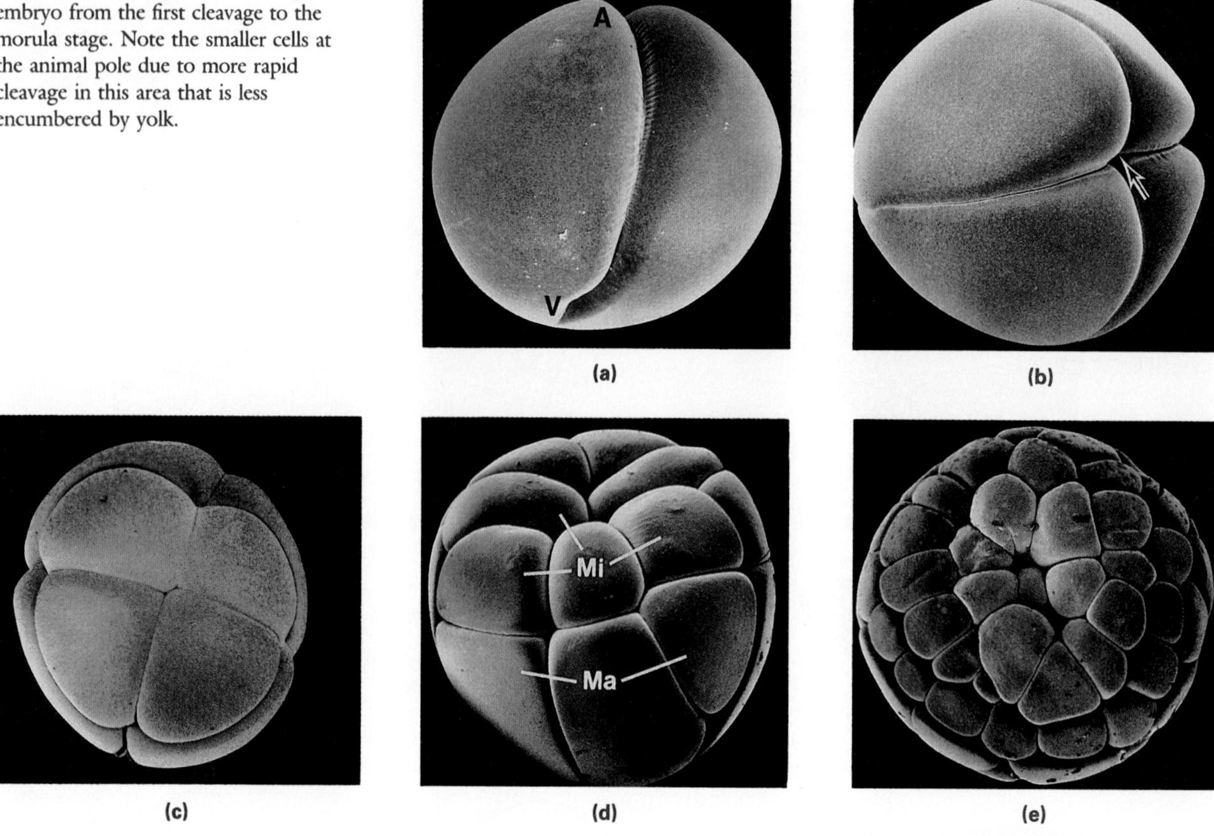

Embryonic Regulation

It must be admitted that at this stage of development, it is hard to see just how this peculiar ball is going to become a frog, but stranger things have happened. Interestingly, even at this early stage (the second day), the fate of some cells has been sealed for hours. In other words, they are firmly set on particular paths that lead to the formation of specific cell types.

However, other cells in the embryo have not yet specialized, and these can presumably take any of a number of developmental routes. They are so "flexible" that if they are transplanted to other parts of the embryo, they will form whatever type of cells their new neighbors happen to be and become something entirely different from what they would have formed had they not been moved. As each hour goes by, though, more and more cells become committed as their changes take them past the point of no return. This specialization is known as the differentiation of cells. It is just this differentiation that gradually, cell by cell, results in the emergence of a new individual.

Keep in mind, also, that these kinds of changes will first build a tadpole. The tadpole will grow for a time, then a dramatic reorganization will begin, transforming the tadpole into a frog.

BIRDS AND THEIR MEMBRANES

Although a bird's egg is very large, the early embryo is very small; in fact, the zygote is at first only a disklike **germinal spot.** Because a bird embryo rests on the bulky and inactive yolk, cell division is limited to the tissue comprising the germinal spot. Cleavage divides the germinal spot into a two-layered embryo called the **blastodisc.** The blastocoel forms as a cavity between the outer cell layer (the ectoderm) and the lower cell layer (the endoderm). The blastopore, in the chicken embryo called the primitive streak, does not develop as a crescent, but as a slit along what will be the body axis (Figure 16.5). As a result, gastrulation in birds is somewhat different than in species with less yolk. The mesoderm is formed as cells

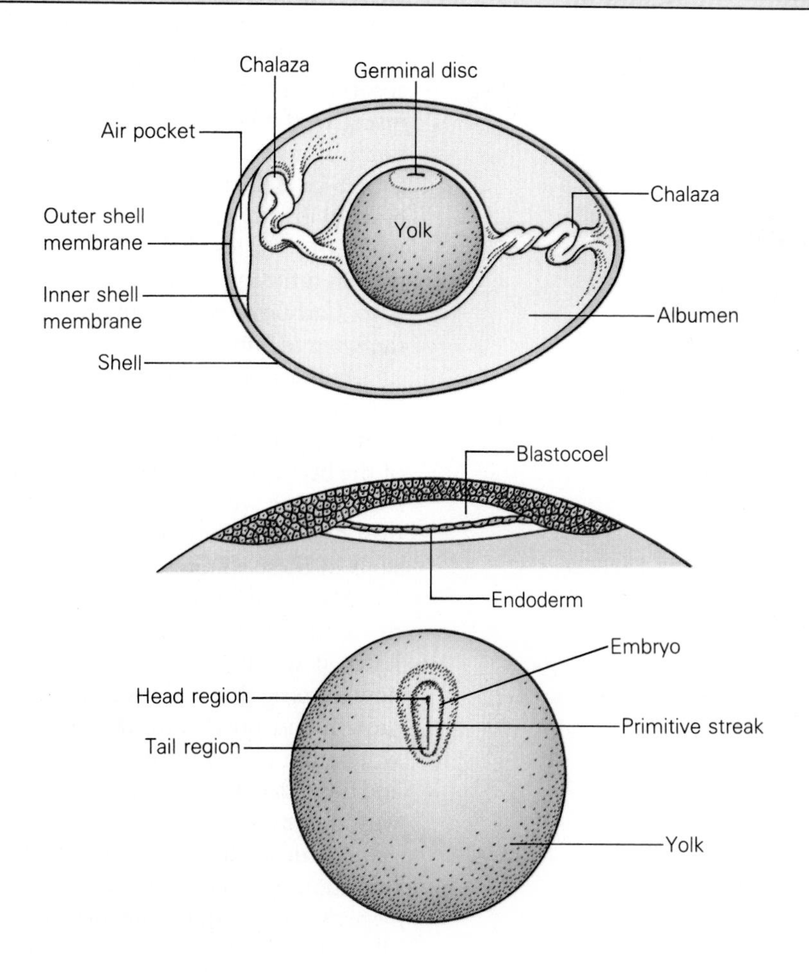

FIGURE 16.5

Top, a cutaway view of a hen's egg with a developing embryo (the germinal disk). The yolk will provide food for the embryo, food that will be transported through embryonic vessels as membranes grow and enclose the yolk. The yolk and embryo are held in place by the chalazae and cushioned by the albumen (or egg white). Later, membranes will come to lie pressed against the inside of the shell, picking up oxygen from the environment and losing CO_2 to it. Center, the blastoceol stage of a developing chick. Compare the developmental stage with the sequences in Figure 16.3. Lower, the embryo is still only a flattened, elongated disk sitting atop the yolk. Developing membranes grow outward from the embryo. The primitive streak is an indentation above the notochord that indicates the axis along which the spinal cord will develop.

FIGURE 16.6

The basic tube-within-a-tube structure of vertebrates and most invertebrates. This general body plan is permitted by the events at gastrulation. Note the number and variety of organs that essentially develop as pouches from the embryonic gut.

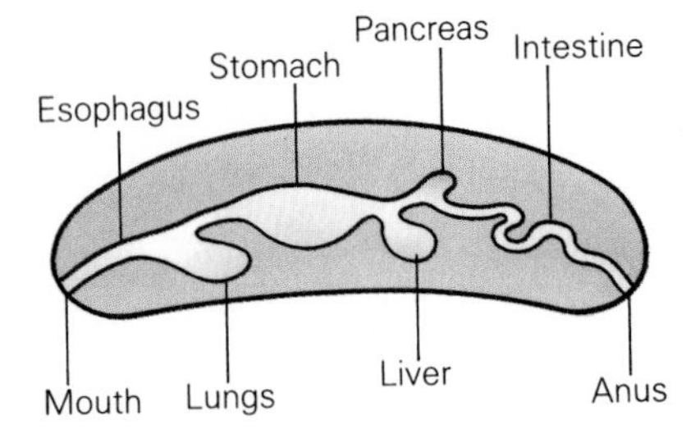

from the upper ectodermal layer roll along the midline of the embryo, forming a new layer of cells within the former blastocoel.

After the spinal cord is formed, the developing head and tail areas grow so rapidly that they overlap the tissue below. Each area is composed of an outer layer of ectoderm and an inner layer of endoderm. The two layers then form pockets, one at the head end and one at the tail end. These thinned areas then rupture and form the mouth and anus. The characteristic tube-within-a-tube structure of vertebrates (Figure 16.6) is now complete.

In the meantime, a membrane called the **yolk sac**, which is composed of endoderm and mesoderm, grows outward from the embryo over the yolk. Blood vessels develop in the membrane and will soon be able to carry food from the yolk to the embryo. Another membrane, called the **allantois,** begins to grow out of the hind area of the developing gut, forming a receptacle for nitrogenous wastes (Figure 16.7). The waste-laden allantois will be left behind when the chick hatches.

As other extra-embryonic membranes develop, they arc upward over the back of the embryo and grow toward each other, finally fusing and enclosing the embryo in a new, four-walled membrane. The inner two layers will become the **amnion** and the outer two the **chorion,** each consisting of both a layer of mesoderm and a layer of ectoderm.

The amnion fills with a slippery fluid that acts as a protective shock absorber and as a lubricant preventing the appendages of the embryo from fusing to the body. The mesoderm on the outside of the allantoic membrane later fuses with the mesoderm on the inside of the chorion, forming a three-layered **chorioallantoic membrane** (see Figure 16.7). This membrane will come to surround the albumen, the egg-white that lies just under the porous shell. This membrane is then in a position to pick up oxygen diffusing in through the porous shell and transport it to the embryo. Carbon dioxide from the respiring embryo can leave the egg through the same membrane.

As the embryo continues to develop, it eventually is connected to the yolk only by a thin stalk through which food passes. In time, the anterior, or head, part of the spinal cord begins to bulge, forming the large lobes of the brain. The mesoderm on either side of the embryo forms blocks of somites that will soon form the vertebrae and large trunk muscles. The mesodermal membranes, in fact, were formed from broad, lateral extensions of these somites.

About this time, a large blood vessel underneath the embryo begins to twitch irregularly and then to pulsate more and more rhythmically. Later, it will loop and fuse to itself, forming the heart. Tiny bulges are now forming that will become the limbs. The brain has continued its rapid growth and now outpockets on either side to form the great orbs that will be the eyes. Meanwhile, the endoderm begins to form pockets here and there that will become the highly complex glands and organs associated with digestion.

A crucial change occurs on the eleventh day. This is a time of rapid transition, marked by a great eruption of enzymes. Many embryos whose systems are not quite functioning properly die at this time. After this, if all goes well, the various systems of the organism will coordinate their activities and begin to function in even greater harmony.

FIGURE 16.7

Later development of membranes in the chick embryo. The head of the embryo is at the right of both figures. Note that as the yolk mass decreases, the membrane complex becomes more extensive. As extra-embryonic membranes grow upward and over the embryo, finally to fuse, they form a four-walled membrane. The inner two form the protective amnion. The outer two form the chorion. The allantois first appears as a saclike extension near the rear of the embryo and is a storage site for embryonic waste. Later the outer layer of the allantois will press against the chorion to form the chorioallantoic membrane across which gases can diffuse.

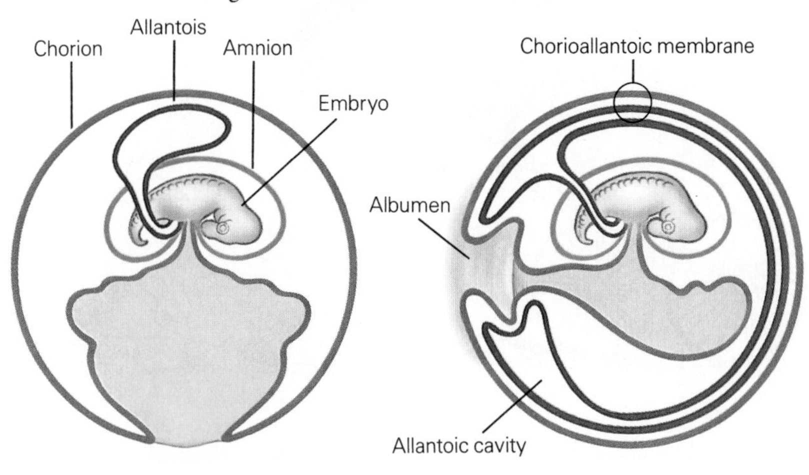

After about three weeks, the chick will begin the first feeble pecks that will eventually enable it to break free of its protective shell. When it finally emerges, it will begin to react to strange and fascinating new experiences. Some responses will be innate, others will develop as the result of experience. As it grows, it will constantly adjust to the wonders of its world, adapting, changing, and learning. Then someone will eat it.

FIGURE 16.8

Pregnancy in humans follows developmental routes common to many animals.

HUMAN DEVELOPMENT

We have seen some of the events pertaining to development and control in two other groups of animals. We should keep in mind, though, that many of the principles we have considered apply to a wide range of animals, including us (Figure 16.8). So, that brings us to us. As you are undoubtedly aware, it is quite easy to plunge into any number of controversial areas by this simple shift in species. There is an understandable tendency for us to assume a certain "specialness" and to minimize the idea of ties with other, less "noble," species. Nonetheless, many commonalities remain, beginning with fertilization.

Descriptions in bus-station novels notwithstanding, the prospective mother has no way of knowing if fertilization has occurred—at least not at first. If there are sperm cells thrashing around in her genital tract from about 48 hours before ovulation to about 12 hours after, the odds are

ESSAY 16.3

Why Your Children Will Not Look Exactly Alike

In the developing female embryo, there are about seven million eggs in the ovaries by the twentieth week. By puberty, many of these have been lost (though the means of selection is unknown) so that there are by then only about 400,000 eggs. At about the age of sexual maturity (usually around 12 to 16), there may only be 300,000 eggs. Of these, only one will escape a ruptured follicle every 28 days. This goes on for about 40 years, so a total of 400 eggs are released. Each woman could actually produce a maximum of about 40 children in a lifetime (the record is a remarkable 69). So, of millions of potential ova, only 25 to 40 can contribute their genes to the next generation. The probability of any one developing to maturity is therefore remote, and the probability of ova with identical sets of genes approaches nil.

Males, after puberty, normally produce millions of sperm each day and the production may continue throughout life. The chromosomes of each sperm and egg have undergone meiosis and crossing over, and there is almost no likelihood of two identical sperm being produced. Of millions of sperm in each ejaculation, only one will fertilize an egg. One reason so many sperm are necessary is that the female reproductive tract is essentially a hostile environment, and most sperm will quickly perish there. Also, a mucous plug may block the cervix, and enzymes carried by sperm are necessary to break it down to open the way for other sperm. These same enzymes act to break down the follicle cells surrounding the egg so that a sperm can get through to fertilize it. If no particular type of sperm (in terms of genetic constituents) has an advantage, you can see that the likelihood of identical individuals being produced from the same parents is very slim. In fact, the odds of your having children with identical genetic constituents have been calculated at about 1 in 14 trillion. Some parents have found this reassuring.

good that pregnancy will result. (The whole story of copulation, pregnancy, and birth may be a bit hard to believe, but let's assume it's true and go on.) As soon as the egg is touched by the head of a sperm (Essay 16.3), the egg begins to pulsate violently, uniting the 23 chromosomes of the sperm with the same number of its own. From this single fertilized cell, now only about 1⁄175 of an inch in diameter, a baby weighing several pounds and composed of trillions of cells will appear about 266 days later. Then, much later, it may vote.

For convenience, we will divide the 266 days, or nine months, of pregnancy into three periods of three months each and consider these trimesters separately.

Human Development

Humans are indeed a remarkable form of life. However, we are only *one* form of life. We almost intuitively try to set ourselves apart from the other forms, but we seem to base the differences primarily on the stages that occur after our first few years of life and what we presume lies in store after life ends. As these photographs show, even if we do differ from other forms of life in both accomplishment and destiny, we are clearly tied to the others in our earliest days. The drama of the beginnings of human life is no more complex, fascinating, and poignant than are the first stirrings of life in other creatures. Every guru was once a gastrula, and perhaps a clearer understanding of our individual beginnings will help us to more comfortably take our place in the grand parade of life.

This remarkable photograph captures the moment of fertilization of a human egg. Shown faintly in the darkened center are the chromosome-laden male and female nuclei approaching each other. At the top are two polar bodies formed during meiosis. One might wonder how much of our behavior and even our values is determined at this moment. It seems strange, but we really don't know just what those chromosomes carry. We know the limits of height are more or less determined genetically. How about the limits of empathy? In any case, about 36 hours after this event, the zygote will undergo its first cleavage, all the while wending its way slowly toward the uterus that will be its next home.

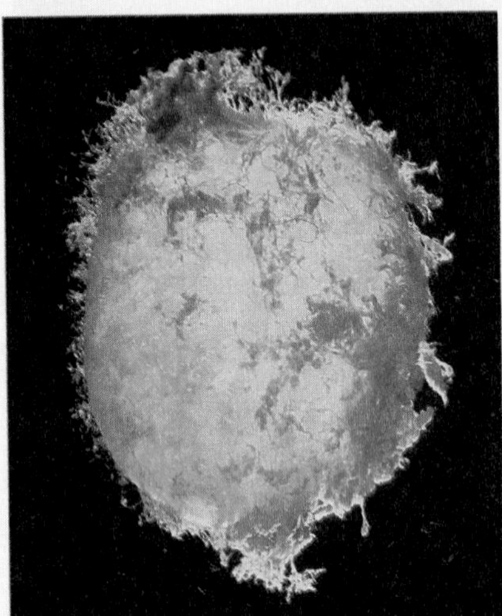

At the beginning of the second week, the embryo has developed extensive membranes that lie in close contact with the mother's tissues (left, top). The delicate projections of the chorion have penetrated the mother's tissues in all directions as they take nutrients and oxygen from her blood and deposit their own metabolic wastes for her system to carry away.

Well into the third week (bottom left) the chorionic membrane has continued to penetrate the mother's endometrium. The chorion, shown radiating outward, carries blood vessels that lie closely intertwined with those of the mother, but the two do not join. The balloonlike structure is the yolk sac. The human embryo need carry only enough food to last through its first few weeks since food will later be derived from the mother's blood.

The embryo at the fourth week (above). It lies protected in its amniotic sac. The dark eye is prominent and the enormous brain lies tucked against the embryonic heart. The embryo by this time has already developed primordial cells that will form its own gametes. By now the tubular heart has begun its first timorous beats.

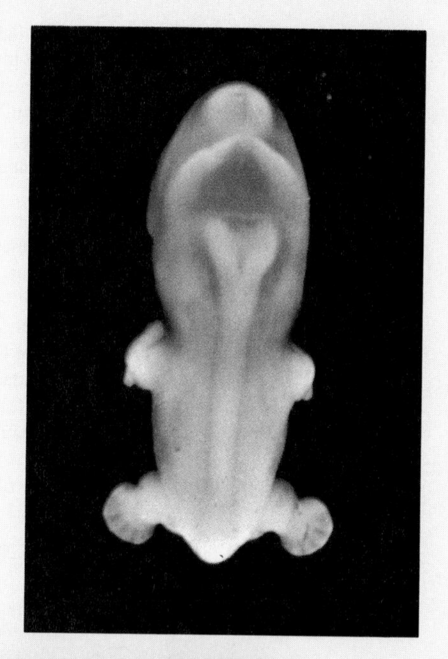

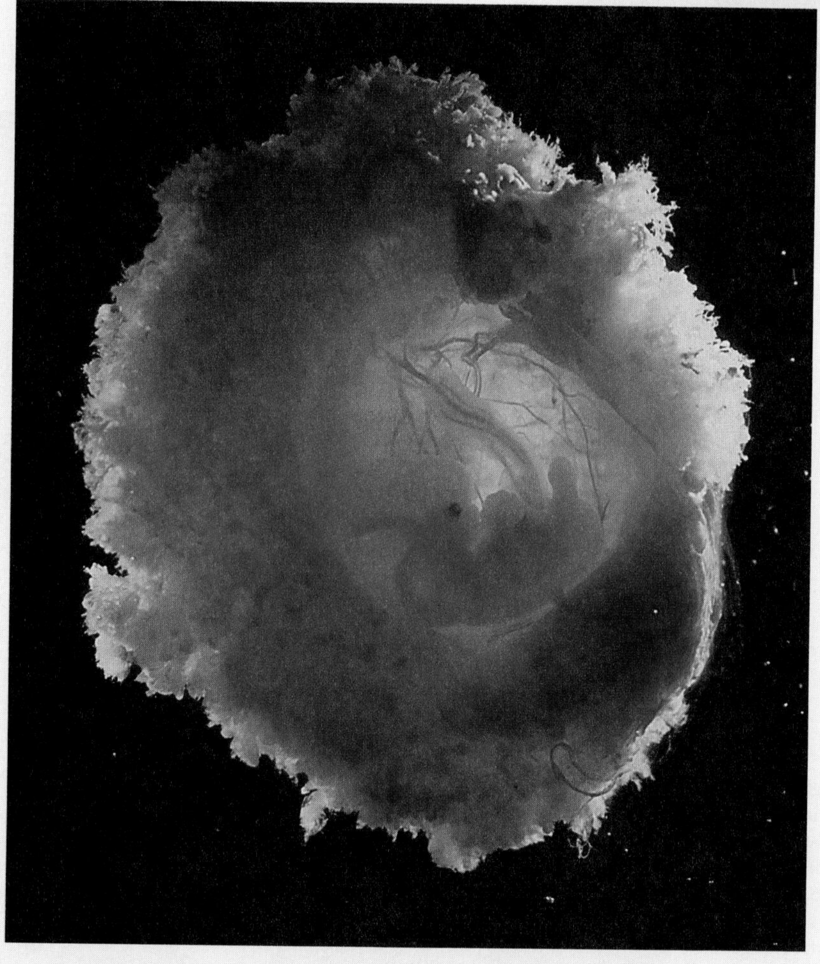

The human embryo is here shown at 42 days with the surrounding membranes removed (above). It is about half an inch (or sixteen millimeters) long. This is a dorsal view showing the enormous head (which helps direct the growth of the rest of the body) and the spinal cord extending to the rump. Notice the paddlelike appendages. Fingers and toes are already apparent as the tissue between them dies as a result of a mysterious chromosomal timing mechanism. Also at this time (left) the extensive vascular system is clearly visible leading from the projections of the embryonic chorion to the embryo itself. The organ below the eye is the now looped heart. The embryo is still so water-laden that its tissues are virtually transparent.

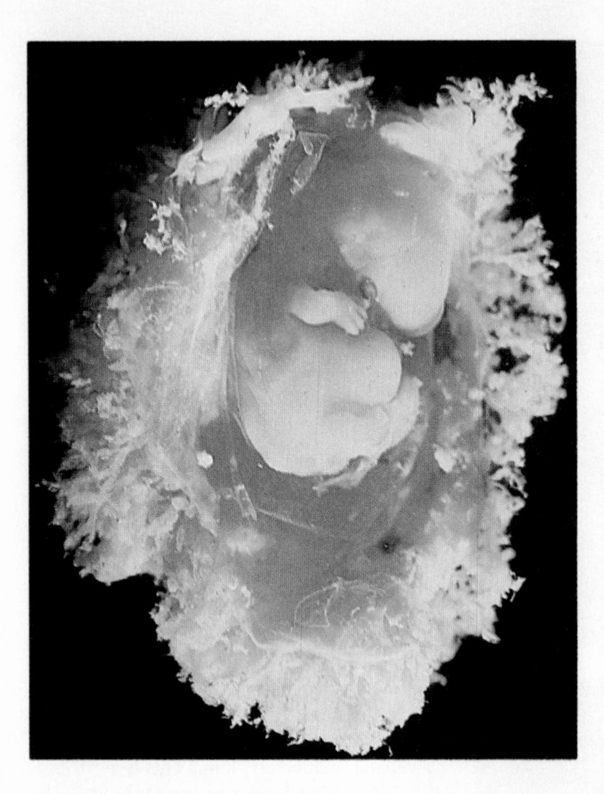

At six weeks (left), with the amnion removed, the fingers are apparent and the bulbous brain still dominates the embryo. Notice the "tail" tucked under the abdomen. (It is destined to disappear.) The tiny pit above the arm will become the ear.

At about this time, the embryo is extremely vulnerable to all sorts of chemical agents. An increasing number of drugs have been found to produce congenital abnormalities, for example, in the growth of limbs. Even X rays, at this time, may endanger the development of the embryo. Certain diseases are also particularly dangerous. For example, if a mother contracts German measles during the fourth to twelfth week of her pregnancy, the result may be deformities in the offspring's eyes, heart, and brain.

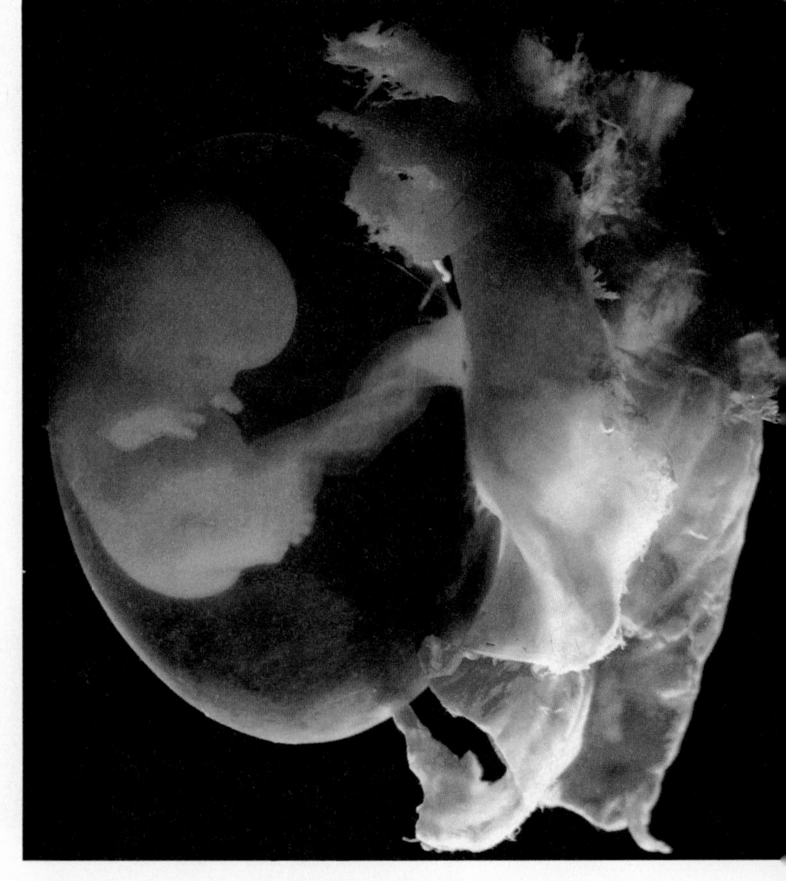

At about seven weeks (right) the embryo, afloat in its amniotic fluid, is clearly anchored to its placenta by the twisted umbilicus through which great blood vessels pass. The abdomen is swollen due to the rapid growth of the liver, the main blood-forming organ at this time.

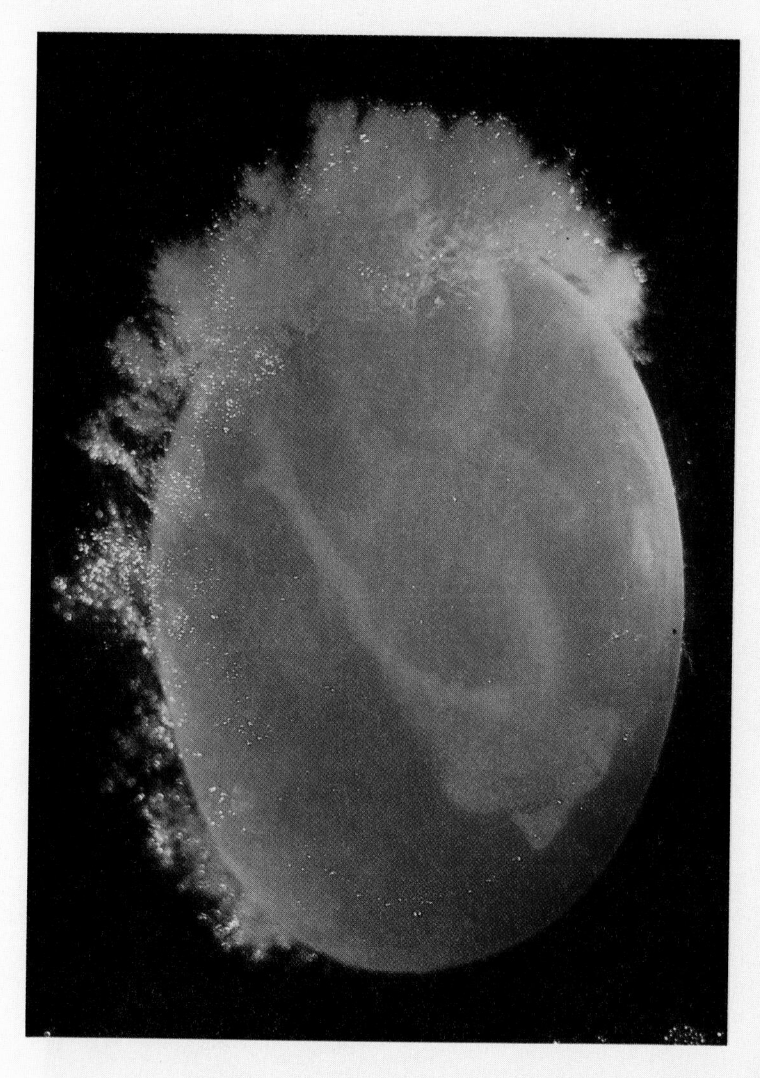

The eight-week embryo is shown here (left) in front view. The organs are now more or less complete after a rapid period of growth and development. From here on, development will primarily consist of refinements of existing structures. The skeletal system is among the last to form, but bones are now evident in the arms and legs.

At nine weeks (below) lids have begun to grow down over the eyes, and the outer ear begins to form. Because the plates of the skull have not fused, the head is rather flexible. During this third month, the fetus may begin to move, wave its arms and legs, and may even suck its thumb. It is now beginning to fill its amniotic space and will assume the typical upside-down fetal posture.

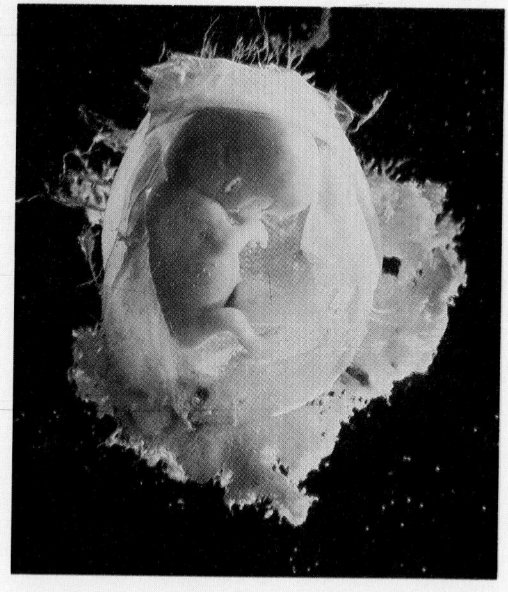

At ten weeks (right) the skeleton is well along in its development. The long bones begin developing independently, growing from areas near their ends. They will join, forming joints, later. In fact, the joints may not be firmly abutted by the time the baby is born. The head is still disproportionately large and will remain so, to a decreasing degree, through childhood. Notice the coiled umbilicus lying near the cuboid ankle bones showing as dark spots. The wrist bones have not yet begun to form and the jaw structure is weak indeed.

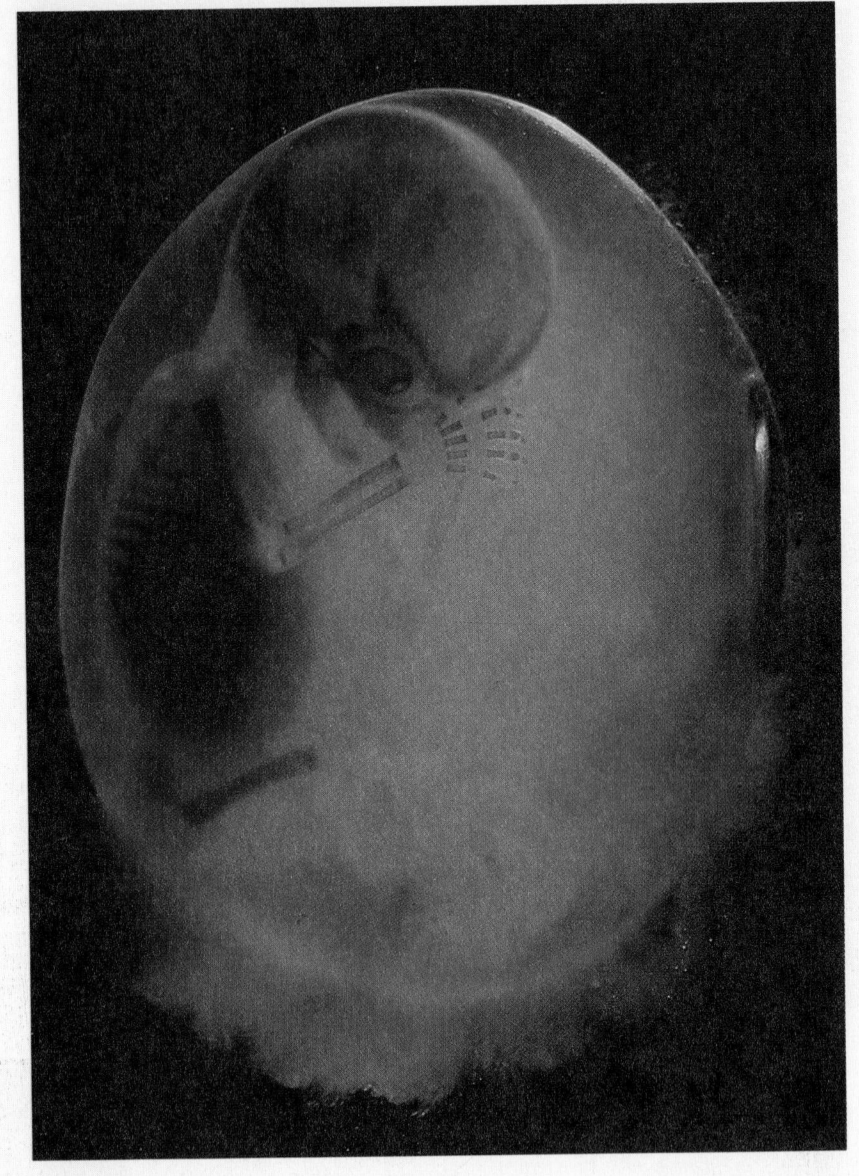

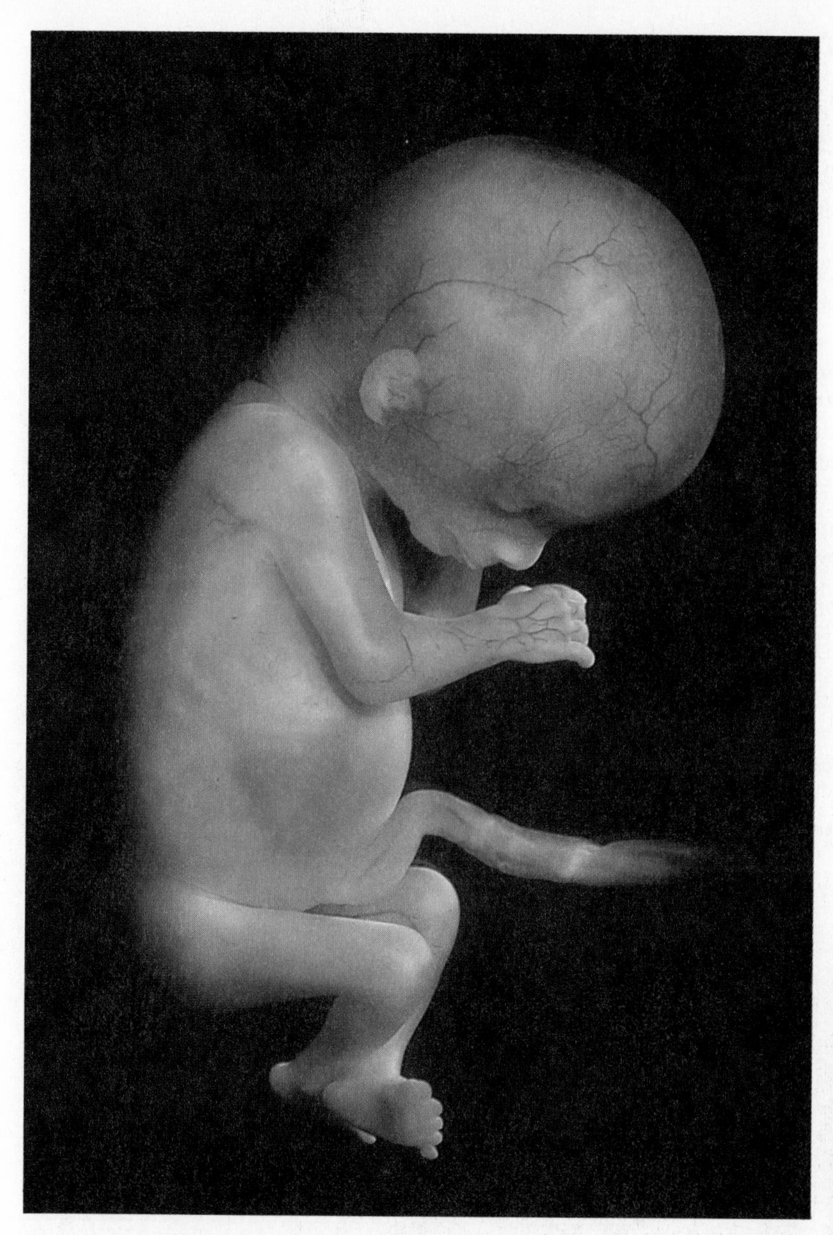

At 14 weeks (left) the fetus is fist-sized. Ribs and blood vessels are visible through the translucent skin. The vigorous movements of the fetus can now be felt by the mother. The delicate skin is actually covered with a cheesy protective coating. Refinements such as fingerprints and fingernails have not yet developed.

By the end of five months (below) the fetus is covered with fine, downy hair and its head may have already started to grow its own crop. It has already started the lifetime process of discarding old cells and replacing them with new ones. The heart is beating now at a rate of 120 to 160 times per minute.

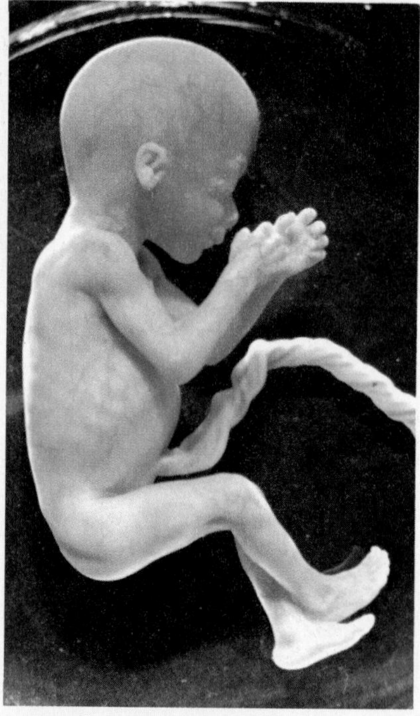

The First Trimester

In those delicate days of the first trimester, the misshapen embryo begins the coordinated changes that will lead to its final form. During this period, the embryo is particularly susceptible to any number of influences on its development. In fact, human embryos often fail to survive this first critical stage. Prospective mothers must be very careful at this time. Later in the pregnancy the fetus will be more resilient to many threats that find it susceptible now.

The first few divisions of a human embryo produce very similar cells with roughly the same potentials. In other words, at this stage the cells are, theoretically anyway, interchangeable. There are very few cells produced in those first days. In fact, only 16 such cells will exist 72 hours after fertilization. (How many divisions will have taken place?) The cells in early embryos grow progressively smaller because, since there is no input of food to fuel growth, each cell divides before it reaches the size of the cell that has produced it. The rate of mitosis is so great that by the

FIGURE 16.9

Each month an egg is released from the ovary. Fertilization generally occurs in the upper oviduct and by the time the embryo implants in the uterus perhaps a week later it will have reached the blastocyst stage.

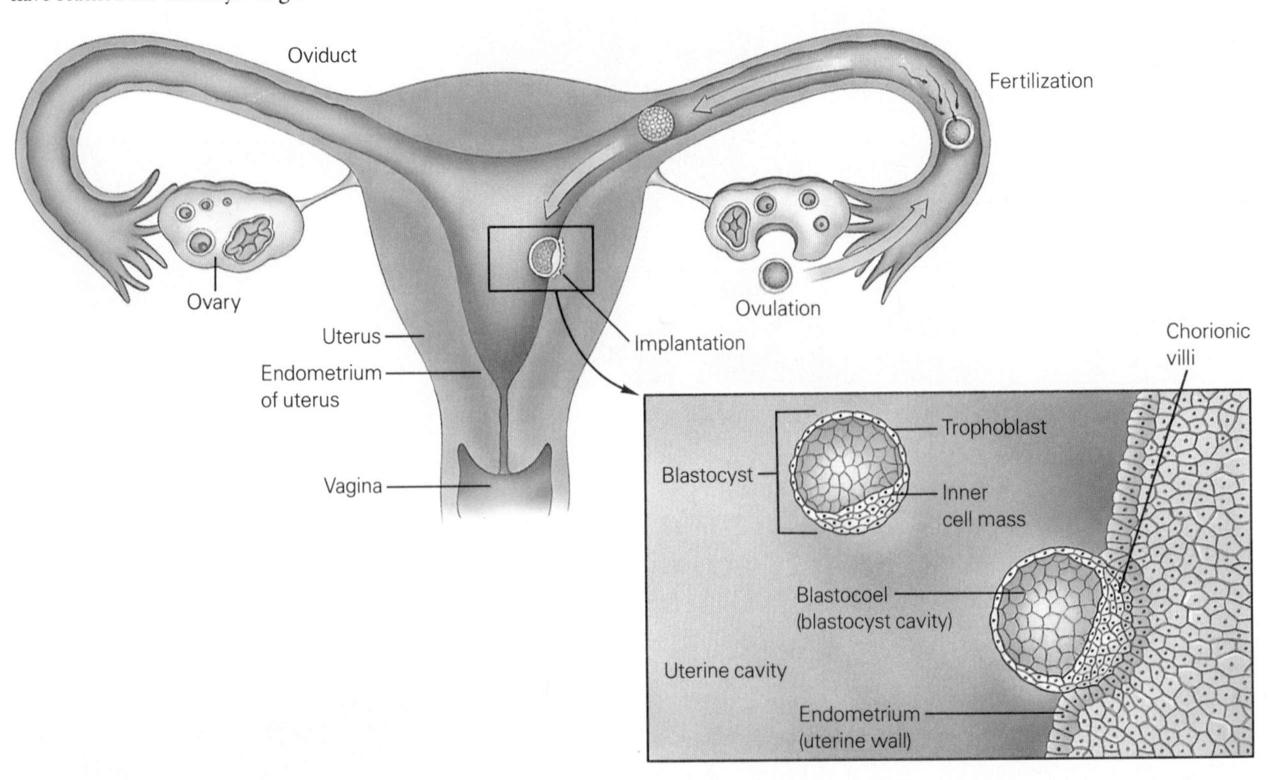

end of the first month, whereas the embryo will be only ⅛ inch long, it will consist of millions of cells.

The First Month

The first cleavages take place in the oviduct before the embryo reaches the uterus. In fact, after the embryo reaches the uterus, a period of a few days, it does not implant for about another three days, even though the uterine wall is already swollen and ready to receive it. The result is that by the time it does implant, perhaps a week after fertilization, it has already reached the hollow ball, or **blastocyst,** stage. This blastocyst is lopsided, somewhat like a class ring. One side of the blastocyst is thin-walled and forms the **trophoblast.** Later, this will produce the complex membranes that surround the embryo. The thicker-walled side (the side of the ring with the setting), forms the **inner cell mass.** It will become the embryo proper (Figure 16.9).

At the time of implantation, the trophoblast secretes a digestive enzyme that breaks down some of the blood vessels in the uterine wall around it. The embryo sinks into this cavity and comes to lie in a pool of the mother's blood. The enzymatic breakdown of the uterus produces some bleeding that is sometimes mistaken for menstruation. (You might like to remember that.) The embryo continues secreting these enzymes for about a week as it sinks deeper into the blood-engorged uterus.

The embryo's early days, then, are spent in a pool of bloody, glycogen-laden fluid that provides it with sustenance. By about 12 days, the injured uterine wall will have repaired itself, and the embryo, buried deep with the flesh of the uterine wall, silently continues its mysterious changes.

By this time, the trophoblast has begun extending fingerlike projections called **chorionic villi** into the uterine wall, where they probe deeply into the mother's tissues, which are also undergoing drastic changes (see Figure 16.9). Where the embryonic and maternal tissues press together, the **placenta** is formed. Through this placenta will pass food and oxygen from the mother and waste and carbon dioxide from the fetus (Figure 16.10, page 464).

Surprisingly, after a month, the embryo is still smaller than a pea (Figure 16.11, page 465), but it has nevertheless undergone momentous changes. The anterior end of the spinal cord has begun to develop the bulges and lobes that will be the brain. The mesoderm alongside the spinal column has now divided into somites that will give rise to much of the lower layer of skin, as well as bone and voluntary muscles. Three pairs of arterial arches have appeared in the neck region. In humans, these disappear, but in fish they become part of the gills. The opening that will be the mouth has broken through, although no nourishment will enter for some time yet. Even large sightless eyes have begun to form. By the end of the third week, the weak, developing heart begins its first twitching pulses.

Also during this first month, undifferentiated cells roaming the yolk sac and chorion begin to form blood cells. These, by day 24, will be pumped by a tubular and incomplete heart through unfinished blood vessels. Nonetheless, they are able to pick up oxygen that has diffused across the placenta from the mother's blood. Near the end of the first

FIGURE 16.10

The chorionic villi are fingerlike projections that provide the means of assuring that sufficient nutrients are exchanged between the mother and the fetus.

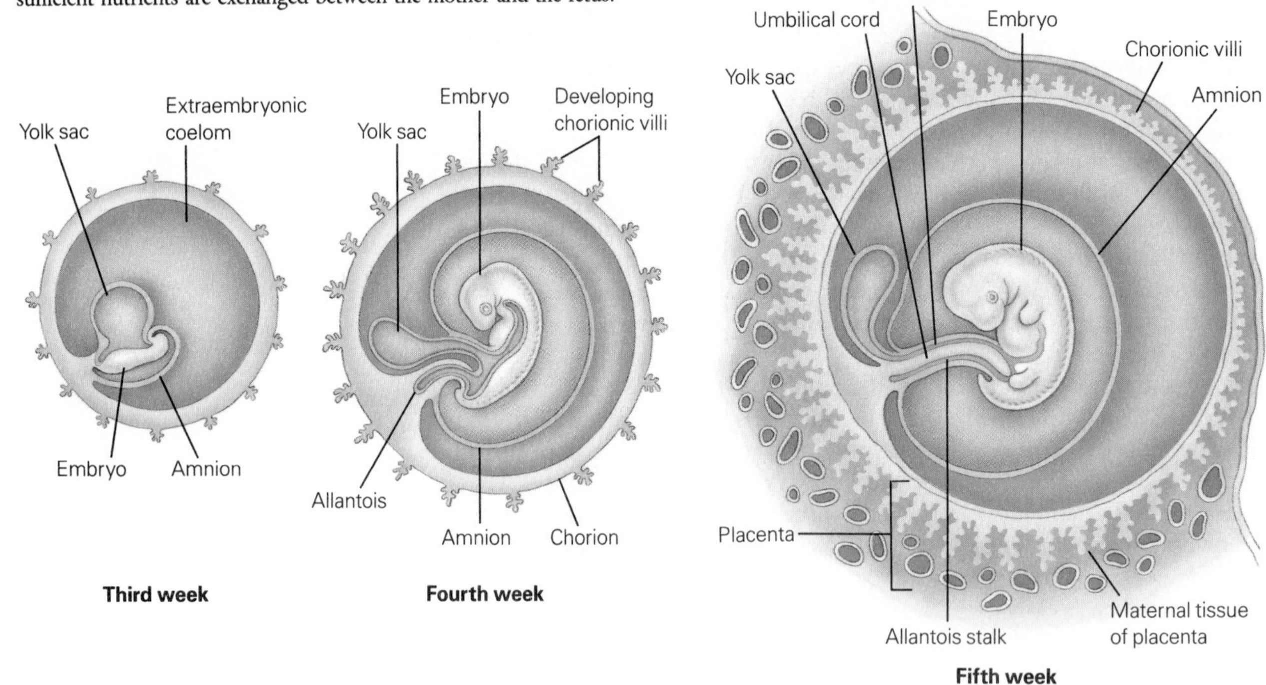

month, primitive kidneys have formed, but they will function for only a few days before they are replaced by a more specialized permanent kidney. By now, the first cells of some endocrine glands are apparent. Tiny buds are visible that will eventually form the arms and legs.

The Second Month

In the second month, the features of the embryo become more recognizable. Bone begins to form throughout the body, first in the jaw and shoulder areas. The head and brain are now developing at a much faster rate than the rest of the body. Ears appear. Open, lidless eyes stare blankly into the amniotic fluid. The tubular heart has looped back on itself, setting the stage for its final four-chambered form. The circulatory system continues to develop, and blood is continuously pumped through the umbilical cord and out to the chorion where it receives life-sustaining nutrients and deposits the poisons it has removed from the developing embryo. The nitrogenous wastes and carbon dioxide filter into the mother's bloodstream, where they will be circulated to her own kidneys and lungs for removal. About day 46, the reproductive organs begin to form, either as testes or ovaries, and now, for the first time, the sex of the embryo is apparent. Near the end of the second month, fingers and toes begin to appear on the flattened paddles that have formed from the limb buds. By this time, the embryo is about two inches long and has the general appearance of a human. From now on it is called a **fetus.**

The Third Month

The fetus continues to grow and change during the third month, and now it may begin to move. It may breathe the amniotic fluid in and out of bulblike lungs, and from time to time it swallows. Even at this stage, fetuses begin to show individual differences, especially in their facial expressions. Some frown a lot; others tend to smile or grimace. It would be interesting to correlate this early behavior with the personality traits that develop after birth.

At the end of the first trimester, the reproductive and excretory organs show marked development as the embryo's urine appears in the amniotic fluid, from where it is filtered into the mother's blood to be removed by her kidneys. Bone continues to form throughout the body, much of it replacing scaffoldings of cartilage. The bones arise independently of each other and grow outward to meet other bones at what will be the joints. All the organ systems have been formed by this time, but if the fetus is removed from its mother it will die.

The Second Trimester

In the second trimester the fetus grows rapidly, and by the end of the sixth month it may be about a foot long, but it will weigh only about a

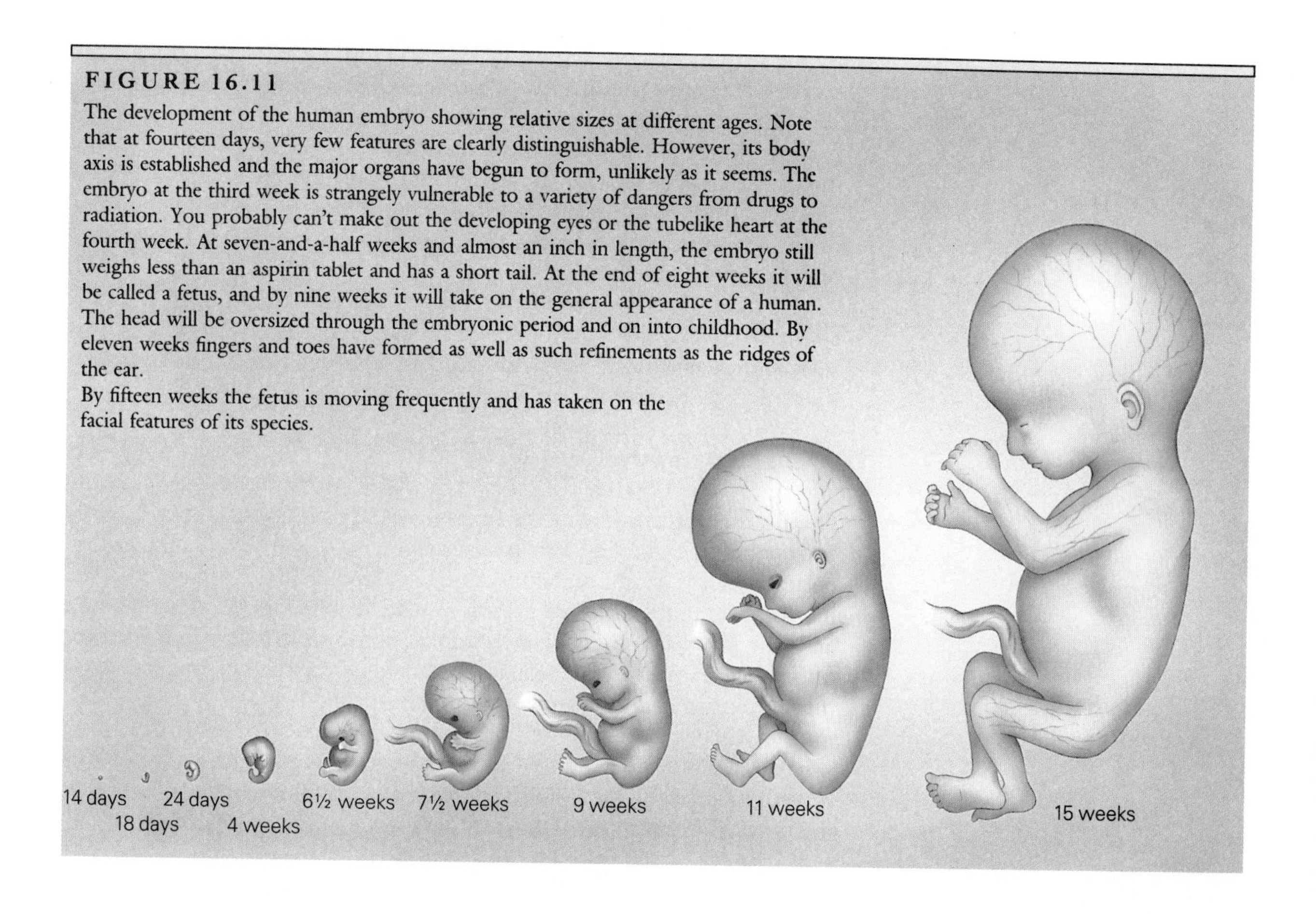

FIGURE 16.11

The development of the human embryo showing relative sizes at different ages. Note that at fourteen days, very few features are clearly distinguishable. However, its body axis is established and the major organs have begun to form, unlikely as it seems. The embryo at the third week is strangely vulnerable to a variety of dangers from drugs to radiation. You probably can't make out the developing eyes or the tubelike heart at the fourth week. At seven-and-a-half weeks and almost an inch in length, the embryo still weighs less than an aspirin tablet and has a short tail. At the end of eight weeks it will be called a fetus, and by nine weeks it will take on the general appearance of a human. The head will be oversized through the embryonic period and on into childhood. By eleven weeks fingers and toes have formed as well as such refinements as the ridges of the ear.

By fifteen weeks the fetus is moving frequently and has taken on the facial features of its species.

pound and a half. Whereas the predominant growth of the fetus during the first trimester was in the head and brain areas, during the second trimester, the rapid growth of the body begins to catch up with the head.

The Fourth and Fifth Months

By the fourth month, the fetus is moving vigorously, kicking and thrashing in its amniotic fluid—movements clearly felt by the mother. Interestingly, it must sleep now, and the thankful mother can also get some rest. As time passes, it becomes increasingly sensitive to more types of stimuli. For example, by the fifth month the eyes are sensitive to light, although there is still no hearing. Other organs, such as the lungs, seem to be complete, but are still nonfunctional. The digestive organs are present but cannot digest food. The skin is well formed, but it cannot adjust to any temperature changes. By the end of the fifth month, the skin is covered by a protective, cheesy paste consisting of wax and sweatlike secretions mixed with loosened skin cells. The fetus is still incapable in nearly all instances of surviving alone.

The skeleton has been developing rapidly during the second trimester, with some bones arising anew from undifferentiated embryonic cells, and others forming through the gradual replacement of cartilage cells by bone cells. Now the mother must supply large amounts of calcium and other bone constituents as building materials for the fetal skeleton.

The Sixth Month

By the sixth month, the fetus is kicking and turning so constantly that the mother often must time her own sleep periods to coincide with its schedule. The distracting effect has been described as feeling somewhat similar to being continually tapped on the shoulder, but not exactly. The fetus now moves so vigorously that its movements can be seen clearly from the outside. To add to the mother's distraction, the fetus may even have periods of hiccups. It is now so large and demanding that it places a tremendous drain on the mother's reserves.

At the end of the second trimester, the fetus is clearly human, but it resembles a very old person because its skin is loose and wrinkled at this stage. In the event of a premature birth around the end of this trimester, the fetus may be able to survive.

The Third Trimester

During the third trimester, the fetus grows until it fills its available space and is no longer floating free in its amniotic pool, and, even in the greatly enlarged uterus, its movement is restricted. In these last three months, the mother's abdomen becomes greatly distended and heavy, and her posture and gait may be noticeably altered in response to the shift in her center of gravity. (However, occasionally a markedly overweight woman may go through much of this stage without realizing she is pregnant at all.) The mass of tissue and amniotic fluid that accompanies the fetus ordinarily weighs about twice as much as the fetus itself. Toward the end of this period, milk begins to form in the woman's mammary glands, which in the previous trimester may have undergone a sudden surge of growth.

At this time, the mother is at a great physical disadvantage in several ways. About 85 percent of the calcium she eats goes to the fetal skeleton, and about the same percentage of her iron intake goes to the fetal blood cells. Much of the protein she eats goes to the brain and other nerve tissues of the fetus.

Some interesting questions arise here. If a woman is unable to afford expensive protein-rich foods during the third trimester, can it affect the brain development and intelligence of her offspring? On the average, poorer people in this country show lower I.Q. scores. Are they poor because their intelligence is low, or is intelligence low because they are poor? Is there a self-perpetuating nature about either of these alternatives?

During the third trimester, the fetus grows quite large. It requires more food each day, and it produces more poisonous wastes for the mother's body to carry away. Her heart must work harder to provide food and oxygen for two bodies. She must breathe, now, for two individuals. Her blood pressure and heart rate rise. The fetus and the tissues maintaining it form a large mass that crowds her internal organs. In fact, the fetus pressing against her diaphragm may make breathing difficult for her in these months. Several weeks before delivery, however, the fetus will change its position, dropping lower in the pelvis (a process called *lightening*), and it thus relieves the pressure against the mother's lungs.

The "finishing" of the fetus proceeds rapidly in the last three months. Such changes are reflected in the survival rate of babies delivered by cesarean section (an incision through the mother's abdomen). In the seventh month, only 10 percent survive; in the eighth month, 70 percent; and in the ninth month, 95 percent survive.

Interestingly, there is a change in the relationship of the fetus and mother the last trimester. In the first trimester, measles and certain other infectious diseases would have affected the embryo. However, during the third trimester, the mother's antibodies confer an immunity to the fetus, a protection that may last through the first weeks of infancy.

About 255 to 265 days after conception, the life-sustaining placenta begins to break down. Parts of it begin to shrink and change and the capillaries begin to disintegrate. The fetal environment becomes rather inhospitable and premature births at this time are not unusual. At about this time, the fetus slows its growth and changes its position so that its head is directed toward the bottom of the uterus. Its internal organs undergo some final changes that will soon enable it to survive in an entirely different kind of world. So far, its home has been warm, sustaining, protected, and confining. It is not likely to encounter anything quite so secure again.

Birth

The signal that there will soon be a new customer on the planet is the onset of **labor,** a series of uterine contractions that at first appear at about half-hour intervals and gradually increase in frequency. Meanwhile, the sphincter muscle around the cervix dilates, and, as the periodic contractions become stronger, the baby's head pushes through the extended cervical canal to the opening of the vagina (Figure 16.12).

FIGURE 16.12

The appearance of a new human on our planet is normally, in our society, an occasion of great joy. We usually don't dwell on the fact that the infant's traits are a result of a chance union of a specific egg and a specific sperm so many months earlier. Another sperm would have resulted in an entirely different individual.

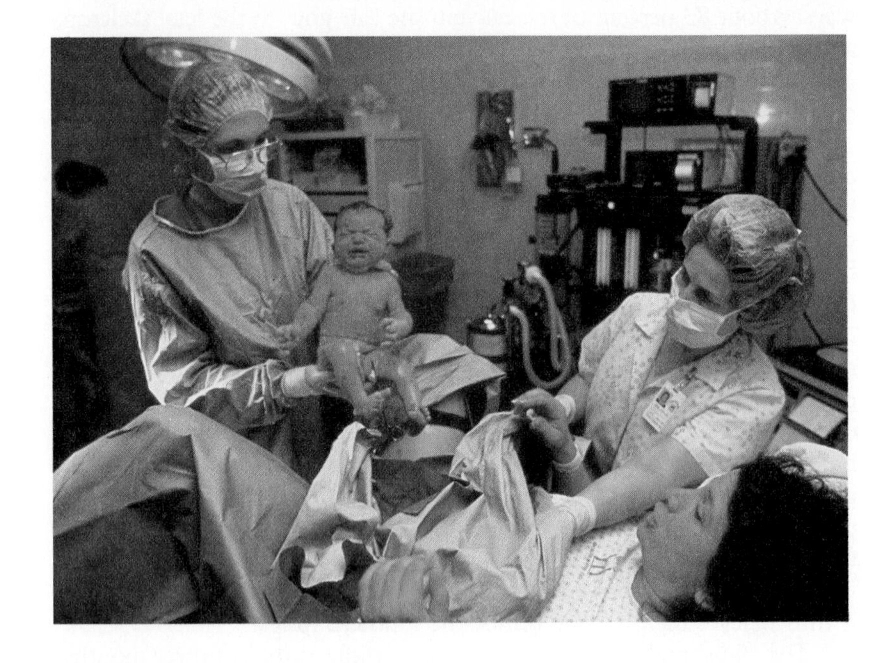

Once the baby's head emerges, the pattern of uterine contractions changes, becoming milder but more frequent. The small shoulders come into view, and then the body appears. With a rush, the baby slips into a new world. A few minutes later, the umbilical cord that had connected the fetus to its life-sustaining placenta is tied off and cut. The contractions continue until the placenta is expelled as the **afterbirth.** The mother recovers surprisingly rapidly, in humans and wild mammals as well. In many other mammals, the mother immediately chews through the umbilicus and eats the afterbirth. She thus regains nutrients and prevents the tissue from advertising to predators the presence of a helpless newborn.

Cutting the umbilicus halts the only source of oxygen the infant has known. With no oxygen, carbon dioxide quickly builds up in the infant's blood. The change triggers a breathing center in the brain that causes a nerve impulse to be fired to the diaphragm. The contraction of the diaphragm causes the baby to gasp its first breath. An exhaling cry means that the baby is breathing on its own.

In American hospitals, a newborn baby is given the first of the many tests it will encounter in its lifetime. This one is called the *Apgar test series,* in which muscle tone, breathing, reflexes, and heart rate are evaluated. The physician then checks for skin lesions and evidence of hernias. If the infant is a boy, it is checked to see whether the testes have properly descended into the scrotum. A footprint is then recorded as a means of identification, since the new individual, despite the boasts of proud parents, does not yet have other distinctive features. (There have been more than a few cases of accidental baby-switching. I'm convinced it happened to me.)

FIGURE 16.13

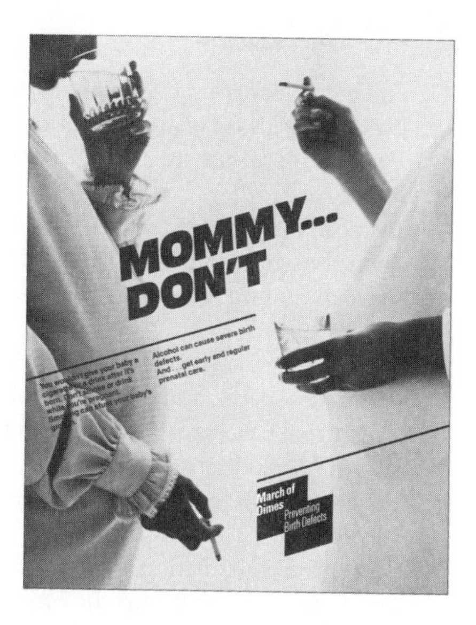

A poster from the March of Dimes Birth Defects Foundation.

Miscarriage

Unfortunately, not all pregnancies result in the delivery of a healthy offspring. In fact, the traditional wedding veil was originally worn to shield the bride from evil spirits that might doom her first pregnancy to miscarriage. The route between conception and birth is indeed fraught with risk, perhaps largely because of the very complexity of living things. After all, a phenomenal array of physical and chemical interactions occur within a developing body, and, if somewhere along the line a single component fails in its function or its timing, the embryo may be lost. Should the problem arise early in the pregnancy, the embryo may be broken down—digested—and reabsorbed by the mother's body, perhaps without her ever knowing she was pregnant. It is interesting that if failure should occur later in the pregnancy, the mother's body often mysteriously seems to recognize that something has gone wrong, and the fetus is aborted as a "miscarriage." About half of all miscarriages involve abnormal fetuses.

A miscarriage can arise from a number of factors. For example, it may occur if any embryo does not implant correctly, or if it invades the uterine wall too soon in its development. The mother's activities also have effects. Alcoholism or heavy smoking may cause premature birth of otherwise healthy babies (Figure 16.13). Drugs, legal or otherwise, may impair the normal development of the embryo. A weak cervix may not be able to support the weight of a fetus and may produce a spontaneous abortion. Other factors may also interfere with a successful pregnancy, such as disease, stress, or malnutrition. In fact, so many things can go wrong that one might well marvel at the successful completion of a pregnancy and the birth of a healthy child.

SUMMARY

1. Although development includes all the processes from conception to death, only those from conception to birth will be considered here.
2. Fertilization results in an embryo with all the genetic information it will ever have.
3. In most cases, the penetration of the egg by one sperm makes the egg unresponsive to other sperm. The two sets of chromosomes join, and a series of mitotic divisions, called cleavage, divide the embryo into smaller cells.
4. During gastrulation, cells migrate and form the three germ layers (ectoderm, mesoderm, and endoderm). Each germ layer contributes to specific structures. Ectoderm forms the outer layers, sense organs, and nervous system. Mesoderm forms muscle, bone, cartilage, blood, heart, blood vessels, gonads, and kidneys. Endoderm contributes largely to the lungs and certain digestive structures.
5. Soon after gastrulation, the notochord forms and causes the overlying ectoderm to develop into the central nervous system. First the ectoderm thickens and forms the neural plate. The edges thicken and become the neural folds, which fuse to create a tube that will become the spinal cord.
6. Cell specialization increases as development proceeds. Although the developmental fate of cells in some animals is flexible, with few exceptions mammalian cell differentiation is permanent. Developmental changes occur in a specific sequence determined by signals that turn sets of genes on or off.
7. Undifferentiated tissue (mesenchyme) moves into areas of the embryo that are stressed, strengthening those areas. In a process called induction, one type of tissue influences the development of another. For example, the mesodermal layer that forms the notochord induces the overlying ectoderm to form the spinal cord.
8. The amount of yolk, a food source, in the egg is correlated with an organism's life history. Human eggs have little yolk, frog eggs have more yolk, and bird eggs have a great deal.
9. During frog development, cleavage occurs more slowly in the yolky vegetal pole than in the animal pole. Cleavage results in a ball of cells called the morula. When a cavity called the blastocoel forms, the ball is called the blastula. In gastrulation, cells migrate inward to form a new cavity, the archenteron, which opens to the outside through the blastopore. Gastrulation forms three germ layers, the ectoderm, the mesoderm, and the endoderm.
10. Due to the large amount of yolk in a chicken egg, cleavage occurs only in the germinal spot. The blastopore is a slit opening along the future body axis. Germ layers form as the ectoderm splits away from the underlying endoderm and the cells that will form the mesoderm roll under along the midline.
11. A membrane called the yolk sac grows over the yolk. The blood vessels form within the mesoderm of this membrane. Another membrane, the allantois, forms a receptacle for nitrogenous wastes. The outer layer of

the allantois fuses with another membrane, the chorion, to form the chorioallantoic membrane, which will eventually function in gas exchange. After about three weeks the chick hatches.

12. Human development begins with fertilization and the union of chromosomes from the sperm and the egg. Development takes about 266 days and is divided into three trimesters.
13. During the first trimester, the embryo is particularly sensitive to many influences. Cleavage begins in the oviduct as the embryo moves toward the uterus. About a week after fertilization, at the blastocyst stage, the embryo implants in the uterus. At this time the embryo is composed of a thin-walled trophoblast, which will form the embryonic membranes, and the inner cell mass, which will become the embryo.
14. Initial nutrition of the human embryo is from a pool of bloody glycogen-rich fluid. After about 12 days chorionic villi form. Soon the placenta forms from embryonic and maternal tissues and delivers nutrients from the mother's blood, removes wastes from fetal blood, and functions in gas exchange.
15. During the first trimester, all major structures and organs begin to form. At the end of the second month the embryo is called a fetus. The second trimester is marked by rapid growth and the beginning of movement. During the third trimester, the fetus grows to fill the available space. Development is completed and the fetus assumes a head down position.
16. Labor is a series of uterine contractions that push the baby through the dilated cervix and vagina to the outside world. The umbilical cord is then tied and cut. The placenta is expelled as the afterbirth. The Apgar test is used to check the newborn's condition.
17. Miscarriage is the loss of the fetus. It may be caused by a number of factors, including malformation; improper implantation; unfavorable maternal activities such as the use of alcohol, drugs, or cigarettes; a weak cervix; disease; stress; or malnutrition.

KEY TERMS

afterbirth (468)
allantois (452)
amnion (452)
animal pole (447)
archenteron (447)
blastocoel (447)
blastocyst (463)
blastodisc (451)
blastopore (447)
blastula (447)
chorioallantoic membrane (452)
chorion (452)
chorionic villi (463)
cleavage (440)
ectoderm (441)
endoderm (441)
fertilization (440)
fetus (464)
gastrulation (441)
germ layer (441)
germinal spot (451)
inducer (445)
induction (445)
inner cell mass (463)
labor (467)
mesenchyme (445)
mesoderm (441)
morula (447)
neural folds (441)
neural groove (441)
neural plate (441)
notochord (441)
placenta (463)
trophoblast (463)
vegetal pole (447)
yolk (446)
yolk sac (452)
zygote (440)

FOR FURTHER THOUGHT

1. Alcohol consumption constricts arteries including the artery in the umbilical cord that unites the fetus to the placenta. With this in mind, explain why alcohol consumption by a pregnant woman is an unwise practice.
2. Pregnant women are often advised to consume calcium-rich products such as milk. Explain why calcium intake is especially important in the third trimester.

FOR REVIEW

True or false?

1. ___ Human eggs contain more yolk than frog eggs.
2. ___ The first embryonic cell division produces a solid ball of cells called a blastula.
3. ___ Mesenchyme moves into areas of embryonic stress and differentiates into supportive tissue.
4. ___ The notochord induces the overlying ectoderm to form the spinal cord.
5. ___ The first cleavage of the human zygote takes place in the oviduct.

Fill in the blank.

6. A membrane called the ___ receives wastes from the chick embryo.
7. The ___ pole of a frog egg contains a large supply of inert yolk.
8. Three distinct cell types called germ layers are formed during ___.
9. Food and oxygen pass from the mother to the embryo through the ___.
10. ___ line either side of the spinal column and give rise to vertebrate skin, bone, and voluntary muscle.

Choose the best answer.

11. The amnion is a membrane that
    A. contains a protective lubricating fluid.
    B. surrounds the albumin.
    C. contains the embryo's food supply.
    D. forms a receptacle for nitrogenous wastes.
12. The ___ that forms during gastrulation will become the frog digestive tract.
    A. morula
    B. blastocoel
    C. archenteron
    D. blastopore
13. Which of the following is a germ layer that forms the outermost layer of skin?
    A. endoderm
    B. mesenchyme
    C. ectoderm
    D. mesoderm

14. The embryo implants in the uterus
    A. before fertilization.
    B. during the first trimester.
    C. during the second trimester.
    D. during the third trimester.
15. A miscarriage may result if the
    A. embryo doesn't implant properly.
    B. embryo implants too quickly.
    C. mother is malnourished.
    D. all of the above.

PART FIVE

Systems and Their Control

Once the principle of homeostasis and internal balance is established, we discuss each of the body's systems, considering a wide array of representative groups, and focusing on humans. Social issues from diet to AIDS are considered, culminating in a review of the effects of drugs.

CHAPTER 17

Support and Muscular Systems

Overview

Support Systems

The Variety of Contractile Systems

Objectives

After reading this chapter you should be able to:

- Describe the various types of skeletal support systems and their functions.
- Describe the functions and physical properties of cartilage and bone.
- Differentiate between the appendicular and axial skeleton and describe four types of joints that connect them.
- Describe the contractile systems of a cnidarian, a roundworm, and an arthropod.
- Describe the structure and function of the three types of vertebrate muscles.
- Explain what is meant by antagonistic muscle pairs and describe how these work to move bones.
- List the components of a skeletal muscle fiber and describe how each functions during muscle contraction.

There are few things that one can be sure of these days, but here's one: There is no such thing as an amoeba large enough to eat your car. The largest amoeba in the world can't even eat your cat. Amoebas are small, and they are destined to stay that way essentially because of one reason: They are composed of a single cell. There are a great many repercussions to this seemingly insignificant trait. For example, you may recall that as an object becomes larger, its volume increases at a proportionately greater rate than its surface area. Thus, a huge amoeba would have an unsatisfactory surface-to-volume ratio. In other words, its cell membrane would be too small to accommodate the cell's mass. Put another way, the cytoplasm would be served by an inadequate structure with which to communicate with the environment.

Even if oxygen were able to move into the amoeba's body, once inside, the only way it could reach the cytoplasm at the center of the beast would be by diffusion, perhaps aided by the cytoplasm slowly shifting due to the amoeba's sluggish movements. The movement of carbon dioxide and oxygen through the amoeba indeed would be slow and inefficient. In addition, the unsupported weight of a huge amoeba would create such great internal pressures that the creature would likely rupture, and you can imagine the mess. Because of such problems, your car is safe and so is your cat.

While there are few creatures on earth that could or would eat your car, there are a number that could legitimately threaten your cat. Many of these are quite large, and we can assume they are composed of many cells. Since evolution has produced so many kinds of multicellular creatures, we can assume the condition has certain marked advantages. So let's look at multicellular animals now and see how those cells are arranged so as to help the creature succeed in its world. We can first remind ourselves that groups of cells existing together, and with the same functions, are called **tissues.** Groups of tissues with the same function are called **organs,** and groups of organs acting together are called **organ systems.** In this chapter, we will concentrate on the skeletal (supportive) and muscular (contractile) tissues and systems. Later we will consider other kinds of systems, keeping in mind that they do not act independently, but in a highly coordinated manner as the organism responds to its environment.

It would seem that bones and muscles would be of such interest to people that we would, as a matter of course, know a great deal about them. But most people probably don't know the answers to some of the most fundamental questions about these great tissues. For example, how many bones are there in the human body? Why can't the knee flex forward? How do muscles grow? What is a bone bruise? What are fallen arches? Why is our back so strong and our belly so weak? If you are allowed to grow old, why should you worry about a broken hip? What we cover here should allow you to deduce the answers to such questions and to gain a better perspective on how we are supported and protected by our skeletons and how that body framework moves.

SUPPORT SYSTEMS

Trees are able to stand erect in spite of their great weights because of the combined strength of their hardened cell walls. Animals such as ourselves, though, lack cell walls. Still, we do not flounder about, collapsing here and there in quivering heaps, except perhaps on very special occasions. Generally, animals maintain their form and structure because of various sorts of support systems. The most familiar of these are **skeletal systems,** formed as special cells secrete a fluid that then hardens to become the supportive material. In many species, including humans, the skeletons are on the inside **(endoskeletons)** but certain other kinds of animals, including arthropods such as insects, and mollusks such as clams, wear their skeletons on the outside **(exoskeletons).**

All skeletons provide support and assist locomotion by providing something firm for muscles to pull against. Some skeletons, such as exoskeletons and vertebrate endoskeletons, also protect internal organs. Let's next consider some of the various forms of skeletal systems.

Types of Skeletal Systems

FIGURE 17.1
A crab, left, after completing molting.

One of the most primitive of skeletal systems is the **hydrostatic skeleton,** in which rigidity is maintained as muscular contraction compresses body fluids. Hydrostatic skeletons are found in certain soft-bodied animals such as cnidarians, flatworms, and annelids.

Another simple skeletal system is found in the sponges (see Figure 13.2). They require little support, of course, because the body is largely supported by the buoyancy of the seawater. Their skeletons consist largely of either tiny hardened needlelike structures called **spicules** or "spongy" protein fibers called **spongin** scattered throughout the tissue. The 10,000 species of sponges can be divided into groups according to the type of spicules they produce.

Species with exoskeletons include lobsters and snails as well as insects and clams. Exoskeletons efficiently protect internal parts (just as do our skull and rib cage). Exoskeletons, however, present special problems of growth that are solved in different ways by different species. Snails and clams simply secrete extension to their shells as they grow, so their shells simply grow with them. Lobsters, crabs, and crayfish, however, discard their old shells in a process called molting and grow new, larger ones. A molting animal usually retreats to a protective place because it is particularly vulnerable at this time.

Before discarding its exoskeleton, the animal withdraws the valuable minerals from its shell, causing the shell to soften. The weakened shell, stressed by the force of the growing animal, then splits down the back. The animal crawls out, soft and vulnerable, its new shell incomplete. It then swells by absorbing quantities of water as it grows, and continues to secrete a new shell into which it redeposits its hoarded minerals (Figure 17.1).

All vertebrates possess an endoskeleton, a skeleton inside their bodies. In the group of vertebrates that includes the sharks and rays, the skeleton consists entirely of cartilage, similar to that composing the bulb of the nose (reach over and squeeze your neighbor's just to get the idea). Even in most other vertebrates, parts of the skeleton are formed of cartilage during embryonic development, but these are largely replaced by bone as the body develops. Some parts of the skeleton, however, form directly as bone.

Types of Connective Tissue

Connective tissue is a supportive tissue of mesodermal origin that includes bones, tendons, ligaments, fat, and blood. Table 17.1 summarizes the types and functions of connective tissues. Connective tissue occurs throughout the body and generally functions as binding material—for example, holding glands in position or binding bone to muscle or bone to bone. Generally, cells comprise only a small part of connective tissue; most of the mass consists of substances secreted by the cells. These secreted, nonliving substances are called **matrix.** The specific qualities of each type of connective tissue are largely dependent on the nature and composition of the matrix.

Bone is heavy material with a matrix comprised mainly of calcium and phosphorus salts, which make the bone hard, and an elastic protein called **collagen** (Figure 17.2). Collagen, the most abundant material in most connective tissue, is a tough, fibrous tissue that gives the bone some degree of flexibility. The mineral salts are embedded in the collagen fibers, which are wound around themselves like the strands in a rope. The larger bones are hollow, and the tissue inside these hollow cavities is called **marrow.** Yellow marrow consists mostly of inactive fat cells, while red marrow is active and forms red blood cells as well as some white blood cells.

Most of compact bone is organized into **Haversian systems,** repeated concentric units of bones. At the center of each Haversian system is a **Haversian canal,** through which run the blood vessels servicing the bone. The bone cells lie in openings in the bone that are connected to each other by tiny canals **(canaliculi).** Thin extensions of the bone cells pass through these openings and permit the cells to physically contact each other, facilitating the passage of materials between cells and between the cells and the blood vessel (Figure 17.3 a, b).

TABLE 17.1
Connective Tissues

TYPE	PURPOSE
Connective tissue proper	
Dense connective tissue	
Tendons	Connect muscle to tissue
Ligaments	Bind bone to bone
Loose connective tissue	Support, elasticity
Fat (adipose) tissue	Store energy; insulate, pad organs
Cartilage	Flexible support; absorbs shock, maintains shape
Bone	Maintain form and structure; firm support; protect internal organs; form blood cells (red marrow)
Blood	Transport nutrients, oxygen, hormones, wastes; immune defense; stabilize body temperature and pH

FIGURE 17.2

The organization of compact bone. Inset, section of one of the body's large bones.

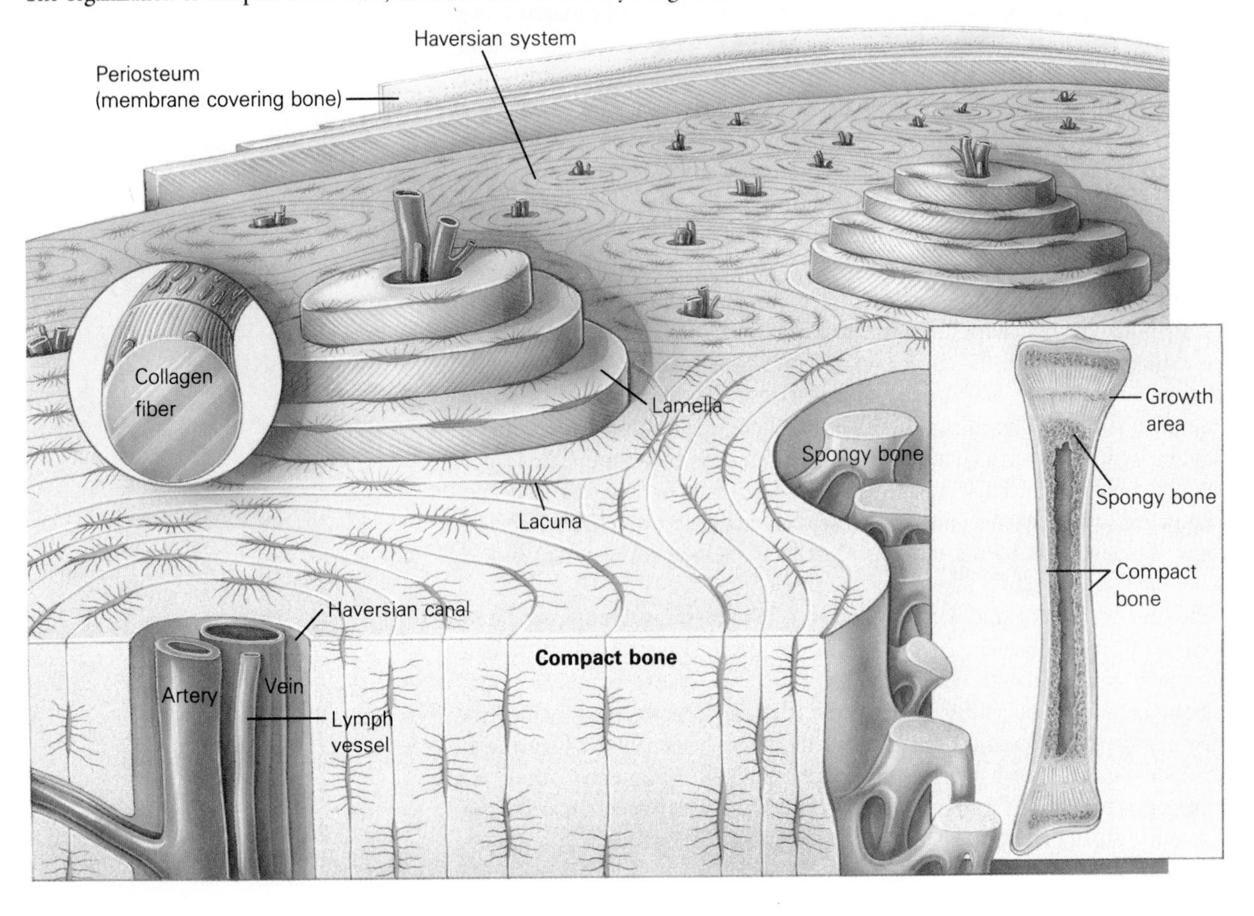

Bone is a living tissue that is broken down and rebuilt all through life. Certain bone cells are specialized to withdraw calcium and phosphate salts from the matrix. The bone would simply dissolve due to their enzymatic activity if it weren't for the fact that new bone is constantly being deposited by other bone cells that are specialized to secrete these salts, forming new matrix. Without this turnover, for some reason, bones grow weak and brittle. The process of deposition slows in older people and this is responsible for some of the increasing fragility of their bones. The constant breakdown and deposition of bone also permits new growth when the bone is most needed. For example, it tends to grow most rapidly in places that are more severely stressed. Certain cultures have used this principle to mold particular configurations in skeletal structures. As an example, in certain African tribes, the heads of women are bound in a tight metal ring that drastically changes the head shape. It is also interesting

FIGURE 17.3

In (a), a Haversian system of a bone. Through the large opening in the center, the Haversian canal, run the blood vessels that serve the bone. The bone-forming cells lie in concentric rings around the canal and communicate with each other by the tiny canals, or canaliculi, radiating from them. In (b), cartilage cells. For some reason, these usually occur in pairs, and they secrete the fibrous mat that is the matrix.

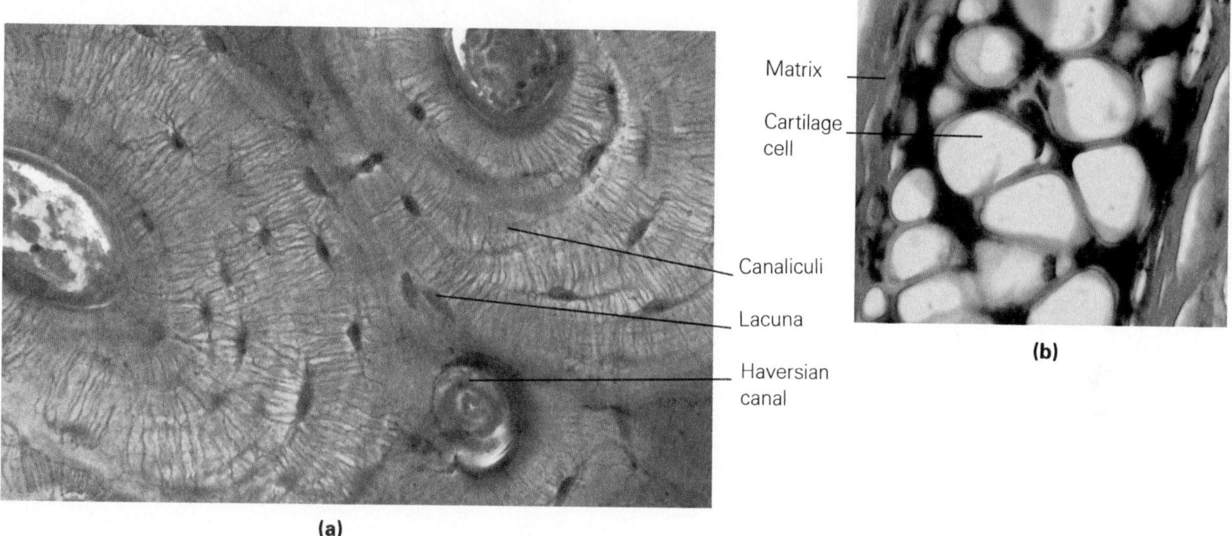

that experts can determine the muscular development of a person by skeletal examination only. Larger muscles place greater stress on the bones where they are attached, and the bone thus enlarges to accommodate the tension. So, as a general rule, the larger the area of attachment, the larger the muscle.

Cartilage is another connective tissue that is important in some support systems. Cartilage cells, like those of bone, are located in small spaces within the matrix they secrete. The characteristics of cartilage as a supporting tissue are those of its matrix: firm, flexible, resilient, and smooth. This makes cartilage well suited for certain sites within the vertebrate skeletal system. For example, it is found in parts of the rib cage, where its flexibility allows the movement needed to inflate the lungs. Because cartilage can withstand pressure and still return to its original shape, it should not be surprising to learn that the pads between the individual bones of the backbone are made of cartilage. Cartilage also forms smooth coverings on the ends of adjoining limb bones that allow the bones to slide easily past one another.

Tendons and **ligaments** are specialized types of fibrous connective tissue, each with a specific function. Ligaments are fairly elastic and serve to bind one bone to another. Tendons are not very elastic, but they are flexible and cordlike, connecting muscle to muscle or muscle to bone. (If you're a very active person, sooner or later you're probably going to have a knee problem. The complex arrangement of ligaments of the knee is shown in Figure 17.4, next page.)

FIGURE 17.4

The ligaments of the knee form a very complex array of supporting and binding structures. The entire system is very vulnerable to twisting injuries, as can be attested by former great skiers and football players who are now virtual cripples because of injury to this delicate area.

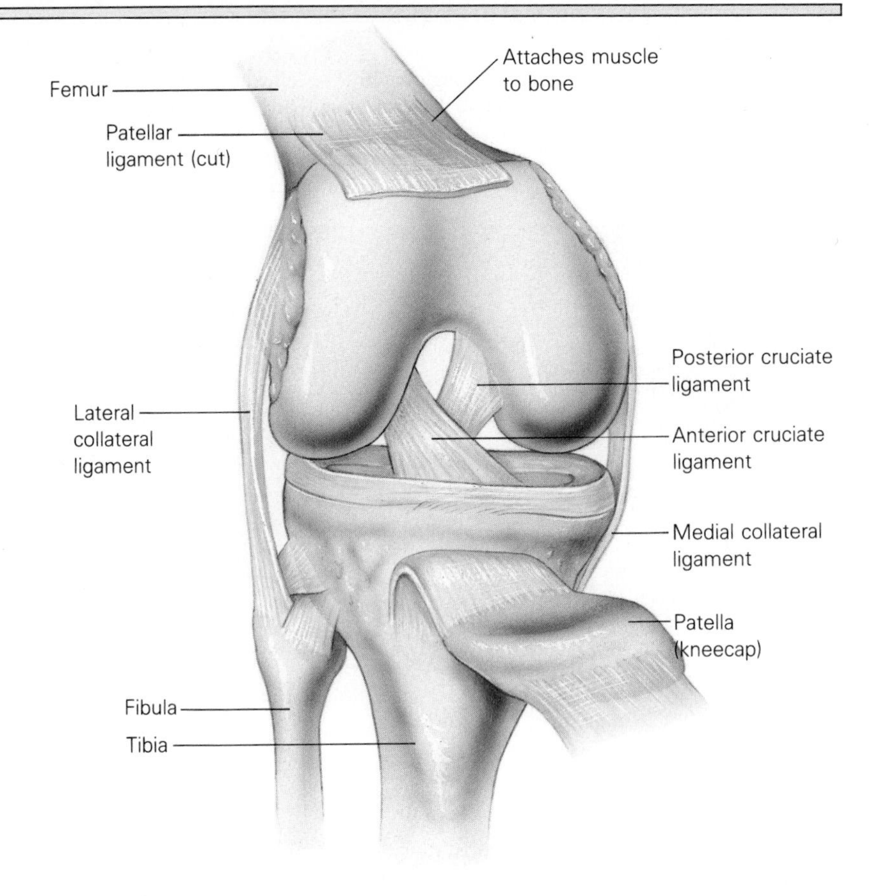

The Human Skeleton

The adult human skeleton consists of 206 bones, although that statement is a bit arbitrary because different people may have different numbers of bones.* In humans, as well as in other vertebrates, the skeleton is generally divided into two major groups called the axial and appendicular skeletons (Figure 17.5).

The **axial skeleton** is the part of the skeleton that forms the axis of the body—lengthwise—running along the direction of the backbone. It includes the skull, the spinal column, the sternum (breastbone), and the ribs. The **skull** consists of the fused bones of the braincase, or cranium, and the bones of the face. The **spinal column** is made up of 33 separate vertebrae, which differ according to their position along the spine (Figure 17.6).

*The number can vary for several reasons. For example, some people have small ribs extending from the cervical (neck) area; people can have various numbers of floating ribs (those not attached to the sternum, or breastbone); and some bones (such as those of the skull) start out separately and fuse with age.

FIGURE 17.5

The human skeleton. The axial skeleton is shaded orange and the appendicular skeleton beige. Cartilage is shown in blue. An astute observer can tell a great deal about an unfamiliar animal's habits simply by looking at the skeleton. Would shortening the upper leg and lengthening the lower leg make humans faster or slower runners?

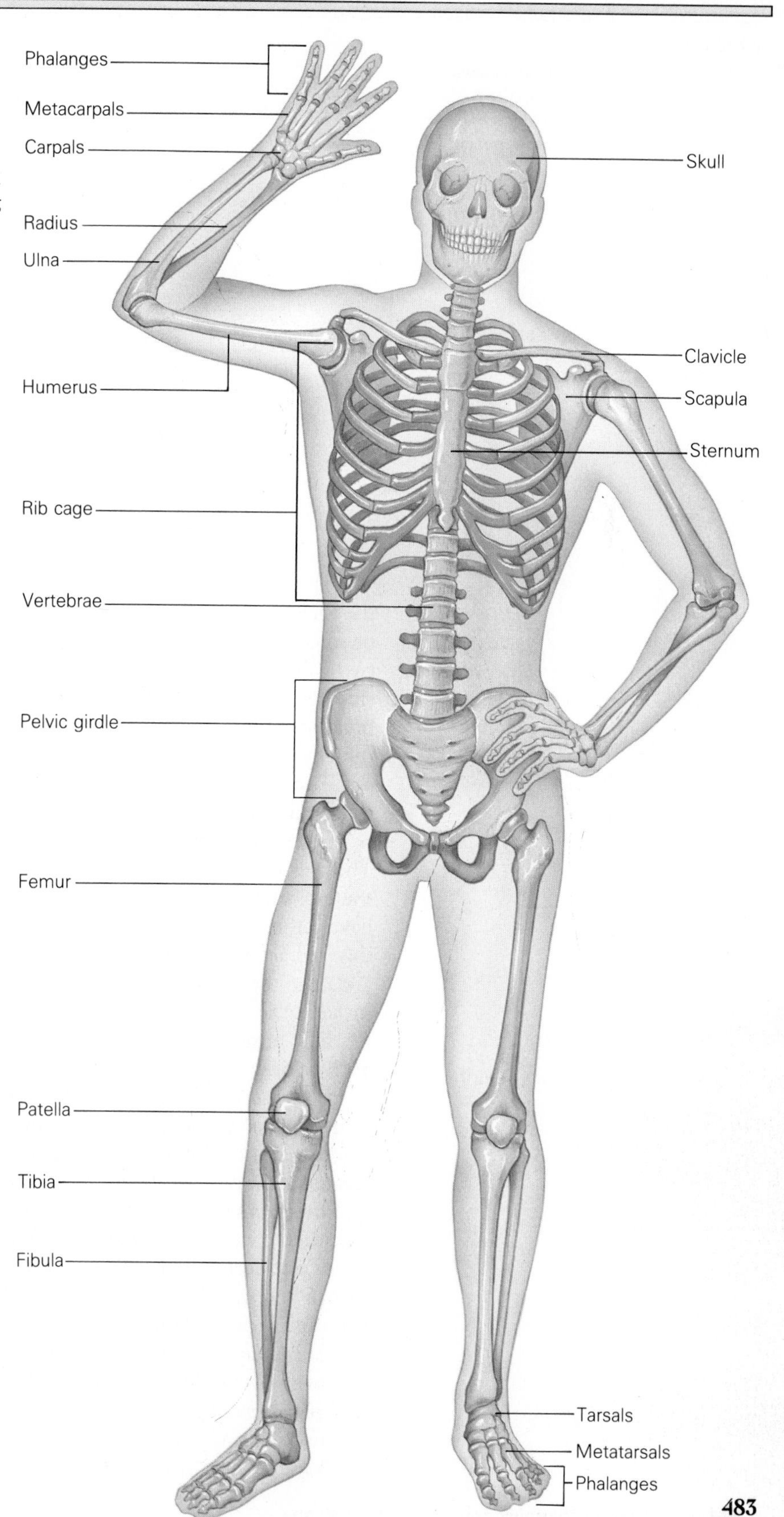

FIGURE 17.6

A lumbar (lower back) vertebra (shown from above) showing the neural arch through which the spinal cord passes. Inset, side view of the lumbar vertebrae.

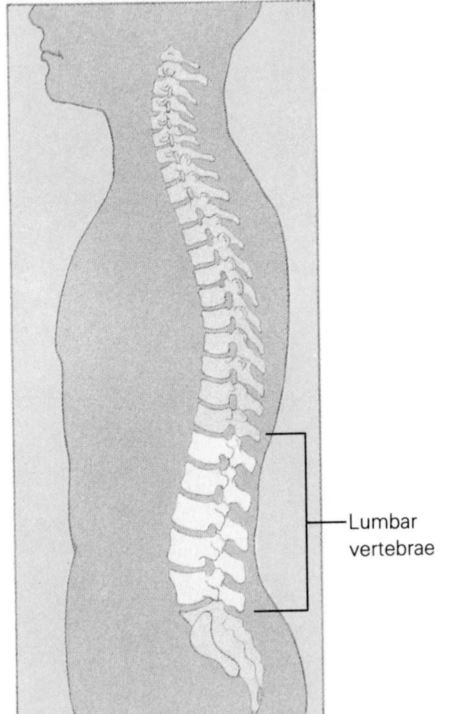

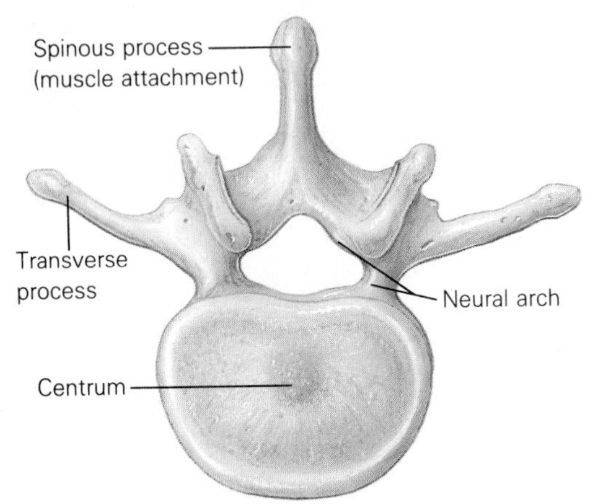

The ribs are attached dorsally (toward the back—see Figure 17.7) to the vertebral column and ventrally (toward the belly surface) to the sterum or to cartilage that is, in turn, attached to the sternum, except for the last two pair (or so), called floating ribs, which are not attached vertrally. The rib cage is slightly movable, which enables it to raise and lower during breathing, and it is strong and flexible enough to absorb powerful blows, thereby protecting the heart, lungs, and other vital organs.

The **appendicular skeleton** includes the arms and hands, the legs and feet, and two girdle complexes. The pectoral girdle consists of two collar bones (clavicles) and two shoulder blades (scapulas) meeting the sternum. The pelvic girdle is composed of two large, heavy, coxal bones that form the hips and provide support for the upper body. Many fish, by the way, have rather primitive pectoral and pelvic girdles. Fossil evidence indicates that the arms and legs of land animals evolved from the fins of early fish, and that these girdles enlarged to support the ancient creatures as they invaded the terrestrial realm.

The arms and legs are comprised of upper and lower parts that join at the elbow and knee respectively. The bone of the upper arm (humerus) is connected to two bones in the lower arm (radius and ulna). Similarly, a single bone in the upper leg (femur) is joined to two bones in the lower leg (tibia and fibula). Both the arms and legs terminate in a complex arrangement of several bones. The arms end with the wrist bones (carpals) and the legs with the ankle bones (tarsals). These bones are able to slide past each other, permitting a rather wide range of movement for both the hands and feet. The hands and feet themselves are made up of the metacarpals and metatarsals, respectively. Both the hands and feet terminate in five digits (or phalanges), the fingers and toes. Although the hands and feet of humans are structurally similar, they differ greatly in their dexterity and maneuverability. In fact, humans are remarkable for a highly specialized hand existing on the same body with a much less talented foot. Both the human arms and legs are rather primitive structures as vertebrate limbs go. A much more specialized limb, for example, is that of the horse, which,

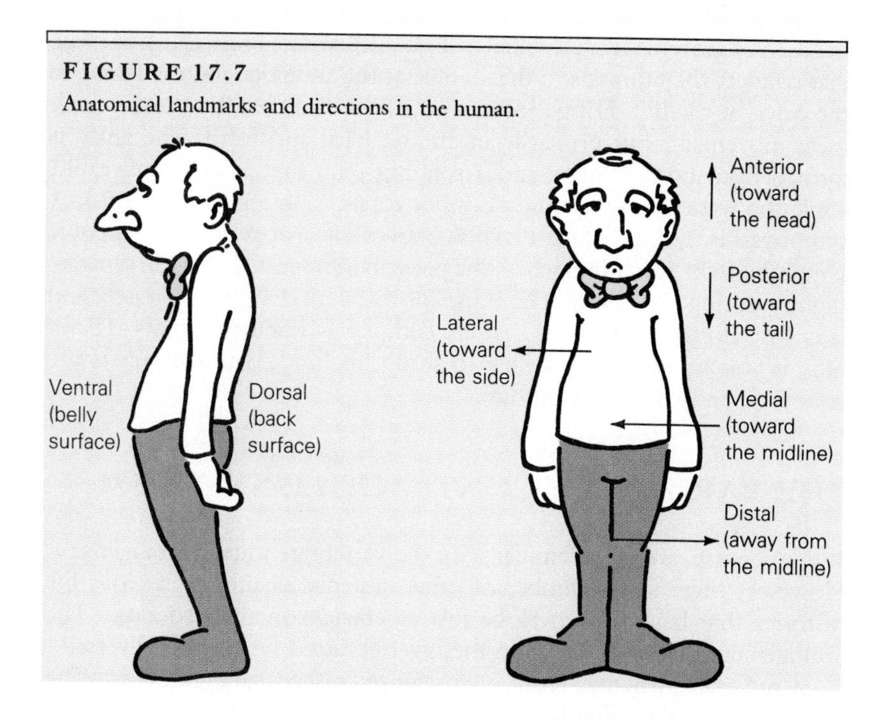

FIGURE 17.7
Anatomical landmarks and directions in the human.

with its shortened upper leg, is adapted for speed. However, specialization has its price, so we find that the legs of horses are not particularly well suited for climbing trees, to the relief of picnickers in the shade.

Joints

Bones are able to move with respect to each other at **joints,** the points of articulation (Figure 17.8). Some bones move freely, while others are only slightly moveable, and yet others are solidly fused together. The greatest

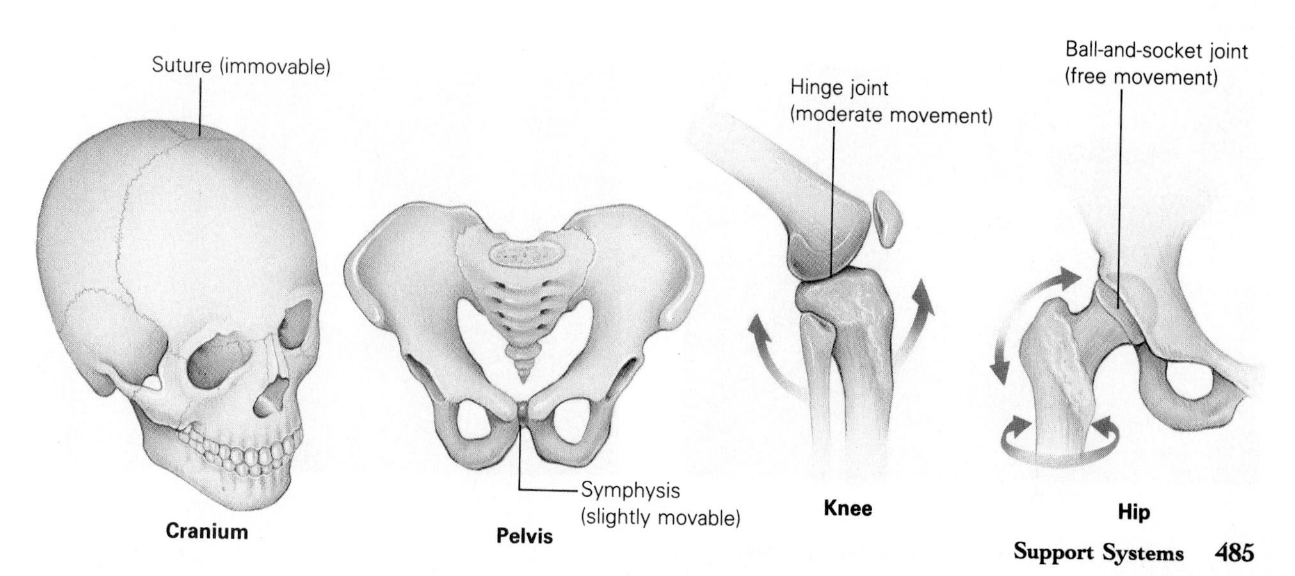

FIGURE 17.8
Representative joints in the human body.

freedom of movement is provided by **ball-and-socket joints,** such as those that connect the humerus to the scapula at the shoulder and the femur to the pelvis at the hip. **Hinge joints** at the knee and the elbow also permit great movement, but primarily in an arc, back and forth. The ankle is formed from both a hinge and a rotating joint and is quite moveable, while the wrist is a hinge and does not rotate (although it seems to). A **symphysis** is a joint that permits little movement. An example is the pubic symphysis, which is found where the pelvic bones meet. These can separate slightly during childbirth. The bones of the skull develop separately and later fuse along **suture lines** that are completely immobile joints. Therefore, if you try to teach a fellow to shrug his head, you will only waste your time and risk annoying him.

THE VARIETY OF CONTRACTILE SYSTEMS

We, of course, are most familiar with the vertebrate muscle arrangement. However, there are a number of other systems adapted to various life histories that have proven to be just as efficient in their contexts. The arrangement of muscle tissue in the invertebrate is remarkably diverse.

Cnidarians, you may recall, are organized rather like a hollow sac. The group includes the jellyfish, hydra, and sea anemones (Figure 17.9). Since cnidarians lack mesoderm, they cannot develop true muscles. However, their double-walled bodies contain two sets of rather efficient contractile fibers. The fibers in the outer wall (which was formed from ectoderm) extend along the length of the animal, and those in the inner wall (formed from endoderm) run circularly. Because anemones are not restricted by some sort of rigid skeleton, they are able to perform some rather remarkable contortions.

FIGURE 17.9
Cnidarians lack mesoderm, hence they do not form true muscle. They can move slowly by contracting their outer ring of contractile tissue formed from ectoderm and an inner radially arranged set formed from endoderm.

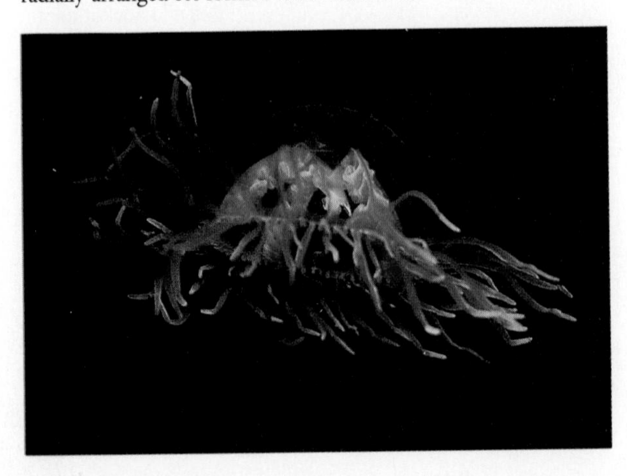

FIGURE 17.10

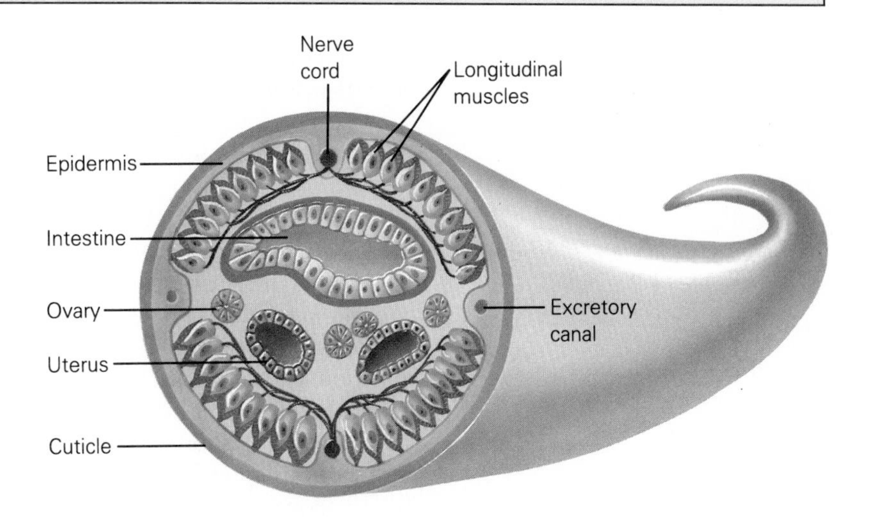

Cross-section of a roundworm, *Ascaris,* a parasite of humans. What kind of movement is possible with only longitudinal muscle fibers?

In the roundworms (Figure 17.10), we see true muscles and so we are now among animals with three germ layers. (Muscles arise from mesoderm.) Whereas the roundworm has only longitudinal muscles, the flatworms and segmented worms also have circular muscles and some species even have diagonal muscles (those that run at an angle between the dorsal and ventral body walls) and dorsoventral muscles (running between the dorsal and ventral walls).

Arthropods, the joint-legged group that includes the insects, are supported by an exoskeleton. This system entails a distinct arrangement of contractile fibers, quite different from any we've considered so far. This is because the muscles lie *within* the skeleton. The joints are also unusual in that they are simply thin, flexible places in the exoskeleton. A muscle may extend across a joint, or it may lie entirely within a single segment of an appendage and insert on the next appendage by a long, thin filament.

The musculature of animals with exoskeletons lends them remarkable strength for their size. You have undoubtedly noticed tiny ants carrying loads of food many times their weight or dragging along the corpses of huge insects. Their strength is largely due to the mechanical advantage inherent in the arrangement of their muscles (Figure 17.11). However, the same physical principles that make them so strong for their size also limit their size. Because of some complex biomechanical principles, an insect the size of a human would have greatly diminished relative strength. Thus, if you should encounter a 150-pound insect on your way home some night, don't be afraid. You're stronger than it is.

The insect muscle-skeleton arrangement is also responsible for the very rapid wingbeats that produce the lovely whine of the mosquito. Physiologists once puzzled over how mosquitoes and other insects could move their wings so fast. It would seem that the nerves that activate the flight muscles simply couldn't fire so rapidly. However, it has now been found that a single muscular contraction can set up a reverberation of the insect's exoskeleton as it is bent and then rebounds, like a tuning fork. As

FIGURE 17.11

A comparison of the musculature of an insect (a) and a human (b). Note, particularly, the relationship of muscles and joints. Insects are much stronger than humans for their weight, so why haven't we evolved internal musculature? What are its disadvantages?

(a)

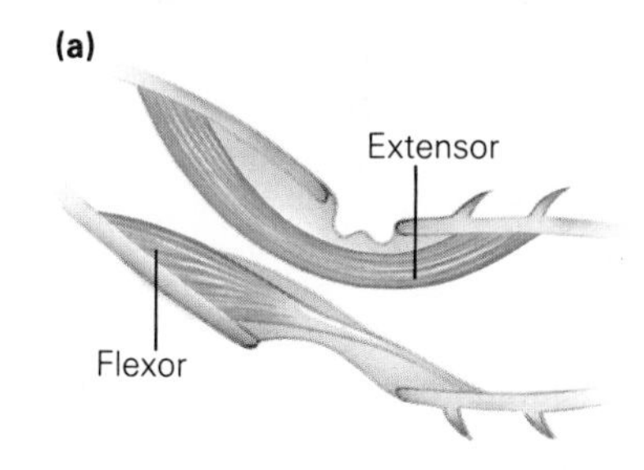

(b)

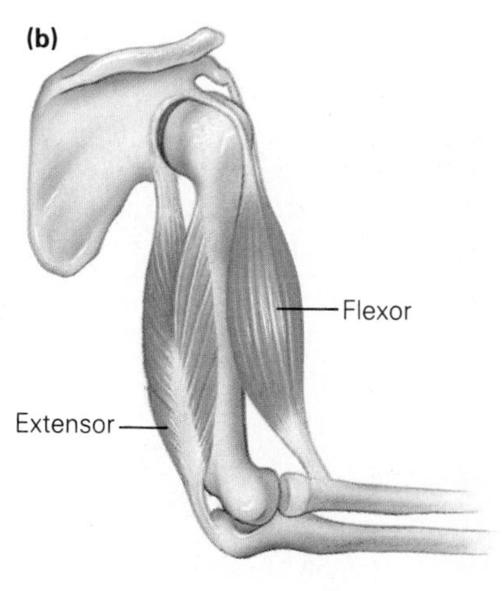

the thorax of the exoskeleton vibrates, it moves the wings. Thus, a single muscle contraction flexes the exoskeleton and produces multiple wingbeats. Let's go to the group, now, with which the mosquito has developed a particularly unwelcome association.

Types of Muscles in Humans

Human muscles can be divided into three groups according to their structure and the nerves that activate them. **Smooth muscle** is found in a number of internal structures, such as the walls of the digestive tract, around some blood vessels, and in certain internal organs. **Cardiac muscle** is found in the walls of the heart. Both of these are involuntary, since they function without conscious control. Thus, you don't have to lie awake nights keeping your heart beating. There is some fascinating evidence that "involuntary" responses can be voluntarily controlled to a degree, as we will see in Chapter 24. **Skeletal muscles** are the voluntary muscles. (Figure 17.12 shows examples of each muscle type.)

The skeletal muscles have several nuclei in each cell or **muscle fiber,** whereas each smooth muscle and cardiac muscle cell has only one nucleus. In skeletal muscles, the nuclei lie just under the membrane of the fiber. The cells of skeletal muscles are also distinctive in that they may be very long, some even extending the entire length of the muscle. Microscopic examination shows that both skeletal and cardiac muscles have alternating light and dark bands, called **striations** that are produced by the regular and repeating arrangement of contractile proteins. The three muscle types are compared in Table 17.2.

Skeletal Muscles

Typically, when we think of a "muscle" we conjure up images of skeletal muscle. A single such muscle is actually a bundle of millions of individual

FIGURE 17.12

Three types of muscle tissue. At left is smooth muscle, unstriated and spindle-shaped (although the shape is not apparent here). The nuclei are elongated and prominent. At center is striated, or skeletal, muscle. The striations are due to the regular and repeating arrangement of contractile proteins, as shown in Figure 17.15. At right is cardiac muscle. Note the striations.

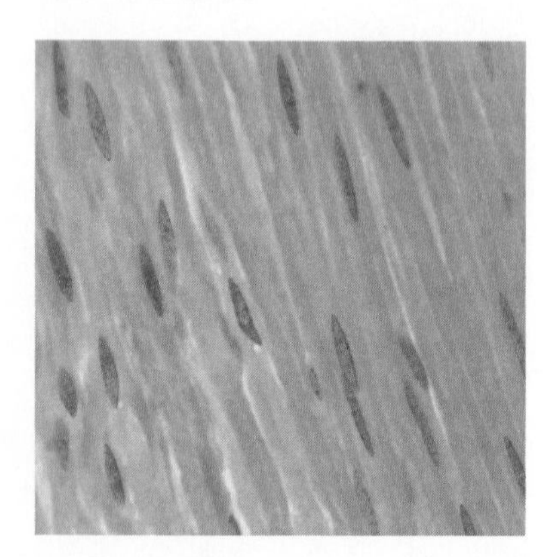

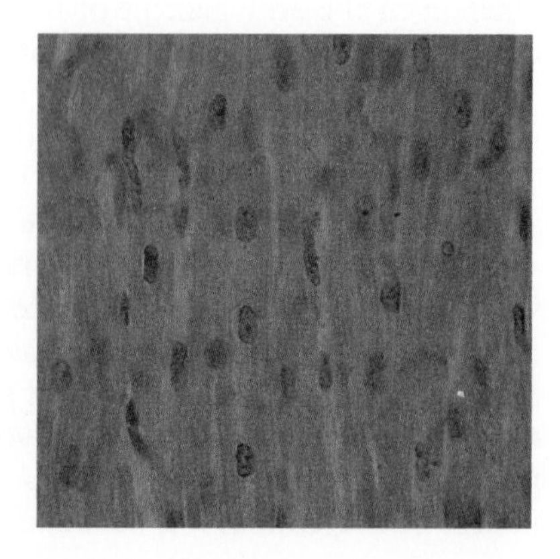

TABLE 17.2
Comparison of Muscle Types

	SKELETAL	SMOOTH	CARDIAC
Location	Attached to skeleton	Walls of viscera, around arteries	Wall of heart
Shape of fiber	Elongated, cylindrical	Elongated, spindle-shaped	Elongated, cylindrical, branched
Number of nuclei	Many	One	One
Position of nuclei	Peripheral	Central	Central
Cross striations	Present	Absent	Present
Speed of contraction	Rapid	Slow	Intermediate
Ability to remain contracted	Least	Most	Intermediate
Common type of control	Voluntary	Involuntary	Involuntary

muscle fibers bound together by connective tissue. (This connective tissue is what makes some meat so tough.) The muscle is surrounded by a tough sheet of whitish connective tissue called the **fascia.** The slippery fascia minimizes friction as the muscles move past each other.

Each end of most skeletal muscles is attached by cordlike tendons to different bones. When a muscle of this type contracts, it shortens and thickens, drawing the areas of attachment closer together, and thereby moving the bones. (Note, then, that muscles "pull" and do not "push.") Ordinarily one of the bones moves less than the other and, in some cases, is completely stationary. The bone that moves less provides the area of attachment for the end of the muscle called the **origin;** the more moveable bone provides the surface for the muscle's **insertion.**

Consider an example. If you ask a fellow to show you his muscle, if he is a gentleman he will display his biceps brachii (Figure 17.13). Note that the origin of this muscle is at the joint of the humerus and scapula (biceps means "two-headed," so note also the double origin). The muscle inserts on the radius, the bone of the forearm that terminates behind the thumb. (As the forearm is rotated, the radius describes an arc, hence its name.) As the biceps is flexed, the lower arm forms an increasingly smaller angle with the upper arm. When the biceps is fully shortened, the muscle mass increases in diameter and the bulge can then be admired by those interested in such things.

Antagonist Muscles

On the other side of the humerus, opposite the biceps brachii, is another large muscle. This one originates on the scapula and at two places on the humerus and is called the triceps brachii. It inserts on the elbow, which is actually the part of the ulna (the bone of the forearm on the same side as the little finger) that extends past the joint. As the biceps is contracted, the triceps must relax and lengthen. When the triceps is contracted, extending the arm, the biceps must then relax and lengthen. Thus the biceps decreases the angle between the upper and lower arms. Muscles that decrease the angles between bone are called **flexors.** The triceps increases the angle and so is called an **extensor.** Most skeletal muscles have these kinds of opposing muscles that reverse their action. Such pairs are called **antagonist muscles.**

FIGURE 17.13

The attachment of the upper arm muscles. The biceps brachii lies on the forward part of the humerus. On the back is its antagonistic muscle, the triceps brachii. Shortening of the biceps has what effect on the lower arm? Note that the origins are attached to bones that will move less than the bones with the insertions when the muscle is shortened.

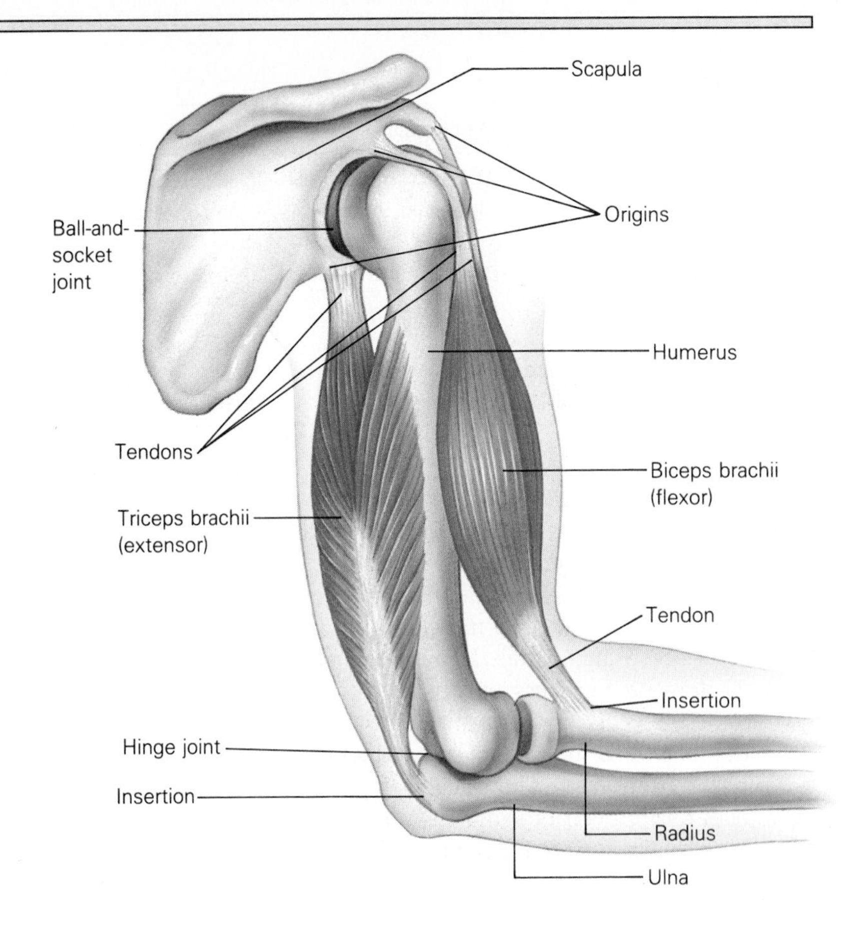

In some cases, antagonistic muscles must partially contract simultaneously in a highly coordinated manner and, thereby, keep a certain tension between them. For example, standing requires simultaneous contractions of opposing leg muscles. The flexor (in the back of the leg) and the extensor muscle (in the front) pull against each other and keep the leg straight and the knees locked. The tension, of course, must be highly regulated to keep us from leaping about and sprawling in such as manner as to arouse the curiosity of our fiancée's parents. Walking, by the same token, is not a simple matter; it requires an amazing degree of intricate coordination between a number of opposing muscles. Balance is maintained by constantly changing tensions as our weight shifts from one leg to the other.

The larger superficial muscles of the human body are shown in Figure 17.14. Look this over, and try to find each of these muscles on yourself. Contract the muscle and see what it does, but not while waiting for a job interview. Does anything about the arrangement surprise you? Why is the gluteus maximus of humans so much larger than that of other species? What muscles are involved in your shrug?

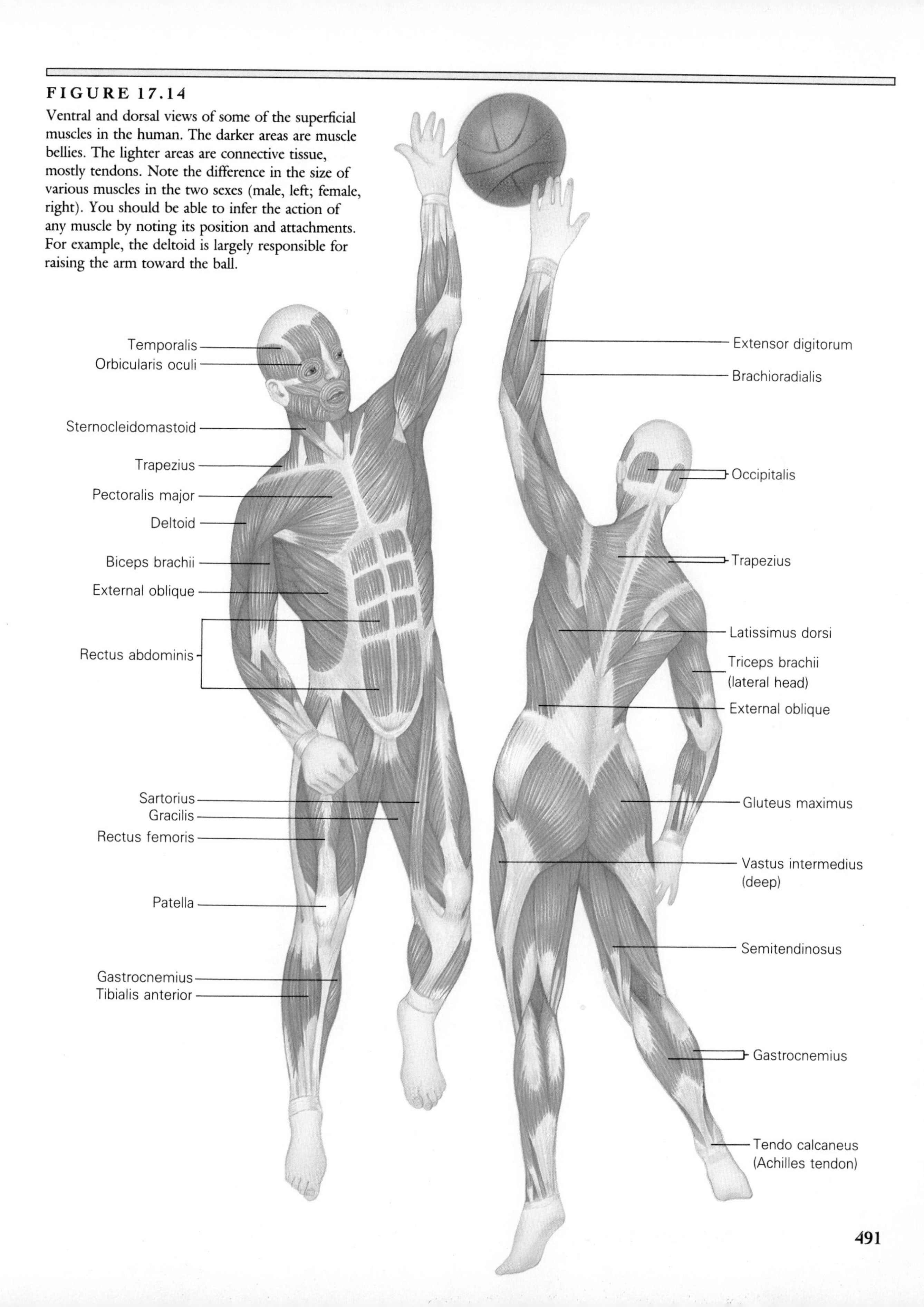

FIGURE 17.14

Ventral and dorsal views of some of the superficial muscles in the human. The darker areas are muscle bellies. The lighter areas are connective tissue, mostly tendons. Note the difference in the size of various muscles in the two sexes (male, left; female, right). You should be able to infer the action of any muscle by noting its position and attachments. For example, the deltoid is largely responsible for raising the arm toward the ball.

Muscle Contraction

The striated muscle fiber, or cell, has several nuclei, as was mentioned. It is also surrounded by a rather tough cell membrane. Closer examination (Figure 17.15) reveals that each fiber is made up of yet smaller fibers. These are composed primarily of protein and are called **myofibrils.** The cross banding, or striations, of the myofibrils gives the fiber its striated appearance. Since muscle tissue is very active, we should not be surprised

FIGURE 17.15

How skeletal muscles are believed to work. The sequence shows increasingly enlarged areas of muscle. Also illustrated here is the sliding filament hypothesis of how such voluntary muscles work at the subcellular level. (a) and (b) indicate what happens when a muscle contracts. Notice that the length of the A-band remains constant. The fibers of the H-zone give rise to tiny "bridges," which the I-fibers can move along in a progressive fashion. The converging I-bands, then, decrease the length of the H-zone as the muscle shortens. The I-bands are believed to be composed of actin, the H-zone of myosin.

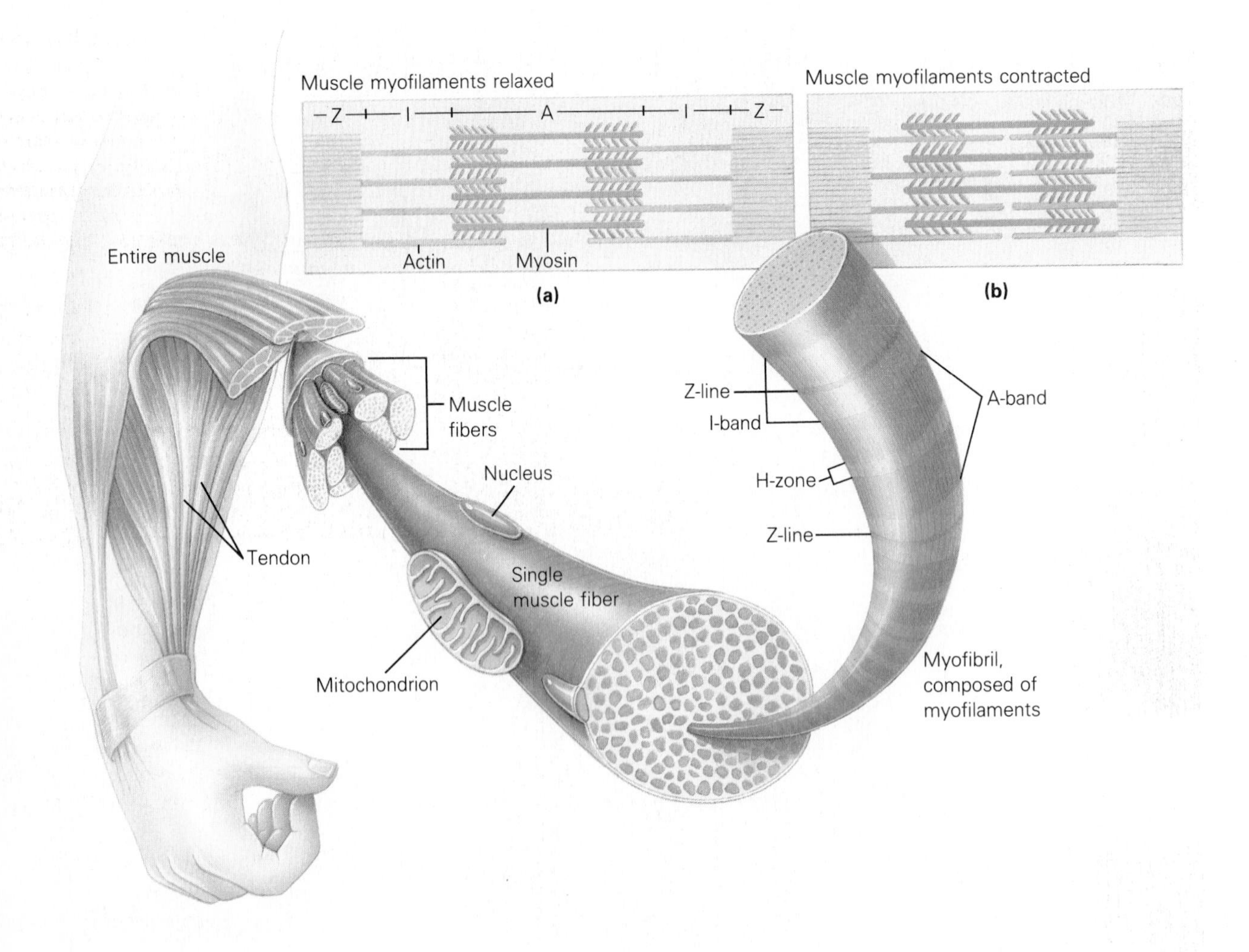

that it contains a rather large number of mitochondria, those tiny organelles so intimately involved in the production of energy. The mitochondria lie snugly against the myofibrils, the actual contractile units.

As we see in Figure 17.15, the tiny myofibrils are made up of still smaller units, called **myofilaments.** There are two kinds of myofilaments; the thicker ones are about 10 nanometers in diameter, and the thinner ones are about half that size. We can now see that the banded appearance of the myofibril is due to how two kinds of myofilaments are arranged with respect to each other.

A myofilament consists primarily of one of two proteins, **actin** and **myosin.** The thinner filaments are composed of actin and the thicker ones of myosin. Note that the various areas along the fiber are designated by letter. The thick myosin filaments, we see, are found only in the darker A-band, while the I-band contains only the thin actin filaments, which may extend into the A-band.

According to the **sliding filament theory,** muscle contraction is due to actin and myosin fibers sliding past each other. The basic idea is diagrammed in Figure 17.15 a and b. It seems clear that the sliding movement is due to the activity of myosin. Myosin filaments have projections called **cross bridges.** These attach to the actin fibers and pull them toward the center of the contractile unit while the myosin itself remains in place. The cross bridges attach, pull, release, reach, and attach again to the actin, thereby pulling it along in a rachetlike fashion.

The myosin cross bridges can only attach to actin in the presence of calcium ions (Ca^{++}) and ATP. The calcium ions react with protein molecules in the actin filament, opening sites along the filament to which the cross bridges of myosin can attach.

After the bridges have attached and pulled the actin along, the myosin takes on the role of an enzyme and breaks down ATP to ADP, releasing energy that can then be used to break the cross bridge, freeing it to reach further along the actin fibril, attach, and pull again. This all sounds rather tedious and plodding, but in actuality it happens with astonishing swiftness. The sequence may occur hundreds of times each second.

The efficiency and swiftness of muscle contractions illustrates the conservative efficiency of living systems that have developed over time to meet certain environmental needs. The environment has traditionally been a demanding, unforgiving, and unprejudiced place, but life has had the time to meet its challenges. We are well adapted for living on this planet, and one prevailing theme has been conservation, not only of energy, but of time.

SUMMARY

1. Groups of cells that exist together and have the same function are called tissues. Groups of tissues with the same function are called organs. Groups of organs that act together are called organ systems.

2. Exoskeletons provide support from outside the body and endoskeletons provide support from within. Muscles of soft-bodied animals, such as hydra, flatworms, and earthworms, contract against body fluids, creating pressure that forms a hydrostatic skeleton. Sponge skeletons consist of spicules or spongin scattered throughout tissue.
3. Exoskeletons present problems for growth. Snails and clams grow by secreting extensions to their shells. Lobsters, crabs, and crayfish discard their old skeletons and grow new ones.
4. Vertebrates have endoskeletons. Endoskeletons of sharks and rays are made of cartilage. In most other vertebrates, a skeleton of cartilage forms within the embryo, but most of the cartilage is later replaced by bone.
5. Connective tissues are supporting tissues. The nature of the tissue depends on the composition of the matrix, a nonliving material secreted by cells that comprises most of the tissue. The matrix of bone is hardened by salts of calcium and phosphorus, and has some flexibility due to a protein called collagen. Large bones have cavities filled with marrow. Whereas yellow marrow consists mostly of inactive fat cells, red marrow forms blood cells.
6. Bone is composed of Haversian systems. In the center of each Haversian system is a canal containing the blood supply for the bone cells that are arranged around it in concentric circles. The cells, which lie in openings in the matrix, extend through tiny passages called canaliculi, permitting physical contact between cells so that materials can pass between the blood vessel and cells.
7. Cartilage is a supporting connective tissue that is firm, flexible, resilient, and smooth.
8. Ligaments are composed of elastic connective tissue and bind bones together. Tendons are flexible cords that bind muscle to muscle or muscle to bone.
9. The human skeleton is divided into two parts: the axial and the appendicular. The axial skeleton forms the long axis of the body and includes the skull, spinal column, sternum, and the ribs. The appendicular skeleton includes bones of the arms, legs, pectoral and pelvic girdles.
10. Bones can move with respect to one another at joints. Different types of joints allow different degrees of movement. Ball-and-socket joints provide the greatest amount of movement. Hinge joints permit back and forth motion. There is little flexibility in a symphysis, such as the pubic symphysis, and the bones of the skull are fused in immobile joints.
11. Although they lack true muscle, cnidarians have two layers of contractile fibers. The outer layer extends longitudinally and the inner layer runs circularly. The roundworms have true muscles that run longitudinally. Flatworms and segmented worms have longitudinal muscles and some species also have muscles running diagonally or dorsoventrally. Muscles in arthropods are attached to the inside of the exoskeleton causing movement at joints, which are thin flexible parts of the exoskeleton.
12. There are three types of human muscle: smooth, cardiac, and skeletal. Involuntary muscle includes smooth muscle, which is found in the walls of many internal structures, and cardiac muscle, which is found in the heart. Skeletal muscles are under voluntary control. Skeletal and cardiac muscle have striations.
13. Each skeletal muscle is a bundle of millions of muscle cells (fibers) bound by connective tissue and surrounded by tough, slippery connective tissue

called fascia, which allows muscles to move past one another easily. Most skeletal muscles are attached to bones by tendons. When a muscle contracts, the bone providing the surface of the muscle's insertion moves closer to the more stationary bone, the one providing the surface for the muscle's origin.

14. Each skeletal muscle cell contains several nuclei and many smaller myofibrils. Each myofibril is comprised of even smaller myofilaments. There are two types of myofilaments, each composed primarily of a different protein: a thin one composed mainly of actin and a thick one composed mainly of myosin. The arrangement of myofilaments gives muscle its striated appearance.
15. According to the sliding filament theory, a muscle shortens (contracts) when the actin filaments slide toward one another, past the myosin filaments. The movement occurs because projections from the myosin filaments, called cross bridges, attach to actin filaments and pull them toward the center of the contractile unit. They then release the actin, reach out, and reattach to actin, pulling it along in a rachetlike fashion. Myosin cross bridges attach to actin only when calcium ions are present to open attachment sites along the actin filament. The energy in ATP is needed to break the cross bridge, freeing it to attach farther along the actin molecule.

KEY TERMS

actin (493)
antagonist muscle (489)
appendicular skeleton (484)
axial skeleton (482)
ball-and-socket joint (486)
bone (479)
canaliculi (479)
cardiac muscle (488)
cartilage (481)
collagen (479)
connective tissue (479)
cross bridge (493)
endoskeleton (478)
exoskeleton (478)
extensor (489)
fascia (489)
flexor (489)
Haversian canal (479)
Haversian system (479)
hinge joint (486)
hydrostatic skeleton (478)
insertion (489)
joint (485)
ligament (481)
marrow (479)
matrix (479)
muscle fiber (488)
myofibril (492)
myofilament (493)
myosin (493)
organs (477)
organ systems (477)
origin (489)
skeletal muscle (488)
skeletal system (478)
skull (482)
sliding filament theory (493)
smooth muscle (488)
spicule (478)
spinal column (482)
spongin (478)
striation (488)
suture line (486)
symphysis (486)
tendon (481)
tissues (477)

FOR FURTHER THOUGHT

1. Arthritis is a disease that affects the most active movable joints. Of the following, which would you LEAST expect to be a candidate for arthritis? Joints in: shoulder bones, wrist bones, skull bones, or knee bones.
2. Explain why a spicule support system would be inappropriate for a large land animal.
3. If you should break your femur in a skiing accident, where will the doctor place the cast?

FOR REVIEW

True or false?

1. ___ Tendons connect muscles to bones and other muscles.
2. ___ Hydrostatic skeletons cover the outside of insects.
3. ___ The widest range of movement is provided by hinge joints.
4. ___ The contractile system of arthropods lies within their skeleton.

Fill in the blank.

5. Vertebrates have internal support systems called ___.
6. Bones, blood, and ligaments are some types of ___ tissue.
7. ___ are points of articulation between bones that enable them to move.
8. ___ muscles occur in pairs and have opposing actions.
9. ___ and ___ are two proteins that give the myofibril its striated appearance.

Choose the best answer.

10. Arm and leg bones are part of the:
    A. pelvic girdle
    B. appendicular skeleton
    C. axial skeleton
    D. pectoral girdle
11. Smooth muscle
    A. is involuntary
    B. is found in walls of the heart
    C. is striated
    D. moves bones

CHAPTER 18

Homeostasis and the Internal Environment

Overview

Homeostasis and the Delicate Balance of Life

Feedback Systems

Homeostasis and the Regulation of Temperature

Homeostasis and the Regulation of Water

Objectives

After reading this chapter you should be able to:

- Define homeostasis and state its relationship to positive and negative feedback systems.
- Differentiate between endotherm and ectotherm and explain how each regulates temperature to maintain homeostasis.
- Name three main nitrogenous waste products and describe the mechanisms by which freshwater, marine, and terrestrial animals eliminate their nitrogenous waste and regulate water balance.
- Name the structures in the human excretory system and give the functions of each.
- Describe the formation of urine in the human kidney.
- Describe the role of hormones in controlling the concentration of urine.

About 200 years ago, Dr. Charles Blagden, then secretary of the Royal Society of London, proved that he was one of the most persuasive people on earth. He talked some friends into joining him, a small dog, and a steak in a room in which the temperature had been raised to 126°C (260°F). In fact, he managed to persuade his friends to stay in there for 45 minutes. (The dog and the steak had no choice.) At the end of this time, the men and the dog emerged unharmed, but the steak was cooked!

In addition to demonstrating his polemic powers, Blagden also showed that the bodies of animals are able to compensate for extreme physiological conditions and, in particular, that some living things can control their internal temperatures in the face of extreme external conditions. Since this early experiment, it has been found that animals can regulate a host of other internal physical and chemical states. The ability to hold things within certain limits should not be unexpected since the delicate processes of life would not be possible under wildly fluctuating conditions.

HOMEOSTASIS AND THE DELICATE BALANCE OF LIFE

Homeostasis is the tendency of living things to maintain a stable internal environment (from the Greek *homios:* same, and *stasis:* standing). The term is a bit misleading, however, because no living thing strictly maintains a constant internal environment. One reason is that maintaining such rigid constancy would place a great demand on the organism's metabolic machinery. Another reason is that some change must occur in order for things to stay the same. That is, as the environment changes, the body's internal processes must also shift in order to counteract the outside changes and keep the internal conditions stable. So cells do change as they constantly monitor, metabolize, and adapt. The result is a "steady state"—keeping the internal environment within certain limits—and that is what homeostasis is all about.

FEEDBACK SYSTEMS

Generally, the steady state is maintained by feedback mechanisms. Feedback occurs when the product of an action influences that action. Biologically, the most important kind of feedback action is called **negative feedback.** Negative feedback occurs when an increase in a system's product causes a slowdown of that system, or when a reduction of the product stimulates the system. One kind of "cruise control" on an automobile engine (a device that keeps the car moving at a certain speed) works because as the motor runs faster, valves are closed, causing the car to slow down. As the car slows, the valves are reopened, and the car accelerates, keeping its speed within certain limits. Also, because of negative feedback mechanisms, you don't have to constantly nibble and fast to keep your blood sugars at the proper level. When blood sugars are low, the liver simply breaks down some of its stores of glycogen and releases glucose into the

FIGURE 18.1

Two types of feedback systems. In (a), a rather constant water level is maintained as long as there is enough rain. As the amount of water increases, more is allowed to run out; as it decreases, the drain is plugged. This, then, is a typical negative feedback system. In (b), as the water level rises, less is allowed to escape, so the water rises out of control. As it falls, even more is released until the barrel is empty. This is a positive feedback system, a type usually not associated with the normal functioning of living things.

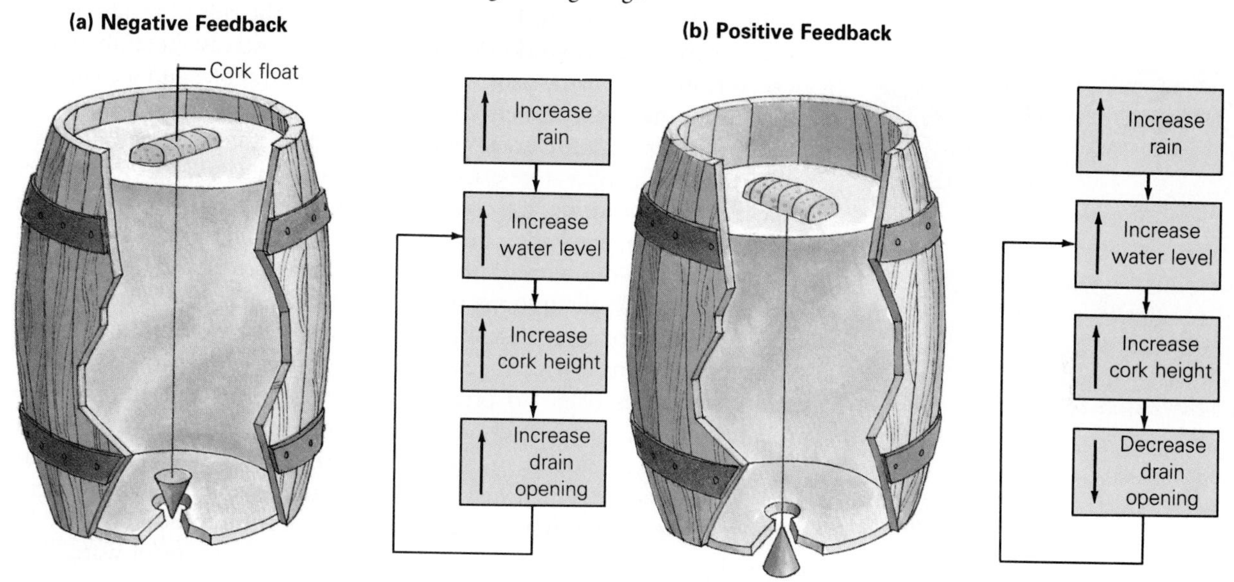

blood. As blood sugars rise, the liver, increasingly, breaks down less glycogen, and instead converts the sugar to glycogen for storage.

Positive feedback works on the opposite principle: the product of a system increases the activity of that system. Using the example of an automobile, with positive feedback, accelerating the motor would tend to open the carburetor and cause the engine to run even faster. Such an engine might be revved to such limits that it would explode. Obviously, then, homeostasis operates primarily through the more delicate mechanisms of negative feedback, but this is not to say that living things are not sometimes subjected to positive feedback.

Positive feedback mechanisms in humans are sometimes associated with severe health problems. For example, if a person's temperature begins to rise above the normal 37°C, the body will activate corrective devices such as sweating and the opening of peripheral blood vessels (producing a heat-dissipating "flush"). However, at some point (usually at about 42°C), the negative feedback system breaks down, and a positive feedback begins. The high temperature begins to cause an increase in metabolic activity, which raises the heat, which increases metabolic activity, which can kill the unfortunate soul. (Positive feedback is the basis of the famed "vicious circle.") To better understand the principles of the two forms of feedback control, consider the operation of two simple storage tanks depicted in Figure 18.1.

FIGURE 18.2

The pupfish is a denizen of hot, briny desert springs. In natural habitats, they are able to live under a temperature range of 38°F (3.3°C) to 108°F (42.2°C). The physiological system that permits such abilities is not completely understood.

HOMEOSTASIS AND THE REGULATION OF TEMPERATURE

Many of the processes of life can occur only within a narrow range of temperatures, and the mechanisms by which this stabilization occurs may serve as a good example of homeostasis. First, we should note that an increase in temperature of only a few degrees can cause great leaps in the rate of chemical reactions. In nonliving material, the rule of thumb is: the rate of chemical reactions doubles for every 10°C increase in temperature. However, in living cytoplasm, such rules may not apply, and a change in temperature may have less effect.

Why is temperature so important to living things? Because of the severe effects of temperature extremes. At about −1°C to −2°C, the water in cells freezes, causing ice crystals to form that may rupture delicate membranes. Also, with water tied up as ice, the remaining cell constituents may become so concentrated that they are unable to function properly. The result of such disruptions may be death. At the other extreme, the upper temperature limit that life can withstand is apparently largely set by the temperature at which the hydrogen bonds holding proteins in their tertiary structures begin to break, thus unwinding (or denaturing) the protein. Because of the effects of temperature, most animals live in places that are not much colder than freezing or warmer than about 40°C (Figure 18.2). (Exceptions to such rules include certain algae, fungi, and bacteria that are able to live in hot springs at temperatures up to 80°C, and some species that cannot survive if "chilled" to the boiling point of water.)

Ectotherms

The **ectotherms,** or the so-called cold-blooded animals, include virtually all the vertebrates except birds and mammals (*ecto:* outside; *therm:* temperature). The ectotherms, therefore, include fish, amphibians, and reptiles. Ectotherms do not physiologically regulate their body temperatures to any great extent. Although they do produce metabolic heat, they have no efficient means of conserving it or of increasing or decreasing its production. If it is necessary to change their body temperatures, they may do so behaviorally (Figure 18.3).

In general, the saltwater fish have no great problems because the oceans rarely change in temperature more than a few degrees. However, freshwater fish living in the shallows are much more at the mercy of the elements. Furthermore, their efforts to thermoregulate behaviorally can put them at risk. As they move from place to place, they may find themselves in danger of becoming landlocked. This means they risk overheating or drying out if they are unable to return to deeper water. Of course, freshwater fish that remain in deeper water suffer no such threat, but because of the low temperature at these depths, their metabolic rate is slow and so must be the pace of their lives.

Among land dwellers it's a different story. They face unbuffered temperatures and dry air with its rapidly changing temperatures. Somewhat surprisingly, amphibians have been able to adapt to such arid environs. One way they adjust is by simply moving overland to places with more agreeable temperatures. Some species may also bury themselves and

FIGURE 18.3

A variety of animals from the Tokay gecko (a) to the sooty tern chick (b) regulate their temperatures behaviorally—by seeking shade or panting, for example.

(a)

(b)

thereby escape the drying air. Other species that are exposed to extreme heat and drought have the ability to **estivate**—that is, to enter a form of summer "stupor" until reactivated by cool temperature and moisture. Toads in certain arid parts of Australia, in fact, have been known to remain buried for as long as two years while waiting for a good rain. The estivating toads have low metabolic rates, just as do hibernating animals, and therefore their food and oxygen requirements are low.

Land reptiles strongly rely on behavioral regulation, as do amphibians; they simply move to areas that are more appropriate to their needs. As with many other ectotherms, they also allow their body temperatures to drop with the cool of night. This results in a certain sluggishness in the morning because their metabolic rates have slowed accordingly. If they are going to catch any food, however, they've got to warm up, and they quickly do so by utilizing the warmth of the sun. Basking ectotherms are able to absorb the heat of the sun even when the air temperature is near freezing. As the day wears on, the animals may seek shade or turn to face the sun, which reduces the surface area exposed to its rays. During the hottest part of the day, some species may retreat underground, reappearing only in the cooler afternoon (Figure 18.4).

FIGURE 18.4

A desert-dwelling lizard. Note the long appendages that not only dissipate heat quickly, but enable the lizard to move quickly to find cooler areas. Even desert lizards cannot stand the direct desert sun for long, and so they hunt in the shade, being most active at dusk and dawn.

Endotherms

Birds and mammals comprise the **endotherms** (*endo:* inside), the so-called warm-blooded creatures. They have evolved physiological methods of keeping their body temperatures within very narrow limits. Their bodies can carry out biochemical activities more efficiently by having specialized to within a range of temperatures that is conducive to biochemical reactions.

Birds and mammals are endotherms with their ancestry rooted in a common reptilian past. Partly because of this heritage, they have a number of traits in common, including traits related to temperature control. First,

FIGURE 18.5

Mammals and birds are endotherms. The yak is an amazingly tough creature that thrives in the frigid arctic regions. It is insulated against the inclement weather by a shaggy coat that sheds each spring. The black-capped chickadee is also insulated, but by feathers that it can raise and lower, trapping more or less air as it regulates its insulation.

they were well-insulated creatures, with their fur and feathers (Figure 18.5). Second, they had a more efficient four-chambered heart.* Theirs was a powerful organ (described in Chapter 19) able to pump enough fuel and oxygen throughout the body to stoke the cells' metabolic furnaces. Remember, the cost of struggling against the environment to maintain a constant internal temperature would have been metabolically expensive and would have required increasingly efficient physiological mechanisms.

Since in endotherms, metabolic heat (a by-product of certain chemical reactions) travels from the inside out, bodies tend to be warmer toward the inside. Thus, if you should touch something that is 37°C (your body's internal temperature), it will feel warm because your skin is cooler than that. So, there is some variation from place to place in your body, and there is even some variation from time to time. (Your temperature usually falls a bit in the wee hours of the morning and rises in the early afternoon.) Because of the greater constancy in the deeper tissues, if you measure someone's temperature with a rectal thermometer, you will get not only his undivided attention, but a more precise reading as well.

How does your body "know" its own temperature? The temperature is monitored by a delicate thermostat in the **hypothalamus** (an ancient part of the brain—see Chapter 22). Certain receptor cells monitor the temperature of the blood reaching the brain. If the blood is too warm, the hypothalamus initiates a chain of events that brings down the temperature. This is done in different ways in different animals, and both

*Since some modern reptiles, notably alligators and crocodiles, have four-chambered hearts, theorists have suggested that the dinosaurs may have had such an advantage as well.

physiological and behavioral responses may be involved. As body temperatures rise, humans and most other large animals sweat, and the evaporation cools them. Dogs pant and rapidly move cooling air over their large tongues and through their lungs. (They may also lie down under a tree—a behavioral response.) Cats are stimulated to lick themselves and are cooled by the evaporation of the saliva. In some animals, including humans, peripheral blood vessels dilate, bringing warm blood to the surface of the skin where heat can be dissipated (Table 18.1).

As the blood cools, the surface vessels contract, reducing heat loss over the skin's surface. The hypothalamus then signals the pituitary to direct the thyroid glands, located in the throat area (Chapter 21), to release a hormone (called thyroxine) that increases metabolic rate. In addition, if the temperature of the blood in the brain should drop to dangerous temperatures, the hypothalamus will direct the adrenal glands to secrete epinephrine (adrenalin), thereby increasing the metabolic rate and raising the body's temperature. If the temperature continues to drop, shivering begins, causing the surface muscles to work and to increase their metabolism, thereby producing more heat and using more energy. Have you noticed that you eat more in the winter? Why do you suppose that is?

Occasionally the body's thermostat is "turned up," and fever results. No one really knows how fever occurs, but infection seems to trigger the release of **pyrogens** ("heat causers") that ultimately raise the body's temperature. Apparently, it does this by resetting the body's thermostat in the hypothalamus to a higher temperature, making the normal temperature feel too cold. Blood vessels in the skin contract ("You look pale; do you feel well?"), and shivering may begin as the rate of metabolism increases. All these things can act together to bring on higher temperatures. Finally,

TABLE 18.1
Mammalian Responses to Temperature Stress

STIMULUS	RESPONSE	EFFECT
Hot temperature	Sweating or panting	Evaporative heat loss
	Dilation of peripheral blood vessels	Heat lost from warm blood at body surface
	Changes in behavior	Increase heat loss or reduce heat generated
Cold temperature	Constriction of peripheral blood vessels	Reduce heat loss at body surface
	Increased muscle activity (shivering)	Increase heat production
	Secretion of thyroxine from thyroid gland and epinephrine from adrenal cortex increases metabolic rate	Increase heat production
	Changes in behavior	Reduce heat loss or increase heat production

the temperature temporarily stabilizes at its new, higher setting. When the fever breaks, the opposite responses occur. The skin becomes flushed and sweating occurs until enough heat is lost to restore the body to normal temperature.

Here I should add that there is some recent evidence suggesting that for those who are ill, aspirin or other fever-reducers may do more harm than good. It has been suggested that a slight rise in the body's temperature can render it inhospitable to temperature-sensitive viruses and bacteria. Physical exercise can also raise the body's temperature and, some say, thereby stave off certain illnesses by rendering the body unsuitable for habitation by pathogens.

Just as your body's temperature is kept within narrow limits, so are other physiological factors. In fact, every activity involving growth and maintenance is controlled by complex regulatory mechanisms. One of the clearest examples of such control involves water balance. You have probably heard that the body to which you are attached is composed mostly of water, and physiologists have glibly quoted rather precise numbers to tell you just what the percentage is. How can they be so sure? Do they include the bodies that have just completed marathons, or those that have been drunk for three days? Not normally, but still they are confident because they are aware of the delicate and intolerant mechanisms of the kidneys.

HOMEOSTASIS AND THE REGULATION OF WATER

Since most of the physical and chemical processes of life involve water in some way, animals have developed very precise means of regulating the water within their bodies. That regulation, we will now see, involves the body's mechanisms for getting rid of excess nitrogens, the by-product of protein metabolism.

All animals have the problem of getting rid of metabolic wastes, or "cell garbage," in the form of excess nitrogens that are left over when proteinaceous foods are metabolized. In this process, the proteins are stripped of their nitrogens and the remainder of the molecules are converted into whatever is needed, such as glucose. This leaves the nitrogens, which are then able to combine with other elements, forming poisonous nitrogenous by-products.

The nitrogen atoms that are stripped from the proteins are not utilized in either the anaerobic or aerobic energy cycles. Once free, they usually pick up three hydrogens—and NH_3 is **ammonia,** a deadly poison. The problem, then, is how to dispose of the ammonia. However, ammonia is so poisonous that it must be highly diluted in order to be handled by the body, and in water-conserving species, any such dilution would result in the loss of too much water. The problem is solved by the addition of components of carbon dioxide to the ammonia, forming **urea,** a less toxic molecule that the body can handle in higher concentrations. A moderately dilute urea is then passed out of the body in the **urine.**

Many terrestrial species, notably the egg-laying birds and the insects, have a particular need to conserve water, so they have developed the means to convert ammonia to an insoluble nitrogenous product called **uric acid.** The uric acid is excreted almost as a dry paste in some species. The major waste products of metabolism in many fishes (ammonia), birds (uric acid), and mammals (urea)* are:

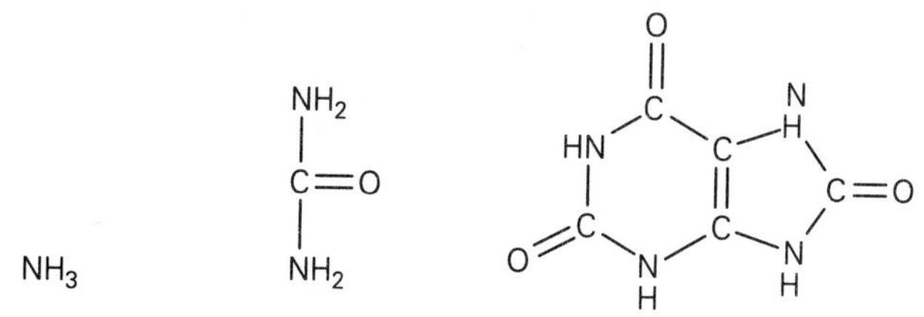

NH_3	$H_2N{-}C({=}O){-}NH_2$	Uric acid structure
Ammonia	Urea	Uric Acid

Solutions to the Water Problem

One of the great problems in the development of excretory systems was how to flush metabolic wastes from the body without losing too much water. The evolutionary solutions, we shall see, have been quite diverse.

Vertebrates have evolved a number of ways of ridding the body of metabolic wastes and maintaining proper internal water levels. For example, freshwater fish, like freshwater protozoa, live in a hypotonic medium; that is, the concentration of particles in their body fluids is higher than that in the surrounding water. Hence, water tends to move into their bodies by osmosis. One means by which they rid themselves of excess water is through a great number of tiny blood vessels in their kidneys. These blood vessels are coiled into tiny clumps that form such circuitous routes that the blood remains in one area for an extended time. These clumps of blood vessels are called glomeruli (singular, **glomerulus**). We'll consider the structure of the glomeruli shortly, but suffice it to say here that, in general, they provide a large surface area through which fluids leave the blood to be collected in the kidney. The metabolic waste in freshwater fish is also highly dilute. This is to be expected since they have no problem finding enough water to flush out their wastes.

Saltwater fish, interestingly enough, have the same problem as desert animals: conserving water. Seawater is a hypertonic medium; that is, its concentration of solutes is higher than the fluids in the marine animal's body. Hence, these animals tend to lose water by osmosis. As a result, many have developed smaller glomeruli, so that there is less opportunity for water to filter out as it passes through the kidney.

Saltwater fish replace the lost water by drinking seawater. They rid themselves of the excess salts from seawater by excreting some across the gills and some in the urine. Other salts are never absorbed and pass out in the feces. In other animals, specialized gills or other structures may also eliminate nitrogenous waste as blood passes through them.

*Actually, one can't strictly group vertebrates according to their excretions. For example, crocodiles, with an abundance of water, excrete both ammonia and uric acid (why would this suggest that they probably evolved from a terrestrial ancestry?). And tortoises, which normally produce urea, can shift to uric acid excretion when water becomes scarce (an example of how natural selection builds in not only restrictions, but also options).

FIGURE 18.6

Regulation of water and salt balances in a representative group of animals. Water enters the freshwater fish by osmosis. Great amounts of fluid cross its glomeruli daily, and it excretes large amounts of water in its urine. The saltwater fish loses water by osmosis, so it saves the remainder by filtering little fluid through its kidneys and urinating little. The shark gains water by osmosis and passes large amounts through its kidneys, urinating copiously. Land mammals drink water and lose some by evaporation. They save water by reabsorbing much of what passes through the kidneys and urinate only a little hypertonic urine. Look at the cases individually and compare their problems and solutions. Why do the fish solve their problems three different ways? What can you say about their cytoplasm?

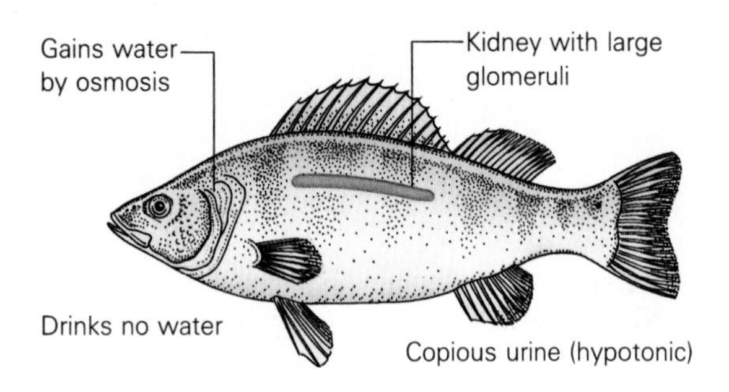

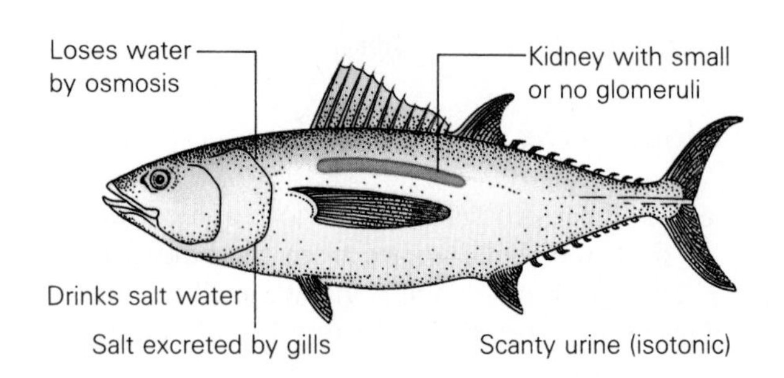

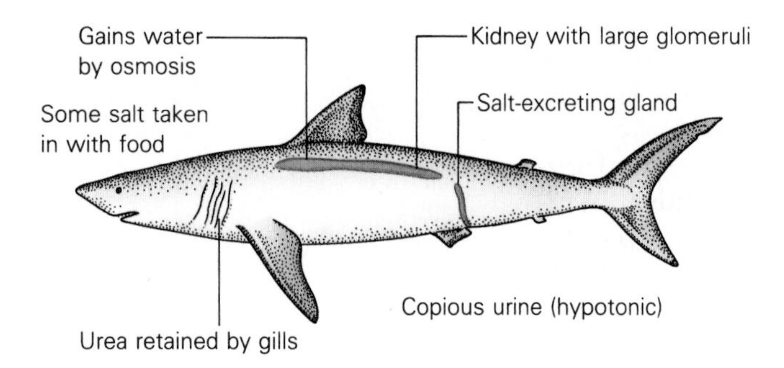

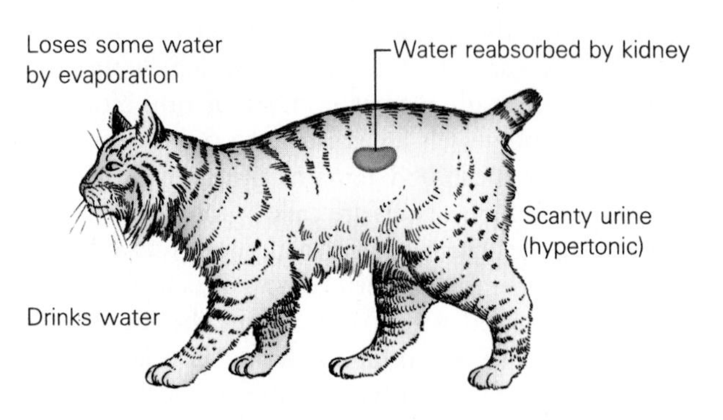

Most sharks and rays in the ocean have a different strategy: they retain some urea in their blood. This raises the osmotic pressure of their body fluid so that it is closer to that of seawater. Thus they sidestep many problems of the excretion of both water and metabolic wastes. (Various solutions to different water and nitrogen problems are summarized in Figure 18.6. See also Essay 18.1.)

Invertebrate Excretory Systems

Different species may dispose of nitrogenous metabolic wastes and regulate water balance in different ways. In protistans, such as the amoeba, nitrogenous wastes simply diffuse across the cell membrane and out into the surrounding water. However, since protoplasm is hypertonic to pond water, water constantly enters the body by osmosis threatening to cause

ESSAY 18.1

Special Solutions to the Water Problem

A quick look around the animal world reminds us that living things solve the problems of life in all sorts of ways. Some of the problems are monumental, indeed, especially those relating to conserving water. For example, some insects are able to survive for years enclosed in jars of dry pepper. They do this by utilizing metabolic water and excreting almost dry uric acid.

Few other species face such severe problems, but water conservation nonetheless remains a problem for many forms of life. Perhaps the most famous water conserver is the camel. A camel can tolerate dry conditions by adopting a number of tactics. At night it drops its body temperature several degrees so that bodily processes slow down. In the heat of day, it doesn't begin to sweat until its body temperature reaches about 105°F. In addition, the camel can lose twice as much of its body water (40 percent) without ill effect than can most other mammals. Interestingly, the thick coat of the camel acts as insulation to keep the heat out. Also, when the camel does drink, it can hold prodigious amounts of water (in its stomach, not in its fatty hump).

The camel has a number of physiological mechanisms that enable it to survive in an arid habitat.

The camel's desert colleague, the tiny kangaroo rat, rarely drinks at all, and it lives on dry plant material. It survives because it doesn't sweat and is active only in the cool of the night. Its feces are almost dry, and its urine is highly concentrated. Most of its water loss is through the lungs. Fats produce more metabolic water than other foods, so it prefers fatty foods. On the other hand, proteinaceous foods, such as soybeans, produce a lot of nitrogenous waste for which a lot of water is needed—so, if a kangaroo rat is fed only soybeans, it will die of thirst.

Some seabirds, marine lizards, and turtles have salt removal glands. The huge tears in the eyes of sea turtles have nothing to do with the realization that we are poisoning the oceans. The tears are flowing from special salt-removing glands that remove sodium chloride from the blood.

the organism to swell and rupture. (You may recall our earlier discussion of osmosis.) Some protistans must constantly expel water by the continual pumping of a contractile vacuole (Figure 18.7). In the case of freshwater protists, then, the problems of metabolic waste and internal water regulation are solved by different mechanisms. In the animals, the two problems may be solved by the same mechanisms.

As an example of a single mechanism adjusting the body's level of both metabolic wastes and water content, consider the excretory system

FIGURE 18.7

Freshwater protists excrete water that moves in osmotically by collecting it in a contractile vacuole and pumping it back out again. Here, a vacuole fills and then contracts, expelling water through the membrane.

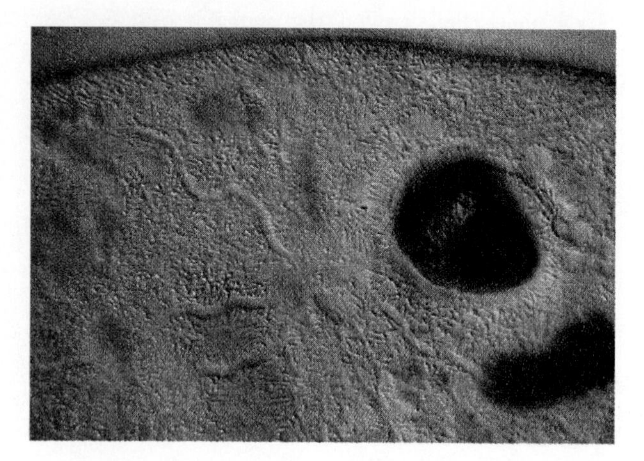

FIGURE 18.8

Earthworm nephridium. Fluids enter the system through a ciliated funnel. Wastes are withdrawn into capillaries surrounding the nephridia, to be excreted through the external pore.

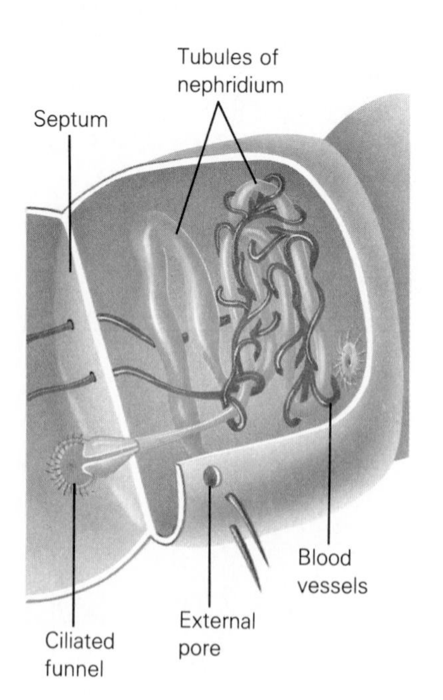

of the earthworm. Each segment of an earthworm has a pair of coiled tubes called **nephridia** (singular, ***nephridium***), connecting the body cavity to the outside (Figure 18.8). The tubes opening inside the body cavity are rimmed with beating cilia that circulate the fluids there. Capillaries lie tightly coiled around the outside of the nephridia, and as the fluid moves into and through the tube toward the outside, water, salts, and minerals are drawn back into the capillaries. The concentrated waste, then, is excreted through the external pore. The amount of water the animal excretes with the wastes (primarily urea and ammonia) depends on how much water has entered the body cavity. Worm urine, in case you've ever wondered, is very dilute, and the amount excreted each day equals about 60 percent of the worm's body weight.

The Human Excretory System

Humans, as do the other mammals, have two kidneys (although we can live with one). These lie in the dorsal area (at the back) and extend slightly below the protective rib cage (which is one reason blows to the kidney area are so dangerous). Actually, they lie behind the membrane that lines the abdominal cavity. Each kidney contains about a million **nephrons.**

Each nephron is composed of several parts, each with a very specific role in urine formation. The *glomerulus* is the blood vessel branching from the renal artery that carries the blood into the nephron. It is highly convoluted and surrounded by a cuplike structure, the **Bowman's capsule,** which accepts the filtered fluid (filtrate) from the glomerulus. The fluid then moves through the **proximal convoluted tubule** (near Bowman's capsule), making a U-turn via the **loop of Henle,** through the **distal convoluted tubule** (away from Bowman's capsule) before entering the

FIGURE 18.9

The human kidney. Note the position of the kidneys in the body. The lower part actually extends below the protective rib cage. Also note the vast maze of blood vessels penetrating the kidney. Blood is brought into the kidney under considerable force, and many of its constituents are filtered through the glomeruli. From there they enter a tubular system from which many of the constituents (including some water) pass back into the bloodsteam. Those that are not recollected are passed into increasingly larger tubes until they enter the urinary bladder as urine. Here the urine is held until, in long lines halfway to the box office, it demands to be released back to the environment. Urine is largely germ-free and has been used to wash wounds under field conditions when the available water was known to be impure.

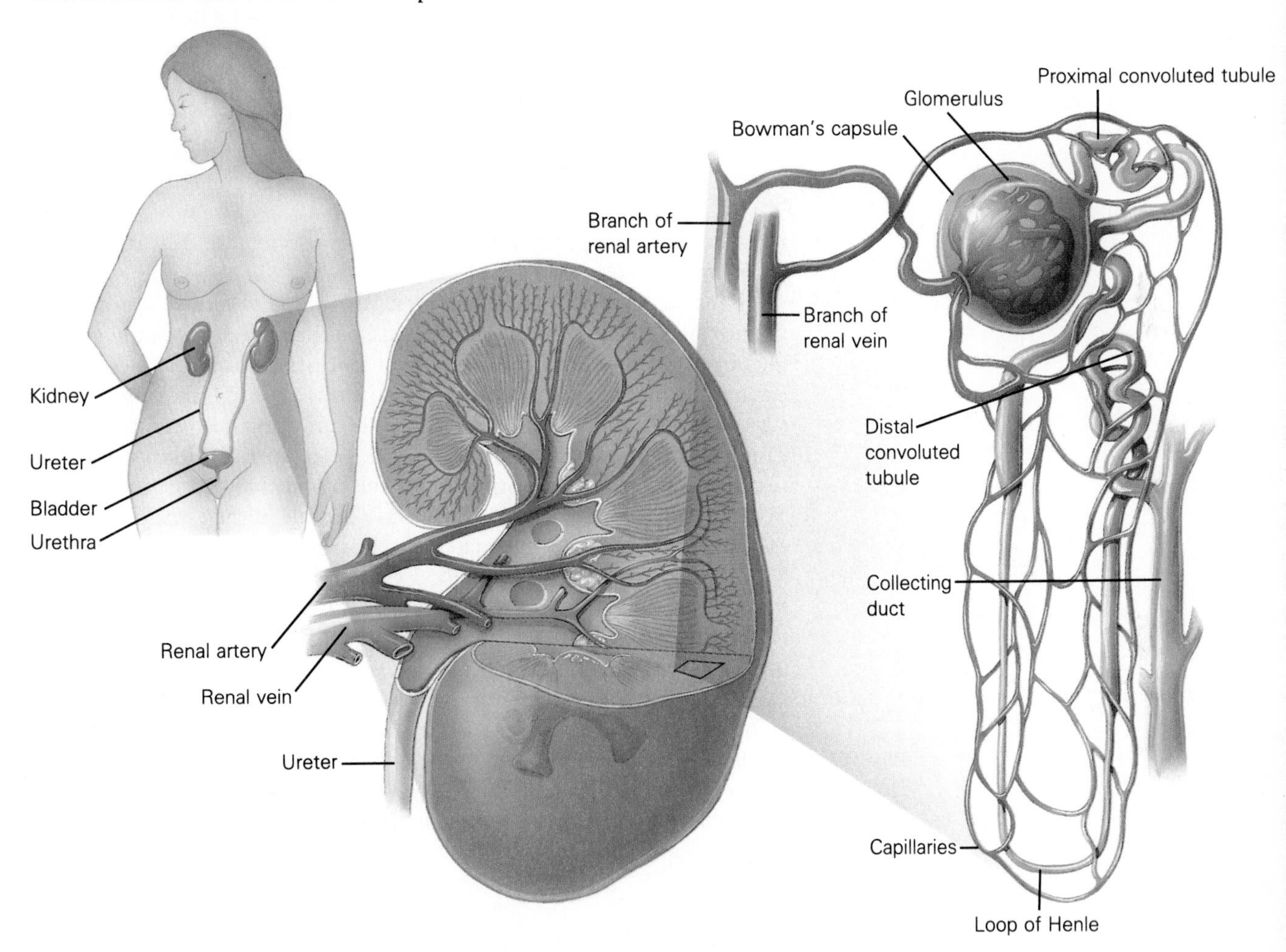

collecting duct, from where it flows to the **ureter,** the **urinary bladder,** and out the **urethra.** The structure of these units is shown, along with the position of the kidney itself, in Figure 18.9.

The Formation of Urine

The basic mechanism of urine formation involves filtering all but the very large substances out of the blood and then adjusting the filtrate by reabsorbing substances needed by the body back into the blood, and by actively secreting some substances from the blood into the filtrate. Thus we see

TABLE 18.2
Processes in Urine Formation

PROCESS	DESCRIPTION
Filtration	All but cells and large molecules (e.g., proteins) are filtered through glomerular capillaries into Bowman's capsule.
Reabsorption	"Valuable" materials (e.g., water, salts, nutrients) are selectively reabsorbed from the filtrate back into the blood.
Secretion	Specific chemicals (e.g., hydrogen ions, potassium ions, and certain drugs) are secreted from the blood into the filtrate.

that there are three processes involved in urine formation: **filtration, reabsorption,** and **secretion** (Table 18.2).

Let's see how this happens. First, the blood is filtered through the glomerulus into Bowman's capsule, pushed through pores in the glomerulus by a pressure gradient. The pressure exists because the blood vessel leaving the glomerulus is smaller than the one entering, making the pressure within the glomerulus higher than the pressure in Bowman's capsule. The glomerulus works as a mechanical sieve, so this filtration is not a selective process; any constituents of the blood that can fit through the pores in the glomerulus enter Bowman's capsule in the same proportions as found in the blood. Some materials, however, simply can't pass through. Most larger particles, such as blood cells and most proteins, cannot pass through the glomerular walls. However, in addition to waste products, the filtrate entering Bowman's capsule contains valuable materials, including water, salts, and nutrients.

As the filtrate moves through the nephron tubule, its contents are adjusted. The blood vessel leaving the glomerulus forms a network of capillaries that surrounds the nephron tubules. Usable materials are reabsorbed from the filtrate back into the capillaries, both through diffusion and by active transport. Certain unwanted materials are actively secreted from the capillaries into the tubules. The material remaining in the tubules is urine.

The Concentration of Urine

The conservation of water is accomplished by the concentration of urine. Although water conservation actually takes place primarily in the collecting ducts, it depends on the activities of the loop of Henle (Essay 18.2).

The principal role of the loop of Henle is to create a dense salt concentration in its surroundings, the kidney medulla. With a million nephrons participating, the salt concentration in the medulla becomes considerable. A high salt concentration means a low water concentration—much lower than the water concentration inside the nephron, so that a natural osmotic gradient forms. Simply stated, where salt goes, water will follow. But how does the loop accomplish this? The answer lies in its peculiar hairpin shape and some very capable cells in the ascending limb of the

ESSAY 18.2

Pressure Gradients in the Kidney Tubule

You will be happy to learn that the mechanism of the kidney is a lot more complicated than it seems. You will also, of course, demand to know the details. You recall that in the human nephron, body fluids filter through the capillary wall into Bowman's capsule and from there through the proximal convoluted tubule (here straightened), the long loop of Henle, the distal convoluted tubule, and then into the collecting duct. As the urine passes through the ascending loop of Henle (on the right), cells lining the loop pump salt out of the urine into the tissue surrounding the loop. In addition, urea moves out of the collecting ducts into the medulla, adding a second solute to the salty fluid there. It then enters the loops of Henle to return to the collecting ducts. This constant circulation means that the loop of Henle and the adjacent collecting duct are always bathed in a salt and urea solution, and it is this solution that sets up the osmotic gradient that withdraws water and solutes from the urine. The ascending loop is apparently impermeable to water since no water passes out with the salt. The descending limb of the loop of Henle and the collecting duct, however, is permeable to water, and water in the urine reaching it freely flows out in response to the high solute concentrations in the surrounding tissue. This water is picked up by the blood vessels that come from the glomerular region, and the hypertonic urine passes to the bladder. So there you are.

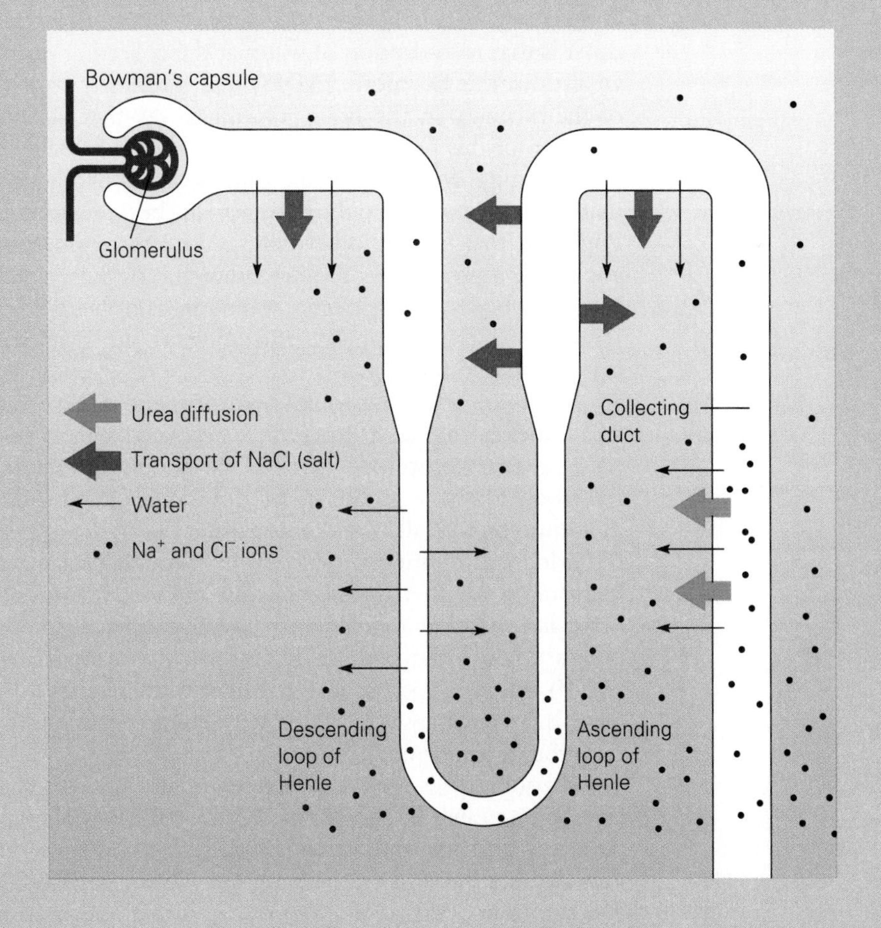

loop. As the crude filtrate makes its way through the loop, cells in the ascending limb actively transport salt out of the filtrate into the fluid surrounding the loop.

We see then, that the continuous cycling of salt in and out of the loop produces the necessary osmotic gradient. Further, active transport also moves some of the salt into the nearby capillary bed, thus extending the gradient. As a result, water continually leaves the descending limb of the loop, moves through the salty medulla, and enters the surrounding capillaries.

Interestingly, the osmotic gradient we've been describing gets a boost in a seemingly odd way. Urea, which is concentrating in the collecting duct, moves out, thereby adding another solute to the salty medulla. It reenters the nephron at Henle's loop, returning to the collecting duct and thereby forming its own cycle. So urea, the primary nitrogenous waste, also plays a role in water retention.

Have you ever noticed how often your rowdy, beer-drinking friends must urinate once the evening gets underway? You may have heard them comment that more seems to be going out than was coming in. The next day they may complain of being thirsty. In your stern lecture to them you should say something about the diuretic effects of ethyl alcohol. You will find them particularly interested when you say that the reason for their thirst is that reabsorption of water at the collecting ducts is controlled by an **antidiuretic hormone (ADH)** that is released from the posterior lobe of the pituitary in the brain. Go on to say that alcohol suppresses the secretion of this hormone, resulting in less water being reabsorbed from the collecting ducts. The urine thus becomes hypotonic, or more dilute than the body fluids, and the body actually becomes somewhat dehydrated. Point out that, on the other hand, ADH secretion also decreases when you drink a great deal of water, since the osmotic concentration of the blood decreases (so drinking water can temporarily "thin" the blood). Drinking water, you can say, also results in the formation of a urine that is hypotonic to the blood.

Then you will want to tell them that the ADH works in opposition to another hormone, **aldosterone,** produced by the adrenal glands. Go on to say that aldosterone increases the amount of sodium reabsorbed into the blood from the filtrate in the distal convoluted tubule. This increase in sodium osmotically draws more water out of the tubules, concentrating the urine and diluting the blood. If, through bleary eyes, they want to know more, say that aldosterone also increases potassium excretion in the urine and therefore regulates the concentration of two critical ions in the body's fluids. Then you can remind them that **atrial natriuretic factor (ANF)** increases sodium in the urine and that water follows osmotically. Refer them to Chapter 21. They will appreciate this information, especially early the next morning.

If your friends are so appreciative that they buy you a sailboat, you may be interested in another bit of practical physiology. If you discover, far at sea, that for some strange reason the boat sinks, you should know that a person on a life raft cannot get the water he needs by eating fish, contrary to popular rumor. The reason is that human urine cannot exceed a salt concentration higher than about 2.2 percent. Fish fluids are not actually this concentrated, but fish are high in protein. This means that in order to get rid of the excess nitrogen, you would have to excrete a lot more water than you could get from eating the fish. Of course, you would do better eating fish than drinking seawater, since the salt concentration in seawater is about 3.5 percent. Some shipwreck survivors claim that the spinal fluid of fish is a good supply of water, but all things considered, you would be better off to have remembered to take water along.

SUMMARY

1. Homeostasis is the tendency of living things to maintain an internal environment within certain limits. This steady state is usually maintained by negative feedback mechanisms, which occur when a system's product slows the system. In positive feedback mechanisms, which are less common, the system's product increases the activity of the system, generating a constantly accelerating system.
2. An example of homeostasis is temperature regulation. If living things become very cold, the water within cells may freeze, disrupting cellular constituents. High temperatures disrupt proteins.
3. Lacking physiological means of regulating body temperature, ectotherms (e.g., fish, amphibians, and reptiles) regulate body temperature behaviorally. The difficulties of temperature regulation are minimal for saltwater fish because the temperature of the water changes little. Temperature regulation is more difficult for freshwater fish. Amphibians move to areas with suitable temperature and may escape heat by estivation (summer hibernation). Reptiles warm themselves in the heat of the sun and seek shade as temperatures rise.
4. Endotherms (e.g., birds and mammals) regulate body temperature physiologically. They have feathers or fur for insulation and a four-chambered heart to efficiently pump oxygen-carrying blood to fuel metabolism for maintaining a constant body temperature. Body temperature is set in the brain's hypothalamus. Among mammals, the body may be cooled by behavior and by evaporation due to sweating, panting, or licking the fur. Heat loss is reduced by constriction of surface blood vessels. Thyroxin (and epinephrine) elevate body temperature by increasing metabolic rate.
5. All animals must rid themselves of metabolic wastes, such as those containing nitrogen from the breakdown of proteins. Natural selection has chosen different nitrogen-containing waste products in species with different water conservation needs. Ammonia, a common waste product among fish, is poisonous and requires a great deal of water to flush it from the body. Because urea, a common waste product among mammals, is less toxic, it can be excreted with less water loss. Uric acid is a common waste product of birds and insects, species with a particular need to conserve water.
6. Animals have the problem of regulating internal water while flushing metabolic waste from the body. Freshwater fish rid themselves of both excess water gained by osmosis and metabolic waste through large glomeruli (clumps of blood vessels) in their kidneys. Saltwater animals tend to lose water by osmosis. The kidneys of many saltwater animals have water-conserving specializations such as small glomeruli and methods of removing excess salt taken in with sea water. Sharks and rays retain urea

in their blood, thereby increasing osmotic pressure and reducing the tendency to lose water to the sea.

7. Different species remove waste and regulate water balance by various mechanisms including diffusion, contractile vacuoles, and excretory systems. In some species, problems of metabolic waste and internal water regulations are solved by the same mechanisms and in others, such as protists, they are solved by different mechanisms.
8. The excretory system of earthworms, coiled tubules called nephridia, adjust the level of metabolic waste and water content. As cilia draw body fluid through the tubules toward the outside, water salts and minerals are drawn back into capillaries surrounding the tubules. The amount of water excreted depends on the amount present in the body fluid.
9. Each of the million or so nephrons in the human kidney consists of a glomerulus (a tuft of blood vessels), Bowman's capsule, and tubules. All but large particles such as proteins and blood cells are filtered from the blood through the thin-walled glomerulus. The filtrate, which contains waste products, water, salts, and nutrients, enters the cuplike Bowman's capsule that surrounds the glomerulus and begins to flow through the proximal convoluted tubule, the loop of Henle, and the distal convoluted tubule. Blood vessels stem from the glomerulus and surround these tubules. Many usuable materials move by diffusion or by active transport from the filtrate within the tubules into the blood vessels. A few substances are secreted into the urine. Material remaining in the tubules, primarily urea and water, flows on to the collecting ducts leading to the ureter and finally to the urinary bladder. The urine leaves the body through a tube called the urethra.
10. Water balance is regulated by thirst and the hormones aldosterone, antidiuretic hormone (ADH), and atrial natriuretic factor (ANF). Aldosterone, secreted by the adrenal glands, causes sodium to be returned to the blood from the filtrate. This draws more water into the blood and concentrates the urine. ADH is released from the posterior pituitary gland. The more ADH secreted, the more concentrated the urine. ADH secretion varies with the osmotic concentration of the blood, which is altered by dietary factors such as water and protein, and is inhibited by alcohol. ANF increases sodium and water in the urine.

KEY TERMS

aldosterone (512)
ammonia (504)
antidiuretic hormone (ADH) (512)
atrial natriuretic factor (ANF) (512)
Bowman's capsule (508)
collecting duct (509)
distal convoluted tubule (508)
ectotherms (500)
endotherms (501)
estivate (501)
filtration (510)
glomerulus (505)
homeostasis (498)
hypothalamus (502)
loop of Henle (508)
negative feedback (498)
nephridia (508)
nephron (508)
positive feedback (499)
proximal convoluted tubule (508)
pyrogen (503)
reabsorption (510)
secretion (510)
urea (504)
ureter (509)
urethra (509)
uric acid (505)
urinary bladder (509)
urine (504)

FOR FURTHER THOUGHT

1. On a chilly morning, would you expect to find a snake basking in the sun or in the protective shade of the bushes? Why?
2. If you place a freshwater fish in your saltwater aquarium, you will shortly find it floating belly up. How might its excretory system have contributed to its untimely demise?

FOR REVIEW

True or false?

1. ___ Body temperature is monitored and regulated by the hypothalamus.
2. ___ Ectotherms regulate their internal temperatures primarily by physiological means.
3. ___ The antidiuretic hormone controls water reabsorption in kidney tubules.
4. ___ Because the sea is a hypertonic medium, saltwater fish tend to lose body water through osmosis.

Fill in the blank.

5. ___ is the tendency of living things to maintain a stable internal environment.
6. Blood sugar levels are maintained by ___ feedback mechanisms.
7. "Warm-blooded" animals are called ___.
8. The human body detoxifies and eliminates poisonous ammonia by ___.

Choose the best answer.

9. An estivating animal would have:
   A. high metabolic rate
   B. increased food requirements
   C. low oxygen requirements
   D. both A and B
10. When human blood temperature is sensed to be too high, the body responds by:
    A. dilating peripheral blood vessels
    B. releasing thyroxine
    C. contracting surface blood vessels
    D. increasing metabolic rates
11. Which of the following correctly matches an animal with its major waste product?
    A. mammals; urea
    B. birds; uric acid
    C. fish; ammonia
    D. all of the above
12. Choose the correct statement.
    A. Blood is delivered to the kidney through the ureter.
    B. Filtration through the glomerulus is a highly selective process.
    C. Usable products are reclaimed from the kidney's tubules by diffusion or active transport processes.
    D. The loop of Henle connects the kidney to the bladder.

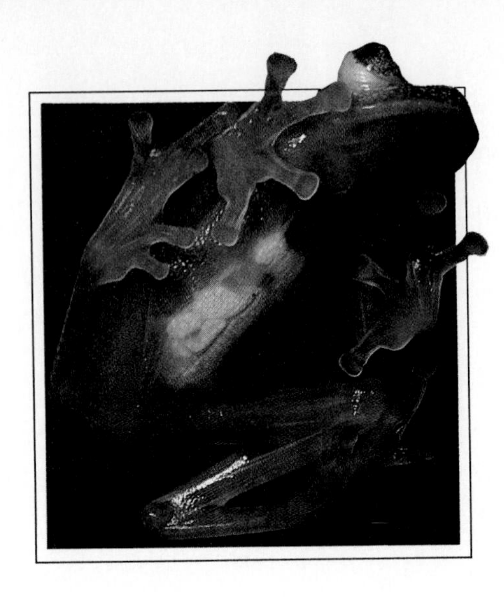

CHAPTER 19

The Respiratory, Circulatory, and Digestive Systems

Overview

Respiratory Systems
Circulatory Systems
Digestive Systems

Objectives

After reading this chapter you should be able to:

- Compare the mechanisms by which aquatic and terrestrial animals obtain and deliver oxygen.
- Describe the mammalian respiratory system and explain how humans ventilate their lungs.
- Compare the circulatory systems of small and large animals.
- Discuss the structural variation in representative vertebrate hearts and describe how blood moves through each.
- Explain what causes the vertebrate heart to beat.
- Trace the path of nutrients through a vertebrate digestive system and compare this mode of digestion to that of representative invertebrates.
- List the structures in the human digestive system and describe how each functions during digestion.

A whole host of problems rained down on the early struggling life on a swiftly changing planet. There was the devastating weather, the withering sunlight, competition at every turn, and there were predators. One solution to survival in such a place was the specialization and efficiency that comes with multicellularity. With multicellularity came larger size, and with larger size came a whole host of new challenges. One of those challenges involved acquiring certain critical molecules and then moving them to where they were needed within the organism. Here, then, we will take a look at three systems that helped solve these sorts of problems.

RESPIRATORY SYSTEMS

Life on the planet evolved in a sea of gases. Eventually, early life forms began to change that atmosphere by "polluting" it with a deadly gas. In time, that gas accumulated until it posed a real threat to the continuance of life (Figure 19.1).

The gas corroded all sorts of hopeful experiments; the air seemed to rot, digest, and break down everything it touched. But life ebbed and flowed, this way and that, in mindless shifts and evolutionary eddies, and in time, there appeared forms that not only could survive the corrosive gas we now call oxygen, but could actually utilize it in metabolic pathways—and so we breathe.

Breathing is often equated with **respiration,** but actually, breathing is only one of three processes involved in respiration. Breathing is one method of **external respiration,** the exchange of gases between the atmosphere and the blood. The next process, **internal respiration,** involves the exchange of gases between the blood and the cells. **Cellular respiration** (Chapter 6), the use of oxygen and the production of carbon dioxide by the cells, is the third process. Here, we are primarily concerned with external respiration.

FIGURE 19.1

The conditions under which life evolved have changed over time. The life-sustaining oxygen that now graces our planet was once a serious threat to most forms of life.

The Various Ways Species Get Oxygen

As life evolved, oxygen was first delivered by diffusing across the moist membranes of living cells, a slow process, indeed, and only effective for very short distances. In time, more effective mechanisms appeared, usually involving the development of an increased membrane area over which oxygen could move into the body and more effective ways for transporting the oxygen around within the body.

Because life is so splendidly varied across the far reaches of the planet, and because so many life forms depend on oxygen, it is not surprising that living things have developed a variety of means to get oxygen. We will consider only a few representative cases, but we will again see that natural selection can take many paths to the same adaptive end.

Aquatic Animals

Small aquatic animals have relatively little problem in getting oxygen to their cells. It simply diffuses in through their body surfaces. Planarian flatworms, for example, have no organs for taking in oxygen, but their thin, flattened shape provides them with a large surface area in relation

FIGURE 19.2
Skin breathing involves gases passing directly through the body wall, as we see in amphibians and invertebrates, such as this planarian, a flatworm.

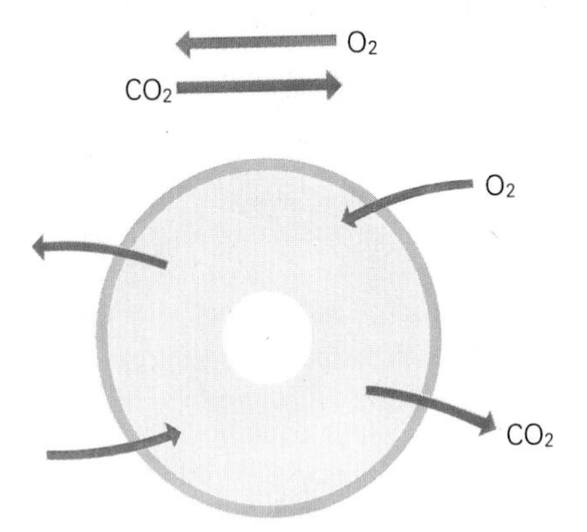

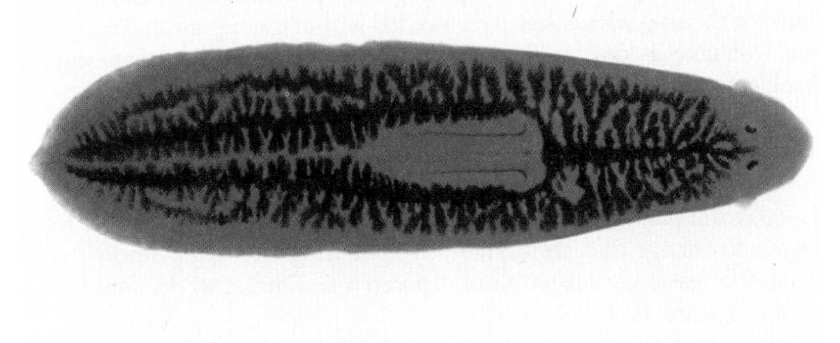

to their mass. Diffusion of oxygen from the water into their bodies gives them all the oxygen they need for their slow-paced lives (Figure 19.2).

Larger aquatic animals have proportionately less surface area relative to their mass; hence, they need some specialized system of obtaining and delivering oxygen. Consider the mud puppy (Figure 19.3), a kind of salamander that keeps its gills (a larval characteristic) as an adult. (Most salamanders as adults breathe through their skin and mouth lining.) As

FIGURE 19.3
Some aquatic animals have external gills, thin-walled structures protruding from the body through which gases can pass.

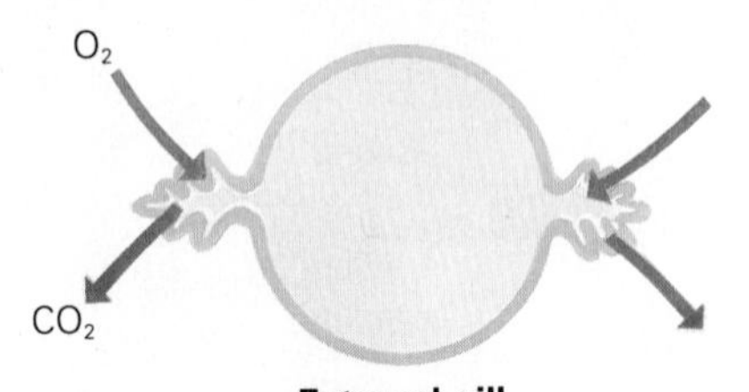

they move about on the bottoms of ponds and lakes, their gills protrude into the water and extract oxygen from it. The gills are very thin-walled with a large surface area and are permeated by tiny vessels; thus, oxygen can enter the body fluid easily and be carried quickly to the rest of the body.

Although fish are much more complex creatures than are mud puppies, they also acquire oxygen through gills. The gills of fish are also essentially comprised of thin protrusions or shelves (lamellae) that provide a very large surface area (Figure 19.4). The cell membranes are very thin and readily permit the passage of gases across them. Lying in close proximity to these cell membranes are tiny, thin-walled capillaries where carbon dioxide produced in the body diffuses into the water, and oxygen from the water enters the blood. The oxygen is then transported to the rest of the body. Remember that there are normally higher carbon dioxide levels within the body and higher oxygen levels outside it, and it is these differences that determine the direction of flow of the dissolved gases according to the principles of diffusion. Thus, the higher level of carbon

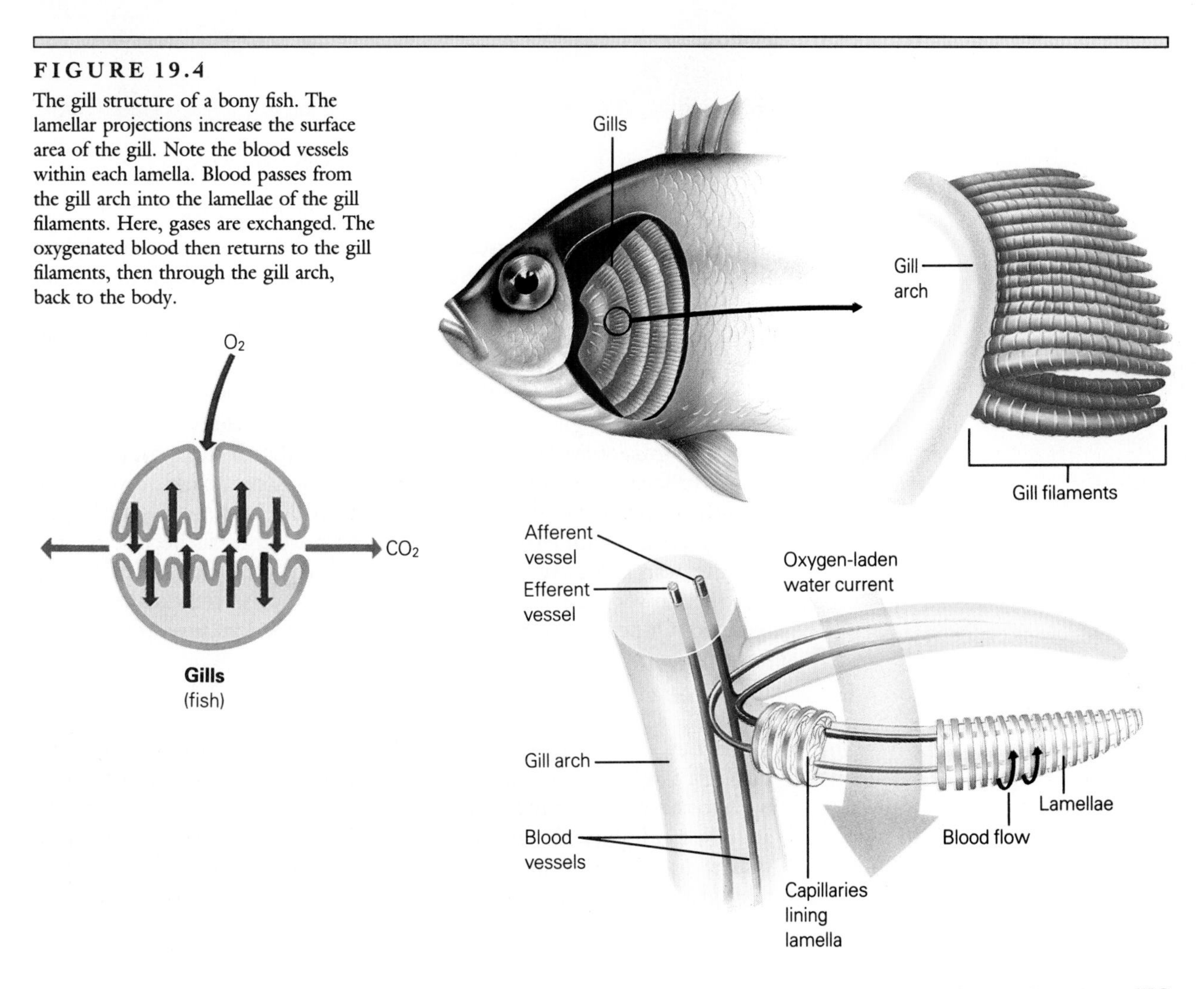

FIGURE 19.4
The gill structure of a bony fish. The lamellar projections increase the surface area of the gill. Note the blood vessels within each lamella. Blood passes from the gill arch into the lamellae of the gill filaments. Here, gases are exchanged. The oxygenated blood then returns to the gill filaments, then through the gill arch, back to the body.

FIGURE 19.5

It was once assumed that sharks must constantly stay in motion to keep oxygen-laden water flowing over their gills. However, divers have recently reported that some species of shark sleep, lying dormant on the bottom. Thus, their physiology must be much more complex than we had thought.

dioxide within the body results in carbon dioxide moving outward, and the lower level of oxygen in the body means oxygen moves inward from the surrounding water.

The fish pumps water over its gills as it enters the mouth and leaves from under the gill covering. Some fish, such as sharks, lack the muscles for pumping water along this pathway and must swim to keep water flowing over their gills. It was long thought that all sharks were destined to swim virtually every minute of their lives—that if they stopped, they would suffocate. Whereas this may be true for some sharks, divers have recently found sharks sleeping motionless on the bottom (Figure 19.5). No one yet quite understands how this can be, but it has been suggested that they can supersaturate their blood with oxygen before sleeping and that they lower their metabolic rate at such times so that less oxygen is required. Or perhaps they simply sleep in oxygen-laden currents with their mouths open.

Insects

Terrestrial insects generally solve the oxygen problem by breathing through a **tracheal system.** The system is comprised of tiny tubes that open to the air and permeate their body tissue. The openings to the system are called **spiracles.** As shown in Figure 19.6, oxygen enters the tracheal system directly from the air and, aided by the insect's bodily movement, moves through the tracheae (plural) deep into the body tissue. Each trachea

FIGURE 19.6

The respiratory system of an insect. Note the spiracles in the animal's side through which air enters, eventually moving into a highly branched tracheal system and, in some species, into large air sacs. The insect's movement hastens the flow of air through these channels.

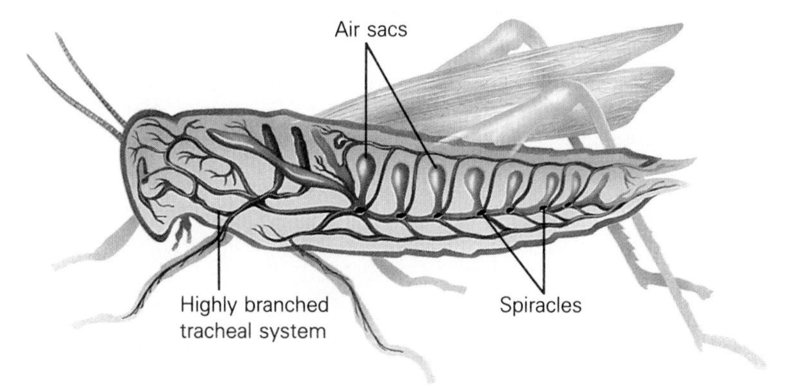

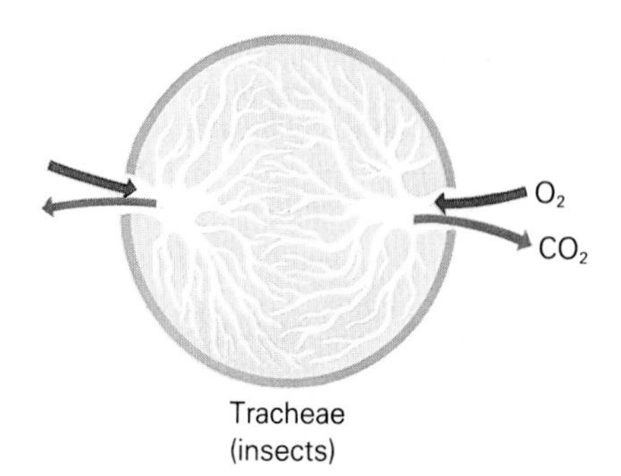

ends amidst groups of cells or in expanded **air sacs.** The tracheal network is so extensive that no cell lies far from an oxygen source.

The insect's system of getting oxygen has probably placed a limit on how large insects can become. This is because a tracheal system cannot deliver oxygen over a great mass. Actually, a tracheal system can work only for organisms no larger than, say, above five centimeters. Thus, an evolutionary direction taken eons ago may be responsible for the fact that we don't have to contend with 8-pound mosquitoes.

Mammals

Among mammals, external respiration takes place by means of special pouches called **lungs.** External respiration is accomplished as oxygen-ladened air is brought into the lungs by **ventilation,** or **breathing.** Air passes from the large **trachea** into the branched **bronchi** (singular, *bronchus*) and on into increasingly smaller **bronchioles** that terminate in the tiny saclike **alveoli,** which is where gas exchange occurs. The alveoli are so numerous that they give a spongelike quality to the lung (Figure 19.7). The total area of the alveoli in human lungs, by the way, is about equal to the area of a tennis court—a large exchange surface indeed. A list of human respiratory structures and their functions is found in Table 19.1. (Essay 19.1 describes the effects of smoking.)

The surface of the lungs is moist, as is the respiratory surface of any animal. The moistness is necessary because oxygen must dissolve before it can cross these delicate membranes of the alveoli and enter the bloodstream. The blood transports the oxygen to the body's tissue where it diffuses into the cells. In the cells, the oxygen has the humble but critical role of picking up spent hydrogen atoms from the electron-transport chain, thereby producing metabolic water.

Meanwhile, the carbon dioxide that has been produced in metabolic reactions throughout the body diffuses from the cells into the capillaries,

FIGURE 19.7

The human respiratory system. The contraction of the diaphragm creates a partial vacuum in the chest and causes oxygen-filled air to rush in to fill the void. The air is warmed and moistened as it travels through the upper air passages. It passes through increasingly smaller passages until it reaches the very thin-walled, saclike alveoli, which allow oxygen to pass into the bloodstream and carbon dioxide to pass into the lungs, from where it is expelled.

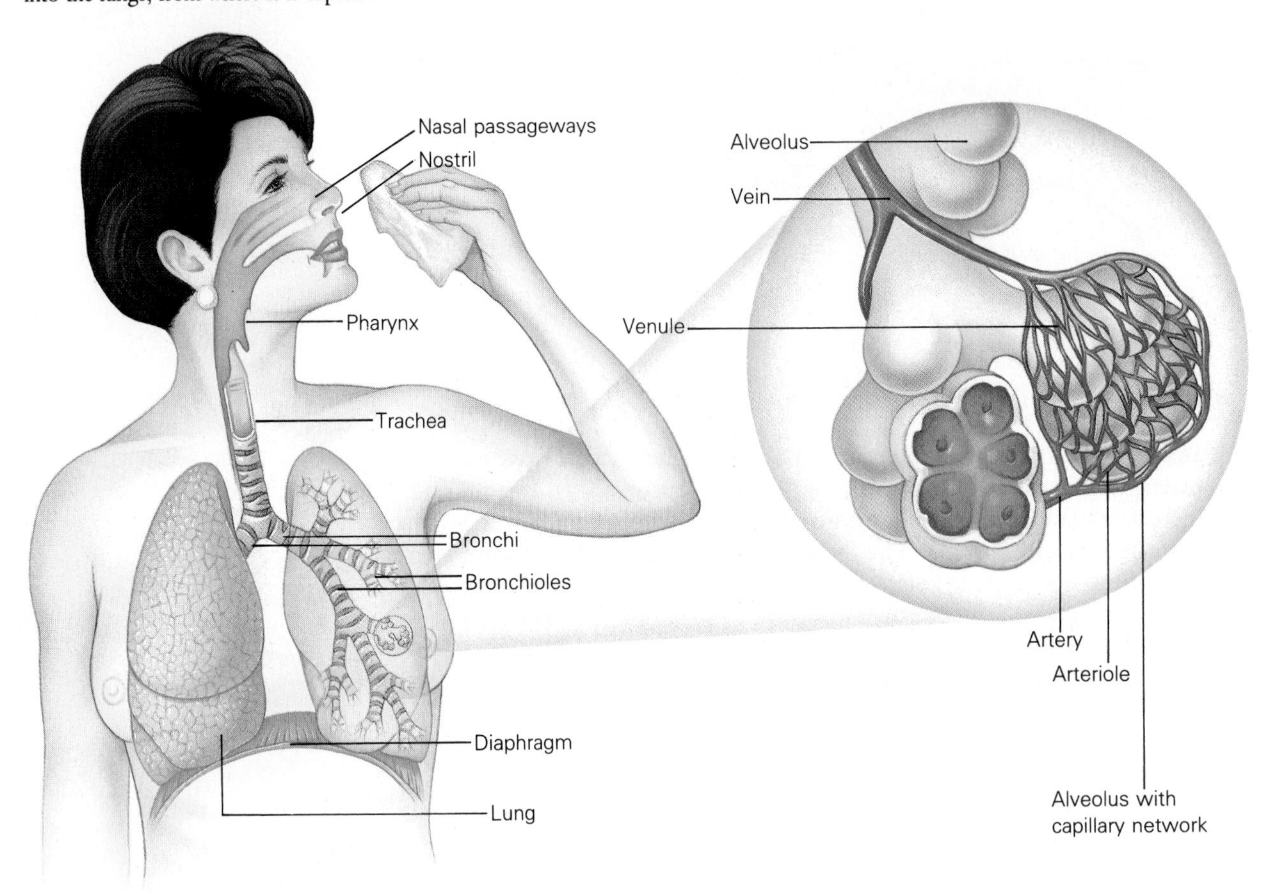

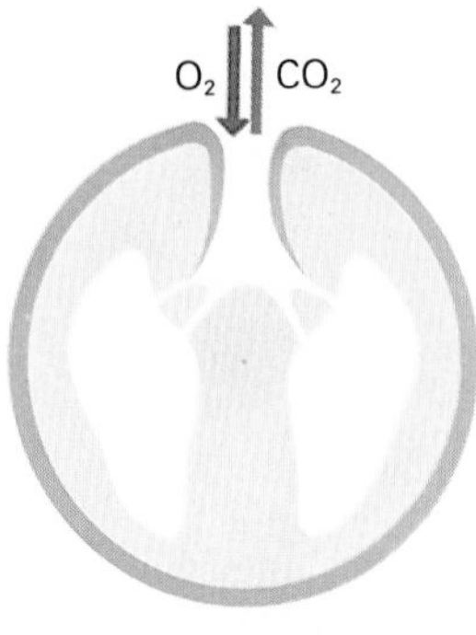

TABLE 19.1
Human Respiratory Structures

STRUCTURE	FUNCTION
Nose	Filters, warms, and moistens incoming air.
Pharynx	Adjustable passageway for air and food.
Trachea	Passageway.
Bronchi	Passage of air to each lung.
Bronchioles	Passage of air to alveoli.
Alveoli	Gas exchange.

ESSAY 19.1

The Joy of Smoking

*I*t is interesting to watch an old film in which the lead characters smoke. Generally, lead characters today don't smoke. Smoking is increasingly characteristic of poorer and less-educated people, or those who started before the hazards were known. (Interestingly, it seems to me that there are regional differences. You don't see many people smoking in Florida, Colorado, and California, while you may see it more often in Missouri, Illinois, and Pennsylvania.)

Smoking is hard to explain even on the basis of smell and cost, but on the basis of health it is incomprehensible. The American Cancer Society tells us that a person aged 25 who smokes two packs of cigarettes a day will live about 8½ years less than a nonsmoker. Furthermore, the end of the smoker's life may be marked by extreme pain due to lung cancer, as well as cancer of the mouth, bladder, pancreas, and esophagus. Even in the absence of cancer, the smoker may be severely debilitated by emphysema when thickened bronchioles and alveolar walls cause trapped air to permanently inflate the lungs. The inefficiency of the damaged lungs may affect both the heart (increasing the risk of coronary heart disease and heart attacks) and brain (bringing on behavioral changes, including sluggishness and irritability). Pregnant women who smoke increase the risk of stillbirths, and those who deliver tend to have smaller, sicker children.

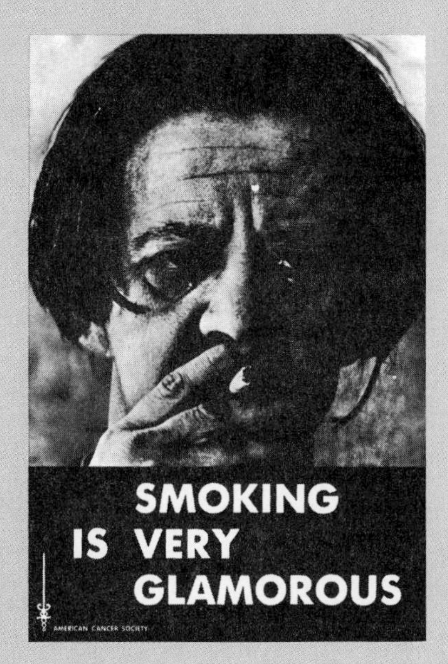

Fortunately, if you quit in time, the damage is largely reversible. Within a year you are markedly at less risk for coronary heart disease and 10 to 15 years later, you are at no greater risk for premature death and coronary heart disease than a nonsmoker. The coughing may stop after a few weeks and, unless the damage is too great, lung function is likely to improve over time. The bottom line is simple: if you smoke, quit.

and from there it is carried to the lungs. In the lungs, CO_2 diffuses through the tiny vessels out of the blood and into the interior of the lungs. From there it is exhaled into the environment.

In humans as well as other mammals, air is drawn into the lungs when the chest cavity is expanded, just as air is sucked into an opening bellows. The expansion of the cavity is due primarily to the contraction of two kinds of muscles. One, the diaphragm, is a broad sheet of muscle that separates the chest cavity from the abdominal cavity. When relaxed, the diaphragm is dome-shaped, like an inverted soup bowl. When contracted, the diaphragm flattens and lowers, thereby increasing the chest cavity from top to bottom. The contraction of other muscles, those between the ribs, causes the rib cage to be lifted upward and outward, thereby enlarging the chest cavity from back to front. The resulting increase in size of the lung cavity causes a pressure difference between the cavity and the atmosphere, causing air to move into the lungs.

FIGURE 19.8

Breathing in humans occurs as the diaphragm contracts and flattens, and as the ribs raise, causing the lungs to fill with air. As the ribs lower and the diaphragm relaxes and rises, air leaves the lungs.

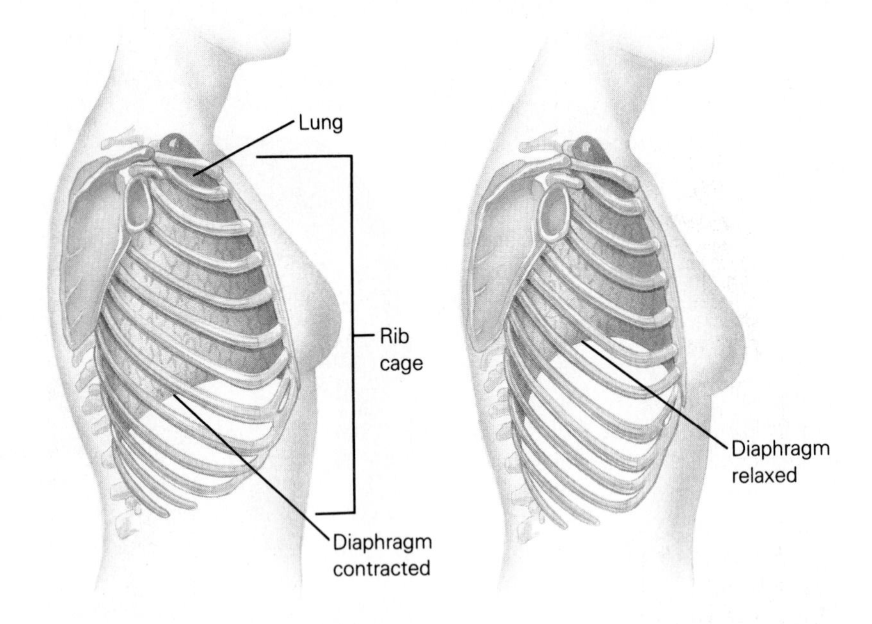

FIGURE 19.9

The primary respiratory control center is located in the medulla of the brain. Breathing rate is adjusted primarily by changes in the level of carbon dioxide in the blood. The medulla, itself, is very sensitive to carbon dioxide levels, but it also receives input from sensory receptors in the carotid arteries, which lead to the head and to the aorta, the large artery leaving the heart.

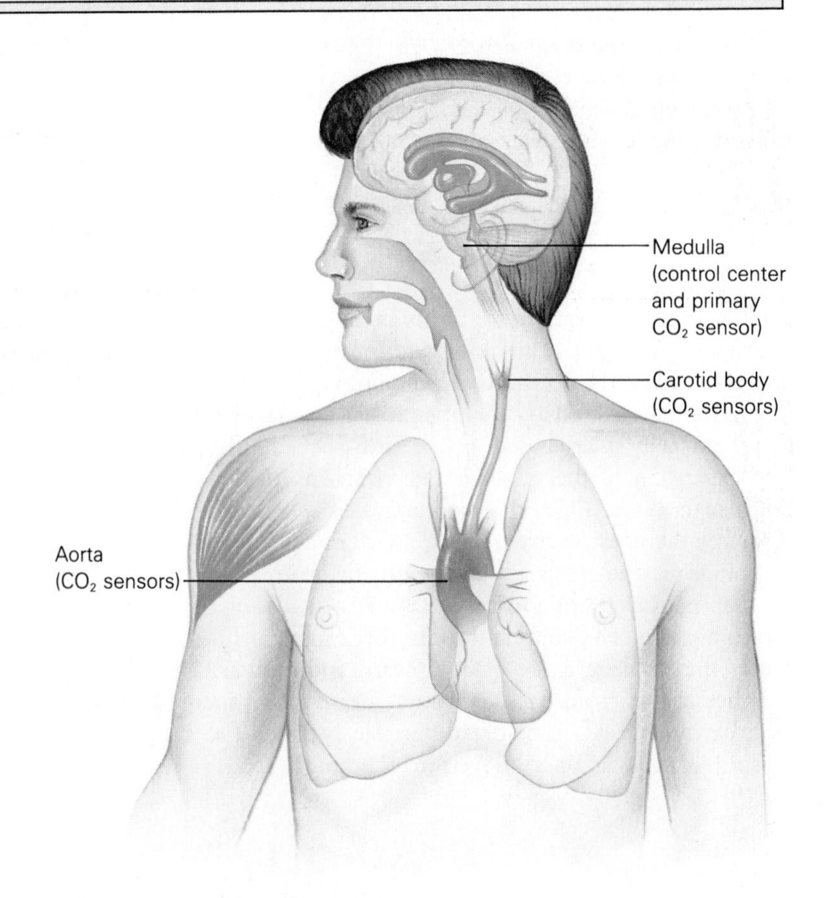

As the diaphragm and rib muscles relax (Figure 19.8), air is allowed to move out of the lungs, helped along by the natural elasticity of the lungs themselves.

Fortunately, we continue to breathe, even when we are asleep or thinking about other matters—a great convenience for those with plans. The continued breathing is due to the activity of certain cells within the medulla, located at the brain stem. These, in turn, are influenced by other neural centers, particularly in the aorta. The medulla, the sensor in the aorta, and the sensor in the carotid artery (called the carotid body) all respond to increasing CO_2 levels in the blood. When they detect rising CO_2, the medulla sends an impulse to the breathing muscles and we breathe faster (Figure 19.9).

CIRCULATORY SYSTEMS

Not only must oxygen and carbon dioxide be transported throughout the body, but a host of other substances must move through the body as well. These include nutrients, various wastes, and hormones. Thus, the development of an efficient circulatory system has been critical. Of course, some species have less need of such a system, and those that have them have developed a variety of mechanisms for moving substances about.

Circulation in Small Animals

A number of factors may influence the type of circulatory systems in living things. For example, size may be important. In one-celled organisms and small muticellular species, there may be no need for a circulatory system at all since materials can easily diffuse from one part of the body to another and a large part of the total membrane area of the body is exposed to the outside environment.

The problem becomes more critical in increasingly larger animals, and it has been met in a variety of ways. Cnidarians have a body wall with only two cell layers that surrounds a large saclike gut, the **gastrovascular cavity,** that is connected by a single opening to the external world. This gut functions not only in circulation, but also in digestion. Once absorbed into the animal, materials can be distributed by an unusual type of circulatory system. Some of the cells in cnidarians are amoeboid. That is, they can crawl and squirm their way through the body's tissues. When they reach the inner layer of the pouchlike animal, they surround and engulf food particles just as an amoeba does. Then they withdraw into the body with the captured nutrients and distribute the food.

Flatworms also have saclike guts with a single opening used for food intake and the elimination of waste. Since the body is flattened, no cell is very far from the gut where digestion takes place; thus, nutrients can easily diffuse to the areas where they will be utilized. Small animals, we see, have little need for the complex systems of their larger colleagues. But this section is supposed to be about circulation, so let's move to the larger creatures.

FIGURE 19.10

In the crustacean's and insect's open circulatory system, blood is pumped through vessels to open sinuses through which it gradually makes a return to the heart. Before its return trip in the crustacean, it is shunted through the gills for oxygenation.

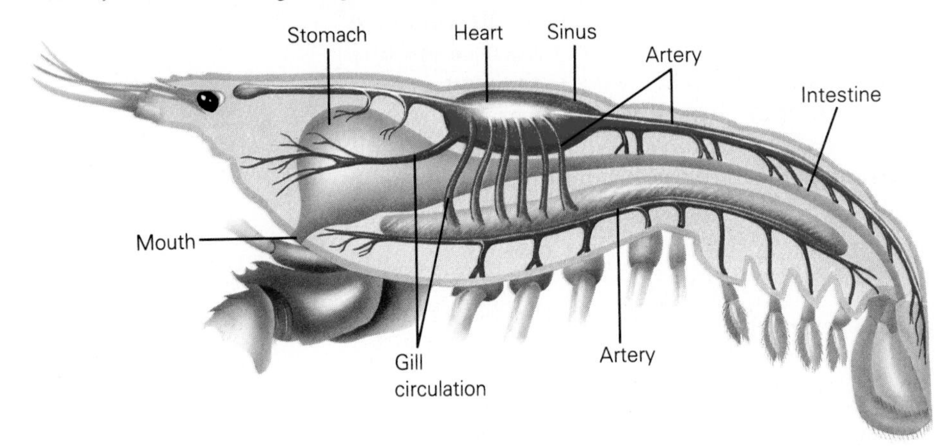

Circulation in Larger Animals

We can begin by noting that circulatory systems may be classified according to whether they are *open* or *closed.* In **open circulatory systems,** the blood (actually, a generally colorless fluid called *hemolymph*) does not remain enclosed in vessels. At certain places, it leaves the vessels to seep through the tissues. For example, in mollusks, such as clams and snails, and in arthropods, such as insects and crayfish (Figure 19.10), a tubular heart pumps hemolymph through vessels for a way, but then it empties into open cavities, or **sinuses.** From there it seeps or percolates through tissues and then collects again in a large sinus around the heart. When the heart relaxes, the hemolymph flows in through tiny, one-way pores in the heart wall. When the heart contracts, the pores close, and hemolymph is pumped out through the vessels. In open circulatory systems, it slowly seeps through the tissues, so the process is generally too inefficient for larger animals in which the more rapid movement of channeled blood would be favored.

In **closed circulatory systems,** blood remains enclosed in vessels (with a few inevitable exceptions) and can therefore move more swiftly with its life-giving burden. Among the most primitive of the closed systems are those of the annelids, the segmented worms. In the earthworm, for example, the heart consists of a series of five tubular rings, the **aortic arches,** that accepts blood flowing forward along a large dorsal vessel, and by contracting, forces blood to flow backward through a large ventral vessel (Figure 19.11).

FIGURE 19.11

The closed circulatory system of the earthworm. The dorsal vessel sends blood downward through the pulsating tubular rings of the aortic arches and then backward through the ventral vessel. Note the branches of the major vessels that lead into other parts of the body. The blood remains enclosed in vessels, thus the system is "closed."

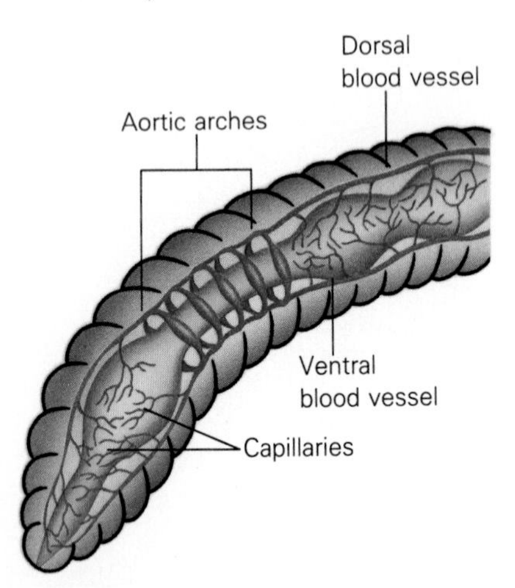

The Vertebrate Vessels

A large muscular heart pumps blood into large muscular **arteries,** which in turn branch into increasingly smaller **arterioles** and finally into **capil-**

ESSAY 19.2

The Incredible Blood-Brain Barrier

One of the most vexing problems in medicine is demonstrated by the fact that physicians can't treat many brain diseases because the brain won't allow their medicines in, yet a smoker may feel a rush within two or three seconds of taking a puff of a cigarette. The nicotine is allowed in, the medicine isn't.

Furthermore, cocaine readily moves into the brain, as does alcohol. Once there, of course, such molecules act to enhance our charm and wit. But why is it they can get into the brain when so many other kinds of molecules can't?

The reason is, they are fat-soluble. That is, they can be dissolved in fatty or oily substances. Those that are water-soluble, as are many kinds of medicinal drugs, can't get across the capillaries of the brain, so they are carried in the bloodstream to the area of the brain and right out again.

Since water-soluble materials generally cross capillaries easily, why can't they do so in the brain? The capillaries of the brain, it turns out, are constructed differently. Whereas other capillaries are somewhat loosely joined and leaky, those of the brain are tight-knit and overlapping, sealing the brain against many of the assorted molecules circulating through the rest of the body.

The blood-brain barrier does not entirely exclude water-soluble molecules, however. For example, the brain needs a continual supply of glucose, a water-soluble molecule. It also needs amino acids, proteins, and iron—all water soluble. These molecules, however, cross via very specialized means. The glucose, for example, squeezes through protein-lined pores that form a channel perfectly shaped to allow their passage. Other substances, it is believed, are captured by special proteins that protrude into the capillary and are brought into the brain and released. Yet others bind to carrier molecules in the blood, which then move into the brain tissue where the carrier is cleaved from its load enzymatically.

It is not surprising that the blood-brain barrier exists in all vertebrates. After all, the brain is a complex organ and its delicate processes must be carefully shielded from the hodgepodge of natural and artificially introduced materials that may be circulating in the bloodstream at any time.

laries. The walls of the capillaries are so thin that oxygen, carbon dioxide, nutrients, and metabolic waste are able to pass through them. (Essay 19.2 describes some special problems getting materials into the brain.) From the capillary beds the waste-laden, oxygen-depleted blood flows into larger vessels called **venules,** and then into increasingly larger **veins** that return the blood to the heart. (Arteries, then, carry blood from the heart, and veins return it to the heart.) In humans, the length of the entire system is estimated to be between 50,000 and 60,000 miles, 70 percent of which is capillaries.

Figure 19.12 shows the major blood vessels of the human body. Where are the pulmonary vessels? What is the large network below the heart on the right side of the body? You might now wonder why the good guys survive so many shoulder wounds in so many bad Westerns. According to the figure, where is a better place to be shot? Some of these vessels,

FIGURE 19.12

The major blood vessels of the human body. The arterial (red) and venous systems (blue) are connected by capillaries. Generally, the arteries (A.) are thicker-walled than the veins (V.), because they are under intense pressure by blood being pumped from the heart. The pressure is dissipated in the immense system of capillaries, some so tiny that red blood cells must move in jerky movements, single file, between the cells. Obviously, blood in the veins of the lower extremities does not have the force to move back up the legs. This blood is forced along largely by muscular contractions.

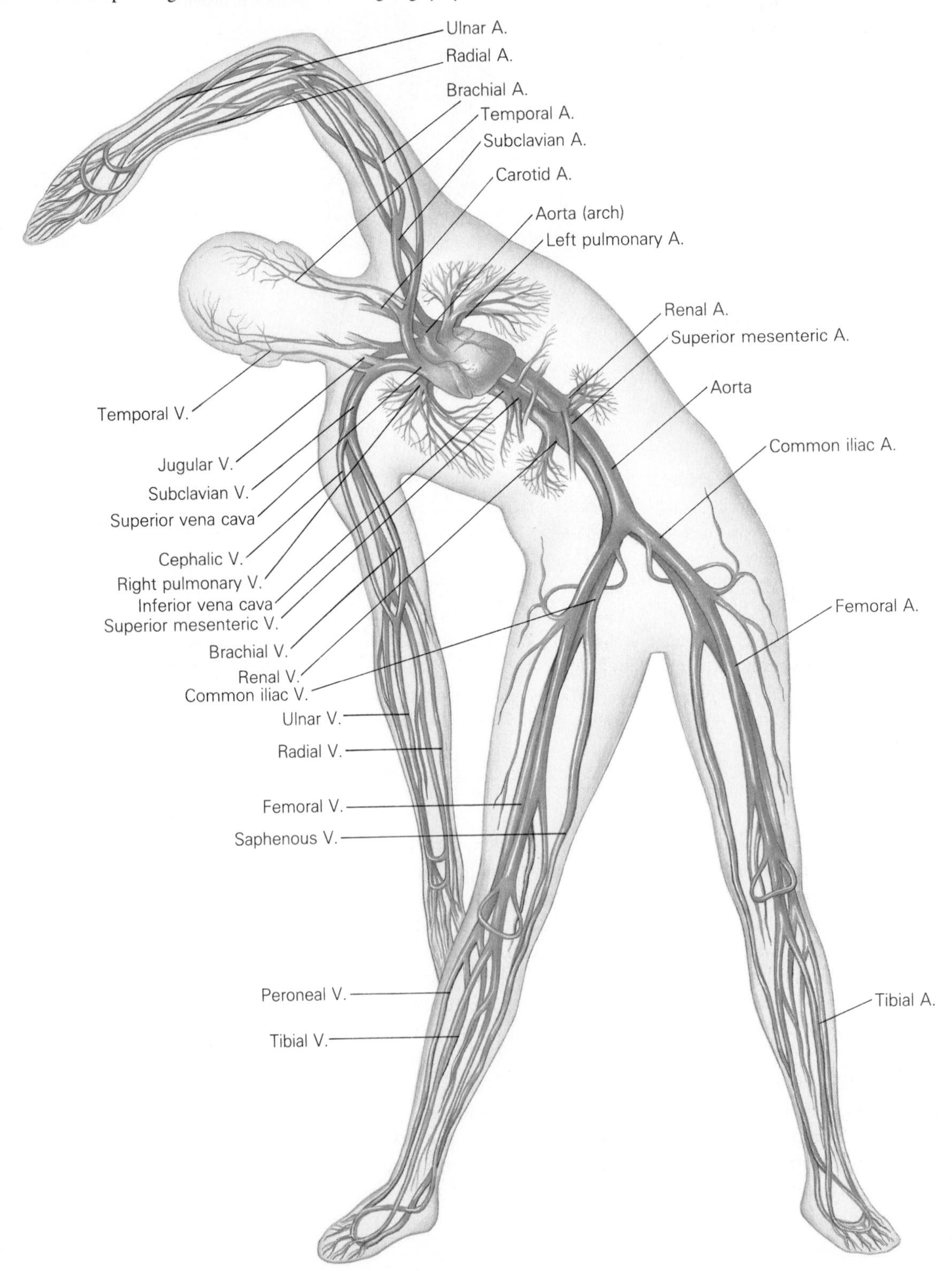

especially the larger ones, are likely to be found in about the same place in any human body. This element of predictability is undoubtedly appreciated by surgeons. But other vessels are extremely variable, as you can demonstrate with an indelible pen by tracing the veins in the hands of the people sitting around you.

Arteries and Blood Pressure

Arteries carry blood away from the heart. They are not simple conduits, however; one of their major functions is to maintain blood pressure. The great vessel that receives blood from the left side of the heart, the aorta, receives the full impact of the heart's powerful surges. The sudden swell of blood during **systole** (contraction of the heart) expands the elastic walls of the aorta. As the blood moves from the last contraction of the heart onward, the elastic walls of the expanded aorta contract automatically. Therefore, during **diastole** (between contractions, when the heart is filling), blood pressure remains high because of the force of the contracting aorta on the remaining blood in the vessel. However, there is some drop in pressure as the ventricles fill. This is called the **diastolic pressure.** A typical **systolic pressure** is 120 mm Hg; a typical diastolic value, 80 mm Hg. The blood pressure in this case would be "120 over 80" (Figure 19.13). (Millimeters of mercury—mm Hg—is a standard way of expressing blood pressure.)

Blood

The liquid portion of the blood is a straw-colored fluid called **plasma.** The plasma contains water, proteins, hormones, nutrients, wastes, and various materials, as well as several types of cells and cell fragments.

FIGURE 19.13

With the sphygmomanometer, a cuff fills with air until it collapses the brachial artery. As the air is slowly released, the movement of blood into the artery is detected by a stethoscope, giving the systolic pressure. When the sound of the blood disappears, the diastolic pressure is registered.

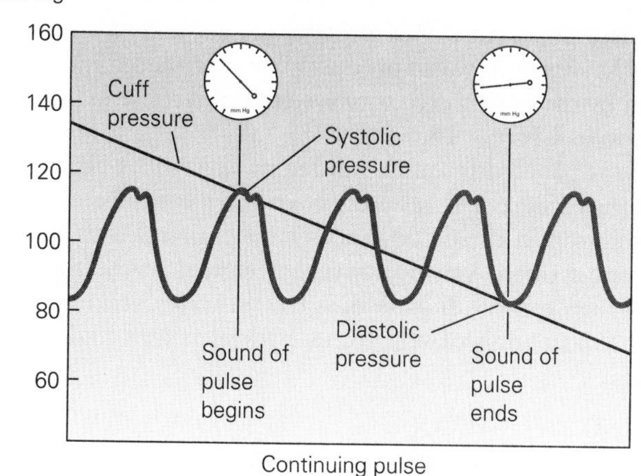

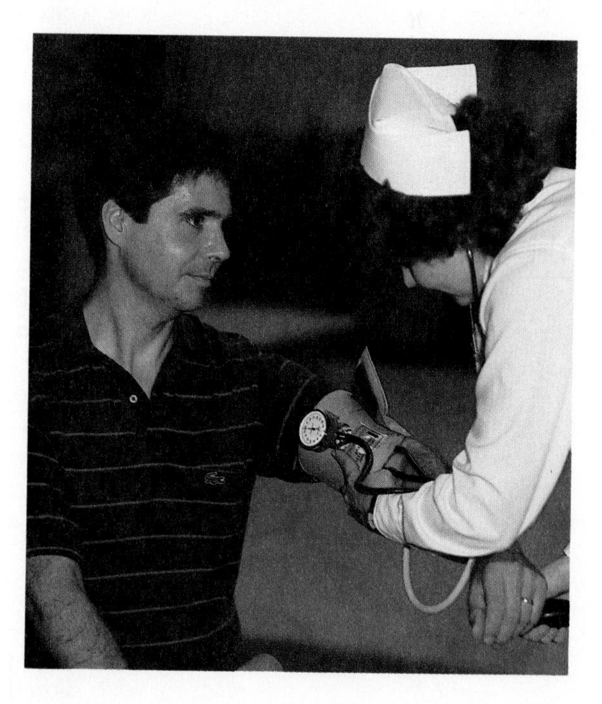

FIGURE 19.14

Red blood cells are rounded and biconcave, pinched in on both sides.

The red blood cells, or **erythrocytes** (Figure 19.14), function to transport oxygen. The red blood cells are so numerous that the blood appears red. The red color comes from a protein-iron compound called **hemoglobin** that fills each red blood cell. Hemoglobin is able to combine reversibly with oxygen. The result is that human blood is able to carry far more oxygen than could be dissolved in plasma alone. Hemoglobin readily combines with oxygen in body parts where the oxygen concentration is high (such as in the lungs). Where oxygen is low (such as in the tissues), hemoglobin tends to give up its oxygen. Hemoglobin is comprised of four subunits, each of which has a protein chain (globin) and *heme* (a complex carbon ring structure that contains iron). Each of these subunits can hold one molecule of oxygen.

Human red blood cells live about 120 days. When their time is up, large white blood cells called macrophages engulf and digest them. No one knows how the macrophages identify these old cells.

The plasma also carries cells in addition to erythrocytes, such as **leukocytes,** or white blood cells, which function mainly in combating invasion or infections, and cellular fragments called **platelets,** which function in clotting.

Another of the blood's jobs is to transport wastes, primarily carbon dioxide and various compounds. This process begins with a series of chemical reactions that result in waste products being produced in a form that can be handled by the body's excretory systems, as we saw in the case of nitrogenous waste in Chapter 18.

Carbon dioxide is handled in quite a different way—it is carried by the blood to the lungs where it is released. Some of the carbon dioxide is transported in solution in the blood, and some combines with hemoglobin. However, most of the carbon dioxide is carried in the form of bicarbonate ions. When carbon dioxide dissolves in water, it forms carbonic acid, which almost immediately forms hydrogen ions and bicarbonate ions. That is:

$$\underset{\text{Dissolved carbon dioxide}}{CO_2} + \underset{\text{Water}}{H_2O} \leftrightarrow \underset{\text{Carbonic acid}}{H_2CO_3} \leftrightarrow \underset{\text{Hydrogen ions}}{H^+} + \underset{\text{Ionic bicarbonate}}{HCO_3^-}$$

The speed at which the carbon dioxide changes to carbonic acid is increased within the red blood cells by the activity of the enzyme **carbonic anhydrase.** The bicarbonate ions that form from the carbonic acid diffuse out of the red blood cells and are carried to the lungs. In the capillaries of the lungs, the process is reversed. Carbonic anhydrase rapidly converts carbonic acid into carbon dioxide and water.

The Vertebrate Heart

The muscular pump called the **heart** varies widely in invertebrates and is related to the life-style of the organism. There is less variation in vertebrates, but there is an important progression regarding the number of chambers.

The fish heart has only two chambers (Figure 19.15). One is a thin-walled **atrium** that accepts blood after it has been circulated though the body. Contraction of the atrium forces blood into the other chamber, the **ventricle.** The atrium doesn't have to be very strong because it doesn't take much work to pump blood only to the next chamber. The ventricle, however, must generate tremendous pressure, since it must force the blood through the large artery called the **aorta** into the tiny mesh of the gill capillaries where gases are exchanged, and then through the rest of the body. From the major tissues of the body it reenters the veins. The fish ventricle, then, must be able to pump blood through two capillary beds—one in the gill and one in the body. Friction is high in these beds due to the enormous surface area of the interior of the tiny vessels of such an expansive network. Not surprisingly, the ventricle is extremely thick and muscular.

The amphibian heart is a bit more complex in that it has two atria. One receives oxygenated blood (blood high in oxygen) from the lungs and the other receives deoxygenated blood (high in carbon dioxide) from the rest of the body. The two are mixed in the single ventricle, so the system is not particularly efficient; but it is efficient enough for the rather lethargic frog (lethargic till you try to catch one).

Reptiles go one step better and have two atria and one ventricle, but the ventricle is partially divided by an incomplete wall. (The wall may be complete in crocodilians.) Because of the incomplete ventricular septum (wall), there is some mixing of blood from the two sides of the heart. Still, this system is more efficient than the frog's, since one side of the heart pumps blood through the lungs and the other side pumps it through the rest of the body.

In birds and mammals, the right and left sides of the heart are completely separated. With an efficient circulatory system, it is easier for these endothermic ("warm-blooded") animals to maintain their body temperature by moving the blood to the heat-generating cells and distributing the heat to the rest of the body when environmental temperatures are low.

What does the separation of the two halves of the heart have to do with efficiency? The two halves are essentially two hearts, one serving the lungs, the other serving the body (Figure 19.16). When CO_2-laden blood enters the right atrium from the large veins (the *superior vena cava* from above and the *inferior vena cava* from below), it is pumped from the atrium to the right ventricle, which sends the blood through the *pulmonary arteries* to the lungs. Here the blood picks up oxygen and releases carbon dioxide

FIGURE 19.15

Various kinds of vertebrate hearts, ranging from the two-chambered to the four-chambered. The simple to complex arrangement is assumed to trace the evolutionary development of the "warm-blooded" heart (that of birds and mammals). Note the increased efficiency in separating oxygenated and deoxygenated blood. These schemes are highly generalized and exceptions do exist. For example, among reptiles, the crocodile heart is more strongly divided than that of snakes. In birds and mammals, the developing tubular heart forms a loop, resulting in the atria coming to lie headward. In our illustration, however, the heart appears as if the looping has not occurred.

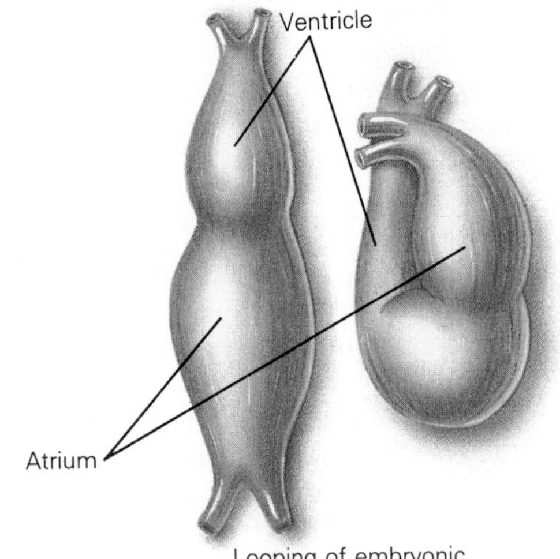

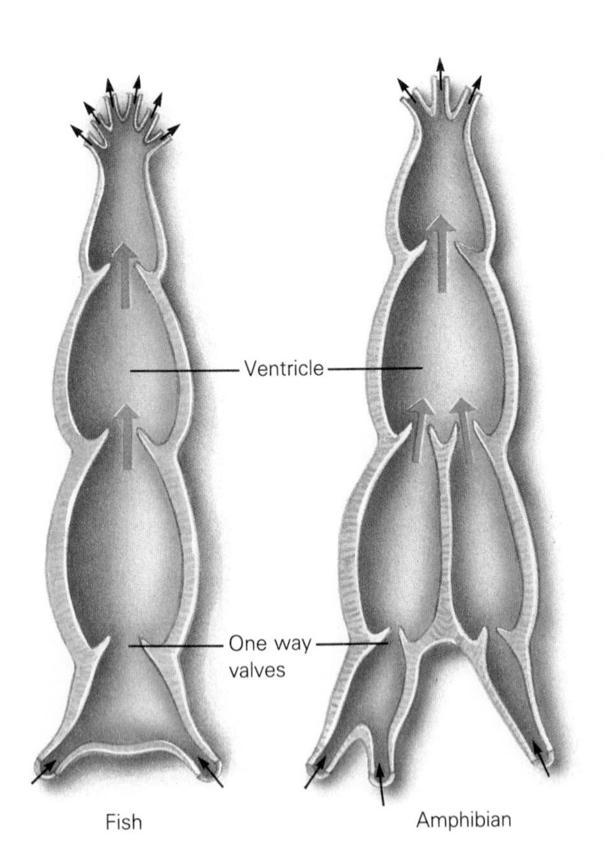

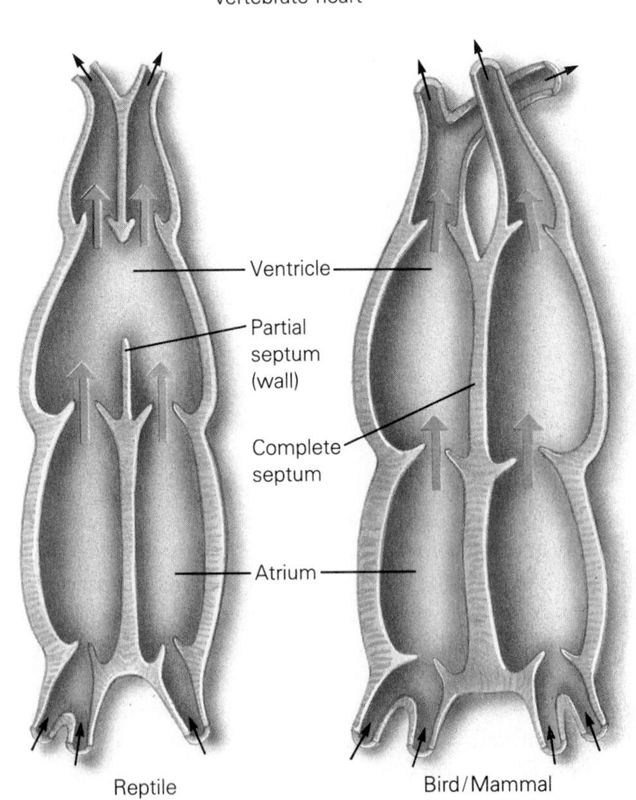

before returning to the left atrium via the *pulmonary veins*. The left atrium pumps the blood into the left ventricle, which then contracts to send the blood into the large aorta, which immediately branches before looping downward, carrying the blood on the first leg of its long journey through the body, bearing its gift of oxygen. The left ventricle, then, is very thick-walled and muscular; hence, the left side of the heart is larger. For this

FIGURE 19.16

The flow of blood through the human heart. Notice the coronary arteries that service the heart itself. Blood from the right atrium (deoxygenated and deep red) is pumped to the right ventricle and from there through pulmonary arteries to the lungs. It returns oxygenated and bright red to the left atrium and from there to the left ventricle to be pumped to the body via the aorta. Notice that the two atria contract simultaneously, as do the ventricles.

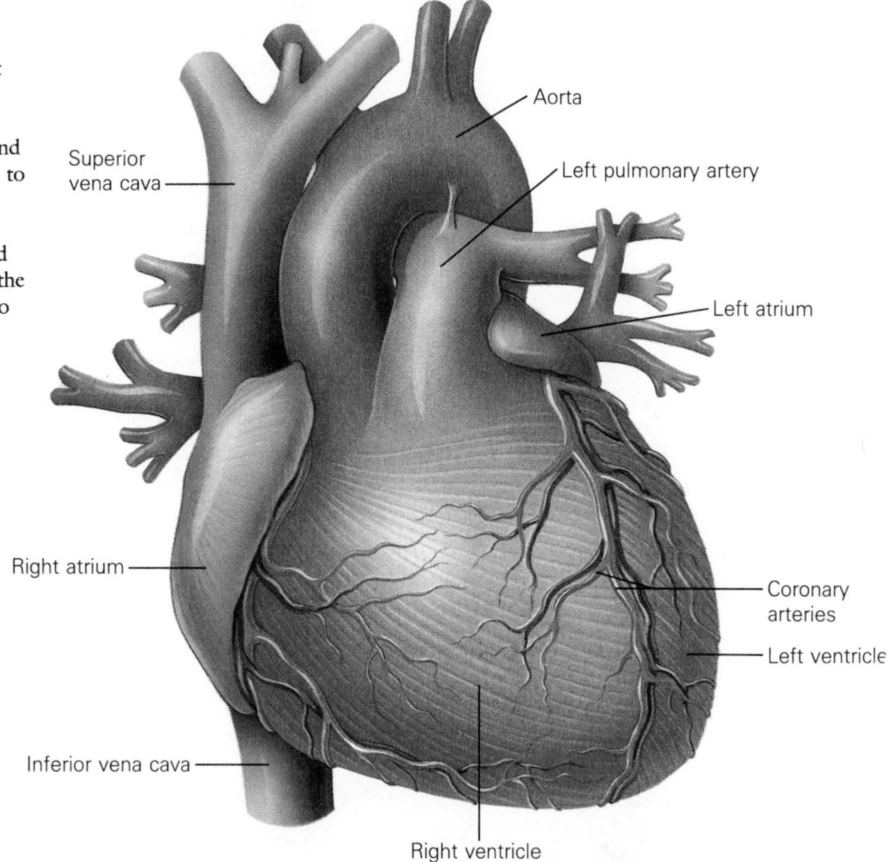

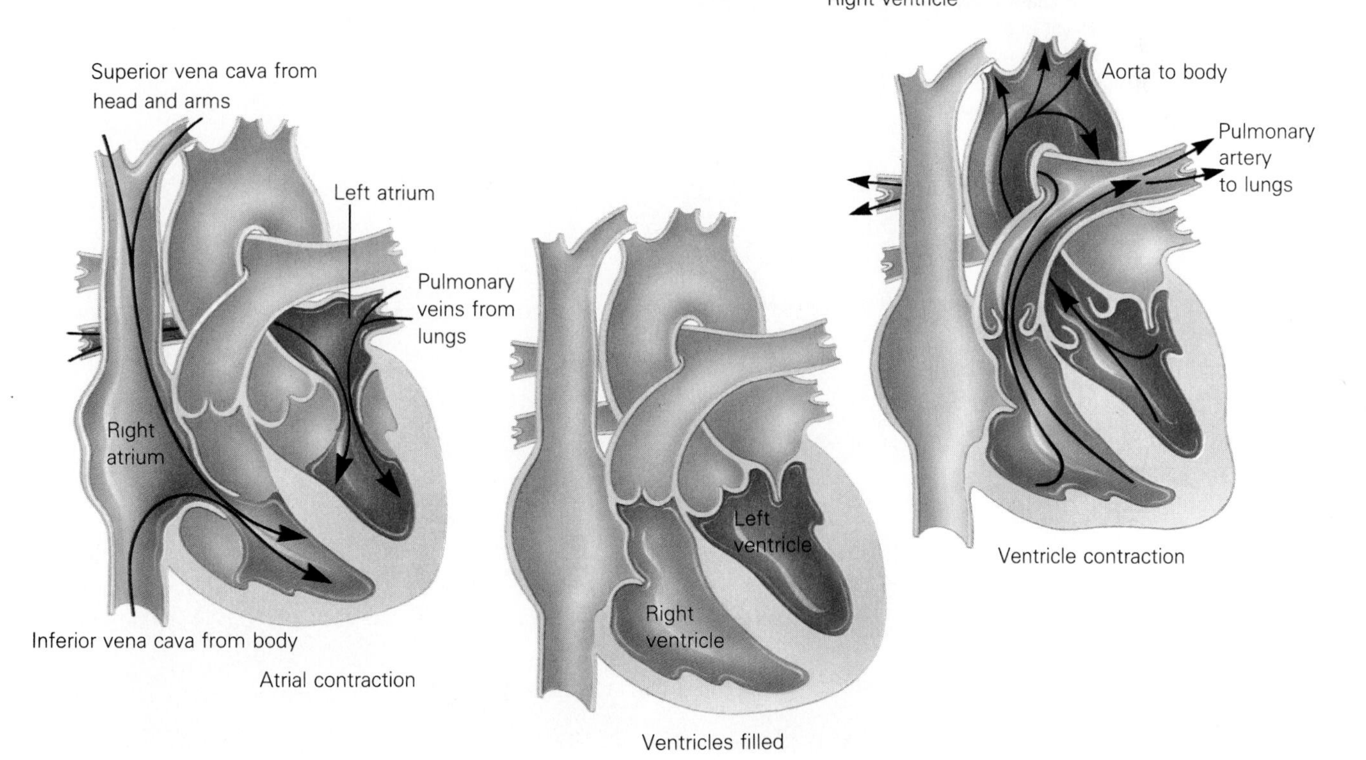

reason, you may have the notion that your heart is on the left side. It isn't. It's right in the middle of your thoracic cavity and is about the size of your fist. Thus, when you pledge allegiance to the flag, your hat is actually over your left lung.

Control of Heartbeat

A tiny heart, excised from some laboratory animal, lies in a pool of osmotically correct fluid in a petri dish. The newest generation of biology students gathers around, wide-eyed, as the tiny heart beats away. No blood; no nerves; no body—yet it pumps. What does the demonstration tell them (about the heart, not the instructor)? It tells them that the beating of the heart is intrinsic (internal) to the heart itself. (Its rate, they will learn, can be influenced by other, external, factors.) "What, then, are the internal mechanisms? How can an isolated heart continue to beat?"

The heart activity results from specialized cardiac cells, called **nodal tissue** that initiate and distribute the impulses throughout the heart. Special junctions between the muscle cells allow those impulses to spread easily from one muscle cell to the next. The impulse begins at an aggregation of nodal tissue called the **sinoatrial (SA) node** (or pacemaker), located in the wall of the right atrium near the vena cava (Figure 19.17). The impulse moves through the atrial muscle to the **atrioventricular (AV) node,** located at the base of the atria near where the ventricles begin. From

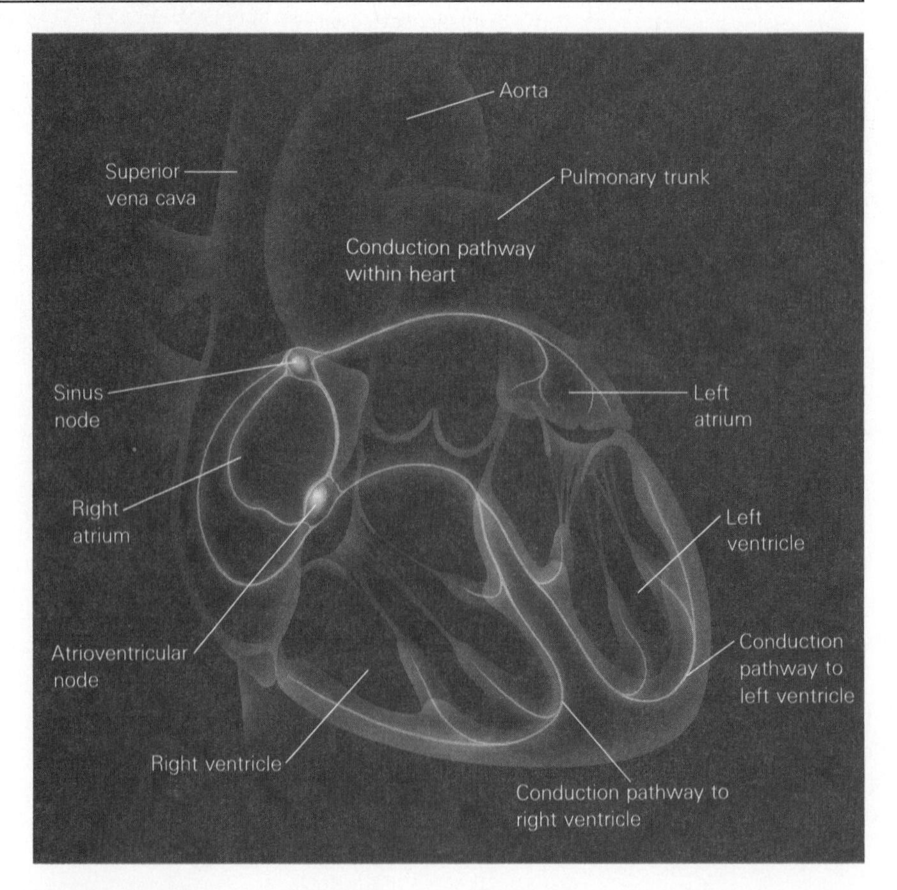

FIGURE 19.17

Heart rate is directly controlled by neural tissue (shown lighter), as the SA node initiates an impulse which immediately is fired to the AV node which, in turn, sends impulses to the ventricles over Purkinje fibers, causing the powerful muscles to contract in a specific sequence.

ESSAY 19.3

Heart Attack

We have all heard the old story about the faint-hearted guard dog who, upon being told "Attack," had one. *Attack* is now a common word in our vocabulary, as well it should be since heart attacks kill over half a million Americans each year. The incidence of heart disease among Americans is one of the highest in the world; about 30 million people are affected. (Some countries have a higher rate, as do the French, or a lower one, as do the Japanese.) But what *is* a heart attack? Technically, it is the result of a *myocardial infarction*—that is, a blockage of the arteries that feed the heart. (The lighter areas in the photo indicate blockage.)

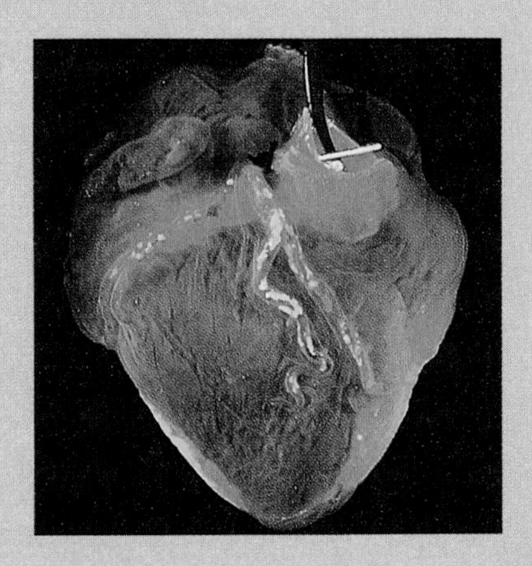

When such an artery is blocked, the oxygen-starved muscles of the heart begin to die. Depending on how much of the heart is damaged and how badly, the results can vary from almost complete recovery to death. Blockage of the coronary vessels that feed the heart is usually caused by one of three things: a clot lodged in the vessel, a prolonged contraction of the walls of the vessel, or atherosclerosis.

Atherosclerosis is the result of the buildup in blood vessels of a number of substances, such as fat, fibrin (formed in clots), parts of dead cells, and calcium. These substances reduce the elasticity of the vessel, and by decreasing the diameter of the vessel, they raise blood pressure, just as you would raise the pressure in a garden hose by holding your thumb over the end. No one knows what causes atherosclerosis, but a number of things can speed its development, such as smoking cigarettes and eating animal fat and cholesterol. Other factors include age, hypertension, diabetes, stress, heredity, and gender (males have more atherosclerosis).

The warnings of a heart attack are often (but not always): (1) a pain that spreads along the shoulder, arm, neck, or jaw; (2) sudden sweating; (3) a heavy pressure and pain in the center of the chest; and (4) nausea, vomiting, and shortness of breath and a powerful feeling of dread. The symptoms may come and go.

People who have not developed a strong and efficient cardiovascular system through exercise are particularly susceptible to *angina pectoris* ("chest pain"), which occurs when the heart fails to receive enough blood, particularly during times of stress or exercise. It should not be confused with a true heart attack. The pain may be relieved by stopping the unusual exercise or by reducing the levels of stress. Blood flow to the heart can be increased by a program of exercise or by surgically inserting vessels from other parts of the body (a coronary bypass). Certain chemicals, such as nitroglycerin, also dilate the heart's vessels and increase the circulation of blood there.

Another form of heart attack results in a phenomenon called *sudden death,* which does not sound good at all. The death may be due to chaotic and uncoordinated contractions of the ventricles. The contractions do not move blood along and, after a few spasms, the heart may stop entirely. Many people afflicted in such a way mysteriously fall dead in their tracks. Some, however, could have been saved if they had been helped in time.

In fact, victims of any form of heart attack stand a much greater chance of surviving if they are treated immediately. In many metropolitan areas, citizens are being trained in cardiopulmonary resuscitation (CPR) to help restore circulation in such emergencies (See Essay 19.4). Their efforts continue the flow of blood to the brain, where sensitive tissues die quickly without oxygen. In Seattle, Washington, with an extensive citizen training program, passersby have performed about one-third of the city's resuscitations. Their success rate is higher than that of professionals because they usually reach victims sooner.

ESSAY 19.4

CPR

In earlier editions of this book, I simply referred to CPR (cardiopulmonary resuscitation) as a critical lifesaving technique. It has since occurred to me that the procedure is so effective and has saved so many lives, that perhaps I should describe the basic procedure. (I am aware that it can be argued that such information does not belong in a biology text. But that's one of the advantages of writing your own book—you can say what you want.) So I'm going to describe the steps here, just in case you have not received formal training.

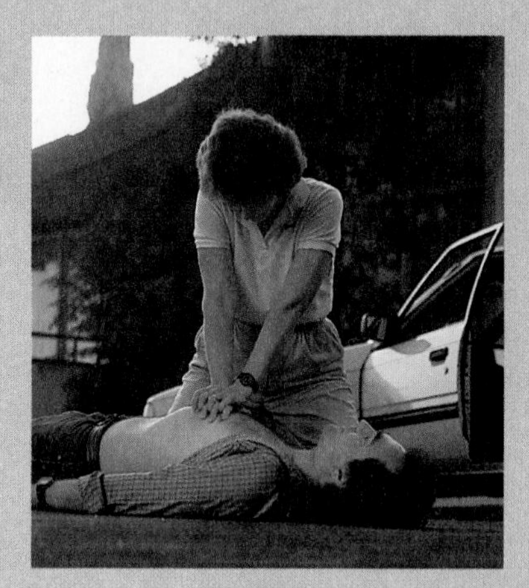

1. Check for cardiac arrest (unconsciousness and lack of pulse). Call or send for help. Roll the victim onto his back.
2. Open the airway by lifting the neck and tilting the chin upward.
3. Check for breathing; clear foreign objects from the mouth.
4. If there is no breathing, pinch the victim's nose and press your open mouth against his. Breathe rapidly into his mouth four times without allowing the victim to exhale completely after each ventilation.
5. If the victim begins breathing and has a pulse, discontinue rescue breathing. If breathing and pulse are absent, begin chest compressions. *(The magic numbers will be 15 and 2.)*
6. Kneel beside the victim. Find a spot two finger-widths above the tip of the breastbone. Put the heel of one hand over the other and place them on this spot.
7. Thrust down, depressing the breastbone about an inch and a half. Thrust rhythmically fifteen times, about once a second.
8. Lean over quickly, breathe twice into the victim's mouth (as in step four).
9. Repeat the cycle of 15 compressions and 2 breaths until help arrives. Check periodically for pulse. Stop chest compressions if pulse appears.

here, the impulse spreads through the ventricles along **Purkinje fibers.** (See Figure 19.17 for a summary of the control of a heartbeat, and Essay 19.3 on heart attack.)

In recent years, people with irregular heartbeats have been fitted with artificial pacemakers, devices that run on batteries or tiny nuclear-powered sources. The power source is implanted beneath the skin of the belly or the shoulder and electrodes are threaded through veins into the right ventricle. The pacemaker then initiates regular patterns of impulses that cause the heart to contract rhythmically. (See Essay 19.4 on performing CPR.)

The Lymphatic System

When we think of a circulatory system, we usually think of blood. However, there is a circulatory system of another kind, the one that circulates **lymph.** This one, logically enough, is called the **lymphatic system** (Figure 19.18). Lymph ultimately comes from the blood. The liquid portion of

FIGURE 19.18

The human lymphatic system, sometimes referred to as the "other" circulatory system. Lymph nodes are scattered throughout the body, but concentrated in specific areas. These nodes harbor large numbers of one kind of white blood cell, the lymphocytes, which help defend the body against disease organisms trapped there. Lymph nodes tend to swell and become sore if they are involved in fighting an infection near them, thus they may signal infections that might otherwise go unnoticed. The axillary lymph nodes are often affected by cancer of the breast and are often removed with the breast and its underlying muscles. Mastectomies of this sort are currently being re-examined to see whether they are merited or effective in most cases.

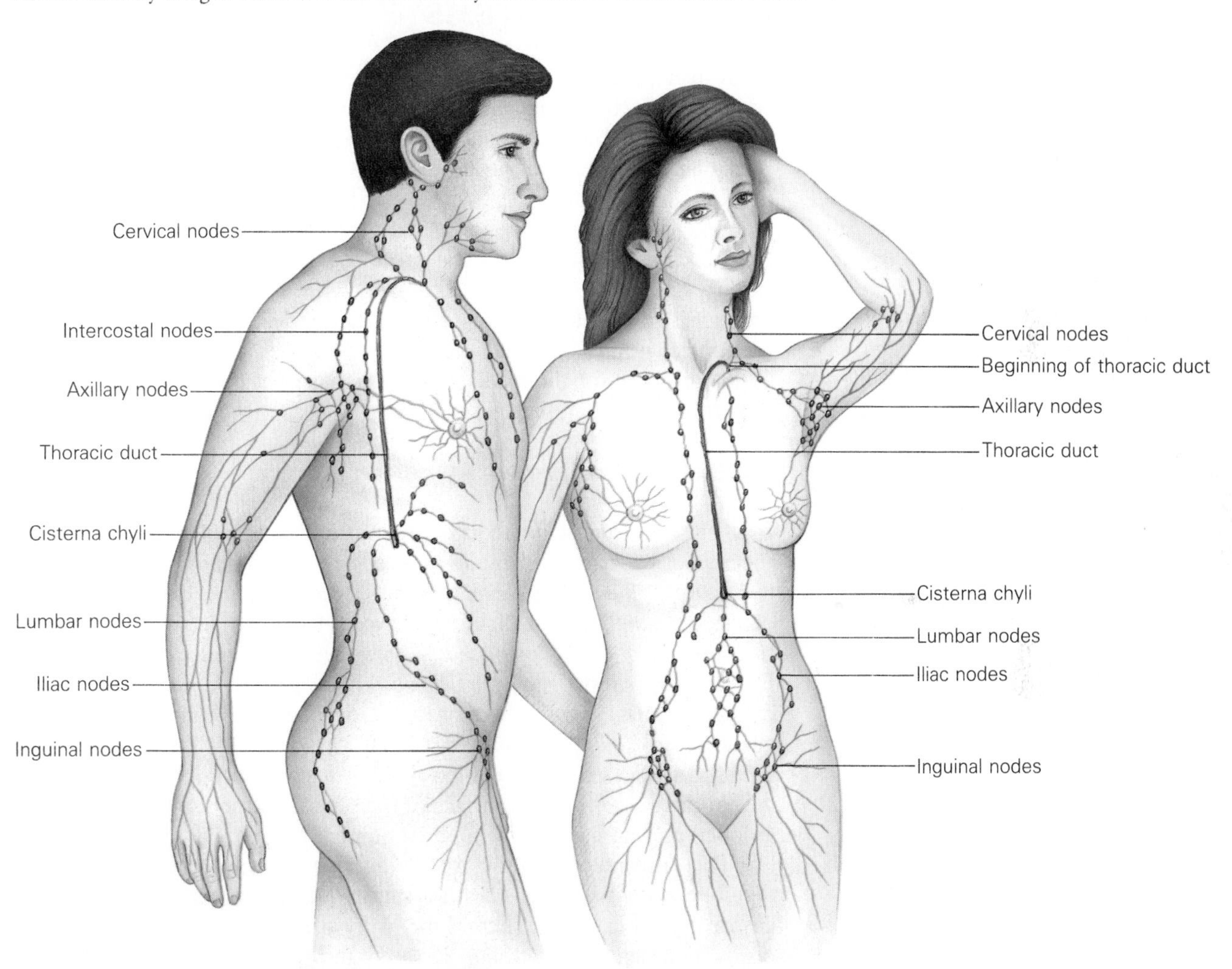

the blood, plasma, filters out through the walls of the tiny capillaries forming lymph. Here it bathes the cells, moves between them and finally winds up in small channels called **lymph capillaries.** The lymph may then move into larger channels, which may eventually bring the fluid to one of the many **lymph nodes** scattered throughout the body. In the nodes it is filtered, thus removing any cellular debris or bacteria. Frogs and some other vertebrates have "lymph hearts" that move the lymph along, but mammals rely chiefly on muscular movements to squeeze the vessels and circulate the fluid. (Again, we see that people are adapted to being on the move.)

White blood cells called **lymphocytes** are found in great numbers in the lymph nodes. As we will see in the next chapter, the lymphocytes are important in defending the body against disease organisms. When such a battle is raging, the lymph nodes enlarge (a sign of infection). As the lymph moves along, it flows into increasingly larger channels, or **lymph ducts,** until, finally, it enters large veins near the heart, rejoining the blood.

DIGESTIVE SYSTEMS

There is a tired old cliché about our bodies being internal combustion engines, with food being the fuel that runs them. However, the idea can still be squeezed for elements of truth. The point, more precisely stated, is that energy is required to do the work that must be done to minimize entropy, or disorganization, within our bodies. The energy for this work is found in food and is released after the food is broken down into its constituents (digestion) and the product transported to the cells (circulation), where it enables the cells—which are actually the engines, if the analogy is to be valid at all—to burn the food for energy (metabolism). Of course, not all ingested material is "burned"; some of it is used as building blocks for the repair and maintenance of the machinery and the growth of the organism.

Digestive Arrangements

In a nutshell, **digestion** reduces the molecular size of nutrients so that they can pass through cell membranes to be distributed to body cells. Since the process is so fundamental, it may seem that the basic scheme for digestion is the same in all animals, except for a few details. This may be true, but these "details" vary widely, as we see in Figure 19.19. An amoeba, for example, simply moves its flexible body around a food particle and engulfs it (phagocytosis), so that the particle becomes contained in a **food vacuole.** Acidic digestive enzymes are then secreted into the vacuole until the food is digested. Any indigestible particles are brought, still enclosed within the vacuole, to the surface of the protist and squeezed out through an opening that appears in the membrane.

The cnidarians and the free-living flatworms both have pouchlike digestive systems—that is, with a single opening to the outside—called gastrovascular cavities. Whereas the hydra stings its prey with special cells on its tentacles and then hauls the prey into its digestive cavity, the flatworm is able to extrude its pharynx, a long, flexible tube, to capture food. In both these animals, digestion begins in the gastrovascular cavity, but before the process is completed, the particles of partly digested food are engulfed by cells of the cavity wall and digestion is completed within these cells. Undigested particles are excreted from the cells and expelled through the single opening. These animals, then, digest food by both extracellular and intracellular processes. (Which of these two processes typifies human digestion?)

In more complex animals, the digestive system is a tube rather than a pouch, with one opening through which food enters and another through which it exits—a much more civilized arrangement, to be sure.

FIGURE 19.19

Digestive systems in a variety of animals. Here, a protist, an amoeba, has an unspecialized arrangement, ingesting food through any part of its surface, simply ejecting waste through its membrane. The hydra and flatworm have saclike digestive tracts, food exiting over the same route it entered. The earthworm has an essentially tubular system with a food-grinding gizzard to prepare the food for digestion. The tubular system of the salamander is embellished by a number of glands and organs that alter the food as it passes. Here, notice that, technically, food always remains outside the body until its digestive products pass through the intestinal membranes. Thus, "I would like to get on the outside of a hamburger," is a proper statement.

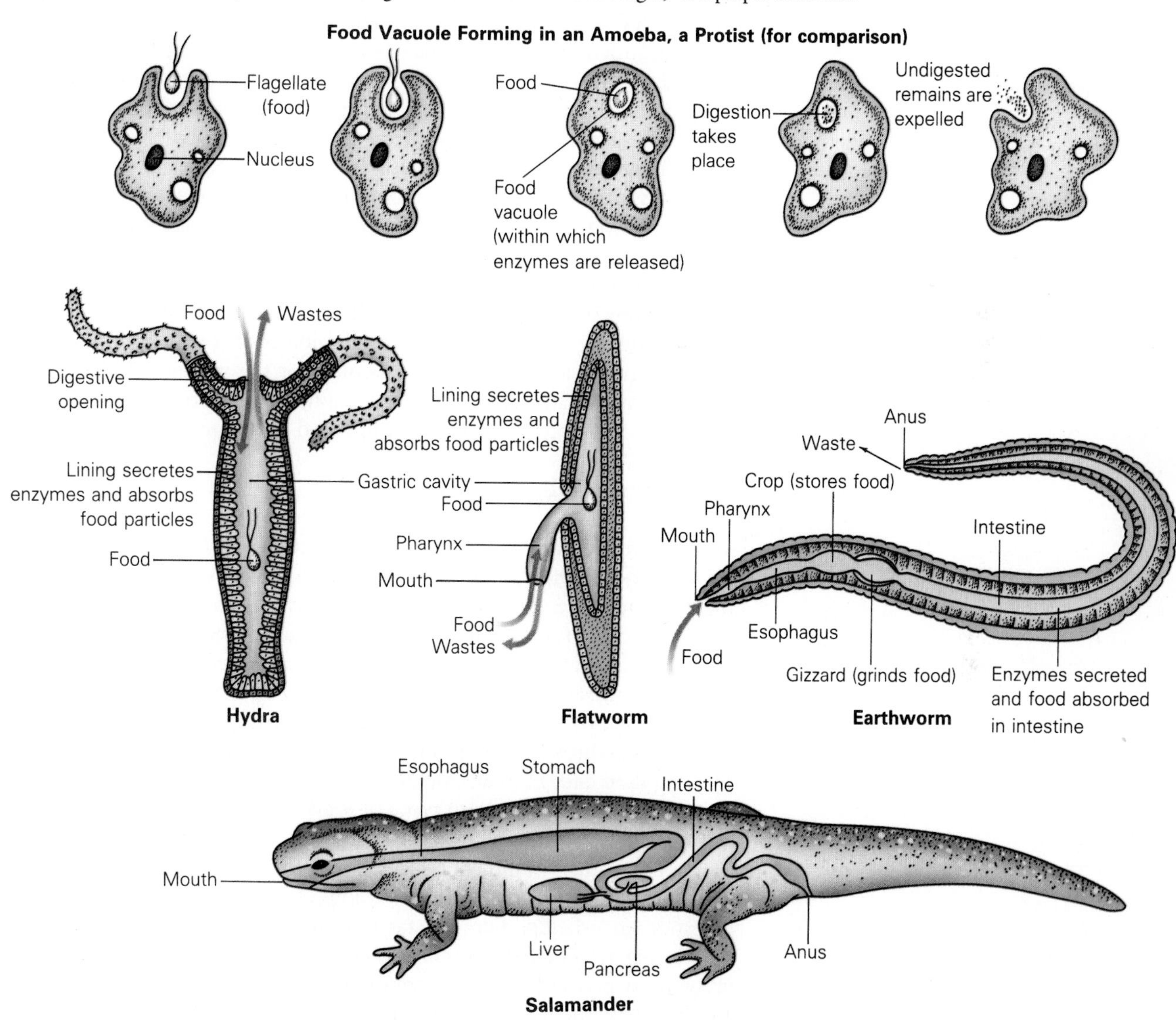

Although both humans and earthworms have tubelike digestive systems, you will notice in the earthworm several structures that are absent in humans, such as the crop and gizzard (also present in birds). The **crop** is where food is stored, and the muscular **gizzard** is where food is ground against small stones that have been ingested. There is now no reason for you to lie awake night after night wondering how an earthworm chews.

As another fascinating aside, earthworms have amoeboid cells in their bodies that ingest particles moving through the digestive tract as the worm literally eats its way through the soil. These cells then migrate through the body and come to rest just under the skin, giving the earthworm a color similar to the soil in which it lives and causing the early bird to come up short.

In vertebrates, such as the salamander, we find elaborations on the basic tubular system. Food enters a well-defined mouth where some basic processes of digestion may begin. The food is swallowed by voluntary action and then moves down the smooth-walled **esophagus** by a wave of involuntary muscular contractions of the tube. The contractions resemble a ring being slid along the tube. The process is called **peristalsis** and may appear along the entire gastrointestinal tract. The food then enters the **stomach,** where digestion continues. From there it passes to the intestine, where it is further broken down by the action of intestinal secretions along with those from complex glands of the pancreas and liver. When the food is finally digested, the products move through the intestinal wall into the blood or lymph vessels that line it. These nutrients are then carried by the blood to the tissues where they will be used. The indigestible particles move on through the intestine, to be eliminated through the **anus** or the **cloaca.** (The cloaca is a common opening for the intestine, kidneys, and reproductive organs. It is found in amphibians, reptiles, and birds—a primitive plumbing arrangement "improved upon" in mammals.)

The Human Digestive System

In the human digestive system (Figure 19.20 and Table 19.2), the breakdown of food particles begins in the mouth, where chewing breaks food apart and increases the surface area on which enzymes can act. Chewing also has another function in those animals that eat plants, since it ruptures the cellulose-laden cell walls that cannot be broken down by digestive enzymes. Since animal cells lack cell walls, most animal tissue *can* be entirely broken down by digestive enzymes. This is why cats don't have flattened molars with which to grind food, and why you never see them lying around chewing their cud. Humans, being omnivores ("all eaters") have teeth that are adapted to handling both types of food.(See Essay 19.5 on the Heimlich maneuver.)

In the human mouth, the food is broken up and lubricated by saliva (this is not my favorite part of biology, either). Also, some chemical digestion may begin there as starches are changed to disaccharides, or double sugars. The food is then swallowed and moves to the stomach by involuntary peristaltic contractions of the smooth-walled esophagus, which also moistens the food by secretory glands in its lining.

The stomach is a muscular sac that churns the food as it secretes mucus, hydrochloric acid, and enzymes that begin the digestion of proteins. The food is meanwhile sealed in the stomach by two sphincters, or rings of muscles, one at either end of the stomach. After the mixing is completed, the lower sphincter opens and the stomach begins to contract repeatedly, squeezing the food into the small intestine. A fatty meal, by the way, slows this process and makes us feel "full" longer. This is also why we're hungry again so soon after a low-fat Chinese dinner.

FIGURE 19.20

The human digestive system. Once food enters the esophagus, it is moved along the digestive tract largely by involuntary muscular movements. Various enzymes and other juices are added along the way to break down the food molecules so that by the time they reach the small intestine, they can be absorbed into the bloodstream. Undigested matter then moves to the large intestine where it waits for the 7:05 out.

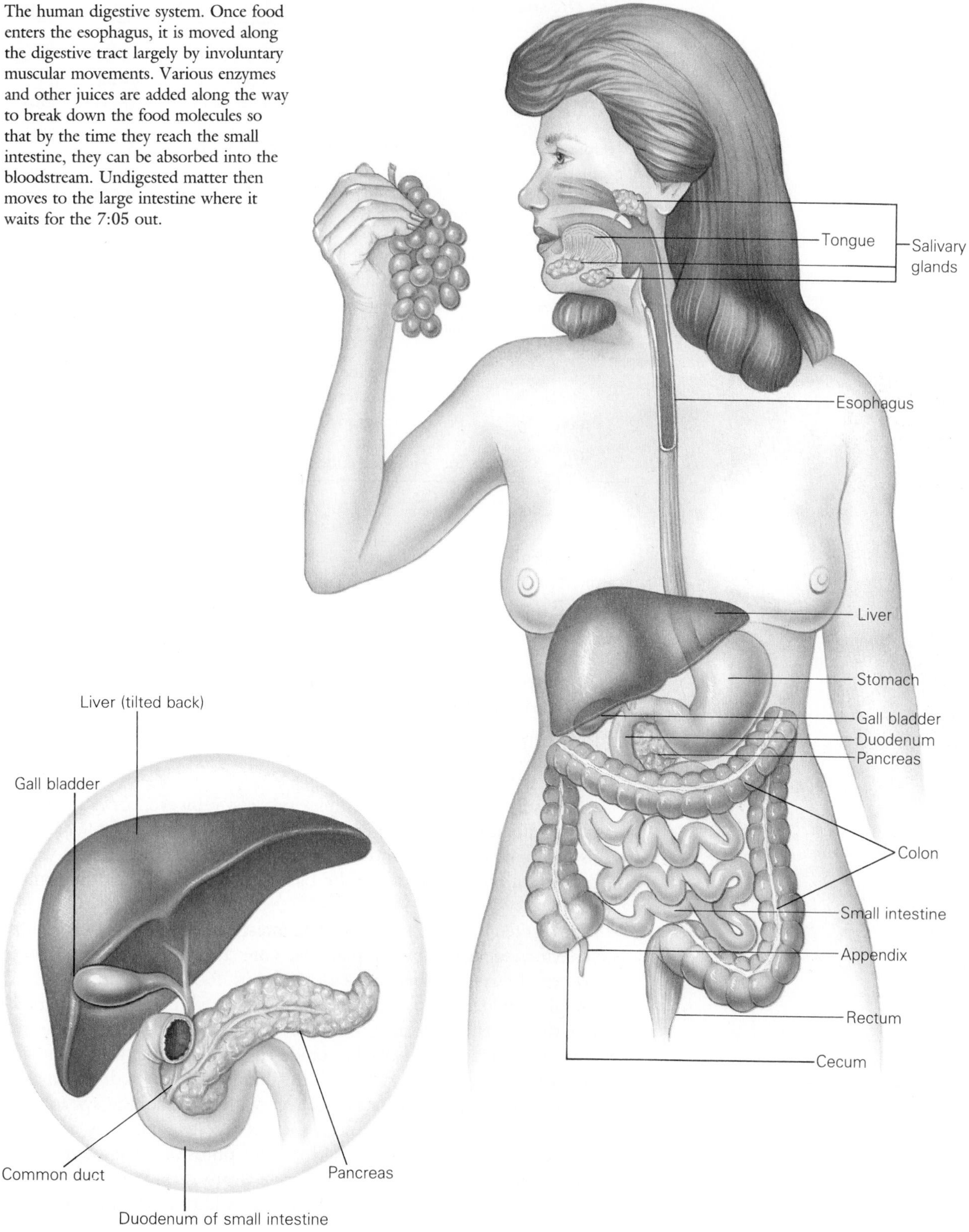

TABLE 19.2
The Digestive System

STRUCTURE	FUNCTION
Mouth	Mechanically breaks down food during chewing; saliva from salivary glands moistens and binds food and begins the breakdown of starch.
Esophagus	Chute between mouth and stomach.
Stomach	Short-term storage; churns food mixing it with enzymes and HCl and liquefying it; begins digestion of proteins and some fat.
Small intestine	Completes digestion of all nutrients using enzymes produced in small intestine and in the pancreas; absorbs digestive products.
Large intestine	Absorbs water; forms feces; houses vitamin-producing bacteria.
Accessory organs	
Liver	Secretes bile for keeping fat in small droplets.
Gall bladder	Stores bile before releasing it to the small intestine.
Pancreas	Produces digestive enzymes that are released into the small intestine; secretes sodium bicarbonate to neutralize the acidic secretions of the stomach.

The **small intestine** is a long convoluted tube in which digestion is completed and through which most nutrient products enter the bloodstream. Its inner surface is covered with tiny, fingerlike projections called **villi,** which increase the surface area of the intestinal lining. Furthermore, the surface area of each villus is increased by about 3000 tiny projections called **microvilli** (Figure 19.21). Within each villus is a minute lymph vessel, called a **lacteal,** which is surrounded by a network of blood capillaries. While the digested products of certain fats move directly into the rather permeable lymph vessel, the products of protein and starch digestion move into the blood capillaries.

Fat absorption is a rather complex process. However, once the products of fat digestion enter the lymphatic system, they are quickly restored as fats and coated with fats and proteins that make the fat soluble in water. In fact after a meal high in fat, the blood may take on a startling milky appearance.

The first ten inches or so of the small intestine comprise the duodenum. The enzymes produced by the **pancreas** enter the gut in this area. These are powerful enzymes that break down many foods including proteins. One might wonder, then, why doesn't the pancreas digest itself? Largely because the protein-splitting enzymes are stored in an inactive form in the pancreas and are activated by the duodenal environment. Well then, the sharp-witted reader will ask, why isn't the duodenum digested? As a matter of fact, it would be if it were not "alive" with a membrane system that actively excludes harmful substances, such as digestive enzymes, from entering the cells. This is also why, in those unfortunate enough to be "wormy" (and many people are), when a worm dies, it is immediately digested—the body's ultimate vengeance. In addition, the intestine pro-

FIGURE 19.21

Microvilli on cells that cover the villi (a), markedly increase the absorptive surface of the small intestine, which is quite long and convoluted (b), providing a great area of exchange.

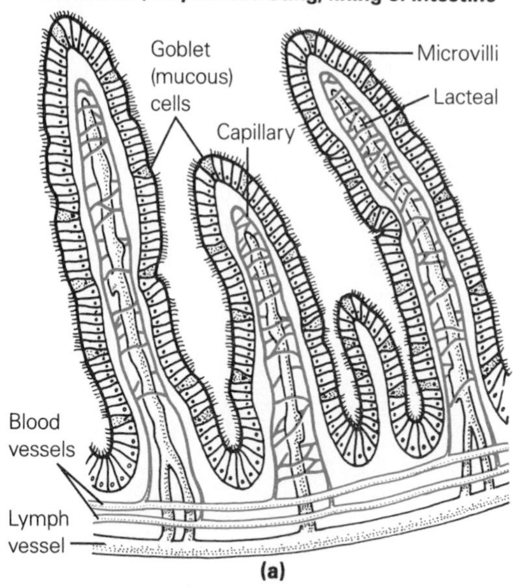

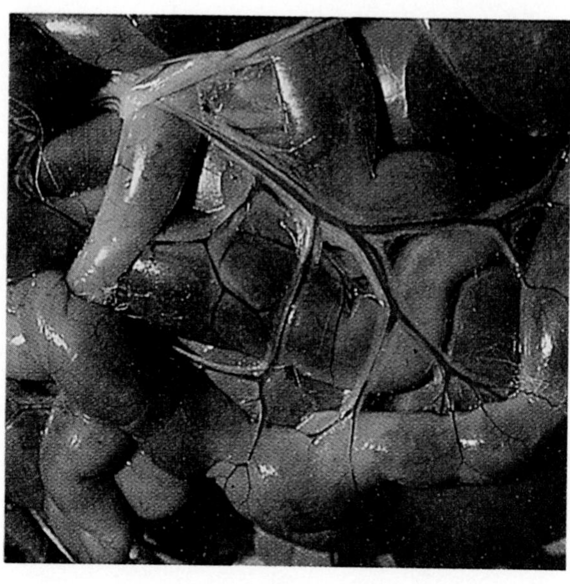

tects itself by secreting a protective shield of mucus, while the pancreas secretes a high level of sodium bicarbonate that helps to neutralize the disruptive acids entering the intestine from the stomach. When the system fails or extreme amounts of acid are produced by the stomach, perhaps due to emotional stress, ulcers (localized lesions of the digestive tract) can result.

Another substance secreted into the duodenum is **bile,** which is stored in the **gall bladder** after it is produced by the **liver.** Bile contains certain sodium salts that act as detergents, breaking up the large fat droplets into smaller ones so that the pancreatic enzymes can work on them.

Once the products of digestion move into the tiny capillaries within the villi, the nutrient-laden blood is carried to the **hepatic portal vein,** which leads to the liver (a portal vessel is one that lies between two capillary beds). The blood then filters through tiny vessels and sinuses in the liver. If the blood contains excess carbohydrates, some are removed by the liver and stored as glycogen. If the blood is low in carbohydrates, stored glycogen is broken down into its glucose subunits, which are then released into the bloodstream. Thus, proper glucose levels are maintained in the blood despite variations in food intake. Such shifts in blood glucose levels can result in various changes, including behavioral ones. For example, as the body's reserves of glucose drop below the amount needed to maintain a constant blood level, one has the sensation of hunger and is motivated to go out and find some glucose (in a variety of forms from rabbits to apples).

ESSAY 19.5

Heimlich Maneuver

Human evolution, unfortunately, has resulted in the opening of the trachea and esophagus being closer together than is the case in many other species. The result is a marked propensity for food to "go down wrong"—that is, for an occasional food particle to move into the air passages. This happens when the epiglottis is not completely closed during swallowing. In this case, the glottis spasmodically contracts and causes choking. In some cases, the food is simply coughed up, but in extreme cases the victim is completely unable to breathe. More than eight Americans die this way each day.

However, a rather simple action can save many of these lives. It is called the Heimlich Maneuver and it works as follows: (1) Stand behind the victim. (2) Wrap your arms around the waist. (3) Make a fist with one hand, knuckles directed upward and inward against the victim. (4) Place the knuckles between the rib cage and the navel. (5) Cup the other hand over the fist. (6) Quickly press inward and upward against the victim's abdomen. (7) Repeat if necessary.

If the victim is lying on his back, kneel, place your knees by the hips and with the heels of your hands (one on top of the other) press upward with a quick thrust. (Repeat if necessary.)

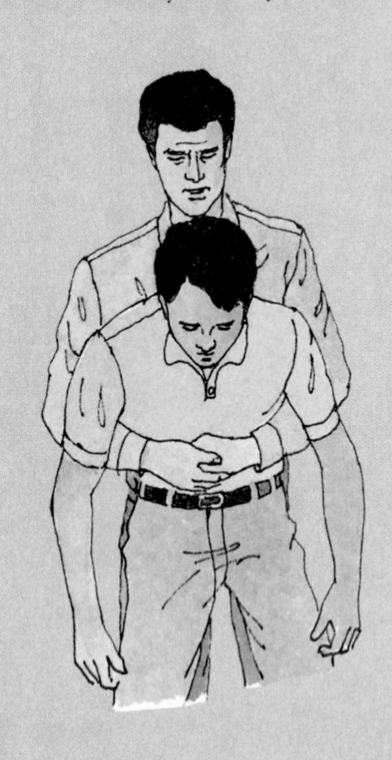

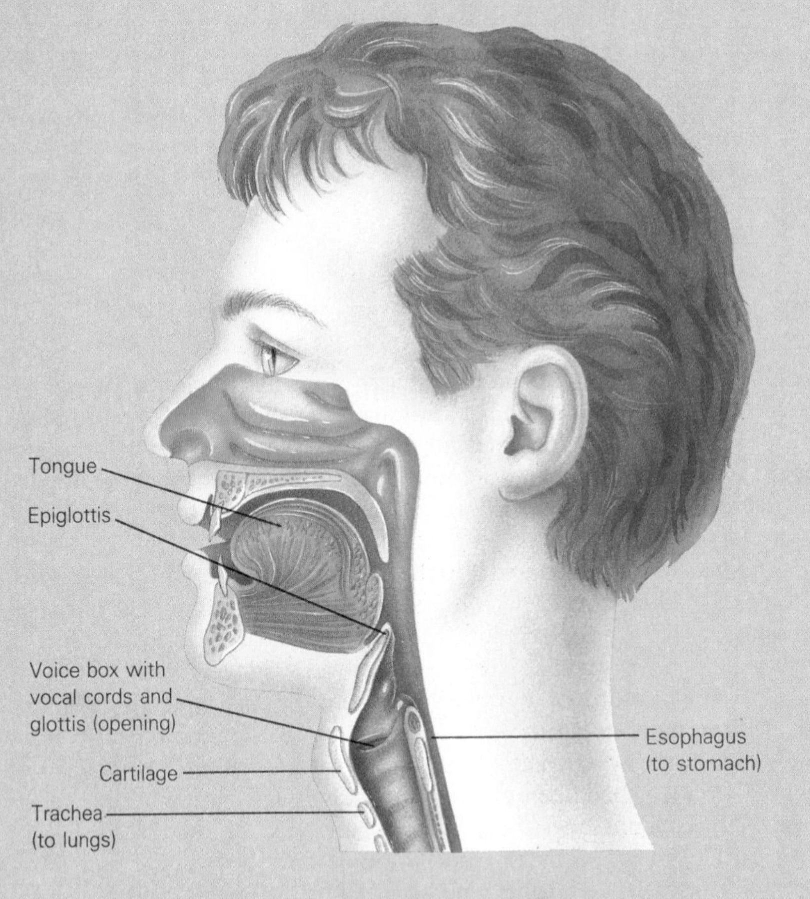

As food passes from the duodenum through the rest of the small intestine, the jejunum and the ileum, further digestion and absorption occurs (as we see in Table 19.3). The remaining matter, mostly composed of undigested foods, bacteria, and water, then move on into the **large intestine.** The inner surface of the large intestine is rather smooth, and no digestion and no absorption of digested food occurs here. However, it is a veritable hotbed of *E. coli* bacteria. The *E. coli* produce a waste that is high in certain vitamins. (The presence of these bacteria in the environment is an indication of fecal contamination, an important factor in the spread of some diseases, such as infectious hepatitis.) The large intestine also extracts water from the solid water product, or **feces.** This is why constipation results if one puts off nature's call and why diarrhea robs the body of both water and nutrients. Finally, let us note that the feces are stored in the **rectum** before elimination through the anus.

We end our discussion on that poetic note. We will next consider the immune system, but we must keep in mind that although we are considering each system separately, each is but a part of a constellation of processes. Ultimately, the various systems must be considered in terms of their precisely coordinated and balanced interactions if we are to begin to understand this complex phenomenon of life.

TABLE 19.3
Digestive Enzymes and Their Functions

SOURCES AND ENZYME	SUBSTRATE	PRODUCT
SALIVARY GLANDS		
Salivary amylase	Starch	Disaccharide
STOMACH LINING		
Pepsin	Protein	Polypeptides
Rennin	Casein (milk protein)	Insoluble curd
Gastric lipase	Triglyceride	Fatty acids + glycerol
PANCREAS		
Proteolytic enzymes	Protein	Peptides
Ribonuclease	RNA	Nucleotides
Deoxyribonuclease	DNA	Deoxynucleotides
Pancreatic amylase	Starch	Glucose
Pancreatic lipase	Triglyceride	Fatty acids + glycerol
Carboxypeptidase	Protein	Shorter peptide and one free amino acid
INTESTINAL LINING		
Peptides	Peptides	Amino acids
Nuclease	Nucleotide	Pentose + nitrogen base
Maltase, Sucrase, Lactase	Disaccharides	Monosaccharides

SUMMARY

1. Breathing is one method of external respiration, getting oxygen to the blood. Internal respiration is the exchange of gases between the blood and cells. The use of oxygen in cellular metabolism is cellular respiration.
2. Small aquatic animals, such as planarians, have enough surface area in relation to their mass for an adequate amount of oxygen to diffuse to all cells. Larger aquatic animals, such as certain salamanders and fish, have gills, which are thin-walled projections that create a large surface area for the diffusion of oxygen into and carbon dioxide out of the blood vessels that permeate the gills. Fish move oxygen over their gills either by pumping it with muscles or by swimming. Insects breathe through a tracheal system comprised of tiny tubes connecting the outside air with the body tissue.
3. External respiration of mammals occurs in pouches called lungs. During ventilation (breathing), oxygen-laden air moves through the trachea to the smaller bronchi, to the branching bronchioles, and finally to tiny alveoli of the lungs.
4. Air is brought into the lungs when the chest cavity is made larger by the contraction of the diaphragm, a large muscle separating the abdominal and chest cavities, and of the muscles between the ribs. When these muscles relax, the elasticity of the lungs helps push air out.
5. A circulatory system moves oxygen, carbon dioxide, nutrients, wastes, and hormones through the body. One-celled organisms distribute materials by diffusion. Cnidarians and free-living flatworms have a saclike gut with a single opening, called a gastrovascular cavity. Digestion begins in this cavity and is completed within cells that line the cavity.
6. In an open circulatory system, found in mollusks and arthropods, blood (hemolymph) leaves the vessels and seeps into open cavities called sinuses. In a closed circulatory system, the blood remains within vessels. However, in vertebrate closed circulatory systems, the blood flows through sinuses in the liver and bone.
7. Humans have a closed circulatory system with a large muscular heart that pumps blood through arteries to arterioles and to thin-walled capillaries where the exchange of materials between the blood and cells occurs. Capillaries join to form venules, which form veins that return blood to the heart. The arteries help maintain blood pressure.
8. Nitrogenous wastes are removed from the blood by the excretory system. Carbon dioxide dissolves in blood and forms carbonic acid, creating hydrogen ions and bicarbonate ions. The ions move to the lungs where carbon dioxide is reformed and exhaled.
9. The liquid portion of the blood, plasma, contains water, proteins, hormones, wastes, and other materials. The red blood cells contain the red pigment hemoglobin that combines with oxygen to transport it to the cells. White blood cells (leukocytes) combat invasion and infection. Platelets function in blood clotting.
10. The heart is a muscular pump that moves the blood. Its form varies among invertebrates. In vertebrates there is a progression in the number of cham-

bers. A fish heart has two chambers: an atrium that pumps blood to the ventricle, from the ventricle to the aorta, from the aorta to the capillary bed in the gills, and another in the body. The amphibian heart has two atria (one receiving oxygenated blood from the lungs and another receiving deoxygenated blood from the body) and a single ventricle. Although reptiles have two atria and one ventricle, the ventricles of most reptiles are not completely divided. In birds and mammals, the right and left sides of the heart are completely separated.

11. The lymphatic system is a second circulatory system that circulates lymph, which is part of the fluid constituent of blood that has been filtered through the tiny capillaries to bathe the cells. The lymph is then collected in lymph capillaries that may bring it to lymph nodes where bacteria and cellular debris is filtered out. Lymphocytes are a type of white blood cell abundant in lymph nodes that attack and destroy bacteria.
12. Food supplies energy for cellular activity and building blocks for maintenance and growth. Digestion reduces the molecular size of nutrients so they can be absorbed into the bloodstream and delivered to cells for use.
13. An amoeba engulfs a food particle by phagocytosis and digests it within a food vacuole. The hydra and flatworm have pouchlike digestive systems with a single opening. The hydra captures prey with its tentacles, and a flatworm uses its extrusible pharynx. In both organisms, digestion begins extracellularly in a cavity and is completed within cells that absorb partially digested particles from the cavity.
14. In more complex animals, the digestive system is a tube with two openings. Earthworms have a crop for storing food and a gizzard for grinding food. In vertebrates, such as a salamander, digestion begins in the mouth. Food then moves through the esophagus and into the stomach where digestion continues. Next the food enters the intestine, where it is broken down by secretions from the intestine and the pancreas and the products are absorbed into blood vessels. Undigested material is eliminated through the anus or cloaca (a common opening for the intestines, kidneys, and reproductive systems found in amphibians, reptiles, and birds). Food is moved through the digestive tube by a wave of contractions called peristalsis.
15. In humans, digestion begins in the mouth where food is chewed (broken apart physically), moistened with saliva, and where the digestion of starches begins. Food is swallowed and moves through the esophagus to the stomach. The food is held in the stomach by two sphincters and is churned and mixed with hydrochloric acid, enzymes, and mucus. In the small intestine, digestion is completed, and the products are absorbed into the capillaries or lymph vessels within fingerlike projections on the intestinal wall called villi. The villi, and many projections from their surface called microvilli, greatly increase the surface area of the small intestine.
16. Digestive enzymes from the pancreas enter the initial region, the duodenum. The pancreas does not digest itself because enzymes are stored in an inactive form. The duodenum is not digested because it secretes a protective layer of mucus and because the pancreas secretes sodium bicarbonate, which helps neutralize the acid from the stomach.
17. Bile, which breaks fat into small droplets and makes it more accessible to pancreatic enzymes, is produced by the liver, stored in the gall bladder, and works in the small intestine.

18. After they are absorbed, the products of digestion are carried to the liver by the hepatic portal vein. As the blood filters through the sinuses of the liver, glucose levels are adjusted.
19. The remaining materials enter the large intestine, which is composed of the colon and the rectum. Water is absorbed from this material and forms feces, which are stored in the rectum before leaving the body through the anus.

KEY TERMS

air sacs (521)
alveoli (521)
anus (540)
aorta (531)
aortic arches (526)
arteriole (526)
artery (526)
atrioventricular (AV) node (534)
atrium (531)
bile (543)
breathing (521)
bronchi (521)
bronchiole (521)
capillary (526)
carbonic anhydrase (531)
cellular respiration (517)
cloaca (540)
closed circulatory system (526)
crop (539)
diastole (529)
diastolic pressure (529)
digestion (538)
erythrocytes (530)
esophagus (540)
external respiration (517)
feces (545)
food vacuole (538)
gall bladder (543)
gastrovascular cavity (525)
gizzard (539)
heart (531)
hemoglobin (530)
hepatic portal vein (543)
internal respiration (517)
lacteal (542)
large intestine (545)
leukocytes (530)
liver (543)
lung (521)
lymph (536)
lymphatic system (536)
lymph capillaries (537)
lymph duct (538)
lymph node (537)
lymphocyte (538)
microvilli (542)
nodal tissue (534)
open circulatory system (526)
pancreas (542)
peristalsis (540)
plasma (529)
platelets (530)
Purkinje fibers (536)
rectum (545)
respiration (517)
sinoatrial (SA) node (534)
sinus (526)
small intestine (542)
spiracle (520)
stomach (540)
systole (529)
systolic pressure (529)
trachea (521)
tracheal system (520)
vein (527)
ventilation (521)
ventricle (531)
venule (527)
villi (542)

FOR FURTHER THOUGHT

1. Of the respiratory systems studied, which ones are not suited for use by larger individuals?
2. Explain why the walls of the human heart's atria are very thin-walled in comparison to the thicker-walled ventricles.
3. The respiratory surface of the common earthworm is its outer layer of skin. If the earthworm is left out in the sun and its skin dries out, it quite literally suffocates. Why?

FOR REVIEW

True or false?

1. ___ Oxygen diffuses into the blood of fish through thin-walled capillaries in their gills.
2. ___ Oxygen enters the mammalian bloodstream through the walls of the alveoli.
3. ___ In open circulatory systems, the heart pumps blood into cavities called sinuses.
4. ___ The chemical digestion of starch begins in the mouth.
5. ___ Most digested products enter the bloodstream through the surface of the stomach.

Fill in the blank.

6. Insects breathe through a network of tubes called a ___ .
7. ___ and ___ are vessels that carry oxygen-depleted blood from capillary beds back to the heart.
8. ___ is the involuntary muscular contraction that moves food through the vertebrate digestive tract.
9. ___ are fingerlike projections that increase the surface area of the small intestine.

Choose the best answer.

10. Small aquatic animals acquire oxygen through:
    A. spiracles
    B. a tracheal system
    C. ventilation
    D. simple diffusion
11. When the atrium contracts, it sends blood directly to the:
    A. veins
    B. ventricle
    C. aorta
    D. capillary beds
12. Which of the following possess a simple two-chambered heart consisting of one atrium and one ventricle?
    A. reptiles
    B. birds
    C. fish
    D. amphibians
13. Some digestive enzymes
    A. enter the gut through mesenteric arteries.
    B. are produced by the pancreas.
    C. are activated in the transverse colon.
    D. are secreted by the gall bladder.
14. The primary function of the large intestine is:
    A. extraction of water from feces
    B. digestion of food
    C. absorption of nutrients
    D. secretion of enzymes

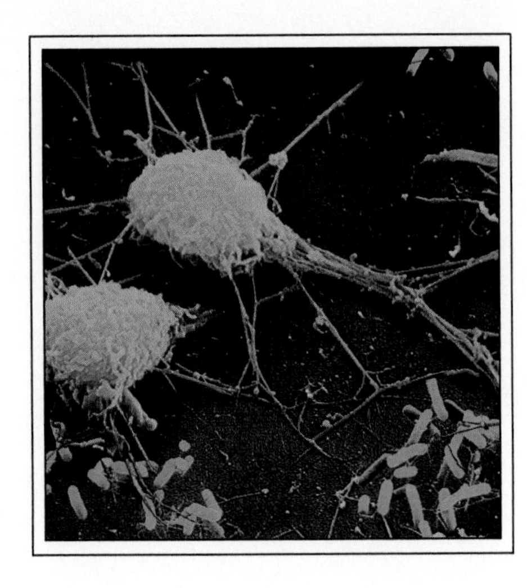

CHAPTER 20

The Immune System

Overview

The Nonspecific Responses
Specific Responses
Interferon: An Exciting New Promise
AIDS: A Devastating New Problem
Mind and Body

Objectives

After reading this chapter you should be able to:

- Name the nonspecific immune responses and state how each defends the body against infection.
- List five major types of white blood cells and describe their roles in human defense responses.
- Describe the structure and function of an antibody.
- Describe how each type of B- and T-lymphocyte functions during specific immune responses.
- Explain what happens during autoimmunity reactions.
- Describe the acquired immune deficiency syndrome, its symptoms, related diseases, and modes of transmission.

The earth is really not a very friendly place. You can get yourself killed here. And the dangers are of many stripes, some much more apparent than others. We occasionally hear a story about someone encountering travellers in an alien spaceship that has landed on our planet. If such beings did land here, their greatest concern need not be a trigger-happy farmer. Their greatest risks may be of a far subtler sort, such as an agonizing corrosion from our oxygen. But in addition to our deadly atmosphere, they might find themselves exposed to innumerable chemical and microbial agents, many of which are able to penetrate the bodies of living things and disrupt their delicate internal balances, bringing life to an end. Of course, we live in this deadly sea, and most of us, most of the time, are able to withstand the dangers. After all, we evolved on the planet, and our presence here attests to the fact that natural selection has endowed us with certain defenses. Primary among such defenses is our **immune system,** our bodies' defenses against disease, poisons, and foreign proteins. In some cases our bodies' immune responses to one kind of infection can be variable and confusing, as we see in our responses to Lyme disease (Essay 20.1).

There are many immune mechanisms in the human body, but these can be divided into two basic lines of defense. The **nonspecific responses** and the **specific responses.** The nonspecific responses are a very general sort of defense that works the same against all invaders (Table 20.1). The specific responses are a selective defense, programmed to work against only certain invaders.

THE NONSPECIFIC RESPONSES

In a sense, the skin not only holds you in but keeps others out. So the first line of defense is the body covering. Remember, the basic vertebrate plan is a tube within a tube (see Figure 16.5). The outer tube is covered with skin and the inner tube is essentially lined with a protective mucous membrane, both of which are effective defenses against intrusion.

The skin is quite an effective barrier, fortified by a tough layer of insoluble **keratin.** The skin is covered with fatty acids, salts, and enzymes that present a very inhospitable environment for many bacteria. Other bacteria, though, do quite well on the skin while generally doing us no

TABLE 20.1
Nonspecific Defense Responses

BARRIERS OF THE BODY COVERING	INFLAMMATORY RESPONSE
1. Intact skin covered with acids, salts, enzymes.	1. In damaged or invaded tissues, blood vessels dilate.
2. Ciliated mucous membranes lining parts of the respiratory tract.	2. Seepage from blood vessels causes local swelling, and also carries with it into the tissues proteins that fight infections.
3. Exocrine gland secretions in surface epithelium.	3. Phagocytes arrive at affected tissues and engulf invaders.
4. Acidic fluid in the stomach, basic in the intestine.	4. Clotting mechanisms result in tissue repair.
5. Microbes that usually inhabit the skin, gut, and vagina.	
6. Lysozyme in sweat, tears, saliva.	
7. Cleansing action of fluids.	

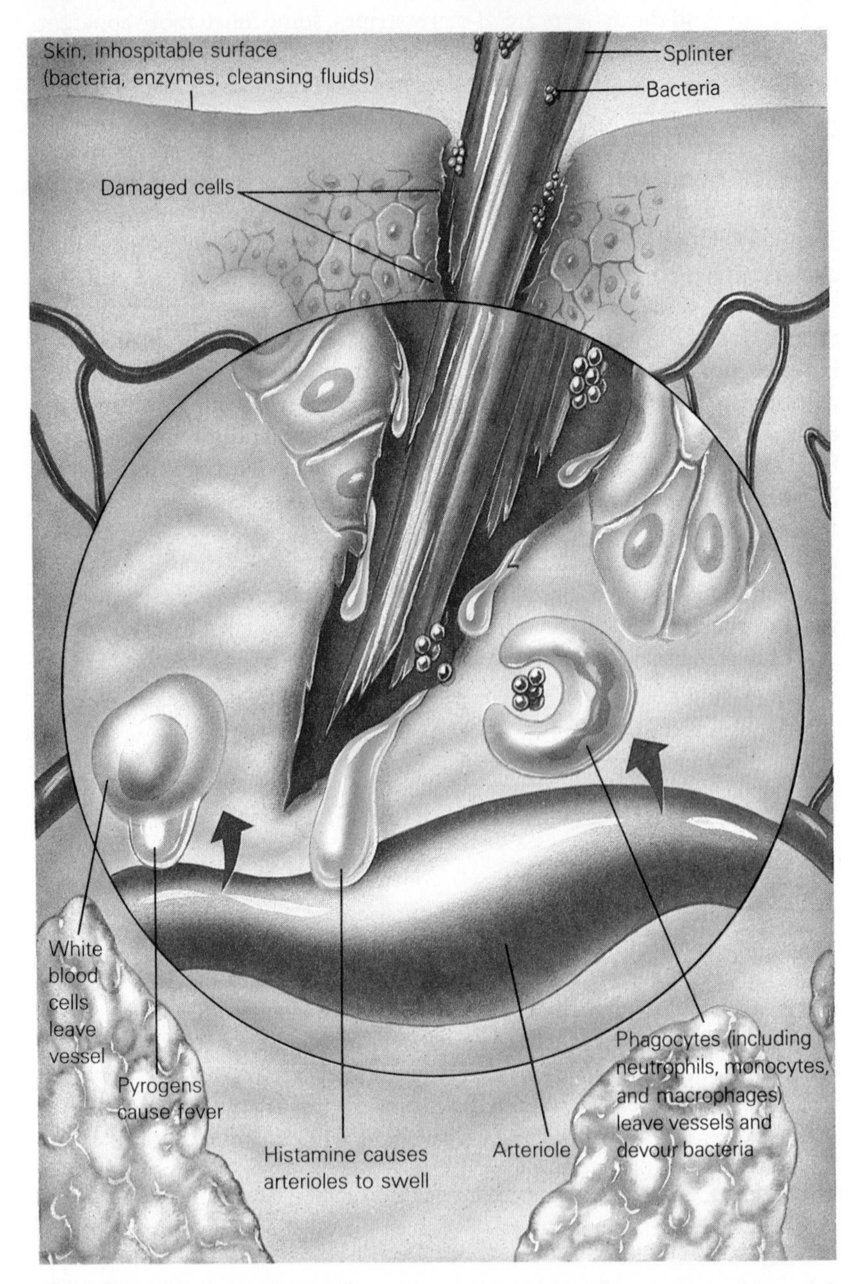

FIGURE 20.1

Nonspecific responses. Here, defense reactions are against the trauma and infection caused by a splinter.

harm. Their presence, though, means competition for any new bacterial colonizers that can survive the environment. Furthermore, sweat, tears, and saliva also contain the enzyme **lysozyme** that can rupture the walls of some bacteria. All three of these fluids can wash away potential invaders.

The inner body covering of mucous tissue lines the gut, the respiratory tract, and the reproductive tract. The mucus itself entraps invading microorganisms and either sweeps them away by the action of beating cilia or holds them until they can be engulfed by roaming white blood cells. In the gut, the highly acidic stomach contents followed by the very basic fluids of the upper intestine kill many forms of microorganisms that enter with the food. The vagina also protects itself by promoting the growth of acid-producing bacteria. In the urinary tract bacteria find it hard to get a toehold for the obvious reason: they are swept away.

Although the body's covering presents a formidable barrier to invading organisms, they do routinely make their way into our bloodstream. (It is disconcerting to learn how many germs can ride in on one splinter.) Once any organism makes its way into the body's interior, it triggers an **inflammatory response,** marked by a reddening area that becomes warmer and increasingly tender. It begins as the cells at the site of the infection immediately begin to secrete **histamine,** which dilates tiny arterioles bringing more blood to the injured area. The increased blood flow delivers defensive substances and cells, and rinses away toxic waste products of the invading organism and dead cells. It also causes the soreness, redness, and swelling, encouraging us to pamper the sensitive area until the infection is beaten back. (Antihistamines have the opposite effect. See Essay 20.2 for a discussion of the body's overreactions to certain stimuli.)

In some situations, the nonspecific immune response is not localized to a single site, but is systemic. In such a case, the entire body reacts, for example by producing fever. Fever is triggered by either toxins produced by the invading organism or by **pyrogens,** chemicals released by certain white blood cells as they respond to an invasion. Pyrogens essentially set the body's thermostat to a higher level (see Chapter 18). The increased temperature can make the body inhospitable to many kinds of invading microorganisms. See Figure 20.1 for a summary of nonspecific responses.

At least five kinds of blood cells are involved in human responses. Figure 20.2 shows the major kinds of blood cells and their origins in the body. Three are **phagocytes—eosinophils, neutrophils,** and **monocytes** engulf any invaders in the bloodstream. **Lymphocytes** produce both cells and proteins that interact in both nonspecific and specific responses. The **basophils** secrete histamine that intensifies the inflammatory response (Table 20.2).

TABLE 20.2
White Blood Cells and Their Functions

CELL TYPE	FUNCTION
PHAGOCYTES	
Neutrophil	Participates in early stages of defense against microorganisms.
Monocyte	Arrives at site after neutrophils, transforms into macrophages; engulfs foreign materials, presents antigens to lymphocytes, stimulates lymphocyte proliferation.
Eosinophil	Responds to allergies and parasitic infections.
LYMPHOCYTES	
Cytotoxic T-cell	Destroys virus-infected and cancerous cells.
Helper T-cell	Stimulates B-cell and cytotoxic T-cell proliferation.
Suppressor T-cell	Slows down immune response.
B-cell	When activated by foreign molecules, produces plasma and memory cells.
Plasma cell	Secretes antibodies.
Memory B-cell	Responds to antigens during secondary response.
Natural killer cell	Directly destroys virus-infected cells and cancerous cells.
BASOPHILS	
	Release histamine in inflammatory response (as do damaged body cells).

FIGURE 20.2

The white blood cells (leukocytes) are far less numerous than red cells. Whereas there are about 5 million red blood cells per cubic millimeter in the adult, there are only 5000–9000 white cells. These are of five known types. Neutrophils and monocytes are phagocytic, engulfing foreign matter, bacteria, and cellular debris from cells destroyed by infection. Lymphocytes are undifferentiated cells that can take a number of immunological directions. Eosinophils increase in number in the presence of foreign proteins, but no one knows why. Basophils secrete a histamine (which causes damaged areas to swell with fluid). Neutrophils and lymphocytes make up about 95 percent of the white cells.

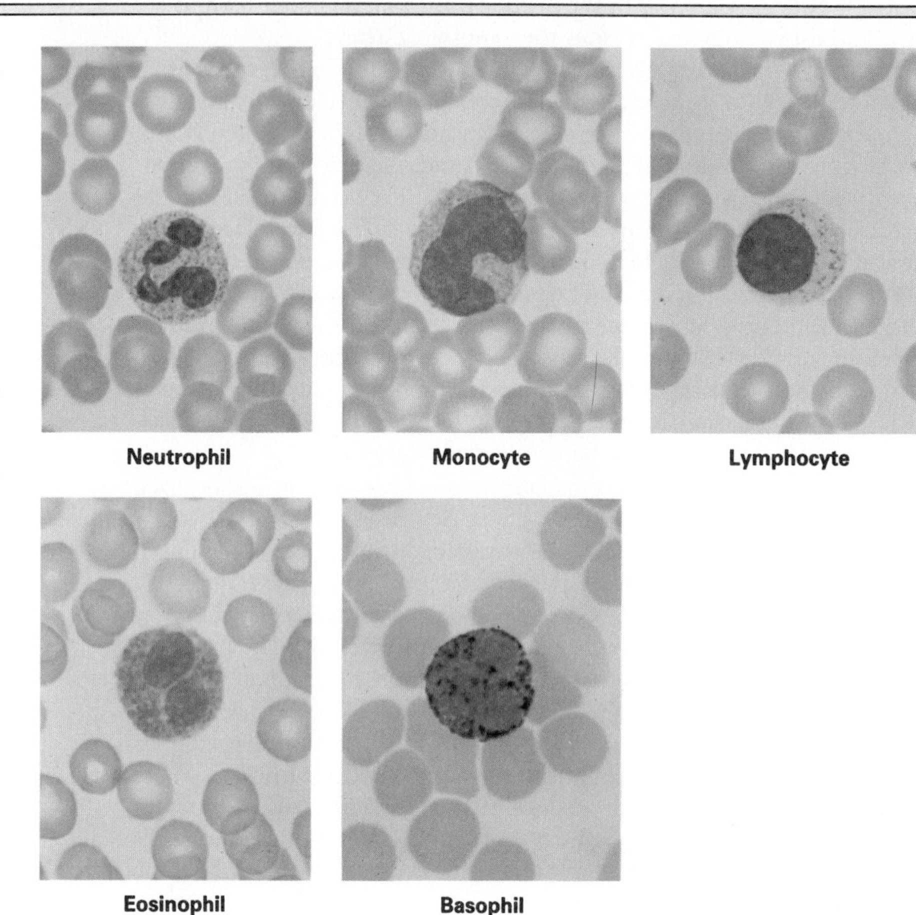

FIGURE 20.3

A macrophage ingesting a red blood cell.

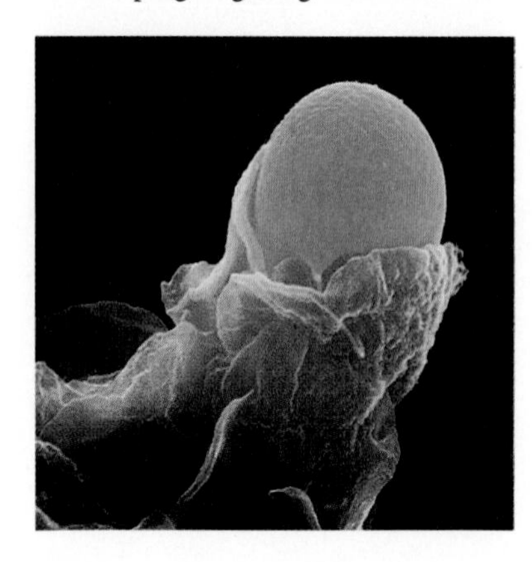

The eosinophils primarily respond to allergies and parasitic infections, but here we will concentrate primarily on the neutrophils and monocytes. The neutrophils are the expendable, frontline soldiers. Hordes of them (perhaps 100 billion) are produced each day and they are the first to arrive at the site of any invasion. They don't survive long, but they may overwhelm an invader by their sheer numbers.

The monocytes arrive next and once they encounter the invader they begin to undergo remarkable changes, growing and swelling until they become huge **macrophages** (*macro:* large). These cells are veritable eating machines that may live for years (Figure 20.3). In spite of their relatively large size, their action is remarkably swift—they can engulf a foreign particle in less than 1/100 second. (The macrophages will also play an important role in the specific responses.) Interestingly, if they come across a particularly large invader—too large for one cell to handle—several of them may merge, their membranes fusing, until they form a giant macrophage that then proceeds to engulf the invader.

The **natural killer cell** (or **NK cell**) is active in the nonspecific response. NK cells are formed from large, granular lymphocytes. They roam the body, constantly checking the body's own cells. When they encounter cancerous cells, or cells harboring viruses, they immediately attack those cells, rupturing their membranes.

ESSAY 20.1

Lyme Disease

As if you didn't have enough to worry about, now there's Lyme disease. (If you haven't heard of it, raise your hand.) What it is, is a disease caused by a spirochaete, a corkscrew-shaped bacterium, *Borrelia burgdorferi,* and carried by a tick, *Imodes dammini* (at least in the Northeast and Midwest). The tick has three stages in its life cycle. In its larval stage, it mainly infects the white-footed mouse. In its nymph stage, it infects a range of mammals, including dogs, raccoons, and humans. In its adult stage, it mainly infects white-tailed deer. The problem is, in the nymph stage, it is very small, about the size of a comma. (It may look like a tiny moving freckle.) Thus, it is easy to overlook.

The disease was first identified in 1975 in Lyme, Connecticut, and has since spread swiftly throughout the United States. In fact, there may have been 50,000 cases in 1989 alone. We say "may have been" because Lyme is the great imitator, mimicking a number of other diseases. Physicians may have a difficult time in diagnosing Lyme (although new procedures are being implemented).

The first sign of the disease may be flu-like symptoms, including headache, fever, weakness, and stiff joints. If antibiotics, such as tetracycline or doxycycline, are administered immediately, the victim is likely to recover. However, as time passes without diagnosis and proper treatment, a debilitating arthritis may appear (from which the victim may never recover). In addition, the spirochaete may invade the brain and spinal cord, producing dizziness, incoherency, visual problems and numbness (suggesting multiple sclerosis). The victim may also begin to have seizures. In pregnant women, Lyme can cross the placenta and cause fetal damage.

In general, the warning signs are a "bull's-eye" rash that appears days to weeks after a bite, followed by fever, chills, fatigue, and headache. The joints may swell and ache and in some cases the heart may beat arrhythmically, with the legs growing very weak. Facial paralysis and numbness may also appear.

The best way to avoid the disease is to avoid the tick. Keep your pets tick-free, stay out of brushy and grassy areas, wear long clothing (tuck your pants into your socks). Use a repellent that contains DEET and permanone, and check yourself for moving freckles.

SPECIFIC RESPONSES

In the fourteenth century, European cities were crowded, dirty, and filled with travellers. Conditions were right for a flea-borne pathogen, *Pasteurella,* to sweep repeatedly through the population causing bubonic plague, or Black Death. Within a few years, one quarter of the population of Europe had been killed. A few infected people managed to survive each onslaught and it was noticed that, for some reason, they were immune to the disease from that time on.

We see the same principle when schoolchildren come down with chickenpox. Once they've had the disease, that's it, they don't catch it again. We now know they're safe because the immune system has been activated against that disease. Such responses are part of the specific responses, when the body is programmed to be activated against a specific invader.

ESSAY 20.2

Allergy: An Overreaction

Certain seasons of the year, when pollen is produced, bring anxiety and apprehension to many people. Others must, at all times, avoid certain wines, cheeses, or oysters. Yet others cannot be around cat dander or dust. The reason is these people are allergic to something associated with these conditions. (You won't be pleased to learn that household dust often contains tiny mites that are kicked up into the air by vacuuming or sweeping, and then drawn into the delicate respiratory tract by breathing the dust.)

The body has ways of cleansing itself of intruders or incompatible substances, such as by violent rushes of air or profuse production of cleansing fluids. And so we sneeze and cough and our eyes and noses run when we encounter some irritant.

In some cases, though, the body overreacts to one of these irritants. This extreme sensitivity may occur after the body has been repeatedly exposed to the antigen in the irritant so that the memory B-cells are always standing by. For example, certain kinds of pollen can bring on the reddening eyes, the sneezing and snuffling that many of us associate with spring. The reaction is definitely extreme for the minor threat presented by the pollen. So what triggers such a violent allergic response? We don't know why the response is so extreme, but we do know, generally, what happens.

The pollen lands on stationary granulated cells called mast cells, which then explosively release their granules. These granules contain histamine, which causes capillary walls to become leaky. Fluid then leaks from the capillaries into the tissue spaces, causing the tissue to swell. The respiratory tract is often the first to encounter any airborne irritant and the swelling tissues are accompanied by coughing, wheezing, sneezing, and a profusely runny nose—usually during a job interview. The result is that the irritant is removed, but the removal mechanism has been extreme, both physically and socially.

Allergic reactions can also produce a dangerous condition called anaphylactic shock. This happens when large areas of the body produce an allergic reaction, as when an individual allergic to bee stings or penicillin receives those antigens and they are rapidly transported throughout the body. In such a case, capillaries throughout the body become leaky and blood pressure suddenly drops, reducing blood flow to the brain and heart. Death can result unless epinephrine (adrenaline) is quickly administered, thereby constricting blood vessels and stopping the leakage.

Two kinds of lymphocytes play a critical role in the specific responses. These are the **B-cells** and the **T-cells.** Let's briefly set the stage for their roles here. We will see that the B-cells are specialized to do two things. One type of B-cell produces **plasma cells** that make antibodies to combat the invader. Antibodies are secreted into the blood, so the B-cells are said to be involved in **humoral immunity** (*humor:* fluid). The second type, memory B-cells, forms a residual force that continues in the body long after the invasion is past, ready to mount a rapid attack should that particular invader show up again. (Other kinds of cells, including helper T-cells, also form memory cells, ensuring a swift and effective response to a second invasion.)

We will also see that the roles of the T-cells are also quite specific. The T-cells must contact the invader in order to attack it, so the T-cells are said to be involved in **cellular immunity.** We will consider three basic kinds of T-cells: helper T-cells, cytotoxic T-cells, and suppressor T-cells. The helper T-cells will interact with other cells to enhance the immune response. The **cytotoxic T-cells** (*cyto:* cell; *toxic:* poison) identify invading cells and rupture their membranes, and the suppressor T-cells help call off the body's defenses.

We will take a closer look at how the B-cells and T-cells work shortly, but first let's consider antibodies and their role in defending the body.

The Antigen-Antibody Response

The cells of the immune system recognize invading organisms and abnormal body cells by certain molecules, called antigens, that they bear on their cell surfaces. **Antigens** are foreign molecules that elicit an immune response in the host organism. When a host is invaded by an antigen-bearing body, it forms antibodies against that antigen. **Antibodies** are proteins produced by the host plasma cells that identify and help destroy antigen-bearing cells.

There are several general classes of antibodies. Typically, they are composed of two identical long **heavy chains** and two shorter, identical **light chains,** arranged in the form of a Y (Figure 20.4). The ends of the arms of the Y are highly variable. That is, the molecules of the arms can take any of millions of different configurations (and so it is called the **variable region**). The rest of the molecule can take only a few different forms (and so it is called the **constant region**). The molecules are placed into their classes according to the configuration of the constant regions. In a sense, the variable region determines whether an antigen is attacked and the constant region determines how any attack is handled.

Basically, antibodies attack antigens in three general ways. First, those with multiple binding sites (with more than one Y, such as IgM) can link groups of antigens together, making the immobile masses easier for the body to deal with (and for phagocytes to devour). Second, antibodies may attach to various sites on a single invader, essentially coating it and marking it for attack by phagocytes. Third, antibodies can trigger a set of reactions that ruptures the membrane of an invading cell.

FIGURE 20.4
Antibodies are proteins that consist of two light and two heavy polypeptide chains, connected by disulfide linkages. Each chain has two general regions common to a number of antibodies and two highly specific antigen recognition sites that bind only to specific antigens.

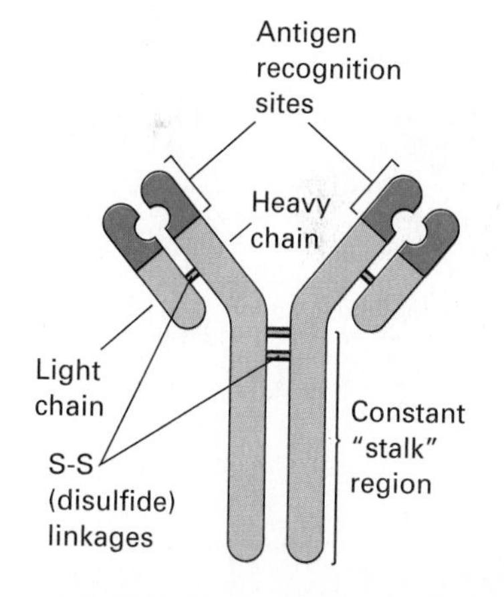

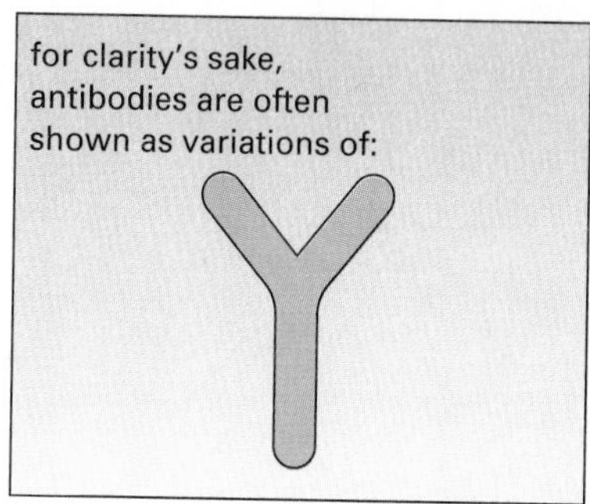

FIGURE 20.5
When a specific antigen is encountered, the recognition regions of the antibody molecules attach to specific binding sites on the antigen, eventually forming an immobile mass that can be engulfed by phagocytes. In other cases, the antibody may simply destroy the antigen.

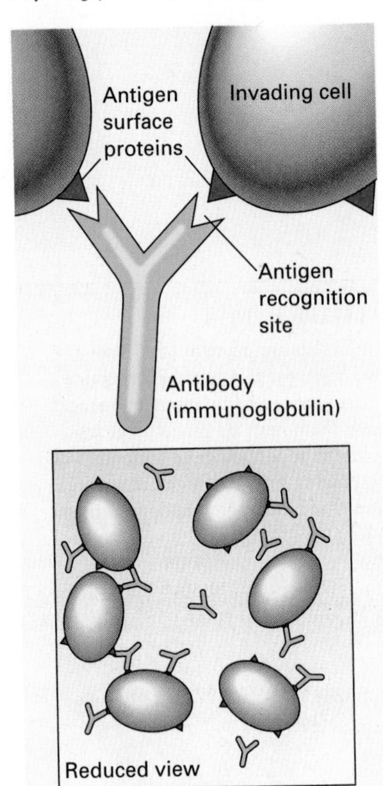

Programming the Lymphocytes

Macrophages, we know, are descended from white blood cells called monocytes, that roam the body attacking invaders with a lightning swiftness. But they have another role as well. When they ingest foreign bodies and dismantle them with their powerful digestive juices, they take the bits and pieces of the victim—pieces that contain the antigen—and wear them on their own membranes (almost like a headhunter carrying an enemy skull). As they move about the body with the grisly trophies studding their membranes, they encounter multitudes of lymphocytes. Among these are helper T-cells, a few of which have a precise **antigen recognition site** with a configuration that matches the molecular structure of a single antigen, much like the match of a lock and key. Sooner or later the macrophage encounters a helper T-cell whose "antigen recognition site" (Essay 20.3) precisely matches the antigen embedded in the macrophage's membrane (Figure 20.5). When this happens, the matching antigen and antigen recognition site lock together (chemically, much as enzymes and substrates lock together). This union arouses the helper T-cells to stimulate rapid cell divisions of cytotoxic T-cells and of B-cells that have also recognized and bound to the foreign antigen. The frenzy of cell division in both kinds of cells is triggered by **interleukin II,** a chemical released by the helper T-cell. Next we will consider the roles of the cytotoxic T-cells and the B-cells, but keep the big picture in mind as we wind our way through a few details: each lymphocyte produced after activation by a helper T-cell bears the antigen recognition site that matches the antigen borne by the macrophage. In this way, an army of lymphocytes is formed, all programmed to attach to any invader bearing that particular antigen. This immune defense is summarized in Figure 20.6.

The Role of the Cytotoxic T-Cells

The cytotoxic T-cells immediately act against invaders. As soon as they are formed, they begin to roam the body, approaching one cell after another. If the cells are normal and healthy (that is, if they are not cancerous and if they don't carry the targeted antigens) the cytotoxic T-cell goes on its way. But if it encounters cancerous cells or a cell that harbors viruses, the infected cell will be destroyed. Cytotoxic T-cells then, attack the body's own infected or abnormal cells. As we will see, the next line of defense attacks not the body's own cells, but extracellular viruses, bacteria, and other invaders.

The Roles of the Helper T-Cells and the B-Cells

B-cells have antigen-recognition sites, just as do the T-cells, and so they are able to recognize and attack any particles or invading cells bearing matching antigens. By themselves though, they are only able to hold them, waiting for a little assistance from helper T-cells. We have already seen that when a helper T-cell encounters a B-cell that has attached to an antigen, the helper T-cell activates this B-cell, causing it to divide rapidly.

FIGURE 20.6

In the ongoing combat against invaders, macrophages bearing captured cell-surface antigens seek out lymphocytes (helper T-cells) with matching membrane surfaces. Once a match is made, the macrophage induces the lymphocyte to begin to reproduce its specific line. These cells then trigger the production of cytotoxic T-cells or B-cells, which then transform into plasma cells that secrete antibodies.

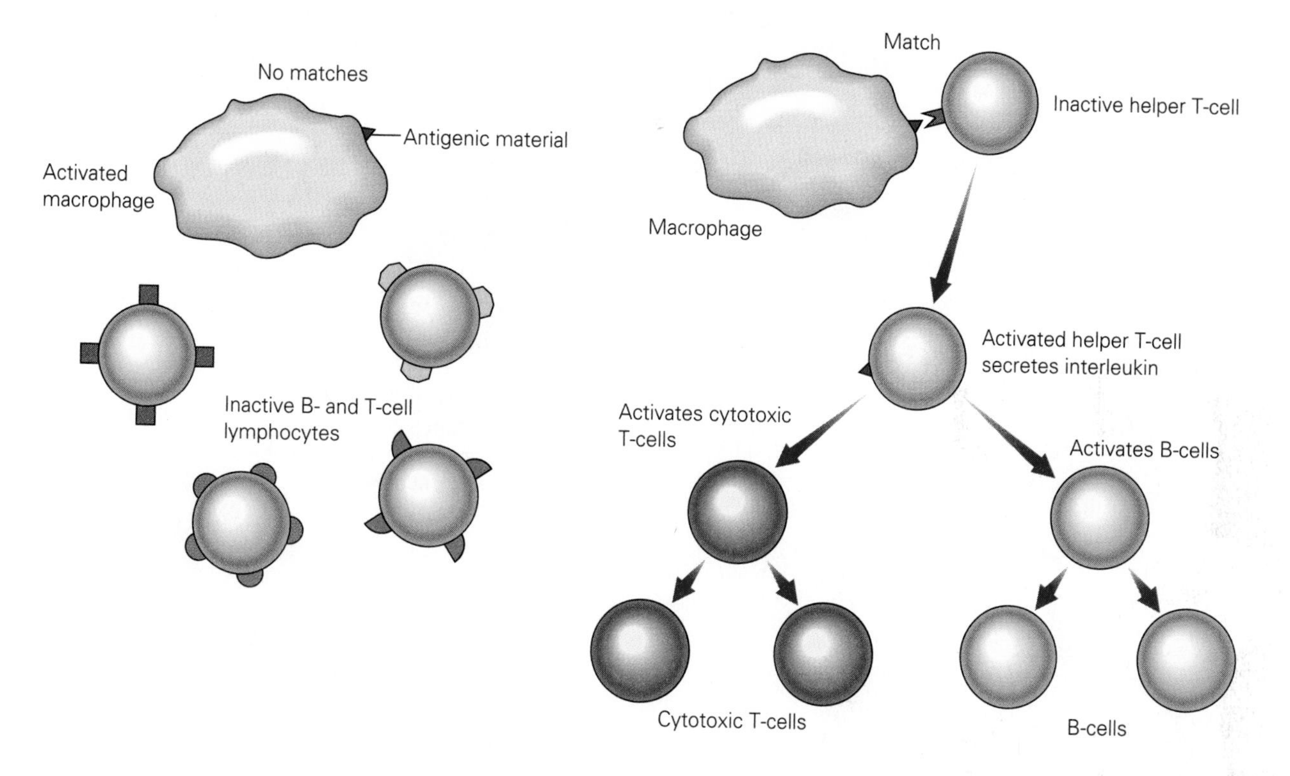

The B-cell produces two kinds of cells: *plasma cells* and **memory B-cells.** The plasma cells live only a few days but during that time they constantly produce antibodies against the antigen that their parent B-cell discovered. Memory B-cells live much longer, perhaps for years. If that same antigen should ever invade the body again, the memory B-cells are waiting for them. The antigen is quickly discovered and the B-cells (with help from the helper T-cells) immediately begin producing new plasma cells and memory B-cells. The memory response is so effective that we catch certain diseases only once. Any later invasion is beaten back before it can get started. As the invasion subsides the defense is called off by the class of lymphocyte called **suppressor T-cells.** (The specific response is summarized in Figure 20.7.)

These defenses can be divided into two stages: the **primary response** occurs when a foreign "nonself" substance is encountered for the first time. The sequence that follows the one we have just described (the helper T-cell joining the complex and prompting the proliferation of B-cells that then manufacture antibodies and of cytotoxic T-cells) takes time and so

FIGURE 20.7
Specific responses.

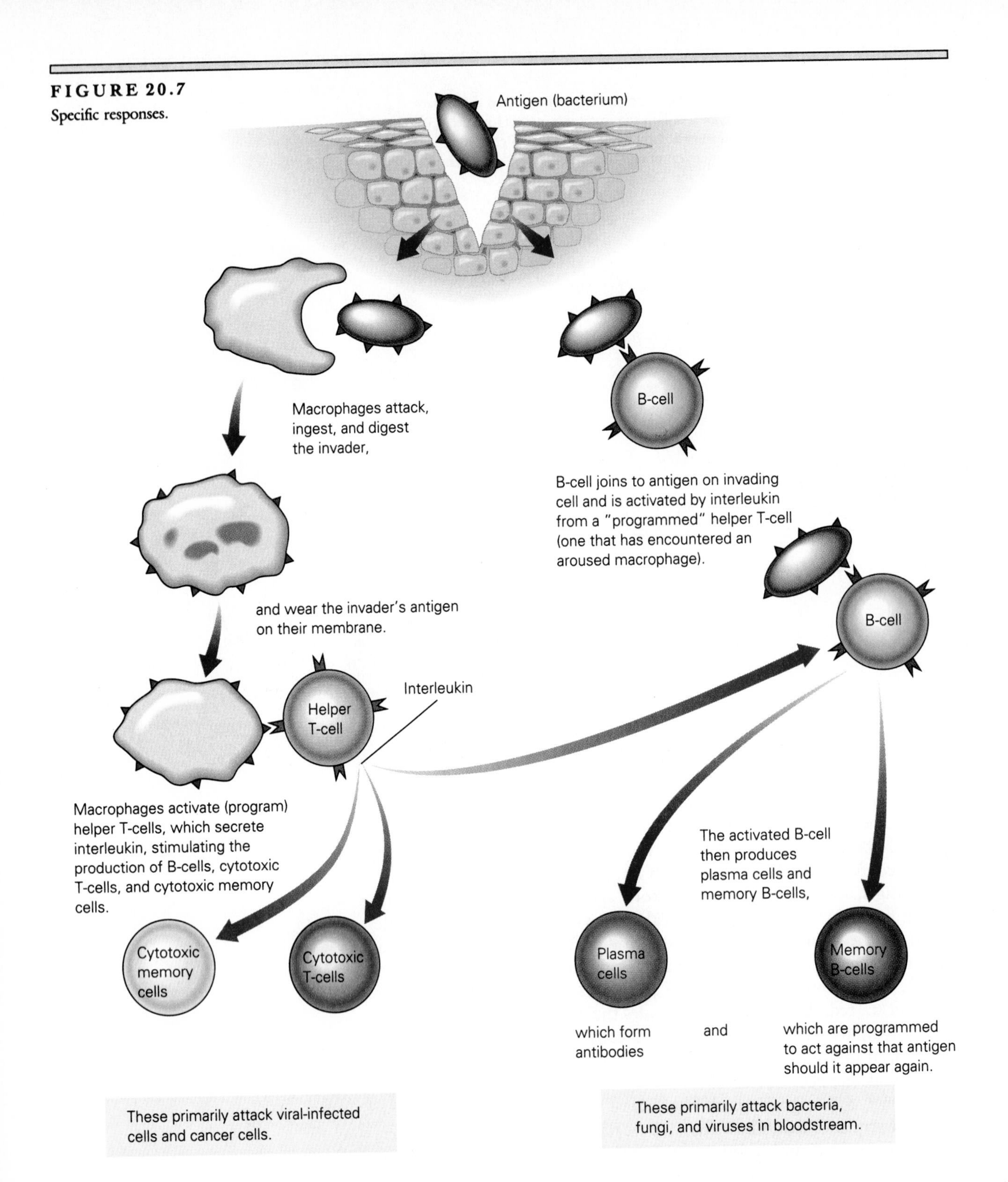

ESSAY 20.3

How African Crocodiles Defend Against Texan Rabbits

If you take hemoglobin from a rabbit that lives in Houston and inject it into a crocodile from the Nile Valley, the crocodile will form antibodies against the rabbit hemoglobin. This means that the crocodile has immune cells that can recognize rabbit blood, even from Texan rabbits. How can this be? Does this imply that the crocodile (or any animal, for that matter) has cells with recognition sites against everything? Furthermore, does this mean that there are enough antibody-coding genes to go around so that any antigen can be matched? Are there enough genes to code for millions upon millions of potential antigens? The answer is yes and no.

Actually, there are only about 300 DNA segments that code for the variable regions of antibodies, clearly not enough to provide an antibody response against every possible antigen. However, these regions are extremely unstable. They break apart and rejoin endlessly, providing a virtually infinite number of combinations (in humans, about 18 billion). When you consider that even this diversity can be increased by single-gene mutations in the antibody-coding DNA, there is indeed enough variation to produce antibodies against just about everything.

the primary response is a bit slow. The **secondary response** is the sequence that is triggered by the army of memory cells produced by the primary response. It is, indeed, so swift and powerful that it can stop a second infection before it can get started.

Vaccinations often involve the injection of antigens (usually associated with dead, altered, or weakened infectious organisms). The body is not at risk, but the lymphocytes detect the agent and mount an immune response (often with few symptoms). Memory cells are thus produced against that antigen and so the body is immune to any later invasion by infectious agents bearing that antigen.

Tolerance and Autoimmunity

All the cells of the body bear their own molecular "markers" embedded in their membranes. These are different for each person and so, in a sense, each cell bears the individual's genetic signature. As the cells of the immune system roam the body, the body's own cells are identified by these markers and so they are not attacked. This acceptance by the body's immune cells is called **tolerance.** Such tolerance means that the B- and T-cells do not begin their immune sequences in response to the body's own molecules. By the same token, those very markers act as antigens when one individual's cells are presented to the immune system of another individual. The immune response of the recipient is, in this way, responsible for the rejection of organ transplants.

In some cases, unfortunately, the immune system *can* turn against its own body, with disastrous results. When this happens, either T- or B-cells begin to recognize the body's cells as antigens and to form clones that act against them. The reaction is called **autoimmunity** (*auto:* self).

It is not known what causes autoimmune reactions and there may be several causes. As an example, antibodies may cross-react with one's own cells. In such a case, antibodies produced against strep throat can, for some reason, begin to act against heart muscles, causing rheumatic fever. As another example, there is evidence that the body's immune system learns the characteristics of "self" at some early embryonic stage. Tissues that are not presented to the immune system at that time are not learned. The cornea of the eye has no blood vessels and so is not presented to the immune system during development. Later in life, though, should the cornea be injured so that its tissue is exposed to the immune system, the lymphocytes will be activated and the cornea attacked, causing the eye to become white and opaque, often causing blindness.

The most common autoimmune disease (affecting about 30 million people) is a form of arthritis that is usually manifest as a crippling inflammation of the joints, but can also affect the spleen, heart, lungs, and blood vessels. Autoimmunity can also cause lupus erythematosus (which affects the kidney), pernicious anemia, and thyroiditis. Since no one is sure what turns the B- and T-cells against one's self, the most promising avenues of treatment at the present are the removal of the lymphocytes by filtering and the administration of immunosuppressant drugs. (What might be an unfortunate side effect of either of these treatments?)

INTERFERON: AN EXCITING NEW PROMISE

The term **interferon** comes from "interference phenomenon," which refers to a group of antiviral substances manufactured by the cells of most vertebrates in response to viral attack. Interferon causes cells to become resistant to attacks by other viruses. As we saw in Chapter 11, an attacking virus tends to alter a cell's own replicating mechanisms, using them to make, instead, more viruses that can infect other cells. Interferon helps to block this deadly geometric increase. It does not act against specific viruses, but will inhibit *any* viral attack. Interferon from one cell can help other cells to resist viral attack, but interferon from one species cannot increase resistance to viruses in another species.

When it recently became known that interferon could be manufactured by recombinant techniques, hopes in the medical community soared. Interferon seemed to be the answer to everything from cancer to the common cold. But the promises were apparently premature; interferon was simply not the magical cure-all we hoped. However, research is progressing, and there are promising signs. In one experiment, not one of 11 volunteers given interferon in a nasal spray caught cold after being exposed to cold viruses, while in control groups, 8 of 11 people exposed and given plain water spray did catch cold. Interferon has also reduced tumor size in a number of patients who did not respond to other treatment. In one case, two of three separate cancers discovered in one man completely disappeared after treatment with interferon.

Interferon does have side effects. In some people, it triggers irregular heartbeats. It may also complicate liver or kidney problems. In high doses, interferon can cause mental confusion, change brain waves, and can bring on seizures. Nonetheless, it is still considered a potentially useful substance and may become a superb form of treatment for some ills, when we learn more about it.

AIDS: A DEVASTATING NEW PROBLEM

Not long ago, an accused murderer was led into a courtroom by a sheriff's deputy who was wearing rubber gloves. The jurors facing the accused did not include the 14 people who had asked to be excused because of the medical condition of the accused. The unusual circumstances arose because the defendant was guilty of having AIDS. The "rubber gloves treatment" is disconcertingly routine in other areas, even those in which AIDS is extremely unlikely (Figure 20.8).

The fear of AIDS now probably surpasses the fear of flying. Sexual behavior in the United States has changed, it has been said, not by messages from the pulpits so much as by messages from the Centers for Disease Control. Not only is casual sex avoided by most informed people, but people with AIDS or those in high risk groups are often shunned, even by those in the health services community. The argument regarding the

FIGURE 20.8

Some of the public responses to the AIDS scare have taken unexpected turns. Some boxing commissions, for example, require referees to wear rubber gloves in the ring.

FIGURE 20.9
AIDS manifests itself in numerous ways, all marked by a deteriorating immune system.

contagion of the **human immunodeficiency virus (HIV),** which causes AIDS, seems to be more vigorous outside the medical community, however, because most researchers in the area seem to agree that the virus is not particularly contagious if certain simple safeguards are taken. But what is AIDS? What is the problem? And what are the safeguards? And if it's not so contagious, why are so many people dying?

AIDS is an acronym for **acquired immune deficiency syndrome.** Essentially, it acts by suppressing the victim's immune system (see Essay 20.4). Presently it is thought that macrophages may be the first cells attacked by the HIV virus, but eventually the helper T-cells are killed. This leaves the body unable to mount a defense against disease organisms that it would normally fight off with ease. People with AIDS are, therefore, susceptible to virtually any disease. In fact, the appearance of rare diseases such as Kaposi's sarcoma (a skin cancer) and pneumocystic pneumonia frequently occur with the virus. Early signs of AIDS include a series of lingering, simple colds, "night sweats," persistent fever, swollen glands, and coughing. (Immediately upon learning this, of course, everyone detects just those symptoms in themselves.) More serious conditions follow, including at least three forms of cancer and destruction of the lungs and brain (Figure 20.9).

By some accounts, the first case of AIDS in the United States appeared in 1979, followed by a half-dozen cases reported in Los Angeles in 1981. In early 1989, the World Health Organization estimated that over one new case of AIDS was developing each minute, worldwide, with 1 million new cases of AIDS expected to be reported by 1993. (The figures have since been revised upward.) It is estimated that between 6 and 10 million people are presently affected with the virus but do not yet show symptoms (which may not appear for years after infection). Many people carry antibodies to the virus that causes AIDS, showing that they have been exposed to it, and some may carry the virus in its early stages without developing the symptoms. It is thought that such people may be able to transmit the virus, nonetheless. Furthermore, one-quarter of a group of high-risk men who had tested negative for the antibody were found to be carrying the virus.

The syndrome, once full blown, is believed to be incurable and to virtually always cause death within a few years (fewer than 14 percent of victims survive past three years). Because so much is unknown about AIDS, much of what is known is misconstrued, often by sensationalist media. Who, then, is at risk, and how does the syndrome progress?

AIDS was once thought to be confined to homosexual men and intravenous drug abusers, the two highest-risk groups in the United States. The fact is, all sexually active people are at risk. In Africa, in fact, AIDS is commonly spread through heterosexual intercourse. The agent, a virus (Essay 20.3), is transmitted in the blood and semen, but is also found in sweat, tears, and mucus. The primary means of contagion is believed to be anal intercourse, when the delicate tissues of the bowel are likely to be injured, allowing the virus to enter the blood through broken vessels. Vaginal intercourse with an infected man is less risky for the woman because the vaginal wall is normally not abraded during the act. The drug abusers may pass the virus along by sharing infected needles. One problem here is that those in the drug subculture are among the least informed people in our society and so they often continue their practices out of sheer ignorance. Some addicts have taken to dipping their needles in bleach before each use, but even this precaution is not entirely effective (Figure 20.10). As with the use of condoms (below) the procedure for cleaning a contaminated syringe must be carefully attended to. There have been cases of people contracting AIDS through medical blood transfusions, but careful screening and processing of blood is reducing this risk. The risk of contracting the virus through heterosexual vaginal intercourse is extremely low but on the rise. One problem is the dependence on condoms as a protection. Unfortunately, some types of condoms are not completely effective. For example, those made of animal membranes, rather than rubber, do not block the passage of viruses. Also, unfortunately, some men simply do not know how to use them safely. (The condom-sheathed penis must be removed immediately after ejaculation.) There are also other means of entrance to the heterosexual population. Some women have even been infected by artificial insemination. And, sadly, AIDS can be contracted by the fetus while still in the uterus.

The geographical source of AIDS is not entirely established, but some researchers believe the virus is a mutant of a strain that infects the African green monkey, a species that lives in close contact with humans in West Africa. The condition is widespread in certain areas there. Certain French-speaking and AIDS-ridden nations of West Africa have developed exchange programs with Haiti, a favorite vacation area of American homosexuals, and AIDS may have spread to America in this way. One problem with tracing the movement, sources, and modes of transmission of the virus is that people tend to be less than honest about their sexual behavior. For example, almost all the hundred or so AIDS victims in the American military claim to have contracted the disease from prostitutes. Of course to say otherwise would be grounds for prosecution and discharge. At present, the American Centers for Disease Control maintain that AIDS is not likely to sweep through the general population, but is

FIGURE 20.10
AIDS is especially prevalent among needle-sharing addicts, probably among our society's least informed citizens.

ESSAY 20.4

AIDS

The infectious agent of AIDS was discovered in 1984 independently by French and American researchers (with some researchers dissenting, saying that the disease is not viral at all). The culprit, a virus, is called HIV (for human immunodeficiency virus). The virus attacks the helper T-cell, penetrating the cell and releasing a single strand of RNA and an enzyme, reverse transcriptase, within it. The enzyme enables the RNA to make a double strand of complementary DNA, which then joins the helper T-cell's DNA. After this, the helper T-cell may continue on as if nothing had happened. Most helper T-cells, after all, lead quiet lives, never being activated. In some cases, though, the infected helper T-cells begin a round of replication, perhaps triggered by an infection to which it is called to react. Perhaps such a defensive event prompts the viral DNA to take over the helper T-cell's genetic machinery. In any case, at some point the viral DNA causes the cell to enter a vigorous round of production that produces not new helper T-cells, but new viruses. The helper T-cell is killed and a host of new viruses is released into the bloodstream to attack other helper T-cells. Finally, there are so few helper T-cells in the blood that no effective immune response can be mounted against any attack whatsoever. The virus does not limit its attack to helper T-cells. Recently, investigators have discovered that it may attack macrophages first. In some cases, it attacks brain cells, bringing on the dementia often associated with AIDS.

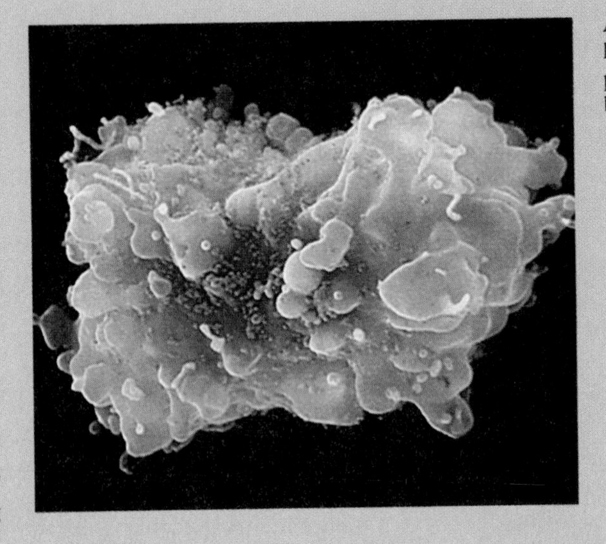

AIDS virus attacking lymphocyte. The viral particles are shown as blue.

The Micro-anatomy of an AIDS Virus (HIV)

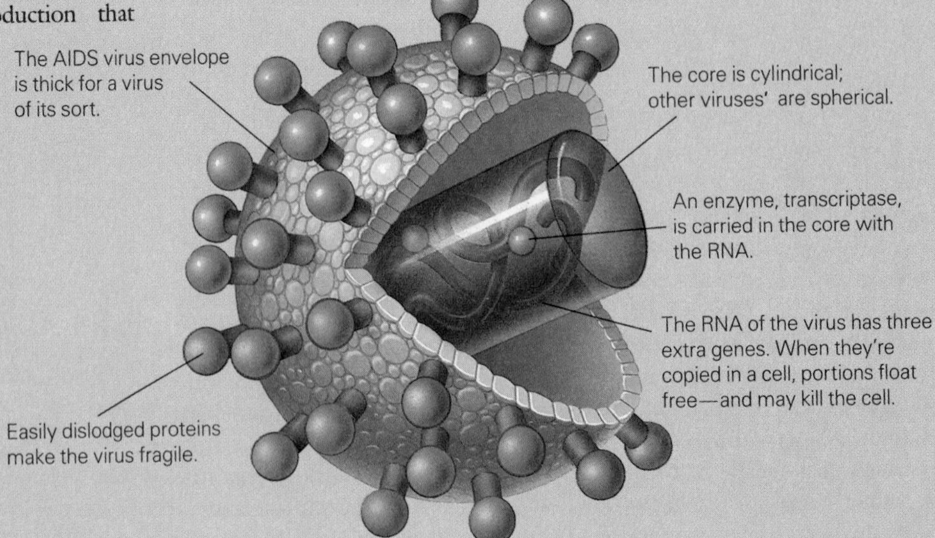

1. Upon entering the bloodstream, the AIDS virus seeks out T-cell lymphocytes, the white blood cells that serve as master controls for the body's immune system.
2. Using the various chemical markers on its surface, the AIDS virus binds easily to a receptor on the surface of the T-cell.
3. The virus then enters the T-cell, in the process shedding its protective protein coat and exposing its core, which contains RNA and the enzyme transcriptase.
4. The enzyme in the virus helps translate the invading RNA into DNA. The DNA is treated by the host cell as its own, and inserted into its chromosomes.
5. The AIDS virus DNA may remain dormant for weeks, months, even years inside the usurped T-cell before it begins to cause disease.
6. Once activated, the viral DNA directs the T-cell to make copies of the virus. The new viruses bud from the T-cell's surface. Eventually the host cell dies.

likely to continue to be transmitted through its present means. In West Africa, the condition affects men and women in roughly equal numbers, but it has been pointed out that heterosexual anal intercourse is common there (often after ritual clitorectomy).

The problem of AIDS all too clearly illustrates the challenges presented to our immune systems. Not only must our own bodies fend off the usual agents that have attacked our delicate systems over evolutionary time, but they must stand ready to meet the new challenges that can be expected in the changing world of living things.

MIND AND BODY

We've discussed the immune system here in very mechanical terms (A triggers B which with C causes D). It's all very tidy, even when we point out those areas that we don't really understand. It appears, though, that if we are to even truly understand the immune system, we're going to have to come to grips with how it is affected by the mind. The very consideration of this topic is met with great resistance in some scientific circles, largely because of the "pop psychology" atmosphere that often surrounds such musings. We hear things like "You're as young as you think you are"—this is often said to octogenarians. (Or, "He thinks he's dead and I hope he's right because we buried him.") Nonetheless, serious researchers are giving increasing credence to the notion that our physical well-being and, in particular, our immune system, is influenced by our state of mind.

FIGURE 20.11
The loss of a loved one can repress immune systems in ways we do not fully understand.

People have long noticed a relationship between mood and illness. Depressed people seem to get sick more often than happy, cheerful people. Bereaved people also tend to be ill more often than others do (Figure 20.11). In a study designed to test this relationship, a group of men were tested, all of whom had wives dying of breast cancer. One month after the death of each wife, the husbands' white blood cell (WBC) count dropped drastically and with their immune systems thereby depressed, the men began to fall ill. The WBC count generally returned to normal after about a year. Elderly people who had been forced to relocate also showed a reduction in their white blood cells for a time. Even astronauts (the picture of health), after being subjected to the stress of manned flight, showed lower WBC counts.

The link between the mind and the immune system has even been shown in other species. In an experiment, mice were fed chocolate milk and then injected with an immune system suppressor. After several such experiments, the mice would get sick from infection at the scent of chocolate milk, their immune systems in disarray. The link between the mind and the body's immunity was building in the minds of some scientists.

With evidence of such a link, the next step would be to control one's mind in order to stay healthy, and many people are attempting to do just that. As an example, the recurrence of herpes has been linked to the mind. Episodes are often triggered by either emotional or physical stress. Some sufferers, though, report that by relaxation, humor, and optimism (perhaps induced hypnotically) they can thwart the onset of the episodes even after the first symptoms have appeared.

The mental set most conducive to good health seems to be a positive, optimistic attitude. Author Norman Cousins, who claimed to have beat heart disease with, among other things, music and Woody Allen videos, subscribes to an approach that he calls a "joyous belief in an outcome" (Figure 20.12).

But how could mind influence immunity? What could the mechanism be? The answer isn't clear yet, but scientists have discovered that neuropeptides, once believed to be restricted to the brain, are found throughout the body. Furthermore, they are chemically related to substances that help regulate the immune system. Some researchers suggest that neuropeptides are the link between the immune system and the brain.

Most of us wish to live long and well. But that implies a continuing existence on an essentially hostile planet. Furthermore, that hostility may be increasing, largely because of our own behavior. Not only are our increasing numbers on the planet threatening our individual access to resources that contribute to good health, but each additional person might be considered a potential reservoir for some mutant threat. Crowding, of course, goes hand-in-hand with contagion, and we grow more crowded daily. In addition, we must rely on new technologies to help us solve our immediate problems, and the earth is becoming permeated with technology's by-products. We are forced to stand against a tide of chemical agents that are totally new to the environment, and that tide rises daily. Indeed, our immune systems, our ability to withstand, may soon be tested in ways we can only imagine.

FIGURE 20.12

Humor, optimism, and a positive outlook somehow enhance the operation of the immune system. Some people are convinced that laughter is, indeed, the best medicine.

SUMMARY

1. Nonspecific defense mechanisms include barriers to invasion, the inflammatory response, fever, phagocytosis, and direct attack by natural killer cells.
2. Barriers that keep microorganisms from entering the body include body surfaces such as the skin and the sticky ciliated mucous membranes lining the inner body surfaces, secretions such as lysozyme and mucus, and acidic environments.
3. An organism entering the body triggers an inflammatory response. Certain cells at the site secrete histamine, which dilates arterioles in the area and causes redness and swelling. Fever makes the body inhospitable to certain invaders.
4. Eosinophils, neutrophils, and monocytes are three types of phagocytic white blood cells that engulf invaders. Eosinophils respond to allergies and parasitic infections. Neutrophils are the first defenders at the site of invasion. The monocytes grow and produce large long-lived macrophages that quickly engulf many invaders.
5. Another white blood cell, the natural killer cell, ruptures the membranes of cancerous body cells or cells harboring viruses.
6. Specific responses are directed against a specific disease. Two types of lymphocytes, B-cells and T-cells, are important in specific responses. When activated by a helper T-cell, B-cells produce plasma cells that make antibodies to combat the invader and long-lived memory cells that can quickly defend against a subsequent invasion of the same organism. Helper T-cells also stimulate the proliferation of cytotoxic T-cells and memory T-cells. Cytotoxic T-cells identify and rupture invaders.
7. Antigens are foreign molecules that trigger the host's immune response. Antibodies produced by the host identify and help destroy antigen-bearing cells. Antibodies are composed of two identical short chains and two identical long chains and are arranged in a Y-shape. The ends of the arms, called the variable region, determine whether an antigen is attacked. The rest of the antibody is relatively constant and determines how the invader is attacked. Three general ways antibodies attack invaders are by linking invaders together, by marking the invaders for attack by phagocytes, and by triggering reactions that rupture the invading cell.
8. After a macrophage destroys an invading organism, it places an antigen-bearing piece of the invader on its own membrane. Then this macrophage roams the body until it finds a helper T-cell with an antigen-recognition site matching the structure of the invader's antigen displayed on its membrane. The antigen and the helper T-cell's antigen-recognition site lock together, triggering the production of interleukin II, which stimulates the production of B-cells and cytotoxic T-cells. Each lymphocyte bears the antigen-recognition site. Cytotoxic T-cells then recognize and destroy body cells infected with the original type of invader. Interleukin also triggers the proliferation of B-cells, which form short-lived plasma cells that secrete antibodies specific for the antigen discovered by their parent cell, and long-lived memory B-cells that will quickly form more plasma cells upon subsequent invasion by organisms bearing the original antigen.

9. The primary response occurs when a "nonself" antigen is first encountered. It begins with the actions of the helper T-cell and is slow. The secondary response is triggered by memory cells and is rapid.
10. A vaccination is the injection of an antigen, triggering the production of memory cells against it. When an organism bearing that antigen invades the body, the response is rapid and vigorous.
11. An individual's cells are marked by genetically determined molecules. Tolerance is the immune system's acceptance of cells with these markers. When the immune system recognizes the markers on the surface of the body's own cells as foreign antigens, it will act against these cells in a reaction called autoimmunity. Autoimmune diseases may result when antibodies produced against a disease organism cross-react with body cells, or when the immune system fails to recognize the body's cells as self.
12. Interferon, a group of antiviral substances produced by most vertebrate cells in response to viral attack, causes resistance to attack by other viruses. It is species-specific. It may prove useful in combating certain viral diseases.
13. AIDS (acquired immune deficiency syndrome) is caused by the HIV (human immunodeficiency virus). It suppresses the victim's immune system by killing the victim's helper T-cells, and the victim becomes susceptible to any disease. Many AIDS victims develop Kaposi's sarcoma or pneumocystic pneumonia, diseases that are otherwise rare. Many people in the U.S. infected with the AIDS virus do not show symptoms, but are thought to be able to transmit the virus. Once full blown, AIDS is incurable and almost always fatal. Most victims are homosexual men or intravenous drug users.
14. The functioning of the immune system is influenced by one's state of mind. It is suppressed by stress. The link between the mind and the immune system may be neuropeptides.

KEY TERMS

acquired immune deficiency syndrome (AIDS) (564)
antibody (557)
antigen (557)
antigen recognition site (558)
autoimmunity (562)
basophil (553)
B-cells (557)
cellular immunity (557)
constant region (557)
cytotoxic T-cells (557)
eosinophil (553)
heavy chains (557)
histamine (553)
human immunodeficiency virus (HIV) (564)
humoral immunity (557)
immune system (551)
inflammatory response (553)
interferon (562)
interleukin II (558)
keratin (551)
light chains (557)
lymphocyte (553)
lysozyme (552)
macrophage (554)
memory B-cell (559)
monocyte (553)
natural killer cell (NK cell) (554)
neutrophil (553)
nonspecific response (551)
phagocyte (557)
plasma cell (557)
primary response (559)
pyrogen (553)
secondary response (561)
specific response (551)
suppressor T-cell (559)
T-cells (557)
tolerance (562)
variable region (557)

FOR FURTHER THOUGHT

1. IgM is an antibody that functions in the primary response. Would you expect IgM's numbers to increase or decrease after you were vaccinated?
2. Multiple sclerosis is a disease in which an individual's antibodies attack the myelin sheath of its own nerve fibers. What type of immune response is this? What is the perceived antigen?

FOR REVIEW

True or false?

1. ___ Neutrophils are phagocytes that participate in the early stages of an inflammatory response.
2. ___ Macrophages with antigens embedded in their membrane may attach to any lymphocyte they encounter.
3. ___ The immune response is slowed down by the action of suppressor T-cells.
4. ___ Interferon is a substance that increases virus replication.
5. ___ By bleaching needles, intravenous drug users can completely eliminate the possibility of contracting AIDS.

Fill in the blank.

6. During an inflammatory response, certain cells at the site of an infection secrete ___, which dilates tiny arterioles.
7. ___ are Y-shaped molecules that destroy foreign invaders.
8. ___ is a substance secreted by helper T-cells that prompts B-cells to proliferate.
9. In the ___ reaction, lymphocytes turn against the body's own cells and recognize them as antigens.
10. ___ is a group of antiviral substances produced by the cells of most vertebrates.

Choose the best answer.

11. ___ develop from monocytes and destroy foreign bodies by ingesting them.
    A. Natural killer cells
    B. Macrophages
    C. Plasma cells
    D. Helper T-cells
12. ___ are cells that produce plasma cells and memory cells.
    A. Cytotoxic T-cells
    B. Phagocytes
    C. Suppressor T-cells
    D. B-cells
13. Memory cells:
    A. confer long-term immunity to infections.
    B. have a short life span.
    C. provide the first line of defense during an initial invasion of the body.
    D. only participate in nonspecific responses.
14. AIDS is a condition that:
    A. is transferred in blood and semen.
    B. is contracted primarily by vaginal intercourse.
    C. always displays symptoms in its early stages.
    D. all of the above.

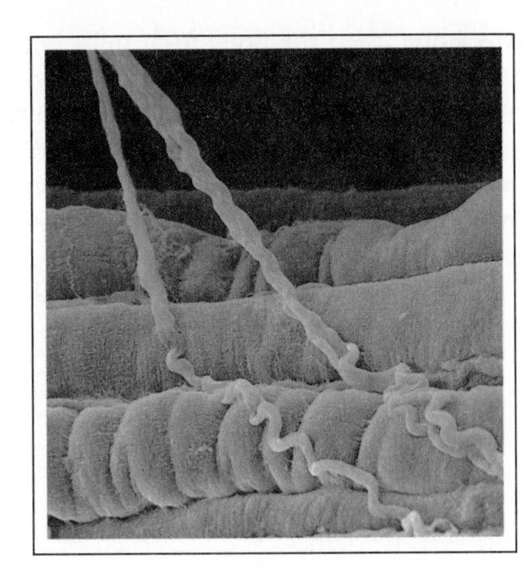

CHAPTER 21

Hormones and Nerves

Overview

Hormones
Human Hormones
The Relationship of Hormones and Nerves
Nerves

Objectives

After reading this chapter you should be able to:

- Describe some regulatory functions of invertebrate hormones.
- Describe the major human endocrine glands, the hormones they produce, and their effects.
- Describe the two basic mechanisms of hormone action.
- Describe the three types of neurons, their impulse pathways, and physical characteristics.
- Explain how an impulse is generated in and conducted along a neuron.
- Discuss the role of neurotransmitters in synaptic transmission.

If life can be thought of as a symphony, then the body itself must be the orchestra. The orchestra is a physical thing, composed of many parts, each contributing to the music of the others, and all indispensable. Each part, however, must act in concert with the rest; each must contribute at just the right time and for the proper duration to produce symphonic harmony. Such coordination is possible only with some means of communication among the various instruments.

In the body, this coordination is possible because of two major means of communication: chemical messengers called *hormones* and electrochemical devices called *nerves*.

HORMONES

A bear steps into a clearing in the Alaskan wilderness and is startled by a flurry of wings rushing into its face. A ptarmigan sweeps away, and then rushes the bear again, feathers abristle. The bear turns away, looking for peace and quiet, and the mother bird returns to her brood lying pressed against the forest floor. That very night in New York City, a woman sits up suddenly in her hotel bed. She is instantly awake and alert. Then she realizes what had awakened her—the sound of a baby crying in the next room. In both cases, the female may have been responding, in a very basic sense, to the presence of tiny amounts of a chemical in her bloodstream. The hormone oxytocin indeed seems to trigger maternal behavior in females, especially those who have already been mothers.

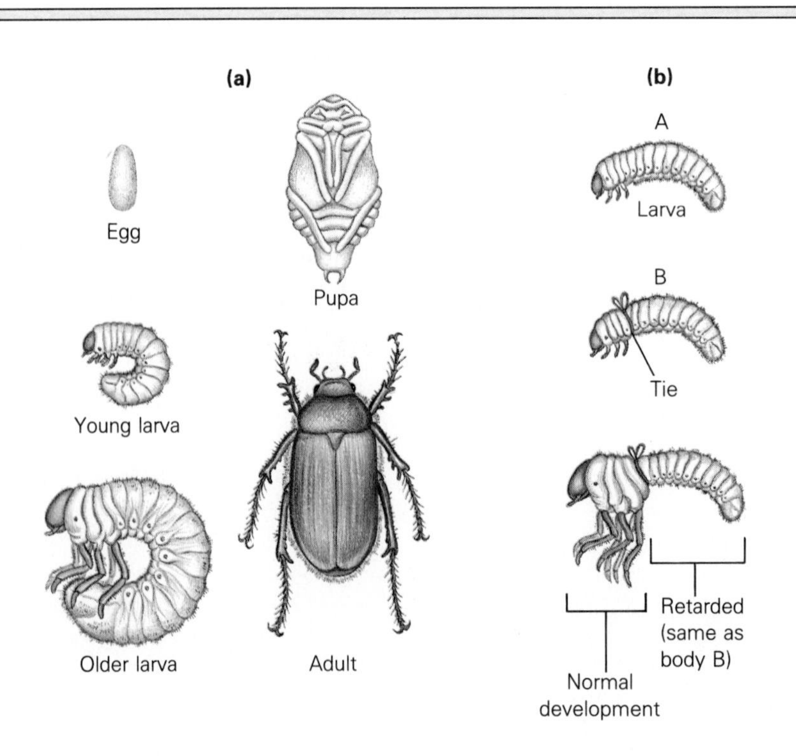

FIGURE 21.1
(a) Development in a beetle. Not all insect species go through the sequence of egg, larva, pupa, adult, but whatever their pattern, each stage is brought on at a specific time by a different hormone. (b) An experiment showing the importance of hormones in insect development. When growth hormones produced in the head are kept from reaching other parts of the body by tying off those parts, the hormone-deprived areas do not mature.

By the same token, some scientists argue that certain men have committed violent crimes because they are relentlessly driven by an overabundance of testosterone, a hormone known to be associated with aggression. Information about such behavioral changes is being added to the more conventional view of hormonal influences, those relating to development and homeostasis, the areas traditionally associated with hormone studies. The list of hormones and their roles continues to grow, regarding human and nonhuman species. Only a few years ago, the list of human hormones numbered about 20. Today, the number is believed nearer to 200.

A **hormone** (from the Greek *horman,* to excite) is a chemical that is produced in one part of the body and carried by the blood to another part of the body where it ultimately influences some process or activity. Hormones are produced by **endocrine** (ductless) **glands** that are found in various parts of the body. (Since hormones are released directly into the bloodstream, there is no need for ducts.)

Although hormones are secreted into the blood and are delivered to virtually all the cells of the body, a particular hormone affects only specific cells, those of its **target organ.** The effect is limited to the cells of the target organs because only they have receptors that recognize and bind to that hormone.

Invertebrate Hormones

Hormones, or chemical messengers, as they are sometimes called, basically function in regulating various bodily processes. Some keep those processes within their proper daily limits, as does the antidiuretic hormone associated with urine formation (Chapter 18), while others may cause permanent changes, such as growth or sexual maturation. They may also function in emergency situations by causing the body to prepare for stress. And they may help adapt the organism to its environment in other ways, as when amphibians or reptiles change colors to match their surroundings. Hormones also play a part in nature's chemical warfare. Some plants, for example, manufacture substances that mimic the growth hormones of insects, and thus they disrupt normal growth of insect larvae that feed on them.

Hormonal regulation is common throughout the animal world, in invertebrates as well as vertebrates. (Their presence in many rather "primitive" animals, such as cnidarians, indicates that this kind of control probably evolved rather early in the history of life.) In some invertebrates, hormones play rather complex roles. For example, in the sea slug, *Aplysia,* the hormone that instigates reproductive behavior inhibits movement so that reproduction can occur.

Both insects and crustaceans have remarkably well-developed endocrine systems. To illustrate, hormones have been found to regulate the change in the body organization of insects as they pass through their larval and pupal stages. The importance of hormones in insect development is illustrated in Figure 21.1. The crustaceans have a particularly extensive battery of chemical messengers. As an example, they have hormones for reproduction, growth, development, water balance, pigments in the exoskeleton, and eye pigments.

An enormous amount of research attention has been focused on hormones, but the work can be quite difficult. One of the greatest problems in doing hormone research is acquiring the hormones in the first place. Animal bodies contain very little hormone substances because only tiny amounts are needed to cause great changes. For example, a woman produces only about a teaspoonful of the female hormones called estrogens in an entire lifetime. A source of the estrogen called estradiol has been pig ovaries. However, more than two tons of pig ovaries are required to extract only a few milligrams of the hormone. (Perhaps recombinant DNA techniques can play a role here.)

HUMAN HORMONES

That delicate balancing act called homeostasis is strongly reliant on the constellation of endocrine glands as their powerful secretions ebb and flow through our sensitive bodies. We will consider each gland separately, but we must keep in mind that any is meaningless alone. Their interaction is critical and life's picture is not complete if any of these stars in this constellation should fade.

In the human endocrine system (Figure 21.2), the **pituitary** has been called the "master gland," but this flattery is misleading, since it is regulated by some of the glands it influences and since many of its activities are also regulated by a part of the brain called the **hypothalamus.** The pituitary does have a number of important functions, however, and it indeed does affect some other glands. Structurally, it is composed of two major parts, the anterior and posterior lobes, with a central midlobe area.

The pituitary is located at the geometric center of the skull, where it lies shielded by heavy layers of surrounding bones. If you point one finger directly between your eyes and stick the other in your ear, you'll point right to it, as well as gain the attention of other people on the bus. The pituitary is only about the size of a bean, but its size belies its physiological complexity. In all, the pituitary secretes nine known hormones. The anterior pituitary alone secretes seven hormones, six of which influence other endocrine glands. The seventh stimulates body growth.

The growth hormone is called **somatotropin,** and it acts by stimulating bone and muscle growth. If too little somatotropin is present during childhood, the result is a pituitary dwarf. Too much somatotropin results in a giant. Excessive somatotropin in an adult increases the size of only those bones that can still respond to the hormone—the jaw and the bones in the hands and feet. This condition is called **acromegaly** (Figure 21.3).

When the **thyroid** gland is stimulated by the pituitary, it produces **thyroxine,** the hormone that controls the body's metabolic rate. Deficiency of thyroxine in childhood results in a characteristic physical appearance and mental retardation called **cretinism.** It is now possible to recognize thyroid deficiency early in life, so that thyroxine can be administered artificially to promote normal development. Excess thyroxine causes hyperactivity and low body weight. The thyroid also produces **calcitonin,** which inhibits the release of calcium from bone into the bloodstream. It operates with the **parathyroid hormone** to regulate levels of calcium ions in the bloodstream. Parathyroid hormone is produced by the **parathyroid**

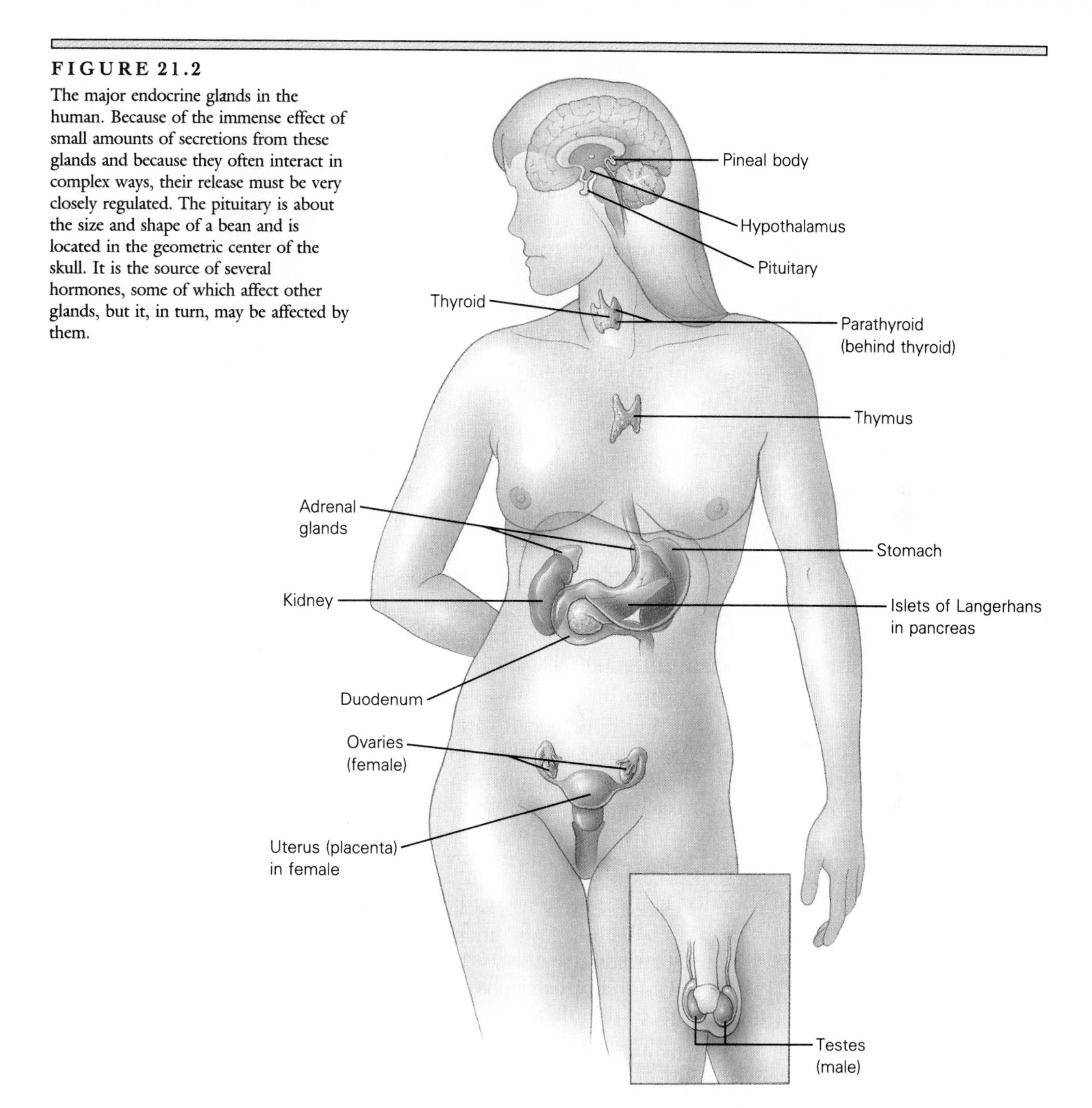

FIGURE 21.2
The major endocrine glands in the human. Because of the immense effect of small amounts of secretions from these glands and because they often interact in complex ways, their release must be very closely regulated. The pituitary is about the size and shape of a bean and is located in the geometric center of the skull. It is the source of several hormones, some of which affect other glands, but it, in turn, may be affected by them.

glands, the smallest endocrine glands, located behind or within the thyroid. Parathyroid hormone increases the absorption of calcium ions from the intestine and reduces its excretion from the kidneys. It can also help withdraw calcium from the bones. It thus raises the calcium concentration in the blood. Calcium ions are important in muscle contraction and the conduction of signals between nerve cells.

FIGURE 21.3

Sudden increases in the production of growth hormone after maturity produces growth in only certain body parts, such as those of the face and hands (a). The condition is called acromegaly. Two improbable brothers-in-law (b) illustrate giantism and dwarfism. Their respective heights are seven feet eight inches, and three feet even. Both conditions arise from abnormal pituitary secretions.

(a)

(b)

The primary hormone produced by the testes is **testosterone** and those produced by the ovaries are **estrogen** and **progesterone.** These hormones are responsible for those secondary sex characteristics we all love and admire. In males, testosterone causes deepening of the voice, broadened shoulders, hairy chests, facial hair, muscular development, sperm production, and increased production of red blood cells (thus, "red-blooded he-man"?). In females, estrogen causes widening of the hips, changes in the distribution of body fat, and development of the breasts. Progesterone stimulates the development of the uterine lining and causes further breast development during pregnancy. There are also recent findings on the effects of the sex hormones on behavior in ways no one had expected (Essay 21.1). Also see Essay 21.2 for a discussion of human pheromones.

Ad- means "upon" and *renal* refers to the kidney, so guess where the **adrenal glands** are. The adrenals have two distinct parts, the outer cortex and the inner medulla. About 50 steroids have been isolated from the adrenal cortex of various mammals, and it is believed that some of these may be used in the production of hormones, but it is not yet known just what role, if any, these steroids play. One hormone produced in the cortex is **cortisol** (manufactured when the cortex is stimulated by adrenal corticotropic hormone (ACTH), from the anterior pituitary). Cortisone is well known as an agent that reduces inflammation. **Aldosterone** is also produced in the adrenal cortex and functions in the regulation of salt and

ESSAY 21.1

Hormones and Test-Taking

Hormones may affect our behavior in ways we are only beginning to suspect. Consider this: the level of sex hormones in the blood of women can affect their reasoning, coordination, and verbal ability. Psychologists Elizabeth Hampson and Doreen Kimura tested women when estrogen levels were high (about 10 days before menstruation) and low (near the beginning of the menstrual cycle). They tested them in areas in which women are generally better than men (such as manual dexterity and verbal facility) and in areas in which men generally do better (such as spatial reasoning, or perhaps mentally rotating a figure). They found that when estrogen is high, women increased their scores by about 10 percent on those tests in which women normally excel but, strangely, on tests on which men generally do better, their scores fell by about 10 percent. The scores were the same for post-menopausal women receiving estrogen replacement therapy, as their estrogen levels were altered artificially.

In earlier work on rats, it was found that differences in exposure to sex hormones during embryonic development would alter the organization of the brain, producing behavioral differences (such as in how male and female rats solve mazes, for example). However, there is, as yet, no explanation for the differences that arise in individual humans subjected to different levels of sex hormones.

Tests on men are more difficult to devise because there is no comparable monthly cycle. However, testosterone levels in men are higher in the morning than in the evening, so this change may permit the development of tests on the effects of the male hormone.

water in the body. It works in the cells of the kidneys' distal convoluted tubules where it increases their rate of sodium reabsorption. (When sodium is reabsorbed, you may recall, water must follow because of the osmotic gradient that is set up, so aldosterone helps the body recover water that is moving through the kidney.) The heart is now known to produce an important hormone, called **atrial natriuretic factor (ANF)**, which acts antagonistically to aldosterone by increasing the excretion of sodium, as well as water, from the kidney. The interplay of the two hormones is important in the regulation of body fluids (and, hence, blood volume). The adrenal cortex also produces some male sex hormones. This function is greatly diminished in women, but if it should be prompted, for example, by a tumor, it will cause such characteristic masculine qualities as beard growth.

The adrenal medulla secretes the hormones **norepinephrine** and **epinephrine** (sometimes called **adrenaline**), as we will see shortly. Other endocrine glands, the hormones they produce, and their principal effects are listed in Table 21.1.

How Hormones Work

There are two general mechanisms by which hormones exert their effects. One pathway involves the formation of second messengers, chemicals that work inside the cell to bring about the hormone's effects. The other involves turning on genes that code for specific proteins.

TABLE 21.1
Major Vertebrate Endocrine Glands and Their Hormones

GLAND	HORMONE	PRINCIPAL ACTION
Anterior lobe of pituitary	Thyroid-stimulating hormone (TSH)	Stimulates thyroid.
	Follicle-stimulating hormone (FSH)	Stimulates ovarian follicle.
	Luteinizing hormone (LH)	Stimulates testes to develop and secrete testosterone in male; stimulates ovulation and the formation of corpus luteum in female.
	Growth hormone (somatotropin)	Stimulates growth of bones and muscles.
	Adrenocorticotropic hormone (ACTH)	Stimulates adrenal cortex.
	Prolactin	Stimulates secretion of milk; parental behavior.
Midlobe of pituitary	Melanocyte-stimulating hormone	Regulates color of skin in reptiles and amphibians.
Posterior lobe of pituitary	Oxytocin	Stimulates contractions of uterine muscle; milk release.
	ADH or vasopressin	Increases water reabsorption in kidneys.
Thyroid	Thyroxine	Increases metabolism.
	Calcitonin	Inhibits the release of calcium from bone.
Parathyroid	Parathyroid hormone	Controls calcium metabolism.
Testes	Testosterone	Stimulates production of sperm and secondary sex characteristics.
Ovary	Estrogens	Stimulates female secondary sex characteristics.
	Progesterone	Prepares and maintains uterine lining for pregnancy.
Adrenal medulla	Epinephrine (adrenaline) Norepinephrine	Augments actions of sympathetic nervous system.
Adrenal cortex	Cortisol, cortisone-like hormones	Contols carbohydrate metabolism; raises blood glucose level; reduces inflammation.
	Aldosterone	Controls salt and water balance; promotes sodium reabsorption.
Pancreas	Glucagon	Stimulates breakdown of glycogen into glucose; elevates blood sugar level.
	Insulin	Lowers blood sugar level; increases formation and storage of glycogen.
Heart	Atrial natriuretic factor (ANF)	Stimulates the secretion of sodium and water from the kidneys.

Second Messenger Systems (Peptide Hormones)

Most of the hormones that work by second messengers are peptides, small nitrogen-containing molecules. They include some of the hormones we have already discussed, such as antidiuretic hormone (ADH) and oxytocin, as well as the enkephalins and endorphins, both of which have a role in reducing pain (Essay 21.3).

Unlike the steroid hormones, which have a direct effect on genes, the hormones that work through second messengers never enter the cell. Rather, they bind with a receptor on the membrane from where they initiate a cascade of reactions within the cell. The general mechanism was discovered by E. W. Sutherland in the early 1960s, for which he was awarded the Nobel Prize.

ESSAY 21.2

Human Pheromones

We are generally aware that humans can communicate by smell, but we usually associate such communication with negative signals. A man reeking of cheap cologne is as likely to be shunned at a social gathering as is his unwashed colleague (see Essay 23.2).

However, we are beginning to learn that the unwashed fellow may be standing there, steeped in his own chemicals, having a totally unexpected effect on others in the room. This is because humans do, indeed, secrete subtle pheromones that can alter the behavior and physiology of others in remarkable ways that are not yet well understood.

Research on human pheromones was greatly stimulated by psychologist Martha McClintock who, in 1970, found that women living together in a college dormitory tended to menstrually cycle together. Then George Preti and Winnifred Cutler carried the research further. They swabbed underarm secretions from men on the upper lip of women with unusually long or unusually short menstrual cycles. Within three months all the women were cycling at about 29 days, considered the optimum. The conclusion was that "male essence" contains something that promotes reproductive health in women.

Then Cutler and Preti, wondering why women living together cycle together, placed underarm sweat from women under the noses of ten women with normal cycles. After three months they were all cycling roughly in unison with the women who donated the sweat. (Women who were treated with plain alcohol showed no change at all.)

Pheromones are well known in other mammals such as rats and pigs. In pigs the chemical has been identified as androstenol. When a sow in heat detects the pheromone, she immediately assumes the mating stance. Fortunately for our social fabric, pheromones apparently do not elicit such immediate behavior in humans. Instead, to the degree that pheromones influence our behavior at all, they probably do so subtly, nudging us from within.

In this system the hormone itself is called the **first messenger.** The hormone attaches to a receptor on the target cell and the hormone-receptor complex binds to an enzyme embedded in the membrane, adenylate cyclase. This binding activates adenylate cyclase which then causes ATP to give up two phosphates and to form a molecule called cyclic AMP (adenosine monophosphate). Cyclic AMP serves as a **second messenger,** causing dramatic changes in the cell by triggering a specific series of events

ESSAY 21.3

Prostaglandins, Endorphins, and Runner's High

Because we live in a complex and changing world, we have developed complex and changing bodies. Many of those changes are due to the body's incredible sensitivity to certain kinds of molecules produced by the body itself. Some of these molecules are hormones but some are not. Although hormones are produced in one place and carried by the bloodstream to another place where they exert their action, there are a number of chemicals that act in the tissues where they are produced. We ran across some of these in our discussion of the immune system. You may recall that certain cells in connective tissue secrete a substance called histamine that triggers the inflammatory response around an infection (Chapter 20). The histamine is secreted into extracellular spaces and therefore may be regarded as a hormonelike substance.

Another group, called the prostaglandins (first found in the prostate gland), is comprised of about 16 different molecules. One kind dilates the blood vessels and the respiratory tubules, increasing the amount of oxygen reaching the muscles. Another functions in regulating the menstrual cycle and in stimulating uterine contractions during childbirth. Yet another brings on fever in response to infection. Fever was once believed to be a harmful side effect of infection, but we now know that it is one of the body's defense mechanisms. Nonetheless, many people still run for the aspirin at the first sign of fever. Aspirin reduces fever, apparently by inhibiting formation of prostaglandins.

In the 1970s, researchers found sites in the cells of the brain that selectively bind to opiates. But, they wondered, why would the body produce sites that enhance the use of such drugs? It turns out that these sites provide surfaces that bind opiatelike molecules produced within the body. These are called enkephalins ("within the head") and endorphins ("internal morphine"). Both kill pain better than does morphine, in fact, and may produce a slight euphoria. Since trauma can trigger their release, they may have a role in natural pain suppression after injury. Long distance running (which can definitely be traumatic) may trigger these opiate relatives, producing the fabled response called "runner's high."

(the biochemical cascade, see Figure 21.4). Since the basic scheme was outlined, a number of variations have been discovered, including other second messengers.

Direct Gene-Activation Systems (Steroid Hormones)

In the early 1960s, researchers came across an intriguing bit of information. They found that estrogen, manufactured in the ovaries and circulated by the bloodstream throughout the body, in female rats was accumulating in the cells of the reproductive tract. Furthermore, the estrogen was localized in the nuclei of those cells. When they looked for the hormone in other organs, such as the spleen and pancreas, they found no such accumulation. This finding led to the discovery of how estrogen works.

Estrogen, as well as other steroid hormones including the male sex hormone testosterone, is derived from cholesterol and is composed of groups of carbon rings that look a bit like remnants of chicken wire after some barnyard breakout. Steroids are responsible for the development of secondary sex characteristics, and so we should not be surprised to find them in the reproductive tract, but, the question arises, why did they accumulate in the nuclei of the cells there?

FIGURE 21.4

A peptide hormone, the first messenger, binds to a receptor on the membrane of the target cell. This hormone-receptor complex interacts with a nearby adenylate cyclase enzyme, causing it to trigger the formation of cyclic AMP from ATP releasing 2 phosphates (2 P_i) into the cell. The cyclic AMP acts as a second messenger, setting into action a cascade of biochemical reactions that produces the effects of the hormone.

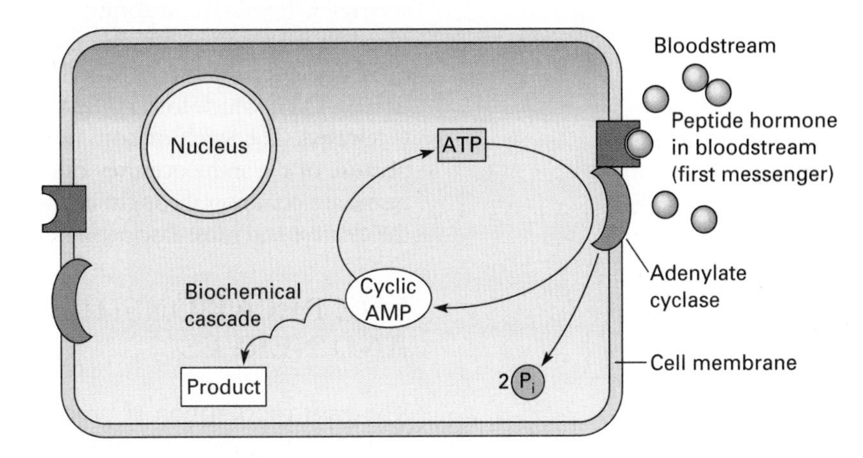

It turns out that steroids function in the nuclei by altering the expression of genes. When a steroid hormone reaches a cell, it moves across the plasma (cell) membrane and into the cytoplasm. If the cell is a target cell, the hormone joins with a receptor molecule located in either the cytoplasm or the nucleus. In the nucleus, the hormone-receptor complex binds to specific receptor sites on a chromosome and directs the synthesis of specific messenger RNA. The mRNA then leaves the nucleus and moves into the cytoplasm to make new proteins that are responsible for causing the hormone's effects (Figure 21.5).

Feedback in Hormonal Systems

Hormone production is regulated, in part, through a negative feedback system in which the product of a process slows the process. As an example, consider the interaction of the tropic hormones (those that regulate other

FIGURE 21.5

A steroid hormone enters the cytoplasm and joins with a receptor molecule. The hormone-receptor complex then enters the nucleus and binds to a special site on the chromosome. There it directs the transcription of certain kinds of mRNA, which then move out to the cytoplasm where they direct the formation of specific proteins, which bring about the hormone's effects.

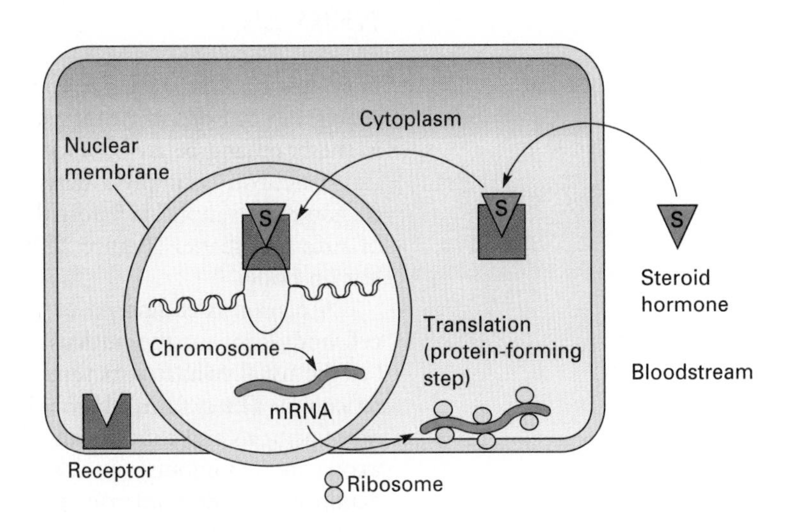

endocrine glands) and their target glands. The anterior pituitary gland secretes a thyroid-stimulating hormone (TSH) that causes the thyroid gland to secrete the hormone thyroxine. As the thyroid responds and the thyroxine level rises in the blood, the thyroxine itself suppresses the secretion of TSH, which lowers the level of thyroxine. As a result, more TSH is released. The levels of both hormones are thus kept within tight limits because of the influence they exert on each other. Such feedback mechanisms are common throughout the endocrine system and furnish some of the clearest and most fascinating examples of homeostasis in living things.

THE RELATIONSHIP OF HORMONES AND NERVES

The great coordination of processes within living things is due, in large part, to hormones and nerves. However, these cannot be considered separate processes, each acting independently of the other. On the contrary, the two systems are inextricably bound together at a number of levels. We can appreciate their interdependence by considering three basic ways they are related.

First, they are *structurally* related. A number of endocrine glands are formed from nervous tissue. For example, the posterior pituitary forms from an extension of the brain, whereas the anterior pituitary is formed from the embryonic mouth.

Second, the two systems are *functionally* related. For example, when the nervous system detects an emergency situation, the adrenal medulla may release epinephrine (adrenaline) into the system, enabling the individual to react with an astonishing vigor. (The story of a little old lady lifting a car off her trapped husband appears periodically.)

Third, the two systems are *chemically* related. A number of chemicals are used to send messages between nerve cells and to act as hormones. Again, we can refer to epinephrine. The same chemical released by the adrenal medulla is used as a signal between certain kinds of nerve cells.

NERVES

Let's now turn our attention to the nervous system. We will first consider nerves and neurons, their individual cells, and then see how they function in the brain and behavior. First, a few definitions are in order.

A **neuron** is simply a nerve cell—although you may reach the conclusion that "simply" is hardly the word for it. Neurons come in a variety of sizes and shapes (Figure 21.6), and there are billions of them in the human body.

A neuron is composed of a cell body, dendrites, and the axon. The **cell body** contains the nucleus of the cell and most of its cytoplasm and has the usual cellular structures, such as the ribosomes and endoplasmic reticulum. The neuron's highly branched **dendrites** receive stimuli from other neurons and conduct impulses toward the cell body. The elongated **axon** conducts impulses away from the cell body. The profusely branching dendrites provide numerous points of contact with other neurons. Most neurons have only one long axon, but in some cases, may branch.

FIGURE 21.6

Four neurons found in human beings show the diversity of these cells. (a) and (b) show clear differences in cell bodies, axons, and dendrites. (c) is a motor neuron with an axon that runs from the nervous system to the effector (in this case, a muscle). (d) is a sensory neuron that runs from the receptor to the spinal cord. Note that the sensory neuron has no true dendrites. The nodes in the myelin sheath are where one Schwann cell ends and another begins. A single nerve cell may be nine feet long, such as those that run from a giraffe's hip region down to its hind leg.

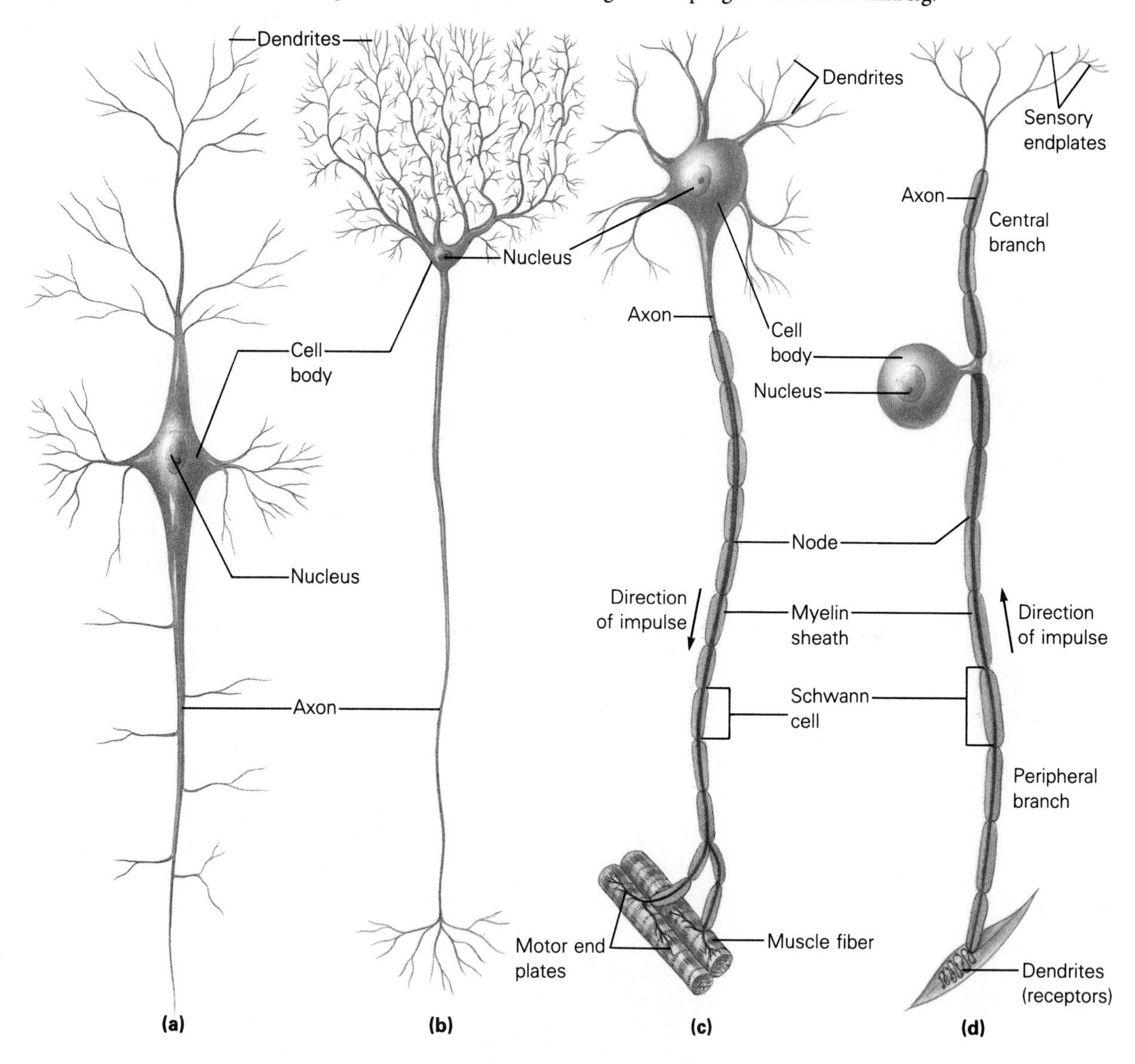

We can differentiate between dendrites and axons on the basis of their interaction with other cells. In general, dendrites can be stimulated by other cells, while axons cannot; conversely, axons can stimulate other cells. (As we might expect, there are certain exceptions to these rules.)

Some vertebrate axons are enveloped in a fatty **myelin sheath.** The myelin sheath serves to speed up the transmission of impulses along the axon by what is called **saltatory propagation.** It is the myelin that gives

FIGURE 21.7

In myelinated neurons, action potentials occur only at the nodes. The neural impulse leaps from node to node down the axon. This type of transmission, known as saltatory propagation, is considerably faster and requires less energy in terms of ATP than transmission in nonmyelinated neurons. Because of saltatory propagation, myelinated neurons transmit impulses up to 20 times faster than the fastest nonmyelinated neurons.

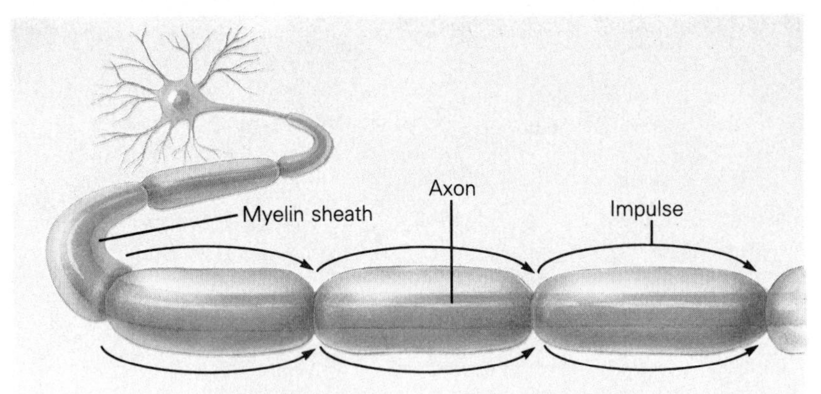

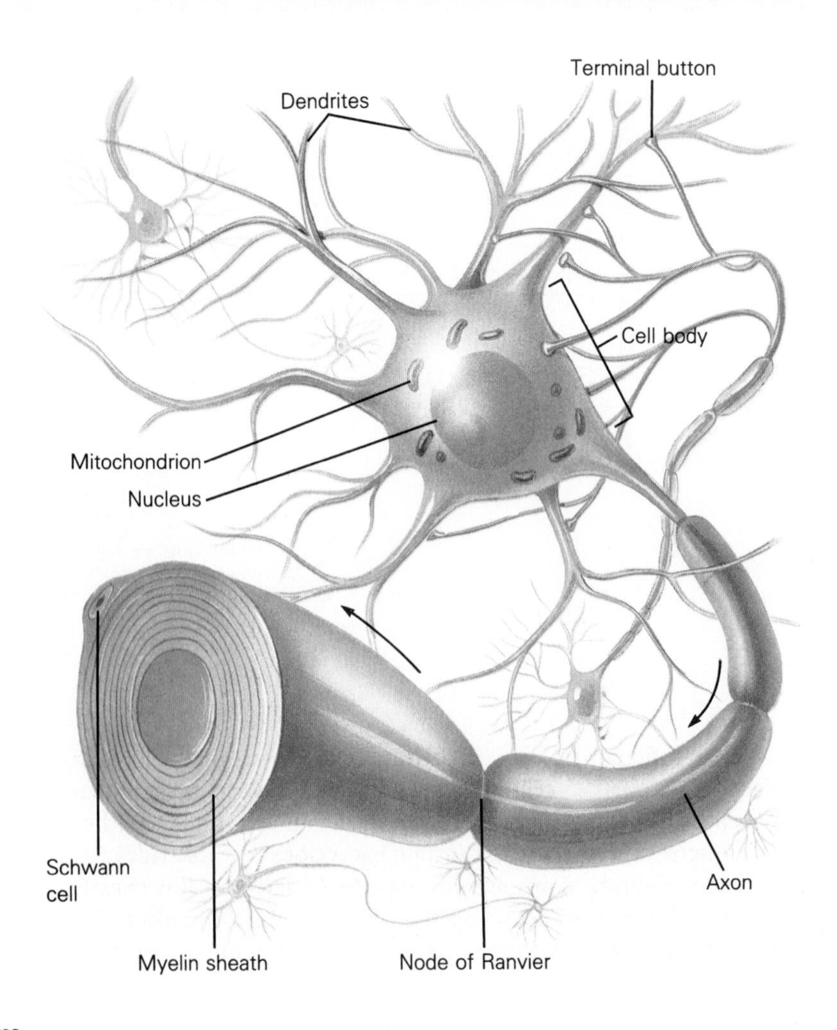

the tissues of the brain and spinal cord a whitish appearance. Nonmyelinated fibers are usually gray. The white myelinated fibers lie on the outside of the spinal cord, but in the brain they retreat to the inside, leaving gray matter visible on the outside. (Thus, "gray matter" has come to refer to the brain.) In most cases, only myelinated fibers outside the brain and spinal cord are capable of regeneration if they are severed (as long as the cell body is not destroyed). Fibers within the brain and spinal cord, myelinated or not, do not normally regenerate. Figure 21.7 summarizes the general structure of myelinated fibers.

Nerves are bundles of axons and dendrites. They are white, glistening strands. Although a nerve may contain many parallel neurons, each neuron can transmit an impulse independently of the others.

Another kind of cell in the vertebrate nervous system is the **glial cell.** Glial cells outnumber nerve cells by ten to one in the nervous system. Their branches often permeate the nerves, weaving between the neurons. They have a number of roles; some serve as binding and support, keeping the neural material bound together; they may provide the neural cells with nourishment from the blood vessels nearby, and channel their waste to these blood vessels; and some may insulate neurons so that transmission along one does not necessarily influence others nearby.

Impulse Pathways

If you step on a tack and are a normally sensitive and intelligent person, you may wish to take your foot off it. For you to be aware of your predicament, however, the condition of your foot has to register in your brain, which is some distance away. This message travels over neurons as an **impulse.** Let's start at the bottom.

To begin with, the tack stimulates certain **receptors** in the sole of your foot. Receptors come in a variety of types, and each type specializes in a particular sensation, such as cold, pressure, or pain (Chapter 23). In some cases, bare nerve endings function as receptors. In any case, the stimulation of a receptor results in the transfer of that impulse to an **afferent** (sensory) **neuron,** which is one that carries impulses *toward* the central nervous system. The impulse may be relayed from one neuron to the next until it enters the spinal cord and is transferred to one or more **interneurons** that carry the message to the neurons that send impulses to the brain, where the message,"Sharp pain in foot!" is generated. If the foot has not already been lifted by a reflex action (which we'll discuss later), you may say to yourself, "I really must remove my foot from the tack," whereupon an impulse is sent from the brain, down the spinal cord, and out along the proper **efferent** (motor) **neuron;** that is, one that carries the impulse away from the spinal cord toward the **effector.** The effector here would be the muscle that raises the leg.

The Mechanism of the Impulse

Whereas neurons conduct impulses from one part of the body to another, the term "conduct" is perhaps misleading. This is because neurons are far more than simple conductors; each neuron not only conducts its impulses, but generates them as well.

Impulse conduction in neurons has been compared to electrical impulses in, say, a copper wire, but the analogy is not a good one. An electrical current in a copper wire diminishes with time and distance, but neural impulses, once started, do not diminish along the length of the neuron. They are as strong at the farthest branch of the axonal tree as they were at their origin in the dendrite. Therefore, instead of being like a current of electricity in a copper wire, the nerve impulse is more like a line of falling dominoes. Each domino triggers the fall of the next, but once triggered, each domino falls with the same energy. And just as dominoes must be set back up before they can repeat their performance, so must the potential energy of a neuron be restored before it can be fired again. Let's see how this happens in a neuron.

When a nerve fiber is at rest, certain chemicals are found in greater abundance within the cell, while others are more concentrated outside the cell. Furthermore, some of these chemicals are ionic, or electrically charged. For example, the resting neuron is literally bathed in sodium ions (Na^+), but these ions are relatively scarce inside the cell. This difference results in a net positive charge outside the cell. The cytoplasm of the neuron, however, is negatively charged with respect to the outside. This is because, in spite of the positively charged ions such as potassium (K^+) in the cell fluid, they are overmatched by the negatively charged proteins found there.

The electrical potential—that is, the difference between the net electrical charges of the cell's interior and its surroundings—that is set up by this peculiar distribution of charged particles is called the cell's **resting potential.** Typically, this potential difference is about −70 millivolts, or about 5 percent as much electrical energy as is found in a regular flashlight battery.

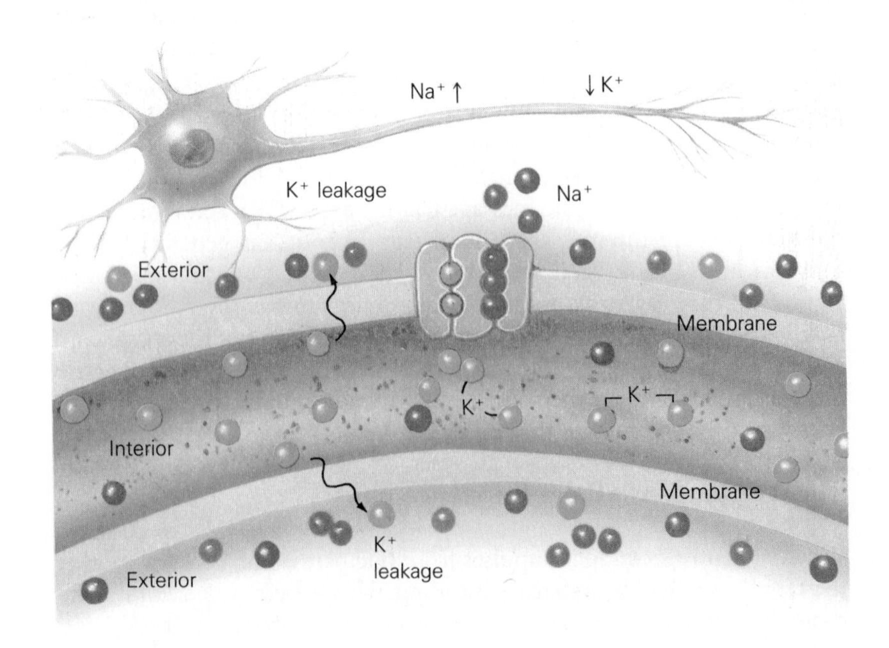

FIGURE 21.8
When not conducting neural impulses, a neuron maintains gradients of sodium ions (Na^+, shown as blue) and of potassium ions (K^+, shown as green). Na^+ are more concentrated outside the cell and K^+ are more concentrated within. These gradients represent a considerable amount of potential energy that will be released when a neural impulse is generated. The sodium/potassium gradients are partially produced by special membranal carriers known as sodium-potassium pumps. In each action of the pumps, three sodium ions are pumped outside and two potassium ions are pumped in. The channels are opened by the release of energy from ATP, which engages in special receptors in the complex.

The difference in the ion concentrations inside and outside the cell is established and maintained by the permeability characteristics of the membrane and the cell's *sodium-potassium pump* (described shortly). The membrane is not very permeable to sodium ions, but it lets potassium ions through relatively easily. Therefore, the positively charged potassium ions accumulate inside, attracted by the negatively charged proteins within. Any sodium ions that leak inside are pumped right back out again by the **sodium-potassium pump,** an assemblage of proteins that simultaneously pumps sodium ions out and potassium ions in.

Now, when an impulse travels along a neuron, the first thing that happens is that the cell membrane suddenly allows sodium ions to rush into the interior of the neuron so that they rush in faster than they can be pumped out. (Both sodium and potassium cross the membrane through special **ion channels.**) However, the change doesn't occur simultaneously along the entire length of the cell; it is sequential. Beginning at the point of stimulation, a wave of inrushing sodium ions sweeps along the length of the cell. Since areas along the membrane are no longer separating positively and negatively charged particles, the neuron is said to be **depolarized** at these areas. This depolarization sweeps along the neuron at a regular rate of speed, depending on the size and type of neuron. When depolarization reaches a certain level, the membrane becomes less permeable to sodium.

As the wave of depolarization passes, a wave of repolarization begins. The **repolarization** occurs as positively charged potassium ions rush out of the cell. This shift sets up a net positive charge outside again. Although the outflow of potassium ions restores the initial electrical potential, the distribution of ions is different, as you can see. The original distribution is restored by the sodium-potassium pump, which ushers sodium ions out and potassium ions back in. The pump uses ATP as its source of energy. For each molecule of ATP used, the pump moves three sodium ions out of the cell and two potassium ions into the cell (Figure 21.8). Thus, the membrane potential of the cell is restored. The description may sound a bit tedious, but the whole process takes place in an instant. Most neurons can be depolarized and repolarized hundreds of times a second.

The inrush of sodium ions at any point along the membrane results in a momentary net positive charge inside the cell at that point, so that there, the polarity is reversed. In other words, for an instant the outside becomes slightly negatively charged and the inside slightly positively charged. Then, as potassium ions leave, the original charge difference is restored. The change in the distribution of charges that sweeps along a neuron as it carries an impulse is called an **action potential** (Figure 21.9).

Two important qualities mark neural impulses. One, as was mentioned, is that an impulse does not increase or decrease in strength as it moves along a neuron. The other is that impulses operate on an "all-or-none" basis. This means that if a stimulus reaches **threshold** (the lowest intensity stimulus that can trigger an action potential in that neuron), the cell fires, and it fires at only one intensity no matter what the strength of the stimulus. This "go" or "no go" principle has been likened to firing a gun. The velocity of the bullet is not changed by how hard you pull the trigger.

FIGURE 21.9

The action potential occurs as a wave of depolarization and repolarization sweeps along the axon. The inside of the cell gains a positive charge. The charge goes from about −70 to +20 millivolts as ion concentrations inside and outside the cell change. Note that the impulse travels, here, from left to right. First, the membrane becomes permeable to sodium, which rushes in and reverses the polarity of the membrane (depolarization). Behind this, potassium rushes out of the "leaky" membrane, restoring the polarity (repolarization). The two ions are restored to their former sides of the membrane by the operation of sodium-potassium pumps.

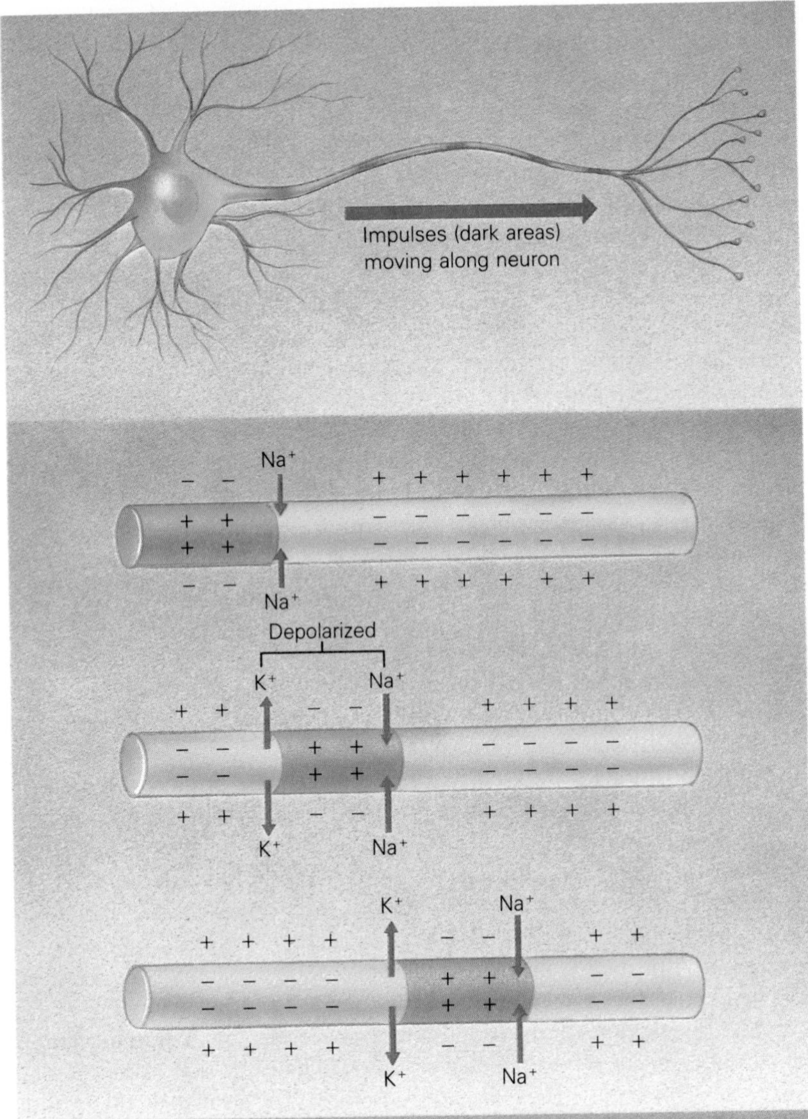

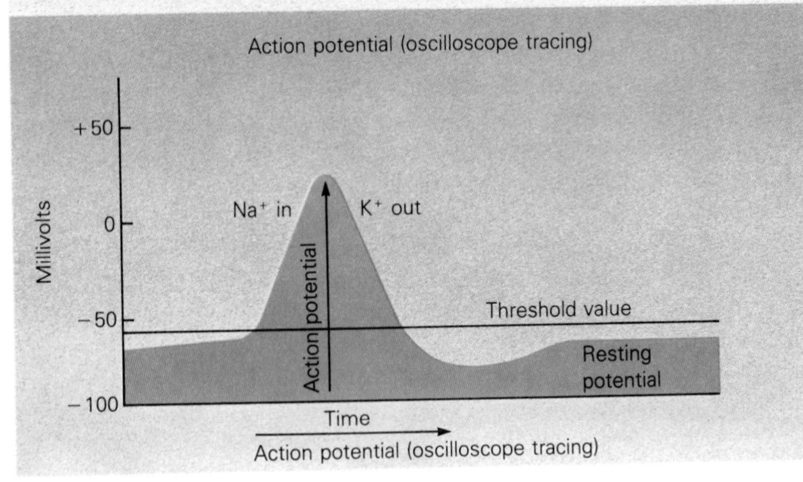

If neurons operate in this all-or-none manner, how is it that we can detect various intensities of stimuli? If you put your hand on a stove, how do you know whether it is simply nice and warm or dangerously hot? Actually, we can make such discriminations in a number of ways. First, the more intense the stimulus (the hotter the stove), the higher the rate of impulses that pass along the neuron (although there are maximum limits). Also, different neurons in the same nerve may have different threshold levels. If some are easily stimulated and others harder to fire, then the stronger the stimulus, the more neurons will be stimulated. Hence, the brain can interpret either the *frequency* of the impulses or the *number* of neurons stimulated to tell you whether you should remove your hand from the stove before you begin to see smoke. Again, fortunately, the process doesn't take as long as the explanation.

The Synapse

Now let's consider how an impulse travels from one neuron to the next. Actually, we are concerned with how one neuron activates another so that the impulse continues along a chain of nerve cells. The story is an interesting one that centers on certain remarkable secretions.

We begin by noting that the junction of two neurons—the point at which the axon of one neuron transmits the impulse to the dendrite of another neuron—is called a **synapse.** Whereas, theoretically, an impulse can travel in either direction along a neuron, it can cross the synapse in only one direction.

When an impulse traveling along a neuron reaches the axon tip, it causes the release of a chemical, called a **neurotransmitter,** into the space between itself and the dendrites of the next neuron. The axon tip is enmeshed in the highly branched dendrites of the next neuron so that there is a large total surface area over which the two neurons communicate. Any neurotransmitter secreted by the axon endings of one neuron, then, can be expected to quickly affect the next neuron. And this is just what happens.

The neurotransmitter is stored in small packets (or vesicles) in the tips of the axon endings (Figure 21.10). When an impulse reaches the end of the axon, some of these packets move toward the membrane and rupture, thus releasing their transmitter substance into the synapse. The chemical quickly diffuses across the gap and joins with a special receptor on the next cell. This causes changes in the receiving neuron that may result in an action potential.

A neuron may have several thousand synapses and these may have quite different effects on the receiving neuron. For example, some are excitatory and some are inhibitory. Those that cause the neuron to fire are called **excitatory neurons;** those that inhibit firing are called **inhibitory neurons.** The effects of the two are compared in a simplified fashion in Figure 21.11. The delicate interplay between excitatory and inhibitory neurons results in a remarkably precise coordination of neural firing patterns.

After a few impulses have traveled down a neuron, causing release of the neurotransmitter into the synaptic space, it might seem that the chemical would build up in the synaptic area and cause the next neuron to continue firing in the absence of a real stimulus. But do not despair.

Neurotransmitters are quickly removed from the synapse, some by special enzymes. For example, the enzyme that breaks down for the common neurotransmitter acetylcholine is called acetylcholinesterase. Others are simply taken back into the tip of the axon and reused.

Humans sometimes accidentally come into contact with substances that inactivate these important neural enzymes, with horrible consequences. For example, certain insecticides, such as Malathion or Parathion, which are sprayed on crops, have the advantage of being short lived, unlike DDT, which remains intact in the environment long after it is applied. Within a few days after spraying, the short-lived chemicals are believed to

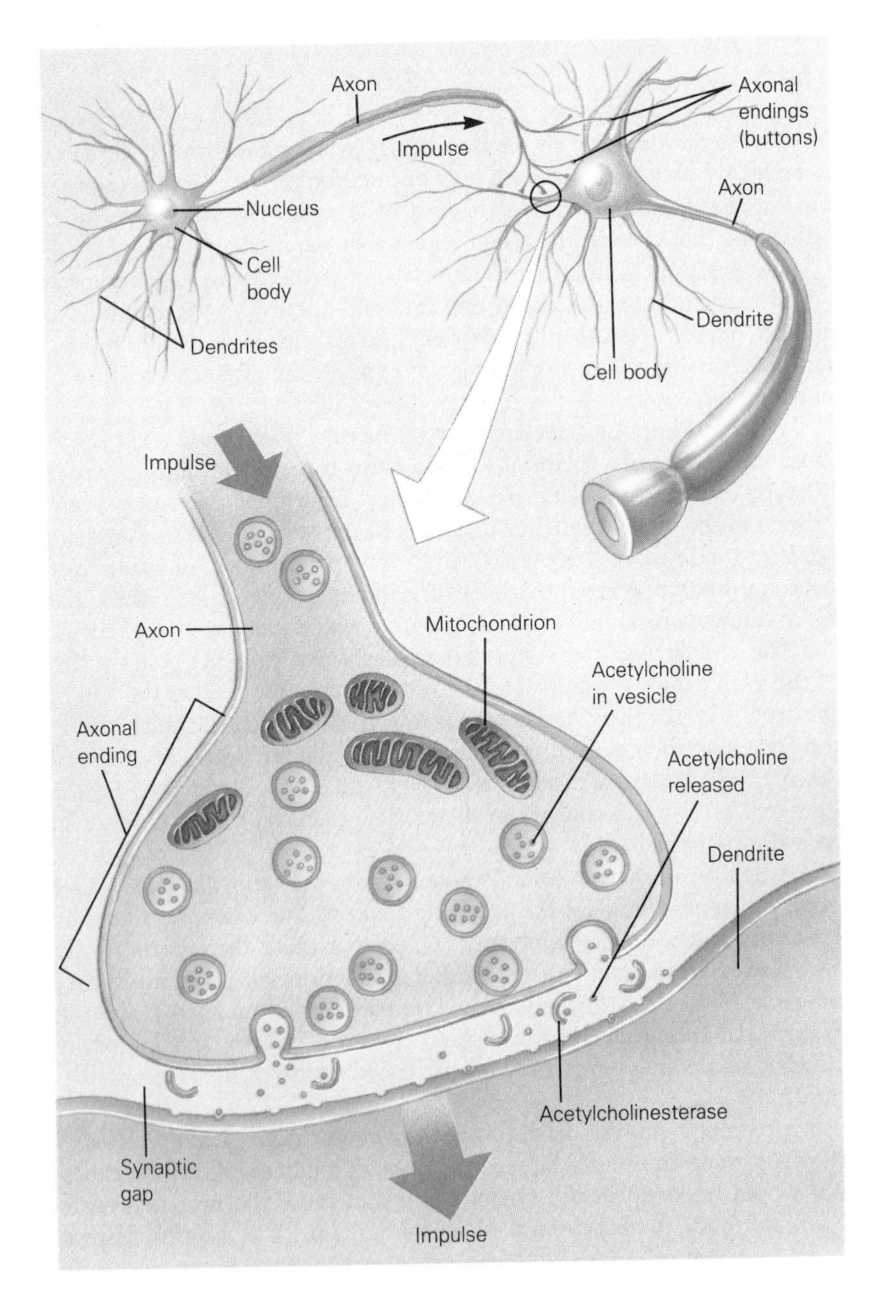

FIGURE 21.10

Synaptic transmission. The axon endings of one neuron lie very near the dendritic endings of the next neuron. As the impulse reaches the end of its axon, it stimulates the release of a neurotransmitter, such as acetylcholine. The process requires the expenditure of energy, thus the axonal endings contain many mitochondria. The neurotransmitter is released through the terminal portion of the axon into the space between the axon and the dendrite of the next neuron. It diffuses across the gap and binds to a receptor on the membrane of the dendrite, thus stimulating that dendrite and perhaps initiating an impulse. The neurotransmitter is quickly destroyed by an enzyme, such as acetylcholinesterase, so that the second neuron does not continue firing in the absence of a real impulse.

FIGURE 21.11

Excitement and inhibition among neurons. Here, two excitatory neurons and one inhibitory neuron synapse with the dendrites of another neuron. The darker axons are firing. In (a) the neuron is at rest with no axonal action; in (b) one excitatory neuron fires, but it does not stimulate to threshold levels; in (c) both excitatory neurons fire and the second neuron is stimulated to fire; in (d) the inhibitory neuron is fired at the same time so that the second neuron is not raised above the threshold level even if both excitatory neurons fire; and in (e) the inhibitory neuron firing alone will, of course, fail to excite the second neuron.

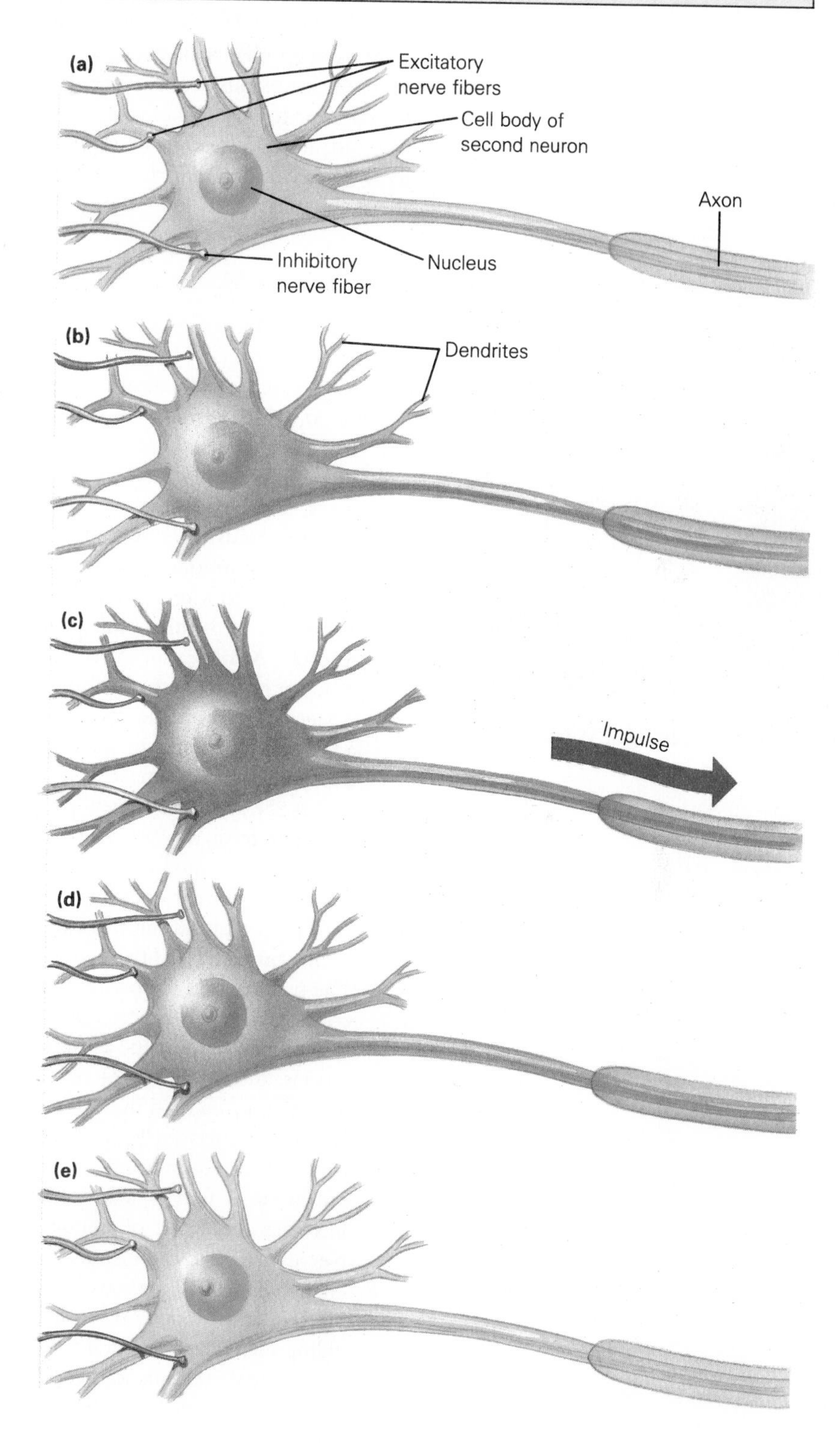

be relatively harmless (Figure 21.12). However, while they are active, they can block the effects of acetylcholinesterase. The enzyme functions not only to keep chains of neurons from firing blindly in response to accumulating transmitter substance, but also to keep acetylcholine from diffusing over to nearby neurons and causing them to fire as well. With this in mind, what do you think might be the effect on a farmworker exposed to the insecticides? Already there have been several deaths attributed to these chemicals.

FIGURE 21.12
Short-lived insecticides such as Malathion have largely replaced the use of DDT in the United States. Whereas DDT was assimilated into fatty tissue where it could cause long-term problems, Malathion destroys the enzymes that break down neurotransmitters causing impulses to fire wildly and indiscriminately, and bringing on convulsions and possibly death.

SUMMARY

1. A hormone is a chemical secreted by an endocrine gland directly into the blood, which carries it to the part of the body where it has an effect.
2. Hormonal regulation evolved early and is common among invertebrates and vertebrates. Hormones regulate body processes, prepare the body for stress, and help adapt the organism to its environment. The major endocrine glands, their hormones, and functions are listed in Table 21.1.
3. A peptide hormone (first messenger) functions from outside a target cell by binding to a receptor on the cell membrane. This complex binds to an enzyme in the membrane, adenylate cyclase, which then converts ATP to cyclic AMP. Cyclic AMP acts as a second messenger by triggering a series of biochemical reactions within the cell.
4. A steroid hormone functions from within cells. It binds to a receptor in a target cell's cytoplasm. This complex enters the nucleus where it directs the synthesis of specific mRNA. The mRNA moves to the cytoplasm and directs the synthesis of new protein.
5. Negative feedback systems regulate the production of many hormones. An example is the interaction between tropic hormones and their target glands. The anterior pituitary gland secretes thyroid-stimulating hormone (TSH) that causes the thyroid to secrete thyroxine. A drop in thyroxine levels stimulates TSH secretion, whereas a rise slows TSH secretion.
6. The nervous and endocrine systems are related structurally (some endocrine glands are composed of nervous tissue), functionally (the endocrine system may support the actions of the nervous system under certain conditions, such as an emergency), and chemically (the same chemical messengers may be used in both systems).
7. Neurons, or nerve cells, consist of a cell body, dendrites that receive stimuli from other neurons, and an axon that conducts impulses away from the cell body. A fatty myelin sheath surrounds some vertebrate axons and increases the speed of impulse transmission by saltatory propagation. Nerves are bundles of axons and dendrites.
8. Glial cells in the nervous system may function in a number of ways: binding and support, nourishing neurons, insulating neurons, and providing materials for neural transmission.
9. An external stimulus is detected by a receptor that is specialized for a particular sensation. The impulse is transferred to an afferent (sensory) neuron, which carries it to the central nervous system. The impulse is conducted from the central nervous system toward an effector in an efferent (motor) neuron.

10. In a resting neuron, there are more sodium ions outside the cell than inside and more potassium ions inside than outside. The inside of the resting neuron has a negative charge due to negatively charged proteins. This distribution of ions creates a resting potential of about −70 millivolts. The resting neuron's membrane is not very permeable to sodium ions. If sodium ions leak into the cell they are removed by the sodium-potassium pump.
11. A nerve impulse begins when the membrane suddenly becomes permeable to sodium ions at some point along the membrane and these ions enter the cell through ion channels, depolarizing the membrane. A wave of inrushing sodium ions sweeps along the length of the cell. At the point along the membrane where sodium ions enter, the interior of the neuron becomes temporarily positive. However, the membrane is immediately repolarized as potassium ions leave the cell. The change in distribution of charges that sweeps along the neuron is called an action potential.
12. Any stimulus that reaches threshold initiates an impulse that does not lose strength as it travels along the cell. The brain interprets the intensity of a stimulus as either the frequency of impulses or the number of neurons stimulated.
13. The point where one neuron meets another is called a synapse. At the axon tip, an impulse triggers the release of a neurotransmitter from small storage packets. The neurotransmitter crosses the gap between neurons. If the chemical comes from an excitatory neuron, it makes it more likely that the receiving neuron will fire. If the chemical comes from an inhibitory neuron, it makes the firing in the receiving neuron less likely. The neurotransmitter in the gap is quickly removed.

KEY TERMS

acromegaly (576)
action potential (589)
adrenal gland (578)
adrenaline (579)
afferent (sensory) neuron (587)
aldosterone (578)
atrial natriuretic factor (ANF) (579)
axon (584)
calcitonin (576)
cell body (584)
cortisol (578)
cretinism (576)
dendrite (584)
depolarized (589)
effector (587)
efferent (motor) neuron (587)
endocrine gland (575)
epinephrine (579)
estrogen (578)
excitatory neurons (591)
first messenger (581)
glial cell (587)
hormone (575)
hypothalamus (576)
impulse (587)
inhibitory neurons (591)
interneurons (587)
ion channels (589)
myelin sheath (585)
nerve (587)
neuron (584)
neurotransmitter (591)
norepinephrine (579)
parathyroid gland (576)
parathyroid hormone (576)
pituitary (576)
progesterone (578)
receptor (587)
repolarization (589)
resting potential (588)
saltatory propagation (585)
second messenger (581)
sodium-potassium pump (589)
somatotropin (576)
synapse (591)
target organ (575)
testosterone (578)
threshold (589)
thyroid (576)
thyroxine (576)

FOR FURTHER THOUGHT

1. The neurons in invertebrate nervous systems are unmyelinated. Based on this information alone, would you expect impulse conduction to be faster in vertebrate or invertebrate nervous systems?
2. Curare is a poison that acts by binding to the receptor sites of dendrites that are normally stimulated by the neurotransmitter acetylcholine. How does this affect nerve transmission?

FOR REVIEW

True or false?

1. ___ The adrenal cortex is the endocrine system's "master gland."
2. ___ Estrogen is an example of a steroid hormone.
3. ___ Dendrites receive input from other neurons and conduct the information toward the cell body.
4. ___ Nerve impulses retain the same intensity as they travel along the length of the neuron.
5. ___ A synapse separates the dendrite from its cell body.

Fill in the blank.

6. Hormones are produced by ductless ___ glands.
7. ___ is a hormone produced in the adrenal cortex that is an antiinflammatory agent.
8. ___ results from excess somatotropin production in adulthood.
9. ___ is the name for the speedy transmission of impulses that occurs in myelinated nerve axons.

Choose the best answer.

10. ___ is secreted by the adrenal cortex and regulates salt and water concentrations in the kidney.
    A. Estrogen
    B. Aldosterone
    C. Testosterone
    D. Progesterone
11. Peptide hormones
    A. promote development of secondary sexual characteristics.
    B. exert their effects via a second messenger within the cell.
    C. must enter target cells to join receptor molecules.
    D. are produced in the ovaries.
12. Choose the incorrect statement.
    A. The lowest intensity of a stimulus that causes a neuron to fire is called the threshold value.
    B. Neurotransmitters are released from the tip of the axon.
    C. Glial cells carry impulses across the synapse.
    D. Efferent neurons carry impulses from the spinal cord to an effector.

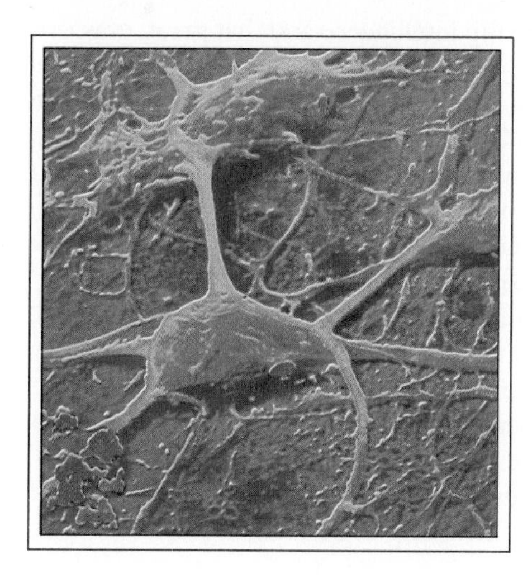

CHAPTER 22

The Nervous System

Overview

The Evolution of Nervous Systems

The Central Nervous System

The Human Brain

The Peripheral Nervous System

Mindbenders

Objectives

After reading this chapter you should be able to:

- Describe the organization and complexity of representative invertebrate nervous systems.
- Describe the neural pathways of the central nervous system and explain how reflex arcs operate.
- Discuss some trends in brain development among vertebrates.
- List the major parts of the brain and their functions.
- Explain how the cerebral hemispheres process information.
- List the divisions of the peripheral nervous system and desribe how each acts.
- Give examples of psychoactive agents, their effects, and potential for dependence.

Humans almost seem to worship intelligence and that great, gray orb from which it stems—the brain. However, a certain irony arises here because we don't even know what intelligence is. People have been wrestling for years to define it and measure it. The result has been sharp disagreement and a complete lack of consensus. And when we try to broach the idea of intelligence in other species, the conversation becomes a shambles. Yet we believe that intelligence, whatever it is, is good—at least for us. And we believe that it somehow resides in the brain.

In our consideration of the brain and other neural structures, we will first review various types of nervous systems, from simple to complex. We could say from "lower" to "higher," except that the terms lead to unfortunate misunderstandings. "Higher," in the minds of many people, implies any characteristic similar to those of humans, such as a large "thinking" center. The implication of this usage is that all species tend to evolve toward humanlike characteristics, including higher intelligence. Nothing could be further from the truth.

THE EVOLUTION OF NERVOUS SYSTEMS

Clues to the evolutionary development of nervous systems can be gleaned from a cross-species survey from simple to complex (Figure 22.1). One of the simplest nervous systems is that of cnidarians such as the freshwater *Hydra.* It consists simply of a two-dimensional net of interconnecting neurons, or nerve cells, spread throughout the outer body layer. The entire surface of the animal is about equally covered. There is no part that controls the rest; no nerve center that functions in the regulation or coordination of the nerve net. Thus, if one part of the animal is stimulated, the entire body responds and shows awareness of the stimulus.

The flatworm has a somewhat more specialized nervous system (Figure 22.1b). The neurons are arranged in two longitudinal nerves connected by transverse nerves, producing a "ladder" as opposed to the *Hydra's* "net." Your keen eye will undoubtedly also have noted the aggregation of nerves in the head region. These are the head ganglia, which are composed of clumps of neural cell bodies. As these become relatively larger and more complex in other species, they eventually are referred to as the brain.

As a brief aside, you may have wondered why the brain is located in the head. There have been exceptions, such as the huge, herbivorous dinosaur, the brontosaurus, which had a second "brain" at the base of its tail to help direct its immense body as it browsed in prehistoric lakes. But the evolutionary reason for nerve centers in the anterior, or head, region may have been to permit quick analysis of the environment into which the animal would be moving. These centers would in all probability have come to be associated with the specialized receptors we refer to as the senses. And as we know, the senses of sight, sound, smell, and hearing are commonly located in the head region. Thus, the environment into which the animal is moving may be quickly assessed. If the brain and these

FIGURE 22.1

Examples of animal nervous systems. (a) The *Hydra,* with its nerve net and no coordinating center; (b) the planarian, or flatworm, with its longitudinal nerves connected by transverse nerves and a nerve concentration in the head; (c) the earthworm, with its single ventral nerve cord and well-defined cerebral ganglia; and (d) the frog, with its dorsal hollow nerve cord and the well-developed brain protected by bone. The trend is toward condensing the nervous system into a longitudinal arrangement (arising with the development of bilateral symmetry). A segmented body innervated by a segmented nervous system sets the stage for specialization along the nerve length. The nerve cord is dorsal and the brain is more complex with the segmentation less regular.

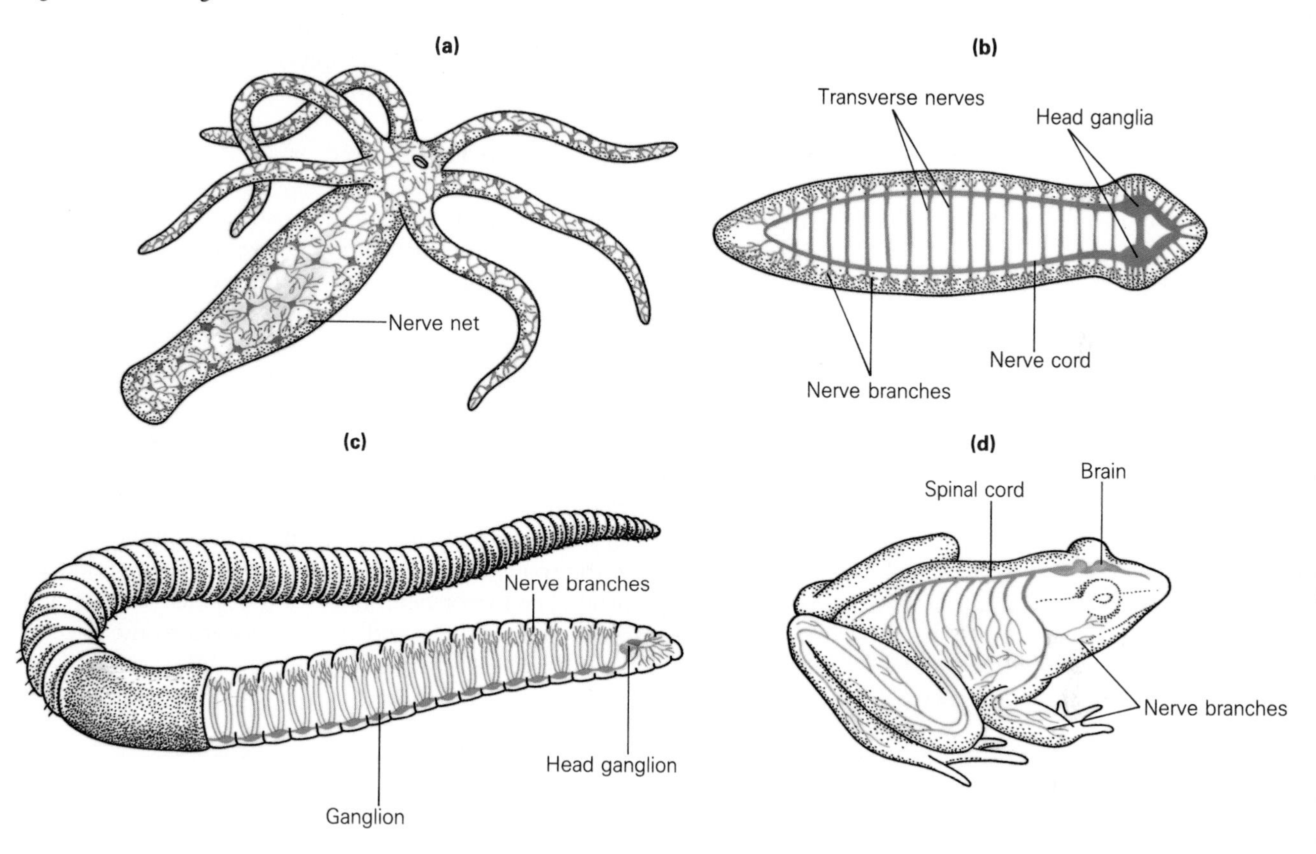

special receptors were located at the posterior end, the animal might find itself in an inhospitable environment by the time it realized its predicament.

A somewhat more complex nervous system is found in the earthworm (Figure 22.1c). The earthworm has a single longitudinal nerve, but it shows vestiges of a paired arrangement in that it is two-lobed, much like two cords pressed together. The nerve is ventral (toward the belly surface) with the heart and digestive tract lying dorsal (toward the back surface) to it. Note the distinctness of the cerebral ganglia and the obvious nodes along the nerve cord, each with paired nerves reaching into a segment of the body. The frog nervous system (Figure 22.1d) is relatively primitive for that of a vertebrate, but it can be used to illustrate the basic neural plan in vertebrates. In this group, we find a distinct brain and spinal cord.

THE CENTRAL NERVOUS SYSTEM

The brain and spinal cord together form the **central nervous system.** In all vertebrates, the longitudinal nerve (the spinal cord) is dorsal, hollow, filled with fluid, and protected by bone. The anterior end is marked by a brain, an elaboration of the primitive ganglionic mass of ancient forebears. The vertebrate brain shows marked specialization; that is, different parts of it are associated with very specific functions. The central nervous system of vertebrates shows traces of the paired and segmented neural arrangements of their distant ancestors. For example, the brain is two-lobed and paired nerves extend from it and from the spinal cord. Table 22.1 summarizes parts of the human nervous system.

The Spinal Cord and the Reflex Arc

If you pride yourself on being a "thinking animal," you may be a little disappointed to realize that your spinal cord can often receive information from the body's receptors, process it, and initiate the proper response before your brain even "knows" what has happened. The neural connections that permit this are called a **reflex arc.**

Physicians often like to tap the tendon below the knee, making the leg jump. This was once thought to be for the physician's amusement, but now we know that it is to test the patient's reflex arc. In the reflex arc shown in Figure 22.2, the message is generated in a special receptor called a **stretch receptor** that responds when it is elongated. The message is then transmitted to a sensory neuron, which enters the spinal column through the **dorsal root** of the spinal nerve. The sensory neuron then excites the proper motor neuron. The motor neuron leaves the spinal cord through the **ventral root** of the spinal nerve and travels outward to the effector,

TABLE 22.1
Parts of the Human Nervous System

CENTRAL NERVOUS SYSTEM		PERIPHERAL NERVOUS SYSTEM	
Brain	Integration; association; thought; directs most behavior	Somatic Nervous System	Primarily involved in sensations and actions of which we are conscious
Spinal Cord	Carries messages to and from brain; center for spinal reflexes	Cranial nerves	Service the head region (the vagus services the body region)
		Spinal nerves	Service the neck and body region
		dorsal root	Houses sensory neurons
		ventral root	Houses motor neurons
		Autonomic Nervous System	Controls involuntary activities of internal organs
		Sympathetic nervous system	"Fight or flight" reactions
		Parasympathetic nervous system	Returns the body to "normal" state after emergency; controls activity of organs under nonstressful conditions

FIGURE 22.2

A simple reflex arc is shown in the familiar knee-jerk response. The tendon that attaches the large muscle in the upper leg to the kneecap (patella) is stretched by a light blow from a mallet. Receptors in the muscle sense this change and excite sensory neurons, which conduct an impulse to the spinal cord. Within the spinal cord, the sensory neuron excites the proper motor neuron. The motor neurons immediately signal the muscle to shorten and the lower leg snaps forward if the neuromuscular systems are in order. The brain is not involved, although it is made aware of the situation after the fact. The rapid shortening is a mechanism that restores the proper state of contraction to the muscle, even if a bit vigorously.

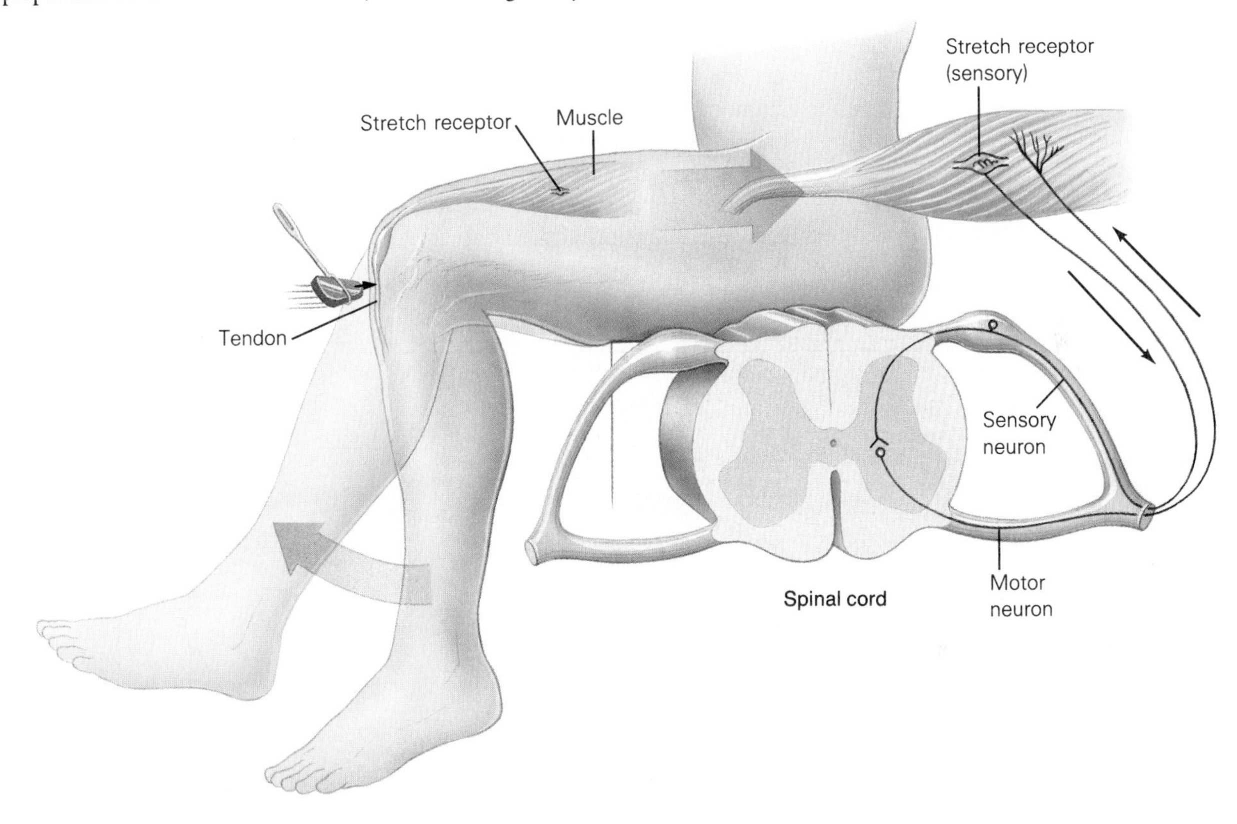

a muscle group. In most other reflex arcs the impulse is transmitted from sensory to motor neurons by an intermediate neuron called an **inter-neuron.**

Note that impulses from one side of the body can cross the spinal cord so that effectors in the other side of the body are stimulated. Direct neural routing from the spinal cord in a reflex arc saves time because the distance the impulse has to travel from receptor to effector is shorter. Furthermore, no time is spent in deliberating over the decision.

Since not all actions are spinal reflexes, the spinal cord also conducts messages to and from the brain. Sensory neurons entering the spinal cord synapse with **ascending neurons** that bring the message to the brain. In this way, even if the response has already been accomplished reflexively, the brain is informed of the change. The brain's messages are conducted downward over **descending neurons.**

The Vertebrate Brain

Actually, there is no such thing as the vertebrate brain, because the vertebrates include widely diverse groups, each of which is highly specialized and distinctive. In the midst of such diversity, however, it is possible to detect trends that give some clues to the general pattern of brain development in animals with backbones.

Figure 22.3 illustrates relative differences in parts of the brain from fish to reptile to bird to mammal. The ***medulla*** is the part of the brain that connects directly with the spinal cord allowing messages to be transmitted between the brain and the rest of the body. In addition, it controls vital functions such as breathing and heart rate. The ***cerebellum*** is associated with sensory motor coordination for locomotion and posture. The ***cerebrum*** is the "gray matter," the thinking part of the brain. The cerebrum is an exceedingly complex integrating center containing areas for both receiving sensation and initiating voluntary movement. (Regarding the

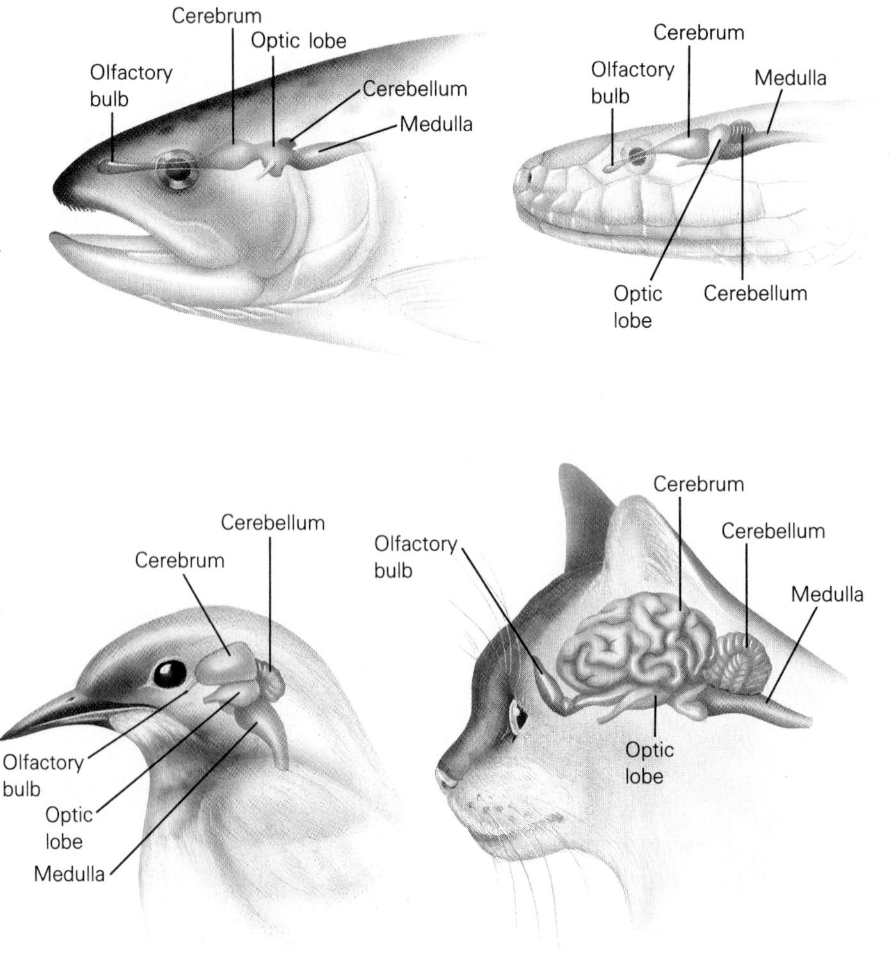

FIGURE 22.3
A comparison of the brains of a fish, a reptile, a bird, and a mammal. Notice the diminutive size of the cerebrum relative to the olfactory bulb in the fish. Also note the relative mass of the lower brain (here, the cerebellum and medulla) compared to that of the cerebrum. Reptiles have a somewhat larger, but still smooth, cerebrum and a reduced olfactory area. Note that the cerebrum of birds is distinctly larger than that of fish and reptiles, compared to the other parts of the brain. The olfactory lobe is generally reduced and the optic lobe is larger. The brain of the cat is dominated by the convoluted cerebrum. The cerebellum, involved in coordination, is well-developed in the cat, as is the olfactory bulb. It is important to realize that all areas of intelligence do not necesarily increase as one moves toward cerebration. Fish, for example, can learn some things easier than reptiles can, even though reptiles are generally more behaviorally adaptable and generally believed to be more intelligent.

brain, the higher centers are those most recently evolved—"advanced"—and usually refer to the cerebrum. The lower centers are more ancient—"primitive"—and generally refer to areas nearer the medulla.) We will discuss all these structures shortly, but there are a few preliminary points you might find interesting. We might note, for example, that not only is there an increase in general brain size (in relation to body size and, particularly, to spinal cord size) as we go from fish to mammal, but there is also an increase in the size of the cerebrum in relation to other parts of the brain. However, there are notable exceptions to this trend. For example, the olfactory lobe doesn't follow this pattern. Olfaction has to do with the sense of smell, so which of the animals in Figure 22.3 do you suppose would rely more on a sense of smell? (Remember to consider the olfactory lobe in relation to total brain size.) Other sensory lobes could also be singled out, such as the optic lobe, which has to do with vision. Which animal do you suppose would have a larger optic lobe with respect to its brain mass, an eagle or an elephant?

The fact that the cerebrum is larger as we move to more recently evolved (advanced) animals does not imply that as one moves up the evolutionary family tree, each species is "smarter" than the ones below it. There may be such a trend toward higher intelligence within certain groups, but there are also many exceptions. For example, the octopus, a mollusk like the snail, is more intelligent than many vertebrates. In fact, the octopus brain is similar to the mammalian brain in terms of its complexity and organization into specialized regions.

The evolutionary development of the mammalian cerebrum, which culminates in the human brain, undoubtedly has been one of the most crucial events in the history of life on earth. Such a statement is admittedly somewhat grandiose, but its validity becomes apparent when we consider the impact of the human species on the fragile life-support system that so thinly covers the planet. It would be interesting, therefore, to know how such a brain came to be. What spurred its development? No one is certain, but one prevailing idea is that an increasingly developed cerebrum aided in the survival of one small seemingly insignificant group of ancestral reptiles that lived in an exceedingly dangerous world dominated by the great dinosaurs. This group, which was to give rise to mammals, branched off from other reptiles 180–200 million years ago.

These small premammals were certainly no match for the speedy, incredibly powerful, and voracious dinosaurs. Since they couldn't outrun or outfight the "ruling reptiles," it is believed that a premium was placed on their mental agility. In order to survive, the premammals were forced to outthink the dinosaurs. The dimmer members among them would rapidly have fallen to predators and, thus, would have failed to leave their "dim genes" in the next generation. So, with their very existence at stake, the premammals must have rapidly increased their mental agility, starting the cerebrum on its way toward dominating the brain of at least one line of animals.

The rapid development of the cerebrum and increasing reliance on intelligence would have been accompanied by other changes among the early mammals. For example, since learning requires experience, the brainier animals would have had to develop a life pattern that gave them a chance to learn *before* they were exposed to the dangers of the world. On

FIGURE 22.4
Birds generally have not had to stay and outwit their terrestrial predators. Instead they just leave and avoid the interaction.

this basis, some believe, parental care developed. Even today the offspring of "learning animals" stay with their parents for extended periods. (How old were you when you left home?) During this time, they are cared for and protected by their parents until they gain enough experience to cope with the world. Also, by associating with parents or others of their kind, they can learn from them. (Most of the more brainy species are social animals, although there are notable exceptions.)

Even in a high-risk world, however, increased braininess is not the only course open to natural selection. As an example of another evolutionary route, consider the "strategy" of birds. The noted psychologist Oscar Heinroth is said to have commented, "Birds are so stupid because they can fly." The idea is that a bird doesn't normally "outfox" its enemies through intellectual maneuvering. It simply flies away (Figure 22.4). Even in escaping airborne predators, such as hawks, most birds do not rely on their cunning. Instead, they employ a few stereotyped behavior patterns, such as attacking in mass, aggregating to confuse predators, taking rather specific evasive maneuvers, or giving warning cries. Birds just never have had to develop great mental capacities, so don't be misled by the discerning frown on an eagle's face.

THE HUMAN BRAIN

Now we come to the rather interesting notion of the human brain talking about itself. Perhaps it is because of this strange twist that we encounter so many endless accolades about this great organ (Figure 22.5). Let's begin with some basic descriptions.

The human brain is divided into three parts: the hindbrain, the midbrain, and the forebrain. The **hindbrain** consists of the medulla, the cerebellum, and the pons. The hindbrain is sometimes called the "old brain" because it evolved first. These structures still dominate the brain of some animals, as we have seen. The **midbrain,** logically enough, is the area between the forebrain and hindbrain and connects the two. The **forebrain,** or "new brain," consists of the two cerebral hemispheres and certain internal structures (Figure 22.6). Table 22.2 summarizes the structure and function of the parts of the brain.

FIGURE 22.5
Under the wrinkled exterior of the human brain resides an incredibly complex array of neurons.

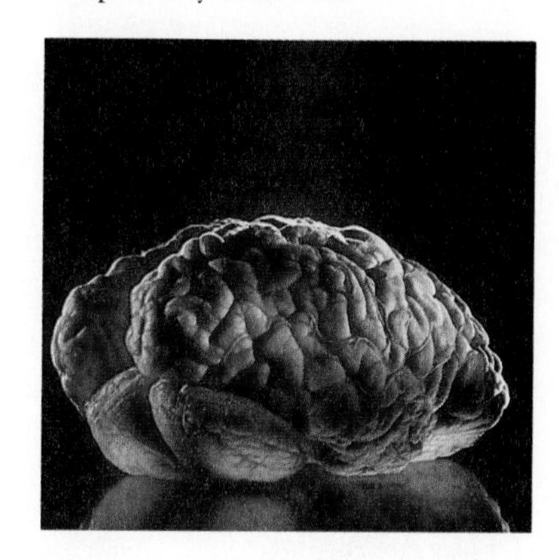

The Hindbrain

The Medulla

As a rough generality, the more subconscious, or mechanical, processes are directed by the more posterior parts of the brain. For example, the hindmost part, the **medulla,** is specialized as a control center for such basic functions as breathing, digestion, and heartbeat. In addition, it is an important center of control for certain charming activities such as swallowing, vomiting, and sneezing. As we have already seen, it connects the spinal cord and the more anterior parts of the brain.

TABLE 22.2
Structure and Function of Human Brain

STRUCTURE	FUNCTION
HINDBRAIN	
Medulla	Control of subconscious activities such as breathing, digestion, heartbeat, swallowing, vomiting, and sneezing; connects the spinal cord with the brain
Cerebellum	Controls balance, equilibrium, and coordination
Pons	Connects the cerebellum and the cerebral cortex and other brain regions
MIDBRAIN	Connects the hindbrain and forebrain; receives sensory input from the eyes
FOREBRAIN	
Thalamus	The "great relay station"; connects various parts of brain
Reticular system	Arousal of certain brain areas; filters impulses from sensory neurons
Hypothalamus	Regulates heart rate, blood pressure, body temperature, and the pituitary; controls drives such as hunger, thirst, and sex
Cerebrum	
Cerebral cortex (outer Cerebrum)	Gray matter; conscious thought; memory, intelligence; speech; association among senses
occipital lobe	Processes visual information
temporal lobe	Auditory reception and some visual information
frontal lobe	Regulates precise voluntary movement and the use of language and speech (includes the prefrontal lobe, which sorts information and orders stimuli)
parietal lobe	Receives information from the skin and processes information about body position
White matter (inner Cerebrum)	Nerve tracts allowing communication between parts of cortex; e.g., corpus callosum connects the two hemispheres and helps them communicate

FIGURE 22.6

The human brain in surface view (a) and in sagittal section (b). Note that there is enough regularity in the convolutions that some have been named.

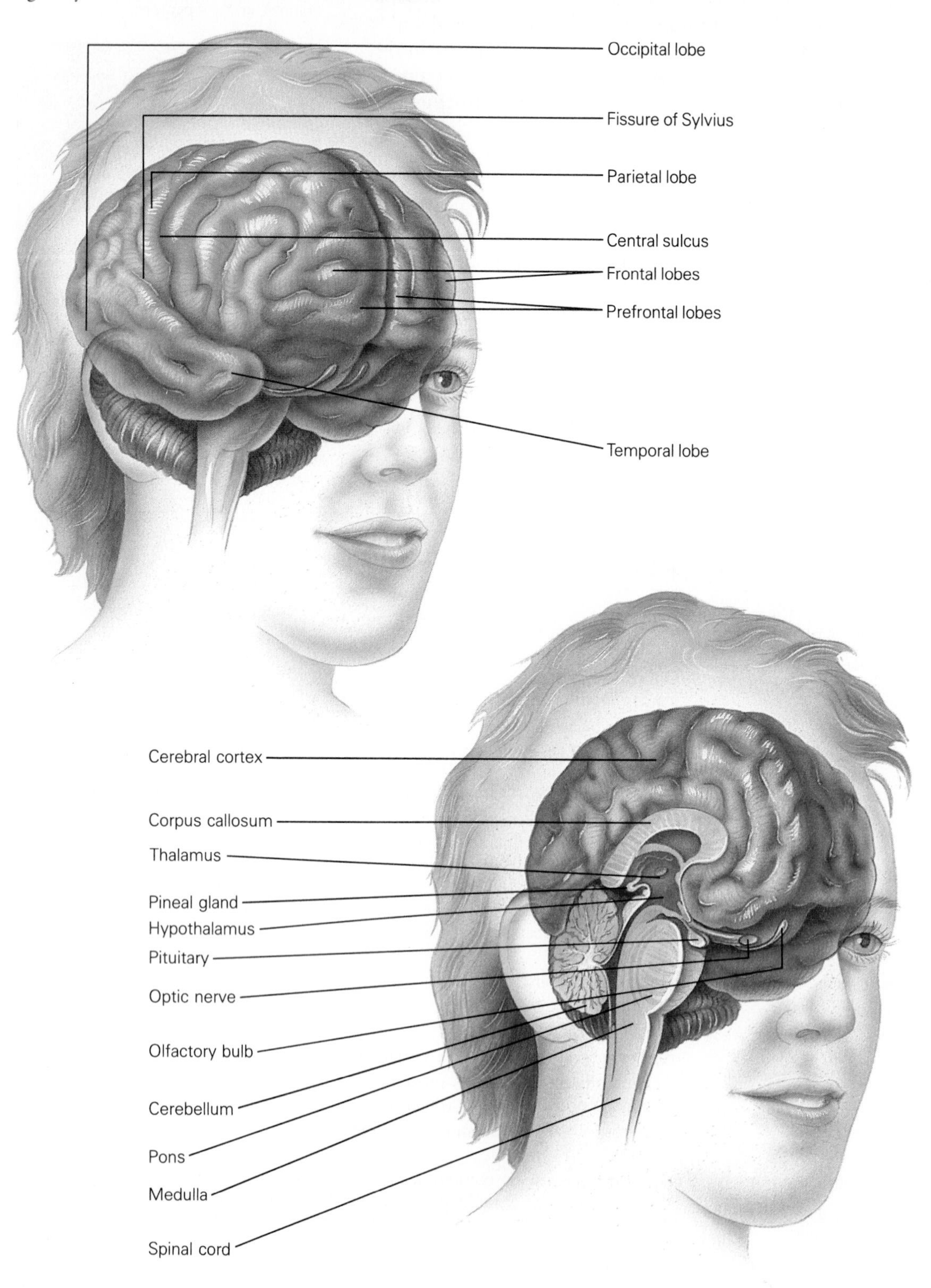

The Cerebellum and Pons

Above the medulla and more toward the back of the head is the **cerebellum,** which is concerned with balance, equilibrium, and coordination. Do you suppose there might be differences between athletes and nonathletes in this part of the brain? (Apparently there are, but the differences are slight.) Do you think this "lower" center of the brain is subject to modification through learning? (Can you improve your coordination through practice?) The **pons,** which is the portion of the brainstem just above the medulla, acts as a bridge connecting certain parts of the brain. For instance, it connects the cerebellum and the cerebral cortex, accenting the relationship between the cerebellar part of the hindbrain and the more "conscious" centers of the forebrain.

The Midbrain

The *midbrain* connects the hindbrain and forebrain by numerous tracts. In addition, certain parts of the midbrain receive sensory input from the eyes and ears. In vertebrates, sound is processed here before being sent to the forebrain. The midbrain has a more complex role in fishes and amphibians than in reptiles, birds, and mammals because in the latter group, many of its functions are taken over by the forebrain.

The Forebrain

The Thalamus and Reticular System

The thalamus and hypothalamus are located at the base of the forebrain. The **thalamus** is rather unpoetically called the "great relay station" of the brain. It consists of densely packed clusters of nervous cells that presumably connect the various parts of the brain—between the forebrain and the hindbrain, between different parts of the forebrain, and between parts of the sensory system and the cerebral cortex.

The thalamus contains a peculiar neural structure called the **reticular system,** an area of interconnected neurons that are almost feltlike in appearance. The reticular system runs through the medulla, pons, and thalamus and extends upward to the cerebral cortex. The role of the reticular system is still a bit mysterious, but several interesting facts are known about it. For example, it bugs your brain. Every afferent and efferent pathway to and from the brain sends side branches to the reticular system as it passes through the thalamus. So all the brain's incoming and outgoing communications are "tapped." Also, these reticular neurons are rather unspecific. That is, the same neuron may respond to stimuli from, say, the hand, foot, ear, or eye. It has been suggested that the reticular system serves to activate the appropriate parts of the brain upon receiving a stimulus. In other words, the reticular system activates the part of the brain that is needed for the particular task at hand. If there is no incoming stimuli—no need for conscious activity—the system sends fewer signals, quieting the brain. You may have noticed it is much easier to fall asleep lying on a soft bed in a quiet, darkened room than on a pool table in a disco. With the quietness, the reticular system receives fewer messages and

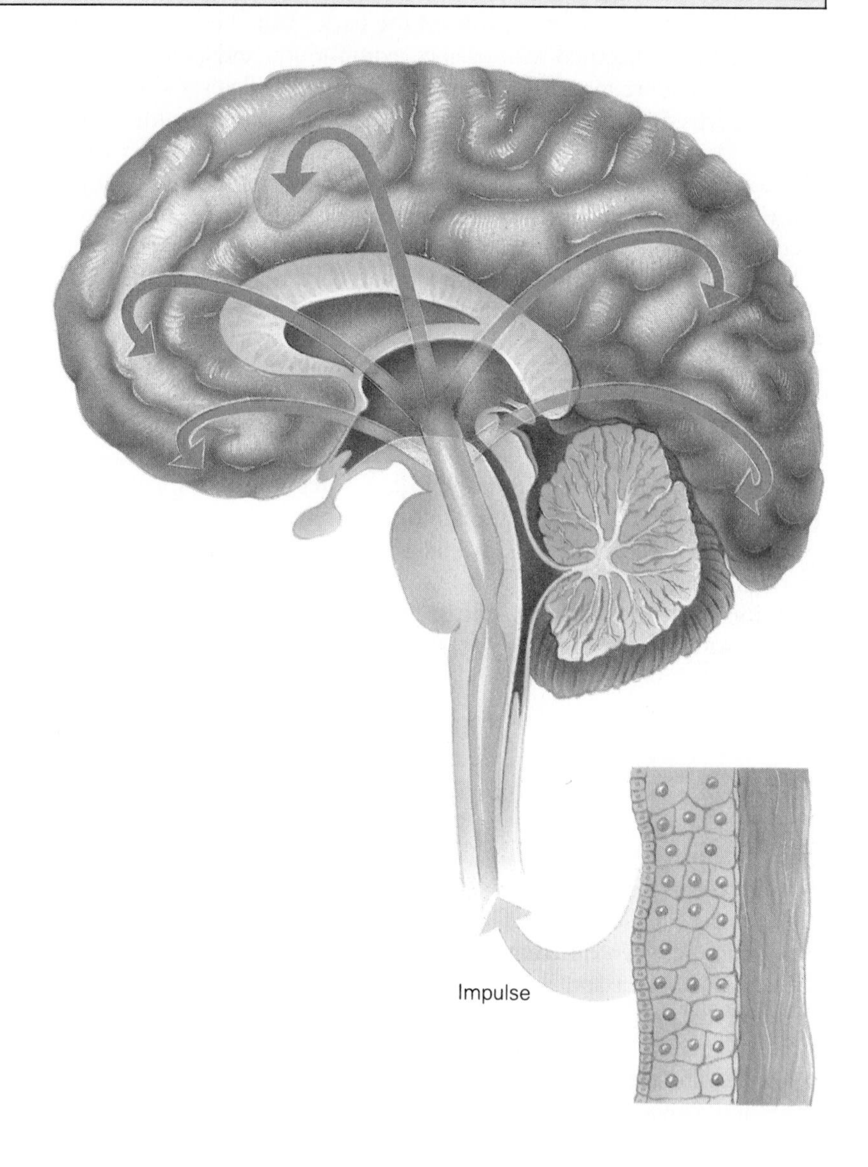

FIGURE 22.7

The reticular system of the human brain. The impulse originating at the lower right passes through the reticular system, with its untold millions of neurons. The smaller arrows indicate that, in this case, the entire cerebrum has been alerted, but a specific part (the shaded area) is the target of most of the impulses. It is likely, then, that this area of the cerebrum will be required to deal with whatever initiated the stimulus in the first place.

the brain is lulled rather than aroused. On those nights when you have the "big eye" and just can't sleep, the cause may be continued (possibly spontaneous) firing of reticular neurons (Figure 22.7).

The reticular system may also regulate which impulses are allowed to register in your brain. When you are engrossed in a television program, you may not notice that someone has entered the room. But when you are engaged in even more absorbing activities, it might take a *general* stimulus on the order of an earthquake to distract you, whereas the *specific* stimulus of a turning doorknob would immediately attract your attention. Such filtering and selective depression of stimuli apparently takes place in the reticular system.

The Hypothalamus

The **hypothalamus** is a small body, densely packed with cells. It helps regulate the internal environment as well as some aspects of behavior. For example, the hypothalamus helps to control heart rate, blood pressure, and body temperature. It also plays a part in the regulation of the pituitary gland, as we learned earlier. And it controls such basic drives as hunger, thirst, and sex. So now you know what to blame for all your problems. Experimental electrical stimulation of various centers in the hypothalamus can cause a cat to act hungry, angry, cold, hot, benign, or horny.

The Cerebrum

For many people the word *brain* conjures up an image of two large, deeply convoluted gray lobes. What they have in mind, of course, is the outside layer of the two cerebral hemispheres, the dominant physical aspect of the human brain. The **cerebrum** is present in all vertebrates, but it assumes particular importance in humans. In some animals, it is essentially an elaborate refinement that implements behavior that could be performed to some degree without it. It has a far greater importance in other animals.

For example, if the cerebral cortex of a frog is removed, the frog will show relatively little change in behavior (*cortex:* rind; the cerebral cortex is the outer layer of the forebrain in which a great deal of the active neural tissue is found). If the frog is turned upside down, it will right itself; if it is touched with an irritant, it will scratch; it will even catch a fly. Also, sexual behavior in frogs can occur without the use of the brain—but we'll try not to extrapolate from that. Rats, on the other hand, are more dependent on their cerebral cortex. A rat that is surgically deprived of its cerebrum can visually distinguish only light and dark, although its body movement seems unimpaired. A decorticated cat (that is, with its cerebral cortex removed) can meow, purr, swallow, and move to avoid pain, but its movements are sluggish and robotlike. A monkey whose cerebral cortex has been removed is severely paralyzed and can barely distinguish light and dark. In humans, the destruction of the cortex causes total blindness and almost complete paralysis. Although such persons can breathe and swallow, they soon die.

It seems apparent that the cerebrum is more than just the center of "intelligence," and that from an evolutionary standpoint, as the cerebrum enlarges, more and more of the functions of the lower brain are transferred to it.

Hemispheres and Lobes

The human cerebrum consists of two hemispheres, the left and the right, each of these being divided into four lobes. At the back is the **occipital lobe,** which receives and analyzes visual information.

The **temporal lobe** is at the side of the brain. It roughly resembles the thumb on a boxing glove, and it is bounded anteriorly by the fissure of Sylvius. The temporal lobe shares in the processing of visual information, but its main function is auditory reception.

The **frontal lobe** is right where you would expect to find it—at the front of the cerebrum, just behind the forehead. This is the part that people hit with the heel of the palm when they suddenly remember what they forgot. One part of the frontal lobe is the center for the regulation of precise voluntary movement. Another part functions importantly in the use of language, and damage here results in speech impairment.

The area at the very front of the frontal lobe is called the **prefrontal area,** if you follow that. Whereas it was once believed that this area was the seat of intellect, it is now apparent that its principal function is sorting out information and ordering stimuli. In other words, it places information and stimuli into their proper context. The gentle touch of a mate or the sight of a hand protruding from the bathtub drain might both serve as stimuli, but they would be sorted differently by the prefrontal area. Up until a few years ago, parts of the frontal lobe were surgically removed in efforts to bring the behavior of certain aberrant individuals more into line with what psychologists had decided was the norm. The operation was called a frontal lobotomy, and it resulted in passive and unimaginative individuals. Fortunately, the practice has been largely discontinued, largely because chemical treatments now meet the same objectives.

The **parietal lobe** lies directly behind the frontal lobe and is separated from it by the central sulcus. This lobe receives stimuli from the skin receptors, and it helps to process information regarding bodily position. Even if you can't see your feet right now, you have some idea of where they are thanks to neurons in the parietal lobe. Damage to the parietal lobe may produce numbness and may cause a person to perceive his or her own body as wildly distorted and to be unable to perceive spatial relationships in the environment.

By probing the brain with electrodes, it has been possible to determine exactly which area of the cerebrum is involved in the body's various sensory and motor activities. We can see the results of such mapping in the rather grotesque Figure 22.8. The pictures are distorted not out of any appreciation of the macabre, but to demonstrate that the area of the cerebrum devoted to each body part is dependent not on the size of the part, but on the relative number of neurons serving the area. Thus, we have the greatest number of sensory receptors in the face, hands, and genitals, but the greatest amount of control only in the face and hands (as you may have already discovered).

Also note that the sensory and motor areas are not randomly scattered through the cortex, but that proximity in brain areas reflects the proximity of the parts of the body they control. Thus the index finger control area lies near the thumb control area, and the elbow area lies closer to the finger area than does the shoulder area. Probing has also revealed that memories are stored in very specific places as described in Essay 22.1.

Two Brains, Two Minds?

The best way to begin a banal conversation a few years ago was, "What sign are you?" Today, though, sophisticates may lead with, "Are you right-brained or left-brained?" Some of the findings of one of the most fascinating branches of neural research has indeed filtered into the public consciousness.

The question is based on information that has been accumulating since the middle of the last century when A. L. Wigam, a British physician,

FIGURE 22.8

The figure at the top indicates the relative amount of the cortex devoted to receiving sensory input from (left) or the motor control of (right) various body parts. The view of the brain at the bottom shows function maps of the cerebral cortex with sensory areas for skin and muscles (left) and motor areas (right). The sensory section is taken posterior to the fissure of Rolando, the motor section anterior to it. Note that a large part of the human brain is devoted to face, hands, and genitalia. Also, the eyes and hands are given more motor than sensory space in the brain in spite of our great dependence on them for sensory information. This is because the visual sensory information from the eye is handled by a different part of the brain. Our lips and genitalia have a rather surprising amount of brain tissue allocated for their sensory input. Perhaps we have so much of the brain devoted to facial areas because the face is so important in the subtle processes of human communication.

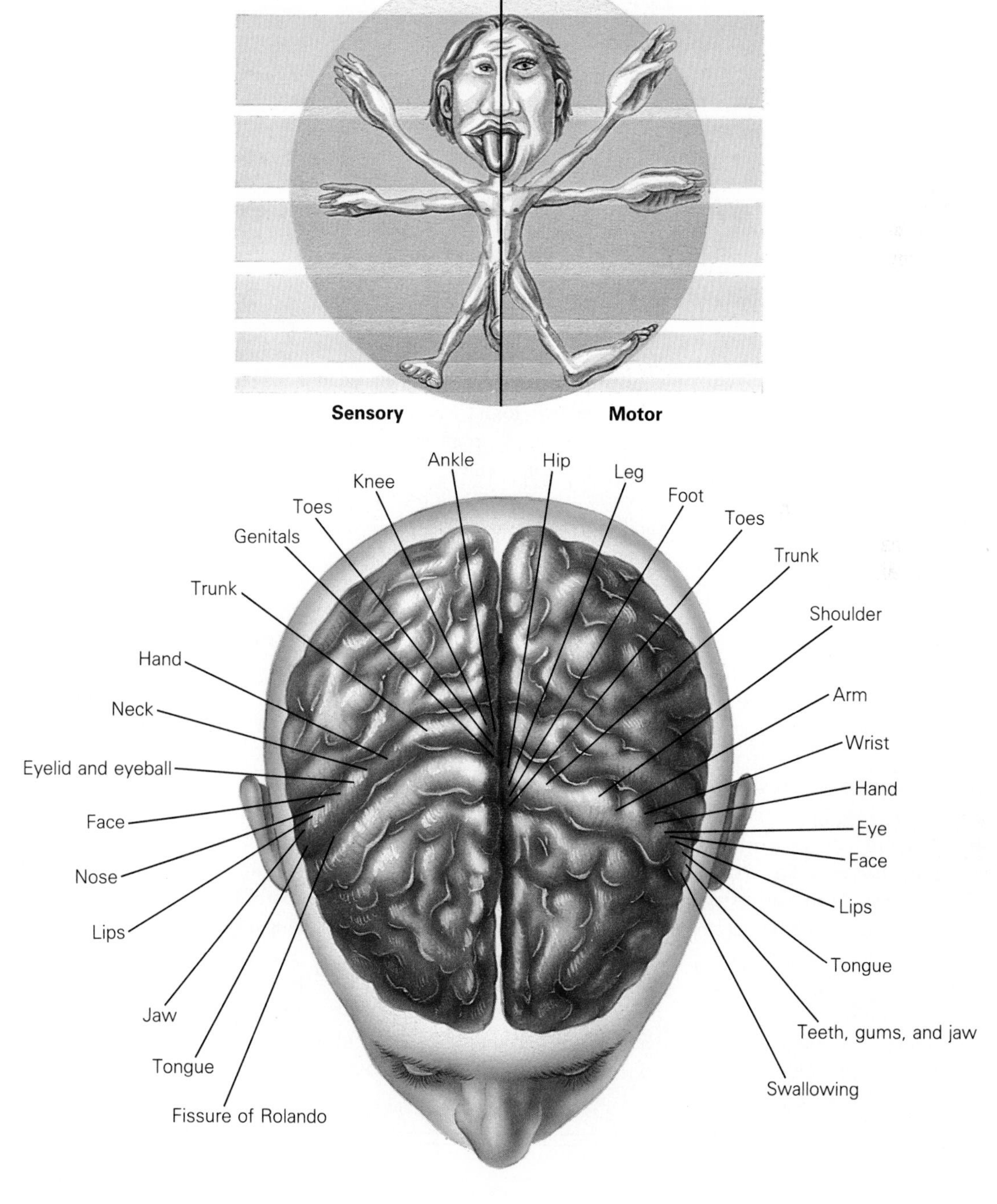

ESSAY 22.1

Penfield's Mapping

Wilder Penfield, of the Montreal Neurological Institute, was one of the group of researchers who fascinated the scientific world by stimulating the human brain with electrodes. Most of Penfield's work was done on epileptic patients in the course of surgical attempts to correct their condition. The attacks are caused by uncontrolled neural discharges from the injured area sporadically surging through the brain. Penfield's surgery was an attempt to identify and remove the damaged area without damaging other areas of the brain. The patients were conscious throughout the operation (the brain cannot feel pain) and could respond to questions when various parts of their brains were stimulated. By carefully positioning the electrodes and recording the patient's response, Penfield was able, in effect, to "map" the brain, to determine which parts are associated with specific functions. He found that such maps are, in a sense, similar to human faces. That is, the general structures are much the same from one person to the next, but the details are highly individual.

One of the most startling and controversial of Penfield's findings was that the human brain may "remember" virtually everything it has ever experienced, and that these memories can be elicited in very specific terms. We all have what are called "photographic memories" to an extent. Whatever stimuli we encounter indeed register with us—every aspect of them. For most of us, it is believed that these impressions actually last less than a second. The line of research begun by Penfield, however, suggests that perhaps much of the stimuli con-

stantly barraging our subconscious minds are indelibly stamped there and, under certain conditions, can be recalled. Penfield found, for example, that stimulation of certain areas could cause patients to "relive" certain experiences. One patient *heard* an old familiar strain of music, as if it were being played in the operating room, when a certain point of the brain was stimulated. When the electrode was removed and reapplied to the same place, the music did not pick up where it had left off but started over again from the beginning.

Another patient saw himself sitting with relatives and friends in their house in South Africa. The event had actually occurred years before. Another patient "watched" a play she had seen years before, and another "heard" again Christmas carols being sung in her old church in Holland.

The events elicited by Penfield's probings are more than simple vivid recollections. According to Penfield, the patients seemed actually to relive the experiences. Penfield has described the storage of such memories as being similar to recording by a tape recorder. The recorder is able to run only forward, and it has only one speed, normal. In other words, electrode stimulation cannot elicit events in any sequence other than the one in which they happened, and when elicited, these remembered events are reexperienced at the same rate they occurred. As long as the electrode is held in place, the experience of a former day goes forward. It cannot be held still, turned back, or crossed with other periods.

If Penfield's work is clearly substantiated, it is exciting to imagine its possible uses of such information. On long train rides (remember trains?), or in boring company, we could re-read a favorite book. There would be little need for testing pure "memory" as is so often done in classroom examinations, since each exam would, in effect, be an open-book test. (In a hypnosis experiment that I performed, a high school girl was able to "see" her notes on the blackboard during an examination after having read them only once. She considered turning herself in for "cheating" but thought better of it.) In cases of terminal disease, instead of facing death in apprehension and racked with pain, perhaps selective brain stimulation could enable a dying person to "relive" happier youthful days, to experience again long hikes with loved ones, robust optimism, and the companionship of old friends, now gone, as life dwindles away.

performed autopsies on men who had led somewhat normal lives with only half a brain. That is, one hemisphere had been destroyed by accident or other trauma. The question then arose, since we only need half a brain, then why do we have two? At first it was assumed that this was another case where nature had built in redundancy, or backup, in a critical system.

Brain research has revealed a fascinating fact: the hemispheres are not duplicates at all, but structures with quite different specializations. To oversimplify, the left hemisphere is the center of logical, stepwise reasoning, of mathematics and language. It processes information in a fragmentary, sequential manner, sorting out the parts of questions and dealing with each quite rationally. (*Star Trek's* Spock was definitely left-brained.) The right brain, on the other hand, is the center of awareness for music

FIGURE 22.9

The right and left sides of the brain control, in a general way, different abilities. These localized effects are less pronounced in women and left-handers.

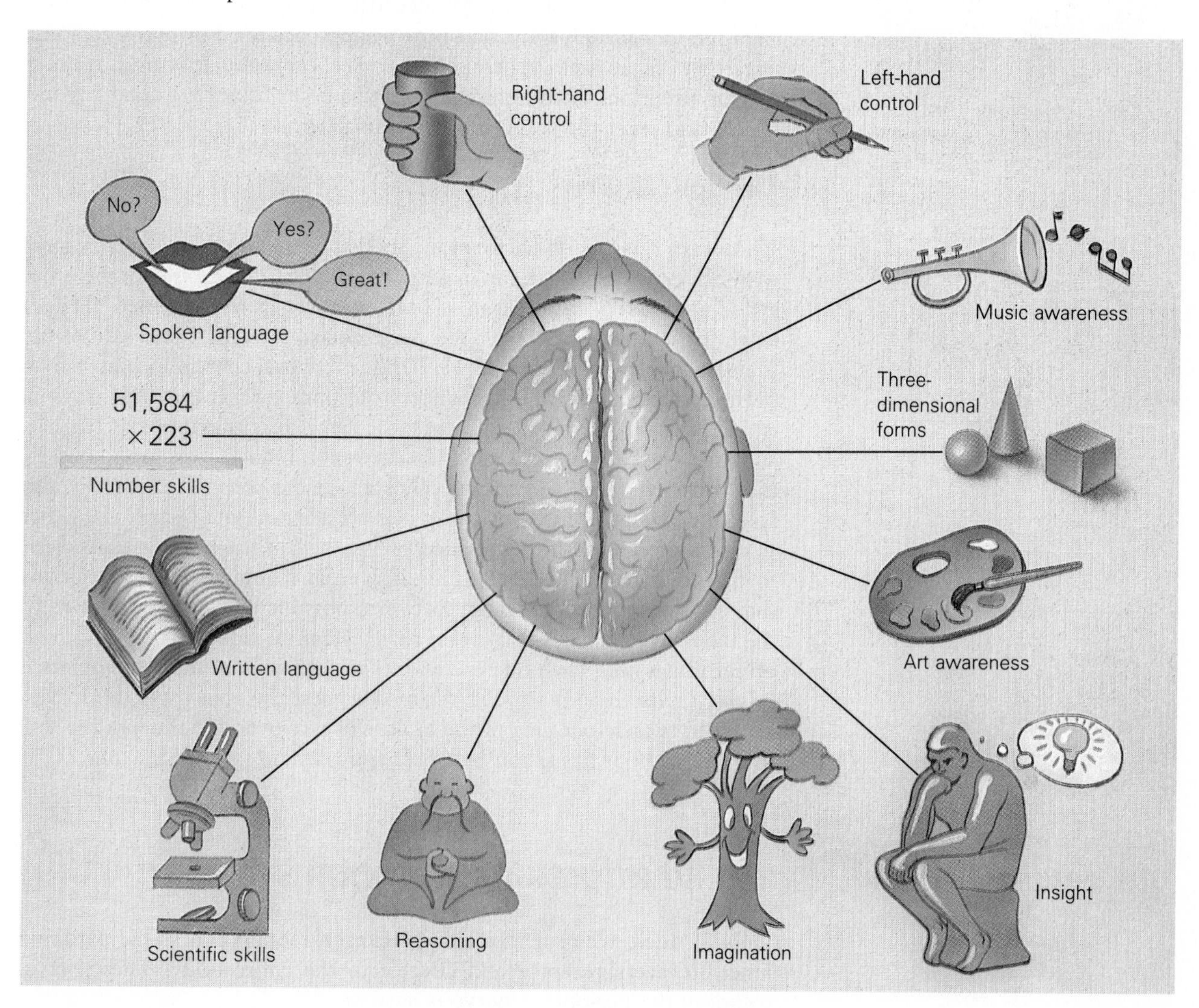

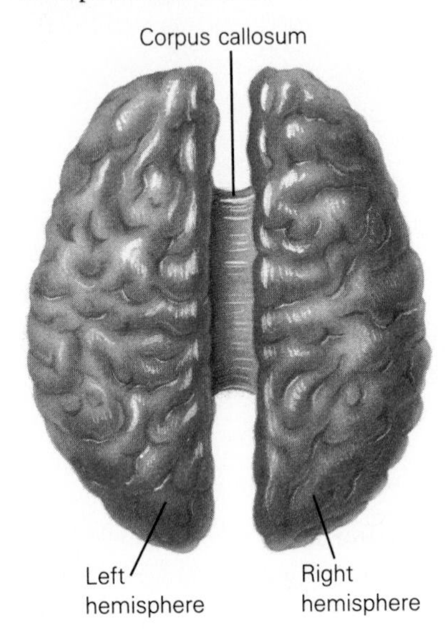

FIGURE 22.10

The corpus callosum is a broad tract of nerve fiber that connects the two hemispheres of the brain.

and art. Imagination swells from this lobe and it sees things in their entirety (holistically), often solving problems through insight, as it compares relationships (Figure 22.9).

The flexibility of the two hemispheres has been shown when one is damaged and the other takes over its role. Such flexibility, by the way, may be greater in left-handers than in right-handers. It seems that the brain centers of southpaws are generally more diffuse, less localized, with functions more equally dispersed between the two hemispheres. Brain damage to left-handers may produce different symptoms than right-handers with similar injuries.

The two halves of the brain are connected by a great, broad tract of nerve fibers, about 4 inches wide, called the **corpus callosum** (Figure 22.10). Information from each half can be communicated to the other half via the corpus callosum. Thus, special abilities of the two parts of the brain can be integrated and we are able to solve problems, perform tasks, and appreciate life's offerings by a grand union of complex and differing abilities.

Or maybe not. Evidence indicates that usually one hemisphere is dominant and inordinately influences how we approach life. Furthermore, some researchers argue that the hemispheres somehow compete with each other for our attention. Such arguments, at this point, quickly extend beyond science and enter the realm of the philosopher.

The Split Brain

If we learn a visual discrimination (such as correct and incorrect shapes) with one eye covered, we can make the same discrimination with the other eye. Anatomically, the reason is because fibers from the inner (medial) halves of each eye cross over the **optic chiasma** to the other side of the brain, as shown in Figure 22.11. Thus, the visual centers in both halves of the cerebrum receive information from both eyes. If the optic chiasma is split, the right eye can still make the same discriminations as the left, and vice versa (Figure 22.12). However, the brain halves are joined by that flattened band, the corpus callosum. If the corpus callosum is also split, then things learned by using one eye remain unknown to the other side of the brain. Nothing learned by one half of the brain is transferred to the other half. This means, in effect, that mammals have two brains that can act independently. In fact, it is possible to train both halves of the brain separately. A split-brain monkey can be trained to approach an object if it is seen with one eye and to withdraw from the same object if it is seen with the other eye. If the monkey sees the object with both eyes, however, usually one side of the brain will take over and the monkey will respond without hesitation by either approaching or withdrawing.

THE PERIPHERAL NERVOUS SYSTEM

Pairs of thick, white nerves emerge from the brain and spinal cord and innervate every receptor and effector in the entire body. These nerves comprise the **peripheral nervous system.**

FIGURE 22.11

The optic chiasma is formed where tracts of visual nerves cross in the brain. The two visual centers of the brain each receive visual input from both eyes. The images from the inner areas of the two eyes cross at the optic chiasma to innervate opposite sides of the brain.

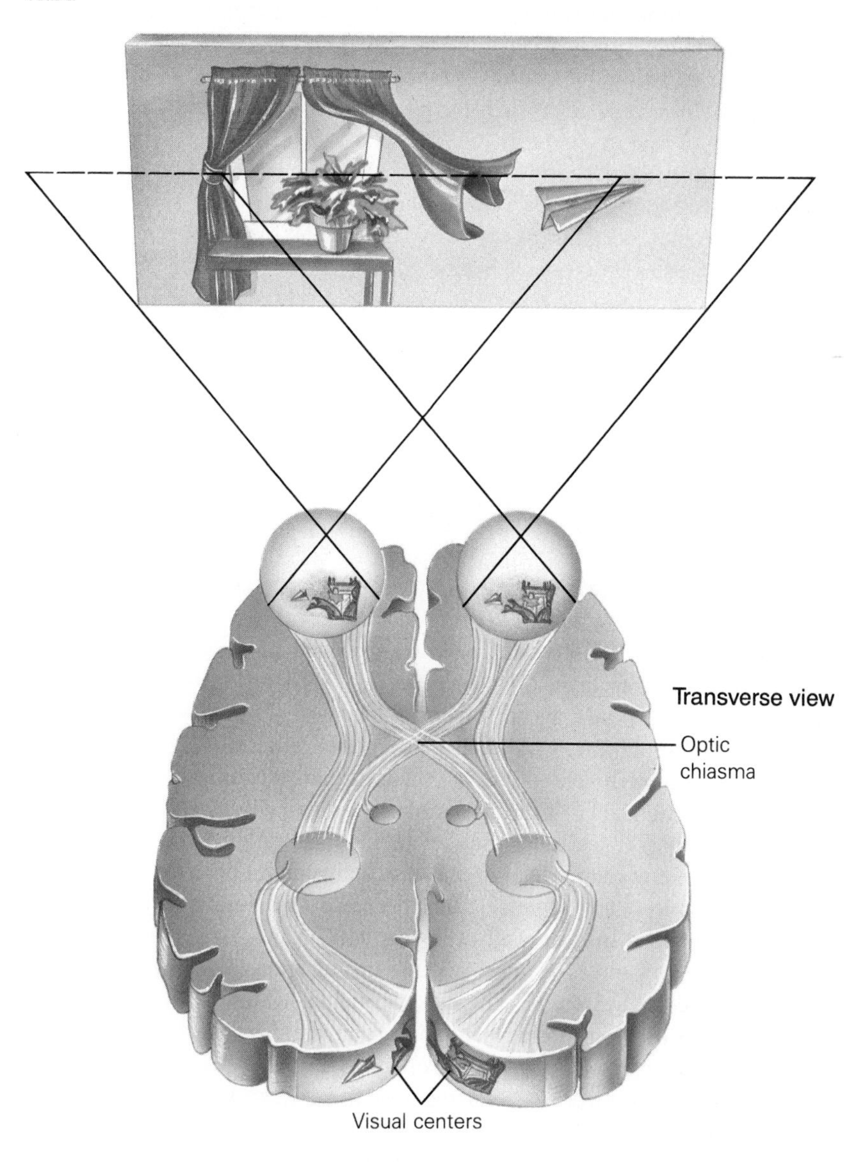

Those nerves (12 pairs in humans) extending from the brain are called the **cranial nerves.** Further down, the spinal cord gives rise to thick **spinal nerves** (31 pairs in humans). Each nerve is formed from the union of a dorsal and ventral nerve root that emerge directly from the spinal cord.

FIGURE 22.12

The major connection between the halves of the brain is the corpus callosum. If the corpus callosum is cut, the individual will continue to behave normally, but experimentation reveals that the halves of the brain can function as separate and roughly equivalent units, although each has its own special qualities. If the optic chiasma is also cut, images seen by one eye can't be transferred to the opposite cerebral hemisphere. When the right eye is covered, the subject can learn tasks and make discriminations using only the left eye. But if the left eye is then covered, the subject can no longer perform the same tasks.

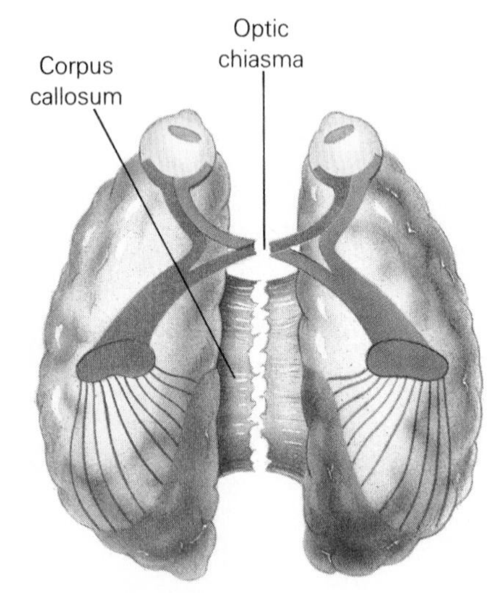

The dorsal root of the spinal nerve (comprised of sensory neurons) bulges with a huge ganglion at about the level where it enters the spinal column. The ganglion houses the cell bodies of all the sensory neurons entering the spinal column. The cell bodies of the neurons comprising the ventral root of the spinal nerve (comprised of motor neurons) lie embedded in the spinal column, so these nerves do not have ganglionic bulges. The two great nerves fuse just outside the spinal column and travel together for a way before giving rise to increasingly smaller nerves—nerves that will ultimately branch into delicate neurons that reach every part of the body.

The peripheral nervous system consists of the somatic nervous system and the autonomic nervous system. The somatic nervous system carries the impulses that we are most conscious of, the commands to our voluntary muscles and the conscious sensations from all parts of our bodies. The autonomic nervous system is more concerned with our unconscious and involuntary internal workings.

The Autonomic Nervous System

The autonomic nervous system is formed from a special set of peripheral nerves that serve the heart, lungs, digestive tract, and other internal organs. It can be divided into the **sympathetic** and **parasympathetic nervous systems,** which act antagonistically (Figure 22.13). If the sympathetic system works to speed up certain body processes, then the parasympathetic system works to slow them down, and vice versa.

The sympathetic nervous system is activated in what has been called "fight-or-flight" reactions. It also plays a role in reproductive behavior, but perhaps we should avoid further alliteration. The fight-or-flight syndrome becomes apparent in certain emergency situations. For example, if a bear rushes into the room where you are quietly reading, your body will react "sympathetically." The pupils of your eyes will dilate, the better to see the bear; blood vessels to the skeletal muscles will increase in diameter, bringing oxygen-laden blood to these structures; the better to effect your escape. Peripheral blood vessels will decrease in diameter, so blood loss will be minimized in case the bear swats you on your way out; heart rate will increase, bringing oxygen to your running muscles; your blood pressure will rise; blood sugar will rise; bronchial tubes will open, getting more oxygen to the muscles; your hair may stand on end; and digestion will almost stop, since your blood is needed elsewhere. The blood supply to your brain may decrease, causing you to faint, in which case you will hope the bear is just there as part of some effort to help curb forest fires. If upon awakening, you should discover the bear was only your roommate in a bear suit, your parasympathetic system will take over and reverse all these responses. You may then wish to activate the sympathetic system of your roommate. Figure 22.14 describes the interaction between the nervous and hormonal systems involved in such responses.

Whereas the parasympathetic nervous system brings about its effects with the neurotransmitter **acetylcholine,** the sympathetic nervous system causes its effects with the transmitter substance **norepinephrine,** (noradrenaline). You might wonder why epinephrine couldn't simply be released from the adrenal glands in an emergency situation, to travel in the

FIGURE 22.13

The autonomic nervous system. The sympathetic components are shown to the right and the parasympathetic components to the left of the central nervous system (in the center). Many internal organs are innervated by both sympathetic and parasympathetic nerves, the two systems having opposite effects.

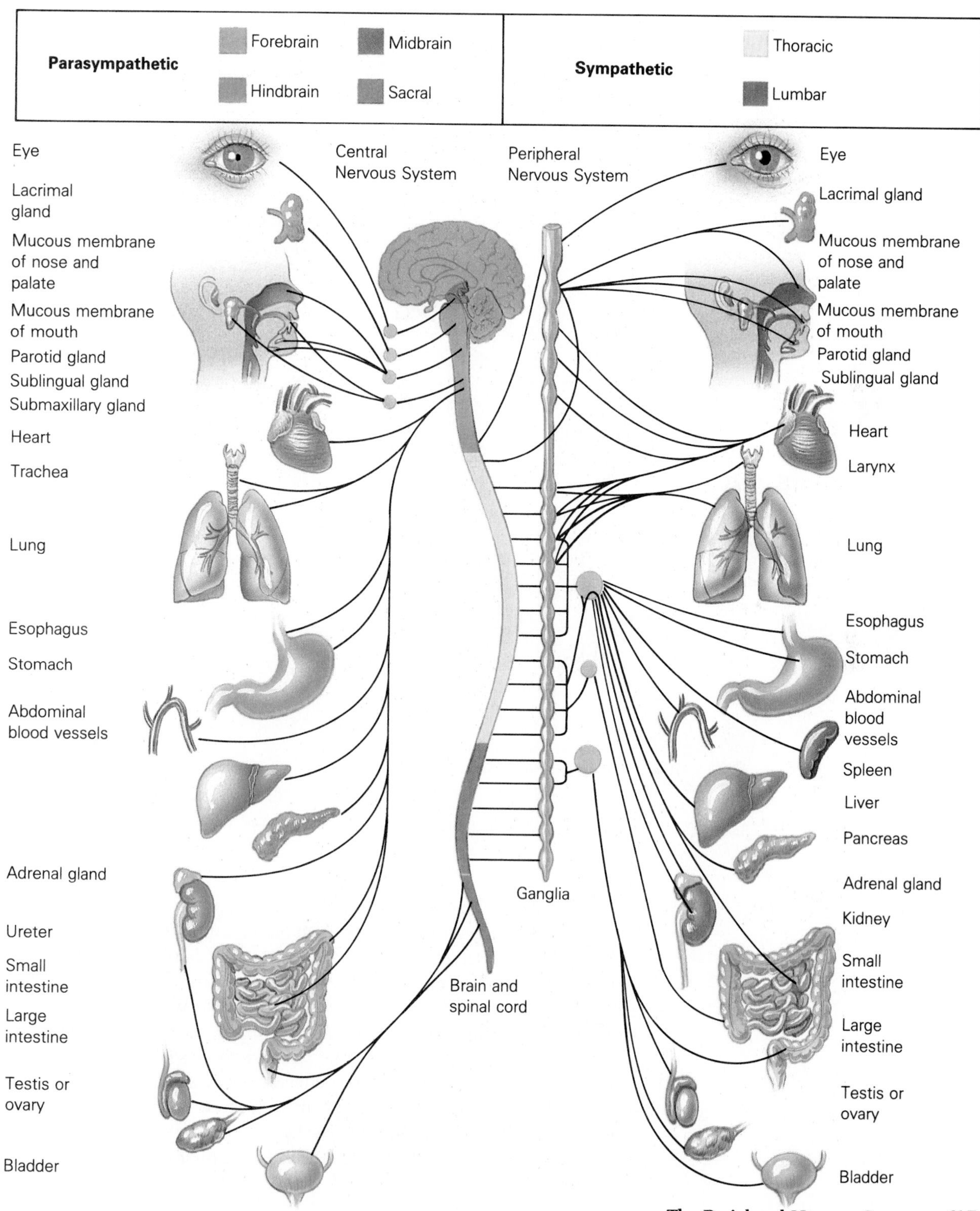

FIGURE 22.14

The sensory system detects the presence of a bear and relays that information to the cerebrum where the bear is recognized as such. The hypothalamus and sympathetic nervous system is then activated, which, in turn, stimulates the adrenal medulla to secrete norepinephrine, thereby enhancing powerful responses in the specific targets of the sympathetic nervous system.

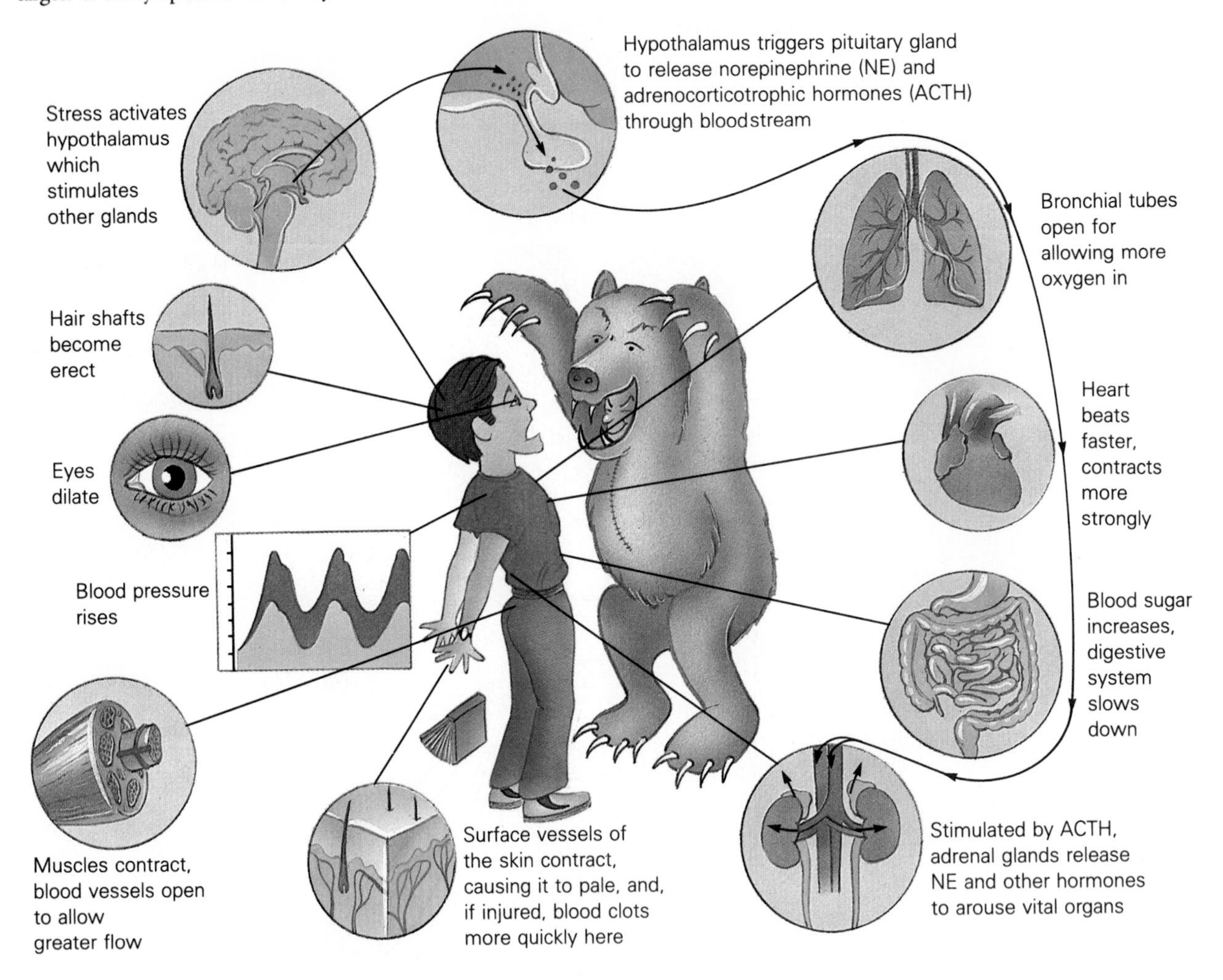

blood and elicit these same changes throughout the body—why it is necessary to develop another, but similar, emergency system. Actually, the adrenal gland may secrete epinephrine into the bloodstream in such a stress situation, but the two emergency reactions (a *general* one, as the adrenal gland alerts the entire system, and a *specific* one, in which certain nerves activate only specific areas) illustrate the important difference between hormonal and neural regulation. The sympathetic system releases its norepinephrine *directly* into the proper effector. Only small amounts are released, so although the response is immediate, it is of short duration. Greater amounts of epinephrine may be secreted by the adrenal gland, and while these take longer to reach the effector, their effect is more long-lasting. As was pointed out earlier, then, neural regulation is more immediate and short-term than hormonal regulation.

Autonomic Learning

The autonomic nervous system is usually described as "involuntary." It is assumed that we do not normally exercise conscious control over its functions. While it is true that the autonomic system can, and normally does, function in the absence of conscious control, there is evidence that some conscious control is possible. In other words, we may be able to learn to influence some of our autonomic reactions.

L. V. Dicara and N. E. Miller of Rockefeller University were able to teach rats to increase or decrease heart rate, blood pressure, intestinal contractions, blood vessel diameter, and even rate of urine formation. They did this by monitoring the animals' normal patterns and fluctuations in these parameters over a period of time. As natural variations occurred, the ones in the desired direction were rewarded by electrical stimulation of the "pleasure center" of the brain; variations in the other direction were punished by a slight but unpleasant shock. For example, if the experimenter wanted the rat to learn to slow its heart rate, when the heart slowed naturally, the animal would be rewarded; when it accelerated naturally, punishment would follow. Soon the heart beat more slowly. The researchers produced similar results in humans and have even taught some to control their blood pressure.

The possible applications of autonomic learning are quite varied and fascinating. For example, some of the amazing feats accomplished by some practitioners of yoga, such as when they drastically lower their metabolic rate, may be due to autonomic learning (Figure 22.15). In other cases, autonomic learning can aid in survival against the elements. It is known, for example, that mountain people of the Himalayas, such as the famed Sherpa guides, are able to withstand extreme cold. They may show no effects from sleeping barefoot in the frigid mountains. It has been suggested that such feats are possible because they have "learned" to withdraw body fluids from their extremities so that cell membranes cannot be ruptured by the formation of ice cyrstals. "Firewalking" has been taught in the United States as a means of instilling confidence and self awareness (Figure 22.16). No one is sure just how this is possible on a physiological basis. Some have suggested that autonomic learning has not received the attention it deserves in the Western world. However, a phenomenon known as *biofeedback* has received a good deal of attention, especially among harried professionals who see themselves caught up in the frenzy of the rat race. Here, people are provided with information that allows them to monitor their autonomic processes and control them. For example, when given information on their brainwaves, they may learn to consciously generate those waves that indicate restfulness and peace (Figure 22.17).

We have considered the nervous system in the simplest of terms, but even so, the complexity of this great central nervous system becomes apparent. As more of its mysteries are solved even more intriguing questions are revealed. It gives up its secrets slowly, but researchers continue to probe at the central nervous system with the greatest tool of all, their own.

FIGURE 22.15
Yogis may work their wonders by using autonomic learning to alter the condition of their tissues. Western physiologists have not yet explored these possibilities as fully as they might.

FIGURE 22.16

Many people have learned to walk across glowing coals as a demonstration of their self-confidence. No one is yet sure why they are not burned but some have suggested that autonomic learning is involved.

Many of the brain's precise mechanisms remain a mystery. And while we often do not understand just *how* the brain is affected by various chemicals, we are often keenly aware of their results. So, let's briefly consider some of the more common means of chemically altering brain function and behavior.

FIGURE 22.17

Biofeedback techniques allow the subject to monitor his or her brainwaves and thus to increase the likelihood of generating waves of a specific type, such as those associated with relaxation. Here, an entire football team reduces stress before a game with techniques learned through biofeedback training. In the graphs we see the distinct differences produced in the brainwaves of a person who is (a) wakeful and (b) at rest.

MINDBENDERS

For some reason, a lot of people don't seem to like the minds they were born with. A great deal of our energy, it seems, is spent in finding ways to "bend" our minds and change our moods. And (in spite of popular beliefs) the search is not a new one. With all the profound problems facing our ancestors, they almost immediately set about learning to make booze. Also, we find hallucinogens were important in many of our earliest known cultures. The search for mindbenders continues today, however, in ways our ancestors could only imagine. (The ancient Greeks, for example, were not big on sniffing transmission fluid.) So some of the things we will mention here have been with us for ages, while others are so new that we really don't know how they act or what their long-term effects are.

We can begin by noting that "drugs" are **psychoactive agents.** That is, they can alter mood, memory, attention, control, judgment, time-and-space sense, emotion, and sensation. Fortunately, probably none do all of these at once. Most cause effects that can be placed along a continuum between *stimulation* (an excitatory state) and *depression* (a state of reduced mental activity—see Figure 22.18). Here, we will review a few general principles about a few of the major groups. Not unexpectedly, we will

FIGURE 22.18

A continuum of drug action. The effects are shown at left, the various agents at right. The neutral area is drug free.

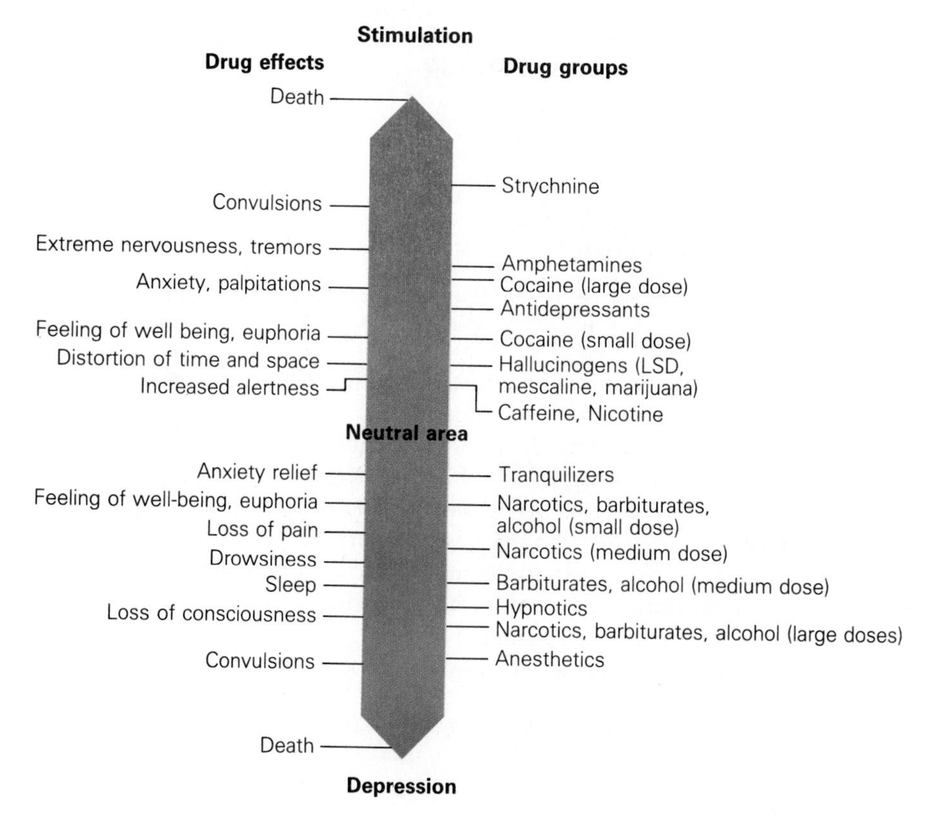

TABLE 22.3
Some Mindbending Drugs

DRUG	ACTION	EFFECTS/RISKS	DEPENDENCE	TOLERANCE
Tobacco	Stimulant	Causes lung cancer and emphysema; heart and blood vessel damage	Physical and psychological	—
Caffeine	Stimulant	Increases alertness; increases heart rate and blood pressure; high doses cause irritability and nervousness	Some degree of psychological	With large amounts
THC (marijuana/hashish)	Low dose is similar to sedative; high dose similar to hallucinogen	Smoke causes lung damage; mild euphoria; THC accumulates in fatty tissues	No physical; possibly slight psychological	No
Alcohol	Depressant	Euphoria; loss of motor coordination (high dose); slows reflexes; damage to liver, stomach, intestines; damage to nerve cells; damage to heart and skeletal muscle	Physical and psychological	Develops quickly
Opiates	Depressant	Euphoria; drowsiness and sleep; muscle relaxation	Strong physical	—
Cocaine	Stimulant	Euphoria; increased energy; increased heart rate and blood pressure	Physical and psychological (especially with crack)	Little
Amphetamines	Stimulant	Increased energy; alertness; decreased appetite	Physical and psychological	Yes
Barbiturates	Depressant	Reduced anxiety; sleep	High potential for physical	Yes, increases rapidly
Phencylidine	Anesthetic; psychedelic	Euphoria; relaxation psychosis; violence; coma	?	?
Methaqualone	Depressant	Reduced anxiety; relaxation, drowsiness, and sleep	Physical	Increases rapidly
Psychedelics	Psychedelic	Hallucinations; distortion of time, space, and sensation; "bad trips"	No physical; possible psychological	Yes

find that all have potential for abuse.* Table 22.3 lists a few mindbending drugs and their effects on the human body.

Abuse may arise because of our tendencies to overdo the drug or to overly rely on the drugs. Use of some drugs can cause *dependence* or **addiction.** There are two major forms of drug dependence. **Physical dependence** occurs when the drug is necessary simply to maintain bodily comfort. It's disuse causes sometimes agonizing discomfort (called **withdrawal symptoms**). **Psychological dependence** exists where a drug is necessary for mental or emotional comfort. Withdrawal symptoms here can be as severe as they are for physical dependence.

*Even the *term* "drug" is abused. It technically refers to a narcotic (*narco:* sleep), or depressant, but largely due to its erroneous application by the federal government, it now refers to any psychoactive agent.

Tobacco

Tobacco is the dried leaf of the plant *Nicotiana tabacum*. It is usually rolled, shredded, or flaked and then burned. The smoke is inhaled, allowing its products to cross the thin-walled alveoli of the lungs and to enter the bloodstream. Over 6800 different chemicals are found in tobacco smoke, many of them carcinogens (cancer-causers). There is evidence that the major psychoactive product *nicotine* is carcinogenic. In large doses, nicotine may also cause cramps, vomiting, diarrhea, dizziness, confusion, and tremors. It can cause respiratory failure and death, the ultimate lesson. The poison is particularly dangerous to nonsmokers who have not developed a tolerance to nicotine. For them, 75 milligrams of nicotine could threaten life (that is, the amount in about 3.5 packs of cigarettes).

The development of the smoking habit is curious because the first attempts can be ghastly. Perhaps, though, its continuance is even more curious because smoking has been clearly linked to lung cancer, emphysema, heart and circulatory ailments, and birth defects. (It also contributes to premature aging of the skin.) Nevertheless, we continue to be susceptible to advertising touting the blessings of smoking (Figure 22.19).

FIGURE 22.19

Advertisers have had our number for a long time. They are well aware that most of us are very conscious of the image we project, as we behave in the way that is most carefully tailored to fit our notion of what that image should be. Thus they can sell us products that they tell us will make us appear more "male," more "female," wealthier, more sophisticated, or whatever. In a sense, they're right, because others who are subjected to the same advertising are persuaded that the use of the advertised product is chic. Sometimes the price of being chic is waking up with the peculiar suspicion that a rabbit slept in your mouth.

Smoke can permanently paralyze the tiny cilia that sweep the breathing passages clean and can cause the lining of the respiratory tract to thicken irregularly. The body's attempt to rid itself of the toxins in smoke may produce a deep, hacking cough in the person next to you at the lunch counter. Console yourself with the knowledge that these hackers are only trying to rid their bodies of nicotines, "tars," formaldehyde, hydrogen sulfide, resins, and who knows what. Just enjoy your meal.

Smoking may cause physiological dependence on the products in the smoke. Withdrawal may produce a variety of unpleasant reactions, and some people are apparently unable to stop, no matter what the results. The American Cancer Society estimates that about 90 percent of all lung cancer (which has a low cure-rate and which is now the number one cause of cancer deaths) is due to smoking and that if smoking ceased, the incidence of all cancers in the U.S. would fall about 25–30 percent. (You should be aware that so-called "smokeless tobacco," such as snuff or chewing plugs, has also been linked to cancer, especially of the mouth.)

Caffeine

Caffeine is a stimulant, affecting the central nervous system. It works by stimulating nerve cell metabolism. Caffeine is a component of coffee, tea, chocolate, and many colas. Such drinks were not always so popular. In fact, at one time in the Near East (Figure 22.20) coffee drinkers were put to death—perhaps, some would say, a fate not worse than having to start the day without coffee.

Caffeine increases alertness and decreases fatigue and boredom. It also speeds the heart rate, increases blood pressure, increases urine formation, and dilates some blood vessels while contracting others. In small to moderate amounts (two to four cups), it may improve performance in boring or repetitive tasks, but it does not help in more complex intellectual tasks,

FIGURE 22.20
Caffeine is a popular stimulant in the Near East, where most users take it in outdoor establishments as a strong, hot tea.

such as reading or doing long division. However, it may help keep you awake so you can perform those tasks, since it inhibits sleep. In higher doses, it causes nervousness, irritability, and a "jangled" feeling. Very high doses can cause convulsions, but you would have to drink about 100 cups before you run a risk of dying. (By then, you probably would have already talked yourself to death.)

People who consume very large amounts of coffee, say 10 to 20 cups a day, may develop **caffeinism.** The symptoms are insomnia, high blood pressure, increased body temperature, racing heart, and chills. (Caffeine is also known to encourage the development of breast cysts in women.)

One develops **tolerance** for caffeine so that increasingly higher doses are needed to produce the same effect. Withdrawal symptoms include headaches and irritability. Withdrawing coffee drinkers are generally not considered dangerous.

Marijuana and Hashish (THC)

Marijuana (also known as dope, pot, grass, reefer, killer weed, or Mary Jane) is a form of Indian hemp *(Cannabis sativa)* (Figure 22.21), and was cultivated in the United States during World War II to produce fibers for ropes after the supply of hemp from the Philippines was cut off. The wild progeny of those plants has driven crusading law enforcement officers up the wall, and the attempt to control the drug continues to absorb enormous amounts of public money.

The active ingredient in marijuana is a group of chemicals called tetrahydrocannabinols (THC). Whereas marijuana users generally smoke the leaves (or eat them in brownies), THC is highest in a preparation called **hashish** (or hash), which is the concentrated resinous exudate col-

FIGURE 22.21

Marijuana plants were once a common weed. The species, however, has undergone tremendous artificial selection to produce plants high in THC.

FIGURE 22.22
An early governmental warning of the hazards of marijuana use. Such exaggerations have led to a general skepticism regarding such warnings.

lected from the female flowers of marijuana plants. Marijuana induces a mild euphoria, sometimes expressed as a happy or giggly mood. It may also produce mild hallucinations (if there is such a thing), forgetfulness, and it may reduce mental agility. Its effect varies among individuals and may be partly dependent upon the setting in which it is used.

Some of the statements, official and otherwise, about marijuana have sometimes been ill-conceived, incorrect, and irrational (Figure 22.22). Because of the ludicrous nature of earlier official warnings about pot, there has been a tendency to reject all warnings regarding drugs. Perhaps this is unfortunate because marijuana can, indeed, produce problems. For one thing, THC is fat-soluble and cannot be flushed out of the body by the kidneys. Instead it collects in the fatty tissues, such as the brain and reproductive organs. THC is slowly released as the fat is metabolized.

Marijuana smoke is a powerful lung irritant and with regular use can cause not only bronchitis, but, apparently, precancerous changes in lung cells. (We must keep in mind that heavy marijuana use is a recent trend and it was only after 60 years of heavy cigarette smoking that its devastating effects became apparent.)

Other studies show that marijuana temporarily reduces sperm production and may cause the production of abnormal sperm. In female monkeys, it can disrupt ovulation and cause abnormal cycles. THC can cross the placenta in some laboratory animals and is associated with a higher rate of miscarriages. (It should definitely be avoided by pregnant women.)

Marijuana is generally not considered addictive, and there seems to be little "tolerance," so doses need not be continuously increased. There may be psychological "dependence" on the drug because of what the user regards as its rewarding effects.

Cannabis sativa has been cultivated in the Near East for centuries. There, its THC is extracted to produce the resinous hashish. Hashish gets its name because it was used by the hashshashin, a Moslem terrorist group whose notorious violence was thought to be a result of addiction to the drug. However, more careful studies of the hashshashin (the source of the word assassin), indicate that hashish was actually given as a reward for their murderous deeds—perhaps to produce the "visions of glory" promised by their leader, who, by the way, was a classmate of the poet Omar Khayyam.

Alcohol

Adolescent cynics have, for years, noted that while their parents were almost rabid in their opposition to other drugs, alcohol was often quite acceptable to them. The kids have a point. Alcohol is, indeed, a drug, even in the technical sense. And it is probably far more harmful than some of the drugs alcoholics fear. The most immediate sign of its harmful effect is in the form of the hangover (Essay 22.2).

No matter what the drink—beer, wine, or Singapore slings—the active ingredient is *ethanol* (CH_3CH_2OH). Ethanol, in spite of its reputation, is not a stimulant, it is a depressant. It may stimulate in a sense, however, because it can depress inhibitions and release the clever fellow within us all. It is also not an aphrodisiac, although it can in smaller

ESSAY 22.2

The Hangover

Egyptian hieroglyphics depict both priests and physicians administering to victims of hangovers. If you've had one, you may prefer the former to the latter. And after all this time no one knows how to remedy the problem. (But some Puerto Ricans may recommend rubbing the underarm with lemon.)

A hangover is a seemingly interminable but temporary chemical imbalance in the body, caused by alcohol acting as an anesthetic on the central nervous system. Usually the results involve a dilation of blood vessels in the brain; the movement of water, potassium, and other ions from the cells outward to the intercellular spaces; depletion of magnesium from the kidneys; and inflammation of the stomach lining. In addition, the sleep that follows is strangely devoid of the rapid eye movement (REM) that is characteristic of the most restful sleep.

The results are insatiable thirst, upset stomach, fatigue (many byproducts of heavy physical exertion appear in the blood), grouchiness, and perhaps remorse (depending on who saw you).

The morning-after drink ("the hair of the dog that bit you") provides superficial relief while slowing down recovery, and coffee stimulates the already exhausted nervous system and prevents needed sleep.

Hangovers, however, have produced more philosophers than all the world's great books, the conventional wisdom being, "There ain't nothing, NOTHING, worth a hangover."

amounts reduce anxieties. In larger doses, it interferes with the sexual act in a most frustrating manner. As Shakespeare observed, it "provokes the desire, but it takes away the performance."

People develop a tolerance for alcohol rather quickly. That is, as the body "handles it," more and more is required to produce the same effect. Essentially, the "handling" is mainly done by the liver, which can oxidize about an ounce of alcohol per hour. (Five to ten percent is excreted by the lungs and kidneys.) Finally, though, the liver may become so damaged that it can detoxify very little alcohol.

Alcohol is very addictive, for some people more than others, and its use can produce severe withdrawal symptoms, the most drastic of which are delirium tremens, or DTs. Long-term use can damage the central and peripheral nervous systems, the liver, the stomach, and the intestines, but the effects themselves vary greatly from one individual to the next (Figure 22.23). One alcoholic's mind may go, while a drinking companion may

FIGURE 22.23
Probably the most widely used drug in the United States is ethanol, and among young people, its most common source is probably beer (with an alcohol content of about 3 to 5 percent, as opposed to wine with about 13 percent, and whisky with about 40 to 50 percent).

only die of cirrhosis of the liver. Interestingly, heavy "binge" drinking may be safer than daily drinking of smaller amounts, perhaps because the layoffs give the liver time to recover.

Opiates

Opiates fall within the realm of what are referred to as "hard drugs." They are technically narcotics in that they depress the nervous system, and they can relieve pain and produce sleep or a stupor. The group includes heroin, opium, and morphine. Users rapidly develop tolerance and continued use often results in addiction. Consequently, the user not only needs the drug continually, but in increasingly higher doses. Because of the cost of opiates, addicts often resort to crime in order to finance their habit.

Injection of opiates produces a "rush," or sudden pleasurable sensation. (In beginners, it may also produce severe nausea.) The rush is followed by a great sense of well-being, accompanied by a marked decrease in physical drive. An accompanying drowsiness produces the nodding you can see in almost any New York subway. Withdrawal symptoms following abrupt discontinuation of the drug are usually very violent. With heavy addiction, withdrawal may produce such an intense shock to the system that death results.

Cocaine

Cocaine (toot, coke, lad, girl) is another of those recently fashionable drugs that we don't know enough about. However, much of what we do know points to an insidious and potentially very dangerous substance. But it was not always regarded so. In fact, it was once sold in the United States

FIGURE 22.24

Photo of Coca-Cola ad touting that it contains cocaine.

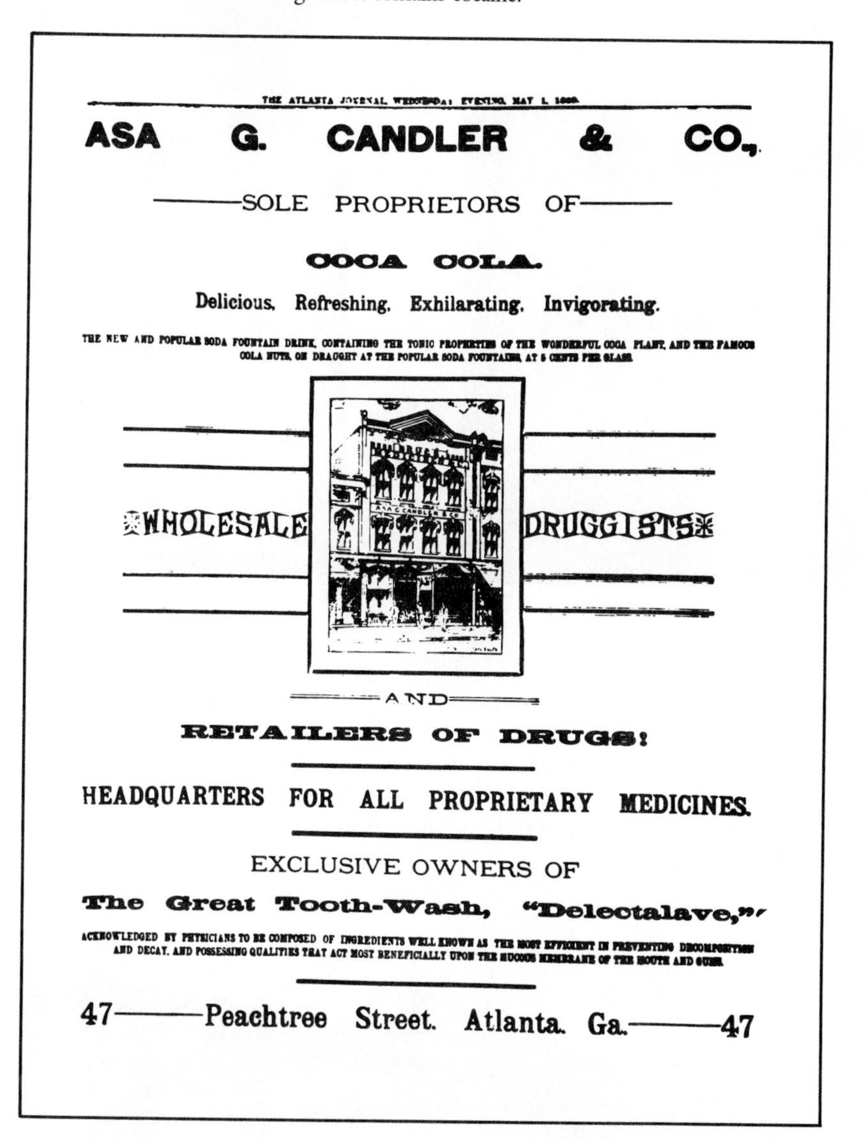

under its own name or used to lace other products, such as Coca-Cola (Figure 22.24). It seems to stimulate neural activity in the brain by enhancing the activity of certain neurotransmitters.

Cocaine is an alkaloid extract usually derived from two species of the coca plant, a South American shrub of the genus *Erythroxylon* (Figure 22.25). Historically, the leaves were chewed by pre-Columbian South American Indians, and the practice continues in that area to this day. The effect of chewing the leaves is a generally elevated intensity and an increase in apparent energy. Today, however, chemical extraction techniques can produce a white, crystalline powder that is usually "snorted" (inhaled through the nose), "shot" (injected into veins), or "based" (where the

FIGURE 22.25

Coca is one of the most lucrative crops in South America. Many pounds of leaves are necessary to produce a single gram of cocaine. In earlier times the cocaine was sold as a syrup.

alkaloid is chemically freed and smoked—"free basing"). Snorting cocaine produces the least drastic effects; basing produces the most powerful. In all cases there is a euphoria, mood elevation, and general stimulation. The result is talkativeness and a general intensity that may cause other people to tiptoe away. As it wears off, a depression sets in that compels the user

FIGURE 22.26

Crack cocaine has become the United States' most widely used illegal drug. "Crack houses" have sprung up in neighborhoods of the rich and poor, black and white, urban and rural.

to seek another "hit." The comedown from crack, a highly addictive and apparently inexpensive based form that is usually smoked (Figure 22.26), is so severe that the user will do anything to avoid it.

The long-term effects of regular cocaine use can be severe. The user may focus on the drug while other facets of life are neglected. The drug was once considered physiologically nonaddictive, but new evidence indicates that it can be powerfully addictive. There is little tolerance, so increasingly larger doses are not required.

Cocaine also has an artificially elevated expense. Whereas it costs only a few cents to manufacture a gram of cocaine, the street price may be $60 to $150. Furthermore, there may be little of the drug left in the street powder by the time it reaches the user. As a rule, it is "stepped on" (cut) by each person through whose hands it passes.

Amphetamines

Amphetamines, often loosely referred to as "speed" or "uppers," include a number of commercial drugs such as Benzedrine, Dexedrine, and Methedrine. Their chemical properties are similar to those of epinephrine—that is, they cause great bursts of energy that can overcome feelings of fatigue. Students and truck drivers have been known to take low doses in order to stay awake during midnight cramming sessions and long hauls, respectively. Weight watchers also use them to decrease appetite. Because amphetamines are effective in improving performance on rigorous physical tasks, they are sometimes used by athletes. Low doses do not impair skills or judgment.

A derivative called "crystal meth" (methamphetamine) is currently undergoing a resurgence on the street. The addition of a methyl group to the active molecule apparently makes for a smoother high. These people with an intense interest in the short term may use "speed balls," a mixture of amphetamines and heroin.

The greatest abuse of amphetamines is by injection of the drug. This produces an initial rush, followed by a feeling of vigor and euphoria that may last several hours. After this, however, come the dues. They appear in the form of aching, discontentment, and irritability. To delay this letdown, the user may boost himself or herself with another injection. The high may thus last for days, during which time the user usually fails to eat or sleep. The end of this period may be followed by exhaustion, severe depression, paranoia, aggressiveness, extreme irritability, and emotional overreaction.

Users develop a tolerance for amphetamines, and cessation after prolonged use may produce withdrawal symptoms, although they are less severe than those associated with opiate withdrawal.

Barbiturates

Barbiturates, or "downers," are sold under a variety of trade names, including Nembutal, Seconal, Luminal, and Amytal. They are all **sedative-hypnotics** that act on the cerebral cortex, midbrain, and brainstem areas. Their effect is to reduce anxiety and induce drowsiness and sleep. These results are accompanied by loss of muscular coordination and slurring of speech, similar reactions to those induced by alcohol (Figure 22.27).

FIGURE 22.27

Admitted to Bellevue Hospital as a result of an overdose of barbiturates. Notice the signs of general poor health in this man, such as his thinness and the condition of his skin. He is in a stupor at this point, but has been tied down because he can be expected to be violent should he ever awaken.

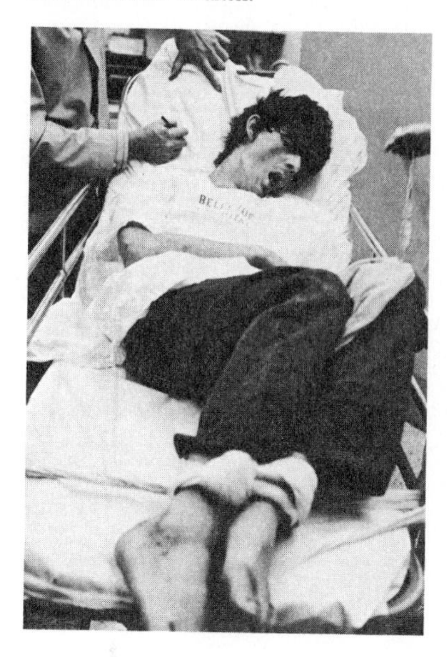

Barbiturates are highly addictive and rapidly produce tolerance. Withdrawal symptoms may be as severe as with opiate or alcohol withdrawal. The heavy barbiturate user is likely to be confused, obnoxious, stubborn, and irritable. In contrast to the placid disinterest of the opiate user, barbiturate users may be particularly aggressive and violent.

Barbiturates and alcohol acting together are particularly dangerous. The combination may cause death by suppressing the breathing centers. In addition, because each drug causes mental confusion, accidental deaths in those who mix them are all too common.

Phencylidine

Phencylidine, also called PCP or angel dust, is one of the more dangerous drugs to make the rounds in recent years, finding its place among certain abysmally ignorant young people. Unfortunately, many people use this powerful animal tranquilizer without realizing it, since it is used to "lace" marijuana and even cocaine. Its users may feel euphoric and extremely relaxed, but the side effects of the drug include extreme violence, psychosis, confusion, and, at high doses, coma.

Methaqualone

Methaqualone (Quaalude, or lude) is a synthetic barbiturate. When it was introduced in 1963, it was believed to be nonaddictive and was prescribed freely. However, all this has changed; it is indeed addictive. It often causes the user to be relaxed, uninhibited, and receptive (as a result it has been touted as an aphrodisiac). When taken with alcohol, one may fall into a stupor ("lude out") and lose control of movement. This is accompanied by a feeling of "pins and needles" in the extremities and around the mouth.

Psychedelics

Psychedelics are comprised of a group of drugs that produce hallucinations and various other phenomena that very closely mimic certain mental disorders. These drugs include lysergic acid diethylamide (LSD), mescaline, peyote, psilocybin, and various commercial preparations such as Sernyl and Ditran.

Of these, LSD is probably the best known. Although its use has apparently diminished since its heyday in the late 1960s, it seems to be making a comeback in some special circles. LSD is synthesized from lysergic acid produced by a fungus (ergot) that is parasitic on cereal grains such as rye. It usually produces responses in a particular sequence. The initial reactions may include weakness, dizziness, and nausea. These symptoms are followed by a distortion of time and space. The senses may become intensified and strangely intertwined—that is, sounds can be "seen" and colors "heard." Finally, there may be changes in mood, a feeling of separation of the self from the framework of time and space, and changes in the perception of the self. The sensations experienced under the influence of psychedelics are unlike anything encountered within the normal range of experiences. The descriptions of users therefore can only be puzzling to nonusers (Figure 22.28). Some users experience bad trips or "bummers," which have been known to produce long-term effects. Bad trips can be terrifying experiences and can occur in experienced users for no apparent reason.

Our nervous systems are the products of natural selection, and so they are specialized to help us deal successfully with the part of the world with which humans must interact. We see, though, that various aspects of our psyches can be satisfied, and perceived "needs" can be met, by a wide range of chemicals that our natural bodies are not prepared to deal with. Some of these chemicals have been around so long that they, and their dangers, have become familiar—even accepted. Others, though, are new, and we await data on the effects of their long-term use. And we are an inventive species, so we must assume that other "mindbenders" lie just over the horizon—new chemicals, new experiences, new behaviors, and new dangers.

FIGURE 22.28

A portrait done by a person under the influence of LSD shows the distortion and sensory confusion the chemical can produce.

SUMMARY

1. The hydra's nervous system is a net of neurons with no control center. Flatworms have head ganglia in the head from which two longitudinal nerve cords extend. These cords are connected by transverse nerves. A brain and sensory structures in the head end allow analysis of the environment into which the animal will be moving. The earthworm has a bilobed ventral nerve cord with distinct cerebral ganglia and nodes in each segment.
2. The central nervous system of vertebrates is composed of a brain with different parts specialized for different functions and a dorsal hollow spinal cord that is protected by bone.
3. A spinal reflex arc can often receive information, process it, and initiate a response. Because the information does not have to go to the brain for deliberation, the response is quicker. In simple reflex arcs, sensory information from a receptor is carried to the spinal cord over a sensory neuron and enters the spinal cord over the dorsal nerve root. It synapses directly with a motor neuron which carries the impulse through the ventral nerve root to the effector, usually a muscle. In more complex reflex arcs, the impulse is transmitted from a sensory neuron to an interneuron and from an interneuron to a motor neuron. Impulses from sensory neurons can be transmitted to neurons that cross the spinal cord so that effectors on the other side of the body are stimulated. They may also be carried to the brain by ascending neurons.
4. There is an evolutionary trend among vertebrates toward larger total brain size and larger cerebrum in relation to the rest of the brain. As the cerebrum increases in size, more functions of the lower brain are transferred to it.
5. The human brain consists of the hindbrain, the midbrain, and the forebrain. Functions of parts of the human brain can be found in Table 22.2.
6. The cerebral hemispheres process information differently. The left hemisphere is specialized for the logical stepwise reasoning of mathematics and language. The right hemisphere sees things in their entirety and compares relationships. It is specialized for art and music. If one hemisphere is damaged, the other takes over its role. Information can be shared between the hemispheres through the corpus callosum.
7. The peripheral nervous system is comprised of pairs of nerves that emerge from the brain and spinal cord and innervate every receptor and effector in the body. Humans have 12 pairs of cranial nerves and 31 pairs of spinal nerves. Each spinal nerve is formed from the union of a dorsal nerve root, containing sensory neurons, and a ventral nerve root, containing motor neurons. The peripheral nervous system is composed of the somatic and the autonomic nervous systems. Although the somatic system carries impulses that we are generally conscious of, the autonomic nervous system is concerned with unconscious and involuntary internal activities.
8. The autonomic nervous system can be divided into the sympathetic and parasympathetic nervous systems, two branches with antagonistic actions. The sympathetic nervous system directs the fight-or-flight reactions that adapt the body to deal with an emergency. The parasympathetic nervous system is active when there is no emergency. The parasympathetic nervous

system causes its effects with transmitter acetylcholine, and the sympathetic nervous system causes its effects with transmitter norepinephrine (noradrenaline).

9. Although the autonomic nervous system is considered to govern involuntary activities, animals may learn to influence some autonomic reactions.
10. "Drugs" are psychoactive agents that alter state of mind. Most can be placed on a continuum between stimulants and depressants and have a potential for abuse. Dependence (addiction) can be physical or psychological and cause withdrawal symptoms. Information on mindbending drugs can be found in Table 22.3.

KEY TERMS

acetylcholine (616)
addiction (622)
ascending neuron (601)
caffeinism (625)
central nervous system (600)
cerebellum (607)
cerebrum (609)
corpus callosum (614)
cranial nerve (615)
descending neuron (601)
dorsal root (600)
forebrain (605)
frontal lobe (609)
hashish (625)
hindbrain (605)
hypothalamus (609)
interneuron (601)
marijuana (625)
medulla (605)
midbrain (605)
norepinephrine (616)
occipital lobe (609)
optic chiasma (614)
parasympathetic nervous system (616)
parietal lobe (610)
peripheral nervous system (614)
physical dependence (622)
pons (607)
prefrontal area (610)
psychedelic (632)
psychoactive agent (621)
psychological dependence (622)
reflex arc (600)
reticular system (607)
sedative-hypnotic (631)
spinal nerve (615)
stretch receptor (600)
sympathetic nervous system (616)
temporal lobe (609)
thalamus (607)
tolerance (625)
ventral root (600)
withdrawal symptoms (622)

FOR FURTHER THOUGHT

1. Explain why an individual with a destroyed reticular system enters into a permanent coma.
2. What neural routing mechanism allows you to quckly remove your hand from a dangerously hot stove?
3. Which division of the autonomic nervous system is operating when you escape from a would-be attacker?

FOR REVIEW

True or false?

1. ___ The medulla controls basic functions such as breathing, digestion, and heart beat.
2. ___ The reticular system determines which impulses are allowed to register in the brain.

3. ___ The left hemisphere of the cerebrum is the center for musical and artistic abilities.
4. ___ The dorsal nerve root is composed of sensory neurons.
5. ___ Withdrawal symptoms only occur in individuals with physical addictions.

Fill in the blank.

6. The central nervous system consists of the ___ and the ___ .
7. The ___ is divided into temporal, parietal, occipital, and frontal lobes.
8. The ___ connects the two halves of the brain.
9. The ___ is a pathway through which the spinal cord processes information from the body's receptors and generates an appropriate response without the involvement of the brain.
10. Cocaine stimulates neural activity by ___ .

Choose the best answer.

11. Balance and coordination are controlled by the:
    A. medulla
    B. cerebellum
    C. midbrain
    D. thalamus
12. Visual stimuli are received and analyzed in the ___ lobe of the brain.
    A. occipital
    B. temporal
    C. frontal
    D. prefrontal
13. The parasympathetic nervous system
    A. speeds up all body processes.
    B. elicits changes in the body by releasing adrenaline.
    C. is part of the central nervous system.
    D. restores body functions to normal levels.
14. Which of the following drugs act by depressing the central nervous system?
    A. opiates
    B. cocaine
    C. amphetamines
    D. caffeine
15. Cirrhosis can result from long-term abuse of:
    A. narcotics
    B. barbiturates
    C. alcohol
    D. none of the above

CHAPTER 23

The Senses

Overview

Thermoreceptors
Tactile Receptors
Auditory Receptors
Chemoreceptors
Proprioceptors
Visual Receptors

Objectives

After reading this chapter you should be able to:

- Describe five main types of sensory receptors and the stimuli to which they respond.
- Compare and contrast the function and distribution of vertebrate and invertebrate receptors.
- Name the structures in the vertebrate ear and explain how each functions to convert sound waves to neural impulses.
- Identify four different tastes and indicate where each is registered.
- Describe the function of rods and cones in the vertebrate eye.

What kind of world is this and how do we know? It seems obvious that whatever we know, we know through our senses. Immediately, however, this answer brings us into difficulties. This is partly because our senses monitor only certain aspects of the environment, so what about those things we can't detect? Aren't they, too, part of our world?

If you were to ask a variety of animals to describe what the earth is like, there would be little consensus. A fly might describe swirling eddies, delicious surfaces, and a kaleidoscopic world of shimmering mosaics. A bee would see deep violet colors in flowers that we call yellow (Figure 23.1), while the red rose would appear to be black, lacking color at all. A dog might describe a gray, drab world dominated by surges of sounds, accented by thundering odors. A tapeworm might speak vaguely of a warm, wet world preceded by harsh light and withering dryness as it lay for months near death. The point is, animals are sensitive to very limited aspects of their environment, and so are we. So we must keep in mind that as we describe our senses, our means of detection, we are dealing with very limited instruments. We are aware of only a very small part of our surroundings. (But we can be sure that this is the part that has been important to our survival.)

The second problem in describing the senses is that sensory abilities may differ widely from one individual to the next. This area is rife with anecdotes and untested claims, but there are many substantiated cases of remarkable and unusual keenness of vision, hearing, touch, and so on. What are we to make of people who seemingly can "feel" the color beneath their fingers? Of the ability of twins in different rooms to silently communicate? Of people who can read newspapers across the room or see the moons on other planets? Here, we're dangerously close to leaving the realm of acceptable scientific topics and, of course, above all things, we want to be acceptable. So let's just admit that we are aware of only a small

FIGURE 23.1

These appear to us as yellow flowers. However, color is partly a function of receptive abilities, and our receptors fail to register the deep violets that the insect sees and that are only apparent to us by use of a special filter.

FIGURE 23.2

Heat sensors in the pit viper (the prairie rattlesnake, *Crotalus viridis*). Pit vipers detect their prey through a pair of heat-sensing devices located in the depressions near the eyes. Each pit consists of an outer chamber that ends in a thin membrane covering an inner chamber below. Extending over the membrane is a highly branched, heat-sensing nerve. The recessed structure of the pits permits the snake to zero in on its prey, just as though it were using its eyes. By moving the head back and forth, it determines the exact location of the prey by the intensity of neural impulses from the membrane. There is some evidence that the snakes also assess the size of the target by the heat generated.

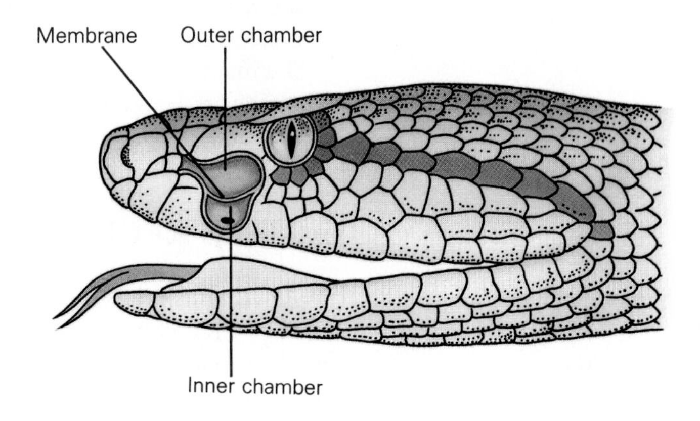

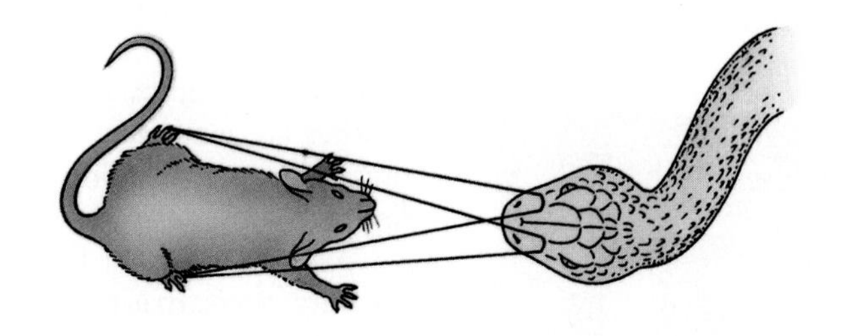

part of our environment, and that there may be great differences in the sensory abilities of people, and move on to consider some of the basic mechanisms of perception. We will review only a few basic types of **receptors,** those neural structures that are capable of responding to environmental stimuli, in a few representative groups of animals.

THERMORECEPTORS

Thermoreception is the ability to sense heat. Such an ability is important because the delicate chemical processes of life can normally be conducted only within certain temperature ranges. A heat-sensitive animal is able to adjust itself in its environment so as to position itself within those ranges.

Not much is known about heat detection in invertebrates, and it is assumed that most of them lack thermoreceptors. Some, however, do have these sensors. For example, cockroaches have heat receptors on their legs, with which they can locate optimal places to live, like your house. In addition, sensitivity to temperature may be important in helping parasites locate their warm-blooded hosts.

The mechanisms of heat detection in vertebrates are better understood. Some species, such as the pit vipers (Figure 23.2), have been intensively studied, and we have been able to determine how they locate prey by heat detection. In many mammals, including humans, temperature is registered by warm receptors, cold receptors, and pain receptors (which register the extremes of temperature). Interestingly, the receptors stop responding when the temperature is constant, but remain responsive if the temperature changes. Thus, standing in a shower that is turning cold is much more startling than standing in a cold shower. Thermal receptors quickly adapt to too cold or too hot so that the temperature becomes bearable.

Thermal receptors are not uniformly distributed over the body. For example, the face and hands are not very sensitive to temperature changes, while the lips and mouth are extremely sensitive.

TACTILE RECEPTORS

Tactile receptors respond to touch. They fire whenever their shape is altered or distorted, and they trigger extremely sensitive, fast-firing neurons. In some cases, bristles, whiskers, or hairs extend from the tissues around these receptors so that objects are perceived by these feelers before they contact living tissue.

Many invertebrates have such sensory feelers, as we see in jumping spiders (Figure 23.3). Web-building spiders have hairy legs that react to vibrations set up by trapped insects. Tiny hairs on the abdomen of cockroaches are extremely sensitive to light air currents—such as those produced by a descending human foot. Many invertebrates use touch in finding food and mates and in avoiding predators.

FIGURE 23.3
Jumping spiders are extremely sensitive to touch. They have a tremendous number of sensory hairs that cover their bodies. When a hair is bent or moved, it activates a neuron that transmits an impulse.

Vertebrates have four kinds of tactile receptors, those that register pressure, those that respond to light touch, free nerve endings, and nerve fibers wrapped around hair follicles. The pressure receptors are located deepest under the skin. They are essentially encapsulated nerve endings called **Pacinian corpuscles.** Near the surface of the skin are **Meissner's corpuscles,** which are believed to respond to light touch (Figure 23.4). In mammals, certain parts of the body are especially sensitive to touch. A sleeping dog can immediately be aroused if you touch the hairy area just under its tail (preferably with a stick). In humans, the most sensitive parts include the hairy areas and the genitals, as you may have noticed. In all primates, the most sensitive areas also include the lips, the area around the eyes, and uncalloused fingertips. (Can you see the adaptive advantages to heightened sensitivity in these areas?)

AUDITORY RECEPTORS

Audition, or hearing, involves the detection of sound, usually a distant stimulus. Most invertebrates lack specific receptors for detecting the vibrating molecules of air that produce sound. However, many are sensitive to the vibration of the air, water, or soil in which they live.

Insects are an exception among the animals without backbones in that some of them can hear quite well. Some, such as grasshoppers and crickets, have **tympanal membranes** that respond to sound much as does the human eardrum. One of the best stories regarding the evolution of hearing in insects involves that of noctuid moths (Essay 23.1).

In general, the hearing structures of land vertebrates include an **auditory canal** and a **tympanic membrane** (eardrum) that vibrates from one to three moveable middle ear bones. The vibrations of the bones then stimulate receptors that carry impulses to the hearing centers of the brain.

FIGURE 23.4

Specialized touch receptors in the skin of humans. Meissner's corpuscles are located close to the surface and register light touch. These are most numerous in the fingertips and around the lips. Pacinian corpuscles are in deeper skin locations, and their complex end bulbs register pressure. Generalized sensory neurons (no distinct structures) surround the hair follicles of mammals and are stimulated by hair movement. (Try to move a single hair on your forearm without feeling it.)

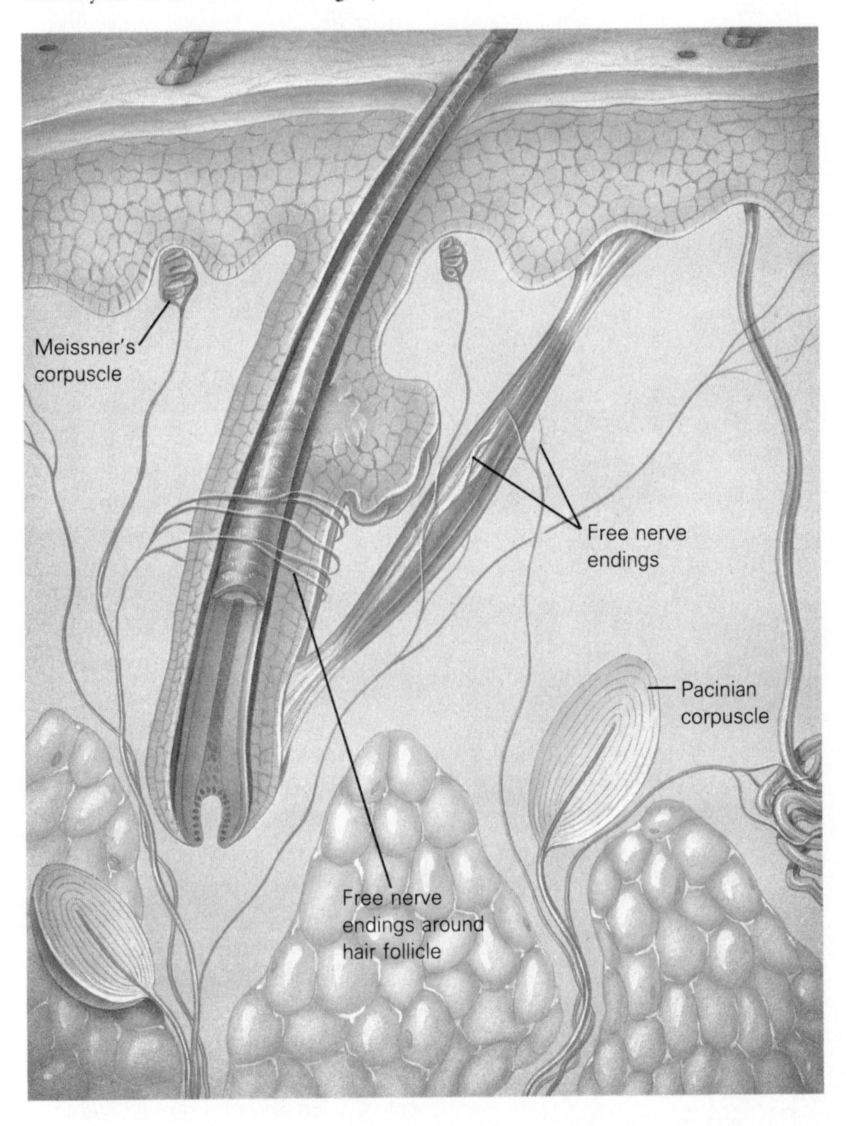

The auditory apparatus of mammals is somewhat distinct among the vertebrates. For one thing, most mammals have an **external ear.** This includes the *pinna* and the *auditory canal*. The pinna is the part that can be moved, as we see when a dog focuses on some sound. (Humans have largely lost the ability to move their ears, but those who can manage it are in great demand at social events.) The auditory canal channels sound toward the eardrum, which vibrates in synchrony with the soundwaves.

ESSAY 23.1

Moth-Bat Coevolution

Noctuid moths have two tympana, one under each wing on either side of the thorax (the insect midsection), and each tympanum has only two receptor cells. Kenneth Roeder found that one, called the A1 cell, is sensitive to low-intensity (weak) sounds. The other, the A2 cell, responds only to loud sounds. Surprisingly, neither kind of receptor is very good at distinguishing frequencies (high versus low notes)—a sound of 20,000 Hertz (cycles per second) elicits the same neural action potential as one of 40,000 Hertz (a much higher sound). As any sound becomes louder, however, the A1 cell fires more frequently and with shorter lag time between receiving the stimulus and firing. Also, the A1 cell shows a greater firing frequency in response to pulses of sound than to continuous sounds; it fires increasingly slower if subjected to a continuous sound. And it just so happens that the bats that prey on noctuid moths emit pulses of sound.

In a sense, the moth has beaten the bat at its own game. Its very sensitive A1 cell is able to detect bat sounds long before the bat is aware of the moth. The moth cannot only detect the distance of the bat, but it can tell whether the bat is coming nearer, since the sound of an approaching bat would grow louder. In addition, the moth is able not only to detect the distance of the bat, but also its direction. The mechanism is simple. If the bat is on the left side, the receptors in the left tympanum of the moth will be exposed to the sounds, while the receptors on the right will be shielded. Therefore, the left receptor fires sooner and more frequently than the right if the bat is on the left. If the bat is directly behind, both neurons will fire simultaneously. Thus, the moth can determine the distance and direction of the bat. But what about its altitude?

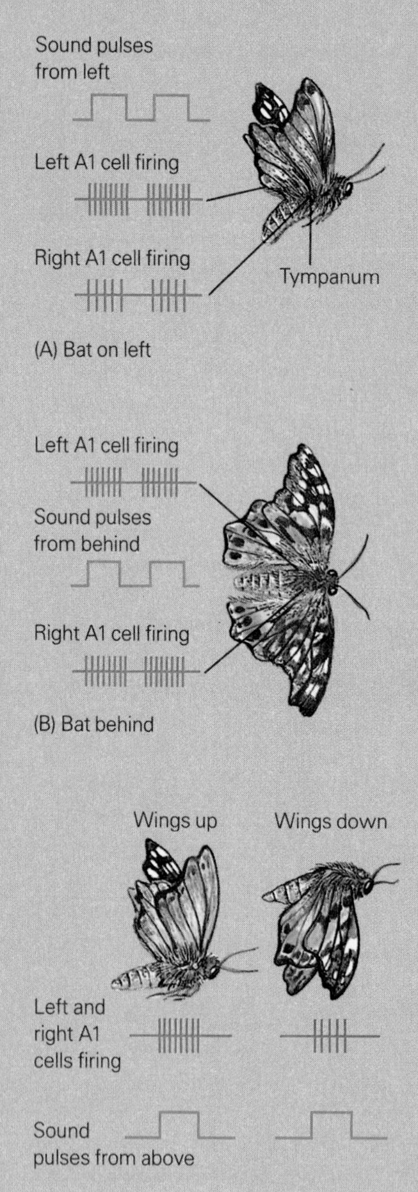

If the bat is above the moth, the loudness of the bat's sound will depend on the position of the moth's wings. When the moth's wings are up, the ears are exposed, and the bat's cries are loud. However, when the moth's wings are down, they cover its ears and the bat's cries are muffled. In contrast, if the bat is below the moth, the loudness of the bat's cries will not be influenced by the position of the moth's wings. The moth, then, decodes the incoming data, probably in its thoracic ganglion (from which the auditory neurons emerge) so that it pretty well has the bat pinpointed.

What does it do with this information? If the bat is some distance away, the moth simply turns and flies in the opposite direction, thus decreasing the likelihood of ever being detected. The moth probably turns until the A1 cell firing from each ear is equalized. When the bat changes direction, so does the moth.

Bats fly faster than moths, though, and if a bat should draw to within 2.5 m (8 feet) of the moth, the moth's number is up—at least if it tries to outrun the bat. So it doesn't. If the bat and moth are on a collision course (that is, if the moth is about to be caught), the sounds of the onrushing bat will become very loud. At this point, the A2 fiber begins to fire—the signal of imminent danger. These messages are relayed to the moth's brain, which then apparently shuts off the thoracic ganglion that had been coordinating the antidetection behavior. Now the jig is up and the moth changes tactics. Its wings begin to beat in peculiar, irregular patterns or not at all. The insect itself probably has no way of knowing where it is going as it begins a series of unpredictable loops, rolls, and dives. But it is also very difficult for the bat to plot a course to intercept the moth. The erratic course may take the moth to the ground where it will be safe since the echoes of the earth will mask its own echoes.

The noctuid moth's evolutionary response to the hunting behavior of the bat serves as a beautiful example of the adaptive response of one organism to another. Also, it shows clearly that the sensory apparatus of any animal is not likely to respond to elements that are irrelevant to its well-being. It is not important for moths to be able to distinguish frequencies of sound, but it is important that they are sensitive to differences in sound volume. Anyone who tried to train a moth to respond to different sound frequencies could only conclude that moths are untrainable.

The relationship of sound pulses from a hunting bat and auditory neural firing in the hunted moth. (a) When a hunting bat, emitting its high-pitched sounds, approaches a noctuid moth from the side, the receptors on that side fire slightly sooner and more rapidly than those on the shielded side. (b) When the bat is behind the moth, the moth's receptors on both sides fire with a similar rapid pattern. (c) When the bat is above the moth, the moth's auditory receptors fire when its wings are up, but not when its wings cover the receptors on the down stroke.

FIGURE 23.5

When sound waves vibrate the human eardrum (tympanic membrane), they set in motion three tiny leverlike bones: the malleus, incus, and stapes. The stapes, attached to the oval window, sets fluids in motion within the snail-shaped cochlea. (b) The cochlea is actually a U-shaped tube, divided by the basilar membrane. (c) Sensory cells of the membrane are embedded in the gelatinous tectorial membrane. The sound impulses pass inward over one surface of the basilar membrane, turn a corner, and pass outward over the opposite surface of the membrane to be dissipated at the round window. Different regions of the basilar membrane are sensitive to different sound frequencies.

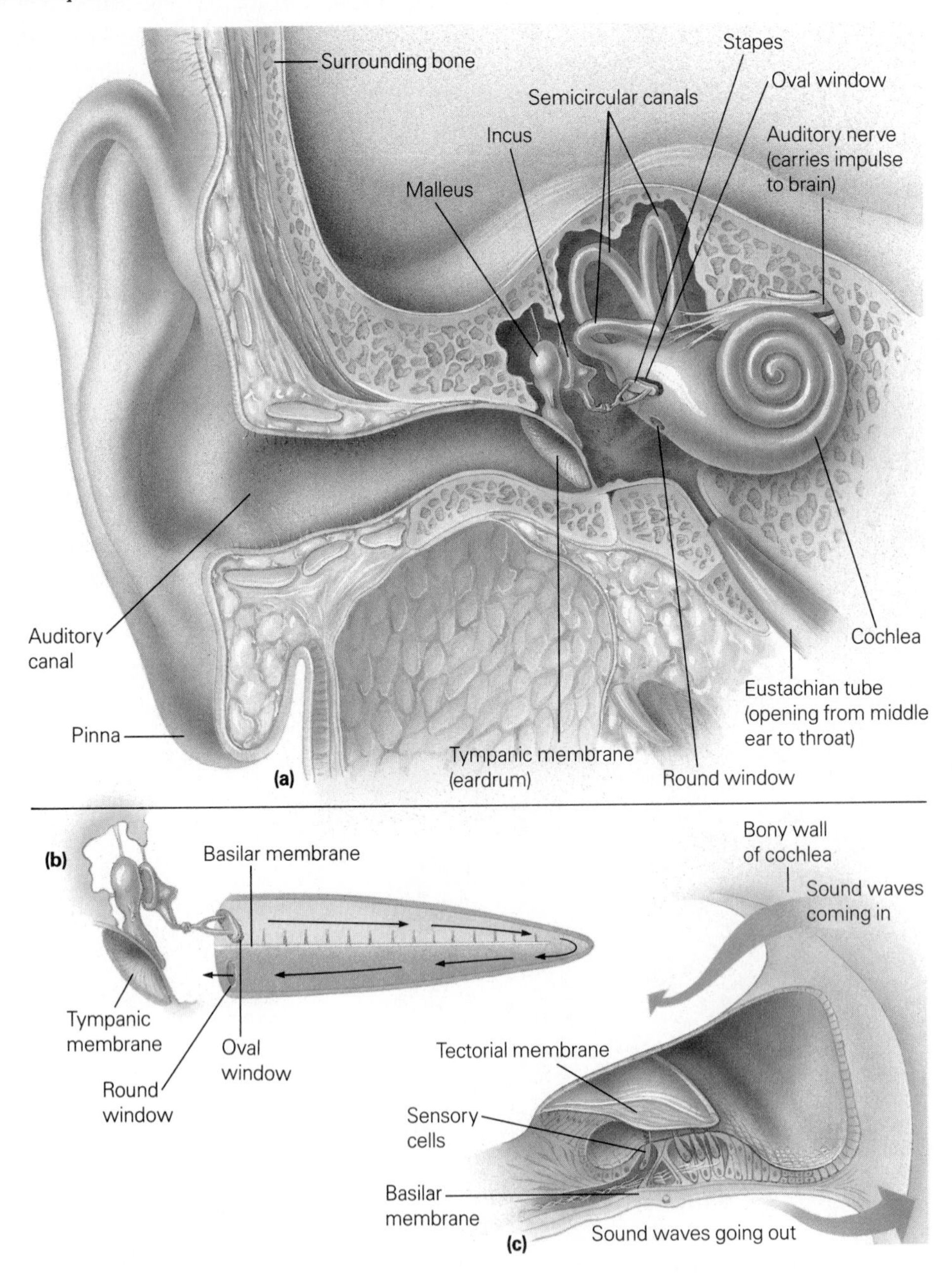

The vibrating eardrum causes the bones of the **middle ear** to vibrate. Whereas most other vertebrates have only one bone in the middle ear, mammals have three: the so-called hammer **(malleus),** anvil **(incus),** and stirrup **(stapes).** Sound vibrations move the hammer, which is pressed against the vibrating eardrum. The movement is transferred to the anvil, which then vibrates the stirrup (Figure 23.5). The stirrup vibrates against the inner ear's **oval window,** which, in turn, sets up movements of the fluid inside the long, coiled **cochlea.** The cochlea is divided along its length into three long chambers. The base of the central chamber forms the **basilar membrane** from which arise hair cells with modified cilia. The sound vibrations move the fluid in the upper chamber of the cochlea, which then presses against the fluid in the middle chamber and moves the basilar membrane. The hair cells extending from it are enmeshed in an underlying, gel-like membrane, the **tectorial membrane.** As the basilar membrane moves, it activates hair cells that stimulate the auditory nerves that lead to the brain. Some hearing impairment is directly related to the loss of the hair cells, and they can be lost by exposure to loud noise, such as amplified rock music. (In fact, partial loss of hearing is considered a badge of honor among some rock musicians and their fans. Others associated with the business are alarmed by the trend.) (See Figure 23.6.) The basilar membrane itself is narrower and more rigid near the eardrum, becoming much wider and thinner further away. The pitch of a sound is registered according to which hairs are stimulated. High pitches stimulate hairs in the narrower regions of the basilar membrane near the oval window; low pitches register at the wider tip of the coiled tube, the part lying in the innermost part of the snail-like coil. Loudness is apparently detected by the number of neurons stimulated and the frequency with which an impulse is generated in each of these neurons. Table 23.1 summarizes the structure and function of the human ear.

FIGURE 23.6
Many rock musicians and some members of their audience have permanently impaired hearing due to the loss of hair cells brought on by loud music.

TABLE 23.1
Structures of the Ear

STRUCTURE	DESCRIPTION	FUNCTION
EXTERNAL EAR		
Pinna	Part of ear outside the head	Catch and direct sound
Auditory canal	Tube between pinna and eardrum	Channel sound to middle ear
MIDDLE EAR		
Tympanic membrane (eardrum)	Vibratory membrane	Vibrates in response to sound
Hammer (malleus) Anvil (incus) Stirrup (stapes)	Three small bones in middle ear	Transmit vibrations of eardrum to inner ear, increase force of vibration
INNER EAR		
Utricle, saccule, and semicircular canals	Two fluid-filled chambers and three bony canals at right angles to one another	Equilibrium
Cochlea	Fluid-filled, snail-shaped bone containing hair cells; auditory nerve starts here	Houses organ of hearing, which generates nerve impulses

ESSAY 23.2

That Wonderful You

You may have noticed that there are certain areas of the human body that are not discussed very much. One of these neglected areas is the armpit, but paradoxically, a neglected armpit will generate discussion. This is because the armpit has a number of large sweat glands that produce the scent that is so distinctly you. Its scent-laden secretion is a mixture of watery sweat and a thick, dark yellow oil. The secretion itself actually has a musky, not unpleasant fragrance. However, its finer qualities are rarely appreciated because, once it is trapped in underarm hair, it is acted upon by bacteria that quickly decompose it into something with an aura reminiscent of goat. Most people tend to give up projecting their own odor preferring to be odor-free, or perfumed, rather than risk the result of bacterial action on their scent.

Underarm secretions are largely steroids, the chemical group that includes the sex hormones estrogen and testosterone. Many people are intensely sensitive to such smells. Perhaps not surprisingly, men (especially those between 18 and 45) produce large amounts of underarm secretions while women generally produce only trace amounts (although there are exceptions). On the other hand, women have scent glands around the nipples. These may be important in recognition and bonding between a mother and nursing infant.

Other sweat glands of the body also secrete an oily mixture that mark us individually and this is why dogs are used to catch us when we escape from prison. (Dogs can perceive odors at up to hundreds of millions lower concentrations than humans.) The dense aggregations of sweat glands on the soles of the feet may release half a fluid ounce of sweat per day. If even one-thousandth of this penetrated the seams of the shoes, millions of molecules would be left behind with each footstep, making tracking us quite simple for a loud-mouth dog. Because each of us bears our own essence, we can be easily distinguished from each other by these specialists.

CHEMORECEPTORS

The ability to detect the presence of chemicals is called **chemoreception,** and it varies widely in sensitivity throughout the animal world. If you encounter a smelly dog, you should keep in mind that your opinion of his odor pales before his perception of you. We can assume that either dogs don't mind our scent or they're just being polite (Essay 23.2).

Both **olfaction** (smell) and **gustation** (taste) are examples of chemoreception. The mechanisms are essentially similar, but olfaction usually involves distant stimuli and gustation registers those in which the source is in contact with the receptors.

The most remarkable chemoreceptive abilities are found among the insects. In this group, there are taste receptors on various parts of the body, including mouthparts, antennae, and forelegs (since some are known to eat what they walk on). Some can even taste with their egg-laying organs, an ability that enables them to lay eggs only on certain kinds of plants. Generally, however, most insect olfactory receptors lie at the ends of minute tubules that branch throughout the insect body. Molecules that diffuse into a tubule become dissolved in the insect's body fluids and can bind to receptors on the membranes of the tiny sensory cells. (One of the most amazing olfactory abilities is found in the atlas moth, Figure 23.7.)

Chemoreception in vertebrates usually involves moving chemicals into specialized sacs or tubes that are lined with receptors. The chemicals must first be dissolved in fluid before they can cross the membranes of the receptors, so these sacs and tubes are usually moist.

Among vertebrates, mammals have the best sense of smell, with the best smellers being the carnivores and rodents. However, some mammals, such as the toothed whales, have no sense of smell at all. The sense of smell is also rather poorly developed in us primates. Chimpanzees, for example, smell about like you do (no offense). In humans, the olfactory receptors in the nasal passage are connected directly to a slender, forward extension of the brain, called the **olfactory bulb** (Figure 23.8). Refer back to Figure 22.3 and compare the relative sizes of the olfactory bulb in various animals. (It is most reduced in humans—and largest in what group?)

FIGURE 23.7

The oversized antennae of the atlas moth, *Artacus atlas*. The receptors in the antennae are extremely sensitive to a chemical attractant released by females. Even one molecule can generate an impulse.

FIGURE 23.8

The olfactory receptors in the human nose connect to the olfactory bulb of the brain. The receptors are able to distinguish a wider variety of stimuli than the taste buds. The olfactory neurons are part of the nasal epithelium dispersed with other cells. Each neuron has numerous olfactory hairs that protrude from the epithelium. Each "hair" is a modified cilium. Interestingly, the olfactory neurons are easily damaged by a blow to the head, rendering the individual unable to smell.

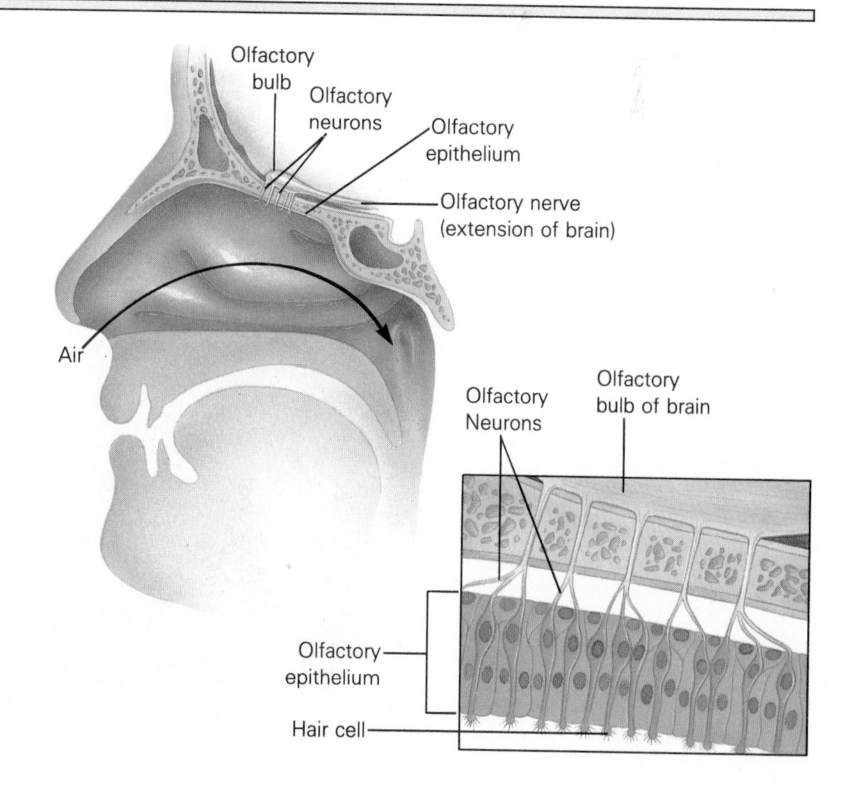

FIGURE 23.9

Taste buds on the tongue respond most strongly to one of four flavors: bitter, salty, sweet, and sour. The photo, taken with a scanning electron microscope, reveals the flattened columns that contain the taste buds.

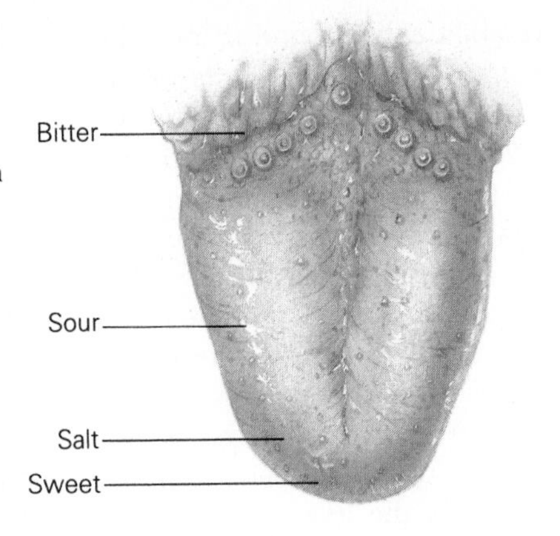

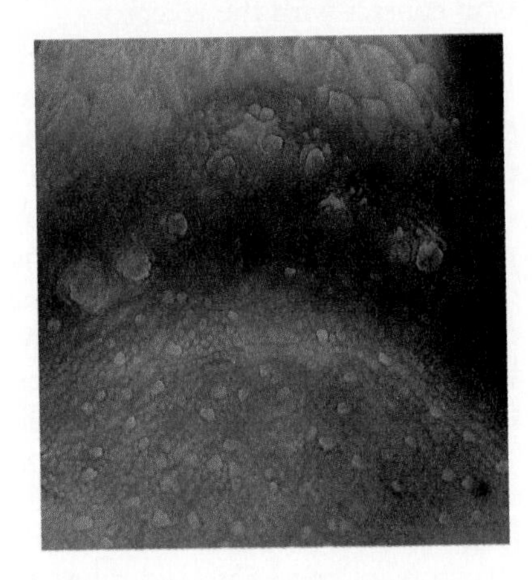

It is generally acknowledged that there are four basic tastes: sweet, sour, salty, and bitter. In humans, these are located in specific areas of the tongue (Figure 23.9). Salt can be tasted over the entire surface of the tongue, but sweet registers on the tip, sour on the sides, and bitter on the back. Research indicates that most cells can respond to three or four tastes, but those of each group are particularly sensitive to a single taste. One might wonder, if a taste receptor can be activated by more than one kind of molecule, how does the brain respond properly to each taste? It turns out that each kind of chemical produces a distinctive pattern of neural firing in the receptors, and the brain deciphers each pattern. The tasting abilities of some individuals are highly developed, due to either heightened sensitivity of receptors or a well-trained ability to recognize certain tastes. Wine tasters are among the most highly trained chemodetectors (Figure 23.10).

FIGURE 23.10

Wine tasters are among the finest chemodetectors, although their attempts to describe their findings can be an exercise in imagination. Sassy and impertinent, indeed.

ESSAY 23.3

Fooling Mother Nature's Sweet Tooth

In West Africa people can eat lemons just as people here eat oranges. The very idea can make our jaws pucker. How can they stand it? Do they like the sour taste? It turns out, there is no sour taste. This is because, before they ate the lemon, they chewed a few miracle fruits, small red berries, about the size of olives, from a plant, *Synsepalum dulcificum*, that grows there. Because of these berries, these people, in effect, are not eating sour lemons, they're eating *sweet* lemons. In fact, they can eat all sorts of local sour foods because chewing the miracle fruit first makes anything sour seem sweet for up to three hours afterwards.

It turns out that the active ingredient is a glycoprotein, a protein with sugar molecules bound to it. It is believed to work because the protein component can attach to a site close to where sweet receptors are located. The acidity of the sour food then alters the glycoprotein so that its sugar components are released. These then bind to the sweet receptor sites, stimulating them, and masking the sour taste.

Another substance, gymnemic acid, found in the leaves of an Indian plant, *Gymnema sylvestre,* can take the sweet taste from any food. "It can make sugar taste like sand and sugar solution taste like tap water." The mechanism for generating the insensitivity is not entirely understood, but those who saw its potential usefulness as a dietary aid may have been disappointed to learn that it also masks a variety of nonnutritive sweeteners, such as cyclamates, as well. And then there are those who would undoubtedly argue that the greatest taste deceiver of all is ketchup.

Chemoreception has been extremely important in the evolutionary history of humans. In essence, it gives us information about the environment before we are forced to interact with it. It lets us know what is good and desirable as well as those things that are to be avoided. For example, sweet is the taste of carbohydrates (ripe fruit tastes sweet). Unripe fruit, however, may be sour, and we have little tolerance for that taste. This means that we are not likely to eat unripe fruit but will probably wait until it has matured and its food value is higher. (Our taste for sweets can be fooled, as we see in Essay 23.3.) Bitter, we might mention, is the taste of a number of powerful poisons and can trigger a gag reflex at the back of the tongue.

We probably have a great deal to learn about the role of chemoreception in human life. For example, recent research has focused on the role of smell in sexual attractiveness. There is evidence that masking human odor with colognes might be counterproductive in increasing one's appeal. It has been suggested that the hair on some parts of our bodies may be useful in trapping our odors and aiding in subtle communcation with the

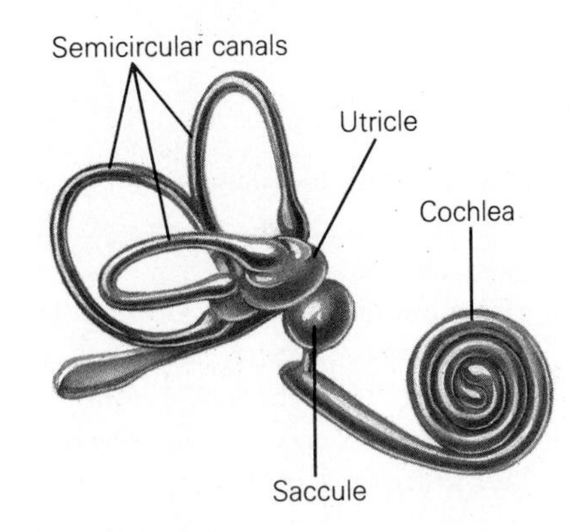

FIGURE 23.11

Semicircular canals. The middle ear is very sensitive to the body's position and movement. The semicircular canals lie at right angles to each other so that bodily movement in any direction shifts the fluid in at least one of them. The saccule and utricle are fluid-filled cavities in which grains are embedded in a jellylike matrix. Movement of these grains sends information to the brain regarding a change in the position of the head with respect to gravity.

opposite sex. (One might think that the communication had best be *very* subtle.) In addition, preliminary experiments show that people can generally distinguish between males and females on the basis of both breath and body scent. The groundwork has been laid for what should be some fascinating lines of research.

PROPRIOCEPTORS

Proprioceptors sense the position of the body or of various parts of the body. They tell you whether you're lying down or standing up and where your hands are. In a society such as ours, these are good things to know. This is an area that perhaps has not received enough research attention, but that may be because it normally works so efficiently.

Proprioception is common in many invertebrates and is achieved by receptors that respond to pressure, stretching, and bending. In some insects, these sensors are at the base of hairs, while in crustaceans, such as lobsters, the receptors are within the muscles. Among vertebrates, certain kinds of proprioceptors are also concentrated deep within the muscles. They are stimulated as the muscle stretches and places pressure on them. It is apparent that proprioception is particularly important among the more active or athletic vertebrates. Clumsy monkeys would tend to fall out of trees, and so we find that monkeys have, in fact, many proprioceptors.

Vertebrates also have very precise equilibrium proprioceptors located in the inner ear. In most species, there are three fluid-filled loops, or **semicircular canals,** opening to two chambers (Figures 23.5 and 23.11). Each canal is filled with fluid and contains sensory hairs. Any change in the animal's position results in movement of the fluid, which then stimulates the hairs. Since the canals lie at right angles to each other, movement in any direction can be detected. The chambers are called the **utricle** and **saccule,** and they contain small granules that shift when the body moves, stimulating the sensory hairs and providing information on the position of the head with respect to gravity.

VISUAL RECEPTORS

Humans are highly visual animals, and so we have a keen interest in sight. However, compared to some animals, especially birds, we don't see well at all. There is indeed a great deal of variation in visual ability among animals, and, not surprisingly, the ability is correlated with a species' simple *need* to see.

But how exactly is any visual receptor stimulated? And what stimulates it? In essence, it is sensitive to a particular part of the electromagnetic spectrum we call light. Light's wavelength ranges from about 430 nm (nanometers) to 750 nm (see Appendix B), but no animal can see more than part of this range. The shorter wavelengths, such as X rays and gamma rays, can't be detected by any animals, nor can the very long ones, such as radio waves. The detectable waves are absorbed by special **visual pigments** that then transform the wave energy into a neural stimulus.

In vertebrates, the light-sensitive part of the eye is the **retina.** (The parts of the eye are summarized in Table 23.2 and labelled in Figure 23.12a.) It is composed of two kinds of cells, **rods** (specialized for black and white vision) and **cones** (specialized for color) (Figure 23.12b). The rods hold large amounts of a pigment called **visual purple.** When this pigment is activated by light, it "bleaches" and the permeability of the rod changes. The change in permeability causes electrochemical changes that

TABLE 23.2
Structures of the Eye

STRUCTURE	DESCRIPTION	FUNCTION
Pupil	Open center of iris	Entrance for incoming light
Iris	Colored part of eye	Regulates the amount of light that enters eye
Cornea	Transparent dome of tissue at front of eye	Bends light rays to help focus them on retina
Lens	Semispherical transparent body of tissue	Adjustable focusing of light rays onto photo-receptors
Aqueous humor	Clear fluid between lens and cornea	Transmits and bends light; pressure of fluid helps maintain shape of eye
Vitreous humor	Jellylike substance within chamber behind lens	Transmits and bends light; pressure of substance helps maintain shape of eye
Retina	Tissue containing rods and cones	Sensory area; receives light and generates nerve impulses
Fovea	Tiny pit on retina with a high density of cones	Most sensitive part of retina
Optic nerve	Bundle of nerve fibers leaving eye	Carries signals from retina to brain

FIGURE 23.12

The eye is actually a rather tough structure (a). The chamber in front of the lens contains a fluid called aqueous humor; the large chamber behind is filled with vitreous humor. The white of the eye, the sclera, is modified in the front to form a transparent window, the cornea; the colored part of the eye is the iris. The lens is focused by the muscles that support it. The sensory area, the retina, is composed of rods and cones that send impulses to the brain over the optic nerve. Where the nerve enters the retina there are no receptors (the "blind spot"). The most sensitive part of the retina is a tiny pit with a very high density of receptors, the fovea. When threading a needle, we usually turn our heads in such a way that the image falls on the fovea.

The rods and cones of the retina (b). The rods can detect light at low levels; the cones can detect different colors.

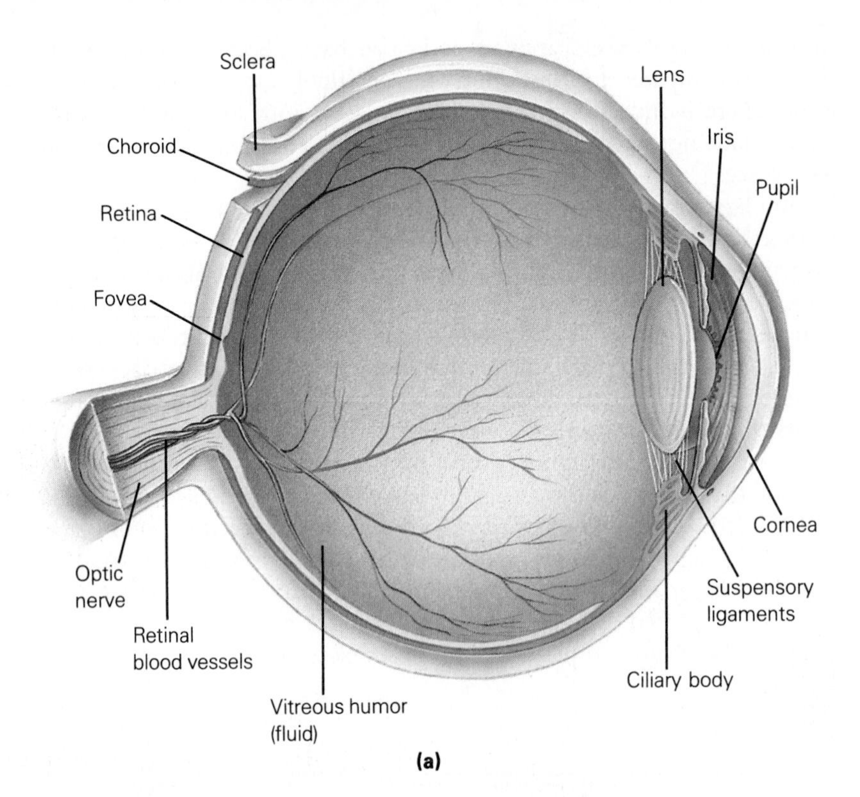

(a)

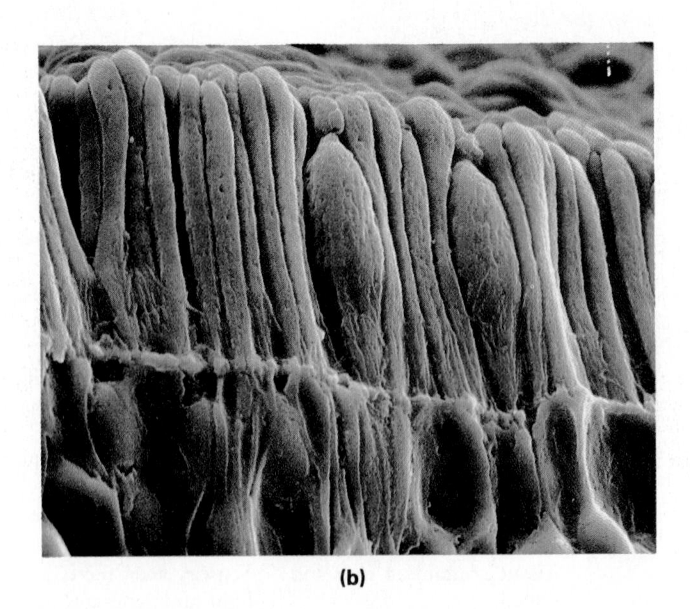

(b)

may lead to action potentials (impulses) that will be sent to the brain where they will be deciphered and integrated.

Whereas rods are sensitive to all wavelengths of visible light, cones respond only to specific wavelengths—that is, to specific colors. Humans and other primates have three kinds of cones that respond either to red, green, or blue. The multitude of colors we see depends on the interplay between these three.

The real question is, of course, do bulls see red? The answer is, not very well, if at all. (Cats may be able to barely detect red, but they cannot be depended upon to charge red capes.) In fact, real color vision is found only among some species of insects, fishes, reptiles, and birds, and among mammals. (In this last group, the primates are the color specialists.)

Since life is an opportunistic phenomenon, living things must have ways of detecting their opportunities (and minimizing their risks). This detection, we see, is the result of a dazzling interplay among a host of specialized receptors that comprise the senses. Next we will see just how animals use this information to get along in their special parts of the world.

SUMMARY

1. Animals know their world through their senses. Animals are sensitive to limited aspects of their environment. Each species is sensitive to stimuli that are important to survival.
2. Receptors are neural structures capable of responding to environmental stimuli.
3. Thermoreception is the ability to sense heat. It allows an animal to adjust itself in its environment so that its body temperature remains within an acceptable temperature range. Only a few invertebrate species have thermoreceptors. In many mammals, temperature is registered by warm receptors, cold receptors, pain receptors, and unspecialized free nerve endings. Thermoreceptors adapt quickly. They are unevenly distributed over the body.
4. Tactile receptors respond to touch. They fire when their shape is distorted. They are common among invertebrates. In vertebrates, Pacinian corpuscles respond to pressure and Meissner's corpuscles respond to touch. In addition, there are free nerve endings and nerve endings that are wrapped around hair follicles.
5. Auditory receptors detect sound. Most invertebrates cannot hear. However, some insects have tympanal membranes that respond to sound. In mammals, the external ear catches sound and funnels it to the eardrum. The resulting vibrations in the eardrum are transferred through the bones of the middle ear: the hammer (malleus), anvil (incus), and stirrup (stapes). The movement of the bones causes the oval window to vibrate and this causes fluid in the cochlea to move. Movement of the fluid causes the basilar membrane to vibrate and push hair cells on its surface into the overlying tectorial membrane, thereby initiating impulses. The pitch of the sound is registered according to which hairs are stimulated. Loudness is encoded in the number of neurons responding to the sound and the frequency of impulses.

6. Chemoreception is the detection of chemicals. Examples of chemoreception are olfaction (smell) and gustation (taste). Chemicals dissolve in fluid and cross the membranes of receptors. In mammals, olfactory receptors in the nasal passage are connected to the brain's olfactory bulb. There are four basic tastes—sweet, sour, salty, and bitter, each located in a specific region of the tongue. Although receptors generally respond to more than one taste, each is particularly sensitive to one. Chemoreception provides information about the environment before interaction with it occurs.
7. Proprioceptors sense the position of various parts of the body. Proprioceptors may be sensors at the base of hairs or receptors within muscles. Vertebrate proprioceptors associated with equilibrium are the fluid-filled semicircular canals that open into the utricle and saccule. Change in body position causes the fluid to move, which in turn causes small granules to move and stimulates sensory hairs in the utricle and saccule.
8. Visual receptors have visual pigments that absorb particular wavelengths of light and transform that wave energy into a neural impulse. The light sensitive part of the eye in vertebrates, the retina, contains rods (specialized for black and white vision) and cones (specialized to detect color).

KEY TERMS

audition (640)
auditory canal (640)
basilar membrane (645)
chemoreception (646)
cochlea (645)
cones (651)
external ear (641)
gustation (646)
incus (645)
malleus (645)
Meissner's corpuscles (640)
middle ear (645)
olfaction (646)
olfactory bulb (647)
oval window (645)
Pacinian corpuscles (640)
proprioceptor (650)
retina (651)
rods (651)
saccule (650)
semicircular canals (650)
stapes (645)
tactile receptors (640)
tectorial membrane (645)
thermoreception (639)
tympanal membrane (640)
tympanic membrane (640)
utricle (650)
visual pigment (651)
visual purple (651)

FOR FURTHER THOUGHT

1. What type of receptor might enable a gymnast to remain upright after performing a series of flips? Where are these receptors located?
2. The piece of hot pizza that you may comfortably hold in your hand burns your mouth when you take a bite. Explain this in terms of thermoreceptors.

FOR REVIEW

True or false?

1. ___ Meissner's corpuscles are tactile receptors that respond to light touch.
2. ___ High pitches of sound are registered by the hairs in the narrower regions of the basilar membrane.
3. ___ Each chemoreceptor in the tongue can distinguish only one of the four basic tastes.
4. ___ Cones in the retina respond to all visible light wavelengths.

Fill in the blanks.

5. ___ are receptors that detect temperature changes.
6. Pacinian corpuscles are ___ receptors that register ___ .
7. Equilibrium is maintained by ___ located in the semicircular canals of the ear.

Choose the correct answer.

8. Which of the following are used by humans to register temperature?
   A. warm receptors
   B. pain receptors
   C. cold receptors
   D. all of the above
9. The stirrup sends sound vibrations directly to the:
   A. oval window
   B. malleus
   C. eardrum
   D. auditory canal
10. ___ are receptors that function in olfaction and gustation.
   A. Proprioceptors
   B. Chemoreceptors
   C. Thermoreceptors
   D. None of the above.

PART SIX

Behavior and the Environment

We consider, first, how certain behaviors have evolved and how they help the animal more closely adapt to its world. Then we consider that world, first by describing its physical nature, then by seeing how populations inhabiting that world behave. We close by focusing on the human condition, and how specific technologies, as well as our ethical codes, have harmed the planet, always with an eye to what we must do if we are to reverse these trends.

CHAPTER 24
Animal Behavior

Overview

Ethology and Comparative Psychology
The Development of the Instinct Idea
Learning
How Instinct and Learning Can Interact
Orientation and Navigation
Social Behavior
Cooperation
Altruism
Sociobiology and Society

Objectives

After reading this chapter you should be able to:

- State how ethologists and comparative psychologists study animal behavior.
- Describe the components of instinctive behavior.
- Give examples of three major types of learning and explain the relationship between learning and innate behavior.
- Explain how birds orient and navigate.
- Describe how animals interact through aggression and cooperation.
- Give examples of altruistic behaviors and explain why they are adaptive.
- Describe how sociobiologists view human behavior and explain why sociobiology is controversial.

Those people who study animal behavior professionally must dread those times when their cover is blown at a dinner party. The unfortunate souls are sure to be seated next to someone with animal stories. The conversation will invariably be about some pet that did this or that, and nonsense is the *polite* word for it. The worst stories are about cats. The proud owners like to talk about their ingenuity, what they are thinking, and how they "miss" them while they're at the party. Those cats would rub the leg of a burglar if he rattled the Friskies box.

The stories about dogs are sometimes not so outlandish. Dogs probably do miss their owners; it's probably easier to know what's crossing their minds (Figure 24.1); and some do tend to protect the house. Even serious scientists have investigated the stories of dogs that, after having been left behind, found their owners after travelling hundreds of miles to places the dogs had never been. Still, even some of the dog stories leave you in danger of falling face down into the soup.

FIGURE 24.1
Some dog owners feel they can tell what's on their pet's mind, as with this smiling, contented fellow.

Why should there be so much misunderstanding about animals? After all, we humans have lived in intimate association with other animals for thousands of years. Not only have animals been important in our cultures, but our evolutionary histories are inextricably linked as well. Yet, somehow, we seem to have learned very little about them. One reason may be that they often seem to be so much like us, and it is easy to assume that they are just somewhat eager but rather ineffective versions of humans. Not so, and our clichés about angry bees, fierce lions, wise elephants, proud eagles, cowardly hyenas, and loving cats may have impeded our attempts to truly understand animal behavior at any meaningful level.

Our goal here, then, will be twofold. First, we will try to take a close look at the structure of animal behavior. Then we will see how the various behaviors help the animals to fit more precisely into their corner of the world. Essentially, we will be asking what animals do and why they do it.

ETHOLOGY AND COMPARATIVE PSYCHOLOGY

The modern study of behavior has taken two distinct routes in this century—one European, one American. Only in recent years have the grand ideas begun to merge, blur, and fuse into an encompassing discipline.

The European approach was called **ethology.** Its goal was to understand behavior by studying its cause, development, evolution, and function through the observation of animals in the wild, or under somewhat natural conditions. It was the ethologists, particularly Niko Tinbergen and Konrad Lorenz (Figure 24.2), who developed the idea upon which the modern concept of instinct is based.

The American approach was called **comparative psychology.** Its goal was to understand behavior by studying animals in the laboratory under carefully controlled conditions. Its focus, at mid-century when the concept of instinct was being developed, was learning, and its primary animal was the Norway rat.

Unfortunately, the two groups were constantly at odds, especially in the early years. However, as each field matured and as researchers learned

FIGURE 24.2
Two of the founders of modern ethology, Konrad Lorenz (a) and Niko Tinbergen (b).

(a)

(b)

more about each other's work, the lines became blurred. Ethologists came into the lab, and comparative psychologists learned where birds go in winter.

THE DEVELOPMENT OF THE INSTINCT IDEA

It was once believed that human behavior results entirely from learning, but that other species respond to "instincts"—patterns that are indelibly stamped into their nervous systems at birth. But the term instinct was never precisely defined and, because of this, the entire concept fell into disrepute. Many behavioral scientists began to hesitate even to use the word. The problem was compounded by the fact that while ethologists were seeking to clarify the concept, the word instinct found its way into the general population. It was handled casually, twisted around, and tossed about. It was used to explain everything from a baby's grip to homemaking to swimming.

Many people seem to think that instinct is any innate behavior—something that appears naturally in an animal and is not due to learning. However, this is only part of the definition of an instinct. Instead, what we will learn here is that the perception of something in the environment (a *releaser*) triggers a reaction in a center in the central nervous system (the *innate releasing mechanism*) that then causes the performance of the instinctive act, sometimes composed of very stereotyped movements (called *fixed action patterns*).

Releasers

According to ethologists, instinctive behavior is triggered by certain very specific signals from the environment. Obviously, the signal is perceived upon activating a receptor, such as a visual receptor. Environmental factors that evoke, or release, instinctive patterns are called **releasers.** The releaser itself may be only a small part of any appropriate situation. For example, fighting behavior may be released in territorial male European robins, not only by the sight of another male, but even by the sight of a tuft of red feathers at a certain height within their territories (Figure 24.3). Of course, such a response is usually adaptive because tufts of red feathers at that height are normally on the breast of a competitor. The point is that the instinctive act may be triggered by only *certain parts* of the environment.

FIGURE 24.3
A male European robin in breeding condition will attack a tuft of red feathers placed in his territory.

Innate Releasing Mechanisms

The exact mechanism by which releasers work isn't known, but one idea is that there are certain neural centers called **innate releasing mechanisms (IRMs),** which, when stimulated by impulses set up by the perception of a releaser, trigger a chain of actions. It is these actions that comprise instinctive behavior. The essence of the IRM is that it is genetically encoded, and once it is triggered by a releaser, it results in the performance of a behavior called the fixed action pattern.

Fixed Action Patterns

It can't be denied that animals are born with certain behaviors that are indelibly stamped into their behavioral repertoire. Birds build nests by using peculiar sideways swipes of their heads, with which they jam twigs into the nest mass. And all dogs scratch their ears the same way, by moving their rear leg outside the foreleg (Figure 24.4). Such precise and identifiable patterns, which are innate, independent of the environment, and characteristic of a given species, are called **fixed action patterns** (Figure 24.5).

FIGURE 24.4
The scratching movements of dogs, as well as many other vertebrates, are considered fixed action patterns.

FIGURE 24.5
Simplified diagram of how a fixed action pattern can be triggered. The releaser is perceived by some sort of receptor, which triggers the IRM to activate certain muscles, thereby producing an instinctive movement that usually involves fixed action patterns.

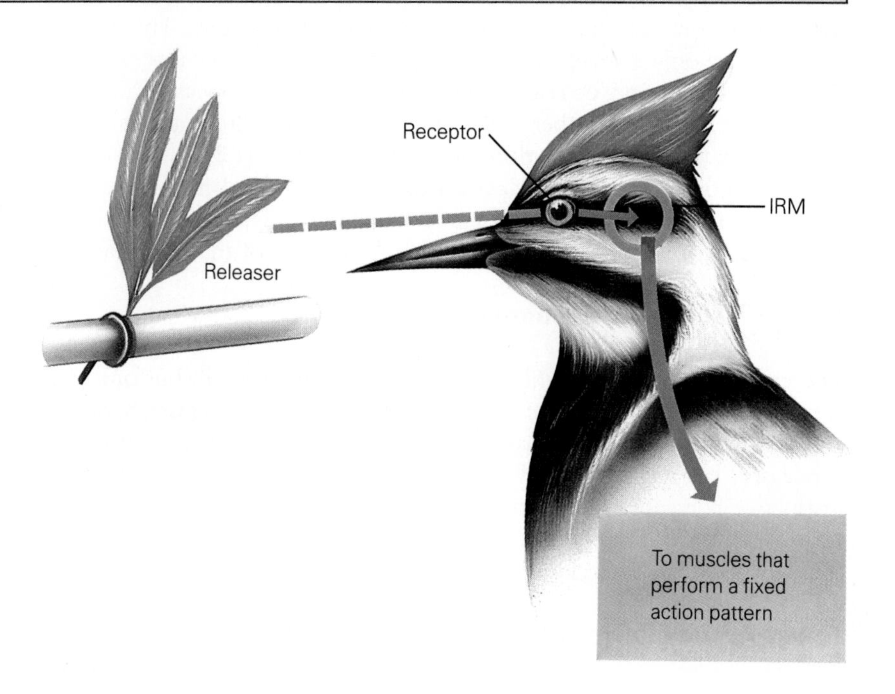

LEARNING

The very word *learning* is revered among humans. We sacrifice and excuse a great deal at its altar. (In fact, it's what you're supposed to be doing at this very moment.) We even like to see it in other species ("I tell you, Chester, that dog's *smart*."). But what is learning? How does one learn? And what does "smart" mean?

The problem is that we haven't learned much about learning in spite of a great deal of research effort. Not only are we vague on how it occurs in humans, but each species seems to have its own learning propensities, and each may learn in its own way, so establishing general principles is exceedingly difficult.

By the way, this is one of the problems in attempting to compare the "I.Q." of various species. We tend to consider the one possessing abilities closest to ours the more intelligent. But in spite of perhaps being able to learn more tricks, a pig is not necessarily smarter than a horse and a cat is not dumber than a bird because it cannot learn to chirp. Nor is a cat smarter than a dog because it sneaks up on its prey. Cats, after all, can't run very far, but dogs can. Each animal, we see, behaves according to the calling of its species, and so comparisons of intellectual ability are likely to be fruitless.

Having said that, some species are clearly more intelligent than others. Tapeworms, for example, have never been considered particularly profound. But then, why should they be? They live in an environment that is soft, warm, moist, and filled with food. The matter of leaving offspring is also simplified; they merely lay thousands and thousands of eggs and leave the rest to chance—sheer blind luck.

In contrast, there are species for which intelligence or the ability to learn is critical. Chimpanzees live in complex and variable environments. They must be able to cope with a variety of conditions. It is no coincidence that they are long-lived, highly social, and mature slowly. These are the traits that are associated with high learning ability. They have time to learn and have a social system that enables them to learn from each other (Figure 24.6).

FIGURE 24.6
Chimpanzees are highly social creatures that live in a complex, variable, and changing world. Intelligence is important under such circumstances.

There is increasing evidence that many species are able to learn more than has been assumed. Bees, for example, aren't usually noted for being very bright. But in experiments in which food is moved a certain distance further away from a beehive each day, bees seem to extrapolate. After the first few moves, they predict where it will be and fly directly to it. Some birds, it is reported, drop tiny white feathers onto the surface of a pond and then capture whatever little fish come to investigate. Squirrels are usually considered smarter than birds, but they have memory problems. They find very few of the nuts they bury in the summer. However, some birds that also bury nuts find almost every one, even months later. The point is that intelligence is both difficult to define and measure. As we stressed earlier, generally, animals are able to learn those kinds of things that are important for them to learn. (Memory is certainly involved in learning and, by an odd twist, *forgetting* can also be important in the adaptiveness of learning; Essay 24.1.)

Although there may be a number of fundamentally distinct ways animals learn, we will consider only three major types: habituation, classical conditioning, and operant conditioning.

Habituation

Habituation is, in a sense, learning *not* to respond to a stimulus. The first time an animal encounters a stimulus, it may respond vigorously. But if the stimulus is presented over and over without consequence, the response to it gradually lessens and may finally disappear altogether (Figure 24.7).

FIGURE 24.7
As animals become accustomed to a stimulus, habituation may occur.

ESSAY 24.1

The Advantage of Forgetting

There are all sorts of strange abilities associated with memory in humans. *Savants,* a special class of retardates, have very low I.Q.s, but some are able to accomplish incredible mathematical feats, such as multiplying 2 five-figure numbers in their heads. Others can immediately tell you the day of the week on which Christmas day fell in 1492—or any day in any year (although this is probably not a memory feat). A normally intelligent Russian man made his living giving stage performances as a *mnemonist.* He would sometimes memorize lines of 50 words. Once, in just a few moments he memorized the nonsense formula:

$$N \cdot \sqrt{d^2 \cdot \frac{85^3}{vx}} \cdot \sqrt[3]{\frac{276^2 \cdot 86x}{n^2 v \cdot 264}} \cdot n^2 b$$

$$= sv \cdot \frac{1624}{32^2} \cdot r^2 s$$

Fifteen years later, upon request, he repeated the entire formula without a single mistake.

And what about those people with "photographic memories"? They exist, and they are called eidetikers. Proof of their abilities has been demonstrated with stereograms. These are apparently randomized dot patterns on two different cards; when they are superimposed by use of a stereoscope, a three-dimensional image appears. One person was asked to look at a 10,000 dot pattern with her right eye for one minute. After ten seconds, she viewed another "random" dot pattern. She then recalled the positions of the dots on the first card and conjured up the image of a T.

So, if such abilities are possible for our species, why haven't they been selected for so that by now we can all, more or less, perform such feats? On the surface, the advantages seem enormous. The reason we can't is, in part, because there are serious drawbacks to remembering everything. Both the Russian mnemonist and the eidetiker could look at a barren tree, "recall" its leaves, and when they looked away, be confused over whether the tree was leafy or not. The mnemonist could watch the hands on a clock and not notice they had moved, remembering (or "seeing") only where they had been. Also, what about all those insignificant events it is of no advantage to remember? The energetically expensive neural apparatus would be wasted retaining such information for recall (assuming such retention takes more energy than the storage of material we normally can't recall). And the mnemonist had trouble with discerning time lapse. He recalled everything so well that it seemed to him as if every event of his life had just occurred.

Obviously, the key here is that the reason we can't remember as well as the people in these examples is that we, as a species, have not found it necessary or useful. We generally don't need to recall every stone on the path to the place where we found food yesterday; we only need to remember the location of the path. In fact, remembering too much about our physical environment might mean that changes in our environment would not be adjusted to as quickly as if we were never quite sure what to expect. Perhaps it is best that we can't rely too strongly on our memories.

Habituation is not necessarily permanent, however. If an animal habituated to some stimulus does not encounter it for a period of time, it may respond if the stimulus later reappears.

Habituation may be quite important in the lives of many animals. For example, a bird must learn not to waste energy by taking flight at the sight of every skittering leaf. A coral-inhabiting fish may come to accept its neighbors, but will immediately attack a strange fish (Figure 24.8). The stranger might be seeking to displace the resident fish, and so the response is adaptive. It may also help to explain why animals respond to the sight of a furtive predator they see only rarely, while ignoring the harmless species they see more often. Habituation is often not given the attention it deserves in studies of learning, perhaps because it seems so simple. But it may well be one of the more important learning phenomena in nature.

FIGURE 24.8
Coral fishes establish "neighborhoods."

Classical Conditioning

Classical conditioning was first described by the famed Russian biologist Ivan Pavlov (Figure 24.9). In classical conditioning, the response to a normal stimulus comes to be elicited by a substitute stimulus. In his experiments, Pavlov found that dogs would salivate at the sight of food. He then began to switch on a light five seconds before food was dropped onto the feeding tray. After doing this a few times, he presented the light without the food and found the dogs would continue to salivate. On the

FIGURE 24.9
The Russian biologist Ivan Pavlov and the apparatus he devised to demonstrate classical conditioning. Upon presentation of a light, meat powder would be blown into the dog's mouth, causing it to salivate. Later it came to salivate at the sight of a light alone. The salivation, then, was conditional on the light. Note in the first graph that the dog salivated at maximal levels after only eight trials. When the experiment was reversed and food no longer followed the light, the dog stopped salivating after only nine trials.

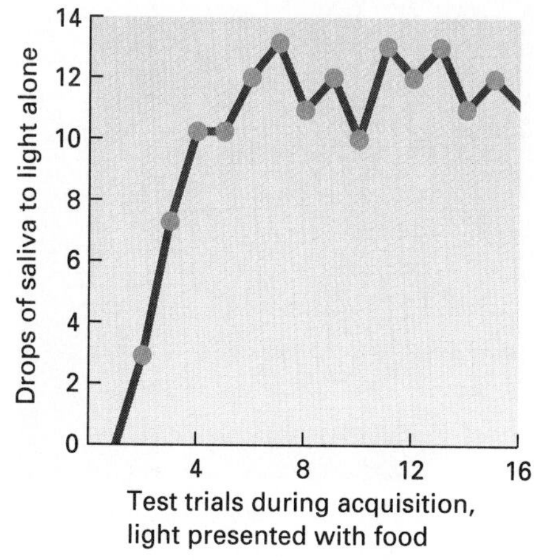

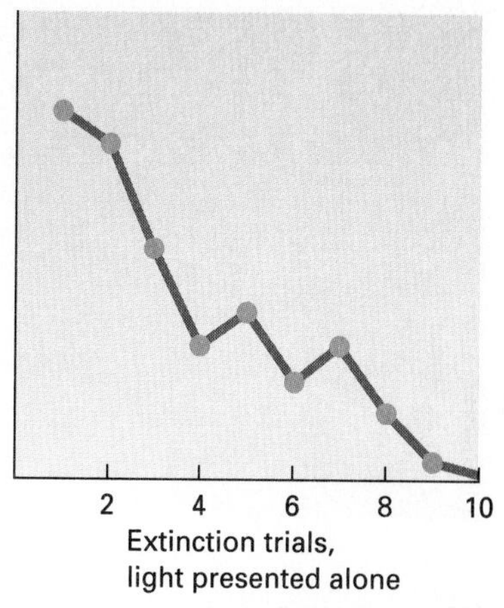

basis of numerous experiments, Pavlov found that the number of drops of saliva elicited by the light alone was in direct proportion to the number of previous trials in which the light had been followed by food. Because the normal stimulus (food in this case) is necessary in the training process, it is said to "reinforce" the response.

Through a process called **extinction,** the response may be lost when it is no longer appropriate. In a sense, Pavlov's dogs learned that the light was a signal that food would appear, so salivation was an appropriate response. When a dog that had come to associate light with food was shown the light over and over again without following it with food, salivation quickly began to decrease in response to the light and finally stopped altogether. The dog had learned that the light was no longer a reliable signal for food.

Operant Conditioning

Operant conditioning differs from classical conditioning in several important ways. Whereas in classical conditioning the reinforcement such as food, follows the stimulus, in operant conditioning, the reinforcement follows the behavioral response. Also, in classical conditioning the experimental animal has no control over the situation. In Pavlov's experiment, all the dog could do was wait for lights to go on and food to appear. There was nothing it could do one way or the other to make it happen. In operant conditioning, the animal's own behavior determines whether or not the reinforcement appears.

In the 1930s, the noted psychologist B. F. Skinner demonstrated operant conditioning by employing a device now called a **Skinner box** (Figure 24.10). An animal placed inside a Skinner box must learn to press

FIGURE 24.10

B. F. Skinner, one of the most important twentieth-century psychologists, and a "Skinner box," which is used to demonstrate operant conditioning.

a small bar in order to receive a pellet of food from an automatic dispenser. Skinner found that when an experimental animal (usually a hungry rat) was first placed in the box, it ordinarily began a random investigation of its surroundings. When it accidentally pressed the bar, a food pellet was delivered. The animal did not immediately show any signs of associating the two events, bar pressing and food, but in time it began to hang around near the bar. As more and more food pellets appeared immediately after the bar was pressed, the animal's behavior became less random until finally it learned to press the bar to obtain food. Eventually, it spent most of its time just sitting and pressing the bar. Skinner called learning through such a sequence operant conditioning. (The essential differences between classical and operant conditioning are shown in Figure 24.11).

The relative importance of each type of learning to animals in the wild isn't known at this point. It is likely that most adaptive, or beneficial, behavior patterns arise in nature as interactions of several types of learning.

HOW INSTINCT AND LEARNING CAN INTERACT

Over the years, a great deal of argument has focused on one simple question: Is a certain behavior innate or learned? Now, the question is no longer regarded as valid. The supposition that a particular behavior stems

FIGURE 24.11

In classical conditioning, a desirable commodity, such as food, comes to be associated with an irrelevant signal until the irrelevant signal alone can elicit an involuntary response normally associated with the commodity. Here, the animal learns passively. In operant conditioning, the animal can act when given a signal, but only one action is rewarded. This action then comes to predominate.

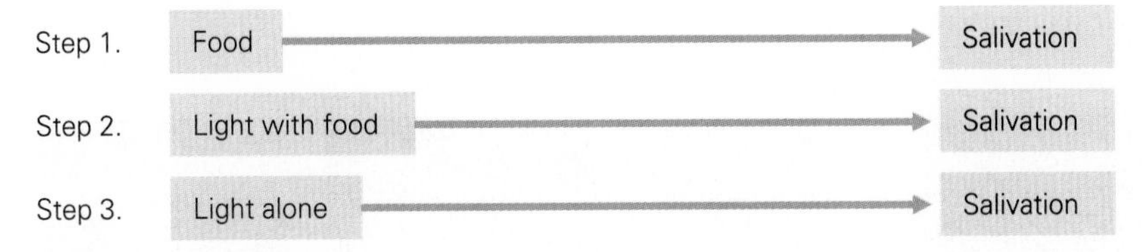

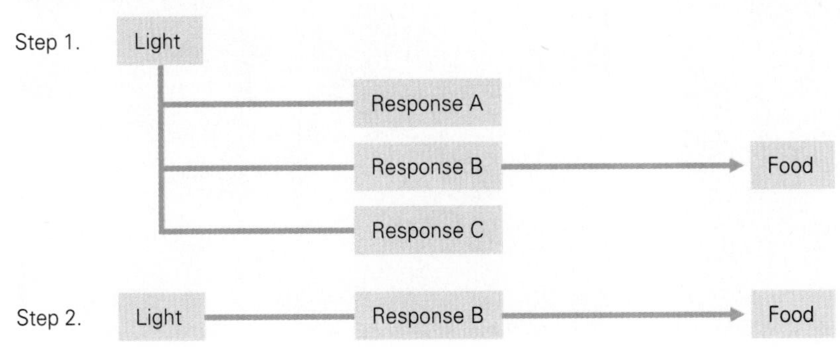

entirely from one source or another neglects the myriad ways in which behavioral components interact. In a sense, it's like asking whether the area of a triangle is due to its height or its base. A better question is: How do innate and learned patterns *interact* to produce an adaptive behavior?

Consider the development of flight in birds. Flight is usually considered a largely innate pattern. Obviously, a young bird must be able to manage it pretty well on the first attempt or it will crash to the ground as surely as would a launched mouse. It was once believed that the little fluttering hops of nestling songbirds were incipient flight movements, and that the birds were, in effect, learning to fly by practicing before they left the nest. But then someone performed some experiments in which a group of nestlings were allowed to flutter and hop, while others were reared in boxes which prevented any such movement. Then, at the time the young birds would have normally begun to fly, both groups were released. The restricted birds flew just as well as the ones that had practiced!

It is apparent, then, that flight behavior is largely an innate, or unlearned, pattern. However, generally, young birds do not fly as well as adults (Figure 24.12). With flight, as well as with many other forms of behavior, an innate pattern can be improved upon by learning through practice.

In other studies, the innate and learned components of a behavior have been more clearly differentiated. If you have ever watched a squirrel open a nut, you may have been impressed with its speed and efficiency. Red squirrels, for example, first cut a groove where the shell is thinnest, along the growth lines on the flat sides of the nut. Then they bite into the groove and break the shell open. In one study, baby squirrels were reared in an environment where they were never exposed to solid bits of food that they could handle with the paws. Interestingly, when later given hazelnuts, they performed the correct gnawing actions and even inserted

FIGURE 24.12

Although young birds have the innate ability to fly, they can improve with practice. Here, an adult eagle has landed gracefully, followed by a younger bird that crashes headfirst into the ground.

their incisors into the grooves correctly, demonstrating that these patterns are largely innate. The problem was that the young squirrels still had trouble actually opening the nuts. They gnawed all over the surface of the nut and started several grooves without completing any of them. After only a few such trials, however, they began to make their grooves at the thinnest part of the shell and easily opened the nuts. So most of the squirrels quickly became proficient at opening hazelnuts as their innate gnawing pattern was modified through learning, thereby producing the complex adaptive result.

FIGURE 24.13

Konrad Lorenz, in a famous photograph, leading a group of goslings that had imprinted on him.

Imprinting

On your annual visits to the farm, you may have seen young ducklings waddling along after their mother, perhaps on the way to the pond. It's a quaint sight, but you're there to see where milk comes from and so you think no more about it. However, if you had visited a farm in southern Germany some years ago, you might have seen a more unusual sight—a column of young ducklings following a white-haired Austrian down to the pond (Figure 24.13). The man was Konrad Lorenz, a future Nobelist, and the following behavior of the ducklings was the product of an experiment he devised.

What he did was to let them see him moving around and making noises a few hours after they had hatched. If they had seen and heard him after that, they would have treated him like any other human. But Lorenz was the figure they encountered during their **critical period,** a window of time when the young are particularly sensitive to certain aspects of their environment. At this time, the ducklings, and the young of many other species as well, learn the traits of whatever is around them. Normally, of course, this would be their mother, but these ducklings developed from eggs that were artificially incubated and so the figure that they saw during their critical period was Konrad Lorenz, and as a result they learn to regard him as one of their own.

Lorenz called this kind of learning **imprinting,** and he defined it as learning that occurs over a defined, relatively brief period of time in which the animal learns to make a specific response to certain aspects of its environment.

A great deal of research has been done on imprinting in the last few decades and we have learned that it is difficult to draw hard and fast rules about its development. It is regarded, though, as a curious interaction of learned and instinctive patterns. In the case of the ducklings, they learned the general characteristics of the stimulus for the innate following response.

Many animals also learn species identification during this critical period; that is, they learn the image of an appropriate mate. As they approach their first breeding season, they seek out an individual with traits generally similar to those of the individual they had followed soon after hatching. If they are raised by parents of another species, they focus on individuals with traits similar to those of their foster parents when it is time to breed. Lorenz once had a tame jackdaw (a European crow) that he had hand-reared, and it would try to "courtship feed" him during mating season. On occasion when Lorenz turned his mouth away, he would receive an

FIGURE 24.14

A crane imprinted on humans. Tex, the only female whooping crane at the International Crane Foundation breeding area in 1982, has been hand-reared and therefore had imprinted on humans. She rejected the mate provided for her, but could be enticed to lay eggs (artificially fertilized) by "dancing" with humans. She preferred Caucasian men of average size with dark hair.

earful of worm pulp! The story of Tex, the dancing whooping crane, provides another example of this type of learning (Figure 24.14).

Imprinting is especially important in the development of song in many species. For example, for male white-crowned sparrows to be able to sing the song of their species, they must hear the song at a particular time during a brief "sensitive period" early in life. If they are exposed to the song after this period, they will produce abnormal songs, lacking the finer details. At the period during which they must hear the song, they are not yet even able to sing.

White-crowned sparrows are wide ranging, and the species forms subpopulations that tend to breed among themselves and sing their own local dialects of the basic song. If young birds from different populations are isolated at hatching and reared in soundproof chambers, they all begin to sing the same rather basic song. Thus, they are apparently born with the basic pattern; the local embellishments of each population are learned later.

Interestingly, isolated hatchling white-crowned sparrows can learn the recorded dialect of any subpopulation of the species while in their learning periods. But if they are exposed to the song of another species, they will not learn that song and will sing as if they had been reared in a soundproof cage. Finally, if a young white-crowned sparrow is deafened after it has heard the proper song but before it has had a chance to sing, it will sing only garbled passages. It apparently must hear its own song in order to match it to its inborn "template." The development of song in white-crowned sparrows, then, serves as an example of the interaction of learned and innate patterns in producing an adaptive response.

ORIENTATION AND NAVIGATION

South Americans probably wonder where the robins go every spring. The robins, of course, are "our" birds; they simply vacation in the south each winter. Furthermore, they fly to very specific places in South America and will often come back to the same trees in our yards the following spring. The question is not why they would leave the cold of winter so much as how they find their way around. The question perplexed people for years, until, in the 1950s, a German scientist named Gustave Kramer provided some answers, and, in the process, raised new questions.

Kramer initiated important new kinds of research regarding how animals orient and navigate. **Orientation** is simply facing in the right direction; **navigation** involves finding one's way from point A to point B.

Early in his research, Kramer found that caged migratory birds became very restless at about the time they would normally have begun migration in the wild. Furthermore, he noticed that as they fluttered around in the cage, they often launched themselves in the direction of their normal migratory route. He then set up experiments with caged starlings and found that their orientation was, in fact, in the proper migratory direction—except when the sky was overcast. At these times, there was no clear direction to their restless movements. Kramer surmised, therefore, that they were orienting according to the position of the sun. To test this idea, he blocked their view of the sun and used mirrors to change its apparent position. He found that under these circumstances, the birds oriented with respect to the position of the new "sun." They seemed to be using the sun as a compass. Of course, this was preposterous. How could a stupid bird navigate by the sun when we lose our way with road maps? Obviously, more testing was in order.

So, in another set of experiments, Kramer put identical food boxes around the cage, with food in only one of the boxes. The boxes were stationary, and the one containing food was always at the same point of the compass. However, its position with respect to the surroundings could be changed by revolving either the inner cage containing the birds or the outer walls, which served as the background. As long as the birds could see the sun, no matter how their surroundings were altered, they went directly to the correct food box. Whether the box appeared in front of the right wall or the left wall, they showed no signs of confusion. On overcast days, however, the birds were disoriented and had trouble locating their food box (Figure 24.15).

FIGURE 24.15

Kramer's orientation cage. The birds can see only sky through the cage roof. The apparent direction of the sun can be shifted with mirrors.

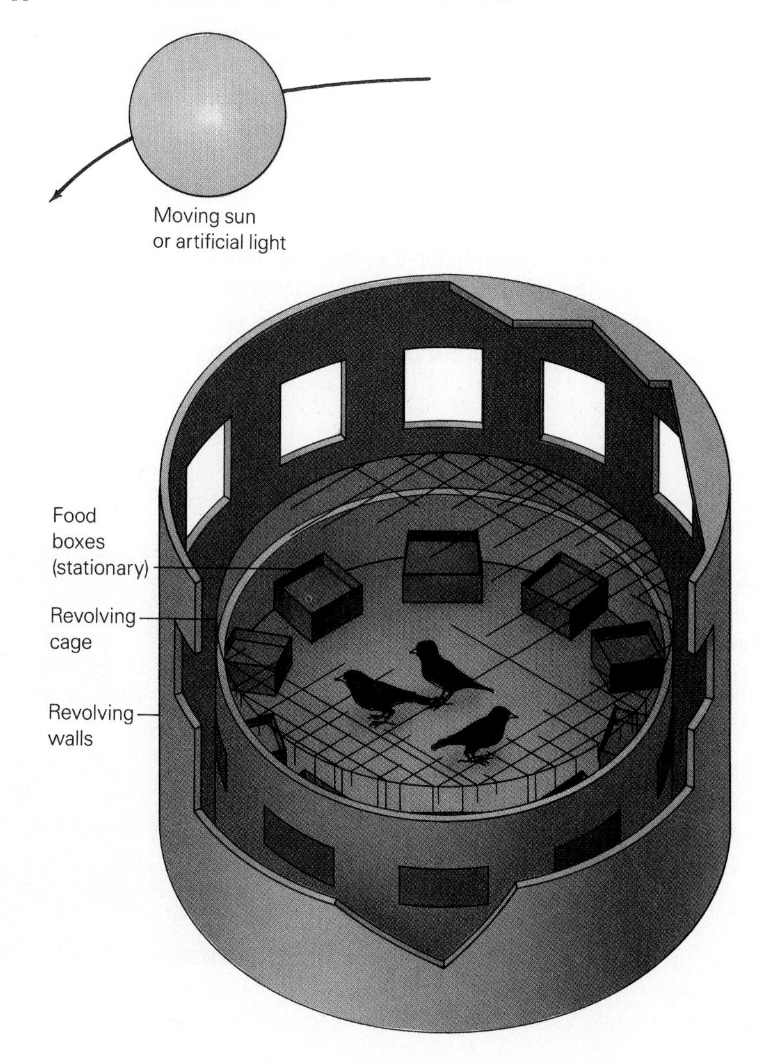

In experimenting with artificial suns, Kramer made another interesting discovery. If the artificial sun remained stationary, the birds would shift their direction with respect to it at a rate of about 15 degrees per hour, the sun's rate of movement across the sky (Essay 24.2). Apparently, the birds were assuming that the "sun" they saw was moving at that rate. When the real sun was visible, however, the birds maintained a constant direction as it moved across the sky. In other words, they were able to compensate for the sun's movement. This meant that some sort of biological clock was operating—and a very precise clock at that.

What about birds that migrate at night? Perhaps they navigate by the night sky. To test the idea, caged night-migrating birds were placed on the floor of a planetarium during their migratory period. A planetarium is essentially a theater with a domelike ceiling onto which a night sky can be projected for any night of the year. When the planetarium sky matched the sky outside, the birds fluttered in the direction of their normal migration. But when the dome was rotated, the birds changed their direction to match the artificial sky. The results clearly indicated that the birds were orienting according to the stars.

There is accumulating evidence indicating that birds navigate by using a wide variety of environmental cues. Other areas under investigation include magnetism (Figure 24.16), landmarks, coastlines, sonar, and even smells. The studies are complicated by the fact that the data are sometimes contradictory and the mechanisms apparently *change* from time to time. Furthermore, one sensory ability may back up another.

SOCIAL BEHAVIOR

The famous student of animal behavior, Jane Goodall (Figure 24.17), who has spent much of her life among the chimpanzees of East Africa's Gombe Stream Preserve, once said, "One chimpanzee is no chimpanzee at all." Her point was that researchers should not attempt to study chimpanzee

FIGURE 24.16

A pigeon with electrical coils on its head. On overcast days, birds carrying coils with reversed polarity tended to home in opposite directions, indicating that pigeons can detect slight magnetic fields and that magnetism can influence their orientation.

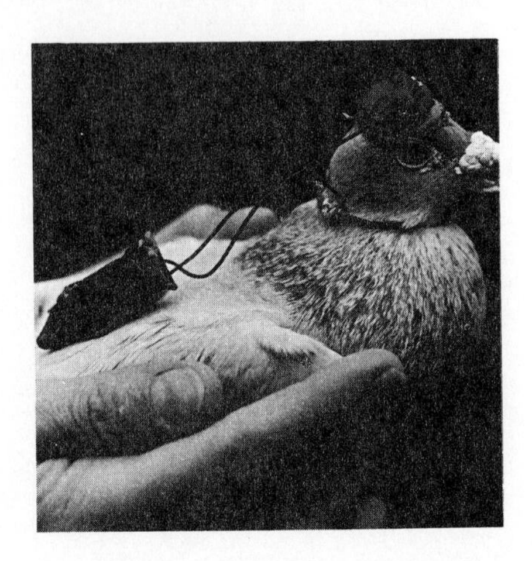

FIGURE 24.17

Jane Goodall, probably the world's greatest authority on chimpanzee behavior, with a friend.

behavior by observing a simple chimpanzee in a cage, because an isolated chimpanzee will behave quite abnormally. Chimpanzees, she noted, are highly social creatures that interact in extremely intensive and complex ways. If you want to know what chimpanzees are like, according to Goodall, then you must watch them when they are with other chimpanzees.

The same statement might be made of any of a number of other creatures, including us. What would a termite be like without other termites? Or a human without other humans? Many of the earth's animals are indeed highly social species, and they interact with each other in subtle and complex ways. Yet, there are some underlying themes. We will look at some of these principles here. In particular, we will consider how animals can aggress against each other, and how they also cooperate. We will also see why we should not attach values to either form of interaction since both are adaptive devices that help the animal survive and reproduce in its own world.

Aggression

The old image of "Nature, red in tooth and claw" has recently been superseded by the popular notion that humans are the only animals that regularly kill members of their own species, while other animals get along with their own kind. Is this true? Let's see what is actually going on out there.

First, we should note that **aggression** is belligerent behavior that normally arises as a result of competition. An animal shows aggression mainly toward other individuals that tend to utilize the same resources. Thus, it is likely to be aggressive toward those most like itself. Those most like itself are, of course, of the same species and sex. And most aggressive

ESSAY 24.2

Biological Clocks

It's the night before a big test. You are trying to study, but your pet hamster, active this night as usual, has a squeaky running wheel. The sound is so distracting that you decide to put your furry friend in the closet. A few nights later, the squeak coming from the closet reminds you that the hamster hasn't seen the sun, or anything else, for three days. Although it had no environmental time cues, such as a light-dark or temperature cycle, the hamster was still active at night, the customary time.

How did the hamster do it? How did it know when it was night? Most researchers assume many living things can measure the passage of time by an internal or biological clock.

Hamsters are not the only organisms with a biological clock. Indeed, the behavior and physiology of most organisms from protists to humans are rhythmic. In fact, rhythms are so common that rhythmicity should be considered a fundamental property of life.

The prevalence of biological rhythms is not surprising when we remember that life evolved in a cyclic environment. Behavior fluctuates in a repeating pattern so that any pattern ideally occurs at the appropriate time of day, in keeping, say, with the state of the tides, phase of the moon, or the season of the year. The rhythmicity on earth reflects the rhythmic movements of certain heavenly bodies, such as the earth, the moon, and the sun. The relative movements of the earth, moon, and sun cause regular changes in such things as light, temperature, geomagnetism, barometric pressure, humidity, and cosmic radiation. Because these environmental changes have been so regular and so predictable, evolution has been able to adjust behavior and physiology to match these cycles.

In nature the fluctuation in behavior is so perfectly tuned to the environmental day/night cycle that you might think that daily changes in light or temperature generate biological rhythms. Most evidence does not support this position, however. If you recorded the activity of your hamster while it was in the perpetual darkness of your closet, you would notice that,

interactions occur within species, between members of the same sex. Predatory behavior, by the way, is not aggressive. A cheetah is about as aggressive toward an antelope as you are toward a hamburger.

Fighting

The most blatant form of aggression is fighting. We can discount the old films of leopards and pythons battling to the death; such fights simply aren't likely to happen. What does a python have that a leopard would risk its life for, or vice versa? Fighting is more likely between competitors.

although bouts of activity regularly alternate with rest, the length of this activity cycle is slightly different than 24 hours. In other words, in constant conditions, daily rhythms are "about a day" in length, or circadian (*circa:* about; *diem:* day). If the hamster were responding to environmental cues, it would stay on a 24-hour cycle.

Although biological rhythms continue without environmental cues, they are not completely independent of such cues. For example, light-dark cycles will set, or entrain, the rhythm so that its period length matches that of the environment. This would be analogous to daily resetting a watch that runs fast.

One might ask, why have a clock control the rhythm instead of having a direct response to the predictable environmental change? There are several answers to this question. For example, some animals must be able to *anticipate* critical changes in their surroundings so that they have adequate time to prepare. A fiddler crab scurrying along the beach must return to its burrow before the tide returns or the waves will wash it away. Other animals use clocks to synchronize their behavior to an event that they cannot sense directly. This is the case for honeybees that travel to distant patches of flowers to gather nectar. Different types of flowers open their petals at different times of the day. The bees' clocks allow them to time their nectar gathering forays so that they arrive when the flowers are open. They might visit morning glories early in the day but want to visit four o'clocks in the afternoon.

Clocks can be used not only to determine the time of day, but how long it has been since some event. For animals such as the birds and the bees, the second function of a clock, measuring the passage of time, is particularly important because it is essential for sun compass orientation (see text). A homing pigeon flying south would be required to keep its path of flight at a 45° angle to the right of the sun at 9 A.M., but would have to change that angle by about 15° an hour as the sun moved across the sky.

What makes the biological clock tick? We don't know. However, because rhythms exist in single cells and protists, we conclude that the clock must be intracellular. Cellular processes that may play a role in the timing process are protein synthesis on the cytoplasmic ribosomes, transport of ions across the plasma membrane, or perhaps proton transport in the mitochondrion.

If a single cell can tell time, does every cell in a multicellular organism have its own "wristwatch"? Apparently so. Isolated tissues often remain rhythmic. For example, it has been found that if the heart of a hamster is removed and kept alive in a tissue culture, it will continue to beat more rapidly at night than during the day. Even a single heart cell will display a daily rhythm.

If there are many clocks in an animal, then they must be set to the same time or there would be internal chaos. Indeed, animals may have master clocks that synchronize the timepieces in individual cells. In mammals such as the rat, the master clock seems to be in a region of the brain called the suprachiasmatic nucleus. When neurosurgeons destroy this tiny group of cells, the rat's running activity, its drinking patterns, and several of its normal hormone rhythms disappear.

Judith Goodenough.

Therefore, interspecific (between species) fighting is much less likely than is intraspecific (within species) fighting. This is because the strongest competitors are likely to be of the same species. No matter who the combatants are, however, fighting is usually a dangerous matter (Figure 24.18).

Animals of different species do sometimes fight, of course. For example, golden-fronted and red-bellied woodpeckers that fail to find enough in common to breed, find enough similarities to exclude each other from their respective territories by fighting (Figure 24.19). And lions may attack and kill African cape dogs at the site of a kill. The lions don't eat the dogs;

FIGURE 24.18

Fighting can be a dangerous activity among some species.

FIGURE 24.19

As we saw in our discussion of evolution (Chapter 10), golden-fronted (a) and red-bellied (b) woodpeckers are remarkably similar. However, they do not interbreed. They obviously have very specific ways of identifying their own species for reproductive purposes, while they also behave as if they recognize each other as strong competitors.

(a)

(b)

they just exclude them. Under certain circumstances, such as at a kill, powerful predators such as bears and cougars may fight (Figure 24.20). In fact, interspecific aggression may be more common than we assume among animals (Figure 24.21).

Fighting is usually a last resort for members of the same species. Generally, the would-be combatants begin by displaying to one another—posturing, making sounds, or giving off other signals that may intimidate the weaker opponent. If one of the adversaries assesses its opponent as stronger, it will signal its submissiveness. In this case, there will be no

FIGURE 24.20

Cougar and black bear fighting. Such animals normally avoid each other. The confrontation may arise over food, such as at a kill.

FIGURE 24.21

In some cases, fighting may occur between species that rarely interact at any level. Here, a raccoon approached a hare in the darkness and in the scuffle, the hare leaped over the raccoon and administered a severe drubbing with its powerful back legs to the raccoon's head. Actually, they may have known each other. There is evidence that animals of different species that share the same area may come to recognize each other individually and may react to each other on an individual basis.

FIGURE 24.22

Male rattlesnakes fighting. The triangular heads show that their poison sacs are full. Each could kill the other, yet they do not bite. The fight is more a test of will and strength as the snakes press against each other, belly to belly. Finally, the weaker individual yields and his head is pushed to the ground by the stronger animal. He then retreats, and no one dies.

battle. However, if the contestants judge themselves to be equals on the basis of these displays, they may come to blows.

The precise methods of fighting vary widely, but however animals fight, most species have means of avoiding injury to each other (Figure 24.22). Such avoidance has several benefits. First, no one is likely to get hurt. Although the opponent is permitted to continue its existence, the possibility of having to compete with him again is less risky than serious fighting.

Fighting between dangerous combatants is usually a stylized ritual and relatively harmless. For example, horned antelope may gore an attacking lion, but when they fight each other, the horns are never directed toward the exposed flank of the opponent (Figure 24.23). Such stylized fighting enables the combatants to establish which is the stronger animal and, once dominance is established, the loser is usually permitted to retreat.

All-out fighting may occur between animals that cannot injure each other seriously, such as hornless female antelope (Figure 24.24). It may also occur between animals that are so fast that the loser can escape before serious injury, as is the case in house cats.

Why don't antelopes gore each other? An incurable romantic might assume that it is because they don't want to hurt each other. In all likelihood, what the two antelope "want" has little to do with it. The fact is, they *can't* hurt each other. When the system works, an antelope could no

FIGURE 24.23

Male impalas fighting. Although the horns of the medium-sized antelope are formidable weapons, neither impala will attack the vulnerable flank of the other. Instead, a harmless pushing contest ensues as the tips of the ridged horns are engaged. These animals effectively employ horns and hoofs against species other than their own, but when confronted with a member of their own species, they are genetically constrained to behave in very circumscribed ways.

FIGURE 24.24

Hornless females of the Nilgai antelope have no inhibitions against attacking the flank of a competitor, but their butts are quite harmless. Keep in mind that "harmless" is used only in the immediate sense. The butt itself may not be dangerous, but if it establishes dominance so that the loser is deprived of commodities, the result can have far-reaching implications. An individual deprived of food, after all, is more likely to fall prey.

more gore an opponent than fly! Perhaps the sight of an opponent's exposed flank inhibits butting behavior. Conversely, the sight of an opponent head-on might release the stereotyped fighting behavior.

There are species, by the way, in which combatants do fight to the death. If a strange rat is placed in a cage with established rats, the group may chase and sniff at the newcomer carefully for a period, but eventually, they will attack it repeatedly until it is dead. Guinea pigs and mice may also fight to the death. The males of a pride of lions are likely to kill any strange male they find within their hunting area, and a pack of hyenas may kill any member of another pack that they can catch (Figure 24.25). Even roving bands of male chimpanzees may attack and kill a male from another band. The chimpanzee findings were quite surprising since it had long been assumed that chimpanzees were essentially peaceful animals. Nonetheless, fights to the death are rare in most species.

FIGURE 24.25
Hyenas may kill any stranger they find on their territory.

COOPERATION

Cooperation may seem to be at the opposite end of the behavioral spectrum from aggression—aggression is not nice, cooperation is nice. But we will see that both aggression and cooperation might be termed "enabling devices." After all, they both function in enabling individual animals to survive and reproduce. Furthermore, just as humans are not unique in their aggressive behavior, neither are they alone in cooperating with each other.

Cooperative behavior occurs both within and between species. As an example of interspecific cooperation, consider the relationship of the rhinoceros and the tickbird that may be found clinging to the rhino's thick hide. The little bird gets free food while the rhinoceros rids itself of ticks and harbors a wary little lookout. The highest levels of cooperation, however, are found among members of the same species.

As an example of intraspecific cooperation, consider the behavior of porpoises. Porpoises are air-breathing mammals, much vaunted in the popular press for their intelligence, and their behavior often seems to support the claim (Figure 24.26). Groups of porpoises will protectively circle a female in the process of giving birth, driving away any predatory sharks that might be attracted by the blood. They have also been known to carry a wounded comrade to the surface where it can breathe. Their behavior in such cases is highly flexible and is influenced by prevailing environmental conditions. Such flexibility indicates that their behavior is not a blind response to innate genetic influences.

Cooperation among mammals is probably most commonly found in their defensive and hunting behavior. For example, the adult musk oxen of the Arctic form a defensive circle around the young at the approach of danger, standing shoulder to shoulder with their massive horns directed outward (Figure 24.27). The defense is effective against all predators except humans, since the beasts try to maintain this stance while they are

FIGURE 24.26

Porpoises are intelligent and often cooperative creatures. Here, two young swim beneath their mothers.

FIGURE 24.27

Musk oxen live above the Arctic circle and are preyed upon by wolves. If attacked, they immediately form a circle with the adults facing outward and the calves inside. The behavior is an effective defense since the oxen are not very large and, individually, could be brought down.

shot one by one. Wolves (Figure 24.28), African cape dogs, jackals, and hyenas often hunt in packs and cooperate in bringing down their prey. In addition, the hunting animals may bring food to those mates or young that were unable to participate in the hunt.

FIGURE 24.28

Wolves often cooperate to bring down prey, but then the dominant individuals feed first. The subordinates, at right, cooperate by waiting their turn.

It might seem that intelligent and purposeful animals, such as mammals, would be expected to show the highest levels of cooperative behavior. It is a bit surprising, therefore, that cooperation is most highly developed in certain lowly insects (Figure 24.29). The behavior is generally considered to be genetically programmed, highly stereotyped, and usually not greatly influenced by learning.

FIGURE 24.29

Leaf cutter ants of the tropics carefully excise sections of leaves, which they then take to the colony where the plant material is used to grow fungi on which the ants feed.

In honey bees (Figure 24.30), the queen lays the eggs and all the other duties are performed by the workers, sterile females. Each worker has a specific job at any given time. For example, newly emerged workers prepare cells in the hive to receive eggs and food. Then, in a day or so, their "brood glands" develop, and they begin to feed larvae. Later they begin to accept nectar from field workers and to pack pollen loads into cells. At about this time, their wax glands develop and they begin to build combs. Some of these "house bees" may become guards, patrolling the area around the hive. Eventually, each bee becomes a field worker, or forager. She flies afield and collects nectar, pollen, or water, according to the needs of the hive. These needs are indicated by the eagerness with which the field bees' different loads are accepted by the house bees.

If a large number of bees with a particular duty are removed from the hive, the normal sequence of duties can be altered. Young bees may shorten or omit certain duties and begin to fill in where they are needed. Other bees may revert to a previous job that is now required again.

The watchword in a beehive is *efficiency*. In the more "feminist" species, the drones (males) exist only as sex objects, reproductive partners. Once the queen has been inseminated, the rest of the drones are quickly killed off by the workers. They are of no further use. The females themselves live only to work. They tend the queen, rear the young, and maintain and defend the hive. When their wings are so torn and tattered that they can no longer fly, they either die or are killed by their sisters. But the hive goes on.

FIGURE 24.30
Honeybees live cooperatively in amazingly complex social systems.

ALTRUISM

We don't wish to shatter anyone, but, well . . . most of the Lassie stories aren't really true (Figure 24.31). Consider what would happen to the genes of any dog that was given to rushing in front of speeding trains to save baby chickens. The reproductive advantages would be considerable

FIGURE 24.31
Lassie may have gone around saving chickens and people, but if she did, she would have expended time and energy that could have been used in having puppies. So where are her genes now?

FIGURE 24.32

Pregnancy may be considered altruistic in that the offspring gain at the mother's expense. However, this is her means of perpetuating her own genes. So is it altruistic?

to chickens, but dogs with those tendencies might be selected out of the population by the action of fast trains. In contrast, the genes of a dog that spent her energy, not in chivalrous deeds, but in seeking out a mate, would be expected to increase in the population. So what kinds of dogs are likely to predominate in the next generation?

If an animal is going to engage in "unselfish" deeds, its best reproductive bet lies in those deeds that advance the genes of other members of its own species. And as we will see, it is here that we are most likely to see selfless behavior.

Altruism may be defined in a biological sense as an activity that benefits another organism at the individual's own expense. It seems to be common among animals, but is it? A more difficult question is, if it is at all common, how did it evolve?

It is easy to see how certain forms of altruism are maintained in any population. For example, pregnancy, in a sense, is altruistic (Figure 24.32). The prospective mother is swollen and slowed. Much of her energy goes to the maintenance of the developing fetus. In labor she is almost completely incapacitated and is in marked danger. Pregnancy is clearly detrimental to her. So why do females so willingly take the risk? It may help to understand the enigma if we remember that the population at any time is composed of the offspring of individuals who have made such a sacrifice. Thus, the females in the population may, to one degree or another, be genetically predisposed to make such a sacrifice themselves.

Kin Selection

Altruism, as we've considered it so far—that is, as a means of increasing one's reproductive output—doesn't explain why an animal would put itself at risk to help the offspring of *another* individual. Why would a bird feed the young of *another* pair, or why would an African hunting dog regurgitate food to almost *any* puppy in the group? Why, also, would a ground squirrel that may have no offspring of its own give a warning cry at the approach of a predator, alerting other ground squirrels at the risk of attracting the predator's attention?

To answer such questions we must look past the answers that first come to mind. It may seem cynical, but we must start with the premise that ground squirrels don't give a hoot about each other. A ground squirrel that issues a warning call isn't thinking, "I must save the others." At least, there is a simpler explanation for its behavior.

Keep in mind that the biologically "successful" individual is the one that maximizes its reproductive output. One way of accomplishing this is for the organism to behave in such a way as to leave as many individuals carrying *its own type of genes* as possible in the next generation. This would explain parental care. However—and this may not be so readily apparent—an individual can also leave its type of genes in the next generation by helping a *relative's* offspring to survive.

Kin selection is the process by which an individual increases its kinds of genes in the population by helping relatives. Remember, relatives have genes in common. For example, an individual shares genes in common with a cousin, although fewer of course, than with a son or a daughter. Hence, there is, theoretically, a point at which an individual could leave more copies of its genes by saving its nieces and nephews (provided there were enough of them) rather than its own offspring. From the standpoint of reproductive output (the number of copies of one's genes that make it into the next generation), the organism would be better off leaving a hundred nephews than one son.

To illustrate, suppose a gene for altruism appears in a population (notice that this sets up the mechanism for the continuance of the behavior, but you can ask yourself if a behavioral tendency could also be passed along culturally in a verbal and social species such as our own). As you can see from Figure 24.33, altruistic behavior would most likely be maintained in groups in which the individuals are related (that is, in which they have some kinds of genes in common). The behavior might be expected, then, in populations in which there is little mixing with outside groups—in other words, where there is a high probability that proximity indicates kinship, or in populations in which individuals have some way of recognizing kin. In many species, by the way, individuals can recognize their kin (Essay 24.3).

Keep in mind that *no conscious decision* on the part of the altruist is necessary. It simply works out that when conditions are right, those individuals that behave altruistically increase their kinds of genes in the population, including the "altruism gene." Nonrelatives would be benefited by the behavior of altruists, of course, but individuals near enough to receive the benefits of an act are likely, to some degree, to be relatives.

It has been determined mathematically that the probability of altruism increasing in a population depends on how closely the altruist and the recipient are related (Figure 24.34). In other words, the advantage to the

FIGURE 24.33

In this population, A and A′ are nonaltruists. They behave in such a way as to maximize their own reproductive success, but do nothing to benefit the offspring of other individuals. In another segment of the population (B and B′), a gene for altruism has appeared that results in individuals benefiting the offspring of others in some way. It can be seen that, assuming the altruistic behavior is only minimally disadvantageous to the altruist, generations springing from B and B′ are likely to increase in the population over those from A and A′. It should be apparent that the altruistic behavior should be more common where B and B′ are most strongly related, so that B shares a maximum number of genes in common with the offspring of B′ and vice versa. The idea is that B, for example, can increase its own reproductive success by caring for the offspring of a relative with whom it has some genes in common. After all, reproduction is simply a way of continuing one's own kind of genes.

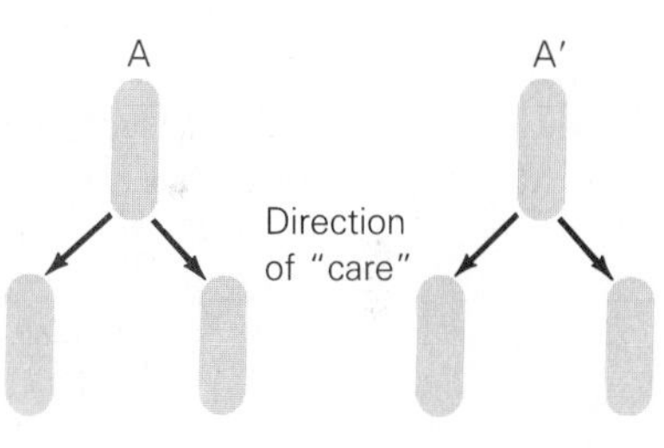

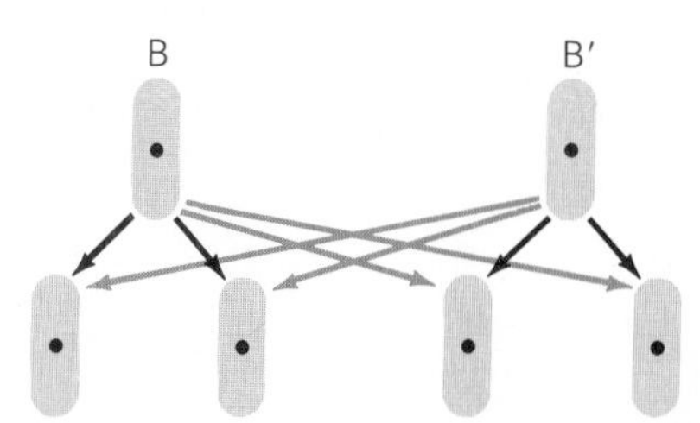

FIGURE 24.34

A male hamadryas baboon will sometimes threaten a predator, thereby placing himself at risk for the sake of the group.

ESSAY 24.3

How to Recognize Kin

A cartoon once depicted an old gentleman sauntering out the door after a grand Sunday dinner with the two startled hosts exclaiming, "*My* uncle! I thought he was *your* uncle!" The implication was that he would have been treated differently if they had known he was unrelated. And so it is in the animal world. In many species, animals behave differently toward kin.

Andrew Blaustein and his coworkers have pointed out that any such recognition is likely to be based on familiarity. In other words, an animal comes to recognize those individuals it grew up with and to treat them as if they were related. If animals disperse soon after birth or hatching, though, there may be no opportunity for them to become familiar with each other and so other means of identification must be employed. One such means is through maternal "labelling." Individuals could acquire, for example, a specific scent while in the uterus or in the same clutch of eggs. Later they would be programmed to treat individuals with that label as if they were relatives. A third way is perhaps the most remarkable of all. It is theoretically done through "genetic markers." In this case, an individual carries a gene that enables it to recognize others with the same gene.

What does it mean to treat others as if they were relatives? For one thing, it means you don't eat them. Cannibalism is startlingly common among many kinds of animals, and so an individual needs to be able to recognize those bearing its same kinds of genes. A second advantage of recognizing kin is to avoid inbreeding (see Chapter 7). Thirdly, animals may warn or protect those to whom they are related when danger threatens, and they may help them in conflicts or in acquiring commodities, as we're discussing in this chapter.

recipient must increase as the kinship becomes more remote, or the behavior will disappear from the population. For instance, altruism toward siblings (brothers and sisters) that results in the death of the altruist will be selected for if the net genetic gain is *more* than twice the loss; for half-siblings, four times the loss; and so on. To put it another way, an altruistic animal would gain reproductively if it sacrificed its life for more than two siblings, but not for fewer, or for more than four half-siblings, but not for fewer, and so on. Therefore, we can deduce that in highly related groups, such as a small troop of baboons, a male might fight a leopard to the death in defense of the troop. By the same token, a ground squirrel will give a warning cry when the chance of attracting a predator to itself is not too great and because the average neighbor is not too distantly related.

Kin selection helps to explain the extreme altruism shown by social insects, such as honey bees. In some species, the queen is inseminated only once, so all the workers in the hive are sisters. Because of the system of sex determination, honeybee sisters are likely to have more genes in common than you would with your sister. Furthermore, the workers share more genes with their sisters than they would with their own offspring! Why? Male honeybees are haploid and females are diploid. Therefore, sisters share three-fourths of their genes, not just half as they would with their offspring. (Can you explain the difference?) In such a system, then, the workers would leave more copies of their genes by helping to raise sisters than by reproducing themselves. So almost any sacrifice is worth any net gain to the hive and to the queen.

The development of altruistic behavior is an admittedly somewhat esoteric topic. However, it seems important to consider even our most cherished and most despised behaviors in the context of their evolutionary history.

Reciprocal Altruism

In a brilliant essay, Robert Trivers expanded our understanding of the evolution of altruism by suggesting that in certain species, notably humans, altruism (outside of parental behavior) depends on the expectation of reciprocation: "I'll scratch your back if you'll scratch mine." **Reciprocal altruism** is selfless behavior that is extended when it is likely that the favor will be returned. It is an evolutionary strategy with some complex rules. Help (altruistic acts) is given to others—even offered to strangers—with the expectation that some kind of help will be reciprocated at another time.

Reciprocal altruism would be expected in those groups that are highly social (so the altruist is likely to encounter the beneficiary again) and relatively intelligent (so that individuals are recognized and their behavior can be remembered and repaid). Some evidence of intraspecific reciprocal altruism has been reported in troops of social mammals, such as hunting dogs and higher primates, but the evidence is not very strong for such behavior in any animals but humans (Figure 24.35). Trivers implies, in

FIGURE 24.35

In a formalized reciprocal altruism, society at large pays a small cost in order to offer help. Here, relief is distributed to hurricane victims at public expense. Implicit is the understanding that each member of society can expect such public assistance in a crisis.

fact, that reciprocal altruism is the key to human evolution. The complexity of such behavior, entailing as it does memory of past actions, the calculation of risk, the foreseeing of the probable consequences of present actions, the possibility of advantageous cheating, and the need to be able to detect such cheating, all require a level of intelligence that is beyond most species. In the opinion of some anthropologists, it is exactly for the management of these elaborate social interactions that the human brain—and the conscious human mind—evolved.

SOCIOBIOLOGY AND SOCIETY

In the mid-1970s, an idea that had been around for decades was revived, reviewed, reanalyzed, refined, reconsidered, and brought to public attention by the noted biologist, Edward O. Wilson of Harvard (Figure 24.36). Because it was a generally familiar concept and was stated very carefully in a professional format, the scientific community was surprised by the turmoil that followed. The turmoil has not yet subsided.

Sociobiology is the study of the effects of natural selection on social behavior. The idea is quite a simple one and has been around long enough that no one is really shocked by its revival; the whole idea might have generated little discussion except for the reaction of a group of people who believe that the concept is based on a politically dangerous premise. They immediately set about to attack it, and their attack was so vigorous and well publicized that it drew a great deal of attention to an idea they didn't want to receive attention.

The defenders of the idea quickly rallied, and the arguments grew vitriolic as both sides firmed their position and stiffened their resolutions.

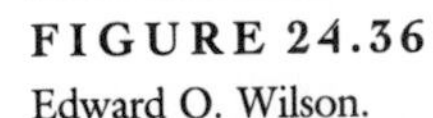

FIGURE 24.36
Edward O. Wilson.

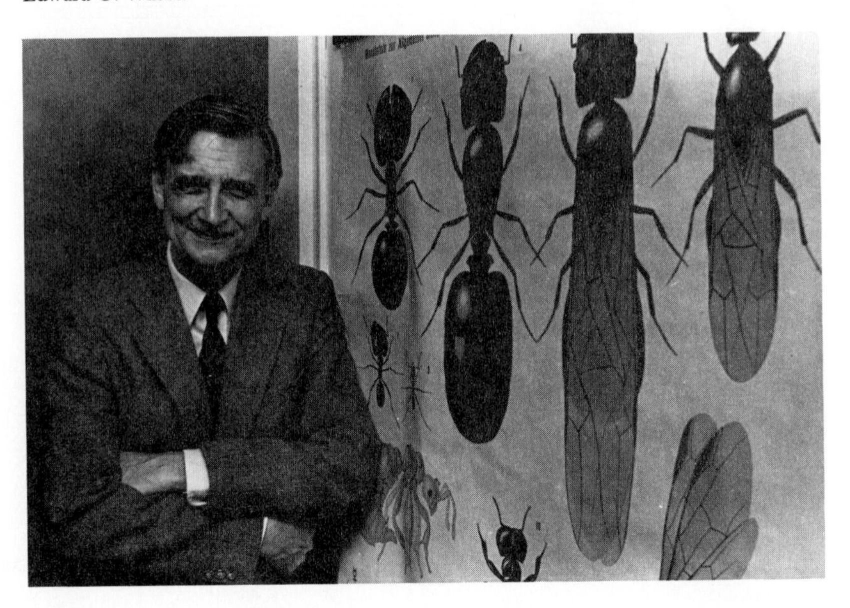

It was an emotional argument, and some people apparently entered the fray with more opinions than information, and rather strange things were said. Furthermore, semantics became a large part of the problem because of sloganeering and loose definitions and because there was so little dialogue between the two groups.

So what was the problem? The opponents of the idea feared that a sociobiological explanation of human behavior would lead to a revival of the notion of **biological determination.** This is the idea that we are primarily the product of our evolution and that we therefore respond to the primitive calling of our heritage. As a result, our social behavior cannot be changed in any fundamental way and it is a waste of time to try. The opponents felt that the acceptance of sociobiology would mean resignation to the status quo. Furthermore, they said it supports such undesirable patterns as racism and sexism. They preferred to believe that our social patterns are molded by culture and learning and that we can change any undesirable trait through education, incentive, and social programs (Figure 24.37).

Sociobiologists strongly deny supporting biological determination and certainly denounce racism and sexism, but they believe that their approach holds real promise for building a better society. Perhaps it does seem a bit irreverent to some to suggest that human behavior, to any great degree, is programmed through evolution and that the forces of natural selection mold our own species as well as others. But the sociobiologists

FIGURE 24.37

Social programs such as this, where inmates at the State School in Giddings, Texas are being trained to be productive citizens, are offered in order to try to change the fabric of society.

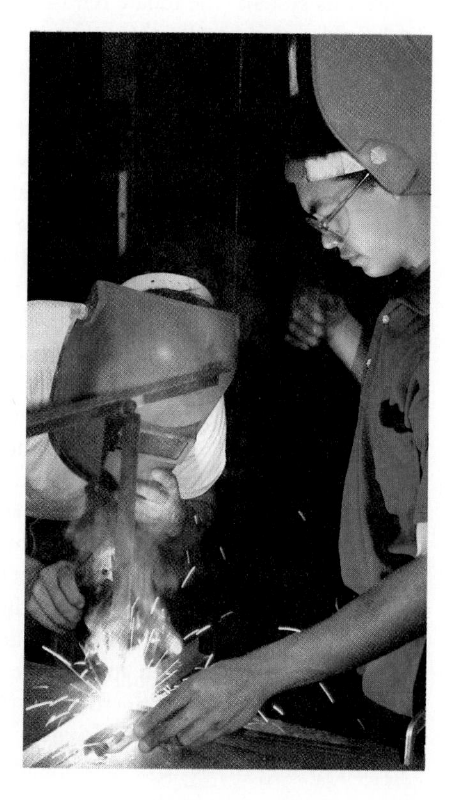

argue that if our behavior is, to *any* degree, genetically controlled or influenced, then we should know it. They note that we can't hope to find solutions if we have ignored the role of our biological heritage. We must, they say, try to understand our social problems at every level and to use every tool we have to improve our lot. (We didn't develop the ability to fly by simply denying the existence of gravity.)

Sociobiology, in its refurbished form, is quickly maturing as data are now appearing from long-term studies and as new researchers approach the problem from different angles. The next few years should be interesting as sociobiologists tighten their premises, precisely define their terms, and present us with new approaches to a weathered idea.

SUMMARY

1. The theory of instinct was formulated by Konrad Lorenz and amended by Niko Tinbergen. Ethology, a school of animal behavior founded in Europe during the 1930s, attempted to understand animal behavior by studying its cause, development, evolution, and function. At the same time, American comparative psychologists were in their labs studying learning processes in animals, typically the Norway rat.
2. An instinct is a species specific, genetically based behavior called a fixed action pattern that is performed when the perception of something in the environment (a releaser) triggers a response in a center in the central nervous system (the innate releasing mechanism).
3. Animals are generally able to learn the kinds of things that are relevant to their lifestyles. There are often species differences in learning ability. Three of many possible different types of learning are considered. Habituation is learning not to respond to a frequently encountered stimulus. It is not permanent. Habituation may save energy by eliminating responses to common harmless stimuli while allowing responses to rare stimuli. In classical conditioning, a response to a normal stimulus comes to be elicited by a substitute stimulus when the two occur together many times. The conditioned response can be lost if the normal stimulus, which reinforces the response, repeatedly fails to follow the conditioned stimulus. In operant conditioning, the animal is rewarded (reinforced) after performing some behavior. The frequency of the behavior increases when it is repeatedly rewarded. Adaptive behavior patterns in nature arise from interactions of several types of learning.
4. Innate and learned patterns interact during the development of a behavior, producing an adaptive response. Innate behavior patterns can be improved by learning. Some species learn certain behaviors only if they are exposed to the relevant stimulus during a brief critical period. This type of learning, called imprinting, is important in the development of the following response of young chicks, ducklings, and goslings, and in some species, in learning the image of an appropriate mate or learning to sing the species' song. For example, white-crowned sparrows must hear their species' song during the sensitive period in order to sing it properly, but they cannot learn to sing the song of another species.

5. Orientation is simply facing in the right direction. Navigation is finding one's way from point A to point B. Migrating birds such as starlings orient themselves by using the sun as a compass. They use a biological clock to compensate for the sun's movement across the sky. Night migrating birds orient themselves by using the stars as a guide. Other cues, which may change over time, may also be involved. One cue may serve as a backup for another.
6. Aggression is belligerent behavior that normally arises from competition. It is most commonly shown toward members of the same species and sex. Members of many species avoid fights by threatening one another through displays. When the weaker individual assumes a submissive posture, no fight will occur. Fights between dangerous animals are usually stylized rituals that establish which combatant is dominant.
7. Cooperative behavior occurs both within (intraspecific) and between (interspecific) species. Cooperation among mammals of the same species often occurs in protective, defensive, and hunting behaviors. Cooperation is most highly developed among insects such as the honeybee.
8. Altruism is an activity that benefits another organism at the individual's own expense. It occurs most commonly among members of the same species. Some forms of altruism, such as pregnancy, increase one's own reproductive output. Other altruistic deeds increase the reproductive output of other individuals.
9. Kin selection, the process by which an individual increases its types of genes in a population by helping relatives, may explain the continuance of certain altruistic behaviors. This would be most likely to occur in populations in which there is a high probability that proximity indicates kinship or in which individuals can recognize kin.
10. Reciprocal altruism is an evolutionary strategy in which help is given to another with the expectation that the favor will be returned. Reciprocal altruism would be expected in highly social groups of intelligent individuals, such as humans.
11. Sociobiology is the study of the effects of natural selection on social behavior. Opponents to the idea feared that it would revive the notion of biological determinism. Proponents argue that our only hope of finding solutions to our social problems lies in understanding our social behavior at every level.

KEY TERMS

aggression (673)
altruism (684)
biological determination (689)
classical conditioning (665)
comparative psychology (659)
critical period (669)
ethology (659)
extinction (666)
fixed action pattern (661)
habituation (663)
imprinting (669)
innate releasing mechanism (IRM) (661)
kin selection (684)
navigation (670)
operant conditioning (666)
orientation (670)
reciprocal altruism (687)
releaser (661)
Skinner box (666)
sociobiology (688)

FOR FURTHER THOUGHT

1. Certain squirrels will begin to disregard danger calls from other squirrels if they repeatedly prove to be false alarms. What behavioral mechanism is at work here?
2. Certain wasps are solitary individuals that come together only to mate. Would you expect members of this species to exhibit altruistic behavior? Explain.

FOR REVIEW

True or false?

1. ___ In operant conditioning, a reward (reinforcement) follows a behavioral response.
2. ___ Navigation is an innate behavior.
3. ___ A fixed action pattern is learned.
4. ___ Harmless stylized fighting rituals are not considered to be acts of aggression.
5. ___ The cooperative behavior of bees in a hive is an example of reciprocal altruism.
6. ___ Altruistic acts are most often performed for members of the same species.

Fill in the blank.

7. ___ is the study of animal behavior that was founded by Lorenz and Tinbergen.
8. ___ are environmental signals that trigger fixed action patterns.
9. ___ cooperation is a mutually helpful relationship between two different species.
10. In ___, an individual increases its own kinds of genes in a population by helping relatives.
11. Biological determinism is ___ .

Choose the best answer.

12. Which of the following is associated with instinctive behavior?
    A. fixed action patterns
    B. releasers
    C. innate releasing mechanism
    D. all of the above
13. In classical conditioning experiments
    A. Skinner boxes are used.
    B. a reward (reinforcement) follows a behavioral response.
    C. food is never used as a reward.
    D. animals are conditioned to respond to a substitute stimulus.
14. Altruistic behavior
    A. benefits only nonrelatives.
    B. is not characteristic of social animals.
    C. may endanger the altruistic individual.
    D. decreases survival chances of a group.

15. In ____ an individual exhibits selfless behavior in hopes of obtaining future favors.
    A. reciprocal altruism
    B. kin selection
    C. altruism
    D. interspecific cooperation
16. Acts of aggression are usually directed at
    A. members of a different species.
    B. animals that utilize the same resources.
    C. individuals of the same species and sex.
    D. both B and C.
17. Choose the incorrect statement.
    A. Caged birds can never orient properly.
    B. When a bird is using the sun to orient, its biological clock compensates for the movement of the sun.
    C. Some birds navigate by the stars.
    D. Birds orient more accurately on overcast days.

CHAPTER 25

Biomes and Communities

Overview

Ecosystems and Communities
Habitat and Niche
Succession
The Web of Life
The Land Environment
The Water Environment
Coastal Areas

Objectives

After reading this chapter you should be able to:

- List the components of an ecosystem.
- Distinguish between a habitat and a niche and describe ways in which different species coexist within a habitat.
- Describe how communities change through primary and secondary succession.
- List the types of consumers within a food chain and explain why their feeding patterns are best described as food webs.
- Explain ways in which species extinction may affect the balance of life at different trophic levels.
- Describe the characteristics of six major types of terrestrial biomes.
- Describe freshwater, marine, and coastal environments and their communities.

If museums weren't generally large, modern artists like Jackson Pollock would be out of luck. A Pollock canvas viewed from close range is a confused, chaotic collection of drips, blobs, and smears. When the viewer can stand back so that the canvas is viewed as a whole, however, the various parts begin to fit and the organization appears.

So it is with life on this planet. Almost any aspect viewed up close seems to stand alone, a part of no grand scheme. But when one steps back, allowing the various parts to play together on the mind, the wholeness and completeness of it all becomes more apparent. The parts begin to be seen as one grand panorama.

Before the large picture could be seen, however, the smaller parts had to come into focus. Specialists have provided us with these narrow views, and those generalists called ecologists have begun to put them into place to form the larger picture. (**Ecologists** are those biologists who are interested in **ecology**, the study of the interrelationship of the organism and the environment.) So let's take a look at the grand scheme of life and see how its components interact in the formation of it all.

ECOSYSTEMS AND COMMUNITIES

We should begin our discussion of the interaction of life with a few basic definitions. First, an **ecosystem** is a group of interacting living things, together with all the environmental factors with which they interact. These "environmental factors" can be living, like predators, or nonliving, like rocks. An ecosystem is considered an independent unit (although its independence would be difficult to prove), and light is usually its only energy source. Ecosystem is a handy term that can be applied at just about any level, from the earth itself to a tiny pool of microorganisms.

Ecosystems can usually be broken down into smaller units called communities. A **community** is an assemblage of interacting living things that forms an identifiable group. So, ecosystems include both living and nonliving factors; communities include only living factors. For example, there are sage desert communities and there are hardwood forest communities, each with its own specific interacting organisms.

Communities, in turn, are composed of **populations,** which are interbreeding groups of the same species. Within a population, individuals (Figure 25.1) interact with the environment in a certain, theoretically circumscribed way. The sum of those interactions, in essence, is the niche. So there you are. Now let's examine the interactions that describe an organism's niche.

HABITAT AND NICHE

It has been said that if the habitat is an organism's address, the niche is its profession. That's nice, but vague. Let's see if we can be a little more precise. The **habitat** is the place where an organism is found, and it can be described in several ways. For example, an animal may live in a desert habitat, or more specifically, in a briny desert pool. Furthermore, it may live in a certain part of that pool—its *microhabitat*. Wherever an organism

FIGURE 25.1

The makeup of ecosystems. Individuals comprise populations, which make up communities. Communities, physical factors of the environment, make up ecosystems.

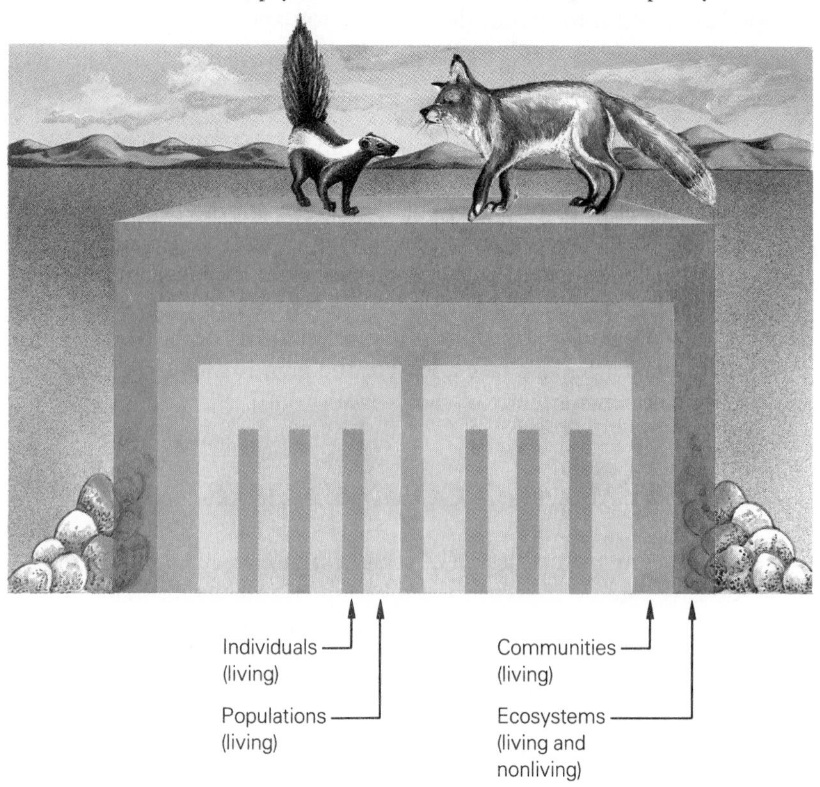

lives, however, it interacts with its surroundings in myriad ways. The interaction involves the environment influencing the organism, which, in turn, influences the physical environment. The organism may also interact with the living things around it. The sum of all the interactions between the organism and the living and nonliving components of its environment, along with the organism's own requirements describe an organism's **niche.** Since the niche involves every aspect of an organism's ecological, physiological, and behavioral interactions with the environment, it is often simply described as an organism's role in nature. As you might imagine if you are beginning to gain some insight into the "scientific mind," there is some disagreement over precisely how these terms should be further defined, but let's not become involved.

It has become axiomatic that two species cannot occupy the same niche indefinitely. If two species were to find themselves in such a situation, it is generally predicted that one would be superior to the other under these conditions and would eventually replace the less fortunate species. Where species do coexist, then, we can assume that they are interacting with the environment in different ways (occupying different niches). This means that when you walk through the woods and see various species of small seed-eating birds in the same area, they are utilizing the habitat differently. This principle was nicely demonstrated by the ecologist Robert H. MacArthur, who showed that five species of American warblers that feed in spruce forests and seem to be occupying the same niche are actually in different niches because each species feeds in different parts of the tree (Figure 25.2). Interestingly, the feeding zones overlap more in times of food abundance. (Why do you suppose this is?)

Animals may also divide up a habitat in other ways. For example, they may utilize the resources at different times of the day or at different times of the year, or they may utilize different commodities within the same part of the habitat. Table 25.1 describes various aspects of a niche.

Partitioning a habitat obviously reduces the level of competition between the species. Generally, it is thought **competition** occurs where organisms utilize resources that are in short supply, or when they harm each other while seeking the same resource that is not in short supply. (Probably the latter cases are much more rare in nature.)

There is currently a vigorous disagreement among biologists regarding the evolutionary role of competition in nature. Part of the problem is that many of our ideas are based largely on intuition or supported by rather tenuous evidence. However, never deterred, we will assume that competition is important and plunge ahead.

TABLE 25.1
Selected Examples of Various Aspects of Niche*

	PLANTS	ANIMALS
BIOLOGICAL FACTORS		
Nature and availability of food sources	Day length; amount of available light; length of growing season	Food source; location of food; time of day for feeding
Reproduction	Reproductive season; method of seed dispersal	Reproductive season; time of day of mating; method of attracting mate
Growth	Rate and requirements	Rate and requirements
Behavior	Interactions with other species	Activity cycles; patterns of movement; interactions with other species
PHYSICAL FACTORS		
Temperature: tolerable and necessary extremes		
Moisture: tolerable and necessary amounts		
Salinity: tolerable and necessary extremes		
Mineral requirements: necessary minimum		
pH: optimum and tolerable extremes		
Oxygen availability: optimum and tolerable extremes		

*These are some of the things that separate one species from another.

FIGURE 25.2
The feeding zones of five species of North American warblers in spruce trees. The darker areas indicate where each species spends at least half its feeding time. By exploiting different parts of the tree, the species reduce their competition for food and thus they can occupy the same habitat. Studies such as this have been done on many species.

FIGURE 25.3
Mt. St. Helens immediately after its eruption (a) and nine months later (b). Many ecologists were surprised at how rapidly the devastated area began its biological recovery.

(a)

(b)

SUCCESSION

If you were able to watch specific places over the earth for a long period of time you might notice that in some places communities of life changed from one time to the next while in other places they didn't change much at all. Those places that change may be involved in **succession,** the orderly sequence of species, structure, and energy flow in a specific area over time. Those areas that are not changing may have gone through successional stages and have now reached a **climax stage,** where the species, structure, and energy flow will remain relatively stable as long as the environment does not change markedly.

There are two fundamental kinds of succession, primary succession and secondary succession. **Primary succession** involves the early progression of life forms where no community previously existed. (Put another way, primary succession starts from scratch.) So we find primary succession after the introduction of life on places such as rocky outcroppings, newly formed deltas (formed as soil is deposited by rivers), sand dunes, volcanic islands, and lava flows. Among the most dramatic recent examples of primary succession was the recovery of the devastated slopes of Mount St. Helens, a volcano that took the world by surprise when it erupted in 1984. The eruption was dramatic and startling, and so has been the recovery of life on its slopes (Figure 25.3).

Primary succession usually begins when small, hardy, drought-resistant species called **pioneer organisms** invade the lifeless area. On rocky outcroppings, for example, lichens may be the first to take hold, held fast by their tenacious, water-seeking fungal component, while the algal component, with its chloroplasts, provides food (Figure 25.4). The lichens pry into every tiny crevice, enlarging the cracks and enabling debris to accumulate. When the lichen meets its own fate, its decomposing body

FIGURE 25.4

An example of primary succession, as lichens grow on bare rock.

adds to the nutrients of the accumulating soil. Soon, then, the rock will play host to grasses and mosses as they follow the pioneers onto the stage in this emerging pageant of life.

As plant roots probe the rocky crevices they exert a remarkably powerful pressure in the tiny cracks, widening them as tiny parts of the rock break away. As insect and decomposer populations begin to appear with the onset of the plants, the lichens that made it all possible begin to give way to the interlopers that, themselves, will one day lose the competitive battle to yet another species (Figure 25.5).

Secondary succession involves the sequence of communities as life recovers in disturbed areas. Here, then, life once existed and soil is in place, so the rate of community change can be much more rapid than in

FIGURE 25.5

As lichens die and decompose, they set the stage for the appearance of yet other species.

FIGURE 25.6

An example of secondary succession. Here, fast-growing species begin to take over a cleared forest.

primary succession. The classic example of secondary succession is sometimes called "old field" succession when a farm or a cleared lot is abandoned, weedy grasses may be the first to take over. The first tree species to appear on the heels of weeds and grasses are often fast-growing softwoods, such as pine. As they grow and their branches begin to shade the soil, shade-tolerant weeds replace the initial invaders. Following these may be slower growing and stronger hardwoods, which as a climax stage is reached, may be interspersed with the softwoods, or may replace them entirely. Eventually, the lines between the surrounding community and the area in succession begin to fade. (Figure 25.6 shows secondary succession in a cleared forest.) Secondary succession can be relatively brief in some cases and remarkably extended in other cases. In grasslands, for example, a climax community can appear on cleared land after only 20 to 40 years. In the fragile Arctic tundra, though, recovery can take centuries.

THE WEB OF LIFE

Life on this planet is tied together in an intricate web in which energy is exchanged between those who eat and those who are eaten. The sun's energy is captured by the **producers,** commonly, photosynthetic organisms. From there energy passes to various consumers, organisms that must rely on others for the organic compounds they require.

Since some consumers eat producers, others eat other consumers, and some eat both, the flow of energy through the consumers involves several trophic (food) levels. Thus we have **primary consumers,** the herbivores that feed directly on producers; **secondary consumers,** carnivores that feed on primary consumers (Figure 25.7); and so on through **tertiary, quaternary,** and even higher consumer levels (with increasingly rare representatives). This general sequence of who eats whom is called a **food chain.**

A moment's reflection will suggest to you that most consumers cross trophic levels. For example, there are few true carnivores (certain sharks and some flies are examples). And then consider humans. At how many trophic levels do we feed? Do we have a green salad with our sirloin steak? When we eat the salad, we are herbivores (primary consumers). The steak comes from a herbivore, of course. So, when we eat the steak, we are carnivores (secondary consumers). What if we have a tuna salad? Tuna eat other carnivores. This would make us a tertiary consumer. Thus the same organism can eat from very high and low on the food chain. Because of such complexities, feeding patterns in a community are better represented by a **food web,** a less linear but more accurate concept.

FIGURE 25.7

A carnivore is a secondary consumer that feeds on primary consumers such as these grazers.

FIGURE 25.8

The animals of a hypothetical food web. Can you tell which organisms serve as food sources for the various species? This picture is greatly simplified—for example, there are also omnivorous birds such as crows, and mice may sometimes eat bird food. The predatory fox may also have to share its food with lynx and wolves.

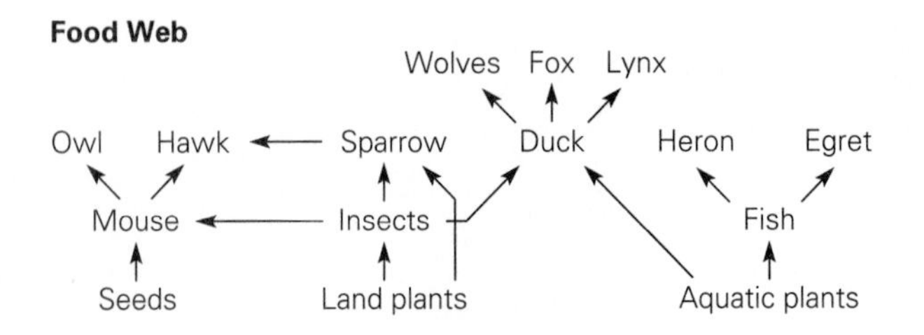

Figure 25.8 shows a hypothetical and greatly simplified food web. Notice that simpler food chains can be identified within the food web. Consider the diagrams that illustrate some of the feeding relationships depicted in the drawing. Can you add additional links to either of the diagrams?

When we consider trophic relationships as complex and weblike, we can see that decomposers play a great role in providing nutrients for a variety of foragers. **Decomposers** are great in number but small in size. This becomes apparent when we realize that they are primarily composed of fungi and bacteria. Decomposers generally feed by secreting digestive enzymes into their food and then absorbing the breakdown products. They are somewhat unusual in that, unlike animals, they readily digest the cellulose-laden corpses of plants and the nitrogenous wastes of animals. In so doing, they release sulfates, nitrites, nitrates, and other mineral ions. Without decomposers the world would be a far different place—a corpse-strewn, mineral-deficient wasteland (Figure 25.9).

It should be clear that the processes of life are indeed complex, interactive, and responsive. Because life is such an interactive process, it is important to keep in mind that if one part changes, the effect can ripple through other parts, sometimes in unexpected ways. Table 25.2 describes relationships within the food web.

FIGURE 25.9

The material that once made up this hyena is being recycled by decomposers.

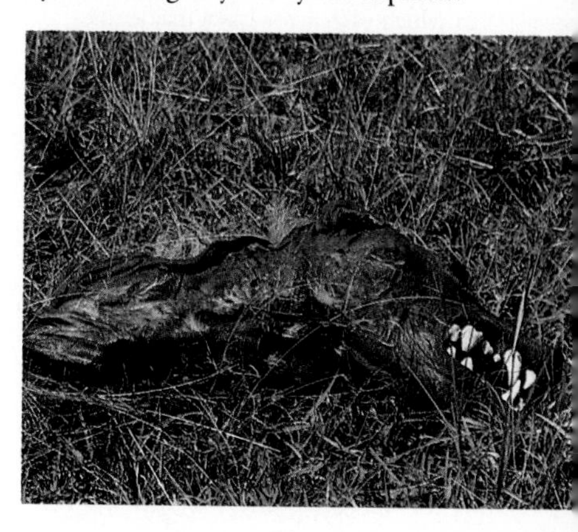

Extinction and the Web of Life

Let's consider an example of how interaction among living things can have far-reaching effects in maintaining the balance of life on the planet. We can begin with the premise that ecologically, simplicity means instability, and that in an unstable ecosystem, extinction rates can be expected to rise. Again we must deal in speculation to some degree because much of the essential data just isn't in.

Refer again to the food web in Figure 25.8. Suppose the only carnivorous (meat-eating) mammal in this food web were the fox. If some

TABLE 25.2
Relationships Within Food Web

LEVEL	ENERGY SOURCE	EXAMPLES
Producers		
Photosynthesizers	Sunlight	Algae; plants
Chemoautotrophs	Oxidation of inorganic substances	Nitrifying bacteria
Primary Consumers (Herbivores)	Primary producers	Caterpillars, grasshoppers, deer
Secondary Consumers[1]	Herbivores	Spiders, fox, lynx, wolf
Tertiary Consumers	Carnivores	Emperor penguin
Decomposers[2]	Organic material	Bacteria

[1] Secondary consumers are carnivores that feed on herbivores, which, in turn, are fed on by tertiary consumers.

[2] Omnivores eat both animals and plants.

fox disease were to sweep the area, what would be the repercussions on the rest of the life there? We can see how the numbers of herbivores, such as rabbits, squirrels, mice, and some birds, might rise. These in turn might put new pressures on their plant food and, in so doing, destroy the habitat of many other species while depleting their own food supply. The resulting initial increase in the numbers of herbivores would provide more food for owls and hawks, which could in turn be expected to increase their own numbers. The resulting abundance of such predators might then reduce the numbers of other animals, such as insect-eating birds and toads. What might then happen to the numbers of snakes? And the numbers of insects?

Now then, suppose the foxes had to share their food with lynx and wolves, as well as with hawks and owls. In this case, a rampant fox disease would have markedly less effect on the ecosystem. Of course, elimination of the foxes would result in some changes in the system, since the fox has a somewhat different niche than the wolf or the hawk. But perhaps the system would continue largely as before until the fox population was restored. The point here is that a simpler system—for example, one with a single "top" predator—is inherently easier to upset.

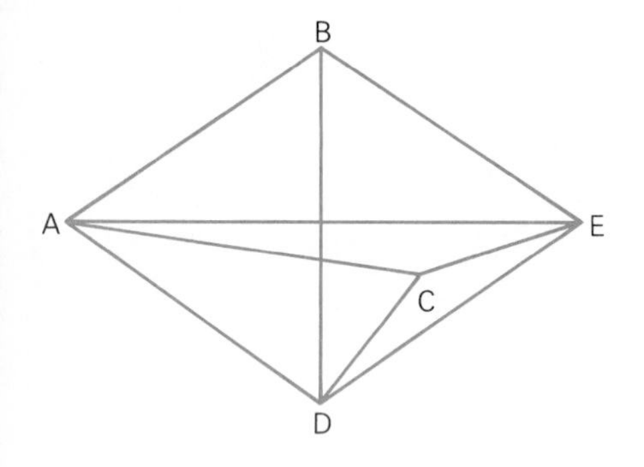

FIGURE 25.10

Interdependency of five theoretical species. This is a relatively simple system in which each actor has a major role. Because of their interdependency, the elimination of any could threaten the existence of all.

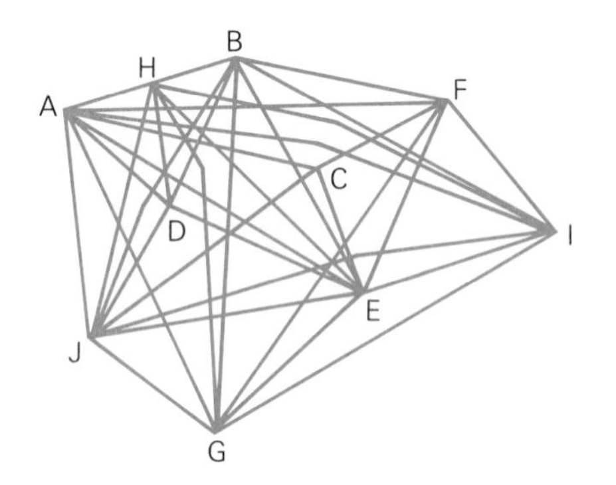

FIGURE 25.11

Interdependency of ten theoretical species. The system becomes vastly more complex by simply doubling the number of species. In such a case, the elimination of any one may not be crucial, since the complexity can act as a buffer. The point is, biological systems may become more unstable by becoming simplified and one way to simplify them is to reduce the numbers of their component species.

Let's draw on a more abstract example to show the effect of the extinction of a single species in an interdependent system. Figure 25.10 shows five species that depend on each other. Of course, any such dependency may be a variety of types and levels. For example, fleas depend on dogs for living quarters and transportation, as well as for food. Starlings depend on old woodpecker holes for nests. A dog may put the tree to a different, more casual use, and the hawks that see starlings and woodpeckers as food have absolutely no use for dogs or their fleas.

Now, suppose that in this group of species that depend on each other in a variety of ways, species C somehow becomes extinct. Since A, E, and D depend on C, we might reasonably expect their numbers to be reduced. B, however, depends on A, E, and D. Even a slight reduction in the numbers of all three might have such an impact on B that it might follow C into oblivion. This would mean that the animals on which A, E, and D were dependent would be reduced by *half*. How might this affect their fate? If A is more susceptible to the loss than E or D, its numbers might then drop quickly. What effect might this have on the two remaining species?

Now, suppose that instead of a five-species system, we have a ten-species system, as in Figure 25.11. Here the system is much more complex, although it is still far simpler than any actual biotic system would be. It is easy to see, however, that the loss of any single component might be more easily absorbed in this system than in the simpler one. Thus, the more complex a system is, the more inherent stability it has.

Extinction and Us

Are there any lessons here for us? Should we care if thousands of species of animals have disappeared from the earth in the last 50 years? Should we care that some scientists calculate that until 2010, one species an hour will meet extinction? Should we be concerned that there are literally thousands of species whose very existence is presently endangered—largely

FIGURE 25.12

Extinction is unlike love, they say, in that extinction is forever.

because of our activities? Extinction, after all, is the natural end of populations. Species are born, they mature, and they die. Some live a long time, perhaps millions of years; some die more quickly. We have hastened the extinction of many species we know about (Figure 25.12) and we have undoubtedly sealed the fate of others. In fact there are undoubtedly many unobtrusive species that have lived among us during our time on earth, but that have disappeared as a result of our activities without our ever having known they existed.

ESSAY 25.1

Subdue the Earth

*I*ndian farmers are incessantly waging war on rats. It has been calculated that the progeny of one pair of rats, in a year's time, can consume enough grain to feed five people. The killing is thereby justified. But who has calculated that five people can, in a year's time, consume enough grain to feed the entire year's progeny of a pair of rats? The question is patently ridiculous. But it hasn't been calculated only because rats can't calculate. Perhaps part of our problem lies here. We see ourselves as more "worthwhile" than other animals. Obviously, our worthiness is subjectively derived since it is we who are doing the measuring and it is to our own best interest to place a higher premium on ourselves and, to carry it further, on those most like us. The mechanism is self-protecting and perhaps justifiable. But can it lead to problems? Will it tend to make us, in very subtle ways, believe that we stand apart from the other species? Can such a belief, expressed or not, cause us to behave as if we are not subject to natural laws? That we are *better,* somehow *different?*

And by the same token, do we *lose* something by elevating ourselves? Do we feel uncomfortable being apart? Why is a deer's footprint in the snow "beautiful," but a man's bootprint "disfiguring"? Why is a beaver dam natural, but a human dam unnatural? After all, we, too, evolved here. We are natural. We are a part of it all, and perhaps, with a greater understanding of our role, we may someday more comfortably take our place in the family of life.

It is hard to explain the rationale of many of us who are concerned about such matters. I have never seen a sei whale, yet I don't want them to become extinct. Moreover, I felt this way long before I understood anything about how they might be an important part of an ecosystem. Possibly such feelings merely reflect the cultural attitude that it is "nice" to wish other living things well; thus, the attitude is rewarded. I feel nice.

There are, of course, more rational reasons for mourning the extermination of any species. For one thing, the kind of attitude that encourages or sanctions the destruction of other species (Essay 25.1) constitutes an intrinsic threat to our own well-being. If such an attitude exists, we ourselves might fall victim to it. Living things (including us) might be expected to fare better where there is reverence for life. The extinction of other species could also threaten us indirectly by simplifying the system of which we are a part or by destroying parts of the ecosystem upon which we directly rely. For example, if we continue to poison the oceans because we are willing to believe only a few bottom dwellers are affected, we might eventually overstep some critical threshold and trigger the wholesale death of plankton, thus finding ourselves without a major source of the world's food and with our oxygen supplies dwindling.

This is all very rational, but none of this precipitated my gut response when I learned that some major whaling nations intended to resist the international ban on whaling. I simply found it very sad. For some reason, I wanted those whales to continue to share the planet. If you had the same reaction and you don't ever expect to see or eat a whale or use its oil, you might try to analyze the roots of your own response.

THE LAND ENVIRONMENT

It seems safe to say that no two organisms interact with their environments in identical ways. If they were asked, two roundworms living in a raccoon's intestine would probably describe their world differently, depending on their precise point of attachment and their individual perceptive and reactive tendencies. Two wildebeest living on an African plain would not only disagree with the roundworms about the nature of the world, but they would certainly have different experiences and might be expected to perceive their grassland environment quite differently. Even identical twins see the world from opposite sides of a baby carriage. Description, we see, is based on perception and interpretation, thus generalizations may not be entirely valid. So now we will generalize.

Biomes

We tend to think of our earth as a ponderous place that unfailingly provides its denizens with those things necessary for life. We often seem to forget, however, that life exists only in a thin film that veils the surface of this immense ball—a delicate shell wherein the wondrous forces of sunlight and water interact to permit life. This fragile film is responsive to a number of influences and, hence, is highly variable from one place to another. Furthermore, each place is likely to be unstable, so that it changes with time. Nonetheless, the different kinds of "places" in which life exists on the earth at present can be roughly categorized based on their present physical and biological properties. We should keep in mind, however, that these are merely arbitrary divisions of the great, complex, and intergrading areas of the earth.

FIGURE 25.13

The world's biomes. Find the major biomes listed in Table 25.3.

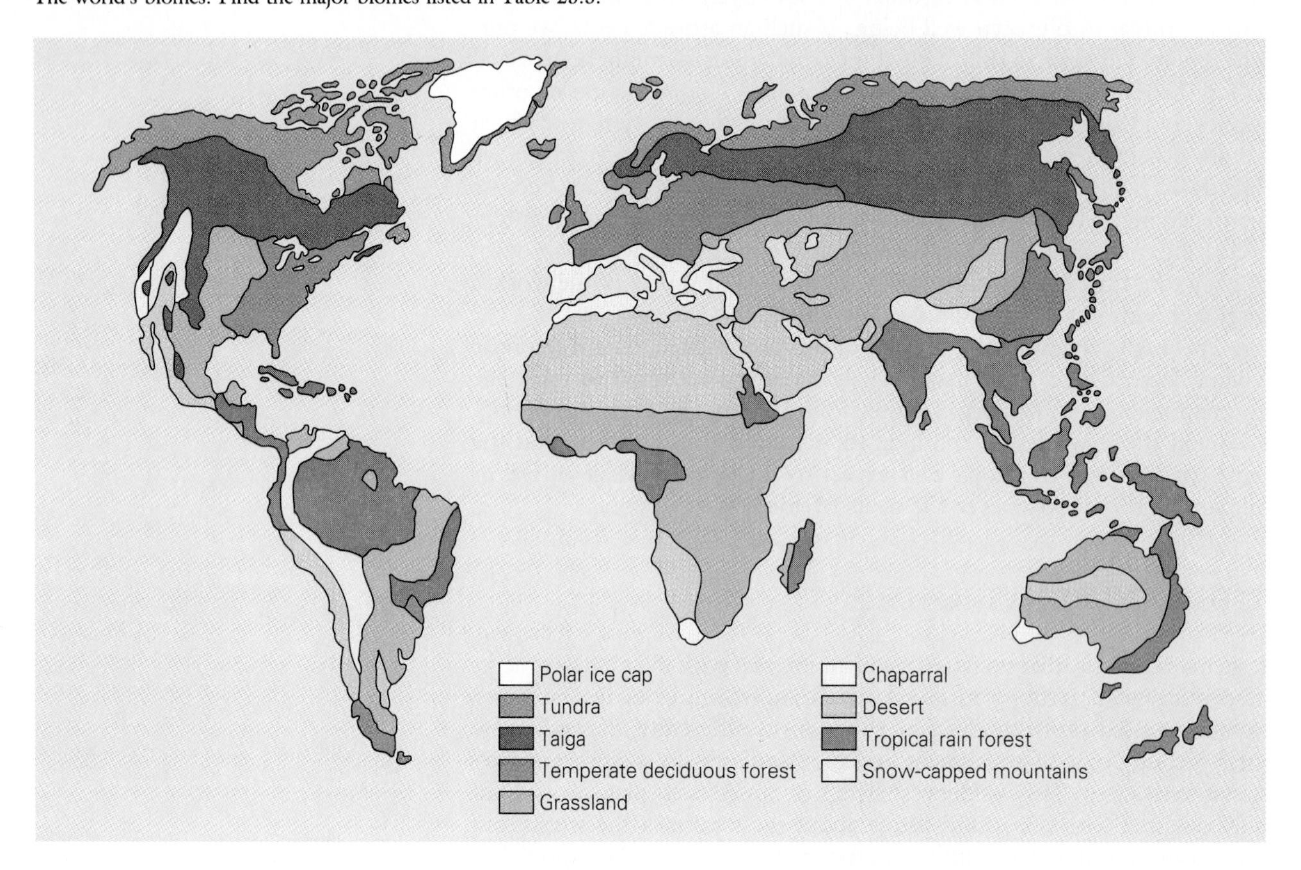

By using imagination, we can divide the earth's land into several kinds of regions called **biomes** (Figure 25.13 and Table 25.3), areas of the earth that support specific assemblages of plants. As would be expected, certain kinds of animals occupy each type of biome, since different species of animals are dependent on different sorts of plant communities for food, shelter, building materials, and hiding places.

The makeup of the plant community is dependent on factors such as soil conditions, available water, weather, day length, and competition, but two climatic factors, available water and temperature, are the main determinants (Figure 25.14). These two factors tend to vary in a somewhat systematic fashion over the earth's surface. Temperature generally decreases as one moves away from the equator and toward the poles or as one goes up in altitude. This uneven heating of the atmosphere is also indirectly involved in determining the pattern of rainfall. Warm moist air rises at the equator. As it does, it cools and loses moisture as rain. The rising air mass moves away from the equator, cooling as it moves. When it reaches roughly 30 degrees north or south, the air has cooled and begins to descend. By then, most of the moisture has been released. In addition to explaining the heavy tropical rainfall, the behavior of the equatorial air cells also explains why many of the earth's deserts form in belts just above and below the equatorial forests (Figure 25.15).

TABLE 25.3
The Major Biomes

BIOME	CLIMATE	VEGETATION	ANIMALS	NOTES
Temperate deciduous forest	Temperate, with seasonal changes; rainfall: 75–130 cm	Hardwoods: oak, maple, beech, poplar, hickory	Deer, turkey, squirrels, chipmunks	Eastern United States, central Europe
Grassland	Tropic or temperate; rainfall: 25–100 cm	Grasses, small bushes, shrubs, possibly bamboo	Prairie dog, gophers, herds of grazing animals such as antelope	Example is American prairies, porous soil
Desert	Hot in daytime, cold at night; rainfall: less than 25 cm	Cacti, succulents, plants taking advantage of sudden rainfall	Reptiles, arthropods, birds, small mammals	The shortage of water necessitates adaptations in plants and animals
Tropical rain forest	Hot, with rainy seasons; rain fall: 200–400 cm	Broad-leaved evergreens; multileveled canopy; epiphytes; sparse understory except where light breaks through canopy	Insects, birds, primates	Amazon and Congo Basins, also Southeast Asia; extreme abundance and diversity of both plant and animal species
Tundra	Covered by ice and snow through most of the year; either arctic (northern) or alpine (high elevation; rainfall: less than 25 cm)	Lichens, herbs, mosses, low-lying shrubs	Ptarmigan, hare, lemming, ground squirrels (and their predators), musk ox, caribou	Northernmost biome, slow to recover from disturbances
Taiga	Long, cold, wet, winters; short summers	Conifers: pine, spruce, fir, hemlock; understory absent or sparse	Porcupines, moose, bear, rodents, hares, wolverines	Northern hemisphere

FIGURE 25.14

The interaction between temperature and precipitation in the formation of the earth's biomes.

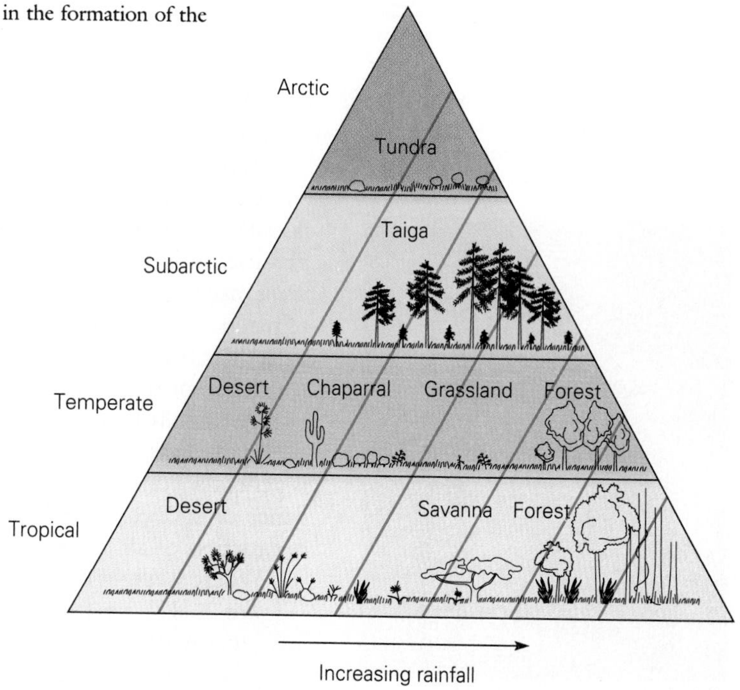

Arctic
Tundra
Taiga
Subarctic
Temperate
Desert
Chaparral
Grassland
Forest
Tropical
Desert
Savanna
Forest
Increasing rainfall

FIGURE 25.15
At left, the prevailing wind directions produce the biomes we see at right. Notice the heavy rainfall in the equatorial regions.

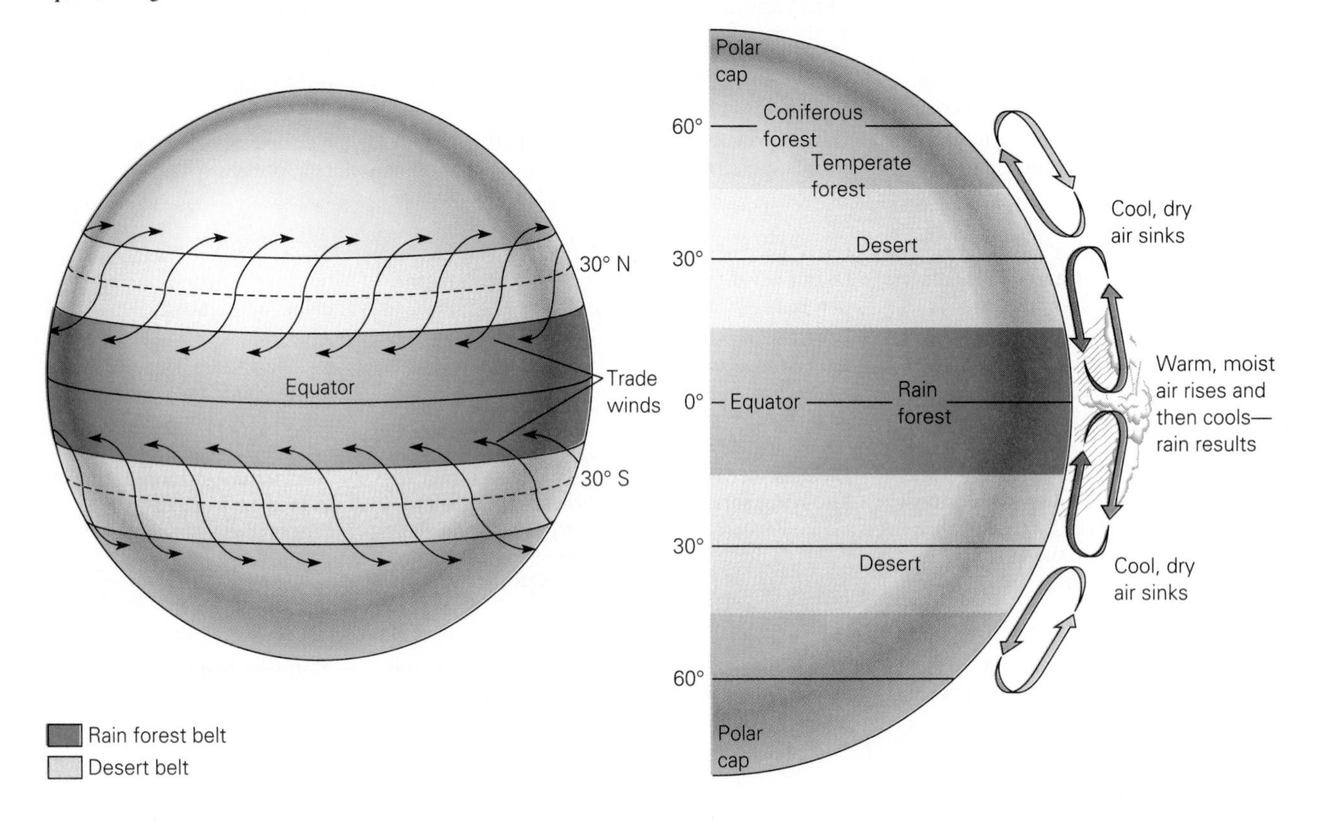

FIGURE 25.16
The interior of a tropical rain forest is often dark and quite open, permitting this Waorani hunter of Ecuador's Amazon basin to pursue monkeys and parrots with a blowgun and poisoned darts.

We will consider only the six largest types of biomes, but there are others, as we see in Figure 25.13. We will begin at the equator and work our way through the temperate regions toward the poles. Then, we will consider the grassland and desert biomes that occur as water becomes increasingly scarce.

Tropical rain forests (Figure 25.16) are found mainly in the Amazon and Congo Basins and in Southeast Asia. The temperature in this biome doesn't vary much throughout the year. Instead, the seasons are marked by variation in the amount of rainfall throughout the year. In some areas, there may be pronounced rainy seasons. These forests support many species of plants. Trees grow throughout the year and reach tremendous heights, with their branches forming a massive canopy overhead. The forest floor, which can be quite open and easy to travel over, may be dark and steamy. Forests literally swarm with insects and birds. Animals may breed throughout the year as a result of the continual availability of food. Competition is generally considered to be very keen in such areas because of the abundance of species. (See Essay 25.2 for a discussion of the results of the destruction of all types of tropical forests.)

Where sunlight does manage to filter through the leafy canopy, **jungles** (Figure 25.17) are formed. These are the densely vegetated and tangled areas that movie companies love to hack their way through. Jungles often grow along river banks, and so nineteenth century water travellers who

ESSAY 25.2

The Destruction of Tropical Forests

Tropical forests occupy only about 7 percent of the earth's surface. Yet these incredibly productive areas probably harbor 50–80 percent of the earth's species. We can't come up with a firm figure for a very simple reason; we don't know how many species live on the planet. To date, we have identified about 1.7 million species, but there may be up to 30 million species of *insects* in the tropical forests *alone*. Because of the great richness of tropical forests, their destruction marks the demise of a great many kinds of living things, some we've never known.

How bad is the destruction? Every minute more than 100 acres of tropical forests are devastated or seriously degraded. Every day more than 240 square miles are wiped out. An area the size of Great Britain is destroyed each year. And the carnage goes on.

Why don't we stop? The reasons stem from a shortsighted view of economics, by professional and peasant alike. Those who should know better, such as those in control of governments (including that of Brazil), feel they must answer to the people's needs. The "people" (often wealthy consortiums) need the land, they say, for expansion, mining, farmland, or whatever. So the governments open up the areas, resolutely determined not to be swayed by the whining of foreign environmentalists. Interestingly, a recent study showed that the profits from the forests themselves are greater than that from the cleared land. The harvest from forests (in edible fruits and latex) was more than double the value of the land used for cattle pastures. Yet the demand for pasture is high, largely due to the wheedling of American fast-food hamburger chains who want to serve poor quality beef raised there in sandwiches with cute names. (The Rainforest Action Network calculates that beef raised on deforested tropical land lowers the average price of a hamburger by about five cents). The destruction is often encouraged by lending practices from organizations, like the World Bank, that may guide or prompt the development of wild areas.

It is indeed a shame to see such wild areas fall before the chainsaw, the axe, and the tractor, but the problem is greater than that. By destroying so many species, we reduce the genetic variation on the planet. This variation is essentially a storehouse we can call on for special genes or combinations of genes to help us solve our mounting agricultural problems. When leaf blight struck the U.S. corn crop in 1970, we were able to add a new, resistant combination of genes to our stock taken from variants of a wild Mexican strain. In addition, the tropical forests absorb much of the earth's carbon dioxide and provide, in turn, much of the earth's oxygen. Without this effect, the earth's climate can warm and great climatic changes can be set in effect. And then there are the potential new foods, medicines, and other products that are bound in the fragile tropical forests, products that we never had a chance to develop. Recently, in one three-acre plot that was being cleared to extend a jungle runway, four species of trees were discovered. Nonetheless, some experts have calculated that over the next 15 years we can expect the extinction of 100 to 200 species a day.

FIGURE 25.17

A Waorani hunter pursuing a capybara into the thick jungle growth along the Cononaco River of Ecuador. Early river travellers believed the entire rain forest was this densely overgrown.

FIGURE 25.18

Temperate deciduous forests are marked by distinct seasons. The winter, when limbs are bare, is quite different from the luxurious greenery of summer.

didn't want to get out and traipse around with the snakes once gained a mistaken impression of tropical rain forests, and their descriptions colored literature for years.

Temperate deciduous forests (Figure 25.18) once covered most of the eastern United States and all of Central Europe. The dominant trees in these forests are hardwoods. The areas characterized by such plants are subject to harsh winters, times when the trees shed their leaves, and warm summers that mark periods of rapid growth and rejuvenation. Before the new leaves begin to shade the forest floor in the spring, a variety of herbaceous (nonwoody) flowering plants may appear. These wildflowers are usually perennials, plants that live and produce flowers year after year. In the early spring, they don't have time to manufacture the food needed to grow and bloom suddenly. Instead, they draw on food produced and stored in underground parts during the previous year. Rainfall may average 75 to 130 centimeters or more each year in these forests, and is rather evenly distributed throughout the year.

People who live in temperate deciduous biomes often consider the seasonal changes as both moving and fascinating. They describe a certain joy that swells within them each spring and a secret pensiveness that overcomes them in the fall as the days darken and the forests become more silent. (Perhaps we are exceeding technical descriptions here, but these are my favorite places.)

Taiga (pronounced "tie-gah," Figure 25.19) is quite unmistakable; there is nothing else like it. It is confined almost exclusively to the Northern

hemisphere and is identified by the great coniferous forests of pine, spruce, fir, and hemlock that extend across North America, Europe, and Asia. Some of these trees are the largest living things on earth (Essay 25.3).

Taiga is marked by long, cold, wet winters and short summer growing seasons. The forest is interrupted here and there by extensive bogs, or muskegs, which are the remains of large ponds (Figure 25.20). The forest floor is usually covered by a carpet of needles. In the dim light at ground level, there may be mosses, ferns, and a few flowering plants. One may move silently on the muffling needles through the Canadian taiga observing a host of mammals, including porcupines, moose, bear, rodents, hares, and wolverines.

Tundra (Figure 25.21) is the northernmost land biome. It is covered throughout most of the year by ice and snow. This biome is most prevalent in the far north (arctic tundra), but it may also appear at high elevations in other parts of the world (alpine tundra). For example, in the United States, it may be seen in the high Rocky Mountains (Figure 25.22). Tundra appears in places where summer usually lasts two to four months, just long enough to thaw a few feet of the soil above the permafrost, or permanently frozen soil. Thaw brings soggy ground, and ponds and bogs appear in the depressions. The plant life consists mostly of lichens, herbs, mosses, and low-lying shrubs and grasses, as well as a few kinds of trees, such as dwarf willows and birches. Such plants obviously must be hardy, but their hardiness disguises their fragility. Once disturbed, these areas take very long periods to restore themselves.

FIGURE 25.19
The great evergreen forests of the taiga cover large areas of the northern hemisphere.

FIGURE 25.20

Where bogs and marshes form, the taiga is interrupted. These areas are often referred to as ***muskegs***. They represent a perpetual tug-of-war between the aquatic and terrestrial environment, as plants continually invade the marshes, some failing, others becoming established.

FIGURE 25.21

After the snow melts in the brief arctic summer, the tundra often reveals a variety of plants, many with lovely delicate hues that belie their rigorous existence.

Although there are far fewer different species of animals on the tundra than in the forests, the life on the tundra is surprisingly abundant for such a raw place. One finds not only ptarmigan, hare, ground squirrels, and lemming (along with their predators, such as owls, weasels, wolves, and foxes), but also large animals, including musk ox and caribou.

FIGURE 25.22

Colorado's Rocky Mountain tundra.

ESSAY 25.3

The Redwoods

It was 390 feet tall. Nothing on earth could match it. It had stood as a slender sapling in the cool coastal air, perhaps moving slightly in a light breeze, on the very day Caesar finally decided to move against Britain. But all that happened a long way from the area that would be called California. Great leaders were born as the tree grew. And they died as the tree became stronger and taller. Wars came and went, as well as plagues and famine. There were great celebrations and deep mournings here and there over the earth. The tree lived through it all.

As the ages passed, the tree continued to grow. No one marked the time when its crown reached above all the others, because it was only one of a vast forest of such trees. In time, however, it *was* noticed. Even before the start of the twentieth century, the straight, tall, and insect-resistant trees had caught the eyes of lumber companies. As the population of California grew, the trees began to be cut. Some citizens tried to establish a national park to save some of them, but the lumber industry blocked it in Congress. In what has been called "one of the greatest swindles of all time," they arranged to change nearly all the redwood lands from public to private ownership.

After World War II, California experienced a population surge of unprecedented dimensions. Factories and homes were being built feverishly and lumber was needed. Redwood was ideal for lawn furniture and tomato stakes, too. The conservationists, led by the Sierra Club, pushed again for parks to be set aside. By 1960, there were only two areas left that were relatively unscarred and of park caliber. While conservationists dickered among themselves over which area was the best, the lumber industry was busy in Sacramento and Washington. They effectively muddied the waters by cynically proposing a number of other sites. The easily confused public became confused. The lumber industry was confident. One official boasted that it takes five years to get a national park bill through Congress and in five years there wouldn't be anything worth fighting for. As the arguments continued, the trees kept falling.

Then, in the summer of 1965, the 390-foot giant was discovered—by the lumber companies. They brought in their chain saws. They worked quickly and the great tree was felled, cut into 20-foot lengths, and hauled away. In fact, every tree in the redwood stand at the junction of Bond and Redwood Creeks was brought down—right in the very heart of the Sierra Club's proposed park site. The crash of the great giant was drowned out by the cheerful ring of the cash register.

FIGURE 25.23
Grasslands support a surprising number of animals.

FIGURE 25.24
Deserts are fascinating places with great temperature variation because the air is unbuffered by water. This puma tends to conserve water by behaving so as to minimize loss. For example, it may avoid direct sunlight except when hunting and may spend most of its time in higher, cooler areas.

Grasslands (Figure 25.23) occur in many parts of the world and are exemplified in the United States by the American prairie. Grasslands are characterized by grasses (not surprisingly), small bushes, and, in some parts of the world, thickets of bamboo, which is a type of grass. The soil in such areas is usually porous. Trees may line streams and rivers. Rainfall averages between 25 and 100 centimeters each year, but may be erratic in its timing. Grasslands are found in both tropic and temperate zones. Grasslands appear to be simple places, so it is often surprising just how many animals they can support, such as the large herds of grazing animals in the African savannah.

Deserts (Figure 25.24) are characteristically hot in the daytime and cold at night. The 25 centimeters or less of rain that fall each year usually comes in sudden downpours, so that much of it runs off, sometimes causing flash floods and marked erosion; the rest quickly evaporates. Directly after a rainfall, the annual plants (those that live for a single growing season) of the desert take advantage of the moisture and explode in an orgy of growth and seed production. Other desert plants meet the water-shortage problem in other ways, either by storing water in their tissues or by reaching far underground to reach ground water with their tap roots (of course). The most common animals of deserts are reptiles, arthropods, birds, and small mammals. They all must beat the heat and conserve water. One way they do this is by moving about mainly by night and by reducing water loss through specialized excretory systems.

THE WATER ENVIRONMENT

The earth's water environments may be classified roughly as fresh water or marine, although not all bodies of water fall neatly into one category or the other. For example, Lake Pontchartrain, near New Orleans, is brackish, or a mixture of salt and fresh water. So are **estuaries,** the places where rivers flow into seas. Fresh water has about 0.1 percent salt; seawater has about 3.5 percent salt; and, as we will see, each has its importance in the earth's drama.

Freshwater Bodies

Rivers and Streams

Sometimes definitions are deceptively simple, as the definition of rivers and streams falls into this class. Technically, rivers and streams are bodies of fresh water that move in one direction. As we see in Figure 25.25 the headwaters, the small streams from which they form, are usually narrow and steep. Since these streams are often springfed and at higher elevations, they also may be intensely cold. The falling, turbulent water tends to be mixed and thoroughly aerated. The communities in these areas are composed of relatively few species, such as trout (species that need the cold temperatures and oxygenated waters). A few invertebrates are found among the algae and mosses that cling tenaciously to the rocky banks. Farther along, the riverbeds tend to become flatter and rivers wider and slower. Here, the waters carry more suspended material and may become

FIGURE 25.25
Moving water often encourages abundant growth along its banks.

The Water Environment

Most of the earth's surface is covered with water, both salty and fresh. It thus constitutes an important part of the biosphere, but we often neglect to consider its importance because in its naturally occurring form, it is essentially disruptive to much of our tissue, and our efforts to examine it are hindered in numerous ways. It puckers our skin, clogs our ears, is rarely a comfortable temperature, and is usually hard to see through. So if we're not drinking or washing, we generally avoid it unless we are sports enthusiasts or specialists, thus we probably imagine we know more about its role in nature than we really do. Yet our lives depend on it and hence we are usually cheered by the sight of clear mountain brooks or long, winding rivers. In fact, we are so dependent on water that most of our great cities have developed along the water's edge.

The rocky coast of Olympic National Park, Washington (left). The incessant pounding of the water has carved out these striking monuments, which serve as attachment for a number of organisms, a particular type of animal dominating each vertical zone.

A mud flat (right), a saltwater area that is covered at high tide and exposed at low tide. The soil is often hard and tightly packed. This is a very difficult place to adapt to because it is subjected to repeated drying and soaking. Predators stalk these areas in search of exposed prey. Such flats may occur at estuaries, the river currents producing the prominent rills seen here. The mud flat estuary harbors a variety of organisms if the river has not been poisoned upstream.

A cypress swamp in winter (above). Notice the water-swollen trunks and the "knees"—roots that have risen above the waterline into the oxygen-laden air. The water in such swamps moves very slowly and hence is low in oxygen.

Tules, or *bulrushes,* which thrive on overflowed land in the American Southwest (right). Large tracts have been overgrown with these plants.

A glacial lake in Banff National Park, Alberta. Such lakes may be exceptionally deep and cold with a predictable turnover rate as their upper reaches are heated or cooled by the air temperature. One of the most beautiful of the world's mountain lakes is, or was, Tahoe, a name now synonymous with artifice and glitter as the struggle between developers and environmentalists continues.

The Channel Islands off the California coast (below). This is Anacapa Island. It is rather small and, as a rule, smaller islands harbor fewer kinds of plants and animals. Whereas Anacapa lacks forests, trees abound on Santa Cruz, her larger neighbor in the distance.

A salt marsh estuary in South Carolina (above). Salt marshes are exceedingly critical places in that the many tiny life forms living here initiate food chains that ultimately involve larger sea creatures. They are vulnerable in that they don't look important and so developers have gained easy access to them.

This sandy beach in the Bahamas supports few life forms, probably because the shifting sands provide very poor attachment and shallow waters tend to surge strongly.

Rivers of ice such as these harbor few forms of life (many of them microscopic).

Highly oxygenated cold waters can be home to a variety of life, including vertebrates such as trout.

FIGURE 25.26

Crater Lake, left, was produced by volcanic activity, while Lake Harris in Florida was produced by gradual uplifting of the surrounding land.

more murky, relegating the penetration of light to the upper levels and shallow areas where algae may thrive. The fish species here need less oxygen, so one may find the slower, less active, bottom-feeding species, such as catfish, carp, and—lurking above them—bass. As the river widens and becomes even slower, suspended materials collected upriver may settle out, forming a nutrient-rich sediment that can support relatively large and complex communities.

Lakes and Ponds

As water moves along the channels cut by rivers and streams, it may pause in its inexorable cycle to rest for a time in still bodies called lakes (and on a smaller scale, ponds).

Lakes and ponds differ not only in size, but usually in depth. Ponds are often shallow and unstable, and therefore can be risky places to live. Lakes are grander places, and, compared to other major features of the earth, they may be of relatively recent geological origin. Most occur in northerly or alpine regions as products of the last glacial retreats (10,000 to 12,000 years ago). They are also produced through volcanic activity, as was the famed Crater Lake in Oregon, and through gradual uplifting of land, as were many of the shallow, acid lakes of Florida (Figure 25.26). The deepest lake in the world is much older than most other lakes: Lake Baikal in the Soviet Union at 1750 m (5742 ft, or well over a mile) was formed during the Mesozoic Era, roughly 240 million years ago.

Due to the peculiar vulnerability of freshwater bodies, human activity has, in many cases, drastically altered their character. The Great Lakes, for

FIGURE 25.27

As a result of human activities the edible lake trout (top) and whitefish (bottom) are now almost nonexistent in the Great Lakes. On the other hand, the parasitic lamprey (attached to the trout) and the destructive alewife (center) have become common. One reason for this change is that the use of detergents and chemical fertilizers have increased the amount of phosphate and nitrate that is washed into the water. Bacteria can break these down but only with the use of great amounts of oxygen. Algae and other plants flourish under such conditions, but as they die, they return their nitrates and phosphates to the water and cause bacteria to use ever more oxygen. Finally, the lake cannot support animals that require a lot of oxygen, such as trout, but the new conditions are fine for alewives.

example, have been altered by human activities so that they can no longer support great numbers of edible lake trout and whitefish (Figure 25.27). However, parasitic lamprey and the destructive alewife began to flourish. (Recently there has been some progress, in some respects, in cleaning up the Great Lakes.)

One way humans can drastically alter lakes is by speeding up a natural process called **eutrophication** ("good food"). This is a normal aging process of lakes, but the use of detergents and chemical fertilizers has increased the amounts of phosphate and nitrate that are washed into the water. These join other forms of human waste in our fresh waters and speed up the aging process and accelerate the production of organic matter in lakes.

FIGURE 25.28
A dying pond, filling in by the natural process of eutrophication.

With increased plant growth, the lake begins to fill in, and as the plants die and sink to the bottom, the lake grows shallower. Eutrophication can cause ponds to choke with growth and, finally, to fill in completely (Figure 25.28).

The Oceans

The oceans of the earth have fascinated humans since "men have gone down to the sea in ships." Not only are sailors attracted to the sea, but so are beach lovers, poets, and practically everyone else with a shred of romanticism about them. The oceans remain as mysterious as they are compelling. Nonetheless, we are free to make our attempts at describing them in some sort of organized way.

First, we should be aware that this is, indeed, a watery planet. Oceans not only cover about three-fourths of the earth's surface, but if that surface was smoothed out, the planet would be covered entirely by water. The average depth of the ocean is about three miles, but there are places where the water is seven miles deep. So the ocean is deeper than the tallest mountains are high (Figure 25.29).

It has been suggested that the oceans will one day provide us with most of our food. However, we presently take only three to five percent of our food from the seas, and even if we were able to double that in the next decade, our food problem would not be solved.* The problem of increased reliance on food from the sea is complicated by the fact that the deep ocean waters, like the deserts, are not very productive. In fact, many parts of the ocean are almost completely devoid of life.

Even the desolate areas of the oceans are not simply still bodies of water, broken only at the surface by waves. In fact, throughout the oceans, the deeper waters are continuously in motion as great silent currents hold sway. These deep currents are primarily the result of surface winds, in particular the trade winds (Figure 25.15). The "trades," once so important to commercial sailors, are prevailing winds—that is, winds that hold rather steadily from the same direction because of the difference in air temperature between the pole and the equator, coupled with the effects of the earth's rotation. The great ocean currents caused by these winds and the earth's rotation spin in wide circles, clockwise in the northern hemisphere, counterclockwise below the equator. One such ocean river is the famed Gulf Stream, which flows along the Atlantic coast as far out as Bermuda and then swings eastward to northern Europe.

Where the currents are deflected upward by the mountains on the ocean floor, cold water from the ocean depths wells up to the warmer surface. This cold water carries with it ages of accumulated sediment from the floor, sediment that is rich in nutrients. The ancient debris then acts as a fertilizer and promotes the bloom of life where light dances through the ocean's surface. The nutrients carried by upwellings are utilized by minute chlorophyll-bearing organisms called **phytoplankton,** which in turn serve as a food source for tiny, drifting animals called **zooplankton.**

*(This is because with the world's population increasing at its present rate, doubling the intake from the sea would add only three to five percent more food to a population that had increased over eighteen percent.)

FIGURE 25.29

The marine environment is divided into the neritic province (from the shallows to where the continental shelf falls off into the steeper continental slope) and the oceanic province (from that line outward). Both provinces, in turn, are divided into lighted (euphotic) zones and unlighted (aphotic) zones. Most marine life is concentrated in the neritic province.

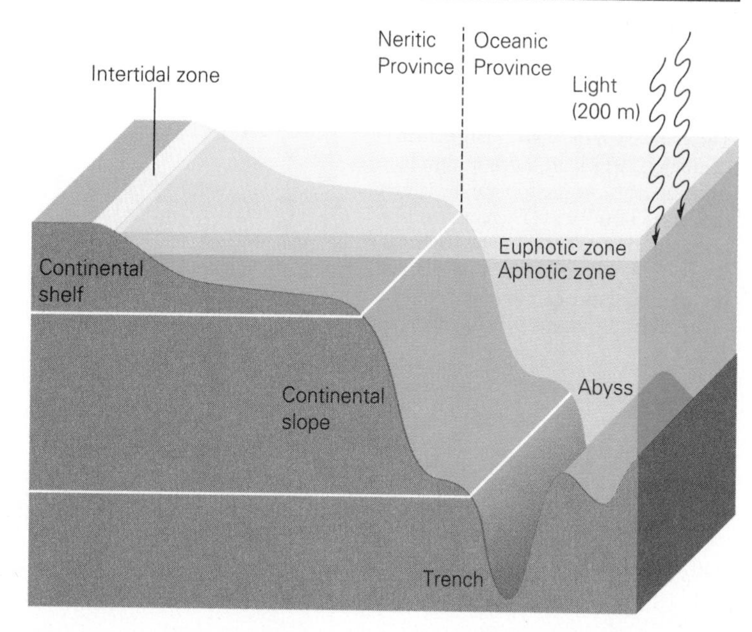

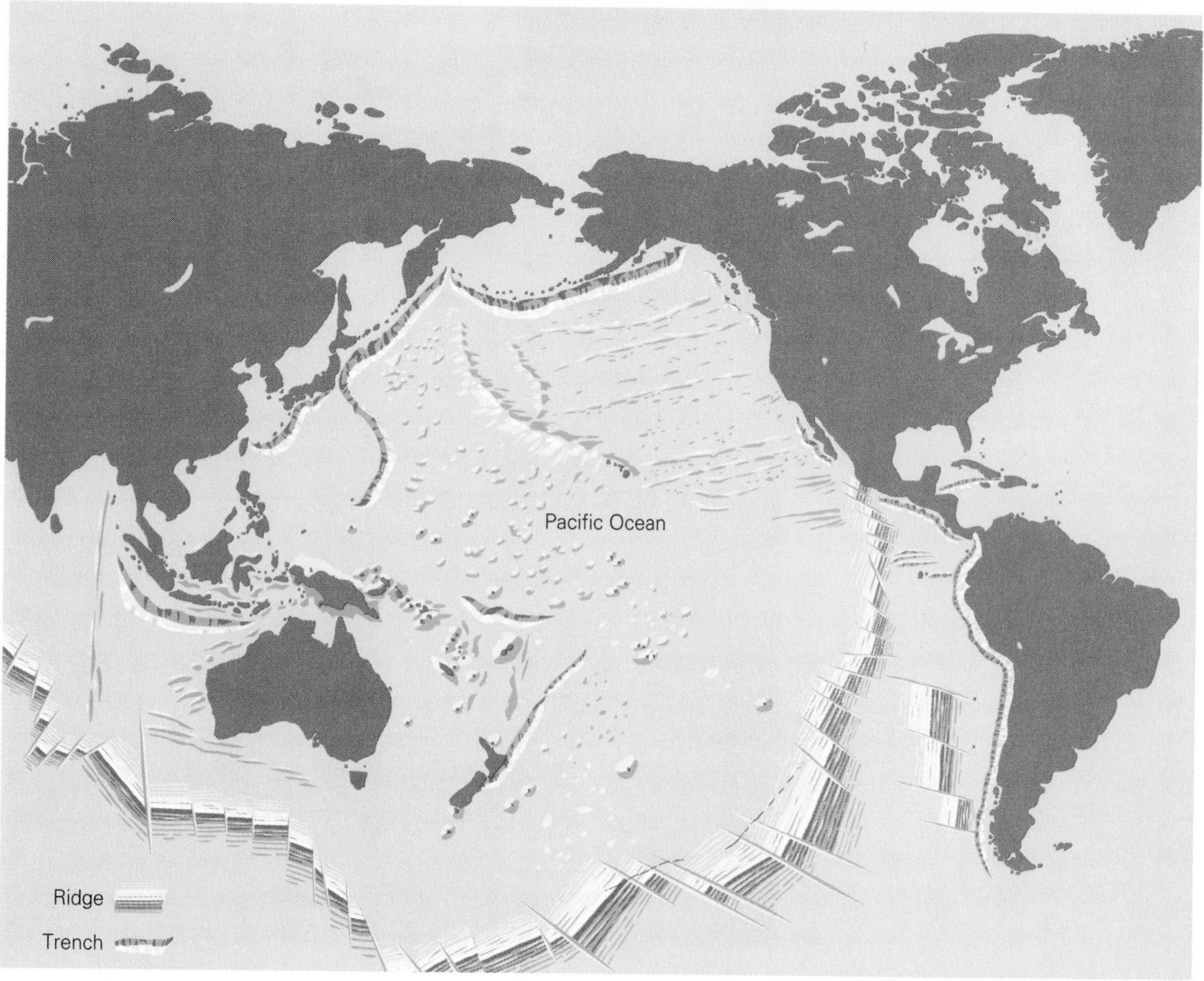

FIGURE 25.30

The food pyramid of the ocean. The tiny phytoplankton at the base (not drawn to scale) capture the energy of the sun. These are eaten by animals larger than themselves, which are eaten in turn by larger animals. At the top are the largest carnivores of the sea. (There are no large herbivores in the open oceans.) What is the position of man in the food chain of the sea? Note that each level is comprised of far fewer organisms that the one below it.

These minute organisms, barely visible to the naked eye, form the basis of the ocean's food pyramid (Figure 25.30). One cubic foot of seawater may contain over 12 million phytoplankton. These food producers, however, can live only near the ocean's surface, since the sunlight necessary for them to carry on photosynthesis cannot penetrate below about 600 feet.

Not only are phytoplankton important as primary food producers in the oceans, but they also manufacture much of the earth's oxygen. Thus the importance of maintaining viable oceans is apparent. Some people are increasingly concerned as we continue to use the world's oceans as a dump. The sight of oil companies' new supertankers, which carry dangerous, partly refined petroleum across the fragile oceans, may also cause frowns of consternation. We were reminded of the potential for disaster when one such tanker went aground in Alaska (Figure 25.31).

People who have descended to the ocean's depths in pressurized bathyscaphes have reported that they can sometimes see light at depths as great as 2000 feet. The light that reaches these depths is pale blue, since the reds and oranges of the light spectrum have been filtered out by the water above. Many of the fish that live there are reddish in color when they are viewed in the light at the surface. In their natural depths, they appear dark and shadowy and difficult to see, since there is no red light for their pigment to reflect and their red color absorbs what blue light there is.

FIGURE 25.31

On Good Friday, March 24, 1989, the captain of the *Exxon Valdez* reportedly steered his ship onto an incorrect course and went below deck. At 12:04 A.M., the ship rammed Bligh Reef. The reef was well known. In fact, it was named after the first European to anchor off it, Captain William Bligh, later of the *Bounty,* in 1778. The result of the collision was that more than 10 million gallons of thick North Slope crude oil spilled into the pristine waters of Prince William Sound. The fishermen who had so carefully protected the beautiful and productive sound stood ready to help immediately, as did the conservation society, Greenpeace. However, Exxon moved slowly, as did the federal government, and the devastating oil sludge soon covered a distance equal to the coast of California, killing innumerable seabirds and mammals. No one knows when the coast will recover.

FIGURE 25.32

Viperfish, an example of a deep-sea fish. Such fish, while not noted for their personality, have enormous jaws and teeth. Thus, when they encounter another animal in their sparsely populated depths, there is a good chance that they can manage to eat it. There are luminescent spots along its sides. Such lighting is apparently useful in signaling between members of the species. This animal, fortunately, is only about a foot long.

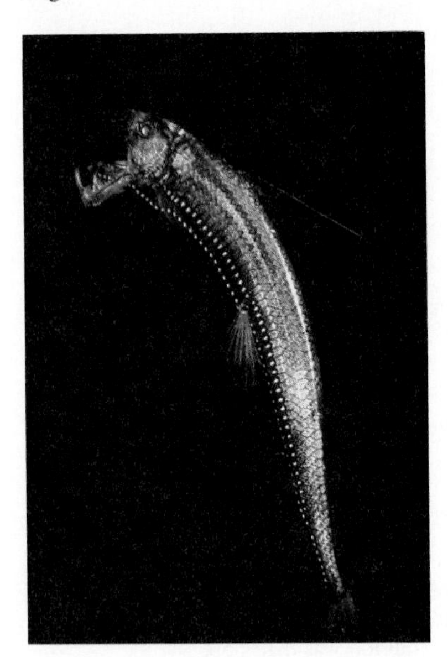

It was once believed that nothing could live below 1800 feet because the pressure there was too great for any form of life to withstand it. However, about the time Darwin wrote on the origin of species, an underwater cable broke in the Mediterranean and was hauled up for repairs from a depth of about 6000 feet. Astonishingly, it was covered with all sorts of living things. So notions concerning the effects of pressure had to be revised. Now it is believed that living things can exist at any depth as long as they are able to develop pressure inside their bodies equal to the pressure outside. This pressure can be so great that fish brought up alive from depths below 2000 feet have literally exploded when they reached the surface.

The deepest reaches of the ocean must be peculiar places indeed. We are just beginning to learn something about them, and the new information is fascinating. It seems that the deepest areas, untouched by the currents above them, are calm places. So far, we know of only a few kinds of animals that live there, and because of the unusual demands placed on them, many are quite weird (Figure 25.32). Other life forms at these depths include a few kinds of fungi, species that do not need light. Among the more interesting new findings on the ocean floor are the volcanic vents where bizarre and unexpected communities derive their energy from water being heated by the earth's core, as we saw in Chapter 5.

The ocean bottom itself is soft and composed of the chalky and glassy corpses of tiny organisms that have rained down steadily onto its surface throughout the ages. Animals that live on the bottom often have long appendages to keep them from sinking into this primal ooze. Because of the utter stillness of the deepest ocean bottom, very delicate and fragile creatures can live there, but we are probably aware of only a small part of the life in such places (Figure 25.33). Some animals, it seems, escape our attention by burrowing into the bottom. Divers have seen many strange holes and burrows that were made by no known creature.

COASTAL AREAS

The land-ocean interface has been called "the seashore." Those romanticists who like to write about it would compose new eulogies if they knew the full impact of the seashore on their lives. And who knows, perhaps coastal developers might even become a bit more cautious.

But why are these such important places? For one thing, life is much more abundant in coastal waters than in any other part of the ocean. The edges of the continents extend beyond the shore for anywhere from 10 to 150 miles as **continental shelves,** which fall off into deeper areas of the continental slope, as we saw in Figure 25.29. At these relatively shallow depths, usually less than 200 feet, sunlight easily penetrates and gives rise to a wide variety of plants. The plants, of course, provide many species of animals with food, shelter, and surfaces for attachment. Coastal areas are generally composed primarily of either rock, sand, or mud. And each type of area has its own special qualities.

Rocky coasts are perhaps the most dramatic and, to coastal navigators, the most awesome of the coastal areas, their jagged formations formed by years of pounding surf that often hides immense reefs below the surface.

FIGURE 25.33

The ocean bottom is often a still and quiet place that can support very fragile forms of life, such as these grass eels.

These coastal areas boast a remarkable array of plant and animal life that often is found in quite distinct zones. For example, the zone dominated by the periwinkle snail is clearly delineated from the barnacle zone, which is, in turn, strongly demarcated from the seaweed zone. The zones are not occupied exclusively by a single species; they are simply identified according to the most prevalent species occupying them.

Sandy seashores have relatively few forms of life, which is perhaps a good thing for timorous bathers. You have probably waded out into the surf and were relieved that you didn't step on some mysterious and toothy bottom dweller. The reason they are so scarce is that wave action in such shallow areas causes a constant shifting of the sandy bottom, depriving the less mobile species of a fixed surface to which they can attach. Any animal that lives in such shifting areas must be able to get around quickly, like crabs, or be able to dig and withstand burial, like clams. (Unfortunately, stingrays can do both). Animals that live in tidal areas and are not mobile probably encounter exceedingly severe conditions at low tide when they are subject to the parching sun and to terrestrial foragers such as shorebirds.

From an aesthetic standpoint, **mud flats** are probably the least appealing part of the seashore, hence perhaps the general lack of interest in protecting them. They aren't especially spectacular as scenery, and they are poor places to spread a towel. Also, they tend to smell peculiar (to say the least). Mud flats do not harbor as many life forms as rocky coasts, it is true, but they support much more life than do the other kinds of coastal areas.

Mud flats are submerged at high tide and exposed at low tide. Bottom dwellers, snails, insects, shrimp, crabs, fish, and birds abound in these areas. Each square yard of the mud flat can support thousands of such individuals. Many sea animals begin their lives in mud flats, only later moving out to take their place in the mysterious pageant of the open ocean. Hence, the life of the ocean itself is, in a very real sense, dependent on the preservation of the unsightly flats. Unfortunately, mud flats and saltwater marshes have attracted the eye of commercial developers who are aware of the public's lack of interest in them. Up and down the coasts, our bayous, marshes, and bays are being filled in for use as industrial sites, "waterfront" housing contracts, and high-rise apartment complexes and condominiums. San Francisco Bay alone is now only two thirds of its original size, much of it having been replaced by long stretches of paved, chromed, and neon-lit areas designed to attract the tourist dollar.

We have described some of the areas and habitats of the planet, and considered some of the ways species interact. We have asked such questions as: How can one species influence the well-being of another? In what ways do species interact with their environment? Does passage of any species from the earth affect us in any real way? As we have seen, there aren't specific answers to such questions, but we do already know enough to make some reasonably intelligent decisions about how we should, in our own best interest, interact with the life around us.

SUMMARY

1. Ecology is the study of the interrelationships among organisms and their environment. Whereas an ecosystem includes all the interacting living and nonliving components that form an independent unit, a community includes only the living organisms. A population is a group of interbreeding members of the same species. The habitat is the place where an organism lives. An organism's immediate surrounding is its microhabitat. The organism's requirements and the ways in which it interacts with both the living and nonliving components of its environment is its niche.
2. Two species cannot occupy the same niche indefinitely. If two species coexist, it is assumed that they are occupying different niches. Partitioning the habitat reduces the level of competition. Competition occurs where organisms use scarce resources or harm each other when seeking the same resource.
3. An orderly sequence of species, structure, and energy flow in a community is called succession. The final or climax stage of a succession does not change markedly. Primary succession occurs when pioneer organisms invade an area where no life previously existed. Secondary succession involves the sequence of communities as life recovers in a disturbed area.
4. Energy flows through the ecosystem from producers (commonly photosynthetic organisms) to primary consumers (herbivores) to secondary consumers (carnivores) to tertiary and quaternary consumers (carnivores). This general sequence is called a food chain. Since organisms often eat at several levels in this food chain, the feeding relationships are more accurately described as a food web. Decomposers get energy from dead organisms and release materials needed for new life.
5. Extinctions are more common in simple food webs than in more complex ones. Ecosystems usually become more stable as the number of interacting species within them increases.
6. Human activity often causes extinctions. In addition to emotional reasons for being concerned about extinctions, there are some rational reasons. Irreverence for life might be directly harmful to humans because we are part of the ecosystem.
7. The earth's land can be divided into several kinds of regions called biomes, which are defined according to the plants they support. The nature of the plant community is dependent on factors such as soil conditions, water,

weather, day length, competition, as well as the nature and abundance of plant eaters and their predators. The nature of the plant community influences the kinds of animals that inhabit the biome. Six of the major biomes are considered: temperate deciduous forest, grasslands, deserts, tropical rain forest, tundra, and taiga.

8. Tropical rain forests experience consistent temperature and heavy rainfall. Some may have rainy seasons. There is an extreme variety of organisms. Competition is keen. Trees grow to great heights and form canopies where their branches meet. Where sun filters through the canopy, jungles form.
9. The dominant trees in the temperate deciduous forest are oaks, maples, beech, poplar, and hickory. These trees shed their leaves during the harsh winter and grow rapidly during the warm summers. Many perennial plants appear in the spring before the new tree leaves shade the forest floor. Annual rainfall is 75 to 130 centimeters.
10. The taiga, or coniferous forest of the north, has long, cold, wet winters and short summer growing seasons. Muskegs (extensive bogs) are common. Due to the carpet of needles, there are few plants on the forest floor. Animals include porcupine, moose, and bear.
11. Both the arctic tundra in the far north and the alpine tundra at high elevations are characterized by a covering of ice and snow, except during the short (2- to 4-month) summer when the soil above the permafrost defrosts so bogs and ponds form. Lichens, herbs, mosses, short shrubs, grasses, and a few trees are found. Animals include ptarmigan, hare, and lemmings. Recovery from disturbance is slow.
12. Grasslands are characterized by grasses, small bushes, shrubs, and in some parts of the world, bamboo. Trees may be found near streams and lakes. Annual rainfall is 25 to 100 centimeters.
13. Deserts have hot days and cold nights. Annual rainfall is less than 25 centimeters. Rain occurs in sudden bursts and then evaporates. Plants are adapted to grow and reproduce at times of water availability, to store water in their tissues, and to tap deep underground water sources. Animals must beat the heat and conserve water.
14. Because they are small, bodies of freshwater are less stable than seawater. The natural process of aging, eutrophication, is speeded up when human activity adds nitrates and phosphates from fertilizer and detergent to freshwater, thereby stimulating plant life. Dead plants fill in the lake and their decomposition depletes oxygen.
15. The oceans, which cover three-fourths of the earth's surface, are an average of three miles deep. Due to the trade winds at the surface, the deep water is in constant motion. Upwelling, which occurs where currents are deflected upward by undersea mountains, brings nutrients from the sediment upward, thereby stimulating the growth of phytoplankton. The phytoplankton are important as primary food producers and manufacture much of the world's oxygen. Phytoplankton are consumed by zooplankton and form the basis of the ocean's food pyramid.
16. The continental shelves, edges of continents, extend 10–150 miles into the ocean. Life abounds in these shallow areas. The life along a rocky coast occurs in zones that are identified by the most prevalent species. Sandy coastlines have little life because they lack a fixed surface to which the organisms could attach. Mud flats harbor less life than the rocky coasts, but more than sandy coasts. Many animals begin life in the mud flat and later enter the open sea.

KEY TERMS

biomes (708)
climax stage (698)
community (695)
competition (697)
continental shelf (730)
decomposers (703)
deserts (716)
ecology (695)
ecosystem (695)
estuary (717)
eutrophication (725)
food chain (701)
food web (701)
grasslands (716)
habitat (695)
jungles (710)
mud flats (731)
niche (696)
phytoplankton (726)
pioneer organisms (698)
populations (695)
primary consumers (701)
primary succession (698)
producers (701)
quaternary consumers (701)
rocky coasts (730)
sandy seashores (731)
secondary consumers (701)
secondary succession (699)
succession (698)
taiga (712)
temperate deciduous forests (712)
tertiary consumers (701)
tropical rain forests (710)
tundra (713)
zooplankton (726)

FOR FURTHER THOUGHT

1. What biome would be found atop a high mountain in a tropical region?
2. If you wish to prevent eutrophication of lakes, why might you choose a washing detergent that is low in phosphate content?

FOR REVIEW

True or false?

1. ___ Muskegs are large bog areas found in the taiga.
2. ___ Grasslands usually form in areas with less than ten inches of annual rainfall.
3. ___ Estuaries contain only freshwater.
4. ___ Ocean environments provide more stable habitats than do fresh-water bodies of water.

Fill in the blank.

5. A smaller section of a habitat in which one species resides is called its ___ .
6. The ___ are responsible for ocean currents such as the Gulf Stream.
7. ___ are the primary producers in the ocean.
8. ___ is the term that describes the natural aging process of lakes.

Choose the best answer.

9. Of the following, which is the largest arbitrary division that is characterized by the plant life it supports?
   A. ecosystem
   B. niche
   C. biome
   D. habitat
10. ___ is a biome that is characterized by hardwood and perennial plants.
    A. Tropical rain forests
    B. Alpine tundra
    C. Taiga
    D. Temperate deciduous forest
11. Diverse and abundant life forms are least likely to be found
    A. in deep ocean waters.
    B. in continental shelf waters.
    C. in mud flats.
    D. along rocky coastlines.

CHAPTER 26

Population Dynamics

Overview

Populations, Ethics, and Necessity

How Populations Change

Controlling Populations Through Reproduction

Controlling Populations Through Mortality

The Advantage of Death

Objectives

After reading this chapter you should be able to:

- Discuss why we should be concerned with species extinction.
- Identify the types of population growth represented by J-shaped and S-shaped population growth curves and list several ways populations crash.
- Explain why most species never achieve their biotic potential.
- Describe some reproductive strategies employed by various species.
- Differentiate between density-dependent and density-independent mechanisms.
- List types of abiotic and biotic control mechanisms and state how each regulates population size.

It is a bit startling to hear population ecologists tell us that about a million species will become extinct over the next 20 years. Many of these, of course, will pass from the earth without our ever knowing they existed. Nonetheless, they will be gone.

We are also learning that the demise of many of our fellow passengers on this small planet is due to human activity. Put simply, we are killing them off. "But *why*?" someone asks. "Why do we do it?"

There are a lot of answers to that question, but most fall into three major groups. First, we kill them out of ignorance. We release toxins that we have created chemically, perhaps as by-products or waste. In many cases, we're simply not aware of the effects of our activities. Second, we often simply out-compete the other species. Elephants may need the same land that we have decided to farm. The elephants are pushed into inferior habitats and, in hard times, many may perish. You can apply the principle to any species. Third, we may purposely kill other species. Few are hunted to extinction these days, but many are hunted to levels at which their populations cannot be sustained. When we know about them we generally place them on our "endangered species list" and the hunting, for the most part, stops.

Probably relatively few species are threatened by hunting, however. Most, by far, are threatened by our ignorance or because they compete with humans, many because they live on land that humans increasingly need.

POPULATIONS, ETHICS, AND NECESSITY

It is sometimes argued that we shouldn't worry about *other* species while humans—people—need the land. However, there are at least a couple of ways that one might justify worrying about them. For one thing, these species may have something we can use, especially considering our emerging abilities in genetic engineering. Such species may serve as genetic reservoirs. Their demise may spell the end of unusual or intriguing combinations of genes that might have proven very useful later on. Second, it is argued that the other species of the earth may be our only companions in the entire universe and that they have a *right* to exist. But perhaps that argument is a bit esoteric for a Latin American farmer who has eight mouths to feed and one on the way, and who wants to clear more land. However, a recent incident illustrates just how valid some of the more esoteric arguments can be, even to the pragmatic farmer.

FIGURE 26.1
The plant *Zea diploperennis,* related to domestic corn.

On a rather ordinary hillside in Mexico, botanists discovered a new plant species. It was in the same genus as domestic corn and was named *Zea diploperennis* (Figure 26.1). These few thousand plants were of great interest because they turned out to be perennial; thus, they do not require replanting each year. Furthermore, they are immune to a number of viruses that infect corn, and they are able to grow in wet areas where corn can't. This is all particularly interesting because there is some promise of being able to cross the wild strain with domestic corn. (It is of such things that agricultural revolutions are born.) Botanists had a devil of a time, however, in convincing the farmer who worked the land not to plow these plants under to make room for his usual crops (Figure 26.2).

FIGURE 26.2

Because of rapidly increasing human populations, land is at a premium throughout many areas of the world. In Latin America, native populations of plants are being destroyed at an alarming rate in order to make room for traditional farming efforts.

This eleventh-hour discovery of a remarkable species can only suggest how many we have already unknowingly sent on their way to extinction. We are reminded of the fact that 95 percent of the genetic varieties of wheat native to Greece have become extinct in the last 40 years. We may thus have lost forms with a genetic resistance to whatever disease might next strike the world's wheat crops.

As a personal aside, in the past few years I have travelled into the Amazon jungle to visit shamans of several remote and vanishing cultures in order to bring out samples of the various plants they use as medicines. I have now collected some 150 barks, leaves, saps, and roots that these witch doctors use as curatives. Interestingly, in two tribes, just miles apart, neither group could even recognize many of the plants used by the other group. In many cases, Amazon plant species are found in pockets (often due to historical factors, such as ancient watersheds). So by destroying even a small area, or displacing a group of Indians from an area with which they are closely associated (and then providing for their needs through some agency, if at all), we may lose precious resources through local extinction and ignorance.

A rather bizarre story illustrates just how perilous the existence of some species may be. It began with a 1959 report of an animal that some have referred to as an aquatic dinosaur of Loch Ness proportions. Fishermen on Lake Tele in Zaire reported that they found the beast and then killed it with spears because it was disturbing their fishing. It was then

FIGURE 26.3

About the turn of the century, scientists reported the existence of a *Palaeotragus* in the Congolese rain forest. The creature was believed to have been extinct for millions of years.

cut up and eaten. They complained about how long it took to prepare it because its neck and tail were so long. Dismayed scientists have offered a reward for any skeletal remains, but nothing has turned up. You may recall that the first coelacanths caught in modern times (Chapter 13) were left to rot on deck or were eaten. As a final reminder of how little we know of our fellow species, a giraffelike creature that lived 20 million years ago, called the palaeotragus (Figure 26.3) turned up alive and well in the Congolese rain forest about the turn of the century.

Extinctions, however, are admittedly only one aspect of population changes. Not all species are tail-spinning into the black hole of oblivion. In fact, most species that we know about seem to be doing quite nicely. I know that in my house, there is no shortage of the huge Florida roaches they call palmetto bugs. One even inexplicably managed to find its way into a guest's shorts at a recent gathering here. (The bugs are still around, but I haven't seen much of my friend lately.) Bluejays and squirrels fight at the feeder, the possum visits nightly, and a big female raccoon (named Rosemary Cooney), always a bit testy, rambles around in blithe evidence that her species is doing fine. And so it is with most of the species that we know of.

Some species are not just holding their own, they're doing better than that. Kudzu, a large leafy vine, is choking off the trees in the Carolinas; local scientists tell us that fire ants are proliferating; "killer bees" (a particularly aggressive strain of honeybee) are moving up from the South; our Florida waterways are becoming choked by water hyacinths (Figure 26.4), partly because the gentle manatee (sea cow) that feeds on them has become endangered (Figure 26.5) (its numbers reduced in large part by the slicing propellers of joyriding boaters who insist on their rights to use the waterways as they like). Many species are indeed on the rise—their numbers growing, their densities increasing, and their ranges expanding.

FIGURE 26.4

Water hyacinths choke many southern waters.

We see, then, that while some populations are dwindling, others are holding their own, and yet others are soaring to new limits. So let's see if we can discover some of the underlying principles of population change. The implications of the findings are enormous, as we will see. It is no secret that human population growth is quite simply out of control.

FIGURE 26.5

Manatees, the gentle giants of Florida waterways, are threatened largely by human carelessness and callousness.

HOW POPULATIONS CHANGE

The subject of **population dynamics** (how populations change) is often a focal point of conversation, from scientific meetings to cocktail parties, perhaps because we *are* becoming increasingly aware of our own population problems. Some scientists insist on telling us that we are breeding ourselves to extinction. "Ourselves?" someone archly asks. "Extinction, you say? My! Care for some coffee?" Since some people are not particularly concerned, we might wonder if it is really time for concern, perhaps for drastic measures. Or is it time for that coffee? Just what *do* the numbers mean? It is difficult to assess the data and understand the implications without some understanding of the basic principles of population dynamics.

Population Growth

The basic story of population growth can be told by two simple, sweeping lines on a graph; they are called the **J-shaped curve** and the **S-shaped curve.** Let's first consider the J-shaped curve.

If we were to place a few bacteria in a suitable medium, or a few sheep on an uninhabited but hospitable island, their numbers would increase in such a way as to produce a J-shaped curve (Figure 26.6). Since their new environment provides them with the necessities of life, their growth will be limited primarily by how fast they can reproduce, rather than by environmental factors.

Note that the rate at which new individuals are added to the population accelerates with time. (Rate is a measure of change over a given time.)

FIGURE 26.6

The J-shaped curve. Time is plotted against population size. In such a case, there is nothing operating to restrict growth, so the population is achieving its biotic (or reproductive) potential. The characteristic curve is produced because the number of reproducing individuals is continually increasing.

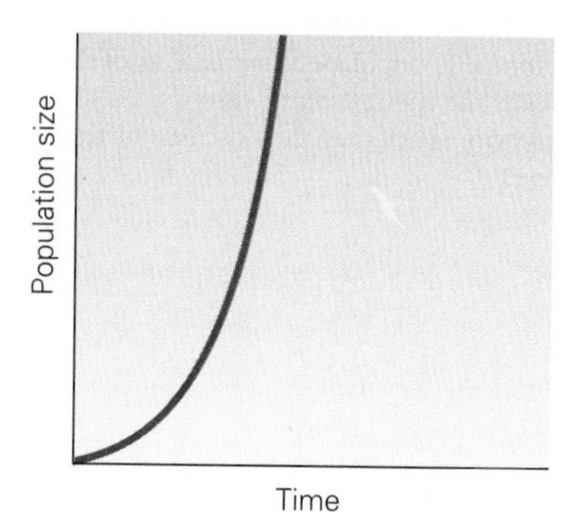

FIGURE 26.7

The S-shaped curve generated when a population approaches the environment's carrying capacity.

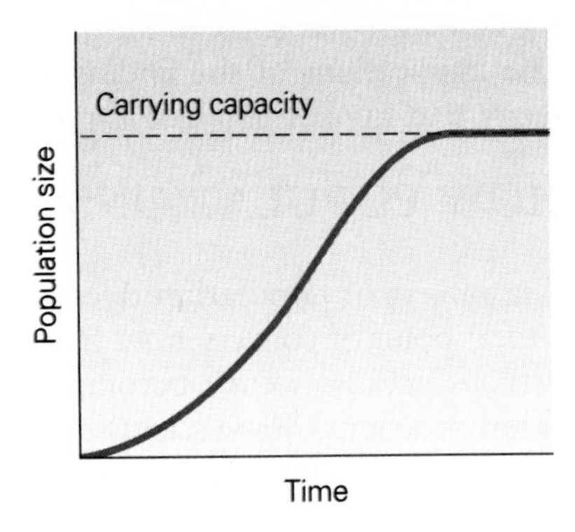

This is true even if individuals in the population continue to reproduce at the same rate because the number of reproducers keeps growing. The distinctive curve is produced because of this change in rate. If each individual in the original group leaves two offspring, the population will remain fairly small for several generations (one bacterium divides to leave two, these leave four, which leave eight—all low numbers). But after a few rounds of reproduction, the numbers begin to skyrocket. (What would be the result after only ten generations? Twenty generations?) The reproductive capacity of any population, when it is unrestricted, is called its **biotic potential,** and the J-shaped curve illustrates a population approaching its biotic potential.

The full biotic potential of any species, however, is not likely to be reached because of the inherent limitations imposed by the environment. **Environmental resistance** includes those factors in the environment that act to reduce a population's increase. Environmental resistance increases as a population approaches the environment's **carrying capacity,** the number of individuals that an environment can sustain over a long time. Generally, populations will finally increase to the environment's carrying capacity and there they will oscillate a bit, not rising too far above, or falling too far below, the carrying capacity (Figure 26.7), so, generally, as a

FIGURE 26.8

Normally, populations increase until they reach the environment's carrying capacity. They may then oscillate at this level.

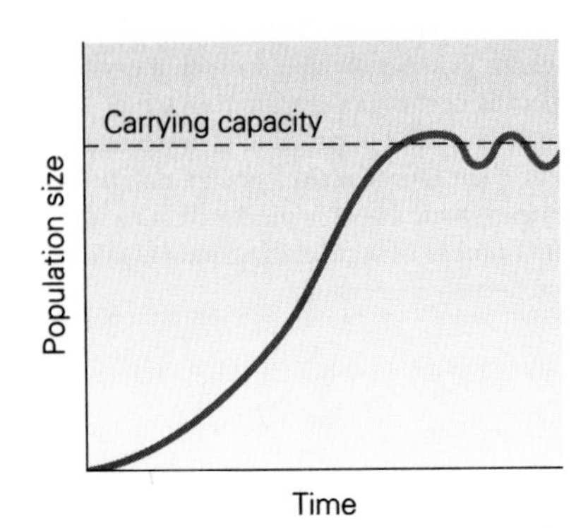

FIGURE 26.9

The theoretical relationships between biotic potential, environmental resistance, and carrying capacity. The biotic potential of a population is rarely, if ever, achieved due to the problems of living in a real world (the problems are collectively called environmental resistance). The characteristic growth curve is produced because a small population grows slowly, since a few individuals can generate only a small number of offspring. Once the breeding population reaches a critical size, however, it skyrockets (still well below its biotic potential) to be slowed only by the effect of its own numbers. Theoretically, it should stabilize around the carrying capacity of the environment.

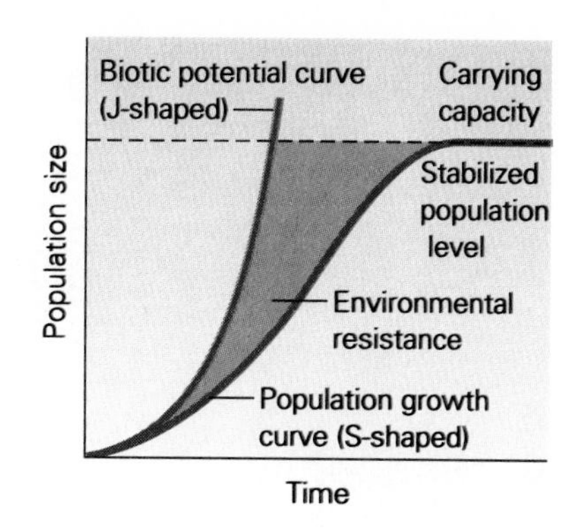

population approaches the carrying capacity of its environment, its rate of growth slows down and its numbers approach the carrying capacity more slowly, finally to remain at about the level that the environment can support. As its numbers begin to level off, they produce an S-shaped curve (Figure 26.8). The relationship between these various factors is summarized in Figure 26.9.

Under certain conditions, a population may rise so sharply that it drastically overshoots the carrying capacity of its environment. When the numbers are too large to be sustained by the environment, the resulting environmental overuse may cause a sudden drop in population numbers (a **crash,** Figure 26.10).

In some cases, populations surge to great numbers and then crash to very low levels. The surges and crashes of some populations are to be expected, as with the desert flowers that suddenly appear in abundance after a rainfall and that disappear just as suddenly as the desert soil becomes parched once again, their descendancy entrusted to their seeds. The spring

FIGURE 26.10

Population size may suddenly drop for a number of reasons, often due to depletion of resources. Such drops are called crashes.

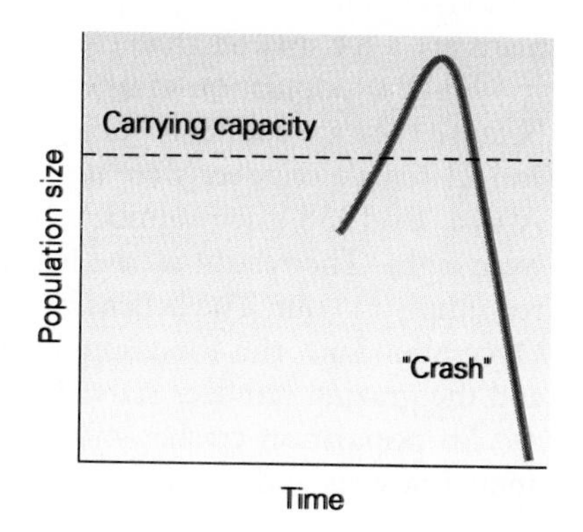

FIGURE 26.11

Elephants eating acacia trees. Unknowingly, they may be destroying the future of their population by eating too much of the foliage that supports them and the soil.

woodlands are home to countless noisy insects that survive through the summer to perish in the hastening cold. These sorts of surges and crashes are predictable and part of the annual life cycles of many species.

In other cases, population crashes are not part of any established pattern. These crashes, in fact, can be catastrophic and, in many cases, they are caused by the population drastically overshooting the carrying capacity of the environment. For example, a bacterial population being grown in a petri dish on a laboratory shelf will rise sharply. However, it will eventually run out of food, even while it befouls its environment with its own wastes. Its numbers will then drop precipitously. Since the carrying capacity has also been lowered, the population cannot recover. As another example of population crashes due to permanent change in carrying capacity, oversized herds of elephants may eat all the young plants in their habitat that, if left alone, would have eventually provided more food and for longer periods, perhaps for generations afterward. Elephants, then, can lower the carrying capacity of their habitat, perhaps permanently. If they strip away so much foliage that they markedly increase erosion, the carrying capacity may become lower still, perhaps to the point that it is no longer suitable for this species at all (Figure 26.11). Fortunately, humans are a lot smarter than elephants.

Finally, population crashes can occur when resources are depleted because they are suddenly subjected to heavy, new uses. In 1891, 4 male and 21 female reindeer were introduced to one of the Pribilof Islands off Alaska. By 1938, their numbers had reached about 2000. By 1950, there were eight. The reason is, the lichens and other materials on which they fed simply ran out. The lichens had grown for centuries—but very slowly. Once abundant, the food supply could not replenish itself fast enough and the grazing reindeer starved (Figure 26.12).

So population crashes may be part of the normal pattern (as with annual flowers and many insects); they may occur when natural popula-

FIGURE 26.12

Numbers of reindeer after introduction to an island. A sudden, new exploitation of a resource can deplete the resource and cause the population of the exploiter to crash.

tions befoul their environment (as with bacteria) or outstrip their traditional resources (as with elephants); or they may occur when resources are subjected to heavy new uses (as with reindeer).

Now we will take a look at just how the environment may influence population numbers. Essentially, we will find that populations can theoretically be regulated in two ways: (1) by the number of offspring produced and (2) by the mortality of the individuals in the population.

CONTROLLING POPULATIONS THROUGH REPRODUCTION

The biotic potential of most species, even the slowest breeding ones, is surprisingly high. Charles Darwin estimated that a single pair of elephants would leave over 19,000,000 descendants in only 750 years. The fact that the entire world was not teeming with elephants indicated to him that some elephants were not reproducing and that the reproducers were somehow "selected" by the environment.

Whereas elephants are very slow breeders, other species are more prolific. For example, the startling reproductive potential of the housefly for a single year is shown in Table 26.1.

Among factors determining the rate at which an organism reproduces is energy. For example, simple energy demands would not allow an elephant to become pregnant several times a year (Figure 26.13). There is just no way that a female could find enough food to produce that many offspring. And what would happen to the helpless baby elephants already born as new brothers and sisters appeared on the scene?

TABLE 26.1
Projected Populations of the Housefly Musca domestica *for One Year*

GENERATION	NUMBERS IF ALL SURVIVE
1	120
2	7,200
3	432,000
4	25,920,000
5	1,555,200,000
6	93,312,000,000
7	5,598,720,000,000

Source: Adapted from E. V. Kormondy, *Concepts of Ecology* (New Jersey: Prentice-Hall, Inc., 1969), p. 63.

If energy is a prime factor in determining reproductive output, it could exert its influence at two levels, ultimately and proximately. At the ultimate (evolutionary) level, natural selection could set the reproductive abilities of an organism according to the natural history of that line. Those that have traditionally had great resources and few demands might develop the physiological abilities to make quite grand attempts at reproduction. At the proximate (immediate) level, reproductive attempts might be dictated by the prevailing conditions. Thus, there are species that produce, say, ten offspring in a good year, but in hard times, their lack of food might cause them to attempt to produce fewer.

Ecologists sometimes speak of the "reproductive strategy" of a species. This "strategy" is not some consciously derived tactic for leaving offspring. It simply refers to the means that have evolved for any species as the most effective way to leave offspring. The means, however, may differ from one

FIGURE 26.13
Elephant reproduction must be carefully programmed because of the great demands the infant places on the mother.

species to the next depending on a number of factors in their natural histories. We can illustrate the point by considering two extremes, tapeworms and chimpanzees. Most species, though, fall somewhere between these and, therefore, have "mixed" strategies, as we will see in a final group, birds.

Tapeworms

The tapeworm living so comfortably in the idyllic environs of your intestines is lucky to be there (although its luck is the inverse of yours). It is lucky because the life of a tapeworm is fraught with risk and its success is largely dependent on chance. The odds are very much against its finding a host. In order for that tapeworm to accompany you on your dates and visit France with you in the summers, it had to successfully survive some rather unlikely events (Figure 26.14). First, it was passed, as an egg, from

FIGURE 26.14

The life cycle of a tapeworm. An infected human may suffer damage to the brain or leg muscles. However, the probability of a man failing to cook the part of a pig in which a tapeworm larva was encysted after the pig had eaten food contaminated by human feces, which had come from another human with a tapeworm, is low. So how does a tapeworm ever reproduce? It simply lays thousands and thousands of eggs and leaves the rest to chance.

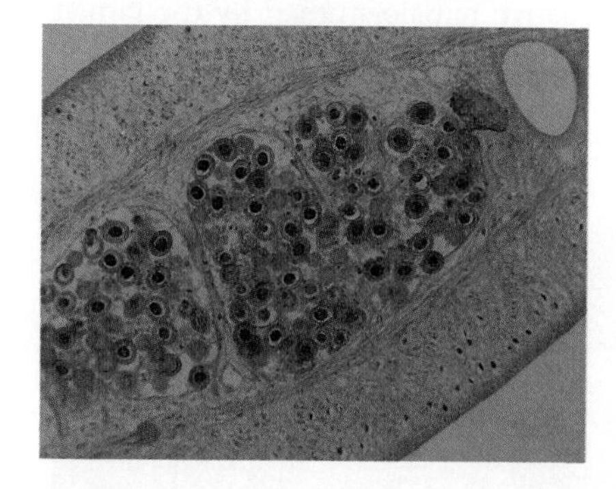

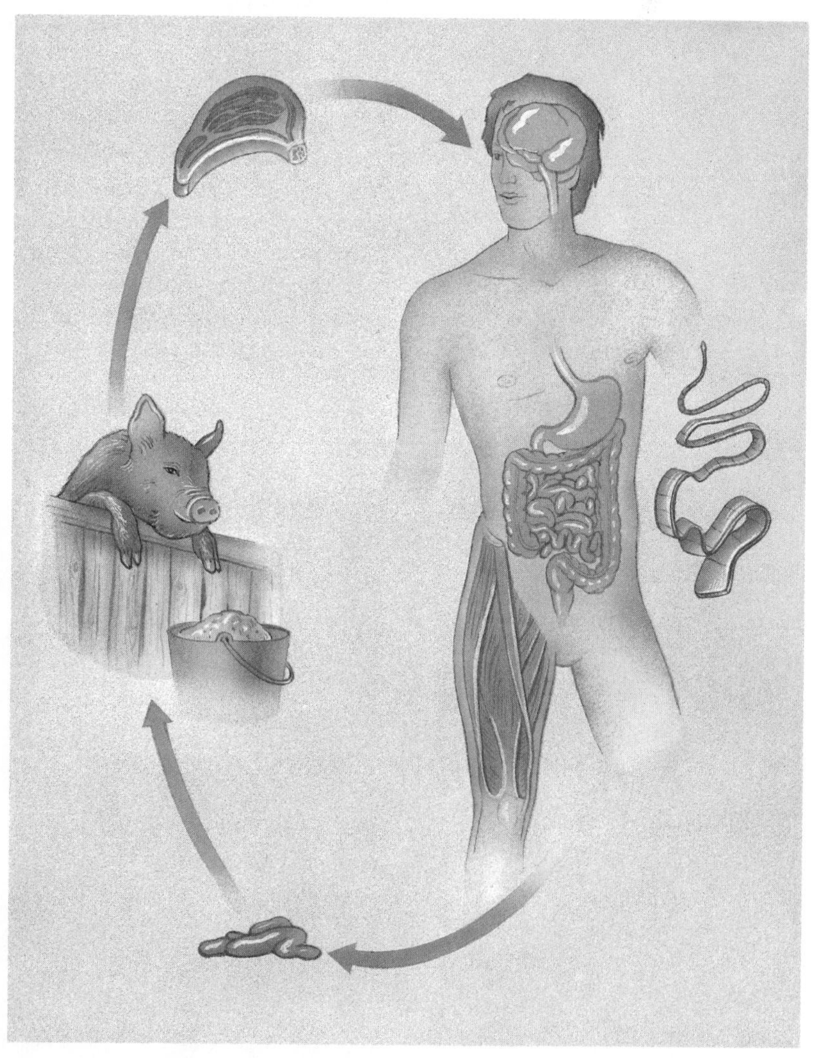

some previous human host. Then it lay around in the sun and rain until it was eaten by a pig. Not only must it survive the elements, but it must also be found by a pig, so already the odds are against it. Once inside the pig, it begins to grow and change—its appearance becoming wormlike. It then moves through the pig's intestinal wall and on through the bloodstream until it reaches the pig's muscle. Here it forms a protective capsule around itself and waits for a human to eat the pig's flesh without cooking it too well. If this should happen, the worm attaches to the intestine of its human host, grows into an adult, and begins to lay eggs.

There is obviously an overwhelming probability that not all these conditions will be met, and thus most eggs will not become adult tapeworms. So the tapeworm must have some means (a "strategy") of overcoming these odds. Through natural selection, the tapeworm has become capable of self-fertilization, a particular convenience when an individual is not likely to ever encounter another of its own species. It is also able to produce prodigious numbers of eggs. Its whole life is devoted to egg production. It exists only to lay eggs, eggs, EGGS! A few of these, with luck, will wend their way through the complex maze that is the tapeworm's life cycle. So tapeworms beat the odds against their reproducing by putting very little energy into any one offspring, but laying so many eggs that some make it.

Chimpanzees

Now let's look at a species that is not in the numbers games. The chimpanzee has developed a different reproductive strategy. It reproduces much more slowly. In fact, during the time a female chimpanzee is sexually receptive, she may copulate with almost any male in the group, but once she has become pregnant, she may not become sexually receptive again for years. Flo, the aging but sexually active female studied by the famed

FIGURE 26.15

The reproductive strategy of intelligent and social creatures such as the chimpanzee involves carefully tending and training few offspring, giving them time to adapt to their complex world.

British field researcher Jane Goodall, did not become sexually receptive for five years after delivering a baby. Of course, this meant she had plenty of time to attend to her offspring.

In the first months, Flo carried her baby everywhere and diligently guarded it against danger. Later, it was permitted brief forays on its own, but never far from her vigilant eye. The curious but wary baby would scurry back to her at any sign of real or imaginary danger. During its first few years, the young chimpanzee gradually became able to care for itself and to associate less and less with its mother. Finally, Flo was free to mate again and to rear another baby. Among chimpanzees, we see, reproductive success does not involve producing large numbers of offspring. Instead, few offspring are produced and are given very careful attention until they are able to be independent. How did such a strategy evolve?

Chimpanzees have historically existed in a precarious and complex world. They have traditionally fallen prey to the big cats (whose numbers have now been drastically reduced by human hunters out to bag a member of some species "before they're all gone"). Furthermore, chimpanzees live in a seasonal world where various kinds of foods come and go. Their food includes seasonal fruits, leaves, shoots, baby baboons, small monkeys, and insects (including ants, which they may pick up by using tools made of broken twigs)—quite a varied menu with a different skill involved in acquiring each item. Also, chimpanzees must move around in order to exploit food sources as they become available. Each night they build a new nest of bent branches, twigs, and leaves. Through all of this, they are highly interactive, each animal recognized by the others and treated according to rank, sex, age, or whatever. Chimpanzees, we see, are constantly dealing with new, complex, changing, and sometimes threatening situations.

To cope with such a physical and social environment, the chimpanzee has developed a high intelligence and a system of extended parental care (Figure 26.15) that gives the young chimpanzee time to gain experience under the watchful eye of its mother as it learns about its complex world. Such extended parental care, however, involves a great deal of time and energy, and so chimpanzees are able to produce only a few offspring during their lifetimes. In this species, selection has resulted in the production of only a few infants that are then carefully attended to. So chimpanzees, like the tapeworm, seem to rear as many young as possible, but in accord with the demands of their particular kind of environment and history. Thus we see that the number of offspring attempted by members of any species is likely to be set by natural selection, and the "strategy" of natural selection will depend on a number of historical and environmental variables.

FIGURE 26.16

The common swift lays different numbers of eggs depending on resource availability.

Birds

Bird studies have revealed reproductive strategies that provide a rather clear picture of how the environment can influence the number of offspring produced by any species. We will see that birds lay the number of eggs that will enable them to rear as many young as possible. In most cases, this is the number of young the parents can *feed* successfully.

Females of the common swift usually lay one to three eggs (Figure 26.16). Apparently, the number has not yet stabilized through selection (stabilizing selection) or it is undergoing a shift due to some kind of new

FIGURE 26.17

The ground-nesting gulls, such as the herring gull (a), have been forced to assume a different reproductive strategy than the cliff-nesting kittiwake (b). The nest of the ground-nesting species is vulnerable to marauding predators, so they compensate by laying more eggs than they will be likely to rear. The ground nesters, for example, lay three eggs, whereas they can probably rear only two young, because at least one of the eggs is likely to fall to a predator. The kittiwake, on the other hand, nests on cliffs safe from predators. Since all their young are likely to survive, they lay only two eggs.

(a)

(b)

ecological pressures (directional selection). In years in which food is abundant, the number of young that reach feathering age was found to be 1.9 in those nests in which two eggs had been laid and 2.3 for nests with three eggs. When eggs were added to a number of nests so that the total was four, an average of only 1.4 young survived until feathering age. (Suppose the food supply were to remain relatively abundant and constant. At what number would the eggs laid by females of this population stabilize?)

The number of offspring attempted (eggs laid) by some species may also be influenced by the vulnerability of the young to predators. For example, some species of gulls nest on accessible beaches, and here the nest and young run a great risk of being discovered by prowling foxes or other nest marauders. Whereas these gulls can usually feed only two young successfully, there is a strong likelihood that at least one of the young will fall to some predator, so three eggs are laid. Since one is likely to be taken, the number will be adjusted to that which results in maximum reproductive success. On the other hand, the kittiwake, a species of gull that nests on cliff sides, normally lays only two eggs. A fox trying to get at these eggs is likely to break its neck. So two eggs are enough; less energy is expended by the parents, the chicks are safe from predators, and both young are likely to reach adulthood (Figure 26.17).

In some species of birds the broods are small, even when the parents could successfully feed more young. One reason is that fewer eggs result in a safer nest. Conspicuous eggs, the activity of young, and the coming and going of parents might advertise the location of the nest and attract predators. Fewer eggs require less conspicuous activity around the nest.

In rather complex ways, time may also influence the number of eggs laid. For example, a given level of food can theoretically raise a small brood quickly or a large brood slowly. In areas with short nesting seasons, such as the polar regions, time is of the essence because food is abundant for only a very brief period. In such areas, there would be greater advantage in raising a small brood quickly than in portioning out the same amount of food among a larger brood that might not mature fast enough to escape the hastening winter.

The mechanism by which some birds regulate the number of eggs they lay may simply be one of energetics. After all, food provides the energy and substance to manufacture eggs. Thus, if food is scarce, the female is able to form fewer eggs. (A reduction in the number of eggs may actually mean that more young will be reared, since each one will receive a larger proportion of whatever food there is.) The following season, if food is more abundant, she may initiate the formation of more eggs.

CONTROLLING POPULATIONS THROUGH MORTALITY

Just as the numbers that are born in a population are influenced by environmental factors, so are the numbers that survive. It is easy to see how population size can be regulated by death, but it is less obvious that death usually occurs through one of two fundamental means.

Abiotic Control and Density-Independence

Death can be met by a number of means, as we all know. But these can basically be categorized according to whether they are nonliving (**abiotic;** *a:* without; *bios:* life) or living **(biotic).** First we will consider abiotic population control, that due to nonliving influences.

One of the greatest causes of death among many forms of life is the weather. We see this when normal seasonal changes bring on winter hardships that decimate populations (Figure 26.18). Such cycles can be hard on populations, but bad weather is especially dangerous when it is irregular or unusual. Almost yearly, we hear of the devastating effects of drought somewhere on the planet. Perhaps less spectacular, but just as deadly, is unseasonably cold weather. Many birds perish in some years because they begin their northward migration in the spring only to be caught by a late cold snap. Natural selection cannot effectively prepare organisms for novel, unusual, or unexpected conditions. Severe weather, then, is an example of abiotic control. A drought is not alive.

Population-depressing factors that are not influenced by density (the numbers of individuals in a given area), such as severe weather, are **density-independent.** That is, such effects are independent of the population's density. In a severe drought, the parching sun is not influenced by how many plants are struggling in the field below. It kills them all. In an area saturated with DDT, insects die whether there are few or many. Their numbers mean nothing unless they somehow tend to protect or shelter each other.

FIGURE 26.18

These deer have died as a result of the harsh effects of winter.

The awesome impact of the human species on the planet's life-support system may soon bring the reality of density-independent population control into sharp focus. For example, many nations, including ours, routinely use the oceans as a dump for poisonous wastes that are not easily manageable on land. (Humans dump about 90 percent of their wastes into the earth's waters.) Is there some threshold, some point at which the sea's chain of life will no longer stand the insults? When will we cross that point of no return? We can't say, because we haven't a clue what that point is, but we keep dumping.

Even the soothing rain has been changed so that it has become a real threat in areas (Essay 26.1). If we continue to befoul our environment so

FIGURE 26.19

Rats and their fleas have drastically reduced human populations by helping transmit bubonic plague.

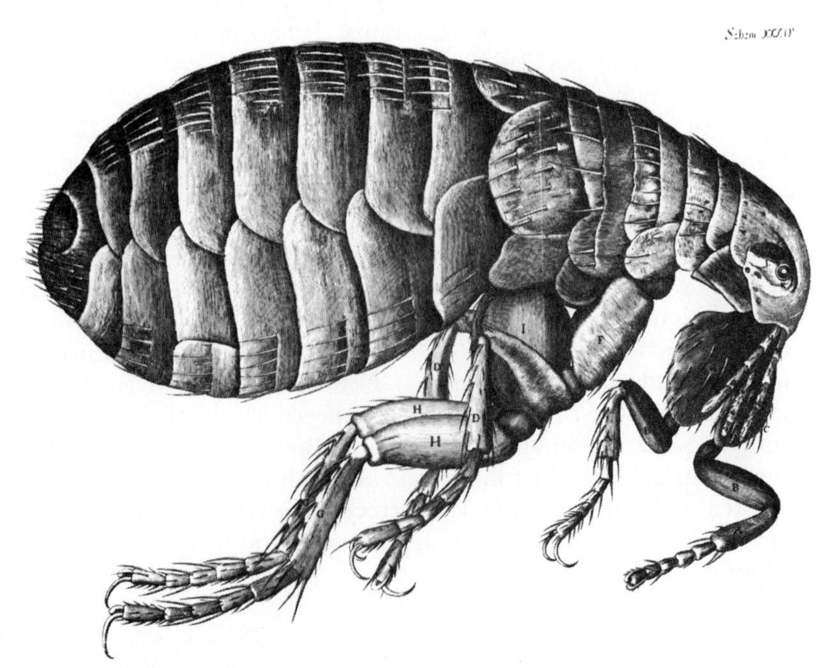

ESSAY 26.1

What Have They Done to the Rain?

In the 1970s, it became clear that the rain was changing. In fact, in some areas the gentle raindrops were downright dangerous. The rain was becoming a dilute mixture of acids. It was first noticed in Scandinavia, then in the northeast United States and southeast Canada, then in Northern Europe and Japan.

Rainwater, of course, had always been slightly acidic because the water dissolved atmospheric carbon dioxide, forming carbonic acids. But now the rain was showing alarming concentrations of the more dangerous sulfuric acid and nitric acid. Where were they coming from? They were the result of accumulations of nitrous oxides and sulfuric oxides in the atmosphere. The nitrous oxides, it turned out, were from power plant and automobile emissions; the sulfuric oxides, mainly from power plants and smelters. Dissolved in the water of cloud formations, they formed nitric acid and sulfuric acid, then fell to earth to bathe our forests and cities and to fill our lakes with the corrosive mix.

The relative proportions of the two acids in our rain depends on where one lives. In the northeastern United States, the acidity is primarily due to sulfuric acid; in California, to nitric acid. So we do have a choice.

The rain has caused the reduction and even the elimination of fish in many of our lakes. The rain apparently doesn't kill the fish, it just keeps them from reproducing. So no young fish are found as the old ones gradually go the way of all flesh. In fact, about 700 lakes in southern Norway are now *entirely devoid* of fish, and our own northeastern lakes are following one by one. As our Adirondack lakes reach pH levels of 5 (not uncommon), 90 percent have no fish whatever. They are also curiously devoid of frogs and salamanders.

Entire patches of forests worldwide are sickening and dying (see photos) as ecologists busily try to find out just what effects the rain is having.

In 1990, the results of a study were released, indicating that in the United States, the effect of acid rain in nature was not as bad as we thought, partly because relatively few lakes are affected, partly because the effect on such lakes is reversible. The effects in cities, though, is much more profound.

Interestingly, the solution is clear to everyone. We simply need to reduce the levels of our effluent from power plants, smelters, and automobiles. Most of the technology exists, but its implementation would be too expensive for the polluters to willingly bear, and our drinking water becomes increasingly unusual.

callously and routinely, we may indeed begin to more clearly understand the meaning of density-independent population control.

Biotic Control and Density-Dependence

Biotic population control refers to living influences on population size. Biotic controls are widely varied and often more complex than abiotic control mechanisms. For example, the organism that causes bubonic plague can reduce populations (Figure 26.19) and so can a tiger. But biotic influences can operate in subtler ways than simply by killing outright. As an example of indirect influences, a territorial bird drives a competitor into an area where there is less food. When winter comes, the underfed competitor may be more likely to succumb to the rigors of the season. Thus, the territory holder has indirectly brought about the reduction of the population.

Biotic controls on populations are likely to be **density-dependent.** Here, the effect of the control mechanism depends on the density of the population. For example, with density-dependent control, an increase in population density increases the effects of mechanisms that reduce the density. Then, as the density falls, the pressures that had reduced the population begin to lessen, permitting the number to increase again. (You may recognize the familiar negative feedback system here.) Let's now consider a number of biotic, density-dependent effects on population, beginning with predation.

Predation

Interestingly, a predator usually doesn't eliminate its prey, but it may devastate its competitors. For example, when the vigorous and intelligent

(a)

(b)

FIGURE 26.20
The vigorous and intelligent dingo (a) proved too great a competitor for the Tasmanian devil (b), where they overlapped on the Australian continent.

dingo, a wild relative of the dog, was introduced to Australia, it didn't kill off the primitive prey it found there. But its hunting prowess proved so superior to that of its competitors, the Tasmanian devil and Tasmanian wolf, that they disappeared from the Australian continent (Figure 26.20). In Tasmania, where the dingo was not introduced, the Tasmanian devil still survives. The question is, why didn't the dingo wipe out its prey?

There is no simple answer to the question of why predators don't wipe out their prey that applies to all populations at all times. One reason may be prey-switching. As a particular prey species becomes scarce, its predator often switches to a more abundant type of prey, one that is perhaps easier to locate and capture. Another reason may be that predators often find it easier to capture the very young or old or the sick individuals in a population. The stronger and healthier ones are likely to escape and live to reproduce.

Whereas the delicate and interacting density-dependent mechanisms may work quite well in the wild, our own species can provide us with remarkable examples of how the system can fail. Many of the great whales have been hunted to the brink of oblivion over the past few decades as modern whaling methods have reduced personal risk and increased profits. Furthermore, instead of relaxing the pressure on threatened species and allowing the whale populations to recover, the fleets initially increased their efforts. Of course, as whales decreased in number, the price of whale products went up and the whalers can be expected to respond even more feverishly to the merchant's pleading (Essay 26.2). Even with new international sanctions and agreements, some countries, notably Iceland, will continue "limited" killing (Figure 26.21). (Admirably, Shoney's, Long John Silver's, and Red Lobster have suspended purchases of Icelandic fish. Guess who hasn't. Ask the managers.)

FIGURE 26.21
Whaling nations continue to find ways around international bans on hunting whales.

ESSAY 26.2

Greenpeace and the Bartenders

Greenpeace is an organization dedicated to the preservation of the sea and its great mammals, notably whales, dolphins, and seals. Its ethic is nonviolent but its aggressiveness in protecting our oceans and the life in them is becoming legendary. In their roving ship, the *Rainbow Warrior*, Greenpeace volunteers have relentlessly hounded the profiteering ships of any nation harming the resources Greenpeace deems to be the property of the world community. Whales, they believe, belong to us all and have a right to exist no matter what the demand for shoe-horns, cosmetics, and machine oil. (In 1985, the *Rainbow Warrior* was sunk in a New Zealand harbor by French military saboteurs just before it was to sail into the South Pacific to protest French nuclear testing there, killing one member of the group.)

Greenpeace volunteers routinely place their lives in danger in many ways, such as by riding along the backs of whales in inflatable zodiacs, keeping themselves between the animal and the harpoons of ships giving chase. They have pulled alongside Dutch ships to stop the dumping of dangerous toxins into the sea. They have placed their zodiacs directly in the paths of ships disrupting delicate breeding grounds of the sea with soundings and have forced some to turn away or even abandon their efforts. They have confronted hostile sealers on northern ice floes to try to stop them from bludgeoning the baby seals on the birthing grounds, skinning them on the spot, and leaving the mother sniffing at the glistening red corpse of her baby as its skin is stacked aboard the ship on the way to warm the backs of very fashionable people who frequent places where the bartender knows their names. (The mother seal would be very happy to know that the skin of her baby had nearly impressed some bartender.)

Parasitism

Parasitism exists when an organism, the parasite, exists on or in the body of another organism, the host, deriving benefit and doing harm, but usually without directly killing the host (Figure 26.22). The relationship between parasite and host is, in a sense, much more delicate than that between predator and prey, because it is to the parasite's advantage for the host to live. Also, the offspring of the parasite will have greater likelihood of success if the host itself lives to reproduce.

Whereas parasites are harmful, in certain cases, the parasite and host are so well adapted to each other that they actually exist quite nicely together. As a rule, the newer parasite-host relationships result in more damage to the host. In older associations, the two have had time to adapt to each other, "fine tuning" their relationship.

FIGURE 26.22

In some cases, the parasite kills the host, as we see here with the sphinx moth larva having been parasitized by a wasp. These wasp larvae will kill the moth larva when they hatch.

In some other cases, such adaptation never has a chance to take place because the initial effects of the parasite are so devastating, to the detriment of both. In 1904, the sac fungus *Endothica parasitica* was accidentally introduced into North America from China. In China, the parasite was held in check by a variety of control mechanisms that do not exist in North America. Sadly, the majestic American chestnut tree proved defenseless to the fungus, and by the late 1940s, the great tree that had dominated our Appalachian forests had been mostly reduced to vegetatively reproducing shrubs, with virtually no self-sustaining populations (Figure 26.23). Fortunately, researchers are still working to save the species and there have been recent successes that may restore the tree to our forests.

FIGURE 26.23

The American chestnut was nearly wiped out by an introduced fungus. Here, a new plant has been grown by new techniques that may restore the tree to our forests.

Competition

We have seen that populations may be held in check by what might be called ecological factors such as competition brought about by low food supply. The role of competition in the world of life has recently been hotly debated in academic circles. However, no one can deny that it exists.

Competition, after all, results when two individuals utilize the same resource. In some instances, each individual in a population scrambles to get as much as possible of the available resource. If the resource is scarce, all members of the population may grow and reproduce more slowly. In other species, behavioral phenomena called territories and hierarchies serve to divide the resource unequally.

Territoriality involves the defense of any area. Territorial species generally simply chase away competitors from the territory (Figure 26.24). In this way, the winner is likely to end up living in a better area than the loser. One area, for example, may hold more commodities, such as nest sites, hiding places, or food. If some areas hold more food than others, say, in times of food shortage, those individuals with the best territories are more likely to survive. Territories may be important in other, perhaps related ways as well. For example, in some species of birds, only those

FIGURE 26.24

Sandhill cranes may engage in fighting over territories that hold resources.

males with the better territories are able to attract females. A female will not mate with a male with an inferior territory no matter what other qualities he has. If there aren't enough good territories to go around, then, not all individuals will be able to reproduce, and thus the population number is lowered.

Hierarchies, or "pecking orders," also result in certain animals having freer access to commodities than do others. In hierarchies, each animal must be able to recognize others individually and to respond to each according to its rank. Hierarchies are established in various ways in different species. They may be established through fighting, ritualistic displays (Figure 26.25), or even play. Hierarchies are adaptive in that they reduce conflict within groups as a result of each animal yielding when confronted by a higher-ranking individual. In stressful times, such as periods of food shortage, then, the higher-ranking animals are more likely to survive than are subordinates because of their freer access to critical resources.

FIGURE 26.25

In hierarchies, the subordinate signals its submission to the dominant animal.

In both territorial and hierarchical systems (which are not, by the way, mutually exclusive), harsh conditions are more likely to kill the lower-ranking animals, the "have-nots," thereby increasing the odds of survival for the "haves." This doesn't mean that the have-nots accept their role for the "good of the species." Quite the contrary. It is to their advantage to stay alive at all costs. However, a low-ranking individual may have better luck by searching for unclaimed commodities than by challenging a dominant, and probably more powerful, individual. The subordinate is likely to lose the conflict and the commodity as well, while risking being battered and weakened for its efforts.

Disease

We should have little trouble believing that disease can reduce the size of populations. (We will see in the next chapter how bubonic plague drastically reduced the human population, and we are aware that all sorts of less dramatic diseases claim thousands of lives daily.) Disease is density-dependent, as a general rule, for several reasons. The more closely a group is packed together, the more easily many diseases are transmitted. Also, the more individuals there are in a population, the greater the number of potential reservoirs in which mutant strains of disease microorganisms can develop. If crowding results in some individuals being weakened, disease will have an even greater impact on the population.

In some cases, disease and predation may act together in reducing populations. For example, a two-week-old caribou fawn can outspring a full-grown timber wolf. However, caribou are subject to a hoof disease that quickly lames them. It is these lamed animals that wolves are likely to attack and cull from the herd. The "sanitizing" effect of predators is now well established. As evidence of such an effect, in those places where wolves have been poisoned to "protect" migrating caribou, the hoof disease has spread unchecked and, in only a few seasons, has wiped out entire herds.

THE ADVANTAGE OF DEATH

Have you ever wondered why you must die? Just the realization of that fact can be a real nuisance. So let's ask why death occurs at all (Figure 26.26). Isn't it logical that natural selection would have tended to extend life spans so that by now only disease or accident could claim us? Why are we slated to start falling apart, one system after the next, until we

FIGURE 26.26

Death in most cultures is marked by elaborate ritual. At left, a Buddhist monk's cremation in South Korea. At right, graveside effigies in Indonesia.

simply can't go on? Why must we suffer the indignity of repairing one bodily breakdown after another, prisoner in a patchwork body until we finally expire, looking like the product of a quilting bee? Why can't all our systems hold? (Or all go at the same time so that one can at least be a handsome corpse?) If "life" is better than "death," then why is there not marked selection for longevity? Why are there delicate insects that don't even have mouths with which to feed themselves, but which must instead spend whatever energy their frail bodies possess in finding mates before their few precious hours of life are gone? Why do humans have so much trouble surviving past "threescore years and ten"? Why death?

People with all sorts of perspectives on life have ruminated on the "meaning" of death. Various notions have been proposed by philosophers, theologians, novelists, dramatists, poets, drunks, and others of bad habit, and none of the explanations seem to coincide. Death to some is a fearful thing, while others welcome it as the gateway to a better life, but without modern conveniences. Some aboriginal societies deal with death casually and fearlessly by treating it as a kind of natural extension of life. Admittedly, the dullest dead person knows more about the meaning of death than all the living philosophers combined. But since it is impossible to prove us wrong, let's join the fray and consider death from a biological point of view. We might first account for its genetic advantages. (Note that last word. Let's see if it makes sense.)

The biological explanation of the advantage of death becomes apparent if we keep three things in mind. First, evolution at its basic level is a means whereby certain kinds of genes (not individuals) are perpetuated. Second, the earth changes, and third, it is a limited place.

To put these principles into the perspective of our argument we must adopt the rather irreverent notion that individuals are simply temporary caretakers of their genes. The evolutionarily more successful types of genes will be those found within individuals who propelled them into the next generation. None of today's *individuals* will be represented in the next generation, but some of today's *genes* will be. The point is that in the cool, unrepentant arithmetic of evolution, the individual is expendable.

Such "wastefulness" can be seen to be evolutionarily advantageous if we keep in mind the next two points—that is, the earth changes and it is limited. We have already noted that the variability of offspring is one way an individual might be able to leave its genes in a changing world. Put simply, an individual might not be malleable (changeable) enough to survive if conditions should change too drastically. But among that individual's variable offspring might be those that had just the traits necessary to continue on under the new conditions. However, since the earth is a finite place with limited resources, a parent that lingered on to compete with its offspring might damage the chances of success for those offspring. So, strange as it seems, an individual's best reproductive (or evolutionary) bet might be to reproduce its kinds of genes in various combinations and then to get out of the way so as not to interfere with them. Of course there are those who might say "To heck with kids. I want to live." No matter, that individual is here because countless earlier generations were programmed to make the sacrifice and thus that rebellious soul's time, too, will come.

FIGURE 26.27

Do we really behave primarily in such a way as to leave our genes behind after we die? What happens to the genes of people who don't?

By considering death as a mechanism, or tool, a way to perpetuate one's kind of genes, we might be left with the idea that an organism is simply a vehicle that genes use in order to reproduce themselves. Of course, we would prefer to feel that our lives have a bit more meaning than that, that we are more than carrying cases for our gonads. Nonetheless, there is a line of logic that suggests that we are simply our genes' way of making more genes. Therefore, we primp, work, save, worship, rationalize, love, lust, avoid, hate, defend, retreat, and sunbathe (Figure 26.27). We have a variety of explanations for all the things we do. But perhaps what we are really doing is behaving in such a way that our genes end up in the next generation. If our primping helps us to reproduce our genes, primping it is. According to this notion, we are really acting at the behest of our genes, and we are given whatever latitude might most effectively enable them to be reproduced and passed on. Of course, we must ask, are we really just great lumbering robots, controlled by our genes, responding to these constant, wheedling messages that are delivered by other mindless intermediates called enzymes?

The question is definitely a bit tortured and overextended, but there is an elegant simplicity to its assumptions, once we can accept the irreverence, just for the sake of entertaining an unfamiliar concept.

We've tossed around some rather weighty concepts fairly loosely here. Perhaps, however, we've managed to ask some rather fundamental questions. Perhaps, also, we have suggested new ways to investigate old problems. This is one of the roles of science in our increasingly complex world.

SUMMARY

1. An alarming number of species is becoming extinct. Reasons for concern about the extinction of species include the loss of a product that may be of future use and that other species have a right to exist. The population size of some species is constant. Still other species are increasing or decreasing in numbers.
2. Organisms growing in an environment with unlimited resources have a rapid rate of increase that is described by a J-shaped growth curve. The unrestricted reproductive capacity of a population is its biotic potential.
3. A population rarely reaches its full biotic potential due to environmental resistance, which is those environmental factors that reduce a population's increase. As a population reaches the carrying capacity of the environment, environmental resistance increases and growth slows. This type of growth is described by an S-shaped growth curve.
4. If a population greatly overshoots its carrying capacity, the environmental overuse may cause the population numbers to crash. Populations of some species, such as annual plants and insects, undergo regular cycles of surges and crashes. Some crashes are followed by recovery. Others are more permanent.
5. A way that population growth is regulated is by reproductive rate. Although biotic potential varies among species, it is surprisingly high. One factor determining the reproductive rate is energy.
6. Natural selection sets a reproductive strategy in accord with a species' environment and history so that reproductive success is maximized. In a species such as the tapeworm, in which an individual's success depends on chance, a common reproductive strategy is to produce a huge number of potential offspring but invest little in each. In contrast, a species such as the chimpanzee, in which an individual's success is increased by learning and protection from parents, produces fewer young but invests a great deal of time and energy in each one. A comparison of bird species reveals that birds lay the number of eggs that will enable them to successfully raise as many young as possible.
7. Population size may be reduced by death. Because nonliving (abiotic) factors, such as severe weather or substances that poison or destroy the environment, are not influenced by the number of individuals in the population, their effects are described as density-independent. Living (biotic) population controls such as predation, parasitism, competition, and disease are likely to depend on the density of the population (density-dependence) and tend to stabilize population size. Predators do not usually wipe out their prey species because a drop in prey number is followed by a decrease in the size of the predator population. A parasite derives benefit from and harms its host without actually killing it. The population-regulating effects of competition can be seen in territoriality and hierarchies. An individual that successfully defends a territory is more likely to survive and reproduce than is the loser because the defended area holds nest sites, hiding places, or food. Hierarchies in which individuals recognize and respond to each according to rank also result in certain animals having freer access to commodities than others. The effects of disease are

density-dependent because disease spreads more easily at higher population densities and because greater numbers of individuals in a population provide more potential reservoirs of mutant strains of disease organisms.

8. Death may be a mechanism of perpetuating the "best" kinds of genes.

KEY TERMS

abiotic (749)
biotic (749)
biotic potential (740)
carrying capacity (740)
crash (741)
density-dependent (752)
density-independent (749)
environmental resistance (740)
hierarchy (756)
J-shaped curve (739)
parasitism (754)
population dynamics (739)
S-shaped curve (739)
territoriality (755)

FOR FURTHER THOUGHT

1. Alligators provide no care for their young while elephants provide several years' support for their young. Discuss the relative numbers of offspring produced by each. Who has more offspring over the course of a lifetime?
2. A certain population produces a J-shaped curve when its numbers are graphed. Has this population reached its carrying capacity?

FOR REVIEW

True or false?

1. ___ A population growing with unlimited resources produces an S-shaped curve when their numbers are graphed.
2. ___ Nesting birds tend to produce more eggs in times of short food supply.
3. ___ When a predator's numbers decline, the numbers of its prey may begin to rise.
4. ___ Species that invest little energy in the care of their offspring produce only a few offspring during their lifetime.

Fill in the blank.

5. ___ is the number of individuals that the environment can support over a long period of time.
6. ___ are organisms that live on or in the body of another organism.
7. Populations in which individuals are "ranked" are called ___ .

Short answer.

8. Describe the reproductive strategy of the chimp.

Choose the best answer.

9. Severe weather that controls population numbers is an example of ___ control.
   A. abiotic, density-dependent
   B. abiotic, density-independent
   C. biotic, density-dependent
   D. none of the above

10. A population that exceeds its environmental resources is most likely to:
    A. crash
    B. lower the environmental resistance
    C. exceed its biotic potential
    D. none of the above
11. Which of the following is an example of a biotic method of population control?
    A. parasitism
    B. predation
    C. disease
    D. all of the above
12. The unrestricted reproductive capacity of a population is called its:
    A. carrying capacity
    B. ultimate level
    C. biotic potential
    D. proximate level

CHAPTER 27

Human Populations

Overview

Early Human Populations

The Advent of Agriculture

Population Changes from 1600 to 1850

Population Changes After 1850

The Human Population Today

The Future of Human Populations

Objectives

After reading this chapter you should be able to:

- Briefly outline the nature and growth of human populations through the seventeenth century.
- List the events that account for major increases in population.
- Describe how demographers determine a population's rate of natural increase.
- Relate crude birth rates and crude death rates to zero population growth and doubling time.
- Compare and contrast the growth rates among the less developed countries.
- Describe how age structure pyramids are used to predict population growth.

"Return with us now to the days of yesteryear."* In fact, let's go back to between one and two million years ago to a far different planet than the one we call home today. Should we appear on that ancient earth, we probably wouldn't be able to recognize the place where our house sits today, and the array of plants and animals we would see would be strange indeed. One kind of animal, though, might have a faint air of familiarity about it. This animal we would know as a primate as it ambled in small bands over the countryside, stopping now and then to grab at some morsel of food. Who could have guessed that virtually every other kind of life, as well as the face of the earth itself, its air and waters, would yield before something as seemingly insignificant as the descendants of that hunched figure sitting in the shade, picking at the soft parts of an insect?

EARLY HUMAN POPULATIONS

We have no idea how many individuals comprised the human species in our earliest days, and we don't know much more about our numbers in more recent times. In fact, we know little about human populations before about 1650 B.C. However, we do have educated guesses. These are often based on what we know about modern groups that seem to be similar to those of earlier times. There are other lines of evidence as well. We know, for example, that agriculture was almost unknown before about 8000 B.C. (possibly a bit earlier in Malaysia and the Middle East). Before that time, most humans must have lived by hunting and gathering their food. Considering the inefficiency of such methods, and assuming a land area of about 58 million square miles, we can infer that in 8000 B.C. the earth could have supported no more than about five million people (Figure 27.1).

It is important to realize that 90 percent of all the people who have ever lived on the earth have been hunters and gatherers, only 6 percent subsisted mainly by agriculture, and only a very small percentage have lived as industrial people. But this small industrial group has played an inordinate role in bringing us to our present situation. Another way to place this group, our group, in its proper perspective is to remind ourselves that about 99 percent of humankind's time on earth has been spent as hunters and gatherers. Obviously, then, our minds, attitudes, values, bodies, and behavior over the eons must have been geared for a different kind of life than we now have. This new kind of life has fallen upon us so suddenly that within the last 1 percent of the period that our species has been on the planet, only 1 percent of us continue to subsist as hunters and gatherers.

Hunters and gatherers have relatively little impact on the environment. They utilize a variety of plants and animals, and as one falls into short supply, they simply move on or switch to other resources, thus relieving the pressure on the scarcer species. Most of the food of hunters and gatherers, by the way, is taken by gathering. In modern groups, the men

*How many of you know where that statement comes from?

FIGURE 27.1
Over most of human history our numbers have been low and we have interacted intimately with our environment.

usually bring home the larger, swifter (or more dangerous) protein-laden meat (Figure 27.2), but the women, quietly gathering plant material, contribute the far greater part of the diet.

In time, the early hunters and gatherers became more efficient and their hunting began to pose more of a real threat to other species. Weapons and tools improved, and men began to hunt together in a more coordinated manner and to bring down larger game. Language would have had to become more sophisticated in order to maximize the results of group hunting ("Tomorrow, half of us will hunt here." And, in time, another concept: "Except Charlie; his wife won't let him go.") It was probably not a great step from someone saying, "You get his attention and I'll hit him with a stick," to the formation of rudimentary leadership and thence to political maneuvering. (Anthropologist Lionel Tiger has suggested that the need for coordination in hunting led to the overwhelming masculine domination of politics.)

FIGURE 27.2

Among the most primitive of today's hunters and gatherers is the Waorani tribe of the Conanaco River of Ecuador. While the men hunt pigs, monkeys, and parrots, and clear land with neolithic axes they find on the forest floor, the women plant the few crops that, with what they find growing wild, form the bulk of their simple diets. The jungle can sustain only small groups, and these must move on every few years to new hunting areas.

We find early indirect signs of man's increasingly proficient hunting ability (Figure 27.3). For example, after the ice ages ended about 10,000 years ago, about 70 percent of the North American mammals became extinct. Their decline correlates suspiciously with the arrival of humans in those areas. In fact, many remains of such animals have been found with human weapons nearby. Furthermore, as hunters grew more cunning, they apparently began to rely more on trickery and imagination. In one place near Soulutre, France, the fossilized remains of over 100,000 horses have been found where they were apparently stampeded over a cliff by primitive hunters. As people began to use fire, they could have made the habitat even more to their liking; some researchers believe that many of the world's great grasslands were created partly by the land being repeatedly burned over by early man. The point is that humans began very early to make sweeping changes in their environment, and the extent of these changes was apparently limited only by their primitive technology.

It is believed that these early human populations formed separate, but interbreeding, groups, possibly divided into small bands. From studies of present-day hunters and gatherers, it has been calculated that the average size of such groups would have been about 200 to 500 men, women, and children, of which about half were of breeding age. Population size was probably relatively stable and controlled by density-dependent factors. (Keeping in mind the food sources, can you tell why? See Chapter 26.) It is also generally believed that the number and size of the groups approached the carrying capacity of the environment.

FIGURE 27.3

Even the art of early humans depicted a keen interest in the game animals of the time, as we see in this ancient cave painting.

THE ADVENT OF AGRICULTURE

About 8000 B.C., the earth's carrying capacity rose to a new level through the advent of agriculture. People began to roam less and to decrease their dependence on the uncertain presence of wild game. By learning to grow and store their food, they added a relatively stable element to their food supply. Stored food meant that the lean seasons no longer marked the deaths of so many children and old people. The result of the shift toward agriculture was a slow, steady increase in the size of human populations.

How could the shift to agriculture have raised the environment's carrying capacity? We have already noted that the ability to store foods, such as grain, meant a more dependable food supply and a greater likelihood of surviving the harsh seasons. But, in addition, agriculture meant a greater reliance on plant food, and so humans began to shift their diet to the "producer" level. As we will see in the next chapter, in a food chain, it is more efficient to eat plants than animals because energy is lost in converting plant food to animal food (see Essay 28.2). All this is not to say that the agricultural revolution has been the only factor that has changed the carrying capacity of our environment. The advent of toolmaking also had its marked effect, and we will learn shortly of the phenomenal impact of the recent scientific-industrial revolution.

POPULATION CHANGES FROM 1600 TO 1850

Somewhere around the seventeenth century, the world's human population suddenly began to rise (Figure 27.4). There have been increases before, such as those associated with the advent of tool-making or the beginning of agriculture. But these were gentle rises; the population surge of about three hundred years ago was much more dramatic. Such an abrupt rise has heralded devastating population crashes in other species, and this increase might have had the same results if it had not been associated with an increase in the environment's carrying capacity. And what brought this about?

The sharp rise in the population of Europe between about 1650 and 1750 is attributed to innovations in agriculture and the beginning of the exploration and the exploitation of the Western Hemisphere. At about that time, life in Europe was dismal almost beyond belief. The cities were dangerous, garbage-filled, and overrun with rats. Sewers ran freely through the streets, their delicate gurgling belying their burden. The stench was devastating. Life in the country was not much better. Outside the cities were scattered hamlets, little more than isolated slums inhabited by people of numbing ignorance. It is hard to imagine how they could be helped. Where to start? The accompanying rise in Asian populations at the same time is harder to explain. India, for example, was in turmoil following the fall of the Mongol Empire. The simultaneous rise in the Chinese population might have been a result of a new political stability brought by the Manchu emperors, who also initiated new and better agricultural policies after the fall of the Ming Dynasty in 1644. Not much is known about Africa in this period, but its population is estimated to have been about 100 million in 1850—about the time that European medicine and technology were introduced. Between 1750 and 1850, the population increase

FIGURE 27.4
Human population growth during the last half million years. If the old Stone Age were drawn to scale, it would extend about 18 feet to the left. Notice the very gradual rise as agriculture began about 8000 years ago.

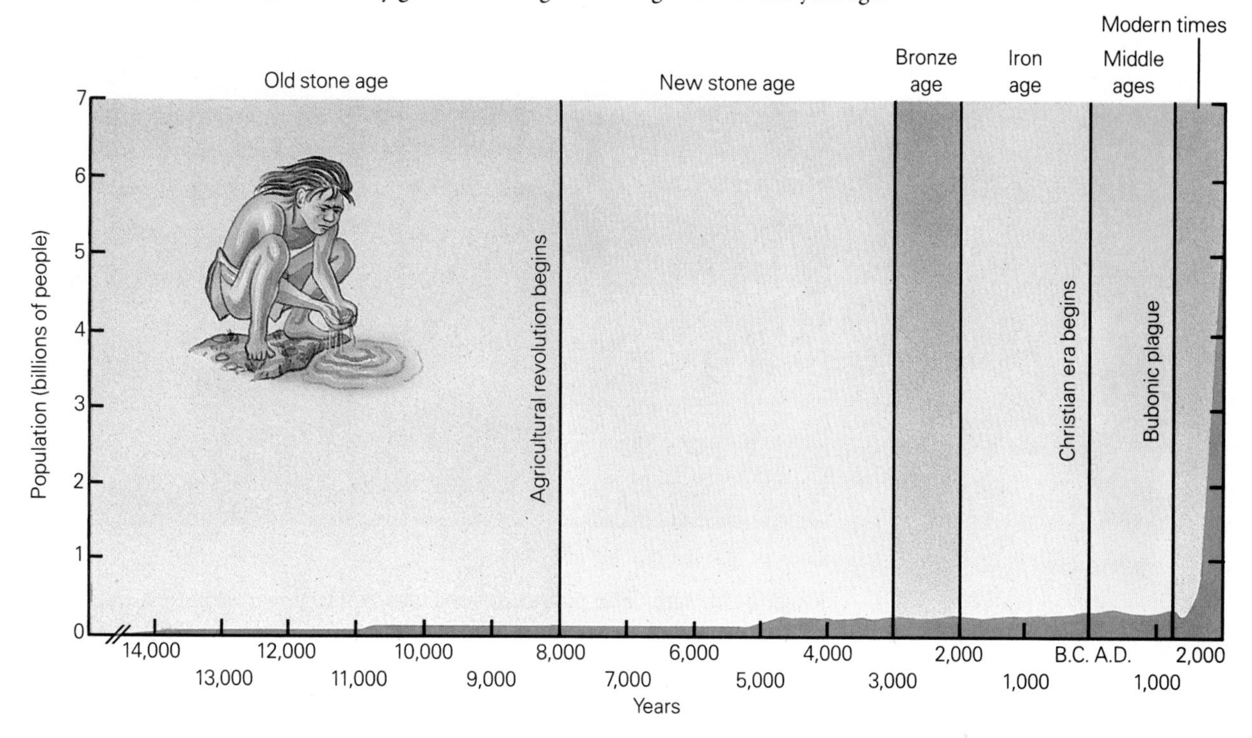

in Asia was only about 50 percent, far less than in Europe. European advances at this time included new agricultural techniques and better sanitation.

POPULATION CHANGES AFTER 1850

Beginning around 1850, European population growth began to accelerate. The growth rate increased due to new agricultural techniques. But perhaps more important was the decrease in the death rate attributed to better sanitation and advances in medicine, such as the development of a smallpox vaccine (Figure 27.5).

By the beginning of the twentieth century, the general situation improved even further, a result of even newer farming techniques that reduced the risk of crop failure. In addition, there were improvements in transportation that meant food could be more easily distributed and imported to locally stricken areas. Also, about this time the role of bacteria in infection became better known, and so more lives were saved. The death rate in Europe dropped from about 24 per 1000 to 20 per 1000 in just those 50 years.

By the beginning of the twentieth century, **demographers** (those who study human population changes) had identified a rather surprising trend. Their records showed that industrialization was generally followed by a

FIGURE 27.5

About the middle of the last century, the European population began to surge, probably because of medical advances and improved sanitation.

lower birth rate. The phenomenon, which has since received a great deal of attention, has never been entirely explained. However, it is suggested that in agricultural societies, children are a source of labor and a form of social security. They provide assistance before they leave home and they care for their aged parents later. In an industrial society, it is suggested, they are less likely to aid in production and are simply "more mouths to feed." Also, mobility, so important in industrial societies, is reduced by the presence of children.

Some researchers believed that simple development leading to industrialization was a sure form of population control. But now it seems that the notion was perhaps a bit simplistic. Industrialization is often accompanied by a reduced population growth, but there are enough exceptions and stipulations that it is no longer held as a "truism."

About the time of World War II, many poorer countries saw a dramatic reduction in death rates, apparently due to imported public health programs, including new drugs, from the technologically more advanced countries. (Among the most important imports was DDT, which was used to control malaria.) Such imported "death control" produced history's most sweeping increases in human populations. It is important to realize that the changes in death rates in these countries were not accompanied by the social and institutional changes that went hand in hand with the far slower reduction of death rates in the technologically more advanced countries.

THE HUMAN POPULATION TODAY

By any standard, the human population today is growing at a phenomenal rate. To illustrate, consider this: there are 30 more humans living on the earth now than there were ten seconds ago when you began reading this paragraph. By tomorrow at this time there will be 250,000 more people living on this planet, and by next year the human population will have grown by 90 million. Unfortunately, most of them can look forward to hard, desperate lives of poverty. And the situation is likely to worsen.

Paradoxically, the worsening conditions will continue in spite of some encouraging numbers. Some small encouragement had appeared by the end of the 1970s. Demographers calculated that the **rate of natural increase** of the human population was slowing. (The rate of natural increase is the difference between **crude birth rates** and **crude death rates.** See Table 27.1). This meant that the number of births each year was becoming closer to the number of deaths. This was, in a sense, good news. The

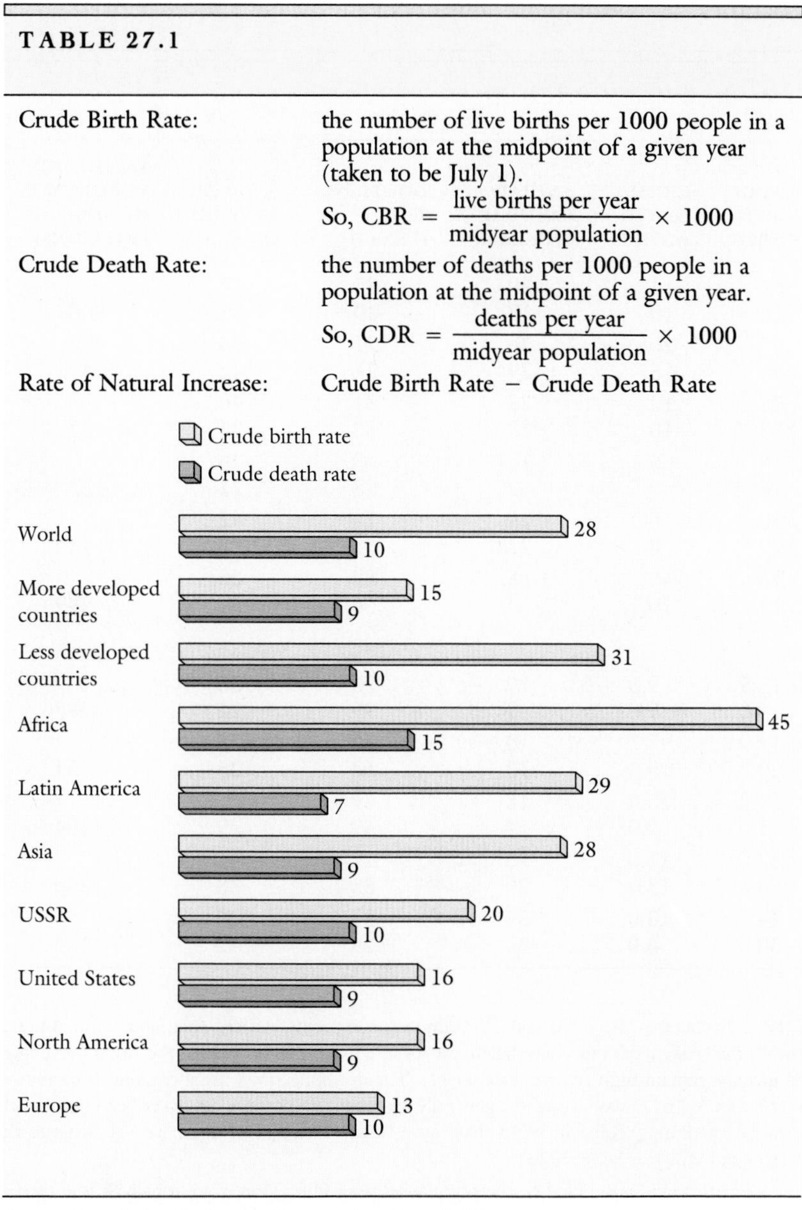

TABLE 27.1

Crude Birth Rate:	the number of live births per 1000 people in a population at the midpoint of a given year (taken to be July 1). So, $CBR = \frac{\text{live births per year}}{\text{midyear population}} \times 1000$
Crude Death Rate:	the number of deaths per 1000 people in a population at the midpoint of a given year. So, $CDR = \frac{\text{deaths per year}}{\text{midyear population}} \times 1000$
Rate of Natural Increase:	Crude Birth Rate − Crude Death Rate

Source: Population Reference Bureau, data from 1989.

problem is that *rate* implies change over a given period of time and a slower rate of natural increase, some said, could be likened to a busload of people heading for a cliff at 30 miles per hour. As their rate slows, they may be moving at only 15 miles per hour as they go over the edge. Obviously, with both the bus and the human population, the solution is to slow the rate to zero and then to reverse direction. (With populations, this would involve the death rate exceeding the birth rate.)

The rate of natural increase varies from one global area to another (Table 27.2). The poorer, less industrialized nations are referred to as **less developed countries (LDCs),** of which there are 142, while those more affluent, industrialized countries are called **more developed countries (MDCs),** of which there are 33. The more developed countries include the United States, Western Europe, Russia, and Japan. The less developed countries are generally notoriously poor and include nations in Asia, Africa,

TABLE 27.2
World Population Data: 1970 and 1990

REGION	YEAR	TOTAL (MILLIONS)	CRUDE BIRTH RATE	CRUDE DEATH RATE	RATE OF NATURAL INCREASE*	DOUBLING TIME (YEARS)	% BELOW 15 YEARS OF AGE	ESTIMATED POPULATION IN 2020 (MILLIONS)
World	1970	3632	34	14	20	35	37	
	1990	5320	28	10	17	40	33	8053
Africa	1970	344	47	20	26	27	44	
	1990	660	44	15	29	24	42	1497
Asia	1970	2045	38	15	23	31	40	
	1990	3111	28	10	18	38	35	4629
North America	1970	228	18	9	11	63	20	
	1990	277	16	9	7	98	21	327
Latin America	1970	283	38	9	29	24	42	
	1990	455	29	8	22	32	38	711
Europe	1970	462	18	10	8	88	25	
	1990	499	13	10	4	266	21	499
NATIONS OF SPECIAL INTEREST								
United States	1970	205	17.5	9.6	10	70	30	
	1990	248	16.0	9.0	7	99	22	297
Soviet Union	1970	243	17.9	7.7	10	70	28	
	1990	291	20.0	10.0	10	68	26	354
People's Rep. of China	1970	760	34.0	15.0	18	39	?	
	1990	1120	21.0	7.0	14	49	29	1404
India	1970	554	42.0	17.0	26	27	41	
	1990	832	33.0	13.0	20	35	38	1309
Mexico	1970	50.7	44.0	10.0	34	21	46	
	1990	89	30.0	6.0	24	29	42	106

Source of Data: Population Reference Bureau.

There are some interesting numbers here. Note the differences in the 1970 and 1990 figures, comparing Africa, Asia, and Latin America with Europe and North America. Note how drastically the crude birth rates have fallen, except in Russia. But because of the larger percentage drop in the crude death rates, the rates of natural increase remain high (where, especially?). The doubling times are increasing (but are still far too short) everywhere but in Africa and Russia. (Russia, with her vast area and a generally inefficient work force has never been particularly concerned with overpopulation, but that may now be changing.) Keep in mind that those below 15 years of age have yet to enter the reproductive ranks. How old were you in 1970? How old will you be in 2020?

*This shows the rate of natural increase as the number of individuals added to the population per 1000. More traditionally it is shown as percent (per 100).

and Latin America. Unfortunately, those countries with the highest birth rates are the poorest and least able to provide for their citizens.

With the approach of the 1990s, the world population was estimated at about 5.3 billion. The rate of natural increase was still disturbingly high, but was becoming lower. The doubling time (the time necessary for the population to double at its current rate) was increasing (Essay 27.1), but was still just over 40 years (how old will your children be?). No one is sure why the population surge was slowing down but several explanations have been offered. In particular, demographers noted the changing attitudes of women regarding their roles in society, changes in the number of children couples wanted, advances in birth control, and liberalized abortion laws in developed countries.

At first glance at Table 27.2, the news from Latin America seems encouraging. The rate of natural increase has, indeed, fallen significantly. However, the encouragement is dampened by the fact that the crude birth rate is 3½ times the crude death rate and the doubling time is similar to that of Africa, a desperate continent, indeed.

Assessing the problem in Asia is more difficult. The problem is that many areas are so primitive or so secretive that world demographers just don't have the basic information they need. We do know, however, that half the people on earth are Asian and half of these are Chinese. Because of China's highly publicized program to limit families to one child (Figure 27.6) the birth rate there has declined somewhat. Partly as a result of this trend, the doubling time in Asia has increased to just under 40 years. Similar efforts at controlling family size in India have been less successful. A program encouraging sterilization never quite caught on and the size of the Indian population is again growing rapidly.

The small, if temporary, successes in controlling the Asian populations, unfortunately, will soon be swamped by another problem looming on the horizon. That is, about a third of the Asian population is under the age of 15 and have yet to enter the breeding population.

FIGURE 27.6
The Chinese government has become keenly aware of the country's population problem. One-child families are encouraged by billboards such as this. Should this encouragement fail, prospective mothers are subjected to more drastic measures, such as being put under intense community pressure to undergo abortion. The Chinese government, finding the policy too stringent, has relaxed the pressure and the population has begun an upward surge. Now, in fact, only 19 percent of Chinese families have only one child.

THE FUTURE OF HUMAN POPULATIONS

Unfortunately, we can't be very confident about predicting future trends in human populations. The sad truth is that almost every major trend of the past few decades has caught even the most sophisticated demographers by surprise. They didn't even anticipate the baby boom that led to the yuppie generation.

However, one thing is clear. Even if we assume the rosiest of projections for the next few decades, the human population will continue to spiral upward, virtually out of control. Our numbers are already placing unprecedented demands on the planet. Forty thousand babies die of starvation each day in third world countries, even as many of these countries are arming themselves to the teeth. Something must be done and the bad news is, it will be, one way or another. Our best hope, it seems, is in the kind of activity generated by information, not hunger. So let's continue, now taking a look at population structure. Remember, we are not engaged in some intellectual exercise stemming from sterile academia. This is real and you are a part of it.

ESSAY 27.1

Doubling Times

It is hard to imagine the growth of the human population on the earth. Perhaps it will help to consider "doubling times." Would you work for someone who paid you a penny your first day, two cents the second, four cents the third, and kept doubling your wages each day for one month? You *should*. Consider this: If this page is say, 1250th of an inch thick and you start doubling it, after only 8 doublings it would be an inch thick, after 12 doublings, thicker than a football field is long. If we doubled the page 42 times, the stack would reach to the moon, and after 50 times we would reach the sun, about 93,000,000 miles away.

Some countries have a much firmer grip on their population problems than others, as evidenced by their doubling times (see table).* To go from five million people on earth (the present number in only three of New York City's five boroughs) in 8000 B.C. to 500 million in A.D. 1650 took six or seven doublings over a period of 9,000 to 10,000 years. During that time, the human population doubled on an average of about every 1500 years. A glance at the table will show that, all other things being equal, in only about 40 years, we will need two cars, two schools, two roads, two wells, two houses, and two cities throughout the world for every one that presently exists. And that will only maintain our status quo as far as material goods are concerned. The problem is that not all things can remain equal over that time. Can India or Guatemala double its food? Can we double the population of the United States? Do we want to? Can California stand another Los Angeles? Where will the water come from? If we double the number of oil wells, can we double the earth's oil?

Doubling Times of the Human Population*

DATE	ESTIMATED WORLD POPULATION	TIME REQUIRED FOR POPULATION TO DOUBLE
8000 B.C.	5 million	1,500 years
A.D. 1650	500 million	200 years
A.D. 1850	1,000 million (1 billion)	80 years
A.D. 1930	2,000 million (2 billion)	45 years
A.D. 1990	5,300 million (5.3 billion)	40 years
A.D. 2010	8,000 million (8 billion)	?

* Doubling times can be calculated from annual percentage growth rate by simply dividing the growth rate into 70. The growth rate of MDCs is 0.6 percent; of LDCs, it's 2.1 percent. Do the arithmetic.

Population Structure

A powerful tool for predicting changes in population growth is the analysis of population structures. One way to show these is through **age structure pyramids.** In Figure 27.7, we can see that populations can be described according to what percentage of their numbers are in various age groups. In Mexico, we see, the population is increasingly expanding in the younger groups. This is partly because of poor health care for older people and a generally high mortality rate throughout life, but it is mainly because of a strong social emphasis on larger families. (Siring children may be considered a sign of virility in these male-dominated societies.) A population such as this one, where a large proportion of the population is still too young to reproduce, is likely to grow in future years as the children reach reproductive age. Recall from the last chapter, that the rate at which new individuals are added to the population increases dramatically as the number of reproducers increases.

In the United States and Sweden, the age structures are more uniform, partly due to better health care and fewer children per family. You can see that Sweden's age structure is more stable than that of the United States, partly due to their long history of socialized medicine. Sweden's age structure indicates that the number of individuals reproducing is likely to remain constant in the near future. If each couple continues to have the same number of children, we would expect that Sweden's population size will also remain fairly constant. Notice the bulge in the age structure pyramid of the United States that represents the baby boomers who are now of reproductive age. The bulge of baby boomers is moving through the population age structure, as would a great egg eaten by a snake. What impact will this have when the bulge reaches the top of the pyramid? How

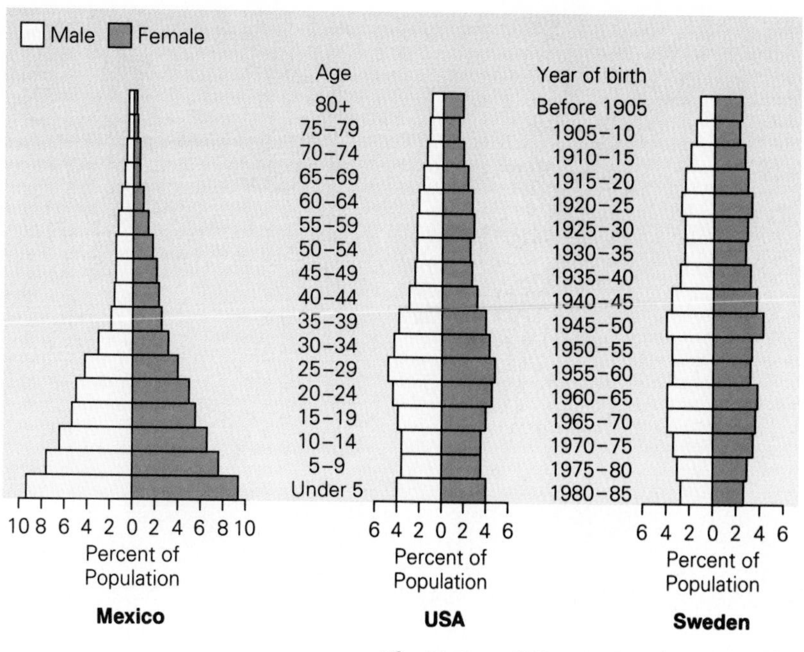

FIGURE 27.7

Age structure pyramids break down the population into five-year age groups, revealing much about past history and permitting predictions of future trends to be made. Left, expanding nation, e.g., Mexico. Middle, moderately stable nation, e.g., United States. Right, long-term stable nation, e.g., Sweden. Source: Population Reference Bureau.

will we care for so many unproductive old people? Also, note that the pyramids reflect a difference in age structures of males and females (Essay 27.2).

There are important social implications in such pyramids. Note that the reproducers are also the workers in the population. In which kind of nation would programs such as Medicare, or other systems of care for the elderly, be more important? Who pays into Social Security programs and who draws from them? Who feeds the young? Would voting tendencies tend to differ in old and young populations? Would housing starts differ? Housing designs? Clothing designs? How about pressures on natural resources or the likelihood of belligerence toward neighboring countries? The point is that many facets of any society reflect the age structure of its population.

There is obviously a great deal of concern in the scientific community over the impact of rising human populations. Present densities, worldwide, are shown in Figure 27.8. But the reaction is not universal dismay. Some scientists believe we can absorb great increases in our numbers with only

FIGURE 27.8

Human population densities around the world. Notice that most of the planet has very low densities. Look at central and southern South America, Northern Africa, southwest Africa, Greenland, and Australia and guess why. Also notice the areas of highest density. What other traits do they have in common?

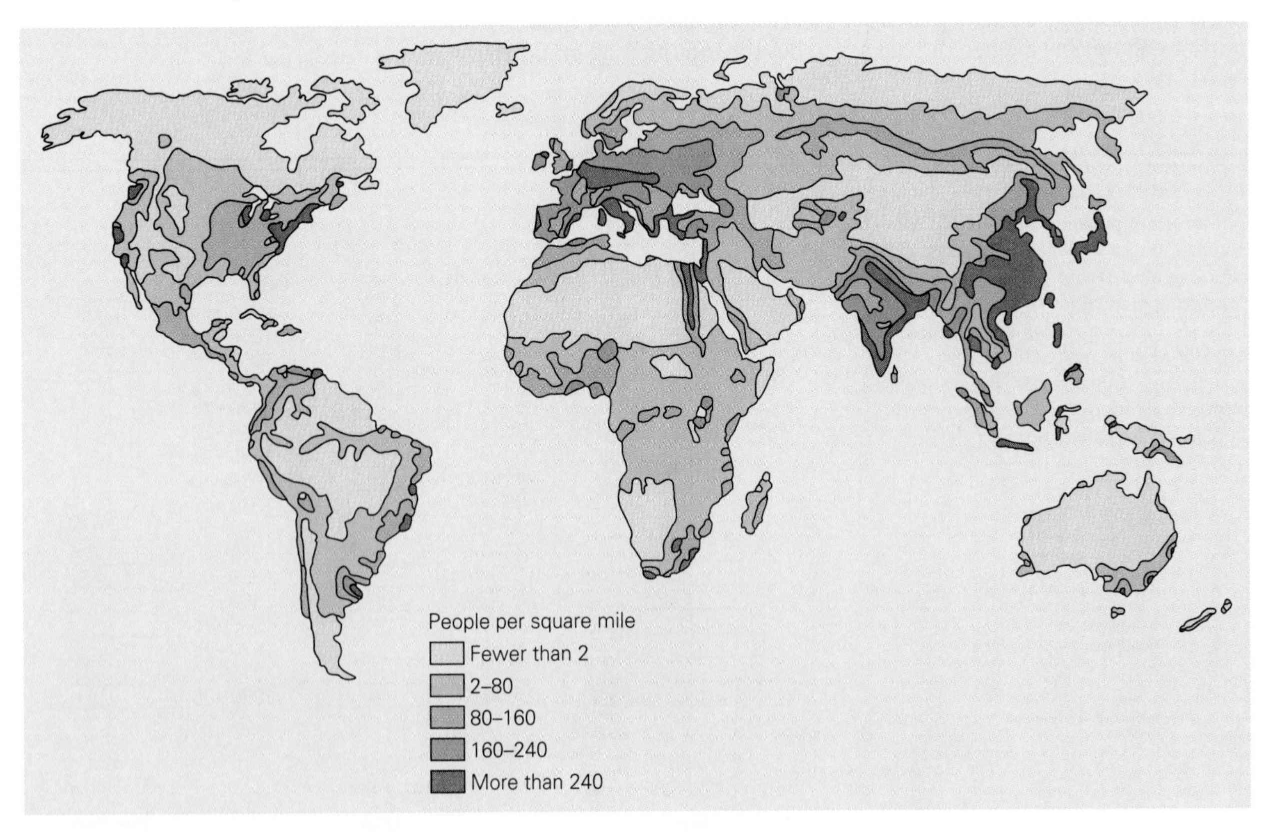

ESSAY 27.2

Why Do Women Live Longer?

People are living longer than ever, but for some reason, women are living longer than men. A baby boy born in 1984 can expect to live to be about 71, a baby girl, about 78.5. This is indeed a wide gap, and no one really knows why it exists. (Some men have claimed it's because women aren't married to women and some women claim that anyone who can put up with a man has to be tough.)

The greater longevity of women has been known for centuries. It was, for example, described in the seventeenth century. However, the difference was smaller then—the gap is growing.

A number of reasons have been proposed to account for the differences. The gap is greatest in industrialized societies, so it has been suggested that women are less susceptible to the strains of "the rat race" that produce such debilitating conditions as heart disease and alcoholism. Sociologists also tell us that women are encouraged to be less adventurous than men (and that they are more careful drivers).

It was once suggested that working women are more likely to smoke and, as they entered the work force, the age-gap would begin to close. (Smoking is related to earlier deaths.) Now, however, we see more women smoking and they still tend to live longer although their lung cancer rate is climbing sharply. (One recent study indicated that if neither sex smokes, they have identical life spans.)

One puzzling aspect of the problem is that women do not appear to be as healthy as men. That is, they report far more illnesses. But when a man reports an illness, it is more likely to be serious.

Men may die earlier because their health is more strongly related to their emotions. For example, men tend to die sooner after losing a spouse than women do. Men even seem to be weakened by loss of a job. (Both of these are linked with a marked decrease in the effectiveness of the immune system.) Among men, death follows retirement with an alarming promptness.

Perhaps we are searching for the answers too close to the surface of the problem. Perhaps the answers lie deeper in our biological heritage. After all, the phenomenon is not isolated to humans. Females have the edge among virtually all mammalian species. Furthermore, the differences begin at the moment of conception; there are more male miscarriages. After birth, more baby boys than baby girls die.

Another biological explanation involves differences in the sex hormones. Estrogen, for example, may help protect against heart disease. Also, the female hormones apparently render the immune system more efficient.

a few policy changes. Others profess an impressive faith in technology and suggest that the answer lies in increasing the carrying capacity of the earth through agricultural, technological, and medical "revolutions." It has recently been estimated that the human population could rise to eight to ten billion people—*temporarily*. From that point, the numbers would have to drop to more sustainable levels in order to avoid permanent ecological disaster.

We are an adaptable and innovative species and there is no doubt that we successfully withstand all sorts of traumas, including great population increases. A dog can get along just fine underwater, too. But not for long. How long can we withstand the relentless pressures of increasing numbers of other people? Even if we can find ways to endure it for long periods of time, the question is, do we want to? How old will *you* be when the earth has six billion people? (This will occur before the year 2000.) At the current rate, how old would you guess your children will be when there are 14 billion people on earth?

SUMMARY

1. Before 8000 B.C., humans were exclusively hunters and gatherers and could not have numbered more than five million. Hunters and gatherers have little effect on the environment. The ability to make tools, such as more efficient weapons, and communication through language would have improved hunting success. Improved hunting techniques and the use of fire led to sweeping environmental changes.
2. Early human populations probably formed interbreeding bands of 200 to 500 individuals. The number and size of the groups is thought to have been close to the carrying capacity of the environment.
3. The advent of agriculture and the ability to store food raised the carrying capacity because it provided a dependable food supply through harsh seasons and because it shifted the diet of humans to a greater dependence on the "producer" level. The human population rose to 200 to 300 million by the time of Christ and to 500 million by 1650.
4. About 300 years ago, the size of human population began to rise sharply for different reasons in different parts of the world. In Europe the population increase between 1650 and 1750 is attributed to new agricultural techniques. The increase between 1750 and 1850 resulted from new agricultural techniques, better sanitation, and advances in medicine.
5. Between 1850 and 1900, the death rate in Europe continued to decrease due to advances in industry, agriculture, and medicine. Industrialization is usually followed by a decrease in birth rate. At the time of World War II, poorer countries experienced a decrease in death rate.
6. The human population today is growing at a phenomenal rate, but it is beginning to slow down. By the end of the 1970s, the number of births each year was becoming closer to the number of deaths (i.e., the rate of natural increase was slowing). However, until the rate slows to zero and reverses, the population problem is not solved. The rate of increase is generally greater in poorer developing nations. Among the reasons for the slowing rate of population growth are the changing attitudes of women's roles in society, changes in the number of children couples want, advances in birth control, and liberalized abortion laws in industrialized societies.
7. Age structure pyramids, which show the percentage of a population in various age groups, are one way to show population structure. They are useful for predicting changes in population size. They have marked social implications.

KEY TERMS

age structure pyramid (775)
crude birth rate (771)
crude death rate (771)
demographer (769)
less developed countries (LDCs) (772)
more developed countries (MDCs) (772)
rate of natural increase (771)

FOR FURTHER THOUGHT

1. Refer to the age structure pyramids in Figure 27.7. From the information given, which country will have the greatest growth rate in 11 years?
2. How do crude birth rates and crude death rates relate to a population's doubling time? (Refer to Table 27.2; compare 1970 United States figures and 1970 Mexico figures.)

FOR REVIEW

True or false?

1. ___ The greatest percentage of mankind's time on earth has been spent in hunting and gathering.
2. ___ The carrying capacity of the earth decreased with the advent of agriculture.
3. ___ Individuals in an industrial society tend to produce more children than those in an agricultural society.
4. ___ In order to attain zero population growth, the birth rate of a population must exceed the death rate.

Fill in the blank.

5. ___ study human population trends or changes.
6. The rate of natural increase of human populations is determined by ___ and ___ .
7. ___ are representations of age groups within a population.

Choose the best answer.

8. The rise in the European population between 1650 and 1750 resulted from:
   A. agricultural advances
   B. better transportation
   C. advances in medicine
   D. none of the above
9. Of the following, which has the highest birth rate?
   A. United States
   B. Asia
   C. Russia
   D. Western Europe
10. The population structure of the United States
    A. is fairly uniform in age structure.
    B. is composed mostly of younger groups.
    C. contains a small proportion of reproducers.
    D. none of the above.

CHAPTER 28

Resources, Energy, and Human Life

Overview

Renewable Resources: Focus on Food and Water
Nonrenewable Resources
Recycling — and Around It Goes
Energy
Encouraging Conservation and Increasing Efficiency

Objectives

After reading this chapter you should be able to:

- Discuss the major food resource problems facing human populations today and list some possible ways the world's food supply may be increased.
- Compare the dietary efficiencies of eating different types of animals.
- Describe some reasons for severe water shortages.
- Give examples of nonrenewable resources.
- Explain why recycling is an energy-efficient and environmentally sound process and name some materials that are currently recycled.
- Describe several energy sources, citing the advantages and disadvantages of each.

All the known field mice in the universe must live on whatever this planet can provide for them. They run little risk of depleting their planet's resources, however, because their needs are few: a warm hole, a little food, and a little social acceptance. It's a good thing for them that their requirements are so few, because they have only the resources that their planet can provide. Should they deplete or foul those resources, they would be in a bit of a fix, because *that's all there is*. The point is not too subtle, is it? We share the planet with the field mice. And our resources, too, are limited. But their gentle touch on the great, delicate globe is far from our sledgehammer impact.

In this chapter, we will review the state of some of those resources necessary to life and try to see just how our impact is influenced by something very ephemeral and hard to pin down: our values. We will approach the question of whether our behavior and values could or should be based on our access to the earth's resources and energy. And how have we treated our heritage?

We will consider the earth's resources to be in one of two categories, renewable and nonrenewable. **Renewable resources** are those that do not exist in set amounts—they can be reused or replenished at least as fast as we use them. If we cut a pine tree, we can grow another one (assuming that the topsoil didn't wash away because we cut the first one). **Nonrenewable resources** are those that exist in set amounts on the planet and cannot be replenished. Once used, they are gone, or they are changed so much that they are difficult to recover in their original form.

RENEWABLE RESOURCES: FOCUS ON FOOD AND WATER

Our heritage, or bequeathment, it could be said, is a generally bountiful earth. We heirs, then, should perhaps consider the earth a *home*—not an opportunity, nor a commodity, nor a bargaining chip, but a place that both blesses us and threatens us as a natural part of things. Strangely, it is the very bounty of a plentiful earth that has been at the root of some pressing problems for our species. How could this situation have come about?

First, the earth's human population, we've seen, is expanding at an alarming rate, and each new face demands its coupons and catalogues. We are placing a severe and increasing demand on our earthly goods, bequeathed by ancestors who lived in a more sparsely populated and far different kind of world. Our numbers are increasing, but many of the earth's resources aren't. Part of the problem is as simple as that. In fact, the addition of 250,000 more people to the world every day means that we will have to produce as much food in the next 30 years as we have produced since the dawn of agriculture, about 10,000 years ago.

The second problem with our bequeathment is that the earth's resources are unequally divided (Figure 28.1). The Arabs got the oil, the South Africans got the chromium, Americans got the forests, and the Japanese got the little cars.

FIGURE 28.1
The earth's resources are unequally distributed. The result is that various nations control the lion's share of certain commodities. For example, here we see the gold mining operations in South Africa. This country is blessed with approximately 55 percent of the world's supply of gold. The country also has unusual amounts of chromium and diamonds.

The unequal distribution of the earth's resources should provide us, one would think, with a marvelous opportunity to achieve world peace through the mutual dependence of trade. It seems we would simply trade, for example, our food for Arab oil. But instead of this unequal distribution being a great blessing, it seems as if it has set the stage for international confrontations. Perhaps it is time to apply our knowledge of the biology of our species to solving such problems. We must try to understand how such biological principles as competition, territoriality, carrying capacity, cooperation, altruism, and so on, apply to our own group. After all, we are the product of the weeding of genes of countless generations by their environment. We have not been exempt from the blind and unprejudiced effects of natural selection.

So let's now review the state of our environment, our home, and let's see how the resources are holding out. We must always keep in mind that the pressure on our resources increases each day as our population grows.

Food and the Present Crisis

FIGURE 28.2

In the search for a better life, many people seek health through better diets. Some are drawn to labels like "organically grown" or "natural," although the terms have no scientific meaning and can be misleading.

It may be hard for most Americans, who define hunger in terms of a late lunch, to understand true hunger. In fact, we seemingly have food to waste, as witnessed by our great attention to trivial detail regarding the earth's foods. We prefer our food this way or that; we hold some critical standard for our tomatoes. We argue the merits of red meats and fish and green vegetables. We pick our way through an abundance of offerings and we waste a great deal as we select only the tip or heart of this plant or that.

Interestingly, many culinary fetishists are abysmally ignorant about nutrition. Their preference for the "natural," for example, may lead them astray. Some actually believe that rose hips provide a better source of vitamin C than do the chemically synthesized pills. It's just not so. Nor is organically grown food better for you than food grown with chemical fertilizers (Figure 28.2). The point is that the dilettantes whose energy is now spent in cultivating their tastes may soon be doing the same to their soil. The food imbalance throughout the world is too great, carrying capacities shift too slowly, and the human population is growing too fast to allow us to continue business as usual. It is possible that in your lifetime (and maybe even sooner), social changes will bring our eating habits more in line with the realities of the worldwide condition.

The problem of food might be more easily solved if it weren't for the fact that people have a problem in utilizing the food that is available, especially the newer kinds of food. That is, we tend to be very conservative and traditional. It is very difficult to introduce a new food to a culture; no matter how nutritious the food is, people just won't eat it. For example, there are over 80,000 species of plants known to be edible. But only 16 plants yield almost all the calories and three-fourths of the protein that humans receive from plants. (And most of that nourishment comes from only four plants: corn, wheat, potatoes, and rice.) The rest of the edible species are almost ignored, even in the face of hunger and starvation. The hungry, by the way, can be divided into two groups: the **undernourished,** those who don't receive enough food, and the **malnourished,** those who don't get the right kinds of food.

FIGURE 28.3

Children stricken with marasmus (a) and kwashiorkor (b). Marasmus is due to the mishandling of scarce food supplies, usually overdiluting babies' formulas, while kwashiorkor results from a lack of protein. Both conditions can permanently impair the victims, both physically and mentally.

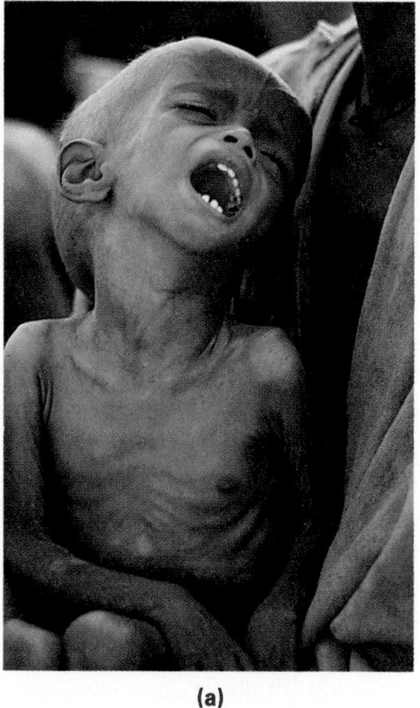

(a)

(b)

The Global Implications of Food Resources

A good diet is particularly important to children, pregnant women, and nursing mothers. Because these women essentially must eat for two, the risk of malnutrition is obvious. Children are especially vulnerable to two very dangerous nutritional diseases, marasmus and kwashiorkor. Marasmus is caused by a diet low in both calories and protein, and its results are a thin body with a bloated belly, wrinkled skin, and a startling appearance of old age. It is associated with drinking overdiluted formulas from unsterile bottles. Kwashiorkor results from protein deficiency and, when breast milk is the major source of protein, it often appears when a child is weaned—usually displaced at its mother's breast by a newborn sibling. Here, the entire body bloats, but especially the belly as it fills with fluid (Figure 28.3). In both cases, the children may suffer severe and permanent mental deficiencies because the brain grows very rapidly in the early years and requires high protein levels in the diet (Essay 28.1). Both diseases could be prevented by extending nursing, but nursing mothers require a good diet that may not be available.

The worldwide distribution of food is very uneven. Not only are there poor, hungry countries, and rich, well-fed ones, but the same country may have rich and poor parts. Southern Brazil is rich and fertile, with affluent and well-fed citizens, while people in the less-fertile northern part of the country are more likely to be undernourished. The problem also exists in the United States. As late as the mid-1980s, it was estimated that 10 to 15 percent of American citizens, mostly in impoverished areas such as northern ghettos and western deserts, were malnourished, while others, not far away, were sniffing at their plates and inquiring about the sauces.

ESSAY 28.1

How Hunger Kills

What happens when someone starves to death? What does starvation actually do to people? Unfortunately, the question is not a purely academic one since starvation is common in certain areas of the earth. The first thing that begins to happen when a person is deprived of food is that he loses weight. After all, the body continues to need the energy stored in chemical bonds to run its machinery. If those chemicals are not provided by food, the body takes them from itself, that is, from its own tissues.

Most Americans have a surprising store of food in their own bodies. In fact, the average U.S. citizen weighing 155 pounds has about 35 pounds of stored fat and can go about three months without food. An obese person with 175 pounds of stored fat can theoretically go without food for a year.

When someone goes without food, the first stored food he draws on is the readily available stored glycogen (starch) in his liver and muscle. When these are depleted he begins to use stored fat. Fat metabolism releases ketones, producing a truly remarkable bad breath. At about this time, the starving body begins to break down protein by stripping the nitrogen away and producing glucose, a necessary molecule for brain function. Later, the brain will change, becoming able to use ketones as well as glucose. So the body shifts back to raiding mainly fat stores, saving the proteins until last. Finally, the proteins, too, are broken down. The body's muscles begin to wither, including the heart muscle, a perilous condition indeed.

On the average, people are considered to be starving when they have lost about 33 percent of their body weight. When they have lost 40 percent, death is almost inevitable. As starvation progresses, various organs begin to lose their efficiency. The liver, kidneys, and endocrine system may cease to function properly. Lack of glycogen may affect the brain so that the starving person becomes listless and confused, often seemingly unable to understand his or her plight. Kwashiorkor and vitamin-deficiency diseases can set in. Soft bones may appear in children who lack vitamin D. Lack of thiamine may cause memory lapses. Lack of niacin causes skin inflammation, diarrhea, and insanity just before death.

Children are often permanently affected since improved diet cannot straighten bones or build normal brain cells, but adults can often approach starvation and then largely recover, as we know from the rescue of concentration camp inmates. Interestingly, no matter how well such people recover, most of them die sooner than those who suffered no such trauma.

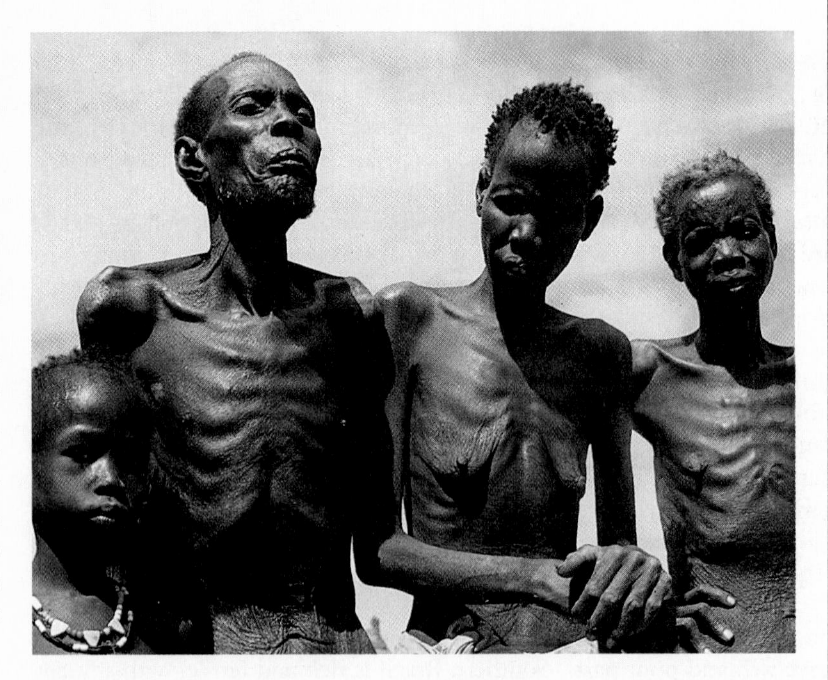

The complexity of the food problem again becomes apparent when we realize that the global food problem involves more than simply finding means to feed people. Food also has its political implications. It is rapidly becoming a bargaining chip, as is the case with oil. For example, in 1972, the Soviet Union began a program to upgrade the diet of its citizens by supplying both more meat and grain products. Because large amounts of grain are needed to produce small amounts of meat (see Essay 28.2), their grain requirements increased dramatically (at the same time that the world's weather took a sharp turn for the worse). The Soviets began negotiating with the United States to buy our surplus grain. At first, the United States refused to sell the grain to the Soviets because of tensions between the two countries, and American farmers lost a lucrative market for their produce. Now, however, the Soviet Union has become a valued agricultural customer, and so we find an opportunity to bolster real world peace through interdependency based on biological as well as political and industrial needs.

FIGURE 28.4

The great Ethiopian famine that began in the early 1980s was triggered by drought, but its basis is much more complicated than that. Changes in farming, herding, and demographics interacted in complex ways to promote the condition.

Some Global Realities

The more developed countries use a disproportionate amount of the world's food, but their behavior cannot be called simple looting because they also grow most of the world's food. Furthermore, they tend to grow more food with each passing year. Interestingly, the poorer and less industrialized developing countries, too, have been growing more food. In fact, from 1950 to 1990, the less developed countries steadily increased their food production, but they actually became hungrier because their populations increased at more than twice the rate of the more affluent nations. In fact, the more developed countries increased per capita food production nearly three times faster than the rest of the world. (The American population increased by a third while we doubled our food production.) The less developed countries, however, did not suffer as greatly as one might think because they had resources that they were able to trade for food. Of course, this meant that instead of the less developed countries receiving the technology they need so desperately, they were forced to trade for food, just to try to feed their booming populations.

It has been calculated that the difference between more developed countries and less developed countries in per capita consumption of calories* and proteins from plant products is only about 13 percent. From plants alone, the poor receive an average of 2016 calories per person per day, near the minimum average requirement. However, the difference in consumption of the more expensive animal proteins was quite striking; the richer countries consumed about five times as much as the poorer ones.

The most tragic story of recent years concerns northeast Africa, particularly Ethiopia (Figure 28.4), where a drought of several years has brought large-scale starvation and death, particularly affecting some six million people in the north (Essay 28.3). Foreign interests took advantage

*Calorie is a measure of the energy in a unit of food. It is the amount of heat (from the chemical energy in food) necessary to raise the temperature of a kilogram of water by 1°C.

ESSAY 28.2

The Food Pyramid

The relative energy efficiencies in using plant and animal food can be shown by use of a "food pyramid." We can begin to illustrate this principle with the premise that there is a certain amount of energy stored in the bodies of plants. However, when an animal, say a cow, eats the plant, only part of the plant energy is made available to the cow. If we, in turn, eat the cow, only part of the energy stored in the cow's flesh is made available to us. The transfer of energy, we see, is not 100 percent efficient. Thus, much of the food stored in living things is wasted. But the wastage varies depending on how far one's food supply is removed from the level of the primary producer, the plant.

In the food pyramid produced in the study described in Figure A, we see that of the great amount of energy stored in the tissues of the plants, less than half (8833/20,810) can be utilized by animals. After all, plants have their own things to do; they aren't in the business of supporting animals. Thus, they turn much of the energy of sunlight, not into food reserves, but into structural materials or metabolic machinery that is of no use to the grazing animal. Furthermore, the plant uses much of the food it produces for its own energy needs. At each step in the pyramid, energy is lost for use by the level above. So, only a portion of the energy a herbivore gets from plants is stored in its own tissues and thus made available to the carnivores that eat it. Much of the energy stored by organisms at any food level is made indigestible, or is used in metabolic activities, or is lost as heat or in other ways. It is simply not made available to the next consumer level.

FIGURE A

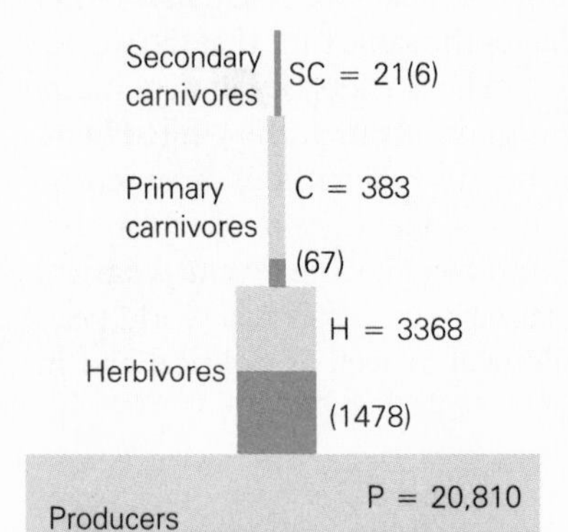

(The red indicates the proportion available to the next level.)

Another reason that animals do not recover all the energy in plant tissue is that animals are not 100 percent efficient at recovering the energy stored in the things they eat. Thus much of the energy that *is* available to them is not utilized. The result of all this is apparent in Figure A. There we see how many producers are necessary to maintain even a few herbivores and even fewer carnivores. Obviously a massive amount of food energy must be produced by plants to maintain only a few secondary carnivores (carnivores that eat carnivores).

With this in mind, then, would you say it is more efficient for humans to eat the fish that ate the fish that ate the fish that ate the algae, *or* is it more efficient to eat the algae? Should we feed grain to beef cattle or should *we* eat the grain? (See Figure B.) The diet of North Americans places *four times* the demand on agricultural resources than do the poorer diets of the less fortunate, because of our interest in eating from higher levels of the food pyramid. With such wastage involved, can you see why steaks are so expensive (politics and market manipulation aside)? Nonetheless, in spite of costs, each American eats about 200 pounds of meat each year. (We seem to be shifting to fish and poultry with increasing studies that link red meat consumption with health problems.)

of the turmoil to advance their own political causes and to fuel the fires of civil war. Roads were closed and much of the relief commodities that were sent could not be distributed. Thus, natural and political causes join to produce a devastating effect in one of the earth's most impoverished areas.

Projections of Food Supply

By 1995, more than 75 percent of all people will live in less developed countries, including China. People in these areas will produce 90 percent

FIGURE B

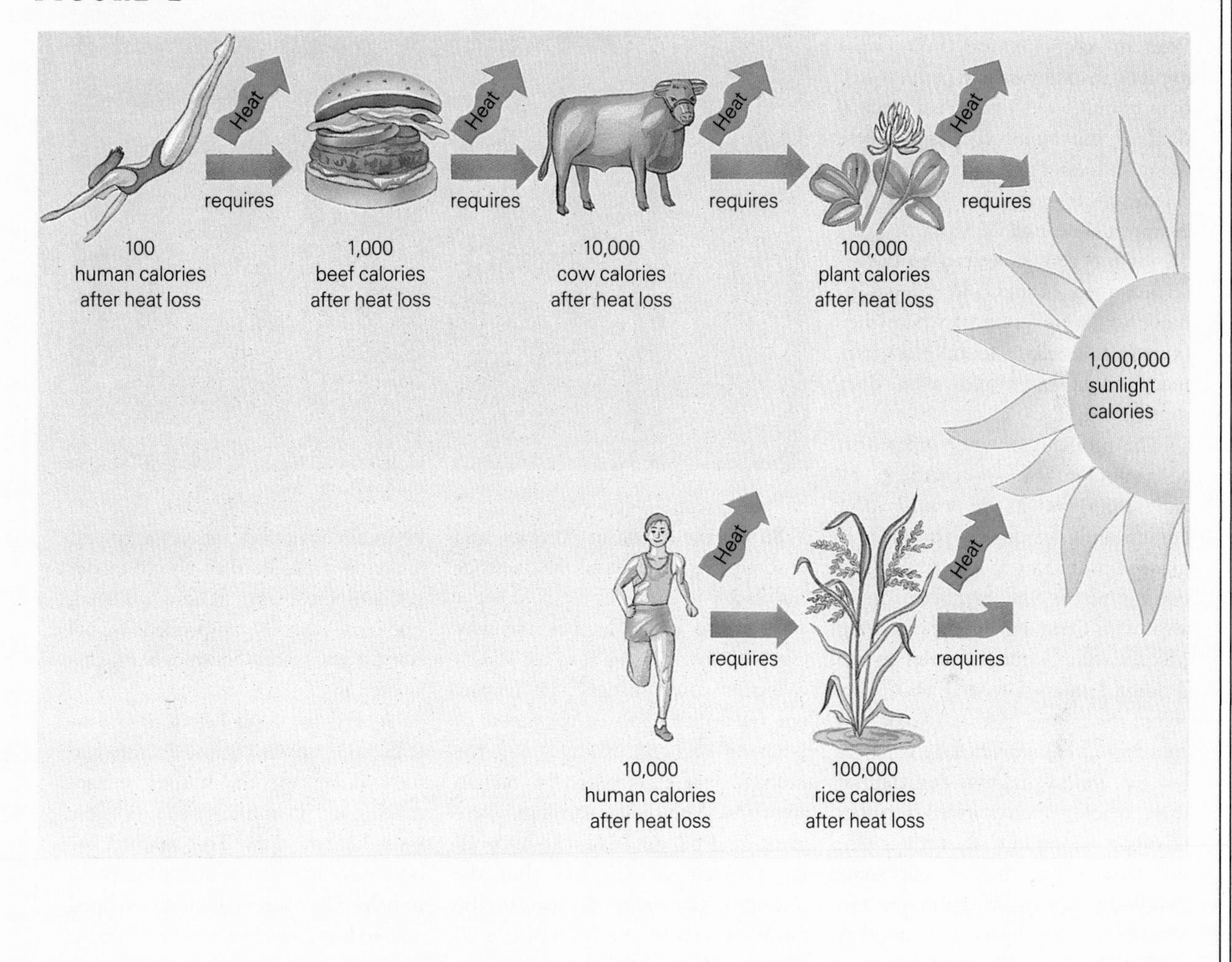

of the new mouths to feed. And worldwide, each year about 90 million people will be *added* to the planet.

Some researchers have suggested that in spite of such problems, a determined and careful worldwide agricultural program has the ability to produce sufficient food, mostly grain, to improve per capita consumption of food by a very limited amount. Of course, production of food is one thing and getting that food to hungry mouths is another. We must always consider the ability of the hungry nations to buy and distribute that food or the willingness of other nations to do it for them.

ESSAY 28.3

The Great African Famine

In late 1984, the plight of a hungry Africa was suddenly thrust upon Western consciousness. We were abruptly made aware that perhaps well over a million Africans had already died of starvation, thousands were dying daily, and many of the rest were on the move in search of food. For example, almost all of Mozambique's 14 million people seemed to be deserting their homes and villages to move to "relief" centers where they hoped food could be found, but where many of them would meet their deaths.

The problem was not isolated to Mozambique. Much of Africa, by early 1985, was in the death-grip of famine and starvation. The area of greatest problem was a swath of land cutting across the northern desert area, with Chad and Ethiopia joining Mozambique as the countries in the greatest danger. A second swath cuts across Southern Africa. Some 30 countries in all are officially listed as hungry, and in the most desperate of these, whole populations are in danger of being wiped out. By early 1985, the famine had helped kill some 200,000 people in Mozambique and 300,000 in Ethiopia (with another million Ethiopians in grave danger).

Africa has had famines before, of course. The last big famine, in fact, occurred in the mid-1970s (when 300,000 died). And, whereas some experts said such a thing could never happen again, the famine of the mid-1980s was the worst the continent has ever seen. In Mauritania in the last 20 years, more than three-quarters of the grazing land has been lost to encroachment by the Sahara. The rainfall in 1983 was the lowest in 70 years, and in some areas, almost all the livestock has died.

How did all this happen and why was it so slowly brought to the West's collective consciousness? Of course, one reason it happened is because of a natural disaster: drought. But the problem was exacerbated by human activities. The first warnings were given in 1982, describing an impending problem in Ethiopia, but the warnings were ignored, not only by the West, but by the Ethiopian government itself. Ethiopia was being torn by civil war and the government elected to spend its money on arms rather than food. Also, the government did not want to dampen a $200 million celebration of its Marxist takeover. In late 1984, a British Broadcasting Corporation film alerted the world to the condition that, by then, had spread to other black African countries. A massive relief effort was begun, but it was severely stymied by local corruption, mismanagement, and civil strife. Further, the countries suffering most from famine had been, not coincidentally, embroiled in civil war for years, their energies being dissipated in conflict.

As arriving food began to be distributed, in some camps, doctors had to walk among the hungry masses making an "X" on the foreheads of the most healthy ones. This marked the ones most likely to be able to respond to help. The others couldn't be helped and no food could be wasted on them.

The rains returned in 1985, but hunger continues. The land had been overworked and overgrazed and its carrying capacity has been permanently lowered. Civil strife has not ended and a great deal of the national effort continues to be used to adjust people's political thinking. Problems of transportation and health dissemination are about as bad as ever, and some experts expect the whole thing to happen again.

Another tragedy is taking place in the Sudan where the Moslem government in Khartoum has been fighting animal worshipers and Christians in the south since 1955, leaving over three million people displaced and unable to feed themselves. In 1984–1985, over 150,000 people died due to the complications caused by a drought. A bumper 1989 harvest was not stored, but was sold to finance the war effort and the country has begun what appears to be another extended drought.

It is important to keep in mind that the world produces enough calories to feed its human inhabitants minimally. Thus, (ignoring protein, vitamin, and mineral needs), there is enough to go around—barely. The fact that people are starving means the problem is partly one of distribution and allocation. The rich have more than their share, the poor have less.

But there are problems other than production and distribution. For example, there is the problem of storage. At least 10 percent of the world's available food is destroyed by pests, waste, and spoilage somewhere between the farm and the consumer. The wastage is greater in less developed countries, the very ones that need the food the most.

In the best of worlds, it would seem that there would be a net flow of food from richer areas to poorer ones, in an effort to alleviate the plight of the hungry. But is this the case in *our* world? It is true that the developed nations send tons of food to underdeveloped countries each year, much of it in the form of protein. (The gifts are well publicized.) It is less well known, however, that the less developed countries send tons of food to more developed countries each year, much of it in the form of *higher quality* protein than they receive. Much of our shrimp, for example, comes from India, and protein-deficient Latin America sends us fish that we use to feed finicky cats with cute names. (As the food crisis worsens, we may have to place strong restrictions on the ownership of pets. Each day, the 500,000 owned dogs in New York City testify to good diets by depositing 90,000 gallons of urine and 150,000 pounds of feces on the city's streets, about half of which ends up on someone's shoes.) Much of the protein that we import from more developed countries is also fed to our poultry and livestock.

FIGURE 28.5

Much of the earth's rain forest is destroyed each year for farming and ranching.

Farming the Earth's Jungles

It was once suggested that the food problem could be greatly alleviated simply by opening up the great jungles of the world to cultivation. Brazil even set up an agricultural colony in the Amazon Basin. The program failed dismally because some very basic rules of agriculture ecology were ignored. The most fundamental of these was the fact that tropical soils are desperately poor in nutrients. This may be surprising, since tropical rain forests seem to be so lush and rich. There are great stores in tropical foliage, to be sure, but when the foliage falls to earth and decays, the nutrients are quickly washed from the soil by torrential rains or reabsorbed into new foliage. Furthermore, if trees are cut and burned (Figure 28.5), many of those nutrients go up in smoke. The agricultural crop plants lack adaptations for the quick absorption of nutrients, and so they lie vulnerable to the rains.

The difficulty of farming such land is compounded by the problem of trying to clear an area in the face of almost overnight encroachment by rapid jungle growth. And then there is the problem of trying to till the

FIGURE 28.6

The soil in tropical rain forests is notoriously poor. However, it can be made productive for growing food by using the slash and burn agricultural technique employed by primitive tribes, as in this small clearing. Indigenous Indians will plant a few crops here, utilize whatever meager nutrients are available in the soil, then move on. Because the plot is small, it is soon reclaimed by the jungle.

jungle soil. In about 5 percent of the jungle, when such soil is exposed to sunlight and oxygen, it forms laterite, a hard, rocklike substance. The beautiful and enduring temples of Angkor Wat in Cambodia are largely made of laterite. Also, the crops that do well in temperate regions, such as wheat, oats, and barley, cannot compete with the plants that grow naturally in the tropical rain forest.

This is not to say that tropical rain forests *cannot* be made to produce food for human populations, but the problem must be approached much more carefully than it has been thus far. We know, for example, that some tropical soils, with very heavy use of fertilizers, can produce up to three crops of grain a year. Unfortunately, there are ten times more insect pests and plant diseases in the tropics. We're going to have to make concentrated efforts to develop new farming techniques and to learn more about how indigenous peoples have been able to successfully cultivate such areas in the past (Figure 28.6 and Essay 28.4).

Domestic Animals as Food

As you may recall, each time food is transferred to a higher level in the food pyramid, most of the energy it possessed in the preceding level is lost (see Essay 28.2). It is therefore much more efficient to eat from low in the pyramid. However, the animal protein from high in the pyramid is advantageous for a number of reasons.

The most obvious is that since we are animals, the flesh of other animals provides us with the eight or ten essential amino acids in proportions that are relatively close to our needs. To get all the amino acids we need from plant tissue, we would have to choose the plants carefully so that one provides what the other lacks.

The choice of which animals to raise for food is important. For example, goats are a good source of milk, but the goat is a destructive grazer, best suited for marginal areas that have no other use. In their overgrazing, goats have contributed to ecological problems by stripping the soil of its protective plant layer. Sheep also tend to overgraze, and they are not very popular as food. Also, because of their heavy wool coats, they are not well suited to tropical areas. Hogs, on the other hand, can be reared in a wide range of habitats and are extremely efficient in converting plant protein to animal protein. The trouble is that their dietary habits make them direct competitors with humans for the available plant food. In fact, like humans, hogs are omnivores, that is, they eat about anything—animals as well as plants. However, whereas they compete very directly with humans, they can also subsist on human refuse (Figure 28.7).

Cattle, because of their complex digestive arrangement, are able to survive on plants that are not food sources for humans. Since they ordinarily don't wander from their grazing areas, they don't require constant attention. Moveover, they travel well, are large enough to be used as work animals, and provide leather as well as beef. In addition, cow's milk is a food staple in many places, whereas people don't tend to get too excited about pig's milk.

Perhaps the most efficient and least costly kind of animal to be raised for food is poultry. More flesh is produced per pound of plant tissue eaten than is produced by any of the other animals just mentioned (Figure 28.8). Chicken is relatively inexpensive because the less efficiently a pound is produced, the more it will cost. (In our frantic efforts to impress each other, we like to feed our friends. But do we offer them beans? No, beans come from too low on the food pyramid. Their conversion efficiency makes them inexpensive and no one will think well of our provisions. However, they appreciate an offering from higher in the food chain. Maybe we should surprise them with something much higher, say, a nice vulture.)

FIGURE 28.7

Hogs are raised as food in many parts of the world because they can eat what humans eat.

FIGURE 28.8

These jungle fowl of Southeast Asia are being raised for their meat. Both American poultry ranchers and Asian farmers recognize the efficiency of raising poultry for food.

ESSAY 28.4

Miracle Crops and the Green Revolution

In the late 1960s, there was a great fanfare heralding the impending "green revolution." New kinds of grains had been developed that yielded plants heavy with seed. For example, one strain of dwarf rice, IR-8, produced over twice as many rice grains per plant as the conventional strains. The new miracle crops do, indeed, hold great promise for the world's populations. However, they have their problems.

First of all, the new strains require special handling; farmers must be carefully trained in new agricultural methods. However, there has traditionally been a reluctance of people steeped in their traditions to adopt new ways. They won't eat new foods and they won't change their farming practices. There is also a problem in underdeveloped countries of providing trained technicians to teach the farmers. In some cases, the technology itself is useless. How can controlled-temperature farming be implemented in areas that have no electricity?

Miracle crops also must be heavily fertilized and fertilizer, of course, costs money. Then there is the problem of the fertilizer runoff. As rains wash fertilizer from the fields into bodies of water, the nutrients may cause the water to eutrophy, or "age," prematurely. Nutrient-rich water may also be subject to great blooms of microscopic algae that can interfere with the balance of life and cause massive fish kills. The "red tides" of the oceans, when the waters turn strangely red and fish die from toxins produced by microorganisms, are the result of such blooms. An increase in agricultural production, then, may reduce the catch of fish in offshore waters.

Wheat, rice, and corn account for about three-fourths of the total yield of grain worldwide. Since grains are concentrated sources of energy and are easily stored, they probably represent a good focus of continued research effort. And, in fact, one new hybrid of wheat and rye, *triticale,* shows great promise (see photo).

Ecologists have warned of the consequences of relying too heavily on artificial strains of grains. After all, they are rather homogenous, thus the low variation leaves little on which natural selection can operate. One of the attendant dangers is that wild populations of insects or disease organisms, replete with their wide genetic variation, might, through their own processes of natural selection, quickly accommodate themselves to any single strain and wipe it out in short order. Thus, in the long run, we might be better off paying the price of lower productivity for increased genetic variation in our crop species.

This point was dramatically illustrated by an unusual event in 1969. Specially developed high-yield corn plants were attacked by a new type of fungus called the southern corn blight. By 1970, the epidemic was so severe that the United States corn crop was reduced by 15 to 20 percent (a loss calculated at more than a billion dollars). The tragedy jolted many agriculturalists into a new appreciation of how food supply is a function of biological realities. The lesson was a simple one: extreme uniformity of our food crops makes them particularly vulnerable to the appearance of new kinds of pathogens or predators.

Some global planners believe that we can indeed boost food production

Fishing

Perhaps fishing should be included in the discussion of nonrenewable resources (those that are not naturally replenished, such as oil or ore), because, in a very real sense, we do not replace the fish we take. Our fishing practices are more akin to mining. The fishing industries are diligently studying fish behavior, developing technical arsenals, and taking more fish as they continue their drive to meet the demands of our hungry numbers (Figure 28.9). The world catch could easily be increased in two ways. One, we discard 20 percent of any catch by keeping potentially useful species that are now thrown overboard as trash fish. Two, the installation of better refrigeration would keep much of the catch from on our planet by basically continuing the green revolution begun in the sixties. It was then that specially bred food plants with exceptionally high yields were disseminated throughout the world. The greatest progress occurred in cereals, especially wheat and rice. Both were short, with upright leaves, and both were highly responsive to fertilizers. In fact, without liberal doses of chemical fertilizers, the plants produced only a little better than traditional varieties. Because of these new crops, India's wheat production rose from about 11 million metric tons in 1964 to 34 million in 1979. The yields of other crops increased comparably, but from 1960 to 1983, the use of chemical fertilizers has increased 18-fold. Pesticide usage has also increased, and these chemicals must go somewhere, so a large part has run off into bodies of fresh water on the continental shelf. Even in areas where miracle crops still hold promise, there is an acknowledged trade-off for immediate results.

Triticale (right) is a crossbreed of wheat (left) and rye (center).

going bad at sea. Some have suggested the use of unconventional species such as lanternfish and the tiny Antarctic arthropods called krill. (Krill has an awful taste, so we need to figure out some way to prepare it.)

The management of the oceans is complicated by what biologist Garrett Hardin refers to as the "law of the commons." "Commons" have historically been areas that belonged to no one, but are available to all. In Victorian England, a common might be an area where everyone could graze sheep. The concept worked well if all the shepherds voluntarily limited their herds so that the commons would not be overgrazed. If someone added a few extra sheep, however, they would have a marketing advantage over the others while not doing too much harm to the grazing

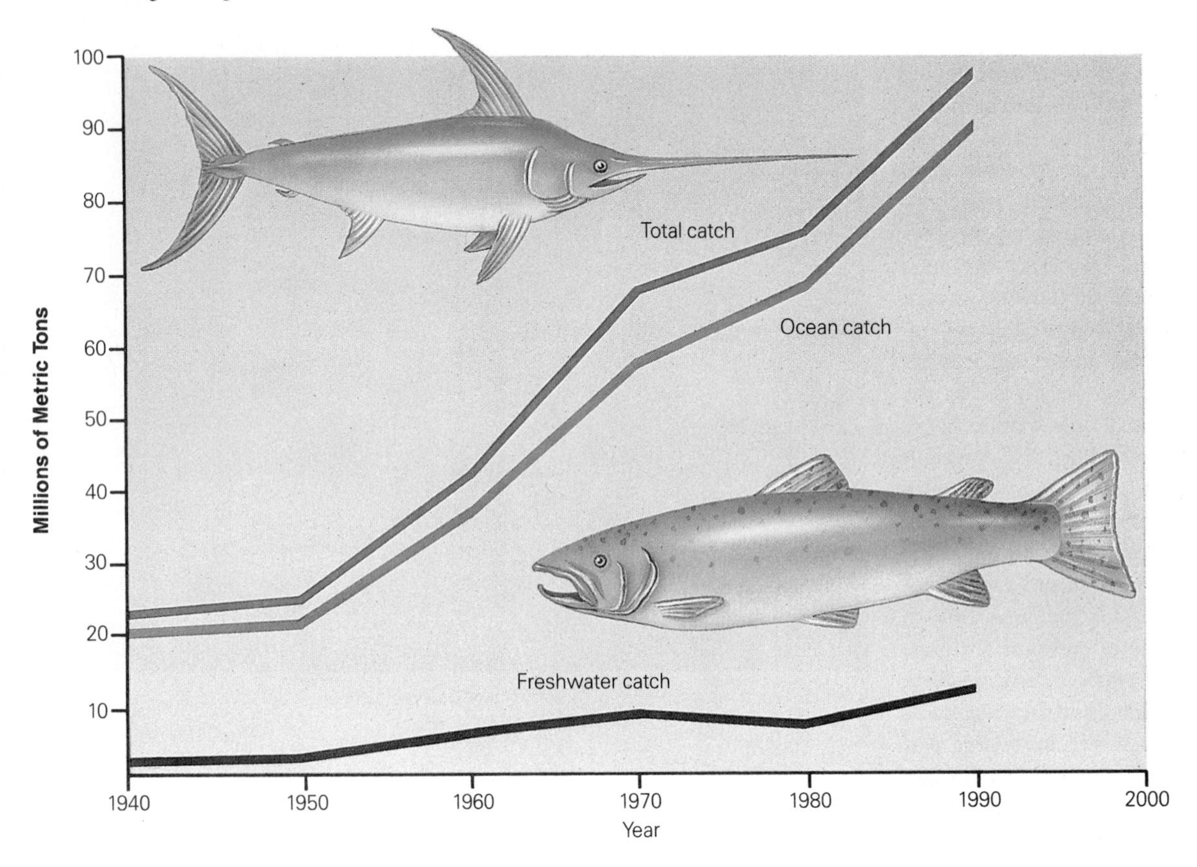

FIGURE 28.9

Our increasingly efficient fishing technology has resulted in more fish being taken each year. We must now ask how long it can go on.

areas. But what were the others to do? Obviously, they would add sheep also, each hoping the others would not notice. Soon the grazing area would be lost to all.

The sea is a "common" today, but for the most part, nations are not behaving so as to ensure optimal productivity. For example, the decision to add another vessel to a nation's fishing fleet may be marginally productive to that country, but in the long run the action may contribute to a diminished catch because of overfishing. Yet if that nation chooses not to add another vessel, it may find itself on the short end as other nations continue to add vessels. Any nation caught in this game comes up short both ways. So how should such decisions be made? What will be the long-range effects of any alternative? What should be our priorities? Our logic? Can you think of some better way of apportioning the world's resources?

Solutions to the Hunger Problem

There are a number of largely untested but fascinating ideas on ways to increase food supplies. For example, it is known that single-celled organisms high in protein can be grown on petroleum. If these organisms could be produced in large enough amounts and purified economically, they could conceivably provide us with high-grade protein. Efforts are presently under way to develop new plants, through recombinant techniques, that can "fix" their own nitrogen from the air. Scientists are also busy trying to produce yet new grains with higher quality proteins (that is, with a better balance of amino acids). A nutritious mixture of corn and cotton seed meal enriched with vitamins A and B, called incaparina (Figure 28.10), has been available in Central America for over 15 years but unfortunately has not been accepted by the people there. As mentioned earlier, this is one of the great problems of food scientists—how to get people to eat new kinds of food. Historically, the hungriest people in the world are those that recognize the fewest items as food. In fact, even people in developed countries are warned to stock familiar foods for emergencies because even such "sophisticates" have been known to starve rather than try something new. (I would probably wait days before opening a can of broccoli.)

FIGURE 28.10

Incaparina—although nutritious—has not been met with great enthusiasm.

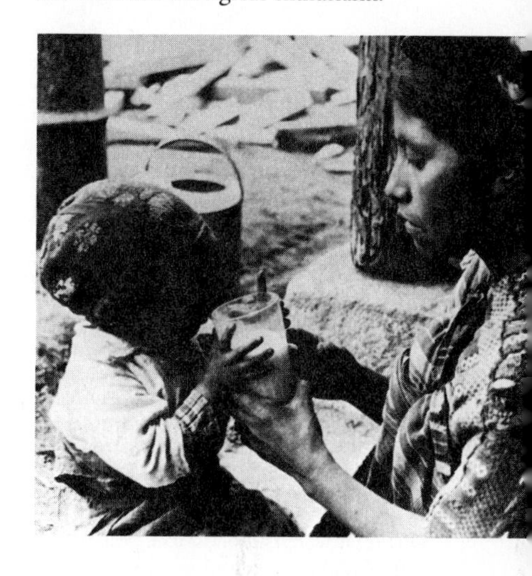

A variety of other ideas to increase food production are being considered, such as culturing algae, ranching various rodents or antelope, converting weeds to cattle feed, extracting protein from plants or small fish, converting wood to cattle feed, and using sewage to grow edible slime. (That last one should be a big hit.)

It is interesting that of the thousands of species with which we share this planet, so few have been treated as food. In fact, it turns out that modern humans, with that vast ability and imagination, have discovered virtually *no* new crop, plant, or domesticable animal. Captain Bligh was directed to take the *Bounty* to bring back a "new" plant that had long been cultivated in the South Pacific. "New" kinds of fish meals are still made of fish. And the miracle grains are simply variations of plants that have been with us for thousands of years. More recently, a promising new "species," triticale, is a hybrid of wheat and rye (Essay 28.4).

One is tempted to ask whether our energies *should* be directed toward frantically trying to provide a minimal diet to teeming millions of people as their numbers grow increasingly faster (Figure 28.11). Should this be

FIGURE 28.11

Ethical questions will arise as more and more people enter an increasingly drained planet. The average family in Bangladesh is reported to have seven children. Should relief agencies encourage or ignore such behavior?

the role of our scientists and the lot of modern humans? Or should we begin to put strong pressures on the swell of our numbers? Should we stop asking how we can manage to keep more and more people barely alive and begin to ask how we can improve the quality of life for those already living? How can this be done? What would be the political, sociological, and moral results?

Water and the Coming Crisis

The planet has a great deal of fresh water, but relatively not much is available for human use (Figure 28.12). We know where most of it is (most of it—97 percent—is in the oceans). Only 3 percent of the earth's water is fresh water. We use it in a variety of ways—some personal, most on a grander scale. For example, most of the fresh water available in the

FIGURE 28.12
Our total water supply. Most fresh water is frozen in glaciers and ice caps. A miniscule amount (0.003% of the total) exists as vapor in the atmosphere.

FIGURE 28.13

This dry reservoir at Kensico, New York, poignantly illustrates the earth's continuing problem with a dependable supply of fresh water.

Perhaps such problems would be of less magnitude if we realized just how much water we actually use. For example, each American uses an average of almost 2000 gallons of water each day. A bath takes 30 to 40 gallons, and a five-minute shower, 25 gallons. Less obvious use includes an incredible 40 gallons of water that goes into the production of a single egg, 75 gallons into one pound of flour, 150 gallons into a loaf of bread, 2500 gallons into a pound of beef, 230 gallons into a gallon of bourbon, 280 gallons into a Sunday newspaper, and 100,000 gallons into a car.

United States is used in agriculture, with the next largest allocation going for steam generation (mostly to cool electric power plants), a trend that is expected to continue at least until the end of the century.

Our use of water in this country is prodigious and is still growing by leaps and bounds. In 1900, for example, Americans used 40 billion gallons of water each day (an average of 530 gallons per person). Now, however, we use over 500 billion gallons a day, an average of almost 2000 gallons per person (Figure 28.13). On the other hand, a person in an underdeveloped country uses about 12 gallons per day.

Many ecologists predict severe shortages within a few years. The problem is pressing in several areas of the United States even now, including the western agricultural states, the northeast, Florida, and Louisiana. The problem is exacerbated because not only is water running short (Figure 28.14), but much of the remainder is becoming dangerously polluted and unusable.

Because of localized water problems, it has been suggested that we shift great waterways in the sparsely populated northern areas of the United States to the south and west where agriculturalists are running short. In fact, in North America, the amount of water available per capita is dwindling at an alarming rate. There are several reasons for the problem. First and foremost, of course, is our increasing population. Beneath the densely populated cities of the northeast, many critical underground water supplies are simply drying up. But, in addition, industries continue to dump dangerous chemicals into our waterways or let them seep into the water table from rusting barrels above, often rendering the water permanently unusable. In some coastal areas, the aquifers (underground supplies of fresh water) are being steadily encroached upon by salt water as the fresh water is pumped out.

FIGURE 28.14

Changing water supply in the continental United States. Source: U.S. Water Resources Council.

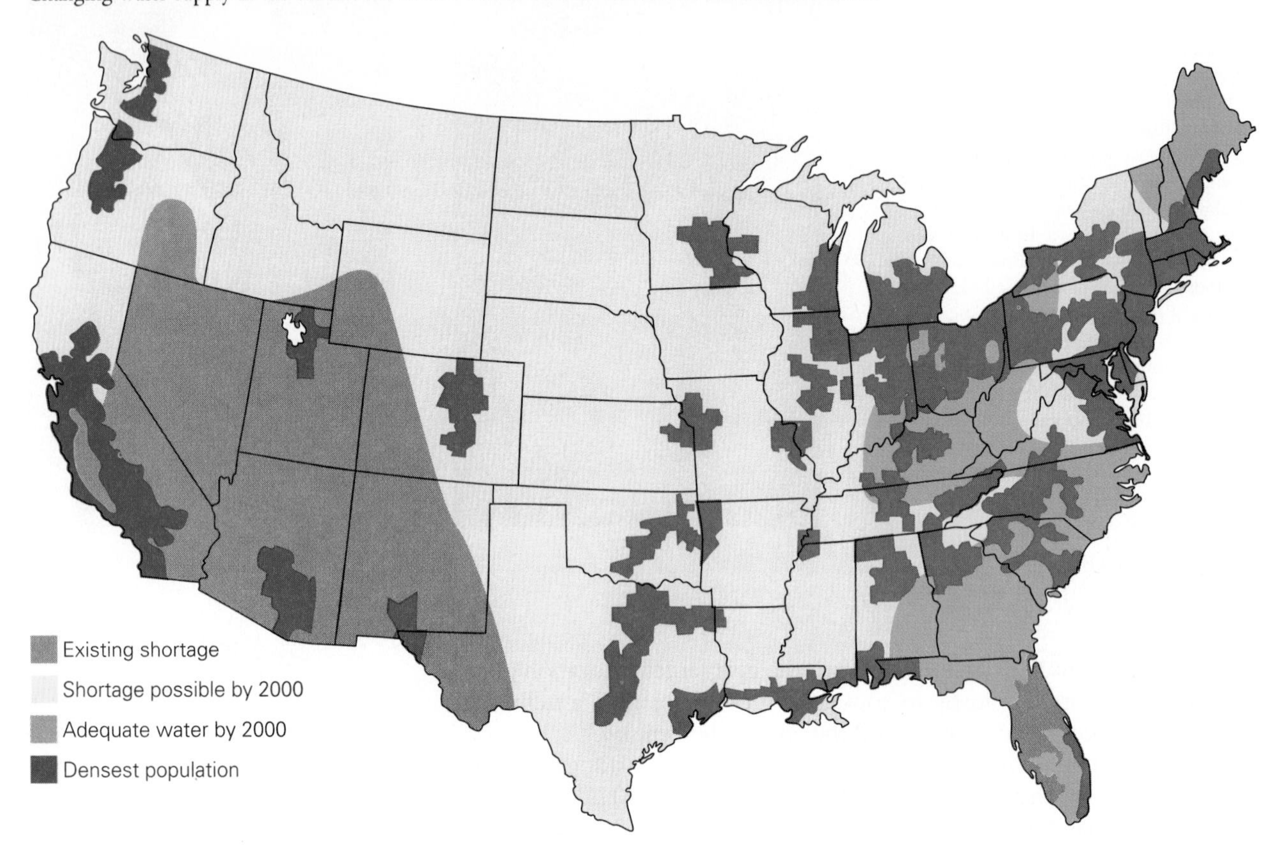

In the west (the area that provides us with most of our vegetables, fruit, and grain), the major resources of irrigation are drying up. For example, the great Ogallala aquifer, an underground reservoir stretching from Texas to Nebraska, is being depleted twice as fast as the waters are being replaced by natural water cycles. Some wells are already over a mile deep and much of the aquifer is expected to be dry by 2020. The water table in the west is dropping by a disconcerting three feet each year. Above ground, the Colorado River, servicing seven states, is a salty trickle by the time it reaches Mexico. Worldwide, the withdrawal of water for irrigation is expected to double by 2025 primarily because of the increasing population of less developed countries.

There are two approaches to managing our water resources. One is to increase the supply and the second is to decrease loss. We can increase supplies only indirectly by moving and allocating water more wisely. We can decrease loss by not polluting the available water and by decreasing opportunities for evaporation. In some countries, for example, farming is no longer permitted in dry areas. We may be asked to switch our eating habits to foods that require less water to grow. We may be asked to largely give up meats because grain to feed the animals takes a great deal of water to grow. Also, we will eventually be asked to pay very dearly for water.

(I, for one, don't want to pay more for anything, and I certainly don't want anyone making any more rules. But the interstate highway near my house just recently collapsed, forming a huge sinkhole, because the water underneath it had been pumped out. I don't like restrictions, but I don't like the earth caving in around me, either. We simply may have to accept more restrictions as our situation becomes increasingly dire.)

NONRENEWABLE RESOURCES

Nonrenewable resources exist in a fixed amount. When we're out, we're out; they cannot be replaced. Examples include aluminum, iron, bauxite, and copper. Normally, a nonrenewable resource is considered depleted when 80 percent of its estimated supply is gone. The reason is, at that level, continued extraction is likely to be too difficult or unrewarding. There is nickel in Alaska, for example, that is not being mined because it's on remote islands. (My friend, Rip Van Winkle, and I discovered big deposits of it back in our prospecting days. As far as we know, it's still just sitting there.)

The line between renewable and nonrenewable resources is a fine one. For example, water is considered cyclic in that it evaporates and falls again as rain and thus is "renewable." However, much water is tied up in various manufactured and natural products and is not easily retrievable. And we know that water may become so polluted that it is no longer usable. Thus, in a sense, it is at least partly nonrenewable. Coal and oil must also be considered nonrenewable because, if they do form, they form slowly over geological time.

We must remember that these resources do not occur uniformly over the earth. Hence, their use may be manipulated by nationalistic policies. Consider the imbalances in metals. Asia is rich in tin, tungsten, and manganese. In fact, over half the world's tin is located in Indonesia, Malaysia, and Thailand. North America, on the other hand, is poor in these elements, but well-endowed with molybdenum. Gold, of course, occurs mostly in South Africa, so does platinum, but not much silver is found there. Over half the world's nickel is in New Caledonia and Cuba, and most mercury is found in Spain, Italy, and the Sino-Soviet areas. Again, a situation that seemingly would encourage cooperation and trade seems, too often, to give rise to aggression and conflict.

It is increasingly apparent that we can expect problems if we continue to take the earth's available resources on a "devil take the hindmost" basis. Each commodity is limited and we do not increase their supplies simply because we find a new deposit. Such thinking is short-sighted, but locating new sources does solve problems on a short-term basis. Problems arise because of our ignorance about the long run. We simply don't know the extent of most of the earth's resources. Some will give out much sooner than expected, and others are probably much more abundant than we now believe. So what should be our policy? Do we gamble? What are the stakes?

Geologist Preston Cloud points out that we can make the best predictions for those substances that are found associated with specific rock layers. These include the fossil fuels—coal, natural gas, and oil—and, to a degree, iron and aluminum. The reserves, grades, locations, and recoverability of many other critical metals are much harder to estimate. Thus the problem of developing long-term policies regarding their use is magnified. Since we may not know how much of certain commodities are left, should we continue to drain each known deposit, confident that others will be found because they *must* be found?

RECYCLING — AND AROUND IT GOES

FIGURE 28.15

The "good life" for many people is based quite simply on more goods and services. As they climb that long ladder of success, however, they utilize disproportionate amounts of commodities and energy. Such artificial demands place a great burden on our reservoir of natural resources and create great amounts of waste that must be handled by the expenditure of yet more commodities and energy.

The odds are that the can you've got in your hand is made from recycled aluminum. In fact, just over half the aluminum cans produced in the United States are recycled from old cans turned in at over 5000 recycling centers nationwide. At about a penny a can, those who sold cans at these centers last year made about $100 million. Still, of the 80 billion aluminum cans made in the United States each year, almost half are simply discarded. In fact, if the cans Americans throw away in a single year were laid end to end, they would reach to the moon and back about eight times. It turns out that Americans will recycle these cans *if* they are paid a minimum of 5 cents per can. Less money means fewer recycled cans.

Recycling aluminum takes only 5 percent of the energy necessary to mine and process aluminum ore, and it produces only about 5 percent of both air and water pollution. Still, overall, Americans recycle only about 30 percent of their aluminum.

Aluminum and iron are found in almost 95 percent of all metal produced. Using scrap iron, instead of virgin iron from the earth, requires 65 percent less energy and produces 85 percent less air pollution and 75 percent less water pollution. The use of scrap iron also means that coal need not be burned to process the ore. Yet our antiquated steel industry still devours enormous amounts of ore and pollutes our air and water and incinerates our coal, turning out iron while scrap iron lies rusting all over our countryside. Why? Largely because mining and energy industries are given subsidies (from your taxes), huge tax breaks, and other governmental incentives that encourage them to go after virgin ores from the soil. Recycling industries, on the other hand, get few tax breaks and very few governmental incentives.

We've really done a job with plastics. Most of the plastics we make simply don't rot. Ever. Those that do typically take 200 to 500 years to break down. Yet we recycle less than 4 percent of our plastics (our fastest growing waste product). Some companies boast that their plastic products are biodegradable, but many environmentalists call the claim a sham. Most are only partly degradable. That is, they break down into smaller pieces of plastic, and many may release poisonous products as they do so. Plastic recycling is becoming lucrative and, by the year 2000, it is estimated that about 45 percent of all plastic waste produced in the United States could be recycled.

A powerful lobby of aluminum, steel, and glass manufacturers, working with supermarkets, brewers, and bottlers, have blocked laws requiring a deposit on bottles in all but 11 states. Yet such laws not only result in up to 70 percent less litter, but massive savings in energy and natural resources. Still people generally won't return a bottle for less than a nickel.

Strange as it seems, one of our biggest solid waste problems involves disposable diapers. Americans discard 18 million disposable diapers a year.

Each diaper will take about 500 years to decompose in American landfills, and it costs taxpayers $360 million each year to dispose of them.

ENERGY

Energy, you recall, is simply the capacity to do work. Just as plants must acquire energy from the sun and animals must derive energy from plants, humans must also utilize energy to be able to rearrange trees and oil and iron in order to build houses so that we may be shielded from disruptive elements in the environment and have a good place to read the newspaper. However, with each new house, or each new commodity of any sort, not only are resources required, but also energy to rearrange the resources.

Of course, some of the demands placed on energy supplies are "legitimate" in that they relate to very basic needs such as food production or housing. We may be reaching the point, however, where it will be necessary to distinguish such essential energy uses from nonessential demands (Figure 28.15). Do we really need elaborate neon-lit billboards, automatic can openers, six straight pins with each new shirt, oversized electronic amplifiers, electric toothbrushes, and powerful cars to carry one person to work? (What nonessential item most violates your own sensibilities? To which do you object the most?)

Now let's take a look at our energy resources and some of the demands being placed on them. Figure 28.16 can help us put the data in perspective by indicating the changes in energy consumption of humans over evolutionary time.

FIGURE 28.16
Estimated average daily per capita energy use at various stages of human cultural evolution.

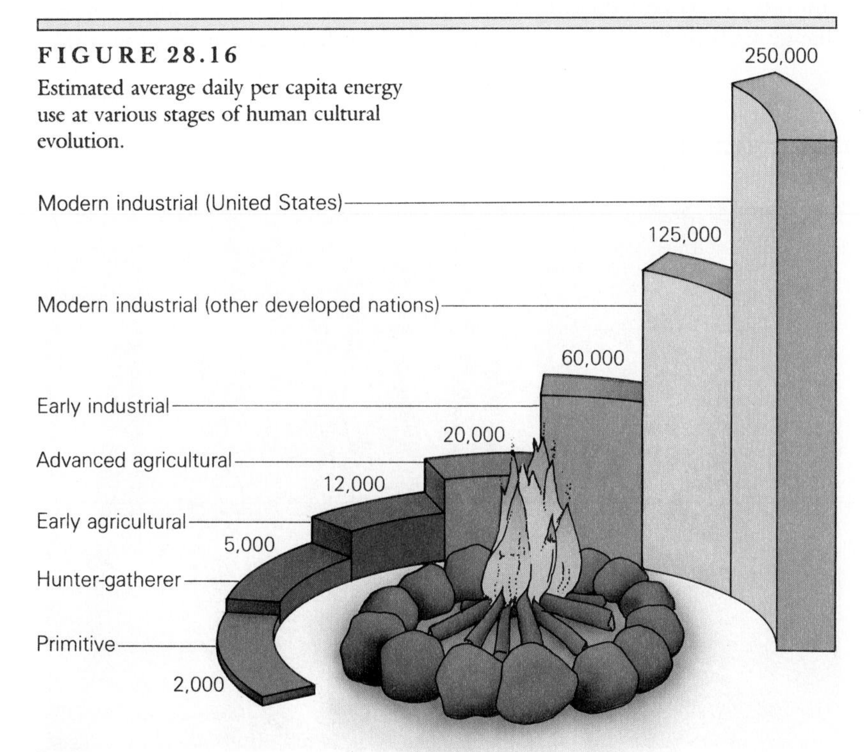

We should first make it clear that, at present, we are not running out of even the current forms of energy. The United States, for example, has vast deposits of coal. However, locating the coal is one thing; getting it out is another. There are heavy costs to our environment both in producing energy and in consuming it, and these costs must be carefully weighed in calculating the net benefit we receive. We will see again the enormous impact of the industrialized countries, particularly the United States, on energy resources, because our attitudes and policies often have such wide-ranging effects worldwide.

Energy from Fossil Fuels

Fossil fuels (oil, coal, and natural gas) are comprised of the partially decomposed remains of ancient organisms. As their bodies fell into the many shallow lakes that covered the earth in those days, great masses of once-living material accumulated and their great weight impeded the natural processes of decay. The partially decomposed masses produced liquified or gaseous material that held great stores of undissipated energy. Today, the fossil fuels (oil and gas) comprise an increasingly large part of the world's energy sources, especially among developed nations, such as the United States. However, human needs are placing inordinate demands on these reserves. In fact, it has been estimated that humans use in one year an amount of fossil fuel that it took nature about a million years to produce. In the United States, that oil is sold at a discounted price (Essay 28.5).

Whereas we have plenty of coal, as well as other fuel sources—such as sunlight, wind, water, geothermal energy, waste energy (from farm and forest), and uranium—in the late 1980s, almost 90 percent of our energy came from gas and oil (Figure 28.17). It seems clear that some sort of change is in order (said the woman as her husband drove into the wall).

FIGURE 28.17
Americans feverishly pump out every drop of oil they can locate, ignoring the simple truth that it cannot be replenished.

ESSAY 28.5

The True Cost of Gasoline

As anyone who has travelled much can tell you, in the United States gasoline is dirt cheap. Americans are generally distressed when they have to pay $20 or $30 to fill up the tank, but Europeans or the Japanese would consider it a bargain (Figure A). So why is gasoline so cheap here? Put simply, it's because we don't pay the real price of gasoline. We pay a discounted, fairy tale price (Figure B).

The price of gasoline in Europe or Japan is two or three times higher than in the United States because their leaders insist that gasoline prices reflect such things as smog, lung disease, trade imbalance, global warming, highway construction, and the cost of defending Middle East oil fields. With the higher prices, consumers are forced to consider the full impact of their gasoline consumption.

The American government, as always, sensitive to the pleading of American big business, is loath to increase the price of fuel. Each small increase in gasoline tax is discussed endlessly. But if America is to take its place among the future leaders of the world, we have to force our consumers to pay for their behavior by factoring in the environmental, economic, and geopolitical costs of oil.

FIGURE A

Gasoline prices (dollars per gallon)

0, 1, 2, 3, 4

U.S.

Canada

W. Germany

France

Japan

Proportion of taxes shown in red

FIGURE B

We are not yet at the wall, but our options narrow as time passes. Furthermore, in addition to becoming concerned about the availability of each energy source, we should consider the consequences of exploiting each potential source. For example, although we have plenty of coal, its use will enhance the greenhouse effect (Essay 28.6) and the problems of acid rain (Essay 26.1).

ESSAY 28.6

The Greenhouse Effect

If it weren't for the greenhouse effect, life as we know it could not exist on the earth. It is the greenhouse effect that keeps the temperature of the planet warm and rather stable as our thin layer of atmosphere traps the heat from the sun and holds it against the earth's surface (Figure A).

In recent years, however, the greenhouse effect has simply become so powerful that some scientists believe it has become a threat to life. The reason for the change is that the atmosphere has begun to accumulate certain molecules that trap the sun's energy so effectively that the temperature of the earth has begun to rise (Figure B). The most important of these molecules is carbon dioxide, which has been steadily increasing in the atmosphere in recent decades (Figure C). It is believed that carbon dioxide, released in large amounts from the burning of fossil fuels (80 percent of the rise in CO_2 is from this source) as well as the burning of forests (20 percent), is responsible for about 60 percent of the global warming trend. CO_2 emissions, by the way, are increasing at about 4 percent a year.

FIGURE A

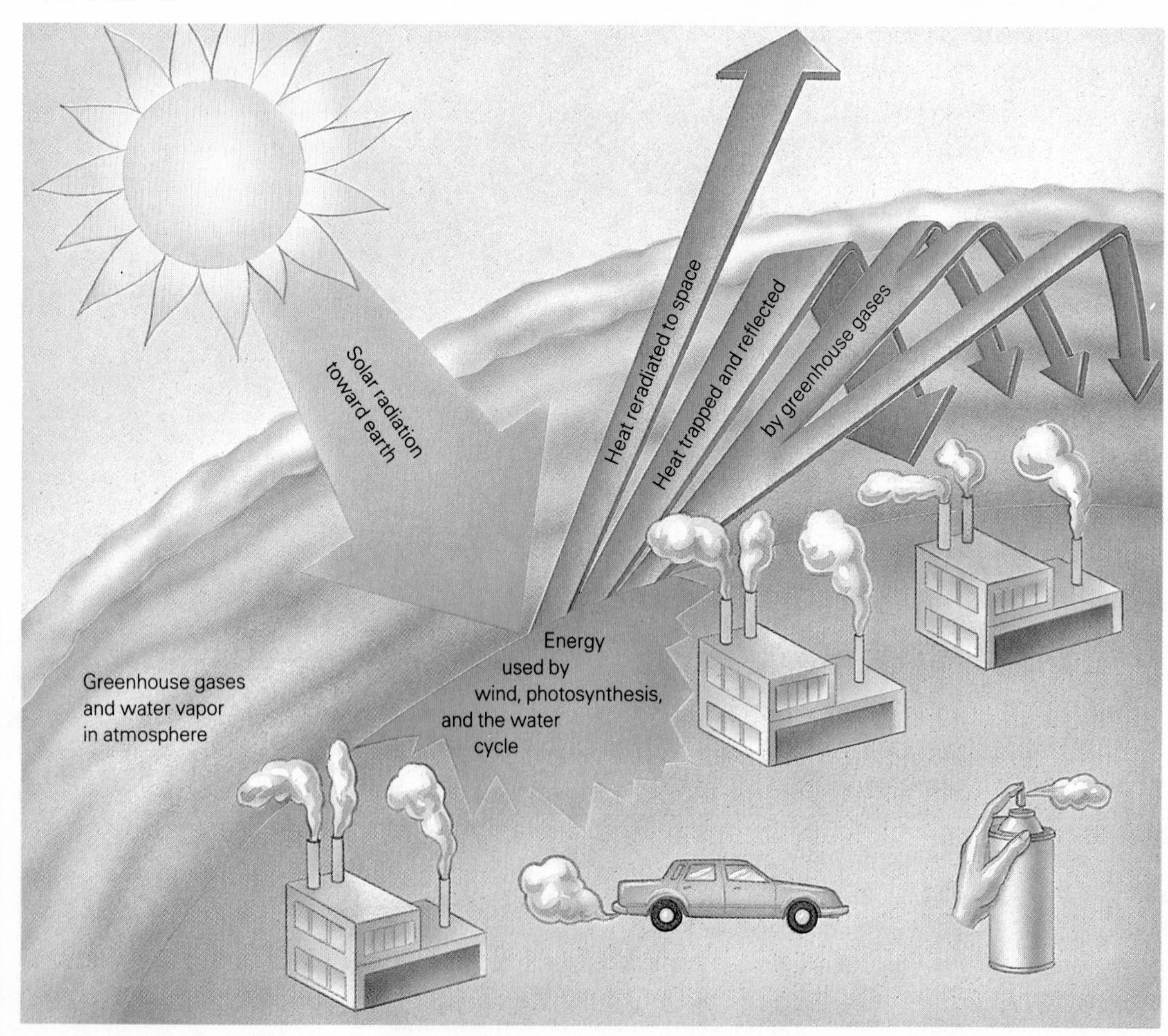

Other heat-trapping molecules have also increased their concentration in the atmosphere, particularly chlorofluorocarbons, methane, and nitrous oxides.

Current predictions forecast, if greenhouse gases continue to rise at the current rate, that between 2030 and 2050, the average temperature on earth will rise by about 3° F (1.5° C) to 18° F (4.5° C). If the temperature should rise by 7° F (4° C), scientists predict massive changes in weather and ocean currents. The result would be drastic shifts in agricultural areas as Iowa became a desert and Alberta became a breadbasket. The Gulf stream would not reach Europe and that area would experience intensely cold weather.

The higher temperatures would also expand the oceans and raise sea levels as the polar ice caps began to melt. Low-lying coastal areas, now home to much of the earth's human population, would be flooded. Great storms would sweep the earth and hurricanes would hit farther north. Diseases could be expected to increase without the periodic winter diebacks of many pathogenic fungi and other disease-causing organisms. The list goes on.

Interestingly, one of President Bush's advisors in 1990 said the effect of global warming is vastly overstated. He said it's only about like moving from Washington to Atlanta.

FIGURE B

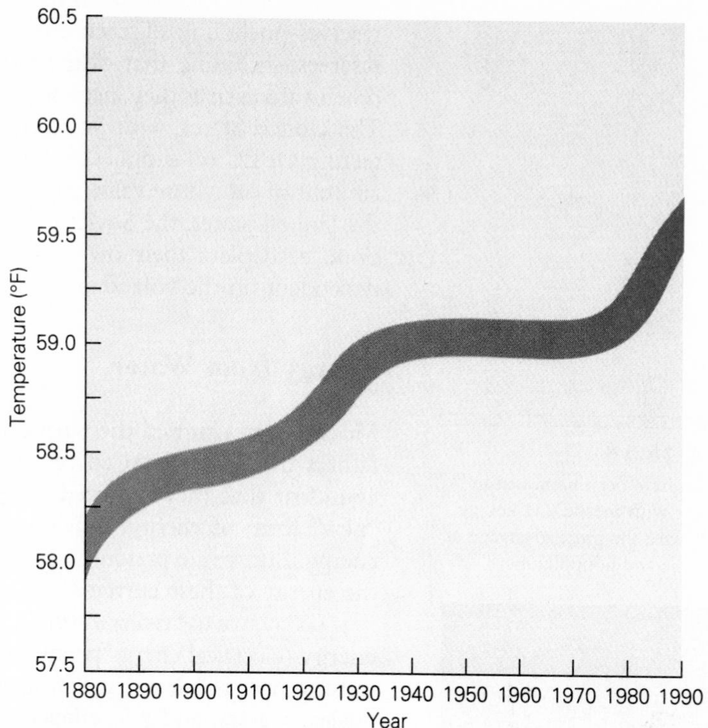

FIGURE C

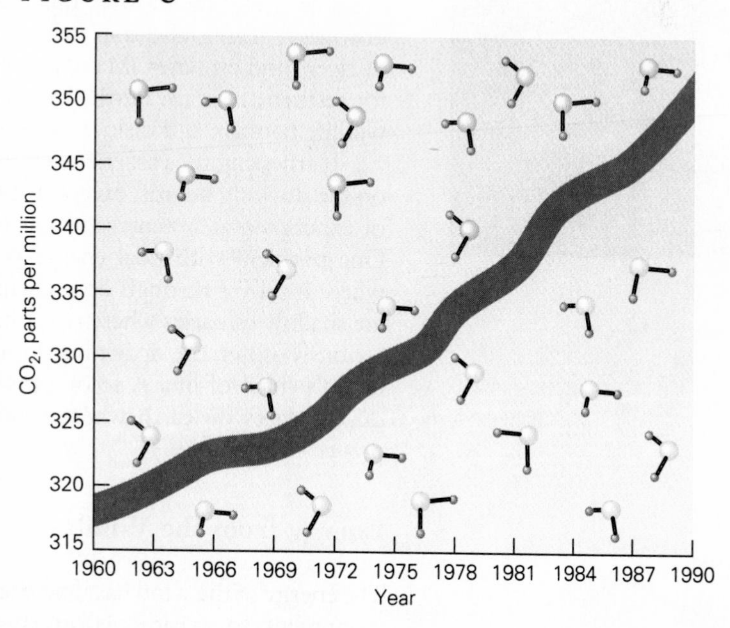

The Geopolitics of Fossil Fuel

Because the industrial world relies so much on fossil fuels, particularly oil, whoever controls the oil controls, to perhaps an alarming extent, the politics of the more industrial nations. Oil means power, plain and simple. So who has the oil? Mainly, the Arabs. In fact, their international agency (the one that sets their prices), the Organization of Petroleum Exporting Countries (OPEC), controls three-fourths of the world's proven crude oil reserves (including all recent major finds). Estimates of the non-OPEC oil reserves, including that of the Soviet Union, have recently been revised downward even as they increase production from their dwindling supplies. The United States, with 3 percent of the world's oil, spends billions protecting OPEC oil supplies and, in fact, the country each year imports an amount of oil whose value equals about a third of the national deficit. As the United States, the Soviet Union, and other non-OPEC countries continue to deplete their own oil supplies, they will become increasingly dependent on the volatile Middle East.

Energy from Water

Moving water turned the wheels that ground the corn that fed our forefathers that gave them energy to have the descendants that became so abundant that they returned to the moving water in their search for a "new" form of energy. Obviously, moving water is not a new form of energy. But we are presently seeking, through a variety of means, to utilize the energy of these currents.

FIGURE 28.18
Even the tides have been harnessed to provide people with inexpensive energy. Such sources have the great advantage of being renewable and nonpolluting.

Today, we use rivers to turn great turbines that produce **hydroelectric energy** (electrical energy produced by the energy of moving water). Such water power is "clean" in that it doesn't pollute the atmosphere or create nuclear wastes, and it is efficient. However, it can create problems. For example, water falling over high dams can pick up nitrogen and kill fish downriver. Dams also trap sediment and thereby alter the natural courses of rivers and estuaries. Many people just do not want to dam wild rivers for aesthetic reasons. Furthermore, dams may destroy both farmlands and wildlife habitats and dislocate people.

Harnessing the vast energy of moving tides is an alternative still largely on the drawing boards, except in England and France where several kinds of experimental systems utilizing tides are in operation (Figure 28.18). One problem with tidal energy is that the water has the greatest force where it moves through deep channels, and most of the natural channels are shallow estuaries where rivers meet oceans. Any disruption here could seriously upset the spawning grounds of some important links in the ocean's chain of life. A second problem is that there are only about two dozen places on earth where conditions are suitable for harnessing the power of tidal flow.

Energy from the Wind

The energy of the wind has long been harnessed, as witnessed by the stilted monuments to an earlier effort, spindly structures standing braced against the winds that once served them. But the principle utilized by these old windmills is being reconsidered. In some cases, not much change is in order. Even old-style windmills work well when the wind blows, and the technology to use the energy is already fairly well developed. Now, however, we have "wind farms"—huge tracts where a number of modern windmills produce energy that can be stored and sold commercially (Figure 28.19). Problems arise because there are relatively few places where winds are dependable, such as in mountain passes and along coastlines. (The winds generally must average 14 to 24 miles per hour.) The most promising places include the Pacific Northwest, the Rocky Mountains, and the mountains of North Carolina. So far, more than 70 percent of the electricity produced by wind worldwide comes from three windy mountain passes of California. Wind power has a low environmental impact, but the new windmills are noisy and unsightly and interfere with radio and television transmission.

FIGURE 28.19
A wind farm. These unsightly and surprisingly noisy windmills produce electrical energy efficiently and without heavy pollution.

Energy from the Earth

It may have occurred to those who saw the awesome eruption of Mt. St. Helens that there is energy under the earth. In fact, one of the greatest explosions of all time was the 1883 eruption of an undersea volcano called Krakatoa between Java and Sumatra. It darkened the sky for months and changed sunsets around the world. This is **geothermal energy**—energy from the heat of the earth's interior.

Wells can be drilled to tap this geothermal energy where the deposits are close enough to the earth's surface. Presently, about 20 countries are tapping geothermal deposits, with most of the energy being extracted in the United States where it is converted to electricity (Figure 28.20).

Geothermal deposits exist in three major forms: dry steam, wet steam, and hot water. Dry steam lacks water droplets (you didn't know there was such a thing, did you?) and is the easiest to use. The three largest fields currently in production are in Italy, Japan, and about 90 miles north of San Francisco.

FIGURE 28.20
Geothermal energy units at El Centro, California utilize the heat energy of the earth's interior.

The wet steam deposits are more common but harder and more expensive to utilize. The largest of these is in New Zealand. There are four small wet steam plants in the United States.

Hot water deposits are more common than either of the steam deposits, but difficult to use because the water is in the form of salty brine, which tends to corrode the machinery and to present problems with disposal.

The geothermal deposits should be considered nonrenewable and can be expected to be depleted in 100 to 200 years, although the heat from the water can be extracted and the water pumped back into the earth. Geothermal energy is cheaper to produce than either fossil fuel or nuclear energy and it does not produce CO_2.

Energy from the Sun

The earth is constantly bathed in **solar energy,** the energy of the sun. Solar energy has, from the beginning, driven much of the life on earth. It has warmed the planet's surface and it has powered the photosynthetic machinery of the earth's algae and plants. Yet, somehow, in the modern

FIGURE 28.21

Sunlight falls on collecting panels and the solar energy is then used for heating water and space and, in some cases, for generating electricity.

search for energy, the most obvious energy of all has been oddly neglected. Some say that is precisely because it is *too* accessible. It's available almost everywhere and it is conceptually difficult to package and sell. Because of its historically insignificant influence on geopolitics, solar energy has traditionally been a concern of tree-huggers and inventors working alone.

All that's changing now. Or at least it seems to be. President Jimmy Carter started the ball rolling as he sought to decrease American reliance on Middle East oil, but President Ronald Reagan brought it to a halt and brought us back to Business as Usual. We all know where that got us. Now even Washington, where Congress, under pressure from the oil, coal, and nuclear energy lobbies recently eliminated tax credits for using low cost, energy-efficient solar energy systems, might be ready to act in the public interest. Just to get an idea of how informed your representatives are, ask your congressman, first chance, about the difference between active solar systems and passive solar systems. But first read a little further.

Active Solar Systems

In **active solar systems,** solar energy is concentrated and stored to heat space and water. This is usually done in individual homes (Figure 28.21) where the sun's rays heat water or antifreeze. The heated fluid is then stored in an insulated tank. Water is pumped past it to be heated or heat radiating from it is blown by fans into the rooms of the house.

As more nations worldwide began to increasingly rely on active solar systems, in the United States between 1984 and 1988 sales of such systems dropped by 75 percent. Do you find it interesting that the states with the greatest potential use of solar energy (Figure 28.22) are also the greatest oil producers?

Passive Solar Systems

In **passive solar systems,** sunlight is captured directly and used for low temperature space and water heating. The heat is not collected and concentrated; sunlight simply heats the space and water. Houses that use passive space-heating systems normally allow the sun in through insulated windows that hold the heat in. With some designs, the natural movement of warm and cool air distributes the heat (Figure 28.23). It is usually difficult to add passive solar systems to existing houses. The system is largely a built-in design feature.

Photovoltaics

The energy of the sun can also be used to create electricity through photovoltaics (solar cells). The problem with using solar energy to produce electricity is that large areas of land must be covered with collecting devices because sunlight is such a diffuse and dispersed form of energy.

Photovoltaics was actually conceived in 1839, but today it involves using sunlight to energize two layers of silicon. The sunlight disrupts the crystalline arrangement and unsettled electrons from one crystal move to the other, generating a slight electrical current. The lost electrons are replaced by attaching the donor crystal to some metal, such as a copper wire.

FIGURE 28.22

The areas of greatest oil and gas production also have the greatest sunlight levels.

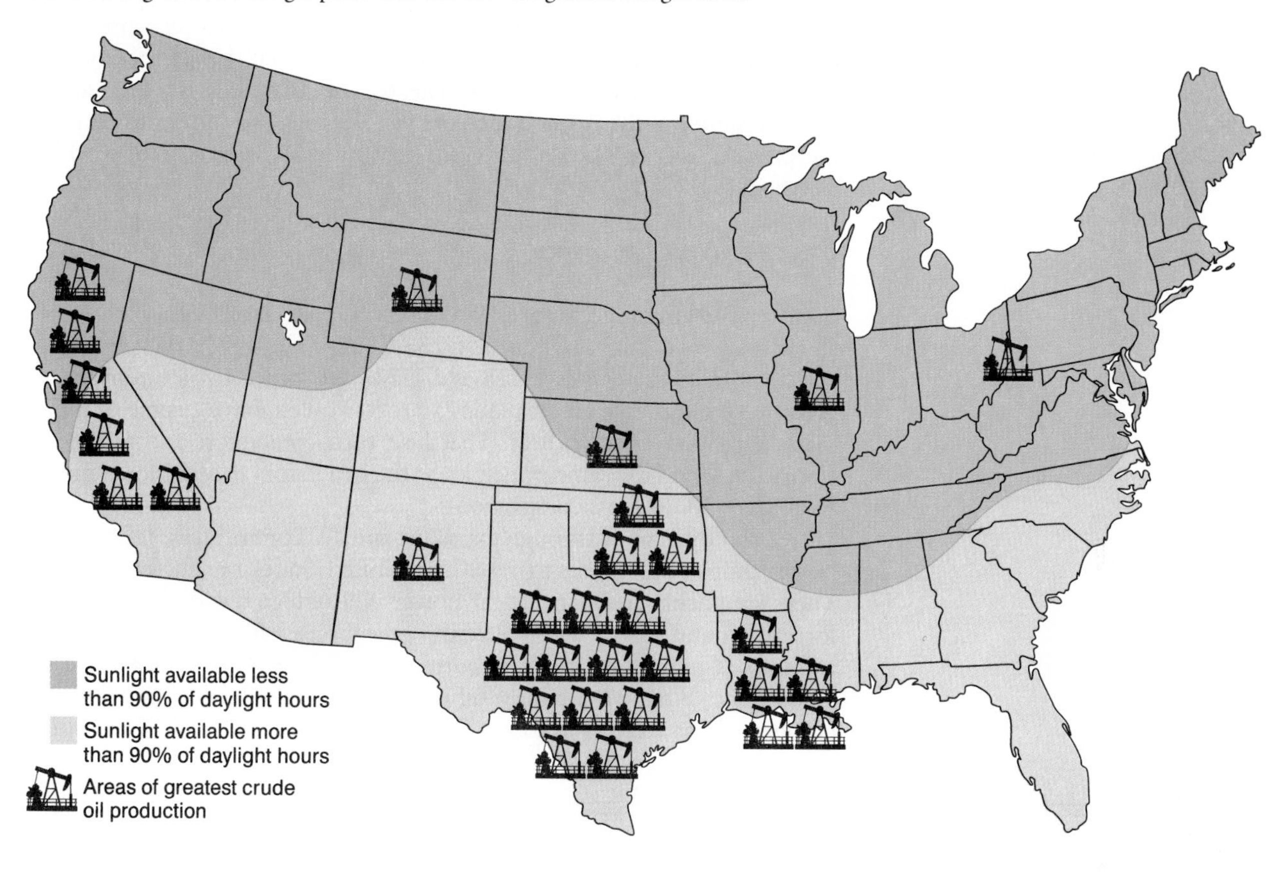

FIGURE 28.23

In passive solar systems such as this one, sunlight warms air and water, which are then distributed by natural expansion, contraction, and convection. Here, heated air moves out into the rest of the house.

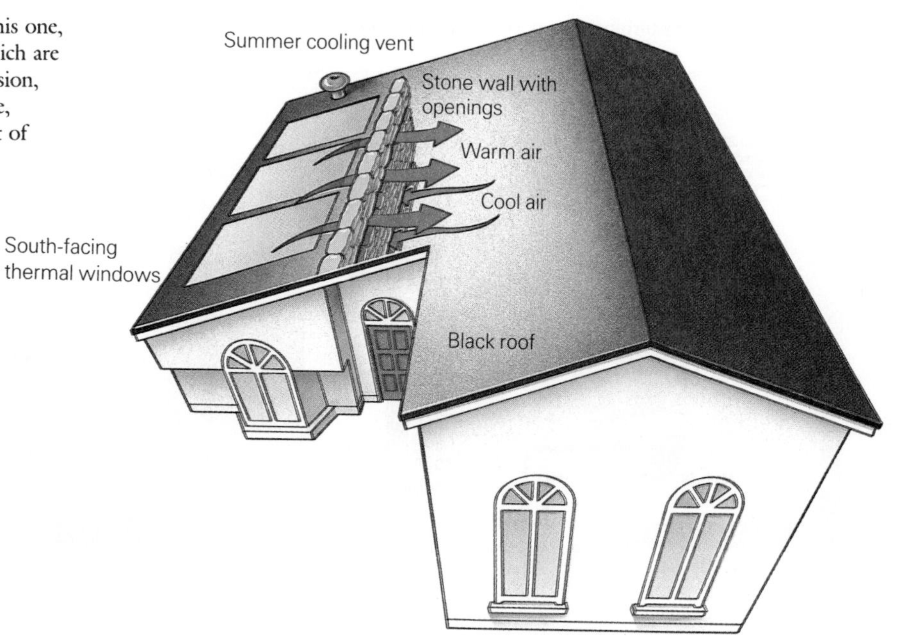

At present, electricity produced by photovoltaics is very expensive. But since this sort of energy can be collected, localized, stored (in battery-like "fuel cells"), and then sold, we now see increasing research in the area.

Although the use of solar cells is still curtailed by their expense, new production techniques are being developed that will lower the cost of producing electricity in this way. The use of solar cells is expected to increase significantly through the next few decades. According to current projections, solar cells could supply 20 percent to 30 percent of the world's electricity by 2050.

Energy from the Atom

Four types of uranium occur on the earth: U238 (about 99.3 percent of all naturally occurring uranium), U235, U234, and U239. When molecules of uranium, such as U238 and U235 are struck by neutrons, they become very unstable and explode (a process called *fission*), some of their mass being converted to heat. That heat turns generators and produces electricity (using the same principles as the generators that use fossil fuel–see Figure 28.24 for a comparison).

Currently, mainly through magazine and TV commercials, the nuclear power industry is trying to resell the United States on nuclear power. Their arguments are that nuclear power will reduce our dependence on foreign oil and that it will reduce the greenhouse effect by lowering the amounts of pollution caused by burning fossil fuels. Neither argument is a good one. First, we use most oil to make gasoline for our cars. Only 3 percent of our electricity comes from burning oil. So, even if we built

FIGURE 28.24

Compare the fossil fuel power plant in (a) with the nuclear power plant in (b). Notice that both use steam to turn turbines that generate electricity. In the nuclear reactor, uranium bars comprise the fuel. As the uranium undergoes fission, it heats water and steam is produced, which turns turbines and operates electric generators.

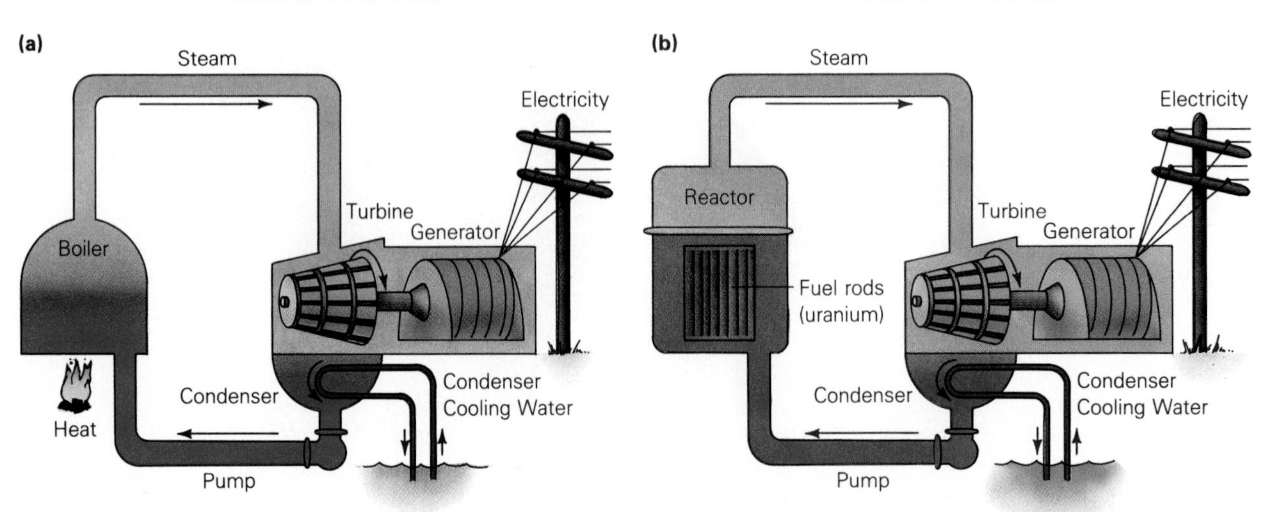

FIGURE 28.25

Features of a well-insulated home.

nuclear plants to provide electricity, we would still need to import oil. Second, it has been calculated that, in order to reduce the world's greenhouse effect by only 2 percent, we would have to build a large reactor *every day* for the next 37 years. Think about *that* one.

ENCOURAGING CONSERVATION AND INCREASING EFFICIENCY

Some astute observers have noted that when we find ourselves in an energy crunch, as when Iraq invaded Kuwait and threatened the Saudi oil fields in 1990, the first response of our national leaders is to act to ensure our continued access to Middle East oil. As the White House outlines what the nation must prepare for at such times, someone invariably asks, "What about encouraging conservation?" To which the usual reply is, "Oh yeah, that too." As a nation we act as if the idea had never occurred to us. (It is interesting that those who consider themselves conservatives are generally the most reluctant to support conservation. Conservatives who think we should conserve are usually branded moderates and booted from the room.)

But for the sake of argument, let's suppose that we should conserve energy in the United States (and as a global community). Where do we begin? We begin by using less energy. This can be accomplished by building our houses to maximally utilize what the environment offers us. For example, the overhangs over our windows should extend just far enough to block the summer midday sun, but to allow in the warming rays of the sun in winter. (In the United States, the sun is lower in the sky in winter, but you knew that.) We should build our houses to take advantage of morning sun, or shade trees, or prevailing breezes, or whatever else would make us comfortable without increased expenditure of energy. We should increase our insulation and make doors and windows airtight (see Figure 28.25). Our homes should be well insulated and we may want to build

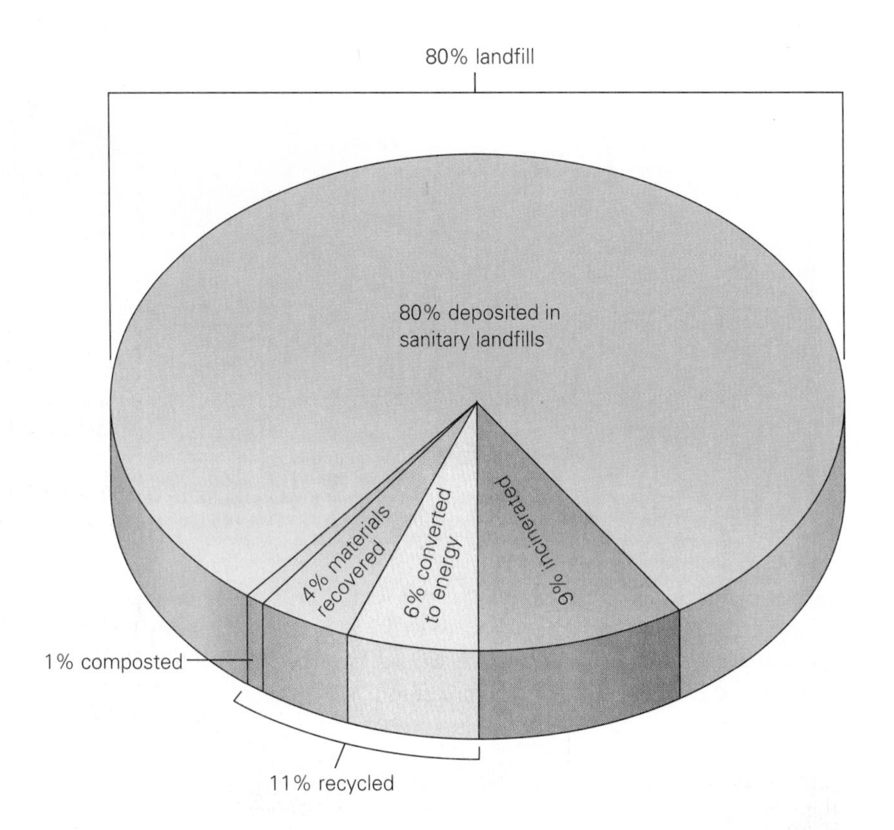

FIGURE 28.26

The various fates of the waste we generate.

them in convenient neighborhoods from which we do not have to travel far on a daily basis. If we do travel, we will want to use public transportation when possible or travel in small, efficient cars.

Second, we can develop a new set of habits. The bicycle is catching on again in industrialized countries. Put simply, we should stay out of cars as much as possible. *Let* our houses be cool in winter and warm in summer. This business of keeping our buildings frigid in the summer (so we have to wear a sweater to the movies) is a sure sign of what might be called "national dumbness." We've got to get smarter. Go ahead and wear your sweaters in winter. That's what winters are for.

Included in our newly emerging American behavior should be a powerful recycling ethic. If your community doesn't recycle, initiate a campaign to get it started. (Remind them that there's *money* in recycling.) At present, most solid waste is put in sanitary landfills (Figure 28.26). In sanitary landfills, the waste is spread out in a thin layer, then covered daily with a layer of soil. There is no burning and the layering discourages rodents and insects. While sanitary landfills beat old fashioned dumping, much of that material could be recycled at great energy savings.

Third, we should encourage, perhaps through revised loan practices and tax incentives, the research and development necessary to produce better devices. These devices would include automobile engines, appliances, insulation materials, biodegradable plastics, and nonpolluting energy production.

We obviously must view ourselves as custodians of our precious resources and not as exploiters, seeking ways to get at the earth's goods before someone else does. However, as our numbers grow, new demands must be met, and thus we must find new sources of commodities as we place greater pressures on the old ones. The race for goods has produced, in many cases, a "devil take the hindmost" ethic. We can hope that we will develop a more careful approach to the use and distribution of the earth's resources. But any such change must begin soon because now the race is on.

SUMMARY

1. Renewable resources are those that can be replenished or reused. Nonrenewable resources exist in set amounts and cannot be replenished. Two problems with the earth's resources are they are not increasing with the demand for them placed by the increasing size of the human population and they are unevenly distributed. The unequal distribution of resources has political implications.
2. An imbalance of food availability exists between and within countries. The hunger problem is worsened by food preferences that prevent people from exploiting different nutritious food sources. Undernourished individuals do not receive enough food. Malnourished individuals don't get the right kinds of food. Children are vulnerable to the nutritional diseases marasmus and kwashiorkor that may result in permanent mental deficiency and could be prevented by extended nursing. Marasmus is caused by a dietary deficiency of both calories and protein, and kwashiorkor is caused by a dietary insufficiency of protein.
3. The more developed countries produce most of the world's food, but they also use a disproportionate share of it. Although less developed nations are increasing their rate of food production, this increase is not keeping up with the increase in population. Although there is only about a 13 percent difference in consumption of calories and plant proteins between more developed and less developed nations, more developed nations consume about five times more animal proteins than do less developed countries.
4. Even though the world may produce enough calories to barely feed all the people, there are problems of inequality, distribution, and storage. Attempts to increase food production by farming the tropical rain forests have failed.
5. Animal protein provides all the essential amino acids. A variety of plants must be eaten to supply these. The choice of which animal to raise for food must consider factors such as cost, degree of competition with human needs, and ability to provide useful by-products. The seas are being farmed at a rate that does not provide for optimal productivity. Many new technologies for developing new food sources to feed the hungry are being investigated.
6. Fresh water is used in a variety of ways. Severe shortages are predicted for the near future because the limited supply of fresh water is being stretched to meet the needs of a growing population and because much of the available supply is being made unusable by pollution.

7. Recycling conserves aluminum and iron, which are nonrenewable resources, lessens the solid waste problem, saves energy, and reduces air and water pollution.
8. Some uses of energy are necessary, but others are unessential. The costs to the environment in producing and consuming energy must be weighed in calculating the benefit we gain. Fossil fuels, partially decomposed remains of ancient life, provide most of our current energy. The supply of fossil fuels is being depleted. The use of fossil fuels is contributing to the greenhouse effect and the problem of acid rain.
9. Hydroelectric energy (electrical energy from the energy of moving water) is an old form of energy whose potential is being rediscovered. Although it does not pollute and is efficient, water falling over dams may pick up enough nitrogen to kill fish, and dams alter the natural course of rivers by trapping sediment. Dams may destroy both farmlands and wildlife habitats and dislocate people. The ability to harness the energy of moving tides has not yet been fully developed. Although the estuaries would be the most efficient place to harness tidal energy, this would disrupt the spawning grounds of oceanic food chains.
10. Wind is an old source of energy that has been redeveloped so it can be stored and sold commercially. Its usefulness is limited because there are only a few places where the winds are dependable.
11. Geothermal energy (energy from the heat of the earth's interior) can be tapped by drilling wells. Geothermal deposits exist primarily as dry steam, wet steam, or hot water. These nonrenewable deposits will be depleted in 100 to 200 years.
12. Solar energy is energy from the sun. Active solar systems concentrate the sun's energy to heat space and water. In passive solar systems, sunlight directly heats space and water. Solar energy can create electricity through photovoltaics (solar cells). Although currently an expensive process, new technology is expected to increase the use of photovoltaics.
13. Fission of uranium (U235) atoms produces heat that can turn generators to produce electricity.

KEY TERMS

active solar systems (808)
geothermal energy (807)
hydroelectric energy (806)
malnourished (782)
nonrenewable resources (781)
passive solar systems (808)
renewable resources (781)
solar energy (807)
undernourished (782)

FOR FURTHER THOUGHT

1. If a less developed country that is in need of food is well suited to either raise sheep or grow wheat crops, which should it choose?
2. Which of the following energy sources can operate without generating carbon dioxide in the process? Photovoltaics, wind energy, atomic energy, fossil fuels.

FOR REVIEW

True or false?

1. ___ Malnourishment results from not eating the right kinds of foods.
2. ___ The consumption of animal foods is greatest in more developed countries.
3. ___ Most of the United States' fresh water supply is used in agriculture.
4. ___ Fossil fuel resources are equally distributed throughout the world.

Fill in the blank.

5. Trees, water, and oil are all examples of ___ resources.
6. ___ energy is electric energy that is produced by moving water.

Choose the best answer.

7. Kwashiorkor is a disease that
   A. is caused by a diet low in calories.
   B. results from protein deficiency.
   C. occurs before a child is fully weaned.
   D. cannot be prevented.
8. Which of the following animals provides the least costly way to eat higher on the food chain?
   A. cow
   B. pig
   C. goat
   D. chicken
9. Solutions to the hunger problem include triticale, a
   A. single-celled organism grown on petroleum.
   B. mixture of corn and cottonseed.
   C. fish protein.
   D. hybrid of wheat and rye.
10. Photovoltaics are used to harness ___ energy.
   A. geothermal
   B. solar
   C. wind
   D. atomic

CHAPTER 29

Bioethics, Technology, and Environment

Overview

The Ethics of Doormats
Environmental Pollution
Hidden Decisions
The Future

Objectives

After reading this chapter you should be able to:

- Describe some issues addressed by bioethics.
- Define *pollutant* and list four ways in which pollutants are harmful.
- Identify the major air pollutants; list the sources and effects of each.
- Describe ways in which water is polluted.
- Describe environmental problems associated with pesticide use.
- List some sources of radiation and describe the risks associated with nuclear power.

In the blink of a geologic eye, it seems, the human species has become incredibly powerful. We can move mountains, straighten rivers, cause extinctions, produce new species, and change the rain. We are now in a position to do great damage or great good. With this ability, some say, comes a heavier responsibility than any species of any age has ever borne. The consideration of this new responsibility that has fallen upon us has produced a philosophical area that has come to be called **bioethics,** the ethics of life. Ethics itself implies inherent right and wrong, and so bioethics deals with the moral values of our behavior toward life on the planet.

Bioethical questions are of many types, approaching many subjects dealing with life. But they all have one thing in common: They are difficult, if not impossible, to answer. Yet because we have become so powerful, so probing, and have touched so many areas of life with our technology, the questions must be asked and we must continue to seek answers. Perhaps just asking the questions has value.

The bioethical questions researchers must address are as varied as the researchers themselves, but let's mention a few that have been asked. Should we send food to countries whose population is spiralling out of control, thereby encouraging more childbearing? Should we build a dam to create electrical energy even if it means the extinction of yet another species? Should abortion be encouraged or even allowed? Should we spend public money to spare the lives of severely handicapped infants who may have to spend their lives in pain? Do we have the right to visit pain on the other species? (See Figure 29.1 and Essay 29.1.) The list could go on and on, but perhaps we have identified the problem.

Here, then, let's see where some of our "advances" have led us. We will delve into the untidy problem of pollution of the environment to identify some specific problems resulting from our traditional ethics. However, as we go, let's focus on how a shift in values, or point of view, might have spared us these problems and see how we might make changes now that would better guarantee our futures.

FIGURE 29.1
This raccoon, caught in a steel leg trap is—in great pain—awaiting the trapper's club. Its coat will be sold to make coats for humans.

THE ETHICS OF DOORMATS

It has been said that we have seen more technological advances in the last ten years than in all our previous history. That may be true, but what does this information tell us? It could be cited with equal enthusiasm by an industrialist or an environmentalist. The former might happily note the increased efficiency of production while the latter might be concerned that our technological abilities have outstripped our wisdom, our ability to deal with our creations. We might certainly, in our quieter moments, wonder if Henry David Thoreau was right. Perhaps we have busied ourselves with trivialities and forgotten about the real business of living. Thoreau once refused a doormat for his cabin at Walden Pond (Figure 29.2) because he figured he wouldn't have time to sweep it. He needed that time, he said, to sit or walk in the woods. Living. *Experiencing!*

Apparently, some people still enjoy the simple life, but even for them, things are becoming more complicated and they are being forced to deal with their own bioethical decisions or those of others. Every year, for example, people heave their backpacks to their shoulders and head for the

FIGURE 29.2
Walden Pond.

ESSAY 29.1

Whose Rights Are They, Anyhow?

After a lifetime of living among the wild chimpanzees, Jane Goodall has become deeply involved in a new cause, the treatment of caged chimpanzees. In particular, she is asking for better treatment for those chimpanzees used in the United States and Europe in medical experimentation. For example, she notes that they should not be kept in isolation since they are such social animals. If it doesn't interfere with the study, they should at least be allowed to interact with each other.

Goodall is a symbol, willingly or not, for a number of people who believe that animals should not be used in research at all. They argue that the animals themselves have rights and that it is unethical for humans to violate those rights just because they can. Some among these people have, recently, become increasingly militant and have, in some cases, resorted to threats and violence.

Medical researchers argue that experimentation with animal models must continue because this is the best way to learn about specific human problems and their treatment. They wryly point out that their research on animals might well someday save the lives of the children of those opposed to the research.

Some of the most blatant misuses of animals are being discontinued, such as the use of whale products in cosmetics and the testing of the effects of new cosmetics by dropping the ingredients into the eyes of immobilized rabbits.

Animal rightists are universally and vehemently opposed to furs for human apparel. They cite the clubbing of baby seals and the anal electrocution of ranch mink as cruel and unnecessary. In particular, they are opposed to leg traps in which the animals are held in agony for hours or days until the trapper returns to club them to death. Those who favor the right to wear furs insist on that right and declare that the furs keep them warm and comfortable, while wool does not.

If one agrees that it is possible to go too far in the abuse of animals for human safety, health, or whim, then where does one draw the line? The search for that line is an exercise of the ethical approach to our behavior toward our fellow animals.

mountains (Figure 29.3). Even there, though, they may run into scores of people on narrow trails who are also trying to reestablish their links with a more natural world. In fact, even the remote lakes of the high Sierra Nevada mountains are becoming seriously polluted by the wastes from so many backpackers in the area. Not far away, trees may be falling, their fibers that had once returned mountain rains to the cool air destined to become part of a gaily colored toothpaste box. The box, of course, is designed to be discarded (Figure 29.4) and to join those huge heaps of refuse whose disposal presents such problems of waste management. It's easy to forget that the sodden box had once been a tree.

What, then, is "quality of life"? Do we trade trees for toothpaste boxes? Is the quality of our lives to be measured by the waste we create? In a sense it is, since more affluent nations indeed create more waste.

Let's now see how meeting our needs and our demands may exact a higher cost than we anticipated. As goods are provided, there are side-effects. One kind of side-effect may be the degrading of our environment.

FIGURE 29.3

Backpackers often can't escape these days.

ENVIRONMENTAL POLLUTION

A **pollutant** is a substance whose presence in the environment is harmful. Pollutants not only take the form of chemicals or garbage, but also such things as noise and heat. Here, we will focus on the pollution produced by human activity, although we must be aware that nature can produce some of the same substances. For example, there is some natural oil seepage from undersea deposits, and it has been going on for thousands of years. Also, volcanoes (Figure 29.5) are among the greatest short-term air polluters on the planet. In fact, our present environment has been molded by such natural occurrences as fires, droughts, oil seepage, earthquakes, volcanoes, and even radiation throughout the earth's history. The life on the planet has adapted to such events, and these are not the things that will concern us here. The kind of pollution that concerns so many scientists today, however, is that which, in kind or degree, is new to the earth and its creatures.

There are four primary ways that these new pollutants can be dangerous or disruptive. First, they may alter critical aspects of our environment so rapidly that some forms of life will not have time to adapt to the changes before they are destroyed. Second, since thousands of known pollutants have never been tested for their risks, they may act subtly and slowly in our bodies, setting the stage for a tragedy down the line. Third, pollutants may have a threshold effect. That is, they may show no effect

FIGURE 29.4

Each year Americans throw away one and a half billion pens, two billion disposable razors, over two million tires, and enough aluminum to build the country's entire airline fleet every three months, even as we continue to ravage the land and sea in the search for more raw materials to replace these items.

FIGURE 29.5

We sometimes consider pollution as human-made. However, those steamy crevasses that belched sulfurous smoke when the earth was young were polluters, as are today's volcanoes that spew chemicals and ash into the air. Our task is to see that we minimize the human contribution to the world's pollution.

until they have accumulated to some critical level. Fourth, pollutants can act synergistically, interacting in such a way that their combined effects are far greater than they would have had individually. For example, both sulfuric acid and ammonium sulfate are rather dangerous pollutants alone, but together they form a far more dangerous combination.

Air Pollution

If you live in a large city, you might be particularly concerned about air pollution because you can *feel* it. It can burn your eyes, nose, and throat on the worst days. It may even make days darker in a literal sense. Air pollution, for example, frequently cuts by 40 percent the amount of sunlight reaching Chicago. A huge stone obelisk from Egypt, a gift to New York City, has deteriorated more in the few years it has stood in Central Park than in the hundreds of years it stood in the desert.

The greatest source of air pollution in the United States is the automobile. However, fuel burned in stationary sources, such as in heating units for buildings and industrial furnaces, is also largely responsible for our foul air, as are cigarettes and cleaning solvents. (Table 29.1 summarizes the effects of some air pollutants.) The city with the worst air pollution in the world, though, is not in the United States. This dubious title belongs to Cubatao, Brazil, outside of São Paulo where a third of the people living downtown have respiratory disease. On some days, breathing the air outside can trigger vomiting. Even the mayor lives somewhere else.

We should realize, also, that air pollution is not just the problem of cities, nor is it restricted to industrialized countries. Meteorologists have found a thin veil of pollutants that hangs over the entire earth. (In recent

TABLE 29.1
Adverse Effects of Specific Air Pollutants

POLLUTANT	EFFECTS
Carbon monoxide	Reduces blood's oxygen carrying ability; stresses heart; causes headaches, dizziness, and nausea
Nitrogen oxides	Causes visible leaf damage; retards plant growth; irritates lungs; irritates eyes and nose; with hydrocarbons, produces photochemical smog; corrodes metals; damages materials
Sulfur oxides	Causes short- and long-term leaf damage; damages many types of trees; destroys plant photosynthetic pigments; irritates eyes, nose, and lungs; erodes statues; corrodes metals; damages fabrics, leather, and paper
Hydrocarbons	Slows plant growth; causes abnormal leaf and bud development; may cause cancer
Particulate matter	Aggravates lung and heart conditions; corrodes metals; forms grime on objects in environment; obscures vision

ESSAY 29.2

Temperature Inversions

Under certain conditions, the polluted air of cities can be amplified by temperature inversions. Normally, air temperature decreases with altitude, but when a layer of warm air moves above the cooled air, it limits the normal upward flow of the air from below. As a result, polluted air nearer the ground is not allowed to escape into the upper atmosphere. This means the foul air is trapped and the unfortunate people below must breathe their own pollutants instead of releasing them into the air for other citizens to breathe. When temperature inversions occur over cities, the number of deaths from respiratory ailments often rises sharply.

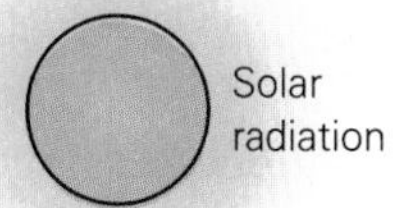

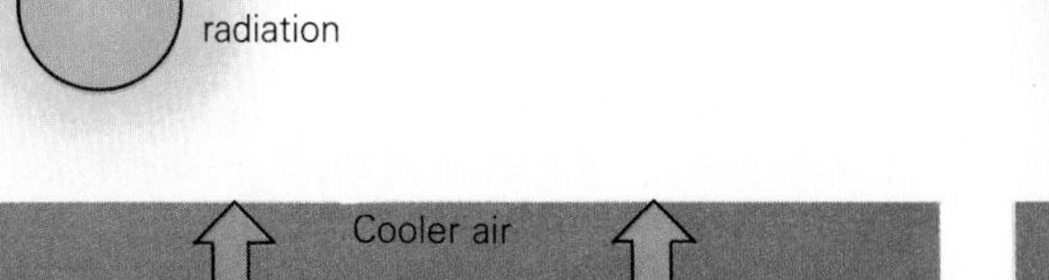

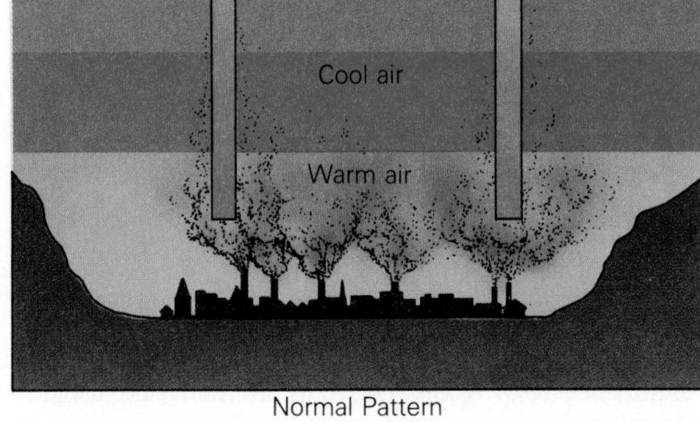

Normal Pattern

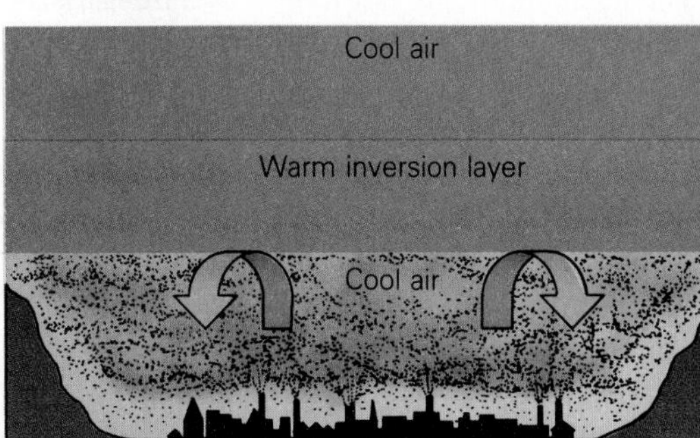

Thermal inversion

years, they have repeatedly detected polluted air over the North Pole.) Generally, however, the foulest air is found in the industrialized countries, where it is particularly concentrated as smog over large cities (Table 29.2). Under certain conditions, the concentration of air pollutants in cities can be magnified to a startling degree (Essay 29.2). And then we have recently become aware that certain air pollutants are destroying the protective ozone in the upper atmosphere (Essay 29.3), and others may be dangerously warming the earth (see Essay 28.6).

Carbon Monoxide

Carbon monoxide is the most prevalent air pollutant. Its source is the incomplete combustion of fossil fuels. It is biologically important because of its tendency to combine with the blood's hemoglobin in the place of oxygen. The effect is to cut down the blood's supply of oxygen to the tissues, thus causing the heart to have to work harder to oxygenate the body. The increased demands on the heart place a severe strain on people with heart or respiratory ailments. Looked at in another way, carbon

ESSAY 29.3

Holes in the Sky

Now let's explore the relationship between underarm deodorants and the death of the oceans. The propellant in underarm sprays is, in many cases, a class of molecules called chlorofluorocarbons (CFCs). These are essentially carbon molecules to which are attached chlorine and/or fluorine atoms. Chlorofluorocarbons are used in a variety of manufactured products, such as air conditioning, refrigeration, insulating foams (the type that keep our hamburgers warm in fast-food places), plastics, and industrial solvents.

The problem is, these molecules are very stable. So after you spray under your arms, or after the insulated fast-food box begins to disintegrate, these long-lived little molecules are released into the air. Because they're light, they eventually, perhaps a few years later, end up in the upper atmosphere.

Paradoxically, these molecules would be safer for life if they stayed closer to earth mingling with living things. The truth is, though, they threaten life precisely because they drift upward away from it. The reason is because at an altitude of 15 miles, the CFCs break down the ozone layer. Ozone is O_3, formed by the sun breaking down atmospheric O_2 molecules, allowing them to rejoin as ozone. The chlorine in the CFCs attacks the ozone, breaking it back down into its components. There isn't much ozone up there to begin with. At sea level, all of it together would form a layer over the earth about as deep as a pencil lead is thick.

The ozone, though, is critical to life on earth. Primarily, it functions by blocking destructive ultraviolet light from the sun. Those rays are destructive on three primary bases. First, they increase the risk of skin cancer, particularly among light-skinned people, such as those who invented chlorofluorocarbons. Second, they depress the immune systems of humans, setting the stage for a host of illnesses. Third, they destroy algae, the first step in the ocean's food chains.

Ozone depletion was first discovered over the Antarctic. In fact, two thirds of the springtime ozone over the Antarctic is now missing since the British began the measurements some years ago. In the northern latitudes where most people live, the ozone levels have declined by several percent since 1969. Now, an ozone hole is reported developing over the Arctic.

The manufacturers of CFCs have been reluctant to take action to reduce the levels of these chemicals over the earth. In fact, the DuPont Company, which makes about $600 million a year on CFCs, took out ads in newspapers saying that the danger to the ozone layer was improved. The company has now agreed to phase out the manufacture of CFCs, but the phasing out will not be complete until the year 2000. Since the United States makes only 30 percent of the CFCs, the effort will have to be global, demanding more cooperation than one usually finds among industrial nations.

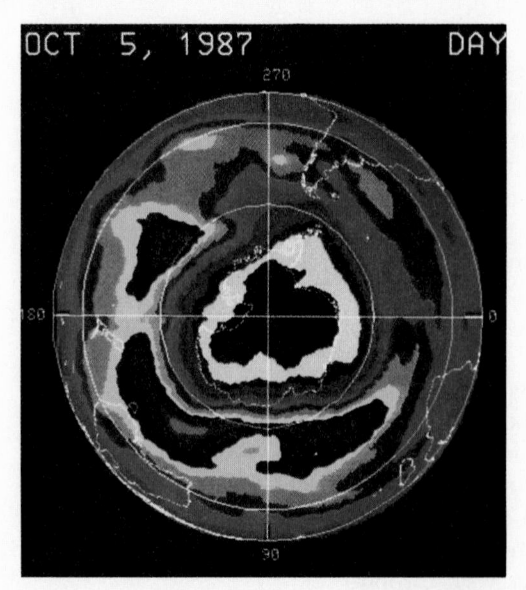

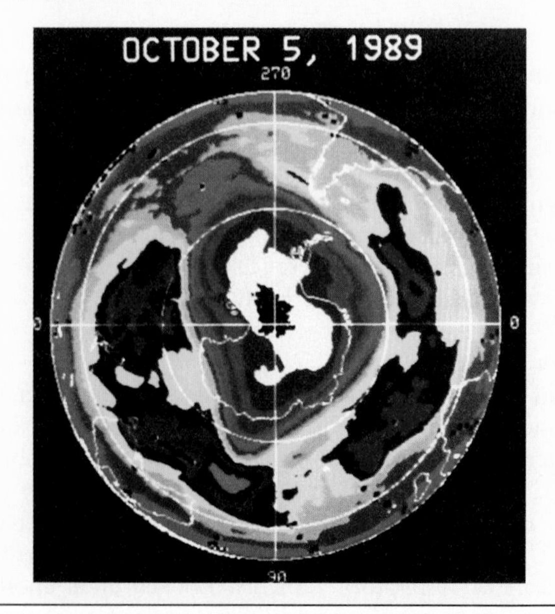

TABLE 29.2
Basic Types of Smog

CHARACTERISTIC	INDUSTRIAL SMOG	PHOTOCHEMICAL SMOG
Typical city	London, Chicago	Mexico City, Los Angeles
Climate	Cool, humid air	Warm, dry air
Chief pollutants	Sulfur oxides, particulates	Ozone, aldehydes, nitrogen oxides, carbon monoxide
Main sources	Industrial and household burning of oil and coal	Motor vehicle gasoline combustion
Time of worst smog	Winter months (especially in the early morning)	Summer months (especially around noontime)

Industrial smog operates in the form in which it is released. Photochemical smog becomes more active when it is hit with sunlight.

FIGURE 29.6
Traffic jams are inconvenient and dangerous.

monoxide inhalation can have the same effect as loss of blood. In fact, spending eight hours in an atmosphere with eighty parts per million of carbon monoxide has the same effect as losing over a *pint* of blood. It may interest you to know that in a traffic jam, the air may contain nearly 400 parts per million of carbon monoxide (Figure 29.6). The people breathing this air may experience headache, loss of coordination, nausea, abdominal cramping, and even partial blindness. (Carbon monoxide is also a major component of cigarette smoke.)

Nitrogen Oxides

Nitrogen oxides are produced in a number of ways, primarily from gasoline engines, power plants, and industry. It has been pointed out that nature produces ten times more oxides of nitrogen than humans do, but the problem is that human products are concentrated in urban areas. The nitrogen oxides dissolve in water forming nitric acid, one of the components in acid rain. Although nitrogen dioxide irritates lungs and withers plants, the greatest danger arises when it combines with hydrocarbons in the presence of sunlight to produce **photochemical smog** (Figure 29.7). This smog is particularly reactive and is one reason pantyhose and automobile tires don't last long in the city.

FIGURE 29.7
Photochemical smog hanging over Mexico City. Many of the chemicals released from traffic and industrial sites are made more reactive by the energy of sunlight, and, thus, their corrosive effects are increased.

Sulfur Oxides

Sulfur oxides, produced mainly by burning coal, are particularly severe during heavy smog. Sulfur oxides can combine with water and produce sulfuric acid, which damages not only lungs and plants, but can dissolve marble (Figure 29.8), iron, and steel. Sulfuric acid is the major contributor to acid rain. Sulfur oxides are very harsh and may bring on bouts of coughing, wheezing, and choking. The most serious damage is done when sulfur oxides interact with ammonia and metallic salts in the air. They are believed to figure importantly in the increased incidence of bronchitis, asthma, and emphysema, which are so frequently suffered by people in high smog areas. (Emphysema is a progressive lung disease, generally considered fatal.) Globally, sulfur oxides may not present a great problem because rain washes them from the air, but locally, they may be the most dangerous single pollutant in the atmosphere.

FIGURE 29.8

The smiles are gradually fading from many of the ancient carved figures in Venice. Sulfur dioxide in the air mixes with soot to form a film over the marble and to gradually change it to calcite and gypsum, which then crumbles away.

Hydrocarbons

Hydrocarbons, as you know, are a diverse and variable group of elements, some of which are implicated in the development of respiratory problems and some cancers. They may form from natural processes, such as plant decay, but most synthetic hydrocarbons come from the incomplete combustion of fossil fuels. More than half of the hydrocarbons that are dumped into the air each year come from our gasoline-powered vehicles. Hydrocarbons may cause cancer. In addition, they slow the growth of plants and interfere with plant development.

Particulate Matter

Particulate matter is composed of particles large enough to be filtered. Certain industrial processes release particulate matter as dusts, minerals, plant products, or other substances. Particles are also released from automobile clutches and brake pads and from tires as they are worn down. Some of these particles may contain chemically active groups of sulfates, nitrates, fluorides, or ammonia. Others are essentially metallic, often containing cobalt, iron, lead, or copper. Some can cause cancer and aggravate lung or heart conditions. Some also can interfere with plant photosynthesis.

Water Pollution

The fresh waters of the United States are generally considered, on an intuitive basis, to be the property of all, a kind of commons. Yet those waters are routinely used for cooling, cleansing, and as dumps for industrial

ESSAY 29.4

Sewage Treatment

When our population was smaller, sewage could be safely emptied into moving rivers where bacteria and fungi (decomposers) could digest it. Those large organic molecules from excretion, fallen leaves, and other byproducts of life were broken down into their elemental constituents. These constituents could then be used as building blocks by primary producers in the recycling of life's materials. When the numbers of waste producers along such rivers became too high, the natural processes of decomposition proved too slow, and sewage treatment plants were devised. These plants are, in effect, places where the natural processes are speeded up within the confines of a restricted area.

Ideally, a two-stage sewage treatment plant collects the wastes discharged into our waters, gets rid of the waste, treats the water, and pumps it back into our supply. First, the largest pieces of waste are screened out. Then the water is pumped into settling tanks where finer material sinks to the bottom as sludge. The supernatant (fluid over the sludge) is pumped into tanks where air is bubbled through it to provide oxygen for bacteria and fungi, which break down the organic molecules. Sludge is again allowed to settle out and the supernatant is chlorinated to kill the bacteria. Then it is ready to drink again.

Unfortunately, many of our cities lack facilities for complete processing of sewage. This means that partly treated or untreated wastes are pumped into our waterways. But even those plants with adequately treated sewage present certain environmental problems. For example, the effluent from the best sewage plants is high in phosphorus, nitrogen, and certain other elements. These can act as fertilizers by enriching the water into which they are dumped. (Their effect is added to that of phosphate from detergents, which accounts for about half the phosphate overload.) The result can be cloudy, smelly water covered with algae scum. Many species of algae that respond so well to inorganic fertilizers are not used as food by zooplankton and thus do little to initiate a food chain. Also, it has been determined that where pollutants raise the overall "productivity" of water, in terms of mass, they decrease the diversity of species present. Certain cyanobacteria produce poisons that can kill cattle and cause rashes and vomiting in humans.

Advanced sewage treatment to remove pollutants after secondary treatment is possible. However, except in Scandinavia, the advanced techniques are rarely used because the plants cost about twice as much to build and are expensive to operate.

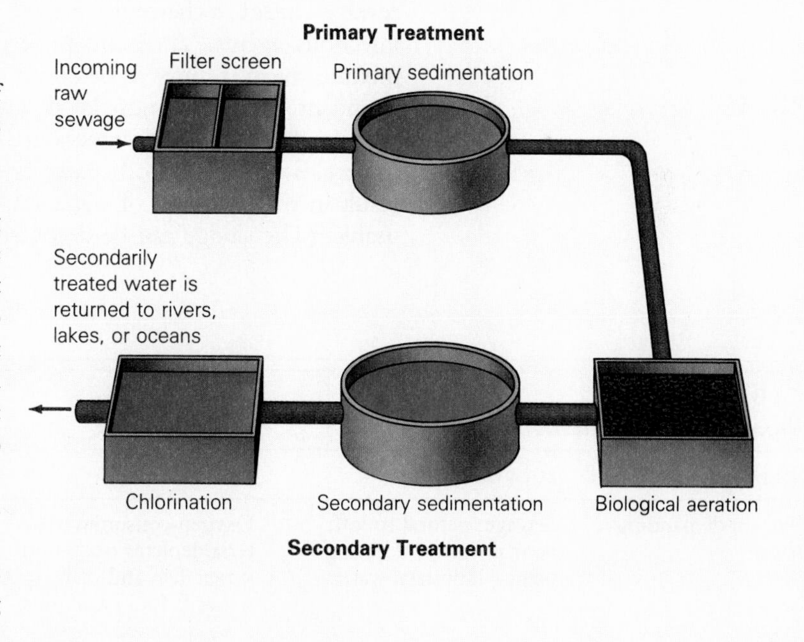

and municipal effluents. By law, when water is utilized for such purposes, it must return to the waterways in a pure, unpolluted form. However, we are aware that, in many cases, our waterways are treated as open sewers and our underground water tables (see Chapter 28) as invisible rugs under which we sweep our poisonous wastes (Essay 29.4 and Table 29.3).

Sewage

It has been said that the tap water of New York City contains many interesting ingredients—not the least of which is chow mein. A more precise, but no less fascinating statement is that in some places in the United States, the water one drinks has already passed through the bodies of seven or eight other good people. Under the best of conditions, that water will have been filtered and treated with chlorine to kill dangerous microorganisms. (Chlorine has been questioned as a water additive because chlorine can interact with organic material to produce potentially carcinogenic hydrocarbons. Also, people who drink chlorinated water have a 13 to 93 percent greater chance of developing colon and bladder cancers than those who drink untreated water. But the consensus is that, considering the risk of impure water, chlorination is by far the lesser of two evils.)

In many locales in the U.S., the water is simply not safe to drink, as evidenced by the continuing outbreaks of infectious hepatitis in the United States. Infectious hepatitis is believed to be caused by a virus carried in human waste, usually through a water supply that is contaminated by sewage. There is some disturbing evidence that this virus may be resistant to chlorine, especially in the presence of high levels of organic material. Despite our national pride in walk-in bathrooms, sewage treatment for many communities in the United States is inadequate, and waste that has been only partially treated is regularly discharged into waterways.

One of the biggest sewage problems is how to properly dispose of **sewage sludge,** a charming mixture of settled solids, toxic chemicals, and infectious agents. There are three basic ways to dispose of this gooey mixture: burn it, bury it, or dump it in the ocean. Burning sites are not abundant since they must be downwind of communities and the burn must be done at a high temperature to completely break down most toxic molecules that would otherwise be released in the smoke. Burying can result in the pollution of water tables and render surrounding land unusable. (The sludge can be decontaminated and used as fertilizer, an ex-

TABLE 29.3
Major Water Pollutants

POLLUTANT	SOURCES	EFFECTS	CONTROL METHODS
Oxygen-demanding wastes	Sewage; natural runoff; animal wastes; decaying plants; industrial wastes	Oxygen-consuming bacteria depletes oxygen in water; fish and plants destroyed; foul odors; livestock sickened	Better treatment of waste water; reduce agricultural fertilizer runoff
Disease-causing wastes	Sewage; animal wastes	Outbreaks of diseases, including typhoid, infectious hepatitis, cholera, and dysentery	Treat waste water; reduce agricultural and ranching runoff
Acids	Mining (especially coal); industrial wastes	Kills plant life, some animals; increases solubility of other harmful minerals	Control surface mining and smoke stack effluent; treat wastewater

pensive process.) Ocean dumping is simply illegal except for New York and New Jersey. These communities dump just over the edge of the continental shelf, about 100 miles out. The problem is, the sludge doesn't dissipate as quickly as expected, and some of it washes back to contaminate beaches (Figure 29.9).

FIGURE 29.9
Sewage sludge accumulates in waterways before reaching the ocean.

Chemical Pollution of Water

Perhaps the most dangerous water pollutants are from inorganic sources, because many of them are so new that we know almost nothing of their long-term effects. However, we do know of the existence of over five million chemicals, about 45,000 of which are used commercially, with literally hundreds more being added to the list each year. We also know that well over 250 synthetic chemicals were isolated from the drinking water of only 80 American cities.

The dangers of releasing chemicals we know nothing about is underscored by evidence that some familiar compounds can be more dangerous than we thought. For example, we learned only around 1950 that nitrates, a common constituent of agricultural fertilizers and water supplies, could alter hemoglobin so as to impair its oxygen-carrying capacity. Even later, it was found that certain bacteria in the digestive tract can convert nitrates to highly dangerous and carcinogenic nitrites. Nitrate water pollution is especially dangerous in certain areas of California, Illinois, Wisconsin, and Missouri.

Heat Pollution of Water

A rather subtle and often ignored form of water pollution is *heat*. Water from our rivers and lakes absorbs heat, for example, when the water is used as a coolant or lubricant in manufacturing or energy production (including nuclear energy). Actually, almost half the water used in the United States is for cooling electric power plants (Figure 29.10). The problem is that when water is returned to its source at this higher temperature, it may alter the forms of life in the water. Furthermore, a temperature rise of only a few degrees is sufficient to cause such changes. Some species may perish, while others begin to thrive under these new conditions, and the organisms affected may range from the producers to the highest level of consumers. Thus, delicate balances, which have been established as the result of long periods of evolution, can be seriously disrupted. The effect is less important in large, fast-moving rivers where the heat is quickly dissipated, but smaller, slower rivers or lakes may show more marked effects.

FIGURE 29.10
Cooling towers at an electrical power plant.

But why is the addition of only a little heat so critical? How does it exert its effects? Basically, heat pollution works in two ways: (1) by increasing the rate at which aquatic organisms consume oxygen, while, at the same time, lowering the oxygen-carrying capacity of the water, and (2) by increasing the rate of evaporation of the water, thus raising the concentration of pollutants left behind. The effects of the heat can be far-reaching, such as in destroying spawning grounds and increasing the susceptibility of fish to diseases. In some areas, enterprising souls are using the heat to grow commercially valuable fish and shellfish. For example, oysters are being cultivated in the warmer water in New York and Japan.

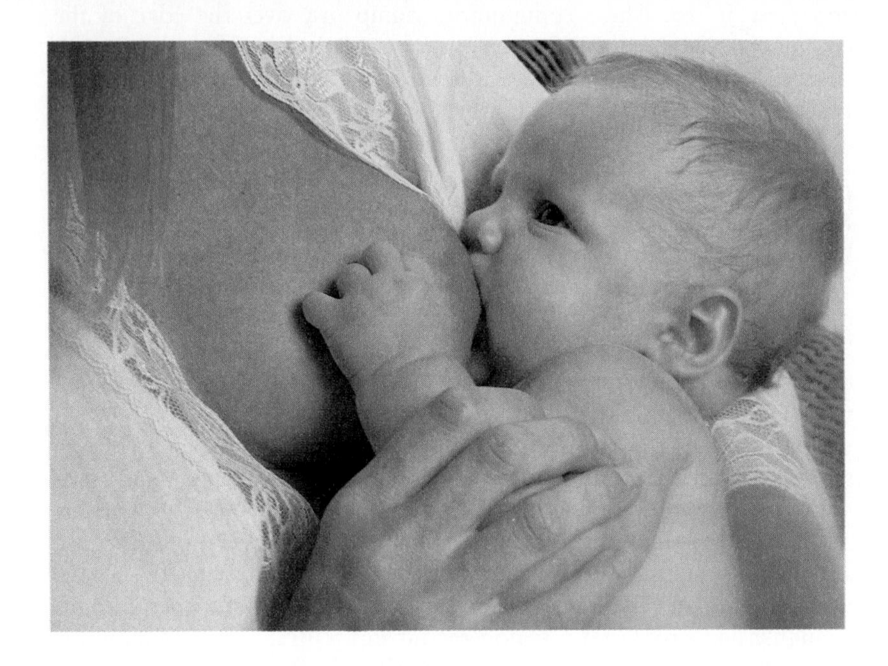

FIGURE 29.11
We often eat from the top of the food chain and thereby encounter food that may, itself, have concentrated DDT in its tissues. DDT is soluble in fat and is often stored in human fatty tissue, and can make its way into human milk.

Pesticides

Pesticides are biologically rather interesting substances. Most have no known counterpart in the natural world, and most didn't even exist 30 years ago. Today, however, a metabolic product of DDT, called DDE, may be the most common synthetic chemical on earth. It has been found in the tissues of living things from polar regions to the remotest parts of the oceans, forests, and mountains. Although the permissible level of DDT in cow's milk, set by the U.S. Food and Drug Administration, is 0.05 parts per million, it often occurs in human milk in concentrations as high as five parts per million and in human fat at levels of more than twelve parts per million (Figure 29.11).

Pesticides, of course, are products that kill pests. But what is a pest? Biologically, the term has no meaning. The Colorado potato beetle, for example, was never regarded as a pest until it made its way (carried by humans) to Europe, where it began to seriously interfere with potato production. Perhaps this episode best illustrates a definition of a **pest**: It is something that interferes with humans.

It seems that the greatest pesticidal efforts have been directed at insects (Figure 29.12) and, clearly, much of it has been beneficial. The heavy application of DDT since World War II has caused sharp decreases in malaria and yellow fever in certain areas of the world. But DDT and other chlorinated hydrocarbons have continued to be spread indiscriminately any place in which insect pests are found. The result, of course, is a kind of (is it artificial or natural?) selection. The problem is that some insects had a bit more resistance to these chemicals than did others. These resistant ones then reproduced and, in turn, the most resistant of their offspring continued the line. The result is that we now have insects that can almost bathe in these chemicals without harm.

FIGURE 29.12
Two insects that have triggered the release of vast amounts of insecticide. The malaria mosquito (left) and the tsetse fly that carries sleeping sickness.

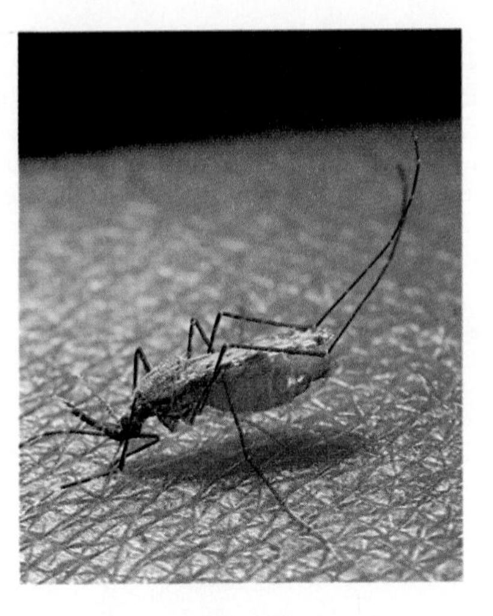

There are also other risks involved in such wide use of insecticides. For example, most are unselective in their targets; they kill virtually *all* the insect species they contact. Many insects, of course, are beneficial and may form an important part of large ecosystems. Also, chemical insecticides move easily through the environment and can permeate far larger areas than intended. Another particularly serious problem with pesticides is that many of them persist in the environment for long periods. In other words, the chemicals are very stable and it is difficult for natural processes to break them down to their harmless components. Newer chemical pesticides are deadly in the short run, but quickly break down into harmless by-products.

The tendency of DDT to be magnified in food chains has been particularly disastrous for predators that feed high on the food pyramid. This is because as one animal eats another in the food chain, the pesticide from each level is added to the next. Thus, species high on the food chain, the predators, tend to accumulate very high levels of these chemicals (Figure 29.13). In this light, recall that humans are often the top predator in food chains. The effects of accumulated DDT on predatory birds have been substantial. Reproductive failures in peregrine falcons, the brown pelican, and the Bermuda petrel have been attributed to ingesting high levels of DDT. The problem is that the pesticide interferes with the birds' ability to metabolize calcium. As a result, they lay eggs with shells too thin to support the weight of a nesting parent.

Radiation

Radiation is energy in the form of fast-moving particles or waves. That radiation can contain enough energy to dislodge electrons from the atoms they hit, turning those atoms into potentially disruptive free radicals. Life on earth evolved under ionizing radiation in one form or another, and, in fact, we still live in a veritable sea of radiation. One source of radiation is the earth's crust, where radioactive substances undergo spontaneous nuclear disintegration, emitting both high-speed atomic particles and penetrating electromagnetic rays (similar to X rays). These types of radiation

FIGURE 29.13

Once DDT enters the food chain, it becomes increasingly concentrated as it moves along from one link to the next. Whereas its concentration may be very low in such organisms as algae, fish that eat algae tend to store most of the DDT they eat in their tissues. Other fish that eat those fish, then, encounter an increased concentration of DDT in their food supply and, in turn, store most of it in their tissues. They eventually pass these boosted amounts along to their own predators. Each level, then, ends up with more DDT in its tissues than the species on which it feeds. The numbers in the figure indicate the parts per million of DDT in the tissues of various species that have been tested in the chain. Where would humans enter the system?

are more prevalent in some regions of the earth than in others. In addition, we are constantly bombarded by fast-moving atomic fragments as our atmosphere is struck by cosmic rays, atomic particles from space. Other radioactive substances, such as one called potassium 40, normally circulate through living systems. Such naturally occurring radiation is collectively referred to as background radiation. We do not know, at present, the degree to which background radiation is responsible for mutations resulting in either cancer or birth defects in the human population, but we do know that such ionizing radiation can produce these conditions under experimental conditions. A much more serious problem is the radiation we produce ourselves through such things as medical radiation, nuclear testing, and nuclear reactors (Essay 29.5).

ESSAY 29.5

Radiation

The nuclei of some atoms are unstable. Of the more than 320 isotopes that exist in nature, about sixty are unstable, or radioactive. In addition, humans have created about 200 more. Radioactive isotopes of any element are called radioisotopes. When a radioisotope decays (or explodes), particles and rays are emitted from its nucleus.

Radioisotopes decay at predictable rates. These rates are usually expressed as the element's half-life. Half-life refers to the time that must expire in order for half the atoms in any amount of the isotope to decay. Half-lives may be very long. The half-life of natural uranium (U238) is 4,500,000,000 years! As a rule of thumb, the time it takes for a substance to become non-radioactive is considered to be 20 of its half-lives.

When radioisotopes decay, the products may strike other atoms in such a way as to tear electrons away from them, leaving them with a positive charge, producing free radicals. Genetically, the danger lies in the molecules of DNA being altered by free radicals, thereby producing mutations. Such changes in DNA can result in a variety of damage to the body (including cancer), or if the change occurs in the DNA of a gamete, the result can produce abnormal offspring. Also, the genetic effects of radiation are accumulative in the exposed individual and his or her offspring. A "snowball effect" operates so that an increased radiation level, which has little effect in the first few generations, if maintained, may become magnified in following generations. The figure shows the effects of radiation on various parts of the body.

Radioactivity of any substance can only be reduced by allowing the radioisotopes to decay naturally. The process cannot be slowed or hastened.

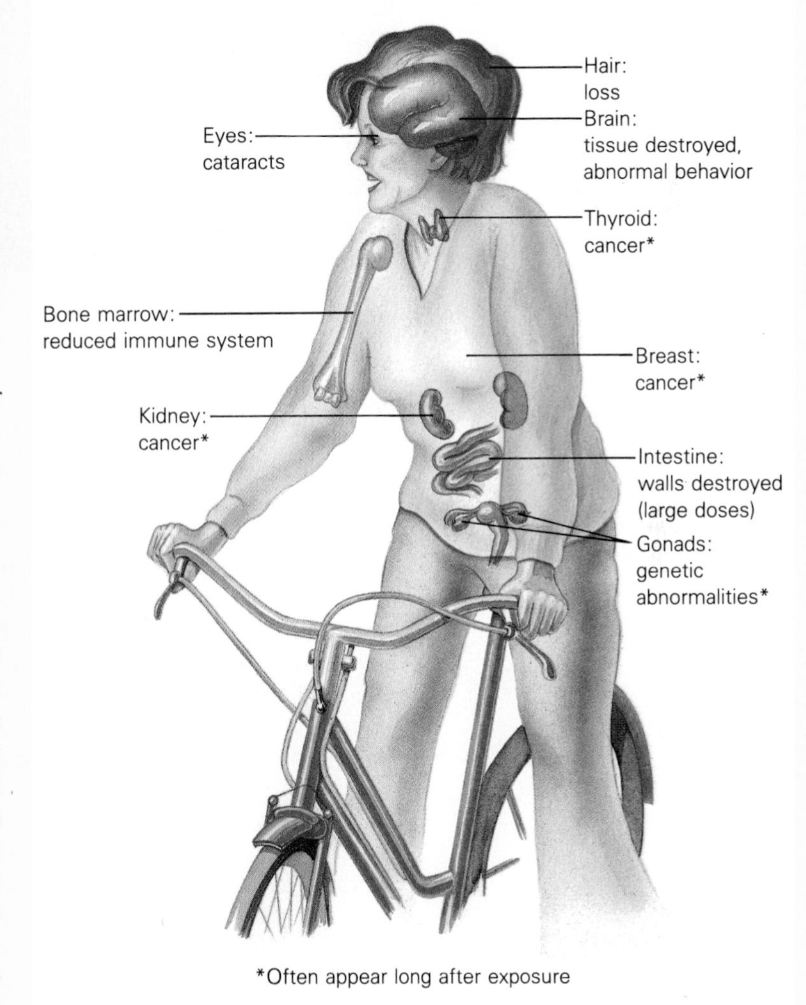

FIGURE 29.14

The worst nuclear accident in the United States occurred at Three Mile Island in the Susquehanna River, Pennsylvania. One of the cooling water pumps failed, immediately activating backup systems. However, valves on emergency water pumps had been accidentally left closed, and a faulty valve (scheduled for replacement) failed to work. Operators miscalculated the amount of water in the cooling system and prematurely shut off the water supply, uncovering the nuclear core for several hours. The housing of the fuel rods oxidized, producing an enormous hydrogen bubble that experts feared could explode. The bubble had not been anticipated. No one yet knows how close we came to a complete meltdown at Three Mile Island.

Pollution from Nuclear Reactors

Americans generally do not favor nuclear power. The interesting thing is they generally don't have scientifically based reasons—they seem to feel that the science is beyond them; they often just have an uneasiness based on something they might have seen on "60 Minutes." Or sometimes the comment may be, "What if the damn thing blows up?" Although modern nuclear reactors are unlikely to produce an explosion of the type produced by nuclear bombs, the worst nuclear disaster in history was due to a mysterious explosion in 1957 at a place called Kyshtym in the Soviet Union. Today the names of 30 towns have been removed from Soviet maps.

A more likely threat is a meltdown, when the nuclear core overheats and melts. As the fissionable material joins, uncontrolled, the heat produced would theoretically not be containable and nuclear material would be released into the environment. Even today we don't know how close we came to a meltdown at Three Mile Island (Figure 29.14), the worst American disaster so far. A far greater tragedy occurred at a nuclear plant at Chernobyl, in the Soviet Union's Ukraine in 1986 (Essay 29.6). The risks associated with nuclear power, however, are not always so spectacular. Some are of a far subtler nature. For example, radioactivity could be released into the environment from activities related to mining and processing nuclear fuel, from the transportation and recycling of the fuel, and from storage of the radioactive wastes. We frequently hear of steam or gas leaks from the reactors themselves. And even the safest reactors normally leak small amounts of radiation into their immediate environment. (The problem with this is that there are no "safe" levels of radioactivity—

ESSAY 29.6

The Chernobyl Meltdown

The first warning came at 9:00 AM on Monday, April 28, 1986. Technicians at a nuclear plant 60 miles north of Stockholm began to see alarming blips across their computer screens. The blips meant one thing: high levels of radiation. The technicians assumed a serious leak at their own plant and began a frantic search for the problem. They found nothing, but the radiation levels kept increasing both at the plant and in the surrounding countryside. They concluded the problem was not with their own facilities and immediately cast a suspicious eye to their powerful neighbor to the south, the Soviet Union.

For days, breezes had been blowing from the south and into Scandinavia, but when the Swedes demanded an explanation, the Soviet response was a stony silence. Later that day, the Soviets broke their silence and blandly said that nothing had happened. On the news that night, an expressionless newscaster read a four-sentence statement admitting that there had been a nuclear accident at the Chernobyl nuclear power plant (80 miles north of Kiev) damaging a reactor and that "measures were being taken."

The Soviets played down the incident, denying reports that thousands had been killed, but admitting two deaths. As the story unfolded, largely through the intense questioning of other countries and American satellite photographs, it became clear that the reactor had experienced a dreaded meltdown. A meltdown occurs when the cooling system fails and the radioactive core units overheat to the melting point. As the molten mass accumulates on the reactor floor, there is a risk of burning through the floor and into the ground below, contaminating the water table (a melt-through). At Chernobyl, the problem was compounded by the type of encasement around the fuel rods. It was graphite, and graphite can burn, producing temperatures over 5000 degrees Fahrenheit. The Soviet technicians then flooded the area with water, but it was not only too late, it was the wrong move. The steam reacted with the uranium fuel, the graphite, and the zirconium sleeves that house the rods, producing a flammable gas that blew up the reactor and released billowing clouds of radioactive gases into the atmosphere (see map).

In the days that followed, the Soviets grudgingly released bits of information telling of an increasing death toll, and mass evacuations of large parts of what had been one of the Soviet Union's greatest agricultural areas.

We still do not know the full effects of the accident on Soviet citizens (or those in other parts of the world) because low doses of radiation may take years to exact their toll. Only two months before the disaster, the Soviets had calculated that a meltdown of this type could statistically be expected only every 10,000 years. The Chernobyl reactor had been operating three years.

only "acceptable" levels. Theoretically, a single radioactive molecule could eject particles that could knock an electron off a molecule within a living cell, thereby triggering cancer.)

We could probably greatly reduce the risks associated with nuclear power by simply exercising more care and common sense. There are numerous published accounts that attest to our carelessness, however. For example, it has been revealed that the Diablo Canyon nuclear power plant in California was built on an earthquake fault line. Of course it was girded for that risk (inasmuch as one can gird for a large earthquake). Incredibly, however, the blueprints were somehow reversed and the earthquake supports were put in backwards. Furthermore, the mistake was not noticed for four years. At the Comanche Peak Plant in Texas, supports were constructed 45 degrees out of line. At the Marble Hill in Indiana, the concrete surrounding the core was found to be full of air bubbles. At the WNP-2 plant in Washington state, the concrete contained air bubbles and pockets of water as well as shields that had been incorrectly welded. At the San Onofre plant in California, a 420-ton reactor vessel was installed backwards and the error was not detected for months. In 1981, the Nuclear Regulatory Commission inspected 43 plants that were under construction (Figure 29.15) and rated seven "below average" and 36 "average." None were rated even "above average."

FIGURE 29.15

Sites of operating and proposed nuclear reactors in the United States. Only Washington State has reactors with graphite and without a containment building (similar to that at Chernobyl). Some sites have more than one reactor.

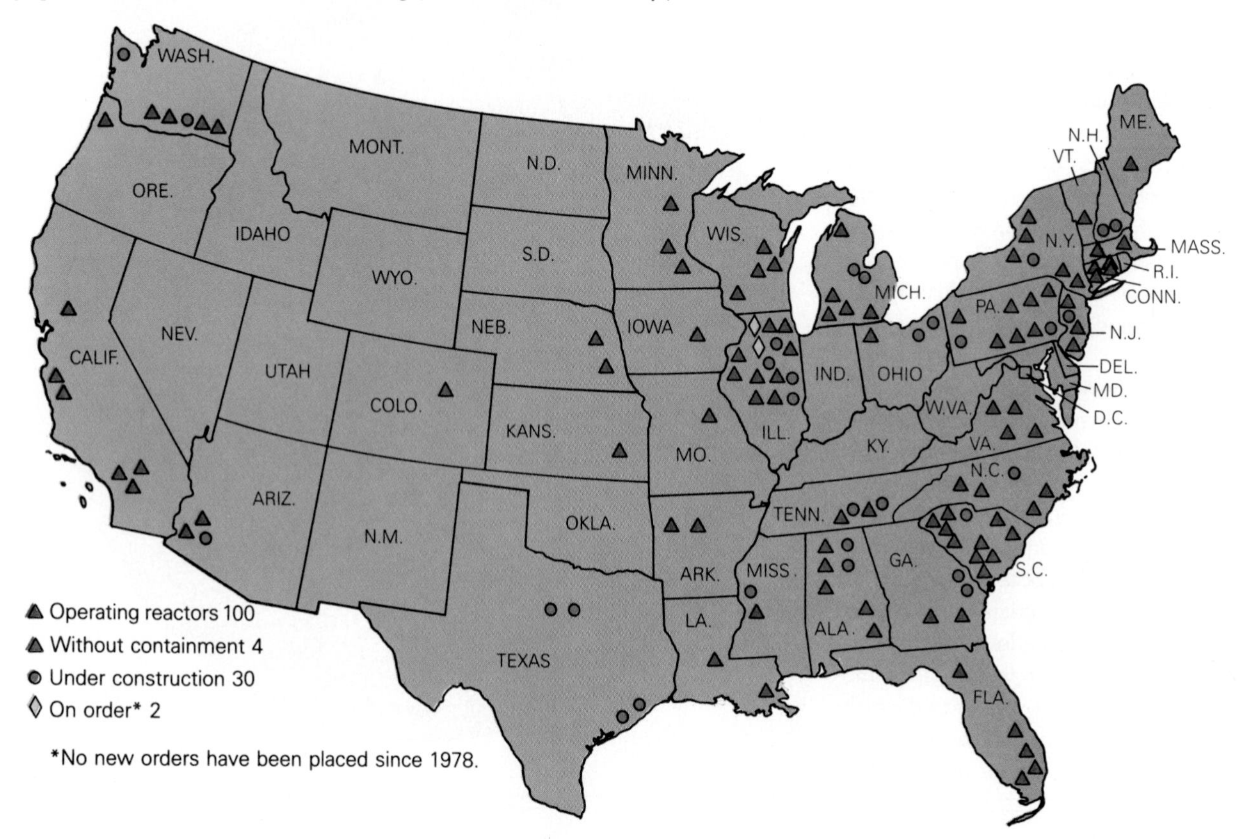

Completely apart from the possibility of accidents, there is the unsolved problem of what to do with the radioactive wastes generated in the course of normal nuclear plant reactions. The problem is a tough one since such wastes can only be rendered safe by the passage of time. The waste radioactivity is generated in the fuel system of the reactors because only a part of the fuel is fissionable and, for technical reasons, not all of the fissionable elements are spent. Much of the spent fuel materials removed from the reactor can be reused. However, some of the radioactive fuel in the spent elements cannot be recovered, and this material adds to the radioactive waste.

We have already generated over 10,000 tons of nuclear waste, with another 47,000 tons expected by 1995. Most of the waste is in the form of fuel rods which are, for now, stored in baths filled with a solution of neutron-absorbing boric acid. The problem is that these are only temporary repositories and, unless new space is found, existing plants must begin closing for lack of space. The rods can be reprocessed, but the technology could lead to the spread of fuel for nuclear weapons (Essay 29.7).

The reprocessing also leaves "high-level liquid waste" that must be stored. The prevailing idea at the moment is to dry the liquid and mix it with molten glass that, when it hardens, can be stored in stainless steel containers. The problem then is what to do with the tanks. Suggestions have ranged from burying them in the Antarctic to sending them into space, but for various reasons these plans have been abandoned. There are now plans to use Sedan Crater at Yucca Flat in Nevada as a storage site for nuclear waste.

The site is scheduled to be completed by 2003, but no one expects it to be on schedule. Even when it is complete, nuclear waste will have to be shipped across 45 states, 6000 shipments a year if by truck, 830 if by train. There are those who wonder what will happen if those trucks or trains have an accident in a heavily populated area or upwind of one.

Finally, there is the unpleasantry of the threat of terrorists. The U239 (plutonium) would undoubtedly be the target of those who would use the material to make nuclear bombs. (The technique was described in detail by a college student a few years ago in a paper he wrote for class, using information legally gathered in this country. The government was very upset with him. Also, the federal government admitted a few years ago that substantial amounts of nuclear fuel could not be accounted for. They added that they did not think it had been stolen.)

HIDDEN DECISIONS

Perhaps in these pages you have come to learn more about the complexities of this great living system of which you are a part. Perhaps, also, you have come to appreciate just how inextricably linked are the various aspects of life. Because life's grand processes are complex and interrelated, it should be apparent now that life is easily touched. A move here, a decision there, and a part of our biosphere heads off in some new direction. Sometimes we know about it; sometimes, however, we do not.

We are by now informed enough about life's processes to avoid unnecessary radiation, to choose the most healthful foods, and to deplore extinctions. When the choice is ours, we choose life. But sometimes we

ESSAY 29.7

Nuclear Winter

The most sadly brutal and devastating effects of a nuclear exchange may not be due to the blast itself or the ionizing radiation that follows. The worst killer may have a different face. Imagine a darkened world blanketed by a constant winter. This is the scene that some scientists* now warn us about. They say that one effect of even a limited exchange of missiles would be nuclear winter.

According to their computer analysis, nuclear explosions at a few target cities would send clouds of smoke and ash high into the atmosphere, where it would hang suspended for long periods, blocking the sun's radiation and throwing the world into a dark and starving winter. Originally, they estimated that these conditions might be restricted to the northern hemisphere, lasting perhaps six months to a year. The early calculations suggested that debris would be lofted perhaps five miles high, but would eventually be washed from the sky by cleansing rains.

However, new data are emerging that paint an even more dismal picture. It seems that the great clouds may be lifted as high as ten miles, from where they could spread their darkening quilt over the entire planet, bringing winter even to the tropics and stopping photosynthesis virtually everywhere. Further, such clouds would rise far above the cleansing effects of weather.

Some scientists disagree with all such projections, saying they encompass too many assumptions and that even newer data show the effects would be far less drastic than first estimated. It is also argued that rains would probably soon cleanse the skies and that the temperature changes would be only moderate. While the argument continues, a great number of people are hoping it will never be resolved.

*The best known is a group comprised of Richard Turco, O. Brian Toon, Thomas Ackerman, James Pollack, and Carl Sagan (TTAPS, they call themselves).

are presented with choices without our knowing it. Precisely because life is a sensitive, responding, and far-flung phenomenon, almost anything we do affects a living system in some way. And we are beginning to learn that changing any part of it can have far-reaching and often unexpected effects. It is important to understand that we can change the nature of life on the planet and our place in its pageant without realizing we were involved in such things at all. It is the role of the educated person to uncover these hidden decisions and bring them to light so that we as a society can begin to understand the cascading impact of our behavior.

As an example, we can influence the environment by our vote. Sometimes this is apparent, such as when we elect some staunch environmentalist (or a developmentally minded realtor). In other cases, though, our political choices may have less apparent, but just as important, biological effects. For example, if elected officials subsidize farmers with public funds, do they encourage people to stay on the land and out of cities? Is this good? Is there a greater sense of community among people in dense populations, or do such conditions encourage anonymity and reduce the sense of responsibility? What happens to an environment in which people do not feel individually responsible?

The United States is proud of its system of public education. Every child in this country has the right, even the obligation, to be educated to some degree. The schools exist, and they are funded. Education is obviously good, so what lawmaker wants to oppose it? However, some say our system of public education has an unexpected side effect—that it subtly encourages people to have children by guaranteeing that the cost of their children's education will be paid for by public money. What are the biological effects of such encouragement?

Even our system of taxation can have biological repercussions. In some years, single people get tax breaks, and in other years, the nod goes to married couples. If marriage is encouraged by government policies such as preferential tax schedules, and if (as is the case) earlier marriages tend to produce more children, then legislators again can subtly influence population growth and social patterns, often without realizing it.

Do tax deductions for home mortgages encourage building? Does home building affect forests? What are the effects of building booms on national parks and outdoor recreation? On land use policies? On sewage? On water supplies? On road building and its handmaidens? What are the social effects of governmental encouragement of home ownership as opposed to apartment rental?

The list of our hidden decisions is endless. The point of all this is that not only *can* we make biological decisions, we *do* make biological decisions. And we do it almost casually on a daily basis. Obviously, we must make such decisions. We have come too far to try to leave well enough alone. However, we must become aware of the far-reaching effects of our manipulations. In other words, our decisions must begin to make biological sense. They must become conducive not only to perpetuating life, but to increasing our quality of life. Perhaps the answer lies in educating our lawmakers in the biological implications of their decisions and policies. Or perhaps the answer lies in first educating ourselves and then electing lawmakers from among us who are well aware of what they're doing and who can confidently expect the support of an enlightened public.

THE FUTURE

We were once the future. We were among the generations our ancestors worked for, hoped for, and fretted about. They wanted to make the world safe for us, and they tried to prepare the way for our coming. The future, however, has a way of becoming the present. And now here we are, the much-loved progeny of people who never knew us. We are indeed reaping what they sowed, and it works both ways. In some ways, we benefit greatly from their concern and foresight. In other ways, we are paying the bill for their binge. However, perhaps they can be forgiven for their unwise choices because there is no way they could have imagined such a world as this.

Now it is our turn. And we, too, are blinded by the present light. We can't see ahead either. But we are to be forgiven less easily for our follies. We must bear greater responsibility than do our ancestors. The reason is quite simple: We know more than they did. We are aware of our present global situation. We know more about our world than any generation that preceded us. We know, in a general way, how bad it is and how good it can be. We have enough information to enable us to understand, more than any other generation, the implications of our decisions. We can see more clearly than ever where we're going and what any charted course will lead us to. Whereas we can't predict what the future holds, we can influence that future to a degree unimagined by our ancestors. Had they had this ability, the world might be a far different place today. But with this ability, born of sheer information, comes an added responsibility. We have but a few more years of this odd interlude in history in which we will be able to ignore our biological imperatives, to fail to recognize that we are a recent upstart among the venerable and ancient species, and that our future is not insured. It is important that as this period of borrowed time runs out, we become aware that our limitations are those of fragile cytoplasm, that we see the urgency of a steady-state global economy, an economy no longer based on a mortgaged future. Will it be said by our descendants that we foreclosed their future and demeaned their existence? Or will they stand in admiration of our strength, dignity, and wisdom at a time when decisions were hard?

SUMMARY

1. Bioethics deals with the moral values of our behavior toward life on the planet. Meeting the needs and demands of an affluent society generates tremendous waste. Pollution is a side effect of this process.
2. A pollutant is a substance whose presence in the environment is harmful. Although some pollution occurs through natural events, pollution from human activity is new in kind or degree and is of concern.
3. Four ways that pollutants can be harmful are: (1) by altering critical aspects of the environment in ways that lead to the destruction of some forms of life, (2) by setting the stage for future effects, (3) by having no effect until reaching a critical level (i.e., a threshold effect), and (4) by acting synergistically with other pollutants so that the combined effects are greater than individual effects.
4. Burning fuel, primarily in automobiles but also in other ways, is the source of air pollution. Although air pollution is present everywhere, it is greatest in the cities of industrialized countries. The most common air pollutant, carbon monoxide, cuts the blood's ability to carry oxygen to the tissues, thereby placing strain on the heart and respiratory system. Nitrogen dioxide irritates lungs, withers plants, and combines with hydrocarbons in the presence of sunlight to produce destructive photochemical smog. Perhaps the most dangerous of the air pollutants are the sulfur oxides, which come mainly from burning coal, and can react with water to form sulfuric acid. Sulfuric acid can cause respiratory illnesses, damage plants, and can dissolve marble, iron, and steel. Hydrocarbons, primarily from the incomplete combustion of fossil fuel, may be carcinogenic and impair plant functioning. Released into the air in a variety of ways are particles of many substances, some of which cause cancer, aggravate lung or heart conditions, or interfere with photosynthesis.

5. Fresh water is polluted as it is used for cooling, cleansing, and dumping for industrial and municipal effluents. Drinking water in the United States is usually filtered and treated with chlorine to kill dangerous microorganisms. If water is contaminated with human waste, it may contain the virus responsible for infectious hepatitis. Drinking water may contain thousands of industrial chemicals, some with known detrimental effects and some with unknown, but potentially harmful effects. When water is used as a lubricant or coolant, it absorbs heat. If it is returned to its source at a higher temperature, it may alter forms of life in the water by simultaneously increasing oxygen needs of organisms and lowering the oxygen content of the water, and by increasing the rate of evaporation, thereby concentrating the remaining pollutants.
6. Pesticides are synthetic chemicals intended to kill pests, organisms that interfere with humans. The heavy use of DDT since World War II has sharply reduced the incidence of malaria and yellow fever. However, its widespread use has led to selection of insects resistant to its effects. Furthermore, it kills beneficial insects. Older chemical pesticides are stable over long time periods and accumulate in food chains. Pesticide levels are greatest in top predators and have interfered with the breeding success of peregrine falcons, the brown pelican, and the Bermuda petrel.
7. Life on earth is always exposed to naturally occurring radiation such as the radiation that comes from the earth's crust or from the atmosphere in the form of cosmic rays. A more serious problem is the radiation we produce ourselves through medical radiation, nuclear testing, nuclear reactors, and nuclear accidents. Nuclear power plants could possibly supply much of the needed energy for our future, but their use is questioned because of the possibility of accidents with disastrous or subtle consequences and because of the unsolved problem of disposal of radioactive waste.
8. Each day, we can and do make decisions that influence biological systems.

KEY TERMS

bioethics (817)
carbon monoxide (821)
hydrocarbon (824)
nitrogen oxide (823)
particulate matter (824)
pest (828)
pesticide (828)
photochemical smog (823)
pollutant (819)
radiation (829)
sewage sludge (826)
sulfur oxide (823)

FOR FURTHER THOUGHT

1. Why should we humans be concerned about the use of pesticides that kill insects at the bottom of the food chain?
2. Describe "hidden decisions" that affect human life that are not discussed in the text. Consider, for example, the costs of oil on air pollution.

FOR REVIEW

True or false?

1. ___ Hydrocarbons may be produced by natural processes.
2. ___ Chemical pollutants such as DDT are not dangerous to animals at the top of the food chain.
3. ___ Particulate matter is a pollutant found in water.

Short answer

4. Give a few examples of naturally occurring pollutants that are not produced by human activity.

Fill in the blank.

5. The ___ refers to how some pollutants may show effects only after they accumulate to a critical level in our bodies.
6. Photochemical smog results from the combining of ___ with hydrocarbons in the presence of sunlight.
7. Waste particles in the air such as dust, minerals, and plant products are collectively referred to as ___ .

Choose the best answer.

8. Bioethics is an area that is concerned with ___ issues.
   A. pollution
   B. extinction
   C. energy
   D. all of the above
9. ___ is a pollutant that decreases the oxygen supply to our tissues by combining with our blood's hemoglobin.
   A. Carbon monoxide
   B. Nitrogen oxide
   C. Sulfur oxide
   D. None of the above
10. Sulfur oxides
    A. are produced by decaying plants.
    B. are the greatest global pollutant.
    C. are not present in heavy smog.
    D. are produced by burning coal.

APPENDIX A

Classification of Organisms

THE PROKARYOTES

This is one of several common ways of dividing this group.

Kingdom Monera

Single-celled organisms, tough cell wall, no membrane-bounded organelles or organized nucleus. Circular DNA, not joined with protein. Reproduction mostly by fission; some by conjugation. Any flagella stiff and rotating.

Subkingdom Archaebacteria: Unique, proteinaceous cell walls. Cell membrane with branched fatty acids. Mostly anaerobic; includes methanogens, halophiles, and thermophiles.

Subkingdom Cyanobacteria: (formerly called blue-green algae) Photosynthetic. Colonial. Some species fix nitrogen. Membranous photosynthetic lamellae.

Subkingdom Eubacteria: ("true" bacteria) Cell walls of peptidoglycan, membranes with straight-chain fatty acids. Includes many pathogens, free-living reducers, phototrophs, and chemotrophs. Found in coccus, bacillus, and spirochaete forms. Occurs singly, in clusters, or in chains.

THE EUKARYOTES

All other organisms. Membrane-bounded cellular organelles, linear chromosomes join protein. Cell division by mitosis and meiosis, sexual reproduction common. Any flagella or cilia are microtubular. Single-celled, colonial, and multicellular.

Kingdom Protista

Includes photosynthetic, plantlike (algal) and heterotrophic, animal-like (protozoan) forms. Primarily single-celled or colonial.

Phylum Pyrrophyta: (1100 species; the dinoflagellates) Single-celled, flagellated, chitinous cell walls, phototrophic.

Phylum Chrysophyta: (11,500 species; yellow-green and golden-brown algae) Single-celled or colonial, glass walls, phototrophic.

Phylum Protozoa:

- *Class Flagellata* (2500 species; flagellated protozoans) Single-celled, heterotrophic.
- *Class Sarcodina* (11,500 species; amoeboid protozoans) Single-celled, phagocytic heterotrophs; includes marine radiolarians and foraminiferans.
- *Class Sporozoa* (6000 species; nonmotile protozoans) Single-celled, parasitic spore formers.
- *Class Ciliata* (7200 species; ciliated protozoans) Single-celled, extremely complex and diverse heterotrophs.

Phylum Acrasiomycota: (26 species; cellular slime molds) Heterotrophic, individual amoeba that live alone or join to form multicellular plasmodia.

Phylum Myxomycota: (450 species; acellular slime molds) Heterotrophic, form multinucleate feeding plasmodia.

Kingdom Fungi

Multicellular heterotrophs, including reducers and parasites. Extracellular digestion. Mycelial organization that may or may not include walls between cells. Either sexual or asexual spores; primarily haploid with brief diploid stage.

Division Oomycota: (475 species; water molds) Flagellated fungi, many parasitic.

Division Zygomycota: (600 species; bread molds) Mycelium without cell end walls; simple sexual zygospores.

Division Ascomycota: (30,000 species; sac fungi) Extensive mycelium with cell end walls. Complex sexual dikaryotic asci, ascospores produced through meiosis. Many species symbiotic with cyanobacteria or algae, forming lichens.

Division Basidiomycota: (25,000 species; club fungi) Extensive mycelium with cell end walls. Large, complex, dikaryotic basidiocarp in which basidiospores are produced.

Division Deuteromycota: (also, *Fungi Imperfecti;* 25,000 species) Various fungi with no known sexual stage.

Kingdom Plantae

Primarily nonmotile, multicellular organisms with dense cell walls or cellulose. Specialized tissues and organs. Most phototrophic. Alternating generations.

Nonvascular Plants

Division Rhodophyta: (4000 species; red algae) Coastal seaweeds, floridean or carageenan as storage carbohydrates.

Division Phaeophyta: (1000 species; brown algae) Coastal seaweeds and kelps, fucoxanthin pigments, laminarin and mannitol storage carbohydrates. Often large, some with vascular tissue.

Division Chlorophyta: (7000 species; green algae) Single-celled, colonial, and multicellular, pigments mostly chlorophylls and carotenoids, usually aquatic.

Division Charophyta: (250 species; stoneworts) Freshwater alga with apical growth, calcareous cell walls.

Division Bryophyta: (16,000 species; mosses, liverworts, hornworts) Multicellular, nonvascular, terrestrial. Simple aerial spore, motile sperm, predominant gametophyte. Generally small with little supportive tissue.

Vascular Plants

Division Tracheophyta: (vascular plants) Vascular tissue in roots, stems, and/or leaves. Predominant sporophyte; some with distinct generation.

Vascular Plants without Seeds

Class Psilophyta (4 species; whisk ferns) Simple, with few surviving species. Vascular stem, scalelike leaves. Aerial spore, motile sperm, distinct sporophyte and gametophyte.

Class Lycophyta (1000 species; club mosses) Vascular roots, stems, and leaves. Usually with aerial spore, motile sperm, and distinct sporophyte and gametophyte.

Class Sphenophyta (12 species; horsetails) One surviving genus, upright vascular stems, prominent nodes and tiny, scalelike nonphotosynthetic leaves; Aerial spore, motile sperm, and distinct sporophyte and gametophyte.

Class Pterophyta (Filicinae) (11,000 species; ferns) Most with aerial spore, motile sperm, predominant sporophyte, photosynthetic gametophyte, and vascular roots, stems, and leaves.

Vascular Plants with Seeds

Subdivision Spermatophytes: (seed plants) Gametophytes develop within sporophyte tissue, male and female spores produced, and microspores released within pollen grains. Embryo develops within seed that includes stored foods and protective seed coats.

Class Gymnospermae (naked seeds)

Cycads (100 species; the cycads) Palmlike leaves, exposed seeds, wind-dispersed pollen, flagellated sperm within pollen tube.

Ginkgos (1 species; ginkgo) Trees with fan-shaped leaves, exposed seeds, wind-dispersed pollen, motile sperm within pollen tube.

Conifers (500 species) Usually large trees with needlelike or scalelike leaves. Evergreens, exposed seeds borne upon cones. Nonmotile sperm.

Gnetophytes (71 species) Gymnosperms with angiosperm features: xylem vessels, pollen cones. Exposed seeds, nonmotile sperm.

Class Angiospermae (flowering plants) Flowers present, seeds enclosed by fruit, xylem vessels present. Nonmotile sperm.

Dicotyledons (200,000 species; the dicots) Diverse, net-veined leaves, secondary growth common, floral parts in fours, fives, or multiples of these, two cotyledons.

Monocotyledons (50,000 species; the monocots) Diverse, parallel-veined leaves, secondary growth rare, floral parts in threes or multiples of threes, one cotyledon.

Kingdom Animalia

Multicellular, heterotrophic eukaryotes. Specialized tissues, most with organ systems. Most highly responsive. Diploid except for gametes. Fertilization without intervening haploid life cycle. Small, flagellated sperm and large stationary egg typical.

Subkingdom Parazoa: Animals of flagellate origins, simple developmental progression.

Phylum Porifera: (5000 species; the sponges) Forms tissues. Nonmotile adults, filter feeding, skeletal elements of calcium carbonate, silicon dioxide, or spongin. Asexual reproduction by budding, sexual reproduction by fertilization of internalized egg.

Subkingdom Metazoa: Animals of ciliate origin; includes all animals except Porifera.

Radiate, Acoelomate Phyla (radial symmetry, no coelom, diploblastic)

Phylum Cnidaria: (9000 species) Radial body of two cell layers, saclike gastrovascular cavity, tentacles and stinging cells. May alternate between medusa and polyp stages, or only one stage may be present. Three classes: Hydrozoa (hydroids), Scyphozoa (jellyfish), Anthozoa (corals and anemones).

Phylum Ctenophora: (90 species; comb jellies) Radial body of two cell layers. Tentacles with glue cells.

Bilateral, Acoelomate Phyla (bilateral symmetry, no coelom, triploblastic)

Phylum Playthelminthes: (13,000 species; flatworms) Flattened body, branching gastrovascular cavity, dense bodies with many cell layers. Three classes: Turbellaria (free-living planarians), Trematoda (parasitic flukes), and Cestoda (tapeworms).

Bilateral, Pseudocoelomate Phyla (bilateral symmetry, pseudocoelom)

Phylum Nematoda: (12,000 species named—estimated half million unnamed; the roundworms and rotifers) Nematodes (roundworms) include free-living and parasitic species; slender body, pseudocoelom (not completely mesodermally lined), tube-within-a-tube body plan (complete gut).

Phylum Rotifera: ("wheel animals") Free-living, minute, with complex organ systems.

Phylum Nematomorpha: (230 species; horsehair worms)

Phylum Rynchocoela: (650 species; proboscis or ribbon worms)

Bilateral, Coelomate, Protostome Phyla (bilateral symmetry, true coelom, embryologically "mouth-first" animals)

Phylum Mollusca: (47,000 species) Controversial classification—segmentation and coelom may not exist. Diversification through modifications of head, foot, mantle, and radula. Includes seven classes: Aplacophora (solenogasters—wormlike, radula only clear characteristic), Monoplacophora (*Neopalina,* deep-sea form once believed to be extinct), Scaphopoda (tooth shells), Polyplacophora (chitons), Bivalvia (bilvalves: two shells—clams, etc.), Gastropoda (snails, slugs), and Cephalopoda (octopus, squid—rather intelligent, fast, predators, foot subdivided into tentacles). Covering mantle, large brain, keen eyesight.

Phylum Annelida: (9000 species; segmented worms) body subdivided into repeating segments, true coelom, well-developed digestive system, closed circulatory system. Three classes: Oligochaeta (earthworms), Hirudinea (leeches), and Polychaeta (marine worms).

Phylum Priapulida: (9 species; proboscis worms)

Phylum Pogonophora: (100 species; beard worms)

Phylum Sipuncula: (300 species; peanut worms)

Phylum Tartigrada: (350 species; water bears)

Phylum Arthropoda: (800,000 to 1,000,000 species; "jointed-footed" animals) Paired, jointed appendages with chitinous exoskeleton, varied segmentation, wide distribution.

Subphylum Chelicerata: Six pairs of appendages, four pairs being legs, with paired chelicerae (fangs). Three classes. Meristomata (horeshoe crabs), Arachnida (spiders, ticks, scorpions, mites, daddy-longlegs), and Pycnogonida (sea spiders).

Subphylum Mandibulata: Most with three pairs of walking legs, mandibles, compound eyes, antennas, some with wings. Four classes: Crustacea (aquatic with crusty exoskeleton, gills), Chilopoda (centipedes), Diplopoda (millipedes), and Insecta (insects—commonly three pairs of legs, wings at some time, three-part body, specialized mouth parts).

Phylum Onycophora: (70 species; *Peripatus*) Possessing both annelid and arthropod characteristics.

Phylum Brachiopoda: (250 species; lampshells) Appear similar to bivalves, but shell mounted differently, lopophore (ring of ciliated tentacles) present.

Phylum Phoronida: (18 species) Lopophore present.

Phylum Ectoprocta: (4000 species; moss animals or bryozoans) Lopophore present.

Bilateral, Coelomate, Deuterostome Phyla (bilateral symmetry, true coelom, embryologically "mouth second")

Phylum Echinodermata: (6000 species) Spiny, skinned animals, five-part radial symmetry as adults, bilateral larvae, endoskeleton, water vascular system. Five classes: Crinoidia (sea lilies), Holothuroidia (sea cucumbers), Echinoidea (sea urchins, sand dollars), Asteroidea (sea stars, basket stars), and Ophiuroidea (serpent stars, brittle stars).

Phylum Hemichordata: (80 species; acorn worms) Gill slits show relatedness to chordates.

Phylum Chordata: (43,000 species) Gill slits, notochord, postanal tail, dorsal hollow nerve cord all present at some time.

Subphylum Urochordata (1300 species; sea squirts) Chordate characteristics seen mainly in bilateral larva.

Subphylum Cephalochordata (28 species; lancelet) Fishlike body, permanent notochord and gill slits, filter feeder.

Subphylum Vertebrata (41,700 species; the vertebrates) Vertebral column of bone or cartilage, heads well developed, ventral heart, dorsal aorta, two pairs of limbs. Seven classes: the fishes: Agnatha (jawless fishes), Chondrichthyes (cartilagenous fishes: sharks, rays, chimera), and Osteichthyes (bony fishes). Also, Amphibia (frogs, toads, salamanders), Reptilia (reptiles), Aves (birds), and Mammalia (mammals).

APPENDIX B

For Further Thought Answers

CHAPTER 1

1. Darwin's *Origin of Species* would have been rejected by those in the 15th century. This era had little interest in biological investigation, and the church would certainly have censured the ideas expressed in the *Origin of Species* since they radically opposed the prevailing church doctrine.
2. The processes are similar in that the animals that are to breed are selected by an external force and both processes are responsible for eliminating undesirable traits over a period of time. Natural selection is the least efficient process because as individuals with slightly less desirable characteristics produce offspring, the traits will take longer to disappear from the population.

CHAPTER 2

1. Sentence C is teleological. It should be rephrased to read: Sexually mature salmon swim upstream where they spawn.
2. A combination of the two methods was utilized. The specific statement was an observation (inductive) of the death of bacterial cells associated with the fungus. From this observation, certain (deductive) propositions were drawn—specifically, that *Penicillium* was an antibacterial agent.

CHAPTER 3

1. Phosphorus will contain 5 electrons in its outer shell because its first and second shells can hold 2 and 8 electrons respectively. It has 15 protons. (Atomic number is equal to the number of protons.)
2. Atom X is inert since its outermost shell is full with 8 electrons, while atom Y will attempt to fill its outermost shell by reacting with other atoms.
3. Equation 1 is valid. Equation 2 is not balanced and therefore is not valid; there should be two hydrogen atoms on both sides of the equation.

CHAPTER 4

1. Because mitochondria are the cell's powerhouses, and congregate in places where work is going on, we can conclude that this is an active cell requiring great amounts of energy.
2. The passive transport process at work here is diffusion. Initially there will be a greater concentration of scent molecules near the incense stick. The molecules will subsequently move away to areas of lower concentration and fill the room.
3. Water moves across semipermeable membranes from regions of high to low water concentration. Since solution B is 50% solutes, it is only 50% water. Solution A with 25% solutes has the highest water concentration (75%); therefore water will move from A to B.

CHAPTER 5

1. In the process of respiration, glucose is broken down to yield energy; whereas in photosynthesis, glucose is an end product.
2. No. The light-independent reactions are so called because they do not require light energy to proceed. They are powered by the high energy molecules produced by the light-requiring reactions.
3. Though glycolysis produces 4 ATPs, it uses two ATPs to break down glucose. Of the possible 34–36 ATPs made in cellular respiration, glycolysis contributes a net yield of only two.

CHAPTER 6

1. DNA—Adenine, cytosine, thymine
   RNA—Adenine, cytosine, uracil
2. Mitosis—daughter cells will have 36 chromosomes
   Meiosis—daughter cells will have 18 chromosomes

CHAPTER 7

1. a. Crossing a homozygous dominant individual **(RR)** with a recessive individual **(rr)** would produce an F_1 generation with one genotype **(Rr)**.
   b. Since the F_1 genotype is **Rr** and since tongue rolling is controlled by the dominant gene **(R)**, all F_1 offspring (100%) will have the ability to roll their tongues.
2. The probability of crossover is proportional to the distance between two genes. Therefore crossovers are more likely to occur between **C** and **D** since they are separated by the greatest distance.

CHAPTER 8

1. By subjecting bacterial populations

to antibiotics, the resistant plasmid-containing bacteria (those with the genes of interest) are culled from the rest of the population. If the plasmids lost their resistance to antibiotics, this method of sorting through recombinants could not be utilized since they too would succumb to the antibiotic treatment.
2. Though recombinant DNA research is governed by firm rules and though some genetic variants are deliberately weakened, a major concern would be a repeat of the smallpox incident—the release of deadly variants from the research labs into the nearby population.
3. It would be a poor choice since a major problem would be simply identifying which genes are responsible for the disease. Once identified, problems would result from the numbers of genes involved since gene replacement becomes increasingly difficult with increased numbers of genes.

CHAPTER 9

1. Since Stanley Miller's experiment used the gases thought to be present in the early earth's atmosphere (methane, ammonia, water vapor, and hydrogen), it is unlikely that today's atmospheric gases—consisting mostly of nitrogen, oxygen, and carbon dioxide—would produce the same results.
2. Early autotrophs, in the process of photosynthesizing, released oxygen which accumulated in the atmosphere and acted as a deadly gas for some early heterotrophs.

CHAPTER 10

1. Since this population exhibits more than two phenotypes it is polymorphic.
2. They would produce a bell-shaped curve.
3. This population is typical of those produced by stable environments in which a population is clustered around an average condition (brown grasshoppers) with divergent forms (green, purple headed, and purple striped grasshoppers) comprising a smaller percentage of the total population.

CHAPTER 11

1. This organism should be placed in Kingdom Protista. It is excluded from Kingdom Monera because it is eukaryotic and it is excluded from Kingdom Fungi because it is photosynthetic.
2. Ecologically important activities of bacteria include decomposition, recycling minerals and nutrients, and nitrogen fixation.
3. Instruments are autoclaved to kill bacterial endospores that resist boiling.

CHAPTER 12

1. Evolutionary advances in flowers include fewer parts, fusion of parts, and bilateral symmetry. Since the magnolia flower exhibits none of these traits, it is considered an early rather than an evolutionarily advanced flower.
2. They have flattened leaf-like structures that absorb water. These curl up during dry conditions, reducing water loss. Bryophytes also grow in groups that enable them to trap moisture.

CHAPTER 13

1. Since specialized segments require a more complex nervous system to coordinate their activities, you could correctly assume that this animal has a well-developed nervous system.
2. Phylum Chordata is the only phylum whose members exhibit pharyngeal slits.
3. Adults of both animals are bilaterally symmetrical and exhibit cephalization. They both contain true organs and a well-developed muscular system.

CHAPTER 14

1. Lateral growth is inhibited by hormones from the terminal bud. Removal of the terminal bud therefore will increase lateral growth, which will produce a fuller, more bushy specimen.
2. No, this is pollination. Fertilization occurs when the pollen tube penetrates the ovule, releasing the sperm that will fertilize the egg cell.
3. Cytokinins increase growth rates of germinating seedlings.

CHAPTER 15

1. Sperm are contained in a basic or alkaline fluid and are inactivated in acidic environments. The acidic spermicide therefore would be most effective.
2. Because the fallopian plug is a type of sterilization that is essentially irreversible, it is a poor choice of contraceptive for a young woman who may some day wish to become pregnant.

CHAPTER 16

1. Alcohol consumption during pregnancy is unwise partly because it constricts the umbilical artery and therefore restricts the flow of nutrients and oxygen to the fetus.
2. Calcium intake is especially important in the third trimester because it is needed to construct the fetal skeleton.

CHAPTER 17

1. Because the joints of the skull are immovable, they are the least likely candidate for arthritis.
2. A spicule skeleton could not adequately support a large land animal. Spicules are better suited to aquatic animals, who require a less rigid skeleton because they are supported in part by the water.
3. The femur is the upper leg bone; the cast will be placed on the leg.

CHAPTER 18

1. On a chilly morning, a snake (an ectotherm) would attempt to raise its body temperature by seeking a sunny, warm location.
2. Freshwater fish do not possess the water conservation methods—such as a reduced glomerulus or osmotic pressure-raising mechanisms—of saltwater fish. Death would result, therefore, from excess water loss from the fish to the hypertonic seawater through osmosis, and through the fish's larger glomeruli that is adapted to freshwater conditions.

CHAPTER 19

1. Unsuitable respiratory systems for large individuals include diffusion (absorbing oxygen through body walls) and a tracheal system, both of which can deliver oxygen only a short distance.
2. Because the atria do not require much pressure to pump blood into the adjacent ventricles, their walls are thinner than those in the ventricles whose contractions produce much pressure to pump blood to the far reaches of the body.
3. The respiratory surfaces of all animals are moist so that oxygen may dissolve. When the skin of the earthworm dries out, oxygen cannot dissolve and therefore cannot enter the body and be delivered to the tissues, so the earthworm suffocates.

CHAPTER 20

1. The antigens injected in the vaccination would elicit the primary immune response and an increase of antibodies. Since IgM is an antibody, its numbers are expected to increase.
2. This is an autoimmune response. The perceived antigen is the nerve fibers' myelin sheath.

CHAPTER 21

1. Because some neurons of the vertebrate nervous system are myelinated, and impulse conduction is faster in myelinated cells, you would expect conduction to be faster in vertebrate nervous systems.
2. Since the neurotransmitter acetylcholine is blocked (by curare) from neighboring dendrites, impulses from one neuron are unable to stimulate adjacent neurons, and nerve transmission cannot proceed.

CHAPTER 22

1. Because the activity of reticular neural firing is responsible for our state of wakefulness, a destroyed reticular system would be unable to activate brain centers, resulting in a permanent sleep, or coma.
2. The neural routing mechanism is the reflex arc.
3. The sympathetic nervous system is operating in this "fight-or-flight" situation.

CHAPTER 23

1. Proprioceptors located in the inner ear are responsible for the gymnast's equilibrium.
2. The difference in temperature sensitivity between the mouth and the hand is a result of uneven distribution of thermoreceptors. Thermoreceptors in the hand are less concentrated than those in the mouth, hence the hand may tolerate temperatures that the mouth cannot.

CHAPTER 24

1. The behavioral mechanism at work here is habituation.
2. No. Reciprocal altruism is expected to occur in highly social, relatively intelligent animals. The members of this species meet none of these criteria.

CHAPTER 25

1. Alpine tundra is a biome found in high elevations.
2. Because phosphates join with other human wastes to speed up the eutrophication process.

CHAPTER 26

1. Because alligators invest no energy in the care of their young, we would expect that they produce more offspring than elephants, who provide extended care.
2. No. Populations that are approaching the carrying capacity produce a characteristic S-shaped curve.

CHAPTER 27

1. In 11 years, the 0–4 age group will enter the reproducing levels of the pyramid. Mexico has the greatest number of people in this category and hence should have the greatest growth rate.
2. Doubling time is longest for populations whose crude death rate is greater than its crude birth rate. For example, the crude death rate of the United States is over half that of its crude birth rate—its doubling time is longer than that of Mexico, whose crude death rate is only one-fourth its crude birth rate.

CHAPTER 28

1. Wheat crops. Because it is more economical and efficient to eat from low on the food pyramid.
2. Photovoltaics, wind energy, and atomic energy.

CHAPTER 29

1. Because one animal eats another in the food chain, the effects of DDT are magnified as each level is added to the next, with animals high on the food chain (humans) receiving the highest accumulation of this toxic chemical.
2. There is no one answer—but any approach must involve education.

APPENDIX C

For Review Answers

CHAPTER 1

1. F 2. F 3. F
4. The Rev. Thomas Malthus
5. Natural selection
6. He devised a classification system of all living things.
7. Darwin's theory met with mixed reviews but for the most part, it conflicted with the views of a society who believed that all forms of life had arisen through special creation.
8. D 9. B 10. C

CHAPTER 2

1. F 2. T 3. T 4. T
5. variable
6. Living things exhibit cellular structure and movement; they metabolize and grow; reproduce; respond; evolve; and adapt.
7. Scientific knowledge is expanded by developing hypotheses and theories, by controlled experimentation, and by simple observation. Simple observation produces the least reliable results.
8. B 9. D 10. A

CHAPTER 3

1. F 2. F 3. T 4. F 5. F
6. electrons
7. ions
8. Lipids
9. enzymes
10. Hydrogen bonds are weak and short-lived. Although heating ruptures the bonds between water molecules, they quickly reform and thereby resist boiling. Hydrogen bonds in water resist freezing by continually shifting and breaking, preventing the formation of a crystalline structure.
11. C 12. D 13. A 14. B 15. A

CHAPTER 4

1. T 2. F 3. T 4. F 5. T
6. Hydrophilic heads of phospholipids project outward while their hydrophobic tails project inward. Proteins are embedded in this layer, some of which completely traverse the cell membrane.
7. microtrabecular lattice
8. microtubules
9. nucleus
10. endocytosis
11. A 12. B 13. D 14. C 15. C

CHAPTER 5

1. F 2. T 3. F 4. F 5. T
6. energy
7. entropy
8. water, photosystem II
9. phosphorylation
10. glycolysis
11. C 12. C 13. D 14. B 15. C

CHAPTER 6

1. F 2. F 3. T 4. T
5. prophase, metaphase, anaphase and telophase
6. homologues
7. Variations are produced by crossovers, random selection of polar bodies in egg formation and by the random line-up of chromosomes at metaphase that mixes paternal and maternal chromosomes.
8. hydrogen bonds
9. A 10. A 11. B 12. C

CHAPTER 7

1. F 2. T 3. F 4. T 5. F
6. segregation
7. Polygenic inheritance
8. chromosome mapping
9. gene or point
10. Hardy-Weinberg
11. D 12. C 13. D 14. C 15. A

CHAPTER 8

1. F 2. F 3. T 4. F
5. plasmid
6. Gene replacement therapy
7. Vectors
8. clone
9. C 10. D 11. B 12. B
13. D 14. A 15. B

CHAPTER 9

1. T 2. F 3. F 4. T
5. Stanley Miller's experiment formed amino acids. Amino acids are the building blocks of proteins, significant molecules that are associated with life.
6. proteinoids
7. coacervate
8. B 9. D 10. D 11. C

CHAPTER 10

1. F 2. T 3. T 4. T 5. T
6. Polyploidy, hybridization
7. Dimorphic
8. genetic isolation
9. stabilizing selection

10. bottleneck effect
11. C 12. D 13. B 14. C 15. B

CHAPTER 11

1. T 2. F 3. T 4. T 5. T
6. archaebacteria
7. oomycetes
8. conjugation
9. mychorrhizae
10. Ascomycota
11. D 12. B 13. D 14. D
15. A

CHAPTER 12

1. F 2. F 3. F 4. T 5. T
6. Phaeophyta
7. They both store carbohydrates, have cell walls composed mainly of cellulose and contain chlorophyll *a* and *b*.
8. They are the flowering plants.
9. Gymnospermae
10. fronds
11. D 12. B 13. B 14. C 15. C

CHAPTER 13

1. T 2. F 3. T 4. T 5. F
6. Cephalization
7. Mollusca
8. coelom
9. At some time in their life they all possess a notochord, a dorsal hollow nerve chord, and pharyngeal slits.
10. hermaphroditic
11. C 12. D 13. A 14. C 15. C

CHAPTER 14

1. F 2. T 3. T 4. F 5. F
6. stamen; carpel (or pistil)
7. Apical meristems
8. xylem; phloem
9. auxin
10. whorls
11. A 12. D 13. C 14. C 15. A

CHAPTER 15

1. T 2. T 3. T 4. F 5. T
6. binary fission; budding
7. endometrium
8. Large numbers of sperm are necessary since so many die en route: some die of natural causes, some are devoured by female white blood cells and some are destroyed by the acidic chemical environment of the vaginal tract.
9. rhythm method
10. Dilation and curettage, vacuum curettage and salting out.
11. A 12. B 13. C 14. D 15. C

CHAPTER 16

1. F 2. F 3. T 4. T 5. T
6. allantois
7. vegetal
8. gastrulation
9. placenta
10. Somites
11. A 12. C 13. C 14. B 15. D

CHAPTER 17

1. T 2. F 3. F 4. T
5. endoskeletons
6. connective
7. joints
8. antagonist
9. actin; myosin
10. B
11. A

CHAPTER 18

1. T 2. F 3. T 4. T
5. homeostasis
6. negative
7. endotherms
8. Components of carbon dioxide are added to ammonia to form the less toxic urea that is then passed out of the body as urine.
9. C 10. A 11. D 12. C

CHAPTER 19

1. T 2. T 3. T 4. T 5. F
6. tracheal system
7. Venules, veins
8. Peristalsis
9. Villi
10. D 11. B 12. C 13. B 14. A

CHAPTER 20

1. T 2. F 3. T 4. F 5. F
6. histamine
7. Antibodies
8. Interleukin II
9. autoimmune
10. interferon
11. D 12. B 13. D 14. A

CHAPTER 21

1. F 2. T 3. T 4. T 5. F
6. endocrine
7. Cortisone
8. acromegaly
9. Saltatory propagation
10. Sodium/potassium ion exchange pump
11. B 12. C

CHAPTER 22

1. T 2. T 3. F 4. T 5. F
6. brain, spinal cord
7. cerebrum
8. corpus callosum
9. reflex arc
10. Cocaine stimulates neural activity by inhibiting the breakdown of neurotransmitters.
11. B 12. A 13. D 14. A 15. C

CHAPTER 23

1. T 2. T 3. F 4. F
5. Thermoreceptors
6. tactile, pressure
7. proprioceptors
8. D 9. A 10. B

CHAPTER 24

1. T 2. T 3. F 4. T 5. T 6. T
7. Ethology
8. They are innate behavioral patterns that are species specific.
9. pheromones
10. Releasers
11. D 12. D 13. D 14. C 15. A
16. D 17. D

CHAPTER 25

1. T 2. F 3. F 4. T
5. microhabitat
6. tradewinds
7. Phytoplankton
8. Eutrophication
9. C 10. D 11. A

CHAPTER 26

1. F 2. F 3. T 4. F
5. Carrying capacity
6. The chimp produces only a few offspring and provides extended care of that offspring.
7. Parasites
8. hierarchies
9. B 10. A 11. D 12. C

CHAPTER 27

1. T 2. F 3. F 4. F

5. Demographers
6. crude birth rates, crude death rates
7. Age structure pyramids
8. A 9. B 10. A

CHAPTER 28

1. T 2. T 3. T 4. F
5. renewable
6. Hydroelectric
7. B 8. D 9. D 10. B

CHAPTER 29

1. T 2. F 3. F
4. undersea oil seepage, volcanoes and radiation
5. threshold effect
6. nitrogen oxides
7. particulate matter
8. D 9. A 10. D

ANSWERS TO GENETICS PROBLEMS

1. A. P_1: Both purple-flowered and white-flowered parents are homozygous.
   F_1: All individuals are heterozygous
   F_2: About 2/3 of the purple-flowered plants are heterozygous
   About 1/3 of the purple-flowered plants are homozygous.
   All white-flowered plants are homozygous.

   B. P_1: Yellow-seeded plants produce all **Y** gametes.
   Green-seeded plants produce all **y** gametes.
   F_1: All plants produce **Y** gametes and **y** gametes in equal numbers.
   F_2: About 2/3 of the yellow-seeded plants produce **Y** gametes and **y** gametes in equal numbers.
   About 1/3 of the yellow-seeded plants produce **Y** gametes only.
   All the green-seeded plants produce **y** gametes only.

2. A. All gametes are **aG.**
   B. 50% **AG,** 50% **aG**
   C. 25% **AG,** 25% **ag,** 25% **Ag,** 25% **ag**
   D. 50% **Ag,** 50% **ag**
3. A. Heterozygous yellow parent is **Yy.**
   Orange parent is **yy.**
   B. **Yy** produces **Y** gametes and **y** gametes in equal numbers.
   yy produces all **y** gametes.
   C. 50% yellow fruit, genotype **Yy.**
   50% orange fruit, genotype **yy.**
4. A. Red pigeon is **Rr;** brown pigeon is **rr.**
   B. **Rr** produces **R** gametes and **r** gametes in equal numbers.
   rr produces all **r** gametes
   C. 50%
5. The red-fruited parent is heterozygous, **Rr,** because it produces roughly equal numbers of the dominant and recessive traits when crossed with the recessive homozygous parent.

Glossary

PRONUNCIATION KEY

a	act, bat		g	give		o͞o	ooze, fool
ā	cape, way		i	if, big		ô	ought
â	dare, Mary		ī	ice, bite		s	see, miss
ä	alms, calm		j	just, Jerry		u	up, love
e	set, merry		o	ox, box		û	urge, burn
ē	equal, bee		ō	over, no		y	yes, lawyer
ə	like *a* in alone or *e* in system		ŏŏ	book, poor		z	zeal, lazy

A

A1 cell: A sound receptor in nocturnal moths that responds only to weak pulses.
A2 cell: A sound receptor in nocturnal moths that responds only to loud pulses.
Abdomen: 1. In mammals, the body cavity between the diaphragm and the pelvis. 2. In other vertebrates, the body cavity containing the stomach, intestines, liver, and reproductive organs. 3. In arthropods, the posterior section of the body.
Abductor: Any muscle that moves or draws away from the axis of the body or from one of its parts. Compare *adductor.*
Abiotic (ā′bī ot′ik)**:** Characterized by the absence of life.
Abortion: The expulsion or removal of an embryo or fetus before it can survive on its own.
Acetate: A salt or ester of acetic acid, that is, the acid with the hydrogen removed from the acid group.
Acetic acid (ə sē′tik as′id)**:** $C_2H_4O_2$, the ionized form of which joins with coenzyme A to form acetyl-CoA before it enters the Krebs cycle.
Acetylcholine (ə sē′tel kō′lēn)**:** A chemical agent that transmits neural impulses across synapses from one neuron to another.
Acetylcholinesterase (ə sē′tel kō lēn es′te rās)**:** A membrane-bound enzyme that hydrolyzes acetylcholine in the course of the synaptic nerve impulse transmission.
Acetyl-CoA (ə sē′tel kō ā)**:** A key intermediate in metabolism, consisting of an acetyl group covalently bonded to coenzyme A.
Acid: A substance that releases hydrogen ions into a solution. It has a sour taste and unites with bases to form salts.
Acquired immune deficiency syndrome (AIDS): A viral disease that results in severely reduced immunity.
Acromegaly (ak′rə meg′ə lē)**:** A chronic disease characterized by enlargement of the bones of the head, the soft parts of the feet and hands, and sometimes other structures, due to dysfunction of the pituitary gland.
Actin: A cytoplasmic protein and a constituent of muscle, known in both globular and fibrous forms. See also *myofilaments.*
Actinomycete (ak′ tən ō mī sēt′)**:** Any of several rod-shaped or filamentous, aerobic or anaerobic bacteria, certain species of which are pathogenic for humans and other animals.
Action potential: In a neuron, a traveling, depolarizing wave. A short-lived change in membrane potential that produces a neural impulse, or, in a muscle, contraction. Compare *resting potential.*
Active site: That part of an enzyme directly involved in specific enzymatic activity.
Active transport: The transport of a substance across the cell membrane, usually against the concentration gradient, that requires energy.
Addiction: The state of being given up or devoted to a habit or practice or to something that is habit-forming to such an extent that stopping may cause trauma.
Adductor: Any muscle that moves or draws toward the axis of the body or one of its parts. Compare *abductor.*
Adenine (ad′ nēn)**:** A purine, one of the nitrogenous bases found in both DNA and RNA, as well as in ATP and several coenzymes.
Adenosine diphosphate (ə den′ə sēn′dī′fos′fāt)**:** ADP, a product formed

by the hydrolysis of ATP, adenosine triphosphate, with the release of energy.
Adenosine triphosphate (ə den'ə sēn'trī'fos'fāt): ATP, a compound found in all cells, that serves as an energy source as it is broken down into ADP plus phosphate with the release of energy.
Adenyl cyclase (ad'ə nil sī'clamās): An enzyme, usually incorporated into the cell membrane, that is capable of transforming ATP into cyclic AMP and pyrophosphate.
ADP: Adenosine diphosphate, a compound of adenine, ribose, and two phosphate groups.
Adrenal gland (e drēn'əl gland): A vertebrate endocrine gland. The outer area produces steroid hormones, the inner produces epinephrine.
Adrenaline (e dren'əl in): See *epinephrine.*
Afferent neuron: A nerve cell that carries impulses toward the brain.
Afterbirth: The placenta with its associated membranes, which is expelled from the uterus after childbirth.
Age structure pyramid: Also called *age profile* or *population pyramid.* A pyramidal graph of a population, divided into age groups. Each age group is represented by a horizontal bar with that of the youngest forming the base.
Aggression: Hostility, attack, or threat, especially unprovoked, usually against a competitor or potential competitor.
Agnatha (ag'nə thə): The class of vertebrates comprising the lampreys, hagfishes, and several extinct forms, having no jaws or paired appendages.
AIDS: See *Acquired immune deficiency syndrome.*
Albumen: The white of an egg.
Alcoholic fermentation: The anaerobic breakdown of glucose by yeast to form alcohol and carbon dioxide.
Aldosterone (al'dō sti rōn'): A steroid hormone produced by the adrenal cortex, involved in potassium reabsorption by the kidney.
Alga (pl. *algae*): Any photosynthetic member of the kingdom *Protista.*
Alimentary canal: A tubular passage functioning in the digestion and absorption of food in humans and most animals, beginning at the mouth and terminating at the anus.
Allantois (ə lan'tō is): One of the extraembryonic membranes. In birds and reptiles, it serves as a repository for the embryo's nitrogenous wastes.
Allele (e lēl'): One of two or more genes that produce a specific characteristic such as blood type, hair color, and so on.
Allopatric speciation (al'ə pat'rək spē'she ā'shən): The formation of new species from populations that are geographically separated. Compare *sympatric speciation.*
Alpha linkage: The type of bonding between glucose molecules of starch.
Alpine tundra: A biome found at elevations above the tree line, characterized by dense growth of mosses, lichens, dwarf herbs, and shrubs and covered throughout most of the year by ice and snow.
Alternation of generations: The existence in the life cycle of an individual of a haploid (1N) gametophyte stage that alternates with a diploid (2N) sporophyte stage.
Altruism: Behavior that is directly beneficial to others at some cost or risk to the altruistic individual.
Alvarez hypothesis: A hypothesis that proposes that much of the massive extinction of life that accompanied the end of the Mesozoic era was produced by the aftereffects of a gigantic asteroid's collision with the earth.
Alveolus (al vē' ə les) (pl. *alveoli*): One of the tiny air sacs that occur in grapelike clusters in the vertebrate lung, in which carbon dioxide and oxygen are exchanged.
Amino acid (ə mē'nō as'id): An organic molecule that contains at least one carboxyl group, COOH, and one amino group, NH_2.
Amino group (ə mē'nō group): The univalent group NH_2, often ionized as NH_3+, which can form amines.
Ammonia: 1. A colorless, pungent, poisonous, highly soluble gas, NH3. 2. Dissolved in water, ammonium hydroxide.
Amnion (am'nē ən): The innermost extraembryonic membrane of reptiles, birds, and animals.
Amoeba (ə mē'bə): 1. Any protozoan of the large genus *Amoeba,* characterized by lobose pseupods and the lack of permanent organelles or supporting structures. 2. Any ameboid protist such as the ameboid stage of a flagellate or sporozoan.
Amoebocyte (ə mē'bə sīt): In many individuals, an ameboid cell that functions in reproduction, digestion, and so on.
Amphetamine (am'fet'ə mēn): A drug that stimulates the central nervous system. Also called *speed* or *uppers.*
Amphibia (am fib'ē ə): The class of cold-blooded vertebrates (frogs, salamanders, etc.) the larva of which are typically aquatic, breathing by gills, and the adult of which are terrestrial, breathing by lungs and through moist, glandular skin.
Amphioxus (am'fē ok'səs): A lancelet of the genus *Brachiostoma,* having such vertebrate characteristics as a notochord and a dorsal cord of nerve tissue.
Anaphase (an'ə fāz): The third phase of mitosis in which the chromatids separate and move to opposite sides of the cell.
Angina pectoris (an jī'nə pek'tə ris): A syndrome characterized by constricting pain below the sternum, most easily brought about by exertion or excitement and caused by insufficient blood flow to the heart muscle, usually due to coronary artery disease, such as arteriosclerosis.
Angiosperm (an'jē ə spėrm): A plant in which the seeds are enclosed in an ovary; a flowering plant.
Animal pole: That end of an animal zygote that has relatively little yolk and experiences more rapid cell division.
Annelid (an'ə lid): Any segmented worm of the phylum *Annelida,* including earthworms, leeches, and various marine forms.
Annual plant: A plant that completes its life cycle in a single season.
Antagonist muscles: One of a pair of skeletal muscles (or groups of muscles) the actions of which oppose one another.
Anterior lobe: The lobe of the pituitary gland that lies toward the front of the brain.
Anther: In a flower, the pollen-producing organ of the stamen.
Antheridium (an'thə rid'ē em): In plants, a male reproductive organ that produces and stores motile sperm.
Anthozoa (an'thə zō'ə): The class of marine coelenterates that contains the corals sea anemones, sea pens, and so on.
Antibiotic: Any of a large number of substances, produced by various microorganisms and fungi, capable of inhibiting or killing bacteria and usually not harmful to higher organisms (e.g., *penicillin, streptomycin,* etc.).
Antibody: A protein molecule of the immune system that can recognize and bind to a foreign substance or invader (such as a bacterium or virus). See also *immunoglobulin.*
Anticodon (an'ti kō'don): A region of a tRNA molecule consisting of three se-

quential nucleotides that will have a matching codon in *m*RNA.
Antidiuretic hormone (ADH) (an′tē dī′yoo ret′ik hormone): A polypeptide hormone secreted by the posterior pituitary, the action of which is to increase the reabsorption of fluid from the kidney filtrate. Also called *vasopressin*.
Antigen: A large molecule, such as a cell-surface protein or carbohydrate, that stimulates the production of specific antibodies, or that binds specifically with such antibodies.
Antigen recognition region: The separated part of a Y-shaped antibody that matches specific antigens.
Anus (ā′nes): The posterior opening of the digestive tract (gut) through which digestive wastes are eliminated.
Aorta (ā ôr′te): In vertebrates, the principal or largest artery. It carries oxygenated blood from the heart to the body.
Apgar test series: A series of tests given infants at birth to help discover any congenital problems.
Apical meristem (ā′pə kel mər′ə stem): The undifferentiated tissue at the stem and root tip that contributes cells for primary growth.
Apodeme (ap′ə dēm): An ingrowth of the arthropod exoskeleton that serves as the point of attachment for a muscle.
Appendicular skeleton (a′pən dik′yə ler skeleton): In vertebrates, bones of the pectoral and pelvic girdles and of the appendages. Compare *axial skeleton*.
Appetitive behavior (ap′i tī′tiv behavior): A variable, nonstereotyped part of instinctive behavior involving searching (for food, water, a mate) for the opportunity to perform.
Arachnida (ə rak′ni də): The class of arthropods that contain spiders, scorpions, mites, ticks, and others.
Archaebacteria (är′kə bak tir′ē ə): One of the two prokaryote kingdoms (or phyla in some schemes); it contains primitive bacteria such as methanogens (methane producers), thermophiles (heat lovers), and acidophiles (acid lovers). See also *Eubacteria*.
Archegonium (är kə gō′nē əm): The female (egg-producing) reproductive structure in the gametophytes of ferns and bryophytes.
Archenteron (är ken′tə ron′): In embryology, the primitive digestive cavity of the gastrula.
Arctic tundra: A biome characterized by level or gently undulating treeless plains of the far north, supporting dense growth of mosses and lichens, as well as dwarf herbs and shrubs. It is underlain by permafrost and seasonally covered by snow.
Arteriole (är tir′ē ōl): A small artery, usually giving rise directly to capillaries.
Artery: A vessel carrying blood away from the heart, toward a capillary bed.
Arthropoda (är throp′ə də): The phylum containing segmented invertebrates with jointed legs, including insects, arachnids, crustaceans, and myriapods.
Artificial selection: The deliberate selection for breeding by humans of domesticated animals or plants on the basis of desired characteristics.
Artiodactyla (är′tē ō dak′ tə lə): "Even toes"; the order comprised of hoofed, even-toed mammals such as pigs, hippopotamuses, camels, deer, giraffes, sheep, goats, antelope, and cattle.
Ascending colon: The first region of the large intestine.
Ascending neuron: A nerve cell that carries impulses up the spinal cord toward the brain.
Ascomycete (as′kə mī sēt′): A fungus of the class that includes the yeasts, mildews, truffles, and so on, characterized by bearing the sexual spores in a sac, the ascus.
Ascomycota (as′kə mī cot′e): See *Ascomycete*.
Ascus (as′kəs): In ascomycetes, the sac in which meiosis occurs and in which four or eight ascospores are subsequently formed.
Association neuron: A nerve cell that connects an afferent neuron to an efferent neuron or to an ascending neuron.
Asteroidea (as′tə roi′dēə): The class of echinoderms that contains starfishes.
Atherosclerosis (ath′ər ō sklə rō′sis): A form of arteriosclerosis characterized by fatty deposits (plaques) in the inner lining of the arterial walls.
Atom: The smallest indivisible unit of an element still retaining the element's characteristics.
Atomic number: The number assigned to a particular element, determined by the number of protons in its atoms.
ATP (Adenosine triphosphate): A ubiquitous small molecule involved in many biological energy exchange reactions, consisting of the nitrogenous base adenine, the sugar ribose, and three phosphate residues.
Atrium (ā′trē əm): The smaller compartment of the heart that receives venous blood from the body or lungs.
Audition: The act, sense, or power of hearing.
Auditory canal: The open, bony canal from the outer ear to the eardrum.
Autoimmunity: The condition existing when the organism's immune system attacks and destroys one or more of the organism's own tissues.
Autosome (ô′tə sōm): Any chromosome other than a sex (X or Y) chromosome.
Autotroph (ô′tə trōf): Self-feeder, able to manufacture food such as through photosynthesis.
Auxin (ôk′sin): A class of natural or artificial substances that acts as the principle growth hormone in plants.
Aves (ā′vēz): The class containing the birds.
Axial skeleton: In vertebrates, the skull, vertebral column, and bones of the chest. Compare *appendicular skeleton*.
Axon: The extension of a neuron that conducts nerve impulses away from the cell body.

B

Bacillus (bə sil′əs): 1. An aerobic, rod-shaped, spore-producing bacterium of the genus *Bacillus*. 2. Any rod-shaped bacterium.
Backbone: The spinal or vertebral column.
Background radiation: Naturally occuring radiation that is present almost everywhere.
Ball-and-socket joint: A joint allowing maximal rotation and flexion, consisting of a ball-like termination on one part, held within a concave, spherical socket on the other (e.g., the hip joint).
Barbiturate (bär bich′ə rāt): Any of a group of barbituric acid derivatives used in medicine as sedatives and hypnotics; "downers."
Barr body: A dark-staining feature in the nuclei of the cells of female mammals, representing the condensed X chromosome.
Base: Any chemical that releases hydroxyl (OH^-) ions in water or joins with free protons. Bases leave a bitter taste and form salts when they react with acids.
Basidiomycete (bə sid′ē ō mī sēt′): A fungus of the class that includes the smuts, rust, mushrooms, puffballs, and so on, characterized by bearing the spores on a basidium.
Basidiomycota (bə sid′ē ō mī cot′ə): See *Basidiomycete*.

Basidiospore (bə sid'ē ō spōr): An aerial spore produced by meiosis in Basidiomycetes.
Basidium (bə sid'ē əm): The meiotic cell of Basidiomycetes such as mushrooms; it produces basidiospores by budding.
Basilar membrane (bas'ə lər membrane): In the vertebrate ear, a membrane that conducts sound waves.
Basophil (bā' so fil): The rarest kind of leukocyte; granular.
Beta linkage (bā'tə linkage): The linkage between the glucose molecules of cellulose.
Biceps: 1. A muscle on the front of the arm that acts to bend the elbow. 2. The hamstring muscle on the back of the thigh that assists in bending the knee and extending the hip joint.
Biceps brachii (biceps brā'kē ī): The large muscle lying along the front (ventral) side of the humerus.
Bile: A bitter-tasting, highly pigmented, alkaline liquid secreted by the liver, containing bile salts and bile pigments, that functions in fat digestion.
Bimodal curve (bī mōd'el curve): A "two-humped" graphic distribution.
Binary fission (bī'ne rē fission): A form of asexual reproduction; fission (splitting) into two organisms of approximately the same size; cell division in prokaryotes.
Binucleate cell (bī nyoo'klē āt' cell): A cell having two nuclei.
Bioethics (bī'ō eth'iks): The philosophical area that deals with the moral implications of decisions that affect life overall.
Biological determination: The influence of genetics, particularly on the social behavior of an organism.
Biomass (bī'ō mas'): The total weight of all the organisms in a prescribed area.
Biome (bī'ōm): A geographical area that is characterized by relatively similar plant and animal representatives; e.g., a tundra.
Biotic (bī ot'ik): Pertaining to life.
Biotic population control: Population control by living factors, including both intraspecific and interspecific influences.
Biotic potential: The maximum growth rate of a population when it is unrestricted by environmental resistance.
Birth control pill: An oral steroid contraceptive that inhibits ovulation, fertilization, or implantation causing temporary infertility in women.
Blastocoel (blas'tō sēl'): The cavity of a blastula, arising in the course of cleavage.
Blastopore (blas'te pōr): In a gastrula, the opening of the archenteron produced by the involution of cells during gastrulation.
Blastula (blas'che le): An early embryonic stage in mammals consisting of a single layer of cells that forms a hollow ball enclosing a central cavity, the blastocoel.
Bombykol (bom'bi kol): The sex attractant produced by the female silkworm moth.
Bone: The hard connective tissue forming the skeleton of most vertebrates.
Bottleneck effect: In population genetics, the result of a population becoming small, resulting in a random change in gene frequencies.
Bowman's capsule: A curved sac at the beginning of the kidney nephric unit that surrounds the glomerulus.
Brackish: Slightly salty.
Bronchiole (brän'kē ōl'): A small airway in the lung that is a branch of a bronchus and part of the respiratory tree.
Bronchus (brän'kəs) (pl. *bronchi*): Either of the two main branches of the trachea.
Bryophyta (brī'ō fīt'e): The plant group that includes the mosses, liverworts, and hornworts.
Budding: Asexual reproduction seen in yeasts and other organisms where smaller cells bud or grow from a parent cell.
Buffer: A solution of chemical compounds capable of neutralizing both acids and bases, thus resisting changes in pH.
Bulk flow: The movement of water or another liquid brought about by pressure or gravity, generally from areas of greater water potential to areas of lesser water potential.

C

Caffeine: A white, crystalline, bitter alkaloid, usually derived from coffee or tea, which acts as a stimulant and diuretic.
Caffeinism: A condition found in people who consume very large amounts of coffee, characterized by insomnia, high blood pressure, increased body temperature, racing heart, and chills.
Calcitonin (kal si'tō nin): One of the two major hormones of the thyroid gland whose major action is to inhibit the release of calcium from the bone.
Calvin cycle: A pathway of carbon dioxide that occurs during the dark phase of photosynthesis as carbohydrate is produced.
Calyx (kā'liks) (pl. *calyces*): The outermost whorl of floral parts (the sepals), usually green and leaflike.
Cambium (kam'bē əm) (pl. *cambia*): Undifferentiated meristematic tissue in a plant, including *cork cambium, procambium,* and *vascular cambium.*
Canaliculus (kan'ə lik'yə ləs) (pl. *canaliculi*): A small canal or tubular passage, as in bone.
Capillary (kap'ə ler'ē): A small blood vessel (with walls one cell thick) in which exchanges between the blood and tissues occur. Capillaries are located between arterioles and venules.
Carbohydrate (kär'bō hī'drāt): An organic substance containing carbon, hydrogen, and oxygen in the ratio of CH_2O. Includes sugars, starches, cellulose, and so on.
Carbon monoxide: A colorless, odorless, poisonous gas, CO, that burns with a pale blue flame; produced when carbon burns in insufficient air.
Carbonic anhydrase (kär bon'ik an hī'drās): An enzyme that catalyzes the reversible conversion of carbonic acid to carbon dioxide gas and water.
Carboxyl group (kär bok'sil group): The univalent group, =COOH, present in and characteristic of organic acids.
Cardiac muscle: Specialized involuntary muscle of the heart whose fibers are striated and branching.
Carnivora (kär niv'ərə): "Meat eaters"; the order comprising chiefly flesh-eating mammals, such as dogs, cats, bears, seals, weasels, and so on.
Carnivore (kär'nə vōr'): Any flesh-eating organism.
Carotene (kar'ə tēn): A red or orange hydrocarbon, found in most plants as an accessory photosynthetic pigment.
Carotenoid (kə rot'ə noid): Any of a group of red, yellow, and orange plant pigments chemically and functionally similar to carotene.
Carpal (kär'pəl): Any wrist bone.
Carpel (kär'pəl): In flowers, a simple pistil, or a single member of a compound pistil; one sector or chamber of a compound fruit.
Carrying capacity: A property of the environment defined as the size of a population that can be maintained indefinitely.
Catalase (kat'ə lās'): An enzyme that decomposes hydrogen peroxide to molec-

ular water and oxygen, in the reaction $2H_2O_2 \rightarrow 2H_2O + O_2$.
Catalysis (kə tal'i sis): a chemical reaction initiated by catalysts.
Catalyst (kat'list): A substance, such as an enzyme, that increases the rate of a reaction without entering into the reaction.
Cecum (sē'kəm): In vertebrates, a blind pouch or diverticulum of the intestine at the juncture of the small and large intestines.
Cell body: The region of a neuron containing the cell nucleus and most of the cytoplasm and organelles.
Cell membrane: The semipermeable membrane that surrounds all cells and consists of a double layer of phospholipids, with proteins, glycoproteins, and glycolipids interspersed in a mosaic arrangement.
Cell theory: The universally accepted proposal that cells are the functional units of organization in living organisms and that all cells today come from preexisting cells.
Cell wall: The semirigid outside of certain cells, particularly plant cells, that give them a definite shape.
Cellular respiration: The energy-yielding metabolism of foods in which oxygen is used. Also see *respiration* and *glycolysis.*
Cellulose (sel'yə lōs): An insoluble carbohydrate comprised of glucose units; forms cell walls in plants.
Cenozoic (sē' nə zō'ik): Noting or pertaining to the present geological era, beginning about 75,000,000 years ago and characterized by the appearance of mammals.
Central nervous system: In vertebrates, the brain and spinal cord to which sensory impulses are transmitted and from which motor impulses are sent.
Centriole (sen'trē ōl'): A paired organelle near the nucleus of animal cells that functions in cell division.
Centromere (sen'trə mir): A structure on the chromosome to which the spindle fiber attaches when chromatids separate in meiosis and mitosis.
Cephalic ganglia (sə fal'ik gang'glē ə): Neural aggregations at the anterior ends of some invertebrates.
Cephalization (sa' fa la zā shun): The evolutionary tendency toward marked and early development of the head region.
Cephalochordate (sef'ə lə kôr'dāt): "Head cord animal"; a lancelet (Branchiostoma) of a chordate subphylum in which the notochord persists throughout life and extends through what would be the head if it had one.
Cerebellum (sər'ə bel'em): The double walnut-shaped portion of the hindbrain that coordinates voluntary movement, posture, and balance.
Cerebral cortex (sər'e brəl kor'teks): The outermost region of the cerebrum, the "gray matter," consisting of several dense layers of neural cell bodies and including numerous conscious centers, as well as regions specializing in voluntary movement and sensory reception.
Cerebrum (sə rē'brem): The anterior portion of the vertebrate brain; the largest portion in humans, consisting of two *cerebral hemispheres* and controlling many localized functions, among them voluntary movement, perception, speech, memory, and thought.
Cervical cap (sûr'vi kəl cap): A birth control device consisting of a small cap that encloses the cervix.
Cestoda (se stō'də): The class composed of parasitic flatworms, like the tapeworm.
CF_1 particle: An ultramicroscopic structure on the outer surface of the hylakoid, the site of the chemiosmotic phosphorylation in photosynthesis.
Chelicerata (chəli' sə ra'tə): Arthropods with no jaws, no antennae, four pairs of legs, and book lungs, including the spiders.
Chemical reaction: The reciprocal action of chemical agents on one another; chemical change.
Chemiosmosis (kem'ē oz mō'sis): The process in mitochondria, chloroplasts, and aerobic bacteria in which an electron transport system utilizes the energy of photosynthesis or oxidation to pump hydrogen ions across a membrane, resulting in a proton concentration gradient that can be utilized to produce ATP.
Chemiosmotic phosphorylation (kəm'ē oz mot'ik fos'fer ə lā'shən): The production of ATP using the energy of protons passing across a membrane and through F_1 and CF_1 particles.
Chemoreception (kēm' ō ri sep'shən): The ability to detect the presence of chemicals in the environment.
Chemosynthesis (kem'e sin'thə sis): The synthesis of organic compounds with energy derived from inorganic chemical reactions.
Chilopoda (kī'lə pod'ə): The class comprising the centipedes.
Chiroptera (kī rop'tərə): "Hand wings"; the order comprising the bats.
Chlorophyll (klôr'ə fil): The light-absorbing pigment of plants and some bacteria that traps light energy for photosynthesis.
Chlorophyll a: A type of chlorophyll common to all plants.
Chlorophyll b: A type of chlorophyll common in certain land plants and in green algae.
Chlorophyta (klôr'ə fī' tə): The green algae.
Chloroplast (klôr'ə plast): A chlorophyll-containing organelle that functions in photosynthesis in plant cells.
Choanocyte (kō an'ə sīt'): Also *collar cell;* a type of cell in all sponges and certain protists in which a single flagellum is surrounded at its base by a screen of fused cilia that filter food from the water current created by the flagellum.
Chondrichthyes (kon drik'thē ēz'): The class containing the cartilaginous fishes.
Chordate (kôr' dāt): Belonging or pertaining to the phylum *Chordata,* which contains the true vertebrates and those animals having a notochord, like the lancelets.
Chorioallantoic membrane (kōr'ē ō al'ən tō'ik membrane): A highly vascular extraembryonic membrane of birds, reptiles, and some mammals, formed by the fusion of the chorion and the allantois.
Chorion (kôr'ē on): The outermost extraembryonic membrane of birds, reptiles, and mammals. It contributes to the formation of the placenta in placental mammals.
Chorionic villi (kôr'ē on'ik vil' ī): Highly branched and pouched growths of the placenta, where the fetal and maternal blood are separated by a membrane's thickness.
Chromatid (krō'mə tid): One of two duplicated strands in a chromosome.
Chromatin (krō'mə tin): Indistinct, granular chromosomal material seen at interphase; DNA.
Chromosome (krō'mə sōm): Threadlike, condensed chromatin (DNA), visible at cell division only, occurring in pairs with specific numbers for each species.
Chromosome mutation: A massive spontaneous change in DNA, generally breakage involving a whole chromosome that has not been repaired or has been repaired improperly.
Chrysophyta (kris'ə fī'tə): The yellow-green and golden-brown algae ("golden plants"), named for their yellow caroten-

oid pigments although they also possess chlorophyll a and b. This group includes the diatoms.
Chyle (kīl): A milky fluid containing emulsified fat and other products of digestion, formed from the chyme in the small intestine and conveyed by the lacteals and the thoracic duct to the veins.
Chyme (kīm): The semifluid masses into which food is converted by gastric secretion and which passes from the stomach into the small intestine.
Ciliophora (si'li ə fôr'ə): The ciliated protozoans.
Cilium (sil' ē əm) (pl. *cilia*): A hairlike cellular organelle that beats rhythmically on the cell surface.
Citrate: A salt or ester of citric acid; citric acid without the hydrogen on the acid group.
Citric acid: A 6-carbon compound formed in the Krebs cycle; the nonionized form of citrate.
Citric acid cycle: A cyclic series chemical transformation in the mitochondria by which pyruvate is degraded to carbon dioxide, NAD and FAD are reduced to $NADH_2$ and $FADH_2$ and ATP is generated. See also *Krebs cycle*.
Class: A major taxonomic subdivision of a phylum, consisting of subordinate groups known as orders.
Classical conditioning: Learning in which a condition stimulus presented together with an unconditional stimulus initiates a reaction to the condition stimulus.
Classical genetics: The study of genetics using the techniques essentially suggested by Mendel.
Clavicle (klav'ə kel): One of a pair of bones of the pectoral girdle, articulating with the sternum and scapula. Also called *collar bone*.
Cleave: To split or divide, esp. along a natural line.
Climax: An orgasm.
Cline (klīn): A variation in physical characteristics across a geographic area.
Clitellum (kli tel'əm): A thickened, ringlike glandular portion of the body wall of earthworms that secretes mucus to form a cocoon for eggs.
Clitoris (klit'ər əs): An external organ of the female genitalia considered to be homologous to the penis in the male.
Cloaca (klō ā'kə): The common cavity into which the intestinal, urinary, and reproductive canals open in all vertebrates except placental and marsupial mammals.
Clone: A genetically identical organism derived from a single individual by asexual reproduction.
Closed circulatory system: A circulatory system in which the blood elements remain in blood vessels and do not leave to percolate through tissue spaces. Compare *open circulatory system*.
Clotting: The coagulation of blood.
Cnidaria (ni dâ' riä): An animal characterized by only one digestive opening, such as the jellyfish.
CoA: See *coenzyme A*..
Coacervate (kō as'ər vit): An organized kind of droplet that may have been the forerunner of primitive life.
Cocaine: A bitter, crystalline alkaloid, obtained from coca leaves and used medically as a local anesthetic.
Coccus (kok'əs): Any spherical bacterium (principally eubacteria).
Cochlea (kok'lē ə): In mammals, a spiral cavity of the inner ear containing fluid, vibrating membranes, and sound-sensitive neural receptors.
Codon (kō'don): A sequence of three nucleotide bases required to form one amino acid in a protein chain synthesized on the RNA messenger.
Coelacanth (sē'lə kanth): A fish thought to have been extinct since the Cretaceous period, but found in 1938 off the southern coast of Africa.
Coelom (sē'ləm): A principal body cavity, or one of several such cavities, between the body wall and gut, entirely lined with mesodermal epithelium. Compare *pseudocoelom*.
Coenzyme A: A biocatalyst required by certain enzymes to produce their reactions.
Coitus interruptus: Coitus that is intentionally interrupted by withdrawal before ejaculation.
Collagen (kol'ə jən): A common, tough, fibrous animal protein occurring principally in connective tissues.
Collar cell: One of certain flagellated cells in sponges that create a current and ingest food particles from the water.
Collenchyma (kə leng'kəmə): In plants, a strengthening tissue; a modified parenchyma consisting of elongated cells with greatly thickened cellulose walls.
Colloid (kol'oid): A solution in which the molecules are large and suspended rather than dissolved.
Colon: The large intestine from the cecum to the rectum; including the ascending, transverse, descending, and sigmoid regions, and the rectum.
Community: In ecology, an assemblage of interacting populations forming an identifiable group within a biome; e.g., a sage desert community or a beech-maple deciduous forest community.
Companion cell: In plants, a nucleated cell adjacent to a sieve tube member and believed to assist in its functions.
Comparative psychologist: One who studies behavior in animals from a comparative point of view, usually in a laboratory environment.
Competition: In ecology, the utilization by two or more individuals or species of the same limiting resource.
Complement: A group of blood proteins that interact with antibody-antigen complexes to destroy foreign cells.
Complete dominance: The expression of a gene with the complete lack of expression of its allele.
Compound: A chemical substance composed of one kind of molecule.
Concentration gradient: 1. A slow, consistent decrease in the concentration of a substance along a line in space. 2. For any spatial difference in concentration, the direction away from the region of greater concentration.
Condensation: Of chromosomes, the coiling and supercoiling that transforms diffuse chromatin into compact, discrete bodies in mitosis.
Condom: A thin sheath of rubber or animal membrane worn over the penis during sexual intercourse to prevent conception or venereal infection. Also called *rubber* or *prophylactic*.
Cone: 1. A male or female reproductive structure of conifers, consisting of a cluster of scalelike modified leaves and either pollen or ovules or the seeds. 2. One of a class of cone-shaped photoreceptors in the retina that detect color, consisting of a highly modified cilium with specialized pigments for detecting red, green, or blue wavelengths.
Conidiophore (kō nid'ē ə fōr): A specialized branch of the mycelium-bearing conidia.
Conidium (kō nid'ē əm): An asexual spore borne on the tip of a fungal hypha.
Conifer (kō' nə fər): An evergreen gymnospore of the order *Coniferales,* bearing ovules and pollen in cones; included are spruce, fir, pine, cedar, and juniper.
Conjugation: Sexual reproduction in which organisms (usually single-celled) fuse to exchange genetic material; in ciliates, a temporary cytoplasmic fusion in

pairs, accompanied by meiosis and the exchange of haploid nuclei.
Connective tissue: One of the primary animal tissues, composed of scattered cells and abundant extra-cellular substances or interlacing fibers, usually including collagen.
Constant region: The C-terminal portion of an antibody chain, not involved in antigen-specific binding.
Consummatory behavior: A part of instinctive behavior that involves fixed, highly stereotyped (invariable) behavior that brings relief (for example, swallowing food).
Continental drift: The slow movement of the continents relative to one another on the earth's surface. See also *plate tectonics*.
Continental shelf: The part of a continent that is submerged in relatively shallow sea.
Contraception: Any process or method intended to prevent the sperm from reaching and fertilizing the egg or preventing ovulation or implantation. Also called *birth control*.
Contractile vacuole (kən trak'təl vak'yoo ōl)**:** An organelle of many freshwater protists that maintains the cell's osmotic equilibrium by bailing excess water through an active, ATP-powered process. Also called a *water vacuole*.
Control: A standard of comparison in scientific experiment; a replicate of the experiment in which a possibly crucial factor being studied is omitted.
Convergent evolution: A similarity in genetically different organisms resulting from their adaptation to similar habitats.
Copulation (kop'yə lā'shən)**:** Sexual intercourse or coitus.
Coracoid process (kôr'ə koid process)**:** In man and other higher mammals, a reduced bony process of the scapula having no connection with the sternum.
Cork cambium (cork kam'bē əm)**:** The perpetually growing tissue of a plant or stem that forms the bark.
Corolla (kə rol'ə)**:** All the petals of a flower.
Corpus callosum (kôr'pes kə lō'səm)**:** A connecting strip of nerve fibers that coordinates the activities of the cerebral hemispheres.
Corpus luteum (kôr'pəs lu'tē əm)**:** "Yellow body"; a temporary endocrine body that develops in the ovarian follicle after ovulation. It secretes estrogen and progesterone, which maintain the endometrium. It continues secretion if pregnancy occurs, and regresses quickly if it does not occur.
Cortex: 1. In plants, the portion of the stem between the epidermis and the vascular tissue. 2. In animals, the outer layer or rind of an organ, such as the *adrenal cortex*.
Cortisone (kôr'ti sōn)**:** A steroid hormone of the adrenal cortex active in carbohydrate and protein metabolism.
Cosmic ray: A radiation of extremely high penetrating power that originates in outer space and consists of high-energy atomic nuclei.
Cotyledon (kot'l ē'dən)**:** A food-storing structure in dicot seeds, sometimes emerging as first leaves; a food-digesting organ in most monocot seeds; first leaves in a gymnosperm embryo. Also called *seed leaf*.
Covalent bond (kō vā'lənt bond)**:** A bond formed between two atoms as they share electrons.
Cranial nerve: In humans, one of the twelve pairs of major nerves that emerge directly from the brain and pass through skull openings to the body's periphery.
Cranium (krā'nē əm)**:** 1. The skull. 2. The part of the skull enclosing the brain.
Crash: Population crash; a sudden die-off, particularly after the carrying capacity of the environment has been exceeded and resources have been depleted.
Cretinism (krēt'ə niz'əm)**:** A recessive genetic abnormality that results in the inability to produce thyroxine. Affected persons are extremely retarded physically and mentally unless thyroxine is administered from early infancy.
Crinoidea (krin oi'dē ə)**:** The class including the echinoderms with cup-shaped bodies to which are attached branched, radiating arms, such as the sea lilies and feather stars.
Crista (kris'tə)**:** A fold of the inner mitochondrial membrane, containing numerous electron transport systems and F_1 particles.
Crop: A food storage sac located below the esophagus in birds, earthworms, and other animals.
Cross bridges: Extensions of the myosin fibers of muscles that attach to actin fibers and pull the fibers past each other in a ratchet fashion as the muscle contracts.
Crossing over: The exchange of chromatid (DNA) segments by enzymatic breakage and reunion during meiotic prophase.
Crude birth rate: The number of live births per 1000 individuals in the population at midyear.
Crude death rate: The number of deaths per 1000 individuals in the population at midyear.
Crustacea (kru stā'shē ə)**:** The class composed of aquatic arthropods typically having the body covered with a hard shell or crust, such as the lobsters, shrimps, crabs, barnacles, and so on.
Cuticle: A tough, often waterproof, nonliving covering, usually secreted by epidermal cells.
Cyclic AMP (cAMP): The "second messenger," adenosine monophosphate. It is synthesized in response to certain hormones arriving at the cell membrane, stimulating further activity in the target cell.
Cyanobacteria (sī' a nō bak tir' ē a)**:** Photosynthetic bacteria containing chlorophyll, that use water as a source of protons; many are capable of nitrogen fixation.
Cyst: 1. A closed sac, containing fluid, embedded in a tissue. 2. A heavy protective covering of a dormant animal or protist. 3. An encysted organism.
Cytokinesis (sī'tō ki ne'səs)**:** The division of cell cytoplasm following mitosis or meiosis.
Cytokinin (sī'tō kin in)**:** A plant cell hormone, mitogen, and plant tissue culture growth factor that interacts with other plant hormones in the control of cell differentiation.
Cytology (sī tol'ə jē)**:** The scientific study of cells.
Cytopyge (sī'to pij)**:** An excretory structure of paramecia.
Cytosine (sī'to sēn')**:** A pyramidine, one of the four nucleotide bases of DNA and RNA.
Cytostome (sī'to stōm')**:** The mouth of a protozoan.
Cytotoxic T-cell (sī'to tocks'ik killer cell)**:** A type of white blood cell that attacks specific antigens after receiving instructions from a macrophage.

D

D and C: See *dilation and curettage*.
Daughter cell: Either of the two cells created when one cell divides.
Deductive method: A logical process in which specific conclusions are drawn from generalities. Compare *inductive method*.

Dehydration: Chemically, an enzymatic reaction during which water is lost and a covalent bond forms between the reactants.
Deletion: In genetics, the loss of any segment of a chromosome or gene.
Delirium tremens (DTs) (di lēr'ē əm trē'mənz): A violent restlessness due to excessive indulgence in alcoholic beverages, characterized by trembling, terrifying visual hallucinations, and so on.
Demographer (di mog'rəfer): One involved in the science of statistics of populations.
Denaturation: The alteration of a protein so as to destroy its properties, through heating or chemical treatment.
Dendrite: A short, branched extension of a neuron that receives impulses and conducts them toward the cell body.
Density-dependent: Factors affecting populations the severity of which is dependent on the densities of the populations. Often an effect of competition.
Density-independent: Those factors affecting population size that are independent of population density, e.g., temperature, salinity, and meteorites.
Deoxyribonucleic acid (DNA) (dē ok'sə rī'bō nü klē'ik acid): A nucleic acid containing deoxyribose, which is found chiefly in the nucleus of cells. DNA is the genetic message functioning in protein synthesis.
Dependence: The state of needing someone or something for aid, support, comfort, or the like.
Depolarization (dē pō'lər ə zā'shən): In neural transmission, a loss of polarization that is characteristic of the resting state; a shift in the neuron from a polarized condition (positively charged exterior and negatively charged interior) to a nonpolarized condition as a neural impulse is generated. See also *action potential.*
Depression: A condition of general emotional dejection, sadness, and withdrawal.
Depressor: A muscle that draws down some part of the body.
Descending colon: A part of the large intestine that descends to the rectum.
Desert: A region characterized by scanty rainfall, especially less than 25cm (10 in) annually.
Deuterostome (dü'tə rə stōm): A bilateral animal (for example, an echinoderm or chordate) whose anus arises from the first embryonic opening (blastopore) and whose mouth arises later as a second embryonic opening. Compare *protostome.*
Developed country: A country that is industrialized, with high per capita income and high literacy. Compare *developing country.*
Developing country: A nation that is characterized by low industrialization, low per capita income, and low literacy. Compare *developed country.*
Diaphragm (dī'ə fram): 1. A birth control device; a thin rubber cap with a springlike rim that covers the cervix. 2. The large, flattened muscle between the thoracic and abdominal cavities that aids in breathing.
Diatomaceous earth (dī'ə tə mā'shəs earth): Geological deposits consisting largely of the cell walls of diatoms; used in filtration and as an abrasive.
Dicot (dī'kot'): See *dicotyledon.*
Dicotyledon (dī kot'l ēd'n): A flowering plant (angiosperm) of the class *Dicotyledonae,* characterized by seeds with two cotyledons. Also called *dicot.*
Diencephalon (dī'en sef'ə lon'): The posterior section of the forebrain.
Differentiation: In development, the poorly understood process by which a cell or cell line becomes structurally or physiologically specialized.
Diffusion: The movement of particles from an area of greater concentration to lesser as a result of random movement.
Digestion: The hydrolic cleavage (adding of water), through enzyme action, of complete food molecules into the molecular subunits, permitting absorption to occur.
Digit: In terrestrial vertebrates, any of the terminal divisions of the limbs; a finger, thumb, or toe.
Dihybrid (dī hī'brid): The result of a genetic cross in which two different traits are involved.
Dilator (dī lā'tər): A muscle that dilates some cavity of the body.
Dilation and curettage (D and C) (dī lā'shən and kyoor'i täzh): A surgical operation in which the cervix is forcefully dilated and the uterine mucosa scraped with a curette.
Dimorphic (dī môr'fik): 1. Having two forms distinct in structure, coloration, and so on among animals of the same species. 2. Having two different forms of flowers, leaves, and so on on the same plant or on distinct plants of the same species.
Dinoflagellate (din'ə flaj'ə lāt): A flagellated, photosynthetic, marine protist of the group *Dinoflagellata.*
Diploid (dip'loid): Having a double set of genes and chromosomes, one set originating from each parent. Compare *haploid, polyploid.*
Diplopoda (dip'lə pod'e): The class composed of millipedes.
Directional selection: In evolution, that selection that favors one or the other of two alleles in a population so that one or the other accumulates.
Disaccharide (dī sak'ə rīd'): Any of a group of carbohydrates, as sucrose or lactose, that hydrolyze into simpler sugar or monosaccharides.
Disruptive selection: Selection that favors more than one trait, such as the extremes of a condition, resulting in the median condition being less common than the extremes.
Distal convoluted tubule: In the kidney, a portion of the nephron between the loop of Henle and the collecting duct.
Divergent evolution: Evolutionary change away from the ancestral type, with selection favoring newly arising conditions.
DNA: See *deoxyribonucleic acid.*
Dominance, principle of: The principle of certain genes being expressed over their allelic alternative.
Dominant trait: A phenotypic characteristic that is always expressed when a certain allele is present.
Dorsal nerve root: Of the two large nerves extending from the spinal column, the one nearer the back surface, carrying afferent impulses.
Double bond: A covalent bond in which two pairs of electrons are shared between two atoms.
Doubling time: The number of years required for a population to double in size.
DTs: See *delirium tremens.*
Duodenum (dōō'ə dē'nem): The first segment of the small intestine posterior to the stomach.
Duplication: Genetically, a doubling in the number of certain genes, resulting in a mutation.

E

E. coli: A common intestinal bacterium.
Early anaphase (early an'ə fāz): The mitotic stage when sister chromatids have separated and just begun to move to opposite poles.
Echinoderm (i kī'nə dûrm): Any organ-

ism of the marine, coelomate, deuterostome phylum *Echinodermata;* i.e., starfishes, sea urchins, sea cucumbers, and so on.
Echinoidea (ek'ə noi'dē ə): The class composed of the sea urchins, sand dollars, and so on.
Ecologist (i kol'ə jist): A scientist who studies the relations between organisms and their environment.
Ecology (i kol'ə jē): The branch of biology dealing with the interrelationship of organisms and their environment.
Ecosystem (ē'kō sis'təm): In ecology, a unit of interaction among organisms and between organisms and their physical environment.
Ectoderm (ek'tə dûrm'): The outer cell layer in an animal embryo that gives rise to skin, nervous tissue, and so on.
Ectoplasm (ek'tə plaz'em): The outer, clear, thin layer of the cytoplasm of a cell.
Ectotherm (ek'tə therm): Lacking the ability to thermoregulate physiologically. These animals are often called "cold-blooded," but their body temperature is usually that of the surroundings. Compare *endotherm.*
Effector: Any structure that elicits a response to neural stimulation (for example, a muscle or gland).
Efferent neuron: A neuron conducting impulses from the brain or spinal cord to an effector, such as a *motor neuron.*
Egg nucleus: The haploid plant cell that, at fertilization by the sperm nucleus, results in a zygote.
Eidetiker (ī det'i kər): A person with a "photographic" memory.
Ejaculation: The forceful expulsion of semen from the penis.
Electrical energy: A form of energy due to the flow of electrons.
Electromagnetic spectrum: The range of electromagnetic radiation, from low-energy, low-frequency radio waves to high-energy, high-frequency gamma rays.
Electron: A minute, negatively charged particle of an atom, in motion about the nucleus.
Electron acceptor: A substance that receives electrons through reduction.
Electron shell: A theoretical band around an atomic nucleus where electrons of a certain energy level are found.
Electron-transport chain: A series of reactions in which electrons are passed from one compound to the next in an energy-producing process.
Element: A substance consisting of atoms of one kind only, such as iron, gold, sulfur, and so on. There are 102 known elements.
Embryo: The developing stage of any organism.
Embryo sac: In flowering plants, the mature megagamtophyte after division into six haploid cells and one binucleate cell, enclosed in a common cell wall.
Endergonic: Any chemical reaction characterized by the absorption of energy from an outside source.
Endocrine gland: A discrete gland that secretes hormones directly into the blood. Also called a *ductless gland.*
Endocytosis (en' dō sī tō'sis): The process of taking food or solutes into the cell (into vacuoles) by engulfment (a form of active transport). See also *phagocytosis,* compare *exocytosis.*
Endoderm: The innermost cell layers of an embryo that differentiate into such organs as the digestive tract, liver, and pancreas.
Endometrium (en'dō mē'trē əm): In mammals, the tissue lining the cavity of the uterus.
Endoplasm: The granular inner portion of the cytoplasm of a cell, containing numerous crystals and mitochondria.
Endoplasmic reticulum (ER) (endoplasmic re tik'yə ləm): The layers of membrane folded through the cytoplasm of a cell, forming complex inner surfaces, after being covered with ribosomes.
Endoskeleton: A mesodermally derived supporting skeleton inside the organism, surrounded by living tissue, as in vertebrates and echinoderms.
Endosperm: A nutritive tissue of seeds, formed around the embryo in the embryo sac.
Endosperm nuclei: A binucleate cell in the plant embryo sac which, when fertilized, produces the triploid endosperm.
Endospore: A resistant, thick-walled spore formed from a bacterial cell. See also *spore.*
Endotherm: An organism with the ability to metabolically thermoregulate; also called "warm-blooded." See also *ectotherm.*
Energy: The capacity or potential to do work; it exists in several interchangeable forms, including heat, chemical, electrical, magnetic and radiant (electromagnetic).
Energy of activation: The energy boost required to initiate a chemical reaction.
Entropy (en'trə pē): The opposite of "free energy." The degree of randomness in a system; a measure of disorder, which means that as the available energy in a system decreases, the system becomes increasingly disorganized.
Enucleate (i nyoo'klē āt'): Having no nucleus.
Environmental resistance: The sum of environmental factors (e.g., limited resources, drought, disease) that restrict the growth of a population below its biotic potential (maximum possible population size).
Enzyme: An organic catalyst, usually a protein, capable of changing the rate of chemical reaction in cells.
Eosinophil (ē' a sin'a fil): Granular leukocytes with pink-staining granules.
Epidermis: 1. In plants, the outer protective cell layer in leaves and in the primary root and stem. 2. In animals, the outer epithelial layer of the skin.
Epinephrine (ep'ə nef'rən): A hormone with numerous effects produced by the adrenal medulla and by nerve synapses of the autonomic nervous system. Also called *adrenaline.*
Epistasis (i pis'ta sis): The making of a trait ordinarily determined by one gene locus by the action of a gene or genes at anther locus.
Epithelium (ep'ə thē'lē əm): A tissue that consists of tightly adjoining cells that cover a surface or line a canal or cavity and that serves to enclose and protect.
Erythrocyte (i rith'rō sīt): A hemoglobin-filled, oxygen-carrying, circulating red blood cell.
Esophagus: The muscular tube connecting the pharynx to the stomach.
Estivate (es'tə vāt): To pass the summer in a state of dormancy or torpor induced by heat or dryness.
Estrogen: A female hormone; one of the hormones involved in the production of secondary sex characteristics and the menstrual cycle.
Estrus: The mating period of female mammals when they become sexually receptive.
Estuary (es'choo er'ē): The part of the mouth of a river in which the river's current meets the sea.
Ethology (e thol'ə jē): The scientific study of animal behavior, usually under natural conditions.
Eubacteria: "True bacteria"; The better-known prokaryote kingdom, containing many familiar human pathogens and important soil and water bacteria. See also *archaebacteria.*

Eukaryotic cell (yoo'kar'ē ō tik cell): A cell having among its traits a true nucleus and inner membrane structures.
Eutheria (yoo thēr'ē ə): The group composed of placental mammals.
Eutrophication (yoo trof'ə kā'shən): The aging process whereby a body of water supports increasing numbers of organisms, often causing lakes to become marshes and then terrestrial communities.
Evolution: The continuous changes occurring in populations, primarily as a result of adapting to the environmental changes.
Exergonic: Any chemical reaction that releases energy.
Exocytosis (ek'sō sī tō'sis): The process of expelling material from vacuoles through the cell membrane. Compare *endocytosis*.
Exoskeleton: An external skeleton or supportive covering, as in arthropods.
Extensor: A muscle that serves to extend or straighten a part of the body.
External: Pertaining to the outer part, or outside, of the body.
External ear: All parts of the ear external to the eardrum, comprising the external ear canal, the external auditory meatus, and the pinna.
External respiration: The process by which gases reach the exchange surfaces, as in breathing.
Extinction: Behaviorally, the loss of a conditioned response as a result of the absence of reinforcement.

F

F_1 particle: An ultramicroscopic mitochondrial organelle attached to the inner surface of the crista; the site of chemiosmotic phosphorylation.
Facilitated diffusion: The diffusion of molecules across a cell membrane assisted by a reversible association with carrier molecules; it differs from active transport in that no energy is expended and net movement follows the concentration gradient.
Fallopian plug (fə lō'pē ən plug): A birth control process in which silicone rubber plugs are inserted into the Fallopian tubes to prevent the eggs from reaching the uterus.
Family: In taxonomic classification, a grouping smaller than *order* but larger than *genus*.
Fascia (fāsh'ē ə): A heavy sheet of connective tissue covering or binding together muscles or other internal structures of the body, often connecting with ligaments or tendons.
Fat: 1. Any of the compounds consisting of glycerol bound by ester linkages to three fatty acid molecules. 2. Parts of an animal consisting largely of cells distended with triglycerides.
Fatty acids: Constituents of fats and oils; long chains of carbon atoms with hydrogen attached and ending with carboxyl groups.
Feces (fē'sēz): Bodily waste discharged through the anus.
Feedback: The return of part of the output of a system back to the input, which regulates or affects the system's functioning. In *negative feedback*, the amount of output is sensed by the system and acts in an inhibiting manner, slowing activity and thus reducing output. In *positive feedback*, the output stimulates activity, further increasing the flow of output.
Feedback mechanism: A pathway in which an end product influences its own production.
Fermentation: Anaerobic respiration with the usual end product being alcohol and carbon dioxide.
Fetus: An unhatched or unborn individual which has developed past the embryo stage.
Fibroblast (fī'brə blast'): A cell in the connective tissue group that produces fibers and matrix substances such as collagen.
Fibrous connective tissue (fī'brəs connective tissue): A vertebrate connective tissue formed from cablelike fibers running in different directions.
Filament (fil'ə mənt): In flowers, the slender stalk of the stamen on which the anther is situated.
Filicinae (fil i'sin ē): The class of plants that contains the ferns.
First law of thermodynamics: The physical law that states that energy cannot be created or destroyed; later amended to allow for the interconversion of matter and energy.
First messenger: A hormone, as distinguished from a *second messenger*, which it causes to form within a cell.
Fissure of Rolando (fish'ər of rō lan'dō): The groove separating the frontal and parietal lobes of the cerebrum.
Fissure of Sylvius (fish'ər of sil'vē es): The groove separating the frontal, temporal, and parietal lobes of the cerebrum.
Fixed action pattern: In behavior, a precise and identifiable set of movements, innate and characteristic of a given species.
Flagellum (flə jel'əm): A threadlike organelle that extends from the cell surface; it may beat with whiplike motions.
Flexing: The process by which the angle between two joined bones is increased, such as by bending the arm.
Flexor: A muscle which serves to flex or bend a part of the body.
Fluid-mosaic model (fluid mō zā'ik model): A description of the plasma membrane as a phospholipid bilayer stabilized by specifically oriented proteins, with some proteins extending through to both surfaces and other proteins specific for the inner or outer surfaces.
Follicle: A fluid-filled body in the ovary during the preovulatory period containing a maturing oocyte.
Follicle stimulating hormone (fol'ə kəl stimulating hormone): A hormone produced by the anterior pituitary that stimulates the growth and maturation of eggs and sperm; also called FSH.
Food vacuole (food vak'yoo ōl): In animals and protists, an intracellular vacuole that arises by the phagocytosis of solid food materials and in which digestive processes occur.
Foot: In the moss sporophyte, the anchoring base.
Forebrain: 1. The anterior of the three primary divisions of the vertebrate embryonic brain. 2. The parts of the adult brain developed from the embryonic forebrain.
Foreplay: Sexual stimulation of another person, intended as a prelude to sexual intercourse.
Founder effect: The chance assortment of genes carried out of the original population by colonizers (or founders) who subsequently give rise to a large population.
Free energy: The energy that is available to do work.
Free radical: A highly reactive atom or molecule unit with an unpaired electron.
Frond: An often large, finely divided leaf, as seen in ferns and certain palms.
Frontal lobe: An anterior division of the cerebral hemisphere, believed to be a site of higher cognition.
Fruit: The mature, seed-bearing ovary of a flowering plant.
Fucozanthin (fū kō zan'thin): A brown carotenoid pigment characteristic of brown algae.
Functional group: A group of atoms in

an organic molecule that are free to participate in chemical reactions.
Fungus: An organism of the kingdom *Fungi,* including yeasts, mushrooms molds, mildews, part of the lichen symbiosis, rusts, smuts, sac fungi, puffballs, water molds, and sometimes the slime molds.

G

Gametangium (gam'i tan'jē əm): In algae and fungi, the structure (cell or organ) in which gametes are formed.
Gamete (gam'ēt): A male or female sex cell with half the chromosomal material. Fertilization is the union of gametes, producing the full amount of hereditary material.
Gametophyte (gə mē'tō fit): In plants with alteration of generations, the haploid form in which gametes are produced. Compare *sporophyte.*
Gametophytic (gam'i tə fit'ik): Referring to the sexual generation of a plant in which the cells have half the normal chromosome numbers.
Gap junction: A dense structure that physically connects membranes of adjacent cells along with channels for cell-to-cell transport.
Gastric cavity: The digestive area, such as the saclike gut of coelenterates.
Gastrodermis (gas'trō dur'mis): The inner cell layer of the body of a coelenterate.
Gastrovascular cavity: The cavity of coelenterates, ctenophores, and flatworms, which has only one opening to the outside and functions as a digestive cavity and crude circulatory system. Also called *coelenteron.*
Gastrula: An early metazoan embryo consisting of a hollow, two-layered cup with an inner cavity (archenteron) opening out through a blastopore.
Gastrulation (gas'troo lā'shən): The formation of the gastrula stage of an embryo. This is the stage in which the embryo acquires its three germ layers.
Gene: The portion or portions of DNA that produce a recognizable effect or trait; e.g., genes produce enzymes, color pigments, and so on.
Gene frequency: The fractional distribution of alleles for a specific trait in a population.
Gene linkage: The condition of genes being physicially linked on the same chromosome.
Gene mutation: A mutation involving alterations in the genes, occurring during the duplication of DNA.
General fertility rate: The average number of live births per 1000 females in their reproductive years (in the U.S., women 15-44).
Gene replacement therapy: The manipulation of a genome by replacing one allele for another.
Generative nucleus: 1. The nucleus of a sperm. 2. Either of the two nuclei that arise from the generative nucleus of a pollen tube and function in the double fertilization characteristic of seed plants.
Genetic code: The biochemical specification (DNA) for producing a particular genetic trait of the entire species' characteristics.
Genetic drift: Changes in the composition of a gene pool caused by random factors rather than by selection.
Genetic engineering: Modern techniques of gene management, including gene cloning, gene splicing, amino acid, DNA and RNA nucleotide sequencing, and gene synthesis.
Genotype (jen'ə tīp): The assortment of genes that make up the genetic characteristics of an organism.
Genus (jēnəs): A major subdivision of a family, consisting of one or more species.
Geothermal energy: Energy obtained from the internal heat of the earth.
Germ layer: Any of the three layers of undifferentiated embryonic cells formed at gastrulation. See *ectoderm, endoderm,* and *mesoderm.*
Germinal spot: A very early stage of development of the bird embryo.
Gibberellin (jib'ə rel'len): Any of a family of plant growth hormones that control cell elongation, bud development, differentiation, and other growth effects.
Gill: 1. In fungi, the thin, flattened structures on the underside of a mushroom (basidiocarp) that bear the spore-forming basidia. 2. In aquatic organisms, the surface over which external respiration occurs.
Gizzard: A muscular sac in the digestive system that mechanically changes food by mashing it against sand particles; found in birds, earthworms, and so on.
Glial cell (glē'əl cell): One of the numerous nonconducting supporting cells of the central nervous system that may play some role in information storage. Also called *neuroglia.*
Glomerulus (gle mer'e les) (pl. *glomeruli*): A cluster of capillaries within the Bowman's capsule of the kidney.
Glycerol (glis'e rôl): A three carbon alcohol molecule that combines with fatty acids to form fats and oils.
Glycogen (glī'kə jən): A complex chain of glucose units produced by animals.
Glycolysis (glī kol'ə sis): The first phase of respiration where glucose is broken down with a small energy yield.
Golgi body (gōl'jē body): A secretory structure in some cells consisting of a complex system of folded membrane.
Gonad (gō'nad): An ovary or testis.
Granum (grā'nəm): A green granule within chloroplasts, appearing as stacks of thylakoids: contains chlorophyll and carotenoids; the site of the light reaction.
Grassland: One of the natural biomes of the earth, characterized by perennial grasses, limited seasonal rainfall, and a great number of herbivorous mammals, birds, and insects.
Guanine (gwä'nēn'): One of the nitrogenous bases of RNA and DNA.
Gullet: The esophagus or food-receiving tube in an animal.
Gustation (ges tā' shan): The sense of taste; the act of tasting.
Gymnosperm (jim'ne spėrm): A non-flowering seed plant of the group *Gymnospermae,* producing seeds that lack fruit; included are conifers, cycads, gnetophytes, and ginkgos.

H

Habitat: The specific place where an organism lives; the native environment of a plant or animal.
Habituation (he bich'o͞o ā shən): A simple form of learning in which an organism's response to a stimulus diminishes.
Half life: In radioisotopes, the time it takes for half of the atoms in a sample to undergo spontaneous decay.
Haploid (hap'loid): Having a single set of genes and chromosomes. Compare *diploid* and *polyploid.*
Hashish (hash'ēsh): The flowering tops, leaves, and so on of Indian hemp, smoked, chewed, or drunk as a narcotic and intoxicant. Also called *hash.*
Haversian canal (he vûr'shen canal): In bone, a small canal through which a blood vessel runs.
Heart: A hollow, muscular organ which by rhythmic contractions and relaxations keeps the blood in circulation throughout the body.
Heat energy: Energy produced by the random movement of molecules that increases with a rise in temperature.

Heavy chain: One of two pairs of polypeptide chains in an antibody molecule, the other pair being light chains.
Helper T-cell: A cell that activates elements of the immune system.
Heme (hēm): A deep-red, iron-containing pigment found in hemoglobin.
Hemichordate (hem'i kôr' dāt): An animal of the deuterostome phylum *Hemichordata,* such as an acorn worm.
Hemoglobin (hē'mə glō'bən): A protein; a respiratory pigment consisting of one or more polypeptide chains, each associated with a heme (iron-containing) group.
Hemophilia (hē'mə fil'ē ə): A sex-linked condition in which the blood does not clot normally. Also called *bleeder's disease.*
Hepatic portal vein (hi pat'ik portal vein): In all vertebrates, a large blood vessel that collects blood from the capillaries and venules of the esophagus, stomach, and intestine and carries it to the liver, where it once more divides into capillaries.
Herbivore (hür'bə vōr): Any organism that feeds on plants.
Hermaphrodite (her maf'rə dīt): An individual animal or plant with both male and female reproductive parts.
Heterotroph (het'ər ə trōf): An organism that depends on an external source of organic substances for its food.
Heterozygous (het'er ō zī'ges): A mixed genotype; different alleles for a particular trait.
Hierarchy (hī'ə rär'kē): Any system of persons, animals, or things in which one is ranked above another; "pecking order."
High selection pressure: Strong selection against departure from the optimum of a phenotype.
Hindbrain: 1. The most posterior of three primary divisions of the embryonic vertebrate brain. 2. The parts of the adult brain derived from the embryonic hindbrain, including the cerebellum, pons, and medulla oblongata.
Hinge joint: A joint that moves in one plane, like a door hinge.
Hirudinea (hir'oo din'ē ə): The class composed of leeches.
Histamine (his'tə mēn): An amine compound released in allergic reactions that dilates blood vessels and reduces blood pressure, stimulates gastric secretions, and causes contraction of the uterus.
Holdfast: A rhizoid base of a seaweed, serving to anchor it to the ocean floor.
Holothuroidea (hol'e thoo roi'dē ə): The class consisting of sea cucumbers.
Homeostasis (hō'mē ō stā'sis): The tendency of organisms to maintain a stable internal environment.
Homologue (hom'ə log): Either of the two members of each pair of chromosomes in a diploid cell.
Homozygous (hō mō zī'gəs): Having identical alleles for a particular trait.
Hormone (hor'mōn): A chemical messenger transmitted in body fluids or sap from one part of the organism to another, producing a specific effect on target cells and regulating physiology, growth, differentiation, or behavior.
Host: A living organism that supports or sustains a parasite.
Host DNA: The normal DNA of a cell into which can be introduced a foreign strand of DNA.
Hybrid: Offspring resulting from the cross of parents with different characteristics or parents of different species.
Hydra: Any fresh-water polyp of the genus *Hydra*..
Hydration: The adding of water.
Hydrocarbon: Any organic compound that contains simply carbon and hydrogen.
Hydroelectric energy: Electric energy derived from falling water or any other hydraulic source.
Hydrogen bond: A weak attractive force between two molecules where one contains hydrogen.
Hydrolysis (hī drol'ə sis): The breakdown of a compound by the addition of water molecules.
Hydrophilic (hī'drə fil'ik): Water-loving, or having an attraction for water.
Hydrophobic (hī'drə fō'bik): Having little or no affinity for water; water-fearing, or repelling of water.
Hydroxyl group (hī drok'sil group): =OH, consisting of an oxygen and hydrogen covalently bonded to the remainder of the molecule; a constituent of alcohols, sugar, glycols, phenols, and other compounds.
Hydrozoa (hī'drə zō'e): The class comprising the coelenterates such as solitary or colonial polyps and free-swimming medusae.
Hypha (hī'fə): One of the individual filaments that make up a fungal mycelium.
Hypothalamus (hī'pə thal'e məs): The region that lies in front of the thalamus in the forebrain. It regulates the internal environment, pituitary secretions, and some of the basic drives.
Hypothesis (hī poth'ə sis): A proposition set forth as an explanation for a specified group of phenomena, either asserted merely as a provisional conjecture to guide investigation (e.g., working hypothesis) or accepted as highly probable in the light of established facts. See also *theory.*

I

Idiot savant (id'ē ət sa vänt'): A mentally defective person with a highly developed special talent, such as an ability to play music, to solve complex mathematical problems mentally at great speed, and so on.
Ileum (il'ē əm): A region of the small intestine.
Ilium (il'ē em): The hip bone; one of the three pairs of fused bones forming the pelvis.
Immune system: In vertebrates, widely-dispersed tissues that respond to the presence of the antigens of invading microorganisms or foreign chemical substances.
Immunoglobulin (im yoo'nō glob'yoo lən): A protein antibody produced by T-cells in response to specific foreign substances; it consists of four subunits that are joined through disulfide linkages and have specific antigen binding sites.
Impulse: Neural impulse; a wave of excitement (transitory membrane depolarization) transmitted along a neuron and between neurons.
Incus: The middle one of a chain of three small bones in the middle ear of humans and other mammals. Also called *anvil.*
Independent assortment: In genetics, Mendel's Second Law: the inheritance of one pair of alleles in an individual occurs independently of the simultaneous inheritance of a second pair of alleles (except) where gene linkage occurs.
Inducer: In embryology, a substance that stimulates the differentiation of cells or the development of a particular structure.
Induction: The influence on cell differentiation or development that one embryonic tissue has on another.
Inductive method: A logical process in which a generalizing conclusion is proposed that contains more information than the observations or experience on which it is based. The truth of the conclusion is verifiable only in terms of future experience.

Inert (in ürt′): Chemically, showing a lack of bonding activity.
Inferior vena cava (inferior vē′nə kā′və): A large vein discharging blood from all parts of the body below the diaphragm into the right atrium of the heart.
Inflammatory response: A reddening and swelling brought on by an immune reaction.
Inhibitory neuron (in hib′i tōr′ē neuron): A nerve cell that functions by inhibiting the firing of the next nerve cell.
Inhibitory synapse (in hib′i tōr′ē sin′aps): The space between an inhibitory nerve cell and the next nerve cell.
Inhibitory transmitter substance (in hib′i tōr′ē transmitter substance): The neurotransmitter of an inhibitory neuron.
Innate behavior (i nāt′ behavior): Behavior that is inborn, not due to learning.
Innate releasing mechanism (IRM) (i nāt′ releasing mechanism): An ethological term that refers to a neural mechanism that produces a specific behavioral event when triggered by a particular stimulus from the environment.
Inner cell mass: In a mammalian blastocyst, the portion that is destined to become the embryo proper.
Insecta (in sek′tə): The class composed of insects.
Insectivora (in sek′tə vōr′ə): "Insect eaters"; the order composed of insect-eating animals such as moles and shrews.
Insertion: In anatomy, the distal attachment of a tendon or muscle (the attachment on the part to be moved). Compare *origin.*
Insoluble (in sol′yoo bəl): Incapable of being dissolved.
Instinct: Any inherent, unlearned behavioral pattern that is functional the first time it happens and can occur in animals reared in total isolation.
Interbreed: To breed or mate with another individual, as in a single population.
Intercourse: Sexual relations or a sexual coupling, especially coitus.
Interference phenomenon: A cellular substance that interferes with viral replication; interferon.
Interferon (in′ter fēr′on): See *interference phenomenon.*
Interleukin II (in′ ter lo͞o′ ken to͞o): A substance secreted by helper T-cells as part of the immune reaction.
Internal: Pertaining to the inside or inner part of something.
Internal respiration: The energy-producing chemical reactions within cells that utilize oxygen.
Interphase: The period between two mitotic events in the same cell.
Interspecific cooperation: Cooperation between members of different species.
Intraspecific cooperation: Cooperation between members of the same species.
Intrauterine device (IUD) (in′tre yoo′tər in device): A plastic device, sometimes containing copper, that is inserted into the uterus as a means of preventing conception.
Inversion: The transposition of a portion of a chromosome following breakage and repair, altering the relative order of gene loci.
Involution: An inward roll or curve; in gastrulation, the movement of cells toward the blastopore, over the blastopore lip, and into the archenteron.
Ion (ī′on): An atom or molecule that has gained or lost one or more electrons.
Ion gates: Areas along a neuron that open and close, regulating the passage of ions.
Ionic bond (ī on′ik bond): An attractive force between atoms that have different charges as a result of the transfer of electrons between them.
Ionizing radiation: Energetic radiation that produces ions in air and free hydronyl radicals in water, the latter causing induced mutations & tissue damage; i.e., X-rays, gamma rays, and streams of charged particles from the breakdown of radioactive isotopes.
IRM: See *innate releasing mechanism.*
Isotope (ī′sə tōp): An atom with a lighter or heavier nucleus than the average; may exhibit radioactivity.
IUD: See *intrauterine device.*

J

Jejunum (ji jo͞o′nəm): The area of the small intestine that lies between the duodenum and the ileum.
Joint: A point where two bones articulate, often movable.
J-shaped curve: A plot of population growth where growth is rapid, approaching the environment's carrying capacity. Compare *S-shaped curve.*
Jungle: A wild land overgrown with dense vegetation, parts of tropical rain forests penetrated by sunlight.

K

Keratin: A fibrous, insoluble structural protein making up most of the substance of the dead cells of hair, horn, nails, claws, feathers, and the outer epidermis.
Kinetic energy (ki net′ik energy): The energy of motion.
Kinetin (kin ē′tin): A plant hormone, one of the cytokinins.
Kingdom: Any of the five categories (monera, protista, fungi, plants, animals) in which organisms are usually placed.
Kin selection: The selection for traits that benefit those individuals carrying some of the same kinds of genes, thereby helping those kinds of genes to be carried into the next generation.
Krebs cycle: A series of chemical reactions in the mitochondrion by which pyruvate is broken down to CO_2, FAD^+ and NAD^+ are reduced to $FADH_2$ and NADH, and ATP is generated.
Kwashiorkor (kwash′ē or′kôr): A syndrome of severe protein deficiency in human infants and children, including failure to grow, deficiency of melanin pigment, edema, degeneration of the liver, anemia, and retardation.

L

Labia minora (lā′bē ə mī′nə rə): The smaller inner folds of skin that border the vagina.
Labor: 1. The efforts of childbirth. 2. The period during which these pains and efforts occur.
Lacteal (lak′tē′əl): A lymphatic vessel of the intestinal villi.
Lactic acid fermentation: A metabolic pathway beginning with pyruvic acid and forming lactic acid.
Lagomorpha (lag′ə mor′fə): "Rabbit shape"; the mammalian order including hares, rabbits, and pikas.
Lamella (lə mel′ə): A layer, or flattened sheet.
Lancelet: Any small, fishlike, chordate animal of the subphylum Cephalochordata, having a notochord in the slender, elongated body pointed at each end.
Large intestine: Also called *colon* or *bowel;* a division of the alimentary canal, primarily functioning in the resorption of water.
Late prophase: The mitotic stage when chromosomal material is highly condensed and ready to enter metaphase.
Lateral meristem (lat′er əl mer′ə stem): Meristem located along the sides of a part, as a stem or root.
Latimeria (lat′ə mer′ē ə): A coelacanth; a primitive, lobe-finned fish once thought to be extinct.
Leaf: In vascular plants, a lateral out-

growth from the stem functioning primarily in photosynthesis, arising in regular succession from the apical meristem, consisting typically of a flattened blade jointed to the stem by a petiole.
Leucoplast (lü'kə plast'): A colorless body in plant cells that stores proteins, lipids, and starch.
Leukocyte (lü'kə sīt): A vertebrate white blood cell; it aids in resisting infections.
Levator: A muscle that raises a part of the body.
Ligament: A tough, flexible, but inelastic band of connective tissue that connects bones or supports an organ in place. Compare *tendon.*
Light chain: One of two pairs of polypeptide chains in an antibody molecule, the other pair being heavy chains.
Light-dependent reaction: That part of photosynthesis directly dependent on the capture of photons; specifically the photolysis of water, the thylakoid electron transport system, and the chemiosmotic synthesis of ATP and NADPH.
Light-independent reaction: That part of photosynthesis not immediately involved in chemiosmosis, specifically the fixation of CO_2 into carbohydrate from the NADPH and ATP produced by the light reaction (Calvin cycle).
Light microscope: The commonly used microscope that illuminates the viewed object with ordinary visible light.
Lignin: An amorphous substance that helps give wood its rigidity.
Lipid: A type of fatty organic compound.
Liposome (lip' a sōm): A fat-containing cellular organelle.
Loop of Henle: The prominent U-shaped loop in the renal tubule of the mammalian kidney.
Low selection pressure: Weak selection against departure from the optimum of a phenotype.
Lumen: The cavity or channel of a hollow tubular organ or organelle.
Lung: In land vertebrates, one of a pair of compound, saclike organs that function in the exchange of gases between the atmosphere and the bloodstream; or any of several analogous organs in invertebrates.
Luteinizing hormone (LH) (loo'tē ə nīz'ing hormone): A pituitary hormone that causes ovulation and stimulates hormone production in the corpus luteum.
Lymph: The clear yellowish intercellular fluid or plasma in the lymphatic system.
Lymph capillary (limf kap'ə ler'ē): One of the many tiny, blind endings in the lymphatic system, including the lacteals of the small intestine.
Lymph duct: See *thoracic duct.*
Lymph node: A clump of cells in the lymphatic system that produces lymphocytes and acts as a filter for lymph.
Lymphatic system (lim fat'ik system): The system of lymphatic vessels and ducts and lymph nodes that serves to redistribute excess tissue fluids and to combat infections.
Lymphocyte: Any of several varieties of similar-looking leukocytes involved in the production of antibodies and in other aspects of the immune response; they are formed in the lymph nodes.
Lysozyme (lī' sa zīm'): An enzyme that destroys bacteria by dissolving the bacteria cell wall; a natural constituent of eggwhite, tears, and saliva.

M

Macronucleus (mak'rō nü'klē əs): The larger of two types of nuclei found in certain protozoans.
Macrophage (mak'rō fāj): A large, phagocyte; one of the leukocytes.
Malleus (mal'ē əs): The outermost of a chain of three small bones in the middle ear of humans and other mammals; also called *hammer.*
Malnourished: Not receiving the proper kinds of food for growth and development. Compare *undernourished.*
Mammalia (mə mā'lē ə): The class of vertebrates that feeds its young with milk from the female mammary glands, that has the body more or less covered with hair, and that, with the exception of the monotremes, brings forth living young rather than eggs.
Mandibulata (man di'byoo la'ta): The jawed arthropods.
Mantel: The shell-secreting organ in mollusks.
Marasmus (mə raz'məs): Gradual loss of flesh and strength, occurring chiefly in infants, caused by a diet low in both calories and protein.
Marijuana (mar'ə wä' nə): Dried leaves and flowers of the Indian hemp, used in cigarettes as a narcotic.
Marrow: *Bone marrow,* the soft tissue in the interior cavity of a bone; *blood marrow,* vascularized bone marrow in which white blood cells of all types are produced; *fatty marrow,* bone marrow consisting of adipose tissue.
Marsupial (mär sü'pē əl): A mammal of the subclass *Metatheria;* usually the female has a pouch (marsupium); included are the kangaroo, wombat, koala, Tasmanian devil, opossum, and wallaby.
Mastigophora (mas'tə gof'ə rə): The phylum consisting of the flagellates.
Matrix (mā'triks): In the mitochondrion, the enzyme-laden region within the highly convoluted inner membrane; the site of the citric acid or Krebs cycle.
Mechanical energy: The energy produced by one physical object's exerting pressure on another.
Mechanism: The theory that the processes of life are based on the same physical and chemical laws that apply to nonliving phenomena.
Medulla (mi dul'ə): 1. The inner portion of a gland or organ. Compare with *cortex.* 2. *Medulla oblongata:* a part of the brain stem developed from the posterior portion of the hindbrain and tapering into the spinal cord.
Medusa (mə dü'sə): The motile, free-swimming jellyfish form of coelenterate. Compare *polyp.*
Megagametophyte (meg'ə gə mēt'ə fīt): The female gametophyte produced from a megaspore in flowering plants; the *embryo sac* consisting of eight haploid nuclei or cells.
Megaspore mother cell: In the ovule, a large diploid cell that will give rise to the megaspore by meiosis and the degeneration of three of the four haploid nuclei.
Meiosis (mī ō'sis): Two successive nuclear divisions in which the chromosomes are shuffled and reorganized and the numbers of chromosomes per cell are halved.
Meissner's corpuscle (mīz'nərz corpuscle): In mammals, a small touch-responsive neural end organ.
Membrane: A thin, pliable sheet or layer of animal or vegetable tissue, serving to line an organ, connect parts, and so on.
Memory B-cell: A mature, long-lived lymphocyte, specialized in retaining specific antigen information.
Menstrual cycle: The cycle of hormonal and physiological events and changes involving growth of the uterine mucosa, ovulation, and the subsequent breakdown and discharge of the uterine mucosa in menstruation (menses). The cycle averages 28 days. Also called *ovarian cycle.*
Menstruation (men'stroo ā'shən): In nonpregnant females of the human species only, the periodic discharge of

blood, secretions, and tissue debris resulting from the normal, temporary breakdown of the uterine mucosa in the absence of implantation following ovulation. Also called *menses.*
Meristem (mer′i stem′): Embryonic plant tissue; undifferentiated, growing, actively-dividing cells. See *apical meristem.*
Merostomata (mer′ō sto ma′ta): The class of arthropods that includes the horseshoe crabs.
Mesenchyme (mes′eng kīm): An aggregation of cells of mesodermal origin that are capable of developing into connective tissues, blood, and lymphatic and blood vessels.
Mesenteric artery (mes′ən ter′ik artery): Artery between the digestive tract and the liver.
Mesenteric vein (mes′ən ter′ik vein): The vein between the digestive tract and the liver.
Mesoderm (mes′ə dėrm): The middle layer of the three primary germ layers of the gastrula, giving rise in development to the skeletal, muscular, vascular, renal, and connective tissues, and to the inner layer of skin and the epithelium of the coelom (peritoneum).
Mesoglea (mes′ə glē′ə): The loose, gelatinous middle layer of the bodies of sponges and coelenterates, between the outer ectoderm and the inner endoderm.
Mesozoic (mez′ə zō′ik): Noting or pertaining to a geological era occurring between 75,000,000 and 220,000,000 years ago, characterized by the appearance of flowering plants and by the appearance and extinction of dinosaurs.
Messenger RNA (*m*RNA): RNA that carries the genetic code to the ribosomes where it is translated into protein production.
Metacarpal (met′ə kär′pel): One of the usually five bones of the hand or forefoot, between the carpals and the digits.
Metaphase (met′ə fāz): The second phase in mitosis during which the chromosomes line up at the center of the cell along the equator of the spindle.
Metatarsal (met′ə tär′səl): One of the usually five bones of the foot or hind foot, between tarsals and toes, forming the instep in human and part of the rear leg in ungulates and birds.
Metatheria (met′ə thēr′ē ə): The group consisting of marsupial mammals.
Metazoa (met′ə zō′ə): All animals other than *Parazoa* or phylum *Porifera* (sponges).
Methaqualone (meth′ə kwā lōn): An animal tranquilizer that has often been misused by humans; also called *Quaalude.*
Microfilament: A submicroscopic filament in the cytoskeleton, involved in cell movement and shape.
Microhabitat: The immediate environment of a species.
Micronucleus: The smaller of two types of nuclei in certain protozoa.
Micropyle (mī′krə pīl′): In seed plants, a minute opening in the integument of an ovule through which the pollen tube enters.
Microspore: In seed plants, one of the four haploid cells formed from meiosis of the microspore mother cell; it undergoes mitosis & differentiation to form a pollen grain.
Microspore mother cells: In the anther of a flowering plant, the diploid cells that will undergo meiosis to form the haploid microspores.
Microtubule: A cytoplasmic hollow tubule composed of spherical molecules of tubulin, found in the cytoskeleton, the spindle, centrioles, basal bodies, cilia, and flagella.
Microvilli (mī′krō vil′ ī) (sing. *microvillus*): Tiny, fingerlike out-pocketings of the cell membrane of various epithelial secretory or absorbing cells, such as those of kidney tubule epithelium and the intestinal epithelium.
Midbrain: 1. The middle of the three primary divisions of the vertebrate embryonic brain. 2. The parts of the adult brain derived from the embryonic midbrain.
Middle ear: The middle portion of the ear, consisting of the tympanic membrane and an air-filled chamber lined with mucous membrane, which contains the malleus, incus, and stapes.
Midlobe: The part of the pituitary gland connecting the anterior and posterior lobes.
Mitochondrion (mī′tə kon′drē ən) (pl. *mitochondria*): An organelle in the cytoplasm that is sometimes called the powerhouse of the cell because it functions in energy-producing processes.
Mitosis (mī tō′sis): The process by which each cell replicates its chromosomes and then divides into two new cells.
Mnemonist (nē mon′ist): A person with a phenomenal memory.
Mole: The quantity of a chemical substance that has a mass in grams numerically equal to its molecular mass.
Molecule: The smallest unit of a compound, composed of two or more atoms.
Mollusk: Any invertebrate of the phylum *Mollusca,* typically having a hard shell that wholly or partly encloses the soft, unsegmented body, such as the chitons, snails, bivalves, squids, and so on.
Moneran (mə nir′ən): Referring to the bacteria or the prokaryotes.
Monocot (mon′ə kot′): See *Monocotyledon.*
Monocotyledon (mon′ə kot′l ēd′n): A flowering plant of the class *monocotyledonae,* characterized by seeds with only one cotyledon (for example, grasses, palms, orchids, lilies). Also called *monocot.*
Monocyte: A large, phagocytic leukocyte, formed in bone marrow & in the spleen.
Monosaccharide (mon′ō sak′ə rīd): A simple sugar, such as glucose or fructose.
Monotreme (mon′ə trēm): A platypus, echidna, or extinct egg-laying mammal of the order *Monotremata,* subclass *Prototheria.*
Morula (môr′yů lə): An early embryonic stage in which the cells are arranged in a solid ball.
***m*RNA:** See *messenger RNA.*
Mud flat: A mud-covered, gently sloping tract of land, alternately covered or left bare by tidal waters.
Muscle: A contractile tissue, in vertebrates including skeletal, smooth, and cardiac muscle.
Muscle fiber: In skeletal muscle, one of the multinucleate cells that take the form of a long, contractile cylinder.
Muskeg: A bog of northern North America, commonly having spagnum mosses, sedge, and sometimes stunted black spruce and tamarack trees.
Mutation: Any permanent genetic change.
Mycelium (mī sē′lē em): The mass of interwoven hyphae that forms the vegetative body of a fungus.
Mycorrhizae (mī′ kō rī′ za): A mutualistic fungus-root association, with the fungal mycelium either surrounding or penetrating the roots of the plant.
Myelin sheath: A fatty sheath surrounding the axoms of some vertebrate neurons.
Myocardial infarction (mī′ə kär dē əl in färk′shən): The necrosis of the muscular substance of the heart caused by blood deprivation. Also called *coronary thrombosis.*
Myofibril (mī′ə fī′brəl): The slender

protein thread in the skeletal muscle that acts in contraction of the muscle.
Myofilament: The highly organized fibrous proteins of striated muscle, including the thin, movable *actin myofilaments* and the thicker, stationary *myosin myofilaments.*
Myosin (mī'ə sən)**:** A protein involved in cell movement and structure, especially in muscle cells. See also *myofilament.*

N

Natural increase: The net gain determined by the crude birth rate minus the crude death rate.
Natural killer cells: Lymphocytes that attack invading antigens.
Natural selection: The differential survival and reproduction of certain individuals in a population. In the same process, other organisms less suited to the environment are eliminated.
Navigation: The process of finding one's way.
Negative feedback: See *feedback.*
Nematocyst (nem'ə tə sist)**:** One of the minute stinging cells of the coelenterates, consisting of a hollow thread coiled within a capsule and an external hair trigger.
Nematode (nəm'ə tōd)**:** Any unsegmented worm of the phylum or class *Nematoda,* having an elongated, cylindrical body, such as the roundworm.
Nephric unit (nef'rik unit)**:** A division of the kidney including the glomerulus, a Bowman's capsule, the loop of Henle, and the collecting duct.
Nephridium (ne frid'ē əm) (pl. *nephridia)***:** An excretory organ found in annelids.
Nerve: A number of neurons following a common pathway, covered by a protective sheath and supporting tissue.
Neural crest: The ridge of a neural fold, migrating cells of which will give rise to spinal ganglia, the adrenal medulla, the autonomic nervous system, and pigment cells.
Neural folds: In early vertebrate embryology, a pair of longitudinal ridges that arise from the neural plate on either side of the neural groove and that fold over and give rise to the neural tube, which eventually becomes the spinal cord.
Neural groove: An ingrowth of the neural plate of a vertebrate embryo; it eventually forms the spinal cord in mammals.
Neural plate: A thick plate formed by rapid cell division over the notochord on the dorsal side of the embryo; it eventually forms the nervous system in mammals.
Neuron: A cell specialized for the transmission of nerve impulses. It consists of one or more branched *dendrites,* a nerve *cell body* in which the nucleus resides, and a terminally branched *axon.* Also called a *nerve cell.*
Neurophysiology (nyoor'ə fiz'ē ol'ə jē)**:** The branch of physiology dealing with the nervous system.
Neurotransmitter: A short-lived, hormonelike chemical (such as acetylcholine) that, when released from an axonal knob into a synaptic cleft, crosses the space to stimulate the next neuron to transmit a nerve impulse. See also *synapse* and *synaptic cleft.*
Neutron: An atomic nuclear particle that has about the same mass as a proton, but is electrically neutral.
Neutrophil (nyo͞o' trō fil)**:** The most common mammalian phagocytic leukocyte.
Niche: All aspects of the biological and physical environment that relate to the activities of an organism.
Nicotine (nik'ə tēn')**:** A colorless, oily, water-soluble, highly toxic, liquid alkaloid obtained from tobacco.
Nitrogen oxide: An oxide of nitrogen; a common air pollutant in some industrial areas.
Node: 1. In a plant stem, any point at which one or more leaves emerge. 2. In a growing stem tip, a region of potential leaf growth, containing meristematic tissue.
Nonrenewable resources: Resources that exist in set amounts on the planet and that cannot be replaced. Once used, they are gone or so changed that they are difficult or impossible to return to their original state.
Nonspecific response: A general immune response, not directed at a specific invader.
Norepinephrine (nôr'ep'ə nef'rən)**:** A compound that serves as a synaptic neurotransmitter and as an adrenal hormone. Also called *noradrenaline.*
Notochord (nō'tə kôrd)**:** A turgid, flexible rod running along the back beneath the nerve cord and serving as a body axis; it exists in all chordates at some point in development, but is replaced in most vertebrates by the vertebral column.
Nuclear membrane: The double membrane surrounding the eukaryote nucleus.
Nuclear pore: A tiny hole in the nuclear membrane of cells.
Nuclear winter: The hypothetical condition that would prevail after a nuclear exchange, when the sunlight would be largely blocked from reaching the earth, thus lowering global temperatures.
Nucleolus (nü klē'ə ləs)**:** An organelle that lies within the nucleus of the cell; it is high in RNA and protein and forms ribosomal RNA.
Nucleotide (nü klē'ə tīd)**:** A molecule made up of phosphate, a 5-carbon sugar, and a purine or pyrimidine base.
Nucleus: An organelle with a cell bounded by a double membrane that contains most of the DNA of the cell. Also, in the atom, a postively charged mass composed of neutrons and protons.

O

Occipital lobe (ok sip'i tal lobe)**:** One of the four major lobes of the brain. It is involved in the reception and processing of visual information.
Olfaction: 1. The sense of smell. 2. The process of smelling.
Olfactory bulb (ol fak'tər ē bulb)**:** An extension of the brain that receives neurons from the olfactory receptors in the nasal passage.
Oligochaeta (ol'ə gō kē'tə)**:** The class composed of annelids that have locomotory setae sunk directly into the body wall, such as earthworms.
Omnivore (om'nə vōr)**:** An organism that eats both animals and plants.
Open circulatory system: A circulatory system in which the vessels open into intercellular spaces, through which the blood percolates before returning to the heart. Compare *closed circulatory system.*
Operant conditioning: A form of learning in which an animal performs an act to receive rewards.
Ophiuroidea (o fi'yoo roi'dē ə)**:** The brittle sea stars.
Opiate: A drug or medicine containing opium, medically used as a sedative or pain killer.
Optic chiasma (optic kī az' mə)**:** The area where the optic nerves meet and part of each crosses to serve the area of the brain opposite the nerve.
Oral groove: In ciliates, a ciliated fold in the body wall, leading into the cytostome or mouth.
Orbit, electron: The approximate and hypothetical path on which an electron moves around the nucleus of an atom.
Order: A taxonomic subdivision of a

class in plants and animals that may consist of several families.
Organ: A distinct structure that consists of a number of tissues and carries out a specific function.
Organ of Corti: On the basilar membrane in the cochlea, an organ containing the neural receptors for hearing.
Organ system: A number of organs participating jointly in carrying out a basic function of life (i.e., respiration, excretion, reproduction, digestion).
Organelle: A specific structure within a cell, usually bounded by a membrane.
Orgasm: In humans, the climax of sexual excitement typically occurring toward the end of coitus, usually accompanied in men by ejaculation and in women by rhythmic contractions of the cervix.
Orientation (ōr'ē en tā'shən): The ability to determine direction in one's environment.
Origin: The nonmoving, skeletal base to which a muscle or tendon is attached. Compare *insertion.*
Osmosis (os mō'sis): The diffusion of water across membranes, from an area of lower solute concentration to one of greater solute concentration.
Osteichthyes (os'tē ik'thē ēz'): The class composed of bony fishes.
Oval window: In the cochlea, a membrane articulating with the stapes that moves in response to its vibrations, subsequently creating movement in the fluid within.
Ovary: 1. In flowering plants, the enlarged, rounded base of a pistil, consisting of a carpal or several united carpels, in which ovules mature and megasporogenesis occurs. 2. The female gonad in which the eggs are formed.
Oviduct: A tube, usually paired, for the passage of eggs from the ovary to a uterus, or the exterior of the body.
Ovule (ō'vyool): In seed plants, an oval body in the ovary that contains the female gametophyte and consists of the embryo sac surrounded by maternal tissue.
Oxaloacetate (ok sal'ō as'i tāt): One of the intermediate molecules in the Krebs cycle.
Oxaloacetic acid (ok sal'ō ə sē'tik acid): The nonionized form of oxaloacetate (an intermediate in the Krebs cycle).
Oxidation: The loss of electrons in chemical reactions, or the combination of oxygen with another substance.

P

Pacinian corpuscle (pa sin'ē ən corpuscle): An oval pressure receptor containing the ends of sensory nerves, especially in the skin of hands and feet.
Palaeotragus (pā'lē ō trā'gəs): A giraffe-like creature that lived twenty million years ago and still survives in the Congolese rain forest.
Paleozoic (pā'lē a zō'ik): Noting or pertaining to a geological era occurring between 220,000,000 and 800,000,000 years ago, characterized by the appearance of fish, insects, and reptiles.
Paramecium (par'ə mē'shē əm): Any ciliated freshwater protozoan having an oval body and a long, deep oral groove.
Parapatric (par'ə pa'trik): Living in separate but adjacent geographic regions; compare *allopatric* and *sympatric.*
Parapod (par'ə pod): Also *parapodium;* one of the short, unsegmented, paired leglike or finlike locomotive organs borne on either side of each body segment in *Nereis* and certain other polycheate worms.
Parasite: An organism that lives in or on another organism and causes some harm to its host as it derives food from it.
Parasympathetic nervous system: One of the divisions of the autonomic nervous system.
Parathyroid gland: One of four small endocrine glands embedded in or adjacent to the thyroid gland and involved in the regulation of calcium ion levels in the blood.
Parathyroid hormone: Also *parathormone;* the internal secretion of the parathyroid glands, involved in maintaining normal calcium balance.
Parenchyma (pə reng'kə mə): A basic plant tissue type, consisting typically of thin-walled cells and commonly specializing in photosynthesis and storage.
Parent cell: The arbitrarily chosen cell used to initiate studies of cellular descendancy.
Parietal lobe (pə rī'ə tel lobe): One of the four major lobes of the cortex. The detection of body position and sensory input are the major functions.
Partial dominance: Where neither of a pair of alleles is totally dominant and the combined expression of the two alleles in the heterozygote produces an intermediate trait; e.g., in the blossoms of four o'clocks, red × white = pink.
Particle: Ecologically, any small unidentified object; often an invasive irritant or contaminant; chemically, any of the components of an atom.
Passive transport: The movement of a substance through the cell membrane without cellular energy (ATP) involved.
PCP: See *phencylidine.*
Pectoral girdle (pek'tər əl girdle): The bones or cartilage supporting and articulating with the vertebrate forelimb.
Pellicle (pel'i kəl): 1. The semirigid, proteinaceous integument of many protists.
Pelvic girdle: The bones or cartilage supporting and articulating with the vertebrate hind limbs; in humans, consisting of the fused bones of the pelvis.
Penicillium (pen'i sil'ē əm): The mold that produces penicillin.
Penis: The male sex organ.
Peptide (pep'tīd): A compound composed of two or more amino acids.
Peptide bond (pep'tīd bond): The dehydration linkage formed between the carboxyl group of one amino acid and the amino acid group of another.
Peptide chain (pep'tīd chain): Two or more amino acids linked by peptide bonds, most often seen as a partial digestive product of a protein or polypeptide.
Peripheral nervous system: One of the two divisions of the vertebrate nervous system; includes the nervous tissue outside the central nervous system.
Periphery (pə rif'ə rē): The external boundary of any surface or area.
Perissodactyla (per'ə sō dak'tə le): "Odd toes"; vertebrates with odd numbers of parts to their hooves, such as horses.
Peristalsis (per'ə stal'sis): Successive waves of involuntary contractions passing along the walls of the esophagus, intestine, or other hollow muscularized tube, forcing the contents onward.
Permafrost: In arctic and high altitude tundra, the permanently frozen layer of soil and/or subsoil.
Permease (per'mē ās): An enzymelike carrier that functions in facilitated transport of a specific substrate across a plasma membrane.
Peroxisome (pə rok'si sōm): A cytoplasmic organelle involved in the detoxification of peroxides.
Pesticide: A chemical preparation for destroying insects such as flies, mosquitoes, and ants.
Petal: One of the usually white or brightly-colored leaflike elements of the corolla of a flower.

PGAL: Phosphoglyceraldehyde; an end product of photosynthesis.
Phaeophyta (fā o fit'ə): The brown algae.
Phagocyte (fag'ə sīt): Any leukocyte that engulfs particles.
Phagocytosis (fag'ə sī tō'sis): The engulfment of solid materials into the cell and the subsequent pinching off of the cell membrane to form a digestive vacuole.
Pharynx (far'ingks): The organ below the mouth that is the common respiratory and digestive structure leading to the esophagus and larynx.
Phencylidine (PCP) (fēn sī li dēn'): A powerful animal tranquilizer that has a variety of undesirable effects on humans, accompanied by or preceded by euphoria.
Phenotype (fē'nə tīp'): The observable genetic characteristics of any organism.
Pheromone (fer'ə mōn') A substance that is secreted by an organism that changes the behavior of another organism of the same species.
Phloem (flō' em): A complex vascular tissue of higher plants that consists of sieve tubes, companion cells, and phloem fibers, and functions in transport of sugars and other nutrients.
Phosphoglyceraldehyde (PGAL) (fos'fō glis'ə ral'de hīd'): An end product of photosynthesis.
Phospholipid (fos'fō lip'id): A fatlike substance of the cell membrane that contains phosphorus, fatty acids, glycerol, and a nitrogenous base.
Photochemical smog: Smog that becomes more reactive when energized by sunlight.
Photon (fō'ton): A unit of light energy.
Photoreceptor: A receptor of light stimuli.
Photosynthesis (fō'tō sin'thə sis): The process in which plants convert carbon dioxide and water into carbohydrates through the use of solar energy.
Photosystem I: The second in the two photosystems in the electron pathway of photosynthesis in cyanobacteria and chloroplasts, and the one involving the reduction of NADP to $NADPH_2$; believed to be evolutionarily more ancient than photosystem II.
Photosystem II: The first of the two photosystems in the electron pathway of photosynthesis in cyanobacteria and all photosynthetic eukaryotes, and the one involving the photolysis of water.
Phycocyanin (fī'kō sī'ə nin): A blue protein pigment found in algae that contributes to the process of photosynthesis.
Phycoerythrin (fī'kō i rith'rin): A reddish protein pigment found in algae that contributes to the process of photosynthesis.
Phycomycete (fī'kō mī'sēt): A fungus from the class that contains downy mildews and so on.
Phycomycota (fi'kō mī kot'e): See *phycomycete.*
Phylum (fī'lem): The major primary subdivision of a kingdom, composed of classes.
Physical dependence: The state that exists when a drug is necessary simply to maintain body comfort.
Phytoplankton (fī'tō plangk'tən): In the aquatic environment (marine and fresh waters), minute photosynthesizing organisms such as diatoms; the base of the marine food chain.
Pigment: Any coloring matter or substance.
Pinocytosis (Pin'ō sī to'sis): Taking dissolved molecular food materials, such as proteins, into the cell by adhering them to the plasma membrane and invaginating portions of the plasma membrane to form digestive vacuoles; a form of active transport.
Pistil: In flowering plants, the female reproductive structure, composed of one or more carpels and ovaries, and a style and a stigma. See also *carpel.*
Pistillate (pis'tə lit): Having pistils but no stamens.
Pituitary gland: A tiny endocrine gland at the base of the brain.
Placenta (plə sen'tə): Embryonic membranes in the uterus of pregnant mammals through which the embryo transports wastes and receives essential nutrients.
Placental mammals: Mammals that form chorioallantoic placentas; all mammals other than marsupials and monotremes.
Planaria (plə nar'ē ə): The genus of free-swimming flatworms.
Planula (plan'yə lə): The early, ciliated, free-swimming larva of a coelenterate.
Plasma: The fluid matrix of blood. It is 90 percent water and 10 percent various other substances, including plasma proteins, ions, and foods.
Plasma cell: A mature, short-lived B-cell lymphocyte, specialized in secreting antibodies.
Plasma membrane: The external semipermeable limiting layer of the cytoplasm.
Plasmid: A small circle of bacterial DNA that exists outside the single large circular chromosome.
Plasmodesmata (plaz'mo dez ma' tə): The tiny cytoplasmic bridges that extend through cell walls and connect the cytoplasm of adjacent living plant cells.
Plasmodium (plaz mō'dē əm): 1. A motile, multinucleate mass of protoplasm produced by the fusion of uninucleate slime mold ameboid cells. 2. The malarial parasite.
Plastid: A small body in a plant cell.
Plate: Any of the great land and ocean floor masses comprising the surface of the earth.
Plate tectonics: The movement of great land and ocean floor masses (plates) on the surface of the earth relative to one another, occurring largely in the Cenozoic era; also called *continental drift.*
Platelets: Minute, fragile noncellular discs present in the vertebrate blood. Upon injury, they are ruptured, releasing factors that initiate blood clotting and wound healing.
Platyhelminth: Any worm of the phylum *Platyhelminthes,* having bilateral symmetry and a soft, solid, usually flattened body, such as the planarians, tapeworms, and trematodes.
Point mutation: A mutation involving a minor change in a DNA sequence, such as a base substitution, addition, or deletion.
Polar body: A small, functionless egg formed during meiosis.
Pollen: The male microspores of flowering plants consisting of reproductive cells in a protective case.
Pollen grain: The male gametophyte of a seed plant; contains a generative nucleus and a tube nucleus and is enclosed in a hardened, resistant case.
Pollen sac: One of two or four chambers in an anther in which pollen develops and is held.
Pollen tube: A tube that extends from a germinating pollen grain and grows down through the style to the embryo sac, into which it releases sperm nuclei.
Pollen tube cell: The cell directly behind the pollen tube as it grows through the style of a flower before actual fertilization by one of the nuclei that follows it.
Pollutant: Anything that makes another thing unclean, impure, or contaminated.
Polychaeta (polrē kē'tə): The class com-

posed of annelids having unsegmented swimming appendages with many chaetae or bristles.
Polygenic inheritance: Inheritance involving many interacting variable genes, each having a small effect on a specific trait.
Polymorphic: Having, assuming, or passing through different forms or stages.
Polyp: The typical attached, nonswimming form of coelenterate. Compare *medusa*.
Polypeptide: A group of two or more amino acids; a protein fragment.
Polyploid: Having more than two complete sets of chromosomes per cell.
Polyploidy: Multiple sets of chromosomes, often occurring in plants.
Polysaccharide (pol′ē sak′ə rīd)**:** A large molecule of sugar subunits.
Pons: Adjacent to the medulla; it contains nerve cells that relate the spinal motor nerves.
Population: An interbreeding group of organisms. Usually a subset of a species.
Population dynamics: The study of how populations change.
Porifera: The phylum composed of the sponges.
Portal: In circulation, a vessel between two capillary beds.
Positive feedback: See *feedback*.
Posterior lobe: The lobe of the pituitary gland nearer the dorsal surface.
Post-Mendelian genetics: The study of genetics after the classical phase suggested by Mendel.
Potential energy: Energy stored in chemical bonds, in nonrandom organization, in elastic bodies, in elevated weight or any other static form in which it can theoretically be transformed into another form or into work.
Pre-Cambrian (prē kam′brē ən)**:** Noting or pertaining to the earliest geological era, ending 600,000,000 years ago, during which the earth's crust was formed and the first life appeared.
Prefrontal area: The area situated at the anterior of the frontal lobe.
Primary growth: The initial growth or elongation of a plant stem or root, resulting in an increase in length and the addition of leaves, buds, and branches. Compare *secondary growth*.
Primary immune response: The relatively slow response of the immune system upon its first contact with an invading organism or foreign protein. Compare *secondary immune response*.
Primary phloem (primary flō′em)**:** Phloem developed from apical meristem; that is, the phloem of primary growth.
Primary response: The immune response that occurs upon the first exposure to alien matter.
Primary structure: The amino acid sequence of proteins.
Primary xylem (primary zī′lem)**:** In primary growth, xylem produced by procambium rather than vascular cambium.
Primata (prī mā′tə)**:** "First"; the order comprising the primates, such as humans, the apes, monkeys, lemurs, tarsiers, and marmosets.
Proboscidea (prō′bə sīd′ē ə)**:** "Front feeder"; the order that includes elephants.
Procambium (prō kam′bē əm)**:** In plants, the primary tissue that gives rise to primary xylem and primary phloem.
Product: The result of a chemical reaction.
Progesterone (prō jes′tə rōn)**:** An ovarian hormone, produced by the corpus luteum; assists in the preparation of the uterus for implantation of a fertilized egg.
Proglottid (prō glot′id)**:** Any segment of a mature tapeworm, containing male and female organs and being shed when full of mature fertilized eggs.
Prokaryotic cell (prō kar′ē ō′tik cell)**:** A cell lacking a membrane-bounded nucleus and membrane-bounded organelles; it has a single circular chromosome of nearly naked DNA.
Pronator (prō nō′tor)**:** A muscle that turns a surface downward.
Prophase: The first stage of mitosis or meiosis.
Proprioception (prō′prē ə sep′shen)**:** The ability to sense and respond to stimuli from within the body, including the integration of the movement of body parts, balance, and stance.
Proprioceptor (prō′ prē ə sep′ tər)**:** A sensory receptor that respond to changes in body position, muscle tension, or internal chemistry.
Prostaglandin (pros′tə gland′ən)**:** A kind of fatty acid hormone; may play a role in fertilization.
Protein: A specific chain of amino acids; essential in producing cell structures and a variety of enzymes.
Proteinoids (prō′ ta noid)**:** Minute globes of amino acid polymers that form spontaneously when proteins are placed in solution, known to grow and divide in a cell-like manner.
Protista (prə tis′tə)**:** The group of organisms that includes all the unicellular animals and plants.
Protobiont (prō′ ta bī′ ant)**:** Abiotically produced droplets that have chemical characteristics different from their surroundings, believed to be a stage in the development of life.
Proton: An atomic nuclear particle that has a mass similar to a neutron, but with a positive charge.
Protostome (prō′tə stōm)**:** An animal in which the mouth derives from the first embryonic opening (the blastopore). Compare *deuterostome*.
Prototheria (prō′tə thēr′ē ə)**:** The group that includes the monotremes like the platypuses.
Protozoan (prō′tə zō′ən)**:** 1. Any of a large group of protists. 2. Any nonphotosynthetic protist.
Proximal convoluted tubule: In the kidney, part of the nephron between Bowman's capsule and the loop of Henle.
Proximate immediate level: The influences on an individual as a result of individual experience.
Pseudocoelom (sü′də sē′lem)**:** In nematodes and rotifers, the body cavity between the body wall and the intestine that is not entirely lined with mesodermal epithelium; the gut is entirely endodermal and thus not muscularized.
Pseudopod (sü′də pod)**:** Literally, a false foot; a temporary projection from the cell used in ameboid cells for feeding and movement.
Psychedelic (sī kə del′ik)**:** A substance that may produce sensory distortions, bizarre thought patterns, and abnormal behavior.
Psychoactive agent: Any of a number of chemicals that alters mood or perception.
Psychological dependence: The state in which a drug is necessary for mental or emotional comfort.
Pulmonary artery (pul′mə ner′ē artery)**:** A branching artery that conveys deoxygenated blood from the right ventricle to the lungs.
Pulmonary vein (pul′mə ner′ē vein)**:** In birds and mammals, the only vein that carries oxygenated blood, returning it from the lungs to the left atrium.
Punctuated equilibrium: A theory stating that evolution does not proceed in a gradual manner but rather in sudden bursts of activity, followed by very long time intervals during which there is little evolutionary activity.

Punnett square (pun'et square): A graphic device used to predict genetic ratios.
Pus: A thick, yellowish-white accumulation of dead phagocytes, dead or living bacteria, and tissue debris in a fluid exudate.
Pyrogen (pī'rə jen'): Any fever-producing substance.
Pyrrophyta (pir'ə fī'te): "Fire-colored plants"; microscopic, photosynthetic algae with two tinsellike flagella.
Pyruvic acid (pī rü'vik acid): A water-soluble liquid, important in many metabolic and fermentation processes.

Q

Quaternary structure: The interaction of two or more polypeptides through disulfide linkages.

R

Radial symmetry: Body symmetry with structures that radiate out from a central point, as in a starfish.
Radioisotope (rā'dē ō ī'se tōp): An unstable isotope that spontaneously breaks down with the release of ionizing radiation; also called *radioactive isotope.*
Radula (raj'u le): In all mollusks except bivalves, a toothed, chitinous band that slides backward and forward, scraping and tearing food and bringing it into the mouth.
Rate of natural increase: In population dynamics, the difference between the crude birth rate and the crude death rate.
Reactant: Any element, ion, or molecule participating in a chemical reaction.
Receptacle: The end of a floral stalk that forms the base on which the flower parts are borne.
Receptor: A sensory structure that detects a particular form of stimulus in the environment, such as light or sound.
Recessive traits: Traits due to alleles that will be masked in the presence of dominant alleles.
Reciprocal altruism: Benefitting another individual at some cost in the expectation (and probability) that the favor will be returned.
Recombinant DNA (rē kom'bə nənt DNA): The general term for laboratory manipulation of DNA; includes gene splicing and gene cloning.
Rectum: The terminal part of the intestine, used for the temporary storage of feces.
Red marrow: The regions within the ribs, sternum, vertebrae, and hip bones where red blood cells are produced.
Reduction: The addition of electrons or hydrogen atoms to a substance.
Reflex arc: A simple nervous pathway that involves a sensory cell, a connecting neuron, and a muscle cell.
Releaser: A stimulus that acts as a cue, releasing a certain behavior in an animal.
Renal artery (rē'nəl artery): The artery that feeds blood to the glomerulus.
Renewable resources: Resources that do not exist in set amounts; they can be replenished or reused.
Replica: A copy or reproduction.
Repolarization: In the neuron, reestablishment of the resting potential or polarized state following an action potential.
Reproductive isolation: The state of a population or species in which successful mating outside the group is impossible because of anatomical, geographic, or behavioral mating barriers.
Reptilia (rep til'ē ə): "Crawlers"; the class composed of cold-blooded vertebrates such as the turtles, lizards, snakes, crocodilians, and the tuatara.
Respiration: 1. The physical and chemical process by which an organism supplies oxygen to its tissues and removes carbon dioxide. 2. Also *aerobic respiration,* any energy-yielding reaction in living matter involving oxygen.
Resting potential: The charge difference across the membrane of a neuron or muscle fiber while it is not transmitting an impulse. Compare *action potential.*
Restrictive enzyme: In bacteria, an enzyme that recognizes & cleaves a specific, short DNA sequence, protecting the cell from all but a few host-specific viruses.
Reticular system (ri tik'yə lər system): A major neural tract in the brain stem containing neural pathways to other parts of the brain and to the reticular activating system (RAS) and arousal center.
Retina (ret'ə nə): The layer of light sensitive cells in the vertebrate eye.
Reward: Reinforcement; something positive given to strengthen the probability of a response to a given stimulus.
Rhizoid (rī'zoid): 1. A portion of a fungal mycelium that penetrates its food medium. 2. A rootlike structure that serves to anchor the gametophyte of a fern or bryophyte to the soil.
Rhizome (rī'zōm): An underground, horizontal plant stem that is often thickened by deposits of reserve food material, produces shoots above and roots below, and is distinguished from a true root in possessing buds, nodes, and usually scale-like leaves.
Rhodophyta (rō'dō fī'tə): The red algae.
Rhynia (rin'ē e): One of the oldest known fossil plants, flourishing about 400,000,000 years ago.
Rhythm method: A method of contraception whereby copulation is avoided during periods when conception is likely; also called *natural birth control.*
Ribonucleic acid (RNA) (rī'bō nü klē'ik acid): A nucleic acid containing ribose sugar and the nitrogen base uracil; functions in protein synthesis.
Ribosomal RNA (rī'bə sō'mel RNA): The RNA that forms the matrix of ribosome structure; its actions are not clearly known, but it is believed to function in protein synthesis.
Ribosome (rī be sōm): A grainy structure found in great number in cells; important in protein synthesis.
Ribulose diphosphate (rī'byoo lōs dī fos'fāt): A form of ribulose that acts as the carbon dioxide acceptor in the Calvin cycle of photosynthesis.
RNA: See *ribonucleic acid.*
Rocky coast: A seacoast area characterized by abundant rock formations, pounding surf, and numerous life forms that occur in distinct zones.
Rod: One of the numerous long, rod-shaped sensory bodies in the vertebrate retina; it contains many membrane layers bearing visual pigments and is responsive to faint light. Compare *cone.*
Rodentia (rō den'shē ə): "Gnawers"; the order of gnawing or nibbling mammals, including the mice, squirrels, beavers, and so on.
Root: The part of a seed plant that functions as an organ of absorption, anchorage, & sometimes food storage, differing from the stem in lacking nodes, buds & leaves.
Rough ER: Endoplasmic reticulum studded with ribosomes, as opposed to smooth ER.
Roundworm: Any nematode that infests the intestine of humans and other mammals.

S

Saccule (sak'yool): 1. The smaller of two sacs in the membranous labyrinth of the ear; compare *utricle.* 2. A small sac anywhere in the body of a plant or animal.
Saltatory propagation (sal' ta tor' ē

prop′ a gā shən): The rapid transmission of a neural impulse, along myelinated fibers as depolarization occurs in internodes, between the Schwann cells.
Salting out: A method of abortion used in the second trimester in which the fetus is killed with an injection of a concentrated salt solution into the amniotic cavity. Also called *saline abortion.*
Sandy seashore: A coastal zone subject to a constant shifting of the sand bottom; has relatively few life forms.
Saprobe (sap′rōb): An organism that reduces dead plant and animal matter.
Sarcodina (sär′kə din′ə): The class of protozoans that move and capture food by forming pseudopods.
Sarcomere (sär′kə mir): The contractile unit of striated muscle bounded by Z-line partitions; consists of actin filaments bound to the Z-line partitions and myosin filaments regularly interspersed between them.
Saturated: Having accepted as many hydrogens as possible. *Saturated fat,* a triglyceride lacking carbon-carbon double bonds.
Scanning electron microscope: A device for visualizing microscopic objects by scanning them with a moving beam of electrons, recording impulses from scattered electrons, and displaying the image by means of the synchronized scan of an electron beam in a cathode ray (television) tube.
Scapula (skap′yə lə): Either of the two flat, triangular bones, forming a shoulder blade.
Schwann cell: One of the many cells that constitute the myelin sheath, wrapped around the axon of a myelinated neuron.
Sclerenchyma (skli reng′kə mə): A protective or supporting plant tissue composed of cells with greatly thickened, lignified, and often mineralized cell walls.
Scolex (skō′ leks): The hook-bearing head of a larval, or adult, tapeworm, from which the proglottids are produced.
Scyphozoa (sī′fə zō′ə): The class consisting of marine jellyfishes.
Secondary growth: Growth in dicot plants that results from the activity of secondary meristem, producing chiefly an increase in the diameter of stem or root. Compare *primary growth.*
Secondary meristem: The tissue that causes growth in plant girth, the vascular and cork cambiums.
Secondary response: The more rapid production of antibodies and conquest of an invader during a second or subsequent infection. Compare *primary immune response.*
Secondary structure: The pattern of folding of adjacent residues of a macromolecule, usually in the form of helices or sheets.
Second law of thermodynamics: The statement in physics that, left alone, all systems proceed toward entropy (disorganization).
Second messenger: An intracellular chemical compound that transfers a hormonal message from the cell membrane to the nucleus or cytoplasm, after being triggered by the first messenger.
Sedative-hypnotic: A drug that acts on the cerebral cortex to reduce anxiety and induce drowsiness and sleep.
Seed: The fertilized and ripened ovule of a seed plant, comprising an embryo, including one or two cotyledons, and usually a supply of food in a protective seed coat; capable of germinating under proper conditions and developing into a plant.
Seed coat: The outer protective covering of a seed, composed of one or more layers of tissue derived from the parent sporophyte.
Segregation, principle of: The separation of allelic genes in different gametes during meiosis, resulting in the separation of their characters in the offspring.
Selectively permeable: In cellular membranes, the characteristic of permitting some substances to pass through while rejecting others.
Semicircular canal: Any of the three curved tubular canals in the labyrinth of the ear, associated with the sense of equilibrium.
Seminal receptacle: A storage organ for sperm in certain invertebrate females. The sperm is released for fertilization after mating.
Senses: The faculties, such as sight, hearing, smell, taste, or touch, by which organisms perceive stimuli originating from outside or inside the body.
Sepal: The leaklike floral parts surrounding the base of the petals. See also *calyx.*
Sequence: In proteins, the linear arrangement of amino acids in the formation of large protein molecules; defines a protein's primary structure.
Seta (sē′tə) (pl. *setae*): 1. In the moss sporophyte, the stalklike growth that supports the sporangium. 2. The bristlelike, chitinous structure in the body wall of annelids, arthropods, and certain other invertebrates.
Sexual dimorphism: The differences—in size, color, anatomy, etc.—between the sexes.
Shell: One of a number of energetic levels at which electrons move around a nucleus.
Sickle cell anemia: A hereditary disease caused by a form of hemoglobin that is defective in its nature and causes sickle-shaped red blood cells.
Sieve tube: In phloem, a thin-walled tube consisting of an end-to-end series of enucleate, living cells joined by sieve plates; a channel through which sap flows.
Sigmoid: S-shaped. *Sigmoid colon,* an S-shaped part of the large intestine.
Simple sugar: A carbohydrate molecule containing a single sugar group. See also *monosaccharide.*
Single-gene effect: Trait due to the expression of a single gene; relatively rare.
Sinus: A cavity that forms part of an animal body; e.g., any of the several air-filled, mucous-membrane lined cavities of the skull.
Skeletal muscle: Muscle attached to the skeleton, under direct and conscious control, striated with multinucleate unbranched fibers. Also called *voluntary muscle, striated muscle.* Compare *cardiac muscle, smooth muscle.*
Skeletal system: The rigid supportive tissue to which muscles (or contractile units) may be attached.
Skinner box: A device for investigating operant conditioning, named after B. F. Skinner, who invented it.
Sliding filament theory: The widely accepted explanation of skeletal muscle contractions in which actin myofilaments in the sacromere are actively drawn through myosin myofilaments, thus shortening the contractile unit.
Small intestine: In vertebrates, the region of the alimentary canal between the stomach and the cecum; the region in which most food absorption occurs.
Smooth ER: Endoplasmic reticulum that is not studded with ribosomes.
Smooth muscle: The muscle tissue of the glands, viscera, iris, piloerectors, and other involuntary structures; consists of masses of uninucleate, unstriated, spindle-shaped cells, usually occurring in thin sheets. Also called *involuntary muscle.*
Sociobiology: The study of ecological,

evolutionary, and genetic influences on social behavior.
Sodium/potassium pump: A poorly understood molecular entity in the plasma membrane, capable of actively transporting sodium out of the cell and potassium in, at a cost of ATP energy. Also called *sodium pump.*
Solar energy: Energy obtained from the sun.
Solute (säl'yüt): Any substance dissolved in a solvent; substance in solution.
Somatotropin (sō'mə tə trō'pin): A growth hormone.
Somite (sō'mīt): One of the paired mesadermal segments along the neural axis of an embryo.
Sound spectrogram: A visual display of the elements of a sound.
Source DNA: The DNA from which a short strand is taken to be placed into another strand of DNA.
Specialization: The process of narrowing abilities so that fewer roles are performed with greater efficiency; the process of becoming increasingly differentiated.
Speciation (spē'shē ā'shən): An evolutionary process by which new species are formed, normally by the division of one species into two.
Species: The major subdivision of a genus. Individuals of a species are able to breed among themselves under natural conditions.
Specific response: An immune response directed against a specific invader.
Sperm cell: The male gamete.
Spermicide: An agent that kills sperm.
Sphincter: A ring of muscles that controls the movement of materials through passages; e.g., sphincters at either end of the stomach.
Spicule (spik'yool): 1. One of the many tiny calcareous or siliceous pointed bodies embedded in and serving to stiffen and support the tissues of various invertebrates, including sponges and sea cucumbers. 2. Any small, stiff, spikelike or needlelike body part.
Spinal cord: The complex band of neurons that runs through the spinal column of vertebrates to the brain.
Spinal nerve: Any of the many nerves that enter and leave the spinal cord, including both somatic and autonomic. Compare *cranial nerve.*
Spindle: A structure formed in mitosis and meiosis that consists of fine threads radiating from the centrosomes along which the chromosomes appear to move.
Spiracle (spī'rə kəl): An external opening to the respiratory system of terrestrial arthropods.
Spirillum (spī ril'əm): Any of several spirally twisted, aerobic bacteria of the genus *Spirillum,* certain species that are pathogenic for man.
Sponge: A birth control device; a small, absorbent polyurethane sponge that is saturated with a spermicide and inserted into the vagina before intercourse.
Sporangium (spə'ran jē'əm): A hollow structure in which spores are formed.
Spore: An asexual reproductive cell that can develop into an adult.
Sporophyte: In plants having an alternation of generations, a diploid individual capable of producing haploid spores by meiosis; the prominent form of ferns and seed plants. Compare *gametophyte.*
Sporophytic (spōr'ə fit'ik): Referring to the spore-producing, diploid generation in plants that alternates with the sexual or gametophyte generation.
Sporozoa: The class composed of parasitic protozoans.
S-shaped curve: A plot of population growth where growth is rapid at first but then slows when *environmental resistance* is met and levels off at some point near or below the *carrying capacity.* Compare *J-shaped curve.*
Stabilizing selection: Selection against both extremes of a continuous phenotype, favoring an intermediate optimum.
Stalk: A stem, shaft, or slender supporting part of the structure of a plant or animal.
Stamen (stā'mən): The male reproductive structure of a flower, consisting of a pollen-bearing anther and the filament on which it is borne.
Staminate (stam'ə nit): Having stamens but no pistils; an exclusively male flower.
Stapes (stā'pēz): The outer, stirrup-shaped bone of a chain of three small bones in the middle ear of man and other mammals. Also called *stirrup.*
Sterilization: The destruction of the ability to reproduce by removing the sex organs or inhibiting their functions.
Sternum: A median ventral bone or cartilage in land vertebrates, connecting with the ribs, shoulder girdle, or both.
Steroid: A type of lipid consisting mainly of fatty acids attached to complex alcohols.
Stigma: In flowers, the top, slightly enlarged and often sticky end of the style, on which pollen grains adhere and germinate.
Stimulation: The state of being excited or invigorated.
Stoma: (pl. *stomata*): One of the minute pores in the epidermis of leaves, stems, and other plant organs; formed by the concave walls of two guard cells; allows the diffusion of gases into and out of the intercellular spaces.
Stomach: A saclike enlargement of the alimentary canal, as in humans and certain animals, forming an organ for storing, diluting, and digesting food.
Stretch receptor: A sensory receptor that is stimulated by stretching, as in a tendon, muscle, or bladder wall.
Striation (strī ī'shən): A striped condition or appearance.
Strobilation (strob'ə lā'shən): Asexual reproduction by traverse division of the body into segments which break free as independent organisms, occurring in certain coelenterates and flatworms (tapeworms).
Stroma: Matrix of a chloroplast in which the grana are imbedded.
Style: The stalk of the pistil in a flower connecting the stigma with the ovary.
Substrate: A substance that is acted upon by an enzyme.
Sudden death: An unanticipated demise, possibly due to congenital defects.
Sugar: A general term for certain larger carbohydrates.
Sulfur oxide: An oxide of sulfur; a common air pollutant in some industrial areas.
Superior vena cava (superior vē'nə kā'və): The anterior large vein by which blood is returned to the right atrium of the heart of land vertebrates.
Supernormal releaser: An environmental stimulus with exaggerated features that produces an instinctive response; not normally encountered in nature.
Supinator (soo'pe nā'ter): A muscle that rotates the hand or forearm so that the palm is facing upward. Compare *pronator.*
Suppressor T-cell: A type of white blood cell that shuts down the immune response after the risk of infection has passed.
Suture line (so͞o'cher line): In anatomy, immovable joints formed by the articulation of skull bones.
Symbiosis hypothesis (sim'bē ō'sis hī poth'ə sis): The hypothesis that the eukaryotic cell evolved from the mutualistic union of various prokaryotic organisms, one of which gave rise basically to the cytoplasm, nucleus, and motile mem-

branes, a second to mitochondria, a third to chloroplasts and other plastids, and a fourth to cilia, eukaryotic flagella, basal bodies, centriles, the spindle, and all other microtubule structures.
Sympathetic nervous system: A subdivison of the autonomic nervous system that increases energy expenditure and prepares the body for emergency situations.
Sympatric (sim pat'rik)**:** Occupying the same geographical area.
Sympatric speciation (sim pat'rik spē'shē ā'shən)**:** Speciation in populations that are not geographically separated. Compare *allopatric speciation.*
Synapse (sin'aps)**:** The junction between the axon of one neuron and the dendrite or cell body of another; crossed by neural impulses.
Synaptic cleft: The minute space between the synaptic knob of one neuron and the dendrite or cell body of another. Neurotransmitters are released into it when nerve impulses are transmitted between cells.

T

Tactile receptor: A sensory receptor responsive to light touch.
Taiga (tī'gə)**:** A subarctic forest biome dominated by spruce and fir trees; it is found in Europe and North America and at high altitudes elsewhere.
Tapeworm: A parasitic flatworm of the class *Cestoda.*
Tarsal: One of the smaller bones of the ankle, between the talus and the metatarsals.
Taxonomist (tak son'ə mist)**:** Person concerned with the identification, naming, and classification of organisms.
Tectorial membrane (tek tor'ē al)**:** A membrane of the cochlea, overlying and contacting the hair cells of the organ of Corti.
Telophase: The final stage of mitosis or meiosis that includes total separation of chromosomes, cytoplasmic division, and the return of the interphase nucleus.
Temperate deciduous forest: A forest biome of the temperate zone, in which the dominant tree species and most other trees are deciduous and are bare in winter months.
Temperature inversion: A reversal in the normal temperature lapse rate, the temperature rising with increased elevations instead of falling.
Temporal lobe (tem'pər əl lobe)**:** One of the four lobes in the human brain; contains auditory and visual centers.
Tendon: A tough, dense cord of fibrous connective tissue that is attached at one end to a muscle and at the other to that part of the skeleton that moves when the muscle contracts. Compare *ligament.*
Terminal bud: The dormant bud at the stem tip, representing the next season's potential growth.
Territorial behavior: Behavior associated with the defense of a territory, in most territorial species primarily by the male.
Tertiary structure (tėr'shē er'ē structure)**:** The pattern of folding of a polypeptide upon itself, which is generally quite specific for each protein type.
Testcross: The cross of a dominant individual with a homozygous recessive individual to determine whether recessive alleles exist.
Testosterone (tes täs'tə rōn)**:** A male hormone, produced in the testes, important in the sex drive and producing secondary sex characteristics.
Tetraploid (tet'rə ploid)**:** Having four complete sets of chromosomes in each cell.
Thalamus (thal'ə məs)**:** The middle part of the diencephalon through which sensory impulses pass to reach the cerebral cortex.
Thallus (thal'əs)**:** A plant body of a multicellular alga, that does not grow from an apical meristem, shows no differentiation into distinct tissues, and lacks stems, leaves, or roots.
Theory: A proposed explanation whose status is still conjectural and unproven, but highly likely and supported by evidence, in contrast to well-established propositions that are regarded as facts.
Thermoreception: The ability to sense temperature or changes in temperature.
Thoracic duct (thō ras'ik duct)**:** The main trunk of the lymphatic system, passing along the spinal column in the thoracic cavity, and conveying a large amount of lymph and chyle into the venous circulation.
Thorax: 1. In animals, the part of the body anterior to the diaphragm and posterior to the neck, containing the lungs and the heart. 2. The middle of the three parts of an insect body, bearing the legs and wings.
Threshold value: The point at which a stimulus is intense enough to initiate a response.
Thylakoid (thī'lə koid)**:** A saclike, membranous structure in the chloroplasts. Stacks of these form grana.
Thymine (thī'mēn')**:** A pyrimidine, one of the four nitrogenous bases of DNA.
Thyroid (thī'roid)**:** A large endocrine gland in the lower neck region of all vertebrates, the secretion of which regulates the rates of metabolism and body growth.
Thyroxine (thī räk'sin)**:** A thyroid hormone that functions in metabolism.
Tissue: A group of contiguous cells of similar origin, structure, and function. Compare *organ.*
Tobacco: Any plant of the genus *Nicotiana,* whose leaves are prepared for smoking or chewing or as snuff.
Tolerance: The power of enduring or resisting the action of a drug, poison, and so on. With drugs, increasingly greater amounts are needed to produce the same effects.
Total fertility rate: A projection of the average number of children women aged 14–44 will bear.
Trachea (trā'kē ə) (pl. *tracheae*)**:** 1. In land vertebrates, the air passage between the lungs and the larynx, usually stiffened with rings of cartilage. 2. One of the air-conveying tubules in the respiratory system of an insect, millipede, or centipede.
Tracheal system (trā'kē əl system)**:** The respiratory system of insects, composed of thin-walled air conducting tubules opening to spiracles and extending to finer, branched tracheoles, terminating in air sacs.
Tracheid (trā'kē ed)**:** A long, tubular xylem element that functions in support and water conductions. It is distinguished from xylem vessels by having tapered, closed ends and communicating with other tracheids through pits.
Tracheophyta (trā'kē ō fī'tə)**:** The vascular plants.
Trade winds: The nearly constant easterly winds that dominate most of the tropics and subtropics throughout the world, blowing mainly from the northeast in the northern hemisphere and from the southeast in the southern hemisphere.
Transfer RNA (*t*RNA)**:** Lightweight nucleic acid molecules that identify with specific amino acids during protein synthesis.
Transmitting electron microscope (TEM): A device for creating magnified images of small specimens by bombarding them with an electron beam and by subsequent magnetic focusing.
Transverse colon: See *colon.*

Transverse nerve: A laterally branching nerve, such as those connecting the longitudinal nerves in flatworms.
Transverse system: Tubules that carry an aetron potential deep into muscle cells, or the muscles that run from one side of the body wall to another.
Tree fern: Any of various, mostly tropical ferns that reach the size of trees, sending up a straight, trunklike stem with fronds at the summit.
Trematoda (trem'ə tō'də)**:** The class composed of parasitic flatworms having one or more external suckers.
Triceps brachii (trī'seps brā'kē ē)**:** The muscle on the back of the arm, the action of which extends the elbow.
Trichina (tri kī'nə)**:** A nematode, the adults of which live in the intestine and produce embryos that encyst in the muscle tissue, especially in pigs, rats, and humans.
Trichocysts (trik'ə sist')**:** In some ciliates, minute harpoonlike bodies below the pellicle that can be extruded.
Tridacna (tri dak'nə)**:** A genus of giant clams found on reefs in the South Pacific, attaining a diameter of four feet or more.
***t*RNA:** See *transfer RNA*.
Trophoblast (trof'ə blast')**:** The thin wall side of a blastocyst that forms the chorion when implantation occurs in mammals.
Tropical rain forest: A tropical woodland biome that has an annual rainfall of at least 250 cm (98 in) and often much more; it is typically restricted to lowland areas and characterized by a mix of many species of tall, broad-leaved evergreen trees that form a continuous canopy, with vines and woody epiphytes, and by a dark, rather bare forest floor.
Tubal ligation: Female sterilization by cutting and tying the Fallopian tubes.
Tubal occlusion: A birth control method that consists of inserting silicone rubber plugs into the Fallopian tubes, which prevents the egg from reaching the uterus; see also *Fallopian plug*.
Tubule: Any slender, elongated channel in an anatomical structure.
Tubulin: A protein consisting of two dissimilar polypeptides making up the subunit of microtubules.
Tumescent (too͞ mes'ent)**:** Swollen.
Tundra: A biome characterized by level or gently undulating treeless plains of the arctic and subarctic that support dense growths of mosses and lichens, as well as dwarf herbs and shrubs; it is underlain by permafrost and seasonally covered by snow.
Turbellaria (tur'be lar'ē e)**:** A class of platyhelminths or flatworms, mostly aquatic, and having cilia on the body surface.
Tympanic membrane: The thin, clear, tense double membrane of connective tissue & epithelium which divides the middle ear from the external ear; it vibrates with received sound & transmits the impulses to the inner ear.
Tympanal membrane (tim pan'el membrane)**:** A thin, tense membrane of an organ of hearing in an insect.
Tympanum (pl. *tympana*)**:** A resonating covering associated with hearing.

U

Ulna: The bone of the forearm on the side opposite the thumb, or the corresponding bone in the forelimb of other land vertebrates.
Ultimate level: Influences on an individual as a result of the evolutionary experience of the historical line.
Undernourished: Not receiving enough food.
Ungulate (ung'gyə lāt)**:** Having hoofs.
Uracil (yü'rə sil)**:** One of the nitrogenous bases of RNA.
Urea (yü'rē ə)**:** A highly soluble nitrogenous compound that is the principle nitrogenous waste of the urine of animals.
Ureter (yü rēt'ər)**:** The tube that conducts urine from the kidney to the bladder in higher animals.
Urethra (yü rē'thrə)**:** The tube that conducts urine from the bladder to the outside of the body in higher animals.
Uric acid (yü'rik acid)**:** A relatively insoluble purine; a principal nitrogenous excretion product of reptiles, birds, and insects. It is excreted in small quantities as a product of nucleic acid breakdown in mammals.
Urinary bladder (yü'rə ner'ē bladder)**:** A membranous sac in which urine is retained until it is discharged from the body.
Urine (yü'rin)**:** The liquid-to-semisolid matter excreted by the kidneys, in humans, being a yellowish, slightly acid, watery fluid.
Urochordata (yü'rə kor da'tə)**:** A subphylum of the chordates that includes the tunicates, salps, and larvaceae.
Utricle (yü'tri kəl)**:** The chamber of the membranous labyrinth of the middle ear into which the semicircular canals open.

V

Vacuole: A space within a cell, bounded by a membrane.
Vacuum curettage (vacuum kyoor'i tüzh)**:** A means of abortion by which the embryonic mass is removed by suction.
Vagina (ve jī'nə)**:** The female copulatory organ and birth canal.
Vaginal wall: The tough, corrugated inner surface of the vagina.
Van der waals forces: Relatively weak attractive forces acting between nonpolar atoms & molecules, binding together the lipid-soluble portions.
Variable: Experimental variable; the focus on an experiment to be tested and compared with a control.
Variable region: The portion of an immunoglobulin polypeptide concerned with the binding of an antigen, which varies from antigen to antigen.
Vascular cambium: The cylinder of meristematic tissue that in secondary growth produces xylem on its inner side and phloem on its outer side, thus contributing to growth in circumference.
Vascular tissue: Any tissue that contains vessels through which fluids are passed.
Vasectomy (vas ek'tə mē)**:** Male sterilization by cutting and tying the seminal ducts.
Vegetal pole: The lower, more yolk-filled end of a zygote or early cleavage blastula, determining the ventral side in development.
Vein: A vessel that carries blood toward the heart.
Ventilation: Exposure to air in the lungs or gills in respiration; oxygenation.
Ventral nerve root: Any of the large motor nerves extending from the ventral area of the vertebrate spinal nerves.
Ventricle (ven'tri kəl)**:** A cavity of a body part or organ; one of the large muscular chambers of the four-chambered heart.
Venule: A small vein.
Vertebra (vü'tə brə)**:** Any of the bones or segments of the spinal column.
Vertebral column (vür'tə brəl column)**:** The articulated series of vertebrae connected by ligaments and separated by intervertebral discs that in vertebrates forms the supporting axis of the body and of the tail in most forms.
Vertebrate (vür'tə brāt)**:** An animal in the subphylum *Vertebrata,* phylum *Chordata;* an animal with a vertebral column or backbone.
Vessel: In botany, a conducting tube in

a dicot formed in the xylem by the end-to-end fusion of a series of cells (vessel elements) followed by the loss of adjacent end walls and of cell cytoplasm. Compare *tracheid.*

Villus (vil'ə s) (pl. *villi*)**:** A small protrusion of the intestinal wall, greatly increasing the absorbing surface of that organ.

Viral DNA: The DNA that exists within the genome of a virus.

Virus: An infectious, submicroscopic parasite that consists of an RNA or DNA core with a protein coat.

Visible light: Electromagnetic wavelengths longer than about 400nm and shorter than about 750nm, which can serve as visual stimuli to most photoreceptive organisms.

Visual pigment: Light-sensitive pigment of the eye that responds biochemically.

Visual purple: Rhodopsin; a bright-red photosensitive pigment found in the rods of the retina of certain fishes and most vertebrates.

Vitalism: The notion that life has unique mystical properties that are distinct from those ascribed by chemical and physical laws.

W

Water vascular system: A system of vessels in echinoderms that contains sea water and is used as a hydraulic system in the movement of tentacles and tube feet.

Wave: In physics, a progressive disturbance moving from point to point in a medium or space without progress or advance by the points themselves, as in the transmission of sound or light.

Wax: A dense, hard, lipid-soluble ester of a long-chain alcohol and a fatty acid.

Whorl: 1. A group of parts repeated in a circle. 2. Any of the four basic radially repeated groups of flower parts.

Withdrawal symptom: Any of a number of physical and psychological disturbances, such as sweating and depression, experienced by a narcotic addict deprived of a required drug dosage.

X

Xylem: One of the two complex tissues in the vascular system of plants; consists of the dead cell walls of vessels, tracheids, or both, often together with sclerenchyma and parenchyma cells; functions chiefly in water conduction and strengthening the plant. See also *tracheid, vessel.* Compare *phloem.*

Y

Yellow marrow: Yellow, fatty material within the central cavity of long bones.

Yolk: The nutrient portion of the egg.

Yolk plug: A pluglike mass of yolk-filled endoderm cells left protruding from the blastopore of an amphibian embryo after the cresentic blastopore enlarges to form a complete circle.

Z

Z line: In striated muscle, the partition between adjacent contractile units to which actin filaments are anchored.

Zooplankton (zō'ə plangk'tən)**:** The nonphotosynthetic animal life drifting at or near the surface of the open sea.

Zygospore (zī'gə spōr)**:** A diploid fungal or algal spore formed by the union of two similar sexual cells; it has a thickened wall and serves as a resistant resting spore.

Selected Readings

PART 1

Albert, B. et al. 1989. *Molecular Biology of the Cell,* 2d ed. Garland Publishing Co., New York.

Baker, J. J. W. and G. E. Allen. 1968. *Hypothesis, Prediction and Implication in Biology.* Addison-Wesley, Reading, Mass.

Clayton, R. 1981. *Photosynthesis: Physical Mechanisms and Chemical Patterns.* Cambridge University Press, New York.

Conant, J. B. 1951. *Science and Common Sense.* Yale University Press., New Haven, Conn.

Darwin, C. R. 1962. *The Voyage of the Beagle.* Doubleday, Garden City, N.Y.

de Beer, G. 1965. *Charles Darwin: A Scientific Biography.* Doubleday, New York.

deDuve, C. 1985. *A Guided Tour of the Living Cell.* Volumes 1 and 2. Freeman, New York.

Eiseley, L. 1956. Charles Darwin, *Scientific American* 194(2):62–72.

———— 1958. *Darwin's Century.* Doubleday, Garden City, N.Y.

———— 1960. *The Firmament of Time.* Atheneum, New York.

Farago, P. and J. Lagnado. 1972. *Life in Action.* Vintage, New York.

Folsome, C. E. 1979. *Life: Origin and Evolution.* Freeman, San Francisco.

Gingerich, O. 1982. The Galileo Affair, *Scientific American,* July.

Jastrow, R. 1967. *Red Giants and White Dwarfs.* Harper and Row, New York.

Knowles, J. 1987. Tinkering with Enzymes: What Are We Learning? *Science* 236:1252–1258.

Koestler, A. 1972. *The Roots of Coincidence.* Random House, New York.

Lederman, L. 1984. The Value of Fundamental Science, *Scientific American,* November.

Mayr, E. 1970. *Populations, Species and Evolution.* Harvard University Press, Cambridge, Mass.

Miller, S. L. 1955. Production of Some Organic Compounds Under Possible Primitive Earth Conditions, *Journal of the American Chemical Society* 77:2351–2361.

Moore, P. 1981. The Varied Ways Plants Tap the Sun, *New Scientist,* February.

Moorehead, A. 1969. *Darwin and the Beagle.* Harper and Row, New York.

Porter, E. 1971. *Galapagos.* Ballatine, New York.

Smith, H. W. 1961. *From Fish to Philosopher.* Doubleday, Garden City, N.Y.

PART 2

Crick, F. H. C. 1966. The Genetic Code: III, *Scientific American* 215(4):55–62.

Delisi, C. 1988. The Human Genome Project, *American Scientist* 76:488–493.

Dobzhansky, T. 1963. Evolutionary and Population Genetics, *Science* 142:3596.

Eckhardt, R. B. 1972. Population Genetics and Human Origins, *Scientific American* 226(2):94–102.

Feder, J. and W. Tolbert. 1983. The Large-Scale Cultivation of Mammalian Cells, *Scientific American,* January.

Gallo, R. C. 1986. The First Human Retrovirus, *Scientific American,* December.

Hayflich, L. 1980. The Cell Biology of Aging, *Scientific American,* January.

Lawn, R. and G. Vehar. 1986. The Molecular Genetics of Hemophilia, *Scientific American,* March.

Leeson, C. R. et al. 1985. *Textbook of Histology.* Saunders, Philadelphia.

Lewin, B. 1987. *Genes III.* John Wiley and Sons, New York.

Lewin, R. 1983. A Naturalist of the Genome, *Science* 222:402.

Lowey, A. G. and P. Siekevitz. 1969. *Cell Structure and Function.* Holt, Rinehart and Winston, New York.

McKusick, V. A. 1965. The Royal Hemophilia, *Scientific American* 213(2): 88–95.

Menoskz, J. A. 1981. The Gene Machine, *Science 81,* July/August.

Pestka, S. 1983. The Purification and Manufacture of Human Interferons, *Scientific American,* August.

Rensberger, B. 1981. Tinkering with Life, *Science 81,* November.

Shine, I., and S. Wrobel. 1976. *Thomas Hunt Morgan: Pioneer of Genetics.* University of Kentucky Press, Lexington, Ky.

Watson, J. D. 1968. *The Double Helix.* Atheneum, New York.

———— 1970. *Molecular Biology of the Gene.* Benjamin, New York.

PART 3

Alexander, T. 1975. A Revolution Called Plate Tectonics Has Given Us a Whole New Earth, *Smithsonian* 5:30.

Anderson, H. T. 1969. *Biology of Marine Mammals.* Academic Press, New York.

Austad, S. 1988. The Adaptable Opossum, *Scientific American,* February.

Bellairs, A. 1970. *Life of Reptiles.* Vol. II. Universe Books, New York.

Bishop, J. A. and Laurence M. Cook. 1975. Moths, Melanism and Clean Air, *Scientific American,* January.

Buchsbaum, R. et al. 1987. *Animals Without Backbones,* 3d ed. University of Chicago Press, Chicago.

Budker, Paul. 1971. *The Life of Sharks.* Columbia University Press, New York.

Cornejo, D. 1982. Night of the Spadefoot Toad, *Science 82,* September.

Cracraft, J. 1988. Early Evolution of Birds, *Nature* 331:389–390.

Esau, K. 1977. *Anatomy of Seed Plants,* 2d ed. Wiley, New York.

Gilbert, L. 1982. The Coevolution of a Butterfly and a Vine, *Scientific American,* August.

Kaplan, D. R. 1983. The Development of Palm Leaves, *Scientific American,* July.

King, J. L. and T. H. Jukes. 1969. Non-Darwinian Evolution, *Science* 164:788.

Lehner, R. and J. Lehner. 1962. *Folklore and Odysseys of Food and Medicinal Plants.* Tudor, New York.

Marshall, N. B. 1966. *The Life of Fishes.* The World Publishing Co., Cleveland.

Panchen, A. 1988. In Search of the Earliest Tetrapods, *Nature* 333:704.

Payne, K. 1989. Elephant Talk. *National Geographic,* August.

Romer, A. S. 1977. *The Vertebrate Body,* 5th ed. W. B. Saunders, Philadelphia.

Russell, D. A. 1982. The Mass Extinctions of the Late Mesozoic, *Scientific American,* January.

Schultz, A. 1969. *Life of Primates.* Weidenfeld and Nicolson.

Sporne, K. R. 1971. *The Mysterious Origin of Flowering Plants.* Carolina Biological Supply Company, Burlington, N.C.

Stanier, R. Y. and M. Douderoff. 1973. *The Microbial World,* 3d ed. Prentice-Hall, Englewood Cliffs, N.J.

Stanley, S. 1984. Mass Extinctions in the Ocean, *Scientific American,* June.

Wallace, R. et al. 1986. *Biology: The Science of Life,* 3d ed. HarperCollins, New York.

Welty, J. C. 1975. *The Life of Birds,* 2d ed. W. B. Saunders, Philadelphia.

Young, J. Z. 1975. *The Life of Mammals,* 2d ed. Clarendon Press, Oxford.

PART 4

Corner, E. H. H. 1968. *The Life of Plants.* New American Library, New York.

Frazer, J. F. D. 1959. *The Sexual Cycles of Vertebrates.* Hutchinson University Library, London.

Ganong, W. 1987. Review of Medical Physiology, 12th ed. Lange Medical Publications, Los Angeles, Calif.

Gehring, W. 1985. The Molecular Basis of Development, *Scientific American,* October.

Masters, W. and V. Johnson. 1966. *Human Sexual Response.* Little, Brown, Boston.

Michelmore, S. 1965. *Sexual Reproduction.* Natural History Press, Garden City, N.Y.

Money, John and Anke A. Ehrhardt. 1972. *Man and Woman, Boy and Girl: The Differentiation and Dimorphism of Gender Identity from Conception to Maturity.* Johns Hopkins University Press, Baltimore.

Nilsson, L., A. Ingleman-Sundberg, and C. Wirsen. 1986. *A Child Is Born: The Drama of Life Before Birth.* Dell, New York.

Rugh, R. and L. B. Shettles. 1971. *From Conception to Birth: The Drama of Life's Beginnings.* Harper and Row, New York.

Shell, E. R. 1982. The Guinea Pig Town, *Science 82,* December.

Shodell, M. 1983. The Prostaglandin Connection, *Science 83,* March.

Wallace, R. A. 1980. *How They Do It.* Morrow, New York.

Wilson, J. et al. 1981. The Hormonal Control of Sexual Development, *Science* 211(4488), 1278–1285.

PART 5

Caravoli, E. and J. Penniston. 1985. The Calcium Signal, *Scientific American,* November.

Cohen, I. 1988. The Self, the World and Autoimmunity, *Scientific American,* April.

Crawshaw, L. et al. 1981. The Evolutionary Development of Vertebrate Thermoregulation, *Scientific American,* September–October.

Currey, J. 1970. *Animal Skeletons.* St. Martin's Press, New York.

Day, R. H. 1971. *Perception.* Brown, Dubuque, Iowa.

Eastman, Joseph T. and Arthur L. DeVries. 1986. Antarctic Fishes, *Scientific American,* November.

Feder, Martin E. and Warren W. Burggren. 1985. Skin Breathing in Vertebrates, *Scientific American,* November.

Greene, R. 1970. *Human Hormones.* McGraw-Hill, New York.

Griffiths, Mervyn. 1988. The Platypus, *Scientific American,* May.

Jaret, P. 1986. Our Immune System, The Wars Within, *National Geographic,* June 169(6), 702–734.

Morton, J. E. 1967. *Guts.* St. Martin's Press, New York.

Nilsson, L. 1987. *The Body Victorious.* Delacourt, New York.

Scientific American. 1988. *What Science Knows About AIDS.* Freeman, New York.

Shepherd, G. 1988. *Neurobiology,* 2d ed. Oxford University Press, New York.

Thompson, R. 1988. *Introduction to Physiological Psychology,* 2d ed. Harper and Row, New York.

——— 1964. *Desert Animals.* Oxford University Press, New York.

——— 1972. *How Animals Work.* Cambridge University Press, New York.

PART 6

Bright, M. 1984. *Animal Language.* Cornell Univ. Press, Ithaca, New York.

Fackelmann, K. 1989. Avian Altruism, *Science News,* June.

Fine, A. 1986. Transplantation in the Central Nervous System, *Scientific American,* August.

Glickstein, M. 1988. The Discovery of the Visual Cortex, *Scientific American,* September.

Goldstein, G. and A. Betz. 1986. The Blood-brain Barrier, *Scientific American,* September.

Jolly, A. 1972. *The Evolution of Primate Behavior.* MacMillan, New York.

Klopfer, P. H. and J. P. Hailman. 1967. *An Introduction to Animal Behavior: Ethology's First Century.* Prentice-Hall, Englewood Cliffs, N.J.

——— 1973. *Behavior Aspects of Ecology.* Prentice-Hall, Englewood Cliffs, N.J.

Morris, D. 1967. *The Naked Ape*. Dell, New York.

Scientific American. 1979. *The Brain*. W. H. Freeman, New York.

Scientific American. 1985. *Progress in Neuroscience*. W. H. Freeman, New York.

Tiger, L. 1969. *Men in Groups*. Random House, New York.

Tinbergen, N. 1951. *The Study of Instinct*. Oxford University Press, Oxford.

__________ 1953. *Social Behavior in Animals*. Methuen, London.

Wallace, R. A. 1978. *The Ecology and Evolution of Animal Behavior,* 2d ed. Goodyear, Santa Monica, Calif.

Wallace, R. A. 1979. *The Genesis Factor*. Morrow, New York.

Wallace, R. A. 1979. *Animal Behavior, Its Development, Ecology and Evolution*. Scott, Foresman, Glenview, IL.

Wilson, E. O. 1984. *Biophilia*. Harvard Univ. Press, Cambridge.

Wilson, E. O. 1975. *Sociobiology*. Belknap, Cambridge, Mass.

Wingfield, J. et al. 1987. Testosterone and Aggression in Birds, *American Scientist* 75:602–658.

Wood-Gush, D. G. M. 1983. *Elements of Ethology*. Chapman and Hall, London.

PART 7

Calahan, D. 1972. Ethics and Population Limitation, *Science* 175:487–492.

Calhoun, J. R. 1962. Population Density and Social Pathology, *Scientific American* 206(2):139–148.

Carson, R. 1962. *Silent Spring*. Houghton Mifflin, Boston.

Clutton-Brock, T. 1985. Reproductive Success in Red Deer, *Scientific American,* February.

Commoner, B. 1966. *Science and Survival*. Viking Press, New York.

__________ 1971. *The Closing Circle*. Knopf, New York.

Edmond, J. and K. Von Damm. 1984. Hot Springs on the Ocean Floor, *Scientific American,* April.

Ehrlich, P. 1968. *The Population Bomb*. Ballantine, New York.

Ehrlich, P. and A. Ehrlich. 1979. What Happened to the Population Bomb? *Human Nature,* January.

Jackson, D. 1989. Searching for Medicinal Wealth in Amazonia, *Smithsonian,* February.

Kerr, R. 1988. Is the Greenhouse Here? *Science* 239:559–561.

Lappe, F. 1971. *Diet for a Small Planet*. Ballantine, New York.

Lee, D. 1989. Tragedy in Alaska Waters, *National Geographic,* August.

Mares, M. A. 1986. Conservation in South America: Problems, Consequences, and Solutions, *Science* 233:734–739.

Meadows, D. et al. 1972. *Limits to Growth*. Universe, New York.

Odum, E. P. 1983. *Basic Ecology*. Saunders, Philadelphia.

Perry, D. 1984. The Canopy of the Tropical Rain Forest, *Scientific American,* November.

Schoener, T. 1982. The Controversy over Intraspecific Competition, *American Scientist* 70:586–595.

Sun, M. 1988. Costa Rica's Campaign for Conservation, *Science* 239:1366–1369.

Tschirley, F. 1986. Dioxin, *Scientific American,* February.

Credits

PHOTO CREDITS

Unless otherwise acknowledged, all photographs are the property of HarperCollins Publishers. Page abbreviations are used as follows: T = top, C = center, B = bottom, L = left, R = right, INS = inset.

VU = Visuals Unlimited
TS = Tom Stack & Associates
PR = Photo Researchers, Inc.

COVER: Richard K. Laval/Earth Scenes

COVER INSET: Tom McHugh/PR

TITLE PAGE: Michael Lawton/American Pangraphics

ABOUT THE AUTHOR: Jayne T. Austin

PART ONE, CHAPTER ONE

pp. 2-3 Stephen Dalton/PR; **p. 4** William H. Amos; **1.1A** Jonathan Wright/Bruce Coleman, Inc.; **1.1B** Charles Darwin. Watercolor by George Richmond, 1840. By permission of the Darwin Museum, Down House and The Royal College of Surgeons of England; **1.2** Walt Anderson/VU; **1.3** "History of the University of Cambridge," by Combe, vol. 2, 1815. Print Collection, Miriam and Ira D. Wallach Division of Art, Prints and Photographs, The New York Public Library, Astor, Lenox and Tilden Foundations; **1.4** By permission of the Darwin Museum, Down House and The Royal College of Surgeons of England; **1.7** Walt Anderson/VU; **1.8** Historical Pictures Service, Chicago; **1.9** "Galilee devant le Saint Office," Robert-Fleury, Musée du Louvre, Paris. © Photo R.M.N.; **1.10** Science VU/VU; **1.10A** R. Gustafson/VU; **1.10B** David S. Addison/VU; **1.11A** The Bettmann Archive; **1.13A** By permission of the Darwin Museum, Down House and The Royal College of Surgeons of England; **1.13B** Staircase of the Old British Museum, Montague House. George Scharf, the Elder. Courtesy of the Trustees of the British Museum; **1.14** Sir Charles Lyell by L. Dickinson (replica), 1883. Courtesy of the National Portrait Gallery, London; **1.15B** A. Q. White/VU; **1.16** *L'Illustration,* vol. 54, Dec. 18, 1969, p. 388. Drawing by Samuel Luke Fildes, 1925; **p. 18 T** The Library of Congress; **p. 18 B** Culver Pictures

PHOTOESSAY

p. 19 G. Prance/VU; **p. 20 T** Gerald A. Corsi/TS; **p. 20 C** G. Prance/VU; **p. 20 B** Norman Owen Tomalin/Bruce Coleman, Inc.; **p. 21 T** William H. Amos; **p. 21 C** C. P. Hickman/VU; **p. 21 B** Joe McDonald/Bruce Coleman, Inc.; **p. 22 TL** Kenneth Poertner/David R. Frazier Photolibrary; **p. 22 TR** Gerald A. Corsi/TS; **p. 22 BL** Leonard Lee Rue III/Bruce Coleman, Inc.; **p. 22 BR** Ben Goldstein/Don & Pat Valenti; **p. 23 TL** William H. Amos; **p. 23 CL** William H. Amos; **p. 23 CR** Gerald & Buff Corsi/TS; **p. 23 BL** William H. Amos; **p. 23 BR** Joe McDonald/VU; **p. 24 TL** Joe McDonald/VU; **p. 24 TR** Daniel W. Gotshall/VU; **p. 24 C** C. P. Hickman/VU; **p. 24 BL** Daniel W. Gotshall/VU; **p. 24 BR** G. Prance/VU **1.17A&B** Richard Howard; **1.17C** John Gerlach/TS; **1.17D** Tony Freeman/PhotoEdit; **1.18** © Biophoto Associates/PR; **1.19A** The Bettmann Archive; **1.19B** Robert Pickett; **1.20** Thomas Henry Huxley by John Collier, 1883. Courtesy of The National Portrait Gallery, London

CHAPTER TWO

p. 32 S. Nielson/Bruce Coleman, Inc.; **2.1** The Bettmann Archive; **p. 34** Manfred Gottschalk/TS; **p. 36** Marty Snyderman/VU; **2.2A** Marty Snyderman/VU; **2.2B** Peter Veit/Sygma; **2.3A** Alan and Sandy Carey; **2.3B** Marty Snyderman/VU; **2.3C** Dwight R. Kuhn; **2.3D** Yogi Kaufmann/Peter Arnold, Inc.; **2.3E** Agence Nature/NHPA; **2.3F** Brian Parker/TS; **p. 44 TL** Jane Thomas/VU; **p. 44 TR** SIU/VU; **p. 44 CL** Cindy Garoutte/TS; **p. 44 BR** Steve McCutcheon/VU; **2.4** Rick Friedman/Black Star; **2.5** Fred Ward/Black Star

PHOTOESSAY

p. 47 T Diane and Michael McBride; **p. 48 Top row** Charles Mazel; **p. 48 Bottom row** John C. D. Bruno; **p. 49 TL** author photo; **p. 49 TR** Lincoln P. Brower; **p. 49 B** Nicholas DeVore/Photographers Aspen

CHAPTER THREE

p. 52 Leonard Lessin/Peter Arnold, Inc.; **3.1 BL** John D. Cunningham/VU; **3.1 BR** Tim Perkins/VU; **3.1 TL** Gamma Liaison; **3.1 TR** Charles Mason/Black Star; **3.2** Tom McHugh, Steinhart Aquarium/PR; **3.5** Wide World Photos; **3.11 BL** Scott Blackman/TS; **3.11 BR** John Shaw/TS; **3.11 TL** R. Prance/VU; **3.11 TR** Steve McCutcheon/VU; **3.12** Mike Yamashita/Woodfin Camp & Associates; **3.14B** Ken W. Davis/TS; **3.16** Film Stills Archive, The Museum of Modern Art; **3.19A** Stanley Flegler/VU; **3.19B** David M. Phillips/VU; **3.20** Gary R. Zahm/Bruce Coleman, Inc.; **3.22** Robert Villani/Peter Arnold, Inc.; **p. 79** Reuters/Bettmann; **p. 83** Tripos Associates, Inc./Peter Arnold, Inc.

PART TWO, CHAPTER FOUR

p. 88 John D. Cunningham/VU; **pp. 88-89** R. Prance/VU; **p. 90** Dr. Jeremy Burgess/Science Photo Library/PR; **4.1A** From *Micrographia* by Robert Hooke, published in 1665 by the Royal Society; **4.1B** From *Micrographia* by Robert Hooke, published in 1665 by the Royal Society; **4.2A** Robert Brons/Biological Photo Service; **4.2B&C** Supplied by Carolina Biological Supply Company; **4.2D** Fred Hossler/VU; **p. 95** University of Oregon photographs; **4.4A** John D. Cunningham/VU; **4.4B** David M. Phillips/VU; **4.4C** Fred Hossler/VU; **4.4D** Biophoto Associates/Science Source/PR; **4.5B** Joe McDonald/TS; **4.7** Martha J. Powell/VU; **4.8B** CNRI/Science Photo Library/PR; **4.9B** M. Schliwa/VU; **4.10A** M.

Schliwa/VU; **4.11** William E. Barstow; **4.12A** Dr. G. Benjamin Bouck, University of Illinois at Chicago; **4.13A** K. G. Murti/VU; **p. 105** Richard Howard; **4.14A** Don W. Fawcett/VU; **4.14B** Don W. Fawcett/VU; **4.15B** M. Powell/VU; **4.16** David M. Phillips/VU; **4.17A** Randy Moore/VU; **4.17B** C. Allan Morgan/Peter Arnold, Inc.; **4.18** Biophoto Associates/Science Source/PR; **4.19** Don W. Fawcett/VU; **4.24A** D. Newman/VU; **4.24B** Brian Parker/TS; **4.24C** David M. Phillips/VU; **4.24D** Supplied by Carolina Biological Supply Company; **4.25** M. Abbey/VU

CHAPTER FIVE

p. 120 Science VU/VU; **5.1** John D. Cunningham/VU; **5.2A** David R. Frazier Photolibrary; **5.2B** BMPRC-Science VU/VU; **5.4A** Tripos Associates, Inc./Peter Arnold, Inc.; **5.4B** author art; **p. 127** John Shaw/TS; **p. 130** Michael G. Gabridge/VU; **p. 135 TL** D. Foster, WHOI-Science VU; **p. 135 TR** J. Edmond, WHOI-Science VU; **p. 135 BL** Rod Catanach/© 1987 Woods Hole Oceanographic Institute; **p. 135 BR** Dr. Peter Lonsdale, Scripps Institution of Oceanography, University of California, San Diego **5.14** author art; **5.15** Nada Pecnik/VU; **5.16** author art

CHAPTER SIX

p. 151 D. Newman/VU; **6.2** Supplied by Carolina Biological Supply Company; **6.4** Supplied by Carolina Biological Supply Company; **6.6** Hasenkampf/Biological Photo Service; **p. 165** De Keerle, UK Press/Gamma-Liaison; **6.9** Photographer: A. C. Barrington Brown. From J. D. Watson, 1968, *The Double Helix*. New York: Atheneum, p. 215. © 1968 by J. D. Watson; **p. 174 L** Gamma Liaison; **p. 174 R** Springer/Bettmann Film Archive; **p. 175 L** Tom Zimberoff/Sygma; **p. 175 R** Gamma Liaison

CHAPTER SEVEN

p. 181 Jane Burton/Bruce Coleman, Inc.; **7.1** Walt Anderson/VU; **7.2** M. Long/VU; **p. 185** Erika Stone/Peter Arnold, Inc.; **p. 186** John Watney/PR; **p. 186** John D. Cunningham/VU; **7.6A** Bob Shanley/*Palm Beach Post*; **7.6B** Wide World; **7.7** PR; **p. 191** Scala/Art Resource, New York; **7.9** Dr. Murray L. Barr; **7.10** R. Calentine/VU; **p. 196** Hulton/Bettmann; **p. 198** Stanley Flegler/VU; **7.19** Collected papers of G. H. Hardy, The Clarendon Press, Oxford

CHAPTER EIGHT

p. 209 John D. Cunningham/VU; **8.1A** Dr. Tony Brain/Science Photo Library/PR; **8.1B** Tom Broker/Rainbow; **p. 214** Charles C. Brinton, Jr. and Judith Carnahan; **p. 217** Hank Morgan/PR; **p. 218 L** Dan McCoy/Rainbow; **p. 218 R** Douglas Kirkland/Sygma; **8.3** Steve Northrup/Black Star; **8.4** Gamma Liaison; **8.5A** Archive Photos; **8.5B** AP/World Wide; **8.6A** The Bettmann Archive; **8.6B** Gamma Liaison; **8.7** Brad Markel/Gamma Liaison

PART THREE, CHAPTER NINE

pp. 226-27 Carl Roessler/ANIMALS ANIMALS **p. 228** William M. Johnson/VU; **9.1A** S. Maslowski/VU; **9.1B** Allan Roberts; **9.1C** John Forsythe/VU; **9.1D** Joe McDonald/TS; **9.2** UPI/Bettmann; **9.3** Science VU; **9.4** Herb Orth/*Life* Magazine, © 1963 Time, Inc. Painting by Charles Bonestell from *The World We Live In,* p. 10; **9.5** Joe McDonald/VU; **9.6** Roger Ressmeyer/Starlight; **9.7** Manfred Kage/Peter Arnold, Inc.; **9.8** Sidney Fox/Science VU; **9.9** John D. Cunningham/VU; **9.10** Science Photo Library/PR; **9.12A** William H. Amos; **9.12B** Dwight R. Kuhn; **9.12C** John Shaw/TS; **9.12D** Tony Freeman/PhotoEdit; **p. 242 TL** John D. Cunningham/VU; **p. 242 BR** NASA; **p. 243 L** Fred Bavendam/Peter Arnold, Inc.; **p. 245 R** S. M. Awramik, Department of Geological Sciences, University of California, Santa Barbara; **9.13** Ted Whittenkraus/VU; **9.14** Brian Parker/TS

CHAPTER TEN

p. 250 Kjell B. Sandved/VU; **10.1** The Houston Chronicle Library; **10.2** Joe McDonald/VU; **10.3** Alan Oddie/PhotoEdit; **10.4** Lynn M. Stone; **10.6** W. Wisniewski/PR; **10.7** Allan Roberts; **10.9** Breck P. Kent; **10.10** Kjell B. Sandved/VU; **p. 261 L** James L. Castner; **p. 261 R** Alan Oddie/PhotoEdit; **10.12** Kjell B. Sandved/VU; **10.13** Frans Lanting/Minden Pictures; **10.14** Stouffer Productions, Ltd./ANIMALS ANIMALS; **10.15A** Anthony Mercieca/PR; **10.15B** Don & Pat Valenti; **10.16** Leonard Lee Rue III/Bruce Coleman, Inc.; **10.17** Pat & Tom Leeson/PR; **p. 274** John D. Cunningham/VU; **10.23** Wendell Metzen/Bruce Coleman, Inc.; **10.24** Lawrence Berkeley Laboratory, University of California

CHAPTER ELEVEN

p. 281 Manfred Kage/Peter Arnold, Inc.; **p. 284** Robert and Linda Mitchell; **11.3** R. Robinson/VU; **11.4** Dr. Tony Brain & David Parker/Science Photo Library/PR; **p. 287 L** David M. Phillips/VU; **p. 287 R** David R. Frazier Photolibrary; **11.5A** Carl O. Wirsen, WHOI/VU; **11.5B** CNRI/Science Photo Library/PR; **11.5C** George Musil/VU; **11.6A** David M. Phillips/VU; **11.6B** ASM/Science VU; **11.6C** John D. Cunningham/VU; **11.7** Courtesy of D. L. Findley, P. L. Walne and R. W. Holton, of Tennessee, Knoxville. From *J. Phycology* 6:182-88, 1970; **p. 290 L** Arthur M. Siegelman/VU; **p. 290 R** Nathan Benn/Woodfin Camp & Associates; **11.8** George Musil/VU; **11.10** Manfred Kage/Peter Arnold, Inc.; **11.11B** Manfred Kage/Peter Arnold, Inc.; **11.13 TL** David M. Phillips/VU; **11.13 R** Veronika Burmeister/VU; **11.13 BL** Manfred Kage/Peter Arnold, Inc.; **11.14** David M. Phillips/VU; **11.15** Kevin Schafer/TS; **11.16** Glenn Oliver/VU; **11.17** N. Allin and G. L. Barron, University of Guelph; **11.20** James H. Karales/Peter Arnold, Inc.; **p. 304 T** Fred Bruemmer/Peter Arnold, Inc.; **p. 304 B** Don & Pat Valenti; **p. 305 T** William H. Amos; **p. 305 R** L. West/PR; **p. 305 B** James Balog/Bruce Coleman, Inc.; **p. 306 B** Gamma Liaison; **p. 306 T** John D. Cunningham/VU; **11.21A** David M. Phillips/VU; **11.21B** John D. Cunningham/VU; **11.22** Bill Keogh/VU; **11.23** Allan Roberts; **11.24** William J. Weber/VU; **11.26** John D. Cunningham/VU

CHAPTER TWELVE

p. 316 Allan Roberts; **12.2** Daniel Gotshall/VU; **12.3** William H. Amos; **12.4** NHMI, M. DeMocker/VU; **12.5A** L. L. Sims/VU; **12.5B** E. R. Degginger; **12.6** William H. Amos; **12.7A** Robert and Linda Mitchell; **12.7B** Ed Reschke/Peter Arnold, Inc.; **12.8** William H. Amos; **12.10A** Walter H. Hodge/Peter Arnold, Inc.; **12.10B** Ed Reschke/Peter Arnold, Inc.; **12.11** Breck P. Kent/Earth Scenes; **12.12** Coco McCoy/Rainbow; **12.13** Ed Cooper; **12.14A** David L. Pearson/VU; **12.14B** R. F. Ashley/VU; **12.15** Walter H. Hodge/Peter Arnold, Inc.; **12.16** William H. Amos; **12.17** John D. Cunningham/VU; **12.18A** Don & Pat Valenti; **12.18B** William H. Amos; **12.18C** Terry Donnelly/TS; **12.18D** Alan Oddie/PhotoEdit; **12.19A** Rod Planck/TS; **12.19B** Jeff Foott/TS; **12.20 R** Robert and Linda Mitchell

CHAPTER THIRTEEN

p. 336 Michael P. Fogden/Bruce Coleman, Inc.; **13.2** Fred Bavendam/Peter Arnold, Inc.; **13.5** Denise Tackett/TS; **13.6** C. Garoutte/TS; **13.7** Daniel W. Gotshall/VU; **13.11** Triarch/VU; **13.13** William H. Amos; **13.16** Ed Robinson/TS; **13.17** Dr. C.F.E. Roper. Photo by Chip Clark; **13.19** David Hughes/Bruce Coleman, Inc.; **13.20** Larry Lipskey/TS; **13.21A** Hans Pfletschinger/Peter Arnold, Inc.; **13.21B** Don & Pat Valenti; **13.21C** Brian Parker/TS; **13.21D** Gary Milburn/TS; **13.24A** John MacGregor/Peter Arnold, Inc.; **13.24B** Zig Leszczynski/ANIMALS ANIMALS; **13.26** Dave B. Fleetham/TS; **13.28A** Kjell B. Sandved; **13.28B** Oxford Scientific Films/ANIMALS ANIMALS; **13.29** Heather Angel/Biofotos; **13.31A** Ken Lucas/Biological Photo Service; **13.31B** Tom Stack/TS; **13.31C** Patrice Ceisel/VU; **13.32** Allan Roberts; **13.33** James R. McCullagh/VU; **13.35** William H. Amos; **13.36A** Peter Scoones/Seaphot, Planet Earth Pictures; **13.37** Science VU—VU; **13.38A** M.P.L. Fogden/Bruce Coleman, Inc.; **13.38B** Nathan Cohen/VU; **13.38C** C. C. Lockwood/ANIMALS ANIMALS; **13.38D** Hans & Judy Beste/ANIMALS ANIMALS; **13.39A** Gerald & Buff Corsi/TS; **13.39B**

Joe McDonald/VU; **13.39C** Thomas Kitchin/TS **13.39D** Brian Parker/TS; **13.39E** Biological Photo Service; **13.39F** Ben Goldstein/Don & Pat Valenti; **13.40** Dr. Carl Welty; **13.41** Mark Newman/TS; **13.42A** Dave Watts/TS; **13.42B** Waina Cheng/Oxford Scientific Films/ANIMALS ANIMALS; **13.42C** John Cancalosi/TS; **13.43** Dwight R. Kuhn; **13.44** Joe McDonald/TS; **13.45** Joe McDonald/VU; **13.46** Gary Milburn/TS; **13.47** Gary Milburn/TS; **13.48** Merlin D. Tuttle, Bat Conservation International; **13.49** John Gerlach/VU; **13.50** Cris Crowley/TS; **13.51** E.P.I. Nancy Adams/TS

PART FOUR, CHAPTER FOURTEEN

pp. 381-82 M. P. Kahl/Bruce Coleman, Inc.; **p. 382** D. Cavagnaro/Peter Arnold, Inc.; **14.3** Rod Planck/TS; **14.6A** Ben Goldstein/Don & Pat Valenti; **14.6B** Ruth Dixon; **14.6C** Kjell B. Sandved, Smithsonian Institution/PR; **p. 393** Hans Pfletschinger/Peter Arnold, Inc.; **14.8A** Runk-Schoenberger/Grant Heilman Photography, Inc.; **14.8B** David M. Phillips/VU; **14.9** J.N.A. Lott, McMaster University/Biological Photo Service; **14.13A** Greg Vaughn/TS; **14.14A** Biophoto Associates/Science Source/PR; **14.14B** A. J. Karpoff/VU; **14.16** AP/Wide World Photos; **14.17** Robert E. Lyons/VU

CHAPTER FIFTEEN

p. 407 Frans Lanting/Minden Pictures; **p. 409** Don & Pat Valenti; **p. 409** Leonard Lee Rue III/DRK Photo; **15.1** T. E. Adams/VU; **15.2B** Stanley Flegler/VU; **15.3** Allan Roberts; **15.4** Daniel W. Gotshall/VU; **15.5** G. Newberry/VU; **15.6** Nancy Adams/TS; **15.7A** IFA/Peter Arnold, Inc.; **15.7B** author photo; **15.8A** Toni Angermayer/PR; **15.8B** Myrleen Ferguson/PhotoEdit; **15.12** John Walsh/Science Photo Library/PR; **15.13** Lennart Nilsson/Bonnier Fakta, from *Behold Man,* 1974, Little Brown; **15.14** Lennart Nilsson, *The Incredible Machine,* pp. 20-21, © 1986 National Geographic Society; **15.15** D. W. Fawcett/Science Source/PR; **15.16** John D. Cunningham/VU; **15.20** Paul Howell/Gamma Liaison; **p. 428** Freeman/PhotoEdit; **p. 429** Taylor/Gamma Liaison

CHAPTER SIXTEEN

p. 439 Supplied by Carolina Biological Supply Company; **16.4** Supplied by Carolina Biological Supply Company; **16.8** John D. Cunningham/VU

PHOTOESSAY

p. 455 Petit Format/Nestle/PR; **pp. 456-61** CEDRI **16.12** John D. Cunningham/VU; **16.13** John D. Cunningham/VU; **p. 443** William Gage; **p. 444** Manfred Kage/Peter Arnold, Inc.

PART FIVE, CHAPTER SEVENTEEN

p. 474 Akira Uchiyama/PR; **pp. 474-75** Francois Gohier/PR; **p. 476** Rod Planck/TS; **17.1** William C. Jorgensen/VU; **17.3A** Biophoto Associates/PR; **17.3B** R. Calentine/VU; **17.9** Charles Seaborn/Odyssey Productions, Chicago; **17.12A** Dr. R. Kessel/Peter Arnold, Inc.; **17.12B** D. W. Fawcett/VU; **17.12C** Fred Hossler/VU

CHAPTER EIGHTEEN

p. 497 Frans Lanting/Minden Pictures; **18.2** Jeff Foott/TS; **18.3A** Dwight R. Kuhn; **18.3B** Frans Lanting/Minden Pictures; **18.4** John Gerlach/VU; **18.5A** Dwight R. Kuhn; **18.5B** Don & Pat Valenti; **p. 507** Wardene Weisser/Bruce Coleman, Inc.; **18.7** Thomas Eisner, Cornell University

CHAPTER NINETEEN

p. 516 Michael Fogden/Bruce Coleman, Inc.; **19.2B** Kent Wood/Peter Arnold, Inc.; **19.3A** William H. Amos; **19.5** Ron & Valerie Taylor/Bruce Coleman, Inc.; **p. 523** American Cancer Society; **19.10** Lennart Nilsson/Bonnier Fakta, from *Behold Man,* p. 63, 1974, Little Brown; **p. 527** Gerold Lim/Unicorn Stock Photos; **19.13** Martha McBride/Unicorn Stock Photos; **19.14** David M. Phillips/VU; **p. 535** Lou Lainey/*Discover* Magazine, March 1984, Time Inc.; **p. 536** Vladimir Lange/The Image Bank, Chicago; **19.21** G. R. Roberts

CHAPTER TWENTY

p. 550 Secchi, Lecaque, Roussel, UCLAF, CNRI/Science Photo Library/PR; **20.2** John D. Cunningham/VU; **20.3** Lennart Nilsson, *The Incredible Machine,* p. 123, © 1986 National Geographic Society; **p. 555** R. Calentine/VU **p. 556 T** SIU/VU; **p. 556 B** John Gerlach/VU; **p. 561 L** George H. Harrison/Bruce Coleman, Inc.; **p. 561 R** John Cancalosi/TS; **20.8** AP/Wide World Photo; **20.9** Alon Reininger/Contact Press Images/Woodfin Camp & Associates; **20.10** Eddie Birch/Unicorn Stock Photos; **p. 566** Lennart Nilsson/Boehringer Ingelheim; **20.11** Film Stills Archive, Museum of Modern Art; **20.12** John D. Cunningham/VU

CHAPTER TWENTY-ONE

p. 573 Lennart Nilsson/Bonnier Fakta, from *Behold Man,* p. 176, 1974, Little Brown; **21.3A** London Daily Mirror; **21.3B** Wide World Photos; **21.12** Bernard Hehl/Unicorn Stock Photos; **p. 581** John D. Cunningham/VU

CHAPTER TWENTY-TWO

p. 597 Secchi, Lecaque, Roussel, UCLAF, CNRI/Science Photo; Library/PR; **22.4** John D. Cunningham/VU; **22.5** Lennart Nilsson/Bonnier Fakta, from *Behold Man,* p. 162, 1974, Little Brown; **p. 612** Charles E. Schmidt/Unicorn Stock Photos; **22.15** Johnathan T. Wright/Bruce Coleman, Inc.; **22.16** Alexander Tsiaras/Stock, Boston; **22.17** Dan McCoy/Rainbow; **22.19** Gamma Liaison; **22.20** Robert Fried/Stock, Boston; **22.21A** Thomas R. Fletcher/Stock, Boston; **22.21B** Frank Oberle/Photographic Resources, Inc.; **22.22** Historical Pictures Service, Chicago; **22.23** Ron P. Jaffe/Unicorn Stock Photos; **22.24** The Coca-Cola Company Archives; **22.25B** Vera Lentz/Black Star; **22.26** Dan Ford Connolly/Picture Group; **22.28** Dr. Albert Hofman

CHAPTER TWENTY-THREE

p. 637 Merlin D. Tuttle, Bat Conservation International; **23.1** Thomas Eisner, Cornell University; **23.3** Rod Planck/TS; **p. 643** Jane Burton and Kim Taylor/Bruce Coleman, Inc.; **23.6** Theo Westenberger/Sygma; **p. 646** Robert Pearcy/ANIMALS ANIMALS; **23.7** M.P.L. Fogden/Bruce Coleman, Inc.; **23.9B** Lennart Nilsson/Bonnier Fakta, from *Behold Man,* p. 244, 1974, Little Brown; **23.10** Seth Resnick/Stock, Boston; **p. 649** Ian Berry/Magnum; **23.12B** Lennart Nilsson/Bonnier Fakta, from *Behold Man,* pp. 196-97, 1974, Little Brown

PART SIX, CHAPTER TWENTY-FOUR

pp. 656-57 Robert Maier/ANIMALS ANIMALS; **p. 658** David MacDonald/Oxford Scientific Films/ANIMALS ANIMALS; **24.1** Peter Miller/The Image Bank, Chicago; **24.2A** Gamma Liaison; **24.2B** John D. Cunningham; **24.3** BBC Natural History Unit. From *The Discovery of Animals Behavior* by John Sparks. 1982. A Collins Publishers/BBC Co-production; **24.4** Walter Chandoha; **24.6** Lawrence Migdale/PR; **24.7** John D. Cunningham/VU; **p. 664** F. Stuart Westmoreland/TS; **24.8** Michael Irrgang/David R. Frazier Photolibrary; **24.9A** The Bettmann Archive; **24.10A** Rick Friedman/Black Star; **24.10B** Gamma Liaison; **24.12** Stephen J. Krasemann/DRK Photo; **24.13** Thomas McAvoy, *Life* Magazine, © Time Inc.; **24.14** John D. Cunningham/VU; **24.16** Charles Walcott; **24.17** P. Breese/Gamma Liaison; **p. 674** IFA/Bruce Coleman, Inc.; **24.18** John Shaw/TS; **24.19A** Anthony Mercieca/PR; **24.19B** Don & Pat Valenti; **24.20** E. R. Degginger; **24.21** Lynwood Chase/PR; **24.22** Gordon Wiltsie/Bruce Coleman, Inc.; **24.23** Peter Davey/Bruce Coleman, Inc.; **24.25** Terry G. Murphy/ANIMALS ANIMALS; **24.26** Marty Snyderman/VU; **24.27** Erwin & Peggy Bauer/Bruce Coleman, Inc.; **24.28** Thomas Kitchin/TS; **24.29** Edward S. Ross, California Academy of Sciences; **24.30** R. Williamson BES/VU; **24.31** UPI/Bettmann Newsphotos; **24.32** Stephen R. Swinburne/Stock, Boston; **24.34** Gerald Lacz/Peter Arnold, Inc; **p. 686** John Cancalosi/Stock, Boston; **24.35 L** Albert Copley/VU; **24.35 R** M. Richards/PhotoEdit; **24.36** Joe

Wrinn/Harvard University News Office **24.37** Daemmrich/Stock, Boston

CHAPTER TWENTY-FIVE

p. 694 Willard Clay; **25.3A** Nancy Dudley/Stock, Boston; **25.3B** Joe McDonald/TS; **25.4** Willard Clay; **25.5** Willard Clay; **25.6** U. S. Forest Service; **25.7** Norman Myers/Bruce Coleman, Inc.; **25.8** Glenn Oliver/VU; **25.12A** C. Allan Morgan/Peter Arnold, Inc.; **25.12B** Marine World/Africa USA/Brian Parker/TS; **25.12C** Spencer Jones/Bruce Coleman, Inc.; **25.12D** Jack Wilburn/ANIMALS ANIMALS; **p. 706** Marc & Evelyne Bernheim/Woodfin Camp & Associates; **25.16** author photo; **p. 711** Jacques Jangoux/Peter Arnold, Inc.; **25.17** author photo; **25.18** Bill Beatty/VU; **25.19** David L. Pearson/VU; **25.20** Kirtley Perkins/VU; **25.21** John Shaw/TS; **25.22** Willard Clay; **p. 715** Robert Frerck/Odyssey Productions, Chicago; **25.23** Norman Myers/Bruce Coleman, Inc.; **25.24** Len Rue, Jr./VU; **25.25** Willard Clay

PHOTOESSAY

p. 718 Stacy Pick/Stock, Boston; **p. 719 T** Robert Frerck/Odyssey Productions, Chicago; **p. 719 B** Milton Rand/TS; **p. 720 T** Max & Bea Hunn/VU; **p. 720 B** Ken W. Davis/TS; **p. 721 T** Thomas Kitchin/TS; **p. 721 B** Frans Lanting/Minden Pictures; **p. 722 T** Lynn M. Stone; **p. 722 B** Roy Attaway/PR; **p. 723 T** Kenneth Mantai/VU; **p. 723 B** John Gerlach/VU **25.26 L** A. H. Benton/VU; **25.26 R** Max & Bea Hunn/VU; **25.28** Doug Sokell/VU; **25.31A** Michelle Barnes/Gamma Liaison; **25.31B** J. L. Atlan/Sygma; **25.32** J. M. Bassot, H. Chaumeton/Nature; **25.33** Al Giddings/Ocean Images

CHAPTER TWENTY-SIX

p. 735 Joe McDonald/VU; **26.1** John D. Cunningham/VU; **26.2** Robert Frerck/Odyssey Productions, Chicago; **26.3** L.L.T. Rhodes/ANIMALS ANIMALS; **26.4** G. Prance/VU; **26.5** C. C. Lockwood/ANIMALS ANIMALS; **26.11** Walt Anderson/TS; **26.12B** Francisco Erize/Bruce Coleman, Inc.; **26.13** S. Powers/VU; **26.14** Bob Gossington/Bruce Coleman, Inc.; **26.15** Gerry Ellis/The Wildlife Collection; **26.16** David Shale/Oxford Scientific Films/ANIMALS ANIMALS; **26.17A** E. R. Degginger/ANIMALS ANIMALS; **26.17B** J. Alcock/VU; **26.18** Tom J. Ulrich/VU; **26.19** *Micrographia* by Robert Hooke, published in 1665 by The Royal Society; **p. 751 T** David M. Dennis/TS; **p. 751 B** Tom McHugh/PR; **26.20A** Mark Newman/TS; **26.20B** Chip Isenhart/TS; **26.21** Adam Woolfitt/Woodfin Camp & Associates; **p. 754** Sygma; **26.22** John Gerlach/VU; **26.23** Martha McBride/Unicorn Stock Photos; **26.24** C. Allan Morgan/Peter Arnold, Inc.; **26.25** Tom Stack/TS; **26.26 L** Nathan Benn/Stock, Boston; **26.26 R** Gordon Groene/VU; **26.27** John Neubauer/PhotoEdit

CHAPTER TWENTY-SEVEN

p. 763 Paul H. Henning; **27.2** author photo; **27.5** Jean Beraud, *Le Boulevard des Capucins et le Theatre de Vaudeville,* 1889, Paris, Carnavalet, Giraudon/Art Resource, New York; **27.6** Owen Franken/Sygma; **p. 774 L** Peter Miller/The Image Bank, Chicago; **p. 774 R** David Young-Wolff/PhotoEdit; **p. 777** David W. Hamilton/The Image Bank, Chicago

CHAPTER TWENTY-EIGHT

p. 780 Kenneth Mantai/VU; **28.1** Sygma; **28.2** Fref Kong/PhotoEdit; **28.3A** Anthony Suau/Black Star; **28.3B** Joseph P. Shapiro/U. S. News & World Report; **p. 784** Springer Liaison; **28.4** Esaias Baitel/Gamma Liaison; **p. 788** M. Philippot/Sygma; **28.5** Chico Paulo/DDB Stock Photo; **28.6** Richard Lord/The Image Works; **28.7** Irven DeVore/Anthro-Photo File; **28.8** Link/VU; **p. 793** U. S. Department of Agriculture Office of Governmental and Public Affairs; **28.10** The United Nations; **28.11** W. Campbell/Sygma; **28.13 L** Ann Duncan/TS; **28.13 R** Zao-Longfield/The Image Bank, Chicago; **28.15** Robert Brenner/PhotoEdit; **28.17** Wendell D. Metzen/Bruce Coleman, Inc.; **p. 803** American Petroleum Institute; **28.18** Kaz Mori/The Image Bank, Chicago; **28.19** Dagmar Fabricius/Gamma Liaison; **28.20** Tony Freeman/PhotoEdit; **28.21** Bruce W. Wellman/Stock, Boston

CHAPTER TWENTY-NINE

p. 816 Bob Pool/TS; **29.1** John Cancalosi/DRK Photo; **29.2** William Johnson/Stock, Boston; **p. 818** Matteini/Sipa Press; **29.3** Kevin Syms/David R. Frazier Photolibrary; **29.4** Arthur Morris/VU; **29.5** Phil Degginger; **p. 822** Gamma Liaison; **29.6** Fernando Bueno/The Image Bank, Chicago; **29.7** Gerry Souter/Tony Stone Worldwide; **29.8** Osvaldo Bohn Photo Studies, © *Discover* Magazine, 2/86; **29.9** Frank Hanna/VU; **29.10** Wernher Krutein/Gamma Liaison; **29.11** Tom McCarthy/Unicorn Stock Photos; **29.12A** Oxford Scientific Films/ANIMALS ANIMALS; **29.12B** F. S. Mitchell/TS; **29.12C** John D. Cunningham/VU; **29.14** Bill Pierce/Woodfin Camp & Associates; **p. 835** Michael Hayman/PR; **p. 836** Spencer Swanger/TS **p. 838** Michael Hayman/PR

ILLUSTRATION CREDITS

Randee Ladden; Teri J. McDermott; Sandra E. McMahon; Precision Graphics; Rossi & Associates; Kevin A. Somerville; Sarah Forbes Woodward

Index

A

Abert squirrel, 266
Abiotic control, 749
Abortion, 434-35
  population growth and, 773
  society and, 434-35
Acadia National Park, 721
Acetate, 142
Acetylcholine, 616
Acid, 64, 73
  nucleic, 82-84
Acid fermentation, 140-41
Acid rain, 751
Acquired immune deficiency syndrome (AIDS), 563-68
Acquired trait, 26-27
Acromegaly, 576
Action potential, 589
Active transport, molecular motion and, 114-16
Adaptiveness of behavior
  coevolution and, 642-43
  sociobiology and, 688-90
Addiction, mindbenders and. *See* Mindbender
Adenosine diphosphate, 124-25
Adenosine triphosphate
  energy and, 124-25
  glycolysis and, 137-39
  photosynthesis and, 131
  repolarization of nerve cell and, 589
ADH. *See* Antidiuretic hormone
ADP. *See* Adenosine diphosphate
Adrenal gland
  anaphylactic shock and, 556
  human endocrine system and, 578
Advertising of smoking, 623
Afferent neuron, 586
Afterbirth, 468
Age structure pyramid, 775
Agent Orange, 401
Aggression, 673-74
Aging
  bone and, 480
  cellular cycles and, 175
  lakes and, 725-26
  nicotine and, 623
Agnatha (jawless fish), 357, 360-62
Agriculture
  advent of, 768
  African famine and, 785-86, 788
  food pyramid and, 790-91
  genetic engineering and, 219-20
  human populations after 1850 and, 769
  population levels and, 768
  slash and burn, 790
  worldwide program and, 789-90
Air, composition of, 232
Air pollution, 820-24
Alaskan oil spill, 729
Albatross
  waved, 22
Albumen, 452
Alcohol
  carbon and, 71
  mood altering substances and, 626-28
  urine concentration and, 512
Aldosterone, 512, 578-79
Algae, 319, 320-23
  life cycle of, 384-86
  sewage and, 825
Allantois, 452
Allele
  Mendelian genetics and, 186
  multiple, 200
Allen, Woody, 569
Allergy, 556
Allopatric speciation, 266
Alpine tundra, 713
Altruism, 683-88
Alvarez hypothesis, 275
Alzheimer's disease, 221, 222
Amanita phalloides, 307
American chestnut, 755
American goldfinch, 367
Amino acid, 78
  genetic code and, 171
Amino group, 71
Ammonia, 504
Amnion, 452
Amoeba, 295
  digestion and, 538
  size limits of, 477
Amoebocyte, 339
Amphetamine, 631
Amphibian, 364-65, 531
  temperature regulation and, 501
Amytal, 631
Anabolic steroid, 79
Anaerobe, 285-86
Anal intercourse, 565
Anaphase, mitosis and, 153
Anaphylactic shock, 556
Andean condor, 367
Anemia, sickle-cell, 198, 220
Anemone, 342
  contractile systems and, 486
Angel dust, 632
Angina pectoris, 535
Angiosperm (flowering plant), 331-33
Anhinga, 367
Animal (Animalia)
  aquatic, 517-20
  classification of, 285
  cold-blooded, 500-501
  development of, 440-69
    egg membranes and, 451-53
    egg types and, 446-47
    embryonic organization and, 442-45
    frog and, 447-51
    human gestation and, 453-69
  domestic, 790-91
  food pyramid and, 790-91
  kingdom of, 336-74
    Agnatha (jawless fish) and, 357, 360-62

Amphibia (amphibian) and, 364-65
Annelida (segmented worm) and, 350-52
Arthropoda (exoskeleton) and, 352-54
Aves (bird) and, 366-69
Chondrichthyes (cartilage fish) and, 362
Chordata (notochord) and, 357-74
Cnidaria (jellyfish) and, 340-44
Echinodermata (spiny skin) and, 356
Hemichordata (marine worm) and, 357
Mammalia (mammal) and, 369-74, 501-504, 521-25
Mollusca (mollusk) and, 347-49
Nematoda (roundworm) and, 346-47
Osteichthyes (bony fish) and, 362-64
phyla and characteristics of, 337-74
Platyhelminthes (flatworm) and, 344-46
Porifera (sponge) and, 339-40
Reptilia (reptile) and, 365-66
reproduction and. *See* Reproduction
temperature regulation and, 500-504
Animal pole, 447
Animal tranquilizer, 632
Animal-like protitst, 292-97
Annelid (segmented worm), 350-52
Antagonist muscle, 489
Antelope, 678
Anterior lobe, human endocrine system, 576
Anther, 392
Antheridium, 386
Anthozoa, 341
Antibiotic, 288, 290
Antibody, 557
acquired immune deficiency syndrome and, 564
Antidiuretic hormone, 512
Antigen-antibody response, 557
Antitoxin, 287
Anus, 540
Anvil, middle ear, 645
Apgar test series, 469
Apical meristem, 396
Appendicular skeleton, 484
Aquatic animal, 517-20
Aquatic plant, 320-23
Arachnida, 352
Arc, reflex, 600-601
Archaebacteria, 283, 285-86
Archegonium, 386
Arctic tundra, 713
Arm, skeletal system and, 484
Armpit, chemoreceptors and, 646
Arthritis, 562
Arthropod
contractile fibers and, 487
Arthropoda (exoskeleton), 352-54
Artificial insemination, 565
Artificial selection, 25-26
Artificial sun, 672
Artiodactyla, 372
Ascending neuron, 601
Ascomycota, 303-306
Ascus, 306
Asexual reproduction
animals and, 408-10
eubacteria and, 288
Asia, projected population of, 773
Aspirin, 502
Association neuron, 601
Asteroidea, 356
Asthma, smog and, 823
Atherosclerosis, 535
Atlantic Ocean, continental drift and, 267
Atmosphere, earth, 232
Atom
electron and, 55, 57-58
elements and, 54-55
energy from, 810-11
inert, 58
structure of, 54-55
Atomic number, 55
ATP. *See* Adenosine triphosphate.
Auditory receptor, 640-45
Autoimmunity, 561-62
Automobile, acid rain and, 751
Autonomic learning, 619-20
Autonomic nervous system, 616-19
Autosome, 192
Autotrophy, 241-42
Auxin
cytokinins and, 402
plant hormones and, 398-401
Aves (bird), 366-69
Axial skeleton, 482
Axilla, chemoreceptors and, 646

B

Baboon, 374
sexual bonding and, 425
Bacillum, 288
Backbone, 360
Background radiation, 830
Bacteria
human populations after 1850 and, 769
reproduction and, 214-15
sewage and, 825
Band, muscle contraction and, 493
Barbiturate, 631-32
Barr body, 192
Base, acid and, 73
Basidiocarp, 309
Basidiomycota, 307-10
Basidiospore, 310
Basilar membrane, middle ear, 645
Basophil, 553
Bat, 372
Bat-moth coevolution, 642-43
B-cell
lymphocytes and, 559
specific immune response and, 557
Beach, food chain and, 722
Beagle, voyage of, 6-8, 14-25
map and, 8
Bear, 370
Beaver, 370
Bee
cooperation and, 683
kin selection and, 686-87
Beer
mood altering substances and, 626
urine concentration and, 512
Beetle, hormones and, 575
Behavior
adaptiveness of
See Adaptiveness of behavior
development of, 659-91
ethology and comparative psychology and, 659
instinct and, 660-67
interaction of instinct and learning and, 667-70
learning and, 662-67
orientation and navigation and, 670-72
innate, 660
isolation and interbreeding and, 265-66
Bell-shaped curve, 258
Benchuga bug, 29
Bereavement, white blood cell count and, 569
Biceps brachii, 489
Big Bang theory, 232
Bile, 543
Bimodal curve, 259
Binary fission, 408
Bioethics, 817-39
environmental pollution and, 819-35
future and, 837-38
hidden decisions and, 835-37
population dynamics and, 739
simple living and, 817-18
water pollution and, 797, 824-35
Biofeedback, 619
Biological clock, 672, 674
Biological determination, 688-90
Biological diversity, 16
Biologist, careers of, 43
Biome and community, 695-732
coastal areas and, 730-32

ecosystems and, 695
extinction and, 704-705
grazing and, 791
habitat and niche and, 695-96
land environment and, 707-16
succession and, 698-700
trophic levels and, 701
water environment and, 717-30
Biotic potential
density-dependence and, 752
J-shaped curve and, 739, 740
Bird
animal kingdom and, 366-69
brain of, 602
egg membranes of, 451-53
Galapagos Islands and, 16
great frigate, 24
heart of, 531
mud flats and, 731-32
reproductive strategy of, 747-48
Birth
from fertilization to, 439-69
See also Animal, development of
human, 468-69
Birth control pill, 431
Birth rate. *See also* Population growth, human
crude, 771
human populations after 1850 and, 770
U.S. tax structure and, 837
Black Death, 555
Black star, 276
Bladder, urinary
chlorinated water and, 826
kidney and, 508-509
Blagden, Charles, 498
Blastocoel, 447
Blastocyst, 463
Blastula, 447
Blaustein, Andrew, 686
Bligh Reef, Alaskan oil spill and, 729
Blight, miracle crops and, 792
Blindness, color, 195
Blood
acquired immune deficiency syndrome and, 565
circulatory system and, 525, 529-31
See also Circulatory system
types of, 200
Blood cell, white, 552
Blood pressure, arteries and, 529
Blood-brain barrier, 527
Blue-green algae. *See* Cyanobacteria
Body cavity, 346-47
Body-mind connection, 568-69
Bog, 713
Bomb, nuclear, 835
Bonding
chemical, 59-63
covalent, 60
double, 60
hydrogen, 61-63
ionic, 59-60
sexual, 425
Bone
aging and, 480
connective tissue and, 479-81
human embryo and, 458-61, 464, 466
middle ear and, 645
Bony fish, 362-64
Booby, 23
Bottleneck effect, 262
Botulism, 287
Bowman's capsule, 508
Brain
acquired immune deficiency syndrome and, 566, 568
human, 604-14
forebrain and, 605-609
hemispheres and, 609-10
hindbrain and, 605
immunity and, 568-69
left/right halves and, 610-14
lobes and, 609-10
midbrain and, 605
split brain and, 614
structure and function of, 605-609
location of, 598
split, 614
vertebrate, 602-604
Branchiostoma, 359
Brazil's tropical forest, 711
Bread mold, 303
Breast, female, 419
Breeding, animal, 26
Brittle star, 356
variation and, 260
Bronchitis, 823
Brown algae, 321-22
Brown pelican, 367
Bryde's whale, 753
Bryophyta (moss), 323-24
Bud scale, 396
Budding, 410
Buffalo, European, 264
Buffon, George-Louis Leclerc De, 11
Bug, Benchuga, 29
Bulrush, 720

C

Caffeine, 624-25
Calcitonin, 576
Calcium, human pregnancy and, 466, 467
Calderas, Galapagos Islands and, 21
Calories, world production of, 789
Calvin cycle, 131
Cambium, 397-98
Camel, 372
water conservation and, 507
Canal, auditory, 640
Canaliculus, 479
Cancer
acquired immune deficiency syndrome and, 564
bladder, 826
chlorinated water and, 826
radiation and, 830
tobacco and, 523, 623
viruses and, 216
Cannabis sativa, 625
Capillary, lymph, 537
Carbohydrate, 74-75
Carbon, 68-72
Carbon monoxide, 821-23
Carboxyl group, 71
Carcinogen, tobacco as, 623
Cardiac muscle, 488
Cardiopulmonary resuscitation, 535, 536
Carnivore, 370
food pyramid and, 786
ocean and, 728
Carotene, 127
Carpal, human skeleton, 484
Carpel, flowers and, 392
Carrying capacity
African famine and, 788, 789
environmental resistance and, 740
food imbalance and, 781
human population levels and, 767, 768
increase in, 768
Cartilage
connective tissue and, 481
skeletal systems and, 478
Cartilage fish, 362
Castle, W.E., 203
Cat, 370
Catalyst, 66
Cattle, 372
grazing and, 791
Cavity, body, 346-47
Cell
aging of, 175
components of, 99-110
cell walls and, 99-100
centriole and, 102-103
cilia and flagella and, 103
cytoskeleton and, 101
endoplasmic reticulum and, 105, 107
Golgi bodies and, 107-108
lysosomes and, 108
microtrabecular lattice and, 102
microtubule and, 102
mitochondria and, 104
nucleus and, 109-110
plasma membrane and, 100
plastids and, 108-109
ribosomes and, 104-105

vacuole and, 109
cytology and, 93
first divisions of, 447
future studies of, 177
glial, 586
life cycles of, 152-80
DNA replication and, 214-15
double helix (DNA) and, 164-66
meiosis and, 157-64
mitosis and, 152-57
nerves and, 584-86
prokaryotic and eukaryotic, 93-97
respiration of, 133-46, 517
energy and, 133
summary of, 146
size limits of, 477
tetraploid, 268
wall of, 99-100
white blood, 552
Cell biology, 176-77
Cell theory, 92
Cellulose
carbohydrates and, 75
enzymes and, 75
Cenozoic era, 267
Centipede, 352, 354
Central nervous system, 600-604
Centriole, 102-103
Centromere, mitosis and, 153
Cephalochordata, 357
Cerebellum, 602, 605
Cerebrum, 602, 609
Cervical cap, 430
Cestoda (tapeworm), 345
Chagas' disease, 29
Chain
antigen-antibody response and, 557
carbon, 70
double helix and, 164-66
electron-transport
cellular respiration and, 143-46
light-dependent reaction and, 128
food
DDT and, 828-29, 830
dinoflagellates and, 298-99
disruption of, 275
ocean and, 726
salt marsh and, 731
sewage and, 825
Chameleon, 263
Channel, ion, 589
Channel Islands, 721
Cheese, fungi imperfecti and, 310
Chelicerata, 352
Chemical
bonding of atoms and, 59-63
effluent and, 825
environmental pollution and, 827
Chemical fertilizer, 725
Chemical messenger, 574
Chemical reaction, 63-68
Chemiosmosis, 131
Chemiosmotic phosphorylation, 143-46
Chemistry
atoms and elements and, 54-55
carbon and, 68-72
chemical bonding and, 59-63
chemical reactions and, 63-68
CHNOPS and, 54-55
electrons and, 56-59
hormone messengers and, 574
molecules of life and, 68-84
Chemoreceptor, 646-50
Chemosynthesis
bacteria and, 289-91
Galapagos Islands and, 134-35
Chernobyl, 832, 833
Chest pain, 535
Chicken, production of, 791
Chilopoda, 354
Chimpanzee, 374
reproductive strategy of, 746-47
China
population policy of, 773
projected population of, 773
17th century and, 768
Chiroptera, 372
Chitin, carbohydrates and, 75
Chlamydia, 428
Chlorine
ionic bonding and, 59-60
sewage and, 826
Chlorophyll
fungi and, 300
photosynthesis and, 126, 130
Chlorophyta (green algae), 322-23
Chloroplast, 127, 130
CHNOPS, six elements and, 54-55
Cholesterol, 77
Chondrichthyes (cartilage fish), 362
Chordata (notochord), 357-74
Chorioallantoic membrane, 452
Chorion, 452
Chorionic villus, 463
Chromatin, mitosis and, 153
Chromium, South Africa and, 781
Chromosome
cellular reproduction and, 168
criminal behavior and, 193
doubling of, 268
mapping and, 197
mitosis and, 153
mutation and, 199
number of, 157
Chrysophyta, 297-98
Cigarette smoking, 523, 623
Cilia, 103
Ciliophora, 292
Circulatory system, 525-38
animals and, 525-36
heart and, 526-27, 531-36
lymphatic system and, 536-38
Civil war, African famine and, 785-86, 788
Clam, 349
Clap, 428
Classical conditioning, 665-66
Classical genetics, 189-97
Classification of five kingdoms, 283-85
Cleavage, mitosis and, 440
Climax stage, 698
Cline, geographic, 256
Clitellum, 351-52
Clitoris, 418
Cloaca, 540
Clock, biological, 672, 674
Clone
gene machine and, 218
genetic interference and, 210
genetic splicing and, 213
Clostridum botulinum, 286, 287
Clostridum tetani, 286
Cloud, Preston, 799
Cnidaria (jellyfish), 340-44
contractile systems and, 486
digestion and, 538
CoA. *See* Coenzyme A
Coacervate droplet, 237-41
Coal, 802-803
Coastal biome and community, 730-32
Coca-Cola, 629
Cocaine, 628-31
Coccus, 288
Cochlea, 645
Coconut milk, 402
Codon, 171
Coelacanth, 364
Coelom, 347-49
Coenzyme A, 142
Coffee, 624-25
Coitus interruptus, 427
Cold-blooded mammal, 500-501
Collenchyma, 398
Colloidal protein, 237-41
Color blindness, 195
Coloring, protective, 258
Columbus, flatness of earth and, 9
Commons, law of, 793
Community
biomes and, 695-732
See also Biome and community
ecosystems and, 695
Comparative psychology, 659-60
Competition
aggression and, 673-74
habitat and, 697
parasitism and, 755
Compound, molecules and, 54-55
Computer, gene machine and, 218
Conception, human, 421-24
Conditioning
classical, 665-66

operant, 666-67
Condom
acquired immune deficiency syndrome and, 565
contraception and, 427-29
Condor, 367
Cone, pine, 388
Congenital disease, 222
Conidia, 303-306
Conidiophore, 306
Conifer (cone bearer), 327
reproduction and, 388-90
Conjugation
bacteria and, 214-15
paramecia and, 294
Connective tissue, 479-82
Conservation of water resources, 798-99
Constant region, 557
Consumer, trophic levels and, 701
Continental drift, 267
Contraception, 425-35
Contractile system, 486-93
Contractile vacuole, 109, 294
Controlled experiment, 36
Convergent evolution, 269-270
Cooperation, 680-83
Copernicus, 9
solar system and, 33
Copulation, 413
human, 418
Cord, spinal. *See* Spinal cord
Cork cambium, 398
Cormorant, 22
Corn
miracle crops and, 792
other edible species and, 736-37
Corpus callosum, 614
Corpus luteum, 614
Cortisone, 578
Cotyledon, 396
Cousins, Norman, 569
Covalent bond, 60
CPR. *See* Cardiopulmonary resuscitation
Crab
chemosynthesis and, 134-35
mud flats and, 731-32
Sally Lightfoot, 21
Crack (cocaine), 631
Cranial nerve, 615
Crash
population, 741
Crayfish, 478
Cretacious period, 273-74
Cretinism, 576
Crick, Francis, 166
Crime, sex chromosomes and, 193
Crinoidea, 356
Critical period, in imprinting, 669
Crocodile, 365, 561
Crop, earthworm digestion and, 539-40
Cross, test, 188-89
Cross bridge, 493
Crossover
gene linkage and, 193-97
meiosis and, 159
Crude birth rate, 771
Crude death rate, 771
Crustacea, 352
circulatory system and, 526
Crystal meth (methamphetamine), 631
Curtiss III, Roy, 217
Curve
bimodal, 259
S & J-shaped, 739, 740
statistical, 258
Cutler, Winnifred, 581
Cyanobacteria, 289-91
sewage and, 825
Cycad tree, 327
Cycle
biological, 674
cellular. *See* Cell, life cycles of
menstrual, 418-21
Cypress, 327
Cypress swamp, 720
Cystitis, 429
Cytokinesis, 152
Cytokinin, 402-403
Cytology, 93
Cytopyge, 294
Cytoskeleton, 101
Cytostome, 293
Cytotoxic T-cell, 557

D

D and C. *See* Dilation and curettage
Darwin, Charles
acquired versus hereditary traits and, 26-27
biotic potential of species and, 743
genetics and, 182, 251, 252
illness of, 29
inductive reasoning and, 33-37
plant hormones and, 398-400
research methods of, 33-38
theory of evolution and, 14-29
voyage of Beagle and, 6-8, 14-25
youth of, 5-8, 13-14
Darwin, Erasmus, 11
Daughter cell, 152
DDT
DDE and, 828
fatty tissue and, 828
human populations after 1850 and, 770
De Buffon, George-Louis Leclerc, 11
Death
abiotic control and, 749-57
advantage of, 757-59
anaphylactic shock and, 556
barbiturates and, 632
controlling populations through, 749-57
heart attack and, 535
starvation and, 784
Death rate
crude, 771-72
human populations after 1850 and, 769
Death star, 272-77
Deciduous forest, 712
Decomposer
sewage and, 825
trophic levels and, 703
Deductive reasoning, 33-37
Deer, 372
Dehydration, 72
Deletion, chromosome mutations and, 199
Dementia, 566
Demography
African famine and, 785-86, 788
human populations after 1850 and, 769
population structure and, 775
Denaturation, 82
Dendrite, 584
Density of populations, 752-57
Density-dependence
biotic control and, 752
human population levels and, 767
Density-independence, 749
Deoxyribonucleic acid. *See* DNA
Depolarization, neuron, 589
Depression
ethanol and, 626
mood altering substances and, 621
Descent with modification. *See* Evolution
Desert, 716
Detergent, water pollution and, 725
Deuteromycota, 310
Deuterostome, 346
Developed country
global food resources and, 785
percent of earth's population and, 772
Developing country
global food resources and, 785
Diablo Canyon nuclear power, 834
Diaphragm, contraception and, 430
Diatom, 298
Diatomacious earth, 298
Dicara, L.V., 619
Dicotyledon, 396
Differentiation, plant development and, 395
Diffusion, molecular motion and, 110-12
Digestion, 538-545
enzymes and, 67
human, 540-45
Digit (finger), 484
Dilation and curettage, 434
Dimorphism, 251

Dingo, 752-53
Dinoflagellate, 298-99
Dinosaur, 738
Diploidy, 384
Diplopoda, 354
Directional selection, 258
Disaccharide, 74
Disease
congenital, 222
genetic engineering and, 220-22
mortality of populations and, 756-57
Disruptive selection, 259
Distribution of food
global resources and, 782-86
storage and, 789
Divergent evolution, 269-70
Diversity, biological, 16
DNA (Deoxyribonucleic acid), 82-84
acquired immune deficiency syndrome and, 566
antibodies and, 561
double helix and, 164-66
eubacteria and, 288
genetic interference and, 210
radiation and, 831
replication of, 168, 214-15
Watson and Crick and, 166-68
DNA sequencing, 218
Dog, 370
Australian dingo and, 752-53
cooperation and, 681
interbreeding and, 264-65
Domestic animal, 790-91
Dominance
aggression and, 673-74
genetics and, 183-84, 199-200
Dominant trait, 184
Dorsal nerve root
central nervous system and, 600
peripheral nervous system and, 616
Double bond, 60
Double helix, 164-66
The Double Helix, 166
Doubling time, population, 773, 774
Downy mildew, 301-303
Drift, continental, 267
Drift, genetic, 260
Droplet, reproducing, 238-41
Drosophila melanogaster, 190-91, 193, 195, 197
Drought, 785-86
Drug
intravenous, 565
psychoactive agents and, 621-33
See also Mindbender
Drug therapy, 218
Duck-billed platypus, 369
Duplication, chromosome mutations and, 199
Dutch Elm disease, 303

E

Ear, 641
Earlobe, 255
Earth
cooling of, 275
crust of, 267
diatomaceous, 298
early history of, 232-33
energy from, 807
future exploration of, 26
subduing of, 706
Earthworm, 350-52
digestion and, 539-40
excretory system and, 507
nervous system and, 599
Echinodermata (spiny skin), 356
Echinoidea, 356
Ecology, 695-732
See also Biome and community
Economic policy, 837
Ecosystem. *See also* Biome and community
communities and, 695
tidal power and, 806
Ectoderm, 441
Ectoplasm, 295
Ectotherm, 500-501
Ecuador, Darwin's voyage and, 15-25
Education, hidden influence of, 837
Effector, nerves and, 586
Efferent neuron, 586
Effluent, chemical, 825
Egg
bird, 451, 747-48
conifers and nucleus of, 389
types of, 446-47
Eggshell, 829
Eidetiker, 664
Ejaculation, 419
Eldredge, Niles, 270
Electricity
fossil fuel versus nuclear power and, 810-11
origin of life and, 235-36
photovoltaics and, 808-10
wind farm and, 806-807
Electron
atom and, 55, 57-58
behavior of, 56-59
Electron acceptor, 127
Electron microscope, 93, 95
Electron-transport chain, 128
cellular respiration and, 143-46
light-dependent reaction and, 128
Element, atoms and, 54-55
Elephant, 372
slow breeding of, 743
Embryo
chick membranes and, 452
egg yolk and, 446-47
flowers and, 392
organization and development of, 440-42
regulation of frog, 450-51
second to sixth month of human, 464-66
Embryo sac, 392-93
Emphysema, 823
Endergonic reaction, 67-68
Endocrine system, 576
Endocytosis, exocytosis and, 115-16
Endoderm, 441
Endometrium, 417
Endoplasm, 295
Endoplasmic reticulum, 105-107
Endorphin, 582
Endoskeleton, 478
Endosperm, 393
Endospore, 288
Endosymbiosis hypothesis, 105
Endotherm, 501-504
Energy, 120-49
adenosine triphosphate and, 124-33
cellular respiration and, 133
chemical reactions and, 65, 67-68
food pyramid and, 786
forms of, 122
laws of thermodynamics and, 123
per capita use of, 801
photosynthesis and, 125
renewable and nonrenewable resources for, 801-11
Engineering, genetic. *See* Genetic engineering
England, industrial melanism and, 258
Entamoeba histolytica, 294
Entropy, second law of thermodynamics and, 123-24
Environment
evolution and, 256, 258
See also Evolution
future of, 837-38
pollution and, 819-35
stable, 258
variation and, 255-56
Environmental resistance, 740
Enzyme
chemical reaction and, 65-67
chick embryo and, 452
restriction, 210
viruses and, 211
Eosinophil, 553
Epidermis
jellyfish and, 340
plant development and, 395
Epilepsy, 612
Epinephrine
anaphylactic shock and, 556
human endocrine system and, 579

sympathetic nervous system and, 616
Epistasis, 200-201
Eratosthenes, 9
Ergot, 306
lysergic acid diethylamide (LSD) and, 306, 632
Erythroxylon, 629
Escherichia coli
genetic crippling of, 217
human digestive system and, 545
vector and, 212
Esophagus, 540
Estrogen
endocrine system and, 578
female longevity and, 777
menstrual cycle and, 419
Estrus, 413
Ethanol, 626
Ethics. *See* Bioethics
Ethiopia, famine and, 785-86, 788
Ethology, 659
Eubacteria, 283, 286-91
Euglena, 296
Euglenophyta, 295-96
Eukaryote
prokaryote and, 285
Protista and, 292-97
Eukaryotic cell, 93-97
Europe, 17th century and, 768
European buffalo, 264
Eutheria, 370
Eutrophication, 725
Eutrophy, 792
Evergreen forest, 713
Evolution, 251-80
advantages of death and, 757-59
chemoreceptors and, 649
competition and, 697
convergent, 269-70
dead ends and, 339-40
definition of, 251
definition of species and, 264-66
directional selection and, 258
divergent, 269-70
environmental influence on variation and, 256
extinctions and, 263, 272-77, 704-705
human, 28
industrial melanism and, 258
natural selection and
Darwin's research and, 33-37
Darwin's theory of, 25-29
protective coloring and, 258
social behavior and, 688-90
sympatric speciation and, 268-69
variation and, 257-60
sexual reproduction and, 383-84
single-gene inheritance and, 254
small populations and, 260-64
social behavior and, 688-90
speciation and, 266-69
theory of, 14-29
variation in populations and, 252-64
Evolutionary tree
animal kingdom and, 348
hypothetical descendants and, 348
Excretory system, 506-13
Exergonic reaction, 67-68
Exocytosis, 115-16
Exoskeleton
Arthropoda and, 352-54
body support and, 478
muscular strength and, 487
Expanding universe, 231-32
Experiment, controlled, 36
Extensor muscle, 489
Extinction, 263, 272-77, 704-705
classical conditioning and, 666
destruction of species and, 704-705
human population growth and, 736
small populations and, 263
tropical forests and, 711
Exxon Valdez, 729
Eye
color of, 202
human, 651
induction and, 445
Eyespot, 297

F

Facilitated active transport, 114
Facilitated diffusion, 112
Fallopian plug, 434
Family tree, 358
Famine, 784, 785, 788, 789
African, 785-86, 788
potato blight and, 303
solutions to, 794-96
Fascia, 489
Fat
DDT and, 826
molecules and, 75-78
solubility of, 527
starvation and, 788
Fatty acid, 75-76
Feedback
homeostasis and, 498-99
hormonal systems and, 583-84
Feeding zone, 697
Female/male ratio, human, 777
Female reproductive system, 416-18
Femur, 484
Fermentation, 140-41
Fern, 326-27
reproduction and, 386
Fertilization
birth and, 439-69
egg membranes and, 451-53
egg types and, 446-47
embryonic organization and, 442-45
frog and, 447-51
human, 423
gestation and, 453-69
moment of, 422-23
Fertilizer
chemical, 725
miracle crops and, 792
water pollution and, 826
Fetish, food, 782
Fetus
acquired immune deficiency syndrome and, 565
thumbsucking and, 459
Fever
acquired immune deficiency syndrome and, 564
human body and, 502
Fibula, 484
Fighting, 674-79
Filament, 392
Finch
allopatric speciation and, 266
Galapagos Islands and, 16
Fireworm, 411
First law of thermodynamics, 123
First messenger, 581
Fish
acid rain and, 751
bony, 362-64
brain of, 602
cartilage, 362
DDT and, 828-29, 830
heart of, 531
jawed, 360, 360-62
jawless, 357, 360-62
mud flats and, 731-32
respiration and, 519
temperature regulation and, 500
Fishing, 792-94
Fission, 408
Fissure of Rolando, 610
Fitzroy, Captain, 7-8
Fixed action pattern, 661
Flagella, 103
Flatworm, 344-46
digestion and, 538
nervous system and, 598
planarian, 345, 408
Flexor, 489
Flight, 366-69
Flightless grasshopper, 268
Flower
reproduction and, 390-94
structure and diversity of, 390-95
Flowering plant, 331-33, 390-92
sympatric speciation and, 268
Fluid-mosaic model, 100
Fly, fruit, 190-91, 193
Follicle, 419
Follicle-stimulating hormone, 419
Food

human populations after 1850 and, 769
new, 782
oceans and, 726
operant conditioning and, 667
population growth and, 782-96
developing vs. developed countries and, 785
domestic animals and, 790-91
farming of jungles and, 789-90
fishing and, 792-94
global implications of, 782, 783-85
projected world supplies and, 786-89
solutions to hunger and, 794-96
Food chain
dinoflagellates and, 298-99
disruption of, 275
domestic animals and, 790-91
efficiency of, 790-91
ocean and, 726
pesticides and, 829
salt marsh and, 731
sewage and, 825
Food vacuole, 293, 538
Food web, 701
Foot, mollusk, 349
Forebrain, human, 605-609
Foreign DNA, 210
Forest
acid rain and, 751
evergreen, 713
temperate deciduous, 712
tropical rain, 710-12
undersea, 322
Fossey, Dian, 38
Fossil, 244-45
Fossil fuel, 802-803
nuclear power versus, 810-11
Founder effect, 262
Fox, Sidney, 236
Franklin, Rosalind, 166
Free radical, 71
Fresh water, 717-26
Frigate bird, 24
Frog
cerebral cortex and, 609
early development of, 447-51
egg of, 447
leopard, 257
nervous system and, 599
reproduction and, 365
Frond, 326
Frontal lobe of brain, 609-10
Fructose, 72, 74
Fruit, 394
Fruit fly, 190-91, 193
FSH. *See* Follicle-stimulating hormone
Fuel, 802-803
fossil versus nuclear, 810-11
Functional group, carbon, 71
Fungus (Kingdom Fungi), 299-311
classification of, 285
fungi imperfecti, 310
lysergic acid diethylamide and, 306, 632
miracle crops and, 792
sewage and, 825
Fur seal, 23

G

Galactic storm, 276
Galapagos Islands, 21
chemosynthesis and, 134-35
Darwin's voyage and, 15-25
iguana and, 22
Galapagos penguin, 24
Galileo, 9
Gamete, 158
Gametophyte, 317
ferns and, 386
moss and, 386
Garbage, 800
Gastrodermis, 340
Gastrovascular cavity, 340
Gastrulation, 441
early frog development and, 447
tube-within-a-tube structure and, 452
Gene
advantages of death and, 757-59
dilution of, 383
frequency of, 204
interaction of, 200-201
isolation of, 211, 265
linkage of, 193-97
mutations and, 197-99
population dynamics and, 736
radiation and, 831
replacement therapy and, 222
transport of, 212
tropical forests and, 711
Gene machine, 218
Gene splicing, 213, 217-24
Generative cell, 392
Genetic code, 171
Genetic drift, 260-64
Genetic engineering, 216-24
agriculture and, 219-20
debate regarding, 216
gene replacement therapy and, 222-23
population dynamics and, 736
Genetic imprinting, 384
Genetic isolation, 211, 265
Genetic variation, 711
Genetics, 251-77
See also Evolution
advances in, 209-24
genetic engineering and, 216-24
human interference and, 209
advantages of death and, 757-59
classical, 189-97
inherited human traits and, 192
interbreeding and, 264-65
species and, 264-65
sympatric speciation and, 268-69
Mendelian, 182-89, 252
mutations and, 197-99
population, 202-204
Genotype, 186
Geographic cline, 256
Geographic isolation, 265
Germ layer, 441
German measles, 458
Germinal spot, 451
Gestation, human. *See* Human, gestation of
Giant squid, 349
Gibberellin, 401
Gill, 518, 519, 520
Gingko tree, 327
Giraffe, 372
acquired versus hereditary traits and, 27
classification of, 372
Gizzard, 539-40
Glacial lake, 724
Gland
human endocrine, 576
pituitary
aging and, 175
antidiuretic hormone and, 512
menstrual cycle and, 419
salt removal, 507
swollen, 564
Glial cell, 586
Global economy, 838
Global warming, 711, 804
Glomerulus, 508
Glucose
carbohydrates and, 72, 74
light-independent reaction and, 133
Glycerol, 75-78
Glycogen, 74
Glycolysis, cellular respiration and, 137-39
summary of, 146
Gnetophyte tree, 327, 329
Goat, 372
grazing and, 791
Golden-fronted woodpecker, 264
Goldfinch, 367
Golgi, Camillo, 107
Golgi body, 107-108
Gonad, 158
Gonorrhea, 428
Gonyaulax, 299
Goodall, Jane, 672, 747, 818
Goodenough, Judith, 674
Gorilla, 374
Gould, Stephen Jay, 270
Granum, 127

Grasshopper, 355
sympatric speciation and, 268
Grassland, 716
Gray, Asa, 27
Great frigate bird, 24
Great Lakes, 724-25
Green algae, 322-23
life cycle of, 384-86
Green monkey, 565
Green revolution, 792-93
Greenhouse effect, 804
Greenpeace, 754
Gulf Stream, 726
Gull, 24
Gustation, 646
Gymnosperm, 327-33
reproduction and, 388-90

H

Habitat
biome and, 695
niche and, 696
Habituation, 663-65
Hagfish, 360
Hair, 461
Hallucinogen, 306
Hammer, middle ear and, 645
Hangover, 627
Hardy, G.H., 203
Hare, 372
Hashish, 625-26
Haversian system, 479
Hawk, 24
Head, evolution of, 344-46
Head ganglia, 598
Healing of wound, 444
Heart
circulatory system and, 526-27, 531-36
human embryo and, 443, 458-61, 465
Heart attack, 535
Heat pollution, 819
water and, 827
Heavy chain, 557
Heimlich maneuver, 544
Heinroth, Oscar, 604
Helium, 55
Helix, double. *See* Double helix
Helper T-cell
acquired immune deficiency syndrome and, 566
lymphocytes and, 558-61
emichordata (marine worm), 357
Hemispheres, human brain and, 609-610
Hemoglobin
blood circulation and, 525
protein structure and, 82
Hemophilia, 195, 196
Henle's loop, 508, 510, 511
Hepatic portal vein, 543
Hepatitis, 826
Herbivore, 786, 791
Hereditary trait
aging and, 175
Lamarckian theory of, 11-12, 27
Heredity. *See* Gene; Genetic entries
Heroin, 628
Herpes, 429
Hershel, John, 33
Heterotrophy, 241-42
Hierarchy, mortality and, 756
Hindbrain, 605
Hindenburg, 57
Hinge joint, 486
Hippopotamus, 372
Hirudinea, 352
Histamine, 553
HIV. *See* Human immunodeficiency virus
Hog grazing, 791
Holothuroidea, 356
Homeostasis
hormonal systems and, 584
internal environment and, 498-513
definition of, 498
excretory system and, 506-13
feedback systems and, 498-99
temperature regulation and, 500-504
water and, 504-13
Homologue, 157
Homosexual male, 565
Honeymoon cystitis, 429
Hood Island, 22
Hooke, Robert, 91-92
Hooker, Joseph
Darwin's logic and, 34
natural selection and, 27
Hormone, 575
aging and, 175
female longevity and, 777
menstrual cycle and, 419
nerves and, 584
plant reproduction and, 398-403
test taking and, 579
vertebrate endocrine glands and, 576-79
Hormone clock, 175
Horse, 372
Host DNA, 210
Housefly, 743
How They Do It, 408
Human, 374
brain of. *See* Brain, human
criminal behavior and, 193
digestive system and, 539, 540-45
disease and, 220-22
endocrine system and, 576
evolution, 28
gestation of, 453-69
birth and, 468-69
first trimester and, 462-66
miscarriage and, 468-69
second trimester and, 466
third trimester and, 467-68
inherited traits and, 192
kidney and, 508
natural laws and, 706
pheromones and, 581
population growth and, 763-68
See also Population growth, human
reproduction and. *See* Reproduction, human
respiration and, 523
Human condition, 46
Human evolution, 28
Human immunodeficiency virus, 564, 566
Human sexual response, 418-425
Humerus, 484
Humor and human immunity, 569
Hunger
death and, 784
solutions to, 794-96
Hunter-gatherer society, 764-65
Huntington's disease, 221, 222
Huxley, Thomas, 29
Darwin's logic and, 34
Huygens, C., 33
Hybrid, 265
isolation of, 265
Hydra, 341
asexual reproduction and, 410
contractile systems and, 486
digestion and, 534, 538
nervous system and, 598
Hydration, 72
Hydrocarbon
carbon and, 70
environmental pollution and, 824
Hydrochloric acid, 64
Hydrogen
Hindenburg and, 57
structure of atom and, 55
Hydrogen bond, 61-63
Hydrolysis
carbohydrates and, 72
Hydrophilic head, 100
Hydrophobic head, 100
Hydroxyl group, 71
Hydrozoa, 340
Hyena, 681
Hypertonic urine, 511
Hypha, 300
mushroom and, 308-309
Hypogonadism, 222
Hypothalamus
human brain and, 502-503, 609
temperature regulation and, 502-503
Hypothermia, 275
Hypothesis, theory and, 35-38

I

Ice age, 767
Idiot savant, 664
Iguana, 22
Ileum, 545
Illness in women, 777
Immune deficiency
  AIDS and, 568
  gene replacement therapy and, 222-23
Immune system, 551-69
  acquired immune deficiency syndrome and, 563-68
  interferon and, 562-63
  mind-body connection and, 568-69
  non-specific responses and, 551-54
  programming lymphocytes and, 558
  specific responses and, 555-562
Imprinting
  genetic, 384
  and instinct and learning, 669-70
Impulse
  mechanism of, 587-91
  pathway of, 586
Incaparina, 795
Incus (anvil), 645
Independent assortment principle, 187-88
Induction, embryonic regulation and, 445
Inductive reasoning, 33-37
Industrial melanism, 258
Industrial society
  population levels and, 768, 769-70
  water pollution and, 797, 800
Inert atom, 58
Infarction, myocardial, 535
Infection, 770
Infectious hepatitis, 826
Inflammatory response, 553
Inheritance. *See* Gene; Genetic entries
Innate behavior, 660
Innate releasing mechanism, 661
Insect, 354
  chemoreceptors and, 647
  circulatory system and, 526
  hormones and, 575
  mud flats and, 731-32
  pesticides and, 829
  pollen and, 393
  respiration and, 520-21
Insecticide, 592
Insectivora, 370
Insemination, external, 410
Instinct, 660
  behavioral development and, 660-67
  learning and, 667-70
Insulin, structure of, 83
Intelligence
  human pregnancy and, 467
  memory and, 663, 664
  nervous system and, 598
  vertebrate brain and, 603
Intelligence quotient (I.Q.), 662
Interbreeding, 264-65
  sympatric speciation and, 268-69
Intercourse, anal, 565
Interdependency, 703-704
Interferon, 562-63
Internal respiration, 517
Interphase, cell reproduction and, 152, 158
Interspecific aggression, 675
Intestine, human, 540, 542, 545
Intraspecific cooperation, 680-83
Intrauterine device, 430
Intravenous drug, 565
Inversion
  chromosome mutations and, 199
  temperature, 821
Invertebrate
  excretory systems and, 506-508
  hormones and, 575
  tactile receptors and, 640
  tube-within-a-tube structure of, 452
Ion, 56, 59-60
Ion channel, 589
Ionic bond, 59-60
I.Q. (intelligence quotient), 662
Iridium, 275
IRM. *See* Innate releasing mechanism
Iron, human pregnancy and, 467
Island
  founder effect and, 262
  Galapagos. *See* Galapagos Islands
Isolation, interbreeding and, 265-66
Isotope, 55-56
IUD. *See* Intrauterine device
IWC. *See* International Whaling Commission

J

Jackal, 370, 681
Jawed fish, 360-62
Jawless fish, 357, 360-62
Jejunum, 545
Jellyfish, 340-44
  contractile systems and, 486
Johnson, Ben, 78
Joint, 485-86
Jungle, 710, 789-90

K

Kaibab squirrel, 266
Kangaroo, 370
Kangaroo rat, 507
Kaposi's sarcoma, 564
Ketone, 784
Kidney, 508
Kin recognition, 686
Kin selection, 684-87
Kinetic energy, 65
Kingdom
  animal. *See* Animal, kingdom of
  characteristics of five, 282-311
    See also Individual kingdom
Knee, 482
Koala bear, 370
Kramer, Gustave, 670
Krebs, (Sir) Hans, 142
Krebs cycle, 141, 142
Kurosawa, E. 401
Kwashiorkor, 783

L

Labia minora, 418
Labor, human birth and, 468
Lactate fermentation, 140-41
Lagomorpha, 372
Lake
  acid rain and, 751
  glacial, 724
Lamarck, Jean Baptiste de
  acquired versus hereditary traits and, 27
  heredity theory of, 11-12
Lamella, 127
Lamprey, 360
Lancelet, 357, 359
Land biomes and communities, 707-16
Large intestine, 545
Lateral meristem, 397
Laterite, 790
Latimeria, 364
Laughter, human immunity and, 569
Laurasia, 267
Law of the commons, 793
Laws of thermodynamics, 123-24
Leaf, 326-27
Learning
  autonomic, 619-20
  developmental behavior and, 662-67
  instinct and, 667-70
Leech, 352
Left-handedness, 614
Left hemisphere of brain, 613
Leg, skeletal system and, 484
Lemur, 374
Lens, 445
Leonardo da Vinci, 191
Leopard frog
  variation and, 257
Leprosy, 288
Leukocyte
  bereavement and, 569
  nonspecific immune response and, 552
LH. *See* Luteinizing hormone
Lichen, 303, 304-305
Life
  chemistry of. *See* Chemistry
  definition of, 39-43
  meanings of, 229-231

oldest known, 244-45
origin of, 227-249
early earth and, 232-33
expanding universe and, 231-32
hypotheses concerning, 233-35
nonlife and, 229-31
protobionts and, 237-41
Stanley Miller's experiments and, 235
quality of, 817-18
Life cycle of plant, 384-94
Life force, vitalists and, 229
Ligament, 481
Light
human vision and, 651
photosynthesis and, 126-31
Light chain, 557
Light microscope, 91-92, 95
Light spectrum, 127
Light-dependent reaction, 125-31
Lightening, 467
Linné, Carl von, 10-11
Linnaean Society of London, 27
Linnaeus, Carolus, 11
Lipid, 75-78
Lipofusion, 175
Living matter, chemicals in, 53-54
Lizard, 365
Galapagos Islands and, 22
temperature regulation and, 501
Lobe
human brain and, 609-610
human endocrine system and, 576
Lockjaw, 286
Logic, 33-35
Long distance running, 582
Longevity, human, 777
Loop of Henle, 508, 510, 511
Lorenz, Konrad, 669
LSD. *See* Lysergic acid diethylamide
Lude, 632
Lumen, 128
Lung
smog and, 823
smoking and, 624
Luteinizing hormone, 419
Lyell, Charles, 14
natural selection and, 27
scientific logic and, 33
Lyme disease, 555
Lymph, 536
Lymphatic system, 536-38
Lymphocyte, 538
nonspecific immune response and, 553
programming of, 558
specific immune response and, 558
Lysergic acid diethylamide (LSD), 632-33
Lysosome, 108

M

MacArthur, Robert H., 697
Macromolecule, 72
Macronucleus, 294
Macrophage, 554, 558
Maggot fly, 268
Magnetism, 267
Malaria
human populations after 1850 and, 770
Plasmodium and, 297
Malathion, 592
Male
homosexual, 565
reproductive system of, 415-16
shorter life span of, 777
XYY, 193
Male/female ratio, human, 777
Malleus, 645
Malnourished person, 784
Malthus, Thomas, 17, 25
Mammal
animal kingdom and, 369-74
brain of, 603
female longevity and, 777
heart of, 531
placental, 370
population levels and, 767
respiration and, 521-25
temperature regulation and, 501-504
Man/woman ratio, 777
Mandibulata, 352
Mangrove, 20
Mantle, 349
Mapping
chromosome, 197
human brain and, 612
Marasmus, 783
Marijuana, hashish and, 625-26
Marine worm, 357
Marrow, 479
Marsh, 731
Marsupial, 370
Marxism, 788
Mating isolation, 265
McClintock, Martha, 581
Measles, 467
Meat, 785
Mechanism, vitalism versus, 229
Medicine, after 1850, 769-70
Medulla, 602, 605
Medusa, 341
Megaspore mother cell, 388
Meiosis, 157-64
Meissner's corpuscle, 640
Melanism, industrial, 258
Meltdown
Chernobyl and, 832, 833
Three Mile Island and, 832
Membrane
birds and, 451-53
middle ear and, 645
plasma, 100
tympanal, 640
tympanic, 640
Memory
electrical stimulation and, 612
intelligence and, 663, 664
Memory B-cell, 559
Mendel, Gregor Johann, 182
Mendelian genetics, 182-89, 252
See also Gene; Genetic entries
Menstrual cycle, 418-21
Mental deficiency, 783
Mercury, nonrenewable resources and, 799
Merostomata, 352
Mesoderm, 441
Mesoglea, 340
Mesozoic era, 267
Messenger, peptide hormones and, 580
Messenger RNA, 169
Metacarpal, 484
Metaphase
meiosis and, 159
mitosis and, 153
Metatarsal, 484
Metatheria, 370
Methamphetamine, 631
Methane, 69
Methaqualone, 632
Mexico, population structure and, 775
Microfilament, 101
Microhabitat, 695
Micronucleus, 294
Micropyle, 392
Microscope, 91-92, 95
Microsphere, 237
Microspore mother cell, 388
Microtrabecular lattice, 102
Microtubule, 102
Microvillus, 542
Midbrain, human, 605
Midlobe, human endocrine system and, 576
Migration, 670-72
Mildew, 301-303
Milk, grazing and, 791
Mill, John Stuart, 34
Miller, N.E., 619
Miller, Stanley, 235
Millipede, 352, 354
Mind. *See* Brain, human; Nervous system
Mind-body connection, 568-69
Mindbender, 621-33
alcohol and, 626-28
amphetamines and, 631
barbiturates and, 631-32
caffeine and, 624-25
cocaine and, 628-31
marijuana and hashish and, 625-26
methaquelone and, 632

opiates and, 628
phencyclidine and, 632
psychedelics and, 632-33
tobacco and, 623-24
Miracle crop, 792
Miscarriage, 468-69
Mitochondria
cell structure and, 104
matrix of, 143
Mitosis, 152-57, 268
cellular life cycle and, 152
Mnemonist, 664
Modification, descent with. *See* Evolution
Mold, 299, 301-303
Mole, 370
Molecule
chemistry and, 68-84
compounds and, 54-55
motion of, 110-16
Mollusca (mollusk), 347-49
Molluscum contagiosum, 429
Molting, crayfish and, 478
Monea
classification of, 285
kingdom of, 285-92
Monkey, 375
acquired immune deficiency syndrome and, 565
Monocotyledon, 396
Monocyte, 553
Monosaccharide, 74
Monotreme, 369
Mood, illness and, 569
Mood altering substance, 621-33
See also Mindbender
Morgan, Thomas Hunt
chromosome mapping and, 197
classical genetics and, 190-91
Morning after pill, 431
Morphine, 628
Mortality. *See* Death
Morula, 447
Moss, 323-24
reproduction and, 386
Moth
coevolution with bat and, 642-43
industrial melanism and, 258
Mountain gorilla, 38
Mouse, 370
Movement, skeletal muscle and, 492-93
mRNA. *See* Messenger RNA
Mucus, AIDS and, 565
Mud flat, 731-32
food chain and, 731-32
Mule, 265
Multiple allele, 200
Musca domestica, 743
Muscle
bone and, 481
contraction of, 492-93
exoskeletons and, 487
skeletal, 488-90
steroids and, 78
types of human, 488
Mushroom, 299, 307
Muskeg, 713
Mutation
genetics and, 197-99
mortality of populations and, 754-55
Mycelium, 301
Mycorrhiza, 311
Myelin sheath, 585
Myocardial infarction, 535
Myofibril, 492
Myofilament, 493

N

NADP+ (nucleotide of adenine), 129
NADPH, 129
Natural killer cell, 554
Natural selection. *See* Evolution
Needle, AIDS and, 565
Negative charge, electrons and, 59-60
Negative feedback, homeostasis and, 498-99
Neisseria gonorrhea, 288, 428
Nematoda (roundworm), 346-47
Nembutal, 631
Nemesis, 275
Nephridium, 508
Nephron, 508
Nerve, 584-94
cranial, 614
hormones and, 584
impulse of, 587-91
Nerve root, 600, 616
Nervous system, 584-94
central, 600-604
evolution of, 598-99
human brain and, 604-14
mindbenders and, 621-33
peripheral, 614-20
Neural induction, 445
Neuron, 584
afferent, 586
ascending, 601
association, 601
efferent, 586
Neurotransmitter, 591
Neutron
atom and, 55
nuclear power and, 810
Neutrophil, 553
Newton, Isaac, 10, 33
Niche, habitat and, 696
Nickel, nonrenewable resources and, 799
Nicotiana tabacum, 623
Nicotine, 623
Nitrate water pollution, 725, 827
lakes and, 725
Nitrite water pollution, 751
Nitrogen
excretory system and, 504-13
sewage and, 825-26
Nitrogen fixing
cyanobacteria and, 291
solutions to hunger and, 795
Nitrogenous bases, 82-84
Nitrogen oxide, 751, 823
NK cell. *See* Natural killer cell
Node, lymph, 537
Noise pollution, 819
Nonrenewable resource, 799
Nonspecific response, immunity and, 551-54
Nonvascular plant, 323-24
Norepinephrine, 579, 616
Normal distribution, as bell-shaped curve, 258
Norplant, 432
Norway, acid rain and, 751
Notochord, 357
Nuclear bomb, 835
Nuclear power
bioethics and, 832-35
fossil fuel versus, 810-11
waste and, 832-35
Nuclear Regulatory Commission, 834
Nucleic acid, 82-84
Nucleotide, 82-84
Nucleotide sequencing, 218
Nucleus
atom and, 55
cell structure and, 109-10
origin within cell of, 109

O

Obligate anaerobe, 285
Observation, scientific value of, 35
Ocean
biomes and communities and, 726-30
chemosynthesis and, 134-35
continental drift and, 267
depth of, 726, 729-30
fishing and, 792-94
food chain and, 726
miracle crops and, 792
poisoning of, 707
Odor, human chemoreceptors and, 649-50
Oil spill, Alaskan, 729
Oken, Lorenz, 92
Olfaction, chemoreceptors and, 646-47
Olfactory bulb, 647
Olympic National Park, 719
Omnivore, 791
Oomycota, 301-303
Oparin, A.I., 232, 237
Operant conditioning, 666-67
Ophiuroidea, 356
Opiate, 628
Opossum, 370
Optimism, human immunity and, 569

Orangutan, 374
Organism, pioneer, 698
Orgasm, human, 419-21
Orientation
instinct and, 670-72
navigation and, 670-72
Orientation cage, 671
Origin of Species, 14, 28
Osmosis
excretory systems and, 505
molecular motion and, 112-14
Osteichthyes (bony fish), 362-64
Otter, 370
Oval window, middle ear and, 645
Ovary
flowers and, 392
human menstrual cycle and, 419
Overpopulation. *See* Population growth
Oviduct, 423
Ovule, 392
Oxaloacetate, 143
Oxidation, electron and, 57
Oxygen
cypress swamp and, 720
extinction and, 709
lakes and, 725
obligate anaerobes and, 285
origin of life and, 246-47
phytoplankton and, 729
respiratory systems and, 517-25
sickle-cell anemia and, 198
Oyster, 410
Ozone, 246, 822

P

Pacinian corpuscle, 641
Pain, chest, 535
Palaeotragus, 738
Paleozoic era, 267
Pangaeae, 267
Paramecium, 292
Parapodia, 352
Parasite
mortality of populations and, 754-55
viruses as, 211
Parasympathetic nervous system, 616-19
Parathyroid hormone, 576
Parenchyma, 398
Parent cell, 152
Parietal lobe, 610
Park, redwoods and, 715
Particulate matter, 824
Passive transport, 110-13
Pasteurella, 555
Patchy environment, 256
Pathway, impulse, 586
Pauling, Linus, 166
Pavlov, Ivan, 665, 666
PCP, 632
Pea plant experiment, 183-84
Peace, trade and, 782
Peacock, 251
Pectoral girdle, 484
Pelican, 367
Pellicle, 292
Pelvic girdle, 484
Pelvic inflammatory disease, 431
Penfield, Wilder, 612
Penguin
Galapagos Islands and, 24
Penicillin, 310
Peppered moth, 258
Peptide
hormone and, 580
protein and, 80
Peripheral nervous system, 614-620
Perissodactyla, 372
Peristalsis, 540
Pesticide
miracle crops and, 792
water pollution and, 828-29
Pet, world food supplies and, 790-91
Petal, flower, 390
Petroleum, edible, 794-95
Phaeophyta (brown algae), 321-22
Phagocytosis, 115
Phalanges (fingers), 484
Phencylidine, 632
Phenomenology, 42-43
Phenotype, 186
Pheromone, 581
Philosophy of science, 33-46
Phloem, 397-98
Phosphate, 725
Phosphoglyceraldehyde, 132
Phospholipid, 77, 237
Phosphorus, 77, 825
Phosphorylation, 143, 146
Photochemical smog, 823
Photographic memory, 664
Photosynthesis
algae and, 297-99
energy and, 125
visible light spectrum and, 127
Photosystems I & II, 128, 129, 131
Photovoltaic, 808-10
Physics, 33
Phytoplankton, 726
oxygen and, 729
Pig, 372
tapeworm and, 746
truffles and, 306
Pika, 372
Pill, birth control, 431
Pine cone, 388
Pinocytosis, 115
Pioneer organism, 698
Pituitary gland
aging and, 175
antidiuretic hormone and, 512
menstrual cycle and, 419
Placenta, human
first month of gestation and, 463-64
third trimester and, 467
Placental mammals, 370
Planaria, 345
Planarian flatworm, 345, 408
Plankton, 709
Plant (Kingdom Plantae), 317-33
alternation of generations and, 317-20
Bryophyta (moss) and, 323-24
Chlorophyta (green algae) and, 322-23
classification of, 285
definition of, 317
development of, 395-98
edible species of, 792-93, 795
flowering, 331-33, 390-92
food pyramid and, 786
nonvascular, 323-24
Phaeophyta (brown algae) and, 321-22
reproduction and, 382-403
development and, 395-98
hormones and, 398-403
life cycles and, 384-94
sexual, 383-84
Rhodophyta, (red algae) and, 320, 321
species of, 737
sympatric speciation and, 268-69
vascular, 324, 325-27
Plant-like protist, 297-99
Planula, 341
Plasma, 537
Plasma cell, 559
Plasma membrane, 100
Plasmid, 212, 213, 214-15
Plasmodium, 297
Plastid, 108-109
Plate tectonics, 267
Platyhelminthes (flatworm), 344-46
Plug, fallopian, 434
Plutonium, 835
Poison, tobacco, 623
Polar body, cell reproduction and, 164
Politics, global food resources and, 785
Pollen
flowers and, 392
gibberellins and, 401
insects and, 393
Pollen grain, 392
Pollen sac, 392
Pollen tube
conifers and, 390
flowers and, 392
Pollock, Jackson, 695
Pollution
acid rain and, 751
air, 820-24
bioethics and, 819-35
heat, 819
nuclear waste and, 832-35

sewage and, 826-27
water, 824-35
Polychaeta, 352
Polydactyly, 190
Polygenic inheritance, 254
evolution and, 254-55
genetic ratios and, 201
variation and, 254-55
Polymorphic shape, 260
Polyp, Cnidaria and, 341
Polypeptide, 80
Polyploidy, 268
Polysaccharide, 74-75
Pons, human brain and, 605
Population
definition of, 202-203, 695
dynamics of, 735-60
advantages of death and, 757-59
change and, 739
control through mortality and, 749-57
control through reproduction and, 743-49
ethics and necessity and, 736
genetics and, 202-204, 736
small, 260-64
variation in, 252-264
Population growth, human, 763-68
advent of agriculture and, 768
ancient history and, 764
changes since 1850 and, 769-70
changes since 17th century and, 768-69
food and, 781-96
future of, 773
future of planet and, 569
Malthus and, 17, 25
recent history and, 770-73
Redwoods and, 715
U.S. tax structure and, 837
Population Reference Bureau, 775
Porifera (sponge), 339-40
Portal vein, hepatic, 543
Positive charge, electrons and, 59-60
Positive feedback, 499
Positive thinking, 569
Posterior lobe, endocrine system and, 576
Potato blight, 303
Poultry, production of, 791
Power plant
acid rain and, 751
fossil fuel versus nuclear power and, 810-11
tidal, 806
Predation
aggression and, 673-74
disease and, 756-57
fluctuation of populations and, 752-53
Prefrontal areas of brain, 610
Pregnancy
first trimester of, 462-66
second trimester of, 466
third trimester of, 467-68
Pressure, water depth and, 729-30
Preti, George, 581
Prey-predator fluctuation, 752-53
Primary consumer, 701
Primary response, lymphocytes and, 559-61
Primary structure, 80
Primary succession, 698
Primata (primate), 374
Prince William Sound, 729
Principle of dominance, 183-84
Principle of independent assortment, 187-88
Principle of segregation, 184-87
Principles of Geology, 14
Proboscidea, 372
Product, enzymes and, 67
Progesterone
endocrine system and, 578
menstrual cycle and, 419
Proglottid, 345
Prokaryote, 285
Prokaryotic cell, 93-97
Prophase, cell reproduction and, 153, 159
Proprioceptor, 650
Prostaglandin, 582
Protective coloring, 258
Protein
colloidal, 237-41
global food resources and, 783
human pregnancy and, 467
molecules and, 78-82
starvation and, 784
structure of, 80-82
synthesis of, 176
urine concentration and, 512
Proteinoid, 236
Protista, 292-311
classification of, 285
internal water balance and, 506
Protobiont, 237-41
Proton, atom and, 55
Protostome, 346
Prototheria, 369
Protozoan, 292-97
Proximate level, reproductive output and, 744
Przewalski's horse, 372
Pseudocoelom, 347
Pseudomonas aeruginosa, 288
Pseudopod, 295
Psychedelics, 632-33
Psychoactive agent, 621
See also Mindbender
Psychology, ethology and, 659
Pterophyta (fern), 326-27
Public education, 837
Puffball, 307
Pump, sodium/potassium, 589
Punnett, R.C., 203
Punnett square, 202
Pupfish, 500
Pyramid, age structure, 775
Pyramid, food
domestic animals and, 790-91
efficiency of, 790-91
pesticides and, 829
Pyrogen
fever and, 503
nonspecific immune response and, 553
Pyrrophyta, 298-99
Pyruvate, 136, 139, 141

Q

Quaalude, 632
Quaternary consumer, 701
Quaternary structure, 82

R

Rabbit, 372, 561
Radial symmetry, 340-44
Radiation, 829-35
Radioactive waste, 832-35
Radioisotope, 831
Radionucleotide sequencing, 218
Radius, 484
Radula, 349
Rain
acid, 751
African famine and, 785-86, 788
desert and, 716
forest and, 710, 712
grasslands and, 716
Rain forest, 710-12
Rain Forest Action Network, 711
Rainbow Warrior, 754
Rat, 370
autonomic learning and, 619
war on, 706
Rate of increase, population growth, 771
Ray, 362
Reasoning, inductive and deductive, 33-37
Receptacle, flower, 390
Receptor
auditory, 640-45
nerve, 586
tactile, 640
visual, 651, 653
Recessive trait, 184
Reciprocal altruism, 687-88
Recognition, species, 264
Recombinant DNA
drug therapy and, 216
genetic interference and, 210, 212, 213-16, 218

Rectum, 545
Red algae, 320-21
Red marrow, 479
Red tide, 299
miracle crops and, 792
Red-bellied woodpecker, 264
Reduction, electron and, 57
Redwood tree, 327, 715
Reflex arc, 600-601
Regeneration of missing part, 410, 443
Releaser, instinct and, 660-61
Religion, ancient, 9
Renewable resource, 781-96
Replication, DNA and, 214-15
Repolarization, 589
Reproduction, 407-35
advantages of death and, 757-59
artificial selection and, 25-26
asexual, 408-10
control of populations through, 743-49
eubacteria, 288
external, 410
human, 413, 415-18
contraception and, 425-35
female system and, 416-18
male system and, 415-16
menstrual cycle and, 418-21
sexual response and, 418-25
society and, 424-25
J-shaped curve and, 739
kin selection and, 684-87
Malthus theory of, 17, 25
plants and, 382-403
development and, 395-98
hormones and, 398-403
life cycles and, 384-94
sexual reproduction and, 383-84
sexual, 383-84, 410-15
survival and, 257, 383
Reproductive isolation, 265
Reproductive strategy, 744
Reptile, 365-66, 531
brain of, 602
temperature regulation and, 501
Resource
nonrenewable, 799
renewable, 781-96
Respiration, 517-25
cellular, 133-46
energy and, 133
summary of, 146
external, 517
mammal and, 521-25
resting potential, 588
Restriction enzyme, 210
Resuscitation, 535, 536
Reticular system, 607
Retina, 651
Reward, operant conditioning and, 667
Rhea, 5
Rhinoceros, 372
Rhizoid, 326
Rhodophyta (red algae), 320-21
Rhythm, biological, 674
Rhythm method of contraception, 427
Rib, human, 484
Ribosomes, 104-105
Rice
miracle crops and, 792
other edible species and, 792
Right hemisphere of brain, 613
RNA (Ribonucleic acid), 82-84, 169
messenger (mRNA). *See* Messenger RNA
Robin, 661, 670
Rock, oldest known, 244-45
Rocky coast, 730-31
Rocky Mountains, 713
Rod, human eye and, 651
Rodentia, 370
Roeder, Kenneth, 642-43
Root meristem, 396
Roseate spoonbill, 367
Roundworm
animal kingdom and, 346-47
contractile systems and, 487
Runner's high, 582
Russian royalty, hemophilia and, 196
Rust, white, 301-303

S

Sac fungus, 755
Saccule, 650
Sahara, famine and, 788
Saint Anthony's fire, 306
Salamander
digestion and, 540
reproduction and, 365
respiration and, 518-20
Sally Lightfoot crab, 21
Salmon, external insemination and, 411
Salt
removal gland and, 507
urine concentration and, 510, 512
Salt marsh, 731
Saltatory propagation, 585-86
Salting out, abortion and, 434
Saltwater, 505
fish and, 500, 505
Sand dollar, 356
Sandy seashore, 731
Sanger, Frederick, 83
Santa Cruz Island, 20
Sarcodina, 294-97
Sargasso Sea, 322
Saturated fat, 76-77
Scanning electron microscope, 95
Schleiden, Matthias Jakob, 92
Schwann, Theodor, 92
Science
historical view and, 8-14
methods and philosophy of, 33-46
defining life and, 39-43
hypothesis and theory and, 35-38
inductive versus deductive reasoning and, 33-37
scientist as skeptic and, 38-39
social responsibility and, 43-46
phenomenology and, 42-43
Sclerenchyma, 398
Scolex, 345
Scorpion, 352
Scyphozoa, 341
Sea anemone, 486
Sea cucumber, 356
Sea lilie, 356
Sea lion, 23
Sea star, 356
Sea urchin, 356
Seal, fur, 23
Galapagos Islands and, 23
Greenpeace and, 754
Seashore, biomes and communities and, 731
Seawater, 505
Seconal, 631
Second law of thermodynamics, 123
Second messenger, 581
Secondary consumer, trophic levels and, 701
Secondary response, lymphocytes and, 561
Secondary structure, 80
Secondary succession, 699-700
Sedative-hypnotic, 631-32
Seed
conifers and, 390
embryonic development in, 393, 395
Seed coat, 393
Segment
repeating, 350-52
specialized, 352-54
Segmented worm, 350-52
Segregation, Mendelian genetics and, 184-87
Selection
artificial, 25-26
natural. *See* Evolution
stabilizing, 258
Semen
acquired immune deficiency syndrome and, 565
human conception and, 419-21
Semi-circular canal, 650
Seminal receptacle, 351
Senses, 638-54
auditory receptors and, 640-45
chemoreceptors and, 646-50
proprioceptors and, 650
tactile receptors and, 640
thermoreceptors and, 639-40
visual receptors and, 651-53

Sequencing, embryonic regulation and, 443-45
Setae, 351
Sewage, 825-27
Sex
criminal behavior and, 193
determination of, 190-93
human odor and, 649-50
linkage and, 195
society and, 424-25
Sex ratio, 275
Sexual dimorphism, 251
Sexual reproduction, 383-84, 410-15
plants and, 383-84
Sexually transmitted disease, 428-29
Shark, 362
respiration and, 520
Sheep, 372
grazing and, 791
Shoebill, 367
Shrew, 370
Shrew, tree, 374
Shrimp
mud flats and, 731-32
world food supplies and, 789
Sickle-cell anemia
genetic engineering and, 220
genetic mutations and, 198
Side, evolution of, 344-46
Sierra Club, 715
Sight, human, 651, 653
Silicon, solar energy and, 808
Silver, nonrenewable resources and, 799
Single-celled organism, 282
digestion and, 538
petroleum as food and, 794-95
size limits of, 477
Single-gene inheritance, 254-55
Skate, 362
Skeleton
body support and, 477-86
human, 482-85
human embryo and, 458-61, 464, 466
muscle and, 488-93
Skin, Echinodermata and, 356
Skinner, B.F., 666
Skinner box, 666
Slash and burn agriculture, 790
Sliding filament theory, 493
Small intestine, 542
Smell, chemoreceptors and, 646-47
Smelter, acid rain and, 751
Smog, 823
Smoking, 523, 623
human longevity and, 523, 777
Smooth muscle, 488
Snake, 365
Society
abortion and, 434-35
industrial
population levels and, 769-70
water pollution and, 797, 800
scientist's response to, 43-46
sex and, 424-25
sociobiology and, 688-90
Sociobiology, 688-90
Sodium, ionic bonding and, 59-60
Sodium-potassium pump, 589
Soil in tropical rain forest, 711, 789-90
Somatotropin, 576
Song, bird, 670
Source DNA, 210
South America
continental drift and, 267
Darwin's voyage and, 14-17
Soviet Union
Chernobyl and, 832, 833
global food resources and, 785
Sparrow, 670
Specialization, plant development and, 395-96
Speciation, 266-70
Species
definition of, 264-66
genetic varieties of, 736-38
numbers of, 283
Species recognition, 264
Specific response, immunity and, 555-62
Speed (drug), 631
Sperm, human, 415, 422-23
conception and, 422
Sperm nucleus, conifer and, 390
Spermicide, 430
Sphincter muscle, 540
Spider, 352
Spinal cord
induction and, 445
reflex arc and, 600-601
Spindle, mitosis and, 153
Spine, 482
Spiny anteater, 369
Spiny skin, 356
Spirillum, 288
Splicing, gene
cloning and, 213, 217-24
dangers of, 216-27
Split brain, 614
Sponge, 339-40
contraception and, 432
Spoonbill, 367
Sporophyte, 317
ferns and, 386
green algae and, 385
moss and, 386
Sporozoa, 297
Sports, steroids and, 78
Squid, 349
Squirrel, 266, 684, 686
S-shaped curve, 739
Stabilizing selection, 258
Stamen, 392
Staminate, conifer and, 388
Stapes (stirrup), 645
Star, death, 272-77
Starch, 74
Starfish, 356
regeneration and, 410
Starvation, 784
Africa and, 785-86
potato blight and, 303
solutions to, 794-96
STD. *See* Sexually transmitted disease
Stem
gibberellins and, 401
growth area of, 398
Sterilization, human, 432-34
Steroid hormone, 77-78, 582
sports and, 78
Stigma, flower and, 392
Stimulation, mindbender and, 621
Stirrup, 645
Stomach, human, 540
Storage, food distribution and, 789
Stress, embryonic distribution and, 445
Stress, embryonic regulation and, 445
Striated muscle, 492
Stroma, 127
Sturtevant, A.H., 197
Style, flower and, 392
Substrate, 67
Succession
biomes and communities and, 698-700
secondary, 699-700
Sudden death, heart attack and, 535
Sugar, 74-75
Sulfur oxide, 823
acid rain and, 751
Sunlight
artificial, 672
energy from, 121, 807-10
Support system, skeletal, 477-86
human, 482-85
embryo and, 458-61, 464, 466
muscle and, 488-93
Suppressor T-cell, 559
Survival, reproduction and, 383
Survival of the fittest, 257
Sutherland, E.W., 580
Sutton, Walter S., 189
Swallowtail gull, 24
Swamp, cypress, 720
Sweat
acquired immune deficiency syndrome and, 565
chemoreceptors and, 646
Symmetry, radial, 340-44
Sympathetic nervous system, 616-19
Sympatric speciation, 268-69
Synapse
nerves and, 591-94
neural impulses and, 591-94

Synthesis
bacteria and, 286-87
DNA, 218
protein, 176
Syphilis, 428
Szent-Gyorgi, Albert, 229-31

T

Tactile receptor, 640
Tadpole, 451
Taiga, 712-13
Tail, human embryo and, 458, 465
Tapeworm, 345
reproductive strategy of, 345, 745-46
Tapir, 372
Tarsal, 484
Tarsier, 374
Taste, 646
Taxation, hidden influence of, 837
Taxonomist, 284
Tay-Sachs disease, 220
T-cell
acquired immune deficiency syndrome and, 566
cytotoxic, 557
lymphocytes and, 559
specific immune response and, 557
Tears
acquired immune deficiency syndrome and, 565
salt removal gland and, 507
Technology
environmental pollution and, 819-35
future and, 837-38
hidden decisions and, 835-37
simple living and, 817-18
water pollution and, 824-35
Tectorial membrane, 645
Teleology, 42
Telophase, mitosis and, 156
TEM. *See* Transmitting electron microscope
Temperate deciduous forest, 712
Temperature
homeostasis and, 500-504
thermoreceptors and, 640
Temperature inversion, 821
Temporal lobe, 609
Tendon, 481
Terminal bud, 396
Territory
fighting and, 674-79
mortality of populations and, 755
Tertiary consumer, 701
Tertiary structure, 80-82
Test cross, 188-89
Test taking, hormone levels and, 579
Testosterone
human endocrine system and, 578
violent crimes and, 575
Tetanus, 286
Tetrahydrocannabinols, 625
Tetraploid cell, 268
Thalamus, 605
THC. *See* Tetrahydrocannabinols
Theory, hypothesis and, 35-38
Therapy
drug, 218
gene replacement, 222-23
Thermal receptor, 640
Thermodynamics, laws of, 123-24
Thermoreceptor, 639-40
Thermostat, human brain and, 502-503
Thoreau, Henry David, 817
Three Mile Island, 832
Threshold, 589
Threshold effect, 820
Thumbsucking, 459
Thykaloid, 127
Thyroid gland, 576
Thyroid-stimulating hormone, 584
Thyroxine, 576
Tibia, 484
Tide, energy from, 806
Time bomb theory, 175
Tin, nonrenewable resources and, 799
Tinbergen, Niko, 659
Toad, 365
Tobacco, 623-24
Tolerance, autoimmunity and, 561-62
Tongue
chemoreceptors and, 648
single gene and, 648
Tortoise, Galapagos, 15, 16, 21
Touch, 640
Toxin, Greenpeace and, 754
Tracheophyte (vascular plant), 325-27
Trade, 782
Trade winds, 726
Trait, hereditary versus acquired, 26-27
Tranquilizer, 632
Transfer RNA, 169
Transmitting electron microscope, 93, 95
Transport, molecular motion and, 110-16
Transverse nerve, 598
Trash
environmental pollution and, 818, 819
natural resources and, 800
Tree
evolutionary
animal kingdom and, 348
hypothetical descendents and, 348
family, 358
gymnosperms and, 327-33
sympatric speciation and, 269
tropical forests and, 711
Tree shrew, 374
Trematoda, 345
Trichina, 347
Trimester of pregnancy
first, 462-66
second, 466
third, 467-68
Triticale, 795
Trivers, Robert, 687
tRNA. *See* Transfer RNA
Trophic level, 701
Trophoblast, 463
Tropical forest, 710-12
destruction of, 711
farming of, 711
Trout, Great Lakes and, 725
Truffle, 306
TSH. *See* Thyroid-stimulating hormone
Tubal ligation, 433
Tube-within-a-tube structure, 452
Tubeworm, 134-35
Tubulin, 102
Tuinal, 631
Tule, overflowed land and, 720
Tundra, 713
Tunicate, 359
Turbellaria, 345
Turtle, 365
Galapagos, 15
salt removal gland and, 507
Tympanal membrane, 640
Tympanic membrane, 640

U

U235, U238, and U239, 810
Ulna, 484
Ulothrix, 385
Ultimate level, reproductive output and, 744
Umbilicus, 467
Undernourished person, 784
See also Starvation
Undersea forest, 322
Uppers (drugs), 631
Urea, 504, 512
Uric acid, 505
Urinary bladder
chlorinated water and, 826
kidney and, 508-509
Urine
concentration of, 510-512
formation of, 509-510
hypertonic, 511
Urochordata, 357
Utricle, 650

V

Vacuole
cell structure and, 109
food, 293
Vacuum curettage, 434
Vaginal intercourse, 565
Vaginitis, 429
Valdez, Alaskan oil spill and, 729
Vampire bat, 372

van Overbeek, J., 402
Variable region, antibody response and, 557
Variation
  environment and, 255-56
  genetic, 711
  natural selection and, 257-60
  populations and, 252-64
  sexual reproduction and, 384
  single-gene and polygenic inheritance and, 254-55
Vascular cambium, 398
Vascular plant, 324, 325-27
Vascular system, 356
Vasectomy, 432
Vector
  escherichia coli as, 212
  genetic interference and, 210
Vegetal pole, 447
Vein, hepatic portal, 543
Venereal wart, 428-29
Ventral nerve root
  central nervous system and, 600
  peripheral nervous system and, 616
Vertebra, 484
Vertebrata (Vertebrate), 359
  brain and, 602-604
  chemoreceptors and, 647
  chordates and, 360
  early development of, 440-42, 447-51, 458-61
  major endocrine glands and, 576-79
  tactile receptors and, 640
  tube-within-a-tube structure of, 452
Vinegar fly, 190-91
Virchow, Rudolf, 92
Virus
  AIDS and, 566
  genetic inference and, 211
  interferon and, 562-63
Visible light spectrum, 127
Vision, human, 651
Visual purple, 651
Visual receptor, 651-53
Vitalism, mechanism versus, 229
Volcano
  environmental pollution and, 819
  Galapagos Islands and, 21
  magnetism and, 267
Volvox colony, 323
von Linné, Carl, 10-11

W

Walden Pond, 817
Wall, cell, 99-100
Wallace, Alfred Russel, 27
Warm-blooded animal, 501-504
  See also Animal; Mammal
Warming effect, 711, 804
Waste
  environmental pollution and, 819-35
  food pyramid and, 786
  natural resources and, 800
  nuclear power and, 832-35
Water
  acid rain and, 751
  coming crisis and, 796-99
  energy from, 806
  fresh, 717-26
  homeostasis and, 504-506, 507
  mold and, 301-303
  nonrenewable resources and, 796-99
  pollution of, 797
  solubility of, 523
  stability of, 62-63
  vascular system and, 356
Water environment, 717-30
  biomes and communities and, 717-30
Watson, James, 166
Waved albatross, 22
Waxes, 77
Wear-and-tear theory of aging, 175
Weather, population mortality and, 749
Weedkiller, 400
Wegener, Alfred, 267
Weinberg, Wilhelm, 203
Went, Fritz, 400
Whale, 754
Wheat
  genetic varieties of, 737
  miracle crops and, 792
  other edible species and, 737
Wheat rust, 308
White blood cell
  bereavement and, 569
  nonspecific immune response and, 552
White rust, 301-303
White-crowned sparrow, 670
Whitefish, 725
Wigam, A.L., 610
Wilburforce, Lord, 29
Wind, energy from, 806-807
Wine, downy mildew and, 303
Wisent, 264
Wolf, cooperation and, 681
Woman, life span of, 777
Woman/man ratio, 777
Wombat, 370
Woodpecker
  golden-fronted, 264
  red-bellied, 264
  species recognition and, 264
Woodpecker finch, 16
World Health Organization, 564
World peace, 782
World population, projected, 770-71
  See also Population; Population growth
World Population Data, 772
Worm
  marine, 357
  segmented, 350-52
  tube, 134-35
Wound, healing of, 444

X

Xylem, 397-98
XYY male, criminal behavior and, 193

Y

Yellow marrow, 479
Yoga, 619
Yolk
  mass of chicken, 451
  three egg types and, 446-47

Z

Zebra, 372
Zooplankton, 726
  sewage and, 825
Zygomycetes, 303
Zygomycota, 303
Zygospore, 303
Zygote
  bird embryo and, 451
  genetic instructions and, 447
  green algae and, 386
  human gestation and, 440-42
  moss and, 386